Quantum Physics

Florian Scheck

Quantum Physics

Second Edition

Springer

Florian Scheck
Institut für Physik,
Theoretische Elementarteilchenphysik
Universität Mainz
Mainz
Germany

ISBN 978-3-642-34562-3 ISBN 978-3-642-34563-0 (eBook)
DOI 10.1007/978-3-642-34563-0
Springer Heidelberg New York Dordrecht London

Library of Congress Control Number: 2012955238

Printed on acid-free paper

Springer is part of Springer Science+Business Media (www.springer.com)

To the memory of my father,
Gustav O. Scheck (1901–1984),
who was a great musician and
an exceptional personality

Preface

This book is divided into two parts: Part I deals with nonrelativistic quantum mechanics, from bound states of a single particle (harmonic oscillator, hydrogen atom) to fermionic many-body systems. Part II is devoted to the theory of quantized fields and ranges from canonical quantization to quantum electrodynamics and some elements of electroweak interactions.

Quantum mechanics provides both the conceptual and the practical basis for almost all branches of modern physics, atomic and molecular physics, condensed matter physics, nuclear and elementary particle physics. By itself it is a fascinating, though difficult, part of theoretical physics whose physical interpretation gives rise, till today, to surprises in novel applications, and to controversies regarding its foundations. The mathematical framework, in principle, ranges from ordinary and partial differential equations to the theory of Lie groups, of Hilbert spaces and linear operators, to functional analysis, more generally. He or she who wants to learn quantum mechanics and is not familiar with these topics, may introduce much of the necessary mathematics in a heuristic manner, by invoking analogies to linear algebra and to classical mechanics. (Although this is not a prerequisite it is certainly very helpful to know a good deal of canonical mechanics!)

Quantum field theory deals with quantum systems with an infinite number of degrees of freedom and generalizes the principles of quantum theory to *fields*, instead of finitely many point particles. As Sergio Doplicher once remarked, quantum field theory is, after all, the *real* theory of matter and radiation. So, in spite of its technical difficulties, every physicist should learn, at least to some extent, concepts and methods of quantum field theory.

Chapter 1 starts with examples for failures of classical mechanics and classical electrodynamics in describing quantum systems and develops what might be called elementary quantum mechanics. The particle-wave dualism, together with certain analogies to Hamilton–Jacobi mechanics are shown to lead to the Schrödinger equation in a rather natural way, leaving open, however, the question of interpretation of the wave function. This problem is solved in a convincing way by Born's statistical interpretation which, in turn, is corroborated by the concept of expectation value and by Ehrenfest's theorem. Having learned how to describe observables of quantum systems one then solves single-particle problems such as the harmonic oscillator in one dimension, the spherical oscillator in three dimensions, and the hydrogen atom.

Chapter 2 develops scattering theory for particles scattered on a given potential. Partial wave analysis of the scattering amplitude as an example for an exact solution, as well as Born approximation for an approximate description are worked out and are illustrated by examples. The chapter also discusses briefly the analytical properties of partial wave amplitudes and the extension of the formalism to inelastic scattering.

Chapter 3 formalizes the general principles of quantum theory, on the basis of the empirical approach adopted in the first chapter. It starts with representation theory for quantum states, moves on to the concept of Hilbert space, and describes classes of linear operators acting on this space. With these tools at hand, it then develops the description and preparation of quantum states by means of the density matrix.

Chapter 4 discusses space–time symmetries in quantum physics, a first tour through the rotation group in nonrelativistic quantum mechanics and its representations, space reflection, and time reversal. It also addresses symmetry and antisymmetry of systems of a finite number of identical particles.

Chapter 5 which concludes Part I, is devoted to important practical applications of quantum mechanics, ranging from quantum information to time-independent as well as time-dependent perturbation theory, and to the description of many-body systems of identical fermions.

Chapter 6, the first of Part II, begins with an extended analysis of symmetries and symmetry groups in quantum physics. Wigner's theorem on the unitary or antiunitary realization of symmetry transformations is in focus here. There follows more material on the rotation group and its use in quantum mechanics, as well as a brief excursion to internal symmetries. The analysis of the Lorentz and Poincaré groups is taken up from the perspective of particle properties, and some of their unitary representations are worked out.

Chapter 7 describes the principles of canonical quantization of Lorentz covariant field theories and illustrates them with examples of the real and complex scalar fields, and the Maxwell field. A section on the interaction of quantum Maxwell fields with nonrelativistic matter illustrates the use of second quantization by a number of physically interesting examples. The specific problems related to quantized Maxwell theory are analyzed and solved in its covariant quantization and in an investigation of the state space of quantum electrodynamics.

Chapter 8 takes up scattering theory in a more general framework by defining the S-matrix and by deriving its properties. The optical theorem is proved for the general case of elastic and inelastic final states and formulae for cross sections and decay widths are worked out in terms of the scattering matrix.

Chapter 9 deals exclusively with the Dirac equation and with quantized fields describing spin-1/2 particles. After the construction of the quantized Dirac field and a first analysis of its interactions we also explore the question to which extent the Dirac equation may be useful as an approximate single-particle theory.

Chapter 10 describes covariant perturbation theory and develops the technique of Feynman diagrams and their translation to analytic amplitudes. A number of physically relevant tree processes of quantum electrodynamics are worked out in detail. Higher order terms and the specific problems they raise serve to introduce and to motivate the concepts of regularization and renormalization in a heuristic manner. Some prominent examples of radiative corrections serve to illustrate their relevance for atomic and particle physics as well as their physical interpretation. The chapter concludes with a short excursion into weak interactions, placing these in the framework of electroweak interactions.

The book covers material (more than) sufficient for two full courses and, thus, may serve as an accompanying textbook for courses on quantum mechanics and introductory quantum field theory. However, as the main text is largely self-contained and contains a considerable number of worked-out examples, it may also be useful for independent individual study. The choice of topics and their presentation closely follows a two-volume German text well established at German speaking universities. Much of the material was tested and fine-tuned in lectures I gave at Johannes Gutenberg University in Mainz. The book contains many exercises for some of which I included complete solutions or gave some hints. In addition, there are a number of appendices collecting or explaining more technical aspects. Finally, I included some historical remarks about the people who pioneered quantum mechanics and quantum field theory, or helped to shape our present understanding of quantum theory.[1]

I am grateful to the students who followed my courses and to my collaborators in research for their questions and critical comments some of which helped to clarify matters and to improve the presentation. Among the many colleagues and friends from whom I learnt a lot about the quantum world I owe special thanks to Martin Reuter who also read large parts of the original German manuscript, to Wolfgang Bulla who made constructive remarks on formal aspects of quantum mechanics, and to the late Othmar Steinmann from whom I learnt a good deal of quantum field theory during my years at ETH and PSI in Zurich.

The excellent cooperation with the people at Springer-Verlag, notably Dr. Thorsten Schneider and his crew, is gratefully acknowledged.

Mainz, April 2013 Florian Scheck

[1] I will keep track of possible errata on an Internet page attached to my home page. The latter can be accessed via http://wwwthep.uni-mainz.de/site. I will be grateful for hints at misprints or errors.

Contents

Part I From the Uncertainty Relation to Many-Body Systems

Part I
From the Uncertainty Relation to Many-Body Systems

Quantum Mechanics of Point Particles

1

▸ **Introduction** In developing quantum mechanics of pointlike particles one is faced with a curious, almost paradoxical situation: One seeks a more general theory which takes proper account of Planck's quantum of action h and which encompasses classical mechanics, in the limit $h \to 0$, but for which initially one has no more than the formal framework of canonical mechanics. This is to say, slightly exaggerating, that one tries to guess a theory for the hydrogen atom and for scattering of electrons by extrapolation from the laws of celestial mechanics. That this adventure eventually is successful rests on both *phenomenological* and on *theoretical* grounds.

On the phenomenological side we know that there are many experimental findings which cannot be interpreted classically and which in some cases strongly contradict the predictions of classical physics. At the same time this phenomenology provides hints at fundamental properties of radiation and of matter which are mostly irrelevant in macroscopic physics: Besides its classically well-known *wave* nature light also possesses *particle* properties; in turn massive particles such as the electron have both *mechanical* and *optical* properties. This discovery leads to one of the basic postulates of quantum theory, de Broglie's relation between the wave length of a monochromatic wave and the momentum of a massive or massless particle in uniform rectilinear motion.

Another basic phenomenological element in the quest for a "greater", more comprehensive theory is the recognition that measurements of *canonically conjugate variables* are always correlated. This is the content of Heisenberg's uncertainty relation which, qualitatively speaking, says that such observables can never be fixed simultaneously and with arbitrary accuracy. More quantitatively, it states in which way the uncertainties as determined by very many identical experiments are correlated by Planck's quantum of action. It also gives a first hint at the fact that observables of quantum mechanics must be described by noncommuting quantities.

A further, ingenious hypothesis starts from the wave properties of matter and the statistical nature of quantum mechanical processes: Max Born's postulate of interpreting the wave function as an amplitude (in general complex) whose absolute square represents a probability in the sense of statistical mechanics.

F. Scheck, *Quantum Physics*, DOI: 10.1007/978-3-642-34563-0_1,

Regarding the *theoretical* aspects one may ask why classical Hamiltonian mechanics is the right stepping-stone for the discovery of the farther reaching, more comprehensive quantum mechanics. To this question I wish to offer two answers:

(i) Our admittedly somewhat mysterious experience is that Hamilton's variational principle, if suitably generalized, suffices as a formal framework for every theory of fundamental physical interactions.

(ii) Hamiltonian systems yield a correct description of basic, *elementary* processes because they contain the principle of energy conservation as well as other conservation laws which follow from symmetries of the theory.

Macroscopic systems, in turn, which are not Hamiltonian, often provide no more than an effective description of a dynamics that one wishes to understand in its essential features but not in every microscopic detail. In this sense the equations of motion of the Kepler problem are elementary, the equation describing a body falling freely in the atmosphere along the vertical z is not because a frictional term of the form $-\kappa \dot{z}$ describes dissipation of energy to the ambient air, without making use of the dynamics of the air molecules. The first of these examples is Hamiltonian, the second is not.

In the light of these remarks one should not be surprised in developing quantum theory that not only the introduction of new, unfamiliar notions will be required but also that new questions will come up regarding the interpretation of measurements. The answers to these questions may suspend the separation of the measuring device from the object of investigation, and may lead to apparent paradoxes whose solution sometimes will be subtle. We will turn to these new aspects in many instances and we will clarify them to a large extent. For the moment I ask the reader for his/her patience and advise him or her not to be discouraged. If one sets out to develop or to discover a new, encompassing theory which goes beyond the familiar framework of classical, nonrelativistic physics, one should be prepared for some qualitatively new properties and interpretations of this theory. These features add greatly to both the fascination and the intellectual challenge of quantum theory.

1.1 Limitations of Classical Physics

There is a wealth of observable effects in the quantum world which cannot be understood in the framework of classical mechanics or classical electrodynamics. Instead of listing them all one by one I choose two characteristic examples that show very clearly that the description within classical physics is incomplete and must be supplemented by some new, fundamental principles. These are: the *quantization of atomic bound states* which does not follow from the Kepler problem for an electron in the field of a positive point charge, and the *electromagnetic radiation emitted by an electron bound in an atom* which, in a purely classical framework, would render atomic quantum states unstable.

When we talk about "classical" here and in the sequel, we mean every domain of physics where Planck's constant does not play a quantitative role and, therefore, can be neglected to a very good approximation.

Example 1.1 Atomic Bound States have Quantized Energies

The physically admissible bound states of the hydrogen atom or, for that matter, of a hydrogen-like atom, have discrete energies given by the formula

$$E_n = -\frac{1}{2n^2}\frac{Z^2 e^4}{\hbar^2}\mu \quad \text{with} \quad n = 1, 2, 3, \ldots \tag{1.1}$$

Here $n \in \mathbb{N}$ is called the *principal quantum number*, Z is the nuclear charge number (this is the number of protons contained in the nucleus), e is the elementary charge, $\hbar = h/(2\pi)$ is Planck's quantum h divided by (2π), and μ is the reduced mass of the system, here of the electron and the point-like nucleus. Upon introduction of Sommerfeld's *fine structure constant*,

$$\alpha := \frac{e^2}{\hbar c},$$

where c is the speed of light, formula (1.1) for the energy takes the form:

$$E_n = -\frac{1}{2n^2}(Z\alpha)^2 \mu c^2 . \tag{1.1'}$$

Note that the velocity of light drops out of this formula, as it should.[1]

In the Kepler problem of *classical* mechanics for an electron of charge $e = -|e|$ which moves in the field of a positive point charge $Z|e|$, the energy of a bound, hence finite orbit can take any negative value. Thus, two properties of (1.1) are particularly remarkable: Firstly, there exists a lowest value, realized for $n = 1$, all other energies are higher than $E_{n=1}$,

$$E_1 < E_2 < E_3 < \cdots .$$

Another way of stating this is to say that *the spectrum is bounded from below*. Secondly, the energy, as long as it is negative, can take only one of the values of the discrete series

$$E_n = \frac{1}{n^2} E_{n=1}, \quad n = 1, 2, \ldots .$$

For $n \longrightarrow \infty$ these values tend to the limit point 0.

Note that these facts which reflect and describe experimental findings (notably the Balmer series of hydrogen), cannot be understood in the framework of classical mechanics. A new, additional principle is missing that excludes all negative values of the energy except for those of (1.1). Nevertheless they are not totally

[1] The formula (1.1) holds in the framework of nonrelativistic kinematics, where there is no place for the velocity of light, or, alternatively, where this velocity can be assumed to be infinitely large. In the second expression (1.1′) for the energy the introduction of the constant c is arbitrary and of no consequence.

incompatible with the Kepler problem because, for large values of the principal quantum number n, the difference of neighbouring energies tends to zero like n^{-3},

$$E_{n+1} - E_n = \frac{2n+1}{2n^2(n+1)^2}(Z\alpha)^2\mu c^2 \sim \frac{1}{n^3} \quad \text{for} \quad n \to \infty .$$

In the limit of large quantum numbers the spectrum becomes nearly continuous.

Before continuing on this example we quote a few numerical values which are relevant for quantitative statements and for estimates, and to which we will return repeatedly in what follows.

Planck's constant has the physical dimension of an *action*, (energy × time), and its numerical value is

$$h = (6.62606957 \pm 0.00000029) \times 10^{-34}\,\text{J s} . \tag{1.2}$$

The reduced constant, h divided by (2π), which is mostly used in practical calculations[2] has the value

$$\hbar \equiv \frac{h}{2\pi} = 1.054 \times 10^{-34}\,\text{J s} . \tag{1.3}$$

As h carries a dimension, $[h] = E \cdot t$, it is called Planck's *quantum of action*. This notion is taken from classical canonical mechanics. We remind the reader that the product of a generalized coordinate q^i and its conjugate, generalized momentum $p_i = \partial L/\partial \dot{q}^i$, where L is the Lagrangian, always carries the dimension of an action,

$$[q^i\ p_i] = \text{energy} \times \text{time} ,$$

independently of how one has chosen the variables q^i and of which dimension they have.

A more tractable number for the atomic world is obtained from the product of $\hbar$ and the velocity of light

$$c = 2.99792458 \times 10^8\,\text{m s}^{-1} ; \tag{1.4}$$

the product has dimension (energy × length). Replacing the energy unit Joule by the million electron volt

$$1\,\text{MeV} = 10^6\,\text{eV} = (1.602176565 \pm 0.000000035) \times 10^{-13}\,\text{J}$$

and the meter by the femtometer, or Fermi unit of length, $1\,\text{fm} = 10^{-15}\,\text{m}$, one obtains a number that may be easier to remember,

$$\hbar c = 197.327\,\text{MeV fm} , \tag{1.5}$$

because it lies close to the rounded value 200 MeV fm.

Sommerfeld's fine structure constant has no physical dimension. Its value is

$$\alpha = (137.036)^{-1} = 0.00729735 . \tag{1.6}$$

---

[2] As of here we shorten the numerical values, for the sake of convenience, to their leading digits. In general, these values will be sufficient for our estimates. Appendix A.8 gives the precise experimental values, as they are known to date.

Finally, the mass of the electron, in these units, is approximately

$$m_e = 0.511 \text{ MeV}/c^2 . \tag{1.7}$$

As a matter of example let us calculate the energy of the ground state and the transition energy from the next higher state to the ground state for the case of the hydrogen atom ($Z = 1$). Since the mass of the hydrogen nucleus is about 1836 times heavier than that of the electron the reduced mass is nearly equal to the electron's mass,

$$\mu = \frac{m_e m_p}{m_e + m_p} \simeq m_e ,$$

and therefore we obtain

$$E_{n=1} = -2.66 \times 10^{-5} m_e c^2 = -13.6 \text{ eV}$$

and

$$\Delta E(n = 2 \to n = 1) = E_2 - E_1 = 10.2 \text{ eV} .$$

Note that E_n is proportional to the square of the nuclear charge number Z and linearly proportional to the reduced mass. In hydrogen-*like* atoms the binding energies increase with Z^2. If, on the other hand, one replaces the electron in hydrogen by a muon which is about 207 times heavier than the electron, all binding and transition energies will be larger by that factor than the corresponding quantities in hydrogen. Spectral lines of ordinary, electronic atoms which lie in the range of visible light, are replaced by X-rays when the electron is replaced by its heavier sister, the muon.

Imagine the lowest state of hydrogen to be described as a circular orbit of the classical Kepler problem and calculate its radius making use of the (classical) virial theorem ([Scheck (2010)], Sect. 1.31, (1.114)). The time averages of the kinetic and potential energies, $\langle T \rangle = -E$, $\langle U \rangle = 2E$, respectively, yield the radius R of the circular orbit as follows:

$$\langle U \rangle = -\frac{(Ze)e}{R} = -\frac{Z^2 e^4}{\hbar^2}\mu , \quad \text{hence} \quad R = \frac{\hbar^2}{Ze^2\mu} .$$

This quantity, evaluated for $Z = 1$ and $\mu = m_e$, is called the *Bohr Radius* of the electron.[3]

$$a_B := \frac{\hbar^2}{e^2 m_e} = \frac{\hbar c}{\alpha m_e c^2} . \tag{1.8}$$

It has the value

$$a_B = 5.292 \times 10^4 \text{ fm} = 5.292 \times 10^{-11} \text{ m} .$$

Taken literally, this classical picture of an electron orbiting around the proton, is not correct. Nevertheless, the number a_B is a measure for the spatial extension of the hydrogen atom. As we will see later, in trying to determine the position of

[3] One often writes a_∞, instead of a_B, in order to emphasize that in (1.8) the mass of the nuclear partner is assumed to be infinitely heavy as compared to m_e.

the electron (by means of a gedanken or thought experiment) one will find it with high probability at the distance a_B from the proton, i.e. from the nucleus in that atom. This distance is also to be compared with the spatial extension of the proton itself for which experiment gives about 0.86×10^{-15} m. This reflects the well-known statement that the spatial extension of the atom is larger by many orders of magnitude than the size of the nucleus and, hence, that the electron essentially moves *outside* the nucleus. Again, it should be remarked that the extension of the atom decreases with Z and with the reduced mass μ:

$$R \propto \frac{1}{Z\mu} .$$

To witness, if the electron is replaced by a muon, the hydrogen nucleus by a lead nucleus ($Z = 82$), then $a_B(m_\mu, Z = 82) = 3.12$ fm, a value which is comparable to or even smaller than the radius of the lead nucleus which is about 5.5 fm. Thus, the muon in the ground state of muonic lead penetrates deeply into the nuclear interior. The nucleus can no longer be described as a point-like charge and the dynamics of the muonic atom will depend on the spatial distribution of charge in the nucleus.

After these considerations and estimates we have become familiar with typical orders of magnitude in the hydrogen atom and we may now return to the discussion of the example: As the orbital angular momentum $\boldsymbol{\ell}$ is a conserved quantity every Keplerian orbit lies in a *plane*. This is the plane perpendicular to $\boldsymbol{\ell}$. Introducing polar coordinates (r, ϕ) in that plane a Lagrangian describing the Kepler problem reads

$$L = \frac{1}{2}\mu\dot{\boldsymbol{r}}^2 - U(r) = \frac{1}{2}\mu(\dot{r}^2 + r^2\dot{\phi}^2) - U(r) \quad \text{with} \quad U(r) = -\frac{e^2}{r} .$$

On a circular orbit one has $\dot{r} = 0$, $r = R = \text{const}$, and there remains only one time dependent variable, $q \equiv \phi$. Its canonically conjugate momentum is given by $p = \partial L/\partial\dot{q} = \mu r^2\dot{\phi}$. This is nothing but the modulus ℓ of the orbital angular momentum.

For a periodic motion in one variable the period and the surface enclosed by the orbit in phase space are related as follows. Let

$$F(E) = \oint p \mathrm{d}q$$

be the surface which is enclosed by the phase portrait of the orbit with energy E, (see Fig. 1.1), and let $T(E)$ be the period. Then one finds (see e.g. [Scheck (2010)], exercise and solution 2.2)

$$T(E) = \frac{\mathrm{d}}{\mathrm{d}E}F(E) .$$

The integral over the phase portrait of a circular orbit with radius R is easy to calculate,

$$F(E) = \oint p \mathrm{d}q = 2\pi\mu R^2\dot{\phi} = 2\pi\ell .$$

In order to express the right-hand side in terms of the energy, one makes use of the principle of energy conservation,

$$\frac{1}{2}\mu R^2\dot{\phi}^2 + U(R) = \frac{\ell^2}{2\mu R^2} + U(R) = E ,$$

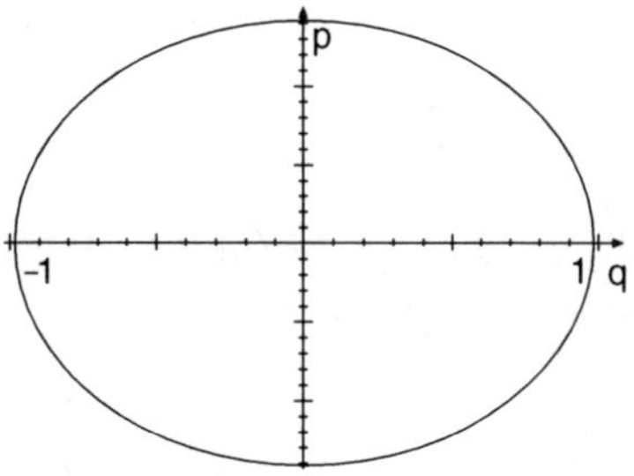

Fig. 1.1 Phase portrait of a periodic motion in one dimension. At any time the mass point has definite values of the coordinate q and of the momentum p. It moves, in a clock-wise direction, along the curve which closes after one revolution

which, upon solving for ℓ, yields

$$F(E) = 2\pi\ell = 2\pi\sqrt{2\mu R^2(E-U)}\,.$$

The derivative with respect to E is given by

$$T(E) = \frac{\mathrm{d}F}{\mathrm{d}E} = 2\pi\frac{\mu R^2}{\sqrt{2\mu R^2(E-U)}} = 2\pi\frac{\mu R^2}{\ell} = \frac{2\pi}{\dot{\phi}}\,.$$

This is correct, though not a surprise! However, making use of the virial theorem $2E = \langle U\rangle = -e^2/R$, one obtains a nontrivial result, viz.

$$T(E) = 2\pi\sqrt{\mu}R^{3/2}/e\,, \quad \text{or} \quad \frac{R^3}{T^2} = \frac{e^2}{(2\pi)^2\mu}\,,$$

which is precisely Kepler's third law ([Scheck (2010)], Sect. 1.7.2).

There is an argument of plausibility that yields the correct energy formula (1.1): First, note that the transition energy, upon division by h, $(E_{n+1} - E_n)/h$, has the dimension of a frequency, viz. 1/time. Furthermore, for large values of n one has

$$(E_{n+1} - E_n) \simeq \frac{1}{n^3}(Z\alpha)^2\mu c^2 = \frac{\mathrm{d}E_n}{\mathrm{d}n}\,.$$

If one postulates that the frequency $(E_{n+1} - E_n)/h$, in the limit $n \longrightarrow \infty$, goes over into the classical frequency $\nu = 1/T$,

$$\lim_{n\to\infty}\frac{1}{h}\frac{\mathrm{d}E_n}{\mathrm{d}n} = \frac{1}{T}\,, \tag{1.9}$$

one has $T(E)\mathrm{d}E = h\mathrm{d}n$. Integrating both sides of this relation gives

$$F(E) = \oint p\mathrm{d}q = hn\,. \tag{1.10}$$

Equation (1.9) is an expression of N. Bohr's *correspondence principle*. This principle aims at establishing relations between quantum mechanical quantities and their classical counterparts. Equation (1.10), promoted to the status of a rule, was called quantization rule of Bohr and Sommerfeld and was formulated before quantum mechanics proper was developed. For circular orbits this rule yields

$$2\pi \mu R^2 \dot{\phi} = hn \, .$$

By equating the attractive electrical force to the centrifugal force, i.e. setting $e^2/R^2 = \mu R \dot{\phi}^2$, the formula

$$R = \frac{h^2 n^2}{(2\pi)^2 \mu e^2}$$

for the radius of the circular orbit with principal quantum number n follows. This result does indeed yield the correct expression (1.1) for the energy.[4]

Although the condition (1.10) is successful only in the case of hydrogen and is not useful to show us the way from classical to quantum mechanics, it is interesting in its own right because it introduces a new principle: It selects those orbits in phase space, among the infinite set of all classical bound states, for which the closed contour integral $\oint p\mathrm{d}q$ is an integer multiple of Planck's quantum of action h. Questions such as why this is so, or, in describing a bound electron, whether or not one may really talk about *orbits*, remain unanswered.

Example 1.2 A Bound Electron Radiates

The hydrogen atom is composed of a positively charged proton and a negatively charged electron, with equal and opposite values of charge. Even if we accept a description of this atom in analogy to the Kepler problem of celestial mechanics there is a marked and important difference. While two celestial bodies (e.g. the sun and a planet, or a double star) interact only through their gravitational forces, their electric charges (if they carry any net charge at all) playing no role in practice, proton and electron are bound essentially only by their Coulomb interaction. In the assumed Keplerian motion electron and proton move on two ellipses or two circles about their common center of mass which are geometrically similar (see [Scheck (2010)], Sect. 1.7.2). On their respective orbits both particles are subject to (positive or negative) accelerations in radial and azimuthal directions. Due to the large ratio of their masses, $m_\mathrm{p}/m_\mathrm{e} = 1836$, the proton will move very little so that its acceleration may be neglected as compared to the one of the electron. It is intuitively plausible that the electron in its periodic, accelerated motion will act as a source for electromagnetic radiation and will loose energy by emitting this radiation.[5] Of course, this contradicts the quantization of the energies of bound states because these can only assume the magic values (1.1). However, as we have assumed again classical physics to be applicable, we may

[4] This is also true for elliptic Kepler orbits in the classical model of the hydrogen atom. It fails, however, already for helium ($Z = 2$).

[5] For example, an electron moving on an elliptic orbit with large excentricity, seen from far away, acts like a small linear antenna in which charge moves periodically up and down. Such a micro-emitter radiates electromagnetic waves and, hence, radiates energy.

estimate the order of magnitude of the energy loss by radiation. This effect will turn out to be dramatic.

At this point, and this will be the only instance where I do so, I quote some notions of electrodynamics, making them plausible but without deriving them in any detail. The essential steps leading to the required formulae for electromagnetic radiation should be understandable without a detailed knowledge of Maxwell's equations. If the arguments sketched here remain unaccessible the reader may turn directly to the results (1.21), (1.22), and (1.23), and may return to their derivation later, after having studied electrodynamics.

Electrodynamics is invariant under Lorentz transformations, it is not Galilei invariant (see for example [Scheck (2010)], Chap. 4, and, specifically, Sect. 4.9.3). In this framework the current density $j^\mu(x)$ is a four-component vector field whose time component ($\mu = 0$) describes the charge density $\varrho(x)$ as a function of time and space coordinates x, and whose spatial components ($\mu = 1, 2, 3$) form the electric current density $\boldsymbol{j}(x)$. Let t be the coordinate time, and $\boldsymbol{x}$ the point in space where the densities are felt or measured, defined with respect to an inertial system of reference **K**. We then have

$$x = (ct, \boldsymbol{x}), \quad j^\mu(x) = [c\varrho(t, \boldsymbol{x}), \boldsymbol{j}(t, \boldsymbol{x})] .$$

Let the electron move along the world line $r(\tau)$ where τ is the Lorentz *invariant* proper time and where r is the four-vector which describes the particle's orbit in space and time. In the reference system **K** one has

$$r(\tau) = [ct_0, \boldsymbol{r}(t_0)] .$$

The four-velocity of the electron, $u^\mu(\tau) = \mathrm{d}r^\mu(\tau)/\mathrm{d}\tau$, is normalized such that its square equals the square of the velocity of light, $u^2 = c^2$. In the given system of reference **K** one has

$$c\mathrm{d}\tau = c\mathrm{d}t\sqrt{1-\beta^2} \quad \text{with} \quad \beta = |\dot{\boldsymbol{x}}|/c ,$$

whereas the four-velocity takes the form

$$u^\mu = (c\gamma, \gamma\boldsymbol{v}(t)) , \quad \text{where} \quad \gamma = \frac{1}{\sqrt{1-\beta^2}} .$$

The motion of the electron generates an electric charge density and a current density, seen in the system **K**. Making use of the δ-distribution these can be written as

$$\varrho(t, \boldsymbol{x}) = e\,\delta^{(3)}[\boldsymbol{x} - \boldsymbol{r}(t)] ,$$
$$\boldsymbol{j}(t, \boldsymbol{x}) = e\boldsymbol{v}(t)\,\delta^{(3)}[\boldsymbol{x} - \boldsymbol{r}(t)]$$

where $\boldsymbol{v} = \dot{\boldsymbol{r}}$. When expressed in covariant form the same densities read

$$j^\mu(x) = ec\int \mathrm{d}\tau\; u^\mu(\tau)\,\delta^{(4)}[x - r(\tau)] . \tag{1.11}$$

In order to check this, evaluate the integral over proper time in the reference system **K** and isolate the one-dimensional δ-distribution which refers to the time components. With $\mathrm{d}\tau = \mathrm{d}t/\gamma$, with $\delta[x^0 - r^0(\tau)] = \delta[c(t-t_0)] = \delta(t-t_0)/c$, and making use of the decomposition of the four-velocity given above one obtains

$$j^0 = ec \int \frac{\mathrm{d}t_0}{\gamma} c\gamma \frac{1}{c}\, \delta(t-t_0)\, \delta^{(3)}[\boldsymbol{x} - \boldsymbol{r}(t_0)] = c\varrho(t, \boldsymbol{x}),$$

$$\boldsymbol{j} = ec \int \frac{\mathrm{d}t_0}{\gamma} \gamma \boldsymbol{v}(t_0) \frac{1}{c}\, \delta(t-t_0)\, \delta^{(3)}[\boldsymbol{x} - \boldsymbol{r}(t_0)] = \boldsymbol{j}(t, \boldsymbol{x}).$$

The calculation then proceeds as follows: Maxwell's equations are solved after inserting the current density (1.11) as the inhomogeneous, or source, term, thus yielding a four-potential $A^\mu = (\Phi, \boldsymbol{A})$. This, in turn, is used to calculate the electric and magnetic fields by means of the formulae

$$\boldsymbol{E} = -\nabla\Phi - \frac{1}{c}\frac{\partial \boldsymbol{A}}{\partial t}, \tag{1.12}$$

$$\boldsymbol{B} = \nabla \times \boldsymbol{A}. \tag{1.13}$$

I skip the method of solution for A^μ and go directly to the result which is

$$A^\mu(x) = 2e \int \mathrm{d}\tau\, u^\mu(\tau) \Theta[x^0 - r^0(\tau)]\, \delta^{(1)}\{[x - r(\tau)]^2\}. \tag{1.14}$$

Here $\Theta(x)$ is the step, or Heaviside, function,

$$\Theta[x^0 - r^0(\tau)] = 1 \quad \text{for} \quad x^0 = ct > r^0 = ct_0,$$

$$\Theta[x^0 - r^0(\tau)] = 0 \quad \text{for} \quad x^0 < r^0.$$

The δ-distribution in (1.14) refers to a scalar quantity, its argument being the invariant scalar product

$$[x - r(\tau)]^2 = [x^0 - r^0(\tau)]^2 - [\boldsymbol{x} - \boldsymbol{r}(t_0)]^2.$$

It guarantees that the action observed in the world point $(x^0 = ct, \boldsymbol{x})$ lies on the light cone of its cause, i.e. of the electron at the space point $\boldsymbol{r}(t_0)$, at time t_0. This relationship is sketched in Fig. 1.2. Expressed differently, the electron which at time $t_0 = r^0/c$ passes the space point $\boldsymbol{r}$, at time t causes a four-potential at the point $\boldsymbol{x}$ such that $r(\tau)$ and x are related by a signal which propagates with the speed of light. The step function whose argument is the difference of the two time components, guarantees that this relation is *causal*. The cause "electron in $\boldsymbol{r}$ at time t_0" comes first, the effect "potentials $A^\mu = (\Phi, \boldsymbol{A})$ in $\boldsymbol{x}$ at the later time $t > t_0$" comes second. This shows that the formula (1.14) is not only plausible but in fact simple and intuitive, even though we have not derived it here.[6]

The integration in (1.14) is done by inserting

$$(x - r)^2 = (x^0 - r^0)^2 - (\boldsymbol{x} - \boldsymbol{r})^2 = c^2(t - t_0)^2 - (\boldsymbol{x} - \boldsymbol{r})^2$$

---

[6] The distinction between forward and backward light cone, i.e. between future and past, is invariant under the proper, orthochronous Lorentz group. As proper time τ is an invariant, the four-potential A^μ inherits the vector nature of the four-velocity u^μ.

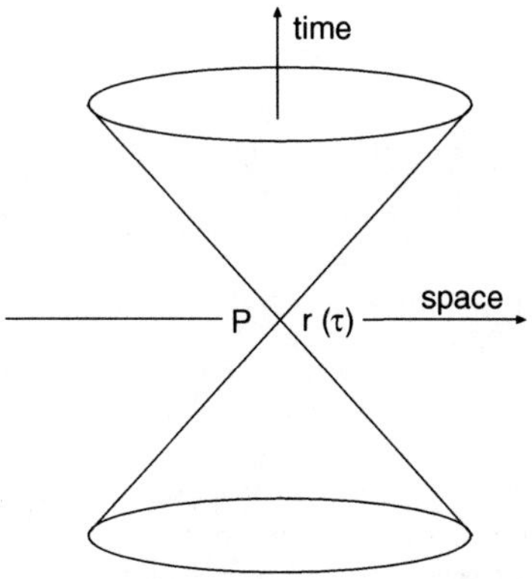

Fig. 1.2 Light cone of a world point in a symbolic representation of space $\mathbb{R}^3$ (plane perpendicular to the ordinate) and of time (ordinate). Every causal action that emanates from the point $r(\tau)$ with the velocity of light, lies on the *upper* part of the cone (forward cone of P)

and by making use of the general formula for δ-distributions

$$\delta[f(y)] = \sum_i \frac{1}{|f'(y_i)|}\delta(y - y_i), \quad \{y_i\text{: } \textit{single} \text{ zeroes of } f(y)\} . \tag{1.15}$$

In the case at hand $f(\tau) = [x - r(\tau)]^2$ and $\mathrm{d}f(\tau)/\mathrm{d}\tau = \mathrm{d}[x - r(\tau)]^2/\mathrm{d}\tau = -2[x - r(\tau)]_\alpha u^\alpha(\tau)$. As can be seen from Fig. 1.3 the point x of observation is lightlike relative to $r(\tau_0)$, the world point of the electron. Therefore $A^\mu(x)$ of (1.14) can be written as

$$A^\mu(x) = e \left. \frac{u^\mu(x)}{u \cdot [x - r(\tau)]} \right|_{\tau=\tau_0} . \tag{1.16}$$

In order to understand better this expression we evaluate it in the frame of reference **K**. The scalar product in the denominator reads

$$u \cdot [x - r(\tau)] = \gamma c^2(t - t_0) - \gamma \boldsymbol{v} \cdot (\boldsymbol{x} - \boldsymbol{r}) .$$

Let $\hat{\boldsymbol{n}}$ be the unit vector in the direction of the vector $\boldsymbol{x} - \boldsymbol{r}(\tau)$, and let $|\boldsymbol{x} - \boldsymbol{r}(\tau_0)| =: R$ be the distance between source and point of observation. As $[x - r(\tau)]^2$ must be zero, we conclude $x^0 - r^0(\tau_0) = R$, so that

$$u \cdot [x - r(\tau_0)] = c\gamma R \left(1 - \frac{1}{c}\boldsymbol{v} \cdot \hat{\boldsymbol{n}}\right) ,$$

while $A^\mu(x) = [\Phi(x), \boldsymbol{A}(x)]$ of (1.16) becomes

$$\Phi(t, \boldsymbol{x}) = \left. \frac{e}{R(1 - \boldsymbol{v} \cdot \hat{\boldsymbol{n}}/c)} \right|_{\text{ret}} , \tag{1.17}$$

$$\boldsymbol{A}(t, \boldsymbol{x}) = \left. \frac{e\boldsymbol{v}/c}{R(1 - \boldsymbol{v} \cdot \hat{\boldsymbol{n}}/c)} \right|_{\text{ret}} .$$

The notation "ret" emphasizes that the time t and the time t_0, when the electron had the distance R from the observer, are related by $t = t_0 + R/c$. The action of

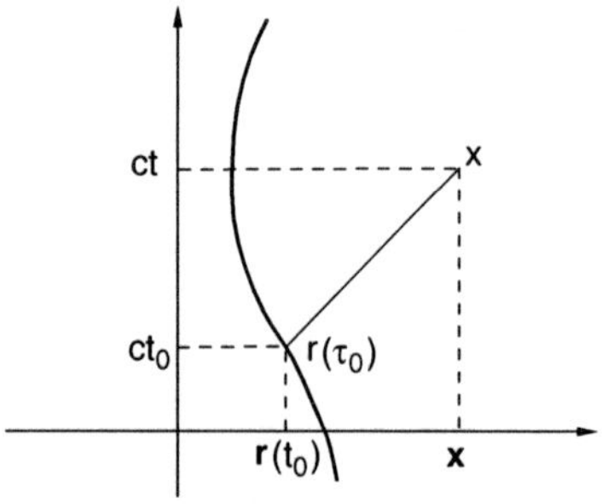

Fig. 1.3 The electron moves along a timelike world line (*full curve*). The tangent to the curve is timelike everywhere which means that the electron moves at a speed whose modulus is smaller than the speed of light. Its radiation at $r(\tau_0) = (ct_0, \boldsymbol{r}(t_0))$ reaches the observer at $\boldsymbol{x}$ after the time of flight $t - t_0 = |\boldsymbol{x} - \boldsymbol{r}(t_0)|/c$

the electron at the point where an observer is located arrives with a delay that is equal to the time-of-flight (R/c).[7]

The potentials (1.16) or (1.17) are called *Liénard-Wiechert potentials.*

From here on there are two equivalent methods of calculating the electric and magnetic fields proper. One is to make use of the expressions (1.17) and to calculate $\boldsymbol{E}$ and $\boldsymbol{B}$ by means of (1.12) and (1.13), keeping track of the retardation required by (1.17).[8] As an alternative one returns to the covariant expression (1.16) of the vector potential and calculates the field strength tensor from it,

$$F^{\mu\nu}(x) = \partial^\mu A^\nu - \partial^\nu A^\mu , \quad \left(\partial^\mu \equiv \frac{\partial}{\partial x_\mu}\right) .$$

Referring to a specific frame of reference the fields are obtained from

$$E^i = F^{i0} , \quad B^1 = F^{32} \quad \text{(and cyclic permutations)} .$$

The result of this calculation is (see e.g. [Jackson (1999)])

$$\boldsymbol{E}(t, \boldsymbol{x}) = e \left. \frac{\hat{\boldsymbol{n}} - \boldsymbol{v}/c}{\gamma^2 (1 - \boldsymbol{v}\cdot\hat{\boldsymbol{n}}/c)^3 R^2}\right|_{\text{ret}} + \frac{e}{c^2} \left. \frac{\hat{\boldsymbol{n}} \times [(\hat{\boldsymbol{n}} - \boldsymbol{v}/c) \times \dot{\boldsymbol{v}}]}{(1 - \boldsymbol{v}\cdot\hat{\boldsymbol{n}}/c)^3 R}\right|_{\text{ret}} \tag{1.18}$$

$$\equiv \boldsymbol{E}_{\text{stat}} + \boldsymbol{E}_{\text{acc}} , \tag{1.19}$$

$$\boldsymbol{B}(t, \boldsymbol{x}) = \left. \left(\hat{\boldsymbol{n}} \times \boldsymbol{E}\right)\right|_{\text{ret}} . \tag{1.20}$$

As usual β and γ are defined as $\beta = |\boldsymbol{v}|/c$, $\gamma = 1/\sqrt{1-\beta^2}$. As above, the notation "ret" stands for the prescription $t = t_0 + R/c$. The first term in (1.18) is a static field in the sense that it exists even when the electron moves with *constant* velocity. It is the second term $\boldsymbol{E}_{\text{acc}}$[9] which is relevant for the question from which

[7] *Retarded, retardation* derive from the French *le retard* = the delay.

[8] This calculation can be found e.g. in the textbook by Landau and Lifshitz, Vol. 2, Sect. 63 [Landau, Lifshitz (1987)].

[9] "acc" is a short-hand for *acceleration*, (or else the French *accélération*).

we started. Only when the electron is *accelerated* will there be nonvanishing radiation.

The flux of energy that is related to that radiation is described by the Poynting vector field

$$S(t, x) = \frac{c}{4\pi} E(t, x) \times B(t, x),$$

this being the second formula, besides (1.14), that I take from electrodynamics and that I do not derive here. In order to understand this formula, one may think of an electromagnetic, monochromatic wave in vacuum whose electric and magnetic fields are perpendicular to each other and to the direction of propagation $\hat{k}$. It describes the amount of energy which flows in the direction of $\hat{k}$, per unit of time.

Imagine a sphere of radius R, at a given time t_0, with the electron at its center. The Poynting vector S is used for the calculation of the power radiated into the cone defined by the differential solid angle $d\Omega$ as follows

$$dP = |S| R^2 d\Omega .$$

The velocity of the electron in the circular orbit that we assumed for the ground state of hydrogen, has the modulus

$$\frac{|v|}{c} = \sqrt{\frac{2E_{\text{kin}}}{m_e c^2}} = \alpha = 0.0073$$

(see (1.6)) and, hence, lies well below the velocity of light. In the approximation $v^2 \ll c^2$ the leading term in (1.18) is

$$E \simeq \frac{e}{c^2} \left. \frac{\hat{n} \times (\hat{n} \times \dot{v})}{R} \right|_{\text{ret}},$$

and, making use of the identity $a \times (b \times c) = b(a \cdot c) - c(a \cdot b)$, the Poynting vector field simplifies to

$$S \simeq \frac{c}{4\pi} E^2 \hat{n} .$$

The power radiated into the solid angle $d\Omega$ is approximately

$$\frac{dP}{d\Omega} \simeq \frac{c}{4\pi} R^2 E^2 \simeq \frac{e^2}{4\pi c^3} |\hat{n} \times (\hat{n} \times \dot{v})|^2 = \frac{e^2}{4\pi c^3} \dot{v}^2 \sin^2\theta .$$

Here θ is the angle between $\dot{v}$ and $\hat{n}$. Integration over the complete solid angle,

$$\int d\Omega \ldots = \int_0^{2\pi} d\phi \int_0^{\pi} \sin\theta d\theta \ldots = \int_0^{2\pi} d\phi \int_{-1}^{+1} dz \ldots , \quad (z = \cos\theta),$$

gives the result

$$P = \int d\Omega \frac{dP}{d\Omega} \simeq \frac{2}{3} \frac{e^2}{c^3} \dot{v}^2 . \tag{1.21}$$

This formula is perfectly suited for our estimates. The exact, relativistic formula deviates from it by terms of the order of v^2/c^2. When there is no acceleration this latter formula gives $P = 0$, too.

On a circular orbit of radius a_B we have

$$|\dot{\boldsymbol{v}}| = \frac{v^2}{a_B} = a_B \omega^2 ,$$

where ω is the circular frequency. The period of the orbit is

$$T = \frac{2\pi}{\omega} = \frac{2\pi a_B}{|\boldsymbol{v}|} = \frac{1}{\alpha}\frac{2\pi a_B}{c} = 1.52 \times 10^{-16}\ \text{s} . \tag{1.22}$$

These data are used to calculate the fraction of the binding energy which is radiated after one complete revolution,

$$\frac{PT}{|E_{n=1}|} = \frac{8\pi}{3}\alpha^3 = 3.26 \times 10^{-6} . \tag{1.23}$$

Thus, after one period the (classical) electron has lost the energy

$$3.26 \times 10^{-6}\ |E_{n=1}|$$

to the radiation field. Referring to (1.22) this means that in a very short time the electron lowers its binding energy, the radius of its orbit shrinks, and, eventually, the electron falls into the nucleus (i.e. the proton, in the case of hydrogen).

Without performing any dedicated measurement, we realize that the age of the terrestrial oceans provides evidence that the hydrogen atom must be extraordinarily stable. Like in the first example classical physics makes a unique and unavoidable prediction which is in marked contradiction to the observed stability of the hydrogen atom.

Quantum theory resolves the failures of classical physics that we illustrated by the two examples described above, in two major steps both of which introduce important new principles that we shall develop one by one in subsequent chapters.

In the first step one learns the quantum mechanics of *stationary systems*. Among these the energy spectrum of the hydrogen atom will provide a key example. For a given, time independent, Hamiltonian system one constructs a quantum analogue of the Hamiltonian function which yields the admissible values of the energy. The energy spectrum of hydrogen, as an example, will be found to be given by

$$\left\{ E_n = -\frac{1}{2n^2}\frac{e^4}{\hbar^2}\mu, \quad (n = 1, 2, 3, \ldots), \text{ and all } \ E \in [0, \infty) \right\} . \tag{1.24}$$

The first group (to the left) describes the bound states and corresponds to the classical finite, circular and elliptic orbits of the Kepler problem. In the limit $n \to \infty$ these energies tend to $E = 0$. Every state n has a well-defined and sharp value of the energy.

The second group (to the right) corresponds to the classical scattering orbits, i.e. the hyperbolic orbits which come in from spatial infinity and return to infinity. In quantum mechanics, too, the states of this group describe scattering states of the electron-proton system where the electron comes in with initial momentum $|\boldsymbol{p}|_\infty = \sqrt{2\mu E}$ along the direction $\hat{\boldsymbol{p}}$. However, no definite trajectory can be assigned to such a state.

In the second step one learns how to couple a stationary system of this kind to the radiation field and to understand its behaviour when its energy is lowered, or increased, by emission or absorption of photons, respectively. All bound states in (1.24), except for the lowest state with $n = 1$, become unstable. They are taken to lower states of the same series, predominantly through emission of photons, and eventually land in the stable ground state. In this way the characteristic *spectral lines* of atoms were understood that had been measured and tabulated long before quantum mechanics was developed.

The possibility for an initial state "i", by emission of one or more photons, to go over into a final state "f" not only renders that state unstable but gives it a broadening, i.e. an uncertainty in energy which is the larger, the faster the decay will take place. If τ denotes the average lifetime of the state and if τ is given in seconds, the line broadening is given by the formula

$$\Gamma = \frac{\hbar}{\tau} = \frac{\hbar c}{\tau c} = \left(6.58 \times 10^{-16}/\tau\right) \mathrm{eV}\,, \tag{1.25}$$

a formula to which we return later in more detail. The ground state, being absolutely stable, is the only bound state which keeps the sharp energy eigenvalue that one found in the first step, in solving the original stationary problem without taking account of the radiation field.

1.2 Heisenberg's Uncertainty Relation for Position and Momentum

Consider a Hamiltonian system of classical mechanics, described by the Hamiltonian $H = T + U$, with U an attractive potential. The fact that after its translation to quantum mechanics this system exhibits an energy spectrum bounded from below, $E \geq E_0$ where E_0 is the energy of the ground state, is a consequence of a fundamental principle of quantum theory: *Heisenberg's uncertainty relations for canonically conjugate variables*. We discuss this principle first on an example but return to it in a more general framework and a more precise formulation in later sections when adequate mathematical tools will be available.

Dynamical quantities of *classical* mechanics, i.e. physical observables in a given system, are described by real, in general smooth functions $F(q, p)$ on phase space. Examples are the coordinates q^i, the components p_j of momentum of a particle, the components ℓ_i or the square $\boldsymbol{\ell}^2$ of its orbital angular momentum, the kinetic energy T, the potential energy U, etc. Expressed somewhat more formally any such *observable* maps domains of phase space onto the reals,

$$F(q^1, \ldots, q^f, p_1, \ldots, p_f) : \mathbb{P} \longrightarrow \mathbb{R}\,. \tag{1.26}$$

For instance, the function q^i maps the point $(q^1, \ldots, q^f, p_1, \ldots, p_f) \in \mathbb{P}$ to its i-th coordinate $q^i \in \mathbb{R}$.

Real functions on a space can be added, they can be multiplied, and they can be multiplied with real numbers. The result is again a function. The product $F \cdot G$ of two functions F and G is the same as $G \cdot F$, with the order reversed. Thus, the set of all real function on phase space $\mathbb{P}$ is an *algebra*. As the product obeys the rule

$$F \cdot G - G \cdot F = 0 \tag{1.27}$$

this algebra is said to be *commutative*. Indeed, the left-hand side contains the commutator of F and G whose general definition reads

$$[A, B] := A \cdot B - B \cdot A \,. \tag{1.28}$$

Expressed in more physical terms, the relation (1.27) says that two dynamical quantities F and G can have well-determined values simultaneously and, hence, can be measured simultaneously. To quote an example in celestial mechanics, the three coordinates as well as the three components of momentum of a body can be measured, or can be predicted, from the knowledge of its orbit in space. This statement which seems obvious in the realm of classical physics, no longer holds in those parts of physics where Planck's constant is relevant. This will be the case if our experimental apparatus allows to resolve volumina in phase space for which the products $\Delta q^i \, \Delta p_i$ of side lengths in the direction of q^i and in the direction of the conjugate variable p_i are no longer large as compared to $\hbar$. In general and depending on the state of the system, observables will exhibit an uncertainty, a "diffuseness". Measurements of two different observables, and this is the essential and new property of quantum theory, may exclude each other. In such cases the uncertainty in one is correlated with the uncertainty in the other observable. In the limiting cases where one of them assumes a sharp, fixed value, the other is completely undetermined.

1.2.1 Uncertainties of Observables

An observable may be known only within some uncertainty, which is to say that in repeated measurements a certain weighted distribution of values is found. This happens in *classical* physics whenever one deals with a system of *many* particles about which one has only limited information. An example is provided by Maxwell's distribution of velocities in a swarm of particles described by the normalized probability distribution

$$\mathrm{d}w(p) = \frac{4\pi}{(2\pi mkT)^{3/2}} e^{-\beta p^2/2m} p^2 \mathrm{d}p \quad \text{with} \quad \beta = \frac{1}{kT} \,.$$

In this expression k denotes Boltzmann's constant and T is the temperature. This distribution gives the differential probability for measuring the modulus $p \equiv |\boldsymbol{p}|$ of the momentum in the interval $(p, p + \mathrm{d}p)$. It is normalized to 1, in accordance with the statement that whatever the value of p is that is obtained in an individual measurement, it lies somewhere in the interval $[0, \infty)$.

More generally, let F be an observable whose measurement yields a real number. The measured values f may lie in a continuum, say in the interval $[a, b]$ of the real axis. In the example above this is the interval $[0, \infty)$. The system (we think here of a many-body system as in the example) is in a given state that we describe by the normalized distribution

$$\varrho(f) \quad \text{with} \quad \int_a^b \varrho(f)\mathrm{d}f = 1 \tag{1.29}$$

In cases where the measured values of F are discrete and belong to the series of ordered real numbers $f_1, f_2, \ldots$, the distribution (1.29) is replaced by a series of probabilities $w_1, w_2, \ldots$,

$$w_i \equiv w(f_i) \quad \text{with the condition} \quad \sum_{i=1} w_i = 1\,, \tag{1.30}$$

where $w_i \equiv w(f_i)$ is the probability to find the value f_i in a measurement of F. The state of the system is defined with reference to the observable F and is described by the distribution $\varrho(f)$ or by the set of probabilities $\{w(f_i)\}$, respectively.

Before moving on we note that this picture, though strongly simplified, shows all features which are essential for our discussion. In general one will need more than one observable, the distribution function will thus depend on more than one variable. For instance, in a system of N particles the coordinates and the momenta

$$\left(\boldsymbol{x}^{(1)}, \ldots, \boldsymbol{x}^{(N)}; \boldsymbol{p}^{(1)}, \ldots, \boldsymbol{p}^{(N)}\right)$$
$$\equiv \left(q^1, q^2, \ldots, q^{3N}; p_1, p_2, \ldots, p_{3N}\right)$$

are the relevant observables which replace the abstract F above, and which are used to define the state. The distribution function

$$\varrho(q^1, q^2, \ldots, q^{3N}; p_1, p_2, \ldots, p_{3N})$$

is now a function of $6N$ variables.

Now, let G be another observable, evaluated as a function of the values of F. For the example of the N body system this could be the Hamiltonian function

$$H(q^1, q^2, \ldots, q^{3N}; p_1, p_2, \ldots, p_{3N})$$

which is evaluated on phase space and which yields the energy of the system. In our simplified description we write $G(f)$ for the value of the observable G at f.

A quantitative measure for the uncertainty of the measured values of G is obtained by calculating the *mean square deviation*, or *standard deviation*, that is, the average of the square of the difference of G and its mean value $\langle G\rangle$, viz.

$$(\Delta G)^2 := \left\langle \{G - \langle G\rangle\}^2 \right\rangle = \left\langle G^2 \right\rangle - \langle G\rangle^2\,. \tag{1.31}$$

The second form on the right-hand side is obtained by expanding the curly brackets

$$\left\langle \{G - \langle G\rangle\}^2 \right\rangle = \left\langle G^2 \right\rangle - 2\langle G\rangle\langle G\rangle + \langle G\rangle^2\,.$$

Depending on whether we deal with a continuous or a discrete distribution of values for F we have

$$\langle G\rangle = \int \varrho(f)G(f)\mathrm{d}f \quad \text{or} \quad \langle G\rangle = \sum_i w_i G(f_i)\,. \tag{1.32}$$

Inserting into (1.31) one obtains the expressions

$$(\Delta G)^2 = \int \left(G(f) - \int G(f')\varrho(f')\mathrm{d}f' \right)^2 \varrho(f)\mathrm{d}f$$

for the continuous distribution, and

$$(\Delta G)^2 = \sum_i w_i \left(G(f_i) - \sum_j w_j G(f_j) \right)^2$$

for the discrete case, respectively.

We summarize this important concept:

Definition 1.1

The *uncertainty*, or *standard deviation*, of an observable in a given state is defined to be the square root of the mean square deviation (1.31),

$$\boxed{\Delta G := \sqrt{\langle G^2 \rangle - \langle G \rangle^2}}\ . \tag{1.33}$$

If the observable F takes only one single value f_0, i.e when

$$\varrho(f) = \delta(f - f_0) \quad \text{or} \quad w_i \equiv w(f_i) = \delta_{i0}\ , \tag{1.34}$$

then the uncertainty (1.33) is equal to zero. In all other cases ΔG has a nonvanishing, positive value. As an example, we calculate the standard deviation of the kinetic energy $T_{\mathrm{kin}} = p^2/2m$ for Maxwell's distribution of momenta given above. Substituting $x = p\sqrt{\beta/(2m)}$ the distribution becomes

$$\mathrm{d}w(x) = \frac{4}{\sqrt{\pi}} x^2 \mathrm{e}^{-x^2} \mathrm{d}x\ .$$

This function is shown in Fig. 1.4. The mean values of T_{kin}^2 and of T_{kin} are calculated as follows

$$\left\langle T_{\mathrm{kin}}^2 \right\rangle = \frac{4}{\beta^2\sqrt{\pi}} \int_0^\infty x^6 \mathrm{e}^{-x^2} \mathrm{d}x = \frac{15}{4}\frac{1}{\beta^2} = \frac{15}{4}(kT)^2\ ,$$

$$\langle T_{\mathrm{kin}} \rangle = \frac{4}{\beta\sqrt{\pi}} \int_0^\infty x^4 \mathrm{e}^{-x^2} \mathrm{d}x = \frac{3}{2\beta} = \frac{3}{2}kT\ .$$

From these expressions the standard deviation of the kinetic energy is obtained

$$\Delta T_{\mathrm{kin}} \equiv \Delta \left(\frac{p^2}{2m} \right) = \sqrt{\frac{3}{2}}\, kT\ .$$

It becomes the larger the higher the temperature.

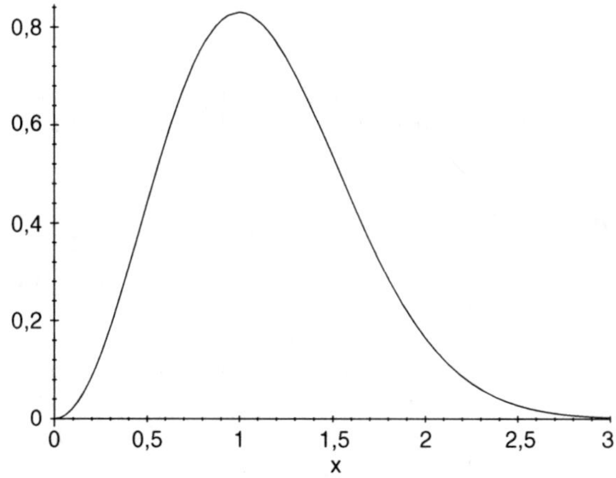

Fig. 1.4 Classical Maxwell distribution of velocities $4/\sqrt{\pi}x^2 \exp(-x^2)$

1.2.2 Quantum Mechanical Uncertainties of Canonically Conjugate Variables

After this excursion to classical mechanics of N body systems we return to quantum mechanics of a *single* particle. In classical mechanics the state of the particle can be characterized, at a given time, by sharp, well-defined values for all its coordinates q^i and all components p_k of its momentum. In quantum mechanics these observables are subject to uncertainties Δq^i and Δp_k, respectively, which obey a fundamental inequality. Let the uncertainty, or standard deviation, be defined as in (1.33) of Definition 1.1 by the difference of the mean value of the square and of the square of the mean value,

$$\Delta q^i = \sqrt{\langle (q^i)^2 \rangle - \langle q^i \rangle^2}\,, \qquad \Delta p_j = \sqrt{\langle (p_j)^2 \rangle - \langle p_j \rangle^2}\,.$$

For the moment we put aside important questions such as: which kind of average is meant here? how should we proceed in calculating the uncertainties in a given state? The uncertainties are the results of measurements on a given quantum mechanical state, and, as such, they are perfectly classical quantities. Yet, the inner dynamics of the system is such that the uncertainties fulfill certain correlated inequalities which limit their measurement in a fundamental way. Indeed, they obey

Heisenberg's uncertainty relation for position and momentum:
Let

$$\{q^i, (i = 1, 2, \ldots, f)\}$$

be a set of coordinates of a Lagrangian or Hamiltonian system with f degrees of freedom. Let

$$p_k = \frac{\partial L}{\partial \dot{q}^k}, \qquad k = 1, 2, \ldots, f$$

be their canonically conjugate momenta. In a given state of the system the results of any measurement will always be such that they are compatible with the following inequalities for the standard deviations of coordinates and momenta:

$$\boxed{(\Delta p_k)(\Delta q^i) \geq \frac{1}{2}\hbar\,\delta^i_k} \,. \tag{1.35}$$

This statement is both strange and remarkable and requires some more comments and remarks.

Remarks

1. For the time being we have in mind only the ordinary coordinates $\boldsymbol{x} = \{q^1, q^2, q^3\}$ and momenta $\boldsymbol{p} = \{p_1, p_2, p_3\}$ of a particle. The uncertainty relation (1.35) is formulated for the more general case of generalized coordinates and their canonically conjugate momenta in a mechanical system with f degrees of freedom. In doing so we assume that the system is such that the Legendre transform is regular (see [Scheck (2010)], Sect. 5.6.1), which is to say that the system can be described equivalently as a Lagrangian system $\{q, \dot{q}, L(q, \dot{q}, t)\}$, or as a Hamiltonian system $\{q, p, H(q, p, t)\}$. The relationship to concepts of classical canonical mechanics is remarkable, but note that the condition (1.35) goes far beyond it. We will return to this in more detail later.
2. Coordinates or momenta in a physical quantum state of a single particle exhibit a distribution, or variation, this being a concept of (classical) statistical physics. This implies, in general, that it is not sufficient to perform a single measurement in a given state of the particle. Rather one will have to perform *very many* identical measurements on that single, given state, in order to determine the distribution of experimental values and to calculate the standard deviation from them.
3. To take an example for the uncertainty relation imagine an experiment which allows to restrict the coordinate q^i to the interval Δ by means of a slit in the i-direction. The corresponding uncertainty in p_i is then at least $\hbar/2\Delta$. The more one localizes the particle in the i-direction the greater the distribution of values in the conjugate momentum. In the limit $\Delta \to 0$ the momentum cannot be determined at all.
4. It is clear from the preceding remark that the state of the particle can by no means be a curve in phase space $\mathbb{P}$. Such a curve would imply that at any given time both coordinates and momenta have definite, sharp values, i.e. that we would have $\Delta q^i = 0$ and $\Delta p_i = 0$. Although repeated measurements of these observables, e.g. in the time interval $(0, T)$, would yield the time averages

$$\overline{q^i} = \frac{1}{T}\int_0^T \mathrm{d}t\, q^i(t)\,, \quad \overline{(q^i)^2} = \frac{1}{T}\int_0^T \mathrm{d}t\, (q^i)^2(t)\,, \quad \text{etc.}$$

the uncertainties would still be zero. This means that in accepting Heisenberg's uncertainty relations we leave the description of the state in phase space. The (quantum) *state* of the particle which is contained in the symbolic notation $\langle\cdots\rangle$ must lie in a larger, and more abstract space than $\mathbb{P}$.

5. Consider the extreme case where the component q^i has the fixed value $\langle q^i\rangle = a^i$ and, hence, where $\langle(q^i)^2\rangle = (a^i)^2$. Since its conjugate momentum p_i, by virtue of the inequality (1.35), is completely undetermined we certainly cannot have $\langle p_i\rangle = b_i$ and $\langle(p_i)^2\rangle = (b_i)^2$. If upon repeated measurements of the i-th coordinate the state answers by "the observable q^i has the value a^i", that same state cannot return one single value b_i in a measurement of the conjugate momentum, otherwise both standard deviations would vanish, in contradiction to the uncertainty relation. This leads to the conjecture that the coordinate q^i and the momentum p_i are represented by quantities $\underline{q}^i$ and $\underline{p}_i$, respectively, which act on the states in some abstract space and which do not commute, in contrast to their classical counterparts. Indeed, we will soon learn that in quantum mechanics they fulfill the relation

$$[\underline{p}_i, \underline{q}^k] = \frac{\hbar}{\mathrm{i}}\delta_i^k\,. \tag{1.36}$$

Objects of this kind may be differential operators or matrices but certainly not smooth functions. For example, one verifies that the following pairs of operators obey the relations (1.36):

$$\left\{\underline{p}_i = +\frac{\hbar}{\mathrm{i}}\frac{\partial}{\partial q^i}, \underline{q}^k = q^k\right\}, \quad \text{and} \tag{i}$$

$$\left\{\underline{p}_i = p_i, \underline{q}^k = -\frac{\hbar}{\mathrm{i}}\frac{\partial}{\partial p_k}\right\}. \tag{ii}$$

In the first example the momentum is a differential operator, the coordinate is a function, i.e. an operator which acts by "multiplication with the function q^i". Indeed, one finds

$$\frac{\hbar}{\mathrm{i}}\frac{\partial}{\partial q^i}[q^k f(\ldots, q^i, \ldots, q^k, \ldots)] - q^k\frac{\hbar}{\mathrm{i}}\frac{\partial f}{\partial q^i}$$
$$= \frac{\hbar}{\mathrm{i}}\delta_i^k f(\ldots, q^i, \ldots, q^k, \ldots)\,.$$

In the second example the momentum is an ordinary function while the coordinate is a differential operator. In this case one has

$$p_i\left(-\frac{\hbar}{\mathrm{i}}\frac{\partial}{\partial p_k}\right)\hat{f}(\ldots, p_i, \ldots, p_k, \ldots)$$
$$-\left(-\frac{\hbar}{\mathrm{i}}\right)\frac{\partial}{\partial p_k}[p_i\hat{f}(\ldots, p_i, \ldots, p_k, \ldots)]$$
$$= \frac{\hbar}{\mathrm{i}}\delta_i^k\hat{f}(\ldots, p_i, \ldots, p_k, \ldots)\,.$$

These remarks to which one will return repeatedly in developing the theory further, raise a number of questions: What is the nature of the states of the system and, if this is known, how does one calculate mean values such as $\langle\cdots\rangle$? What is the nature of the abstract spaces which are spanned by physically admissible states of a system? If coordinates and momenta are to be represented by operators or some other set of noncommuting objects, then also all other observables that are constructed from them, will become operators. What are the rules that determine the translation of the classical observables to their representation in quantum mechanics?

The answers to these questions need more preparatory work and a good deal of patience. Before we turn to them let us illustrate the physical significance of (1.35) by means of three examples.

1.2.3 Examples for Heisenberg's Uncertainty Relation

Example 1.3 Harmonic Oscillator in one Dimension

The harmonic oscillator in one spatial dimension is described by the Hamiltonian function

$$H = \frac{p^2}{2m} + \frac{1}{2}m\omega^2 q^2 \equiv \frac{1}{2}\left[z_2^2 + z_1^2\right]$$

$$\text{where} \quad z_2 := \frac{p}{\sqrt{m}}, \quad z_1 := \sqrt{m}\omega q\,.$$

Suppose the oscillator is replaced by the corresponding quantum system (by replacing q and p by operators). The average of H in a state with energy E is

$$E = \langle H\rangle = \frac{\langle p^2\rangle}{2m} + \frac{1}{2}m\omega^2\langle q^2\rangle \equiv \frac{1}{2}\left[\langle z_2^2\rangle + \langle z_1^2\rangle\right].$$

The obvious symmetry $q \leftrightarrow -q$ and $p \leftrightarrow -p$ suggests that the mean values of these variables vanish, $\langle q\rangle = 0$, $\langle p\rangle = 0$. If this is the case we have

$$(\Delta q)^2 = \langle q^2\rangle, \quad (\Delta p)^2 = \langle p^2\rangle\,,$$

and (1.35) yields the inequality

$$\langle z_2^2\rangle\langle z_1^2\rangle = \omega^2\langle q^2\rangle\langle p^2\rangle \geq \frac{\hbar^2}{4}\omega^2\,.$$

Even if the reader does not accept the conjecture $\langle q\rangle = 0$, $\langle p\rangle = 0$ at this point the estimate just given remains true. Indeed, from (1.31) one concludes that $\langle G^2\rangle \geq (\Delta G)^2$ and, therefore, that

$$\langle q^2\rangle\langle p^2\rangle \geq (\Delta q)^2(\Delta p)^2 \geq \frac{\hbar^2}{4}\,.$$

This is sufficient for the following sequence of inequalities to hold true

$$0 \leq \frac{1}{2}\left(\sqrt{\langle z_2^2\rangle} - \sqrt{\langle z_1^2\rangle}\right)^2 = \langle H\rangle - \sqrt{\langle z_1^2\rangle\langle z_2^2\rangle} \leq E - \frac{\hbar\omega}{2}\,.$$

They show that the energy is bounded from below, $E_0 \leqslant E$ with $E_0 = \hbar\omega/2$. We will see below that this is precisely the energy of the lowest state. Now, if indeed $E = E_0 = \hbar\omega/2$, then

$$\left\langle z_1^2 \right\rangle = \left\langle z_2^2 \right\rangle = \frac{\hbar\omega}{2} ,$$

in agreement with the virial theorem which requires $\langle T_{\text{kin}} \rangle = \langle U \rangle = E/2$ for the case of the oscillator potential.

Thus the energy spectrum of the oscillator is bounded from below because of Heisenberg's uncertainty relation for position and momentum. The lowest state with energy $E = \hbar\omega/2$ and with the properties

$$\left\langle p^2 \right\rangle = \frac{1}{2} m\hbar\omega \quad \text{and} \quad \left\langle q^2 \right\rangle = \frac{1}{2} \frac{\hbar}{m\omega}$$

is marginally compatible with that relation. Since in quantum mechanics the oscillator can never come to a complete rest one says that its motion in the ground state consists in *zero point oscillations*. The energy E_0 of the ground state is called *the zero point energy*. Even in its lowest state both potential and kinetic energies exhibit nonvanishing, though minimal deviations. This is an intrinsic and invariant property of the oscillator.

Example 1.4 Spherical Oscillator

In classical mechanics the spherical oscillator is described by the Hamiltonian function

$$H = \frac{\boldsymbol{p}^2}{2m} + \frac{1}{2} m\omega^2 r^2 .$$

If converted to cartesian coordinates

$$H = \sum_{i=1}^{3} \frac{p_i^2}{2m} + \frac{1}{2} m\omega^2, \sum_{i=1}^{3} (q^i)^2 ,$$

it is seen to be equivalent to the sum of three linear oscillators, all three with the same mass and the same circular frequency ω. Thus, the analysis of the previous example can be applied directly. The standard deviations of the coordinate q^i and the conjugate momentum p_i are correlated, the ones of pairs (q^k, p_l) with different indices $k \neq l$ are not. Therefore, repeating the estimate of Example 1.3 yields the inequality

$$0 \leq E - 3\frac{\hbar\omega}{2} .$$

The lowest state has the energy $E = E_0 = 3\hbar\omega/2$. This system has three degrees of freedom each of which contributes the amount $\hbar\omega/2$ to the zero point energy.

Example 1.5 Hydrogen Atom

An analogous, though rougher, estimate for the hydrogen atom shows that, here too, the uncertainty relation is responsible for the fact that the energy spectrum is bounded from below. Using polar coordinates in the plane of the classical motion the Hamiltonian function reads

$$H = \frac{p_r^2}{2\mu} + \frac{\ell^2}{2\mu r^2} - \frac{e^2}{r},$$

(see [Scheck (2010)], Sect. 2.16), where p_r is the momentum canonically conjugate to r, ℓ is the modulus of the (conserved) orbital angular momentum. Although the mean value of p_r does not vanish we can make use of the property $\langle p_r^2 \rangle \geq (\Delta p_r)^2$ to estimate the mean value of H, for nonvanishing ℓ, as follows

$$\begin{aligned} E = \langle H \rangle (\ell = 0) &= \frac{\langle p_r^2 \rangle}{2\mu} - e^2 \left\langle \frac{1}{r} \right\rangle \\ &> \frac{\hbar^2}{8\mu(\Delta r)^2} - \frac{e^2}{(\Delta r)} . \end{aligned}$$

In this estimate we have used the uncertainty relation $(\Delta p_r)(\Delta r) \geq \hbar/2$ and we have approximated the term portional to $1/r$ by $1/(\Delta r)$. The right-hand side is a function of (Δr). Its minimum is attained at the value

$$(\Delta r) = \frac{\hbar^2}{4\mu e^2} .$$

If this is inserted in the expression above we see that the energy must be bounded from below by at least $E > (-2\mu e^4/\hbar^2)$. The right-hand side is four times the value (1.1) of the true energy of the ground state – presumably because our estimate is not optimal yet. The result shows, however, that it is again the uncertainty relation between position and momentum that prevents binding energies from becoming arbitrarily large.

1.3 The Particle-Wave Dualism

Energy E and momentum $\boldsymbol{p}$ of classical physics are understood to be properties of mechanical bodies, i.e., in the simplest case, of point-like *particles* of mass m. For such a particle these two kinematic quantities are related by an energy–momentum relation which reads

$$E = \boldsymbol{p}^2/2m \quad \text{or} \quad E = \sqrt{c^2 \boldsymbol{p}^2 + (mc^2)^2} \tag{1.37}$$

in the nonrelativistic and the relativistic case, respectively. In contrast to this, a circular frequency $\omega = 2\pi/T$, with T the period, and a wave vector $\boldsymbol{k}$, whose modulus is related to the wave length λ by $k = 2\pi/\lambda$, are attributes of a monochromatic *wave* which propagates in the direction $\hat{\boldsymbol{k}}$. The frequency ω and the wave number k are related by a dispersion relation $\omega = \omega(k)$.

The interpretation of the photoelectric effect and the derivation of Planck's formula for the spectral distribution of black body radiation show that light appears in quanta of energy which are given by the *Einstein–Planck* relation

$$\boxed{E = h\nu} \,. \tag{1.38}$$

This relation is quite remarkable in that it relates a particle property, "energy" E, with a wave property, "frequency" ν, via Planck's constant. The energy of a monochromatic electromagnetic wave is proportional to its frequency. Thus, light, or any other electromagnetic radiation for that matter, besides its well-known wave character, also has particle properties which will be particularly important when the number n of photons of a given energy is small. In such cases the light quanta, or photons, must be treated liked point particles of mass $m_{\text{photon}} = 0$. As we will see later this is a direct consequence of the long–range nature of the Coulomb potential $U_C(r) = \text{const}/r$. According to the second equation of (1.37) the photon must be ascribed a momentum, its energy and its momentum being related by (1.37) with mass zero, viz. $E = c|\boldsymbol{p}|$. On the other hand, when light propagates in vacuum, its frequency and wave length are related by $\nu\lambda = c$, where c is the speed of light. Therefore, the Einstein–Planck relation (1.38) is translated to a formula relating the modulus of the momentum to the wave length, viz.

$$|\boldsymbol{p}|_{\text{Photon}} = \frac{h}{\lambda} \,.$$

This dual nature of electromagnetic radiation on one side, and the diffraction phenomena of free massive elementary particles, on the other, lead Louis de Broglie[10] to the following fundamental hypothesis:

In close analogy to light which also possesses particle properties, all massive objects, and, in particular, all elementary particles exhibit wave properties. To a material particle of definite momentum $\boldsymbol{p}$ one must ascribe a monochromatic wave which propagates in the direction of $\hat{\boldsymbol{p}}$ and whose wave length is

$$\boxed{\lambda = \frac{h}{p} \quad \text{where} \quad p = |\boldsymbol{p}| \qquad \text{(de Broglie, 1923)}} \,. \tag{1.39}$$

This wave length is called *de Broglie wave length* of the material particle.

We comment on this hypothesis by the following

Remarks

1. If Planck's constant were equal to zero, $h = 0$, we would have $\lambda = 0$ for all values of p. The particle would then have no wave nature and could be described

[10] The name is pronounced "Broj", see e.g. *Petit Larousse*, Librairie Larousse, Paris.

exclusively by classical mechanics. Therefore, one expects classical mechanics to correspond to the *limit of short wave lengths* of quantum mechanics. We conjecture that this limiting situation is reached somewhat like in optics: Geometric optics corresponds to wave optics in the limit of short waves. If the wave length of light is very small as compared to the linear dimension d of the object on which it is scattered, optical set-ups such as slits, screens, or lenses, can be described by means of simple ray optics. If, on the other hand, $\lambda \simeq d$, i.e. if the wave length is comparable to typical linear dimensions of the set-up, there will be diffraction phenomena.

2. Quantum effects will be noticeable when the de Broglie wave length λ is of the same magnitude as the linear dimensions d which are relevant in a given situation. As an example, consider the scattering of a particle with momentum $\boldsymbol{p}$ on a target whose size is d. Whenever $\lambda \ll d$, i.e. if $dp \gg h$, classical mechanics will be applicable – though here as a limiting form of quantum mechanics. In other terms, one expects to find classical mechanics as the limit $h \longrightarrow 0$ of quantum mechanics. However, if $\lambda \simeq d$ we expect to find new and specific quantum effects.
3. Of course, the particle nature, in the strict sense of classical mechanics, and the postulated wave nature of matter are not readily compatible. Rather, particle properties and wave properties must be complementary aspects. Both aspects are essential in the description of matter particles. This assertion, although still somewhat vague for the time being, is described by *Bohr's principle of complementarity*.
4. By associating a wave to a particle the uncertainty relation receives an interpretation in terms of wave optics: a monochromatic wave corresponds to a fixed value of $\boldsymbol{p}$. Such a wave is nowhere localized in space. Conversely, if one wishes to construct an optical signal which is localized in some finite domain of space, one will need an appropriate superposition of partial waves taken from a certain spectrum of wave lengths. The smaller, i.e. the more localized this *wave packet* is, the broader the spectrum of contributing wave lengths must be, or, through de Broglie's relation, the larger the range of momenta must be chosen.

1.3.1 The Wave Function and its Interpretation

On the basis of de Broglie's hypothesis we associate to a particle such as the electron a wave function $\psi(t, \boldsymbol{x})$. If this electron has a sharp value of momentum $\boldsymbol{p}$ this wave function will be a plane wave of the form

$$e^{i(\boldsymbol{p}\cdot\boldsymbol{x}/\hbar-\omega t)} = e^{i(\boldsymbol{k}\cdot\boldsymbol{x}-\omega t)}$$

where $\boldsymbol{k}$ is the wave vector, $k = |\boldsymbol{k}|$ is the wave number, and $\omega = \omega(k)$ is a function still to be determined. In accordance with the uncertainty relation such a plane wave is nowhere localized in space and, for this reason, it is not obvious how to interpret it physically. It would be more helpful if ψ were a strongly localized wave phenomenon.

Indeed, we could compare such a wave packet at time t to the position in space that the particle would pass at this time if it were described by classical mechanics. With this idea in mind we write the wave function as a superposition of plane waves

$$\psi(t,\boldsymbol{x}) = \frac{1}{(2\pi)^{3/2}} \int \mathrm{d}^3k\, \widehat{\psi}(\boldsymbol{k}) \mathrm{e}^{\mathrm{i}(\boldsymbol{k}\cdot\boldsymbol{x}-\omega t)} \tag{1.40}$$

by choosing the function $\widehat{\psi}(\boldsymbol{k})$ such that it is concentrated around a central value, say $\boldsymbol{k}_0$. The numerical factor in front of (1.40) is chosen in order to render the Fourier transform between $\psi(t,\boldsymbol{x})$ and $\widehat{\psi}(\boldsymbol{k})$ symmetric. Expanding around the wave vector $\boldsymbol{k}_0$, we have

$$\boldsymbol{k}\cdot\boldsymbol{x} = (\boldsymbol{k}-\boldsymbol{k}_0)\cdot\boldsymbol{x} + \boldsymbol{k}_0\cdot\boldsymbol{x},$$

$$\omega(k) \simeq \omega(k_0) + (\boldsymbol{k}-\boldsymbol{k}_0)\cdot\nabla|_k\, \omega(k)|_{k=k_0} = \omega_0 + \frac{(\boldsymbol{k}-\boldsymbol{k}_0)\cdot\boldsymbol{k}_0}{k_0}\left.\frac{\mathrm{d}\omega}{\mathrm{d}k}\right|_{k=k_0} .$$

In this expansion we set $k \equiv |\boldsymbol{k}|$ and $k_0 \equiv |\boldsymbol{k}_0|$, the gradient with respect to the three components of $\boldsymbol{k}$ is replaced by the derivative with respect to its modulus $k = |\boldsymbol{k}|$, $\nabla_k = (\nabla_k|\boldsymbol{k}|)\mathrm{d}/\mathrm{d}k$, by means of the chain rule. In this approximation (1.40) takes a form that is easily interpreted:

$$\psi(t,\boldsymbol{x}) \simeq \frac{1}{(2\pi)^{3/2}} \mathrm{e}^{\mathrm{i}(\boldsymbol{k}_0\cdot\boldsymbol{x}-\omega_0 t)} A(\boldsymbol{x}-\hat{\boldsymbol{k}}_0 v_0 t)$$

with

$$A(\boldsymbol{x}-\hat{\boldsymbol{k}}_0 v_0 t) = \int \mathrm{d}^3k\, \widehat{\psi}(\boldsymbol{k}) \mathrm{e}^{\mathrm{i}(\boldsymbol{k}-\boldsymbol{k}_0)(\boldsymbol{x}-\hat{\boldsymbol{k}}_0 v_0 t)}$$

and

$$v_0 = \left.\frac{\mathrm{d}\omega}{\mathrm{d}k}\right|_{k=k_0} .$$

The wave function just obtained may be understood as follows: The amplitude A is determined by the distribution $\widehat{\psi}(\boldsymbol{k})$. It moves with velocity v_0. If one dealt with a theory of (classical) waves one would call this the *group velocity* while ω/k would be said to be the *phase velocity*. Relating v_0 to the momentum, by virtue of de Broglie's relation (1.39), one has

$$\left.\frac{\mathrm{d}\omega}{\mathrm{d}k}\right|_{k=k_0} = v_0 = \frac{p_0}{m} = \frac{\hbar k_0}{m} .$$

Upon integration, the functions $\omega(k_0)$ or, somewhat more generally, $\omega(k)$ are seen to be given by

$$\omega(k) = \frac{\hbar \boldsymbol{k}^2}{2m} \quad \text{or} \quad E \equiv \hbar\omega(k) = \frac{\boldsymbol{p}^2}{2m} . \tag{1.41}$$

Thus, we recover the well-known nonrelativistic relation between energy and momentum and, hence, fix the dispersion relation $\omega = \omega(k)$. Note that the connection between particle and wave properties was used several times.

The superposition (1.40) that was chosen, mimics the analogous classical situation. It describes an object localized at time t which moves with the group velocity v_0, equal to the velocity of the classical particle. However, as we shall discover soon, the localization cannot last for long. The well-localized wave packet at time t disperses in the course of time. In constructing the packet we have tacitly assumed that different wave functions can be superimposed linearly – a property which is in agreement with the interference phenomena observed in experiment. The question of *interpretation* of the wave function, for the time being, remains unanswered. This is the question of how to derive measurable, testable predictions from the knowledge of $\psi(t, \boldsymbol{x})$.

In a next step we show that a wave function of the type (1.40) satisfies the differential equation

$$\boxed{\mathrm{i}\hbar\dot{\psi}(t, \boldsymbol{x}) = -\frac{\hbar^2}{2m}\Delta\psi(t, \boldsymbol{x})} \, . \tag{1.42}$$

Proof

As the function $\widehat{\psi}(\boldsymbol{k})$ in (1.40) is localized, the integral exists. Differentiation with respect to time or space coordinates, and integration can be interchanged. Making use of (1.41) for ω, we have

$$\dot{\psi} = -\mathrm{i}\frac{\hbar}{2m}\frac{1}{(2\pi)^{3/2}}\int \mathrm{d}^3k\, \boldsymbol{k}^2\widehat{\psi}(\boldsymbol{k})\mathrm{e}^{\mathrm{i}(\boldsymbol{k}\cdot\boldsymbol{x}-\omega t)} \, .$$

Replacing two of the factors in the integrand as follows,

$$\boldsymbol{k}^2\mathrm{e}^{\mathrm{i}(\boldsymbol{k}\cdot\boldsymbol{x}-\omega t)} = -\Delta_x\, \mathrm{e}^{\mathrm{i}(\boldsymbol{k}\cdot\boldsymbol{x}-\omega t)} \, ,$$

and taking the Laplace operator out of the integral gives the differential equation (1.42).

Having derived (1.42) we have found the Schrödinger equation for the case of force-free motion. This is a *homogeneous, linear* differential equation: It is homogeneous because it does not contain a source term independent of ψ. Linearity means that if $\psi_1(t, \boldsymbol{x})$ and $\psi_2(t, \boldsymbol{x})$ are solutions then also any linear combination

$$\psi(t, \boldsymbol{x}) = c_1\psi_1(t, \boldsymbol{x}) + c_2\psi_2(t, \boldsymbol{x}) \quad \text{with} \quad c_1, c_2 \in \mathbb{C}$$

is a solution. The statement that two different solutions can interfere is an expression of the *superposition principle*. This principle has far reaching observable consequences.

Equation (1.42) is *first* order in the time variable. This means that a given initial distribution $\psi(t_0, \boldsymbol{x})$ fixes the wave field for all times. Being of *second* order in derivatives with respect to space coordinates, this equation cannot be Lorentz covariant (but it is Galilei invariant). This is not really surprising because we obtained (1.41) by making use of the nonrelativistic relation between velocity and momentum.

1.3.2 A First Link to Classical Mechanics

In the spirit of the eikonal approximation in optics we attempt to solve (1.42) by means of the ansatz

$$\psi(t,\boldsymbol{x}) = \psi_0 \exp\left[\frac{\mathrm{i}}{\hbar} S(t,\boldsymbol{x})\right] \tag{1.43}$$

where ψ_0 is a complex constant. The time derivative and the Laplacian applied to this give, respectively,

$$\mathrm{i}\hbar\dot{\psi} = -\frac{\partial S}{\partial t}\psi_0 \exp\left[\frac{\mathrm{i}}{\hbar}S(t,\boldsymbol{x})\right] = -\frac{\partial S}{\partial t}\psi\,,$$

$$\Delta\psi = \left[-\frac{1}{\hbar^2}(\nabla S)^2 + \frac{\mathrm{i}}{\hbar}\Delta S\right]\psi_0 \exp\left[\frac{\mathrm{i}}{\hbar}S(t,\boldsymbol{x})\right].$$

Upon insertion in (1.42) one obtains a differential equation for $S(t,\boldsymbol{x})$,

$$\frac{\partial S}{\partial t} + \frac{1}{2m}(\nabla S)^2 = \mathrm{i}\frac{\hbar}{2m}\Delta S\,.$$

If the function S is such that the term on the right-hand side may be neglected, this is seen to be the Hamilton–Jacobi differential equation for the case of the Hamiltonian function $H = \boldsymbol{p}^2/2m$,

$$\widetilde{H} = H\left(\frac{\partial S}{\partial q^i}, q^k, t\right) + \frac{\partial S}{\partial t} = 0$$

with the well-known formulae

$$p_i = \frac{\partial S}{\partial q^i}\,, \quad Q^k = \frac{\partial S}{\partial P_k}\,, \quad S = S(q,\alpha,t)\,,$$

and with $P_k = \alpha_k = \text{const}$ (see [Scheck (2010)], Sect. 2.35 where this particular canonical transformation is denoted by S^*). In mechanics one learns that the general solution is

$$S(\boldsymbol{x},\boldsymbol{\alpha},t) = \boldsymbol{\alpha}\cdot\boldsymbol{x} - \frac{\boldsymbol{\alpha}^2}{2m}t + \text{const}$$

and that it describes uniform, rectilinear motion, as expected,

$$\boldsymbol{x} - \frac{\boldsymbol{\alpha}}{m}t = \boldsymbol{\beta}\,.$$

The particle trajectories are perpendicular to the surfaces $S(\boldsymbol{x},\boldsymbol{\alpha},t) = \text{const}$. This result is interesting in that it says that these surfaces are the wave fronts of the wave function ψ. In the approximation made above the classical orbits are orthogonal trajectories of the wave fronts $\psi(t,\boldsymbol{x})$.

Remark

The ansatz (1.43) is the starting point for a systematic expansion in powers of $\hbar$, i.e. a series of approximations around the classical limit. This is called the *WKBJ-method* where the short-hand stands for its authors Wentzel, Kramers, Brillouin, and Jeffreys.

1.3.3 Gaussian Wave Packet

For the sake of simplicity we first consider wave functions in one spatial dimension, denoted x. Plane waves are chosen as follows

$$\psi_k(t,x) = \frac{1}{\sqrt{2\pi}} e^{i(kx-\omega t)} = \frac{1}{\sqrt{2\pi}} e^{i/\hbar(px-Et)}$$

with $\omega = \hbar k^2/(2m)$, the normalization being chosen such that

$$\int dx\, \psi^*_{k'}(t,x)\psi_k(t,x) = \delta^{(1)}(k-k') .$$

Appendix A.1 gives a summary of the properties of δ-distributions and indicates how this improper integral is to be understood and how it is evaluated. The wave packet is taken to be

$$\psi(t,x) = \frac{1}{\sqrt{2\pi}} \int dk\, \widehat{\psi}(k) e^{i(kx-\omega t)} . \tag{1.44}$$

Taking the Fourier transform of this ansatz, assuming the integrals to exist, yields

$$\widehat{\psi}(k) = \frac{1}{\sqrt{2\pi}} \int dx\, \psi(t,x) e^{-i(kx-\omega t)} . \tag{1.45}$$

At time $t = 0$ the wave function is assumed to be a Gaussian wave packet, of the form

$$\psi(t=0,x) = \alpha e^{-x^2/(2b^2)} e^{ik_0 x} ,$$

where α is a complex constant that we fix as follows:

$$\alpha = \frac{1}{b^{1/2}\pi^{1/4}} e^{i\varphi_\alpha} .$$

With this choice of the factor α the distribution $\psi^*\psi = |\psi|^2$ is normalized to 1, it has the shape sketched in Fig. 1.5. Its width is

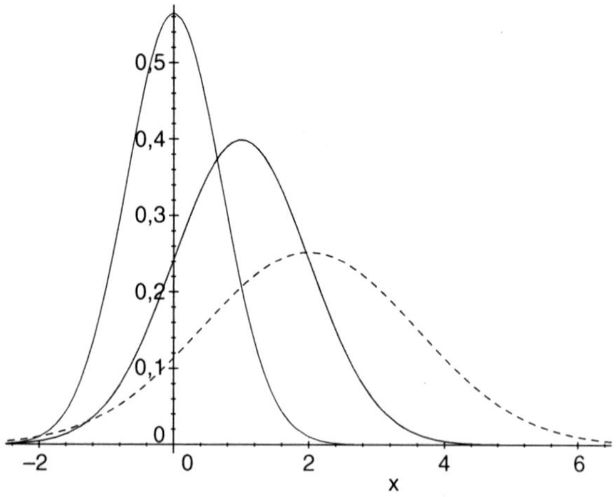

Fig. 1.5 A Gaussian wave packet (1.49) (in one dimension) while moving to the right in this picture, broadens in the course of time. The three curves show the shape of the wave packet as a function of the space coordinate x, at $t = 0$, at $t = \tau(b)$, and at $t = 2\tau(b)$, where $\tau = mb^2/\hbar$

$$\Gamma|_{(0,x)} = 2b\sqrt{\ln 2} \simeq 1.665\, b\,,$$

the index $(0, x)$ being added in order to point out that this is the width at time $t = 0$ and in coordinate space. In order to check the normalization and to calculate the Fourier transform we need the Gauss integral which is calculated as follows

Gauss integral: In a first step calculate the integral

$$\int_{-\infty}^{\infty} \mathrm{d}x\, \mathrm{e}^{-x^2}$$

by taking its square. This yields a double integral over two independent cartesian variables which may be identified with two orthogonal variables x and y. In turn, this double integral is converted to polar coordinates in the (x, y)-plane and takes a form which is integrated in an elementary way, viz.

$$\left(\int_{-\infty}^{\infty} \mathrm{d}x\, \mathrm{e}^{-x^2}\right)^2 = \int_{-\infty}^{\infty}\int_{-\infty}^{\infty} \mathrm{d}x\mathrm{d}y\, \mathrm{e}^{-(x^2+y^2)}$$
$$= \int_0^{2\pi} \mathrm{d}\phi \int_0^{\infty} \mathrm{d}r\, r\mathrm{e}^{-r^2} = \pi\,.$$

Let a be a positive real number, or else a complex number with positive real part, and let b and c be two arbitrary complex numbers. Then one obtains

$$\int_{-\infty}^{\infty} \mathrm{d}x\, \mathrm{e}^{-(ax^2+2bx+c)} = \sqrt{\frac{\pi}{a}}\mathrm{e}^{(b^2-ac)/a}\,. \tag{1.46}$$

The latter formula follows from the one obtained in the first step by completing the argument of the exponential to a complete square,

$$ax^2 + 2bx + c = a\left(x + \frac{b}{a}\right)^2 - \frac{b^2 - ac}{a}$$

and by substituting $u = \sqrt{a}(x + b/a)$.

This result is used to verify that

$$\int_{-\infty}^{\infty} \mathrm{d}x\, |\psi(0, x)|^2 = 1$$

and, likewise, that $\widehat{\psi}(k)$ is also normalized to 1.

Making use of the formula (1.46) one calculates the Fourier transform (1.45) for the distribution $\psi(0, x)$ as given above,

$$\widehat{\psi}(k) = \alpha b \exp\left[-\frac{1}{2}(k - k_0)^2 b^2\right], \tag{1.47}$$

α being the same parameter as before. Inserting this result in (1.44) one calculates the wave function for arbitrary times, making use once more of the Gauss integral. The result is

$$\psi(t,x) = \frac{\alpha b}{\sqrt{b^2 + \mathrm{i}\hbar t/m}} \exp\left[-\frac{x^2}{2(b^2 + \mathrm{i}\hbar t/m)}\right]$$
$$\times \exp\left[\mathrm{i}\frac{k_0 x - \hbar k_0^2 t/(2m)}{1 + \mathrm{i}\hbar t/(mb^2)}\right]. \tag{1.48}$$

This somewhat cumbersome formula is understood better if one calculates its absolute square, viz.

$$|\psi(t,x)|^2 = \frac{|\alpha|^2}{\sqrt{1 + \hbar^2 t^2/(m^2 b^4)}} \exp\left[-\frac{(x - \hbar k_0 t/m)^2}{b^2[1 + \hbar^2 t^2/(m^2 b^4)]}\right]. \tag{1.49}$$

The expression (1.49) in coordinate space, and the corresponding absolute square $|\widehat{\psi}(k)|^2$ in momentum space that follows from it, are now easy to interpret: The tip of the wave packet moves with velocity $\hbar k_0/m$. Inserting $|\alpha|^2 = 1/(b\sqrt{\pi})$ its width is seen to be proportional to

$$b\sqrt{1 + \frac{\hbar^2 t^2}{m^2 b^4}} \quad \text{(in coordinate space)},$$

while the distribution $|\widehat{\psi}(k)|^2$ in momentum space has constant width,

$$\frac{1}{b} \quad \text{(in momentum space)}.$$

Two properties can be read off these formulae:

1. If one wishes to prepare a sharply localized packet in coordinate space, at time $t = 0$, then one must choose b as small as possible. In this case the corresponding distribution in momentum space is necessarily broad. In turn, the packet will have a large spatial extension already at $t = 0$ if its momentum space representation is strongly localized around the central value k_0. Both situations are in accord with Heisenberg's uncertainty relation (1.35).
2. In the course of time the wave packet in coordinate space broadens (independently of whether one extrapolates to the future or to the past of $t = 0$). Writing the factor characterizing the width as follows

$$b\sqrt{1 + \frac{\hbar^2 t^2}{m^2 b^4}} = b\sqrt{1 + \frac{t^2}{\tau^2(b)}} \quad \text{with} \quad \tau(b) = \frac{m}{\hbar} b^2,$$

one sees that the width will double after the time $t = \sqrt{3}\tau(b)$ as compared to what it was at $t = 0$. In Fig. 1.5 I have plotted the quantity (1.49) for $t_0 = 0$, for $t_1 = \tau(b)$, and for $t_2 = 2\tau(b)$, as a function of x, with the choice of parameters $k_0 = 1$, $b = 1$ (in arbitrary units).

It is instructive to estimate more quantitatively the dispersion of the wave packet. For an electron the characteristic time during which the width increases to the double of its initial value, is

$$t = \sqrt{3}\tau(b) \simeq 1.5 \times 10^{-26}\,\text{fm}^{-2}\text{s}\, b^2 .$$

Suppose that the wave packet assigned to an electron, at time $t = 0$, had the width $b = 1$ fm. The width will be twice its initial value after 1.5×10^{-26} s, a very short time indeed.

Let us compare this to a tennis ball with mass $m = 0.1$ kg, taken to be a wave packet of length $b = 6\,\text{cm} = 6 \times 10^{13}$ fm, at time $t = 0$. The time it will take to double its size is

$$\sqrt{3}\tau(b) = 1642\,\text{fm}^{-2}\,\text{s}\, b^2 = 5.91 \times 10^{30}\,\text{s} \simeq 1.9 \times 10^{23}\ \text{years} .$$

Thus, there is no reason to worry about the validity of macroscopic, classical mechanics!

1.3.4 Electron in External Electromagnetic Fields

Classically, an electron subject to external electromagnetic fields is described by the Hamiltonian function

$$H(\boldsymbol{p}, \boldsymbol{x}, t) = \frac{1}{2m}\left(\boldsymbol{p} - \frac{e}{c}\boldsymbol{A}(t, \boldsymbol{x})\right)^2 + e\Phi(t, \boldsymbol{x}) \tag{1.50}$$

(see, e.g., [Scheck (2010)], Sect. 2.16). Here e denotes its electric charge, $\boldsymbol{A}$ and Φ are the vector and scalar potentials, respectively, the electric and the magnetic fields being derived from them by means of the formulae (1.12) and (1.13), respectively. The corresponding differential equation of Hamilton and Jacobi reads

$$\frac{1}{2m}\left(\nabla S - \frac{e}{c}\boldsymbol{A}\right)^2 + e\Phi + \frac{\partial S}{\partial t} = 0 .$$

Furthermore, one has the equations

$$\dot{x}^i = \frac{\partial H}{\partial p_i} = \frac{1}{m}\frac{\partial S}{\partial x_i} - \frac{e}{mc}A^i .$$

It seems reasonable to require that this classical differential equation follow from its quantum analogue, by means of the ansatz (1.43) and by expanding in powers of $\hbar$, very much as in the previous case of force-free motion. One then sees that a possible generalization of (1.42) could be[11]

$$\mathrm{i}\hbar\dot{\psi}(t, \boldsymbol{x}) = \frac{1}{2m}\left(\frac{\hbar}{\mathrm{i}}\nabla - \frac{e}{c}\boldsymbol{A}\right)^2 \psi(t, \boldsymbol{x}) + e\Phi\psi(t, \boldsymbol{x}) . \tag{1.51}$$

[11] The "dot" applied to the function ψ is a short-hand for the partial derivative with respect to time. This generally accepted convention should not give rise to confusion because the coordinates $\boldsymbol{x}$ on which ψ depends too, as such, are not functions of time.

For vanishing external fields, $\boldsymbol{A} \equiv 0$, $\Phi \equiv 0$, it is identical with (1.42). Furthermore, one easily verifies, upon inserting the ansatz (1.43), that it yields the correct Hamilton–Jacobi equation, at the order $\mathcal{O}(\hbar^0)$, when the fields do not vanish.

The following discussion gives further support to the ansatz (1.51). At the same time it takes us closer towards an interpretation of the wave function $\psi(t, \boldsymbol{x})$. We proceed in two steps.

(1) Consider first the case $\boldsymbol{A} \equiv 0$, with only the scalar potential different from zero. The conjectured differential equation (1.51) simplifies to

$$\mathrm{i}\hbar\dot{\psi}(t, \boldsymbol{x}) = \left(-\frac{\hbar^2}{2m}\Delta + e\Phi(t, \boldsymbol{x})\right)\psi(t, \boldsymbol{x}) . \tag{1.52}$$

As we know already that the absolute square $|\psi(t, \boldsymbol{x})|^2$ must be related to localization of the electron in space, it seems plausible that the integral of this positive-semidefinite quantity, weighted with the external potential Φ, is proportional to the potential energy of the electron in the external field. Thus, we assume

$$e\int \mathrm{d}^3x\, |\psi(t, \boldsymbol{x})|^2 \Phi(t, \boldsymbol{x}) = E_{\mathrm{pot}}$$

(where, possibly, a constant of proportionality might have to be inserted). The average force acting on the electron is then obtained from the integral over the gradient field of Φ,

$$\langle \boldsymbol{F} \rangle = -e\int \mathrm{d}^3x |\psi|^2 \nabla\Phi(t, \boldsymbol{x}) .$$

If for $|\boldsymbol{x}| \to \infty$ the wave function tends to zero sufficiently rapidly such that all surface terms vanish at infinity, we shift the operator ∇ over to $|\psi|^2 = \psi^*\psi$, by partial integration, and obtain

$$\langle \boldsymbol{F} \rangle = +e\int \mathrm{d}^3x \left[(\nabla\psi^*)\psi + \psi^*(\nabla\psi)\right]\Phi(t, \boldsymbol{x}) .$$

We show next that this expression is also equal to

$$\frac{\mathrm{d}}{\mathrm{d}t}\left(\int \mathrm{d}^3x\, \psi^* \frac{\hbar}{\mathrm{i}}\nabla\psi\right) = \int \mathrm{d}^3x\, \dot{\psi}^* \frac{\hbar}{\mathrm{i}}\nabla\psi + \int \mathrm{d}^3x\, \psi^* \frac{\hbar}{\mathrm{i}}\nabla\dot{\psi}$$

provided use is made of (1.52) for ψ, as well as its complex conjugate for ψ^*. The terms containing the Laplace operator cancel by virtue of

$$\int \mathrm{d}^3x\, (\Delta\psi^*\nabla_i\psi - \psi^*\nabla_i\Delta\psi)$$
$$= \sum_k \int \mathrm{d}^3x\, \left(-\nabla_k\psi^*\nabla_k\nabla_i\psi + \nabla_k\psi^*\nabla_i\nabla_k\psi\right) = 0 ,$$

using partial integration. The terms containing the scalar potential Φ also cancel except for the one in which the operator ∇ acts on the function Φ. This term reduces to

$$\frac{1}{\mathrm{i}\hbar}e\int \mathrm{d}^3x\, \psi^* \frac{\hbar}{\mathrm{i}}(\nabla\Phi)\psi .$$

Putting these results together we obtain the equation

$$\langle \boldsymbol{F} \rangle = \frac{\mathrm{d}}{\mathrm{d}t} \int \mathrm{d}^3x\, \psi^* \frac{\hbar}{\mathrm{i}} \nabla \psi \equiv \frac{\mathrm{d}}{\mathrm{d}t} \langle \boldsymbol{p} \rangle \,.$$

In the second part of this equation we have tentatively identified the integral with the average of the momentum. This interpretation receives further support by calculating the time derivative of E_{pot}, making use of (1.52), viz.

$$e\Phi\psi = -\frac{\hbar}{\mathrm{i}}\dot{\psi} + \frac{\hbar^2}{2m}\Delta\psi$$

as well as of its complex conjugate:

$$\begin{aligned}
\frac{\mathrm{d}}{\mathrm{d}t} E_{\mathrm{pot}} &= e \int \mathrm{d}^3x \left(\dot{\psi}^* \psi\, \Phi + \psi^* \dot{\psi}\, \Phi \right) \\
&= \frac{\hbar^2}{2m} \int \mathrm{d}^3x \left(\dot{\psi}^* \Delta\psi + (\Delta\psi)^* \dot{\psi} \right) \\
&= \frac{\hbar^2}{2m} \int \mathrm{d}^3x \left(\dot{\psi}^* \Delta\psi + \psi^* \Delta\dot{\psi} \right) \\
&= \frac{\mathrm{d}}{\mathrm{d}t} \left(\frac{\hbar^2}{2m} \int \mathrm{d}^3x\, \psi^* \Delta\psi \right) .
\end{aligned}$$

It is suggestive to interpret the right-hand side as the kinetic energy, but for the sign

$$E_{\mathrm{kin}} = -\frac{\hbar^2}{2m} \int \mathrm{d}^3x\, \psi^* \Delta\psi \,,$$

such that the total energy is represented by

$$E = E_{\mathrm{pot}} + E_{\mathrm{kin}} = \int \mathrm{d}^3x\, \psi^* \left\{ -\frac{\hbar^2}{2m}\Delta + e\Phi \right\} \psi$$

and is constant in time. The operator in curly brackets would then be the analogue of the classical Hamiltonian function. In particular, its first term would take the role of the classical kinetic energy $\boldsymbol{p}^2/(2m)$. This is compatible with the identification of the momentum above because the square of the operator in the integral for $\boldsymbol{p}$ gives

$$\frac{\hbar}{\mathrm{i}}\nabla \cdot \frac{\hbar}{\mathrm{i}}\nabla = -\hbar^2\Delta \,.$$

(2) Next we consider a situation where both $\boldsymbol{A}$ and Φ are different from zero. In view of the relation of the wave function to the observables that we wish to establish, we define the following densities:

$$\varrho(t, \boldsymbol{x}) := \psi^*(t, \boldsymbol{x})\psi(t, \boldsymbol{x}) = |\psi(t, \boldsymbol{x})|^2 \,, \tag{1.53}$$

$$\begin{aligned}
\boldsymbol{j}(t, \boldsymbol{x}) := \frac{1}{2m} \Big\{ \psi^*(t, \boldsymbol{x}) \left(\frac{\hbar}{\mathrm{i}}\nabla - \frac{e}{c}\boldsymbol{A} \right) \psi(t, \boldsymbol{x}) \\
+ \left[\left(\frac{\hbar}{i}\nabla - \frac{e}{c}\boldsymbol{A} \right) \psi(t, \boldsymbol{x}) \right]^* \psi(t, \boldsymbol{x}) \Big\} \qquad (1.54) \\
= \frac{\hbar}{2m\mathrm{i}} \left[\left(\psi^* \nabla\psi - (\nabla\psi^*)\psi \right) - \frac{\mathrm{i}2e}{\hbar c} \boldsymbol{A}\psi^*\psi \right] .
\end{aligned}$$

By construction both the scalar density (1.53) and the vectorial density (1.54) are real. If it is true that the operator $(\hbar/\mathrm{i})\nabla$ represents the (canonical) momentum, then

$$\frac{\hbar}{\mathrm{i}}\nabla - \frac{e}{c}\boldsymbol{A}$$

must be the kinematic momentum.

We start by calculating the divergence of the current density (1.54). We find

$$\begin{aligned}\nabla\cdot\boldsymbol{j} &= \frac{1}{2m}\Big\{-\mathrm{i}\hbar\left[\psi^*\Delta\psi - (\Delta\psi^*)\psi\right] \\ &\quad -\frac{2e}{c}\boldsymbol{A}\left[(\nabla\psi^*)\psi + \psi^*(\nabla\psi)\right] - \frac{2e}{c}(\nabla\cdot\boldsymbol{A})\psi^*\psi\Big\} \\ &= \frac{\mathrm{i}}{2m\hbar}\left\{\psi^*\left(\frac{\hbar}{\mathrm{i}}\nabla - \frac{e}{c}\boldsymbol{A}\right)^2\psi - \left[\left(\frac{\hbar}{\mathrm{i}}\nabla + \frac{e}{c}\boldsymbol{A}\right)^2\psi^*\right]\psi\right\}.\end{aligned}$$

(In the last step two terms proportional to $\boldsymbol{A}^2$ were added and subtracted in order to obtain perfect squares.)

This becomes a remarkable result when we also calculate the time derivative of the scalar density (1.53), by making use of the differential equation (1.52) for ψ and for ψ^*,

$$\begin{aligned}\frac{\partial\varrho}{\partial t} &= \dot{\psi}^*\psi + \psi^*\dot{\psi} \\ &= -\frac{\mathrm{i}}{2m\hbar}\left\{\psi^*\left(\frac{\hbar}{\mathrm{i}}\nabla - \frac{e}{c}\boldsymbol{A}\right)^2\psi - \left[\left(\frac{\hbar}{\mathrm{i}}\nabla + \frac{e}{c}\boldsymbol{A}\right)^2\psi^*\right]\psi\right\}.\end{aligned}$$

Indeed we obtain the continuity equation relating the densities (1.53) and (1.54):

$$\boxed{\nabla\cdot\boldsymbol{j} + \frac{\partial\varrho}{\partial t} = 0}\,. \tag{1.55}$$

On the basis of this result one might be tempted to interpret the density (1.53) as the electric charge density, and the vectorial density (1.54) as the electric current density of the moving charged particle. However, as one realizes after a moment of thought, this interpretation is in conflict with observation: Upon integration over the whole three-dimensional space one would obtain the total electric charge

$$Q_{\text{e.m.}} = \int \mathrm{d}^3x\,|\psi(t,\boldsymbol{x})|^2\,.$$

We know that this must be a multiple of the elementary charge $|e|$. Thus, we would have to require $Q_{\text{e.m.}} = \pm n|e|$. For an electron, more specifically, we would choose the negative sign and $n = 1$. As a consequence, a localized fraction of the integral

$$\int\limits_V \mathrm{d}^3x\,|\psi|^2\,,$$

obtained by integrating over a finite volume V, would represent a certain fraction of that integer charge. This contradicts experiment: A free electron carrying a fraction of its charge has never been observed.

There is further, perhaps even more convincing evidence against this interpretation. As this interpretation is purely classical, all diffraction phenomena of matter waves would be of the same nature as those of classical optics. In particular, interference patterns would always be perfect and complete, no matter how low the intensity of the incoming wave. A single electron which is scattered on two or more slits in a wall, would give rise to a complete interference pattern on a screen behind the slits, though with strongly reduced intensity. This is in contradiction with experiment. What one really observes is a *statistical* phenomenon. Any single electron hits a well-defined point on the screen (which cannot be predicted, though!). The interference pattern appears only after some time, after having a large number of identically prepared electrons scatter in the same experimental set-up.

1.4 Schrödinger Equation and Born's Interpretation of the Wave Function

Return for a while to the differential equation (1.42) which is used to describe the motion of free particles. Setting

$$\psi(t, \boldsymbol{x}) = \mathrm{e}^{-(\mathrm{i}/\hbar)Et}\psi(\boldsymbol{x}) \tag{1.56}$$

it goes over into the differential equation

$$\frac{1}{2m}\left(\frac{\hbar}{\mathrm{i}}\nabla\right)^2 \psi(\boldsymbol{x}) = E\psi(\boldsymbol{x})\,.$$

Plane waves of the form

$$\psi(\boldsymbol{x}) = \frac{1}{(2\pi)^{3/2}}\mathrm{e}^{\mathrm{i}\boldsymbol{k}\boldsymbol{x}} = \frac{1}{(2\pi)^{3/2}}\mathrm{e}^{(\mathrm{i}/\hbar)\boldsymbol{p}\boldsymbol{x}} \tag{1.57}$$

are seen to be solutions of this equation provided

$$E = \frac{\boldsymbol{p}^2}{2m} = \frac{\hbar^2\boldsymbol{k}^2}{2m}\,,$$

i.e. provided E and $\boldsymbol{p}$ obey the energy-momentum relation (1.37) valid for nonrelativistic kinematics. These simple calculations and the considerations of the previous section lead to the conjecture that quantum mechanics assigns differential operators to energy and momentum, respectively, such that

$$E \longleftrightarrow \mathrm{i}\hbar\frac{\partial}{\partial t}, \qquad \boldsymbol{p} \longleftrightarrow \frac{\hbar}{\mathrm{i}}\nabla\,. \tag{1.58}$$

Indeed, by this formal replacement the nonrelativistic energy-momentum relation goes over into the differential equation (1.42).

Consider next the differential equation (1.51) for a charged particle in electromagnetic fields. For simplicity, we study the case $\boldsymbol{A} \equiv 0$ and assume the scalar potential Φ to be independent of time. Inserting once more the ansatz (1.56) gives

$$E = \frac{\boldsymbol{p}^2}{2m} + U(\boldsymbol{x}) \quad \text{with} \quad U(\boldsymbol{x}) = e\Phi(\boldsymbol{x})\,.$$

Again, this is nothing else than the energy-momentum relation in the presence of an external potential. In an autonomous system of classical mechanics this equation describes the conserved total energy which is the sum of the kinetic and of the potential energies. In mechanics it is interpreted in the sense that in any point of the classical orbit $\boldsymbol{x}(t)$ the momentum is adjusted such that $|\boldsymbol{p}| = \sqrt{2m[E - U(\boldsymbol{x}(t))]}$. In quantum mechanics there are no orbits because of the uncertainty relation between position and momentum so that this interpretation can no longer hold. On the other hand, our experience with classical mechanics tells us that the electric potential energy $e\Phi(\boldsymbol{x})$ can equally well be replaced by a more general potential energy $U(\boldsymbol{x})$ describing other forces than the electric ones. In doing so the Eq. (1.52) is generalized to a fundamental differential equation of nonrelativistic quantum mechanics:

$$\boxed{\mathrm{i}\hbar\dot{\psi}(t,\boldsymbol{x}) = \left(-\frac{\hbar^2}{2m}\Delta + U(t,\boldsymbol{x})\right)\psi(t,\boldsymbol{x}) \quad \text{(E. Schrödinger, 1926)}} \,. \tag{1.59}$$

This equation is the *time-dependent Schrödinger equation*. Its right-hand side contains the analogue of the classical Hamiltonian function $H = \boldsymbol{p}^2/(2m) + U(t,\boldsymbol{x})$. Therefore, the Schrödinger equation can also be written as follows

$$\mathrm{i}\hbar\dot{\psi}(t,\boldsymbol{x}) = H\psi(t,\boldsymbol{x}) \quad \text{with} \quad H = \left(-\frac{\hbar^2}{2m}\Delta + U(t,\boldsymbol{x})\right) .$$

Note that H, which is no longer a function on phase space, becomes an operator acting on wave functions $\psi(t,\boldsymbol{x})$.

In those cases where the function U does not depend on time, one inserts (1.56) into (1.59) and obtains the *time-independent Schrödinger equation*

$$\boxed{E\psi(\boldsymbol{x}) = \left(-\frac{\hbar^2}{2m}\Delta + U(\boldsymbol{x})\right)\psi(\boldsymbol{x})} \,. \tag{1.60}$$

These two equations, their generalization to more than one particle and to other degrees of freedom than position and momentum, will be the subject of a detailed analysis in what follows.

In the previous section we argued that the wave function cannot and should not be understood to be a classical wave. Rather, we concluded that, in a sense to be made more precise, it contains statistical information about an individual particle, and, as a consequence, that it cannot yield deterministic predictions for a *single* particle. Only a very large number of events obtained under identical conditions, can be compared to theoretical predictions. The statistical interpretation which is suggested here, is made more precise by the following fundamental postulate:

Postulate

If $\psi(t,\boldsymbol{x})$ is a solution of the Schrödinger equation (1.59), then $|\psi(t,\boldsymbol{x})|^2$ is the probability density for detecting the particle described by this equation in the point $\boldsymbol{x}$ of space, at time t.

This probabilistic interpretation of Schrödinger wave functions was first proposed by Max Born. In view of a wider range of applications to be dealt with below we formulate the postulate in a somewhat more general form:

Born's Interpretation for the Wave Function: $|\psi|^2(t)$ is the probability density to find the system at a given time t in the configuration described by ψ.

Note the important step initiated by this postulate: The quantum dynamics of the particle, or, for that matter, of a more general system, is contained in the wave function ψ which fulfills the Schrödinger equation (1.59) (in this form, or in a form adapted to more general systems). This equation has many features familiar from classical dynamics but, a priori, it tells us little on how to extract physics from its solutions and, in particular, how to derive physical observables from it. The postulate of Born says that $|\psi|^2$ is a probability density. This means that

$$|\psi(t, \boldsymbol{x})|^2 \mathrm{d}^3 x \,,$$

in the case of a single particle, is the probability to find the particle at time t in the volume element $\mathrm{d}^3 x$ around the point $\boldsymbol{x}$ in space.

We encounter here a probability distribution similar to (1.29) of Sect. 1.2.1, though in a radically different context: There, we dealt with a ensemble of very many particles, described by classical mechanics. Our knowledge of the ensemble is incomplete but can be improved at any time, at least in principle. Here, a complex function $\psi(t, \boldsymbol{x})$ is the source of the observable density $|\psi|^2$ that, qualitatively speaking, we cannot penetrate any further. As we remarked earlier, this function is strictly deterministic in the sense that a given initial distribution $\psi(t_0, \boldsymbol{x})$ fixes the wave function for all times (as long as one does not run into possible singularities of $U(t, \boldsymbol{x})$). Nevertheless, in general, it does not allow for a definite prediction for the individual particle (or, more generally, for an individual event). Only after having performed very many measurements on a set of identically prepared particles can one compare to predictions that are obtained by a well-defined prescription from the density $|\psi|^2$.

One discovers here a way of describing physical phenomena which is fundamentally new as compared to classical, non-quantum physics: a statistical description that does not allow, as a matter of principle, to focus and improve indefinitely on the information about the system under consideration up to the point where states are points in phase space. In other terms, unlike in classical mechanics, a point of phase space has no physical meaning in quantum mechanics.

This way of proceeding raises deep questions which go far beyond the familiar framework of classical physics. Therefore, it will need great care and solid preparation to answer them. I advise the reader, if he or she is one of them, to first study in depth the postulates of quantum mechanics, their consequences and their experimental tests, without prejudice. In doing so one will not forget the fundamental questions raised by quantum theory but will establish a more solid ground for pondering them.

The statistical interpretation of the wave function, a quite bold step indeed, clarifies the situation almost at once. In a natural way it leads to a number of new concepts which are decisive for the predictive power of the theory and the description of experiments. If $|\psi(t, \boldsymbol{x})|^2$ is the probability density of Born's interpretation, then the integral

$$\int_V \mathrm{d}^3x \, |\psi(t, \boldsymbol{x})|^2$$

over a closed domain V of $\mathbb{R}^3$ is the probability to find the particle in this volume at time t. As the particle, at any time, must be *somewhere* in space, integrating over the whole space must yield the answer "with certainty", i.e. the probability 1. Thus, it is natural to impose the condition of integrability

$$\int \mathrm{d}^3x \, |\psi(t, \boldsymbol{x})|^2 = 1 \,. \tag{1.61}$$

Wave functions which are interpreted in a statistical sense, must be square integrable. Expressed in mathematical symbols,

$$\psi(t, \boldsymbol{x}) \in L^2(\mathbb{R}^3) \,,$$

where $L^2(\mathbb{R}^3)$ is the space of complex, square integrable functions over $\mathbb{R}^3$.

The current density defined in (1.54) describes then the flux of probability. That is to say, if one calculates the surface integral of the normal component of $\boldsymbol{j}(t, \boldsymbol{x})$ over the surface Σ enclosing the volume,

$$\int_\Sigma \mathrm{d}\sigma \, \boldsymbol{j}(t, \boldsymbol{x}) \cdot \hat{\boldsymbol{n}} \,,$$

one obtains the probability per unit of time for a particle to cross the surface Σ. If the integral is positive the particle left the volume. If it is negative the particle penetrated this closed domain.

The continuity equation (1.55) expresses in mathematical terms the conservation of the probability to find the particle somewhere in space, at any time. This is seen as follows. Consider the time derivative of the probability density $\varrho(t, \boldsymbol{x}) = \psi^*(t, \boldsymbol{x})\psi(t, \boldsymbol{x})$, integrated over the entire space. Making use of the continuity equation (1.55) one has

$$\frac{\mathrm{d}}{\mathrm{d}t} \int \mathrm{d}^3x \, \varrho(t, \boldsymbol{x}) = \int \mathrm{d}^3x \, \frac{\partial}{\partial t} \varrho(t, \boldsymbol{x}) = -\int \mathrm{d}^3x \, \nabla \cdot \boldsymbol{j}(t, \boldsymbol{x}) \,.$$

If the wave function ψ vanishes sufficiently rapidly for $|\boldsymbol{x}| \to \infty$ then the integral on the right-hand side can be converted to a surface integral of the normal component of $\boldsymbol{j}$. This surface being at infinity the surface integral vanishes. Thus, one obtains the conservation law

$$\boxed{\frac{\mathrm{d}}{\mathrm{d}t} \int \mathrm{d}^3x \, \varrho(t, \boldsymbol{x}) = 0} \,. \tag{1.62}$$

The probability of finding the particle *somewhere* in space is independent of time. This means that the particle can neither be created nor can it disappear. If the wave function is normalized at some initial time t_0 it remains normalized for all times. This remark shows why it is important that the Schrödinger equation (1.59) be of *first* order in the time derivative. It is this fact which guarantees the important property (1.62).

Born's interpretation of the wave function clarifies at once the statistical nature of the interference of matter waves. Suppose, for simplicity, that the initial state is prepared, at time $t = t_0$, as a linear combination of two solutions ψ_1 and ψ_2 of the Schrödinger equation,

$$\psi(t, \boldsymbol{x}) = c_1\psi_1(t, \boldsymbol{x}) + c_2\psi_2(t, \boldsymbol{x}) \quad \text{with} \quad c_1, c_2 \in \mathbb{C}\,.$$

The absolute square of this function is given by

$$\begin{aligned}|\psi(t, \boldsymbol{x})|^2 = |c_1|^2|\psi_1(t, \boldsymbol{x})|^2 + |c_2|^2|\psi_2(t, \boldsymbol{x})|^2 \\ + 2\mathrm{Re}[c_1^* c_2 \psi_1^*(t, \boldsymbol{x})\psi_2(t, \boldsymbol{x})]\end{aligned}$$

and describes the probability density for detecting the particle. As compared to an analogous classical situation the third term is new. It indicates that the wave functions ψ_1 and ψ_2 interfere, through coherent superposition. The sum of the individual probabilities of the first two terms can be enhanced or weakened by the interference term. In an extreme situation the interference term cancels the first two terms altogether.

Interference phenomena are familiar from the theory of classical waves. The statistical interpretation of the quantum wave function is new. It says that a single particle, at time $t > t_0$, can be recorded by a given, space-fixed detector with probability between 0 and 1. This probability may even be zero locally, whenever the interference is perfect and destructive. Imagine a particle state prepared at $t = 0$, $\boldsymbol{x} = \boldsymbol{0}$. Suppose that this point is surrounded by a sphere whose surface is equipped homogeneously with devices that allow to detect the particle.[12] At some time $t > 0$ an individual particle will be detected somewhere on the surface of the sphere, in one of the detectors, but it is impossible to predict which detector that will be. The predicted interference pattern will only appear, through the rates at which the detectors will fire, after very many identical measurements. A detector in a maximum of the interference pattern will give the highest rates, another detector in a minimum will give the lowest rate. In the case of complete and destructive interference, a detector in a minimum will never fire because the probability density vanishes at its position.

In quantum mechanics a harmonic time dependence such as in (1.56) means that the state described by the wave function ψ is a *stationary state*. Indeed, all physically relevant densities do not depend on time. In contrast to classical mechanics, wave functions of this type describe time independent states and are not oscillatory solutions.

As we will see shortly the requirement that $\psi(t, \boldsymbol{x})$ be square integrable is the decisive boundary condition for solutions of the Schrödinger equation. For example, the discrete energy spectra which are the analogues of the classical finite orbits, follow from it. We summarize the boundary condition obtained from the physical arguments discussed above, as follows:

---

[12] An arrangement of detectors which covers the entire solid angle is also called a "4π detector".

Born boundary condition: Only square integrable, normalized solutions are admissible and can be interpreted in terms of physics.

In actually solving the Schrödinger equation one often uses the following alternative condition:

Schrödinger boundary condition: In the whole domain of their definition physically realizable solutions must be uniquely defined and bounded.

This condition is not identical with Born's condition. A wave function that satisfies the Born boundary condition is not always bounded.

These two types of boundary conditions are relevant for important parts of quantum physics and will be illustrated by various applications. Nevertheless, their significance must be put in perspective. The careful reader will have noticed that the plane waves (1.57) are not square integrable. Wave functions of this type will be needed for the description of scattering states and must be understood as limiting expressions of normalized wave packets.

Furthermore, there are many quantum processes in which particles are created or annihilated. Examples are provided by the emission of a photon in the transition from an excited atomic state to the ground state, e.g.

$$(\text{H-atom}, n = 2) \longrightarrow (\text{H-atom}, n = 1) + \gamma \,.$$

Another example is pair annihilation of an electron and a positron into two photons

$$\mathrm{e}^- + \mathrm{e}^+ \longrightarrow \gamma + \gamma' \,.$$

In processes of this kind the "conservation of the probability" will still be valid in a generalized form, though certainly not in the simple one described in (1.62).

1.5 Expectation Values and Observables

The probability density (1.53) is a real, measurable, hence *classical* quantity. Although of a very different origin than the densities (1.29) which describe a many particle system in statistical mechanics, it will enter, in much the same way, the calculation of averages of observables. In the simplest case let $F(\boldsymbol{x})$ be an observable which depends on the coordinates only, i.e. which is a real function of $\boldsymbol{x}$ over phase space. The average of this quantity in the state described by the wave function ψ is calculated in the same way as in (1.32):

$$\langle F \rangle_\psi (t) = \int \mathrm{d}^3x \; F(\boldsymbol{x}) |\psi(t, \boldsymbol{x})|^2 \equiv \int \mathrm{d}^3x \; \psi^*(t, \boldsymbol{x}) F(\boldsymbol{x}) \psi(t, \boldsymbol{x}) \,.$$

In the case of a function which depends on $\boldsymbol{x}$ only, the second integral is trivially equal to the first because $F(\boldsymbol{x})$ or $F(t, \boldsymbol{x})$ commute with ψ or ψ^*. However, if the observable also depends on momenta, i.e. on the remaining coordinates in phase space, $F = F(t, \boldsymbol{x}, \boldsymbol{p})$, then more care is needed. Indeed, if the conjecture (1.58) holds true, i.e. if the classical momentum variable is replaced by the nabla operator, then also F becomes an operator,

$$F = F\left(\boldsymbol{x}, \frac{\hbar}{\mathrm{i}}\nabla\right) .$$

In this situation we must take care of the fact that F no longer commutes with ψ and ψ^*, and must make sure that the average is a real, not a complex number. As we will see in a moment, only the second form of the average fulfills this condition.

Because of the basic conceptual differences, as compared to the *classical* statistical mechanics, that we worked out in some detail, the quantum mechanical average has a different formal and physical content. It is called *expectation value* and is defined as follows:

Definition 1.2 Expectation Value

Let $F(\boldsymbol{x}, \boldsymbol{p})$ be a classical physical observable, defined on the phase space of a single particle system. From this function an operator is constructed by replacing $\boldsymbol{p}$ by $(\hbar/i)\nabla$,

$$F\left(\boldsymbol{x}, \frac{\hbar}{\mathrm{i}}\nabla\right) ,$$

in such a way that the quantity

$$\langle F\rangle_{\psi}(t) := \int \mathrm{d}^3x\, \psi^*(t, \boldsymbol{x}) F\left(\boldsymbol{x}, \frac{\hbar}{\mathrm{i}}\nabla\right) \psi(t, \boldsymbol{x}) \equiv (\psi, F\psi) \qquad (1.63)$$

is real. This quantity is said to be the *expectation value* of the observable F in the state ψ. It yields the experimental value of the observable, i.e. the value that one will find, in the sense of statistics, after very many measurements under similar conditions.

Remarks

1. It would be more consistent to mark the symbol that stands for an operator constructed from the classical function F in a special way. For instance, one could write $\underline{F}$ for the operator, and F for the classical function. However, in most cases it will be clear from the context whether we have in mind the function or the associated operator. For this reason I make this typographical distinction only in exceptional cases.
2. In the second form of (1.63) I have used a notation that reminds the mathematically inclined reader of a scalar product. Although of no relevance for the moment this will gain special significance later.

3. The condition of reality $\langle F\rangle_\psi = \langle F\rangle^*_\psi$ means that one must have

$$\int \mathrm{d}^3x\, \psi^*(t,\boldsymbol{x})F\left(\boldsymbol{x},\frac{\hbar}{\mathrm{i}}\nabla\right)\psi(t,\boldsymbol{x})$$
$$= \int \mathrm{d}^3x \left[F\left(\boldsymbol{x},\frac{\hbar}{\mathrm{i}}\nabla\right)\psi(t,\boldsymbol{x})\right]^* \psi(t,\boldsymbol{x})\,.$$

Operators which have this property are called *self-adjoint*. As one verifies by partial integration the simple examples

$$\boldsymbol{x} \;\text{(position)}, \qquad \frac{\hbar}{\mathrm{i}}\nabla \;\text{(momentum)},$$
$$\boldsymbol{\ell} = \frac{\hbar}{\mathrm{i}}\boldsymbol{x}\times\nabla \;\text{(orbital angular momentum)}$$

fulfill this condition. Here is an example:

$$\int \mathrm{d}^3x\, \psi^*\frac{\hbar}{\mathrm{i}}\nabla\psi = -\frac{\hbar}{\mathrm{i}}\int \mathrm{d}^3x(\nabla\psi)^*\psi = +\int \mathrm{d}^3x\left(\frac{\hbar}{\mathrm{i}}\nabla\psi\right)^*\psi\,.$$

For other observables the "translation" of the classical function on phase space to their associated, self-adjoint operators needs a careful discussion. The following section begins this analysis.

1.5.1 Observables as Self-Adjoint Operators on $L^2(\mathbb{R}^3)$

The notion of observable is a familiar one in classical mechanics. In mechanics it is represented by a real function on phase space and describes a quantity that can be measured with some physical apparatus. The dynamics of the system under consideration tells us which observables are relevant for its analysis. In particular, the dynamics tells us how many observables are needed for a complete description of the system. Hamiltonian systems whose dynamics is defined by the knowledge of the Hamiltonian function, are used as landmarks for orientation in the transition to quantum mechanics. As we noticed in the previous section the observables of quantum mechanics which take over the role of their classical analogues, must yield real expectation values. This is indeed the case if these operators are self-adjoint. This property is made more precise by the following definition:

Definition 1.3

An operator F, defined on the space $L^2(\mathbb{R}^3)$ of square integrable functions, is said to be *self-adjoint* if its action on sufficiently many elements φ of this space is well-defined and if

$$\int \mathrm{d}^3x\, \varphi^* F\varphi = \int \mathrm{d}^3x\,(F\varphi)^*\varphi \tag{1.64}$$

holds for all such elements $\varphi \in L^2(\mathbb{R}^3)$. (For further details see later.)

Using the notation of a scalar product, cf. right-hand side of (1.63), this property takes the form

$$(\varphi, F\varphi) = (F\varphi, \varphi) \tag{1.65}$$

whose general significance will be clarified by a later, more detailed mathematical analysis. The property (1.64) entails that every expectation value of the observable F is real,

$$\langle F \rangle_\psi = \langle F \rangle_\psi^* \ .$$

The following list gives a few examples for observables.

$$\begin{aligned}
x^k &\longleftrightarrow x^k , \\
p_k &\longleftrightarrow \frac{\hbar}{\mathrm{i}} \frac{\partial}{\partial x^k} , \\
\frac{\boldsymbol{p}^2}{2m} &\longleftrightarrow -\frac{\hbar^2}{2m} \Delta , \\
\boldsymbol{x} \times \boldsymbol{p} &\longleftrightarrow \frac{\hbar}{\mathrm{i}} \boldsymbol{x} \times \nabla , \\
\boldsymbol{x} \cdot \boldsymbol{p} &\longleftrightarrow \frac{\hbar}{2\mathrm{i}} \{\boldsymbol{x} \cdot \nabla + \nabla \cdot \boldsymbol{x}\} , \\
\boldsymbol{A} \cdot \boldsymbol{p} &\longleftrightarrow \frac{\hbar}{2\mathrm{i}} \{\boldsymbol{A} \cdot \nabla + \nabla \cdot \boldsymbol{A}\} \ .
\end{aligned}$$

The left column shows the classical function on phase space while the right column gives the corresponding self-adjoint operator.

Remarks

1. Of course, for a given operator one must investigate on which elements $\varphi \in L^2(\mathbb{R}^3)$ it is well-defined. In this way one identifies the domain of definition of the operator. We will come back to these questions in more detail later. For the moment the heuristic approach described above may be sufficient
2. Among the examples just given the last two are particularly noteworthy. They show that the product of a vector field $\boldsymbol{v}(\boldsymbol{x})$ and the momentum $\boldsymbol{p}$ must be replaced by $\hbar/2\mathrm{i}$ times the symmetric combination of $\boldsymbol{v} \cdot \nabla$ and $\nabla \cdot \boldsymbol{v}$. It is understood that the gradient acts on all functions to the right in accordance with the product rule. In the second term, for instance, this action is

$$\nabla \cdot \boldsymbol{v}(\boldsymbol{x})\psi(\boldsymbol{x}) = \psi(\boldsymbol{x})[\nabla \cdot \boldsymbol{v}(\boldsymbol{x})] + \boldsymbol{v}(\boldsymbol{x}) \cdot [\nabla\psi(\boldsymbol{x})] \, .$$

   If one retained only the first, or the second, term the operator constructed in this way would not be self-adjoint.
3. The examples given here raise the question of uniqueness in translating classical observables to self-adjoint operators. In essence, the answer to this question is the following: In general, the transition from a real function on phase space which might describe a classical observable, to a self-adjoint operator is not unique. That is to say, it may happen that there is more than one such operator corresponding to the given classical function on phase space. In such cases

a further principle fixing the choice may be necessary. In itself, this statement will not surprise the reader because quantum mechanics is expected to be the embracing theory which contains classical mechanics as a limiting case.
The dynamical variables relevant for point-like particles, as a rule, are polynomials in $\boldsymbol{x}$ and $\boldsymbol{p}$, whose degree is smaller than or equal to 2. For functions of this type there is generally only one way of choosing a self-adjoint operator. Thus, it seems that in cases relevant in practice, the translation leads to a uniquely defined operator.

4. The property of self-adjointness (1.64) or (1.65) implies that for two different elements φ_n, φ_m of $L^2(\mathbb{R}^3)$ one has

$$\int \mathrm{d}^3x\, \varphi_m^* F\varphi_n = \int \mathrm{d}^3x\, (F\varphi_m)^* \varphi_n$$

or, when written in terms of scalar products,

$$(\varphi_m, F\varphi_n) = (F\varphi_m, \varphi_n) = (\varphi_n, F\varphi_m)^* \,.$$

The proof will be given below when further mathematical tools will be available. Note that in general $(\varphi_m, F\varphi_n)$ are complex numbers, in contrast to the expectation values $(\varphi_m, F\varphi_m)$ which are real. We know by now that in physics this second example describes the result of a large number of measurements of the observable F in the state described by φ_m. The possibly complex number $(\varphi_m, F\varphi_n)$, with $m \neq n$, will enter the calculation of the probability of the state φ_n to make a transition to the state φ_m under the action of the observable F.

5. A specific example that we study in the next section will illustrate a general property of the space of functions $L^2(\mathbb{R}^3)$: this space can be described by means of a *basis of functions*

$$\{\varphi_n(\boldsymbol{x}) | n = 1, 2, \ldots\} \,.$$

A basis of this type *spans* the space $L^2(\mathbb{R}^3)$, in a sense known from linear algebra. If this is so the numbers $(\varphi_m, F\varphi_n)$ are the entries of an infinite-dimensional matrix

$$F_{mn} = (\varphi_m, F\varphi_n) \,,$$

which is *hermitian*,[13] which is to say that its entries fulfill the relations $F_{mn} = F_{nm}^*$.

6. All operators listed in the table are *linear*, i.e. with φ_1, and φ_2 elements of $L^2(\mathbb{R}^3)$ and with arbitrary complex constants c_1, c_2, they fulfill

$$F(c_1\varphi_1 + c_2\varphi_2) = c_1 F\varphi_1 + c_2 F\varphi_2, \qquad c_1, c_2 \in \mathbb{C}\,,$$
$$\varphi_1, \varphi_2 \in L^2(\mathbb{R}^3)\,. \tag{1.66}$$

The class of linear operators plays a central role in quantum mechanics. Linear operators are used not only for describing observables (in which case they

[13] Called so after Charles Hermite (1822–1901), the French mathematician.

must be self-adjoint) but also for the analogues of classical canonical transformations (in which case they are unitary). In the context of time reversal, or reversal of motion, we will also encounter *antilinear* operators. These are operators for which (1.66) is replaced by

$$F(c_1\varphi_1 + c_2\varphi_2) = c_1^* F\varphi_1 + c_2^* F\varphi_2 \,. \tag{1.67}$$

Note that the right-hand side contains the complex conjugate c-numbers.

Returning to the Schrödinger equation (1.59) one sees that, indeed, it has the general form

$$\boxed{\mathrm{i}\hbar\dot{\psi}(t, \boldsymbol{x}) = H\psi(t, \boldsymbol{x})}\,, \tag{1.68}$$

where H is the self-adjoint *Hamilton operator*, or *Hamiltonian* for short, which is obtained from the classical Hamiltonian function by the procedure described above. As an example consider the Hamiltonian function (1.50) which describes the electron in external fields. In accord with the first remark above it is replaced by the self-adjoint Hamilton operator

$$\begin{aligned} H &= \frac{1}{2m}\left(\frac{\hbar}{\mathrm{i}}\nabla - \frac{e}{c}\boldsymbol{A}\right)^2 + e\Phi \qquad (1.69)\\ &= \frac{1}{2m}\left(-\hbar^2\Delta - \frac{\hbar}{\mathrm{i}}\frac{e}{c}\nabla\cdot\boldsymbol{A} - \frac{\hbar}{\mathrm{i}}\frac{e}{c}\boldsymbol{A}\cdot\nabla + \frac{e^2}{c^2}\boldsymbol{A}^2\right) + e\Phi \,. \end{aligned}$$

Note, in particular, the replacement of $\boldsymbol{p}\cdot\boldsymbol{A}$ by half the sum of $\boldsymbol{p}\cdot\boldsymbol{A}$ and $\boldsymbol{A}\cdot\boldsymbol{p}$, needed to make H hermitian.

1.5.2 Ehrenfest's Theorem

Let F be a hermitian operator, defined on the space $L^2(\mathbb{R}^3)$ of complex, square integrable functions, and which corresponds to a classical, possibly time dependent, observable $F(t, \boldsymbol{x}, \boldsymbol{p})$. Let $\langle F\rangle$ be its expectation value in some state ψ which is a solution of the Schrödinger equation (1.68) with the Hamiltonian

$$H = \left(-\frac{\hbar^2}{2m}\Delta + U(t, \boldsymbol{x})\right) .$$

By making use of the equations

$$\dot{\psi} = -\frac{\mathrm{i}}{\hbar}H\psi \quad \text{and} \quad \dot{\psi}^* = \frac{\mathrm{i}}{\hbar}H\psi^*$$

the time derivative of the expectation value is expressed in terms of the commutator of the Hamiltonian and the observable as follows

$$\begin{aligned} \frac{\mathrm{d}}{\mathrm{d}t}\langle F\rangle &= \frac{\partial\langle F\rangle}{\partial t} + \int \mathrm{d}^3x\, \left\{\psi^* F\dot{\psi} + \dot{\psi}^* F\psi\right\} \\ &= \frac{\partial\langle F\rangle}{\partial t} + \frac{\mathrm{i}}{\hbar}\int \mathrm{d}^3x\, \psi^*\{HF - FH\}\psi \\ &= \left\langle\frac{\partial F}{\partial t}\right\rangle + \frac{\mathrm{i}}{\hbar}\langle [H, F]\rangle \,. \end{aligned}$$

The second term of the integral initially contains the factor $(H\psi^*)(F\psi)$. The operator H is moved to the right of ψ^*, without further change, because it is self-adjoint.

I insert here an important remark: The same equation applies also to the time derivative of arbitrary matrix elements of F

$$(\varphi_m, F\varphi_n) = \int \mathrm{d}^3x\, \varphi_m^* F \varphi_n\,,$$

so that we may write it as an identity relating operators,

$$\frac{\mathrm{d}}{\mathrm{d}t}F = \frac{\partial F}{\partial t} + \frac{\mathrm{i}}{\hbar}[H, F]\,. \tag{1.70}$$

(This conclusion may need a more careful mathematical analysis, though.) This equation is said to be the *Heisenberg equation of motion* . It has a striking similarity to a well-known equation of classical mechanics,

$$\frac{\mathrm{d}}{\mathrm{d}t}F(t, \boldsymbol{x}, \boldsymbol{p}) = \frac{\partial F(t, \boldsymbol{x}, \boldsymbol{p})}{\partial t} + \{H(t, \boldsymbol{x}, \boldsymbol{p}), F(t, \boldsymbol{x}, \boldsymbol{p})\}$$

([Scheck (2010)], Sect. 2.32), in which $\{\cdot,\cdot\}$ denotes the Poisson bracket

$$\{f, g\} = \frac{\partial f}{\partial p_i}\frac{\partial g}{\partial q^i} - \frac{\partial f}{\partial q^i}\frac{\partial g}{\partial p_i}\,.$$

Apparently, in the process of quantization the Poisson bracket of the Hamiltonian function H with the observable F is replaced by the commutator of the Hamiltonian $\underline{H}$ with the operator $\underline{F}$ according to the rule

$$\{H, F\} \longleftrightarrow \frac{\mathrm{i}}{\hbar}[\underline{H}, \underline{F}]\,. \tag{1.71}$$

The constant $\hbar$ which appears in the denominator is perhaps not too surprising in the light of the following remark: Strictly speaking, the analogy should be formulated the other way around, quantum mechanics being expected to be the more general framework that embraces classical mechanics. Suppose we expand the commutator $[\underline{H}, \underline{F}]$, as well as the operators $\underline{H}$ and $\underline{F}$ themselves in powers of $\hbar$. At order $(\hbar)^0$ we will obtain the commutator of two ordinary functions which, of course, vanishes. The order $(\hbar)^1$, however, will contain first derivatives which do not commute and which are likely to be the ones appearing in the Poisson bracket. The factor $1/\hbar$ drops out and so does the factor i because momenta are replaced by $-\mathrm{i}$ times first derivatives.

There is an application of the equation proven above, viz.

$$\frac{\mathrm{d}}{\mathrm{d}t}\langle\underline{F}\rangle = \left\langle\frac{\partial \underline{F}}{\partial t}\right\rangle + \frac{\mathrm{i}}{\hbar}\langle[\underline{H}, \underline{F}]\rangle\,,$$

which is of special importance for physics:

Ehrenfest's Theorem: The expectation values of position and momentum in a quantum mechanical system which corresponds to a classical Hamiltonian system of mechanics, satisfy the classical equations of motion. In the case of a single particle described by

$$\underline{H} = \underline{\boldsymbol{p}}^2/(2m) + U(t, \underline{\boldsymbol{x}})$$

the equations of motion read

$$\frac{\mathrm{d}}{\mathrm{d}t}\langle \boldsymbol{x}\rangle = \frac{1}{m}\left\langle \underline{\boldsymbol{p}}\right\rangle , \tag{1.72}$$

$$\frac{\mathrm{d}}{\mathrm{d}t}\left\langle \underline{\boldsymbol{p}}\right\rangle = -\left\langle \nabla \underline{U}\right\rangle . \tag{1.73}$$

For the sake of clarity, and as an exception, we distinguished for a while the functions H, F, on phase space, and the operators $\underline{H}$, $\underline{F}$, respectively. From here on we return to the former, less pedantic notation.

Proof of the theorem

Neither the position operator nor the momentum operator depend on time. In calculating the commutators of the Hamiltonian with these two operators we make use, for the former, of the formula

$$[A^2, B] = AAB - BAA = A[A, B] + [A, B]A$$

and obtain

$$[\underline{H}, \underline{x}^i] = \frac{1}{2m}[\underline{p}_i^2, \underline{x}^i] = \frac{1}{2m}2\frac{\hbar}{\mathrm{i}}\underline{p}_i ,$$

so that

$$\frac{\mathrm{i}}{\hbar}[\underline{H}, \underline{\boldsymbol{x}}] = \frac{1}{m}\underline{\boldsymbol{p}} .$$

Taking the expectation value of this result proves the first part (1.72) of the theorem. For the proof of its second part one calculates the commutator

$$[\underline{H}, \underline{p}_i] = -\frac{\hbar}{\mathrm{i}}\frac{\partial U}{\partial x^i}, \quad \text{or} \quad \frac{\mathrm{i}}{\hbar}[\underline{H}, \underline{\boldsymbol{p}}] = -\nabla U .$$

Inserting this result into the expectation value proves the second half (1.73) of the theorem.

1.6 A Discrete Spectrum: Harmonic Oscillator in one Dimension

If a particle moves in an *attractive* potential $U(\boldsymbol{x})$ there may exist bound states. In classical physics this will happen whenever the particle is "caught" in a potential well, i.e. if the function U is locally concave and if the energy is chosen such that the particle cannot escape to spatial infinity. Figure 1.6 shows an example for $U(x)$ in one dimension, though of no particular physical significance, which illustrates what we mean by this.

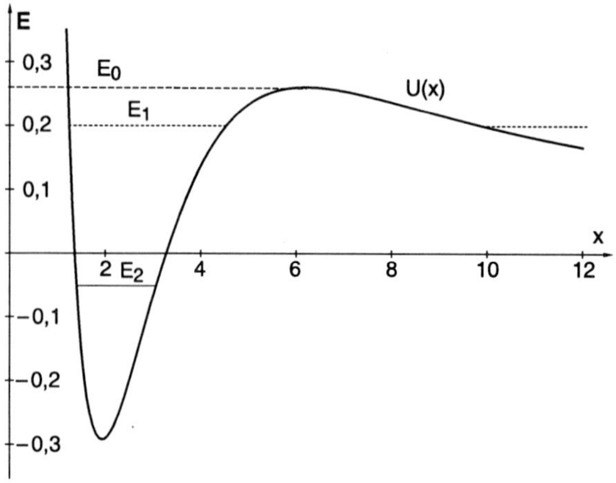

Fig. 1.6 Example of a potential in one space dimension which classically allows both for finite, bound orbits, and for unbound trajectories on which the particle can escape to infinity. Classically, the particle is confined whenever it moves within the potential well on the left of the picture

For the examples of the one-dimensional harmonic oscillator and of the spherical oscillator where

$$U(x) = \frac{1}{2} m\omega^2 x^2 \, ,$$

$$U(r) = \frac{1}{2} m\omega^2 r^2 \, , \qquad (r = |\boldsymbol{x}|) \, ,$$

respectively, the particle is caught for any value of the energy E, and all states are bound states. In the case of the attractive Coulomb potential

$$U(r) = -\frac{\alpha}{r} \quad \text{with} \quad r = |\boldsymbol{x}|, \quad \alpha > 0 \, ,$$

bound states, i.e. finite (classical) orbits occur only for negative energies, $E < 0$. A particle on an orbit with positive energy possesses enough kinetic energy for escaping to infinity and, hence, is not bound.

In intervals of the energy where there are classical finite orbits the corresponding quantum system *may* (but *need not*) have bound states. If such states exist they belong to discrete values of the energy. The reason for this to happen is that bound states must have localized, square integrable wave functions. In general, Born's boundary condition will be satisfied, in the most favourable cases, only for selected, discrete values of the energy. In the cases of the one-dimensional oscillator and of the spherical oscillator all states are bound states, the energy spectrum is found to be a *fully discrete spectrum*.

In the case of the attractive Coulomb potential there are bound states and discrete energies only if $E < 0$, while states with positive energy are not bound and can take any value $E > 0$ of the energy. The corresponding quantum mechanical energy spectrum consists of a discrete part ($E < 0$), and a continuous part ($E > 0$). One says that this system has a *mixed spectrum*.

In examples such as the repulsive Coulomb potential, or the special case $U \equiv 0$, there are no bound states, neither in classical nor in quantum mechanics. The spectrum then is *fully continuous*.

This section deals with a simple but especially important example for a fully discrete spectrum: the harmonic oscillator in one dimension. The hydrogen atom as an important example of a mixed spectrum, will be analyzed in Sect. 1.9.5. The case of the fully continuous spectrum will be illustrated by the study of plane waves in Sect. 1.8.4.

We return to the one-dimensional form of the Schrödinger equation (1.68), insert the Hamiltonian

$$H = -\frac{\hbar^2}{2m}\frac{\mathrm{d}^2}{\mathrm{d}x^2} + \frac{1}{2}m\omega^2 x^2 ,$$

and choose the stationary form of solutions

$$\psi(t, x) = \mathrm{e}^{-(\mathrm{i}/\hbar)Et}\varphi(x) .$$

Equation (1.60) then takes the form

$$-\frac{\hbar^2}{2m}\varphi''(x) + \frac{1}{2}m\omega^2 x^2 \varphi(x) = E\varphi(x) \qquad (*)$$

The constants $\hbar$, m, and ω are combined to a reference energy and a reference length, respectively, as follows

$$\hbar\omega \quad \text{and} \quad b := \sqrt{\frac{\hbar}{m\omega}} .$$

This suggests to replace both the energy and the variable x by dimensionless variables

$$\varepsilon := \frac{E}{\hbar\omega} \quad \text{and} \quad u := \frac{x}{b} ,$$

respectively.[14] As one verifies immediately the stationary differential equation $(*)$ takes the simple form

$$-\varphi''(u) + u^2\varphi(u) = 2\varepsilon\varphi(u) . \qquad (**)$$

For simplicity we keep the same symbol for the unknown function but we write explicitly its dependence on the dimensionless variable u. The problem to be solved is well defined: find all solutions of this ordinary differential equation of second order which are everywhere finite and square integrable, and determine those values of ε for which such solutions exist. Instead of attacking the problem directly we use a seemingly innocent trick which will turn out to be instructive, from the perspective of physics, and will allow for alternative physical interpretations. Define two differential operators

$$a^\dagger := \frac{1}{\sqrt{2}}\left(u - \frac{\mathrm{d}}{\mathrm{d}u}\right) = \frac{1}{b\sqrt{2}}\left(x - b^2\frac{\mathrm{d}}{\mathrm{d}x}\right) , \qquad (1.74)$$

---

[14] This was not possible in the case of the classical oscillator which does not contain the dimensionful constant $\hbar$. Only in the example of the plane mathematical pendulum was there a reference energy, $mg\ell$.

$$a := \frac{1}{\sqrt{2}}\left(u + \frac{\mathrm{d}}{\mathrm{d}u}\right) = \frac{1}{b\sqrt{2}}\left(x + b^2\frac{\mathrm{d}}{\mathrm{d}x}\right) . \tag{1.75}$$

Neither of these operators is self-adjoint. Indeed, while $\mathrm{i}\,\mathrm{d}/\mathrm{d}u$ has this property, $\mathrm{d}/\mathrm{d}u$ without the factor i does not. In turn, partial integration shows that

$$\left(\varphi, \frac{\mathrm{d}}{\mathrm{d}u}\varphi\right) = \left(\left(-\frac{\mathrm{d}}{\mathrm{d}u}\right)\varphi, \varphi\right) .$$

If two operators A and A^* have the same domain of definition $\mathcal{D}$ and if

$$(\varphi, A\varphi) = (A^*\varphi, \varphi) \quad \text{for all} \quad \varphi \in \mathcal{D} ,$$

the operator A^* is said to be the *adjoint* operator of A. In this case one also has $(A^*)^* = A$, i.e. A is the adjoint of A^*. Thus, the operators $A = \mathrm{d}/\mathrm{d}u$ and $A^* = -\mathrm{d}/\mathrm{d}u$ are adjoints of one another. The same property is shared by the pair a and $a^\dagger$. If these operators were matrices the hermitian conjugate of M would be denoted by $M^\dagger$ – hence the notation in (1.74) and (1.75).[15]

The product $a^\dagger a$ is calculated as follows. Using the product rule for differentials in the second term one finds

$$a^\dagger a = \frac{1}{2}\left(u^2 - \frac{\mathrm{d}}{\mathrm{d}u}u + u\frac{\mathrm{d}}{\mathrm{d}u} - \frac{\mathrm{d}^2}{\mathrm{d}u^2}\right) = \frac{1}{2}\left(u^2 - 1 - \frac{\mathrm{d}^2}{\mathrm{d}u^2}\right) .$$

Thus, the differential equation $(**)$ takes the simple form

$$\left(a^\dagger a + \frac{1}{2}\right)\varphi(u) = \varepsilon\varphi(u) . \tag{***}$$

Note that the Hamiltonian now becomes remarkably simple, viz.

$$H = \hbar\omega\left(a^\dagger a + \frac{1}{2}\right) . \tag{1.76}$$

In much the same way one calculates the product $aa^\dagger$ of the two operators in the alternative order. One finds

$$aa^\dagger = \frac{1}{2}\left(u^2 + 1 - \frac{\mathrm{d}^2}{\mathrm{d}u^2}\right) .$$

This and the previous result combined give the important commutator

$$\boxed{[a, a^\dagger] \equiv aa^\dagger - a^\dagger a = 1} , \tag{1.77}$$

supplemented by the obvious relations

$$[a, a] = 0, \qquad [a^\dagger, a^\dagger] = 0 .$$

[15] In the mathematical literature adjoints are usually denoted by a "star", while complex conjugates are marked with an "over-bar". In the literature on quantum physics the star is traditionally used for complex conjugates, the "dagger" for adjoints, the over-bar being needed in the theory of Dirac spinors.

As the next step one shows that if $\varphi(u)$ is a solution of (***) pertaining to the eigenvalue ε then the functions obtained from φ by the action of $a^\dagger$ or of a, are also solutions and pertain to the eigenvalues $\varepsilon + 1$ and $\varepsilon - 1$, respectively. To see this take

$$\left(a^\dagger a + \frac{1}{2}\right)\left(a^\dagger \varphi\right) = \left(a^\dagger (a^\dagger a + 1) + \frac{1}{2} a^\dagger\right)\varphi = \left(\varepsilon - \frac{1}{2} + \frac{3}{2}\right)(a^\dagger \varphi)$$
$$= (\varepsilon + 1)(a^\dagger \varphi)\,.$$

In the first step one replaces $aa^\dagger = a^\dagger a + 1$ by means of (1.77), in the second step we use the Schrödinger equation in the form of (***) and insert $a^\dagger a\varphi = (\varepsilon - 1/2)\varphi$. This shows that the wave function $(a^\dagger \varphi)$ is a solution and pertains to the eigenvalue $\varepsilon + 1$.

In much the same way one verifies that

$$\left(a^\dagger a + \frac{1}{2}\right)(a\varphi) = (\varepsilon - 1)(a\varphi)$$

This shows that $(a\varphi)$, if it does not vanish identically, is also a solution and belongs to the eigenvalue $\varepsilon - 1$. In other terms, by applying repeatedly the *raising operator* $a^\dagger$ to a given solution φ with eigenvalue ε one generates an infinite series of new solutions which belong to the eigenvalues

$$\varepsilon + 1, \varepsilon + 2, \varepsilon + 3, \ldots\,.$$

Alternatively, one may apply the *lowering operator* a to the same solution φ and generate a series of further solutions with eigenvalues

$$\varepsilon - 1, \varepsilon - 2, \ldots\,,$$

except for the case where $(a\varphi)$ vanishes identically. Indeed, one finds that this series stops after finitely many steps downward. The smallest value of ε is $\varepsilon_0 = 1/2$, all eigenvalues are contained in the formula $\varepsilon_n = 1/2 + n$ with $n \in \mathbb{N}_0$, i.e. $n = 0, 1, 2, \ldots$. In order to show this, one proves two properties:

1. All admissible values of ε must be *positive*: Making use of (***) one calculates the integral

$$\int_{-\infty}^{+\infty} \mathrm{d}u\, \varphi^*(u) a^\dagger a \varphi(u) = \left(\varepsilon - \frac{1}{2}\right) \int_{-\infty}^{+\infty} \mathrm{d}u\, \varphi^*(u)\varphi(u)$$
$$= \left(\varepsilon - \frac{1}{2}\right) \int_{-\infty}^{+\infty} \mathrm{d}u\, |\varphi(u)|^2\,.$$

By partial integration, the operator $a^\dagger$ is moved in front of $\varphi^*(u)$ so that the integral becomes

$$\int_{-\infty}^{+\infty} \mathrm{d}u\, [a\varphi(u)]^* [a\varphi(u)] = \int_{-\infty}^{+\infty} \mathrm{d}u\, |[a\varphi(u)]|^2 \geq 0\,,$$

and is found to be positive semi-definite. This is compatible with the previous expression only if the factor $(\varepsilon - 1/2)$ in front of the integral is greater than or equal to zero, i.e. if $\varepsilon \geqslant 1/2$.

2. The eigenfunction pertaining to the lowest eigenvalue ε_0 must fulfill the relation $[a\varphi_0(u)] \equiv 0$: If this were not so then also $(a\varphi_0)$ would be a solution with eigenvalue $\varepsilon_0 - 1$ – in contradiction to the assumption which was that ε_0 should be the lowest eigenvalue.
   As a result we obtain the spectrum

$$\varepsilon_n = \left(n + \frac{1}{2}\right), \quad \text{i.e.} \quad E_n = \left(n + \frac{1}{2}\right)\hbar\omega,$$
$$n \in \mathbb{N}_0 \quad (n = 0, 1, 2, \ldots). \tag{1.78}$$

   Indeed, the lowest state has exactly the minimal energy which was found to be marginally compatible with Heisenberg's uncertainty relation, cf. Sect. 1.2.3, Example 1.3. The remaining part of the spectrum is remarkably simple. All eigenvalues are equidistant, their difference is given by the quantum of energy $\hbar\omega$.

What is the shape of the corresponding eigenfunctions and what are their properties? In order to answer these questions consider first the ground state $(\varepsilon_0 = 1/2, \varphi_0)$ whose wave function follows from the condition

$$[a\varphi_0(u)] = 0, \ \text{i.e.} \ \left(u + \frac{\mathrm{d}}{\mathrm{d}u}\right)\varphi_0(u) = 0\,.$$

One sees that φ_0 must be proportional to $\mathrm{e}^{-u^2/2}$. Returning to the dimensionful variable x, normalizing $|\varphi_0(x)|^2$ to 1, and making use of the Gaussian integral of Sect. 1.3.3, one obtains the result

$$\varphi_0(x) = \frac{1}{b^{1/2}\pi^{1/4}}\mathrm{e}^{-x^2/(2b^2)}\,. \tag{1.79}$$

Note that, generally, wave functions whose arguments are points in d spatial dimensions, i.e. $\boldsymbol{x} \in \mathbb{R}^d$, and which are to be interpreted in the spirit of Born's postulate, must carry the physical dimension $1/L^{d/2}$ where L stands for "Length". Thus, on $\mathbb{R}^3$ the dimension must be $1/L^{3/2}$, while in our one-dimensional example it must be $1/L^{1/2}$.

The higher states are generated from φ_0 by repeated action of $a^\dagger$, i.e.

$$\varepsilon_n = n + \frac{1}{2}: \quad \varphi_n = \text{const}\, \underbrace{a^\dagger a^\dagger a^\dagger \cdots a^\dagger}_{n \text{ times}} \varphi_0$$

Defining the following polynomials, called *Hermite polynomials*

$$H_n(u) := \mathrm{e}^{u^2/2}\left(u - \frac{\mathrm{d}}{\mathrm{d}u}\right)^n \mathrm{e}^{-u^2/2} = \mathrm{e}^{u^2/2}\left(\sqrt{2}\,a^\dagger\right)^n \mathrm{e}^{-u^2/2}\,, \tag{1.80}$$

the eigenfunctions are seen to be given by

$$\varphi_n(x) = N_n \mathrm{e}^{-x^2/(2b^2)} H_n\left(\frac{x}{b}\right)\,.$$

The factor N_n must be determined such that $|\varphi_n(x)|^2$ is normalized to 1. Before turning to the calculation of N_n we collect a few properties of the polynomials defined in (1.80).

Hermite Polynomials:

1. $H_n(u)$ is a real polynomial of degree n, the coefficient of u^n is 2^n. This is seen as follows

$$
\begin{aligned}
e^{u^2/2}\left(u-\frac{d}{du}\right)^n e^{-u^2/2} &= \sum_{m=0}^{n}\binom{n}{m}(-)^m u^{n-m} e^{u^2/2}\frac{d^m}{du^m}e^{-u^2/2} \\
&= \sum_{m=0}^{n}\binom{n}{m}(-)^m u^{n-m} e^{u^2/2}\left((-)^m u^m+\ldots\right)e^{-u^2/2} \\
&= \left[\sum_{m=0}^{n}\binom{n}{m}\right]u^n+\mathcal{O}(u^{n-1}) = 2^n u^n+\mathcal{O}(u^{n-1})\,.
\end{aligned}
$$

2. There is an equivalent definition that one finds in some books on special functions, viz.

$$
H_n(u) = e^{u^2}\left(-\frac{d}{du}\right)^n e^{-u^2}\,.
$$

The equivalence to the formula given above can be verified as follows

$$
\begin{aligned}
e^{u^2/2}\left(u-\frac{d}{du}\right)^n e^{-u^2/2} &= e^{u^2}\left[e^{-u^2/2}\left(u-\frac{d}{du}\right)e^{u^2/2}\right]^n e^{-u^2} \\
&= e^{u^2}\left(-\frac{d}{du}\right)^n e^{-u^2}\,.
\end{aligned}
$$

The first step becomes obvious when one writes the factors of the square brackets $[\ldots]^n$ side by side. In a second step one notes that $u-d/du$ when acting on $e^{u^2/2}$, gives zero so that, by the product rule, only the derivative $-d/du$ survives, acting to the right.

3. The first six polynomials are

$$
\begin{aligned}
&H_0(u) = 1, && H_3(u) = 8u^3-12u, \\
&H_1(u) = 2u, && H_4(u) = 16u^4-48u^2+12, \\
&H_2(u) = 4u^2-2, && H_5(u) = 32u^5-160u^3+120u\,.
\end{aligned}
$$

4. Replacing u by $-u$, one sees that

$$
H_n(-u) = (-)^n H_n(u)\,.
$$

The polynomials of even order are even under space reflection (or parity) $\boldsymbol{\Pi}: x \longrightarrow -x$, polynomials of odd order are odd.

5. The Hermite polynomials are orthogonal to one another in the following, generalized sense:

$$
\int_{-\infty}^{\infty} du\, H_m(u)H_n(u)e^{-u^2} = 0 \quad \text{for all} \quad m \neq n\,. \tag{1.81}
$$

This new kind of orthogonality will be a separate subject in the next section. Here a most direct way of proving orthogonality is to make use of the Schrödinger equation in the form of $(*)$ and to use the hermiticity of H. If $m \neq n$ then also

$E_m \neq E_n$. Multiplying the equation $H\varphi_n = E_n\varphi_n$ by φ_m^* from the left and integrating over all space, one obtains, in a short-hand notation,

$$(\varphi_m, H\varphi_n) = E_n(\varphi_m, \varphi_n) = (H\varphi_m, \varphi_n) .$$

In the second step the self-adjoint operator H is shifted to φ_m^*, by partial integration. As φ_m is a solution with energy E_m, too, and as E_m is real, the right-hand side may be continued as follows:

$$(H\varphi_m, \varphi_n) = E_m(\varphi_m, \varphi_n) .$$

As the energies E_m and E_n are assumed to be different, these equations are compatible only if $(\varphi_m, \varphi_n) = 0$. This is precisely the claim (1.81).
In the next section we prove further properties of Hermite polynomials which also apply to other, similarly defined sets of polynomials, in a more general framework.

6. For many practical calculations and in various applications it is useful to introduce a generating function for Hermite polynomials. Generally speaking, one talks about a *generating function* of the set of polynomials $\{P_n(u), n = 0, 1, \ldots\}$ if there is a function $g(u, t)$ of two variables u and t such that

$$g(u, t) = \sum_{n=0}^{\infty} a_n P_n(u) t^n \tag{1.82}$$

with given constant coefficients a_n. In many cases the polynomials of a given class are defined by giving the generating function and the coefficients.
Here we do not proceed in this way. Instead, we start from the expression given in Remark 2 above which followed from a physical argument, and derive a generating function for Hermite polynomials. As one easily sees

$$\left[\frac{\mathrm{d}^n}{\mathrm{d}u^n} \mathrm{e}^{-(t-u)^2}\right]_{t=0} = (-)^n \frac{\mathrm{d}^n}{\mathrm{d}u^n} \mathrm{e}^{-u^2}$$

so that the Hermite polynomials can be transformed to

$$H_n(u) = \mathrm{e}^{u^2} \left[\frac{\mathrm{d}^n}{\mathrm{d}u^n} \mathrm{e}^{-(t-u)^2}\right]_{t=0} .$$

Function theory teaches us that the n-th derivative of an analytic function at the point z_0 can be expressed by the integral

$$f^{(n)}(z_0) = \frac{n!}{2\pi \mathrm{i}} \oint \frac{f(z)}{(z - z_0)^{n+1}} \mathrm{d}z ,$$

taken over a closed contour which encloses the point z_0 once and with counter-clockwise orientation. Taking $z_0 = 0$ and using $f(z) = \exp[-(z-u)^2]$ yields

$$H_n(u) = \mathrm{e}^{u^2} \frac{n!}{2\pi \mathrm{i}} \oint \frac{\mathrm{e}^{-(z-u)^2}}{z^{n+1}} \mathrm{d}z = \frac{n!}{2\pi \mathrm{i}} \oint \frac{\mathrm{e}^{u^2-(z-u)^2}}{z^{n+1}} \mathrm{d}z .$$

Construct then the series

$$\sum_{n=0}^{\infty} \frac{1}{n!} H_n(u) t^n = \frac{1}{2\pi \mathrm{i}} \sum_{n=0}^{\infty} \oint \frac{\mathrm{e}^{u^2-(z-u)^2}}{z} \left(\frac{t}{z}\right)^n \mathrm{d}z$$

$$= \frac{1}{2\pi \mathrm{i}} \oint \frac{\mathrm{e}^{u^2-(z-u)^2}}{z-t} \mathrm{d}z = \mathrm{e}^{u^2-(t-u)^2} = \mathrm{e}^{2tu-t^2},$$

where in the last step Cauchy's integral theorem was used. This shows that, indeed,

$$g(u,t) = \mathrm{e}^{2tu-t^2}$$

is a generating function for Hermite polynomials.

It remains to determine the normalization factor N_n for arbitrary n. An elegant way of calculating this normalization consists in first choosing the constant in

$$\varphi_n = \text{const}\,(a^\dagger)^n \varphi_0$$

such that φ_n is normalized in the same way as φ_0. Instead of using an explicit notation containing the integral over x, I use the same short-hand notation as in (1.63). In one dimension it reads

$$(\varphi, F\varphi) \equiv \int_{-\infty}^{\infty} \mathrm{d}x\, \varphi^*(x) F \varphi(x)\,.$$

By n-fold partial integration one has

$$(\varphi_n, \varphi_n) = \text{const.}\big((a^\dagger)^n \varphi_0, (a^\dagger)^n \varphi_0\big) = \text{const.}\big(\varphi_0, (a)^n (a^\dagger)^n \varphi_0\big)\,.$$

The expectation value on the right-hand side,

$$(\varphi_0, \underbrace{a\,a \cdots a}_{n}\ \underbrace{a^\dagger a^\dagger \cdots a^\dagger}_{n} \varphi_0)$$

is calculated by inserting repeatedly the commutator (1.77), $aa^\dagger = a^\dagger a + 1$, such as to shift the right-most operator a past all operators $a^\dagger$, until it hits φ_0. This last term vanishes because a annihilates φ_0. The commutation of neighbours yields n times the 1 and there remains

$$n(\varphi_0, \underbrace{a\,a \cdots a}_{(n-1)}\ \underbrace{a^\dagger a^\dagger \cdots a^\dagger}_{(n-1)} \varphi_0)\,.$$

Let then the second operator a migrate, by commutation with its neighbours, as far right as possible. This time one obtains a factor $(n-1)$. This process is continued until all a are taken to the right of all $a^\dagger$. At the end of this procedure there remains the factor

$$n(n-1)(n-2)\cdots 1 = n!\,,$$

showing that

$$\varphi_n = \frac{1}{\sqrt{n!}} (a^\dagger)^n \varphi_0$$

is normalized to 1. Inserting then the formula (1.79) for φ_0 and the definition (1.80) of Hermite polynomials shows that the wave functions

$$\varphi_n(x) = \frac{1}{b^{1/2}} \frac{1}{\sqrt{\pi^{1/2} 2^n n!}} e^{-x^2/(2b^2)} H_n\left(\frac{x}{b}\right) \tag{1.83}$$

are correctly normalized. From this result and from the result (1.81) we conclude that the solutions φ_n are normalized and orthogonal (in the generalized sense), or, for short, that they are *orthonormal*,

$$(\varphi_m, \varphi_n) \equiv \int_{-\infty}^{\infty} dx\, \varphi_m^*(x)\varphi_n(x) = \delta_{mn} .$$

Note that in the case we discuss here the solutions φ_m can be chosen real. Thus, there is no need for the complex conjugation mark on the left function. As one sees immediately, if φ_m is a solution then also all functions

$$\left\{ e^{i\alpha} \varphi_m | \alpha \in \mathbb{R} \right\}$$

are solutions with the same eigenvalue E_m. These alternative eigenfunctions are indistinguishable as far as physics is concerned. No expectation value, i.e. no result of any real measurement, is modified by the phase factor. The more general question as to when solutions of the Schrödinger equation can be chosen real is related to the behaviour of the solutions with respect to time reversal.
The wave function of the lowest state is written down in (1.79). The three solutions of (∗) which follow this one and which are normalized to 1 read explicitly:

$$\varphi_1(x) = \frac{\sqrt{2}}{b^{1/2}\pi^{1/4}} \left(\frac{x}{b}\right) e^{-x^2/(2b^2)} , \tag{1.84}$$

$$\varphi_2(x) = \frac{1}{b^{1/2} 2\sqrt{2}\pi^{1/4}} \left[4\left(\frac{x}{b}\right)^2 - 2 \right] e^{-x^2/(2b^2)} , \tag{1.85}$$

$$\varphi_3(x) = \frac{1}{b^{1/2} 4\sqrt{3}\pi^{1/4}} \left[8\left(\frac{x}{b}\right)^3 - 12\left(\frac{x}{b}\right) \right] e^{-x^2/(2b^2)} . \tag{1.86}$$

Figure 1.7 shows the graphs of φ_0 up to φ_3. This picture shows a striking pattern regarding the number and the position of the zeroes on which we comment in the next section, in a more general context. Figure 1.8 shows the graphs of the probability densities $|\varphi_n|^2$, $n = 0, 1, 2, 3$.

Remark A Representation of the Heisenberg Algebra

The action of the raising and lowering operators on a given wave function φ_n follows from the normalization $(\varphi_0, \varphi_0) = 1$ derived above

$$\varphi_n = \frac{1}{\sqrt{n!}} \underbrace{a^\dagger \cdots a^\dagger}_{n} \varphi_0 .$$

They are found to be as follows

$$a^\dagger \varphi_n = \sqrt{n+1}\, \varphi_{n+1}, \qquad a\varphi_n = \sqrt{n}\, \varphi_{n-1} .$$

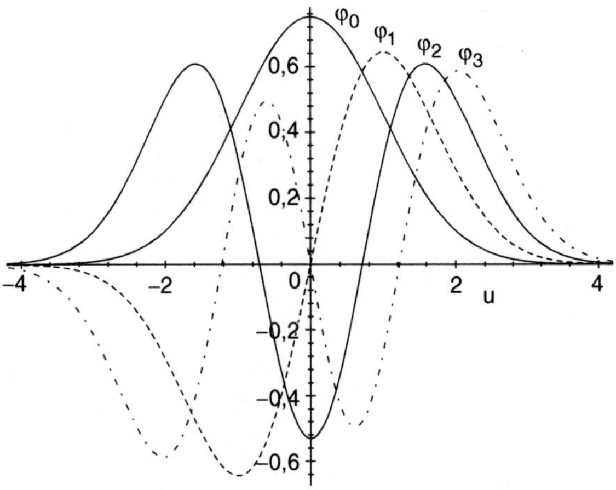

Fig. 1.7 Graphs of the wave functions (1.79), (1.84)–(1.86) pertaining to energy eigenvalues $E_0 = \hbar\omega/2$, $E_1 = 3\hbar\omega/2$, $E_2 = 5\hbar\omega/2$, and $E_3 = 7\hbar\omega/2$ of the harmonic oscillator in one dimension, respectively, as a function of $u = x/b$

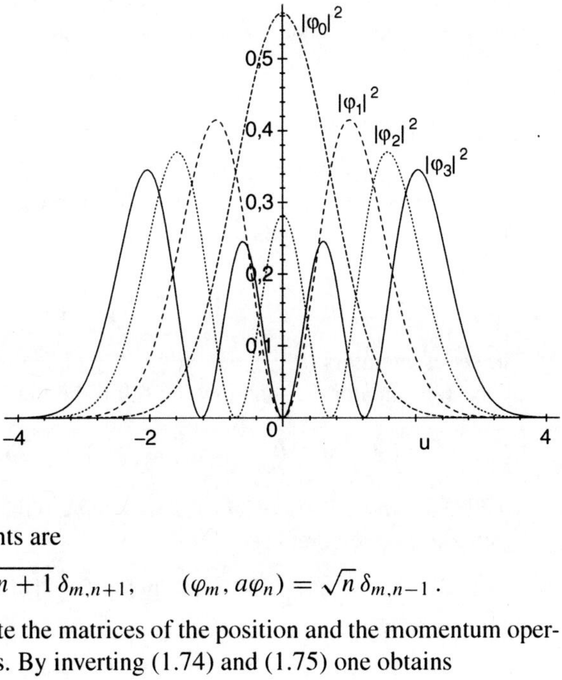

Fig. 1.8 Squares of the wave functions shown in Fig. 1.7. These are the probability densities in the four lowest oscillator states

Thus, their matrix elements are

$$(\varphi_m, a^\dagger \varphi_n) = \sqrt{n+1}\,\delta_{m,n+1}, \qquad (\varphi_m, a\varphi_n) = \sqrt{n}\,\delta_{m,n-1}.$$

It is instructive to calculate the matrices of the position and the momentum operators from these elements. By inverting (1.74) and (1.75) one obtains

$$q \equiv x = \sqrt{\frac{\hbar}{2m\omega}}(a^\dagger + a) = b\frac{1}{\sqrt{2}}(a^\dagger + a),$$

$$p \equiv \frac{\hbar}{\mathrm{i}}\frac{\mathrm{d}}{\mathrm{d}x} = \sqrt{\frac{\hbar m\omega}{2}}\,\mathrm{i}(a^\dagger - a) = \frac{\hbar}{b}\frac{\mathrm{i}}{\sqrt{2}}(a^\dagger - a),$$

where $b = \sqrt{\hbar/(m\omega)}$, as before.

Denoting the matrix representations of q and p by $\{q\}$ and $\{p\}$, respectively, and numbering rows and columns by $n = 0, 1, 2, \ldots$, one finds

$$\{q\} = b\frac{1}{\sqrt{2}}\begin{pmatrix} 0 & 1 & 0 & 0 & \ldots \\ 1 & 0 & \sqrt{2} & 0 & \ldots \\ 0 & \sqrt{2} & 0 & \sqrt{3} & \ldots \\ 0 & 0 & \sqrt{3} & 0 & \ldots \\ \vdots & \vdots & \vdots & \vdots & \ddots \end{pmatrix},$$

$$\{p\} = \frac{\hbar}{b}\frac{\mathrm{i}}{\sqrt{2}}\begin{pmatrix} 0 & -1 & 0 & 0 & \ldots \\ 1 & 0 & -\sqrt{2} & 0 & \ldots \\ 0 & \sqrt{2} & 0 & -\sqrt{3} & \ldots \\ 0 & 0 & \sqrt{3} & 0 & \ldots \\ \vdots & \vdots & \vdots & \vdots & \ddots \end{pmatrix}.$$

We calculate the commutator of p and q by taking the commutator of these matrices. The result has a very simple structure

$$[\{p\}, \{q\}] = \frac{\hbar}{2}\mathrm{i}\begin{pmatrix} -2 & 0 & \ldots \\ 0 & -2 & \ldots \\ \vdots & \vdots & \ddots \end{pmatrix} = \frac{\hbar}{\mathrm{i}}\mathbb{1}.$$

This result is nothing else than a matrix representation of the relation (1.36). Thus, it is equivalent to the commutator in coordinate space

$$\left[\frac{\hbar}{\mathrm{i}}\frac{\partial}{\partial x}, x\right] = \frac{\hbar}{\mathrm{i}}.$$

The set of operators $\{q^i, p_k | i, k = 1, \ldots, f\}$, endowed with the product defined by the commutator, $[\cdot,\cdot]$, and which fulfill the fundamental commutators

$$[q^i, q^k] = 0, \qquad [p_i, p_k] = 0, \qquad [p_i, q^k] = \frac{\hbar}{\mathrm{i}}\,\delta_i^k$$

is called the *Heisenberg algebra*. This algebra is the analogue of the fundamental algebra of Poisson brackets

$$\{q^i, q^k\} = 0, \quad \{p_i, p_k\} = 0, \quad \{p_i, q^k\} = \delta_i^k,$$

well-known from mechanics (see [Scheck (2010)], Sect. 2.31).

The matrices that were computed above belong to what is called a *representation*. These matrices are infinite-dimensional and span a particular representation of the Heisenberg algebra in one space dimension. They are interesting both for a historical and a mathematical reason.

Heisenberg developed his version of quantum mechanics in precisely this form. For this reason, in the early development of quantum mechanics, Heisenberg's approach was called *matrix mechanics*. It was E. Schrödinger who subsequently proved the equivalence of matrix mechanics to the wave mechanics developed by him and by L. de Broglie.

From a mathematical point of view our example is interesting because it illustrates the fact that the Heisenberg algebra has no physically relevant finite-dimensional matrix representations. It can neither be realized by finite-dimensional matrices nor by bounded operators (see, e.g., [Thirring (1981)], [Blanchard and Brüning (2003)]).

1.7 Orthogonal Polynomials in One Real Variable

The properties of Hermite polynomials that we discovered in the context of the harmonic oscillator, are so important and at the same time so general that we wish to place them in a larger framework and derive and discuss them in greater generality. In this context, the notion of generalized orthogonality that appeared in the example (1.81) is of paramount importance:

Definition 1.4 Generalized Orthogonality

Assume the following data to be given

1. an interval $I = [a, b] \subset \mathbb{R}$ on the real axis and
2. a positive-semidefinite function $\varrho : \mathbb{R} \longrightarrow \mathbb{R}$ which is strictly positive on I, and whose growth for large absolute values of x is moderate, or, expressed in symbols,

$$\varrho(x) \geq 0 \quad \forall x \in \mathbb{R}\,, \quad \varrho(x) > 0 \quad \forall\, x \in [a, b]\,,$$
$$\varrho(x)\mathrm{e}^{\alpha|x|} \leq c < \infty \qquad \text{for appropriately chosen } \alpha,\ \forall\, x\,.$$

The function ϱ is called the *density*, or *weight function*. An infinite sequence of polynomials $P_k(x)$, $k = 0, 1, 2, \ldots$, constructed in such a way that

$$\int_a^b \mathrm{d}x\, \varrho(x) P_m(x) P_n(x) = \delta_{mn}\,, \tag{1.87}$$

is said to be orthogonal and normalized with respect to the interval $[a, b]$ and the weight function $\varrho(x)$.

A special feature of this definition is that there exists an explicit procedure of constructing this sequence of polynomials, for given interval I and weight function ϱ. For any pair (I, ϱ) (obeying certain conditions) there is a set of orthogonal polynomials. Once these polynomials are constructed one defines the functions

$$\varphi_k := \sqrt{\varrho(x)}\, P_k(x) \tag{1.88}$$

(which, in general, are no longer polynomials) and concludes that they are orthogonal and normalized, in a generalized sense, viz.

$$(\varphi_m, \varphi_n) \equiv \int_a^b \mathrm{d}x\, \varphi_m^*(x)\varphi_n(x) = \delta_{mn}\,. \tag{1.89}$$

(For the time being we only consider real polynomials and, therefore, real functions. It then makes no difference whether we write φ_m^* or φ_m in the integral. We keep the complex conjugation of the left factor in view of the more general case of complex valued functions.)

Construction of the Polynomials (Gram-Schmidt Method)
Let the symbol (f, g) denote the integral (1.89) over the interval $[a, b]$. Define

$$g_k(x) := \sqrt{\varrho(x)}\, x^k, \qquad k = 0, 1, 2, \ldots \quad \text{and}$$

$$f_0(x) = g_0(x), \qquad f_1(x) = g_1(x) - \frac{(f_0, g_1)}{(f_0, f_0)} f_0(x).$$

One verifies that $(f_1, f_0) = 0$ and that

$$(f_1, f_1) = (g_1, g_1) - \frac{(f_0, g_1)^2}{(f_0, f_0)}.$$

Furthermore, one defines

$$f_2(x) = g_2(x) - \frac{(f_0, g_2)}{(f_0, f_0)} f_0 - \frac{(f_1, g_2)}{(f_1, f_1)} f_1$$

and verifies that $(f_2, f_0) = 0$ and $(f_2, f_1) = 0$. This construction is continued so that, for arbitrary k,

$$f_k(x) = g_k(x) - \sum_{l=0}^{k-1} \frac{(f_l, g_k)}{(f_l, f_l)} f_l(x).$$

One confirms that all previously defined functions with $l = 0, 1, \ldots, k-1$, are orthogonal to f_k, $(f_k, f_l) = 0$.

By construction $(f_n, f_n) > 0$. It follows that the functions

$$\varphi_n(x) := \frac{f_n(x)}{\sqrt{(f_n, f_n)}}$$

are orthogonal and normalized to 1. The weight function, by assumption, is strictly positive on the interval I. Therefore, by dividing by the square root of the weight function, one obtains the orthogonal polynomials that were to be constructed,

$$P_n(x) = \frac{\varphi_n(x)}{\sqrt{\varrho(x)}}. \tag{1.90}$$

The construction just described implies

Lemma 1.1

Let $Q_m(x)$ be a polynomial of degree m. This polynomial can be expressed as a linear combination of the orthogonal polynomials constructed above,

$$Q_m(x) = \sum_{l=0}^{m} c_l P_l(x).$$

For all degrees $n > m$ one has

$$\int_a^b \mathrm{d}x\, \varrho(x) Q_m(x) P_n(x) = 0, \qquad (n > m)\,.$$

Examining once more Fig. 1.7 which shows the first four Hermite polynomials, two properties are striking: the number and the position of the zeroes. The function φ_0 has no zero at all, φ_1 has exactly one zero, φ_2 has two, and φ_3 has three zeroes. Furthermore, these zeroes are entangled: the zero of φ_1 lies between the two zeroes of φ_2, while those of φ_2 lie between the zeroes of φ_3. Even though Fig. 1.7 shows the wave functions of the harmonic oscillator, not the Hermite polynomials proper, this observation applies also to the latter. This remark hints at a general property of all orthogonal polynomials as clarified by the following two theorems. Recall that a real polynomial of degree n has n zeroes. In case some of these are not real, the complex zeroes occur in pairs of complex conjugate values (fundamental theorem of analysis).

Theorem 1.1

The polynomial $P_n(x)$ has exactly n real simple zeroes in the interval $I = [a, b]$.

Theorem 1.2

The zeroes of $P_{n-1}(x)$ separate the zeroes of $P_n(x)$. In other terms, given two neighbouring zeroes of $P_n(x)$, there is exactly one zero of P_{n-1} which lies between the two.

The proof of Theorem 1.1 constructs a contradiction: Consider all real zeroes of *odd* order of $P_n(x)$ (that is, the simple, triple, etc. zeroes) situated at the points α_i, with

$$\alpha_1 < \alpha_2 < \cdots < \alpha_h\,.$$

Assume that h is smaller than n. From these construct an auxiliary polynomial

$$Q_h(x) = (x - \alpha_1)(x - \alpha_2)\cdots(x - \alpha_h)\,.$$

On the whole interval I the product of this auxiliary polynomial and of P_n has the property: either $Q_h(x)P_n(x) \geqslant 0$ or $Q_h(x)P_n(x) \leqslant 0$. Obviously, it does not vanish identically. Therefore, the integral over I,

$$\int_a^b \mathrm{d}x\, \varrho(x) Q_h(x) P_n(x)\,,$$

is either positive or negative but is not equal to zero. This contradicts Lemma 1.1, except if $h = n$. This proves Theorem 1.1.

The proof of Theorem 1.2 rests on two lemmata:

Lemma 1.2

The polynomial $Q_k(\lambda, x) = P_k(x) + \lambda P_{k-1}(x)$ has exactly k real and simple zeroes for all real λ.

The number of real zeroes of $Q_k(\lambda, x)$ either is k, or is smaller than or equal to $(k-2)$. Assume Q_k to have the following real zeroes of *odd* order

$$\alpha_1 < \alpha_2 < \cdots < \alpha_h \quad \text{with} \quad h \le k-2\,.$$

Using these values construct another auxiliary polynomial,

$$R_h(x) = (x-\alpha_1)(x-\alpha_2)\cdots(x-\alpha_h)\,.$$

Again, for the product one concludes that $R_h(x)Q_k(\lambda, x) \geqslant 0$ or $R_h(x)Q_k(\lambda, x) \leqslant 0$, for all x. This contradicts Lemma 1.1 which implies that

$$\int_a^b \mathrm{d}x\, R_h(x)Q_k(\lambda, x) = 0\,.$$

The contradiction is avoided only if $h = k$. This shows that Lemma 1.2 holds true.

Lemma 1.3

There is no point $x_i \in I$, where both $P_k(x_i) = 0$ and $P_{k-1}(x_i) = 0$.

If there were such a point one would have $Q_k(\lambda, x = x_i) = 0$ for all λ. One could then choose

$$\lambda_0 := -\frac{P'_k(x_i)}{P'_{k-1}(x_i)}\,,$$

and would conclude that $Q_k(\lambda_0, x = x_i) = 0$ and $Q'_k(\lambda_0, x = x_i) = 0$. In other terms, this polynomial would have a *double* zero in x_i – in contradiction to Lemma 1.2.

One then proves Theorem 1.2 as follows: Suppose the theorem is not true. Then there must be two zeroes α and β of $P_n(x)$, $P_n(\alpha) = 0 = P_n(\beta)$ with $\alpha < \beta$, such that

$$P_n(x) \neq 0 \quad \text{for all} \quad x \in (\alpha, \beta) \quad \text{and}$$
$$P_{n-1}(x) \neq 0 \quad \text{for all} \quad x \in [\alpha, \beta]\,.$$

For all $x \in [\alpha, \beta]$ the polynomial $Q_n(\lambda, x) = P_n(x) + \lambda P_{n-1}(x)$ then has the zero

$$\lambda_0(x) := -\frac{P_n(x)}{P_{n-1}(x)}\,.$$

Furthermore, one has $\lambda_0(x = \alpha) = 0 = \lambda_0(x = \beta)$, but $\lambda_0(x) \neq 0$ for all $x \in (\alpha, \beta)$. Therefore, the function $\lambda_0(x)$ has the same sign everywhere in the open interval (α, β), and it reaches an extremum in some point $x_0 \in (\alpha, \beta)$. At this point one has

$$\left.\frac{\mathrm{d}\lambda_0(x)}{\mathrm{d}x}\right|_{x=x_0} = 0\,.$$

Consider then the polynomial, vanishing by construction,

$$Q_n(\lambda_0(x), x) = P_n(x) + \lambda_0(x)P_{n-1}(x) = 0\,.$$

Taking its derivative with respect to x and choosing $x = x_0$, one concludes

$$Q'_n(\lambda_0(x_0), x_0) = P'_n(x_0) + \lambda_0(x_0)P'_{n-1}(x_0) = 0\,.$$

Thus, $Q_n(\lambda_0(x_0), x)$ has a zero of *second* order at $x = x_0$. As this contradicts Lemma 1.2 Theorem 1.2 is true.

The set of functions $\{\varphi_n\}$ is not only orthogonal (in the sense of Definition 1.4) and normalized but is also *complete.* This notion is a direct generalization of the notion of completeness for systems of base vectors in finite-dimensional vector spaces. The precise definition is the following:

Definition 1.5

A set of orthonormal functions $\{\varphi_n\}$ is said to be *complete*, if every square integrable function $h(x)$ for which $(\varphi_n, h) = 0$ for all n, vanishes identically. This is equivalent to the statement that square integrable functions $f(x)$ can be expanded in the basis of functions $\{\varphi_n\}$.

The statement made in this definition is proved by means of function theory. If $(\varphi_n, h) = 0\ \forall n$ then

$$\int_a^b \mathrm{d}x\, \sqrt{\varrho(x)}x^n h(x) = 0\,.$$

This is so because x^n can be written as a linear combination of P_0 to P_n. Consider then the complex function

$$F(p) := \int_a^b \mathrm{d}x\, \sqrt{\varrho(x)}h(x)\mathrm{e}^{\mathrm{i}px}\,.$$

As this function is analytic[16] one calculates its derivatives by differentiating the integrand with respect to p,

$$F^{(m)}(p) := \int_a^b \mathrm{d}x\, \sqrt{\varrho(x)}h(x)x^m\mathrm{e}^{\mathrm{i}px}\,.$$

Specializing to the point $p = 0$ one has

$$F^{(m)}(0) := \int_a^b \mathrm{d}x\, \sqrt{\varrho(x)}h(x)x^m = 0 \quad \text{for all} \quad m\,.$$

Thus, $F(p)$ vanishes identically and so does the integrand $\sqrt{\varrho(x)}h(x)$. As ϱ is strictly positive on the interval one concludes that $h(x)$ vanishes identically.

[16] In cases where the interval $I = [a, b]$ extends to infinity, $F(p)$ is defined only for $|\mathrm{Im}\, p| < \alpha$ where α is the parameter that controls the growth of the weight function $\varrho(x)$, cf. Definition 1.4.

Example 1.6

The set of Hermite polynomials $H_n(u)$ is orthogonal on the interval $(-\infty, +\infty)$ and with the weight function e^{-u^2}. Dividing them by $\sqrt{\pi^{1/2}2^n n!}$ renders them normalized to 1. The two theorems on the zeroes are illustrated by Fig. 1.7.

Example 1.7

Choose the interval $(a = -1, b = 1)$ and the weight function $\varrho(x) = \Theta(x - a) - \Theta(x - b)$, i.e. such that it equals 1 on the interval but vanishes outside. The Gram–Schmidt procedure then yields the *Legendre polynomials* $P_l(x \equiv \cos\theta)$ with $0 \le \theta \le \pi$. Traditionally the Legendre polynomials are not normalized to 1. Rather, they are defined in such a way that $P_l(x = 1) \equiv P_l(\theta = 0) = 1$ for all values of l. There are general formulae for Legendre polynomials (such as the formula of Rodrigues). Nevertheless, it may be a good exercise to construct, say, the first six of them explicitly, by means of the Gram–Schmidt procedure. (The reader is encouraged to do so!) One finds

$$P_0(x) = 1,\ P_1(x) = x,\ P_2(x) = \frac{1}{2}(3x^2 - 1),\ P_3(x) = \frac{1}{2}(5x^3 - 3x),$$

$$P_4(x) = \frac{1}{8}(35x^4 - 30x^2 + 3),\quad P_5(x) = \frac{1}{8}(63x^5 - 70x^3 + 15x).$$

Some of their general properties are

$$P_l(1) = 1, \qquad P_l(-x) = (-1)^l P_l(x), \qquad P_{2l+1}(0) = 0.$$

If we supplement them by the factor $\sqrt{(2l + 1)/2}$, the Legendre polynomials are normalized to 1,

$$\int_{-1}^{+1} dx \sqrt{\frac{2l+1}{2}} P_l(x) \sqrt{\frac{2l'+1}{2}} P_{l'}(x) = \delta_{ll'}.$$

Figure 1.9 illustrates the theorems on their zeroes and shows some examples of normalized polynomials $\sqrt{(2l + 1)/2}\, P_l(x)$ in the interval $[-1, +1]$.

Every function $f(\theta)$ which is regular in the interval $0 \le \theta \le \pi$ can be expanded in terms of Legendre polynomials:

$$f(\theta) = \sum_{l=0}^{\infty} c_l P_l(\cos\theta). \tag{1.91}$$

With $x = \cos\theta$, $dx = -\sin\theta d\theta$, and with the normalization described above, the expansion coefficients are given by

$$c_l = \frac{2l+1}{2} \int_0^{\pi} \sin\theta d\theta\, P_l(\cos\theta) f(\theta). \tag{1.92}$$

We will meet further examples of orthogonal polynomials in later sections.

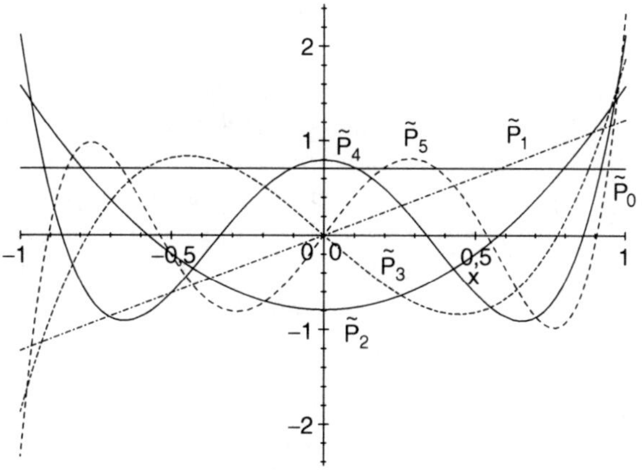

Fig. 1.9 Graphs of the first six Legendre polynomials, normalized to 1, as functions of $x = \cos\theta$, this means that $\tilde{P}_\ell(x) = \sqrt{(\ell + 1/2)}\, P_\ell(x)$ are plotted here

Remarks

1. The results of this section lead to a better understanding of the quantized harmonic oscillator that we studied in Sect. 1.6 in great detail. Since the Hermite polynomials are not only orthogonal but also complete, the wave functions (1.83) provide a *complete* and *orthonormal* system of functions. They span an infinite-dimensional space of functions, the space of square integrable functions $L^2(\mathbb{R})$ over $\mathbb{R}^1$. Every element of this space can be expanded in terms of the φ_n.
2. It is the property of completeness, Definition 1.5, which justifies, a posteriori, calling the integral

$$(\varphi_m, \varphi_n) \equiv \int_{-\infty}^{\infty} \mathrm{d}x\, \varphi_m^*(x)\varphi_n(x) = \delta_{mn}$$

   a scalar product. Indeed, an integral of this type fulfills all conditions imposed on a scalar product: If the functions φ_n are real or can be chosen real (this was the case for the harmonic oscillator) the product is symmetric. If they are complex one has the relation

$$(\varphi_n, \varphi_m) = (\varphi_m, \varphi_n)^* .$$

   The product (φ_m, φ_n) is nondegenerate because any f for which $(\varphi_n, f) = 0$ for all n, vanishes identically.
3. Their is a great deal of analogy to linear algebra of finite-dimensional vector spaces. If we compare the expansion of an element $v \in V$ in a d-dimensional vector space in terms of a basis $\hat{e}_i$, with the expansion of a function $f \in L^2(\mathbb{R})$ in terms of the basis φ_n, i.e

$$v = \sum_{i=1}^{d} c_i \hat{e}_i \quad \text{with} \quad f(x) = \sum_{k=0}^{\infty} c_k \varphi_k(x) ,$$

then the roles of the expansion coefficients and of the bases are seen to be the same. Also the rules by which the coefficients are calculated from the scalar products of v and f with $\hat{e}_i$ and φ_n, respectively, are very similar.
Of course, there are significant differences to the case of finite-dimensional vector spaces when going over to infinite-dimensional function spaces. To quote just one important difference, in the latter case the notion of convergence must be studied carefully. For the moment, however, we emphasize the similarities because they are helpful in visualizing the wave functions of a self-adjoint quantum system.

1.8 Observables and Expectation Values

1.8.1 Observables with Nondegenerate Spectrum

Consider an observable $F(\boldsymbol{p}, \boldsymbol{x})$ which possesses a complete system of orthonormal wave functions obeying the following differential equation,

$$F\left(\boldsymbol{p} = \frac{\hbar}{\mathrm{i}}\nabla, \boldsymbol{x}\right) \varphi_n(\boldsymbol{x}) = \lambda_n \varphi_n(\boldsymbol{x}) . \tag{1.93}$$

The (real) numbers λ_n are called *eigenvalues*, the wave functions $\varphi_n(\boldsymbol{x})$ are called *eigenfunctions* of the observable $F(\boldsymbol{p}, \boldsymbol{x})$, with φ_n being the eigenfunction that belongs to the eigenvalue λ_n.[17] This nomenclature stems from linear algebra. For instance, every real, symmetric $m \times m$-matrix $\{M_{ik}\}$ has m real eigenvalues and eigenvectors both of which are obtained by solving the linear system of equations

$$\sum_{k=1}^{m} M_{ik} c_k^{(n)} = \mu_n c_i^{(n)} \qquad n = 1, 2, \ldots, m .$$

To quote an example from physics recall the inertia tensor $\mathbf{J} = \{J_{ik}\}$ in mechanics of rigid bodies which is a real, symmetric 3×3-matrix. The system of linear equations

$$\sum_{k=1}^{3} J_{ik} \omega_k^{(n)} = I_n \omega_i^{(n)}$$

yields the moments of inertia I_1, I_2, I_3, the eigenvalues of $\mathbf{J}$, while the corresponding eigenvectors $\omega^{(n)}$ define the orthogonal directions of the principal axes of inertia, i.e. those axes for which the angular velocity and the angular momentum have the same direction.

[17] Both expressions are derived from their German names, where "eigen" stands for "proper". In French eigenvalues are called *valeurs propres*, eigenfunctions, or eigenvectors are called *fonctions/vecteurs propres*.

For the sake of clarity, the definitions introduced above are summarized as follows:

- *Orthonormal*: means that the eigenfunctions are mutually orthogonal and, in view of their probabilistic interpretation, are normalized to 1,

$$(\varphi_m, \varphi_n) \equiv \int d^3x\, \varphi_m^*(\boldsymbol{x})\varphi_n(\boldsymbol{x}) = \delta_{mn}\,.$$

- *Nondegenerate*: one says that a spectrum is nondegenerate if all eigenvalues are different, $\lambda_m \neq \lambda_n$ for $m \neq n$. In other terms, this means that for each eigenvalue there is exactly one eigenfunction φ_n. In turn, one will say that eigenvalues are degenerate whenever for fixed n there is a sequence of eigenfunctions

$$\varphi_{n,1}, \varphi_{n,2}, \cdots \varphi_{n,k_n}\,,$$

all of which obey the differential equation (1.93) for the same eigenvalue λ_n. We exclude this case for the moment but return to it, as well as to its physical interpretation in Sect. 1.8.3.

- *Completeness*: One says that a set of orthonormal functions is complete if every square integrable function can be expanded in terms of its elements φ_n,

$$\psi(t, \boldsymbol{x}) = \sum_{n=0}^{\infty} c_n(t)\varphi_n(\boldsymbol{x}) \quad \text{with}$$
$$c_n(t) = \int d^3x\, \varphi_n^*(\boldsymbol{x})\psi(t, \boldsymbol{x}) \equiv (\varphi_n, \psi)(t)\,. \tag{1.94}$$

This series converges in the mean, that is to say, it converges in the following sense

$$\lim_{N\to\infty} \int d^3x \left|\psi(t, \boldsymbol{x}) - \sum_{n=0}^{N} c_n(t)\varphi_n(\boldsymbol{x})\right|^2$$
$$= \lim_{N\to\infty} \left(\int d^3x\, |\psi|^2 - \sum_{n=0}^{N} |c_n|^2\right) = 0\,.$$

Here is an example for an operator with the properties just summarized: Suppose F is the Hamiltonian of a harmonic oscillator in three spatial dimensions,

$$H = -\frac{\hbar^2}{2m}\Delta + \frac{1}{2}m[\omega_1^2(x^1)^2 + \omega_2^2(x^2)^2 + \omega_3^2(x^3)^2]\,,$$

the circular frequencies ω_i being chosen so that they are pairwise relative irrational. Clearly, H can be written as the sum of three one-dimensional operators with different frequencies. For the latter we can use the results of Sect. 1.6. The spectrum of the eigenvalues of H reads

$$E_{n_1,n_2,n_3} = \sum_{i=1}^{3} \left(n_i + \frac{1}{2}\right)\hbar\omega_i\,.$$

By assumption it is nondegenerate.

The (differential) operator $F(\boldsymbol{p}, \boldsymbol{x})$ is supposed to represent an observable and, therefore, must be self-adjoint. Indeed, if it has this property then the eigenvalues are real:

$$(\varphi_n, F\varphi_n) = \lambda_n = (F\varphi_n, \varphi_n) = \lambda_n^*,$$

where we use the same argument as the one following (1.81).

As an example for a further property of self-adjoint operators we prove the relation announced in Remark 3 in Sect. 1.5.1 for $n \neq m$,

$$(\varphi_m, F\varphi_n) \equiv \int d^3x\, \varphi_m^* F\varphi_n = \int d^3x\, (F\varphi_m)^* \varphi_n$$
$$\equiv (F\varphi_m, \varphi_n) . \tag{1.95}$$

Take $\psi = u\varphi_n + v\varphi_m$ with arbitrary complex constant coefficients $u, v \in \mathbb{C}$ and consider G, somewhat more generally than hitherto, a self-adjoint operator defined on the system of functions $\{\varphi_n\}$. This observable G is also defined on ψ and one has the relation

$$\int d^3x\, \psi^* G\psi - \int d^3x\, (G\psi)^* \psi \equiv (\psi, G\psi) - (G\psi, \psi) = 0 .$$

Inserting here the decomposition $\psi = u\varphi_n + v\varphi_m$ yields

$$u^* v \left[\int d^3x\, \varphi_n^* G\varphi_m - \int d^3x\, (G\varphi_n)^* \varphi_m \right]$$
$$+ uv^* \left[\int d^3x\, \varphi_m^* G\varphi_n - \int d^3x\, (G\varphi_m)^* \varphi_n \right] = 0 .$$

As u and v are arbitrary the two expressions in square brackets must vanish independently. This proves the claim.

The eigenfunctions φ_n of the observable F provide a basis of the infinite-dimensional space of square integrable functions. Another observable G can equally well be replaced by its matrix representation in this basis,

$$G_{mn} := (\varphi_m, G\varphi_n)$$

which inherits the property

$$G_{mn} = G_{nm}^* \quad \text{or, in matrix notation,} \quad \mathbf{G} = \mathbf{G}^\dagger .$$

The matrix $\mathbf{G}$ equals the complex conjugate of its transposed. It is said to be *hermitian*. The representation by an infinite-dimensional matrix is equivalent to the representation by a differential operator. The matrix is said to be hermitian, the operator is called self-adjoint. Because of this equivalence the properties "hermitian" and "self-adjoint" are often used as synonyms.

Let us return to the observable F whose eigenvalues and eigenfunctions are assumed to be known from (1.93). Suppose an arbitrary physical state ψ is expanded as in (1.94). The norm of ψ is obtained from

$$\int d^3x\, |\psi(t, \boldsymbol{x})|^2 = 1 = \sum_{n=0}^{\infty} |c_n(t)|^2 .$$

The expectation value of the observable F in the state ψ is then given by

$$\langle F\rangle_\psi = \int \mathrm{d}^3x\, \psi(t,\boldsymbol{x})^* F\psi(t,\boldsymbol{x}) = \sum_{n=0}^{\infty} \lambda_n |c_n(t)|^2 .$$

These formulae, together with Born's interpretation of the wave function suggest the following interpretation:

The set of eigenvalues λ_n of the self-adjoint operator F is the set of possible values that one will find in any particular single measurement of the observable F. If the measurement of F is done on a quantum mechanical state described by the wave function ψ, the probability to find a specific eigenvalue λ_q is given by $|c_q(t)|^2$.

This means in practice that if one performs measurements of F on very many, identically prepared systems characterized by the wave function ψ, then in each individual measurement one will find one of the eigenvalues λ_n. The set of all measurements will follow a distribution of eigenvalues weighted with the probabilities $|c_n|^2$. The mean-square deviation (1.31) of the observable F in the state ψ is calculated from

$$(\Delta F)^2 = \left\langle (F - \langle F\rangle_\psi)^2 \right\rangle_\psi = \sum_{i=0}^{\infty} |c_i|^2 \left(\lambda_i - \sum_{j=0}^{\infty} \lambda_j |c_j|^2 \right)^2 \geqslant 0 .$$

As a result for expectation values, that is, for the outcome of very many individual measurements, this is a *classical* result. It can be compared directly with the general statistical information of Sect. 1.2.1. If we set $w_i \equiv |c_i|^2$ this corresponds to what was called the discrete distribution in Sect. 1.2.1. In particular, we can sharpen the observation made in relation to (1.34): The mean square deviation of the observable F in the state ψ vanishes if and only if ψ is an eigenstate of F. In this case only one of the coefficients, c_k, has the absolute value 1, all others are equal to zero,

$$|c_k| = 1, \qquad c_n = 0 \quad \text{for all} \quad n \neq k .$$

It is instructive to prove this statement as follows: We assume that ψ belongs to the domain on which F is defined, and that the result of the action of F, $\phi = F\psi$ also belongs to that domain. Assume ϕ to be different from zero. We then have

$$\left\langle F^2 \right\rangle_\psi = (\psi, F^2\psi) = (F\psi, F\psi) = (\phi, \phi) \quad \text{and} \quad \langle F\rangle_\psi^2 = (\psi, \phi)^2 .$$

Therefore, the mean square deviation is

$$\left\langle F^2 \right\rangle_\psi - \langle F\rangle_\psi^2 = \big([\phi - (\psi, \phi)\psi], \phi \big) \geqslant 0 .$$

In the left entry of the scalar product the projection of ϕ onto ψ is subtracted from ϕ. The analogous expression in a vector space and with $\boldsymbol{a}, \boldsymbol{b} \in V$ would read (using the abbreviation $\hat{\boldsymbol{a}} = \boldsymbol{a}/|\boldsymbol{a}|$)

$$\big(\boldsymbol{b} - (\hat{\boldsymbol{a}} \cdot \boldsymbol{b})\hat{\boldsymbol{a}} \big) \cdot \boldsymbol{b} \geqslant 0 .$$

In both cases the expressions on the left-hand side of the inequalities vanish precisely if $\boldsymbol{b}$ is parallel to $\boldsymbol{a}$, and if ϕ is proportional to ψ, $\phi = \lambda\psi$, respectively. In the latter case ψ is an eigenfunction of F. Finally, if ϕ vanishes identically, the claim is trivially true.

This section concludes with the following remarks:

Remarks

1. The quantum state is equivalently described by the wave function $\psi(t, \boldsymbol{x})$ over the coordinate space $\mathbb{R}^3$, or by the set of expansion coefficients $\{c_n(t)\}$. The knowledge of all $c_n(t)$ completely determines $\psi(t, \boldsymbol{x})$.
2. Measured values are always of the form $\int \psi^* \dots \psi$, or, as one also says, they are *sesquilinear* in the wave function. Therefore, two wave functions $\psi(t, \boldsymbol{x})$ and $e^{i\alpha}\psi(t, \boldsymbol{x})$, where α is a real number, cannot be distinguished by measurements. The set
$$\underline{\psi} := \{e^{i\alpha}\psi | \alpha \in \mathbb{R}\}$$
is called a *unit ray*. Below, when we will explore the spaces on which the wave functions ψ are defined, as well as the symmetries that act on them in these spaces, unit rays will appear in the context of *projective representations*.
3. A measurement of an observable can be used as a means to *prepare* a quantum state. Like in a filter one selects a specific eigenvalue λ_k at time t_0, and rejects all others. At this time the state of the system is
$$\psi(t_0, \boldsymbol{x}) = \varphi_k(\boldsymbol{x}) .$$
Its time evolution is described by the time-dependent Schrödinger equation (1.59). In particular, let F describe a time independent, conserved quantity. Then F commutes with the Hamiltonian H. If φ_n is an eigenfunction of F then also $(H\varphi_n)$ is an eigenfunction of F,
$$F(H\varphi_n) = HF\varphi_n = \lambda_n(H\varphi_n)$$
and pertains to the same eigenvalue λ_n. As the eigenvalues are nondegenerate one concludes
$$H\varphi_n = E_n\varphi_n$$
with real E_n. As a consequence, the expansion coefficients in (1.94) fulfill the differential equations
$$\dot{c}_n = (\varphi_n, \dot{\psi}) = -\frac{i}{\hbar}(\varphi_n, H\psi) = -\frac{i}{\hbar}(H\varphi_n, \psi) = -\frac{i}{\hbar}E_n c_n .$$
Thus, the time dependence of c_n is harmonic, i.e.
$$c_n(t) = c_n(t_0)e^{-(i/\hbar)E_n(t-t_0)} ,$$
the probabilities $|c_n(t)|^2$ are independent of time.

1.8.2 An Example: Coherent States

Unfortunately the cases that we discussed until now are still somewhat academic because we studied exclusively stationary, stable states about which not much can be measured. This situation will change only when we know how to describe the scattering of two systems on one another, or when we learn to quantize the radiation field and to couple it to hitherto stationary systems such as oscillators, hydrogen atoms etc. It is only by observation of scattering processes, in the first case, or of emission and absorption of γ-rays, in the second case, that characteristic properties of quantum systems can be verified by experiment.

The following example is a little more realistic insofar as it describes a state with a nontrivial evolution in time, i.e. a state that has more than just harmonic dependence on time. Let φ_n be, once more, the basis of eigenfunctions (1.83) of the harmonic oscillator in one dimension, and let

$$\psi(t,x) = \sum_{n=0}^{\infty} c_n(t)\varphi_n(x)$$

be a time dependent state which is constructed following the model of (1.94). The time dependence of the expansion coefficients is harmonic, as shown in Sect. 1.8.1, Remark 3, which is to say

$$c_n(t) = c_n(0)\mathrm{e}^{-(\mathrm{i}/\hbar)E_n t} = c_n(0)\mathrm{e}^{-(\mathrm{i}/2)\omega t}\mathrm{e}^{-\mathrm{i}n\omega t} ,$$

and where we have inserted the formula (1.78) for the energy. Let $z(0) = r\mathrm{e}^{-\mathrm{i}\phi(0)}$ be an arbitrary complex number, written in terms of its modulus and a phase. For the sake of convenience we have chosen a minus sign in front of the phase. Choosing the coefficients at time zero to be

$$c_n(0) = \left(\frac{1}{\sqrt{n!}} z(0)^n\right) \mathrm{e}^{-r^2/2} ,$$

the wave function ψ is seen to be normalized to 1 at time $t = 0$, and, thus, also for all times,

$$\sum_{n=0}^{\infty} |c_n(0)|^2 = \left(\sum_0^{\infty} \frac{1}{n!}(r^2)^n\right) \mathrm{e}^{-r^2} = 1 .$$

If we set

$$z(t) = r\mathrm{e}^{-\mathrm{i}\phi(t)} = r\mathrm{e}^{-\mathrm{i}[\omega t + \phi(0)]} \quad \text{and} \quad \phi(t) = \omega t + \phi(0) ,$$

then ψ reads

$$\psi(t,x) = \mathrm{e}^{-r^2/2}\mathrm{e}^{-\mathrm{i}\omega t/2} \sum_{n=0}^{\infty} \frac{1}{\sqrt{n!}} z^n(t)\, \varphi_n(x) .$$

Whenever $r \neq 0$, obviously, ψ is not an eigenstate of the energy. However, in the limit $r \to 0$ it goes over into the ground state of the harmonic oscillator. In order to explore its physics content we calculate the expectation values of the coordinate x and of the momentum p, as well as their standard deviations (Δx) and (Δp), respectively. Introducing the raising and lowering operators of Sect. 1.6, x and p are given by

$$x = b\frac{1}{\sqrt{2}}(a^\dagger + a)\,, \qquad p = \frac{\hbar}{b}\frac{\mathrm{i}}{\sqrt{2}}(a^\dagger - a)\,,$$

while the action of $a^\dagger$ and of a on the eigenfunctions is

$$(\varphi_m, a^\dagger \varphi_n) = \sqrt{n+1}\,\delta_{m,n+1}\,, \qquad (\varphi_m, a\varphi_n) = \sqrt{n}\,\delta_{m,n-1}\,.$$

The expectation value of x in the state ψ is calculated as follows:

$$\begin{aligned}\langle x\rangle_\psi &= \frac{b}{\sqrt{2}}\mathrm{e}^{-r^2}\left(\sum_{n=0}^{\infty}\frac{z^{*\,n+1}z^n\sqrt{n+1}}{\sqrt{(n+1)!n!}} + \sum_{n=1}^{\infty}\frac{z^{*\,n-1}z^n\sqrt{n}}{\sqrt{(n-1)!n!}}\right)\\ &= \frac{b}{\sqrt{2}}\mathrm{e}^{-r^2}\mathrm{e}^{+r^2}(z^* + z) = rb\sqrt{2}\cos[\omega t + \phi(0)]\,.\end{aligned}$$

The expectation value of p follows from Ehrenfest's theorem (1.72)

$$\langle p\rangle_\psi = m\frac{\mathrm{d}}{\mathrm{d}t}\langle x\rangle_\psi = -\frac{\hbar\sqrt{2}}{b}r\sin[\omega t + \phi(0)]\,.$$

These intermediate results, in themselves, are quite interesting: the expectation values of $\omega\sqrt{m}\,x$ and of $p/\sqrt{m}$ move on a circle with radius $r\sqrt{2\hbar\omega} \equiv \sqrt{2E_{\mathrm{cl}}}$ in phase space and with angular velocity ω,

$$\begin{aligned}\omega\sqrt{m}\,\langle x\rangle_\psi &= r\sqrt{2\hbar\omega}\cos[\omega t + \phi(0)]\,,\\ \frac{1}{\sqrt{m}}\langle p\rangle_\psi &= -r\sqrt{2\hbar\omega}\sin[\omega t + \phi(0)]\,.\end{aligned}$$

The time dependent state $\psi(t,x)$ belongs to the class of *coherent states*. It comes closest to a classical oscillator motion with energy $E_{\mathrm{cl}} = r^2\hbar\omega$. As a quantum mechanical state it does not have a fixed, well-defined energy. The probability to find the eigenvalue $E_n = (n+1/2)\hbar\omega$ in a measurement of the energy is

$$w(E_n) = \frac{r^{2n}\mathrm{e}^{-r^2}}{n!}\,.$$

It is independent of time and has its maximum at

$$n = r^2, \quad \text{i.e. at} \quad E_{n=r^2} = \left(r^2 + \frac{1}{2}\right)\hbar\omega\,.$$

Except for the zero point energy the maximum has the value of the corresponding classical energy.

It is instructive to investigate this coherent state in more detail. We calculate the operator

$$x^2 = \frac{1}{2}b^2(a^\dagger a^\dagger + a^\dagger a + aa^\dagger + aa) = \frac{b^2}{2}(a^\dagger a^\dagger + 2a^\dagger a + 1 + aa)$$

and, from this, the expectation value

$$\begin{aligned}\left\langle x^2\right\rangle_\psi &= \frac{1}{2}b^2\left\{1+\left\langle a^\dagger a^\dagger + 2a^\dagger a + aa\right\rangle_\psi\right\}\\ &= \frac{1}{2}b^2\left\{1+r^2\left[z^{*2}(t)+2z^*(t)z(t)+z^2(t)\right]\right\}\\ &= \frac{1}{2}b^2+2r^2b^2\cos^2[\omega t+\phi(0)]\,.\end{aligned}$$

This calculation is not difficult. As a matter of example, I show a typical intermediate step. The raising operator, when applied twice to ψ, yields

$$a^\dagger a^\dagger \psi = \mathrm{e}^{-r^2/2}\sum_{n=0}^{\infty}\frac{z^n}{\sqrt{n!}}\sqrt{(n+2)(n+1)}\varphi_{n+2}(x)\,.$$

One takes the scalar product of this with ψ, i.e. one multiplies with ψ^* from the left and integrates over x. Making use of the orthogonality of the base functions φ_n, the expectation value is found to be

$$\left\langle a^\dagger a^\dagger\right\rangle_\psi = \mathrm{e}^{-r^2}\sum_{n=0}^{\infty}\frac{z^{*\,n+2}z^n\sqrt{(n+2)(n+1)}}{\sqrt{(n+2)!n!}} = \mathrm{e}^{-r^2}\mathrm{e}^{+r^2}z^{*2}(t)\,.$$

With the results obtained so far the mean square deviation and the standard deviation are found to be

$$(\Delta x)^2 = \left\langle x^2\right\rangle_\psi - \langle x\rangle_\psi^2 = \frac{1}{2}b^2\,,\quad \text{so that}\quad (\Delta x) = \frac{b}{\sqrt{2}}\,.$$

The expectation value of p^2 is calculated in much the same way,

$$\left\langle p^2\right\rangle_\psi = \frac{\hbar^2}{2b^2}\left\{1+4r^2\sin^2[\omega t+\phi(0)]\right\}\,,$$

so that the standard deviation is

$$(\Delta p) = \frac{\hbar}{b\sqrt{2}}\,.$$

This is an interesting result: the product of the standard deviations of x and p has the value

$$(\Delta x)(\Delta p) = \frac{\hbar}{2}\,.$$

This is the minimum allowed by Heisenberg's uncertainty relation (1.35). It equals the product of the uncertainties in the (stationary) ground state of the harmonic oscillator. The coherent state is not stationary, it moves along the classical orbit in phase space and is marginally compatible with the uncertainty relation.

Finally, we calculate the standard deviation (ΔE) of the energy. The expectation values of H and of H^2 are

$$\begin{aligned}\langle H\rangle_\psi &= \frac{1}{2m}\left\langle p^2\right\rangle_\psi + \frac{1}{2}m\omega^2\left\langle x^2\right\rangle_\psi = \left(r^2+\frac{1}{2}\right)\hbar\omega\,,\\ \left\langle H^2\right\rangle_\psi &= \left(\frac{1}{4}+2r^2+r^4\right)(\hbar\omega)^2\,,\end{aligned}$$

respectively, so that the standard deviation is found to be

$$(\Delta E) \equiv (\Delta H) = r(\hbar\omega) .$$

There is an interesting observation in connection with the uncertainty of the energy. Classically, one would calculate the period of the motion from the formula

$$t(x) - t(x_0) = \sqrt{\frac{m}{2}} \int_{x_0}^{x} \mathrm{d}x' \, \frac{1}{\sqrt{E - m\omega^2 x'^2/2}} ,$$

(cf. [Scheck (2010)], Sect. 1.21), where the integral would have to be taken over a complete revolution. In the quantum system the position can be given only with the uncertainty (Δx). Therefore, the calculation of the period must have an uncertainty, too, which follows from this formula when one integrates over the interval $2(\Delta x)$,

$$\begin{aligned} \Delta T &= \sqrt{\frac{m}{2}} \int_{-(\Delta x)}^{(\Delta x)} \mathrm{d}x' \, \frac{1}{\sqrt{E - m\omega^2 x'^2/2}} \\ &= \frac{2}{\omega} \arcsin\left(\frac{(\Delta x)\omega\sqrt{m}}{\sqrt{2E}} \right) \simeq 2\frac{(\Delta x)\sqrt{m}}{\sqrt{2E}} . \end{aligned}$$

The approximation in the last line is applicable when (Δx) is sufficiently small. This quantity was calculated for the coherent state ψ above. Inserting these results and restricting to values of r large as compared to $1/\sqrt{2}$, one finds $\Delta T \simeq 1/(r\omega)$, so that the product of (ΔE) and of (ΔT) becomes

$$(\Delta E)(\Delta T) \simeq r(\hbar\omega) \, \frac{1}{r\omega} = \hbar .$$

The uncertainties of the energy and of the period are correlated by Planck's constant. The better the period is known the larger the uncertainty in the energy. Note, however, that the correlation of the uncertainties of energy and time is of a different nature than the one between position and momentum, the reason being that time plays the role of a *parameter*, and is not an operator.

1.8.3 Observables with Degenerate, Discrete Spectrum

For reasons that will become clear below an observable with a nondegenerate spectrum, like the one developed in Sect. 1.8.1, is rather the exception in physical situations. Consider the example of the harmonic oscillator in $\mathbb{R}^3$ and assume the three circular frequency to be equal. The Hamiltonian reads

$$H = -\frac{\hbar^2}{2m}\Delta + \frac{1}{2}m\omega^2 \sum_{i=1}^{3} (x^i)^2 = -\frac{\hbar^2}{2m}\Delta + \frac{1}{2}m\omega^2 r^2 ,$$

its eigenvalues are

$$E_N = \left(N + \frac{3}{2} \right) \hbar\omega \quad \text{with} \quad N = n_1 + n_2 + n_3 .$$

One easily verifies that although for $N = 0$ there is only one eigenstate, there are three for $N = 1$, six for $N = 2$, ten for $N = 3$, etc. The degree of degeneracy grows rapidly with N.

In classical mechanics the following setting is known: Given a Hamiltonian function $H \equiv F_0$ which has no explicit time dependence, and a set of time independent constants of the motion $F_1, F_2, \ldots$, all of which are in involution. This is to say that the Poisson brackets of the F_i with H, and all brackets of every F_i with every F_j vanish. An example is provided by the two-body system with central potential, viz.

$$H = \frac{\boldsymbol{p}^2}{2m} + U(r)\,, \qquad F_1 = \boldsymbol{P}\,, \qquad F_2 = \boldsymbol{\ell}^2, \qquad F_3 = \ell_3$$

where $\boldsymbol{p}$ and $\boldsymbol{P}$ denote relative and center-of-mass momenta, respectively. This set fulfills the assumption. If there are sufficiently many of such constants of the motion, or, more precisely, if there are f of them, f being the number of degrees of freedom, then the system is integrable (this is one of Liouville's theorems).

In quantum mechanics we expect an analogous situation to be one where a time independent Hamiltonian is given, and where there exist further (time independent) observables $F_1, F_2, \ldots$ whose commutators vanish,

$$[H, F_i] = 0\,, \qquad [F_i, F_j] = 0$$

The physical significance of the first of these relations is that each one of these observables is a constant of the motion, cf. (1.70). Mathematically speaking it says that one can choose the eigenfunctions of the self-adjoint operator H such that they are also eigenfunctions of the observables F_i and vice versa. In other terms, one can find a basis ψ_n which has the property that the matrices $(\psi_m, H\psi_n)$ and $(\psi_p, F_i\psi_q)$ can be brought to diagonal form simultaneously. The second part of the assumption guarantees that it is possible to construct simultaneous eigenfunctions for H and all F_i, or to find a basis in which H as well as the observables F_i are represented by diagonal matrices.

A typical case will be one where the observable F has a discrete, but degenerate spectrum. The operator F then obeys the equation

$$F\left(\boldsymbol{p} = \frac{\hbar}{\mathrm{i}}\nabla, \boldsymbol{x}\right)\varphi_{nk}(\boldsymbol{x}) = \lambda_n \varphi_{nk}(\boldsymbol{x})\,, \tag{1.96}$$

with eigenvalues which have the property

$$\lambda_m \neq \lambda_n \quad \text{for} \quad n \neq m\,,$$

but where there is more than one eigenfunction belonging to the eigenvalue λ_n. If the degree of degeneracy is k_n then there are k_n linearly independent functions

$$\varphi_{n1}(\boldsymbol{x}), \varphi_{n2}(\boldsymbol{x}), \cdots, \varphi_{nk_n}(\boldsymbol{x})$$

which belong to the eigenvalue λ_n. Suppose the set of eigenfunctions φ_{nk} to be orthonormal and complete. The orthogonality and normalization condition reads

$$\int \mathrm{d}^3x\, \varphi^*_{nk}\varphi_{n'k'} = \delta_{nn'}\delta_{kk'}\,.$$

A square integrable wave function defined on the same domain can be expanded in this basis,

$$\psi(t, \boldsymbol{x}) = \sum_{n=0}^{\infty} \sum_{k=1}^{k_n} c_{nk}(t) \varphi_{nk}(\boldsymbol{x}) .$$

The coefficients $c_{nk}(t)$ are obtained by the formulae

$$c_{nk}(t) = \int \mathrm{d}^3x \, \varphi_{nk}^*(\boldsymbol{x}) \psi(t, \boldsymbol{x}) .$$

The normalization of ψ follows from the equation

$$\int \mathrm{d}^3x \, |\psi(t, \boldsymbol{x})|^2 = \sum_{n,k} |c_{nk}(t)|^2 = 1 ,$$

the expectation value of F in the state ψ is given by

$$\langle F \rangle_\psi = \sum_{n=0}^{\infty} \lambda_n \sum_{k=1}^{k_n} |c_{nk}(t)|^2 .$$

We now assume that F commutes with H, $[H, F] = 0$, and that the two operators have the same domain of definition. If φ_{nk} is an eigenfunction of F belonging to the eigenvalue λ_n, then $(H\varphi_{nk})$ is also eigenfunction of F for the same eigenvalue,

$$F\,(H\varphi_{nk}) = \lambda_n\,(H\varphi_{nk}) \; .$$

A way to visualize this situation is the following: The base functions φ_{nk} with fixed n span a subspace which is characterized by the eigenvalue λ_n of F and which has dimension k_n. The state $(H\varphi_{nk})$ is an element of this subspace. Therefore, as the basis is complete, it must be possible to decompose this state in terms of the φ_{nk} with fixed n,

$$(H\varphi_{nk}) = \sum_{k'=1}^{k_n} \varphi_{nk'} H_{k'k} ,$$

where $\mathbf{H} \equiv \{H_{k'k}\}$ is the $k_n \times k_n$ hermitian matrix representation of H in the subspace belonging to λ_n,

$$H_{k'k} = (\varphi_{nk'}, H\varphi_{nk}) = H_{kk'}^* .$$

This finite-dimensional, hermitian matrix is diagonalized by means of a unitary matrix $\mathbf{U}$,

$$\mathbf{U}^\dagger \mathbf{H} \mathbf{U} = \overset{0}{\mathbf{H}} \quad \text{with} \quad \mathbf{U}^\dagger \mathbf{U} = \mathbb{1} ,$$

or, when written in components,

$$\sum_{j,k=1}^{k_n} U_{ji}^* H_{jk} U_{kl} = E_{nl} \delta_{il} .$$

A little calculation shows that in the new basis defined by

$$\psi_{nl}(\boldsymbol{x}) = \sum_{j=1}^{k_n} \varphi_{nj}(\boldsymbol{x}) U_{jl}$$

both F and H are diagonal, i.e.

$$F\psi_{nl} = \lambda_n \psi_{nl} \quad \text{and} \quad H\psi_{nl} = E_{nl}\psi_{nl} .$$

Although somewhat schematic the example shows what is the reason for the degeneracy and the multiple indices of the wave function: The wave functions ψ_{nl} are eigenfunctions of the two commuting operators F and H both of which have (in our example) discrete spectra. The indices on ψ serve to count these spectra.

Here is another example which may serve as an exercise for the reader and which may help in learning some calculational techniques. (We come back to this example below, in the context of problems with spherical symmetry.)

Example 1.8 Spherical Oscillator

At the beginning of this section we mentioned the example of the harmonic oscillator with equal circular frequencies in three space dimensions. In classical mechanics it is described by a spherically symmetric Hamiltonian function, in quantum mechanics by a spherically symmetric Hamiltonian operator. Therefore, one expects the orbital angular momentum to play an important role in determining the states of the system in classical and quantum mechanics, respectively. We treat this important aspect in the framework of a general analysis of orbital angular momentum in Sect. 1.9. The present example is meant to give a first example for commuting observables which are at the root of the degeneracy of the eigenvalues of the Hamiltonian.

We represent the spherical oscillator as the sum of three linear oscillators and make use of the raising and lowering operators (1.74) and (1.75), for each of the spatial coordinates. With the expression (1.76) the Hamiltonian can be written in the form

$$H = \left(\sum_{i=1}^{3} a_i^\dagger a_i + \frac{3}{2} \right) \hbar\omega .$$

The pairs of operators $(a_i^\dagger, a_i)$ refer to the cartesian directions in space. The obey the commutation rules (1.77) for every $i = 1, 2$, or 3, but they commute for different values of the indices, viz.

$$[a_i, a_k^\dagger] = \delta_{ik} , \qquad [a_i, a_k] = 0 , \qquad [a_i^\dagger, a_k^\dagger] = 0 .$$

The eigenfunctions of the Hamiltonian are given by the products

$$\varphi_{n_1 n_2 n_3}(\boldsymbol{x}) = \frac{1}{\sqrt{n_1! n_2! n_3!}} (a_1^\dagger)^{n_1} (a_2^\dagger)^{n_2} (a_3^\dagger)^{n_3} \varphi_0(x^1)\varphi_0(x^2)\varphi_0(x^3) \tag{1.97}$$

with φ_0 as given in (1.79). It is useful to define the following operators:

$$N_1 = a_1^\dagger a_1 , \quad N_2 = a_2^\dagger a_2 , \quad N_3 = a_3^\dagger a_3 , \quad N_{ik} = a_i^\dagger a_k \quad \text{for} \quad i \neq k .$$

All these operators do not change the total energy

$$E_{n_1 n_2 n_3} = \hbar\omega\left(n_1 + n_2 + n_3 + \frac{3}{2}\right)$$

and, thus, commute with the Hamiltonian. As far as the operators N_i are concerned this is obvious. Regarding N_{ik} one may wish to verify

$$[H, N_{ik}] = \hbar\omega[a_i^\dagger a_i + a_k^\dagger a_k, a_i^\dagger a_k] = \hbar\omega(a_i^\dagger a_k - a_i^\dagger a_k) = 0\,.$$

The first three operators N_i are diagonal in the basis (1.97). Indeed, the interpretation of N_i is easy to find out by calculating its action on the state (1.97),

$$a_i^\dagger a_i\, \varphi_{n_1 n_2 n_3}(\boldsymbol{x}) = n_i\, \varphi_{n_1 n_2 n_3}(\boldsymbol{x})\,.$$

It reproduces the wave function with the eigenvalue n_i, and it measures the number of quanta $\hbar\omega$ which are excited in the degree of freedom i. For reasons to become clear later it is called *number operator* for quanta or particles of the species i. The operators N_{ik}, in turn, change the eigenfunctions because they lower n_k by one, while increasing n_i by one. For example, in the basis used here N_{12} has the matrix representation

$$(\varphi_{n_1' n_2' n_3'}, N_{12}\varphi_{n_1 n_2 n_3}) = \sqrt{n_2(n_1+1)}\,\delta_{n_1', n_1+1}\delta_{n_2', n_2-1}\,.$$

This example becomes more conspicuous and better interpretable if one calculates the components of the orbital angular momentum. By the formulae (1.74) and (1.75) one finds

$$\begin{aligned}\ell_3 = x_1 p_2 - x_2 p_1 &= \mathrm{i}\frac{\hbar}{2}\{(a_1^\dagger + a_1)(a_2^\dagger - a_2) - (a_2^\dagger + a_2)(a_1^\dagger - a_1)\}\\ &= \mathrm{i}\hbar\{N_{21} - N_{12}\}\,.\end{aligned}$$

The two other components follow from this by cyclic permutation of the indices 1, 2, 3, so that

$$\ell_1 = \mathrm{i}\hbar\{N_{32} - N_{23}\}\,, \qquad \ell_2 = \mathrm{i}\hbar\{N_{13} - N_{31}\}\,.$$

All three components commute with H, but they do not commute among themselves. For instance, one has

$$[\ell_1, \ell_2] = -\hbar^2[a_3^\dagger a_2 - a_2^\dagger a_3, a_1^\dagger a_3 - a_3^\dagger a_1] = -\hbar^2\{-a_1^\dagger a_2 + a_2^\dagger a_1\} = \mathrm{i}\hbar\ell_3\,.$$

Clearly, the two remaining commutators follow from this one by cyclic permutation, giving

$$[\ell_2, \ell_3] = \mathrm{i}\hbar\ell_1\,, \qquad [\ell_3, \ell_1] = \mathrm{i}\hbar\ell_2\,.$$

Dividing by $\hbar$ one obtains precisely the commutators for the generators of the rotation group in three real dimensions:

$$\left[\frac{\ell_i}{\hbar}, \frac{\ell_j}{\hbar}\right] = \mathrm{i}\sum_{k=1}^{3} \varepsilon_{ijk}\frac{\ell_k}{\hbar}\,.$$

The calculation of $\boldsymbol{\ell}^2$ is a little more involved, one finds

$$\begin{aligned}\boldsymbol{\ell}^2 &= \ell_1^2 + \ell_2^2 + \ell_3^2 \\ &= \hbar^2\{2(N_1 + N_2 + N_3 + N_1N_2 + N_2N_3 + N_3N_1) \\ &\quad - N_{32}^2 - N_{23}^2 - N_{13}^2 - N_{31}^2 - N_{21}^2 - N_{12}^2\}\,.\end{aligned}$$

Inspection of this operator shows that it commutes with H, too, but that the eigenfunctions of H found previously are not eigenfunctions of $\boldsymbol{\ell}^2$. Finally, one shows that $\boldsymbol{\ell}^2$ commutes with every component. The following case is sufficient to show this:

$$\begin{aligned}[\boldsymbol{\ell}^2, \ell_3] &= [\ell_1^2, \ell_3] + [\ell_2^2, \ell_3] \\ &= \ell_1[\ell_1, \ell_3] + [\ell_1, \ell_3]\ell_1 + \ell_2[\ell_2, \ell_3] + [\ell_2, \ell_3]\ell_2 = 0\,.\end{aligned}$$

The conclusion is that a set of *commuting* observables consists of H, of $\boldsymbol{\ell}^2$, and *one* of the three components of the orbital angular momentum, say ℓ_3. This explains, at least to some extent, the degeneracy of the eigenvalues of H, noticed above. In Sect. 1.9 below we will learn how to construct common eigenfunctions of these three operators, in a more concrete setting. Finally, note that this result is completely analogous to the corresponding classical situation: there the functions on phase space H, $\boldsymbol{\ell}^2$, and ℓ_3, are in involution. We just have to replace the commutators by Poisson brackets.

1.8.4 Observables with Purely Continuous Spectrum

Besides the observables with fully discrete spectrum there are also observables whose spectrum is purely continuous, as well as observables which possess a mixed spectrum consisting of a discrete series and a continuous interval. One says that the spectrum is fully continuous if it is not countable, i.e. if the eigenvalue equation of the observable A reads

$$A\left(\boldsymbol{p} = \frac{\hbar}{\mathrm{i}}\boldsymbol{\nabla}, \boldsymbol{x}\right)\varphi(\boldsymbol{x}, \alpha) = \alpha\varphi(\boldsymbol{x}, \alpha) \tag{1.98}$$

and if α can take any value in an interval $I \subset \mathbb{R}$. The momentum operator provides an example, viz.

$$\underline{p} = \frac{\hbar}{\mathrm{i}}\frac{\mathrm{d}}{\mathrm{d}x}\,.$$

For simplicity it is written in one space dimension only and, as an exception, the operator is underlined in order to distinguish it from its eigenvalues p. In this case the eigenvalue equation (1.98) reads

$$\underline{p}\,\varphi(x, p) = p\,\varphi(x, p) \quad \text{with} \quad p \in (-\infty, +\infty)\,.$$

The eigenfunction which belongs to the eigenvalue p is proportional to $\exp(\mathrm{i}px/\hbar)$. Obviously, it is not square integrable. In order to find out something about its normalization we first note that the naïve expression

$$\frac{1}{2\pi}\int_{-\infty}^{\infty} \mathrm{d}x\, \mathrm{e}^{\mathrm{i}(\alpha-\beta)x} = \delta(\alpha-\beta) \tag{1.99}$$

is neither a Riemann nor a Lebesgues integral, and can only be understood as a *tempered distribution*, that is to say, roughly speaking, as a functional $\delta[f]$ which yields a finite, nonsingular, result only when weighted with sufficiently "tamed" functions f.[18] Some important properties of tempered distributions are summarized in Appendix A.1. In particular, it is shown that definitions can be adjusted such that the formal rules are the same as for genuine functions. In particular, Dirac's δ-distribution has the property, for sufficiently smooth functions f,

$$\delta[f] \equiv \int_{-\infty}^{\infty} \mathrm{d}\alpha\, \delta(\alpha-\beta) f(\alpha) = f(\beta)\,.$$

In practical applications one must take care of two of its properties: (a) the normalization factor $1/(2\pi)$ on the left-hand side of (1.99) is essential; (b) the distribution $\delta(z)$ carries a physical dimension whenever its argument z has a dimension. Indeed, one convinces oneself that

$$\text{if}\quad \dim[z] = D \quad\text{then}\quad \dim[\delta(z)] = \frac{1}{D}\,.$$

This is so because, formally,

$$\int_{-\infty}^{\infty} \mathrm{d}z\, \delta(z) = 1\,,$$

must come out without dimension.

The formula (1.99) and the normalization can be obtained by a formal limiting process from true Riemann integrals to distributions. As this can be understood without further knowledge of distributions I insert a digression on the following example:

Example 1.9 Plane Waves in a Limit

The set of functions

$$\left\{ \varphi_m(x) = \frac{1}{\sqrt{a}}\mathrm{e}^{\mathrm{i}(2\pi m/a)x} \,\middle|\, a, x \in \mathbb{R},\ m = 0, \pm 1, \pm 2, \ldots \right\} \tag{1.100}$$

[18] In this book we only use tempered distributions. For this reason we henceforth talk about distributions, for short, and omit the adjective tempered.

form an orthonormal system in the interval $I = [-a/2, +a/2]$. Indeed, we have

$$\int_{-a/2}^{+a/2} \mathrm{d}x\, \varphi_m^*(x)\varphi_n(x) = \frac{\sin(n-m)\pi}{(n-m)\pi} = \delta_{nm}\,.$$

A periodic function $f(x)$ with period a, $f(x+a) = f(x)$, can be expanded in this basis

$$f(x) = \frac{1}{\sqrt{a}} \sum_{m=-\infty}^{+\infty} c_m \mathrm{e}^{\mathrm{i}(2\pi m/a)x} = \frac{1}{a} \sum_{m=-\infty}^{+\infty} g\left(\frac{m}{a}\right) \mathrm{e}^{\mathrm{i}(2\pi m/a)x}\,,$$

where the function g of $y_m := m/a$ is given by

$$g(y_m) = \int_{-a/2}^{+a/2} \mathrm{d}x\, \mathrm{e}^{-\mathrm{i}(2\pi m/a)x} f(x) \equiv \int_{-a/2}^{+a/2} \mathrm{d}x\, \mathrm{e}^{-\mathrm{i}2\pi y_m x} f(x)\,.$$

If one performs the formal transition $a \to \infty$, one has

$$y_m = \frac{m}{a}\,, \qquad y_{m+1} = \frac{m+1}{a}\,, \quad \text{and} \quad y_{m+1} - y_m = \frac{1}{a} \longrightarrow \mathrm{d}y\,.$$

The sum over m goes over into an integral over y,

$$\frac{1}{a} \sum_{m=-\infty}^{+\infty} \longrightarrow \int_{-\infty}^{+\infty} \mathrm{d}y\,.$$

Thus, one obtains

$$f(x) = \int_{-\infty}^{+\infty} \mathrm{d}y\, g(y)\mathrm{e}^{\mathrm{i}2\pi yx}\,, \quad g(y) = \int_{-\infty}^{+\infty} \mathrm{d}x\, \mathrm{e}^{-\mathrm{i}2\pi yx} f(x)\,.$$

Finally, substituting $u := \sqrt{2\pi}\,x$, and $v := \sqrt{2\pi}\,y$, one has

$$f(u) = \frac{1}{\sqrt{2\pi}} \int_{-\infty}^{+\infty} \mathrm{d}v\, g(v)\mathrm{e}^{\mathrm{i}vu}\,, \qquad g(v) = \frac{1}{\sqrt{2\pi}} \int_{-\infty}^{+\infty} \mathrm{d}u\, \mathrm{e}^{-\mathrm{i}vu} f(u)\,.$$

It is tacitly assumed that these integrals exist, i.e. that $f(x)$ decreases sufficiently rapidly at infinity. These formulae represent what is called Fourier transformation in one dimension and its inverse. The base functions (1.100) go over into the functions

$$\left\{ \frac{1}{\sqrt{2\pi}} \mathrm{e}^{\mathrm{i}vu} \,\middle|\, v, u \in \mathbb{R} \right\}.$$

The orthogonality relation

$$\int_{-a/2}^{+a/2} \mathrm{d}x\, \varphi_m^*(x)\varphi_n(x) = \delta_{nm}$$

is replaced by the normalization (1.99) which is to be understood in the sense of a distribution.

The result of this example can be applied directly to the construction of the eigenfunctions of the momentum operator. If the normalization of the eigenfunctions of $\underline{p}$ is chosen to be

$$\varphi(x, p) = \frac{1}{(2\pi\hbar)^{1/2}} e^{(i/\hbar)px}, \tag{1.101}$$

the generalized orthogonality relation reads

$$\int_{-\infty}^{\infty} dx\, \varphi^*(x, p')\varphi(x, p) = \delta(p' - p) \quad \text{"Orthogonality"}. \tag{1.102}$$

The δ-distribution replaces the Kronecker deltas in the scalar product $(\varphi_m, \varphi_n) = \delta_{mn}$ and inherits the physical dimension 1/(dimension of momentum). As the relation is to be interpreted as a functional over the space of the p-variables which, qualitatively speaking, yields nonsingular expressions only upon integration over p or p', on says that (1.101) is *normalized in momentum scale.*

In practice it may happen that one has to deal with eigenfunctions of energy, not of momentum, so that the plane waves must be normalized in the energy scale, instead of the momentum scale. The substitution required for this is illustrated by the following example. Formally, one has

$$\delta(p - p') = \delta[p(E) - p(E')] = \left(\left| \frac{dp}{dE} \right|_{E=E'} \right)^{-1} \delta(E - E').$$

Thus, if one must normalize to $\delta(E - E')$ then the wave function which was normalized according to $\delta(p - p')$ must be multiplied by the square root of

$$\left| \frac{dp}{dE} \right|.$$

With nonrelativistic kinematics $p = \sqrt{2mE}$, the wave function (1.101) becomes

$$\varphi(x, E) = \frac{1}{(2\pi\hbar)^{1/2}} \frac{m^{1/4}}{(2E)^{1/4}} e^{(i/\hbar)px}.$$

It is then *normalized in the energy scale*, viz.

$$\int_{-\infty}^{\infty} dx\, \varphi^*(x, E')\varphi(x, E) = \delta(E' - E).$$

The wave function (1.101) is symmetric in the variables x and p. Therefore, the analogue of the orthogonality relation (1.102) is

$$\int_{-\infty}^{\infty} dp\, \varphi^*(x', p)\varphi(x, p) = \delta(x' - x) \qquad \text{(completeness)}. \tag{1.103}$$

Indeed, this relation is an expression for the completeness of plane waves. This may be verified by means of the analogous relations in the purely discrete case of Sect. 1.8.1. Restricting them, for simplicity, to one coordinate one finds

$$\int_{-\infty}^{+\infty} \mathrm{d}x\, \varphi_m^*(x)\varphi_n(x) = \delta_{mn} \qquad \text{(orthogonality)}\,, \tag{1.104}$$

$$\sum_{n=0}^{\infty} \varphi_n^*(x')\varphi_n(x) = \delta(x'-x) \qquad \text{(completeness)}\,. \tag{1.105}$$

The second relation can be derived in a formal manner as follows: Since the basis φ_n is complete every ψ can be expanded

$$\psi(x) = \sum_{n=0}^{\infty} c_n\varphi_n(x) = \sum_{n=0}^{\infty}\int_{-\infty}^{\infty} \mathrm{d}x'\, \varphi_n^*(x')\varphi_n(x)\psi(x')\,.$$

Interchanging summation and integration, with the understanding that this is a relation between distributions, the equation is consistent only if (1.105) holds true. This justifies calling (1.105) a completeness relation.

It is not difficult to generalize the results of this section to three space dimensions. Plane waves which are normalized in the momentum scale, are given by

$$\varphi(\boldsymbol{x},\boldsymbol{p}) = \frac{1}{(2\pi\hbar)^{3/2}}\mathrm{e}^{(\mathrm{i}/\hbar)\boldsymbol{p}\cdot\boldsymbol{x}}\,.$$

If one prefers to work in the energy scale instead, it is convenient to write the momentum in spherical polar coordinates, i.e. to express $\boldsymbol{p}$ in terms of its modulus $p \equiv |\boldsymbol{p}|$ and polar angles θ_p and ϕ_p. In these coordinates one has

$$\mathrm{d}^3p = p^2\mathrm{d}p\mathrm{d}(\cos\theta_p)\mathrm{d}\phi_p\,, \quad \text{and}$$

$$\delta(\boldsymbol{p}'-\boldsymbol{p}) = \frac{1}{p'p}\delta(p'-p)\delta(\cos\theta_{p'}-\cos\theta_p)\delta(\phi_{p'}-\phi_p)\,.$$

As before, the δ-distribution for the moduli of the momenta is converted to a δ-distribution for the energies by means of the relation $E = \boldsymbol{p}^2/(2m)$.

A third case which occurs frequently is the case of a *mixed spectrum*, i.e. of a spectrum which has both a discrete part and a continuous part. An example of special importance for physics is the spectrum (1.24) of the hydrogen atom. The hydrogen atom belongs to the class of problems with a central field in $\mathbb{R}^3$. This needs, as a preparation, the analysis of orbital angular momentum in quantum mechanics, and the separation into radial and angular motion. These are subjects to which we now turn.

1.9 Central Forces and the Schrödinger Equation

Very much like in classical mechanics the problems with central forces belong to a class of applications of quantum mechanics which is of great theoretical and practical importance. If the forces are continuous functions they can be expressed as (negative) gradient fields of spherically symmetric potentials $U(r)$. Like in the classical case the prime example is a system of two bodies which interact via a central force $\boldsymbol{F} = F(r)\hat{\boldsymbol{r}}$ with $F(r) = -\mathrm{d}U(r)/\mathrm{d}r$, where $\boldsymbol{r} = \boldsymbol{r}_1 - \boldsymbol{r}_2$ is the relative coordinate, and $r = |\boldsymbol{r}|$ is its modulus. The separation into center-of-mass and relative motion is done in the same way as in classical mechanics and will not be repeated here. Besides the force-free center-of-mass motion which is separated from the rest, one obtains an effective *one-body* problem whose Hamiltonian depends on the relative coordinate only and which contains the potential and the reduced mass. For simplicity we denote the latter with the symbol m and investigate the stationary Schrödinger equation (1.60) with

$$H = -\frac{\hbar^2}{2m}\Delta + U(r)\,.$$

The general strategy in solving problems of this kind is much the same as in mechanics. One separates the relative motion into purely radial motion in the variable r, and angular motion in the variables θ and ϕ, making use of the fact that the modulus of the angular momentum as well as one of its projections are conserved. Clearly, because of the Heisenberg uncertainty relations for the variables r, θ, ϕ and their canonically conjugate momenta p_r, p_θ, p_ϕ one can no longer talk about "orbits", and one cannot even claim that the motion takes place in a plane. Nevertheless, there are similarities between the classical and the quantum mechanical cases, even though the results will be technically different and of a different physical significance. We start by studying the orbital angular momentum and show how the Schrödinger equation is reduced to a differential equation in the radial variable alone. The radial differential equation is then solved for three examples of particular importance.

1.9.1 The Orbital Angular Momentum: Eigenvalues and Eigenfunctions

The orbital angular momentum $\boldsymbol{x} \times \boldsymbol{p}$ is an observable with three components whose operator representation in coordinate space is found by means of the substitution rules

$$x^i \longmapsto x^i\,, \qquad p_k \longmapsto \frac{\hbar}{\mathrm{i}}\frac{\partial}{\partial x^k}\,.$$

As the momentum is proportional to $\hbar$, it is convenient to take out a factor $\hbar$ in the definition of the operator of orbital angular momentum, viz.

$$\hbar\boldsymbol{\ell} := \boldsymbol{x} \times \boldsymbol{p} \quad \text{so that} \quad \boldsymbol{\ell} := \frac{1}{\mathrm{i}}\boldsymbol{x} \times \nabla\,. \tag{1.106}$$

Its components and the square of its modulus which then have no physical dimension, are, respectively,

$$\ell_1 = \frac{1}{\mathrm{i}}\left(x^2\frac{\partial}{\partial x^3} - x^3\frac{\partial}{\partial x^2}\right),$$
$$\ell_2 = \frac{1}{\mathrm{i}}\left(x^3\frac{\partial}{\partial x^1} - x^1\frac{\partial}{\partial x^3}\right),$$
$$\ell_3 = \frac{1}{\mathrm{i}}\left(x^1\frac{\partial}{\partial x^2} - x^2\frac{\partial}{\partial x^1}\right),$$
$$\boldsymbol{\ell}^2 = \ell_1^2 + \ell_2^2 + \ell_3^2 .$$

One verifies that these operators are all self-adjoint.

From the commutation rules for coordinates and derivatives

$$\left[x^i, \frac{\partial}{\partial x^k}\right] = -\delta_{ik} = \mathrm{i}^2\delta_{ik}$$

one derives the commutator of ℓ_1 with ℓ_2. One finds

$$[\ell_1, \ell_2] = \left(x^1\frac{\partial}{\partial x^2} - x^2\frac{\partial}{\partial x^1}\right) = \mathrm{i}\ell_3 . \tag{1.107}$$

Clearly, this commutator is continued to two more commutators by cyclic permutation of the indices. Therefore, the general result reads

$$\boxed{[\ell_i, \ell_j] = \mathrm{i}\sum_k \varepsilon_{ijk}\ell_k} \tag{1.108}$$

The symbol ε denotes the antisymmetric tensor in dimension three which has the value $+1$ (-1) if the set (i, j, k) is an *even* (*odd*) permutation of $(1, 2, 3)$, and which vanishes whenever two or more indices are equal. The commutator of the square $\boldsymbol{\ell}^2$ with any of the components was calculated in Sect. 1.8.3. It follows from the formula $[A^2, B] = A[A, B] + [A, B]A$ and is found to be zero,

$$\boxed{[\boldsymbol{\ell}^2, \ell_i] = 0} . \tag{1.109}$$

The results (1.107)–(1.109) have the following physical interpretation: Only the square of the modulus and *one* component can be measured simultaneously, while the remaining two components cannot have sharp values. The convention, up to exceptions, is to choose $\boldsymbol{\ell}^2$ and ℓ_3, or, expressed differently, to choose the 3-axis of the system of reference in a direction singled out by the specific physical system one is studying. This privileged axis is often called *axis of quantization.*

Before turning to the calculation of the eigenvalues and to the construction of the common eigenfunctions of $\boldsymbol{\ell}^2$ and ℓ_3 it is useful to work out a few more commutators. Examples are

$$[\ell_i, x^j] = \mathrm{i}\sum_k \varepsilon_{ijk}x^k , \qquad [\ell_i, p_j] = \mathrm{i}\sum_k \varepsilon_{ijk}p_k .$$

It is important to note that every component of angular momentum commutes with the modulus r of the position vector, as well as with the modulus of the momentum. This is verified, in the first case, by

$$[\ell_i, r] = \frac{1}{\mathrm{i}} \sum_{m,n} \varepsilon_{imn} x^m \left[\frac{\partial}{\partial x^n}, r \right] = \frac{1}{\mathrm{i}} \sum_{m,n} \varepsilon_{imn} x^m \frac{x^n}{r} = 0 .$$

In the last step the contraction of the totally antisymmetric ε-tensor with the symmetric form $x^m x^n$ gives zero. Thus, all components and $\boldsymbol{\ell}^2$ commute with any differentiable function of r. Similarly, the three components and $\boldsymbol{\ell}^2$ commute with any smooth function of $|\boldsymbol{p}|$. This shows that

$$\left[\ell_i, \frac{\boldsymbol{p}^2}{2m} + U(r) \right] = 0 \quad \text{and} \quad \left[\boldsymbol{\ell}^2, \frac{\boldsymbol{p}^2}{2m} + U(r) \right] = 0 .$$

The physical interpretation is that the modulus and the components of orbital angular momentum are constants of the motion whenever the potential is spherically symmetric. However, as the components do not commute with one another there exist common eigenfunctions only for H, $\boldsymbol{\ell}^2$ and *one* of the three components such as, e.g., ℓ_3. Adding the total momentum $\boldsymbol{P}$ of the two-body system to this list, the result is seen to correspond precisely to the classical situation where

$$H = \frac{\boldsymbol{p}^2}{2m} + U(r), \quad \boldsymbol{P}, \quad \boldsymbol{\ell}^2, \quad \text{and} \quad \ell_3$$

are in involution (cf. [Scheck (2010)], Sect. 2.37.2, example (iii)).

Since every component ℓ_i commutes with the variable r, none of them can contain derivatives with respect to r. This observation suggests using spherical polar coordinates

$$x^1 = r \sin\theta \cos\phi , \quad x^2 = r \sin\theta \sin\phi , \quad x^3 = r \cos\theta$$

and to write the operators ℓ_i and $\boldsymbol{\ell}^2$ as differential operators in the angular variables θ and ϕ. We now show that they are given by the following expressions

$$\ell_1 = \mathrm{i} \left\{ \sin\phi \frac{\partial}{\partial\theta} + \cot\theta \cos\phi \frac{\partial}{\partial\phi} \right\} ,$$

$$\ell_2 = \mathrm{i} \left\{ -\cos\phi \frac{\partial}{\partial\theta} + \cot\theta \sin\phi \frac{\partial}{\partial\phi} \right\} ,$$

$$\ell_3 = -\mathrm{i} \frac{\partial}{\partial\phi} . \tag{1.110}$$

$$\boldsymbol{\ell}^2 = - \left\{ \frac{1}{\sin^2\theta} \frac{\partial^2}{\partial\phi^2} + \frac{1}{\sin\theta} \frac{\partial}{\partial\theta} \left(\sin\theta \frac{\partial}{\partial\theta} \right) \right\} . \tag{1.111}$$

In fact, one could have guessed the third of (1.110) because one knows from classical physics that $\hbar\ell_3$ is the variable canonically conjugated to ϕ. Therefore, their commutator (1.36) must be

$$[\hbar\ell_3, \phi] = -\mathrm{i}\hbar .$$

Indeed, this result is obtained if the derivatives with respect to ϕ are written in terms of derivatives with respect to cartesian coordinates, by means of the chain rule,

$$\frac{\partial}{\partial\phi} = \frac{\partial x^1}{\partial\phi}\frac{\partial}{\partial x^1} + \frac{\partial x^2}{\partial\phi}\frac{\partial}{\partial x^2} = -x^2\frac{\partial}{\partial x^1} + x^1\frac{\partial}{\partial x^2} = \mathrm{i}\ell_3 .$$

In much the same way one expresses the partial derivative with respect to θ in terms of cartesian partial derivatives thereby finding the relation

$$\frac{\partial}{\partial\theta} = \mathrm{i}\,(-\ell_1 \sin\phi + \ell_2 \cos\phi) .$$

A further trick is to note that the following operator is zero

$$\boldsymbol{x}\cdot\boldsymbol{\ell} = \ell_1 x^1 + \ell_2 x^2 + \ell_3 x^3 = 0 .$$

Dividing by x^3 and substituting polar coordinates gives the result

$$\tan\theta(\cos\phi\ell_1 + \sin\phi\ell_2) = -\ell_3 = \mathrm{i}\frac{\partial}{\partial\phi} .$$

Thus, there are two linearly independent equations for ℓ_1 and ℓ_2 whose solution gives the first two formulae (1.110).

The proof of the formula (1.111) is made easier if instead of the cartesian components ℓ_1 and ℓ_2 one introduces the linear combinations

$$\ell_\pm := \ell_1 \pm \mathrm{i}\ell_2 = \mathrm{e}^{\pm\mathrm{i}\phi}\left\{\pm\frac{\partial}{\partial\theta} + \mathrm{i}\cot\theta\frac{\partial}{\partial\phi}\right\}$$

These operators are called *ladder operators*. Their product can be written as follows

$$\ell_\pm\ell_\mp = \ell_1^2 + \ell_2^2 \pm \mathrm{i}\,(\ell_2\ell_1 - \ell_1\ell_2) = \ell_1^2 + \ell_2^2 \pm \ell_3 .$$

The product $\ell_+\ell_-$ is obtained by careful differentiation, using the chain rule,

$$\begin{aligned}\ell_+\ell_- &= \mathrm{e}^{\mathrm{i}\phi}\left\{\frac{\partial}{\partial\theta} + \mathrm{i}\cot\theta\frac{\partial}{\partial\phi}\right\}\mathrm{e}^{-\mathrm{i}\phi}\left\{-\frac{\partial}{\partial\theta} + \mathrm{i}\cot\theta\frac{\partial}{\partial\phi}\right\} \\ &= -\frac{\partial^2}{\partial\theta^2} - \mathrm{i}\frac{1-\cos^2\theta}{\sin^2\theta}\frac{\partial}{\partial\phi} - \cot\theta\frac{\partial}{\partial\theta} - \cot^2\theta\frac{\partial^2}{\partial\phi^2} .\end{aligned}$$

The formula (1.111) follows from this result and from the relation

$$\boldsymbol{\ell}^2 = \ell_+\ell_- + \ell_3^2 - \ell_3 .$$

In a next step we derive the eigenvalues and the eigenfunctions of the operators ℓ_3 and $\boldsymbol{\ell}^2$. The interpretation of the commutators (1.108), (1.109) and their relation with the rotation group SO(3) in three real dimensions, as well as with its covering group SU(2), will be dealt with later, cf. Sects. 4.1 and 6.2.1.

Eigenvalues and Eigenfunctions of $\boldsymbol{\ell_3}$: The eigenfunctions of ℓ_3 fulfill the differential equation

$$\ell_3 f(\phi) = -\mathrm{i}\frac{\partial}{\partial\phi} f(\phi) = m f(\phi) .$$

Obviously, they are proportional to $\mathrm{e}^{\mathrm{i}m\phi}$ where m is a real number. Such a function must allow for a quantum mechanical interpretation which is to say that it must be invariant under a complete rotation of the system of reference about the 3-axis, i.e. under $\mathbf{R}_3(2\pi)$,

$$f(\phi + 2\pi) = f(\phi)\,.$$

This requirement of *uniqueness of the wave function* implies that the eigenvalue m of ℓ_3 must be a positive or negative integer, or zero,

$$m = 0, \pm 1, \pm 2, \pm 3, \ldots\,.$$

Alternatively one may argue that $f(\phi)$ is defined on the unit circle S^1, not on the real axis. Therefore, it must be *well-defined* on S^1 which is to say that $f(\phi + 2\pi)$ must equal $f(\phi)$. Whether one uses the *uniqueness* condition for the wave function, or the requirement that it be *well defined* on S^1, leads to the same conclusion: the solutions normalized to 1 on the interval $[0, 2\pi]$ are

$$f_m(\phi) = \frac{1}{\sqrt{2\pi}} \mathrm{e}^{\mathrm{i}m\phi}\,.$$

The normalization follows from the analysis of the functions (1.100) in Sect. 1.8.4. Indeed, for integer values of m and m' one has

$$\int_0^{2\pi} \mathrm{d}\phi\, f^*_{m'}(\phi) f_m(\phi) = \frac{1}{2\pi} \int_0^{2\pi} \mathrm{d}\phi\, \mathrm{e}^{\mathrm{i}(m-m')\phi}$$

$$= \mathrm{e}^{\mathrm{i}\pi(m-m')} \frac{\sin[\pi(m-m')]}{\pi(m-m')} = \delta_{mm'}\,.$$

Regarding the eigenvalue equation of the operator $\boldsymbol{\ell}^2$, (1.111), we try an ansatz whereby the eigenfunction factorizes in a function of θ and another function $f(\phi)$ of ϕ alone,

$$\boldsymbol{\ell}^2 Y(\theta) f(\phi) = \lambda\, Y(\theta) f(\phi)\,.$$

Inserting (1.111) and dividing by the product $Y(\theta)f(\phi)$ gives

$$\frac{1}{f(\phi)} \frac{\mathrm{d}^2 f(\phi)}{\mathrm{d}\phi^2} + \frac{\sin^2\theta}{Y(\theta)} \left[\frac{1}{\sin\theta} \frac{\mathrm{d}}{\mathrm{d}\theta} \left(\sin\theta \frac{\mathrm{d}Y(\theta)}{\mathrm{d}\theta} \right) + \lambda\, Y(\theta) \right] = 0\,.$$

The first term depends only on ϕ, while the second term depends only on θ. This observation explains why the formerly partial derivatives ∂ are now replaced by ordinary derivatives d, and shows that the separation into a function of θ and a function of ϕ is justified. The solutions $f(\phi)$ were derived above, so that there remains a differential equation for $Y(\theta)$. It reads

$$\frac{1}{\sin\theta} \frac{\mathrm{d}}{\mathrm{d}\theta} \left(\sin\theta \frac{\mathrm{d}Y(\theta)}{\mathrm{d}\theta} \right) + \left(\lambda - \frac{m^2}{\sin^2\theta} \right) Y(\theta) = 0\,.$$

Substituting $z := \cos\theta$ and making use of $\mathrm{d}z = -\sin\theta\, \mathrm{d}\theta$, this equation becomes

$$\frac{\mathrm{d}}{\mathrm{d}z} \left((1 - z^2) \frac{\mathrm{d}Y(z)}{\mathrm{d}z} \right) + \left(\lambda - \frac{m^2}{1 - z^2} \right) Y(z) = 0\,. \qquad (1.112)$$

This is a differential equation which is known from the theory of spherical harmonics and which belongs to the class of *differential equations of Fuchsian type*. Their general form is

$$\boxed{\frac{d^2 y(z)}{dz^2} + \frac{\mathcal{P}_0(z-z_0)}{z-z_0}\frac{dy(z)}{dz} + \frac{\mathcal{P}_1(z-z_0)}{(z-z_0)^2}y(z) = 0}\,, \tag{1.113}$$

where $\mathcal{P}_0$ and $\mathcal{P}_1$ are polynomials in $(z - z_0)$ (or Taylor series which converge in the interval of definition). The singularities of its coefficient functions are characteristic for this type of differential equation: The function that multiplies the first derivative of $y(z)$ has a pole of first order in $z = z_0$, the function multiplying the homogeneous term has a pole of second order in the same point. This type of differential equation occurs in many different eigenvalue problems of quantum mechanics. Their solutions can be constructed explicitly in terms of series in the variable $(z - z_0)$.

In the case at stake (1.112) the differential equation reads

$$\frac{d^2 Y}{dz^2} - \frac{2z}{(1-z)(1+z)}\frac{dY}{dz} + \frac{\lambda(1-z)(1+z) - m^2}{(1-z)^2(1+z)^2}Y = 0\,.$$

As is obvious from its explicit form it has Fuchsian singularities both at $z = 1$ and at $z = -1$. These are the boundaries of the interval of definition of $z = \cos\theta$. Thus, in view of their physical interpretation, one must search for solutions which are regular at the two boundaries $z = \pm 1$.

The theory of spherical harmonics shows that this condition can only be met if the eigenvalue λ is of the form

$$\lambda = \ell(\ell+1) \quad \text{with} \quad \ell = 0, 1, 2, \ldots$$

and if m and ℓ fulfill the inequality $m^2 \leqslant \ell^2$. The solutions of (1.112) which correspond to these eigenvalues are regular in the whole interval $[-1, +1]$. In the case $m = 0$ (1.112) coincides with the differential equation of Legendre polynomials (cf. Example 1.7). These polynomials are obtained from the formula of Rodrigues

$$P_\ell(z) = \frac{1}{2^\ell \ell!}\frac{d^\ell}{dz^\ell}\left(z^2 - 1\right)^\ell \quad \text{(formula of Rodrigues)}\,. \tag{1.114}$$

For $m \geqslant 0$ the solutions can be written in terms of derivatives of Legendre polynomials as follows,

$$P_\ell^m(z) = (-)^m(1-z^2)^{m/2}\frac{d^m}{dz^m}P_\ell(z)\,, \tag{1.115}$$

with $P_\ell(z)$ as defined in (1.114). These solutions are called *associated Legendre functions of the first kind*. Obviously, they are no longer polynomials in z.

Collecting the solutions in the two angular variables one obtains the eigenfunctions $Y_{\ell m}(\theta, \phi)$ of $\boldsymbol{\ell}^2$,

$$Y_{\ell m}(\theta, \phi) = \sqrt{\frac{(2\ell+1)}{4\pi}\frac{(\ell-m)!}{(\ell+m)!}}\,P_\ell^m(\cos\theta)e^{im\phi}\,. \tag{1.116}$$

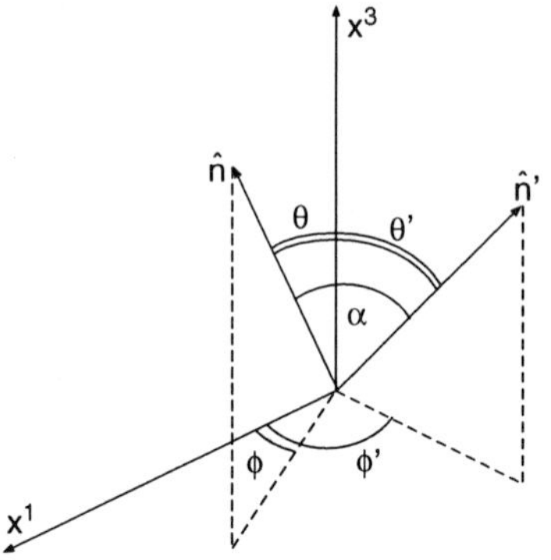

Fig. 1.10 Two unit vectors in $\mathbb{R}^3$ are defined by their polar angles (θ, ϕ) and (θ', ϕ'), respectively. They span the relative angle α. The first two pairs appear on the left-hand side of the addition theorem (1.121) while the angle α is the argument of the Legendre polynomial on the right-hand side

These are called *spherical harmonics* and they have the following properties:

1. The complex conjugate functions are obtained from the relation
$$Y^*_{\ell m}(\theta, \phi) = (-)^m Y_{\ell -m}(\theta, \phi)\,. \tag{1.117}$$
The restriction to $m \geqslant 0$ of (1.115) is circumvented by this symmetry. Equivalently, the index m of (1.115) can be replaced by $|m|$.
2. The spherical harmonics provide a complete system of orthogonal and normalized functions on S^2, the sphere with radius 1 in $\mathbb{R}^3$. Writing $\mathrm{d}\Omega = \mathrm{d}\phi\, \sin\theta \mathrm{d}\theta$ their *orthogonality relation* reads
$$\int \mathrm{d}\Omega\; Y^*_{\ell' m'}(\theta, \phi) Y_{\ell m}(\theta, \phi) = \delta_{\ell' \ell}\delta_{m' m}\,, \tag{1.118}$$
while their *completeness* is expressed by
$$\sum_{\ell=0}^{\infty} \sum_{m=-\ell}^{+\ell} Y_{\ell m}(\theta, \phi) Y^*_{\ell m}(\theta', \phi') = \delta(\phi - \phi')\delta(\cos\theta - \cos\theta')\,. \tag{1.119}$$
3. They are simultaneous eigenfunctions of $\boldsymbol{\ell}^2$ and of ℓ_3, the eigenvalues being
$$\boxed{\begin{aligned} \boldsymbol{\ell}^2 Y_{\ell m} &= \ell(\ell+1) Y_{\ell m}\,, \quad \ell = 0, 1, 2, \ldots \\ \ell_3 Y_{\ell m} &= m Y_{\ell m}\,, \qquad m = -\ell, -\ell+1, \ldots, \ell-1, \ell \end{aligned}}\,. \tag{1.120}$$
The square of the modulus takes the values $\ell(\ell+1)$ with $\ell \in \mathbb{N}_0$, while the 3-component takes one of the $(2\ell+1)$ integer values between $m = -\ell$ and $m = \ell$.
4. Given a direction $\hat{\boldsymbol{n}}$ in space defined by the angles (θ, ϕ), and another direction $\hat{\boldsymbol{n}}'$ with angular coordinates (θ', ϕ'), let α denote the angle spanned by these unit vectors, i.e. $\hat{\boldsymbol{n}} \cdot \hat{\boldsymbol{n}}' = \cos\alpha$ as sketched in Fig. 1.10. Then
$$\frac{4\pi}{2\ell+1} \sum_{m=-\ell}^{+\ell} Y^*_{\ell m}(\theta', \phi') Y_{\ell m}(\theta, \phi) = P_\ell(\cos\alpha)\,. \tag{1.121}$$
This important relation is called the *addition theorem for spherical harmonics*.

We conclude this section with a few remarks and an example.

Remarks

1. Every square integrable function $F(\theta, \phi)$ on S^2 can be expanded in terms of spherical harmonics,

$$F(\theta, \phi) = \sum_{\ell=0}^{\infty} \sum_{m=-\ell}^{+\ell} Y_{\ell m}(\theta, \phi) c_{\ell m}$$

where the expansion coefficients are given by

$$c_{\ell m} = \int \mathrm{d}\Omega \, Y^*_{\ell m}(\theta, \phi) F(\theta, \phi)$$
$$= \int_0^{2\pi} \mathrm{d}\phi \int_0^{\pi} \sin\theta \mathrm{d}\theta \, Y^*_{\ell m}(\theta, \phi) F(\theta, \phi) .$$

2. The action of the operators $\ell_\pm$ on $Y_{\ell m}$ is worked out by means of the formulae (1.115) and (1.114). With

$$z = \cos\theta, \qquad \frac{\partial}{\partial\theta} = -(1-z^2)^{1/2} \frac{\partial}{\partial z}, \qquad \cot\theta = \frac{z}{\sqrt{1-z^2}},$$

and taking account of the normalization in (1.116), straightforward calculation yields the result

$$\boxed{\ell_\pm Y_{\ell m} = \sqrt{\ell(\ell+1) - m(m \pm 1)} Y_{\ell, m\pm 1}} . \tag{1.122}$$

The operator ℓ_+ does not take out of the subspace with fixed ℓ, but it raises the eigenvalue of ℓ_3 by 1. Analogously, the operator ℓ_- lowers the eigenvalue of ℓ_3 by one unit. This is the reason why these operators are called *ladder operators*. Note, in particular, that the chain of eigenstates of $\boldsymbol{\ell}^2$ and of ℓ_3 which are obtained from $Y_{\ell m}$ by repeated action of ℓ_+, does indeed stop at $m = \ell$. Similarly, the descending chain $(\ell_-)^n Y_{\ell m}$ stops at $m = -\ell$.
3. Denoting temporarily the state $Y_{\ell m}$ by the short-hand ψ, one sees that the expectation values of the components

$$\ell_1 = \frac{1}{2}(\ell_+ + \ell_-), \qquad \ell_2 = \mathrm{i}\frac{1}{2}(\ell_- - \ell_+)$$

vanish,

$$\langle \ell_1 \rangle_\psi = \langle \ell_2 \rangle_\psi = 0, \qquad (\psi \equiv Y_{\ell m}) .$$

The expectation values of their squares do not vanish, however. They may be obtained as follows:

$$\left\langle \ell_1^2 + \ell_2^2 \right\rangle_\psi = \left\langle \boldsymbol{\ell}^2 - \ell_3^2 \right\rangle_\psi = \ell(\ell+1) - m^2 .$$

As no particular direction perpendicular to the 3-axis is singled out, one concludes that the expectation values of ℓ_1^2 and of ℓ_2^2 are equal. This implies that

$$\left\langle \ell_1^2 \right\rangle_\psi = \left\langle \ell_2^2 \right\rangle_\psi = \frac{1}{2}[\ell(\ell+1) - m^2] .$$

Nevertheless, it is certainly instructive to do this calculation more directly. One writes ℓ_1^2 in terms of the ladder operators

$$\ell_1^2 = \frac{1}{4}(\ell_+^2 + \ell_-^2 + \ell_+\ell_- + \ell_-\ell_+)$$

and calculates the expectation value of the right-hand side. The operators ℓ_+^2 and ℓ_-^2 give no contribution because they raise/lower $Y_{\ell m}$ to $Y_{\ell,m\pm 2}$ and because the latter functions are orthogonal to $Y_{\ell m}$. The expectation values of the remaining, diagonal, operators $\ell_+\ell_-$ and $\ell_-\ell_+$, follow from (1.122) so that

$$\begin{aligned}\left\langle \ell_1^2 \right\rangle_\psi &= \frac{1}{4}\Big[\sqrt{\ell(\ell+1)-(m-1)m}\,\sqrt{\ell(\ell+1)-m(m-1)} \\ &\quad + \sqrt{\ell(\ell+1)-(m+1)m}\,\sqrt{\ell(\ell+1)-m(m+1)}\Big] \\ &= \frac{1}{2}[\ell(\ell+1)-m^2]\,.\end{aligned}$$

Thus, the standard deviation of ℓ_1 and of ℓ_2 is the same for both,

$$(\Delta\ell_1) = (\Delta\ell_2) = \frac{1}{\sqrt{2}}\sqrt{\ell(\ell+1)-m^2}\,.$$

Note that this deviation is different from zero even when $|m|$ takes its maximal value $|m| = \ell$, i.e. even in a situation when the classical angular momentum is completely aligned along the 3-axis.[19]

4. A description of quantum angular momentum by a vector that takes discrete, quantized directions, should be considered with caution. Indeed, the calculation of the previous remark shows that measurements of a component of angular momentum perpendicular to the 3-axis will yield eigenvalues $+q$ and $-q$ with equal probabilities, q being in the set $q \in \{-\ell, \ldots, +\ell\}$.

Example 1.10

The result (1.122) shows that the ladder operators $\ell_\pm$ have the following matrix representation in the basis of spherical harmonics:

$$(\ell_\pm)_{\ell' m', \ell m} \equiv (Y_{\ell' m'}, \ell_\pm Y_{\ell m}) = \sqrt{\ell(\ell+1)-m(m\pm 1)}\delta_{\ell'\ell}\delta_{m', m\pm 1}\,.$$

These infinite-dimensional matrices decompose into square blocks along the main diagonal, one block of dimension $(2\ell+1)\times(2\ell+1)$ for every eigenvalue of $\boldsymbol{\ell}^2$. Rows and columns are numbered by ℓ and m, taking $\ell = 0, 1, 2\ldots$ in increasing order, and $m = \ell, \ell-1, \ldots, -\ell$, for fixed ℓ, in decreasing order. Thus, rows and columns are marked by

$$\begin{aligned}(\ell, m) = &(0,0), (1,1), (1,0), (1,-1), \\ &(2,2), (2,1), (2,0), (2,-1), (2,-2)\cdots\,.\end{aligned}$$

[19] In the early days of atomic physics this was called the "stretched case".

For example, in the subspace characterized by $\ell = 1$ one obtains

$$(\boldsymbol{\ell}^2)_{1m',1m} = \begin{pmatrix} 2 & 0 & 0 \\ 0 & 2 & 0 \\ 0 & 0 & 2 \end{pmatrix} ,$$

$$(\ell_+)_{1m',1m} = \begin{pmatrix} 0 & \sqrt{2} & 0 \\ 0 & 0 & \sqrt{2} \\ 0 & 0 & 0 \end{pmatrix} , \qquad (\ell_-)_{1m',1m} = \begin{pmatrix} 0 & 0 & 0 \\ \sqrt{2} & 0 & 0 \\ 0 & \sqrt{2} & 0 \end{pmatrix} .$$

The matrix representations of ℓ_1 and ℓ_2 are obtained from these by the formulae of the remarks above. Including the matrix representation of ℓ_3 they are

$$(\ell_1)_{1m',1m} = \frac{1}{2} \begin{pmatrix} 0 & \sqrt{2} & 0 \\ \sqrt{2} & 0 & \sqrt{2} \\ 0 & \sqrt{2} & 0 \end{pmatrix} ,$$

$$(\ell_2)_{1m',1m} = \mathrm{i}\, \frac{1}{2} \begin{pmatrix} 0 & -\sqrt{2} & 0 \\ \sqrt{2} & 0 & -\sqrt{2} \\ 0 & \sqrt{2} & 0 \end{pmatrix} ,$$

$$(\ell_3)_{1m',1m} = \begin{pmatrix} 1 & 0 & 0 \\ 0 & 0 & 0 \\ 0 & 0 & -1 \end{pmatrix} .$$

Striking features of this result are that ℓ_1 is represented by a *real* matrix, all its entries being *positive*, while the matrix ℓ_2 is *purely imaginary*. (ℓ_3 is chosen diagonal. As it is hermitian it is automatically real.) This need not be so in general. These specific properties are a consequence of a choice of phases to which we come back later (so-called Condon-Shortley phase convention).

Given this representation a few more exercises can be done. One confirms that the commutator of the matrices of ℓ_1 and ℓ_2 does indeed yield $\mathrm{i}\,\ell_3$. Calculation of the eigenvalues and eigenfunctions of ℓ_1 in the subspace requires the characteristic polynomial to vanish,

$$\det(\ell_1 - \mu \mathbb{1}) = \det \begin{pmatrix} -\mu & 1/\sqrt{2} & 0 \\ 1/\sqrt{2} & -\mu & 1/\sqrt{2} \\ 0 & 1/\sqrt{2} & -\mu \end{pmatrix} = -\mu(\mu^2 - 1) = 0 .$$

As expected the eigenvalues are $\mu = 1, 0, -1$. The corresponding eigenfunctions are obtained by solving the homogeneous system of linear equations

$$\frac{1}{\sqrt{2}} \begin{pmatrix} 0 & 1 & 0 \\ 1 & 0 & 1 \\ 0 & 1 & 0 \end{pmatrix} \begin{pmatrix} c_1^{(\mu)} \\ c_0^{(\mu)} \\ c_{-1}^{(\mu)} \end{pmatrix} = \mu \begin{pmatrix} c_1^{(\mu)} \\ c_0^{(\mu)} \\ c_{-1}^{(\mu)} \end{pmatrix} , \qquad \mu = 1, 0 \text{ or } -1 .$$

Except for possible phase factors the eigenvectors are

$$\boldsymbol{c}^{(\pm 1)} = \frac{1}{2}(1, \pm\sqrt{2}, 1)^T , \qquad \boldsymbol{c}^{(0)} = \frac{1}{\sqrt{2}}(1, 0, -1)^T .$$

This means that the eigenfunctions of ℓ_1 which pertain to the eigenvalues $\mu = +1$ and $\mu = -1$, respectively, are

$$\psi'_{\ell=1,\mu=\pm 1} = \frac{1}{2}(Y_{11} \pm \sqrt{2}Y_{10} + Y_{1,-1}) ,$$

while the eigenfunction pertaining to $\mu = 0$ is

$$\psi'_{\ell=1,\mu=0} = \frac{1}{\sqrt{2}}(Y_{11} - Y_{1,-1}) .$$

All three of them are normalized to 1, any two of them are orthogonal.

1.9.2 Radial Momentum and Kinetic Energy

The result (1.111) represents the operator $\boldsymbol{\ell}^2$ as a differential operator on the surface of S^2. The expression in curly brackets of (1.111) also appears as part of the Laplace operator if the latter is given in spherical polar coordinates, viz.

$$\Delta = \frac{1}{r^2}\frac{\partial}{\partial r}\left(r^2\frac{\partial}{\partial r}\right) + \frac{1}{r^2}\left[\frac{1}{\sin^2\theta}\frac{\partial^2}{\partial\phi^2} + \frac{1}{\sin\theta}\frac{\partial}{\partial\theta}\left(\sin\theta\frac{\partial}{\partial\theta}\right)\right] .$$

In turn, the Laplace operator is contained in the operator describing the kinetic energy. Therefore, the kinetic energy can be written in the form

$$T_{\text{kin}} = -\frac{\hbar^2}{2m}\left[\frac{1}{r^2}\frac{\partial}{\partial r}\left(r^2\frac{\partial}{\partial r}\right) - \frac{1}{r^2}\ell^2\right] .$$

The similarity to the decomposition of the *classical* kinetic energy into a radial and an angular part

$$(T_{\text{kin}})_{\text{cl}} = \frac{(p_r^2)_{\text{cl}}}{2m} + \frac{(\boldsymbol{\ell}^2)_{\text{cl}}}{2mr^2} \quad \text{(classical)}$$

is remarkable and raises the question whether there exists an operator associated to the radial variable p_r which would yield the first, r-dependent term of T_{kin}.

For the classical radial momentum we could write

$$(p_r)_{\text{cl}} = \frac{\boldsymbol{x} \cdot \boldsymbol{p}}{r} .$$

Replacing naïvely $\boldsymbol{p}$ by $\hbar\nabla/i$ produces an operator which is not self-adjoint. A better try is to start from the classically equivalent, symmetrized expression

$$\frac{1}{2}\left(\frac{\boldsymbol{x}}{r}\cdot\boldsymbol{p} + \boldsymbol{p}\cdot\frac{\boldsymbol{x}}{r}\right)$$

which, upon quantization, becomes

$$\frac{\hbar}{2\mathrm{i}}\left(\frac{\boldsymbol{x}}{r}\cdot\nabla + \nabla\cdot\frac{\boldsymbol{x}}{r}\right) = \frac{\hbar}{2\mathrm{i}}\left[2\frac{\boldsymbol{x}}{r}\cdot\nabla + \left(\nabla\cdot\frac{\boldsymbol{x}}{r}\right)\right] .$$

The two terms are evaluated as follows:

$$\frac{1}{r}\boldsymbol{x}\cdot\nabla=\sum_i \frac{x^i}{r}\frac{\partial r}{\partial x^i}\frac{\partial}{\partial r}=\sum_i \frac{(x^i)^2}{r^2}\frac{\partial}{\partial r}=\frac{\partial}{\partial r}\,,$$

$$\left(\nabla\cdot\frac{\boldsymbol{x}}{r}\right)=\frac{2}{r}\,.$$

Putting these formulae together the new operator is

$$p_r=\frac{\hbar}{\mathrm{i}}\left(\frac{\partial}{\partial r}+\frac{1}{r}\right)=\frac{\hbar}{\mathrm{i}}\frac{1}{r}\frac{\partial}{\partial r}r\,. \tag{1.123}$$

This operator acts on the integrable functions over the interval $r\in[0,\infty)$. Unfortunately it is not self-adjoint either, but it is *symmetric*. The meaning of this term is the following: The original operator is defined on the positive real half-axis $\mathbb{R}_+\backslash\{0\}$. Its adjoint, however, is defined on the whole real axis. The domain of definition of the adjoint differs from the one of the original operator, so that here $\mathcal{D}\subset\mathcal{D}^\dagger$. In Definition 3.6 of selfadjointness (Chap. 3) the requirement will be that the domains of definition must be the same, $\mathcal{D}=\mathcal{D}^\dagger$. This is not the case here. [A more detailed discussion may found, e.g., in [Galindo and Pascual (1990)] vol I, Sect. 6.2.
The commutator of p_r with r is

$$[p_r,r]=\frac{\hbar}{\mathrm{i}}\,.$$

Very much like in classical mechanics it stands for the momentum canonically conjugate to r. Finally, one calculates its square and finds, indeed,

$$p_r^2=-\hbar^2\left(\frac{\partial}{\partial r}+\frac{1}{r}\right)^2=-\hbar^2\left(\frac{\partial^2}{\partial r^2}+\frac{2}{r}\frac{\partial}{\partial r}\right)=-\hbar^2\frac{1}{r^2}\frac{\partial}{\partial r}\left(r^2\frac{\partial}{\partial r}\right)\,.$$

Thus, the decomposition of the kinetic energy into kinetic energy of radial motion and of angular motion that is well known from classical physics, also applies to the corresponding operators of quantum mechanics,

$$T_{\text{kin}}=\frac{p_r^2}{2m}+\frac{\hbar^2\boldsymbol{\ell}^2}{2mr^2}\quad\text{(quantum operators)}\,. \tag{1.124}$$

(Note that the factor $\hbar^2$ shows up explicitly only because the operator of angular momentum (1.106) was defined by extracting $\hbar$.)

Perhaps the most important consequence of this decomposition, from a physical point of view, is the observation that the second term in (1.124) can be interpreted as a potential of the centrifugal force. In a problem with central field, described by the Hamiltonian

$$H=\frac{p_r^2}{2m}+\frac{\hbar^2\boldsymbol{\ell}^2}{2mr^2}+U(r) \tag{1.125}$$

the centrifugal potential will compete with the true, attractive or repulsive, potential $U(r)$ – in close analogy to the classical situation. This is seen very clearly if stationary eigenfunctions of H separate in radial and angular variables,

$$\psi_{\alpha\ell m}(\boldsymbol{x})=R_\alpha(r)Y_{\ell m}\quad\text{or}\quad\psi_{\ell m}(\alpha,\boldsymbol{x})=R(\alpha,r)Y_{\ell m} \tag{1.126}$$

The quantum numbers ℓ and m play the same role as before while α characterizes the radial motion and depends on the nature of the potential $U(r)$. In the first of (1.126) α is a denumerable, discrete quantum number (examples are provided by the spherical oscillator and by the bound part of the hydrogen spectrum). In the second of (1.126) α is a continuous variable (examples are the force-free motion, and the unbound states in the hydrogen atom). The operator $\boldsymbol{\ell}^2$, acting on $Y_{\ell m}$, answers by the eigenvalue $\ell(\ell+1)$, the operator p_r^2 only acts on the radial function $R_\alpha(r)$, while the action of all other terms is just multiplication by real numbers. Dividing the whole equation by $Y_{\ell m}$ one obtains the differential equation

$$-\frac{\hbar^2}{2m}\frac{1}{r^2}\frac{\mathrm{d}}{\mathrm{d}r}\left(r^2\frac{\mathrm{d}R(r)}{\mathrm{d}r}\right)+\left(\frac{\hbar^2\ell(\ell+1)}{2mr^2}+U(r)\right)R(r)=ER(r)\,. \qquad (1.127)$$

This equation describes the radial dynamics which is governed by the *effective* potential

$$U_{\mathrm{eff}}(r)=\frac{\hbar^2\ell(\ell+1)}{2mr^2}+U(r)\,,$$

in close analogy to the corresponding classical situation. For instance, an attractive potential $U(r)$ acts in competition with the repulsive centrifugal potential so that the radial wave functions, for increasing ℓ, are pushed away from small values of r, and, thus, are screened more and more from the influence of the true potential. The problems that will be dealt with in the following sections provide good illustrations of this interpretation.

1.9.3 Force Free Motion with Sharp Angular Momentum

In a situation where there is no (true) potential, $U(r)\equiv 0$, three commuting observables may be chosen. For example, one chooses

1. *either* the set

$$\{p_1, p_2, p_3\}$$

2. *or* the set

$$\{H, \boldsymbol{\ell}^2, \ell_3\}\,.$$

In the first case H is not listed explicitly because its eigenvalues $E=\boldsymbol{p}^2/(2m)$ are fixed as soon as those of all three operators p_i are given.

The two alternatives exclude each other because, even though $\boldsymbol{\ell}^2$ and ℓ_3 commute with $\boldsymbol{p}^2$, they do not commute with the three components p_i. With the choice 1. the plane waves of Sect. 1.8.4 are seen to be simultaneous eigenfunctions of the three observables. A particle of mass m moves with given momentum $\boldsymbol{p}$ along the straight line defined by the direction $\hat{\boldsymbol{p}}$. With the second alternative, choice 2., the particle is in a state with fixed energy, i.e. with fixed modulus $p:=|\boldsymbol{p}|$ of the momentum, and with sharp values $\ell(\ell+1)$ and m for the square of the orbital angular momentum and its component ℓ_3 along an arbitrarily chosen 3-axis, respectively. At first sight this

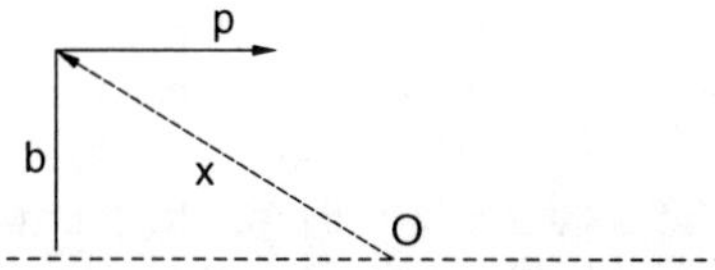

Fig. 1.11 A particle of classical mechanics moving with momentum $\boldsymbol{p}$ at the distance b from the parallel through the origin $\mathcal{O}$, possesses a well-defined orbital angular momentum. This angular momentum lies perpendicular to the plane of the drawing (pointing away from the observer) and has the modulus $|\ell_{\text{cl}}| = b|\boldsymbol{p}|$

seems to be very different from classical kinematics: There, a particle which comes in with momentum $\boldsymbol{p}$ and impact parameter b, has the orbital angular momentum

$$\boldsymbol{\ell}_{\text{cl}} = \boldsymbol{x} \times \boldsymbol{p} \quad \text{with} \quad |\boldsymbol{\ell}_{\text{cl}}| = bp$$

relative to the origin $\mathcal{O}$. This is sketched in Fig. 1.11. One might argue that the origin could be chosen differently and, as a consequence, that the orbital angular momentum is not well-defined. This is true. However, if the plane waves are taken to be the (asymptotic) incoming states in describing scattering on a central potential $U(r)$ then the origin $\mathcal{O}$ is the center of the force so that $\boldsymbol{\ell}$ is a physically well defined observable. Yet, the relationship between impact parameter and orbital angular momentum is not completely lost in quantum mechanics. We will show this by means of the stationary solutions of the radial differential equation (1.127) to whose construction we turn next.

Defining

$$k^2 := \frac{2mE}{\hbar^2}, \qquad \varrho := kr\,, \tag{1.128}$$

the radial equation (1.127) with $U(r) \equiv 0$ turns into the following differential equation in the dimensionless variable ϱ:

$$\frac{1}{\varrho^2}\frac{\mathrm{d}}{\mathrm{d}\varrho}\left(\varrho^2\frac{\mathrm{d}R(\varrho)}{\mathrm{d}\varrho}\right) - \frac{\ell(\ell+1)}{\varrho^2}R(\varrho) + R(\varrho) = 0\,. \tag{1.129}$$

Working out the first term one sees immediately that this equation is of Fuchsian type (1.113) with pole position $z_0 = 0$. Solutions of this differential equation which are to be interpreted as probability amplitudes, must not be singular at $\varrho = 0$ where the coefficient functions have poles of first and second order, respectively. This condition can be tested by trying the ansatz

$$R(\varrho) = \varrho^\alpha f(\varrho) \quad \text{with} \quad f(0) \neq 0 \quad \text{finite}\,.$$

Substitution yields a differential equation for $f(\varrho)$, viz.

$$\begin{aligned}&\varrho^\alpha f'' + 2(\alpha+1)\varrho^{\alpha-1} f' \\ &\quad + \left[\alpha(\alpha-1)\varrho^{\alpha-2} + 2\alpha\varrho^{\alpha-2} - \ell(\ell+1)\varrho^{\alpha-2} + \varrho^\alpha\right] f = 0\,.\end{aligned}$$

Comparing the terms of this equation as $\varrho \to 0$, yields the algebraic condition

$$\alpha(\alpha+1) = \ell(\ell+1)\,,$$

whose solutions are $\alpha = \ell$ and $\alpha = -\ell - 1$.[20] Obviously, in describing scattering states with sharp angular momentum we must choose the first solution which is regular at $\varrho = 0$.

Note that the differential equation (1.129) for $R(\varrho)$, or (1.130) for $Z(\varrho)$ (see just below), are well known from the theory of Bessel functions.

In the mathematical literature on Special Functions [Abramowitz and Stegun (1965)] one finds either the differential equation (1.129) of *spherical Bessel functions*, or a somewhat different form which is obtained from it by the substitution

$$Z(\varrho) = \sqrt{\varrho}R(\varrho) .$$

It reads

$$Z''(\varrho) + \frac{1}{\varrho}Z'(\varrho) + \left[1 - \frac{(\ell + 1/2)^2}{\varrho^2}\right]Z(\varrho) = 0 \tag{1.130}$$

and is called *Bessel's differential equation*.

For lack of space we do not dwell upon the theory of Bessel functions. Rather, we merely quote solutions of (1.129) and describe their relevant properties.

The solutions which are regular at the origin $\varrho = 0$ are called *spherical Bessel functions*, they are given by

$$j_\ell(\varrho) = (-\varrho)^\ell \left(\frac{1}{\varrho}\frac{\mathrm{d}}{\mathrm{d}\varrho}\right)^\ell \frac{\sin\varrho}{\varrho} \tag{1.131}$$

The first three functions read explicitly

$$j_0(\varrho) = \frac{\sin\varrho}{\varrho}, \qquad j_1(\varrho) = \frac{\sin\varrho}{\varrho^2} - \frac{\cos\varrho}{\varrho},$$
$$j_2(\varrho) = \frac{3\sin\varrho}{\varrho^3} - \frac{3\cos\varrho}{\varrho^2} - \frac{\sin\varrho}{\varrho}.$$

Their behaviour in the limit $\varrho \to 0$ is the one expected on general grounds, viz.

$$\varrho \to 0: \quad j_\ell(\varrho) \sim \frac{\varrho^\ell}{(2\ell + 1)!!}, \tag{1.132}$$

the double factorial in the denominator being defined by

$$(2\ell + 1)!! := (2\ell + 1)\cdot(2\ell - 1)\cdots 5\cdot 3\cdot 1 .$$

The asymptotic behaviour for $\varrho \to \infty$ is

$$\varrho \to \infty: \quad j_\ell(\varrho) \sim \frac{1}{\varrho}\sin\left(\varrho - \ell\frac{\pi}{2}\right). \tag{1.133}$$

The examples $\ell = 0$, 1 and 2 given above may be useful in testing the two limits for small and large values of ϱ, respectively.

---

[20] In the theory of differential equations of Fuchsian type the coefficient α is called *characteristic exponent*.

Summarizing our formulae, the eigenfunctions which are common to the operators $\{H, \boldsymbol{\ell}^2, \ell_3\}$, and which are regular at $r = 0$ are

$$\psi_{\ell m}(k, \boldsymbol{x}) = j_\ell(kr) Y_{\ell m}(\theta, \phi) . \tag{1.134}$$

A complete stationary solution of the Schrödinger equation reads

$$\Psi_{\ell m}(k, t, \boldsymbol{x}) = \mathrm{e}^{-(\mathrm{i}/\hbar)Et} j_\ell(kr) Y_{\ell m}(\theta, \phi) ,$$

where $E = \hbar^2 k^2/(2m)$. Its asymptotic form follows from (1.133),

$r \to \infty$:

$$\Psi_{\ell m}(t, k, \boldsymbol{x}) \sim \frac{1}{2\mathrm{i}kr} \left[\mathrm{e}^{\mathrm{i}(kr-(\ell\pi/2)-(Et/\hbar))} - \mathrm{e}^{-\mathrm{i}(kr-(\ell\pi/2)+(Et/\hbar))} \right] Y_{\ell m} . \tag{1.135}$$

As we will see in the analysis of scattering states in Chap. 2, the first term describes outgoing spherical waves while the second describes incoming spherical waves.

The solutions (1.134) are said to be *partial waves* of fixed angular momentum ℓ. They are not eigenfunctions of the momentum operator $\boldsymbol{p}$. To the contrary, as we show below, eigenfunctions of momentum contain *all* values of ℓ. Nevertheless, the relation of the angular momentum to the impact parameter is not lost completely. Indeed, in studying the graph of the spherical Bessel function $j_\ell(kr)$ one discovers that this function, for $\ell \gg 1$, has a pronounced maximum at

$$\varrho = kr \simeq \left(\ell + \frac{1}{2} \right) ,$$

i.e. practically at the point where the relation between ℓ and the impact parameter holds true (cf. [Abramowitz and Stegun (1965)], Sect. 10.1.59). Figure 1.12 shows the function $j^2_{\ell=10}(\varrho)$, while Fig. 1.13 shows its square multiplied by ϱ^2. In this example j_{10} is obtained either from (1.131) or through repeated application of the formula

$$j_\ell(\varrho) = \left(-\frac{\mathrm{d}}{\mathrm{d}\varrho} + \frac{\ell - 1}{\varrho} \right) j_{\ell-1}(\varrho), \qquad \ell \geq 1 .$$

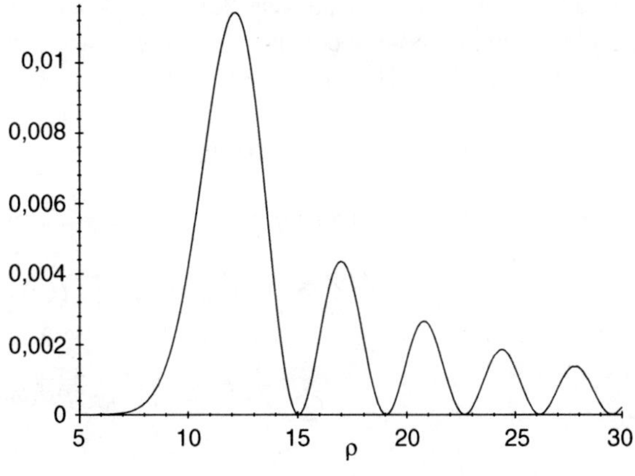

Fig. 1.12 Square of the spherical Bessel function with $\ell = 10$ as a function of $\varrho = kr$

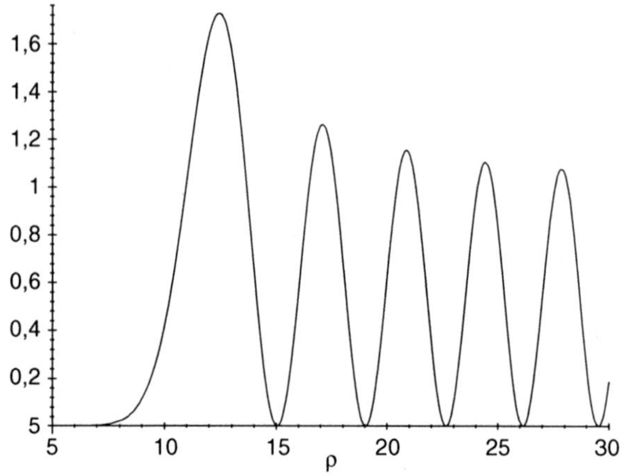

Fig. 1.13 Square of $j_{10}(\varrho)$ multiplied with ϱ^2, as a function of $\varrho = kr$

Thus, it is correct to state that the centrifugal potential displaces the ℓth partial wave from the origin, the more the higher the value of ℓ. In describing scattering from a (true) potential $U(r)$ high partial waves feel its influence less than low partial waves, even if the potential is attractive.

At this point, the reader might ask about the relationship between the simultaneous eigenfunctions of the first set of operators, $\{p_1, p_2, p_3\}$, and those of the second set, $\{H, \boldsymbol{\ell}^2, \ell_3\}$. The answer to this question is contained in an important formula which gives the expansion of the plane wave in terms of partial waves. With $\boldsymbol{p} = \hbar\boldsymbol{k}$, and $k = |\boldsymbol{k}|$, as before, it reads

$$\boxed{e^{i\boldsymbol{k}\cdot\boldsymbol{x}} = 4\pi \sum_{\ell=0}^{\infty} i^{\ell} j_{\ell}(kr) \sum_{m=-\ell}^{+\ell} Y^*_{\ell m}(\theta_k, \phi_k) Y_{\ell m}(\theta_x, \phi_x)} \,. \tag{1.136}$$

The arguments of the first spherical harmonic are the angular coordinates of the vector $\boldsymbol{k}$, the arguments of the second spherical harmonic are those of the vector $\boldsymbol{x}$. The physical interpretation of this formula is that a plane wave contains all partial waves $\ell = 0, 1, 2, \ldots$. Likewise, for every value of ℓ, it contains all values of m unless the momentum vector points in the 3-direction. In this is the case, we have $\theta_k = 0$ and $\phi_k = 0$. The formulae (1.115) and (1.116) show that

$$Y_{\ell m}(\theta_k = 0, \phi_k = 0) = \frac{\sqrt{2\ell+1}}{\sqrt{4\pi}} \delta_{m0} \,.$$

In this case the plane wave and its expansion reduce to

$$\begin{aligned} e^{ikx^3} &= \sum_{\ell=0}^{\infty} i^{\ell} \sqrt{4\pi(2\ell+1)}\, j_{\ell}(kr) Y_{\ell 0}(\theta_x, \phi_x) \\ &= \sum_{\ell=0}^{\infty} i^{\ell} (2\ell+1) j_{\ell}(kr) P_{\ell}(\cos\theta_x) \,. \end{aligned}$$

This is an important result:

Although a plane wave contains all partial waves ℓ, the projection of the orbital angular momentum onto the direction of the momentum is equal to zero for all partial waves, $m_\ell = 0$.

Proof of the formula (1.136)

In a first step take $\boldsymbol{k}$ to point along the 3-direction. In this case axial symmetry implies that the expansion of a plane wave in terms of spherical harmonics only contains contributions with $m = 0$,

$$e^{ikx^3} = \sum_{\ell=0}^{\infty} G_\ell(r) Y_{\ell m=0}(\theta, \phi)\,,$$

whereby the functions $G_\ell(r)$ are to be calculated from

$$G_\ell(r) = \int d\Omega\, Y^*_{\ell 0} e^{ikx^3} = \int_0^{2\pi} d\phi \int_0^{\pi} \sin\theta d\theta\, Y^*_{\ell 0}(\theta) e^{ikr\cos\theta}\,.$$

Rather than working out this integral in all details one resorts to a trick: One calculates the leading term for asymptotically large values of r and compares to the asymptotics (1.133) of the spherical Bessel functions. By partial integration in the variable $z := \cos\theta$ one obtains

$$\begin{aligned} G_\ell(r) &= 2\pi \left[\frac{1}{ikr} \int_{-1}^{+1} dz\, Y_{\ell 0}(z)(ikr e^{ikrz}) \right] \\ &= \frac{2\pi}{ikr} \left[\left[Y_{\ell 0}(z) e^{ikrz} \Big|_{-1}^{+1} - \int_{-1}^{+1} dz\, \frac{dY_{\ell 0}}{dz} e^{ikrz} \right] \right]. \end{aligned}$$

Partial integration of the second term in square brackets generates further inverse powers of r so that, to leading order, only the first term contributes. This term becomes

$r \to \infty$:

$$\begin{aligned} G_\ell(r) &\sim \frac{\sqrt{4\pi(2\ell+1)}}{2ikr}[e^{ikr} - (-1)^\ell e^{-ikr}] P_\ell(z=1) + \mathcal{O}[(kr)^{-2}] \\ &= \frac{\sqrt{4\pi(2\ell+1)}}{2ikr} i^\ell \left(e^{i(kr-\ell(\pi/2))} - e^{-i(kr-\ell(\pi/2))} \right) P_\ell(z=1) \\ &\quad + \mathcal{O}[(kr)^{-2}]\,. \end{aligned}$$

Compare now to the asymptotics of (1.133). The expansion in terms of spherical harmonics being unique, the formula for $\exp(ikx^3)$ given above holds true. One then substitutes

$$Y_{\ell 0}(\theta) = \sqrt{\frac{2\ell+1}{4\pi}} P_\ell(\cos\theta)$$

in this formula. Finally, if $\boldsymbol{k}$ does not point along the 3-axis the formula holds with the replacement $\cos\theta \longmapsto \cos\alpha$, where α denotes the angle between $\boldsymbol{k}$ and $\boldsymbol{x}$. At this point one uses the addition theorem (1.121) thus obtaining the result (1.136). This concludes the proof.

The expansion (1.136) of the plane wave in terms of solutions with sharp ℓ and which are regular at the origin is useful in determining the normalization of the functions (1.134). One has, successively,

$$\begin{aligned}
&\int \mathrm{d}^3x\, \mathrm{e}^{-\mathrm{i}(\boldsymbol{k}'-\boldsymbol{k})\cdot\boldsymbol{x}} = (2\pi)^3\delta^{(3)}(\boldsymbol{k}-\boldsymbol{k}') \\
&= \frac{(2\pi)^3}{kk'}\delta(k-k')\delta(\cos\theta-\cos\theta')\delta(\phi-\phi') \\
&= (4\pi)^2\sum_{\ell\ell'} \mathrm{i}^\ell(-\mathrm{i})^{\ell'} \sum_{mm'} Y^*_{\ell m}(\hat{k}) Y_{\ell' m'}(\hat{k}')\delta_{\ell\ell'}\delta_{mm'} \int_0^\infty r^2\mathrm{d}r\, j_\ell(kr) j_{\ell'}(k'r) \\
&= (4\pi)^2\delta(\cos\theta-\cos\theta')\delta(\phi-\phi') \int_0^\infty r^2\mathrm{d}r\, j_\ell(kr) j_\ell(k'r)\,.
\end{aligned}$$

(The short-hand notation $\hat{k}, \hat{k}'$ stands for the angular coordinates of $\boldsymbol{k}$ and of $\boldsymbol{k}'$, respectively.) In the last two steps the sum over ℓ' and m' was carried out, and the completeness relation (1.119) of spherical harmonics was used. By comparison of coefficients one obtains the important formula

$$\boxed{\int_0^\infty r^2\mathrm{d}r\, j_\ell(kr) j_\ell(k'r) = \frac{\pi}{2kk'}\delta(k-k')}\,. \tag{1.137}$$

Spherical Bessel functions are orthogonal but they are not normalizable in the usual sense. Like plane waves they are normalized, in a more general manner, to δ-distributions in the modulus of the momentum or, equivalently, in the energy scale (cf. Sect. 1.8.4).

This section concludes with a few statements on further solutions of the differential equation (1.129) that are relevant for the theory of scattering.

Solutions of linear, homogeneous, ordinary differential equations of second order such as (1.129) can be expressed as superpositions of two, linearly independent, fundamental solutions. The spherical Bessel function (1.131) being one choice, the function

$$n_\ell(\varrho) = -(-\varrho)^\ell \left(\frac{1}{\varrho}\frac{\mathrm{d}}{\mathrm{d}\varrho}\right)^\ell \frac{\cos\varrho}{\varrho}\,, \tag{1.138}$$

being linearly independent of $j_\ell(\varrho)$, is another one. The set of these functions with $\ell = 0, 1, \ldots$ are called *spherical Neumann functions*. They have the expected behaviour in the limit $\varrho \to 0$,

$$\varrho \to 0: \quad n_\ell(\varrho) \sim -\frac{(2\ell-1)!!}{\varrho^{\ell+1}} . \tag{1.139}$$

At infinity their behaviour is similar to the one of spherical Bessel functions but for a shift by $\pi/2$,

$$\varrho \to \infty: \quad n_\ell(\varrho) \sim -\frac{1}{\varrho}\cos\left(\varrho - \ell\frac{\pi}{2}\right) . \tag{1.140}$$

Another choice of a fundamental system, instead of $\{j_\ell(\varrho), n_\ell(\varrho)\}$, is provided by the *spherical Hankel functions*. They are defined as follows

$$h_\ell^{(\pm)}(\varrho) := (-\varrho)^\ell \left(\frac{1}{\varrho}\frac{\mathrm{d}}{\mathrm{d}\varrho}\right)^\ell \frac{\mathrm{e}^{\pm\mathrm{i}\varrho}}{\varrho} , \tag{1.141}$$

their relation to the former set being

$$j_\ell(\varrho) = \frac{1}{2\mathrm{i}}[h_\ell^{(+)}(\varrho) - h_\ell^{(-)}(\varrho)], \qquad n_\ell(\varrho) = -\frac{1}{2}[h_\ell^{(+)}(\varrho) + h_\ell^{(-)}(\varrho)] .$$

As both Hankel functions contain the spherical Neumann functions their behaviour at the origin is singular. In turn, their asymptotic properties are simple:

$$\varrho \to \infty: \quad h_\ell^{(\pm)}(\varrho) \sim \frac{1}{\varrho}\mathrm{e}^{\pm\mathrm{i}[\varrho - \ell(\pi/2)]} .$$

As this is the behaviour of outgoing and incoming spherical waves, respectively, it is plausible that this basis will play a special role in scattering problems.

1.9.4 The Spherical Oscillator

The spherical oscillator is an example for a problem with central field with a purely discrete spectrum. The spherically symmetric potential reads

$$U(r) = \frac{1}{2}m\omega^2 r^2 ,$$

the differential equation (1.127) for the radial part of the motion is

$$-\frac{\hbar^2}{2m}\frac{1}{r^2}\frac{\mathrm{d}}{\mathrm{d}r}\left(r^2\frac{\mathrm{d}R_\alpha(r)}{dr}\right) + \left[\frac{\hbar^2\ell(\ell+1)}{2mr^2} + \frac{1}{2}m\omega^2 r^2\right] R_\alpha(r) = E R_\alpha(r) . \tag{1.142}$$

Like in Sect. 1.6 it is useful to introduce the reference length b and the dimensionless energy variable ε which were defined by

$$b = \sqrt{\frac{\hbar}{m\omega}} = \frac{\hbar c}{\sqrt{mc^2\hbar\omega}} , \qquad \varepsilon = \frac{E}{\hbar\omega} . \tag{1.143}$$

The variable r is replaced by the dimensionless variable

$$q := \frac{r}{b} .$$

The radial differential equation then goes over into

$$\frac{1}{q^2}\frac{\mathrm{d}}{\mathrm{d}q}\left(q^2\frac{\mathrm{d}R(q)}{\mathrm{d}q}\right)-\left(\frac{\ell(\ell+1)}{q^2}+q^2\right)R(q)=-2\varepsilon R(q)\,.$$

Instead of trying to solve this equation in its full generality it is helpful to first collect the conditions to be imposed on its solutions from a physical perspective. Like in the force-free case, Sect. 1.9.3, every physically interpretable solution must remain finite at $r \to 0$. Like in that example, the substitution

$$R(q)=q^{\alpha}f(q) \quad \text{with} \quad f(0)\neq 0\,,$$

yields the algebraic condition

$$\alpha(\alpha+1)=\ell(\ell+1), \quad \text{i.e. either} \quad \alpha=\ell \quad \text{or} \quad \alpha=-\ell-1\,.$$

For bound states only the first value $\alpha = \ell$ of the characteristic exponent is admissible. We note that this result holds for all other central fields for which $\lim_{r\to 0} r^2 U(r) = 0$. The physical reason for this is that the behaviour for $r \to 0$ is dominated by the centrifugal potential as long as the true potential $U(r)$ is less singular than that at the origin.

Taking out the "centrifugal factor" q^{ℓ}, the above differential equation is modified to

$$f''(q)+2\frac{\ell+1}{q}f'(q)+(2\varepsilon-q^2)f(q)=0\,.$$

(The calculation is the same as in Sect. 1.9.3.) As a striking property of this second form of the radial equation one notices that it remains unchanged by the replacement $q \to -q$. This means that the solutions depend on q^2, not on q. Of course, this property is a consequence of the potential being quadratic in r. This observation suggests to substitute the variable once more by taking

$$z := q^2=\left(\frac{r}{b}\right)^2, \qquad f(q)\equiv v(z)\,. \tag{1.144}$$

Making use of the formulae

$$q=\sqrt{z}, \qquad \frac{\mathrm{d}}{\mathrm{d}q}=2\sqrt{z}\frac{\mathrm{d}}{\mathrm{d}z}, \qquad \frac{\mathrm{d}^2}{\mathrm{d}q^2}=2\frac{\mathrm{d}}{\mathrm{d}z}+4z\frac{\mathrm{d}^2}{\mathrm{d}z^2}$$

the differential equation for $v(z)$ is seen to be

$$v''(z)+\frac{\ell+3/2}{z}v'(z)+\left(\frac{\varepsilon}{2z}-\frac{1}{4}\right)v(z)=0\,. \tag{$*$}$$

At this point one may wish to pause and to ponder a further requirement imposed by physics: the wave function of bound states should be square integrable. This is a strong condition on their asymptotic behaviour at $r \to \infty$, that will be of key importance in the analysis of bound states in the hydrogen atom. For large values of z, the differential equation reduces to the approximate form

$$v''(z)-\frac{1}{4}v(z)\simeq 0\,,$$

which is independent of ℓ and of ε. This equation would be easy to solve if it described the whole problem. Indeed, solutions would be

$$v(z) \simeq \mathrm{e}^{\pm z/2} = \mathrm{e}^{\pm r^2/(2b^2)} .$$

The solution which grows exponentially is not compatible with (∗) because the term in the first derivative of $v(z)$ would be positive, requiring the parameter ε to be negative. As the potential energy is everywhere positive, the total energy, being the sum of the expectation values of kinetic and potential energies, must be positive. This implies that all solutions decrease like $\exp[-r^2/(2b^2)]$, independently of their angular momentum ℓ and of their energy. This result should not be surprising for two reasons: the potential increases quadratically for $r \to \infty$ so that the wave function must decrease in this limit. On the other hand, we know that the spherical oscillator can be decomposed into three linear oscillators with equal frequencies, cf. Sect. 1.8.3, whose wave functions have precisely this property.

If the exponential behaviour at infinity is extracted as well, by substituting

$$v(z) = \mathrm{e}^{-z/2} w(z) ,$$

the newly defined function $w(z)$ should turn out to be something very simple such as, presumably, polynomials in z. Although this is somewhat tedious and, perhaps, tiring for the reader to follow, it is worthwhile to make a last substitution aiming at converting the differential equation for $v(z)$ to one for $w(z)$. The method being rather general, this step of the calculations is useful also for other problems. We note

$$v'(z) = \left[-\frac{1}{2} w(z) + w'(z) \right] \mathrm{e}^{-z/2} ,$$

$$v''(z) = \left[\frac{1}{4} w(z) - w'(z) + w''(z) \right] \mathrm{e}^{-z/2}$$

and insert these formulae, thereby obtaining a differential equation for $w(z)$:

$$z w''(z) + \left(\ell + \frac{3}{2} - z \right) w'(z) + \frac{1}{2} \left(\varepsilon - \ell - \frac{3}{2} \right) w(z) = 0 .$$

This equation is well-known from the theory of Special Functions. Its general form is

$$\boxed{z w''(z) + (c - z)\, w'(z) - a\, w(z) = 0} , \tag{1.145}$$

where c and a are real or complex constants. It is called *Kummer's equation*. As it is of central importance for quantum mechanics we devote a whole section of Appendix A.2 to a summary of its most important properties. There one learns that the solution which is regular at $z = 0$ can be written as an infinite series which reads

$$ {}_1F_1(a; c; z) = 1 + \frac{a}{c} z + \frac{a(a+1)}{2!\, c(c+1)} z^2 + \ldots + \frac{(a)_k}{k!\, (c)_k} z^k + \ldots , \tag{1.146}$$

and where the following abbreviation is used

$$(\lambda)_0 = 1 , \qquad (\lambda)_k = \lambda(\lambda+1)(\lambda+2)\ldots(\lambda+k-1) , \qquad \lambda = a, c .$$

The function defined by the series (1.146) is called the *confluent hypergeometric function*.[21] Among its most remarkable properties we note the following:

1. In the sense of function theory the series given above defines an *entire* function which is to say that it converges for all finite values in the complex plane of the variable z. At infinity, in general, it has an essential singularity. This property is illustrated by the example $a = c$ for which
$$ {}_1F_1(a;a;z) = \sum_{k=0}^{\infty} \frac{1}{k!} z^k = e^z . $$
2. If a equals a negative integer or zero,
$$ -a \in \mathbb{N}_0 , $$
the series terminates after a finite number of terms, and ${}_1F_1(a = -n; c; z)$ is a polynomial of degree n.
3. At infinity there exists an asymptotic expansion in $1/z$ for ${}_1F_1(a; c; z)$ which is important for many applications in quantum mechanics. It is derived in Appendix A.2 where we show that it is
$$ |z| \to \infty, \qquad a \text{ fixed}, \qquad c \text{ fixed} $$
$$ {}_1F_1(a;c;z) \sim \frac{\Gamma(c)}{\Gamma(c-a)} e^{\pm i\pi a} z^{-a} \left[1 + \mathcal{O}\left(\frac{1}{z}\right) \right] + e^z z^{a-c} \frac{\Gamma(c)}{\Gamma(a)} \left[1 + \mathcal{O}\left(\frac{1}{z}\right) \right] . \tag{1.147} $$
The upper sign in the first term applies for $-\pi/2 < \arg z < 3\pi/2$, the lower sign applies for $-3\pi/2 < \arg z < -\pi/2$. The symbol $\Gamma(x)$ denotes the Gamma function, i.e. the generalized factorial, whose salient properties are also collected in Appendix A.2.

Applying this information to the third form of the differential equation of the spherical oscillator, one obtains
$$ a = -\frac{1}{2}\left(\varepsilon - \ell - \frac{3}{2}\right) , $$
$$ c = \ell + \frac{3}{2} \ : \ w(z) = {}_1F_1\left[-\frac{1}{2}\left(\varepsilon - \ell - \frac{3}{2}\right); \ell + \frac{3}{2}; z \right] . $$

Clearly, the second term of the asymptotic expansion (1.147) is potentially dangerous because it grows exponentially and, therefore, may destroy the good behaviour of the radial function noted above. One sees that this catastrophe can only be avoided if the second term is absent altogether, that is, if the factor multiplying it vanishes. The Gamma function has no zeroes for real argument. However, it has first-order poles at

---

[21] The notation "hypergeometric" is meant to remind that it is built following the model of the geometric series; it is called "confluent" because it is the result of the junction, or "confluence", of two first-order poles. This transition from Gauss' hypergeometric function to the confluent hypergeometric function is explained and is carried out in Appendix A.2.

zero and at all negative integers. Therefore, if $\Gamma(a)$ which appears in the denominator, has a pole position at a then the exponentially growing term is absent. The important conclusion is that the radial wave function is square integrable and, thus, amenable to a statistical interpretation, only if $a = -n$ with $n \in \mathbb{N}_0$. This implies that the eigenvalues ε are quantized and must obey the formula $\varepsilon = 2n + \ell + 3/2$.

The result of this analysis is the following. The eigenvalues of the Hamiltonian are

$$E_{n\ell} = \left(2n + \ell + \frac{3}{2}\right) \hbar\omega , \qquad n = 0, 1, 2, \dots . \tag{1.148}$$

The radial functions carry the quantum numbers $\alpha \equiv (n, \ell)$ and are given by

$$R_{n\ell}(r) = N_{n\ell} r^{\ell} \mathrm{e}^{-r^2/(2b^2)} {}_1F_1\left(-n; \ell + \frac{3}{2}; \frac{r^2}{b^2}\right) , \tag{1.149}$$

where $N_{n\ell}$ denotes the normalization. Without delving into its calculation[22] I merely quote the result:

$$N_{n\ell} = (-)^n \frac{1}{b^{\ell+3/2}} \frac{\sqrt{2\Gamma(n + \ell + 3/2)}}{\Gamma(\ell + 3/2)\sqrt{n!}} . \tag{1.150}$$

(The sign $(-)^n$ is physically irrelevant. I have chosen this sign for the purpose of rendering the coefficient of the highest power of r positive.)

Remarks

1. As expected the energy formula contains the term $3\hbar\omega/2$, i.e. one term $\hbar\omega/2$ for each of the three degrees of freedom. This is the zero point energy which is a direct consequence of the uncertainty relation. The oscillator can never have an energy lower than this value.
2. The allowed values of the energy are $E_{n\ell} = (\Lambda + 3/2)\hbar\omega$ with $\Lambda = 2n + \ell$ which shows that for $\Lambda \geqslant 1$ they are multiply degenerate. This degeneracy is due in part to the projection of the angular momentum because for fixed ℓ the states with $m = -\ell, m = -\ell + 1, \dots , m = +\ell$ all have the same energy. (The Hamiltonian does not depend on ℓ_3.) Part of the degeneracy must have a dynamical origin and must be a pecularity of the potential being proportional to r^2. For example, the state which has $\Lambda = 2$ has a sixfold degeneracy because this value is obtained from either $(n = 0, \ell = 2)$ or from $(n = 1, \ell = 0)$. Counting the m-degeneracy there are $5 + 1 = 6$ states which have the same energy.
3. The derivative term in the radial equation (1.142) is written in a manifestly self-adjoint form. Indeed, if $R_{n\ell}(r)$ and $R_{n'\ell'}(r)$ are two different radial functions, one has (noting that these functions are real!)

[22] Integrals with confluent hypergeometric functions, powers and exponentials, are known. They are found in good tables of integrals in the context of Laguerre polynomials.

$$\int_0^\infty r^2 \mathrm{d}r\, R_{n'\ell'}(r) \frac{1}{r^2} \frac{\mathrm{d}}{\mathrm{d}r} \left(r^2 \frac{\mathrm{d}R_{n\ell}(r)}{\mathrm{d}r} \right)$$

$$- \int_0^\infty r^2 \mathrm{d}r\, R_{n\ell}(r) \frac{1}{r^2} \frac{\mathrm{d}}{\mathrm{d}r} \left(r^2 \frac{\mathrm{d}R_{n'\ell'}(r)}{\mathrm{d}r} \right) = 0 .$$

Now, write the radial equation (1.142) once of $R_{n\ell}$, and once more for $R_{n'\ell}$, with possibly different values of n but with the same value of ℓ. Then multiply the first by $R_{n'\ell}$ from the left, the second by $R_{n\ell}$ from the left, integrate over the entire interval $[0, \infty)$, $\int_0^\infty r^2 \mathrm{d}r \ldots$, and subtract the results. What remains in the difference is the term

$$(E_{n'\ell} - E_{n\ell}) \int_0^\infty r^2 \mathrm{d}r\, R_{n'\ell}(r) R_{n\ell}(r) = 0 .$$

If $n' \neq n$, then $(E_{n'\ell} - E_{n\ell}) \neq 0$, and the integral must vanish. This means that the radial functions with the *same* value of ℓ are orthogonal. For *different* values $\ell \neq \ell'$ there remains the term

$$\int_0^\infty r^2 \mathrm{d}r\, R_{n'\ell'}(r) \frac{\hbar^2}{2mr^2} [\ell(\ell+1) - \ell'(\ell'+1)] R_{n\ell}(r) ,$$

the radial functions are no longer orthogonal. In this case the orthogonality is taken care of by other factors of the whole wave function

$$\psi_{n\ell m}(\boldsymbol{x}) = N_{n\ell} R_{n\ell}(r) Y_{\ell m}(\theta, \phi) ,$$

so that one always has

$$\int_0^\infty r^2 \mathrm{d}r \int \mathrm{d}\Omega\, \psi^*_{n'\ell' m'}(\boldsymbol{x}) \psi_{n\ell m}(\boldsymbol{x}) = \delta_{nn'} \delta_{\ell\ell'} \delta_{mm'} .$$

4. In atomic and nuclear spectroscopy states with $\ell = 0$ are called s-states, states which have $\ell = 1$ are called p-states, while states which have $\ell = 2$ are called d-states. Originally, these labels served the purpose of characterizing atomic spectral lines, with "s" standing for "sharp", "p" for "principal", "d" for "diffuse". From $\ell = 3$ on up atomic states are labelled following the alphabet, that is to say, f-states have $\ell = 3$, g-states have $\ell = 4$, h-states have $\ell = 5$, etc. Using this notation the first four radial functions are, from (1.149) and (1.150),

$$E = \frac{3}{2}\hbar\omega : \quad R_{0s}(r) = \frac{2}{\sqrt{\pi^{1/2} b^3}} \mathrm{e}^{-r^2/(2b^2)} ,$$

$$E = \frac{5}{2}\hbar\omega : \quad R_{0p}(r) = \frac{\sqrt{8}}{\sqrt{3\pi^{1/2} b^3}} \left(\frac{r}{b}\right) \mathrm{e}^{-r^2/(2b^2)} ,$$

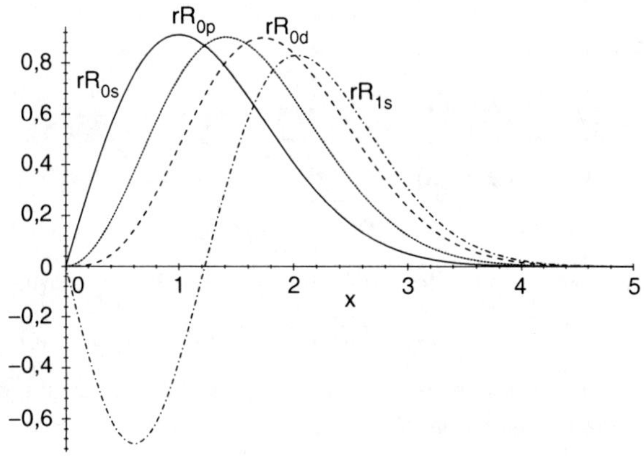

Fig. 1.14 The radial wave functions of the spherical oscillator (1.148), multiplied by r, for the states $0s$, $0p$, $0d$, and $1s$, as functions of r/b

$$E = \frac{7}{2}\hbar\omega : \quad R_{0d}(r) = \frac{4}{\sqrt{15\pi^{1/2}b^3}} \left(\frac{r}{b}\right)^2 \mathrm{e}^{-r^2/(2b^2)} ,$$

$$E = \frac{7}{2}\hbar\omega : \quad R_{1s}(r) = \frac{\sqrt{8}}{\sqrt{3\pi^{1/2}b^3}} \left[\left(\frac{r}{b}\right)^2 - \frac{3}{2}\right] \mathrm{e}^{-r^2/(2b^2)} .$$

Figure 1.14 shows the graphs of these functions. Among these, and in accord with the discussion above, only R_{0s} and R_{1s} are orthogonal.

5. On the basis of our knowledge of the one-dimensional linear oscillator we conclude that the system of eigenfunctions of the spherical oscillator

$$\psi_{n\ell m}(\boldsymbol{x}) = N_{n\ell} R_{n\ell}(r) Y_{\ell m}(\theta, \phi)$$

provides a complete, orthonormal set of square integrable functions over $\mathbb{R}^3$. This is important to know because this base system can be used in expansions of other, square integrable wave functions. Nuclear theory frequently makes use of this fact.

1.9.5 Mixed Spectrum: The Hydrogen Atom

We are now well prepared, through the experience of Sects. 1.9.3 and 1.9.4, to derive in a few steps the energy spectrum and the corresponding wave functions of the hydrogen atom. Let m denote the reduced mass, as before, let r be the modulus of the relative coordinate, $\hat{r} \equiv (\theta, \phi)$ its angular coordinates, and let $\boldsymbol{\ell}$ be the relative orbital angular momentum. Like in classical mechanics the center-of-mass S behaves like a pointlike particle of total mass M which moves freely. Thus, the eigenvalues of the corresponding term in the Hamiltonian

$$(H)_S = \frac{\boldsymbol{P}^2}{2M}$$

are $E_S = \boldsymbol{P}^2/(2M)$, the corresponding wave functions are plane waves. The Hamiltonian of relative motion reads

$$(H)_{\rm rel} = \frac{p_r^2}{2m} + \frac{\hbar^2 \boldsymbol{\ell}^2}{2mr^2} - \frac{e^2}{r} . \tag{1.151}$$

In order to determine simultaneous eigenfunctions of the set of observables

$$H, \ \boldsymbol{\ell}^2, \quad \text{and} \quad \ell_3$$

we start from the factorization form (1.126), by setting

$$\psi_{\alpha\ell m}(\boldsymbol{x}) = R_{\alpha\ell}(r) Y_{\ell m}(\hat{x}) \quad \text{or} \quad \psi_{\ell m}(\alpha, \boldsymbol{x}) = R_\ell(\alpha, r) Y_{\ell m}(\hat{x})$$

for the discrete or the continuous part of the spectrum, respectively. The radial equation (1.127) becomes

$$\frac{1}{r^2}\frac{\mathrm{d}}{\mathrm{d}r}\left(r^2 \frac{\mathrm{d}R(r)}{\mathrm{d}r}\right) - \left(\frac{\ell(\ell+1)}{r^2} - \frac{2me^2}{\hbar^2 r} - \frac{2mE}{\hbar^2}\right) R(r) = 0 .$$

Unlike in the previous examples we first substitute $R(r)$ as follows

$$u(r) := rR(r) .$$

This substitution has no other purpose than to simplify the radial differential equation. Indeed, it contains no longer any first derivative because

$$\frac{1}{r^2}\frac{\mathrm{d}}{\mathrm{d}r}\left(r^2 \frac{\mathrm{d}}{\mathrm{d}r}\frac{u(r)}{r}\right) = \frac{u''}{r} - 2\frac{u'}{r^2} + 2\frac{u}{r^3} + \frac{2}{r}\left(\frac{u'}{r} - \frac{u}{r^2}\right) = \frac{u''}{r} .$$

Furthermore, all radial integrals $\int r^2 \mathrm{d}r$ are replaced by $\int \mathrm{d}r$. The radial equation becomes

$$\frac{\mathrm{d}^2 u(r)}{\mathrm{d}r^2} - \left[\frac{\ell(\ell+1)}{r^2} - \frac{2me^2}{\hbar^2 r} - \frac{2mE}{\hbar^2}\right] u(r) = 0 .$$

As the potential tends to zero as $r \to \infty$, states with *positive* energy can escape to infinity and, therefore, will be similar to the force-free solutions of Sect. 1.9.3. However, the Coulomb potential being of infinite range, these states will be sensibly deformed even for large values of r. States with *negative* energy, in turn, must be fully localized because, if this were not so, the kinetic energy would have negative values at very large radii. Therefore, like in classical mechanics, these states must be bound states. For these reasons we analyze the cases $E > 0$ and $E < 0$ separately. We begin by the latter case:

Bound States: Introducing $B := -E$, the binding energy, $\kappa := \sqrt{2mB}/\hbar$ a wave number, and the dimensionless constant

$$\overline{\gamma} := \frac{me^2}{\hbar^2 \kappa} = \frac{e^2}{\hbar c}\sqrt{\frac{mc^2}{2B}}$$

the variable r is replaced by the dimensionless variable

$$\varrho := 2\kappa r . \tag{1.152}$$

The radial equation then reads

$$\frac{d^2u(\varrho)}{d\varrho^2} - \left[\frac{\ell(\ell+1)}{\varrho^2} - \frac{\overline{\gamma}}{\varrho} + \frac{1}{4}\right]u(\varrho) = 0. \qquad (**)$$

Very much like in the previous examples, in a first step, one analyzes the behaviour of $u(\varrho)$ at the origin, and at infinity. As we substituted $R(r) = u(r)/r$ the solutions regular at the origin must have the behaviour

$$\varrho \to 0: \quad u(\varrho) \sim \varrho^{\ell+1}v(\varrho).$$

In turn, for asymptotically large values of ϱ the differential equation $(**)$ simplifies so that its approximate solutions are

$$\varrho \to \infty: \quad u(\varrho) \sim a(B)e^{-(1/2)\varrho} + b(B)e^{+(1/2)\varrho}.$$

Note that both terms are admissible, a priori, in contrast to the spherical oscillator. The first term which decays exponentially, is welcome, while the second must certainly be absent in order to preserve square integrability. This raises the question of whether there are special values of the binding energy B (remember $E = -B$!) for which the coefficient $b(B)$ vanishes.

The example of the spherical oscillator taught us that it is advisable to extract from the wave function its asymptotic forms both for small and for large values of the radial variable. This is achieved by the ansatz

$$u(\varrho) = \varrho^{\ell+1}e^{-1/2\varrho}w(\varrho).$$

Equipped with the experience of the previous examples, $(**)$ is easily converted to a differential equation for the function $w(\varrho)$. Who is surprised to find once more Kummer's equation (1.145)? Its specific form in the present application reads

$$\varrho w''(\varrho) + (2\ell + 2 - \varrho)w'(\varrho) - (\ell + 1 - \overline{\gamma})w(\varrho) = 0.$$

Comparing to the general form (1.145) we see that a and c are

$$a = \ell + 1 - \overline{\gamma}, \quad c = 2\ell + 2.$$

The solution regular at the origin is

$$w(\varrho) = {}_1F_1(\ell + 1 - \overline{\gamma}; 2\ell + 2; \varrho).$$

The asymptotic representation (1.147) of the confluent hypergeometric function shows that its second term which grows like $e^{+\varrho}$, would destroy the exponential decay of $u(\varrho)$ at infinity unless the factor that multiplies this term vanishes,

$$\frac{1}{\Gamma(a)} = \frac{1}{\Gamma(\ell + 1 - \overline{\gamma})} = 0.$$

This happens precisely when $-a \in \mathbb{N}_0$ or, in detail,

$$\ell + 1 - \overline{\gamma} = -n', \qquad n' = 0, 1, 2, \dots.$$

Unlike the case of the spherical oscillator one defines

$$n := n' + \ell + 1, \quad \text{so that} \quad n = 1, 2, 3, \dots. \qquad (1.153)$$

The integer n is called *principal quantum number*. If $n' \in \mathbb{N}_0$ then $n \in \mathbb{N}$, excluding the zero. It follows from the definition (1.153) that, for given n, the orbital angular momentum ℓ can only take one of the values

$$\ell = 0, 1, \ldots, n-1 .$$

This tells us, in the interplay of the repulsive centrifugal potential and the attractive Coulomb potential, that the orbital angular momentum must not be too large if there are to be bound states.

The eigenvalues of the energy follow from the condition $\overline{\gamma} = n$. They have the remarkable property that they only depend on n, but not on ℓ, viz.

$$E_n \equiv -B_n = -\frac{me^4}{\hbar^2}\frac{1}{2n^2} = -\frac{1}{2n^2}\alpha^2 mc^2 . \tag{1.154}$$

This is indeed the discrete part of the hydrogen spectrum quoted in (1.24). As a new property one sees that the degree of degeneracy is

$$\sum_{\ell=0}^{n-1}\sum_{m=-\ell}^{+\ell} 1 = \sum_{\ell=0}^{n-1}(2\ell+1) = n^2 .$$

In addition to the directional degeneracy which yields a factor $(2\ell + 1)$ there is a further, dynamical, degeneracy which is specific to the $1/r$ potential.

The eigenfunctions of the Hamiltonian, normalized to 1, are as follows

$$\psi_{n\ell m}(\boldsymbol{x}) = R_{n\ell}(r)Y_{\ell m}(\hat{x}) \equiv \frac{1}{r}y_{n\ell}(r)Y_{\ell m}(\hat{x})$$

where

$$y_{n\ell}(r) = \sqrt{\frac{(\ell+n)!}{a_{\mathrm{B}}(n-\ell-1)!\,n(2\ell+1)!}\frac{1}{}}\varrho^{\ell+1}e^{-\varrho/2}{}_1F_1(-n+\ell+1;2\ell+2;\varrho) , \tag{1.155}$$

and where a_{B} denotes the Bohr radius (1.8). Upon insertion of the values of the energy obtained above, the variable ϱ is seen to be equal to $2/n$ times the ratio of r and of the Bohr radius,

$$\varrho = 2\kappa r = \frac{1}{\hbar}\sqrt{-2mE_n}r = \frac{2\alpha mc^2}{n\hbar c}r = \frac{2r}{na_{\mathrm{B}}} .$$

I skip the calculation of the normalization in (1.155). The integrals containing powers, exponentials, and confluent hypergeometric functions which are needed for this calculation, are found e.g. in [Gradshteyn and Ryzhik (1965)].

While the energy only depends on the principal quantum number n the radial wave functions depend on both n and the orbital angular momentum ℓ. Like in the case of the spherical oscillator one notes that two radial wave functions are orthogonal only for equal values of ℓ and different values of n, but not for different values of ℓ. In the latter case, orthogonality of the entire wave function is taken care of by the spherical harmonics.

Here are the normalized radial functions for $n = 1, 2, 3$, using the spectroscopic notation for the orbital angular momentum:

$$R_{1s}(r) = \frac{2}{a_B^{3/2}} e^{-(r/a_B)},$$

$$R_{2p}(r) = \frac{1}{r}\frac{1}{a_B^{1/2} 2\sqrt{6}} \left(\frac{r}{a_B}\right)^2 e^{-(r/2a_B)},$$

$$R_{2s}(r) = \frac{1}{r}\frac{1}{a_B^{1/2}\sqrt{2}} \left(\frac{r}{a_B}\right)\left[1 - \frac{1}{2}\left(\frac{r}{a_B}\right)\right] e^{-(r/2a_B)},$$

$$R_{3d}(r) = \frac{1}{r}\frac{1}{a_B^{1/2} 3\sqrt{5!}} \left(\frac{2r}{3a_B}\right)^3 e^{-(r/3a_B)},$$

$$R_{3p}(r) = \frac{1}{r}\frac{\sqrt{2}}{a_B^{1/2} 3\sqrt{3}} \left(\frac{2r}{3a_B}\right)^2 \left[1 - \frac{1}{4}\left(\frac{2r}{3a_B}\right)\right] e^{-(r/3a_B)},$$

$$R_{3s}(r) = \frac{1}{r}\frac{1}{a_B^{1/2}\sqrt{3}} \left(\frac{2r}{3a_B}\right)\left[1 - \left(\frac{2r}{3a_B}\right) + \frac{1}{6}\left(\frac{2r}{3a_B}\right)^2\right] e^{-(r/3a_B)}.$$

Figure 1.15 shows the first three s-functions $\{r \cdot R_{ns}(r), n = 1, 2, 3\}$, Fig. 1.16 shows their squares $r^2 R_{ns}^2(r)$, as functions of r expressed in units of a_B. In order to interpret these graphs we calculate the expectation values of r^α for the three states and with α an integer, positive or negative, power. One finds the following results

$$\left\langle r^\alpha \right\rangle_{1s} = a_B^\alpha \frac{1}{2^{\alpha+1}}(\alpha + 2)!,$$

$$\left\langle r^\alpha \right\rangle_{2s} = a_B^\alpha \frac{1}{2}(\alpha + 2)! \left(1 + \frac{3}{4}\alpha + \frac{1}{4}\alpha^2\right),$$

$$\left\langle r^\alpha \right\rangle_{3s} = a_B^\alpha \frac{3^\alpha}{2^{\alpha+1}}(\alpha + 2)! \left(1 + \frac{7}{6}\alpha + \frac{23}{36}\alpha^2 + \frac{1}{6}\alpha^3 + \frac{1}{36}\alpha^4\right).$$

For $\alpha = 0$ the right-hand sides are equal to 1, in agreement with the normalization of the wave functions. For $\alpha = 1$ and $\alpha = 2$ these formulae give

$$\langle r \rangle_{1s} = \frac{3}{2}a_B, \qquad \langle r \rangle_{2s} = 6a_B, \qquad \langle r \rangle_{3s} = \frac{27}{2}a_B$$

$$\left\langle r^2 \right\rangle_{1s}^{1/2} = \sqrt{3}\, a_B, \qquad \left\langle r^2 \right\rangle_{2s}^{1/2} = \sqrt{42}\, a_B, \qquad \left\langle r^2 \right\rangle_{3s}^{1/2} = 3\sqrt{23}\, a_B,$$

respectively. As the abscissa of Fig. 1.16 shows the ratio r/a_B one may mark the numbers just obtained on this axis, and thereby interpret the graphs of the (radial) probabilities $r^2 R_{ns}^2$.

Evaluating the results for $\alpha = -1$ one finds for all three examples

$$\left\langle \frac{1}{r} \right\rangle_{ns} = \frac{1}{n^2}\frac{1}{a_B}.$$

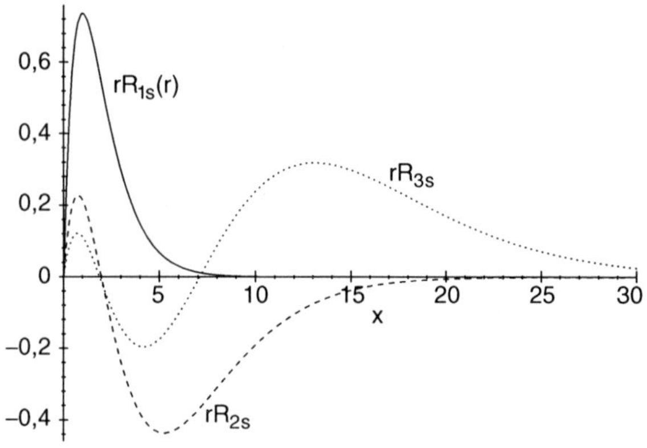

Fig. 1.15 Graphs of the functions $y_{ns}(r) = rR_{ns}$ of the hydrogen atom $n = 1, 2, 3$ as functions of (r/a_B). These states are pairwise orthogonal

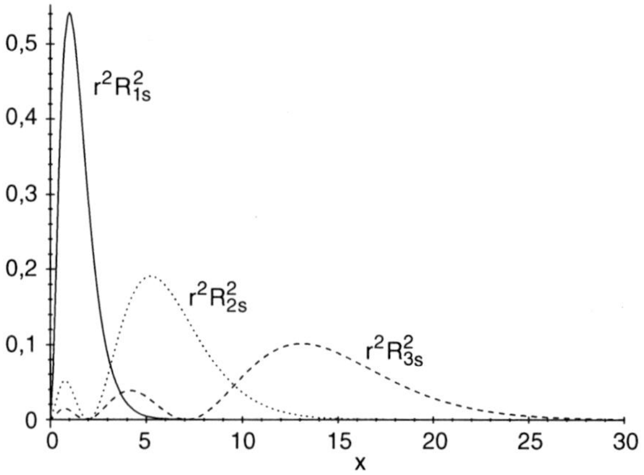

Fig. 1.16 Radial probability densities $r^2 R_{ns}^2(r)$ of the first three s-states of Fig. 1.15, as functions of (r/a_B). The angular part is $Y_{00} = 1/\sqrt{4\pi}$ and, hence, is isotropic. Therefore, these densities, completed by spherical symmetry and multiplied by $1/(4\pi)$, yield the full densities

This is a result one could have guessed in advance. Indeed, it follows from the virial theorem which yields $\langle U(r)\rangle_{n\ell} = 2E_n$ for the case of a $1/r$-potential.[23]

Figures 1.17 and 1.18 show the graphs of the radial functions for the same value $n = 3$ of the principal quantum number and for $\ell = 2, 1, 0$, i.e. the functions $rR_{3d}(r)$, $rR_{3p}(r)$, and $rR_{3s}(r)$, and of their squares, respectively, as functions of r/a_B. (Note that these functions are not orthogonal.)

In contrast to the case of the spherical oscillator this first set of wave functions is *not* complete. What is missing to obtain completeness are the eigenfunctions of the Hamiltonian which correspond to *positive* energies. To these we now turn.

---

[23] Note that I have used the virial theorem of classical mechanics, and have replaced the averages by expectation values as suggested by Ehrenfest's theorem, Sect. 1.5.2. In fact one proves the virial theorem for expectation values directly (cf. Exercise 1.11).

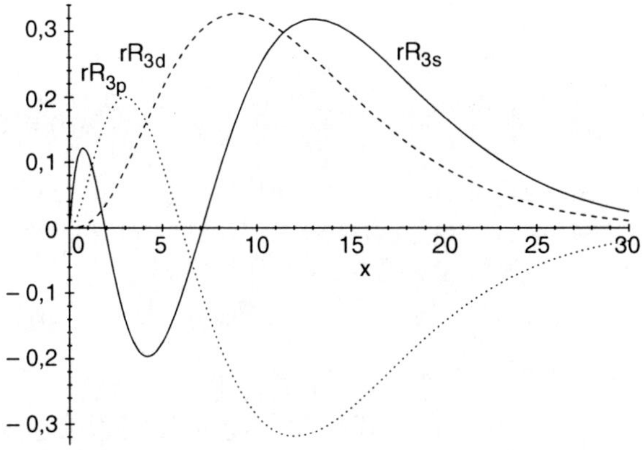

Fig. 1.17 Graphs of the radial eigenfunctions, multiplied by r, of the states $(n = 3, \ell)$, $\ell = 0, 1, 2$, as functions of the variable (r/a_B)

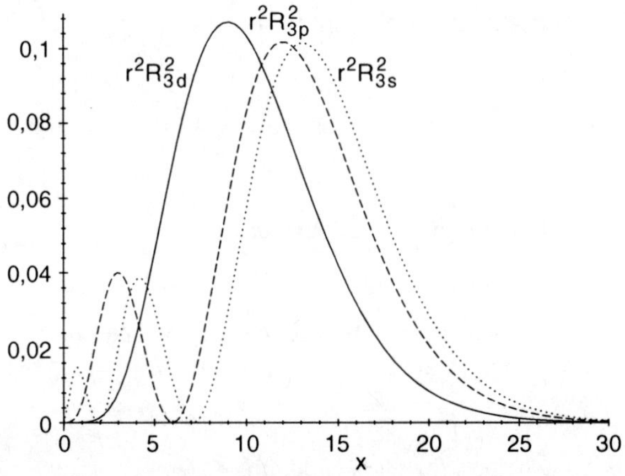

Fig. 1.18 Radial probability densities $r^2 R^2_{3\ell}(r)$ in the states shown in Fig. 1.17. The full spatial densities are obtained by multiplying the wave function of the s-state by $1/(4\pi)$, the one of the p-state by $|Y_{1m}(\theta, \phi)|^2$, and the one of the d-state by $|Y_{2m}(\theta, \phi)|^2$

Eigenstates in the Continuum: If the energy is positive then

$$k := \frac{\sqrt{2mE}}{\hbar}$$

is the wave number that is to be associated to the electron whenever it moves asymptotically far from the origin, the center of the force field. The constant $\overline{\gamma}$ of bound states is replaced by the definition

$$\gamma := -\frac{e^2\sqrt{m}}{\hbar\sqrt{2E}} ,$$

or, somewhat more generally,

$$\gamma := -\frac{me^2}{\hbar^2 k} = \frac{ZZ'e^2}{\hbar v} , \qquad (1.156)$$

where Z and Z' are the charge numbers of the two particles that scatter off one another, $v = \hbar k/m = \sqrt{2E/m}$ being their relative velocity. In the case of hydrogen we have $Z = 1$, $Z' = -1$, hence the choice of the sign in the definition. Taking again $\varrho := 2kr$, the radial equation reads

$$\frac{d^2u(\varrho)}{d\varrho^2} - \left[\frac{\ell(\ell+1)}{\varrho^2} + \frac{\gamma}{\varrho} - \frac{1}{4}\right] u(\varrho) = 0 . \qquad (\varrho = 2kr) \qquad (***)$$

In comparing to (**) note the different sign of the last term in square brackets. While the regular solution starts like $\varrho^{\ell+1}$, as before, this sign change causes the oscillatory behaviour to become $\exp(\pm i\varrho/2) = \exp(\pm ikr)$ at infinity. This suggests the ansatz

$$u(\varrho) = e^{i\varrho/2}\varrho^{\ell+1} w(\varrho) ,$$

thus obtaining the differential equation

$$\varrho w''(\varrho) + (2\ell + 2 + i\varrho)w'(\varrho) + (i(\ell+1) - \gamma)w(\varrho) = 0 .$$

This is almost, but not quite, Kummer's equation (1.145). Closer examination shows, however, that it is sufficient to substitute

$$z := -i\varrho$$

to obtain that differential equation with the constants

$$a = \ell + 1 + i\gamma \quad \text{and} \quad c = 2\ell + 2 .$$

The solution regular at the origin is

$$w(z = -i\varrho) = w(-2ikr) = N_\ell \, {}_1F_1(\ell + 1 + i\gamma; 2\ell + 2; z)$$

where N_ℓ is a constant that remains to be determined. The asymptotics for $r \to \infty$ is of particular interest. It is read off the formula (1.147): Writing

$$\Gamma(\ell + 1 + i\gamma) = |\Gamma(\ell + 1 + i\gamma)| e^{i\sigma_\ell} ,$$

thereby defining what is called the *Coulomb phase* σ_ℓ, one has

$$\begin{aligned}
{}_1F_1 &\sim \frac{\Gamma(2\ell+2)}{\Gamma(\ell+1-i\gamma)}(+2ikr)^{-\ell-1-i\gamma} + \frac{\Gamma(2\ell+2)}{\Gamma(\ell+1+i\gamma)} e^{-2ikr}(-2ikr)^{-\ell-1+i\gamma} \\
&= \frac{\Gamma(2\ell+2)}{|\Gamma(\ell+1-i\gamma)|} \frac{1}{(2kr)^{\ell+1}} e^{-ikr} e^{(\pi\gamma/2)} \\
&\quad \times \left[i^{-\ell-1} e^{i[kr - \gamma\ln(2kr) + \sigma_\ell]} + (-i)^{-\ell-1} e^{-i[kr - \gamma\ln(2kr) + \sigma_\ell]} \right] \\
&= \frac{\Gamma(2\ell+2)}{|\Gamma(\ell+1-i\gamma)|} e^{(\pi\gamma/2)} \frac{2}{(2kr)^{\ell+1}} e^{-ikr} \frac{1}{2i} \\
&\quad \times \left[e^{i[kr - \gamma\ln(2kr) - \ell(\pi/2) + \sigma_\ell]} - e^{-i[kr - \gamma\ln(2kr) - \ell(\pi/2) + \sigma_\ell]} \right] .
\end{aligned}$$

If we choose the normalization N_ℓ as follows

$$N_\ell = \frac{|\Gamma(\ell+1-i\gamma)|}{2\Gamma(2\ell+2)} e^{-\pi\gamma/2} ,$$

the radial function obtains an asymptotic behaviour

$$u_\ell(\varrho = 2kr) = N_\ell \, e^{i\varrho/2}\varrho^{\ell+1} \, {}_1F_1(\ell + 1 + i\gamma, 2\ell + 2, -i\varrho) \qquad (1.157)$$

which is similar to the asymptotics of the free solutions

$$\varrho \to \infty : \quad u_\ell(\varrho) \sim \sin\left(kr - \ell\frac{\pi}{2} - \gamma \ln(2kr) + \sigma_\ell\right).$$

It differs from the asymptotics of the spherical Bessel functions by the constant scattering phase

$$\sigma_\ell = \arg\Gamma(\ell + 1 + \mathrm{i}\gamma)$$

and by the phase $-\gamma \ln(2kr)$, with a logarithmic dependence on r which is characteristic for the $1/r$-dependence of the potential.

The complete wave function with positive energy and with definite values of ℓ and m reads

$$\psi_{\ell m}(E, \boldsymbol{x}) = R_\ell(E, r) Y_{\ell m}(\hat{x}) \equiv \frac{1}{r} u_\ell(E, r) Y_{\ell m}(\hat{x})$$

with, as noted above,

$$u_\ell(E, r) \equiv u_\ell(\varrho),$$

the normalization being chosen according to the needs of the specific situation one is studying.

Remarks

1. The energy spectrum of hydrogen is the classical example for a mixed spectrum. It consists of a countably infinite, discrete, set of values, with a limit point at zero, and of a continuum of positive values starting at $E = 0$. Besides the degeneracy in the projection m of orbital angular momentum the discrete spectrum exhibits a dynamical degeneracy which grows strongly with n. This is sketched in Fig. 1.19. This dynamical degeneracy is lifted as soon as the radial dependence of the spherically symmetric potential deviates from $1/r$. This happens, for instance, if nuclei of hydrogen-like atoms are no longer described by point charges Ze but by charge distributions of finite spatial extension.
2. Only the combined set of wave functions (1.155) with negative eigenvalues, and (1.157) with positive energies are complete. The two parts taken separately, are not. Although rarely done, there may be situations where to use the eigenfunctions of the hydrogen Hamiltonian as a basis for calculations in atomic physics. In such a case one must first normalize the eigenfunctions (1.157) in the energy scale and must use both groups of eigenfunctions. The completeness relation then reads

$$\sum_{n=1}^{\infty} \sum_{\ell=0}^{n-1} \sum_{m=-\ell}^{+\ell} \psi_{n\ell m}(\boldsymbol{x}) \psi^*_{n\ell m}(\boldsymbol{x}')$$

$$+ \int_0^\infty \mathrm{d}E \sum_{\ell=0}^{\infty} \sum_{m=-\ell}^{+\ell} \psi_{\ell m}(E, \boldsymbol{x}) \psi^*_{\ell m}(E, \boldsymbol{x}') = \delta(\boldsymbol{x} - \boldsymbol{x}').$$

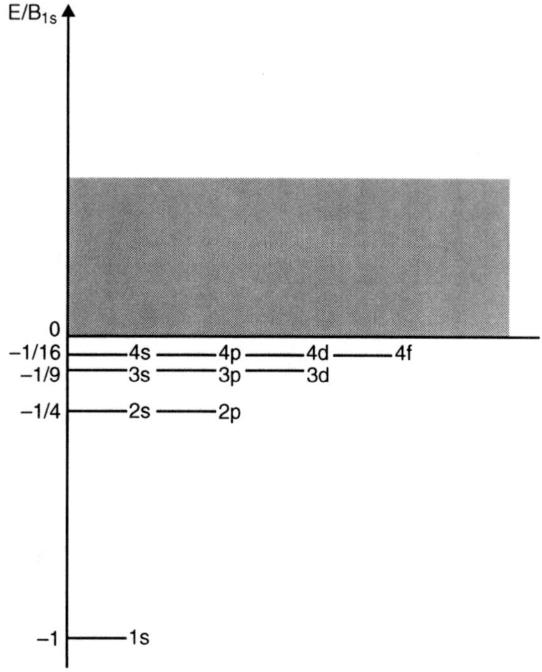

Fig. 1.19 Energy spectrum of the hydrogen atom. Besides the degeneracy due to the magnetic quantum number m which is typical for all central field problems, the discrete values with negative E exhibit a dynamical ℓ-degeneracy: for given principal quantum number n all levels with $\ell = 0$ up to $\ell = n - 1$ have the same energy. The discrete part of the spectrum has the limiting value $E = 0$. At this point the continuum of positive energies starts

3. The confluent hypergeometric functions contained in the eigenfunctions (1.155) of the bound states are polynomials which are identical with the *associated Laguerre polynomials*, up to the normalization,

$$L_{\ell+n}^{2\ell+1}(\varrho) = -\frac{[(\ell+n)!]^2}{(n-\ell-1)!(2\ell+1)!}\, {}_1F_1(-n+\ell+1, 2\ell+2, \varrho) .$$

These polynomials are defined as follows:
Laguerre polynomials:

$$L_\mu(x) = \mathrm{e}^x \frac{\mathrm{d}^\mu}{\mathrm{d}x^\mu}(\mathrm{e}^{-x} x^\mu) = \sum_{\nu=0}^{\mu} (-)^\nu \binom{\mu}{\nu} \frac{\mu!}{\nu!} x^\nu ;$$

Associated Laguerre polynomials:

$$L_\mu^\sigma(x) = \frac{\mathrm{d}^\sigma}{\mathrm{d}x^\sigma} L_\mu(x) .$$

In practical calculations one often uses the associated Laguerre polynomials, instead of the confluent hypergeometric, because there are useful recurrence relations for them which are useful in simplifying integrals.

2 Scattering of Particles by Potentials

Introduction The three prototypes of spectra of self-adjoint operators, the *discrete spectrum*, with or without degeneracy, the *continuous spectrum*, and the *mixed spectrum*, as well as the corresponding wave functions, contain important information about the physical systems that they describe. Yet, from a physicist's point of view, the results obtained until now remain somewhat academic as long as we do not know how to render the information they contain visible by means of concrete experiments. For example, the static spectrum of the Hamiltonian describing the hydrogen atom and the spatial shape of its stationary wave functions, *a priori*, are not observable for us, the macroscopic observers, as long as the atom is not forced to change its state by interaction with external electromagnetic fields or by interaction with scattering beams of electrons. In other terms, the pure stationary systems that we studied so far, must be subject to nonstationary interactions in set-ups of realistic experiments which allow for preparation and detection, before we can decide whether or not these systems describe reality. As this question is of central importance I insert here a first chapter on the description of scattering processes in what is called *potential scattering*, before turning to the more formal framework of quantum theory.

2.1 Macroscopic and Microscopic Scales

In studying classical macroscopic systems our experience shows that it is always possible to perform observations without disturbing the system: In observing the swinging pendulum of an upright clock we measure its maximal elongation, its period, perhaps even the velocity at the moment of passage through the vertical, a stop-watch in our hand, by just "looking" at it and without interfering, to any sizable degree, with the motion of the pendulum. Even extremely precise measurements on satellites or on planets by means of radar signals and interferometry are done with practically no back-reaction on their state of motion. This familiar, almost obvious fact is paraphrased by the statement that the object, i.e. the isolated physical system

F. Scheck, *Quantum Physics*, DOI: 10.1007/978-3-642-34563-0_2,

that one wishes to study, is separated from the observer and his measuring apparatus in the sense that back-reactions of the measuring procedure on the object are negligible. The perturbation of the system by the measurement either is negligibly small, or can be corrected for afterwards. In particular, the system is obviously not influenced by the mere fact of being observed.

There is a further aspect that we should realize: The length scales and the time scales of macroscopic processes are the typical scales of our familiar environment, or do not differ from them much, that is, do not go beyond the realm of what we can imagine based on our day-to-day experience. Examples are the milliseconds that matter in sportive competitions, or the very precise length measurements in micro mechanics.

Matters change when the object of investigation is a microscopic system such as a molecule, an atom, an atomic nucleus, or a single elementary particle:

1. Every measurement on a micro system is a more or less intrusive intervention that can strongly modify the system or even destroy it. As an example, think of an atom described by a wave function $\psi(t, \boldsymbol{x})$. If this atom is bombarded by a relatively "coarse" beam such as, e.g., an unpolarized light beam, one intuitively expects the subtle phase correlations that are responsible for the interference phenomena of the wave function, to be partially or completely destroyed. In fact, the impossibility of separating measuring device and object belongs to the most difficult aspects of quantum theory.
2. Furthermore, the spatial and temporal scales of typical quantum processes, in general, are small as compared to spatial distances or time intervals of experiments set up to detect them. An example may illustrate this: The size of a hydrogen atom is of the order of the Bohr radius (1.8), that is about 10^{-10} m. This is a very small quantity as compared to the distance of the hydrogen target from the source emitting the incoming beam that is used to investigate the atom, as well as from the detector designed to detect the scattered beam. A similar remark applies to the temporal conditions in atoms. Characteristic times of an atom are defined by the transition energies,
$$\tau(m \to n) = \frac{2\pi}{c} \frac{\hbar c}{E_m - E_n} .$$
For the example of the (2p $\to$ 1s)-transition in hydrogen this time is $\tau(2 \to 1) \approx 4 \times 10^{-16}$ s – a time interval which is very short as compared to the time scales in a typical experiment.

The general conclusion from these considerations is that in general we can only observe *asymptotic* states, realized long before or long after the process proper, and at large spatial distances from it. More concretely this means the following: We wish to investigate a quantum system which, when isolated, is a stationary one, by means of a beam of particles. The system is the *target*, the particles are the *projectiles* which are directed onto the target, or which are seen in a detector after the scattering has taken place. The interaction process of the beam and the system happens within a time interval Δt around, say, $t = 0$. It is localized within a volume V of space around the origin $\boldsymbol{x} = 0$ that is characterized by the radius R_0. At time $t \to -\infty$

the beam is constructed in a controlled way at an asymptotically large distance from the target. The beam and the target in this configuration define what is called the *in*-state (abbreviation for "incoming state"). For $t \to +\infty$ the scattered particles, or more generally the reaction products of the scattering process, are identified in the detector which also has an asymptotically large distance from the target. The scattered particles, together with the target in its final state, are in what one calls the *out*-state (abbreviation for "outgoing state").

Other situations where we actually can perform measurements are provided by systems which are stationary by themselves but which are unstable because of interactions with other systems. For example, the 2p-state of hydrogen is unstable because it decays to the 1s-state by emission of a photon, the typical time for the transition being on the order of 10^{-9} s. In this example the *in*-state is the atom in its excited 2p-state, the *out*-state consists of the outgoing photon and the atom in the stable ground state. In this example, too, our information on the (unstable) quantum system stems from an asymptotic measurement, the decay products being detected a long time after the decay process happened and very far in space.[1]

As a general conclusion we note that experimental information on quantum mechanical systems is obtained from asymptotic, incoming or outgoing states. We cannot penetrate the interaction region proper and cannot interfere with the typical time scale of the interaction. In quantum mechanics of molecules, atoms, and nuclei, the important methods of investigation are: scattering of particles, i.e. electrons, protons, neutrons, or α particles, on these systems; excitation and decay of their excited states by interaction with the electromagnetic radiation field.

Scattering processes which can be dealt with by means of the notions and methods developed in Chap. 1, are the subject of this chapter. The interaction with the radiation field requires further and more comprehensive preparation and is postponed to later chapters. Likewise, scattering theory in a physically more general, mathematically more formal framework will be taken up again later.

2.2 Scattering on a Central Potential

We assume that a potential $U(\boldsymbol{x})$ is given which describes the interaction of two particles and which is spherically symmetric and has finite range. Somewhat more formally this means

$$U(\boldsymbol{x}) \equiv U(r) \quad \text{with} \quad r = |\boldsymbol{x}|\,, \qquad \lim_{r\to\infty} [rU(r)] = 0\,. \tag{2.1}$$

[1] Of course, the unstable state must have been created beforehand and one may ask why the preparation procedure is not taken to be part of the *in*-state, or under which condition this becomes a necessity. The answer is a qualitative one: The total decay probability of the unstable state, when multiplied by $\hbar$, yields the energy uncertainty, or width Γ of the unstable state. If $\Gamma \ll E_\alpha$, i.e if the width Γ is small as compared to the energy E_α of the state then this state is *quasi-stable*. The process used to prepare it can be separated from the decay process.

Spherically symmetric potentials such as

$$U(r) = U_0 \Theta(r_0 - r) \quad \text{or} \quad U(r) = g\frac{e^{-r/r_0}}{r} ,$$

the first of which vanishes outside of the fixed radius r_0, while the second decreases exponentially, fulfill this condition. The Coulomb potential

$$U_C(r) = \frac{e_1 e_2}{r}$$

does not. The aim is to investigate the scattering of a particle of mass m on the potential and to calculate the corresponding differential cross section. (Note that in the case of the two-body system m is the reduced mass.) The differential cross section is an observable, i.e. a classical quantity. Its definition is the same as in classical mechanics (cf. [Scheck (2010)], Sect. 1.27): It is the ratio of the number dn of particles which are scattered, in the unit of time, into scattering angles between θ and $\theta + d\theta$, and of the number n_0 of incoming particles per unit of area and unit of time. In other terms, one determines the number of particles that were actually scattered, and normalizes to the incoming flux. In contrast to the classical situation these numbers are obtained from the current density (1.54) (with $\boldsymbol{A} \equiv 0$) describing the flow of probability, via Born's interpretation, not from classical trajectories (which no longer exist).

The correct way of proceeding would be to construct a state coming in along the 3-direction as a wave packet at $t = -\infty$ which clusters around the average momentum $\boldsymbol{p} = p\hat{\boldsymbol{e}}_3$, then to calculate the time evolution of this wave packet by means of the Schrödinger equation, and eventually analyze the outgoing flux at $t \to +\infty$. As this is tedious and cumbersome, one resorts to an intuitive method which is simpler and, yet, leads to the correct results. One considers the scattering process as a stationary situation. The incoming beam is taken to be a stationary plane wave, while the scattered state is represented by an outgoing spherical wave. At asymptotic distances from the scattering center the wave field then has the form

$$r \to \infty : \quad \psi_{\text{Somm}}(\boldsymbol{x}) \sim e^{ikx^3} + f(\theta)\frac{e^{ikr}}{r} , \quad k = \frac{1}{\hbar}|\boldsymbol{p}| , \tag{2.2}$$

whose first term is the incoming beam with momentum $\boldsymbol{p} = p\hat{\boldsymbol{e}}_3 = \hbar k\hat{\boldsymbol{e}}_3$ while the second is the outgoing spherical wave. This ansatz is called *Sommerfeld's radiation condition*

The role of the (in general) complex amplitude $f(\theta)$ is clarified by calculating the current densities of the incoming and outgoing parts. It is useful, here and below, to denote the skew-symmetric derivative of (1.54) by a symbol of its own,

$$f^*(\boldsymbol{x}) \overleftrightarrow{\nabla} g(\boldsymbol{x}) := f^*(\boldsymbol{x})\nabla g(\boldsymbol{x}) - [\nabla f^*(\boldsymbol{x})]g(\boldsymbol{x}) \tag{2.3}$$

where f and g are complex functions which are at least C^1 (once continuously differentiable). For the incoming wave one finds

$$\boldsymbol{j}_{\text{in}} = \frac{\hbar}{2m\text{i}} e^{-ikx^3} \overleftrightarrow{\nabla} e^{ikx^3} = \frac{\hbar k}{m}\hat{\boldsymbol{e}}_3 = v\hat{\boldsymbol{e}}_3 ,$$

with $v\hat{\boldsymbol{e}}_3$ the velocity of the incoming particle.

Using spherical polar coordinates for which

$$\nabla = \left(\frac{\partial}{\partial r}, \frac{1}{r}\frac{\partial}{\partial \theta}, \frac{1}{r \sin\theta}\frac{\partial}{\partial \phi} \right) ,$$

and the following expressions for the gradient of $\psi = f(\theta)\mathrm{e}^{\mathrm{i}kr}/r$

$$(\nabla\psi)_r = \frac{\partial \psi}{\partial r} = \left(-\frac{1}{r^2} + \frac{\mathrm{i}k}{r} \right) f(\theta)\mathrm{e}^{\mathrm{i}kr} ,$$

$$(\nabla\psi)_\theta = \frac{1}{r}\frac{\partial \psi}{\partial \theta} = \frac{1}{r^2}\frac{\partial f(\theta)}{\partial \theta}\mathrm{e}^{\mathrm{i}kr} , \qquad (\nabla\psi)_\phi = 0 ,$$

the outgoing current density is easily calculated,

$$\boldsymbol{j}_{\mathrm{out}} = \frac{\hbar k}{m}\frac{|f(\theta)|^2}{r^2}\hat{\boldsymbol{e}}_r + \frac{\hbar}{2m\mathrm{i}}\frac{1}{r^3} f^*(\theta) \overset{\leftrightarrow}{\nabla} f(\theta)\hat{\boldsymbol{e}}_\theta .$$

The first term, upon multiplication by the area element $r^2\,\mathrm{d}\Omega$ of a sphere with radius r and center at the origin, yields a probability current in the radial direction proportional to $|f(\theta)|^2$. The second term, in contrast, yields a current decreasing like $1/r$ which must be neglected asymptotically. The flux of particles across the cone with solid angle $\mathrm{d}\Omega$ in an outgoing radial direction becomes (for large values of r)

$$\boldsymbol{j}_{\mathrm{out}} \cdot \hat{\boldsymbol{e}}_r r^2\,\mathrm{d}\Omega = \frac{\hbar k}{m}\frac{|f(\theta)|^2}{r^2} r^2\,\mathrm{d}\Omega \qquad (r \to \infty) .$$

Normalizing to the incoming flux, the differential cross section is found to be

$$\mathrm{d}\sigma_{\mathrm{el}} = \frac{\boldsymbol{j}_{\mathrm{out}} \cdot \hat{\boldsymbol{e}}_r r^2\,\mathrm{d}\Omega}{|\boldsymbol{j}_{\mathrm{in}}|} = |f(\theta)|^2\,\mathrm{d}\Omega .$$

This result clarifies the physical interpretation of the amplitude $f(\theta)$: This amplitude determines the differential cross section

$$\boxed{\frac{\mathrm{d}\sigma_{\mathrm{el}}}{\mathrm{d}\Omega} = |f(\theta)|^2} . \tag{2.4}$$

The function $f(\theta)$ is called *scattering amplitude.* It is a probability amplitude, in the spirit of Born's interpretation. The square of its modulus is the differential cross section and, hence, is a classical observable. As we will see soon it describes *elastic* scattering. The total elastic cross section is given by the integral taken over the complete solid angle

$$\boxed{\sigma_{\mathrm{el}} = \int \mathrm{d}\Omega |f(\theta)|^2 = 2\pi \int_0^\pi \sin\theta\,\mathrm{d}\theta\,|f(\theta)|^2} . \tag{2.5}$$

Before continuing with a discussion of methods that allow to calculate the scattering amplitude and the cross sections we add a few complements to and comments on these results.

Remarks

1. The result obtained for the asymptotic form of $\boldsymbol{j}_{\text{out}}$ shows that it was justified to call the two terms in (1.135), Sect. 1.9.3, *out*going and *in*coming spherical waves, respectively.
2. In the two-body problem with central force, the variable r is the modulus of the relative coordinate, m is the reduced mass,
$$m \longmapsto \frac{m_1 m_2}{m_1 + m_2}, \quad \text{and} \quad \theta \longmapsto \theta^*$$
is the scattering angle in the center-of-mass system. Thus, the amplitude $f(\theta)$ is the *scattering amplitude in the center-of-mass system.*
3. The potential $U(r)$ must be real if the Hamiltonian is to be self-adjoint. If this is so, then there is only *elastic* scattering. No matter how it is scattered, the particle must be found somewhere in the final state, or, in the spirit of quantum mechanics, the probability to find the particle *somewhere* in space must be conserved. Therefore, the expression (2.4) describes the differential cross section for *elastic* scattering, the expression (2.5) gives the *integrated elastic* cross section.
However, there are processes where the final state is not the same as the initial state. An electron which is scattered on an atom, may loose energy and may leave behind the atom in an excited state. A photon used as a projectile may be scattered inelastically on the atom, or may even be absorbed completely. In those cases one says that the final state belongs to another *channel* than the initial state. Besides the real potential responsible for elastic scattering, the Hamiltonian must also contain interaction terms which allow to cross from the initial channel to other, inelastic channels. Loosely speaking, in such a situation the total probability is distributed, after the scattering, over the various final state channels. In Sect. 2.6 we will develop a bulk method for describing such a situation without knowing the details of the reaction dynamics.
4. Even though the ansatz (2.2) is intuitively compelling, in a strict sense the ensuing calculation is not correct. The condition (2.2) assumes a stationary wave function where both the incoming plane wave and the outgoing spherical wave are present at all times. In particular, when one calculates the current density (1.54) there should be terms arising from the interference between the incoming and outgoing parts. Instead, in our calculation of the current densities we proceeded as if at $t = -\infty$ there was only the plane wave, and at $t = +\infty$ there will be only the scattered spherical wave. Although this derivation rests on intuition and, strictly speaking, is not correct, it yields the right result. This is so because the plane wave is but an idealization and should be replaced by a suitably composed wave packet. We skip this painstaking calculation at this point and just report the essential result: One follows the evolution of a wave packet which at $t \to -\infty$ described a localized particle with average momentum $\boldsymbol{p} = p\hat{\boldsymbol{e}}_3$. For large positive times and at asymptotic distances there appears an outgoing spherical wave with the shape assumed above. There are indeed interference terms between the initial state and the final state, but, as

these terms oscillate very rapidly, integration over the spectrum of momenta renders them negligibly small. Except for the forward direction, that is for scattering where $\boldsymbol{p}' = \boldsymbol{p}$, the ansatz (2.2), together with the interpretation given above, is correct.

5. It is instructive to compare the quantum mechanical description with the theory of elastic scattering in classical mechanics. The definition of the differential cross section, of course, is the same (number of particles per unit of time scattered into the solid angle $\mathrm{d}\Omega$, normalized to the incoming flux). The physical processes behind it are not the same. The classical particle that comes in with momentum $\boldsymbol{p} = p\hat{\boldsymbol{e}}_3$ and impact parameter b, moves on a well-defined trajectory. It suffices to follow this orbit from $t = -\infty$ to $t = +\infty$ to find out, with certainty, where the particle has gone. In quantum mechanics we associate a wave packet to the particle which, for example, contains only momenta in the 3-direction and which, at $t = -\infty$, is centered at a value $\boldsymbol{p} = p\hat{\boldsymbol{e}}_3$. Values for its 3-coordinate can be limited within what the uncertainty relation allows for. As the particle has no momentum components in the 1- and the 2-directions, or, in other terms, since $\boldsymbol{p}_1$ and $\boldsymbol{p}_2$ have the sharp values 0, the position of the particle in the plane perpendicular to the 3-axis is completely undetermined. At $t \to +\infty$ quantum mechanics yields a probability for detecting the particle in a detector which is positioned at a scattering angle θ with respect to the incoming beam. It is impossible to predict where any individual particle will be scattered. The probability $\mathrm{d}\sigma_{\mathrm{el}}/\mathrm{d}\Omega$ which is defined by the complex scattering amplitude $f(\theta)$ will be confirmed only if one allows very many particles to scatter under identical experimental conditions.

2.3 Partial Wave Analysis

Clearly, the scattering amplitude must be a function of the energy $E = \hbar^2 k^2/(2m)$ of the incoming beam, or, what amounts to the same, a function of the wave number k. Whenever this dependence matters we should write, more precisely, $f(k, \theta)$ instead of $f(\theta)$. The cross section has the physical dimension [area]. Hence, the scattering amplitude has dimension [length]. In the physically allowed region $\theta \in [0, \pi]$ or $z \equiv \cos\theta \in [-1, +1]$, and for fixed k, the amplitude $f(k, \theta)$ is a nonsingular, in general square integrable function of θ.[2] Therefore, it may be expanded in terms of spherical harmonics $Y_{\ell m}(\theta, \phi)$. However, as it depends on θ, by the spherical symmetry of the potential, and does not depend on ϕ, this expansion contains only spherical

[2] Assuming the restriction (2.1) the amplitude f is square integrable, indeed. In the case of the Coulomb potential the amplitude is singular in the forward direction $\theta = 0$, its behaviour being like $1/\sin\theta$, and, hence, is no longer square integrable. Nevertheless, the scattering amplitude can still be expanded in terms of Legendre polynomials. However, the series (2.6) is no longer convergent in the forward direction, and the expression (2.7) for the integrated cross section diverges.

harmonics with $m = 0$, $Y_{\ell 0}$, which are proportional to Legendre polynomials,

$$Y_{\ell 0} = \sqrt{\frac{2\ell+1}{4\pi}} P_\ell(z = \cos\theta) .$$

As a consequence one can always choose an expansion in terms of Legendre polynomials,

$$\boxed{f(k,\theta) = \frac{1}{k} \sum_{\ell=0}^{\infty} (2\ell+1) a_\ell(k) P_\ell(z)} . \tag{2.6}$$

The factor $1/k$ is introduced in order to keep track of the physical dimension of the scattering amplitude, the factor $(2\ell + 1)$ is a matter of convention and will prove to be useful. The complex quantities $a_\ell(k)$ which are defined by (2.6) are called *partial wave amplitudes*. They are fuctions of the energy only (or, equivalently, of the wave number). The spherical harmonics are orthogonal and are normalized to 1. Therefore, the integrated cross section (2.5) is

$$\sigma_{\text{el}}(k) = \frac{4\pi}{k^2} \sum_{\ell,\ell'} \sqrt{(2\ell+1)(2\ell'+1)}\, a_\ell(k) a^*_{\ell'}(k') \int \mathrm{d}\Omega\, Y^*_{\ell' 0} Y_{\ell 0}$$

which, by the orthogonality of spherical harmonics, simplifies to

$$\boxed{\sigma_{\text{el}}(k) = \frac{4\pi}{k^2} \sum_{\ell=0}^{\infty} (2\ell+1) |a_\ell(k)|^2} . \tag{2.7}$$

Like the formulae (2.4) and (2.5) these expressions are completely general, and make no use of the underlying dynamics, i.e. in the case being studied here, of the Schrödinger equation with a central potential $U(r)$.

We now show that the amplitudes $a_\ell(k)$ are obtained by solving the radial equation (1.127) for all partial waves. The arguments presented in Sect. 1.9.3 and the comparison to the analogous classical situation show that this is not only an exact method for studying elastic scattering but that it is also particularly useful from a physical point of view. By assumption the potential has finite range, i.e. it fulfills the condition (2.1). In the corresponding *classical* situation a particle with a large value of angular momentum ℓ_{cl} stays further away from the origin $r = 0$ than a particle with a smaller value of ℓ_{cl}, and the action of the potential on it is correspondingly weaker. Very much like in classical mechanics the quantum effective potential

$$U_{\text{eff}}(r) = \frac{\hbar^2 \ell(\ell+1)}{2mr^2} + U(r) ,$$

for large values of ℓ, is dominated by the centrifugal term. Thus, one expects the amplitudes a_ℓ to decrease rapidly with increasing ℓ, so that the series (2.6) and (2.7) converge rapidly.

If the ℓ-th partial wave is taken to be

$$R_\ell(r) Y_{\ell m} = \frac{u_\ell(r)}{r} Y_{\ell m}$$

the radial function $u_\ell(r)$ obeys the differential equation

$$u_\ell''(r) - \left(\frac{2m}{\hbar^2}U_{\text{eff}}(r) - k^2\right)u_\ell(r) = 0\,, \tag{2.8}$$

(see also Sect. 1.9.5). At the origin, $r = 0$, the radial function must be regular. This means that we must choose the solution which behaves like $R_\ell \sim r^\ell$, or $u_\ell \sim r^{\ell+1}$, respectively, in the neighbourhood of the origin. For $r \to \infty$ the effective potential becomes negligible as compared to k^2 so that (2.8) simplifies to the approximate form

$$r \to \infty: \quad u_\ell''(r) + k^2 u_\ell(r) \approx 0\,.$$

Therefore, the asymptotic behaviour of the partial wave must be given by

$$r \to \infty: \quad u_\ell(r) \sim \sin\left(kr - \ell\frac{\pi}{2} + \delta_\ell(k)\right)\,. \tag{2.9}$$

The phase $\delta_\ell(k)$ which is defined by this equation is called the *scattering phase* in the partial wave with orbital angular momentum ℓ.

The following argument shows that the asymptotic form (2.9) is very natural: On the one hand, the function $u_\ell(r)$ is real (or can be chosen so) if the potential $U(r)$ is real. Under the same assumption the scattering phase must be real. On the other hand, the asymptotics of the force-free solution is known from the asymptotics (1.133) of spherical Bessel functions (which are regular at $r = 0$),

$$u_\ell^{(0)}(r) = (kr)j_\ell(kr) \sim \sin\left(kr - \ell\frac{\pi}{2}\right)\,.$$

If the potential $U(r)$ is identically zero all scattering phases are equal to zero. Therefore, the scattering phases "measure" to which extent the asymptotic, oscillatory behaviour of the radial function $u_\ell(r)$ is shifted relative to the force-free solution $u_\ell^{(0)}(r)$.

There remains the problem of expressing the scattering amplitude, or, equivalently, the amplitudes $a_\ell(k)$ in terms of the scattering phases. In solving this problem, the idea is to write the unknown scattering solution $\psi(\boldsymbol{x})$ of the Schrödinger equation in terms of a series in partial waves,

$$\psi(\boldsymbol{x}) = \sum_{\ell=0}^{\infty} c_\ell\, R_\ell(r)Y_{\ell 0}(\theta) = \frac{1}{r}\sum_{\ell=0}^{\infty} c_\ell\, u_\ell(r)Y_{\ell 0}(\theta) \tag{2.10}$$

and to choose this expansion such that the *in*coming spherical wave $\alpha_{\text{in}}\,\mathrm{e}^{-ikr}/r$ which is contained in it for $r \to \infty$, coincides with the incoming spherical wave contained in the condition (2.2). We begin with the latter: Expanding the plane wave in terms of spherical harmonics (cf. (1.136)),

$$\mathrm{e}^{ikx^3} = \sum_{\ell=0}^{\infty} \mathrm{i}^\ell\sqrt{4\pi(2\ell+1)}\,j_\ell(kr)Y_{\ell 0}\,,$$

and making use of the asymptotics (1.133) of the spherical Bessel functions,

$$j_\ell(kr) \sim \frac{1}{kr}\sin\left(kr - \ell\frac{\pi}{2}\right) = \frac{1}{2ikr}(e^{ikr}e^{-i\ell\pi/2} - e^{-ikr}e^{i\ell\pi/2}),$$

the piece proportional to e^{-ikr}/r can be read off. The *out*going spherical wave, on the other hand, in (2.2), in addition to the term proportional to the scattering amplitude $f(\theta)$, also contains a piece of the plane wave that can be read off from the same expression. This is to say that the Sommerfeld condition (2.2) is rewritten in terms of spherical waves as follows:

$$\psi_{\text{Somm}}(\boldsymbol{x}) \sim -\frac{e^{-ikr}}{2ikr}\left(\sum_{\ell=0}^{\infty} i^\ell\sqrt{4\pi(2\ell+1)}\,e^{i\ell\pi/2}Y_{\ell 0}\right)$$
$$+\frac{e^{ikr}}{2ikr}\left(\sum_{\ell=0}^{\infty} i^\ell\sqrt{4\pi(2\ell+1)}\,e^{-i\ell\pi/2}Y_{\ell 0} + 2ikf(\theta)\right).$$

This is to be compared to the representation (2.10) for the scattering solution for $r \to \infty$ which becomes, upon inserting (2.9) once more,

$$\psi(\boldsymbol{x}) \sim -\frac{e^{-ikr}}{2ir}\left(\sum_{\ell=0}^{\infty} c_\ell e^{-i\delta_\ell}e^{i\ell\pi/2}Y_{\ell 0}\right) + \frac{e^{ikr}}{2ir}\left(\sum_{\ell=0}^{\infty} c_\ell e^{i\delta_\ell}e^{-i\ell\pi/2}Y_{\ell 0}\right).$$

The *in*coming spherical waves in ψ_{Somm} and in ψ are equal provided the coefficients c_ℓ are chosen to be

$$c_\ell = \frac{i^\ell}{k}\sqrt{4\pi(2\ell+1)}\,e^{i\delta_\ell}.$$

For the rest of the calculation one just has to compare these formulae. Inserting the result for c_ℓ into the *out*going part of (2.10), one finds

$$f(\theta) = \frac{1}{k}\sum_{\ell=0}^{\infty}\frac{e^{2i\delta_\ell}-1}{2i}\sqrt{4\pi(2\ell+1)}Y_{\ell 0}$$
$$= \frac{1}{k}\sum_{\ell=0}^{\infty}\frac{e^{2i\delta_\ell}-1}{2i}(2\ell+1)P_\ell(\cos\theta),$$

where use is made of the relation

$$Y_{\ell 0}(\theta) = \sqrt{\frac{2\ell+1}{4\pi}}P_\ell(\cos\theta).$$

Comparison with the general expansion (2.6) yields the following *exact* expression for the amplitudes $a_\ell(k)$ as functions of the scattering phases,

$$\boxed{a_\ell(k) = \frac{e^{2i\delta_\ell}-1}{2i} = e^{i\delta_\ell(k)}\sin\delta_\ell(k)}. \tag{2.11}$$

Applications and Remarks

1. It was indeed useful to define the amplitudes a_ℓ by extracting an explicit factor $(2\ell+1)$ in (2.6). The so-defined amplitudes then have moduli which are smaller than or equal to 1. The second equation in (2.11) is correct because the phase δ_ℓ is real.[3]
2. The result (2.11) for the partial wave amplitudes which follows from the Schrödinger equation, has a remarkable property: The imaginary part of a_ℓ is positive-semidefinite

$$\mathrm{Im}\, a_\ell(k) = \sin^2\delta_\ell \geq 0\,.$$

Calculating the elastic scattering amplitude (2.6) in the forward direction $\theta = 0$, where $P_\ell(z = 1) = 1$, its imaginary part is seen to be

$$\mathrm{Im}\, f(0) = \frac{1}{k}\sum_{\ell=0}^{\infty}(2\ell+1)\,\mathrm{Im}\, a_\ell(k) = \frac{1}{k}\sum_{\ell=0}^{\infty}(2\ell+1)\sin^2\delta_\ell(k)\,.$$

In turn, the integrated elastic cross section (2.7) is found to be

$$\sigma_{\mathrm{el}}(k) = \frac{4\pi}{k^2}\sum_{\ell=0}^{\infty}(2\ell+1)\sin^2\delta_\ell(k)\,.$$

In the case of a real potential there is elastic scattering only, the integrated cross section (2.7) is identical with the total cross section. Comparison of the results just obtained yields an important relation between the imaginary part of the elastic scattering amplitude in the forward direction and the total cross section, viz.

$$\boxed{\sigma_{\mathrm{tot}} = \frac{4\pi}{k}\,\mathrm{Im}\, f(0)}\,. \tag{2.12}$$

This relationship is called *the optical theorem*.[4] Loosely speaking it is a consequence of the conservation of (Born) probability.
3. As we will see in Sect. 2.6 and in Chap. 8, the optical theorem also holds in more general situations. When two particles, A and B, are scattered on one another, besides elastic scattering $A + B \longrightarrow A + B$, there will in general be inelastic processes as well, in which one of them, or both, are left in excited states, $A + B \longrightarrow A + B^*$, $A + B \longrightarrow A^* + B^*$, or where further particles are created, $A + B \longrightarrow A + B + C + \cdots$. As a short-hand allowed final states are denoted by "n". The optical theorem then relates the imaginary part of the

[3] This remark is important because the same analysis can be applied to the case of a complex, absorptive potential. In this case the scattering phases are complex functions. The first part of the formula (2.11) still applies, the second part does not.

[4] Indeed, this term was coined in classical optics. It was known, in a different context, before quantum mechanics was developed.

elastic forward scattering amplitude at a given value of the energy, to the *total* cross section at this energy,

$$\sigma_{\text{tot}} = \sum_n \sigma(A + B \longrightarrow n) .$$

The more general form of the optical theorem (2.12) is

$$\sigma_{\text{tot}}(k) = \frac{4\pi}{k} \operatorname{Im} f_{\text{el}}(k, \theta = 0) ,$$

the quantity k being the modulus of the momentum $\boldsymbol{k}^*$ in the center-of-mass system.

2.3.1 How to Calculate Scattering Phases

The asymptotic condition (2.9) can be interpreted in still a different way that, by the same token, gives a hint at possibilities of obtaining the scattering phases. As the potential has finite range the interval of definition of the radial variable r splits into an inner domain where the (true) potential $U(r)$ is different from zero, and an outer domain where either it vanishes or becomes negligibly small, and where only the centrifugal potential is active. Thus, in the outer domain every solution $u_\ell(r)$ is a linear combination of two fundamental solutions of the force-free case. For instance, these may be a spherical Bessel function $j_\ell(kr)$ and a spherical Neumann function $n_\ell(kr)$, so that

$$u_\ell(k, r) = (kr)\big[j_\ell(kr)\alpha_\ell(k) + n_\ell(kr)\beta_\ell(k)\big] ,$$
$$(\text{for } r \text{ such that } U(r) \approx 0) .$$

One now takes this formula to large values of r, and compares the asymptotic behaviour (2.9) with the asymptotics of the spherical Bessel and Neumann functions (1.133) and (1.140), respectively. Using the addition formula for Sines

$$\sin\left(kr - \ell\frac{\pi}{2} + \delta_\ell(k)\right)$$
$$= \sin\left(kr - \ell\frac{\pi}{2}\right)\cos\delta_\ell(k) + \cos\left(kr - \ell\frac{\pi}{2}\right)\sin\delta_\ell(k)$$

then yields an equation for the scattering phase which is

$$\tan\delta_\ell(k) = \frac{\beta_\ell}{\alpha_\ell} . \tag{2.13}$$

This shows that the differential equation for the radial function must be solved only in the inner domain of the variable r: One determines the solution regular at $r = 0$, e.g. by numerical integration on a computer, and follows this solution to the boundary of the outer domain. At this point one writes it as a linear combination of j_ℓ and n_ℓ, and reads off the coefficients α_ℓ and β_ℓ whose ratio (2.13) yields the scattering phase in the interval $[0, \pi/2]$.

Here are some examples of potentials for which the reader may wish to determine the scattering phases:

1. The spherically symmetric potential well

$$U(r) = U_0 \Theta(r_0 - r)\,; \tag{2.14}$$

2. The electrostatic potential

$$U(r) = -4\pi Q \left(\frac{1}{r} \int_0^r \mathrm{d}r' \, \varrho(r') r'^2 + \int_r^\infty \mathrm{d}r' \varrho(r') r' \right) , \tag{2.15}$$

which is obtained from the charge distribution

$$\varrho(r) = N \frac{1}{1 + \exp[(r - c)/z]} \tag{2.16}$$

with normalization factor

$$N = \frac{3}{4\pi c^3} \left[1 + \left(\frac{\pi z}{c} \right)^2 - 6 \left(\frac{z}{c} \right)^3 \sum_{n=1}^{\infty} \frac{(-)^n}{n^3} \mathrm{e}^{-nc/z} \right]^{-1} .$$

The distribution (2.16) is often used in the description of nuclear charge densities. The parameter c characterizes the radial extension while the parameter z characterizes the surface region. Its specific functional form is also known from statistical mechanics which is the reason why it is usually called *Fermi distribution*. Figure 2.1 shows an example applicable to realistic nuclei where z is very small as compared to c.

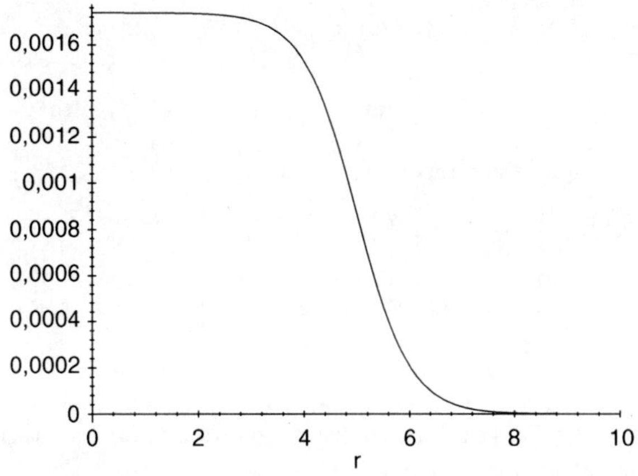

Fig. 2.1 Illustration of the model (2.16) for a normalized charge distribution. In the case shown here, $c \gg z$. The example shows the distribution with $c = 5$ fm, $z = 0.5$ fm. The parameter c is the distance from the origin to the radius where the function $\varrho(r)$ has dropped to half its value at $r = 0$. The two points where it assumes 90 and 10 % of its value in $r = 0$, respectively, are situated at $r_{90} = c - 2z \ln 3$ and at $r_{10} = c + 2z \ln 3$, respectively. They are separated by the approximate distance $t \equiv 4z \ln 3 \approx 4.394z$

3. The Yukawa potential, that we mentioned in the introduction to Sect. 2.2,
$$U_Y(r) = g\frac{e^{-r/r_0}}{r}\,. \tag{2.17}$$
We will show later that this potential describes the interaction of two particles which can exchange a scalar particle of mass $M = \hbar/(r_0 c)$ (c denotes the speed of light). The name is due to H. Yukawa who had postulated that the *strong* interactions of nucleons were due to the exchange of particles with spin 0, the π-mesons, long before these particles were actually discovered. The length $r_0 = \hbar c/(Mc^2)$ is interpreted as the range of the potential. It is the Compton wave length of particles of mass M. These remarks emphasize that the example (2.17) may have a deeper significance for physics than the pure model potentials (2.14), (2.15) and (2.16).

Consider now two potentials of finite range, $U^{(1)}$ and $U^{(2)}$, and compare their scattering phases. For equal values of the energy the corresponding radial functions obey the radial differential equations
$$u_\ell^{(j)\prime\prime}(r) - \left[\frac{2m}{\hbar^2}U_{\text{eff}}^{(j)}(r) - k^2\right]u_\ell^{(j)}(r) = 0\,, \qquad j = 1, 2\,,$$
where the effective potentials differ only by the true, dynamical potentials
$$U_{\text{eff}}^{(2)}(r) - U_{\text{eff}}^{(1)}(r) = U^{(2)}(r) - U^{(1)}(r)\,.$$
Both radial functions are assumed to be regular at $r = 0$. As we are factoring $1/r$, this means that $u_\ell^{(j)}(0) = 0$ ($j = 1, 2$). Compute then the following derivative, making use of the differential equations for $u_\ell^{(1)}$ and $u_\ell^{(2)}$,
$$\begin{aligned}\frac{\mathrm{d}}{\mathrm{d}r}\left(u_\ell^{(1)}u_\ell^{(2)\prime} - u_\ell^{(2)}u_\ell^{(1)\prime}\right) &= u_\ell^{(1)}u_\ell^{(2)\prime\prime} - u_\ell^{(2)}u_\ell^{(1)\prime\prime} \\ &= \frac{2m}{\hbar^2}\left(U^{(2)} - U^{(1)}\right)u_\ell^{(1)}u_\ell^{(2)}\,.\end{aligned}$$
Taking the integral over the interval $[0, r]$, one has
$$\begin{aligned}&u_\ell^{(1)}(r)u_\ell^{(2)\prime}(r) - u_\ell^{(2)}(r)u_\ell^{(1)\prime}(r) \\ &\quad = \frac{2m}{\hbar^2}\int_0^r \mathrm{d}r'\,\left(U^{(2)}(r') - U^{(1)}(r')\right)u_\ell^{(1)}(r')u_\ell^{(2)}(r')\,.\end{aligned}$$
In the limit of r going to infinity, and using the asymptotic form (2.9) as well as its derivative on the left-hand side of this equation, one obtains an integral representation for the difference of scattering phases:
$$k\sin(\delta_\ell^{(1)} - \delta_\ell^{(2)}) = \frac{2m}{\hbar^2}\int_0^\infty \mathrm{d}r\,\left(U^{(2)}(r) - U^{(1)}(r)\right)u_\ell^{(1)}(r)u_\ell^{(2)}(r)\,. \tag{2.18}$$

This formula is useful for testing the sensitivity of high, intermediate, and low partial waves to the potential, by letting $U^{(1)}$ and $U^{(2)}$ differ but little. Instead of the scattering phases themselves one calculates the change in any given partial wave as a function of the change in the potential.

Alternatively one may consider a situation where $U^{(2)}$ vanishes identically, i.e. where the corresponding radial functions are proportional to spherical Bessel functions,

$$u_\ell^{(2)}(r) = (kr) j_\ell(kr) .$$

With $\delta_\ell^{(1)} \equiv \delta_\ell$ and $\delta_\ell^{(2)} = 0$ the integral representation reduces to

$$\sin(\delta_\ell) = -\frac{2m}{\hbar^2} \int_0^\infty r \mathrm{d}r\, U(r) u_\ell(r) j_\ell(kr) . \tag{2.19}$$

The Yukawa potential (2.17) may serve as an illustration of this formula. For the sake of simplicity we assume this potential to be weak enough so that the corresponding radial function can be approximated by the force-free solution,[5]

$$u_\ell^{(Y)}(r) \approx (kr) j_\ell(kr) .$$

With this assumption one has

$$\begin{aligned}
\sin \delta_\ell^{(Y)} &\approx -\frac{2mk}{\hbar^2} \int_0^\infty r^2 \mathrm{d}r\, U_Y(r) j_\ell^2(kr) \\
&= -\frac{2mkg}{\hbar^2} \int_0^\infty r \mathrm{d}r\, \mathrm{e}^{-r/r_0} j_\ell^2(kr) \\
&= -\frac{mg\pi}{\hbar^2 k} \int_0^\infty \mathrm{d}\varrho\, \mathrm{e}^{-\varrho/(kr_0)} J_{\ell+1/2}^2(\varrho) \\
&= -\frac{mg}{\hbar^2 k} Q_\ell \left(1 + \frac{1}{2(kr_0)^2}\right) .
\end{aligned}$$

In this derivation the variable $\varrho = kr$ is introduced, and the spherical Bessel function is written in the standard form of Bessel functions found in monographs on special functions, viz.

$$j_\ell(\varrho) = \left(\frac{\pi}{2\varrho}\right)^{1/2} J_{\ell+1/2}(\varrho) .$$

[5] This approximation is nothing but the first Born approximation that is studied in more detail in Sect. 2.4.1.

The last step makes use of a known definite integral which is found, e.g. in [Gradshteyn and Ryzhik (1965)], Eq. 6.612.3. Here, Q_ℓ is a Legendre function of second kind whose properties are well known.[6] These functions are known to decrease rapidly both for increasing ℓ and for increasing values of the argument ≥ 1, cf., e.g., [Abramowitz and Stegun (1965)], Fig. 8.5.

2.3.2 Potentials with Infinite Range: Coulomb Potential

The Coulomb potential violates the condition (2.1). This means that its influence is still felt when the particle moves at very large distances from the scattering center. This is seen very clearly in the asymptotics of the partial waves, derived in Sect. 1.9.5, which in addition to the constant phase σ_ℓ, contain an r-dependent, logarithmic phase $-\gamma \ln(2kr)$ multiplied by the factor (1.156),

$$\gamma = \frac{ZZ'e^2 m}{\hbar^2 k} . \tag{2.20}$$

The scattering solutions of the Coulomb potential certainly do not obey the radiation condition (2.2). Both the outgoing spherical wave and the plane wave are modified due to the long range of the potential, and the formulae of partial wave analysis derived above, cannot be applied directly. A similar statement holds for any spherically symmetric potential which, though different from the $(1/r)$-behaviour in the inner domain, approaches the Coulomb potential for large values of r. An example is provided by the electrostatic potential corresponding to the charge distribution (2.16) in Example 2 of Sect. 2.3.1. In order to solve this new problem we proceed in two steps:

In the first step we show that the condition (2.2) is modified to

$$r \to \infty : \quad \psi \sim e^{i\{kx^3 + \gamma \ln[2kr \sin^2(\theta/2)]\}} + f_C(\theta) \frac{e^{i[kr - \gamma \ln(2kr)]}}{r} , \tag{2.21}$$

with r-dependent, logarithmic phases both in the incoming wave and in the outgoing spherical wave, and calculate the scattering amplitude $f_C(\theta)$ for the pure Coulomb potential.

In the second step we study spherically symmetric potentials which deviate from the $(1/r)$-form in the inner region but decrease like $1/r$ in the outer region, in other words, which approach the Coulomb potential for $r \to \infty$. In this case it is sufficient to calculate the *shift* of the scattering phases relative to their values in the pure Coulomb potential, and not relative to the force-free case.

Step 1: Although we already know the scattering phases for the Coulomb potential from Sect. 1.9.5 it is instructive to calculate the scattering amplitude directly, using

[6] The function $Q_\ell(z)$ and the Legendre polynomial $P_\ell(z)$ form a fundamental system of solutions of the differential equation (1.112) with $\lambda = \ell(\ell+1)$ and $m = 0$. In contrast to $P_\ell(z)$ the function $Q_\ell(z)$ is singular in $z = 1$, the singularity being a branch point. For all values $|z| > 1$ of the argument $Q_\ell(z)$ is a one-valued function.

a somewhat different method. Indeed, the (nonrelativistic) Schrödinger equation can be solved exactly in a way adapted to the specific scattering situation at hand.[7] With $k^2 = 2mE/\hbar^2$, $U(r) = ZZ'e^2/r$, and with the definition (2.20) for γ, the stationary Schrödinger equation (1.160) reads

$$\left(\Delta + k^2 - \frac{2\gamma k}{r}\right)\psi(\boldsymbol{x}) = 0\,. \tag{2.22}$$

This is solved using parabolic coordinates

$$\xi = \sqrt{r - x^3}\,, \qquad \eta = \sqrt{r + x^3}\,, \qquad \phi$$

and by means of the ansatz

$$\psi(\boldsymbol{x}) = c_\psi e^{ikx^3} f(r - x^3) = c_\psi e^{ik(\eta^2-\xi^2)/2} f(\xi^2)\,,$$

in which c_ψ is a complex number still to be determined. As before, the direction of the incoming, asymptotic momentum is taken to be the 3-direction. Since no other, perpendicular direction is singled out in the *in*-state and since the potential is spherically symmetric, the scattering amplitude does not depend on ϕ. Surprisingly, the differential equation (2.22) separates in these coordinates, too. The variable $\xi^2 = r - x^3$ is denoted by u, first and second derivatives with respect to that variable are written f' and f'', respectively. Then, for $i = 1, 2$:

$$\frac{\partial\psi}{\partial x^i} = e^{ikx^3} f'(u)\frac{\partial r}{\partial x^i} = e^{ikx^3} f'(u)\frac{x^i}{r}\,, \quad \frac{\partial^2\psi}{\partial(x^i)^2}$$
$$= e^{ikx^3}\left[f''\frac{(x^i)^2}{r^2} + f'\left(\frac{1}{r} - \frac{(x^i)^2}{r^3}\right)\right].$$

The derivatives with respect to x^3 give

$$\frac{\partial\psi}{\partial x^3} = e^{ikx^3}\left[ikf + f'(u)\left(\frac{x^3}{r} - 1\right)\right], \quad \frac{\partial^2\psi}{\partial(x^3)^2}$$
$$= e^{ikx^3}\left[-k^2 f + 2ikf'\left(\frac{x^3}{r} - 1\right) + f''\left(\frac{x^3}{r} - 1\right)^2 + f'\left(\frac{1}{r} - \frac{(x^3)^2}{r^3}\right)\right].$$

Inserting these formulae in (2.22) one obtains

$$\left(u\frac{d^2}{du^2} + (1 - iku)\frac{d}{du} - \gamma k\right) f(u) = 0\,.$$

This differential equation is again of Fuchsian type and appears to be very close to Kummer's equation (1.145). The identification becomes perfect if one replaces the variable u by the variable $v := iku$. Indeed, the differential equation then becomes

$$v\frac{d^2 f(v)}{dv^2} + (1 - v)\frac{d f(v)}{dv} + i\gamma f(v) = 0\,.$$

[7] This no longer holds true when the relativistic form of the wave equation is used.

The solution which is regular at $r = 0$ is

$$f(v) = c_\psi\, {}_1F_1(-\mathrm{i}\gamma\,;1\,;v) = c_\psi\, {}_1F_1[-\mathrm{i}\gamma\,;1\,;\mathrm{i}k(r-x^3)]\,.$$

The asymptotics of the confluent hypergeometric function is obtained from the formula (1.147),

$${}_1F_1 \sim \frac{1}{\Gamma(1+\mathrm{i}\gamma)}\mathrm{e}^{\pi\gamma}[\mathrm{i}k(r-x^3)]^{\mathrm{i}\gamma} + \frac{1}{\Gamma(-\mathrm{i}\gamma)}\mathrm{e}^{\mathrm{i}k(r-x^3)}[\mathrm{i}k(r-x^3)]^{-\mathrm{i}\gamma+1}\,.$$

Setting

$$\mathrm{i}^{\mathrm{i}\gamma} = \mathrm{e}^{-\pi\gamma/2}$$

and choosing the coefficient c_ψ as follows

$$c_\psi = \Gamma(1+\mathrm{i}\gamma)\mathrm{e}^{-\pi\gamma/2}\,,$$

the solution ψ takes the asymptotic form postulated above

$$\psi \sim \mathrm{e}^{\mathrm{i}\{kx^3+\gamma\ln[k(r-x^3)]\}} - \frac{\Gamma(1+\mathrm{i}\gamma)}{\Gamma(1-\mathrm{i}\gamma)}\frac{\gamma}{k(r-x^3)}\mathrm{e}^{\mathrm{i}\{kr-\gamma\ln[k(r-x^3)]\}}\,.$$

Inserting $r - x^3 = r(1-\cos\theta) = 2r\sin^2(\theta/2)$ shows that this is the asymptotic decomposition, (2.21), into an incoming, but deformed plane wave, and an outgoing, deformed spherical wave. The complex Γ-function is written in terms of modulus and phase,

$$\Gamma(1\pm\mathrm{i}\gamma) = |\Gamma(1+\mathrm{i}\gamma)|\,\mathrm{e}^{\pm\mathrm{i}\sigma_\mathrm{C}}\,,$$

so that the scattering amplitude for the pure Coulomb potential is seen to be

$$f_\mathrm{C}(\theta) = -\frac{\gamma}{2k\sin^2(\theta/2)}\mathrm{e}^{\mathrm{i}\{2\sigma_\mathrm{C}-\gamma\ln[\sin^2(\theta/2)]\}}\,. \qquad (2.23)$$

This amplitude contains a phase factor that depends on the scattering angle, and which is characteristic for the long range of the Coulomb potential. This phase factor drops out of the differential cross section (2.4) for which one obtains

$$\frac{\mathrm{d}\sigma_\mathrm{el}}{\mathrm{d}\Omega} = \frac{\gamma^2}{4k^2}\frac{1}{\sin^4(\theta/2)} = \left(\frac{ZZ'\mathrm{e}^2}{4E}\right)^2\frac{1}{\sin^4(\theta/2)}\,, \qquad (2.24)$$

where the definition (2.20) and $E = \hbar^2k^2/(2m)$ were used. The result (2.24) is called the *Rutherford cross section*. Note that it agrees with the corresponding expression of classical mechanics (cf. [Scheck (2010)], Sect. 1.27). This important formula was essential in analyzing the scattering experiments of α particles on nuclei, performed by Rutherford, Geiger, and Marsden from 1906 on. These experiments proved that nuclei are practically point-like as compared to typical radii of atoms.

Step 2: Consider now a spherically symmetric charge distribution which is no longer concentrated in a point but is localized in the sense that it lies inside a sphere with a given, finite radius R. One calculates the electrostatic potential by means of the formula (2.15) and notes that for values $r > R$ it coincides, either completely or to a very good approximation, with the pure Coulomb potential but deviates from it for values $r < R$. It should be clear immediately, in the light of the general discussion of

Sect. 2.3, that high partial waves are insensitive to these deviations. The information on the precise shape of the charge distribution is contained in the low and intermediate partial waves. This suggests to design the partial wave analysis for these cases such that it is not the force-free case but the Coulomb potential which is taken as the reference potential. This means that the phase shift analysis should be designed such as to yield the difference

$$\delta_\ell = \delta_{U(r)} - \delta_C$$

between the true phase and the Coulomb phase.

2.4 Born Series and Born Approximation

We emphasize again that the expansion in terms of partial waves is an *exact* method to calculate the cross section for spherically symmetric potentials which, in addition, has the advantage of making optimal use of the information about the range of the potential. In the case of potentials which are not spherically symmetric but may be expanded in terms of spherical harmonics the cross section can also be computed by expanding the scattering amplitude in partial waves. However, this method becomes technically complex and cumbersome, and looses much of the simplicity and transparency it has for spherically symmetric potentials.

The Born series that we describe in this section, does not have this disadvantage. It yields an exact, though formal, solution of the scattering problem by means of the technique of Green functions, and can equally well be applied to potentials with or without spherical symmetry. Its most stringent disadvantage is the fact that beyond first order it is not very practicable and becomes cumbersome. The first iteration, or *first Born approximation*, in turn, is easy to calculate and allows for simple and convincing physical interpretation, but it violates the optical theorem.

The starting point is again the stationary Schrödinger equation (1.160) in the form

$$(\Delta + k^2)\psi(\boldsymbol{x}) = \frac{2m}{\hbar^2}U(\boldsymbol{x})\psi(\boldsymbol{x}), \tag{2.25}$$

where $k^2 = 2mE/\hbar^2$. If one deals with a two-body problem the parameter m is the reduced mass; if one studies scattering of a single particle on a fixed external potential then m is just the mass of that particle.[8] The differential equation (2.25) is solved by means of Green functions, i.e. of functions (more precisely: distributions) $G(\boldsymbol{x}, \boldsymbol{x}')$, which obey the differential equation

$$(\Delta + k^2)G(\boldsymbol{x}, \boldsymbol{x}') = \delta(\boldsymbol{x} - \boldsymbol{x}').$$

[8] The latter case can also be viewed as the limit of the former in which the mass of the heavier partner is very large as compared to the one of the lighter partner.

The well-known relation

$$(\Delta + k^2)\frac{e^{\pm ik|z|}}{|z|} = -4\pi\,\delta(z)$$

shows that the general solution can be given in the form

$$G(\boldsymbol{x}, \boldsymbol{x}') = -\frac{1}{4\pi}\frac{1}{|\boldsymbol{x}-\boldsymbol{x}'|}\left[a e^{ik|\boldsymbol{x}-\boldsymbol{x}'|} + (1-a)e^{-ik|\boldsymbol{x}-\boldsymbol{x}'|}\right].$$

Formally, the differential equation (2.25) then has the solution

$$\psi_{\boldsymbol{k}}(\boldsymbol{x}) = e^{i\boldsymbol{k}\cdot\boldsymbol{x}} + \frac{2m}{\hbar^2}\int d^3x' G(\boldsymbol{x}, \boldsymbol{x}')U(\boldsymbol{x}')\psi_{\boldsymbol{k}}(\boldsymbol{x}') . \tag{2.26}$$

It is formal because the differential equation (2.25) is replaced by an *integral equation* which contains the unknown wave functions both on the left-hand side and in the integrand of the right-hand side. Nevertheless, it has two essential advantages: The constant a in the Green function can be chosen such that the scattering solution fulfills the right boundary condition, which in our case is the Sommerfeld radiation condition (2.2). Furthermore, if the strength of the potential is small in some sense, this integral equation can be used as the starting basis for an iterative solution, i.e. an expansion of the scattering function around the force-free solution (the plane wave).

The correct asymptotics (2.2) is reached with the choice $a = 1$. This is seen as follows: Define $r := |\boldsymbol{x}|$, $r' := |\boldsymbol{x}'|$, and assume the potential $U(\boldsymbol{x}')$ to be localized. As r goes to infinity one has

$$r \gg r' : \quad |\boldsymbol{x}-\boldsymbol{x}'| = \sqrt{r^2 + r'^2 - 2\boldsymbol{x}\cdot\boldsymbol{x}'} \approx r - \frac{1}{r}\boldsymbol{x}\cdot\boldsymbol{x}' .$$

The scattering function takes the asymptotic form

$$r \to \infty : \quad \psi_{\boldsymbol{k}}(\boldsymbol{x}) \sim e^{i\boldsymbol{k}\cdot\boldsymbol{x}} - \frac{2m}{4\pi\hbar^2}\frac{e^{ikr}}{r}\int d^3x' e^{-i\boldsymbol{k}'\boldsymbol{x}'}U(\boldsymbol{x}')\psi_{\boldsymbol{k}}(\boldsymbol{x}') .$$

The reader will have noticed that we defined $k\boldsymbol{x}/r =: \boldsymbol{k}'$ in this expression. Indeed, the momentum of the scattered particle is $\hbar\boldsymbol{k}'$. It moves in the direction of $\boldsymbol{x}/r$ and, because the scattering is elastic, one has $|\boldsymbol{k}'| = |\boldsymbol{k}|$.

By the same token, this result yields a general formula for the scattering amplitude, viz.

$$\boxed{f(\theta, \phi) = -\frac{2m}{4\pi\hbar^2}\int d^3x\, e^{-i\boldsymbol{k}'\cdot\boldsymbol{x}}U(\boldsymbol{x})\psi_{\boldsymbol{k}}(\boldsymbol{x})} . \tag{2.27}$$

(As this equation no longer contains the point of reference $\boldsymbol{x}$, the integration variable $\boldsymbol{x}'$ was renamed $\boldsymbol{x}$.)

This equation is an interesting result. If the potential has a strictly finite range we need to know the exact scattering function only in the domain where $U(\boldsymbol{x})$ is sizeably different from zero.[9]

---

[9] An approximation which makes use of this fact and which is particularly useful for scattering at high energies, is provided by the *eikonal expansion*. The reader will find an extended description of this method in, e.g., [Scheck (2012)], Chap. 5, and illustrated by explicit examples.

If one knew the exact scattering solution this formula would yield the *exact* scattering amplitude. Although this ambitious goal cannot be reached, the formula serves as a basis for approximation methods which are relevant for various kinematic conditions. One of these is the *Born series* which is obtained from an iterating solution of the integral equation (2.26). The idea is simple: one imagines the potential as a perturbation of the force-free solution

$$\psi_{\boldsymbol{k}}^{(0)} = e^{i\boldsymbol{k}\cdot\boldsymbol{x}}$$

such that in a decomposition of the full wave function

$$\psi_{\boldsymbol{k}}(\boldsymbol{x}) = \sum_{n=0}^{\infty} \psi_{\boldsymbol{k}}^{(n)}(\boldsymbol{x})$$

the n-th term is obtained from the $(n-1)$-st by means of the integral equation (2.26), i.e.

$$\psi_{\boldsymbol{k}}^{(n)}(\boldsymbol{x}) = -\frac{1}{4\pi}\frac{2m}{\hbar^2}\int d^3x' \frac{e^{ik|\boldsymbol{x}-\boldsymbol{x}'|}}{|\boldsymbol{x}-\boldsymbol{x}'|} U(\boldsymbol{x}')\psi_{\boldsymbol{k}}^{(n-1)}(\boldsymbol{x}'), \qquad n \geq 1 \tag{2.28}$$

Even without touching the (difficult) question of its convergence, one realizes at once that this provides a method of representing the scattering amplitude as a series whose structure is very different from the expansion in terms of partial waves. While in the latter one expands in increasing values of ℓ, the former is an expansion in the strength of the potential.

2.4.1 First Born Approximation

It is primarily the first and simplest approximation which matters for practical applications. It consists in truncating the series (2.28) at $n = 1$. In this case the full scattering function in the integrand of the right-hand side of (2.27) is replaced by $\psi_{\boldsymbol{k}}^{(0)} = \exp(i\boldsymbol{k}\cdot\boldsymbol{x})$ so that one obtains

$$f^{(1)}(\theta,\phi) = -\frac{2m}{4\pi\hbar^2}\int d^3x\, e^{-i\boldsymbol{k}'\cdot\boldsymbol{x}} U(\boldsymbol{x}) e^{i\boldsymbol{k}\cdot\boldsymbol{x}} .$$

Introducing the momentum transfer

$$\boldsymbol{q} := \boldsymbol{k} - \boldsymbol{k}' \quad \text{with} \quad |\boldsymbol{k}| = |\boldsymbol{k}'| = k\,, \quad \hat{\boldsymbol{q}} = (\theta,\phi)\,,$$

the *first Born approximation* for the scattering amplitude reads

$$\boxed{f^{(1)}(\boldsymbol{q}) = -\frac{2m}{4\pi\hbar^2}\int d^3x\, e^{i\boldsymbol{q}\cdot\boldsymbol{x}} U(\boldsymbol{x})} . \tag{2.29}$$

This formula tells us that in first Born approximation the scattering amplitude is the Fourier transform of the potential with respect to the variable $\boldsymbol{q}$.

The formula (2.29) simplifies further if the potential has spherical symmetry, $U(\boldsymbol{x}) \equiv U(r)$. One inserts the expansion (1.136) of $\exp(i\boldsymbol{q}\cdot\boldsymbol{x})$ in terms of spherical harmonics and notes that by integrating over $\mathrm{d}\Omega_x$ only the term with $\ell = 0$ survives. This follows from the fact that $Y_{00} = 1/\sqrt{4\pi}$ is a constant and that

$$\int \mathrm{d}\Omega_x \, Y_{\ell m}(\hat{\boldsymbol{x}}) = \sqrt{4\pi}\,\delta_{\ell 0}\,\delta_{m0}\,.$$

Thus, one obtains the expression

$$f^{(1)}(\theta) = -\frac{2m}{\hbar^2}\int_0^\infty r^2\mathrm{d}r\, U(r)\, j_0(qr) \tag{2.30}$$

where

$$q \equiv |\boldsymbol{q}| = 2k\sin(\theta/2)\,, \quad \text{and where} \quad j_0(u) = \frac{\sin u}{u}$$

is the spherical Bessel function with $\ell = 0$ (see Sect. 1.9.3). The functional dependence of the scattering amplitude could be written as $f(q)$, or, even more precisely, $f(q^2)$ because the result (2.30) shows that it depends only on the modulus of $\boldsymbol{q}$ and is invariant under the exchange $q \to -q$. As an alternative, one may express q by the scattering angle θ and write the scattering amplitude in the form

$$f^{(1)}(\theta) = -\frac{m}{\hbar^2 k\sin(\theta/2)}\int_0^\infty r\mathrm{d}r\, U(r)\sin[2kr\sin(\theta/2)]\,.$$

We illustrate this result by the following example.

Example 2.1

Let us return to the Yukawa potential (2.17),

$$U_Y(r) = g\frac{\mathrm{e}^{-\mu r}}{r}\,, \quad \text{where} \quad \mu = \frac{1}{r_0}\,.$$

The following integral is obtained in an elementary way

$$\int_0^\infty \mathrm{d}r \mathrm{e}^{-\mu r}\sin(\alpha r) = \frac{\alpha}{\mu^2+\alpha^2}\,.$$

With $\alpha = 2k\sin(\theta/2)$ the formula (2.30) yields

$$f_Y^{(1)}(\theta) = -\frac{2mg}{\hbar^2}\frac{1}{4k^2\sin^2(\theta/2)+\mu^2}\,. \tag{2.31}$$

Two properties of the result (2.31) should be noticed

1. In the limit $\mu \to 0$ (though not allowed), and with $g = ZZ'e^2$ and $\hbar^2k^2 = 2mE$, the amplitude becomes

$$f_C^{(1)}(\theta) = -\frac{ZZ'e^2}{4E}\frac{1}{\sin^2(\theta/2)}\,.$$

This is the scattering amplitude for the pure Coulomb potential, except for the phase factor in (2.23). Its absolute square gives the correct expression (2.24) of the differential cross section.

2. There is a well-known expansion of $1/(z-t)$ in terms of Legendre polynomials and Legendre functions of the second kind, (see [Gradshteyn and Ryzhik (1965)], Eq. 8.791.1)

$$\frac{1}{z-t} = \sum_{\ell=0}^{\infty} (2\ell+1) Q_\ell(z) P_\ell(t) \,.$$

Write the amplitude (2.31) as

$$\begin{aligned} f_Y^{(1)}(\theta) &= -\frac{mg}{\hbar^2 k^2} \frac{1}{2\sin^2(\theta/2) + \mu^2/(2k^2)} \\ &= -\frac{mg}{\hbar^2 k^2} \frac{1}{1 + \mu^2/(2k^2) - \cos\theta} \,, \end{aligned}$$

set $1 + \mu^2/(2k^2) = z$ and $\cos\theta = t$. This yields

$$f_Y^{(1)}(\theta) = -\frac{mg}{\hbar^2 k^2} \sum_{\ell=0}^{\infty} Q_\ell \left(1 + \frac{\mu^2}{2k^2}\right) P_\ell(\cos\theta) \,.$$

One now compares this with the general expansion in terms of partial waves (2.6) and notices that the coefficients of this series are

$$a_\ell = e^{i\delta_\ell} \sin\delta_\ell \approx \sin\delta_\ell = -\frac{mg}{\hbar^2 k} Q_\ell \left(1 + \frac{\mu^2}{2k^2}\right)$$

and that they agree with the example (2.19) in Sect. 2.3.1. Note that here and in that example, the scattering phases δ_ℓ are small.

Remark

The results (2.27) and (2.30) show that the scattering amplitude in first Born approximation is real, i.e. $\mathrm{Im}\, f^{(1)} = 0$. This is in contradiction with the optical theorem. The first Born approximation does not respect the conservation of probability.

2.4.2 Form Factors in Elastic Scattering

The first Born approximation leads in a natural way to a new notion, called *form factor*, which is important for the analysis of scattering experiments. This section gives its definition and illustrates it by a few examples. The question and the idea are the following: Suppose a localized distribution $\varrho(\boldsymbol{x})$ of elementary scattering centers is given whose interaction with the projectile is known. If we know the scattering amplitude for the elementary process, i.e. for the scattering of the projectile off a single, isolated elementary scattering center, can we calculate the scattering amplitude off the *distribution* $\varrho(\boldsymbol{x})$?

The answer to this question is simple if the Born approximation is sufficiently accurate for the calculation of the scattering amplitudes. In this case, the amplitude for the distribution is equal to the product of the elementary scattering amplitude and a function which depends only on the density $\varrho(\boldsymbol{x})$ and on the momentum transfer $\boldsymbol{q}$. We show this by means of an example:

Suppose the projectile is scattered by a number A of particles whose distribution in space is described by the density $\widetilde{\varrho}(\boldsymbol{x}) = A\varrho(\boldsymbol{x})$. This is to say that

$$\int d^3x\, \widetilde{\varrho}(\boldsymbol{x}) = A \quad \text{and} \quad \int d^3x\, \varrho(\boldsymbol{x}) = 1 \,.$$

It is customary to normalize the density $\varrho(\boldsymbol{x})$ to 1, i.e. to take out an explicit factor A. Let the elementary interaction be described by the Yukawa potential (2.17). The potential created by all particles, with their density $\varrho(\boldsymbol{x})$, then is, using $\mu = 1/r_0$

$$U(\boldsymbol{x}) = gA \int d^3x' \frac{e^{-\mu|\boldsymbol{x}-\boldsymbol{x}'|}}{|\boldsymbol{x}-\boldsymbol{x}'|} \varrho(\boldsymbol{x}') \,. \tag{2.32}$$

It obeys the differential equation

$$(\Delta_x - \mu^2) U(\boldsymbol{x}) = -4\pi A \varrho(\boldsymbol{x}) \,. \tag{2.33}$$

In first Born approximation the scattering amplitude $F^{(1)}$ describing scattering on the distribution $\varrho(\boldsymbol{x})$ is given by the formula (2.29), upon insertion of the potential (2.32). The exponential function is replaced by the identity

$$e^{i\boldsymbol{q}\cdot\boldsymbol{x}} = -\frac{1}{q^2 + \mu^2} (\Delta_x - \mu^2) e^{i\boldsymbol{q}\cdot\boldsymbol{x}} \,.$$

The differential operator $(\Delta_x - \mu^2)$ is shifted to $U(\boldsymbol{x})$, by partially integrating twice, and the differential equation (2.33) replaces the potential by the density. One obtains the expression

$$F^{(1)}(\boldsymbol{q}) = A\, f_Y^{(1)}(\theta) \cdot F(\boldsymbol{q}) \tag{2.34}$$

for the scattering amplitude. The first factor is the elementary amplitude (2.31), the second factor is defined by

$$\boxed{F(\boldsymbol{q}) = \int d^3x e^{i\boldsymbol{q}\cdot\boldsymbol{x}} \varrho(\boldsymbol{x})} \,. \tag{2.35}$$

This factor which depends on the density $\varrho(\boldsymbol{x})$ and on the momentum transfer only, is called *form factor*. Its physical interpretation is clarified by its properties the most important of which are summarized here:

Properties of the Form Factor:

1. If it were possible to measure the form factor for *all* values of the momentum transfer then the density would be obtained by inverse Fourier transform,

$$\varrho(\boldsymbol{x}) = \frac{1}{(2\pi)^3} \int d^3q e^{-i\boldsymbol{q}\cdot\boldsymbol{x}} F(\boldsymbol{q})$$

   The density $\varrho(\boldsymbol{x})$ describes a *composite* target such as, e.g., an atomic nucleus composed of A nucleons. If the interaction of the projectile with the individual particle in the target is known (in our example this is a nucleon), the form factor measures the spatial distribution of the target particles.

2. Scattering in the forward direction, $\boldsymbol{q} = \mathbf{0}$, tests the normalization

$$F(\boldsymbol{q} = \mathbf{0}) = 1 \, .$$

If the target is point-like, i.e. has no spatial extension at all, the form factor equals 1 for *all* momentum transfers,

$$\varrho(\boldsymbol{x}) = \delta(\boldsymbol{x}) \longrightarrow F(\boldsymbol{q}) = 1 \quad \forall \boldsymbol{q} \, .$$

3. For a spherical density, $\varrho(\boldsymbol{x}) = \varrho(r)$ where $r := |\boldsymbol{x}|$, the formulae for the form factor simplify further. Like in the derivation of (2.30) one shows that the form factor only depends on q^2 ($q := |\boldsymbol{q}|$),

$$F(\boldsymbol{q}) \equiv F(q^2) = \frac{4\pi}{q} \int_0^\infty r \mathrm{d}r \varrho(r) \sin(qr) \, , \tag{2.36}$$

the density being given by the inversion of this formula:

$$\varrho(r) = \frac{4\pi}{(2\pi)^3 r} \int_0^\infty q \mathrm{d}q \, F(q) \sin(qr) \, .$$

Expanding (2.36) for small values of q, one has

$$F(q) \approx 4\pi \int_0^\infty r^2 \mathrm{d}r \varrho(r) - \frac{4\pi}{6} q^2 \int_0^\infty r^2 \mathrm{d}r \, r^2 \varrho(r) = 1 - \frac{1}{6} q^2 \left\langle r^2 \right\rangle .$$

The first term is independent of q and is equal to 1 (using the normalization as defined above). The second term contains the *mean square radius*

$$\left\langle r^2 \right\rangle := 4\pi \int_0^\infty r^2 \mathrm{d}r \, r^2 \varrho(r) \, , \tag{2.37}$$

which is characteristic for the given distribution $\varrho(\boldsymbol{x})$. If the density is not known, but the scattering amplitude and, hence, the form factor, are measured for small values of q, the mean square radius is obtained from the first derivative of the form factor by q^2,

$$\left\langle r^2 \right\rangle = -6 \, \frac{\mathrm{d}F(q^2)}{\mathrm{d}q^2} \, . \tag{2.38}$$

That is to say, in order to obtain the root-mean-square radius one plots the form factor as a function of q^2, reads off the slope at the origin, and multiplies by (-6).

Example 2.2

The following density is normalized to 1,

$$\varrho(r) = \frac{1}{\pi^{3/2} r_0^3} \mathrm{e}^{-r^2/r_0^2} \, , \qquad 4\pi \int_0^\infty r^2 \mathrm{d}r \, \varrho(r) = 1 \, .$$

It is easy to verify that it leads to the form factor

$$F(q^2) = \mathrm{e}^{-q^2 r_0^2/4} .$$

The mean square radius is found to be

$$\left\langle r^2 \right\rangle = \frac{3}{2} r_0^2 ,$$

so that the form factor and the density can be written equivalently as

$$F(q^2) = \mathrm{e}^{-(1/6)\left\langle r^2 \right\rangle q^2} , \quad \text{and} \quad \varrho(r) = \frac{3\sqrt{6}}{4} \frac{1}{(\pi \langle r^2 \rangle)^{3/2}} \mathrm{e}^{-(3/2) r^2 / \langle r^2 \rangle}$$

respectively.

This example is more than an academic one. Indeed, to a good approximation, the function $\varrho(r)$ describes the distribution of electric charge in the interior of a proton, with

$$\left\langle r^2 \right\rangle_{\text{Proton}} = (0.86 \times 10^{-15}\ \text{m})^2$$

as a typical value for the mean square radius as obtained from electron scattering on protons.

Remarks

1. Although the definition of the form factor is based on Born approximation it has a more general significance. When the potential becomes too strong for Born approximation to be applicable, this will be felt in a deformation of the partial waves (in comparison to the force-free case) which is the stronger the smaller ℓ is. Nevertheless, the information on the distribution of the scattering centers that is contained in the scattering amplitude, is not modified in any essential way. Thus, there are methods which allow to isolate this effect and to define an effective Born approximation which enables one to draw conclusions on the density $\varrho(\boldsymbol{x})$ from the form factor (2.35).
2. The case of the Coulomb potential which is of infinite range must be treated with special care. Although this is mathematically not correct, we let the parameter μ in the formulae for the Yukawa potential tend to zero. This limit gives the correct scattering amplitude, except for the characteristic logarithmic phases, and, hence, the correct differential cross section. This cross section then takes the form

$$\frac{\mathrm{d}\sigma}{\mathrm{d}\theta} = \left(\frac{\mathrm{d}\sigma}{\mathrm{d}\theta} \right)_{\text{point}} F^2(\boldsymbol{q}) , \tag{2.39}$$

   the first factor being the cross section for a point-like target, i.e.

$$\left(\frac{\mathrm{d}\sigma}{\mathrm{d}\theta} \right)_{\text{point}} = \left(\frac{Z Z' e^2}{4E} \right)^2 \frac{1}{\sin^4(\theta/2)} ,$$

   while $F(\boldsymbol{q})$ is the electric form factor (2.35) of the target. The density $\varrho(\boldsymbol{x})$ now is the charge density and is normalized to 1 provided the total charge $Q = Z' e$ is factored.

Example 2.3

Assume the charge distribution of an atomic nucleus to be given by the homogeneous density

$$\varrho(r) = \frac{3}{4\pi r_0^3}\Theta(r_0 - r) .$$

Equation (2.36), by elementary integration, yields the form factor

$$F(q^2) = \frac{3}{(qr_0)^3}[\sin(qr_0) - (qr_0)\cos(qr_0)] = \frac{3}{(qr_0)} j_1(qr_0) ,$$

where $j_1(z)$ is the spherical Bessel function with $\ell = 1$, cf. Sect. 1.9.3. This function has zeroes at the points

$$z_1 = 4.493 , \qquad z_2 = 7.725 , \qquad z_3 = 10.904, \ldots .$$

While the cross section for the point charge has no zeroes, the cross section for scattering on a homogeneous charge density has zeroes at the values of the product $qr_0 = 2kr_0 \sin(\theta/2)$ given above. These are artifacts of Born approximation. Yet, they are not unphysical: An exact calculation of the cross section will differ from Born approximation only by the varying deformation of individual partial waves. As a result, the zeroes are "smeared out", that is, they are replaced by minima of the cross section which are less pronounced than the zeroes of Born approximation and are shifted slightly as compared to those. These minima are called *diffraction minima*, they contain essentially the same physical information as the zeroes obtained in first Born approximation.

2.5 *Analytical Properties of Partial Wave Amplitudes

Up to this point all important concepts of scattering theory were introduced. As far as *observables* are concerned these were carried over from classical to quantum mechanics, all other concepts were new. In particular, we developed some practical methods for calculating scattering amplitudes and cross sections.

Quantum theoretic scattering theory has a number of further aspects which need more comprehensive analysis and which become important in various applications. Among them we list the following

1. Generalization to the case where, besides elastic scattering, also scattering into inelastic channels is possible.
2. The more general expression of the optical theorem announced above, and its relation to the conservation of probability.
3. A formal, operator theoretic description of potential scattering which makes extensive use of the theory of integral equations.
4. A more detailed description of scattering on composite targets, i.e. the calculation of the scattering amplitude for a target composed of elementary constituents, from the amplitude for the individual scattering centers.

5. The analysis of scattering processes, by means of Heisenberg's scattering matrix, in which the projectiles may have relativistic velocities and where the dynamics and the kinematics allow for creation or annihilation of particles.
6. The analytical properties of scattering amplitudes and their consequences for the physics of scattering.

Some of these subjects will be taken up later when the concepts and methods will be ready that are needed to treat them. At this stage I sketch one of these topics which gives a good impression for the richness of quantum scattering theory: the analytic properties of scattering amplitudes for definite values of the angular momentum ℓ (i.e. item 6 on our list). This section which makes use of some function theory, is slightly more difficult than the previous ones and this is why it is marked by an asterisk. This also means that there is no harm if one decides to skip it in a first reading.

The starting point is the differential equation (2.8) for the radial function $u_\ell(r)$. We repeat it here, for the sake of convenience:

$$u_\ell''(r) - \left(\frac{\ell(\ell+1)}{r^2} + \frac{2m}{\hbar^2}U(r) - k^2\right) u_\ell(r) = 0\,, \qquad k^2 = \frac{2mE}{\hbar^2}\,. \tag{2.40}$$

As this equation is homogeneous we can normalize the solution which is regular at $r = 0$ such that

$$u_\ell(r) = r^{\ell+1} f_\ell(r) \quad \text{with} \quad f_\ell(r=0) = 1\,,$$

that is, we take out the centrifugal tail and put the remainder to 1 at the origin.

2.5.1 Jost Functions

In this section we study the analytic properties of the wave function, of the scattering phase δ_ℓ, and of the scattering amplitude, as functions of k^2, taken as a variable in the complex plane. What do we know about the differential equation (2.40)? It is of Fuchsian type, (1.113). The regular solution is free of the singularities that the coefficients of (1.113) have at $r = 0$. As is obvious, the coefficients are analytic functions of the complex variable k^2. The boundary conditions imposed on u_ℓ (the condition at $r = 0$ and the asymptotics for $r \to \infty$) are analytic as well. One can then show that the solutions are analytic functions of k^2. Hence, we write $u_\ell(r) \equiv u_\ell(r, k^2)$, in order to emphasize this property, and formulate their asymptotic behaviour as follows,

$$r \to \infty: \quad u_\ell(r, k^2) \sim \varphi_\ell^{(-)}(k^2)\mathrm{e}^{\mathrm{i}kr} + \varphi_\ell^{(+)}(k^2)\mathrm{e}^{-\mathrm{i}kr}\,. \tag{2.41}$$

The functions $\varphi_\ell^{(\pm)}(k^2)$ are called *Jost functions*.[10] Comparison to the asymptotic formula (2.9),

$$u_\ell \sim \frac{1}{2\mathrm{i}}[\mathrm{e}^{\mathrm{i}(-\ell\pi/2+\delta_\ell)}\mathrm{e}^{\mathrm{i}kr} - \mathrm{e}^{\mathrm{i}(\ell\pi/2-\delta_\ell)}\mathrm{e}^{-\mathrm{i}kr}]$$

[10] After Res Jost (1918–1990), who developed the mathematical foundations of scattering theory in a series of important publications.

yields the relation of the scattering phase to the Jost functions

$$S_\ell(k^2) := e^{2i\delta_\ell} = (-)^{\ell+1} \frac{\varphi_\ell^{(-)}(k^2)}{\varphi_\ell^{(+)}(k^2)} . \tag{2.42}$$

In a physical scattering process, i.e. for real k, the definition (2.41) implies that the Jost functions are complex conjugates of each other. Indeed, with a real potential the radial function $u_\ell(r, k^2)$ is real and, hence,

$$\varphi_\ell^{(+)}(k^2) = [\varphi_\ell^{(-)}(k^2)]^* \qquad (k \text{ real}) .$$

One of the reasons why these functions are important is the fact that one can derive and understand the analytic properties of the function $S_\ell(k^2)$ (just defined) and of the partial wave amplitude $a_\ell(k^2)$ from the Jost functions. The following remarks and examples may serve to illustrate this statement.

When the energy is negative, E, then also $k^2 < 0$ and $k = i\kappa$, with real κ, is pure imaginary. The asymptotic form (2.41) then has a term that decreases exponentially, and a term that grows exponentially. There will be a bound state with energy $E = E_n < 0$ if the wave function is square integrable, i.e. if the second term is absent. This means that

$$\varphi^{(+)}\left(k^2 = -\frac{2m|E_n|}{\hbar^2}\right) = 0 \qquad (k = i\kappa \text{ pure imaginary}) .$$

As a consequence the function $S_\ell(k^2)$ defined in (2.42) has a *pole* at this point. A more detailed analysis of the singularities of the partial wave amplitude a_ℓ will show where this pole is situated.

2.5.2 Dynamic and Kinematic Cuts

Here and below we analyze partial wave amplitudes which already contain the factor $1/k$ of the expansion (2.6), i.e we write

$$f(k, \theta) = \sum_{\ell=0}^{\infty} (2\ell + 1) f_\ell(k) P_\ell(\cos\theta) \quad \text{with} \quad f_\ell := \frac{1}{k} a_\ell ,$$

a_ℓ being defined by (2.11). The amplitude f_ℓ is studied both as a function of k^2 and of k. The convention is that k as a square root of k^2 is always chosen such that it has *positive* imaginary part. Furthermore, the potential in (2.40) is assumed to be the bare Yukawa potential (2.17). As $u_\ell(r, k^2)$ is analytic, $f_\ell(k^2)$ is analytic as long as one stays clear of the negative real half-axis of k^2.

Note that the asymptotic form (2.41) holds under the assumption that the two first terms in the parentheses of (2.40) (centrifugal and true potentials) are negligible against k^2. If $k^2 = -\kappa^2$, i.e. is negative real, the first term in (2.41) decreases like $e^{-\kappa r}$. If it does so more rapidly than this, one can no longer neglect the potential which decreases like e^{-r/r_0}. Therefore, it seems plausible that the Jost function $\varphi_\ell^{(-)}(k^2)$ is no longer defined. Indeed, one can show that this function has a singularity at

$k_c^2 = -\kappa_c^2 = -1/(4r_0^2)$ and is not defined below this point, $k^2 < k_c^2$. Qualitatively speaking, the value κ_c is the point where the exponentially growing part, when multiplied by the term e^{-r/r_0} of the potential, equals the exponentially decreasing term,

$$e^{(\kappa_c - 1/r_0)r} = e^{-\kappa_c r} .$$

The interval $[-\infty, -1/(4r_0^2)]$ of the negative real half-axis in the complex k^2-plane is called *left cut*, or *dynamical cut*.

Let us analyze the dynamical cut by means of first Born approximation. We showed in Sect. 2.4.1 that the amplitude which belongs to the angular momentum ℓ, in first Born approximation, is given by

$$f_\ell(k^2) \equiv \frac{a_\ell}{k} = -\frac{mg}{\hbar^2 k^2} Q_\ell \left(1 + \frac{1}{2(kr_0)^2}\right) .$$

The Legendre function of the second kind $Q_\ell(z)$ is known to be an analytic function of z if the complex plane is cut from -1 to $+1$ along the real axis. It is singular on the cut $[-1, +1]$. As a matter of illustration, we quote the first three Legendre functions of the second kind,

$$Q_0(z) = \frac{1}{2} \ln\left(\frac{z+1}{z-1}\right) , \qquad Q_1(z) = \frac{z}{2} \ln\left(\frac{z+1}{z-1}\right) - 1 ,$$
$$Q_2(z) = \frac{3z^2 - 1}{4} \ln\left(\frac{z+1}{z-1}\right) - \frac{3z}{2} ,$$

whose singularity structure is evident.

The point -1 corresponds to $k^2 = k_c^2 = -1/(2r_0)^2$, while the point $+1$ corresponds to the infinity in the k^2-plane that we choose to lie at $-\infty$. This shows that the amplitude $f_\ell(k^2)$, in first Born approximation, already exhibits the left cut. The position of this cut depends on r_0 and, hence, on the range of the potential.

It is not difficult to compute the discontinuity of $f_\ell(k^2)$ across the cut $[-1, +1]$. In Sect. 2.4.1, Example 2.1, we quoted the expansion

$$\frac{1}{z-t} = \sum_{\ell'=0}^{\infty} (2\ell' + 1) Q_{\ell'}(z) P_{\ell'}(t) = \sum_{\ell'=0}^{\infty} \sqrt{4\pi(2\ell' + 1)} Q_{\ell'}(z) Y_{\ell' 0}(\theta, \phi)$$

with $t = \cos\theta$. This equation can be solved for Q_ℓ by multiplying with $Y^*_{\ell m}(\theta, \phi)$, and integrating over the entire solid angle. Making use of the orthogonality of the spherical harmonics, we have

$$Q_\ell(z) = \frac{1}{\sqrt{4\pi(2\ell + 1)}} \int d\Omega \frac{1}{z - \cos\theta} Y_{\ell 0}(\theta, \phi) = \frac{1}{2} \int_{-1}^{+1} dt \frac{P_\ell(t)}{z - t} .$$

The discontinuity then follows from

$$f_\ell(k^2 + i\varepsilon) - f_\ell(k^2 - i\varepsilon) = -\frac{mg}{\hbar^2 k^2} [Q_\ell(z + i\varepsilon) - Q_\ell(z - i\varepsilon)]$$

and the formula

$$\frac{1}{w \pm \mathrm{i}\varepsilon} = \mathcal{P}\frac{1}{w} \mp \mathrm{i}\pi\delta(w) \tag{2.43}$$

with $w = z - t$ and, as before, $z = 1 - 1/(2(kr_0)^2)$. The formula refers to the evaluation of integrals along the real axis. The pole at $w = 0$ of the left-hand side is shifted to the upper or to the lower half-plane, respectively. The right-hand side contains the principal value[11] and Dirac's δ-distribution. Inserting this formula into the expression for the discontinuity, the contribution of the principal value cancels out and one obtains

$$f_\ell(k^2 + \mathrm{i}\varepsilon) - f_\ell(k^2 - \mathrm{i}\varepsilon) = \mathrm{i}\pi\frac{mg}{\hbar^2 k^2} P_\ell\left(1 + \frac{1}{2(kr_0)^2}\right).$$

As a result we note that the *position* of the left cut is determined by the *range*, its *discontinuity* by the *strength* g of the potential. This is the reason why this cut is called dynamical cut.

The partial wave amplitude f_ℓ exhibits yet another cut whose nature is purely kinematical. This is the reason it is called *kinematic cut*. In order to show this we return to the Jost functions and study them both in the complex k^2-plane and in the complex k-plane.

The asymptotic expansion (2.41) is not defined in the point $k^2 = 0$. Physically speaking, the reason is that at this point the centrifugal term and the potential cannot be neglected as compared to k^2. Let $z := k^2 = r\mathrm{e}^{\mathrm{i}\alpha}$ with r very small, α positive and small, too. After a complete rotation about the origin this becomes $z \longmapsto z' = r\mathrm{e}^{\mathrm{i}(\alpha+2\pi)}$, while its square root becomes $k = \sqrt{z} \longmapsto k' = -k$. At the same time the two Jost functions, taken as functions of k, exchange their roles,

$$\varphi_\ell^{(+)}(k) = \varphi_\ell^{(-)}(-k).$$

It follows from this observation that the point $k^2 = 0$ is a two-sheeted branch point (one also says a branch point of order 1). Let us investigate in somewhat more detail this singularity and the Riemann surface that is related to it. To do so we define

$$\phi_\ell(k) := \varphi_\ell^{(-)}(k^2).$$

This function is analytic in the upper half-plane $\operatorname{Im} k > 0$, cut along the interval $[\mathrm{i}/(2r_0), +\mathrm{i}\infty]$ of the imaginary axis. If k is replaced by $-k$, the solution $u_\ell(r, k^2)$ remains unchanged. However, the two exponential functions in its asymptotic expansion (2.41) are interchanged. From this follows an important relation for the Jost functions,

$$\phi_\ell(k) = \phi_\ell(-k)$$

which shows that the analytic continuation of $\phi_\ell(k)$ to the lower half-plane $\operatorname{Im} k < 0$ is precisely the function $\varphi_\ell^{(+)}(k^2)$. This function has no other singularities in that half-plane.

[11] As a reminder: The principal value is half the sum of the integrals for which the integration path is deformed once above, once below the point 0.

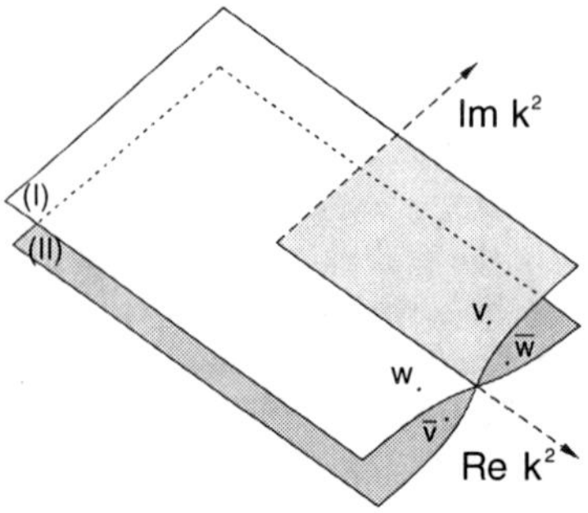

Fig. 2.2 Because of the kinematic cut the origin of the complex k^2-plane is a branch point of order 1. The scattering amplitude with fixed ℓ is defined on a two-sheet Riemann surface. The physical and the unphysical sheets are denoted by (I) and (II), respectively

We conclude that the manifold of the k^2 is a two-sheeted Riemann surface whose sheets, for fixed k^2, are distinguished by the two values of k. This surface is sketched in Fig. 2.2. The two sheets are tangent to each other along the positive real axis. In scattering theory the customary nomenclature is the following:

Sheet (I) with $\operatorname{Im} k > 0$ is called the *physical sheet*, sheet (II) with $\operatorname{Im} k < 0$ is the *unphysical sheet*.

2.5.3 Partial Wave Amplitudes as Analytic Functions

The formula (2.11) for the partial wave amplitude, upon insertion of the definition (2.42), gives the scattering amplitude in terms of the S-matrix,

$$f_\ell(k^2) = \frac{a_\ell}{k} = \frac{1}{2\mathrm{i}k}[S_\ell(k^2) - 1] .$$

Thus, $f_\ell(k^2)$ is seen to be a function of the two Jost functions. Therefore, the analytic properties of $f_\ell(k^2)$ can be deduced from the analytic properties of the Jost functions. By convention and for the sake of clarity let us denote the function (2.42) by $S_\ell(k^2)$ on the first, physical sheet, and by $\overline{S}_\ell(k^2)$ its continuation to the second, unphysical sheet. Consider a point v on the physical sheet (i.e. the square root of $k^2 + \mathrm{i}\eta$),

$$z = k^2 + \mathrm{i}\eta \equiv r\mathrm{e}^{\mathrm{i}\alpha} , \qquad v = \sqrt{z} = \sqrt{r}\mathrm{e}^{\mathrm{i}\alpha/2} ,$$

and its neighbour $\overline{v} = \sqrt{z^*} = \sqrt{r}\mathrm{e}^{-\mathrm{i}\alpha/2}$ (square root of $k^2 - \mathrm{i}\eta$) on the unphysical sheet. The other square root of z, $\overline{w} = -v$, lies on the unphysical sheet, while the other root of z^*, $w = -\overline{v}$, lies on the physical sheet. The four points $\{v, \overline{v}, w, \overline{w}\}$ are shown in Fig. 2.2 on the Riemann surface of complex k^2, and in Fig. 2.3 in the complex k-plane.

The function (2.42), taken on the unphysical sheet, is given by

$$\overline{S}_\ell(k^2) = (-)^{\ell+1}\frac{\varphi_\ell^{(-)}(-k)}{\varphi_\ell^{(+)}(-k)} = (-)^{\ell+1}\frac{\varphi_\ell^{(+)}(k)}{\varphi_\ell^{(-)}(k)} = \frac{1}{S_\ell(k^2)}\,. \tag{2.44}$$

This relation provides the key for analytic continuation of the amplitude f_ℓ from the physical to the unphysical sheet. Indeed, the amplitude on the second sheet is

$$\overline{f}_\ell(k^2) = \frac{1}{2\mathrm{i}k}[\overline{S}_\ell(k^2) - 1] = \frac{1}{\overline{2\mathrm{i}k}}\left(\frac{1}{S_\ell} - 1\right) = \frac{f_\ell(k^2)}{1 + 2\mathrm{i}kf_\ell(k^2)}\,.$$

Thus, $\overline{f}_\ell(k^2)$ is expressed in terms of the same amplitude on the first sheet. (Note that a first minus sign in the transformation is compensated by another minus sign: in performing the analytic continuation the factor $1/k$ goes over into $\overline{k} = -k$.) This analysis shows, in particular, that f_ℓ has the left cut $[-\infty, -1/(2r_0)^2]$ both on the first and on the second sheets.

2.5.4 Resonances

We know from Sect. 2.5.1 that poles on the negative real half-axis in the complex k^2-plane correspond to genuine bound states. We now want to show that poles that appear on the second, unphysical sheet play a physical role, too.

In a first step we show that the function $S_\ell(k^2)$ (studied both in the first *and* the second sheets) takes complex conjugate values for complex conjugate arguments k^2, i.e. that

$$S_\ell[(k^2)^*] = [S_\ell(k^2)]^*\,. \tag{2.45}$$

The differential equation (2.40) has real coefficients. Hence, its solutions obey the relation

$$u_\ell(r, k^2) = \{u_\ell[r, (k^2)^*]\}^*\,.$$

The left-hand side has the asymptotic behaviour (2.41). If $v = \sqrt{k^2 + \mathrm{i}\eta}$ is the root of k^2 then the root of $(k^2)^*$ (which also lies in the first sheet) equals $w = -\overline{v} = -v^*$. Therefore, the asymptotics of the right-hand side is

$$r \to \infty: \quad u_\ell[r, (k^2)^*] \sim \varphi_\ell^{(-)}[(k^2)^*]\mathrm{e}^{-\mathrm{i}k^*r} + \varphi_\ell^{(+)}[(k^2)^*]\mathrm{e}^{\mathrm{i}k^*r}\,.$$

Comparing this to the complex conjugate of (2.41) yields

$$\varphi_\ell^{(\pm)}[(k^2)^*] = [\varphi_\ell^{(\pm)}(k^2)]^*\,,$$

so that the assertion (2.45) holds true. For any pair of two points in the first or the second sheet whose positions are symmetric with respect to the real axis, the function S_ℓ takes complex conjugate values.

Suppose the function $\overline{S}_\ell(k^2)$ has a first-order pole in

$$\overline{v}\,,\quad \text{positive square root of} \quad k^2 = k_0^2 - \mathrm{i}\Gamma/2\,,$$

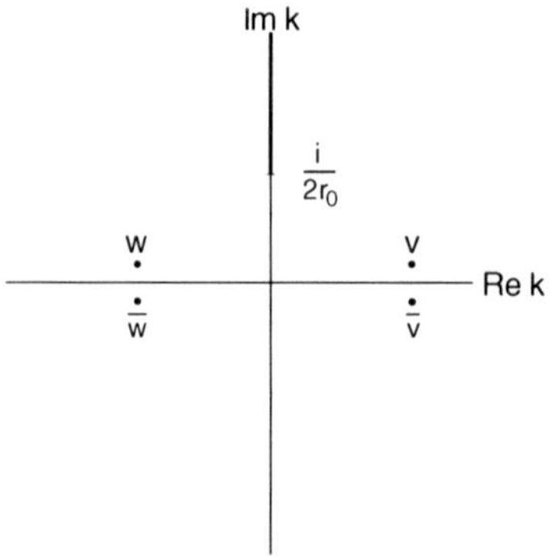

Fig. 2.3 The function $\phi_\ell(k) = \varphi_\ell^{(-)}(k^2)$ is analytic in the cut k-plane. The points marked here are the same as in Fig. 2.2

where k_0 and Γ are real. By the symmetry (2.45) it then has also a pole in

$$\overline{w}\,,\quad \text{negative square root of}\quad k^2 = k_0^2 + \mathrm{i}\Gamma/2\,;$$

both of which lie in the second sheet. The relation (2.44) implies that $S_\ell(k^2)$ has zeroes in the points v and w of the *first* sheet, see Fig. 2.3. If one approaches the point k_0 of the real, positive axis from above, it is clear that the variation of S_ℓ, as a function of k^2, is dominated by the pole in $\overline{v}$ and by the zero in v. Thus, in the neighbourhood of k_0 one can write

$$S_\ell(k^2) = \frac{k^2 - k_0^2 - \mathrm{i}\Gamma/2}{k^2 - k_0^2 + \mathrm{i}\Gamma/2} S_\ell^{(\text{n.r.})}(k^2)$$

where the "nonresonant" function $S_\ell^{(\text{n.r.})}$ is slowly varying. The factor $(k^2 - k_0^2 - \mathrm{i}\Gamma/2)(k^2 - k_0^2 + \mathrm{i}\Gamma/2)$ is a pure phase factor. As the product has the form indicated in (2.42), the second factor must also be a pure phase. Thus, both factors can be written as follows

$$\frac{k^2 - k_0^2 - \mathrm{i}\Gamma/2}{k^2 - k_0^2 + \mathrm{i}\Gamma/2} = \mathrm{e}^{2\mathrm{i}\delta_\ell^{(\text{res})}}\,, \qquad S_\ell^{(\text{n.r.})} = \mathrm{e}^{2\mathrm{i}\delta_\ell^{(\text{n.r.})}}\,,$$

where the "resonant" phase is given by

$$\delta_\ell^{(\text{res})} = \arctan\left(\frac{\Gamma/2}{k_0^2 - k^2}\right). \tag{2.46}$$

The other phase, $\delta_\ell^{(\text{n.r.})}$, is a parametrization of the nonresonant amplitude, and is a slowly varying function of k^2. Inserting this in (2.11) the partial wave amplitude is

$$f_\ell(k) = \frac{1}{k}\mathrm{e}^{\mathrm{i}(\delta_\ell^{(\text{res})} + \delta_\ell^{(\text{n.r.})})} \sin(\delta_\ell^{(\text{res})} + \delta_\ell^{(\text{n.r.})})\,. \tag{2.47}$$

It is not difficult to interpret these results. Suppose first that the nonresonant phase is negligibly small as compared to the resonant phase. If we let k^2 run through an

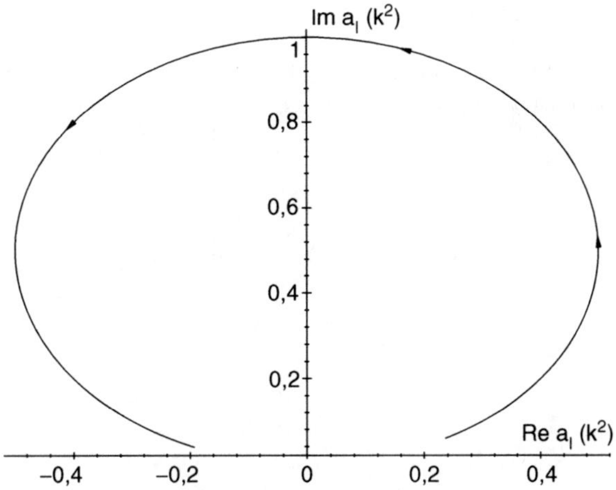

Fig. 2.4 With increasing k^2 the amplitude $a_\ell = kf_\ell$ runs along a curve in the complex plane that crosses the point (0, i) for $k^2 = k_0^2$. In the example shown here, we chose $k_0^2 = 10$, $\Gamma = 4$ (in arbitrary units)

interval I on the real axis that includes the point k_0^2, the phase $\delta_\ell^{(\mathrm{res})}$ runs from nearly zero to values close to π, while at $k^2 = k_0^2$ it takes the value $\pi/2$. The amplitude

$$f_\ell(k^2) \approx f_\ell^{(\mathrm{res})} = \frac{1}{k} e^{i\delta_\ell^{(\mathrm{res})}} \sin\delta^{(\mathrm{res})} = -\frac{1}{k} \frac{\Gamma/2}{k^2 - k_0^2 + i\Gamma/2}$$

runs through the curve sketched in Fig. 2.4, in the complex plane for f_ℓ. For $k^2 = k_0^2$ it is pure imaginary and has the value i/k. The special role of this point becomes clear if we note that the cross section (2.7) for a partial wave ℓ,

$$\sigma_\ell(k^2) = 4\pi(2\ell+1)\left|f_\ell(k^2)\right|^2 = \frac{4\pi(2\ell+1)}{k^2} \frac{\Gamma^2/4}{(k^2-k_0^2)^2 + \Gamma^2/4}$$

is proportional to $|a_\ell^{(\mathrm{res})}(k^2)|^2$, and if we plot this quantity over k^2 in the interval I. Indeed, Fig. 2.5 shows that this function has a sharp maximum at the point $k^2 = k_0^2$. For $k^2 = k_0^2 \pm \Gamma/2$ it takes half the value of the maximum. A graph of this type is called Breit–Wigner curve, or Lorentz curve.

The pole at $(k_0^2, i\Gamma/2)$ leads to a resonance in the partial wave cross section σ_ℓ, the quantity Γ is the width of the resonance. It is not difficult to find out qualitatively how these results change when the nonresonant phase is not small, or when the contributions of other partial waves $\ell' \neq \ell$ to the cross section (2.7) are not negligible. I just mention in passing that there are methods which allow to reconstruct the individual partial wave amplitudes from experimental data and thereby to check whether one of these exhibits a resonance of this type at some value k_0^2 in the physical range. Like bound states, resonances contain physical information on the potential.

2.5.5 Scattering Length and Effective Range

This section is devoted to a discussion of two notions of special importance for scattering at very low energies: the *scattering length* and the *effective range*.

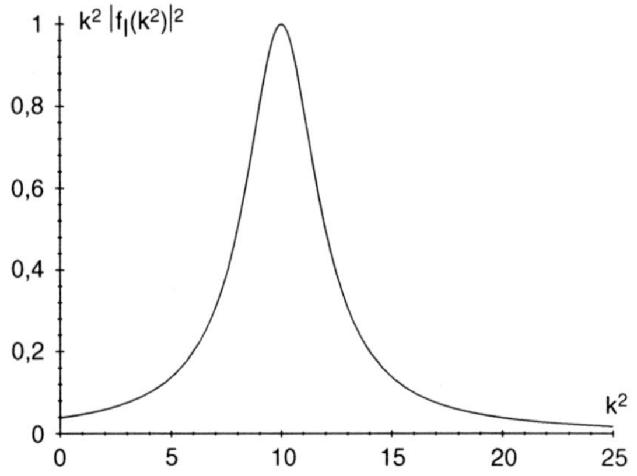

Fig. 2.5 Resonance curve of $|a_\ell|^2 = k^2|f_\ell|^2$ as a function of k^2. The parameters are chosen as in Fig. 2.4, i.e. $k_0^2 = 10$, $\Gamma = 4$

With decreasing energy the modulus k of the momentum tends to zero. With some more mathematics one can show rigorously that when $k \to 0$, the amplitude a_ℓ, (2.11), tends to zero like $k^{2\ell+1}$. I skip the proof of this result but give a plausibility argument based on the formula (2.19) for $\sin\delta_\ell$ and on simple dimensional analysis. The potential $U(r)$, as well as the kinetic energy $\hbar^2k^2/(2m)$ have dimension [energy]. The left-hand side of (2.19) being dimensionless, this must also be true for its right-hand side which can be written in terms of the ratio of potential to kinetic energy, viz.

$$\int_0^\infty (kr)\mathrm{d}(kr)\, u_\ell(r)\, j_\ell(kr)\frac{U(r)}{\hbar^2k^2/(2m)}\,.$$

We already know that for

$$r \to 0: \quad u_\ell(r) \sim r^{\ell+1}\,, \qquad j_\ell(kr) \sim (kr)^\ell\,,$$

and, of course, that the product (kr) carries no dimension. Thus, in order to balance physical dimensions, u_ℓ must behave like $(kr)^{\ell+1}$. This shows that the integral goes to zero for $k \to 0$. More specifically, for small k, it is proportional to $k^{2\ell+1}$. If this is so then, as k tends to zero,

$$k \to 0: \quad f_\ell = \frac{1}{k}\mathrm{e}^{\mathrm{i}\delta_\ell}\sin\delta_\ell \approx \frac{1}{k}\sin\delta_\ell \approx \frac{1}{k}\delta_\ell \sim k^{2\ell}\,.$$

This consideration shows that it is meaningful to define the limit

$$\lim_{k\to 0}\left(\frac{f_\ell(k)}{k^{2\ell}}\right) = \lim_{k\to 0}\left(\frac{\delta_\ell(k)}{k^{2\ell+1}}\right) =: a^{(\ell)}\,. \tag{2.48}$$

The quantities $a^{(\ell)}$ determine the cross section at the threshold, i.e. at very small positive energy. They are called *scattering lengths*. This nomenclature is somewhat inaccurate because only $a^{(\ell=0)}$ really has the dimension of a length L. For $\ell = 1$ the amplitude $a^{(1)}$ is a volume, the more general $a^{(\ell)}$ has dimension $L^{2\ell+1}$.

The expansion in small values of k can be pushed one step further. For this purpose define the function

$$R_\ell(k) := \frac{1 + \mathrm{i}kf_\ell(k)}{f_\ell(k)} = k\frac{1 + \mathrm{i}\mathrm{e}^{\mathrm{i}\delta_\ell}\sin\delta^{(\ell)}}{\mathrm{e}^{\mathrm{i}\delta_\ell}\sin\delta_\ell} = k\cot\delta_\ell \,.$$

The analytical properties of this function, when understood as a function of the complex variable k, are derived from those of the amplitude f_ℓ. For our discussion the most relevant feature is that the function $R_\ell(k)$, in contrast to $f_\ell(k)$, does *not* have the right, or kinematic, cut. This is shown as follows. Starting from the specific form (2.11) of the amplitude one sees that

$$\frac{1}{f_\ell^*(k)} - \frac{1}{f_\ell(k)} = k\frac{\mathrm{e}^{\mathrm{i}\delta} - \mathrm{e}^{-\mathrm{i}\delta}}{\sin\delta} = 2\mathrm{i}k \,.$$

Thus, if one calculates the discontinuity of R_ℓ one finds that it vanishes,

$$R_\ell^*(k) - R_\ell(k) = \left(\frac{1 - \mathrm{i}kf_\ell^*(k)}{f_\ell^*(k)} - \frac{1 + \mathrm{i}kf_\ell(k)}{f_\ell(k)}\right) = k(2\mathrm{i} - 2\mathrm{i}) = 0 \,,$$

the discontinuity of $1/f_\ell$ (which is not zero) is cancelled.

With $k \to 0$, $R_\ell(k)$ behaves like $k^{-2\ell}$. From this fact and the fact that R_ℓ has no cut on the real positive axis one concludes that the product $r^{2\ell}R_\ell(k)$ can be expanded around the origin,

$$k^{2\ell}R_\ell(k) = k^{2\ell+1}\cot\delta_\ell(k) = \frac{1}{a^{(\ell)}} + \frac{1}{2}r_0^{(\ell)}k^2 + \mathcal{O}(k^3) \,. \tag{2.49}$$

The first term contains the scattering length defined in (2.48). The second term contains the new parameter $r_0^{(\ell)}$, called *effective range*. (Note that $r_0^{(\ell)}$ has physical dimension [length] only for s-waves.)

The formula (2.49) is a good approximation for sufficiently low values of k. The following example serves to illustrate these concepts and their usefulness at low energies.

Example 2.4

We consider s-wave scattering in an attractive square-well potential, $U(r) = -U_0\Theta(R - r)$, in which case (2.40) reduces to

$$u''(r) + \left(k^2 + \frac{2m}{\hbar^2}U_0\Theta(R - r)\right)u(r) = 0 \,.$$

The index $\ell = 0$ is dropped, for convenience. In the outside domain, $r > R$, the relation between wave number and energy is, as before, $k^2 = 2mE/\hbar^2$. Inside the radius R we define

$$\kappa^2 := k^2 + K^2 \quad \text{where} \quad K^2 := \frac{2mU_0}{\hbar^2} \,.$$

The inner solution which is regular at $r = 0$, is seen to be $u^{(\mathrm{i})}(r) = \sin(\kappa r)$. For the outer solution we write $u^{(\mathrm{o})}(r) = \sin(kr + \delta)$, with δ still to be determined.

The requirement that the wave function and its first derivative be continuous at $r = R$,

$$\left.\frac{u^{(\mathrm{i})\prime}}{u^{(\mathrm{i})}}\right|_{r=R} = \left.\frac{u^{(\mathrm{o})\prime}}{u^{(\mathrm{o})}}\right|_{r=R} \tag{2.50}$$

yields the condition

$$k \cot \delta = \frac{\kappa + k \tan(kR) \tan(\kappa R)}{\tan(\kappa R) - \sqrt{1 + K^2/k^2} \tan(kR)} .$$

This result is expanded in k^2 and is compared to (2.49). The scattering length and the effective range then are found to be

$$a^{(0)} = -R + \frac{1}{K} \tan(KR) , \qquad r_0^{(0)} = R - \frac{1}{3} \frac{R^3}{(a^{(0)})^2} + \frac{1}{K^2 a^{(0)}} .$$

The scattering amplitude $f_{\ell=0}$ proper is easily expressed in terms of scattering length and effective range. Writing $\sin \delta = 1/\sqrt{1+u^2}$, $\cos \delta = u/\sqrt{1+u^2}$, and $u \equiv \cot \delta$ one has

$$f_0 = \frac{1}{k} \mathrm{e}^{\mathrm{i}\delta} \sin \delta = \frac{1}{u - \mathrm{i}} \approx \frac{a^{(0)}}{1 - \mathrm{i}a^{(0)}k + a^{(0)} r_0^{(0)} k^2/2} .$$

This amplitude has poles on the negative real axis in the complex k^2-plane which correspond to the bound states in this potential. If one wants to know the exact values of their energy one would have to solve the condition (2.50), with $k = \mathrm{i}\sqrt{2m(-E)}$, either analytically or numerically. However, if the term of the denominator containing the effective range is small, the pole, or bound state, is obtained from $1 - \mathrm{i}a^{(0)}k = 1 + a^{(0)}\sqrt{2m(-E)} = 0$. Thus,

$$E \approx -\frac{\hbar^2}{2m(a^{(0)})^2}$$

gives an approximate value for the binding energy.

2.6 Inelastic Scattering and Partial Wave Analysis

Examination of the differential equation for the radial function $u_\ell(r)$ in the form of (2.8) or of (2.40) shows that a real, attractive or repulsive potential leads to real scattering phases. In these cases there is only elastic scattering. If, on the other hand, the initial state can make transitions to other states that differ from it, there will be both elastic and inelastic scattering amplitudes whose absolute square yield the cross sections for the various inelastic channels. Loosely speaking, the elastic final state will be "depopulated" in favour of new, inelastic channels. Whether such channels are "open", and, if so, how many there are, depends on the dynamics of the scattering process and of the energy of the incoming state. For example, an electron which scatters on an atom, can lift this atom to an excited discrete state,

$$e + (Z, A) \longrightarrow (Z, A)^* + e' ,$$

if its energy is high enough as to furnish the finite, discrete difference $E(Z, A)^* - E(Z, A)$.

An exact quantum theoretic description would have to be a *multi-channel* calculation of the transition probabilities into all channels, the elastic one as well as all open inelastic channels, by solving a finite number of coupled wave equations. Depending on the kind and on the complexity of the system on which the scattering takes place, this may be an extensive, technically challenging calculation. If, on the other hand, one is primarily interested in the back-reaction onto the elastic channel, there is a simpler bulk procedure to parametrize the partial wave contributions to (2.6). The key to this method is provided by the optical theorem (2.12) which says that the *total* cross section

$$\sigma_{\text{tot}} = \sigma_{\text{el}} + \sigma_{\text{abs}}$$

is proportional to the imaginary part of the *elastic* scattering amplitude in the forward direction,

$$\sigma_{\text{tot}} = \sigma_{\text{el}} + \sigma_{\text{abs}} = \frac{4\pi}{k} \, \text{Im} \, f_{\text{el}}(k, \theta = 0) \, . \tag{2.51}$$

The elastic cross section, integrated over the whole solid angle, is given by

$$\sigma_{\text{el}} = \int \mathrm{d}\Omega \, |f_{\text{el}}|^2 = \frac{4\pi}{k^2} \sum_{\ell=0}^{\infty} (2\ell+1) |a_\ell(k)|^2 = 4\pi \sum_{\ell=0}^{\infty} (2\ell+1) |f_\ell(k)|^2 \, .$$

The remainder σ_{abs} is the sum of all cross sections into inelastic channels open at the given energy. It represents the absorption out of the elastic channel.

So far, we did not prove the optical theorem in this general form because we did not yet develop all tools needed for its proof. In the theory of potential scattering the theorem follows essentially from the conservation of probability: if the particle can be scattered away from the elastic channel, the probability to find it *somewhere* in one of the kinematically (and dynamically) allowed final states must be equal to 1.

The expansion in partial waves being so useful and well adapted to the physics of scattering, it is suggestive to define total, elastic, and absorption cross sections for each partial wave ℓ,

$$\sigma_{\text{el}}^{(\ell)} := 4\pi (2\ell + 1) |f_\ell(k)|^2 \, ,$$

$$\sigma_{\text{tot}}^{(\ell)} := \frac{4\pi}{k} (2\ell + 1) \, \text{Im} \, f_\ell(k) \, ,$$

$$\sigma_{\text{abs}}^{(\ell)} := \sigma_{\text{tot}}^{(\ell)} - \sigma_{\text{el}}^{(\ell)} \, .$$

Note that we have made use of the optical theorem in the second definition. As $\sigma_{(\text{tot})}^{(\ell)} \geq \sigma_{\text{el}}^{(\ell)}$ there follows an important condition on the partial wave amplitudes:

$$\boxed{\text{Im} \, f_\ell(k) \geq k |f_\ell(k)|^2} \, . \tag{2.52}$$

The condition (2.52) is called *positivity condition*. This condition implies that $f_\ell(k)$ must have the general form

$$f_\ell(k) = \frac{1}{2ik}(e^{2i\delta_\ell(k)} - 1) \tag{2.53}$$

where δ_ℓ is a phase which may be complex, and whose imaginary part must be positive or zero. This is shown as follows:

The polar decomposition of the complex function $1 + 2ikf_\ell$ is written as follows

$$1 + 2ikf_\ell = \eta_\ell e^{2i\varepsilon_\ell} \quad \text{with} \quad \eta_\ell = e^{-2\,\mathrm{Im}\,\delta_\ell}\,, \qquad \varepsilon_\ell = \mathrm{Re}\,\delta_\ell\,.$$

(The factor 2 in the exponent is introduced for convenience in order to facilitate comparison with the results of Sect. 2.3.) One calculates

$$\mathrm{Im}\, f_\ell = \frac{1}{2k}[1 - \eta_\ell \cos(2\varepsilon_\ell)]\,, \quad |f_\ell|^2 = \frac{1}{4k^2}[1 + \eta_\ell^2 - 2\eta_\ell \cos(2\varepsilon_\ell)]\,.$$

The positivity condition (2.52) implies the inequalities

$$1 \geq \frac{1 + \eta_\ell^2}{2} \quad \text{or} \quad 0 \leq \eta_\ell^2 \leq 1\,.$$

This was precisely the assertion: If $\eta_\ell = 1$, then $\mathrm{Im}\,\delta_\ell = 0$; if $\eta_\ell < 1$, then $\mathrm{Im}\,\delta_\ell > 0$. The quantity η_ℓ, which, by definition, is positive semi-definite and which obeys the inequality

$$\boxed{0 \leq \eta_\ell \leq 1} \tag{2.54}$$

is called *inelasticity*. Indeed, qualitatively speaking, the inelasticity is a measure for the amount taken out of the ℓ-th partial wave of elastic scattering, due to absorption.

The result (2.53) with the property (2.54) is quite remarkable: The same form (2.11) for the amplitude $f_\ell = a_\ell/k$ was obtained from the Schrödinger equation, though with *real* potentials and, hence, real scattering phases. In deriving the more general formulae of this section we need no more than the optical theorem (2.51)!

The cross sections of fixed partial wave, defined above, when written as functions of the inelasticity and the real scattering phase, are given by

$$\sigma_{\mathrm{tot}}^{(\ell)} = \frac{2\pi}{k^2}(2\ell + 1)[1 - \eta_\ell \cos(2\varepsilon_\ell)]\,, \tag{2.55}$$

$$\sigma_{\mathrm{el}}^{(\ell)} = \frac{\pi}{k^2}(2\ell + 1)[1 + \eta_\ell^2 - 2\eta_\ell \cos(2\varepsilon_\ell)]\,, \tag{2.56}$$

$$\sigma_{\mathrm{abs}}^{(\ell)} = \sigma_{\mathrm{tot}}^{(\ell)} - \sigma_{\mathrm{el}}^{(\ell)} = \frac{\pi}{k^2}(2\ell + 1)[1 - \eta_\ell^2]\,. \tag{2.57}$$

The interpretation of the results (2.55)–(2.57) is simple:

1. If $\eta_\ell = 1$ there is no absorption at all in this partial wave. The inelastic contribution vanishes, the elastic cross section (2.57) equals the total cross section (2.55),

$$\sigma_{\mathrm{abs}}^{(\ell)} = 0\,, \qquad \sigma_{\mathrm{el}}^{(\ell)} = \sigma_{\mathrm{tot}}^{(\ell)}\,.$$

In this case the scattering phase δ_ℓ is real, and (2.53) can be written in the form known from (2.11),

$$f_\ell(k) = \frac{1}{k} e^{i\delta_\ell} \sin\delta_\ell\,.$$

2. The other extreme case is $\eta_\ell = 0$. The absorption is maximal in the partial wave with angular momentum ℓ. This does not mean that there is no elastic scattering at all! Rather, the elastic and the inelastic cross sections are equal. The results (2.55)–(2.57) which were derived from the optical theorem, give the result

$$\sigma_{\text{abs}}^{(\ell)} = \sigma_{\text{el}}^{(\ell)} = \frac{1}{2}\sigma_{\text{tot}}^{(\ell)} .$$

The scattering amplitude proper is pure imaginary and is equal to

$$f_\ell = \frac{\mathrm{i}}{2k} .$$

3. A case of interest is one where there is a resonance in the ℓ-th partial wave, accompanied by absorption. The resonance curve Fig. 2.4 remains qualitatively similar to the case without absorption. However, it no longer intersects the ordinate in the point i (or i/k), but at a smaller value from which the inelasticity can be read off.

3 The Principles of Quantum Theory

Introduction This chapter develops the formal framework of quantum mechanics: the mathematical tools, generalization and abstraction of the notion of state, representation theory, and a first version of the postulates on which quantum theory rests.

Regarding the mathematical framework quantum mechanics makes extensive use of the concept of Hilbert space, of the theory of linear operators which act on elements of this space, and of more general functional analysis. In themselves these are important and comprehensive fields of mathematics whose even sketchy description would go far beyond the scope of this book. For this reason I adopt a somewhat pragmatic approach introducing all definitions and methods of relevance for quantum mechanics but skipping some of the detailed justifications. Some of the general concepts are made plausible and, to some extent, are visualized by means of matrix representations. Even though these matrices will often be infinite dimensional, this approach allows to adopt, by analogy, methods familiar from linear algebra.

3.1 Representation Theory

Observables, by definition, are classical quantities. In quantum mechanics they are represented by self-adjoint operators. In the physical examples that we studied up to this point, the eigenfunctions of such operators define systems of base functions which are orthogonal, and either square integrable (and hence normalizable to 1) or normalizable to δ-distributions. Regarding the corresponding spectrum of eigenvalues there are three possibilities:

1. The spectrum may be *pure discrete*. Examples are the square of the orbital angular momentum $\boldsymbol{\ell}^2$, and one of its components, say ℓ_3. Both operators are defined on S^2, the sphere with unit radius in $\mathbb{R}^3$, their eigenfunctions $Y_{\ell m}(\theta, \phi)$ are orthonormal and complete.

F. Scheck, *Quantum Physics*, DOI: 10.1007/978-3-642-34563-0_3,

The Hamiltonian of the spherical oscillator is another example,

$$H = -\frac{\hbar^2}{2m}\Delta + \frac{1}{2}m\omega^2 r^2 \,. \tag{3.1}$$

It is defined on $\mathbb{R}^3$, its spectrum and its eigenfunctions were derived in Sect. 1.9.4.

2. The spectrum may be *pure continuous*. Examples are provided by the operator of momentum of a particle $\boldsymbol{p}$, the position operator $\boldsymbol{x}$, and the operator of kinetic energy $\boldsymbol{p}^2/(2m)$.
3. Finally, the spectrum may have *both discrete as well as continuous parts*. An important example is the Hamiltonian describing the hydrogen atom,

$$H = -\frac{\hbar^2}{2m}\Delta - \frac{e^2}{r} \,, \tag{3.2}$$

that we studied in Sect. 1.9.5. More examples are provided by the Hamiltonians for one-particle motion where the potential is $U(r) = -U_0\Theta(R_0 - r)$, i.e. an attractive well of finite radius. Like in the hydrogen atom there are bound states with $E < 0$ and states with $E > 0$ which belong to the continuum.

Let $\psi_\alpha(\boldsymbol{x})$, or, more generally, $\psi_\alpha(t, \boldsymbol{x})$ be the quantum state of a physical system which is characterized by the quantum number(s) α. Indeed, α may stand for more than just one quantum number as exemplified by the discrete bound states of the hydrogen atom where α is a short-hand notation for the triple (n, ℓ, m). The Fourier transform of ψ_α

$$\widetilde{\psi}_\alpha(t, \boldsymbol{p}) = \frac{1}{(2\pi\hbar)^{3/2}} \int \mathrm{d}^3x \, \exp\left(-\frac{\mathrm{i}}{\hbar}\boldsymbol{p}\cdot\boldsymbol{x}\right) \psi_\alpha(t, \boldsymbol{x})$$

is unique. It provides a means of expanding the physical wave function

$$\psi_\alpha(t, \boldsymbol{x}) = \frac{1}{(2\pi\hbar)^{3/2}} \int \mathrm{d}^3p \, \exp\left(+\frac{\mathrm{i}}{\hbar}\boldsymbol{p}\cdot\boldsymbol{x}\right) \widetilde{\psi}_\alpha(t, \boldsymbol{p})$$

in terms of eigenfunctions of the momentum operator

$$\varphi(\boldsymbol{p}, \boldsymbol{x}) = \frac{1}{(2\pi\hbar)^{3/2}} \exp\left(\frac{\mathrm{i}}{\hbar}\boldsymbol{p}\cdot\boldsymbol{x}\right) . \tag{3.3}$$

The fact that these are not square integrable and, hence, not normalizable in the usual sense, plays no special role because completeness can be formulated equally well by means of δ-distributions. The function $\widetilde{\psi}_\alpha(t, \boldsymbol{p})$ is as suitable for describing the state with quantum numbers "α" as was the function $\psi_\alpha(t, \boldsymbol{x})$. Therefore, when considering the wave function $\widetilde{\psi}_\alpha(t, \boldsymbol{p})$ one says one is using the *momentum space representation*, while when using $\psi_\alpha(t, \boldsymbol{x})$, one says that one is working in the *position* or *coordinate space representation*.

Representing the state "α" in coordinate space, the original Born interpretation of $|\psi_\alpha(t, \boldsymbol{x})|^2$ applies, i.e. $|\psi_\alpha(t, \boldsymbol{x})|^2 \, \mathrm{d}^3x$ is the probability for finding the particle at time t in an infinitesimal neighbourhood of the point $\boldsymbol{x}$. By analogy $|\widetilde{\psi}_\alpha(t, \boldsymbol{p})|^2 \, \mathrm{d}^3p$ is the probability to detect the particle at time t with a momentum in a ε-neighbourhood of the point $\boldsymbol{p}$ in momentum space.

Let A be an observable whose spectrum of eigenvalues is assumed to be fully discrete and, for simplicity, whose eigenvalues are not degenerate. The eigenvalues are denoted by a_n, the eigenfunctions are denoted by φ_n, viz.

$$A\varphi_n(\boldsymbol{x}) = a_n\varphi_n(\boldsymbol{x}) .$$

The system of functions $\{\varphi_n\}$ is complete and normalized to 1. If a given state vector $\psi_\alpha(t, \boldsymbol{x})$ is (absolutely) square integrable as well, it can be expanded in the base $\{\varphi_n\}$,

$$\psi_\alpha(t, \boldsymbol{x}) = \sum_n \varphi_n(\boldsymbol{x}) c_n^{(\alpha)}(t) \quad \text{with} \quad c_n^{(\alpha)}(t) = \int \mathrm{d}^3x\, \varphi_n^*(\boldsymbol{x})\psi_\alpha(t, \boldsymbol{x}) .$$

At any time the set of all expansion coefficients $\{c_n^{(\alpha)}(t)\}$ gives a complete description of the state α. In a measurement of the observable A in the state described by ψ_α, the quantity $|c_n^{(\alpha)}(t)|^2$ is the probability to find the eigenvalue a_n of A.

If the spectrum of A is degenerate one must also sum over the base states spanning the subspace of fixed eigenvalue a_n. The Hamiltonian of the spherical oscillator (3.1) provides an example of a purely discrete, degenerate spectrum in which case $a_n \equiv E_{n\ell}$ and

$$\varphi_\nu(\boldsymbol{x}) = R_{n\ell}(r)Y_{\ell m}(\theta, \phi) , \qquad \nu \equiv (n, \ell, m) .$$

In case A has a spectrum that includes both a discrete and a continuous part, the expansion of the wave function reads, in a somewhat symbolic but suggestive notation,

$$\psi_\alpha(t, \boldsymbol{x}) = \sum_\nu d_\nu^{(\alpha)}(t)\varphi_\nu(\boldsymbol{x}) + \int \mathrm{d}\nu\, d^{(\alpha)}(t, \nu)\varphi(\nu, \boldsymbol{x}) .$$

The classical example of this case is the Hamiltonian (3.2) which describes the hydrogen atom.

As a first result we note that the state "α" is represented alternatively by the data

$$\begin{aligned} &\psi_\alpha(t, \boldsymbol{x}) \quad \text{or} \quad \widetilde{\psi}_\alpha(t, \boldsymbol{p}) \quad \text{or} \\ &\{c_n^{(\alpha)}(t)\} \quad \text{or} \quad \{d_\nu^{(\alpha)}(t), d^{(\alpha)}(t, \nu)\} . \end{aligned} \tag{3.4}$$

This observation suggests to introduce a more abstract concept of "quantum state", by stripping it off any specific representation. In turn, in concrete considerations or in practical calculations, this freedom may be used to select the representation which is adapted best to the situation at stake. As a perhaps even more important bonus, this abstract concept of quantum state allows to also treat systems for which there is no classical analogue.

In a certain sense the transformations between different, but equivalent representations are analogous to the canonical transformations in mechanics of Hamiltonian systems. Similar to the situation in mechanics the physical system (that is, its quantum wave function) is invariant. Its representation in one of the concrete bases sketched above is a kind of "choice of coordinates" which may be more or less fortunate but, in any case, should be adapted to the specific problem. Of course, we will have to work out more precisely the transformation of wave functions and of operators from one representation to another. Before we do so, however, we wish to introduce a notation which is particularly useful for practical purposes and, at the same time, takes good account of the abstract nature of a quantum state.

3.1.1 Dirac's Bracket Notation

In many situations it is useful to write a quantum state characterized by the quantum number(s) α in a symbolic notation as $|\alpha\rangle$, independently of its specific representations (3.4). This notation is due to Dirac who called this symbol a "ket", and its dual $\langle\alpha|$ a "bra" – thereby alluding to the word *bracket*, broken up in two pieces, i.e. $\langle\cdots|\cdots\rangle$ into $\langle\cdots|$ and $|\cdots\rangle$. It is instructive to go a little deeper into this notation: As the Schrödinger equation is a homogeneous, linear differential equation, linear combinations of its solutions are again solutions. Thus, a physical state in one of the forms (3.4), is a (generalized) *vector* in a linear vector space over the complex numbers $\mathbb{C}$. The four representations (3.4) have in common the property of *linearity*, and the physical content which is coded in the quantum numbers α. They differ only insofar as the state is given either by a square integrable function over coordinate space $\mathbb{R}^3_x$, or by such a function over momentum space $\mathbb{R}^3_p$, or as a column vector with an infinite number of components. Dirac's notation $|\cdots\rangle$ summarizes the invariant information on the state, its vector nature and its physical content, but stands, in fact, for *all* representations. If $|\alpha\rangle$ is a vector, and $\langle\beta|\alpha\rangle$ is the complex number

$$\int \mathrm{d}^3x\, \psi^*_\beta(t,\boldsymbol{x})\psi_\alpha(t,\boldsymbol{x}) \quad \text{or} \quad \int \mathrm{d}^3p\, \widetilde{\psi}^*_\beta(t,\boldsymbol{p})\widetilde{\psi}_\alpha(t,\boldsymbol{p}) \quad \text{or}$$

$$\sum_{n=0}^{\infty} c_n^{(\beta)\,*} c_n^{(\alpha)} ,$$

depending on the representation one has chosen, this implies that $\langle\beta|$ is a linear form which acts on vectors $|\alpha\rangle$, thereby yielding a complex number. This is to say that $\langle\beta|$ is *dual* to $|\alpha\rangle$. For example, in coordinate space $|\alpha\rangle$ is the wave function $\psi_\alpha(t,\boldsymbol{x})$ while $\langle\beta|$ represents the integral operator

$$\int \mathrm{d}^3x\, \psi^*_\beta(t,\boldsymbol{x}) \;\bullet$$

which acts on the slot marked by a "bullet".

Dirac's notation is a somewhat pragmatic way of writing which is useful for many practical calculations. This is the reason it is widely used in everyday physics. The mathematical literature makes less use of it, probably because it is not unique[1] and could possibly give rise to misunderstanding in some situations. In what follows we shall often, though not everywhere, make use of it. Here we illustrate the "bra" and "ket" language by some simple examples:

[1] For instance, $|n\rangle$ could stand for the base system $\varphi_n(x)$ of stationary eigenfunctions of the one-dimensional harmonic oscillator but could also be any other, fully discrete system with some other Hamiltonian.

Example 3.1

Let $|n\rangle$ characterize the base system that belongs to the fully discrete, initially non-degenerate, spectrum of eigenvalues of some observable A. Then $\langle m|n\rangle = \delta_{mn}$. The expansion of a physical state in terms of this basis which in coordinate space has the form $\psi_\alpha(t, \boldsymbol{x}) = \sum \varphi_n(\boldsymbol{x})\, a_n^{(\alpha)}(t)$, takes the abstract, representation-free form

$$|\alpha\rangle = \sum_n |n\rangle\langle n|\alpha\rangle \,.$$

In this series $\langle n|\alpha\rangle$ is the expansion coefficient corresponding to the state "n". For instance, in coordinate representation one has

$$\langle n|\alpha\rangle = (\varphi_n, \psi_\alpha) = \int \mathrm{d}^3x\, \varphi_n^*(\boldsymbol{x})\psi_\alpha(t, \boldsymbol{x}) = \langle\alpha|n\rangle^* \,.$$

The expansion of a "bra" reads correspondingly

$$\langle\beta| = \sum_m \langle m|\beta\rangle^* \,\langle m| \,.$$

The scalar product of two states in this notation,

$$\langle\beta|\alpha\rangle = \sum_{n,m} \langle m|\beta\rangle^* \langle n|\alpha\rangle\delta_{mn} = \sum_n \langle\beta|n\rangle\langle n|\alpha\rangle \,,$$

is realized by the formula

$$\int \mathrm{d}^3x\, \psi_\beta^*(t, \boldsymbol{x})\psi_\alpha(t, \boldsymbol{x}) = \sum_n a_n^{(\beta)\,*} a_n^{(\alpha)}$$

where the left-hand side holds in coordinate space while the right-hand side holds in the A-representation.

In those cases where the basis belongs to an observable with a discrete, but degenerate spectrum, or to an observable with a mixed spectrum, the sum over n must be replaced by a multiple sum, or by a sum plus an integral, respectively. The common eigenfunctions of $\boldsymbol{\ell}^2$ and of ℓ_3 are an example for the first case and, in Dirac's notation are written as $|\ell m\rangle$. The second case is illustrated by the eigenfunctions of the hydrogen Hamiltonian.

Example 3.2

In the *bra* and *ket* notation the completeness relation takes the symbolic, yet immediately intelligible form

$$\sum_{n=0}^{\infty} |n\rangle\langle n| = \mathbb{1} \quad \text{and} \quad \sum_n |n\rangle\langle n| + \int \mathrm{d}\nu\, |\nu\rangle\langle\nu| = \mathbb{1} \tag{3.5}$$

for the purely discrete and the mixed cases, respectively. If one decided to write the second expression in coordinate space, it would read

$$\sum_n \varphi_n(\boldsymbol{x}) \int d^3x' \, \varphi_n^*(\boldsymbol{x}') \bullet + \int d\nu \, \varphi_n(\nu, \boldsymbol{x}) \int d^3x' \, \varphi_n^*(\nu, \boldsymbol{x}') \bullet$$
$$= \int d^3\boldsymbol{x}' \, \delta(\boldsymbol{x} - \boldsymbol{x}') \bullet .$$

The "bullet" marks an empty slot, in other terms stands in lieu of the wave function on which this expression acts. Obviously, this explicit way of writing is less transparent than the more general, abstract notation.

The completeness relation in a complex, infinite-dimensional vector space is written in an analogous way: Let $\{e_n\}$ with $e_i = (0, \cdots 0, 1, 0, \cdots, 0)^T$ be a system of base vectors (with a 1 as the i-th entry), spanning this space. Then we have

$$\sum_{n=1}^{\infty} |n\rangle\langle n| = \sum_{i=1}^{\infty} e_i e_i^\dagger$$
$$= \sum_i \begin{pmatrix} 0 \\ \vdots \\ 0 \\ 1 \\ 0 \\ \vdots \end{pmatrix} (0 \ldots 0 \, 1 \, 0 \ldots) = \begin{pmatrix} 1 \, 0 \, 0 \ldots \\ 0 \, 1 \, 0 \ldots \\ 0 \, 0 \, 1 \ldots \\ \vdots \; \vdots \; \vdots \; \ddots \end{pmatrix} = \mathbb{1},$$

in close analogy to finite dimensional vector spaces.

Example 3.3

Expectation values or more general, nondiagonal matrix elements of operators are written as $\langle\beta|A|\alpha\rangle$. In coordinate space this expression is the familiar integral over $\mathbb{R}^3$ of Chap. 1. If one considers such a matrix element of the product of two operators A and B, the following, somewhat formal calculation uses the completeness relation for expressing matrix elements of the product in terms of products of matrix elements of individual operators,

$$\langle\beta|AB|\alpha\rangle = \langle\beta|A\mathbb{1}B|\alpha\rangle = \sum_n \langle\beta|A|n\rangle\langle n|B|\alpha\rangle .$$

One often makes use of this way of reducing a product to individual operators.

Example 3.4

The generalization to improper states, i.e. to states which are not normalizable to 1, poses no particular problem. We denote by $|\boldsymbol{x}\rangle$ and by $|\boldsymbol{p}\rangle$ the eigenfunctions of the operator $\boldsymbol{x}$ and the operator $\boldsymbol{p}$, respectively, in Dirac's notation. We then have

$$\langle\boldsymbol{x}'|\boldsymbol{x}\rangle = \delta(\boldsymbol{x}' - \boldsymbol{x}), \quad \langle\boldsymbol{p}'|\boldsymbol{p}\rangle = \delta(\boldsymbol{p}' - \boldsymbol{p}) .$$

The expansion coefficients of a physical state "α" in terms of eigenfunctions of $\boldsymbol{x}$ take the form $\langle\boldsymbol{x}|\alpha\rangle$ and are identical with $\psi_\alpha(t, \boldsymbol{x})$, i.e. the coordinate representation of the wave function. In this light the formula

$$\langle x | p \rangle = \frac{1}{(2\pi\hbar)^{3/2}} \exp\left(\frac{\mathrm{i}}{\hbar} p \cdot x\right) = \langle p | x \rangle^*$$

is immediately understandable: The left-hand side is the coordinate space representation of the wave function $|p\rangle$, the right-hand side is the complex conjugate of the momentum space representation of the eigenfunction $|x\rangle$ of the position operator.

3.1.2 Transformations Relating Different Representations

In changing the representation of states the operators that act on these states must be transformed as well. As a first orientation one should recall the description of finite dimensional vector spaces in linear algebra. Let V be a real vector space of dimension n, $\hat{e} = \{\hat{e}_k\}$, $k = 1, \ldots n$, an orthonormal basis, $\hat{e}_i \cdot \hat{e}_k = \delta_{ik}$. Every *orthogonal* transformation $\mathbf{R}$ takes it to a new orthonormal basis,

$$\hat{e} \longmapsto \hat{f} = \mathbf{R}\hat{e}\,, \quad \mathbf{R}^T\mathbf{R} = \mathbb{1}\,.$$

Observables over a $\mathbb{R}$-vector space are real, symmetric matrices whose action on an arbitrary element $a = \sum_k \hat{e}_k c_k^{(a)}$ is as follows

$$\mathbf{A}a = \sum_k (\mathbf{A}\hat{e}_k) c_k^{(a)} = \sum_{ik} A_{ik} \hat{e}_i c_k^{(a)} \quad \text{with} \quad A_{ik} = \hat{e}_i \cdot (\mathbf{A}\hat{e}_k)\,.$$

With $\mathbf{R}^T = \mathbf{R}^{-1}$ the observable in the new basis is

$$\widetilde{A}_{ik} = \hat{f}_i \cdot (\mathbf{A}\hat{f}_k) = \sum_{pq} R_{ip} R_{kq} A_{pq} = (\mathbf{R}\mathbf{A}\mathbf{R}^{-1})_{ik}\,,$$

that is to say that the operator transforms according to the rule

$$\mathbf{A} \longmapsto \widetilde{\mathbf{A}} = \mathbf{R}\mathbf{A}\mathbf{R}^{-1}\,.$$

This rule is easy to remember: Reading the product on the right-hand side in the order in which the matrices act, i.e. from right to left, $\mathbf{R}^{-1}$ "rotates back" to the old basis, where $\mathbf{A}$ acts as before. Finally the result is taken back to the new basis by the "rotation" $\mathbf{R}$.

Matters are similar in vector spaces over $\mathbb{C}$, the difference being that the orthogonal transformation $\mathbf{R}$ is replaced by a *unitary* $\mathbf{U}$, i.e. $\mathbf{U}\mathbf{U}^\dagger = \mathbb{1} = \mathbf{U}^\dagger\mathbf{U}$, where $\mathbf{U}^\dagger$ denotes the transposed and complex conjugate matrix. As we saw in Sect. 1.8.1 observables are no longer real-symmetric but complex hermitian matrices.

Let us begin by an example: Let Q^k, $k = 1, 2, 3$, be the three operators of position in a cartesian basis. As they commute among each other their action on every solution $\psi(x)$ of the Schrödinger equation in coordinate space is

$$Q^k \psi(x) = x^k \psi(x)\,.$$

One says that Q^k act multiplicatively. Expanding these functions in terms of eigenfunctions (3.3) of the momentum operator one obtains

$$Q^k \int \mathrm{d}^3 p\, \varphi(\boldsymbol{p}, \boldsymbol{x})\widetilde{\psi}(\boldsymbol{p}) = \int \mathrm{d}^3 p\, \varphi(\boldsymbol{p}, \boldsymbol{x}) x^k \widetilde{\psi}(\boldsymbol{p})$$
$$= -\frac{\hbar}{\mathrm{i}} \int \mathrm{d}^3 p\, \varphi(\boldsymbol{p}, \boldsymbol{x}) \frac{\partial}{\partial p_k} \widetilde{\psi}(\boldsymbol{p}) .$$

Here we inserted the relation

$$\exp\left(\frac{\mathrm{i}}{\hbar}\boldsymbol{p}\cdot\boldsymbol{x}\right) x^k = \frac{\hbar}{\mathrm{i}} \frac{\partial}{\partial p_k} \exp\left(\frac{\mathrm{i}}{\hbar}\boldsymbol{p}\cdot\boldsymbol{x}\right)$$

and performed partial integration with respect to the variable p_k. This equation holds for all $\boldsymbol{x}$. By inverse Fourier transformation one concludes that

$$Q^k \widetilde{\psi}(\boldsymbol{p}) = -\frac{\hbar}{\mathrm{i}} \frac{\partial}{\partial p_k} \widetilde{\psi}(\boldsymbol{p}) . \tag{3.6}$$

Thus, in momentum space the operator Q^k is represented by the first derivative with respect to p_k, in analogy to the component P_k of momentum in coordinate space which is realized as first derivative by x^k, see (1.58) and Sect. 1.8.4. Note, however, the difference in signs in (1.58) and in (3.6).

Let A be an observable with a purely discrete, non-degenerate spectrum whose eigenfunctions are $\varphi_n(\boldsymbol{x})$. The state ψ may be expanded in the basis of these eigenfunctions, so that

$$Q^k \psi(\boldsymbol{x}) = Q^k \sum_n \varphi_n(\boldsymbol{x}) c_n = \sum_n x^k \varphi_n(\boldsymbol{x}) c_n .$$

In turn, $x^k \varphi_n(\boldsymbol{x})$ is again a square integrable function and, hence, can also be expanded in the basis φ_n. Denoting the expansion coefficients by $X^{(k)}_{mn}$ one has

$$x^k \varphi_n(\boldsymbol{x}) = \sum_m \varphi_m(\boldsymbol{x}) X^{(k)}_{mn} \quad \text{with} \quad X^{(k)}_{mn} = \int \mathrm{d}^3 x\, \varphi^*_m(\boldsymbol{x}) x^k \varphi_n(\boldsymbol{x}) .$$

This result has the following interpretation: in the A-representation the state $\psi(\boldsymbol{x})$ appears in the form of a (in general infinite dimensional) vector $\mathbf{c} = (c_1, c_2, \ldots)^T$, the position operator Q^k is represented by the matrix $\mathbf{X}^{(k)} = \left\{X^{(k)}_{mn}\right\}$ and we have

$$Q^k \mathbf{c} = \mathbf{X}^{(j)} \mathbf{c} \quad \text{or} \quad Q^k (c_m)^T = \left(\sum_n X^{(k)}_{mn} c_n\right)^T . \tag{3.7}$$

As a result we note that the operator Q^k which describes the k-th cartesian component of the position operator, may appear in quite different forms:

– In coordinate space as the function x^k, acting multiplicatively.
– In momentum space as a differential operator,

$$-\frac{\hbar}{\mathrm{i}} \frac{\partial}{\partial p_k} .$$

– In the space spanned by the eigenfunctions of the observable A as the infinite dimensional matrix

$$X^{(k)}_{mn} = \int \mathrm{d}^3 x\, \varphi^*_m(\boldsymbol{x}) x^k \varphi_n(\boldsymbol{x}) .$$

It is instructive to continue on this example by considering another cartesian component, say P_j, of the momentum operator. In coordinate space it is $(\hbar/\mathrm{i})(\partial/\partial x^j)$, in momentum space it is the function p_j (acting multiplicatively), in the space spanned by the eigenfunctions of A it is the matrix

$$P^{(j)}_{mn} = \int \mathrm{d}^3x\, \varphi^*_m(\boldsymbol{x}) \frac{\hbar}{\mathrm{i}} \frac{\partial}{\partial x^j} \varphi_n(\boldsymbol{x}) .$$

Obviously, the symbols Q^k and P_j denote these operators in *all* representations, which is to say, they are an abstract notation for what is essential about these operators. For instance, Heisenberg's commutation relations, in abstract notation, are

$$\boxed{[P_j, Q^k] = \frac{\hbar}{\mathrm{i}} \delta_{jk} \mathbb{1}, \quad [Q^j, Q^k] = 0, \quad [P_j, P_k] = 0}\,, \tag{3.8}$$

where $\mathbb{1}$ is either the number 1 or the infinite-dimensional unit matrix. When spelled out, their concrete realization is

– in coordinate space: $[P_j, Q^k] = \frac{\hbar}{\mathrm{i}} \frac{\partial}{\partial x^j} x^k - x^k \frac{\hbar}{\mathrm{i}} \frac{\partial}{\partial x^j} = \frac{\hbar}{\mathrm{i}} \delta_{jk}$;

– in momentum space: $[P_j, Q^k] = p_j \left(-\frac{\hbar}{\mathrm{i}} \frac{\partial}{\partial p_k} \right) - \left(-\frac{\hbar}{\mathrm{i}} \frac{\partial}{\partial p_k} \right) p_j = \frac{\hbar}{\mathrm{i}} \delta_{jk}$;

– in "A-space": $\sum_l \left[P^{(j)}_{ml} X^{(k)}_{ln} - X^{(k)}_{ml} P^{(j)}_{ln} \right] = \frac{\hbar}{\mathrm{i}} \delta_{jk} \delta_{mn}$.

It is not difficult to convert any one of these three representations into one of the two others. For example, in transforming from coordinate space to the A-representation one makes use of formulae of the type

$$\begin{aligned} \int \mathrm{d}^3x\, \varphi^*_m(\boldsymbol{x}) \frac{\hbar}{\mathrm{i}} \frac{\partial}{\partial x^j} x^k \varphi_n(\boldsymbol{x}) &= \int \mathrm{d}^3x\, \varphi^*_m(\boldsymbol{x}) \frac{\hbar}{\mathrm{i}} \frac{\partial}{\partial x^j} \sum_l \varphi_l(\boldsymbol{x}) X^{(k)}_{ln} \\ &= \sum_l P^{(j)}_{ml} X^{(k)}_{ln} . \end{aligned}$$

In position space and in momentum space the relations (3.8) refer to the commutator of a function and a differential operator. In the space spanned by the eigenfunctions of A they mean the commutator of two matrices. This statement which basically is simple, sheds light on an important historical step in the development of quantum mechanics. While Erwin Schrödinger treated quantum mechanics of nonrelativistic atomic systems by means of the differential equation that bears his name, Werner Heisenberg, together with Max Born and Pascual Jordan, developed the same theory in the framework of what was called *matrix mechanics*. The two approaches turned out to be just two different representations of one and the same theory, on the one hand what we now call the "coordinate space representation", on the other hand what is called the "A-representation". It was Schrödinger who proved the equivalence of his and Heisenberg's approaches, shortly after the birth of quantum mechanics.

3.2 The Concept of Hilbert Space

In Chap. 1 we studied important examples of self-adjoint Hamiltonians, and introduced the notions of orthogonality in function spaces and of completeness of base systems. In the previous section we studied formally different but physically equivalent representations of operators which describe observables. Here we wish to learn more about the spaces in which physical wave functions live. Central to this endeavour is the concept of Hilbert space. In many respects it corresponds to our conception of finite dimensional vector spaces, in others it is markedly different, due to its dimension being infinite. Of course, a detailed and mathematically rigorous treatment would go far beyond the scope of this book and would lead us astray for a while from the physical aspects of quantum theory that we wish to learn and to understand. Therefore, I restrict this text to a somewhat cursory and, in some respects, qualitative discussion. Those who wish to study these matters in greater depth are referred to the literature in mathematics and mathematical physics.

We start with a few remarks which are meant to clarify what we need for a formulation of the principles of quantum mechanics, and which will help to motivate the subsequent definitions.

Remarks

1. A striking property of Schrödinger's wave functions is, on one hand, that they are defined over our physical time axis $\mathbb{R}_t$ and over the usual physical space $\mathbb{R}^3_x$ where experiments are done, but, on the other hand, that they live in function spaces $\mathcal{H}$ where they have certain remarkable properties (such as, for instance, being square integrable). In a more scholarly language, $\psi_\alpha(t, \boldsymbol{x})$ is defined over $\mathbb{R}_t \times \mathbb{R}^3_x$ but is an element of $\mathcal{H}$. This raises the question of how the wave function ψ_α will react when we perform, e. g., Galilei transformations in space such as translations, rotations, or special Galilei transformations which leave the dynamics (formulated by means of a Hamiltonian) invariant. This question which leads to interesting statements both of principle and practical use, will be taken up extensively below.
   Note, however, that there are systems without classical analogues and to which wave functions are ascribed that have no, or only indirect, relation to the space-time of physics. This situation will be encountered in the description of spin, i.e. of intrinsic angular momentum of particles.
2. A central principle of quantum mechanics is the *superposition principle*, which says that with two different solutions ψ_α and ψ_β of the Schrödinger equation any linear combination $\lambda\psi_\alpha + \mu\psi_\beta$, with $\lambda, \mu \in \mathbb{C}$ two complex numbers, is also a solution. Therefore, the spaces in which wave functions are defined, must be *linear* spaces which is to say they must be vector spaces over $\mathbb{C}$.
3. Recalling Born's interpretation of the wave function, Sect. 1.4, or generalizations thereof, it is clear that the spaces $\mathcal{H}$ must carry a *metric structure*, it must be possible to define, or to measure, the norm or the length of a state ψ. For,

if we ask for the probability to measure the eigenvalue a_n of the observable A in the normalized state ψ, this is equivalent to asking about the scalar product (φ_n, ψ) of the eigenfunction of A corresponding to a_n and of the state ψ. In other words, one is asking for the projection of ψ onto φ_n, i.e. for the angle included between these two functions.

4. Both the metric and the geometric structure are obtained by the correct definition of the scalar product of wave functions (or, more generally, of state vectors). By the same token, a general, formal framework is provided where *expectation values* are well-defined which, as we know, represent physical observables and which are essential for the interpretation of the theory.

3.2.1 Definition of Hilbert Spaces

The previous remarks, hopefully, were helpful in preparing and motivating the following definition:

Definition 3.1 Hilbert Space

(I) A Hilbert space $\mathcal{H}$ is a linear vector space over the complex numbers $\mathbb{C}$. Addition of elements $f \in \mathcal{H}$ and $g \in \mathcal{H}$ exists, $f + g \in \mathcal{H}$, and has the usual properties, i.e. it is associative, there is a null element, for which $f + 0 = f$ for all $f \in \mathcal{H}$, and for every f there is an element $(-f)$ such that $f + (-f) = 0$. Multiplication with complex numbers is well-defined, it is associative and distributive.

(II) On $\mathcal{H}$ a scalar product is defined

$$(\cdot,\cdot): \quad \mathcal{H} \times \mathcal{H} \longrightarrow \mathbb{C}: \quad f, g \longmapsto (f, g),$$

which has the following properties:

The scalar product (f, g) of two elements $f, g \in \mathcal{H}$ is $\mathbb{C}$-linear in its second argument,

$$(f, g_1 + g_2) = (f, g_1) + (f, g_2), \tag{3.9a}$$

$$(f, \lambda g) = \lambda(f, g), \quad \lambda \in \mathbb{C}. \tag{3.9b}$$

The scalar product of an element (f, f) with itself is positive definite. It is zero if and only if f is the null element,

$$(f, f) \geq 0 \quad \forall f, \quad (f, f) = 0 \Longleftrightarrow f = 0. \tag{3.10}$$

When one interchanges its arguments the scalar product takes its complex conjugate value,

$$(g, f) = (f, g)^*. \tag{3.11}$$

(III) The space $\mathcal{H}$ is complete, i.e. every Cauchy series $f_1, f_2, \ldots$ converges to a limit f which is an element of $\mathcal{H}$,

$$f_n \longrightarrow f\,, \quad \text{if} \quad \lim_{n\to\infty} \|f_n - f\| = 0\,. \tag{3.12}$$

(IV) The space $\mathcal{H}$ has countably infinite dimension.

Comments on the axioms (I)–(IV). The properties (3.9a), (3.9b) and (3.11) imply that the scalar product is *anti*linear in its *first* entry, which is to say that

$$(\mu_1 f_1 + \mu_2 f_2, g) = \mu_1^*(f_1, g) + \mu_2^*(f_2, g)\,, \quad f_i, g \in \mathcal{H},\ \mu_i \in \mathbb{C}\,.$$

If the scalar product were real then, by (3.10) and (3.11), it would define a positive definite bilinear form. Being linear in the second entry, but antilinear in the first, it is said to be a *positive definite sesquilinear form*.

Definition 3.2

1. Two elements f and g in $\mathcal{H}$ are said to be *orthogonal*, if their scalar product vanishes,
$$(f, g) = 0 \quad f \text{ and } g \text{ orthogonal}\,. \tag{3.13}$$
2. The scalar product defines a *norm*
$$\|f\| := (f, f)^{1/2} \quad \text{norm of} \quad f \in \mathcal{H}\,. \tag{3.14}$$

Equation (3.13) takes up the concept of orthogonality of functions that we studied extensively in Chap. 1, in a more general framework. In close analogy to finite dimensional vector spaces one proves

$$\text{Schwarz' inequality:} \quad |(f, g)| \le \|f\| \cdot \|g\|\,, \tag{3.15}$$

and the

$$\text{triangle inequality:} \quad \big|\, \|f\| - \|g\| \,\big| \le \|f + g\| \le \|f\| + \|g\|\,. \tag{3.16}$$

If $f_1, f_2, \ldots, f_N$ is a set of orthogonal elements of $\mathcal{H}$ which are normalized to 1, one has *Bessel's inequality*

$$\sum_{n=1}^{N} |(f_n, g)|^2 \le \|g\|^2 \quad \text{for all} \quad g \in \mathcal{H}\,. \tag{3.17}$$

The norm $\|f\|$ is the (generalized) length of the vector $f \in \mathcal{H}$. By Schwarz' inequality (3.15) the ratio $|(f, g)|/(\|f\|\ \|g\|) =: \cos\alpha$ defines the angle comprised by the vectors f and g.

We note in passing that a space which has the properties (I) and (II) only, is called *pre-Hilbert space*.

Axiom (III) makes use of the notion of Cauchy series which may be summarized as follows: A set of functions forms a Cauchy series if for every $\varepsilon > 0$ there is a positive integer N such that

$$\|f_n - f_m\| < \varepsilon \quad \text{for all} \quad n, m > N\,.$$

In fact, it is the requirement (III) which turns a pre-Hilbert space into a full Hilbert space.

The axiom (IV) is not really necessary. In fact, in the mathematical literature a space which fulfills (I)–(III) – irrespective of its dimension – is called a Hilbert space. As a rule, the Hilbert space(s) of quantum theory are infinite dimensional, so axiom (IV) is included to remind us of this observation. In many applications we will deal with finite dimensional Hilbert spaces but note that these are subspaces of a "physical", infinite dimensional Hilbert space. In what follows and in the examples we will distinguish, if necessary, finite dimensional and infinite dimensional situations.

In the context of axiom (III) a certain type of convergence was made use of, called *strong convergence*. We note in passing that, unlike the finite-dimensional case, there are further definitions of convergence in infinite dimensional spaces which are distinct from strong convergence.

The following examples serve to illustrate the definition of Hilbert spaces. In particular, they show to which extent these spaces resemble the vector spaces familiar from linear algebra.

Example 3.5

The set of all infinite dimensional, complex vectors for which the sum of absolute squares of their components is convergent

$$\boldsymbol{a} = (a_1, a_2, a_3, \ldots)^T \quad \text{with} \quad \sum_{n=1}^{\infty} |a_n|^2 < \infty ,$$

is a linear vector space over $\mathbb{C}$ provided addition of elements and multiplication by complex numbers $\lambda \in \mathbb{C}$ are defined as usual, i.e.

$$\boldsymbol{a} + \boldsymbol{b} = \boldsymbol{c} \Longleftrightarrow c_n = a_n + b_n , \quad \lambda \boldsymbol{a} = (\lambda a_n)^T .$$

It is obvious that the condition of convergence is met for $\lambda \boldsymbol{a}$ if it is fulfilled for $\boldsymbol{a}$. It is less obvious for the sum of two elements: One has

$$|a_n + b_n|^2 \le \big|\, |a_n| + |b_n| \,\big|^2 + \big|\, |a_n| - |b_n| \,\big|^2 = 2\left(|a_n|^2 + |b_n|^2\right) < \infty ,$$

and, therefore, $\sum |c_n|^2 < \infty$.

The scalar product

$$(\boldsymbol{a}, \boldsymbol{b}) := \sum_{n=1}^{\infty} a_n^* b_n$$

has the properties (3.9a–3.11), and the result fulfills the condition of convergence. Indeed,

$$|(\boldsymbol{a}, \boldsymbol{b})| \le \sum_n |a_n||b_n| \le \frac{1}{2} \sum_n \left(|a_n|^2 + |b_n|^2\right) .$$

One shows, furthermore, that this vector space is complete. If its dimension were finite, say equal to N, one would refer to the fact that the field of real numbers $\mathbb{R}$ and, hence, also the direct product $\mathbb{R}^N$ of N copies of it, are complete. In infinite dimension things are not so simple. One must consider genuine Cauchy series and show that the limit of any such series is an element of the same vector space. Finally, one defines a countably-infinite base system $\hat{e}^{(i)} = (\ldots, \delta_{ni}, \ldots)^T$ by choosing the i-th entry equal to 1, all others equal to 0. This vector space fulfills all axioms of Definition 3.1 and, therefore, is a Hilbert space.

If one chooses the dimension of this space to be *finite*, $n = 1, 2, \ldots, N$, the condition of convergence is unnecessary. The so-defined N-dimensional vector space is a Hilbert space (fulfilling axioms (I)–(III)). Quantum mechanics often makes use of such spaces, for instance, in describing eigenstates of orbital angular momentum or of spin, though they appear as subspaces of a big Hilbert space with infinite dimension.

Example 3.6

The second example, in a first step, is chosen finite dimensional. Let $M_N(\mathbb{C})$ be the set of all $N \times N$-matrices with complex entries, $N \in \mathbb{N}$. Addition of elements and multiplication by complex numbers,

$$\mathbf{A}, \mathbf{B} \in M_N(\mathbb{C}) : \quad \mathbf{C} = \mathbf{A} + \mathbf{B} \Longleftrightarrow C_{jk} = A_{jk} + B_{jk}$$

makes it a $\mathbb{C}$-vector space. An obvious candidate for the scalar product of $\mathbf{A}$ and $\mathbf{B}$ is the trace of the product of the hermitian conjugate matrix $\mathbf{A}^\dagger = (\mathbf{A}^*)^T$ with $\mathbf{B}$,

$$(\mathbf{A}, \mathbf{B}) := \operatorname{tr}(\mathbf{A}^\dagger \mathbf{B}) = \sum_{j,k=1}^{N} A^*_{jk} B_{jk} .$$

(Note the unusual position of indices of the first factor $\mathbf{A}$ which is due to its being transposed.) Indeed, it fulfills the properties (3.9a–3.11). As the set of complex $N \times N$ matrices $M_N(\mathbb{C})$ is isomorphic to $\mathbb{C}^{N^2}$ and as this multiple direct product of $\mathbb{C}$ with itself has this property, the so-defined vector space is complete. Thus this provides another example of Hilbert space.

In a second step, one may let the dimension N go to infinity. Obviously, one must then restrict the set to those matrices whose trace is finite, and one must examine Cauchy series of matrices more carefully.

Example 3.7

Consider the set of all complex-valued functions $\psi(\boldsymbol{x})$ over three-dimensional space $\mathbb{R}^3$ whose absolute square admits a Lebesgue measure,

$$\int \mathrm{d}^3x \, |\psi(\boldsymbol{x})|^2 < \infty .$$

Take the scalar product of two such functions ψ and χ to be

$$(\psi, \chi) := \int \mathrm{d}^3x \, \psi^*(\boldsymbol{x}) \chi(\boldsymbol{x}) .$$

The following estimate

$$\int \mathrm{d}^3x \, |\psi^*(\boldsymbol{x}) \chi(\boldsymbol{x})| \le \frac{1}{2} \left[\int \mathrm{d}^3x \, |\psi(\boldsymbol{x})|^2 + \int \mathrm{d}^3x \, |\chi(\boldsymbol{x})|^2 \right]$$

shows that this scalar product is well-defined. Addition $\psi + \chi$ and multiplication $\lambda\psi$ by complex numbers make this set a linear vector space over $\mathbb{C}$. Indeed, the absolute square of the sum of two elements is finite because $|\psi(\boldsymbol{x}) + \chi(\boldsymbol{x})|^2 \le 2(|\psi(\boldsymbol{x})|^2 + |\chi(\boldsymbol{x})|^2)$ is finite.

It is more difficult to prove directly the completeness of this space. For our purposes it may be sufficient to hint at the examples of complete systems $\{\varphi_n(\boldsymbol{x})\}$ of orthonormal functions developed in Chap. 1 which were used as bases for an expansion of elements ψ, χ of this space. Hence, it is plausible that all axioms (I)–(IV) are fulfilled. Obviously, this Hilbert space of *square integrable functions over* $\mathbb{R}^3$ is of special importance for wave mechanics. It is denoted by $L^2(\mathbb{R}^3)$.

Example 3.8 (Interplay of Position Space and Hilbert Space)

Here is a simple example which shows the reaction, alluded to above, of elements of Hilbert space $L^2(\mathbb{R}^3)$ to transformations in $\mathbb{R}^3$, i.e. which illustrates the interplay between the space of physics, where we perform measurements, and the space of quantum wave functions. Consider the eigenfunctions of the Hamiltonian for the spherical oscillator of Sect. 1.9.4. The functions $\psi_{n\ell m}(\boldsymbol{x}) = R_{n\ell}(r)Y_{\ell m}(\theta,\phi)$ are defined over $\mathbb{R}^3$ or, more precisely, over $\mathbb{R}_+ \times S^2$: The variable r takes its values on the positive real half-axis, the variables θ and ϕ refer to the unit sphere S^2 in $\mathbb{R}^3$. The knowledge of these coordinates and of the quantum number m implies that a particular frame of reference $\mathbf{K}$ was chosen. Of course, we can change this reference system. As the Hamiltonian refers to a given center of force it would not be reasonable to perform space translations of $\mathbf{K}$. However, a different orientation in space of the frame of reference is a meaningful alternative, thus replacing $\mathbf{K}$ by a new system $\mathbf{K}'$ which is obtained from the original by a rotation $\mathbf{R}$, $\boldsymbol{x} \mapsto \boldsymbol{x}' = \mathbf{R}\boldsymbol{x}$. How does the base function $\psi_{n\ell m}$ react to this rotation? The answer in the most general case is derived below, in Sect. 4.1 where we study angular momentum in quantum theory. For the purpose of this example we consider the simple case illustrated by Fig. 3.1: Let $\mathbf{K}'$ be obtained from $\mathbf{K}$ by a rotation about the 3-axis, by the angle α. Comparing polar coordinates of the new system of reference and those in the old system one has

$$r \mapsto r' = r, \quad \theta \mapsto \theta' = \theta, \quad \phi \mapsto \phi' = \phi - \alpha$$

so that the wave functions in $\mathbf{K}'$ and $\mathbf{K}$, respectively, are related as follows

$$\psi_{n\ell m} \longmapsto \psi'_{n\ell m}(\boldsymbol{x}') = \mathrm{e}^{-\mathrm{i}m\alpha}\psi_{n\ell m}(\boldsymbol{x}) .$$

Writing the basis as one symbol $\boldsymbol{\Psi} = \{\psi_{n\ell m}\}$, one has

$$\boldsymbol{\Psi}(\boldsymbol{x}) \longmapsto \boldsymbol{\Psi}'(\boldsymbol{x}) = \mathbf{U}(\alpha)\boldsymbol{\Psi}(\boldsymbol{x})$$

where the unitary matrix is

$$\mathbf{U}(\alpha) = \mathrm{diag}(1, \mathrm{e}^{\mathrm{i}\alpha}, 1, \mathrm{e}^{-\mathrm{i}\alpha}, \mathrm{e}^{2\mathrm{i}\alpha}, \mathrm{e}^{\mathrm{i}\alpha}, 1, \mathrm{e}^{-\mathrm{i}\alpha}, \mathrm{e}^{-2\mathrm{i}\alpha}, \ldots),$$
$$\mathbf{U}^\dagger(\alpha)\mathbf{U}(\alpha) = \mathbb{1} .$$

This simple result is interesting: An orthogonal transformation $\mathbf{R} \in \mathrm{SO}(3)$ of the system of reference in $\mathbb{R}^3$ induces a *unitary* transformation of the basis in Hilbert

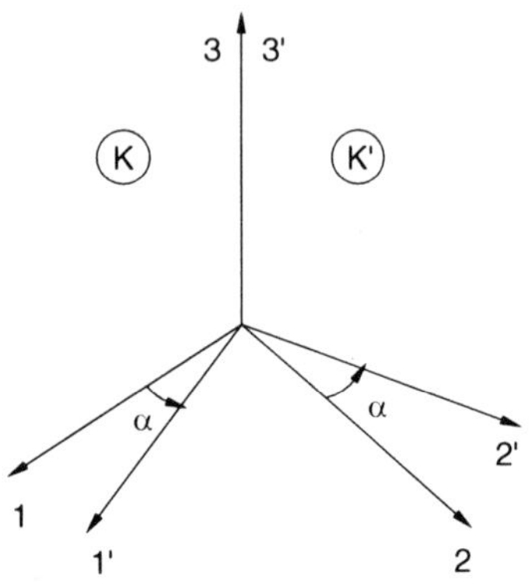

Fig. 3.1 A rotation $\mathbb{R}^3$ in three-dimensional, physical space induces a unitary transformation in Hilbert space. The figure shows the example of a rotation about the 3-axis by the angle α. The state with definite values of ℓ and m reacts to this rotation by the phase factor $\exp(-\mathrm{i}m\alpha)$

space.[2] The result suggested by this elementary example is not surprising if one keeps in mind the following fact: Both systems $\boldsymbol{\Psi}$ and $\boldsymbol{\Psi}'$ are orthonormal and complete; they refer to two systems of reference which are connected by a rotation $\mathbf{R}$, described, say, by Euler angles (ϕ, θ, ψ). Therefore, the transformation between the two systems must be unitary,

$$\psi'_{n\ell m'} = \sum_{m=-\ell}^{\ell} U^{(\ell)}_{m'm}(\phi, \theta, \psi)\psi_{n\ell m} ;$$

only then are the functions $\psi'_{n\ell m'}$ orthogonal and normalized to 1. The unitary matrix of transformation is a function of the Euler angles that will be calculated and analyzed in Chap. 4.

3.2.2 Subspaces of Hilbert Spaces

As an example we consider the Hilbert space $L^2(S^2)$ spanned by the eigenfunctions $Y_{\ell m}(\theta, \phi)$ of $\boldsymbol{\ell}^2$ and ℓ_3. Its dimension is infinite since the quantum number ℓ runs through zero and all naturals. This Hilbert space decomposes into an infinite series of finite dimensional subspaces whose dimension is $(2\ell + 1)$, which are characterized by fixed eigenvalues $\ell(\ell + 1)$ of $\boldsymbol{\ell}^2$, and which are spanned by the base functions $Y_{\ell m}$ with $m = -\ell, -\ell + 1, \ldots, +\ell$. Situations like this occur frequently in quantum theory. As they are important for the understanding of physical quantum systems let us consider the concept of subspace more closely.

Definition 3.3

A subset $\mathcal{H}_i \subset \mathcal{H}$ of a Hilbert space is said to be a *subspace* of $\mathcal{H}$ if
1. $\mathcal{H}_i$ is a sub-vector space of $\mathcal{H}$, and if
2. $\mathcal{H}_i$ is closed in $\mathcal{H}$.

[2] In the example this unitary transformation is diagonal. This will not be so in the general case.

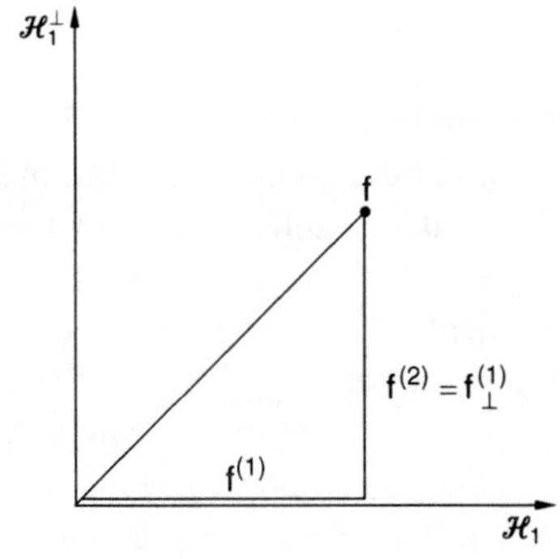

Fig. 3.2 Symbolic representation of the decomposition of an element $f \in \mathcal{H}$ into two orthogonal components $f^{(1)} \in \mathcal{H}_1$, and $f^{(2)}$ in the orthogonal complement $\mathcal{H}_1^\perp$. Note the analogy to the decomposition of a vector over $\mathbb{R}^2$ into two orthogonal components

If the subspace is equipped with the restriction of the metric of $\mathcal{H}$ to $\mathcal{H}_i$, then $\mathcal{H}_i$ is itself a Hilbert space.

Another way of phrasing these criteria is this: Every finite linear combination $\sum \lambda_n \psi_n$ of elements $\psi_n \in \mathcal{H}_i$ is again an element of $\mathcal{H}_i$, and $\mathcal{H}_i$ is closed.

As the "big" Hilbert space has a metric, the orthogonal complement of any subspace (or some other subset W of $\mathcal{H}$, for that matter) is well-defined. It is the set of all those elements of $\mathcal{H}$ which are orthogonal to every element of the subspace (or the more general subset, respectively),

$$W^\perp = \left\{ f \in \mathcal{H} \middle| (g, f) = 0 \text{ for all } g \in W \right\}.$$

The set $W^\perp$ is said to be the *orthogonal complement* of W in $\mathcal{H}$.

If W is a subspace $\mathcal{H}_i$ of Hilbert space the important decomposition theorem applies:

Theorem 3.1 Decomposition Theorem

Every element $f \in \mathcal{H}$ can be written in a unique way as the sum of an element $f^{(i)} \in \mathcal{H}_i$ and an element $f_\perp^{(i)} \in \mathcal{H}_i^\perp$ of the orthogonal complement,

$$f = f^{(i)} + f_\perp^{(i)}.$$

The norm of the component which lies in the orthogonal complement, is given by

$$\| f_\perp^{(i)} \| = \inf_{g \in \mathcal{H}_i} \| f - g \|.$$

It is instructive to highlight the close analogy to the decomposition of an element of Euclidean space $\mathbb{R}^2$ into two orthogonal components: Fig. 3.2 shows this decomposition $f = f^{(1)} + f^{(2)}$ of an element $f \in \mathbb{R}^2$. If we identify the abscissa with the subspace $\mathcal{H}_1 \equiv \mathbb{R}$, the orthogonal complement $\mathcal{H}_1^\perp$ corresponds to the ordinate. We have $f_\perp^{(1)} = f^{(2)}$, the length of this element is the usual geometric distance to the 1-axis.

3.2.3 Dual Space of a Hilbert Space and Dirac's Notation

Returning to Dirac's *bra* and *ket* notation I begin with an example that is known from linear algebra and from mechanics: Let M be a finite dimensional, real, differentiable manifold equipped with a metric $\mathbf{g} = \{g_{ik}\}$. Let $T_m M$ be the tangent space, $T_m^* M$ the cotangent space at the point $m \in M$, $v = \{v^k\}$ and $w = \{w^i\}$ two elements of the tangent space. The metric evaluated with v and w, $\mathbf{g}(w, v)$, gives a real number $\sum_{i,k} w^i g_{ik} v^k$. This means that $\sum_i w^i g_{ik}$, or, written differently, $\mathbf{g}(w, \bullet)$ is a linear form which acts on elements of $T_m M$, that is to say, is an element of the *co*tangent space. It is easy to show that the mapping $w \longmapsto \mathbf{g}(w, \bullet)$ is bijective. When expressed in the language of coordinates this means that to a given tangent vector $v = (v_1, v_2, \ldots)$ there corresponds the unique element $v^* = (\sum_i v^i g_{i1}, \sum_i v^i g_{i2}, \ldots)$ of the cotangent space, and, vice versa, that to each $u^* \in T_m^* M$ there corresponds the tangent vector $u = (\sum_k g^{ik} u_k)$. The vector spaces $T_m^* M$ and $T_m M$ are isomorphic. The isomorphism being established by the metric, one also says that they are *metrically equivalent*. These well-known facts can be translated to Hilbert spaces.

Let $\mathcal{H}$ be a Hilbert space, $\mathcal{H}^*$ the dual space. By definition the dual space contains all linear and continuous functionals $T : \mathcal{H} \to \mathbb{C}$ which when applied to elements of $\mathcal{H}$, yield complex numbers. The property of linearity is immediately clear. Continuity means that the functionals T are bounded, i.e. for all $g \in \mathcal{H}$ there is a finite real number c such that $|T(g)| \leq c \, \|g\|$. Such functionals can be assigned a norm by taking

$$\|T\| = \sup \left\{ |T(g)| \, \middle| \, g \in \mathcal{H}, \|g\| \leq 1 \right\}$$

for all $g \in \mathcal{H}$ whose norm is equal to or less than 1. By analogy to the case discussed above one is led to try the functionals $T_f := (f, \bullet)$ with $f \in \mathcal{H}$ whose action on an arbitrary element $g \in \mathcal{H}$ produces the complex number $T_f(g) = (f, g)$. Obviously, these functionals are linear. By the inequality (3.15) and for all g one has

$$\left| T_f(g) \right| = |(f, g)| \leq \|f\| \, \|g\| \, .$$

Thus, continuity is guaranteed. It is not difficult to show that the norm of T_f equals the norm of f,

$$\left\| T_f \right\| = \|f\| \, .$$

In order to see this consider the action of T_f onto the vector $f/\|f\|$ which is normalized to 1:

$$\begin{aligned} T_f \left(\frac{f}{\|f\|} \right) &= \frac{1}{\|f\|} (f, f) = \|f\| \leq \sup \left\{ \left| T_f(g) \right|, \|g\| \leq 1 \right\} \\ &= \|T_f\| \leq \|f\| \, . \end{aligned}$$

Hence, as $\|T_f\|$ is both less than or equal to and greater than or equal to $\|f\|$, it is equal to $\|f\|$.

Matters are clarified by the following theorem by Riesz and Fréchet:

Theorem 3.2 (Riesz and Fréchet)

For every functional $T \in \mathcal{H}^*$, acting on the elements of the Hilbert space $\mathcal{H}$, there is one and only one element $f \in \mathcal{H}$ such that $T = T_f = (f, \bullet)$ and $\|T\| = \|f\|$.

(For a proof of this theorem see, e. g. [Blanchard and Brüning (2003)].)

The theorem establishes that the dual of a Hilbert space is isomorphic to it. Indeed, the mapping

$$\Gamma : \mathcal{H} \longrightarrow \mathcal{H}^* : f \longmapsto T_f \quad \text{with} \quad T_f(g) = (f, g)$$

is an *isometry*, because $\|\Gamma(f)\| = \|T_f\| = \|f\|$ and, hence, is injective. The Theorem 3.2 shows that it is also surjective.

Notice the close analogy to the example studied above: In the example it is the metric $\mathbf{g}$ which maps the isomorphic vector spaces $T_m M$ and $T_m^* M$ onto one another. Here, the isomorphism between $\mathcal{H}$ and $\mathcal{H}^*$ is effected by the mapping Γ, that is to say, again by means of the scalar product. There is a difference, however: Because of the property

$$\Gamma(\mu_1 f_1 + \mu_2 f_2) = \mu_1^* \Gamma(f_1) + \mu_2^* \Gamma(f_2),$$

the mapping Γ is an *anti-isomorphim*. The isomorphism $\mathcal{H}^* \simeq \mathcal{H}$ provides the justification, a posteriori, of Dirac's bracket notation which was introduced heuristically in Sect. 3.1.1. To every *ket* $|\alpha\rangle \equiv |\psi^{(\alpha)}\rangle$, element of the Hilbert space $\mathcal{H}$, the mapping Γ associates the functional $T_\alpha = \langle\psi^{(\alpha)}| \equiv \langle\alpha|$. The action of T_α on a state $|\beta\rangle$ is $T_\alpha(|\beta\rangle) = \langle\alpha|\beta\rangle$, and is given by the scalar product.

3.3 Linear Operators on Hilbert Spaces

A linear operator $\mathcal{O}$ maps the Hilbert space $\mathcal{H}$, or parts thereof, onto itself, this mapping being $\mathbb{C}$-linear. The action of $\mathcal{O}$ on two vectors $f_1, f_2 \in \mathcal{H}$ being defined, we have

$$\mathcal{O}(\mu_1 f_1 + \mu_2 f_2) = \mu_1 \mathcal{O}(f_1) + \mu_2 \mathcal{O}(f_2), \quad \mu_i \in \mathbb{C} \quad \text{(linearity)}.$$

The definition of an operator must contain, on the one hand, a rule which says how it acts on a given f, and, on the other hand, a domain $\mathcal{D}$ of elements in $\mathcal{H}$ on which it acts. This domain is a subset of $\mathcal{H}$, i.e. a subset $\{f_\lambda \in \mathcal{H}\}$ which is such that every linear combination $\sum c_\lambda f_\lambda$ of its elements belongs to it. Therefore, when talking about a *linear operator* what is really meant is the pair $(\mathcal{O}, \mathcal{D})$ consisting of the operator and its domain of definition $\mathcal{D}$,

$$\mathcal{O} : \quad \mathcal{D} \longrightarrow \mathcal{H} : \quad f \in \mathcal{D} \longmapsto g \in \mathcal{H}.$$

If the domain of definition of $\mathcal{O}$ is dense in $\mathcal{H}$, i.e. if the closure of $\mathcal{D}$ equals $\mathcal{H}$, $\overline{\mathcal{D}} = \mathcal{H}$, the operator is said to be *densely defined*. The set of all nonvanishing elements f of $\mathcal{H}$ which are images of elements $g \in \mathcal{D}$, is called the *range* of the operator. The set of all $g \in \mathcal{D}$ which are mapped to the null element is called the *kernel* of the operator.

The domain of definition of those operators which are relevant for physics, is always related to the concrete physical situation and its description, and we need not go into more academic examples which may be important for the theory of linear operators on Hilbert spaces. Furthermore, we are usually on the safe side if we assume the operators appearing in quantum mechanics to be densely defined.

Definition 3.4 Bounded Operator

An operator $\mathcal{O}$ which is defined everywhere on $\mathcal{H}$ is said to be *bounded* if for all $f \in \mathcal{H}$ the inequality

$$\|\mathcal{O} f\| \leq c\|f\| \tag{3.18}$$

holds where c is a positive constant.

If the operator $\mathcal{O}$ is bounded one defines a norm for it by taking the supremum of $(\mathcal{O}g, \mathcal{O}g) = \|\mathcal{O}g\|^2$ in the set of all states normalized to 1:

$$\|\mathcal{O}\| := \left\{\sup \|\mathcal{O}g\| \,\middle|\, g \in \mathcal{D} \quad \text{with} \quad \|g\| = 1\right\} . \tag{3.19}$$

Operators for which (3.18) does not hold and, hence, which cannot be ascribed a norm, are said to be *unbounded*. Quantum mechanics makes use of both bounded and unbounded operators. For instance, one shows that the position operator x^i is unbounded on $\mathcal{H} = L^2(\mathbb{R}^3)$, the Hilbert space of square integrable functions over three-dimensional space.

Important examples are provided by operators which bear the names of Hilbert and Schmidt. Given a function of two arguments $\boldsymbol{x}, \boldsymbol{y} \in \mathbb{R}^3$ which is square integrable in both arguments, $\int \mathrm{d}^3x \int \mathrm{d}^3y\, K(\boldsymbol{x}, \boldsymbol{y}) < \infty$. Then, if $g(\boldsymbol{x})$ is square integrable, then also

$$f(\boldsymbol{y}) = \int \mathrm{d}^3x\, K(\boldsymbol{y}, \boldsymbol{x}) g(\boldsymbol{x})$$

is a square integrable function. The mapping from g to f, written as $f = \mathbf{K}g$, is an integral operator and is called a *Hilbert-Schmidt operator*.

3.3.1 Self-Adjoint Operators

To operators of quantum mechanics one associates adjoints. For instance, the creation operator $a^\dagger$ in the theory of the quantum oscillator is the adjoint of the annihilation operator a. Whenever an operator describes an observable the adjoint and the original must have the same domain of definition and, hence, are identical. For example, the number operator $N = a^\dagger a$ of the oscillator is an observable and is equal to its adjoint. Matters are made more precise by the following definitions and comments.

Definition 3.5 Adjoint Operator

Given an operator $(\mathcal{O}, \mathcal{D})$ whose domain of definition is dense in $\mathcal{H}$. Consider the scalar products $(f, \mathcal{O}g)$ with $g \in \mathcal{D}$. The set of all f, for which there is an element $f' \in \mathcal{H}$ such that $(f, \mathcal{O}g) = (f', g)$, for all $g \in \mathcal{D}$, defines the domain $\mathcal{D}^\dagger$ of the *adjoint* operator $\mathcal{O}^\dagger$. Let $\mathcal{O}^\dagger f = f'$. Scalar products then follow the rule

$$(f, \mathcal{O}g) = (\mathcal{O}^\dagger f, g) = (g, \mathcal{O}^\dagger f)^* . \tag{3.20}$$

Remarks

1. In the mathematical literature the adjoint operator is marked by an asterisk, i.e. $\mathcal{O}^*$, while complex conjugate numbers are marked by an "over-bar", i.e. $\overline{\lambda}$. I use consistently the standard notation of the physical literature where adjoint operators, as well as hermitian conjugate matrices, are written with the "dagger" symbol $\dagger$, while complex conjugate numbers are written λ^*. It seems reasonable to adhere to this tradition because the "over-bar" will be needed in relativistic quantum field theory of spin-1/2 particles where it has a different meaning.
2. It is important that $\mathcal{D}$ be dense in $\mathcal{H}$. The adjoint operator is uniquely defined and different from zero only if this condition is fulfilled.
3. If the operator $\mathcal{O}$ has an inverse operator $\mathcal{O}^{-1}$ and if the domains $\mathcal{D}(\mathcal{O})$ and $\mathcal{D}(\mathcal{O}^{-1})$ both are dense in $\mathcal{H}$ then one has
$$(\mathcal{O}^\dagger)^{-1} = (\mathcal{O}^{-1})^\dagger ,$$
as expected intuitively.
4. Let $\varphi_i \in L^2(\mathbb{R}^3)$ be square integrable wave functions which are defined on all of $\mathcal{H} = L^2(\mathbb{R}^3)$, and let $\mathcal{O} = \boldsymbol{\mu} \cdot \nabla$ where $\boldsymbol{\mu} = (\mu_1, \mu_2, \mu_3)$ is a triple of complex numbers. The definition (3.20)
$$(\varphi_m, \mathcal{O}\varphi_n) = (\mathcal{O}^\dagger \varphi_m, \varphi_n)$$
and a partial integration in each of the three variables x^i shows that the adjoint is $\mathcal{O}^\dagger = -\boldsymbol{\mu}^* \cdot \nabla$. In particular, if the coefficients μ_k are pure imaginary the adjoint operator and the original are identical.

Definition 3.6 Self-Adjoint Operator

An operator which coincides with his adjoint, $\mathcal{O}^\dagger \equiv \mathcal{O}$, is called *self-adjoint*. In this case one has $\mathcal{D}^\dagger = \mathcal{D}$ and $\mathcal{O}f = \mathcal{O}^\dagger f$ for all $f \in \mathcal{D}$. In particular, for all f and g in the domain of definition $\mathcal{D}$ one has

$$(g, \mathcal{O}f) = (\mathcal{O}g, f) = (f, \mathcal{O}g)^* . \tag{3.21}$$

All expectation values $(f, \mathcal{O}f)$ are real.

These definitions become particularly transparent if a given operator $\mathcal{O}$ admits an "A-representation", i.e. if it is represented by a matrix $\mathbf{O} = \{O_{ik}\}$ (usually infinite dimensional). Its adjoint is obtained by reflection on the main diagonal and by complex conjugation, viz.

$$\mathbf{O}^{\dagger} = (\mathbf{O}^{T})^{*}, \quad (O^{\dagger})_{ik} = O^{*}_{ki} .$$

A self-adjoint operator is represented by a hermitian matrix, $\mathbf{O}^{\dagger} = \mathbf{O}$. Its entries in the diagonal are real, the entries outside the main diagonal are pairwise complex conjugates, i.e. $O^{*}_{ki} = O_{ik}$.

Example 3.9

The self-adjoint operators ℓ_1, ℓ_2, and ℓ_3 describe the cartesian components of orbital angular momentum. In the basis of the states $|\ell m\rangle \equiv Y_{\ell m}$ they are given by the matrices:

$$\begin{aligned}
(Y_{\ell' m'}, \ell_1 Y_{\ell m}) &= \frac{1}{2}\delta_{\ell'\ell}\sqrt{\ell(\ell+1) - mm'}\left\{\delta_{m',m-1} + \delta_{m',m+1}\right\} \\
(Y_{\ell' m'}, \ell_2 Y_{\ell m}) &= \frac{\mathrm{i}}{2}\delta_{\ell'\ell}\sqrt{\ell(\ell+1) - mm'}\left\{\delta_{m',m-1} - \delta_{m',m+1}\right\} \\
(Y_{\ell' m'}, \ell_3 Y_{\ell m}) &= m\,\delta_{\ell'\ell}\delta_{m'm} .
\end{aligned}$$

Exchanging the quantum numbers $\ell' m' \leftrightarrow \ell m$ exchanges the two terms in the curly brackets of ℓ_1 and ℓ_2. The former matrix is real and, hence, symmetric, the latter is pure imaginary. With this change of sign its adjoint is seen to be equal to the original matrix. The matrix describing ℓ_3 is diagonal and real. (Note that these formulae are taken from Sect. 1.9.1 and, specifically, from (1.122).)

Example 3.10

The three Pauli matrices

$$\sigma_1 = \begin{pmatrix} 0 & 1 \\ 1 & 0 \end{pmatrix}, \quad \sigma_2 = \begin{pmatrix} 0 & -\mathrm{i} \\ \mathrm{i} & 0 \end{pmatrix}, \quad \sigma_3 = \begin{pmatrix} 1 & 0 \\ 0 & -1 \end{pmatrix}, \tag{3.22}$$

provide examples of operators acting on elements of a Hilbert (sub-)space with dimension 2. They are used in the description of the rotation group in quantum mechanics and, in particular, of particles with spin 1/2. All three matrices (3.22) are hermitian, $\sigma_i^{\dagger} = \sigma_i$. Their eigenvalues are 1 and -1. The corresponding eigenfunctions are easily determined. Choose the eigenvectors $(1, 0)^T$ and $(0, 1)^T$ of σ_3 as the basis, so that an arbitrary, normalized element of $\mathcal{H}$ reads

$$\alpha\begin{pmatrix} 1 \\ 0 \end{pmatrix} + \beta\begin{pmatrix} 0 \\ 1 \end{pmatrix} \quad \text{with} \quad |\alpha|^2 + |\beta|^2 = 1 .$$

The states with $\alpha = \pm\beta$ and $|\alpha| = 1/\sqrt{2}$ are eigenstates of σ_1 and belong to the eigenvalues 1 and -1, respectively. Likewise, the states with $\alpha = \pm\mathrm{i}\beta$ are the analogous eigenstates of σ_2.

As was shown in Sect. 1.8 the operators which describe *observables* belong to the class of self-adjoint operators. The following statements apply to them:

Theorem 3.3 Eigenvalues and Eigenvectors of Observables

1. The eigenvalues of a self-adjoint operator are *real*.
2. Two eigenvectors which belong to different eigenvalues $\lambda_1 \neq \lambda_2$ are orthogonal.

Proof

1. Let $\mathcal{O}f = \lambda f$, where f is different from the null element. As the squared norm (f, f) is different from zero the relations (3.21) allow to conclude
$$\lambda = \frac{(f, \mathcal{O}f)}{(f, f)} = \frac{(\mathcal{O}f, f)}{(f, f)} = \frac{(f, \mathcal{O}f)^*}{(f, f)} = \lambda^* .$$
2. Let $\mathcal{O}f_1 = \lambda_1 f_1$ and $\mathcal{O}f_2 = \lambda_2 f_2$. Then the chain of equations
$$\lambda_1(f_2, f_1) = (f_2, \mathcal{O}f_1) = (\mathcal{O}f_2, f_1) = \lambda_2(f_2, f_1)$$
holds true. If $\lambda_1 \neq \lambda_2$, the scalar product of f_1 and f_2 must vanish, $(f_2, f_1) = 0$.

The case where there is more than one eigenvector with a given eigenvalue is dealt with in the following definition.

Definition 3.7 Eigenspace

The eigenvectors of a self-adjoint operator $\mathcal{O}$ which belong to the same eigenvalue λ, are elements of a subspace $\mathcal{H}_\lambda$ of $\mathcal{H}$, called *eigenspace* for the eigenvalue λ. The dimension of this subspace is equal to the degree of degeneracy of the eigenvalue λ.

Example 3.11

We remind the reader of two examples that were studied in Chap. 1:
1. The spectrum of the operator $\boldsymbol{\ell}^2$ is $\ell(\ell + 1)$ where ℓ runs through the naturals and zero, $\{\ell\} = (0, 1, 2, \ldots)$. Every fixed value ℓ has $(2\ell+1)$-fold degeneracy. The subspace $\mathcal{H}_\ell$ is spanned by the spherical harmonics $Y_{\ell m}$ with the given ℓ and with m in the set $-\ell, -\ell + 1, \ldots, +\ell$.
2. Except for the ground state, the eigenvalues of the energy of the spherical oscillator are degenerate. The subspace with fixed energy (1.148) is spanned by the eigenfunctions for all triples of eigenvalues (n, ℓ, m) which give this energy, i.e. for which $2n + \ell$ has a fixed value and m has one of the values $-\ell, -\ell + 1, \ldots, +\ell$.

3.3.2 Projection Operators

Operators which project onto subspaces $\mathcal{H}_\lambda$ of the kind considered above, belong to a particularly important class of self-adjoint operators. As we shall see soon their physical interpretation is in terms of "Yes–No" experiments, and they are essential in the general definition of quantum states. Their importance for mathematics lies in the fact that they allow for a rigorous description of the spectra of physically relevant operators, even when these operators are not bounded.

For simplicity, we begin with the example of a Hilbert space spanned by the eigenfunctions of an observable whose spectrum is purely discrete. The notation is chosen such as to remind us of the examples of the previous section, and with the aim of illustrating the new concepts and definitions.

Definition 3.8 Projection Operator

If $\mathcal{H}_\lambda \subset \mathcal{H}$ is a subspace of Hilbert space of dimension K and if $\{\varphi_k\}$, $k = 1, 2, \ldots, K$ is an orthonormal system spanning $\mathcal{H}_\lambda$, the projection of an arbitrary vector $f \in \mathcal{H}$ into the subspace $\mathcal{H}_\lambda$ is defined by

$$P_\lambda f := \sum_{k=1}^{K} \varphi_k(\varphi_k, f) . \tag{3.23}$$

The relation to observables of importance for physics should be obvious: The real number λ may be viewed as the degenerate eigenvalue of an observable $\mathcal{O}$ whose degree of degeneracy is K. For example, the operator P_ℓ describes the projection onto the subspace $\mathcal{H}_\ell$ of fixed ℓ, spanned by the eigenfunctions $\varphi_k \equiv Y_{\ell m}$ of ℓ_3 with $m = -\ell, \ldots, +\ell$. Note, however, that nothing prevents us from choosing other eigenfunctions such as, for instance,

$$\{\psi_m\} = \{\text{eigenfunctions of } \boldsymbol{\ell}^2, \ell \text{ fixed, and of } \ell_\alpha = \ell_1 \cos\alpha + \ell_2 \sin\alpha\} ,$$

the definition (3.23) being independent of the specific choice of the basis. Indeed and more generally, with $\varphi_k = \sum_m \psi_m(\psi_m, \varphi_k)$ one finds

$$P_\lambda f = \sum_{m,m'} \sum_k \psi_m(\psi_m, \varphi_k)(\varphi_k, \psi_{m'})(\psi_{m'}, f) = \sum_m \psi_m(\psi_m, f) .$$

Both sets of eigenfunctions are orthonormal and both span the same space $\mathcal{H}_\lambda$.

Using Dirac's notation, the definition (3.23) takes the form

$$P_\lambda = \sum_{k=1}^{K} |\varphi_k\rangle\langle\varphi_k| = \sum_{k=1}^{K} |\psi_k\rangle\langle\psi_k|$$

which shows again that P_λ does not depend on the basis one has chosen.

Projection operators are self-adjoint. The square of a projection operator equals the operator. One says that projection operators are *idempotent*,

$$\boxed{P_\lambda^\dagger = P_\lambda \quad \text{(a)}, \qquad P_\lambda^2 = P_\lambda \quad \text{(b)}} . \tag{3.24}$$

These assertions are easily proven: (a): With two arbitrary vectors f and g, and by definition (3.23) one calculates

$$(P_\lambda g, f) = \sum_k (\varphi_k, g)^* (\varphi_k, f) = \sum_k (g, \varphi_k)(\varphi_k, f) = (g, P_\lambda f) .$$

(b): As the base functions φ_k are orthonormal, applying the projection operator twice gives

$$P_\lambda(P_\lambda f) = \sum_{k',k} \varphi_{k'}(\varphi_{k'}, \varphi_k)(\varphi_k, f) = \sum_{k',k} \varphi_{k'}\delta_{k'k}(\varphi_k, f) = P_\lambda f .$$

The second equation (3.24) tells us that P_λ has the only eigenvalues 0 and 1. Its physical interpretation is this: If one asks whether a state $f \in \mathcal{H}$ has components with eigenvalue λ of the observable (this is the question whether there is a finite probability to find the eigenvalue λ in a measurement) the answer is "Yes" if the eigenvalue is 1, and "No" if the eigenvalue is 0.

Matters become particularly simple in the case of finite dimensional Hilbert spaces. Let us return to the Example 3.10 of Sect. 3.3.1. As one easily convinces oneself, the operators

$$P_+ = \frac{1}{2}(\mathbb{1} + \sigma_3) = \begin{pmatrix} 1 & 0 \\ 0 & 0 \end{pmatrix} , \quad P_- = \frac{1}{2}(\mathbb{1} - \sigma_3) = \begin{pmatrix} 0 & 0 \\ 0 & 1 \end{pmatrix}$$

are projection operators. They fulfill $P_+^2 = P_+$, $P_-^2 = P_-$, and project onto mutually orthogonal subspaces, i.e. $P_+ P_- = 0 = P_- P_+$, and $P_+ + P_- = \mathbb{1}$. These operators project onto the two eigenvectors, respectively, of the hermitian operator σ_3.

3.3.3 Spectral Theory of Observables

A theorem of central importance in the theory of linear operators on Hilbert space says that every self-adjoint operator can be represented by means of its eigenvalues and of the projection operators which project onto the corresponding subspaces. This representation is called the *spectral representation*. It is uniquely defined and allows for a unified description of bounded and unbounded operators, of operators with purely discrete spectra, with mixed spectra, and with purely continuous spectra alike. As a detailed discussion of the theorems relevant for this topic would go beyond the scope of this chapter, we restrict the discussion to qualitative arguments and a few instructive examples.

The following example shows what the main questions are. Let A be an operator with purely discrete spectrum $\{\lambda_i\}$. Every eigenvalue λ_i defines a subspace $\mathcal{H}_i$ of $\mathcal{H}$ whose dimension equals the degree of degeneracy of the eigenvalue one considers. Denoting the eigenfunctions which belong to a given λ_i by $\varphi_{i,k}$, $k = 1, \ldots, K_i$, the projection operator to $\mathcal{H}_i$ is given by

$$P_i = \sum_{k=1}^{K_i} \varphi_{i,k}(\varphi_{i,k}, \bullet) \equiv \sum_{k=1}^{K_i} |i, k\rangle\langle i, k| ,$$

(the second form using the bracket notation). The subspaces $\mathcal{H}_i$ are pairwise orthogonal. Hence, the sum of two projection operators $P_i + P_j$ with $i \neq j$ is again a projection operator (reader please verify!). As the eigenfunctions of A are complete, the sum of *all* subspaces equals the identity on $\mathcal{H}$,

$$\sum_{i=1}^{\infty} P_i = \mathbb{1} .$$

Thus, one has found a sort of partition of unity in $\mathcal{H}$.

For a given state $f \in \mathcal{H}$ one has

$$f = \sum_{i=1}^{\infty} P_i f, \quad Af = \sum_{i=1}^{\infty} \lambda_i P_i f .$$

This implies, in particular, that the expectation value of the operator A in the state f can be expressed in terms of the spectrum of eigenvalues of A and of the corresponding projection operators

$$\langle A \rangle_f \equiv (f, Af) = \sum_i \lambda_i (f, P_i f) = \sum_i \lambda_i (f, P_i^2 f) = \sum_i \lambda_i \|P_i f\|^2 .$$

In particular, calculating the expectation value of the unity in the state f one has

$$(f, \mathbb{1} f) = 1 = \|f\|^2 = \sum_i (f, P_i f) = \sum_i \|P_i f\|^2 .$$

This formula is easy to interpret if one recalls its analogue in a finite dimensional vector space: The square of the length of a vector equals the sum of the squares of its orthogonal components.

The eigenvalues of operators A whose spectrum is pure discrete, can be ordered, $\lambda_1 < \lambda_2 < \dots$. In all cases of relevance for physics this spectrum is bounded from below, i.e. there is a smallest, finite eigenvalue. All Hamiltonians that we studied in Chap. 1 have this property. This leads one to define a spectral family of projection operators by also ordering the corresponding projection operators P_{λ_i}, and by taking the sum of all projectors for which the corresponding eigenvalue λ_i is smaller than or equal to a given real number,

$$\boxed{E(\mu) := \sum_{i,(\lambda_i \leq \mu)} P_i \quad \text{with} \quad \mu \in \mathbb{R}} . \tag{3.25}$$

As the P_i project onto mutually orthogonal subspaces, the operator $E(\mu)$ is again a projection operator. Its expectation value in a state $f \in \mathcal{H}$

$$\big(f, E(\mu) f\big) = \sum_{i,(\lambda_i \leq \mu)} (f, P_i f)$$

is a real, monotonous, non decreasing function of the real variable μ. In our example it is a step function because every time μ passes an eigenvalue λ_j it increases by a finite amount (unless, by coincidence, f has no component in $\mathcal{H}_j$).

There is a natural ordering relation for projection operators

$$P_j > P_i \,, \quad \text{if} \quad \mathcal{H}_j \supset \mathcal{H}_i \,.$$

It states that P_j is "bigger" than P_i if the subspace $\mathcal{H}_i$ onto which P_i projects, is contained in $\mathcal{H}_j$ as a genuine subspace. Indeed, for all $f \in \mathcal{H}$ one has $(f, P_j f) \geq (f, P_i f)$. The family defined in (3.25) for the observable A has this property: $E(\mu') \geq E(\mu)$ holds whenever $\mu' > \mu$. Furthermore, one has always $\lim_{\varepsilon \to 0} E(\mu + \varepsilon) = E(\mu)$, or, expressed in words, in approaching the real number μ from above, $E(\mu+\varepsilon)$ goes over into the projection operator $E(\mu)$. For $\mu = -\infty$ one has $E(-\infty) = 0$ because at this point the spectrum has not yet begun. On the other hand, at $\mu = +\infty$ the full spectrum is exhausted and, therefore, $E(+\infty) = 1$.

Though we still work here in the framework of the simple example of an observable with discrete spectrum, it is plausible that these concepts can be applied to more general cases of observables with mixed or continuous spectra. Indeed, the properties introduced above are part of the definition of a general spectral family:

Definition 3.9 Spectral Family

A spectral family is a set of projection operators $E(\mu)$ which depend on a real variable μ and have the properties

$$E(\mu') \geq E(\mu) \quad \text{for} \quad \mu' > \mu$$

$$\lim_{\varepsilon \to 0^+} E(\mu + \varepsilon) = E(\mu) \,, \qquad E(-\infty) = 0 \,, \qquad E(+\infty) = 1 \,.$$

The benefit of this concept for quantum theory is twofold: On the one hand, it allows to define the spectrum of eigenvalues of an observable in a way which is applicable equally well to the discrete case, to the mixed case, and to the continuous case. This definition is:

Definition 3.10 Spectrum of Eigenvalues

The spectrum of eigenvalues is the set of all values for which the spectral family is *not constant*.

Indeed, discrete eigenvalues are the points on the real axis where the spectral family is discontinuous. A – possibly piecewise – continuous spectrum occurs where $E(\mu)$ is a continuous function which is not constant and does not decrease.

On the other hand, Definition 3.9 allows to write integrals over the spectra of observables such that the three categories need no longer be distinguished. The expectation value $(f, E(\mu) f)$ of $E(\mu)$ in the state $f \in \mathcal{H}$ is a function which is bounded but not necessarily continuous. As μ increases, this function is either piecewise constant (this happens for μ between two successive discrete eigenvalues), or increases monotonously (in the continuum). As it is confined to the interval between 0 and 1, it is bounded and, therefore, has the properties which are needed in defining Stieltjes integrals such as, e.g.,

$$\int_{-\infty}^{+\infty} \mathrm{d}(f, E(\mu)f) = \|f\|^2 = 1 ,$$

$$\int_{-\infty}^{+\infty} \mathrm{d}(f, E(\mu)f)\mu = (f, Af) \equiv \langle A \rangle_f .$$

Remarks

I do not define this integral in any detail. Rather, I give two examples which are directly related to the physical context and which illustrate the calculus involving Stieltjes integrals.

1. Suppose the real function $g(x)$ (which is meant to be the analogue of $(f, E(\mu)f)$) is piecewise constant on the interval $[a, b]$ of the real axis. Suppose further that it is discontinuous in the points $c_0 = a, c_1, \ldots, c_p = b$ as sketched in Fig. 3.3, and that its value in the interval (c_{k-1}, c_k) is $g(x) = g_k$. Define the differences $\delta_0 = g_1 - g(a)$, $\delta_1 = g_2 - g_1, \ldots, \delta_p = g(b) - g_p$. Finally, let $f(x)$ be a function which is continuous in the interval $[a, b]$. The Stieltjes integral receives contributions only at the points where $g(x)$ is discontinuous. In our example it is given by
$$\int_a^b \mathrm{d}g f(x) = \sum_{i=0}^{p} f(c_i)\delta_i .$$
2. Let $f(x)$ and $g(x)$ be continuous functions in $[a, b]$, with $g(x)$ also differentiable. In principle the definition of the Stieltjes integral requires a series of refinements of the partition of $[a, b]$, and a proof of the convergence of the result. In the first example a further refinement of the interval's partition would not be meaningful because it is defined by the discontinuities of $g(x)$, so that the result would not change. However, if $g(x)$ is continuous the partition can be refined indefinitely. Then, by the mean-value theorem applied to the differences of the values of g, $g(x_{k+1}) - g(x_k) = g'(\xi_k)(x_{k+1} - x_k)$, with ξ_k

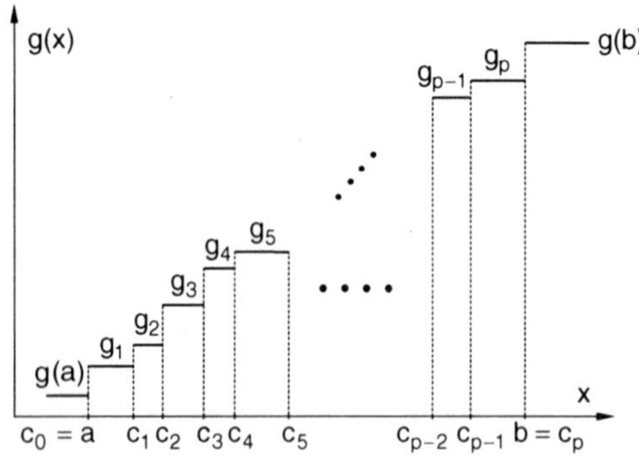

Fig. 3.3 A non-decreasing, piecewise constant function $g(x)$ in the interval $[a, b]$, (arbitrary example) whose Stieltjes integral is taken

an intermediate value between x_k and x_{k+1}, one arrives at the usual Riemann integral,

$$\int_a^b \mathrm{d}g(x)\, f(x) = \int_a^b g'(x)\, \mathrm{d}x\, f(x)\,.$$

The results obtained above in terms of expectation values and of integrals over expectation values, can be written symbolically $\int_{-\infty}^{+\infty} \mathrm{d}E(\mu) f = f$ and $\int_{-\infty}^{+\infty} \mu\, \mathrm{d}E(\mu) f = Af$ or, even more abstractly, in the form of equations for operators,

$$\mathbb{1} = \int_{-\infty}^{+\infty} \mathrm{d}E(\mu)\,, \qquad A = \int_{-\infty}^{+\infty} \mu\, \mathrm{d}E(\mu)\,. \tag{3.26}$$

This abstract formulation receives deeper significance in the light of an important theorem in the theory of linear operators on Hilbert space:

Theorem 3.4 Spectral Theorem

Every self-adjoint operator $(A, \mathcal{D})$ with $\mathcal{D} \subset \mathcal{H}$ admits a uniquely defined spectral family (Definition 3.9), with

$$\mathcal{D} = \left\{ f \in \mathcal{H} \,\middle|\, \int_{-\infty}^{+\infty} \mu^2\, \mathrm{d}(f, E(\mu) f) < \infty \right\},$$

its action on vectors in the domain of definition being

$$Af = \int_{-\infty}^{+\infty} \mu\, \mathrm{d}E(\mu)\, f \quad \text{with} \quad f \in \mathcal{D}\,.$$

Conversely, any operator which is defined by the integral over a spectral family is self-adjoint.

3.3.4 Unitary Operators

A bounded linear operator $A : \mathcal{H}^{(1)} \to \mathcal{H}^{(2)}$ which maps a given Hilbert space $\mathcal{H}^{(1)}$ to itself or to another Hilbert space $\mathcal{H}^{(2)}$ and which conserves the norm,

$$\|Af\|_{\mathcal{H}^{(2)}} = \|f\|_{\mathcal{H}^{(1)}} \quad \text{for all} \quad f \in \mathcal{H}^{(1)}\,,$$

is called an *isometry*.

As one realizes easily, $A^\dagger A$ is the identity $\mathbb{1}_{\mathcal{H}^{(1)}}$ on the initial space $\mathcal{H}^{(1)}$, while $AA^\dagger$ is a projection operator on $\mathcal{H}^{(2)}$: It projects onto the range of A. If the range of A coincides with the entire image space $\mathcal{H}^{(2)}$ one calls the operator a *unitary operator*. As unitary operators are of great importance in quantum theory they are given a symbol of their own, U. They are defined as follows:

Definition 3.11 Unitary Operator

A linear and bounded operator $U : \mathcal{H}^{(1)} \to \mathcal{H}^{(2)}$ which is isometric and surjective, i.e. which conserves the norm and whose range is the whole of $\mathcal{H}^{(2)}$, is called *unitary*.

Unitary operators have a number of properties that we summarize as follows:

1. For any unitary operator U there is an inverse U^{-1} and an adjoint operator $U^\dagger$, both of which are unitary and which fulfill $U^{-1} = U^\dagger$, i.e. the adjoint is the inverse of U.
2. The products of U and its adjoint are the identities on the original space and on the target space, respectively,
$$U^\dagger U = \mathbb{1}_{\mathcal{H}^{(1)}}, \quad UU^\dagger = \mathbb{1}_{\mathcal{H}^{(2)}} .$$
3. If $U : \mathcal{H}^{(2)} \to \mathcal{H}^{(3)}$ and $V : \mathcal{H}^{(1)} \to \mathcal{H}^{(2)}$ are unitary, their product $(UV) : \mathcal{H}^{(1)} \to \mathcal{H}^{(3)}$ is unitary and one has
$$(UV)^\dagger = V^\dagger U^\dagger .$$
4. If the target space can be identified with the original space – this, in fact, is the rule in quantum mechanics – and if φ_n is a countably infinite basis of $\mathcal{H}$, then every unitary operator U has a matrix representation $U_{nm} = (\varphi_n, U\varphi_m)$. These matrices are unitary in the sense known from linear algebra, i.e.
$$UU^\dagger = U^\dagger U = \mathbb{1} \quad \text{or} \quad \sum_i U^*_{im} U_{in} = \delta_{mn} .$$

For the sake of illustration let us consider these properties in the case where the two spaces in the Definition 3.11 can be identified. With this definition and for any pair of elements $f, g \in \mathcal{H}$ the scalar product (f, g) as well as the norms $\|f\|$ and $\|f - g\|$ are invariant under the mapping U,
$$(Uf, Ug) = (f, g), \quad \|Uf\| = \|f\|, \quad \|U(f - g)\| = \|(f - g)\| .$$
As the scalar product, by definition, is nondegenerate, this means that different originals $f \neq g$ have different images $f' = Uf \neq g' = Ug$. The mapping U is surjective. Therefore, it has an inverse U^{-1} and one concludes
$$(Uf, g) = (Uf, UU^{-1}g) = (f, U^{-1}g) \quad \text{for all} \quad f, g \in \mathcal{H} .$$
This shows, indeed, that $U^{-1} = U^\dagger$. One also sees that $(U^\dagger)^\dagger = U$, and, hence, that U is *linear*. Furthermore, the norm of a unitary operator exists and has the value $\|U\| = 1$, cf. (3.19). Finally, one has
$$\begin{aligned}(f, (UV)g) &= (f, U(Vg)) = (U^\dagger f, Vg) = (V^\dagger(U^\dagger f), g)\\ &= ((UV)^\dagger f, g) ,\end{aligned}$$
which proves the property 3.

In a sense unitary operators are generalized rotations. Rotations in the customary physical space are intimately interwoven with a group of unitary transformations in Hilbert space. This will be clarified when studying the rotation group in Chap. 4. Note that we came across a first example in Sect. 3.2.1: Rotations $\mathbf{R}_3(\alpha)$ about the 3-axis in $\mathbb{R}^3$ induce unitary transformations $U(\alpha)$ which belong to a one-parameter group. Indeed, we have $U(\alpha = 0) = \mathbb{1}$, $U(\alpha_2)U(\alpha_1) = U(\alpha_1 + \alpha_2)$ and $U^{-1}(\alpha) = U(-\alpha) = U^\dagger(\alpha)$. With due mathematical scrutiny one shows that for any such unitary operator which can be deformed continuously into the identity $\mathbb{1}$ there is a traceless, self-adjoint operator J such that U can be written as the exponential series

$$U(\alpha) = \exp(-\mathrm{i}\alpha J) \quad \text{with} \quad J^\dagger = J, \quad \operatorname{tr} J = 0 .$$

In analogy to the rotations in $\mathbb{R}^3$ the hermitian operator J is called *generator of infinitesimal unitary transformations*.

Example 3.12

The Pauli matrices (3.22) are distinguished by the fact that they are both hermitian and unitary. Their trace is zero. Exponential series in $(\mathrm{i}\alpha\sigma_k)$ are also unitary matrices. Examples are

$$\mathrm{e}^{\mathrm{i}\phi\sigma_3} = \mathbb{1}\cos\phi + \mathrm{i}\sigma_3 \sin\phi = \begin{pmatrix} \mathrm{e}^{\mathrm{i}\phi} & 0 \\ 0 & \mathrm{e}^{-\mathrm{i}\phi} \end{pmatrix} ,$$

$$\mathrm{e}^{\mathrm{i}\theta\sigma_2} = \mathbb{1}\cos\theta + \mathrm{i}\sigma_2 \sin\theta = \begin{pmatrix} \cos\theta & \sin\theta \\ -\sin\theta & \cos\theta \end{pmatrix} .$$

Note that we have used the fact that all even powers of σ_k are equal to the unit matrix, $(\sigma_k)^{2n} = \mathbb{1}$, while for all odd powers $(\sigma_k)^{2n+1} = \sigma_k$.

3.3.5 Time Evolution of Quantum Systems

A first important example of a unitary operator follows directly from the time dependent Schrödinger equation (1.59). Assume, for simplicity, that the Hamiltonian in

$$\mathrm{i}\hbar\dot{\psi}(t, \boldsymbol{x}) = H\psi(t, \boldsymbol{x})$$

is independent of time. Construct from it the operator

$$U(t, t_0) := \exp\left(-\frac{\mathrm{i}}{\hbar} H(t - t_0)\right) . \tag{3.27}$$

This is a unitary operator which describes the time evolution of a quantum state by a unitary mapping of the initial configuration $\psi(t_0, \boldsymbol{x})$ to the field distribution $\psi(t, \boldsymbol{x})$ at a time earlier or later than t_0,

$$\psi(t, \boldsymbol{x}) = U(t, t_0)\psi(t_0, \boldsymbol{x}) . \tag{3.28}$$

The operator (3.27) itself obeys the Schrödinger equation,

$$\mathrm{i}\hbar\dot{U}(t, t_0) = \mathrm{i}\hbar\frac{\mathrm{d}}{\mathrm{d}t}U(t, t_0) = HU(t, t_0) , \tag{3.29}$$

with initial condition $U(t_0, t_0) = \mathbb{1}$. For an infinitesimal time difference we have

$$\psi(t, \boldsymbol{x}) \approx \psi(t_0, \boldsymbol{x}) + \left.\frac{\mathrm{d}\psi}{\mathrm{d}t}\right|_{t_0} (t - t_0) = \left(\mathbb{1} - \frac{\mathrm{i}}{\hbar} H(t - t_0)\right) \psi(t_0, \boldsymbol{x}) \,.$$

If the time difference $(t - t_0)$ is finite the evolution may be thought of as very many successive infinitesimal steps, making use of Gauss' formula for the exponential function, viz.

$$\lim_{n\to\infty} \left(\mathbb{1} - \frac{\mathrm{i}}{\hbar} H \frac{t - t_0}{n}\right)^n = \exp\left(-\frac{\mathrm{i}}{\hbar} H(t - t_0)\right) \,.$$

Remarks

1. The restriction to time independent Hamiltonians is not really essential. When the Hamiltonian H depends on time the Schrödinger equation $\mathrm{i}\hbar\dot{U} = HU$ applies as before, and the evolution is still described by (3.28). However, the evolution operator no longer is a simple exponential series. It satisfies the integral equation

$$U(t, t_0) = \mathbb{1} - \frac{\mathrm{i}}{\hbar} \int_{t_0}^{t} \mathrm{d}t' H(t') U(t', t_0) \,, \quad \text{with} \quad U(t_0, t_0) = \mathbb{1} \,, \tag{3.30}$$

which is equivalent to (3.29) and which may be solved by an iterative procedure.
2. It is well-known from mechanics that the Hamiltonian function can be interpreted as the generator for infinitesimal canonical transformations which "boosts" the system along its physical orbit. The construction (3.27) and the formula (3.28) show that the Hamiltonian operator of quantum mechanics has a similar interpretation: It boosts the wave function locally.

3.4 Quantum States

Having prepared the ground by assembling the necessary mathematical tools we can now tackle some questions of central importance for physics: the *preparation* and *detection* of quantum states in experiment. We have learnt that states of quantum mechanical systems bear wave properties and, hence, that they can exhibit interference phenomena. Let us first recall what we know about wave phenomena in *classical* physics. In describing classical waves one distinguishes coherent and incoherent situations. Electromagnetic radiation, i.e. visible light, laser beams, radio waves or the like, is realized in rather different forms. For instance, light may be fully polarized, or partially polarized, or not polarized at all, depending on how it was prepared. There is polarization if the wave contains only one polarization component, or if there are fixed phase relations between different components. In turn, if there is no polarization, this means that the components are incoherent, i.e. have no phase correlation at all.

3.4.1 Preparation of States

Quantum mechanics has many similarities to classical wave theory. There are states which are capable of interference without restriction, and, hence, which exhibit the constructive and destructive interference phenomena which are typical for quantum theory. Every state of this kind spans a one-dimensional subspace of Hilbert space. A state of this kind is fixed up to a constant phase and is described by the equivalence class of wave functions

$$\{e^{i\sigma}\psi\}, \quad \sigma \in \mathbb{R}$$

which form a unit ray. This phase degree of freedom is taken care of automatically if one uses the projector onto the corresponding one-dimensional subspace,

$$P_\psi = \psi(\psi, \bullet) \equiv |\psi\rangle\langle\psi| .$$

The expectation value of an observable $\mathcal{O}$ in a fully interfering state is calculated as described in Chap. 1. Assume $\mathcal{O}$ to be bounded and assume φ_n to be an orthonormal base system that spans $\mathcal{H}$. Then

$$\begin{aligned}(\psi, \mathcal{O}\psi) &= \left(\psi, \mathcal{O}\sum_{n=1}^{\infty}\varphi_n(\varphi_n, \psi)\right) = \sum_{n=1}^{\infty}(\varphi_n, \psi)(\psi, \mathcal{O}\varphi_n) \\ &= \sum_{n=1}^{\infty}(\varphi_n, P_\psi \mathcal{O}\varphi_n) = \mathrm{tr}(P_\psi \mathcal{O}) . \end{aligned} \tag{3.31}$$

The sum over n converges absolutely for bounded operators $\mathcal{O}$. In the case of unbounded operators one recurs to the spectral family (3.25) of the operator $\mathcal{O}$, see also (3.26), and defines the Stieltjes Integral

$$\mathrm{tr}(P_\psi \mathcal{O}) := \int \mu \, \mathrm{d}\, \mathrm{tr}(P_\psi E(\mu)) . \tag{3.32}$$

Now, let A be another observable which describes a simple, idealized "source" and let α be one of its eigenvalues. The state ψ is created through a measurement of A by fixing the eigenvalue α. As before, Sect. 3.3.2, the projection operator to the subspace with fixed eigenvalue α of A is denoted by P_α. Then the following alternatives must be considered:

1. The eigenvalue α is not degenerate. In this case $P_\psi = P_\alpha$, the state ψ is equal to the eigenstate φ_α of A, modulo constant phases, which belongs to α.
2. The eigenvalue α is degenerate, its degree of degeneracy is K_α. The corresponding subspace $\mathcal{H}_\alpha$ has dimension K_α, it is spanned by the eigenfunctions $\{\varphi_{\alpha i}, i = 1, \ldots, K_\alpha\}$ (or any other unitarily equivalent base system). The state ψ, or, for that matter, P_ψ is prepared by a measurement of A and by sorting out the eigenvalue α. But how is it to be described?

One might be tempted to try the ansatz

$$P_\psi = P_\chi \quad \text{with} \quad \chi = \varphi_{\alpha j} \quad \text{or} \quad \chi = \sum_{i=1}^{K_\alpha} \varphi_{\alpha i}\, c_i ,$$

where c_i are complex numbers fulfilling the normalization condition $\sum_i |c_i|^2 = 1$. Yet, as one verifies immediately, this approach cannot be correct: The state described by P_χ contains components with well-defined phase relations and, hence, is apt to interferences without restriction. Apparently it contains *more* information than what was prepared by the measurement.

An example may be helpful in illustrating this point. Suppose that we had developed an apparatus which allows to measure the eigenvalue $\ell(\ell+1)$ of the squared orbital angular momentum and that we had applied a filter which is transparent only for the eigenvalue $\ell = 1$. Thus, this *source* prepares a state which is known to lie in the subspace $\mathcal{H}_{\ell=1}$ but about which nothing else is known. Every coherent superposition of the base states Y_{1m}, $\chi = \sum Y_{1m} c_m$, would contain information about the spatial orientation of angular momentum. For example, a state χ with $c_{+1} = 1/\sqrt{2}$, $c_0 = 0$, and $c_{-1} = -1/\sqrt{2}$ would also be eigenstate of ℓ_1 with eigenvalue $\mu = 0$, (cf. the Example 1.10 in Sect. 1.9.1), even though we did not impose this additional information. Similarly, a state

$$\chi = \frac{1}{\sqrt{N}} \sum_{m=-1,0,+1} Y_{1m} c_m \,, \quad N = |c_{-1}|^2 + |c_0|^2 + |c_{+1}|^2 \,,$$

where we choose $|c_{-1}| = |c_0| = |c_{+1}|$, would contain the following phase-dependent information about the expectation values of the components ℓ_1, ℓ_2, and ℓ_3

$$\langle \ell_3 \rangle_\chi = \frac{1}{N}(|c_{+1}|^2 - |c_{-1}|^2) = 0 \,,$$

$$\left\langle \ell_{1/2} \right\rangle_\chi = \frac{\sqrt{2}}{N} \mathrm{Re}/\mathrm{Im}(c_{+1}^* c_0 + c_0^* c_{-1}) \,.$$

This contradicts our intuition. If we set out to prepare a state which carries the quantum number $\ell = 1$ but for which all directions are equivalent, then the expectation values of all three components must be equal and, in fact, equal to zero![3]

This example suggests the solution of our problem: If, indeed, no more than the information "α" is available about the prepared state, it must be described by an incoherent statistical mixture of components, in much the same way as in classical, unquantized physics. This means, that to every substate $\varphi_{\alpha i}$ we must associate a real, positive-semidefinite weight w_i such that

$$0 \le w_i \le 1 \,, \quad \sum_{i=1}^{K_\alpha} w_i = 1 \,. \tag{3.33}$$

The real number w_i is the probability that a given particle is found in the state $\varphi_{\alpha i}$, or, more precisely, in the unit ray $P_{\alpha i} = |\varphi_{\alpha i}\rangle\langle\varphi_{\alpha i}|$. These classical probabilities do

[3] The formulae of this example are written in such a way that one can easily specialize to the eigenfunctions of ℓ_1 or ℓ_2, and check that they yield the correct eigenvalues.

not interfere. The expectation value of an observable $\mathcal{O}$ in a state described in this fashion, is given by

$$\langle \mathcal{O} \rangle_\psi = \sum_{i=1}^{K_\alpha} w_i (\varphi_{\alpha i}, \mathcal{O}\varphi_{\alpha i}) . \tag{3.34}$$

The actual values of the weights w_i depends on the history of the quantum state that was sent through the filter "A" – an important aspect that will be taken up further below.

Let us return once more to the example studied above in which the filter generated the value $\ell = 1$. If there are reasons to assume that in the so-prepared state all directions are equivalent then the weight factors w_{+1}, w_0, and w_{-1} must be equal for all three substates and, by the normalization (3.33), must have the value $1/3$. The expectation value (3.34) then is $\langle \mathcal{O} \rangle_\psi = \sum_m (Y_{1m}, \mathcal{O} Y_{1m})/3$. Indeed, with this choice the expectation values of all components ℓ_k are equal to zero

$$\langle \ell_1 \rangle_\psi = \langle \ell_2 \rangle_\psi = \langle \ell_3 \rangle_\psi = 0 .$$

In the case of ℓ_1 and ℓ_2 this follows from the formulae Sect. 1.9.1, in the case ℓ_3 the contributions of $m = +1$ and of $m = -1$ cancel.

We summarize once more the previous, still preliminary results. A quantum state is sent through the "A"-filter which identifies the eigenvalue α of A and which is transparent only for components which have this property. We define the operator

$$\boxed{W := \sum_i w_i \, P_{\alpha i} \quad \text{with} \quad 0 \le w_i \le 1 , \quad \sum_i w_i = 1} . \tag{3.35}$$

The weight w_i which is real and positive-semidefinite, represents a *classical* probability (not dependent on any interferences) to find the substate with quantum numbers (α, i) in a subsequent measurement. Its value depends on the nature of the state *before* the preparing measurement α. The operator W is called *statistical operator*. It provides the most general description of a quantum state. In case only one single weight is different from zero and, by the normalization (3.35), is equal to 1, say $w_k = 1$, $w_i = 0$ for all $i \neq k$, the state is fully susceptible to interference. For that reason, it is called a *pure state*. Examples for pure states are provided by the wave functions derived in Chap. 1 which were normalizable solutions of the Schrödinger equation. In turn, if at least two weights, say w_k and w_j, are different from zero then the state can exhibit interferences only *within* the components (α, k) and (α, j), respectively, but not between these two. A state of this kind is called *mixed state*, or *mixed ensemble*. In either case the expectation value of an observable $\mathcal{O}$, in the state described by the statistical operator W, is given by the trace of the product of $\mathcal{O}$ and W,

$$\langle \mathcal{O} \rangle = \operatorname{tr}(\mathcal{O} W), \quad \langle \mathbb{1} \rangle = \operatorname{tr}(\mathbb{1} W) = \operatorname{tr} W = 1 .$$

The second formula expresses the normalization of the state.

3.4.2 Statistical Operator and Density Matrix

The concept of state elaborated by the arguments and the examples of the previous section is made more precise by the following postulate:

Postulate 3.1 Description of Quantum States

A quantum state is described by a statistical operator W. This operator is a convex linear combination of projection operators with real, nonnegative coefficients. It is self-adjoint and is normalized to 1, i.e. fulfills the condition $\operatorname{tr} W = 1$. The outcome of measurements of physical observables $\mathcal{O}$ are described by the expectation value

$$\langle \mathcal{O} \rangle = \operatorname{tr}(W \mathcal{O}) . \tag{3.36}$$

The trace of its square W^2 contains the information whether the state is a pure state or a mixed ensemble. If $\operatorname{tr} W^2 = \operatorname{tr} W = 1$ the state is a *pure* state; if $\operatorname{tr} W^2 < \operatorname{tr} W$ and, hence, $\operatorname{tr} W^2 < 1$ the state is a *mixed state*.

Remarks

1. Recall that the trace of an operator is calculated as exemplified by (3.31), (3.32), and (3.34).
2. The convex sum of N objects $\mathcal{O}_i$ is defined by

$$\sum_{i=1}^{N} w_i \mathcal{O}_i , \quad \text{with} \quad \sum_{i=1}^{N} w_i = 1 .$$

It is a weighted sum with positive-semidefinite factors $0 \leqslant w_i \leqslant 1$.

It would be worth a more precise mathematical analysis how to calculate such traces, if they exist, and to justify that if $\operatorname{tr} W^2 < 1$ there is no pure state. Both questions will become intuitively clear when we study a matrix representation of W, instead of the operator itself. Let B be a self-adjoint operator which is defined on the given Hilbert space and whose spectrum is fully discrete. Its eigenfunctions ψ_m provide a basis of $\mathcal{H}$, which means that we can go over to the "B"-representation of the statistical operator W, in the spirit of representation theory (Sect. 3.1),

$$\varrho_{mn} = (\psi_m, W \psi_n) . \tag{3.37}$$

The matrix $\boldsymbol{\varrho}$ obtained in this way is called *density matrix*. Its properties are summarized in the following definition.

Definition 3.12 Density matrix

The density matrix is a matrix representation of the statistical operator in an arbitrary basis of Hilbert space. Its properties are:

1. It is hermitian $\boldsymbol{\varrho}^\dagger = \boldsymbol{\varrho}$, its eigenvalues are real, positive-semidefinite numbers between 0 and 1, $0 \leq w_j \leq 1$, i.e. $\boldsymbol{\varrho}$ is a positive matrix.

2. It obeys the invariant inequality

$$0 < \operatorname{tr} \varrho^2 \leq \operatorname{tr} \varrho = 1 \,. \tag{3.38}$$

3. It serves to characterize the quantum state by the following criteria:
   1. If $\operatorname{tr} \varrho^2 = \operatorname{tr} \varrho = 1$ the state is a *pure state*,
   2. if $\operatorname{tr} \varrho^2 < \operatorname{tr} \varrho = 1$ the state is a *mixed state*.
4. Expectation values of an observable $\mathcal{O}$, in the B-representation, are given by the trace of the product of ϱ and the matrix representation $\mathcal{O}_{pq}$ of the observable,

$$\langle \mathcal{O} \rangle = \operatorname{tr}(\varrho \mathcal{O}) = \sum_{m,n} \mathcal{O}_{mn} \varrho_{nm} \,. \tag{3.39}$$

Let us return to the preparing measurement via the eigenvalue α of A and let us expand the states $\varphi_{\alpha i}$ in terms of the eigenstates of B,

$$\varphi_{\alpha i} = \sum_m \psi_m c_m^{(\alpha i)}$$

where $c_m^{(\alpha i)} = (\psi_m, \varphi_{\alpha i})$. We then find

$$\begin{aligned} \varrho_{mn} &= \sum_i w_i (\psi_m, P_{\alpha i} \psi_n) = \sum_i w_i (\psi_m, \varphi_{\alpha i})(\varphi_{\alpha i}, \psi_n) \\ &= \sum_i w_i \, c_m^{(\alpha i)} c_n^{(\alpha i)*} \,. \end{aligned}$$

Taking the trace gives

$$\operatorname{tr} \varrho = \sum_i w_i \sum_m \left| c_m^{(\alpha i)} \right|^2 = \sum_i w_i = 1 \,,$$

while the trace of the square gives

$$\begin{aligned} \operatorname{tr} \varrho^2 &= \sum_{i,k} w_i w_k \sum_{mn} c_m^{(\alpha i)} c_n^{(\alpha i)*} c_n^{(\alpha k)} c_m^{(\alpha k)*} \\ &= \sum_{i,k} w_i w_k \delta_{\alpha i, \alpha k} = \sum_i w_i^2 \leq \sum_i w_i = 1 \,. \end{aligned}$$

If the basis ψ_m is chosen to coincide with that of the eigenfunctions $\varphi_{\beta j}$ of the "filter" A, then ϱ, though an infinite dimensional matrix, has nonvanishing entries only in the subspace $\mathcal{H}^{(\alpha)}$ which pertains to the eigenvalue α. In this subspace it is diagonal and has the explicit form $\varrho = \operatorname{diag}\left(w_1, w_2, \ldots, w_{K_\alpha}\right)$.

Example 3.13

In the two-dimensional Hilbert space spanned by the eigenvectors of σ_3, (3.22), define

$$\varrho = \begin{pmatrix} w_+ & 0 \\ 0 & w_- \end{pmatrix} = \frac{1}{2}(\mathbb{1} + P\sigma_3) \,,$$

with $w_+ + w_- = 1$ and $P := w_+ - w_-$ so that the number P lies between -1 and 1. The trace of ϱ equals 1 while the trace of ϱ^2 equals $(1 + P^2)/2$. Indeed,

$$\varrho^2 = \frac{1}{2}\left(\frac{1}{2}(1 + P^2)\mathbb{1} + P\sigma_3\right),$$

whose trace receives a factor 2 from the unit matrix, while tr $\sigma_3 = 0$. If $P = \pm 1$ the density matrix describes pure states, if $|P| < 1$ it describes mixed ensembles. In particular, if $P = 0$ the weights of the two base states are equal.

Consider now the observable $\mathcal{O} := \sigma_3/2$ (in Chap. 4 we will learn that it represents the 3-component of spin of a spin-1/2 particle). Its expectation value in the state defined by ϱ follows from $\text{tr}(\varrho\sigma_3) = w_+ - w_- = P$. The states with $(w_+ = 1, w_- = 0)$ and $(w_+ = 0, w_- = 1)$ are pure states. The first of these describes particles which are fully polarized in the positive 3-direction, the second describes particles polarized in the negative 3-direction. A state for which both weights are different from zero describes a particle beam with partial polarization. In the special case $w_+ = w_-$ and, hence, $P = 0$, this beam is unpolarized. In a measurement of the observable $\mathcal{O}$ the probabilities to find the eigenvalues $+1/2$ or $-1/2$ are the same.

At this point it may be useful to return for a while to the discussion of classical probabilities with positive semi-definite weights in Sect. 1.2.1, and to note the differences to the quantum case. As we have learnt how to calculate the expectation values in the components of a statistical ensemble the questions posed towards the end of Sect. 1.2.2 now are answered.

3.4.3 Dependence of a State on Its History

In Sect. 3.4.1 we left the question unanswered of how to obtain the weights w_i which determine the incoherent mixture of eigenstates with quantum numbers α of the "filter" A. The answer to this question which is worked out in this section, leads to new aspects which may be surprising but are typical for quantum theory.

Of course, *before* its preparation by means of the observable A, the system is in a quantum state that may be a pure or a mixed state. In order to cover the most general case we assign a statistical operator $W^{(i)}$ to the initial state ("i" for *initial*) which satisfies the Postulate 3.1. We send this state through the filter α, as described in the two preceding sections, i.e. by shielding all eigenvalues different from α, and construct the statistical operator $W^{(f)}$ ("f" for *final*) which describes the prepared state. Let

$$P_\alpha := \sum_{i=1}^{K_\alpha} P_{\alpha i}$$

be the projection operator to the subspace $\mathcal{H}_\alpha$ which corresponds to the possibly degenerate eigenvalue α of A. With these definitions and notations the relation between the statistical operators *before* and *after* the preparing measurement is given by

$$\boxed{W^{(f)} = P_\alpha W^{(i)} P_\alpha / \mathrm{tr}(P_\alpha W^{(i)} P_\alpha)} \,. \tag{3.40}$$

The numerator of this formula contains the projection P_α to the subspace $\mathcal{H}_\alpha$, to the right and to the left of $W^{(i)}$. The denominator is a real number which is chosen such that $W^{(f)}$ is normalized. The following arguments show that (3.40) does indeed describe the desired preparation:

1. The product $P_\alpha W^{(i)} P_\alpha$ is a self-adjoint operator on $\mathcal{H}$, its trace is real and positive. Therefore, the operator $W^{(f)}$ is self-adjoint. As for all eigenvalues β of A which are different from α, one has $W^{(f)} P_\beta = 0$, its action is different from zero only in the subspace $\mathcal{H}_\alpha$. For states in $\mathcal{H}_\alpha$, on the other hand, we have
$$W^{(f)} \varphi_{\alpha j} = N \sum_{k=1}^{K_\alpha} \varphi_{\alpha k} (\varphi_{\alpha k}, W^{(i)} \varphi_{\alpha j})$$
where the normalization factor
$$N = \frac{1}{\mathrm{tr}(P_\alpha W^{(i)} P_\alpha)} = \frac{1}{\sum_{k=1}^{K_\alpha} (\varphi_{\alpha k}, W^{(i)} \varphi_{\alpha k})}$$
is real and positive.
2. Let $\chi_m = \sum_{i=1}^{K_\alpha} \varphi_{\alpha i} c_i^{(m)}$ be an arbitrary element of $\mathcal{H}_\alpha$, P_{χ_m} the corresponding projection operator. The probability to find the system in this state, is given by $\mathrm{tr}(W^{(i)} P_{\chi_m})$ before, by $\mathrm{tr}(W^{(f)} P_{\chi_m})$ after the preparation. Since the preparation fixes only the eigenvalue α but no further property, these two probabilities must be proportional, with a constant of proportionality which is independent of the element χ_m that was chosen. In other terms, for all $\chi_m, \chi_n \in \mathcal{H}_\alpha$ the condition
$$\frac{\mathrm{tr}(W^{(i)} P_{\chi_m})}{\mathrm{tr}(W^{(i)} P_{\chi_n})} = \frac{\mathrm{tr}(W^{(f)} P_{\chi_m})}{\mathrm{tr}(W^{(f)} P_{\chi_n})}$$
must be fulfilled.
3. The eigenvalues and eigenvectors of $W^{(f)}$ are obtained by diagonalizing the matrix $(\varphi_{\alpha k}, W^{(i)} \varphi_{\alpha j})$, and by multiplying the result by the normalization factor N. For every element $\chi \in \mathcal{H}_\alpha$ one has $(\chi, W^{(f)} \chi) = N(\chi, W^{(i)} \chi) \geq 0$. Thus, the requirement 2 is fulfilled. This last equation says also that $W^{(f)}$ is positive, i.e. that its eigenvalues, the weights $w_j^{(f)}$, are positive semi-definite. Finally, as $\mathrm{tr}\, W^{(f)} = 1$, the sum of the weights is one, $\sum_{j=1}^{K_\alpha} w_j^{(f)} = 1$.

The formula (3.40) is best illustrated by some examples. It allows for various possibilities of preparing quantum states:

1. A state that was pure before the preparation may remain to be a pure state. This happens, for example if the initial state is an eigenstate of A, i.e. if $W^{(i)} = P_{\beta k}$. The filter "α" either confirms this state, or gives zero,
$$W^{(f)} = \delta_{\alpha\beta} W^{(i)} \,.$$

2. Filtering an initially mixed state may produce a pure state. For instance, the filter could choose from $W^{(i)} = \sum_\mu w_\mu P_\mu$ the specific state with quantum numbers (μk) in $\mathcal{H}_\mu$.
3. Conversely, a pure state can be turned into a mixed ensemble. Qualitatively speaking, this will happen if one does measure the filter observable A, that is to say, if one records the measured eigenvalue for each individual event, but allows part or all of the spectrum to go through the filter.

   For this purpose consider two observables E and F which, for simplicity, are assumed to have discrete spectra, but which do not commute. The eigenfunctions of E are denoted by $\varphi_\mu \equiv |\mu\rangle$, those of F by $\psi_a \equiv |a\rangle$. Let the observable F be the filter, and let this filter be exposed to the pure state $W^{(i)} = P_\mu$, eigenstate of E. As $[E, F] \neq 0$, the operators E and F have no common eigenfunctions. Decomposing the given eigenstate of E in terms of eigenfunctions of F, one has

$$P_\mu \equiv |\mu\rangle\langle\mu| = \sum_{a\,a'} c_a^{(\mu)} c_{a'}^{(\mu)*} |a\rangle \langle a'| \,.$$

   The preparation filter F is designed such that it selects a subset Δ of eigenvalues of F without actually measuring the eigenvalue for each individual event. This means that on the right-hand side of the formula (3.40) we must insert the projection operator $P_\Delta = \sum_{a\in\Delta} P_a$. One then calculates

$$\begin{aligned} W^{(f)} &= \frac{P_\Delta W^{(i)} P_\Delta}{\operatorname{tr}(P_\Delta W^{(i)} P_\Delta)} \\ &= \frac{1}{\operatorname{tr}(P_\Delta W^{(i)} P_\Delta)} \sum_{a\in\Delta} \sum_{a'\in\Delta} |a\rangle \sum_{b,b'} c_b^{(\mu)} c_{b'}^{(\mu)*} \langle a|b\rangle \langle b'|a'\rangle \langle a'| \\ &= \frac{1}{\sum_{a\in\Delta} |c_a^{(\mu)}|^2} \sum_{b,b'\in\Delta} |b\rangle c_b^{(\mu)} c_{b'}^{(\mu)*} \langle b'| \,. \end{aligned} \tag{3.41}$$

   The state prepared in this manner still has fixed phase relations. Hence, like the initial state, it is a pure state. This is confirmed by verifying that the trace of $W^{(f)2}$ is equal to 1,

$$\begin{aligned} \operatorname{tr} W^{(f)2} &= \sum_{d=1}^{\infty} \langle d|W^{(f)2}|d\rangle \\ &= \frac{1}{\left(\sum_{a\in\Delta} |c_a^{(\mu)}|^2\right)^2} \sum_{b\in\Delta} \left|c_b^{(\mu)}\right|^2 \sum_{b'\in\Delta} |c_{b'}^{\mu}|^2 = 1 \,. \end{aligned}$$

   In turn, if the preparation is designed such that the eigenvalues of the filter F are measured, one by one, and all those which do not fall in the interval Δ are rejected, then (3.41) must be replaced by

$$W^{(f)} = \frac{1}{\sum_{a\in\Delta} |c_a^{(\mu)}|^2} \sum_{b\in\Delta} |b\rangle \left|c_b^{(\mu)}\right|^2 \langle b| \,. \tag{3.42}$$

   This is a statistical mixture because $\sum_{b\in\Delta} |c_b^{(\mu)}|^4 < (\sum_{a\in\Delta} |c_a^{(\mu)}|^2)^2$ and, therefore, the trace $\operatorname{tr} W^{(f)2} < 1$ is smaller than one. By the measurement of F all phase relations are destroyed, the states $b \in \Delta$ are contained in the final state with the real weights

$$w_b = \frac{|c_b^{(\mu)}|^2}{\sum_{a\in\Delta} |c_a^{(\mu)}|^2}, \qquad b \in \Delta\,.$$

We meet here a characteristic property of quantum mechanics which seems very peculiar from a classical viewpoint: If in the preparation process the actual values of the filter observable F are recorded, then all phase correlations are lost, and the new, final state is a mixed ensemble. In either case, (3.41) and (3.42), at least some information on the state *before* the preparation is conserved. In the first case, through the expansion coefficients $c_b^{(\mu)}$ with $b \in \Delta$, in the second case through the relative weights w_b. The case where only one single eigenstate of F goes through, all others being rejected, is an exception. In this case $W^{(f)} = P_b$, all information on the state of the system before the preparation measurement is lost. That nature does indeed work in this way is confirmed by experiment.

3.4.4 Examples for Preparation of States

Assuming the observables E and F to have discrete spectra is no serious restriction and we drop this assumption in what follows. We study two examples in which the initial state is an eigenstate of $\boldsymbol{p}$, the momentum operator,

$$|\mu\rangle \equiv |\boldsymbol{p}\rangle = \frac{1}{(2\pi\hbar)^{3/2}}\, e^{i\boldsymbol{p}\cdot\boldsymbol{x}/\hbar}\,.$$

The "filter", i.e. the observable serving the purpose of preparation is taken to be $F = \boldsymbol{\ell}^2$. These observables do not commute. However, we know from Sect. 1.9.3 how to relate the eigenfunctions of $\boldsymbol{p}$ with those of $\boldsymbol{\ell}^2$, cf. (1.136).

Example 3.14

Let the filter F be set up such that only one single eigenvalue $\ell(\ell+1)$ is accepted but the corresponding m-values are not discriminated. Before preparation we have

$$W^{(i)} = P_{\boldsymbol{p}} = |\boldsymbol{p}\rangle\langle\boldsymbol{p}| = \sum_{\ell' m'}\sum_{\ell'' m''} d_{\ell' m'} d^*_{\ell'' m''}\, |\ell' m'\rangle\langle \ell'' m''|\,,$$

where the coefficients are obtained from (1.136)

$$d_{\ell m} = \frac{4\pi}{(2\pi\hbar)^{3/2}} i^\ell j_\ell(kr) Y^*_{\ell m}(\widehat{\boldsymbol{p}}) \quad \text{with} \quad k = \frac{1}{\hbar}|\boldsymbol{p}|\,.$$

We must insert $P_\alpha \equiv P_\ell = \sum_{m=-\ell}^{\ell} |\ell m\rangle\langle \ell m|$ in the formula (3.40) for the operator $W^{(f)}$. Then, calculating $W^{(f)}$ as in (3.41) one finds

$$W^{(f)} = \frac{1}{\sum_m |d_{\ell m}|^2} \sum_{m,m'=-\ell}^{\ell} d_{\ell m} d^*_{\ell m'}\, |\ell m\rangle\langle \ell m'|\,.$$

Of course, we already know that this still is a pure state. However, it is instructive to confirm this in the present example in another, more direct way. If the 3-direction is taken along $\boldsymbol{p}$, $\boldsymbol{p} = p\hat{\boldsymbol{e}}_3$, only partial waves with $m = 0$ contribute,

$$Y_{\ell m}(\widehat{\boldsymbol{p}}) = Y_{\ell m}(\theta = 0, \phi) = \sqrt{\frac{2\ell + 1}{4\pi}}\delta_{m0} .$$

In this case we find $W^{(f)} = |\ell\, 0\rangle\langle\ell\, 0|$ which obviously describes a pure state.

Example 3.15

Choose the 3-direction to be along the direction of the momentum, $\boldsymbol{p} = p\hat{\boldsymbol{e}}_3$. Assume that the filter records the values of ℓ but does not shield any of them. When all values of ℓ are collected in this way we have, like in (3.42),

$$W^{(f)} = \frac{1}{\sum_{\ell=0}^{\infty} |d_{\ell\, 0}|^2} \sum_{\ell=0}^{\infty} |d_{\ell\, 0}|^2 |\ell\, 0\rangle\langle\ell\, 0| \quad \text{with}$$

$$|d_{\ell\, 0}|^2 = \frac{4\pi}{(2\pi\hbar)^3}(2\ell + 1) j_\ell^2(kr) .$$

It is illuminating to analyze these formulae more closely, by making use of known properties of Bessel functions. First one notes that $\sum_0^\infty (2\ell + 1) j_\ell^2(kr) = 1$, cf. [Abramowitz and Stegun (1965)] (10.1.50), and, hence, that

$$W^{(f)} = \sum_{\ell=0}^{\infty} (2\ell + 1) j_\ell^2(kr) |\ell\, 0\rangle\langle\ell\, 0| \equiv \sum_{\ell=0}^{\infty} w_\ell |\ell\, 0\rangle\langle\ell\, 0| .$$

Let us fix the product (kr) and ask where w_ℓ assumes a maximum, as a function of ℓ. For values of ℓ which are not too small, the formula (10.1.59) of [Abramowitz and Stegun (1965)] yields the answer: The first maximum of $j_\ell^2(z)$, and, hence, of w_ℓ, occurs at $z \approx (\ell + 1/2)$. Thus, we recover in an approximate way the classical relation between the modulus of the orbital angular momentum and the impact parameter

$$pr = \hbar kr (= \ell_{\text{cl}}) \approx \hbar \left(\ell + \frac{1}{2}\right) .$$

This is the relation that we found in Sect. 1.9.3 in the analysis of free solutions of the radial Schrödinger equation.

3.5 A First Summary

At this point of the development we established a good deal of the foundations on which quantum theory rests. Before turning to further important applications it seems appropriate to halt for a while, and to summarize some of the essential features which are typical for quantum theory. I do this by means of a list of key concepts, each of which is accompanied by a short abstract.

Observables: By definition, observables describe measurable, hence classical variables. To every observable there corresponds a uniquely defined, self-adjoint operator. These operators are defined on Hilbert space, or parts thereof, whose elements are used to describe states of physical systems. Their eigenvalues which are real, correspond to the set of values that one will find in individual measurements.
Quantization: In most cases, canonical mechanics serves as a model for the quantization of classical observables. One postulates Heisenberg commutators for pairs of canonically conjugate variables. The prime example are the commutators (3.8) for coordinates and momenta.
An alternative is provided by path integral quantization to which we turn in Sect. 7.7.
States: In the most general case a quantum state is described by a statistical operator (3.35), expectation values of observables in this state being given by (3.36). An equivalent description is provided by the density matrix, Definition 3.12, which is a matrix representation of the statistical operator. The criterion (3.38) serves the purpose of distinguishing between pure and mixed states. Quantum mechanics makes quantitative predictions only for *ensembles*, i.e. either for very many, identically prepared, systems, or for a very large number of measurements on a single system which is always prepared in exactly the same way.
Preparation Measurements: The statistical operator is fixed by a "filter", i.e. by way of measuring an observable and by a choice of its eigenvalues. The relation between the initial state, described by $W^{(i)}$, and the final state prepared by the filter and described by the statistical operator $W^{(f)}$, is given by the formula (3.40). If the projection operator P_α in (3.40) projects onto a one-dimensional subspace, the prepared state is a pure state. However, while information on this state is optimal, any knowledge on the state *before* preparation gets lost. Only if the prepared state is a mixed ensemble does it still contain (though only partial) information on its past.
Time Evolution: The evolution of a quantum system is determined by its Hamiltonian H. In classical mechanics the Hamiltonian function is the generating function for those infinitesimal canonical transformations which boost the system along its physical orbits. The quantum Hamiltonian, in turn, determines the evolution mapping $U(t, t_0)$ which transports the initial distribution of a Schrödinger field at time t_0 to its present form at time t. The operator $U(t, t_0)$ is a solution of the Schrödinger equation in the form of (3.29), or, equivalently, of the integral Equation (3.30). The time dependence of the expectation values of an observable $\mathcal{O}$ is given by the following prescription. If W denotes the statistical operator which describes a state prepared at time t_0 then, as time goes by, this state evolves under the influence of the Hamiltonian H in such a way that measurements of the observable $\mathcal{O}$, at times $t \neq t_0$, are given by

$$\boxed{\langle \mathcal{O} \rangle_t = \mathrm{tr}[U(t, t_0) W U(t, t_0)^\dagger \mathcal{O}]} \,. \tag{3.43}$$

Remarks

The experience gained in Chaps. 1 and 2 shows that this set of postulates is sufficient for treating a number of important applications of quantum mechanics. Yet, two important questions remain unanswered:

The first of these is about the *completeness* of the description. What we have in mind here is that we do not know yet how many, mutually commuting observables are needed to describe a given physical system. In other words, this is the question how many commuting observables must be known to define a pure state such as a beam of identically prepared electrons. The answer is given by still another postulate that we will be able to state only after having developed the description of the spin of particles.

The second question concerns states of many identical, hence indistinguishable, particles. The answer to this question is contained in a fundamental relation between the nature of the spin, integer or half-integer, of a particle, and the symmetry of many particle wave functions under permutations of identical particles. It will be seen later that in our four-dimensional world there is only the alternative of Bose-Einstein statistics for particles with integer spin, or Fermi-Dirac statistics for particles with half-integer spin. This is the content of the *spin-statistics theorem*, first derived by Fierz and by Pauli, to which we return later in this book (cf. Sect. 4.3.4 and Part Two).

3.6 Schrödinger and Heisenberg Pictures

From all examples studied up to this point we are used to see the time evolution of a quantum system in its wave function that obeys the time dependent Schrödinger equation. Observables $\mathcal{O}$, in contrast, seem to be fixed by the quantization procedure once and forever and, therefore, seem to be independent of time. This is in accord with classical physics: the system moves, as time goes on, but observables are defined by apparatus which is static. If we analyze the quantum mechanical description more closely we discover that this view of matters is only one of several possibilities.

Measurable, hence, testable information is contained solely in expectation values. These, however, are calculated by means of the general formula (3.43). Now, if the trace in (3.43) exists, it is cyclic in the factors of its arguments. For example, we could write

$$\operatorname{tr}\left\{\left[U(t,t_0)WU(t,t_0)^\dagger\right]\mathcal{O}\right\} = \operatorname{tr}\left\{W\left[U(t,t_0)^\dagger \mathcal{O} U(t,t_0)\right]\right\} .$$

Here, I inserted square brackets in view of the following considerations.

The description of the state discussed above is equivalent to defining the product

$$W_t := \left[U(t,t_0)WU(t,t_0)^\dagger\right] \quad \text{(Schrödinger picture)} \tag{3.44}$$

to be the statistical operator which represents the system at time t. Whenever one decides to shift the entire time dependence into the wave function, or, more generally, the statistical operator, one says one is using the *Schrödinger picture.*

It is equally well possible to "distribute" the evolution operator and its hermitian conjugate in such a way that we define

$$[U(t,t_0)^\dagger \mathcal{O} U(t,t_0)] =: \mathcal{O}_t \quad \text{(Heisenberg picture)}\,. \tag{3.45}$$

While the statistical operator (or the wave function) now is independent of time, the time evolution is contained in the operator $\mathcal{O}_t$. The measurable quantities proper remain unchanged by this new definition. If one shifts the entire time dependence to the operators which represent physical observables, one says one is working in the *Heisenberg picture.*

The Schrödinger equation which describes the time evolution, takes a somewhat different form in the two pictures. As one easily verifies, using the Schrödinger picture it reads

$$\boxed{\dot{W}_t = -\frac{\mathrm{i}}{\hbar}[H, W_t]}\,, \tag{3.46}$$

while, when using the Heisenberg picture it becomes

$$\boxed{\dot{\mathcal{O}}_t = \frac{\mathrm{i}}{\hbar}[H, \mathcal{O}_t]}\,. \tag{3.47}$$

Note the characteristic change of sign in these equations. The differential equation (3.47) holds for operators in the Heisenberg picture and is called *Heisenberg equation of motion.* Note its analogy to the equation

$$\frac{\mathrm{d}}{\mathrm{d}t} f(\boldsymbol{q}, \boldsymbol{p}) = \{H, f(\boldsymbol{q}, \boldsymbol{p})\}\,,$$

in classical, canonical mechanics which expresses the change in time of a dynamical quantity defined on phase space, by the Poisson bracket of the Hamiltonian function and the observable, cf. [Scheck (2010)] (2.128). For example, using an "energy representation", i.e. the A-representation in the sense of Sect. 3.1, with A a Hamiltonian with a fully discrete spectrum, one has

$$(\varphi_n, \mathcal{O}_t \varphi_m) = \mathrm{e}^{-\mathrm{i}/\hbar (E_m - E_n)t}(\varphi_n, \mathcal{O}_0 \varphi_m)\,.$$

This yields a matrix representation with typical harmonic time dependence in the transition frequencies

$$\omega_{mn} = \frac{E_m - E_n}{\hbar}\,.$$

It was this description that Heisenberg made use of in developing his matrix mechanics.

Further Comments: Only expectation values are observable and, hence, physically relevant. Therefore, we have the freedom to read the formula (3.43) in still another way. Suppose the Hamiltonian that describes a given system is the sum of a time-independent part H_0 and an additional term H' which explicitly depends on time, $H = H_0 + H'$. Both operators, H_0 and H', are self-adjoint. Inserting the identity

$$\mathbb{1} = \mathrm{e}^{-\mathrm{i}H_0 t/\hbar}\,\mathrm{e}^{\mathrm{i}H_0 t/\hbar}$$

in two positions, and making use of the cyclic property of the trace, one has

$$\langle \mathcal{O} \rangle_t = \mathrm{tr}\left(e^{iH_0 t/\hbar} U(t, t_0) W U(t, t_0)^\dagger \, e^{-iH_0 t/\hbar} \, e^{iH_0 t/\hbar} \mathcal{O} \, e^{-iH_0 t/\hbar} \right) .$$

Define then the modified evolution operator

$$U^{(\mathrm{int})}(t, t_0) := \exp\left(\frac{i}{\hbar} H_0 t\right) U(t, t_0) ,$$

and the modified observable

$$\mathcal{O}_t^{(\mathrm{int})} := \exp\left(\frac{i}{\hbar} H_0 t\right) \mathcal{O} \exp\left(-\frac{i}{\hbar} H_0 t\right) .$$

These operators are seen to obey the differential equations

$$\begin{aligned} \dot{\mathcal{O}}_t^{(\mathrm{int})} &= \frac{i}{\hbar} [H_0, \mathcal{O}_t^{(\mathrm{int})}] , \quad \text{and} \\ i\hbar \dot{U}^{(\mathrm{int})} &= e^{iH_0 t/\hbar} (H - H_0) \, e^{-iH_0 t/\hbar} U^{(\mathrm{int})} = H' U^{(\mathrm{int})} , \end{aligned} \tag{3.48}$$

respectively. The purely harmonic time dependence which is due to the time independent operator H_0, is shifted to the operators. The true time dependence which is due to the operator H', is contained in the modified evolution operator. The formula (3.43) takes the form

$$\langle \mathcal{O} \rangle_t = \mathrm{tr}(U^{(\mathrm{int})} W U^{(\mathrm{int})\dagger} \mathcal{O}^{(\mathrm{int})}) . \tag{3.49}$$

In many applications one takes H_0 to describe an unperturbed, in some cases solvable, system, while H' describes interactions and is often interpreted as a perturbation. With this idea in mind one calls this representation the *interaction picture*.

Space-Time Symmetries in Quantum Physics

4

Introduction The transformations in space and in time which belong to the Galilei group play an important role in quantum theory. In some respect and for some aspects, their role is new as compared to classical mechanics. Rotations, translations, and space reflection induce unitary transformations of those elements of Hilbert space which are defined with respect to the physical space $\mathbb{R}^3$ and to the time axis $\mathbb{R}_t$. Reversal of the arrow of time induces an *anti*unitary transformation in $\mathcal{H}$. Invariance of the Hamiltonian H of a quantum system under Galilei transformations implies certain properties of its eigenvalues and eigenfunctions which can be tested in experiment. This chapter deals, in this order, with rotations in $\mathbb{R}^3$, space reflection, and time reversal. A further and more detailed analysis of the rotation group is the subject of Chap. 6 in Part Two.

4.1 The Rotation Group (Part 1)

Consider a Hilbert space and a countably infinite basis $\{\varphi_\nu(\boldsymbol{x})\}$ thereof. The functions $\{\varphi_\nu(\boldsymbol{x})\}$ are defined over the physical space $\mathbb{R}^3$. Being elements of $\mathcal{H}$ they are orthogonal and normalized to 1. Given a physical wave function $\psi = \sum_\nu \varphi_\nu a_\nu$, the vector $(a_1, a_2, \ldots)^T$ is a specific representation of this state. Every transformation $\mathbf{R} \in SO(3)$, or $\mathbf{R} \in O(3)$, interpreted as a passive transformation in $\mathbb{R}^3$ (i.e. a rotation of the frame of reference) induces a unitary transformation in $\mathcal{H}$ such that

$$\{a_\nu\} \longmapsto \left\{ a'_\mu = \sum_\nu \mathbf{D}_{\mu\nu}(\Theta_i) a_\nu;\ \mathbf{D}\mathbf{D}^\dagger = \mathbf{D}^\dagger\mathbf{D} = \mathbb{1} \right\} . \tag{4.1}$$

As the physical state ψ does not depend on the base used for its expansion, this implies that the base functions are contragredient to the expansion coefficients, that is to say, transform according to $(\mathbf{D}^{-1})^T$.

F. Scheck, *Quantum Physics*, DOI: 10.1007/978-3-642-34563-0_4,

4.1.1 Generators of the Rotation Group

The infinite dimensional matrices **D** depend on the Euler angles $\{\Theta_i\} \equiv (\phi, \theta, \psi)$, or any other parametrization of the rotation in space. They are elements of the unitary, in general reducible, representations of the rotation group in Hilbert space. The eigenfunctions $R_{n\ell}^{\text{h.o.}}(r) Y_{\ell m}(\hat{\boldsymbol{x}})$ of the spherical oscillator, Sect. 1.9.4, provide an example of an orthonormal system which is defined with reference to a coordinate system in $\mathbb{R}^3$. An example for the unitary transformation which is induced by a rotation about the 3-axis was discussed in Sect. 3.2.1.

The elements of $SO(3)$ are continuous functions of the angles and can be continuously deformed into the identical mapping $\mathbb{1}$. Therefore, they can be written as exponential series in three angles and three generators for infinitesimal rotations, see [Scheck (2010)]. If, for example, we choose a Cartesian basis in the physical space, denote the angles by $\boldsymbol{\varphi} = (\varphi_1, \varphi_2, \varphi_3)$, and the generators by $\boldsymbol{J} = (\mathbf{J}_1, \mathbf{J}_2, \mathbf{J}_3)$, where the matrices $\mathbf{J}_k$ are given by

$$\mathbf{J}_1 = \begin{pmatrix} 0 & 0 & 0 \\ 0 & 0 & -1 \\ 0 & 1 & 0 \end{pmatrix}, \quad \mathbf{J}_2 = \begin{pmatrix} 0 & 0 & 1 \\ 0 & 0 & 0 \\ -1 & 0 & 0 \end{pmatrix}, \quad \mathbf{J}_3 = \begin{pmatrix} 0 & -1 & 0 \\ 1 & 0 & 0 \\ 0 & 0 & 0 \end{pmatrix},$$

then a passive rotation in $\mathbb{R}^3$ reads

$$\boldsymbol{x}' = \exp(-\boldsymbol{\varphi} \cdot \boldsymbol{J})\boldsymbol{x} .$$

Thus, the aim is to find an analogous decomposition of the corresponding, induced, transformation **D** in the form of an exponential series in the rotation angles and the generators.

The matrices $\mathbf{J}_i$ are antisymmetric, or, if interpreted as matrices over the complex numbers $\mathbb{C}$, are *antihermitian*, which is to say that they fulfill the relations $\mathbf{J}_k^\dagger = -\mathbf{J}_k$. As we know from mechanics their commutators are $[\mathbf{J}_1, \mathbf{J}_2] = \mathbf{J}_3$, with cyclic permutation of their indices. These antihermitian matrices can easily be replaced by hermitian ones by using, instead of the $\mathbf{J}_k$, the matrices $\mathrm{i}\mathbf{J}_k$, viz.

$$\widetilde{\mathbf{J}}_k := \mathrm{i}\mathbf{J}_k .$$

Their commutators are now $[\widetilde{\mathbf{J}}_1, \widetilde{\mathbf{J}}_2] = \mathrm{i}\widetilde{\mathbf{J}}_3$ with the characteristic factor i on the right-hand side. The commutator of two hermitian matrices is antihermitian, the factor i turns the result into a hermitian matrix.

For reasons that will become clear below, we replace the Cartesian frame of reference in $\mathbb{R}^3$ and its unit vectors $(\hat{\boldsymbol{e}}_1, \hat{\boldsymbol{e}}_2, \hat{\boldsymbol{e}}_3)$ by what are called *spherical coordinates* and *spherical base vectors*. These are defined as follows

$$\boldsymbol{\zeta}_1 := -\frac{1}{\sqrt{2}}(\hat{\boldsymbol{e}}_1 + \mathrm{i}\hat{\boldsymbol{e}}_2), \quad \boldsymbol{\zeta}_0 := \hat{\boldsymbol{e}}_3, \quad \boldsymbol{\zeta}_{-1} := +\frac{1}{\sqrt{2}}(\hat{\boldsymbol{e}}_1 - \mathrm{i}\hat{\boldsymbol{e}}_2). \tag{4.2}$$

One verifies that they fulfill the symmetry and orthogonality relations

$$\boldsymbol{\zeta}_m^* = (-)^m \boldsymbol{\zeta}_{-m}, \quad \boldsymbol{\zeta}_m^* \cdot \boldsymbol{\zeta}_{m'} = \delta_{mm'} . \tag{4.3}$$

This definition is motivated by the following argument: If in the expression for a position vector one transforms the base to the linear combinations (4.2), and makes use of spherical polar coordinates, one sees that

$$\begin{aligned}\boldsymbol{x} &= x^1\hat{\boldsymbol{e}}_1 + x^2\hat{\boldsymbol{e}}_2 + x^3\hat{\boldsymbol{e}}_3\\ &= -\frac{1}{\sqrt{2}}(x^1+\mathrm{i}x^2)\left(-\frac{1}{\sqrt{2}}\right)(\hat{\boldsymbol{e}}_1 - \mathrm{i}\hat{\boldsymbol{e}}_2) + \frac{1}{\sqrt{2}}(x^1-\mathrm{i}x^2)\frac{1}{\sqrt{2}}(\hat{\boldsymbol{e}}_1+\mathrm{i}\hat{\boldsymbol{e}}_2) + x^3\hat{\boldsymbol{e}}_3\\ &= r\left(-\frac{1}{\sqrt{2}}\sin\theta\, e^{\mathrm{i}\phi}\boldsymbol{\zeta}_1^* + \frac{1}{\sqrt{2}}\sin\theta\, e^{-\mathrm{i}\phi}\boldsymbol{\zeta}_{-1}^* + \cos\theta\,\boldsymbol{\zeta}_0^*\right).\end{aligned}$$

Making use of the formulae (1.116) for the spherical harmonics with $\ell = 1$, this becomes

$$\boldsymbol{x} = r\sqrt{\frac{4\pi}{3}}(Y_{11}\boldsymbol{\zeta}_1^* + Y_{1-1}\boldsymbol{\zeta}_{-1}^* + Y_{10}\boldsymbol{\zeta}_0^*),$$

the linear combinations $x^1 \pm \mathrm{i}x^2$ being proportional to $Y_{1\pm1}$, x^3 proportional to Y_{10}. By the symmetry relations of (4.3) and of (1.117) one has

$$\sum_m Y_{1m}\boldsymbol{\zeta}_m^* = \sum_m Y_{1m}^*\boldsymbol{\zeta}_m .$$

The decomposition of an arbitrary vector $\boldsymbol{a}$ over $\mathbb{R}^3$ in terms of a spherical basis has always this same form, viz.

$$\sum_m a_m\boldsymbol{\zeta}_m^* = \sum_m a_m^*\boldsymbol{\zeta}_m .$$

The relation between the basis $\boldsymbol{\zeta}^* \equiv (\boldsymbol{\zeta}_1^*, \boldsymbol{\zeta}_0^*, \boldsymbol{\zeta}_{-1}^*)^T$ and the Cartesian basis $\hat{\boldsymbol{e}} \equiv (\hat{\boldsymbol{e}}_1, \hat{\boldsymbol{e}}_2, \hat{\boldsymbol{e}}_3)^T$, from which we started, is given by a matrix $\mathbf{A}$, $\boldsymbol{\zeta}^* = \mathbf{A}\hat{\boldsymbol{e}}$, which is easily determined. This matrix $\mathbf{A}$ and its inverse $\mathbf{A}^{-1}$ are given by

$$\mathbf{A} = \frac{1}{\sqrt{2}}\begin{pmatrix}-1 & \mathrm{i} & 0\\ 0 & 0 & \sqrt{2}\\ 1 & \mathrm{i} & 0\end{pmatrix}, \quad \mathbf{A}^{-1} = \frac{1}{\sqrt{2}}\begin{pmatrix}-1 & 0 & 1\\ -\mathrm{i} & 0 & -\mathrm{i}\\ 0 & \sqrt{2} & 0\end{pmatrix}.$$

Transforming the generators $\tilde{\mathbf{J}}_k$ to this basis by working out $\mathbf{A}\tilde{\mathbf{J}}_k\mathbf{A}^{-1}$, one finds

$$\mathbf{A}\tilde{\mathbf{J}}_1\mathbf{A}^{-1} = \frac{1}{2}\begin{pmatrix}0 & \sqrt{2} & 0\\ \sqrt{2} & 0 & \sqrt{2}\\ 0 & \sqrt{2} & 0\end{pmatrix}, \quad \mathbf{A}\tilde{\mathbf{J}}_2\mathbf{A}^{-1} = \frac{\mathrm{i}}{2}\begin{pmatrix}0 & -\sqrt{2} & 0\\ \sqrt{2} & 0 & -\sqrt{2}\\ 0 & \sqrt{2} & 0\end{pmatrix},$$

$$\mathbf{A}\tilde{\mathbf{J}}_3\mathbf{A}^{-1} = \begin{pmatrix}1 & 0 & 0\\ 0 & 0 & 0\\ 0 & 0 & -1\end{pmatrix}. \tag{4.4}$$

These matrices are seen to be identical with the ones we found in Example 1.10, Sect. 1.9.1, for angular momentum $\ell = 1$. A characteristic property of this representation is that the 3-component is diagonal (and, hence real), the 1-component is real and positive, the 2-component is pure imaginary. It is this coincidence which motivates the choice of the spherical basis.

Of course, the commutators of the matrices $\widetilde{\mathbf{J}}_k$ are unaffected by this change of basis. What we have achieved, however, is to render the 3-component diagonal. This is different from the Cartesian basis from which we started and where none of the three generators had this property.

From here on we adopt the convention to *choose the generators of the rotation group to be hermitian*. For simplicity we choose the same symbol for them, i.e. $\mathbf{J}_k$ instead of $\widetilde{\mathbf{J}}_k$. Passive and active rotations then read

$$\exp(-\mathrm{i}\boldsymbol{\varphi}\cdot\boldsymbol{J}), \quad \exp(+\mathrm{i}\boldsymbol{\varphi}\cdot\boldsymbol{J}),$$

respectively, with $\boldsymbol{\varphi}\cdot\boldsymbol{J} = \varphi_1\mathbf{J}_1 + \varphi_2\mathbf{J}_2 + \varphi_3\mathbf{J}_3$ in the Cartesian basis. The generators fulfill the commutation relations

$$\boxed{[\mathbf{J}_1, \mathbf{J}_2] = \mathrm{i}\mathbf{J}_3 \quad \text{(and cyclic permutations)}}, \tag{4.5}$$

or, alternatively,

$$\boxed{[\mathbf{J}_i, \mathbf{J}_k] = \mathrm{i}\sum_{l=1}^{3} \varepsilon_{ikl}\mathbf{J}_l}. \tag{4.6}$$

when using the totally antisymmetric ε-symbol in three dimensions, with $\varepsilon_{ikl} = 1$ for all even permutations of $(1, 2, 3)$, $\varepsilon_{ikl} = -1$ for all odd permutations, and zero otherwise.

4.1.2 Representations of the Rotation Group

The action (4.1) in Hilbert space which is induced by a rotation in $\mathbb{R}^3$, also applies to the generators: If $\mathbf{D}(\mathbf{R})$ denotes the *unitary* transformation induced by $\mathbf{R}$, then $\mathbf{D}(\mathbf{J}_k)$ is the *hermitian* matrix which represents the generator $\mathbf{J}_k$ in Hilbert space. It has the same dimension as $\mathbf{D}(\mathbf{R})$, $n = \dim \mathbf{D}$. Written differently we have

$$\mathbf{D}(\mathbf{R}(\boldsymbol{\varphi})) = \exp[\mathrm{i}\boldsymbol{\varphi}\cdot\mathbf{D}(\boldsymbol{J})] = \lim_{k\to\infty}\left(\mathbb{1}_{n\times n} + \frac{\mathrm{i}}{k}\boldsymbol{\varphi}\cdot\mathbf{D}(\boldsymbol{J})\right)^k, \tag{4.7}$$

where we inserted Gauss' formula

$$\mathrm{e}^x = \lim_{k\to\infty}\left(1 + \frac{x}{k}\right)^k$$

for the exponential series, with a matrix valued argument here.

Obviously, the generators fulfill the commutation relations (4.5), or (4.6), in *any* representation. As a matter of example, consider the first commutator (4.5): It is obtained by performing successively two rotations, the first about the 1-axis, the second about the 2-axis, i.e. $\mathbf{R}(\eta\,\hat{\boldsymbol{e}}_2)\mathbf{R}(\varepsilon\,\hat{\boldsymbol{e}}_1)$, and by inverting these transformations in the other, "wrong", order, that is $\mathbf{R}(-\eta\,\hat{\boldsymbol{e}}_2)\mathbf{R}(-\varepsilon\,\hat{\boldsymbol{e}}_1)$, as sketched in Fig. 4.1. The result of these four infinitesimal transformations is found to be a rotation about the 3-axis by the angle $(\varepsilon\eta)$, hence of second order,

$$\mathbf{R}(-\eta\,\hat{\boldsymbol{e}}_2)\mathbf{R}(-\varepsilon\,\hat{\boldsymbol{e}}_1)\mathbf{R}(\eta\,\hat{\boldsymbol{e}}_2)\mathbf{R}(\varepsilon\,\hat{\boldsymbol{e}}_1) = \mathbf{R}(-\varepsilon\eta\,\hat{\boldsymbol{e}}_3).$$

This is verified by calculating the product of the four exponential series to second order in ε and η. All linear and pure quadratic terms cancel, while the mixed term

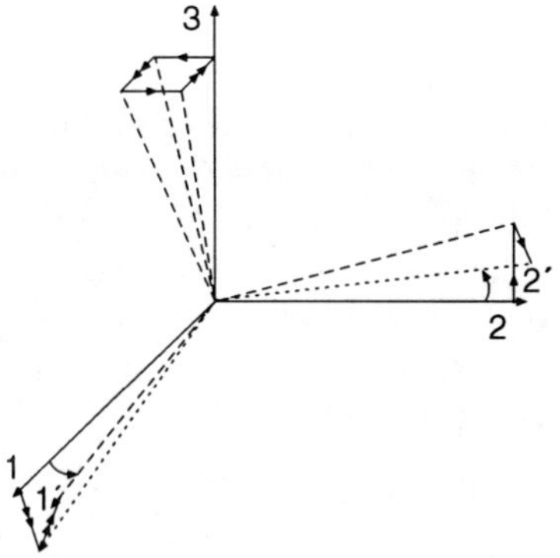

Fig. 4.1 A rotation by the angle ε about the 1-axis, followed by a rotation by the angle η about the new 2-axis, but then inverted in the other, "wrong" order, takes the 3-axis back to its initial position. However, the 1-axis and the 2-axis are taken to $1'$ and to $2'$, respectively. Thus, the result is a rotation by the angle $(\varepsilon\eta)$ about the 3-axis

yields $\varepsilon\eta(\mathbf{J}_1\mathbf{J}_2 - \mathbf{J}_2\mathbf{J}_1)$. In turn, by drawing carefully the figure shown in Fig. 4.1 one sees that there results a rotation about the 3-axis by the angle $(\varepsilon\eta)$ so that one concludes

$$\mathbb{1} + \varepsilon\eta(\mathbf{J}_1\mathbf{J}_2 - \mathbf{J}_2\mathbf{J}_1) = \mathbb{1} + \mathrm{i}\varepsilon\eta\,\mathbf{J}_3 \approx \exp(\mathrm{i}\varepsilon\eta\,\mathbf{J}_3)\,.$$

Thus, one could have derived the commutator (4.5) from this figure without even calculating the matrices $\mathbf{J}_k$ explicitly. Obviously, the same relation must hold also for the representation of the generators in Hilbert space, viz.

$$[\mathbf{D}(\mathbf{J}_1), \mathbf{D}(\mathbf{J}_2)] = \mathrm{i}\mathbf{D}(\mathbf{J}_3) \quad \text{(and cyclic permutations of } 1, 2, 3)\,.$$

As this is so one can simplify the explicit notation $\mathbf{D}(\mathbf{J}_k)$ by replacing it by the symbol $\mathbf{J}_k$ of the generator itself, with the understanding that it stands for *all* representations. This is what we do in the sequel.

The representation (4.4) which was obtained from the analysis of the rotation group in three-dimensional space $\mathbb{R}^3$, and, hence, follows from the very definition of this group, is called the *defining representation*. The one-dimensional representation in which $\mathbf{J}_1 = \mathbf{J}_2 = \mathbf{J}_3 = 0$, is called the *trivial representation*.

Further representations are obtained by the analogy to the orbital angular momentum that we studied in Sect. 1.9.1: The components of orbital angular momentum obey the commutation rules (1.107) which are the same as those for the generators $\mathbf{J}_k$. In Sect. 1.9.1 we showed that one can always choose $\boldsymbol{\ell}^2$ and one component, say ℓ_3, simultaneously diagonal, and that the eigenvalues of $\boldsymbol{\ell}^2$ and of ℓ_3 are given by $\ell(\ell+1)$ and by m, respectively, with $\ell \in \mathbb{N}_0$ and $m = -\ell, -\ell+1, \ldots, \ell$. This allows for two conclusions: One the one hand we found a countably infinite tower of representations in subspaces of Hilbert space whose dimension is $(2\ell + 1)$ (with integer ℓ), and in which the rotation matrices (4.7) and the generators are represented by unitary and hermitian $(2\ell + 1) \times (2\ell + 1)$-matrices, respectively. On the other hand we showed that angular momentum is intimately related to the rotation group. The components of angular momentum generate infinitesimal rotations, their commutators are those of the Lie algebra of the rotation group.

The problem to be solved is now clearly defined. Firstly, one must construct all representations which are compatible with (4.5), and, secondly, all unitary matrices $\mathbf{D}(\mathbf{R}(\boldsymbol{\varphi}))$ which span these representations. The first part of this program can be carried out completely on the basis of the commutators (4.5) only. The second part needs further tools that are developed in Part Two of this book.

With the example of orbital angular momentum in mind, cf. Sect. 1.9.1, one defines the square $\mathbf{J}^2$ and the ladder operators[1] $\mathbf{J}_\pm$,

$$\boldsymbol{J}^2 := \mathbf{J}_1^2 + \mathbf{J}_2^2 + \mathbf{J}_3^2 , \quad \mathbf{J}_\pm := \mathbf{J}_1 \pm \mathrm{i}\mathbf{J}_2 . \tag{4.8}$$

While $\boldsymbol{J}^2$ commutes with all components and, hence, also with $\mathbf{J}_\pm$, the remaining commutators are (reader please verify!)

$$\boxed{[\mathbf{J}_3, \mathbf{J}_\pm] = \pm\mathbf{J}_\pm , \quad [\mathbf{J}_+, \mathbf{J}_-] = 2\mathbf{J}_3 .} \tag{4.9}$$

The operators $\boldsymbol{J}^2$ and $\mathbf{J}_3$ are hermitian, the ladder operators are not. Instead, they are adjoints of each other, $\mathbf{J}_+^\dagger = \mathbf{J}_-$. The following two formulae are quite useful for subsequent calculations,

$$\boldsymbol{J}^2 = \mathbf{J}_+\mathbf{J}_- + \mathbf{J}_3^2 - \mathbf{J}_3 , \quad \boldsymbol{J}^2 = \mathbf{J}_-\mathbf{J}_+ + \mathbf{J}_3^2 + \mathbf{J}_3 , \tag{4.10}$$

they are easily verified by means of (4.5).

We will use the following, equivalent notations for the rotation $\mathbf{R}(\boldsymbol{\varphi})$ in Hilbert space

$$\mathbf{D}(\mathbf{R}) \equiv \mathbf{D}(\mathbf{R}(\boldsymbol{\varphi})) \equiv \mathbf{D}(\boldsymbol{\varphi}) .$$

As $\boldsymbol{J}^2$ commutes with all components it commutes also with $\mathbf{D}(\boldsymbol{\varphi})$,

$$[\boldsymbol{J}^2, \mathbf{D}(\boldsymbol{\varphi})] = 0 .$$

This commutator is important because of the following implication: If $\boldsymbol{J}^2$ is chosen to be diagonal the infinite dimensional matrix $\mathbf{D}(\mathbf{R})$ must have a form where only entries contained in square blocs along the main diagonal are different from zero, as sketched in Fig. 4.2. Each of these blocs pertains to one of the eigenvalues of $\boldsymbol{J}^2$, their dimension is equal to the degree of degeneracy of the eigenvalues of $\boldsymbol{J}^2$. If, in turn, $\mathbf{D}(\mathbf{R})$ has this bloc-diagonal form for all rotations $\mathbf{R}$, this matrix is said to be *irreducible*. The matrices $\mathbf{D}$ span a *unitary, irreducible representation of the rotation group*. This is a consequence of the following lemma:

[1] Note that this definition is not completely congruent with the definition of the spherical basis: the "+" component does not have the characteristic minus sign, and the normalization factor $1/\sqrt{2}$ is absent here.

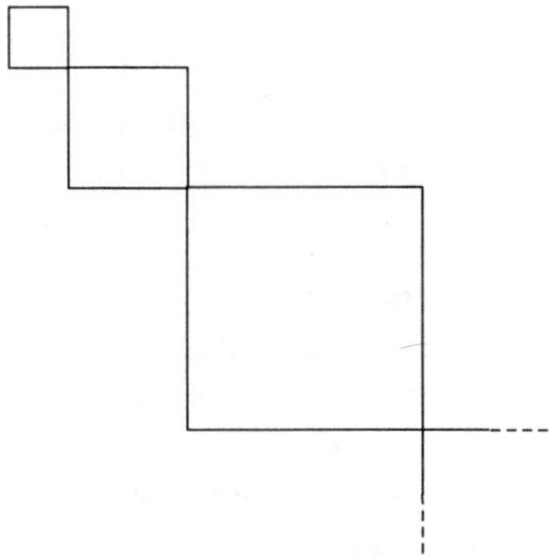

Fig. 4.2 If it is irreducible, the rotation matrix **D** has the structure of a sequence of square blocs along the main diagonal. Only entries within these blocs can be different from zero, all entries outside the blocs vanish, for every rotation in $\mathbb{R}^3$. The blocs themselves cannot be decomposed any further

Schur's lemma. Let $\mathbf{D}(\mathbf{R})$ and $\mathbf{D}'(\mathbf{R})$ be two matrices of dimension n and n', respectively, which are unitary and irreducible, and which depend on the rotations $\mathbf{R} \in SO(3)$. Furthermore, let $\mathbf{M}$ be a matrix with n columns and n' rows, which fulfills the relation

$$\mathbf{M}\mathbf{D}(\mathbf{R}) = \mathbf{D}'(\mathbf{R})\mathbf{M} \quad \text{for all} \quad \mathbf{R} \in SO(3)\,. \qquad (*)$$

Then one of the following two alternatives holds true: Either $\mathbf{M}$ vanishes, $\mathbf{M} = 0$, or $n = n'$ and $\det \mathbf{M} \neq 0$. In the second case $\mathbf{D}(\mathbf{R})$ and $\mathbf{D}'(\mathbf{R})$ are equivalent.

Proof

Multiply the hermitian conjugate of $(*)$, $\mathbf{D}^\dagger\mathbf{M}^\dagger = \mathbf{M}^\dagger\mathbf{D}'^\dagger$ by $\mathbf{D}$ from the left, and by $\mathbf{D}'$ from the right, thus obtaining $\mathbf{M}^\dagger\mathbf{D}' = \mathbf{D}\mathbf{M}^\dagger$. Multiplying this equation from the left by $\mathbf{M}$ and making use, in a second step, of the original equation $(*)$, one obtains

$$\mathbf{M}\mathbf{M}^\dagger\mathbf{D}'(\mathbf{R}) = \mathbf{M}\mathbf{D}(\mathbf{R})\mathbf{M}^\dagger = \mathbf{D}'(\mathbf{R})\mathbf{M}\mathbf{M}^\dagger\,. \qquad (**)$$

As $\mathbf{D}'(\mathbf{R})$ is irreducible by assumption, the product $\mathbf{M}\mathbf{M}^\dagger$ must be a multiple of the $n' \times n'$ unit matrix, $\mathbf{M}\mathbf{M}^\dagger = c\mathbb{1}$. The constant c is real because the product $\mathbf{M}\mathbf{M}^\dagger$ is hermitian. There are then three possibilities:

(a) $n = n'$, $c \neq 0$: In this case $\det \mathbf{M} \neq 0$ and, thus,

$$\mathbf{D}'(\mathbf{R}) = \mathbf{M}\mathbf{D}(\mathbf{R})\mathbf{M}^{-1}\,,$$

which says that $\mathbf{D}$ and $\mathbf{D}'$ are equivalent.

(b) $n = n'$ and $c = 0$: One now has $\sum_k M_{ik}M^*_{jk} = 0$, and specifically for $i = j$ one finds $\sum_k |M_{ik}|^2 = 0$. This means that $\mathbf{M}$ as a whole vanishes.

(c) $n < n'$ (or $n > n'$): In this case one enlarges the $n' \times n$-matrix $\mathbf{M}$ to a square $n' \times n'$ matrix $\mathbf{N}$ by inserting $n' - n$ columns all of whose entries are 0. Then one has $\mathbf{N}\mathbf{N}^\dagger = \mathbf{M}\mathbf{M}^\dagger$ and $\det \mathbf{N} = 0$. This implies $c = 0$, so that $\mathbf{M} = 0$ as before. (The same conclusion is reached if $n > n'$. In this case just interchange $\mathbf{D}$ and $\mathbf{D}'$.) This proves Schur's lemma.

Our aim is to derive the algebraic properties of all representations of the rotation group, starting from the commutators (4.6) or from the equivalent commutators (4.9).

Let $|\beta m\rangle$ be a common eigenstate of the operators $\boldsymbol{J}^2$ and $\mathbf{J}_3$, with β the eigenvalue of the first, and m the eigenvalue of the second operator. As $[\boldsymbol{J}^2, \mathbf{J}_\pm] = 0$, one generates further eigenstates of $\mathbf{J}^2$ by the action of $\mathbf{J}_+$ or of $\mathbf{J}_-$ on $|\beta m\rangle$ which belong to the same eigenvalue β, unless the action of $\mathbf{J}_\pm$ on $|\beta m\rangle$ gives the null vector. The nonvanishing states obtained in this way are again eigenstates of $\mathbf{J}_3$. Indeed, by the first commutator (4.9), one finds

$$\mathbf{J}_3(\mathbf{J}_\pm |\beta m\rangle) = \mathbf{J}_\pm[(\mathbf{J}_3 \pm 1)|\beta m\rangle] = (m \pm 1)(\mathbf{J}_\pm |\beta m\rangle) .$$

The state $\mathbf{J}_+ |\beta m\rangle = \text{const}\, |\beta, m+1\rangle$ is an eigenstate of $\mathbf{J}_3$ and pertains to the eigenvalue $(m+1)$. It is not normalized to 1, though. Likewise $\mathbf{J}_- |\beta m\rangle = \text{const}\, |\beta, m-1\rangle$ is an eigenstate of $\mathbf{J}_3$ for the eigenvalue $(m-1)$. The squared norm of these new eigenstates are easily calculated by means of the formulae

$$\mathbf{J}_\pm \mathbf{J}_\mp = \boldsymbol{J}^2 - \mathbf{J}_3^2 \pm \mathbf{J}_3$$

which follow from (4.10). With $\mathbf{J}_+^\dagger = \mathbf{J}_-$ one obtains

$$\|\mathbf{J}_+ |\beta m\rangle\|^2 = \langle \beta m|\mathbf{J}_-\mathbf{J}_+ |\beta m\rangle = (\beta - m^2 - m)\| \, |\beta m\rangle\|^2 ,$$

$$\|\mathbf{J}_- |\beta m\rangle\|^2 = \langle \beta m|\mathbf{J}_+\mathbf{J}_- |\beta m\rangle = (\beta - m^2 + m)\| \, |\beta m\rangle\|^2 ,$$

from which one deduces the inequalities

$$\beta - m(m+1) \geq 0, \quad \beta - m(m-1) \geq 0 . \qquad (+)$$

The series of eigenvalues $\ldots, m-2, m-1, m, m+1, m+2, \ldots$ of $\mathbf{J}_3$ must be bounded from above and from below because otherwise the inequalities (+) would be violated. With *increasing m* the series terminates if and only if there is a *largest* value $m_{\text{max}} =: j$ for which $\beta - j(j+1) = 0$. Similarly, for *decreasing m* the series terminates if and only if there is a *lowest* value m_{min} such that $\beta - m_{\text{min}}(m_{\text{min}} - 1)$ equals zero. Thus, if we set $\beta = j(j+1)$ we obtain the condition

$$m_{\text{min}}(m_{\text{min}} - 1) = j(j+1) .$$

This condition is met for $m_{\text{min}} = -j$ provided j is positive. The second root of this equation, $m'_{\text{min}} = j + 1$, must be rejected because m_{min} cannot exceed the value $m_{\text{max}} = j$ which is the largest, by assumption.

Finally, one realizes that the series of m values which proceeds in steps of 1, contains both the smallest value $m_{\text{min}} = -j$ and the largest value $m_{\text{max}} = +j$ if and only if j is either an *integer*, or a *half-integer*. In all other cases the increasing series generated by $\mathbf{J}_+$ would miss the decreasing series generated by $\mathbf{J}_-$, and neither of them would ever terminate.

As a result, the eigenvalues of J^2 and of $\mathbf{J}_3$ are

$$J^2 |JM\rangle = j(j+1)\,|j,m\rangle\,, \quad \mathbf{J}_3\,|JM\rangle = m\,|JM\rangle \tag{4.11a}$$

$$j = 0, \frac{1}{2}, 1, \frac{3}{2}, 2, \ldots, \quad m = -j, -j+1, \ldots, j\,. \tag{4.11b}$$

As expected the values of j include the series of integers $0, 1, 2, \ldots$, by now well known from the analysis of orbital angular momentum. In addition, one has a complete set of orthonormal eigenfunctions spanning these representations. What is new and surprising are the *half*-integer values of j among which the so-called *spinor representation*, $j = 1/2$, is particularly important and will be studied further below.

By construction, these representations are unitary and irreducible. They are realized in finite dimensional subspaces of Hilbert space whose dimensions are $d = (2j+1)$. Notations and dimensions for the first three values of j are as follows:

Remarks

1. We return for a moment to the Cartesian components of angular momentum, $\mathbf{J}_1 = (\mathbf{J}_+ + \mathbf{J}_-)/2$, $\mathbf{J}_2 = -i(\mathbf{J}_+ - \mathbf{J}_-)/2$, and make use of a phase convention in which the matrix elements of $\mathbf{J}_1$ are real and positive, while the elements of $\mathbf{J}_2$ are pure imaginary. We then have

$$\begin{aligned}\langle m'|\mathbf{J}_1\,|m\rangle &= \frac{1}{2}\sqrt{j(j+1) - m'm}\,(\delta_{m',m+1} + \delta_{m',m-1})\,,\\ \langle m'|\mathbf{J}_2\,|m\rangle &= -\mathrm{i}(m'-m)\langle m'|\mathbf{J}_1\,|m\rangle\,,\\ \langle m'|\mathbf{J}_3\,|m\rangle &= m\delta_{m',m}\,.\end{aligned} \tag{4.12}$$

Note that the second of these equations is obtained from the commutator $\mathbf{J}_2 = -\mathrm{i}[\mathbf{J}_3, \mathbf{J}_1]$.

2. How do we proceed in case the eigenstates of $\mathbf{J}_3$ contain a further degeneracy, that is to say, in case the common eigenfunctions of J^2 and of $\mathbf{J}_3$ have the form $|\alpha jm\rangle$, where $\alpha = 1, 2, \ldots, k_m$?
In this case one shows, in a first step, that the degrees of degeneracy k_m are all equal, i.e. $k_m \equiv k$ for all $m \in [-j, +j]$. In a second step one notices that the representation can be reduced further into k representations each of which has $(2j+1)$ elements. First step: We have

$$\sum_{\alpha'} \langle \alpha j, m+1|\mathbf{J}_+\,\big|\alpha' jm\big\rangle\big\langle \alpha' jm\big|\mathbf{J}_-\,|\alpha j, m+1\rangle$$

$$= j(j+1) - m(m+1)\,,$$

and summing over α,

$$\sum_{\alpha,\alpha'} \langle \alpha j, m+1 | \mathbf{J}_+ \left| \alpha' jm \right\rangle \left\langle \alpha' jm \right| \mathbf{J}_- \left| \alpha j, m+1 \right\rangle$$
$$= k_{m+1}[j(j+1) - m(m+1)] .$$

The same reasoning applied to the product with the two factors interchanged, yields

$$\sum_{\alpha,\alpha'} \langle \alpha jm | \mathbf{J}_- \left| \alpha' j, m+1 \right\rangle \left\langle \alpha' j, m+1 \right| \mathbf{J}_+ \left| \alpha jm \right\rangle$$
$$= k_m[j(j+1) - m(m+1)] .$$

As the left-hand sides are equal one concludes $k_{m+1} = k_m$. Thus, one obtains the same degree of degeneracy $k \equiv k_m$ for all m.
Second step: Define

$$\langle \alpha j, m+1 | \mathbf{J}_+ \left| \alpha' jm \right\rangle = \sqrt{j(j+1) - m(m+1)}\, U^{(m)}_{\alpha\alpha'} ,$$
$$\left\langle \alpha' jm \right| \mathbf{J}_- \left| \alpha j, m+1 \right\rangle = \sqrt{j(j+1) - m(m+1)}\, U^{(m)\dagger}_{\alpha'\alpha}$$

where $\mathbf{U}^{(m)}$ is a unitary $k \times k$ matrix. Choose then a new basis,

$$|\beta jm\rangle = \sum_{\alpha} V^{(m)}_{\beta\alpha} \, |\alpha jm\rangle \quad \text{with} \quad \mathbf{V}^{(m)} = \mathbb{1}\, \mathbf{U}^{(j-1)} \mathbf{U}^{(j-2)} \ldots \mathbf{U}^{(m)} .$$

The matrix representation of $\mathbf{J}_+$ in the new basis is obtained from the former one by the unitary transformation $\mathbf{V}^{(m+1)}\mathbf{U}^{(m)}\mathbf{V}^{(m)\dagger}$. By the choice of $\mathbf{V}^{(m)}$ this product is seen to be the $k \times k$ unit matrix so that $\langle \beta j, m+1 | \mathbf{J}_+ | \beta' jm \rangle = \sqrt{j(j+1) - m(m+1)}\, \delta_{\beta\beta'}$, with $\beta = 1, 2, \ldots, k$. Analogous formulae are obtained for the remaining operators.

4.1.3 The Rotation Matrices $\mathbf{D}^{(j)}$

The algebraic construction of the preceding section which was based on the *Lie algebra* (4.6) or (4.9) of the rotation group (not on the group proper), yields the eigenvalue spectra (4.11a) of the square of the angular momentum and of one component. There remains the problem to construct the infinite dimensional matrices $\mathbf{D}(\mathbf{R})$, i.e. the unitary transformations in Hilbert space which are induced by the rotations $\mathbf{R}$. The complete solution of this problem is postponed to Part Two. Here, I restrict the analysis and discussion to some general properties as well as to the case of $j = 1/2$ which can be dealt with in an elementary way.

Conventions

1. *Nomenclature*: The matrices $\mathbf{D}(\mathbf{R})$ are called *representation coefficients of the rotation group*. As this name is cumbersome one often calls them simply "rotation matrices" or, more simply, "D-matrices".

2. *Condon-Shortley phase convention*: The representation (4.12) is based on a specific choice of phases which goes back to the construction of the matrices for the ladder operators. This is seen by returning to the construction of representations of the Lie algebra (see preceding section): Indeed, in determining the states $\mathbf{J}_\pm|JM\rangle = c_\pm|j, m \pm 1\rangle$ only the square of the norm $||\,\mathbf{J}_\pm|JM\rangle||^2$ is fixed, the phases of the coefficients $c_\pm$ remain undetermined. The phases that were chosen in (4.12) followed from the decision to choose the matrix elements of the raising and lowering operators real,
$$\langle j'm'|\mathbf{J}_\pm|JM\rangle = [j(j+1) - m(m \pm 1)]^{1/2}\delta_{j'j}\delta_{m',m\pm 1}\,. \tag{4.13}$$
This phase convention which is due to Condon and Shortley, is the generally accepted convention in the theory of the rotation group.
3. *D-Matrices as functions of Euler angles*: It is useful to parametrize the rotation in $\mathbb{R}^3$ by Euler angles. One should note, however, that in quantum theory one uses a definition which is not identical with the traditional choice made in classical mechanics. The two definitions differ in the choice of the axis for the second rotation. For details see [Scheck (2010)], Sect. 3.10. Figure 4.3 shows the choice made in this book: First a rotation about the (original) 3-axis by the angle ϕ, then a rotation about the intermediate 2-axis by the angle θ, and finally a rotation about the (new) 3-axis by the angle ψ.[2]
It follows from Schur's lemma that by diagonalizing $\boldsymbol{J}^2$ the D-matrix decomposes into square blocs along the main diagonal. Each of these blocs belongs to one of the values of j, its dimension is $d = (2j+1)$. By counting the values of j

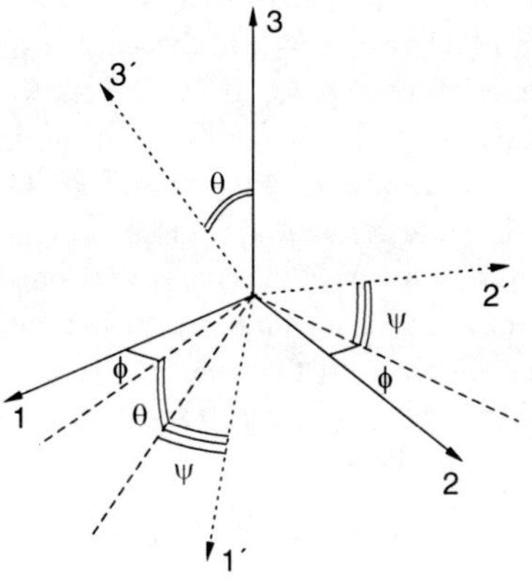

Fig. 4.3 Definition of Eulerian angles which is customarily used in quantum mechanics (a definition that differs from the preferential choice in classical mechanics). The intermediate position of the 2-axis is the axis for the second rotation

[2] In [Scheck (2010)] these angles are denoted by α, β, and γ, respectively, while the ones of the traditional convention of mechanics are denoted by Φ, Θ, and Ψ. The relation between the two choices is given in [Scheck (2010)], Eq. (3.39).

in increasing order and by ordering the values of m in decreasing order, from $m = +j$ to $m = -j$, in every subspace $\mathcal{H}^{(j)}$, the qualitative scheme of Fig. 4.2 is replaced by the more precise decoding

$$\begin{pmatrix} D^{(0)}_{0,0} & 0 & 0 & 0 & 0 & 0 & \cdots \\ 0 & D^{(1/2)}_{1/2,1/2} & D^{(1/2)}_{1/2,-1/2} & 0 & 0 & 0 & \cdots \\ 0 & D^{(1/2)}_{-1/2,1/2} & D^{(1/2)}_{-1/2,-1/2} & 0 & 0 & 0 & \cdots \\ 0 & 0 & 0 & D^{(1)}_{1,1} & D^{(1)}_{1,0} & D^{(1)}_{1,-1} & \cdots \\ 0 & 0 & 0 & D^{(1)}_{0,1} & D^{(1)}_{0,0} & D^{(1)}_{0,-1} & \cdots \\ 0 & 0 & 0 & D^{(1)}_{-1,1} & D^{(1)}_{-1,0} & D^{(1)}_{-1,-1} & \cdots \\ \vdots & \vdots & \vdots & \vdots & \vdots & \vdots & \ddots \end{pmatrix} .$$

For every finite value of j, and parametrizing the rotation by means of Euler angles, the rotation reads

$$\mathbf{D}^{(j)} = \exp(\mathrm{i}\psi \mathbf{J}_3) \exp(\mathrm{i}\theta \mathbf{J}_2) \exp(\mathrm{i}\phi \mathbf{J}_3) . \tag{4.14}$$

As $\mathbf{J}_3$ is chosen diagonal this implies for the matrix elements

$$D^{(j)}_{m'm}(\psi, \theta, \phi) = \mathrm{e}^{\mathrm{i}m'\psi}\, d^{(j)}_{m'm}(\theta)\, \mathrm{e}^{\mathrm{i}m\phi} , \tag{4.15}$$

where $d^{(j)}_{m'm}(\theta) = \langle jm' | \exp(\mathrm{i}\theta \mathbf{J}_2) | JM \rangle$ is a function of the second Euler angle only. In choosing to parametrize the rotation by means of Euler angles, instead of Cartesian angles, the benefit is obvious: The construction of the D-matrices is reduced to the calculation of the matrices $\mathbf{d}(\theta)$.

4. *Phase convention for D-Matrices*: While the choice of phases in the representation (4.12) of the generators is generally accepted, this is unfortunately not so for D-matrices. In consulting one or the other of the many monographs on the rotation group in quantum mechanics, the reader, if she or he is one of them, is advised to check carefully the conventions adopted in the monograph and to work out carefully the relation to his or her own conventions.
   In this book I adopt conventions which are in agreement with what is generally accepted in linear algebra, so that one can easily reconstruct them at any moment. Suppose we expand a *physical* state Ψ in terms of a base system φ_{jm} of eigenfunctions of $\boldsymbol{J}^2$ and $\mathbf{J}_3$,

$$\Psi = \sum_j \sum_m \varphi_{jm}\, a^{(j)}_m .$$

   The functions φ_{jm} are the *basis*, the coefficients a_{jm} are the *expansion coefficients*. A rotation in $\mathbb{R}^3$, $\mathbf{R} \in SO(3)$, induces the unitary transformation

$$a'^{(j)}_{m'} = \sum_m D^{(j)}_{m'm}(\mathbf{R}) a^{(j)}_m \tag{4.16}$$

   of the vectors $\boldsymbol{a} = (a^{(j)}_j, a^{(j)}_{j-1}, \ldots, a^{(j)}_{-j})^T$, that is, of the expansion coefficients. This implies that the basis transforms by the mapping $(\mathbf{D}^{-1})^T(\mathbf{R})$ which is contragredient to $\mathbf{D(R)}$. The physical state stays invariant $\sum_{jm'} \varphi'_{jm'} a'^{(j)}_{m'} = \sum_{jm} \varphi_{jm} a^{(j)}_m$. As $\mathbf{D}$ is unitary, its inverse equals its adjoint. The subsequent transposition takes it to $D^{(j)*}_{m'm}$, so that for every j, one obtains

$$\varphi'_{jm'} = \sum_{m''} D^{(j)*}_{m'm''}(\psi, \theta, \phi)\, \varphi_{jm''} \,.$$

By (4.16) and by adopting the Condon-Shortley convention (4.12) the phases of the D-matrices are fixed uniquely. Many books, though unfortunately not all, treating this topic make use of these conventions.[3]

4.1.4 Examples and Some Formulae for D-Matrices

In the subspace with $j = 0$ we have $\mathbf{D}^{(0)} = 1$. A state vector or an operator with $j = 0$ is a scalar under rotations. Thus, it remains unmodified by a rotation $\mathbf{R} \in SO(3)$.

In the fundamental representation, $j = 1/2$, the formulae (4.12) yield the matrix representations

$$\mathbf{J}_1 = \frac{1}{2}\begin{pmatrix} 0 & 1 \\ 1 & 0 \end{pmatrix}, \quad \mathbf{J}_2 = \frac{1}{2}\begin{pmatrix} 0 & -\mathrm{i} \\ \mathrm{i} & 0 \end{pmatrix}, \quad \mathbf{J}_3 = \frac{1}{2}\begin{pmatrix} 1 & 0 \\ 0 & -1 \end{pmatrix} \tag{4.17}$$

for the components of the angular momentum operator. Except for the factor $1/2$ these are precisely the Pauli matrices (3.22). We calculate the matrix $\mathbf{d}^{(1/2)}(\theta)$ of (4.15) by writing the exponential series and by making use of the fact that all even powers of σ_2 are equal to the unit matrix, all odd powers are equal to σ_2, see Sect. 3.3.4, Example 3.12,

$$\begin{aligned} \mathbf{d}^{(1/2)}(\theta) &= \exp\left(\mathrm{i}\frac{\theta}{2}\sigma_2\right) = \mathbb{1}\cos\frac{\theta}{2} + \mathrm{i}\sigma_2 \sin\frac{\theta}{2} \\ &= \begin{pmatrix} \cos(\theta/2) & \sin(\theta/2) \\ -\sin(\theta/2) & \cos(\theta/2) \end{pmatrix}. \end{aligned}$$

With this result the complete D-matrix for $j = 1/2$ is given by

$$\mathbf{D}^{(1/2)}(\theta) = \begin{pmatrix} \cos(\theta/2)\, e^{\mathrm{i}(\psi+\phi)/2} & \sin(\theta/2)\, e^{\mathrm{i}(\psi-\phi)/2} \\ -\sin(\theta/2)\, e^{-\mathrm{i}(\psi-\phi)/2} & \cos(\theta/2)\, e^{-\mathrm{i}(\psi+\phi)/2} \end{pmatrix}. \tag{4.18}$$

The result (4.18) has a truly remarkable property: Performing a rotation in $\mathbb{R}^3$ by the angle $360° = 2\pi$, by choosing for instance ($\psi = 0, \theta = 0, \phi = 2\pi$), changes nothing in that space. However, the induced transformation $\mathbf{D}^{(1/2)}(0, 0, 2\pi)$ is *not* the identity! Instead, it is $\mathbf{D}^{(1/2)}(0, 0, 2\pi) = -\mathbb{1}$. Thus, only after having performed successively two such complete rotations does one return to the identity, for example $\mathbf{D}^{(1/2)}(0, 0, 4\pi) = +\mathbb{1}$. This peculiar property is characteristic for all *half-integer* values of j, it will be understood better when developing the second part of the

---

[3] There are authors who choose to transform the *basis* by $\mathbf{D}$. The expansion coefficients then transform by $\mathbf{D}^*$. Some authors modify the standard rule for matrix multiplication. A good reason to compare matters very carefully!

theory of the rotation group. We just remark that it has an important implication for the description of indistinguishable particles with half-integer spin but postpone this topic for the moment.

I quote here the general formula for $\mathbf{d}^{(j)}$, without derivation. Details will be worked out in Part Two. One finds the following expression

$$d^{(j)}_{nm}(\theta) = \sum_p (-)^p \frac{\sqrt{(j+n)!\,(j-n)!\,(j+m)!\,(j-m)!}}{(j-n-p)!\,(j+m-p)!\,p!\,(p+n-m)!} \times \left(\cos\frac{\theta}{2}\right)^{2j-n+m-2p} \left(\sin\frac{\theta}{2}\right)^{2p+n-m} . \tag{4.19}$$

The sum over p has a finite number of terms, the smallest and the largest value being determined by the factorials in the denominator. As is well-known, $q! = \Gamma(q+1)$. The Gamma function $\Gamma(z)$ has first order poles at $z = 0, -1, -2, \ldots$. Thus, its inverse $1/\Gamma(z)$ has zeroes in these points on the negative real axis. This means that whenever p is either so large, or so small that one of the terms in round brackets in the denominator of (4.19) becomes negative, the sum terminates.

From the general formula (4.19) one deduces the following symmetry properties of the d-functions

$$\begin{aligned} d^{(j)}_{mn}(\theta) &= (-)^{n-m}\, d^{(j)}_{nm}(\theta) \\ d^{(j)}_{-n,-m}(\theta) &= (-)^{n-m}\, d^{(j)}_{nm}(\theta) \\ d^{(j)}_{n,-m}(\theta) &= (-)^{j-n}\, d^{(j)}_{nm}(\pi-\theta) . \end{aligned} \tag{4.20}$$

If j is an integer, $j \equiv \ell$, there is a close relationship between the D-functions and the spherical harmonics. One finds

$$Y_{\ell m}(\theta, \phi) = \sqrt{\frac{2\ell+1}{4\pi}}\, D^{(\ell)}_{0,m}(0, \theta, \phi) . \tag{4.21}$$

4.1.5 Spin and Magnetic Moment of Particles with $j = 1/2$

It is an empirical fact that the elementary particles observed in nature are characterized not only by their mass m and by a well defined charge but also by an intrinsic angular momentum s, called *spin*, which, in contrast to orbital angular momentum, is independent of the state of motion of the particle. This intrinsic angular momentum is an inner, invariant property of the particle. For example, the electron, the muon, the proton, and the neutron all carry spin $1/2$. This means that they are to be classified in the fundamental representation of the rotation group and, if all other features in the state of motion are kept fixed, that they can have two states, $|1/2, +1/2\rangle$ and $|1/2, -1/2\rangle$, the first of which describes the spin oriented along the positive 3-direction, while the second describes the orientation along the negative 3-direction.

The physical manifestation of the spin $1/2$ of these particles is through the corresponding magnetic moment which is proportional to the *Bohr magneton* $\mu_{\mathrm{B}}^{(i)}$ of the particle,

$$\mu = g^{(i)} \mu_{\mathrm{B}}^{(i)} \frac{1}{2} \quad \text{with} \quad \mu_{\mathrm{B}}^{(i)} := \frac{e\hbar}{2m_i c} . \tag{4.22}$$

Here $g^{(i)}$ is the gyromagnetic ratio whose value, for charged particles, should have the approximate value 2, e is its charge, i.e. $e = -|e|$ for the electron e^- and the muon μ^-, $e = |e|$ for the positron e^+, for the positive myon μ^+, and for the proton. The factor $1/2$ is nothing but m_{max}: Indeed, the self-adjoint *operator* to be associated to the magnetic moment, is given by

$$\boldsymbol{\mu} = g^{(i)} \mu_{\text{B}}^{(i)} \boldsymbol{s} \,, \tag{4.23}$$

and one defines the observable μ as being the largest eigenvalue of the operator (4.23).

The following argument shows that the Bohr magneton is the natural unit for magnetic moments to be associated to elementary particles. We calculate the magnetic moment which is related to the *orbital* motion of an electron bound in its atomic state. The magnetic moment $\boldsymbol{M}$ is given by the space integral of the magnetization density $\boldsymbol{m}(\boldsymbol{x})$ which, in turn, is determined by the electric current density $\boldsymbol{j}(\boldsymbol{x})$, viz.

$$\boldsymbol{M} = \int \mathrm{d}^3x \, \boldsymbol{m}(\boldsymbol{x}) \quad \text{with} \quad \boldsymbol{m}(\boldsymbol{x}) = \frac{1}{2c} \boldsymbol{x} \times \boldsymbol{j}(\boldsymbol{x}) \,.$$

Inserting the expression

$$\boldsymbol{j}(\boldsymbol{x}) = \frac{\hbar}{\mathrm{i}} \frac{e}{2m} [\psi^* \nabla \psi - (\nabla \psi)^* \psi]$$

for the electric current density, one sees that the magnetization density contains the operator $\boldsymbol{\ell}$,

$$\boldsymbol{m}(\boldsymbol{x}) = \frac{e\hbar}{4mc} [\psi^* \boldsymbol{\ell} \psi + (\boldsymbol{\ell}\psi)^* \psi] \,.$$

Thus, the magnetic moment caused by the orbital motion is proportional to the expectation value of $\boldsymbol{\ell}$,

$$\boldsymbol{M} = \frac{e\hbar}{2mc} \langle \boldsymbol{\ell} \rangle \,, \tag{4.24}$$

an expression which contains the Bohr magneton, indeed.

For the rest, classical electrodynamics tells us that this magnetic moment interacts with external magnetic fields $\boldsymbol{B}$ via the term $-\boldsymbol{M} \cdot \boldsymbol{B}$. Therefore, if H_0 denotes the Hamiltonian that describes the atom, the presence of an external magnetic field will modify it to

$$H = H_0 - \frac{e\hbar}{2mc} \boldsymbol{\ell} \cdot \boldsymbol{B}$$

Of course, the intrinsic magnetic moment (related to the spin) interacts with the external field, too. Also, the two moments, the orbital and the spin magnetic moments, interact with each other. The interaction of the spin magnetic moment with the external field is proportional to $\boldsymbol{s} \cdot \boldsymbol{B}$, its interaction with the magnetic moment generated by the orbital motion is proportional to $\boldsymbol{\ell} \cdot \boldsymbol{s}$. Inserting the correct factors the interaction Hamiltonian reads

$$H = H_0 - \frac{e\hbar}{2mc} \boldsymbol{\ell} \cdot \boldsymbol{B} - g \frac{e\hbar}{2mc} \boldsymbol{s} \cdot \boldsymbol{B} + \frac{\hbar^2}{2m^2c^2} \frac{1}{r} \frac{\mathrm{d}U(r)}{\mathrm{d}r} \boldsymbol{\ell} \cdot \boldsymbol{s} \,. \tag{4.25}$$

The second and third terms on the right-hand side reflect the interaction of the orbital and the spin magnetic moments with the field, respectively, the fourth term describes the *spin-orbit coupling* which manifests itself in what is called the *fine structure* of spectral lines. The factor in front of it contains the derivative of the spherically symmetric potential. It follows from the relativistic dynamics of the hydrogen atom.

Remarks

1. The spin of the electron was discovered through the magnetic moment attached to it and the interaction of this moment with inhomogeneous magnetic fields (experiment by Stern and Gerlach). Furthermore, the two values of s_3 play an important role in explaining the structure of the electronic shells of atoms, as well as, more generally, in the relation between the spin of the electron and the statistics that many-electron systems obey (Pauli principle). This led to one of the basic postulates of relativistic quantum theory to which we will return in Part Two. It says that elementary particles are to be classified by irreducible representations of the Poincaré group which correspond to fixed mass and definite spin. The spin is always defined in the *rest system* of the particle where its linear momentum and, hence, also its orbital angular momentum vanish. Qualitatively speaking, one investigates the reaction of the particle at rest to rotations of the frame of reference.
2. The operator $\boldsymbol{\mu}$ which represents the intrinsic magnetic moment of the electron, with respect to rotations in $\mathbb{R}^3$, must be a vector operator, that is to say, it must transform like $\boldsymbol{x}$ or like $\{Y_{1m}|m = -1, 0, +1\}$. In the rest system the only nonvanishing vector operator is the operator of spin $\boldsymbol{s}$. Therefore, the magnetic moment must be proportional to the spin operator, $\boldsymbol{\mu} \propto \boldsymbol{s}$.
3. The g-factor, or gyromagnetic ratio, hints at a deeper physical structure of particles. A relativistic version of quantum mechanics yields, at first, the value $g = 2$. However, one finds that this value can (and will) be modified by the interactions that the particle is subject to. In the case of the electron, or, by analogy, of the muon, this is the interaction with the Maxwell radiation field. In lowest approximation one finds
$$g^{(e)} \approx g^{(\mu)} \approx 2 + \frac{\alpha}{\pi}\,. \tag{4.26}$$
4. The nucleus of the hydrogen atom itself is an elementary particle with spin $1/2$. Thus, it too, carries a magnetic moment related to its spin,
$$\boldsymbol{\mu}^{(p)} = g^{(p)} \frac{|e|\hbar}{2m_{\mathrm{p}}c} \boldsymbol{s}\,, \quad \text{with} \quad g^{(p)} = 5.586\,,$$
whose absolute value is smaller by a factor $g^{(p)} m_{\mathrm{e}}/(g^{(e)} m_{\mathrm{p}}) \simeq 1.5 \times 10^{-3}$ than the absolute value of the electron's magnetic moment. Its interaction with the moment of the electron causes the *hyperfine structure* observed in the spectra of the hydrogen atom.

4.1.6 Clebsch-Gordan Series and Coupling of Angular Momenta

Vectorial addition of two angular momenta $\boldsymbol{j}_1$ and $\boldsymbol{j}_2$ to a resulting, total angular momentum $\boldsymbol{J} = \boldsymbol{j}_1 + \boldsymbol{j}_2$ is well known from classical physics. The analogous operation in quantum mechanics, for irreducible representations of the rotation group, is of great theoretical and practical interest. The problem is the following: Given two unitary, irreducible representations spanned by the states $|j_1, m_1\rangle$ and $|j_2, m_2\rangle$, respectively, the corresponding D-matrices being $\mathbf{D}^{(j_i)}$, $i = 1, 2$. The states formed by the products $|j_1, m_1\rangle\, |j_2, m_2\rangle$ provide a unitary representation, too, but this representation is reducible. This can be seen, for instance, by realizing that these states transform with the product $\mathbf{D}^{(j_1)} \times \mathbf{D}^{(j_2)}$ which, in the general case, does not have the structure of square blocs along the main diagonal which cannot be reduced further. On the other hand, the union of all representations is complete, and it must be possible to expand the product states in terms of irreducible representations, i.e., in a somewhat symbolic notation,

$$\mathbf{D}^{(j_1)} \times \mathbf{D}^{(j_2)} = \sum_J \mathbf{D}^{(J)} . \tag{4.27}$$

This series is called *Clebsch-Gordan series*.[4]

If $|JM\rangle$ denote the eigenstates of the commuting operators $\boldsymbol{J}^2 = (\boldsymbol{j}_1 + \boldsymbol{j}_2)^2$ and $\mathbf{J}_3 = (\mathbf{j}_1)_3 + (\mathbf{j}_2)_3$, this series is written in the form

$$|JM\rangle = \sum_{m_1, m_2} (j_1 m_1, j_2 m_2 | JM)\ |j_1 m_1\rangle\, |j_2 m_2\rangle \ . \tag{4.28}$$

The expansion coefficients $(j_1 m_1, j_2 m_2 | JM)$ are called *Clebsch-Gordan coefficients.* Alternative notations for Clebsch-Gordan coefficients are $C(j_1 m_1, j_2 m_2 | JM)$, or $C(j_1 j_2 J | m_1 m_2 M)$, or, even shorter, $(m_1 m_2 | JM)$. They are the entries of the unitary matrix which maps the orthonormal product basis $|j_1, m_1\rangle\, |j_2, m_2\rangle$ to the new, orthonormal base states $|JM\rangle$. In Part Two we will show that, as a consequence of the Condon-Shortley phase convention (4.12), the Clebsch-Gordan coefficients are *real*, which is to say that the transformation matrix is even *orthogonal.* This implies that the inverse of (4.28) is obtained from the transposed matrix, so that the product states can be written as follows

$$|j_1 m_1\rangle\, |j_2 m_2\rangle = \sum_{J, M} (j_1 m_1, j_2 m_2 | JM)\ |JM\rangle \ . \tag{4.29}$$

By comparing the series (4.28) and its inverse (4.29) one sees that the notation $(m_1 m_2 | JM)$ is the most concise in expressing the change of basis. The values of j_1 and of j_2 are fixed anyway, and their repetition in the coefficients is redundant. Explicit methods that allow to calculate these coefficients are explained in Chap. 6, Part Two.

---

[4] The series (4.27) is named after the mathematicians A. Clebsch (1833–1872) and P. Gordan (1837–1912).

Already at this stage, before even calculating the Clebsch-Gordan coefficients explicitly, one can determine the values of the total angular momentum J and of its 3-component M that are possible for given values of j_1 and j_2. This is achieved by the following arguments:

1. The operator $\mathbf{J}_3$ is the sum of the 3-components of $\boldsymbol{j}_1$ and of $\boldsymbol{j}_2$. Applying this operator to the two sides of (4.28) shows that one must have $M = m_1 + m_2$.
2. It is obvious that the largest value of M is obtained if one chooses $m_1 = j_1$ and $m_2 = j_2$, so that $M = j_1 + j_2$. The corresponding value of J must be $J = j_1 + j_2$: A *smaller* value contradicts the properties (4.11a), (4.11b) of the representations, a *larger* value is excluded because then there would be states with $M > m_1 + m_2$, in contradiction with the assumption.
3. Considering the next lower value $M = j_1 + j_2 - 1$, there are two possibilities to choose the quantum numbers m_i, either $(m_1 = j_1, m_2 = j_2 - 1)$ or $(m_1 = j_1 - 1, m_2 = j_2)$. One linear combination of the corresponding product states belongs to total angular momentum $J = j_1 + j_2$, the corresponding eigenstate $|J = j_1 + j_2, M = j_1 + j_2 - 1\rangle$ is obtained from the state $|J = j_1 + j_2, M = j_1 + j_2\rangle$ by applying the lowering operator $\mathbf{J}_- = (\mathbf{j}_1)_- + (\mathbf{j}_2)_-$. The other linear combination which is orthogonal to the first, must belong to a multiplet which has $J = j_1 + j_2 - 1$.
4. The example $(j_1 = 3/2, j_2 = 1)$ sketched in Fig. 4.4, shows that this process continues: For $M = j_1 + j_2 - 2$ there are three possibilities to choose the pair (m_1, m_2). Two orthogonal combinations of product states with $|j_1 m_1\rangle$ and $|j_2 m_2\rangle$ belong to the two values of J, determined previously. The third linear combination which is orthogonal to the first two, opens a new multiplet which carries $J = j_1 + j_2 - 2$. This construction terminates when the value $j = |j_1 - j_2|$ is reached. For example, assume like in Fig. 4.4 that $j_1 > j_2$. Then, for $M = j_1 - j_2 - 1$ one possible choice is missing (the one with $m_1 = j_1$ and $m_2 = -j_2 - 1$), so that no new multiplet opens up. (Of course, the other case, $j_2 > j_1$, is reduced to the first by interchanging the two angular momenta.)

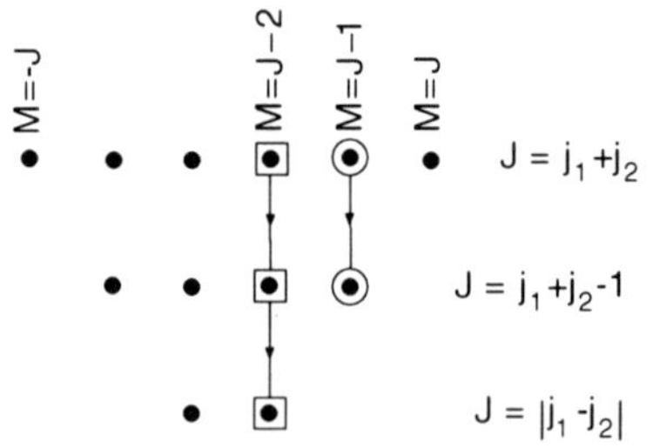

Fig. 4.4 Construction of the coupled states $|JM\rangle$ for the example $j_1 = 3/2$, $j_2 = 1$. Starting from the state $|J, J\rangle$ one constructs first the multiplet which has $J = j_1 + j_2$. With every step in M (to the *left* in the figure) there emerges a new multiplet until one has reached $J = |j_1 - j_2|$

As a result we note the rules:

$$m_1 + m_2 = M\,, \quad j_1 + j_2 - J = n\,, \quad n \in \mathbb{N}_0$$
$$j_1 + j_2 \geq J \geq |j_1 - j_2|\,. \tag{4.30}$$

The first of these repeats the statement that the 3-components must be added; the second and third, taken together, are said to form the *triangle rule for angular momenta.* By writing $J \equiv j_3$ they can be written in an equivalent, symmetric form:

$$\boxed{j_1 + j_2 + j_3 = n\,, \quad j_1 + j_2 \geq j_3 \geq |j_1 - j_2| \ \text{(cyclic)}}\,. \tag{4.31}$$

(Note that we used the fact that $2J$ is always a nonnegative integer.)

Remarks

1. The rules (4.30), (4.31) show a certain analogy to the restrictions that one has to observe in adding ordinary vectors. At the same time, they also take account of the peculiarities of "quantum angular momentum".
2. Every Clebsch-Gordan coefficient which violates these selection rules, vanishes.
3. It is easy to verify that the product basis comprises the same number of independent orthonormal elements as the coupled basis $|JM\rangle$. Assuming $j_1 > j_2$, without restricting the generality of the argument, we find

$$\sum_{J=|j_1-j_2|}^{j_1+j_2} (2J+1) = [2(j_1+j_2)+1] + [2(j_1+j_2-1)+1] + \ldots$$
$$+ [2(j_1-j_2)+1] = (2j_2+1)(2j_1+1)\,.$$

   This is precisely the number of product states.
4. The Clebsch-Gordan coefficients, being real, are the entries of *orthogonal* matrices. More precisely, they obey the orthogonality relations

$$\boxed{\sum_{m_1 m_2} (j_1 m_1, j_2 m_2 | JM)(j_1 m_1, j_2 m_2 | J' M') = \delta_{JJ'} \delta_{MM'}}\,, \tag{4.32}$$

$$\boxed{\sum_{JM} (j_1 m_1, j_2 m_2 | JM)(j_1 m_1', j_2 m_2' | JM) = \delta_{m_1 m_1'} \delta_{m_2 m_2'}}\,. \tag{4.33}$$

5. I quote here two symmetry relations which apply when two of the angular momenta are interchanged. The proof is deferred to Part Two. Taking $J \equiv j_3$ they read

$$\boxed{(j_2 m_2, j_1 m_1 | j_3 m_3) = (-)^{j_1+j_2-j_3} (j_1 m_1, j_2 m_2 | j_3 m_3)}\,, \tag{4.34}$$

$$\boxed{(j_1m_1, j_2m_2|j_3m_3) = (-)^{j_1-m_1}\sqrt{\frac{2j_3+1}{2j_2+1}}(j_1m_1, j_3-m_3|j_2-m_2)} \; . \tag{4.35}$$

6. There are simple cases in which the construction by means of raising and lowering operators sketched above, can be done "by hand". Examples are the coupling of two spin-1/2 states, and the coupling of angular momentum and spin of an electron, ($j_1 = \ell$, $j_2 = 1/2$). (The reader is encouraged to try these cases.)

4.1.7 Spin and Orbital Wave Functions

The eigenfunctions of spin, $J = 1/2$, span a two-dimensional subspace of Hilbert space. The D-matrices are given in (4.18), the matrix representation of the spin operators is given in (4.17). Adopting the conventional notation, $\boldsymbol{s}$ instead of $\boldsymbol{J}$, we have $\boldsymbol{s} = \boldsymbol{\sigma}/2$ where $\boldsymbol{\sigma} = (\sigma_1, \sigma_2, \sigma_3)$ stands for the three Pauli matrices. For the sake of completeness and as a starting point for calculations with spinors I repeat here these matrices and their properties

$$\sigma_1 = \begin{pmatrix} 0 & 1 \\ 1 & 0 \end{pmatrix}, \quad \sigma_2 = \begin{pmatrix} 0 & -\mathrm{i} \\ \mathrm{i} & 0 \end{pmatrix}, \quad \sigma_3 = \begin{pmatrix} 1 & 0 \\ 0 & -1 \end{pmatrix}; \tag{4.36}$$

$$\sigma_i^\dagger = \sigma_i \, , \quad \sigma_i\sigma_j = \delta_{ij} + \mathrm{i}\sum_k \varepsilon_{ijk}\sigma_k \, . \tag{4.37}$$

The second equation in (4.37) summarizes the information that the square of any Pauli matrix is the unit matrix, $\sigma_i^2 = \mathbb{1}$, and that the commutator of two different matrices is given by

$$[\sigma_i, \sigma_j] = 2\mathrm{i}\sum_k \varepsilon_{ijk}\sigma_k \, ,$$

in agreement with the more general form (4.6), (note the factor $1/2$ in the definition $\boldsymbol{s} = \boldsymbol{\sigma}/2$!).

There are three alternative notations for the normed eigenstates of $\mathbf{s}_3$ which belong to the eigenvalues $1/2$ and $-1/2$, respectively. They are

$$(\chi_+, \chi_-) \quad \text{or} \quad \left(\left|\frac{1}{2}, +\frac{1}{2}\right\rangle, \left|\frac{1}{2}, -\frac{1}{2}\right\rangle\right) \quad \text{or} \quad \left[\begin{pmatrix} 1 \\ 0 \end{pmatrix}, \begin{pmatrix} 0 \\ 1 \end{pmatrix}\right] .$$

Eigenstates of $\mathbf{s}_1$, or of $\mathbf{s}_2$, or of the projection of the spin onto an arbitrary direction $\hat{\boldsymbol{n}}$ all have the form

$$\chi = \frac{1}{\sqrt{|a_1|^2 + |a_2|^2}}\left[a_1\begin{pmatrix} 1 \\ 0 \end{pmatrix} + a_2\begin{pmatrix} 0 \\ 1 \end{pmatrix}\right] . \tag{4.38}$$

Combining the orbital wave functions $\psi_\nu(t, \boldsymbol{x})$ of an electron with its spin functions $\chi_\pm$ yields a total wave function $\Psi = (\psi_+, \psi_-)^T$ which has two components. In general, it no longer factorizes in an orbital and a spin part. Rather, it is a linear combination of product states $\sum_{\nu m_s} c_{\nu m_s} \psi_\nu \chi_{m_s}$. For example, this will happen when the orbital wave functions are eigenfunctions of orbital angular momentum while Ψ is meant to describe eigenstates of total angular momentum $\boldsymbol{j} = \boldsymbol{\ell} + \boldsymbol{s}$.

Given two such composite wave functions, say $\Psi^{(i)} = (\psi_+^{(i)}, \psi_-^{(i)})^T$ with $i = 1, 2$, their scalar product is

$$(\Psi^{(1)}, \Psi^{(2)}) = \int \mathrm{d}^3x\, \Psi^{(1)\dagger} \mathbb{1}\, \Psi^{(2)} = \int \mathrm{d}^3x \left[\psi_+^{(1)*} \psi_+^{(2)} + \psi_-^{(1)*} \psi_-^{(2)}\right].$$

The probability density to find the electron at time t at the position $\boldsymbol{x}$, *and* in the eigenstate $|1/2, m_s\rangle$ of $\mathbf{s}_3$ is given by $|\psi_{m_s}|^2$. In turn, the function $|\psi_+|^2 + |\psi_-|^2$ describes the probability density to find the electron in the world point $(t, \boldsymbol{x})$, irrespective of the orientation of its spin in space.

An example for a factorizing state is provided by the plane wave which describes a polarized electron with (linear) momentum $\boldsymbol{p}$,

$$\frac{1}{(2\pi\hbar)^{3/2}}\, \mathrm{e}^{\mathrm{i}\,\boldsymbol{p}\cdot\boldsymbol{x}/\hbar} \chi\,.$$

This example is a special case of the following more general situation.

4.1.8 Pure and Mixed States for Spin 1/2

In the two-dimensional subspace of $\mathcal{H}$, and using the basis of eigenstates of $\mathbf{s}_3 = \sigma_3/2$, define the density matrix

$$\varrho^{(0)} = \begin{pmatrix} w_+ & 0 \\ 0 & w_- \end{pmatrix} = \frac{1}{2}\left(\mathbb{1} + \zeta\,\sigma_3\right) \quad \text{with} \quad \zeta = w_+ - w_-\,.$$

By definition $w_+ + w_- = 1$ (this relation was used above), both numbers being real. The expectation values of the components of the spin operator are found to be

$$\langle \mathbf{s}_1 \rangle = 0 = \langle \mathbf{s}_2 \rangle\,, \quad \langle \mathbf{s}_3 \rangle = \operatorname{tr}\left(\varrho^{(0)} \frac{\sigma_3}{2}\right) = \frac{1}{2}\zeta\,.$$

In calculating these traces use was made of the formulae

$$\operatorname{tr} \sigma_i = 0, \quad \operatorname{tr}(\sigma_i \sigma_k) = 2\delta_{ik}$$

which follow from (4.37). If $w_+ = 1$ and, hence, $w_- = 0$ (or if $w_+ = 0$ and $w_- = 1$), then $\varrho^{(0)}$ describes a *pure* state of complete polarization along the positive (or negative) 3-direction. Any other choice of the real weights corresponds to a *mixed* state in which the particle is only partially polarized, or not polarized at all. In these cases the two spin states do not interfere. Specifically, the choice $w_+ = w_- = 1/2$ describes an unpolarized state, the probabilities to find the spin in positive or negative 3-direction are equal.

Consider a physical state which is a statistical mixture of eigenstates of the operator $\boldsymbol{\ell} \cdot \hat{\boldsymbol{n}}$, where $\hat{\boldsymbol{n}} = (\sin\theta\cos\phi, \sin\theta\sin\phi, \cos\theta)$ is a unit vector in 3-space, and with given weights w_+ and w_-. In a frame of reference whose 3-direction is directed along $\hat{\boldsymbol{n}}$ the density matrix is the same as above, $\varrho|_K = \operatorname{diag}(w_+, w_-)$. The same density matrix expressed with respect to the original frame $\mathbf{K}_0$ is obtained by the rotation relating $\mathbf{K}$ to $\mathbf{K}_0$, i.e. by

$$\varrho|_{K_0} = \mathbf{D}^{(1/2)\dagger}(\psi,\theta,\phi)\; \varrho|_K\; \mathbf{D}^{(1/2)}(\psi,\theta,\phi)\,,$$

where the D-matrix is given by (4.18). When calculating the product $\mathbf{D}^{(1/2)\dagger}\sigma_3\mathbf{D}^{(1/2)}$, the Euler angle ψ is seen to drop out. One obtains the result

$$\varrho|_{K_0} = \frac{1}{2}\left[\begin{pmatrix} 1 & 0 \\ 0 & 1 \end{pmatrix} + (w_+ - w_-)\begin{pmatrix} \cos\theta & \sin\theta\, e^{-i\phi} \\ \sin\theta\, e^{i\phi} & -\cos\theta \end{pmatrix}\right]$$
$$= \frac{1}{2}[\mathbb{1} + (w_+ - w_-)\hat{\boldsymbol{n}}\cdot\boldsymbol{\sigma}]\,. \tag{4.39}$$

Note that the *half* angles which are the arguments of $\mathbf{D}^{(1/2)}$, by the well-known addition theorems for trigonometric functions, are replaced by the *full* angles θ and ϕ.

The result (4.39) is interpreted easily and gives rise to a few interesting comments. First of all, one verifies the general properties of a density matrix (we omit the reference to $\mathbf{K}_0$),

$$\varrho^\dagger = \varrho, \quad \text{tr}\,\varrho = 1\,, \quad \text{tr}(\varrho^2) = \frac{1}{2}[1+(w_+-w_-)^2] = w_+^2+w_-^2 \le 1\,.$$

In doing this calculation one makes use of the second formula in (4.37) to show that $(\hat{\boldsymbol{n}}\cdot\boldsymbol{\sigma})(\hat{\boldsymbol{n}}\cdot\boldsymbol{\sigma}) = \hat{\boldsymbol{n}}^2 = 1$. Of course, use is made of the normalization condition $w_+ + w_- = 1$. If one of the weights equals 1 while the other vanishes, then tr $\varrho^2 =$ tr $\varrho = 1$. Thus, the density matrix describes a *pure* state. In all other cases ϱ describes a statistical mixture.

In some considerations it is useful to define the vector

$$\boldsymbol{\zeta} := (w_+ - w_-)\hat{\boldsymbol{n}} \tag{4.40}$$

and to write the density matrix as follows,

$$\varrho = \frac{1}{2}(\mathbb{1} + \boldsymbol{\zeta}\cdot\boldsymbol{\sigma})\,.$$

For example, calculating the expectation value of the spin operator in the state described by the density matrix ϱ, yields, as expected,

$$\langle \boldsymbol{s}\rangle = \frac{1}{2}\,\langle\boldsymbol{\sigma}\rangle = \frac{1}{2}\text{tr}(\varrho\boldsymbol{\sigma}) = \frac{1}{2}\boldsymbol{\zeta}\,.$$

The polarization points in the direction of $\boldsymbol{\zeta}$. With $|\langle \boldsymbol{s}\rangle|_{\max} = 1/2$, the degree of polarization is equal to

$$P := \frac{|\langle \boldsymbol{s}\rangle|}{|\langle \boldsymbol{s}\rangle|_{\max}} = |\boldsymbol{\zeta}| = \frac{w_+ - w_-}{w_+ + w_-}\,. \tag{4.41}$$

The square of the norm of $\boldsymbol{\zeta}$ is equal to $(w_+ - w_-)^2 = (1-2w_-)^2$ and is equal to or smaller than 1. By the same token, the formula (4.41) yields the relevant observable that is determined by experiment: One measures the number N_+ of particles which are polarized *along* the direction of $\boldsymbol{\zeta}$ as well as the number N_- of particles polarized in the *opposite* direction. One then takes the difference $N_+ - N_-$ of these numbers and normalizes to their sum, viz.

$$P = \frac{N_+ - N_-}{N_+ + N_-} . \tag{4.42}$$

A numerical example may serve to illustrate these results. Suppose one measures the polarization (4.41) to be 40 %. In order to take account of this experimental result one must choose $w_+ = 0.7$, and $w_- = 0.3$. The trace of ϱ^2 equals 0.58 and, hence, is smaller than 1.

4.2 Space Reflection and Time Reversal in Quantum Mechanics

The example of the rotation group showed that symmetries under transformations in space and time play an important role in quantum mechanics and, as compared to classical physics, present some novel aspects. In nonrelativistic quantum mechanics the relevant symmetry group is the Galilei group, in relativistic quantum (field) theory, to be treated in Part Two, it is the Poincaré group, in both cases including reflection in space and reversal of the direction of time. The consequences of the invariance of a given theory under space-time symmetries that follow from the theorem of E. Noether, are best worked out in the framework of second quantization. Although this topic is dealt with later in Part Two, I discuss space reflection and time reversal already at this point of the development.

4.2.1 Space Reflection and Parity

One can show that reflection of the axes in the physical space $\mathbb{R}^3$

$$\boldsymbol{x} \longmapsto \boldsymbol{x}' = -\boldsymbol{x} , \quad t \longmapsto t' = t$$

induces a *unitary* transformation $\mathbf{\Pi}$ in Hilbert space. An important theorem of E. Wigner says that every symmetry S of a quantum system induces in a unique way a transformation of unit rays in Hilbert space

$$\{\psi\} \longmapsto S\{\psi\}$$

which is either unitary or antiunitary. This theorem is spelled out in Part Two. If $\psi_U^{(i)} = U\psi^{(i)}$ is a unitary transformation induced by a symmetry of the physical system, then all transition matrix elements fulfill the relation

$$\langle \psi_U^{(i)} | \psi_U^{(k)} \rangle = \langle \psi^{(i)} | \psi^{(k)} \rangle . \tag{4.43}$$

In the case of *antiunitary* realization of the symmetry the mapping $\psi_S^{(i)} = S\psi^{(i)}$ implies the relations

$$\langle \psi_S^{(i)} | \psi_S^{(k)} \rangle = \langle \psi^{(i)} | \psi^{(k)} \rangle^* = \langle \psi^{(k)} | \psi^{(i)} \rangle \tag{4.44}$$

for all i and k.

If the states transform under space reflection according to

$$\psi(\boldsymbol{x}) \longmapsto \psi'(\boldsymbol{x}') = \boldsymbol{\Pi}\psi(-\boldsymbol{x}),$$

then the observables will have the transformation behaviour

$$\mathcal{O} \longmapsto \widetilde{\mathcal{O}} = \boldsymbol{\Pi}\,\mathcal{O}\,\boldsymbol{\Pi}^{-1}.$$

It is obvious that expectation values are invariant $(\psi, \mathcal{O}\psi) = (\psi', \widetilde{\mathcal{O}}\psi')$. The properties of the operator $\boldsymbol{\Pi}$ are

$$\boldsymbol{\Pi}^2 = \mathbb{1}, \quad \boldsymbol{\Pi} = \boldsymbol{\Pi}^\dagger = \boldsymbol{\Pi}^{-1}. \tag{4.45}$$

This operator is called *parity operator*. It is unitary and self-adjoint. Its eigenvalues are $+1$ and -1. An eigenstate with eigenvalue $+1$ is called a state with *even* parity, an eigenstate with eigenvalue -1 is said to be a state with *odd* parity.

The actions of the parity operator on the operators of position, of momentum, of orbital angular momentum, and of spin are, respectively,

$$\boldsymbol{\Pi}\boldsymbol{Q}\boldsymbol{\Pi}^{-1} = -\boldsymbol{Q}, \quad \boldsymbol{\Pi}\boldsymbol{P}\boldsymbol{\Pi}^{-1} = -\boldsymbol{P}, \tag{4.46}$$

$$\boldsymbol{\Pi}\boldsymbol{\ell}\boldsymbol{\Pi}^{-1} = +\boldsymbol{\ell}, \quad \boldsymbol{\Pi}\boldsymbol{s}\boldsymbol{\Pi}^{-1} = +\boldsymbol{s}. \tag{4.47}$$

The two formulae (4.46) are a direct consequence of the definition of space reflection. The first of the formulae (4.47) follows from the classical expression $\boldsymbol{\ell} = \boldsymbol{Q} \times \boldsymbol{P}$ for the orbital angular momentum: Indeed, as both $\boldsymbol{Q}$ and $\boldsymbol{P}$ are odd, their vector product $\boldsymbol{\ell}$ must be even. Both of (4.47) leave the commutation rules (4.5) invariant. This is remarkable in view of the fact that these rules are nonlinear. Their left-hand side contains two operators, while their right side contains only one! Of course, the result is in agreement with the well-known relationship between $O(3)$ and $SO(3)$ which says that every element of $O(3)$ whose determinant is -1 can be written as the product of an element of $SO(3)$ and of space reflection.

The action of $\boldsymbol{\Pi}$ on a wave function with spin projection m_s is seen to be

$$\boldsymbol{\Pi}\psi_{m_s}(t, \boldsymbol{x}) = \psi_{m_s}(t, -\boldsymbol{x}).$$

The action on an eigenfunction of orbital angular momentum is

$$\boldsymbol{\Pi}R_\alpha(r)Y_{\ell m}(\theta, \phi) = R_\alpha(r)Y_{\ell m}(\pi - \theta, \pi + \phi) = (-)^\ell R_\alpha(r)Y_{\ell m}(\theta, \phi).$$

The sign $(-)^\ell$ arises from the following observations: The mapping $\theta \mapsto (\pi - \theta)$ means that $z \equiv \cos\theta$ is replaced by $-\cos\theta = -z$. Inspection of the formulae (1.114) and (1.115) shows that the factors $(z^2 - 1)^\ell$ and $(1 - z^2)^{m/2}$ remain unchanged, while the derivative $\mathrm{d}/\mathrm{d}z$ is multiplied by (-1), $\mathrm{d}/\mathrm{d}z \mapsto -\mathrm{d}/\mathrm{d}z$. Therefore, the associated Legendre function P_ℓ^m (1.115) obtains the factor $(-)^{\ell+m}$. On the other hand, the factor $e^{im\phi}$ gets multiplied by $(-)^m$. The product of these sign factors gives indeed $(-)^\ell$ and we obtain the important relation

$$\boldsymbol{\Pi}Y_{\ell m}(\theta, \phi) = (-)^\ell Y_{\ell m}(\theta, \phi). \tag{4.48}$$

What is the role of the parity operation $\boldsymbol{\Pi}$ in the dynamics of a system described by the Hamiltonian H? The operator of kinetic energy is proportional to the Laplace operator Δ which is invariant under space reflection. Asking whether or not $\boldsymbol{\Pi}$ is a symmetry of the theory is equivalent to asking whether the interaction has a well-defined behaviour under parity. For example, any spherically symmetric potential as well as the spin-orbit interaction

$$U(r) \quad \text{and} \quad f(r)\boldsymbol{\ell}\cdot\boldsymbol{s}\,,$$

respectively, are *even* with respect to $\mathbf{\Pi}$. A velocity dependent term of the kind

$$g(r)\boldsymbol{\ell}\cdot\boldsymbol{q}\,,$$

with $\boldsymbol{q}$ a momentum or momentum transfer, would be *odd*.

Whenever there is an interaction which is neither even nor odd, e.g. being the sum of an even and an odd part, there will be observables which are odd, and, hence, which signal violation of invariance under parity. Examples for observables of this kind are the spin-momentum correlations. These are observables which are proportional to the (scalar) product of an even and an odd observable such as

$$2\frac{1}{|\boldsymbol{p}|}\langle\boldsymbol{s}\rangle\cdot\boldsymbol{p} =: P_l\,.$$

This observable describes the longitudinal polarization of an electron.[5] Nature makes use of such interactions: The weak interaction with charged currents which is responsible, e.g., for β-decay of nuclei, violates parity. Here, parity violation is even *maximal*, the observable effects are as large as they can possibly be.

If the Hamiltonian which appears in the Schrödinger equation is such that it commutes with the parity operator, $[H, \mathbf{\Pi}] = 0$, the eigenfunctions of H (i.e. the solutions with fixed energy) can be chosen such that they are also eigenfunctions of the parity operator, with eigenvalue $+1$ or -1. For instance, the eigenfunctions of the spherical oscillator, Sect. 1.9.4, and, likewise, the eigenfunctions of the bound states of the hydrogen atom, Sect. 1.9.5, are eigenfunctions also of $\mathbf{\Pi}$. The eigenvalue of $\mathbf{\Pi}$ is fixed by the value of ℓ. All states with $\ell = 0, 2, 4, \ldots$ are even, all states with $\ell = 1, 3, 5, \ldots$ are odd.

This observation is of paramount importance for the discussion of *selection rules*. In a transition from the initial state ψ_i to the final state ψ_f, by the action of an operator $\mathcal{O}$, we have

$$(\psi_f, \mathcal{O}\psi_i) = (\psi_f, \mathbf{\Pi}^{-1}\mathbf{\Pi}\mathcal{O}\mathbf{\Pi}^{-1}\mathbf{\Pi}\psi_i) = (\mathbf{\Pi}\psi_f, \widetilde{\mathcal{O}}\mathbf{\Pi}\psi_i)\,.$$

If ψ_f and ψ_i are eigenfunctions of $\mathbf{\Pi}$ and pertain to the eigenvalues $(-)^{\Pi_f}$ and $(-)^{\Pi_i}$, respectively, and if $\widetilde{\mathcal{O}} = (-)^{\Pi_O}\mathcal{O}$, the transition matrix elements can be different from zero only if

$$(-)^{\Pi_i+\Pi_O} = (-)^{\Pi_f}\,.$$

[5] The precise statement is: If an initial state which is even under $\mathbf{\Pi}$, by the action of some interaction, goes over into a final state which exhibits a correlation of this kind, the parity-odd observable is a measure for the amount of parity violation. In contrast, if the electron in the initial state had already a longitudinal polarization, then there is not necessarily parity violation. What matters is the change of state.

The parity of the initial state multiplied by the parity of the operator must equal the parity of the final state.

Electric multipole transitions in atoms provide important examples. The transition amplitude is proportional to matrix elements of the form

$$\langle n'\ell'm' \big| j_\lambda(kr)Y_{\lambda\mu} \, |n\ell m\rangle$$

where j_λ with $\lambda \in \mathbb{N}$ is a spherical Bessel function, and k is the wave number of the emitted light quanta. A matrix element of this kind must vanish if the parities do not match, that is to say, if the selection rule

$$(-)^\ell(-)^\lambda = (-)^{\ell'}$$

is not fulfilled. Electric dipole transitions have $\lambda = 1$, and the selection rule requires the initial and the final states to have different parities. A 2p-state of hydrogen which by (4.48) is parity-odd, can make an electric dipole transition to the 1s-state which has even parity. This is not possible for a 2s-state because it has even parity.

Of course, there are further selection rules, beyond parity. For example, the orbital angular momenta ℓ, λ, and ℓ' must fulfill the triangle rule (4.30) and one must have $m + \mu = m'$. These remarks and examples illustrate why space reflection takes a fundamentally different and more important role in quantum mechanics of particles than in classical mechanics of point particles.

4.2.2 Reversal of Motion and of Time

Time reversal $t \longmapsto -t$ in the physical spacetime is the prime example for a symmetry transformation which in Hilbert space is represented by an *anti*unitary operator **T**. The reason for this will become clear soon. However, we first give the precise definition of antiunitary operators and collect a few of their properties.

Definition 4.1 Antiunitary Operator

An operator **K** which maps the Hilbert space bijectively onto itself, is said to be *antiunitary* if it has the following properties

1. $\mathbf{K}[c_1 f^{(1)} + c_2 f^{(2)}] = c_1^*[\mathbf{K} f^{(1)}] + c_2^*[\mathbf{K} f^{(2)}], \quad c_1, c_2 \in \mathbb{C}$,
2. $|| f||^2 = ||\mathbf{K}|| f^2 \quad \text{for all} \quad f^{(1)}, f^{(2)}, f \in \mathcal{H}$.

One easily proves the following properties:

Theorem 4.1 Antiunitary Operators

1. With $f, g \in \mathcal{H}$ any two elements one has
$$(\mathbf{K}f, \mathbf{K}g) = (g, f) = (f, g)^* . \tag{4.49}$$
2. The product of two *antiunitary* operators $\mathbf{K}^{(1)}$ and $\mathbf{K}^{(2)}$ is *unitary*.
3. The product of an *antiunitary* operator and a *unitary* operator is again *antiunitary*.

Remarks

1. The relation (4.49) is identical with (4.44), quoted in Sect. 4.2.1, which relates different transition amplitudes. One proves the relation (4.49) by evaluating $\big(\mathbf{K}(c_f f + c_g g), \mathbf{K}(c_f f + c_g g)\big)$ for arbitrary complex numbers c_f and c_g, making use of property 2 in the definition, and by comparing coefficients. Indeed, by property 1
$$\begin{aligned}&\big(\mathbf{K}(c_f f + c_g g), \mathbf{K}(c_f f + c_g g)\big)\\ &\quad = |c_f|^2(\mathbf{K}f, \mathbf{K}f) + |c_g|^2(\mathbf{K}g, \mathbf{K}g)\\ &\qquad + c_f c_g^*(\mathbf{K}f, \mathbf{K}g) + c_f^* c_g(\mathbf{K}g, \mathbf{K}f)\,.\end{aligned}$$
By property 2 this is equal to
$$\begin{aligned}= \big((c_f f + c_g g), (c_f f + c_g g)\big) &= |c_f|^2(f, f) + |c_g|^2(g, g)\\ &\quad + c_f^* c_g(f, g) + c_f c_g^*(g, f)\,.\end{aligned}$$
As c_f and c_g can be chosen at will one concludes $(\mathbf{K}f, \mathbf{K}g) = (g, f)$.
2. As a corollary of the assertion 3 in Theorem 4.1 one notes that every antiunitary operator can be written as the product of a unitary operator and a fixed antiunitary operator $\mathbf{K}^{(0)}$, $\mathbf{K} = \mathbf{U}\mathbf{K}^{(0)}$.
3. The operator $\mathbf{K}^{(0)}$ may be chosen to be "complex conjugation", that is to say, the operator which does no more than to replace every complex number (also called a *c-number*) by its complex conjugate.
4. In classical physics time reversal is equivalent to reversal of the sense of motion. Indeed, the relation (4.44) or (4.49) means that initial and final states are interchanged. Thus, it seems plausible that time reversal is effected by an antiunitary operator.
5. In turn, if we already know that time reversal is represented by an antiunitary symmetry transformation in Hilbert space, then (4.44) implies a simple rule: Time reversal, applied to a transition matrix element, *either* means that initial and final states must be interchanged, *or* that all c-numbers must be replaced by their complex conjugates.

We work out the action of the transformation
$$t \mapsto t' = -t\,, \quad \boldsymbol{x} \mapsto \boldsymbol{x}' = \boldsymbol{x}$$
on the Schrödinger equation. Let
$$\psi(t, \boldsymbol{x}) \longmapsto \psi'(t', \boldsymbol{x}) = \mathbf{T}\psi(t, \boldsymbol{x})\,, \quad (t' = -t)\,.$$
If H has no explicit time dependence, the operator $\mathbf{T}$ must be determined such that the Schrödinger equation remains form invariant, that is,
$$\mathrm{i}\hbar\frac{\mathrm{d}}{\mathrm{d}t'}\psi'(t', \boldsymbol{x}) = H\psi'(t', \boldsymbol{x}) \quad \text{or} \quad -\mathrm{i}\hbar\frac{\mathrm{d}}{\mathrm{d}t}\mathbf{T}\psi(t, \boldsymbol{x}) = H\mathbf{T}\psi(t, \boldsymbol{x})\,.$$
As long as ψ is a scalar, hence a one-component wave function, this requirement is met if the operator $\mathbf{T}$ acts by complex conjugation $\mathbf{K}^{(0)}$,
$$\mathbf{T}\psi(t, \boldsymbol{x}) = \mathbf{K}^{(0)}\psi(t, \boldsymbol{x}) = \psi^*(t, \boldsymbol{x})\,. \tag{4.50}$$

Thus the time reversed wave function satisfies the complex conjugate Schrödinger equation.

If the wave function contains also a spin-1/2, that is to say, if it is a two-component vector $\Psi = (\psi_+, \psi_-)^T$ (cf. Sect. 4.1.7), then $\mathbf{T}$ can be taken to be of the form

$$\mathbf{T} = \mathbf{U}\mathbf{K}^{(0)} \tag{4.51}$$

where $\mathbf{U}$ is a *unitary* transformation still to be determined. Thus, its action will be

$$\mathbf{T}\begin{pmatrix}\psi_+\\ \psi_-\end{pmatrix} = \mathbf{U}\begin{pmatrix}\psi_+^*\\ \psi_-^*\end{pmatrix}.$$

The unitary transformation $\mathbf{U}$ is obtained by the following argument: With $\Psi = (\psi_+, \psi_-)^T$ also $\Psi^* = (\psi_+^*, \psi_-^*)^T$ is a spinor representation of the rotation group. While the former transforms under rotations with $\mathbf{D}^{(1/2)}$, cf. Sect. 4.1.3, the latter, Ψ^*, transforms with $\mathbf{D}^{(1/2)*}$. However, this is consistent only if the relation

$$\mathbf{U}D^{(1/2)}\mathbf{U}^\dagger = \mathbf{D}^{(1/2)*}, \quad \text{i.e.} \quad \mathbf{U}\sigma_j^*\mathbf{U}^\dagger = -\sigma_j$$

is satisfied. The matrix σ_2 is pure imaginary. It obviously commutes with itself, but anticommutes with σ_1 and with σ_3. Therefore $\mathbf{U}$ must be proportional to σ_2. The conventional choice is the following

$$\mathbf{T} = \mathbf{U}\mathbf{K}^{(0)} \quad \text{with} \quad \mathbf{U} = \mathrm{i}\sigma_2 = \begin{pmatrix}0 & 1\\ -1 & 0\end{pmatrix}. \tag{4.52}$$

A look at the formula (4.18) tells us that $\mathbf{U}$, in reality, is a rotation about the 2-axis, by an angle π,

$$\mathbf{U} = \mathrm{i}\sigma_2 = \mathbf{D}^{(1/2)}(0, \pi, 0) = \mathrm{e}^{\mathrm{i}\pi\sigma_2/2}.$$

This result calls for two comments.

1. The same reasoning may be applied to any representation of the rotation group in which the generators are $(2j+1) \times (2j+1)$ matrices $\mathbf{J}_i$, $i = 1, 2, 3$. The rotation $\mathbf{D}^{(j)}(0, \pi, 0)$ takes $\mathbf{J}_1$ and $\mathbf{J}_3$ to their negatives, but leaves invariant $\mathbf{J}_2$, viz.

   $$\mathbf{U}\mathbf{J}_{1/3}\mathbf{U}^{-1} = -\mathbf{J}_{1/3}, \quad \mathbf{U}\mathbf{J}_2\mathbf{U}^{-1} = +\mathbf{J}_2.$$

   With the phase convention (4.12) the 1- and 3-components are real, while the 2-component is pure imaginary. Therefore, for all three of them we have

   $$\mathbf{U}\mathbf{J}_i^*\mathbf{U}^{-1} = -\mathbf{J}_i, \quad i = 1, 2, 3, \tag{4.53}$$

   which hold in all representations. Thus, time reversal is realized by the antiunitary transformation $\mathbf{T} = \mathbf{U}\mathbf{K}^{(0)}$.
2. The matrix $\mathbf{U}$ is real (hence, in fact, orthogonal) and commutes with the operation of complex conjugation. Therefore, applying time reversal twice, one obtains

   $$\mathbf{T}^2 = \mathbf{U}\mathbf{K}^{(0)}\mathbf{U}\mathbf{K}^{(0)} = \mathbf{U}^2\mathbf{K}^{(0)\,2} = \exp(\mathrm{i}2\pi\mathbf{J}_2) = (-)^{2j}\mathbb{1}.$$

   Thus, for integer angular momentum one has $\mathbf{T}^2 = +\mathbb{1}$, while for half-integer angular momentum one finds $\mathbf{T}^2 = -\mathbb{1}$. In particular, in a system containing N particles with spin 1/2 (fermions) one finds

   $$\mathbf{T}^2 = (-)^N\mathbb{1}.$$

If the Hamiltonian of this system commutes with time reversal, this factor has an important consequence: From $H\psi = E\psi$ follows $H(\mathbf{T}\psi) = E(\mathbf{T}\psi)$ which means that if ψ is a solution of the Schrödinger equation then also $\mathbf{T}\psi$ is a solution and has the same energy E as the original. If the number of particles N is *even* one can always manage to have $(\mathbf{T}\psi) = \psi$. However, if N is *odd* the two states ψ and $\mathbf{T}\psi$ are different. This means that the eigenvalues of the Hamiltonian H for a system with an *odd* number of fermions are always degenerate. The degree of degeneracy is *even*, hence, at least equal to 2. This statement is called *Kramer's theorem*.

Under the action of time reversal one obtains relations which are analogous to (4.46) and to (4.47), and which read

$$\mathbf{T}\boldsymbol{Q}\mathbf{T}^{-1} = +\boldsymbol{Q}\,, \quad \mathbf{T}\boldsymbol{P}\mathbf{T}^{-1} = -\boldsymbol{P}, \tag{4.54}$$

$$\mathbf{T}\boldsymbol{\ell}\mathbf{T}^{-1} = -\boldsymbol{\ell}\,, \quad \mathbf{T}\boldsymbol{s}\mathbf{T}^{-1} = -\boldsymbol{s}\,. \tag{4.55}$$

An external electric field is invariant under time reversal, while a magnetic field $\boldsymbol{B}$ goes over into $-\boldsymbol{B}$. The relative sign between $\boldsymbol{E}$ and $\boldsymbol{B}$ is immediately clear if one recalls the Lorentz force $\boldsymbol{F} = e(\boldsymbol{E} + \boldsymbol{v} \times \boldsymbol{B}/c)$ and notes that the velocity $\boldsymbol{v}$ is odd. If H_0 commutes with $\mathbf{T}$, then the Hamiltonian (4.25), as a whole, is invariant under time reversal. Note that we apply time reversal both to the observables of the electron and to the external fields. If instead we had kept the external fields unchanged then the terms H_0 and $\boldsymbol{\ell} \cdot \boldsymbol{s}$ would be even, but the terms $\boldsymbol{\ell} \cdot \boldsymbol{B}$ and $\boldsymbol{s} \cdot \boldsymbol{B}$ would be odd. An electron which crosses a fixed external magnetic field does not follow the same trajectory when $t \mapsto -t$.

Suppose $\mathbf{T}$ commutes with H. The action of time reversal on the evolution operator (3.27) then is

$$\mathbf{T}U(t, t_0)\mathbf{T} = \mathbf{T}\exp\left(-\frac{\mathrm{i}}{\hbar}H(t - t_0)\right)\mathbf{T} = \exp\left(+\frac{\mathrm{i}}{\hbar}H(t - t_0)\right) = U^{\dagger}(t, t_0)\,.$$

This is the quantum theoretic analogue of the classical equivalence between time reversal and reversal of the motion.

4.2.3 Concluding Remarks on T and Π

The rotation group being a subgroup of the Galilei and of the Poincaré groups, plays an important role in both nonrelativistic quantum mechanics and quantum field theory. As this is a continuous group, invariance of the dynamics of a given system under rotations about an (arbitrary) axis $\hat{\boldsymbol{n}}$ implies conservation of the projection of the angular momentum onto that axis – in close analogy to the analogous situation in classical physics (theorem of E. Noether). Indeed, if $\varphi(\boldsymbol{x})$ is an eigenfunction of a given Hamiltonian H, than the transformed, "rotated" wave function

$$\varphi'(\boldsymbol{x}') = \mathbf{D}_{\hat{n}}(\alpha)\varphi(\boldsymbol{x}) = \exp[\mathrm{i}\alpha(\boldsymbol{J} \cdot \hat{\boldsymbol{n}})]\varphi(\boldsymbol{x})$$

is an eigenfunction of H, for all values of the angle of rotation α, if and only if

$$[H, (\boldsymbol{J} \cdot \hat{\boldsymbol{n}})] = 0\,,$$

i.e. if the projection of $\boldsymbol{J}$ onto the given direction commutes with the Hamiltonian. If this holds true for all directions the angular momentum as a whole is conserved.

In contrast to rotations and translations, time reversal and space reflection are *discrete* transformations. Invariance of the dynamics of a physical system with respect to $\mathbf{\Pi}$ or to $\mathbf{T}$ does not lead to new conserved quantities, but implies certain selection rules. These selection rules are a characteristic feature in the quantum world, and are new as compared to classical physics. Parity selection rules in the case of $\mathbf{\Pi}$, and Kramer's theorem in the case of $\mathbf{T}$ provide good examples.

There is another, far reaching aspect which comes in addition to the two discrete transformations studied above: Relativistic quantum physics predicts, somewhat loosely speaking, that for every elementary particle there exists an antiparticle. Particles and antiparticles have the same mass and the same spin but they differ in the sign of all additively conserved quantum numbers.[6] An example for an additive quantum number is the electric charge $q/|e|$, expressed in units of the elementary charge: An electron has charge -1, the positron being its antiparticle, has charge $+1$. Electric charge is universally conserved. Therefore, in any reaction process the sum of all charges in the initial state equals the sum of all charges in the final state.

Conversely, a particle which is identical with its own antiparticle, can have no additively conserved quantum numbers. This is the case for the photon which carries no electric charge and which, indeed, coincides with its antiparticle. Furthermore, one learns that quantum field theory can be formulated in a way which is completely symmetric in particles and antiparticles, and that it is a matter of pure convention whether one calls the electron "particle", the positron "antiparticle", or vice versa. In order to take proper and formal account of the particle-antiparticle relationship one defines one further discrete transformation $\mathbf{C}$, called *charge conjugation*, which replaces every particle (antiparticle) by its antiparticle (particle), without changing any of its other dynamical attributes such as linear momentum, spin, or the like. For example, its action on the state of an electron is

$$\mathbf{C}: \quad \left|e^-, \boldsymbol{p}, m_s\right\rangle \longmapsto e^{i\eta_C} \left| e^+, \boldsymbol{p}, m_s\right\rangle , \tag{4.56}$$

Note that there can be a phase factor $e^{i\eta_C}$ in this mapping.

It turns out that charge conjugation is intimately related to space reflection and to time reversal. A fundamental theorem in quantum field theory, discovered by G. Lüders and W. Pauli, and proven in its most general form by R. Jost, says that a theory which is Lorentz covariant and fulfills certain conditions of locality and causality, is invariant under the *product*

$$\mathbf{\Pi C T} =: \boldsymbol{\Theta} \tag{4.57}$$

[6] In nonrelativistic quantum mechanics there exists an analogue for the particle-antiparticle relation. For example, in a many body system consisting of N fermions in an external, attractive potential the ground state (i.e. the state of lowest energy) is a state where the particles are distributed over the lowest bound states, in accord with the Pauli principle. Excited states of the system contain configurations where one or several particles are taken out of a bound state, creating a "hole", and are lifted to a formerly unoccupied state. The relation of *hole states* to occupied particle states is somewhat like the relation of antiparticles to particles.

of space reflection, charge conjugation, and time reversal [Streater and Wightman (1964)]. Thus, if the theory fulfills these conditions but is not invariant under one of the three discrete transformations, it must violate one of the remaining two. Here is an example which illustrates this important relation.

Example 4.1 Decay of Charged Pions

A positively charged pion π^+, after a mean life of about 2.6×10^{-8} s, decays predominantly into a positive muon and a muonic neutrino,

$$\pi^+ \longrightarrow \mu^+ + \nu_\mu \, .$$

The pion has no spin. Seen from the pion's rest system the muon and the neutrino have opposite and equal (spatial) momenta, as sketched in Fig. 4.5. The plane wave describing the relative motion of the particles in the final state contains all values of the orbital angular momentum ℓ, cf. (1.136), but all partial waves have vanishing projection onto the line along which the particles escape, $m_\ell = 0$. (Recall that we showed this in Sect. 1.9.3!) Take this direction to be the 3-axis (axis of quantization). As the projection $\mathbf{J}_3$ of the total angular momentum (which is the sum of orbital and spin angular momenta) is conserved, and as all m_ℓ vanish, the projections of the spins $m_s^{(\mu)}$ and $m_s^{(\nu)}$ must be equal and opposite. Neutrinos having their spin oriented opposite to their momentum, the m quantum numbers must be the ones shown in Fig. 4.5a.

Applying charge conjugation $\mathbf{C}$ to this process yields the new process shown in Fig. 4.5b: A negatively charged pion goes over into a μ^- with its spin antiparallel to its momentum, and a muonic antineutrino whose spin is antiparallel to its momentum, too. This process was never observed. Experiment tells us that neutrinos occur always with their spin *antiparallel* to the momentum, while antineutrinos come always with their spin *parallel* to the momentum. Another way of expressing the same observation is this: Define *helicity*, or "handedness" to be the projection

$$h := \frac{\boldsymbol{s} \cdot \boldsymbol{p}}{|\boldsymbol{p}|} \, . \tag{4.58}$$

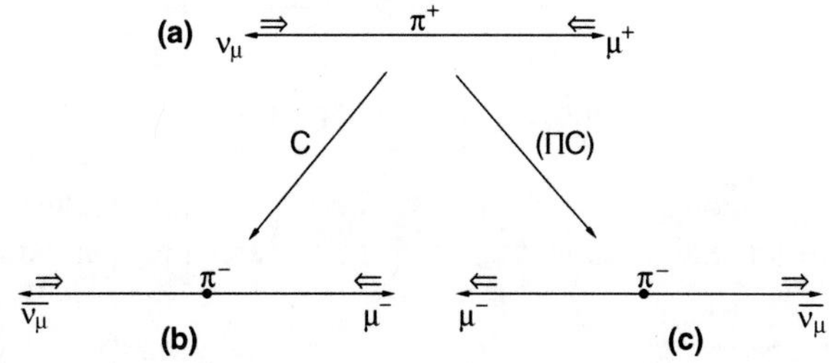

Fig. 4.5 The figure shows the decay $\pi^+ \longrightarrow \mu^+ + \nu_\mu$, in part (**a**), as well as the processes which are obtained from it by the action of charge conjugation $\mathbf{C}$, and of the product of $\mathbf{C}$ and of space reflection $\mathbf{\Pi}$. The process (**c**) is observed in nature, the process (**b**) is not

The empirical observation is that *neutrinos* always come with *negative* helicity, *antineutrinos* with *positive* helicity. Now, while it seems prohibitively difficult to measure this observable directly with neutrinos, this measurement is possible for their charged partners. One deduces the helicity of the neutrino or the antineutrino, respectively, from the conservation of $\mathbf{J}_3$, as described above.

If, in addition, one applies space reflection to the (unobserved) process of Fig. 4.5b, or, equivalently, if one acts with the *product* $\mathbf{C}\,\mathbf{\Pi}$ on the initial process of Fig. 4.5a, then one obtains the decay process shown in Fig. 4.5c. Indeed, this is a process which is seen in experiment and, hence, which is physically allowed!

The mere fact that one observes a nonvanishing spin-momentum correlation of the kind of (4.58), or $\boldsymbol{s}\cdot\boldsymbol{p}/E$, which is odd under $\mathbf{\Pi}$ but even under $\mathbf{T}$, is an indication that the weak interactions which cause the decay do not conserve parity. The experimental finding is that for leptons (electrons, neutrinos, muons, etc.) this correlation even takes the largest possible value. This means that violation of parity invariance must be *maximal*.[7] The example shows, furthermore, that the weak interaction also violates invariance under charge conjugation $\mathbf{C}$ in a maximal way. Only the combined transformation $\mathbf{C}\,\mathbf{\Pi}$ is a symmetry of the interaction.

4.3 Symmetry and Antisymmetry of Identical Particles

The quantum mechanics of a single particle that was the subject of Chap. 1 can be generalized to systems of many particles in a simple manner and in close analogy to classical mechanics. I show this in quite some detail on the example of two-body systems, and then, in a second step, extend the results to systems with a finite number N of particles. As compared to classical mechanics, a fundamentally new aspect will come into play when these particles are identical particles. In micro-physics, as a matter of principle, identical particles are indistinguishable. The analysis of this fact, together with Born's interpretation of the wave function, implies that the total wave function of the N-particle system carries a specific symmetry character with respect to interchange of particles. Furthermore, the type of exchange symmetry it has is related to the spin class of the particles. Particles with half-integer spin have another exchange symmetry than particles with integer spin.

4.3.1 Two Distinct Particles in Interaction

Take the hydrogen atom as an example: it is composed of an electron which has electric charge $-|e|$, and mass $m_e c^2 = 0.511\,\text{keV}$, and a proton which has charge

[7] The precise statement is the following: A particle with nonvanishing mass which is created in β-decay, or in some other related process, has longitudinal polarization (i.e. along the positive or negative direction of its momentum). The degree of polarization assumes the maximal value $P_l = \pm v/c$. If the particle has no mass, then this goes over into $P_l = \pm 1$, the helicity is $h = \pm 1/2$.

$+|e|$, and mass $m_p c^2 = 938.3$ MeV. Hence, we deal with two distinct particles whose properties differ. We consider two particles with masses m_1 and m_2, respectively, which interact via a central force. The central force is conservative, and, hence, may be derived from a potential $U(r)$ with $r := |\boldsymbol{x}^{(2)} - \boldsymbol{x}^{(1)}|$ the modulus of the relative coordinate. A Hamiltonian describing this system then has the form

$$H = -\frac{\hbar^2}{2m_1}\Delta^{(1)} - \frac{\hbar^2}{2m_2}\Delta^{(2)} + U(r)\,. \tag{4.59}$$

The particle index (i) attached to the Laplace operators indicates that derivatives are to be taken with respect to the coordinates $\boldsymbol{x}^{(i)}$.

The *kinematics* of the problem is the same as in classical mechanics: Center-of-mass and relative coordinates are, respectively,[8]

$$\boldsymbol{X} := \frac{1}{m_1 + m_2}(m_1\boldsymbol{x}^{(1)} + m_2\boldsymbol{x}^{(2)})\,, \quad \boldsymbol{r} := \boldsymbol{x}^{(2)} - \boldsymbol{x}^{(1)}\,.$$

The corresponding canonically conjugate momenta are, respectively,

$$\boldsymbol{P} := \boldsymbol{p}^{(1)} + \boldsymbol{p}^{(2)}\,, \quad \boldsymbol{p} := \frac{1}{m_1 + m_2}(m_1\boldsymbol{p}^{(2)} - m_2\boldsymbol{p}^{(1)})\,.$$

If $\mu = m_1 m_2/(m_1 + m_2)$ denotes the reduced mass, then we also have $\boldsymbol{p} = \mu\dot{\boldsymbol{r}}$.

The *dynamics* is different. The transformation

$$(\boldsymbol{x}^{(1)}, \boldsymbol{x}^{(2)}, \boldsymbol{p}^{(1)}, \boldsymbol{p}^{(2)}) \longmapsto (\boldsymbol{X}, \boldsymbol{r}, \boldsymbol{P}, \boldsymbol{p})$$

implies that the two physical particles are replaced by two fictitious but equally independent particles, one of which has the mass $M := m_1 + m_2$ and the phase space variables $(\boldsymbol{X}, \boldsymbol{P})$, while the other has mass μ and the phase space variables $(\boldsymbol{r}, \boldsymbol{p})$. Following the rules developed in Chap. 1 the momenta must be replaced by self-adjoint differential operators as follows

$$\boldsymbol{P} \longrightarrow \frac{\hbar}{\mathrm{i}}\boldsymbol{\nabla}^{(X)}\,, \quad \boldsymbol{p} \longrightarrow \frac{\hbar}{\mathrm{i}}\boldsymbol{\nabla}^{(r)}\,.$$

These operators and the position operators fulfill the commutation relations

$$[P_i, X^k] = \frac{\hbar}{\mathrm{i}}\delta_i^k = [p_i, r^k]\,, \quad [P_i, r^k] = 0 = [p_i, X^k]\,,$$

$$[P_i, P_k] = 0 = [p_i, p_k]\,, \quad [X^i, X^k] = 0 = [r^i, r^k]\,.$$

This prescription is verified by the following calculation. Making use of the chain rule for differentials one shows that

$$\boldsymbol{\nabla}^{(1)} = \frac{m_1}{M}\boldsymbol{\nabla}^{(X)} - \boldsymbol{\nabla}^{(r)}\,, \quad \boldsymbol{\nabla}^{(2)} = \frac{m_2}{M}\boldsymbol{\nabla}^{(X)} + \boldsymbol{\nabla}^{(r)}\,.$$

---

[8] In view of Jacobi coordinates for N particles I choose here $\boldsymbol{r}$ to point from particle 1 to particle 2. Note, however, that in [Scheck (2010)], Sects. 1.7.1 and 1.7.3, I used $\boldsymbol{r} = \boldsymbol{x}^{(1)} - \boldsymbol{x}^{(2)}$ instead.

Squaring these operators, multiplying the result for the first operator by $1/(2m_1)$, for the second by $1/(2m_2)$, and adding, one finds

$$\frac{1}{2M}\left(\nabla^{(X)}\right)^2 + \frac{1}{2}\left(\frac{1}{m_1}+\frac{1}{m_2}\right)\left(\nabla^{(r)}\right)^2 = \frac{1}{2M}\Delta^{(X)} + \frac{1}{2\mu}\Delta^{(r)} .$$

Thereby, the Hamiltonian (4.59) takes the expected form

$$H = -\frac{\hbar^2}{2M}\Delta^{(X)} - \frac{\hbar^2}{2\mu}\Delta^{(r)} + U(r) \equiv H^{(X)} + H^{(r)} . \tag{4.60}$$

It separates into the (force-free) motion of the center-of-mass and an effective one-body problem in the relative motion. Stationary solutions of the Schrödinger equation may be assumed to factor as follows,

$$\Psi(\boldsymbol{X}, \boldsymbol{r}, \text{spins}) = \psi(\boldsymbol{X})\varphi(\boldsymbol{r})\chi(s^{(1)}, s^{(2)}) , \tag{4.61}$$

where χ is a wave function which describes the spin(s) of the two particles. For example, if the total state is an eigenstate of the center-of-mass momentum $\boldsymbol{P}$, then

$$\Psi(\boldsymbol{X}, \boldsymbol{r}, \text{spins}) = \exp\left(\frac{\mathrm{i}}{\hbar}\boldsymbol{P}\cdot\boldsymbol{X}\right)\varphi(\boldsymbol{r})\chi(s^{(1)}, s^{(2)}) .$$

At the same time the stationary Schrödinger equation is decomposed into two additive parts, and the eigenvalue of energy is the sum of the kinetic energy of the center-of-mass and the energy of the relative motion, $E = \boldsymbol{P}^2/(2M) + E_{\text{rel}}$. The full dynamics is contained, apart from free motion of the center-of-mass, in the Schrödinger equation for relative motion,

$$H^{(r)}\varphi(\boldsymbol{r}) = \left(-\frac{\hbar^2}{2\mu}\Delta^{(r)} + U(r)\right)\varphi(\boldsymbol{r}) = E_{\text{rel}}\varphi(\boldsymbol{r}) .$$

Thus, we are taken back to the central field problems for one single particle that we discussed extensively in Chap. 1.

The spin function χ, in turn, for given spins $s^{(1)}$ and $s^{(2)}$ is to be constructed as described previously.

Remarks

As long as all particles are different from each other, the generalization to $N > 2$ is obvious and follows the corresponding rules of classical mechanics. Suppose the external forces are potential forces, and, thus, are described by potentials $U_n(\boldsymbol{x}^{(n)})$. Suppose further that the internal forces are central forces described by potentials $U_{mn}(|\boldsymbol{x}^{(m)} - \boldsymbol{x}^{(n)}|)$. Then a typical Hamiltonian reads

$$H = \sum_{n=1}^{N}\left(-\frac{\hbar^2}{2m_n}\Delta^{(n)} + U_n(\boldsymbol{x}^{(n)})\right) + \frac{1}{2}\sum_{m\neq n=1}^{N} U_{mn}\left(\left|\boldsymbol{x}^{(m)} - \boldsymbol{x}^{(n)}\right|\right) . \tag{4.62}$$

Its stationary eigenstates depend on all coordinates $\{\boldsymbol{x}^{(n)}\}$ and, if occasion arises, on the spins of the participating particles,

$$\Psi = \Psi(\boldsymbol{x}^{(1)}, s^{(1)}; \boldsymbol{x}^{(2)}, s^{(2)}; \ldots; \boldsymbol{x}^{(n)}, s^{(n)}) .$$

The interpretation of this wave function follows from Born's interpretation.

If all external forces are zero it is appropriate to separate the center-of-mass motion and to introduce Jacobi coordinates[9] $\{\boldsymbol{r}^{(i)}, \boldsymbol{\pi}^{(i)}\}$, cf. [Scheck (2010)], Exercise 2.24. In the notation used here, these are

$$\boldsymbol{r}^{(j)} = \boldsymbol{x}^{(j+1)} - \frac{1}{M_j}\sum_{i=1}^{j} m_i \boldsymbol{x}^{(i)}, \quad \boldsymbol{r}^{(N)} = \frac{1}{M_N}\sum_{i=1}^{N} m_i \boldsymbol{x}^{(i)}, \tag{4.63}$$

$$\boldsymbol{\pi}^{(j)} = \frac{1}{M_{j+1}}\left(M_j \boldsymbol{p}^{(j+1)} - m_{j+1}\sum_{i=1}^{j} \boldsymbol{p}^{(i)}\right), \quad \boldsymbol{\pi}^{(N)} = \sum_{i=1}^{N} \boldsymbol{p}^{(i)}. \tag{4.64}$$

In these formulae M_j is the sum of the first j masses, $M_j = \sum_{i=1}^{j} m_i$, the index j running from 1 to $N-1$. It is also of interest to determine the inverse formulae of (4.64). They are

$$\begin{aligned}
\boldsymbol{p}^{(1)} &= -\boldsymbol{\pi}^{(1)} - \sum_{j=2}^{N-1} \frac{m_1}{M_j}\boldsymbol{\pi}^{(j)} + \frac{m_1}{M_N}\boldsymbol{\pi}^{(N)} \\
\boldsymbol{p}^{(2)} &= \quad \boldsymbol{\pi}^{(1)} - \sum_{j=2}^{N-1} \frac{m_2}{M_j}\boldsymbol{\pi}^{(j)} + \frac{m_2}{M_N}\boldsymbol{\pi}^{(N)} \\
\boldsymbol{p}^{(3)} &= \quad \boldsymbol{\pi}^{(2)} - \sum_{j=3}^{N-1} \frac{m_3}{M_j}\boldsymbol{\pi}^{(j)} + \frac{m_3}{M_N}\boldsymbol{\pi}^{(N)} \\
\boldsymbol{p}^{(4)} &= \quad \boldsymbol{\pi}^{(3)} - \sum_{j=3}^{N-1} \frac{m_4}{M_j}\boldsymbol{\pi}^{(j)} + \frac{m_4}{M_N}\boldsymbol{\pi}^{(N)} \\
\vdots &= \quad \vdots \\
\boldsymbol{p}^{(N)} &= \boldsymbol{\pi}^{(N-1)} \qquad\qquad\qquad + \frac{m_N}{M_N}\boldsymbol{\pi}^{(N)}.
\end{aligned}$$

The variables $(\boldsymbol{r}^{(j)}, \boldsymbol{\pi}^{(j)})$ are canonically conjugate variables, and so are the original variables $(\boldsymbol{x}^{(k)}, \boldsymbol{p}^{(k)})$.

4.3.2 Identical Particles with the Example $N = 2$

When the particles are identical and hence indistinguishable, the description of the N-particle system receives a fundamentally new feature. While in the classical,

[9] K.G.J. Jacobi, Crelles Journal für reine und angewandte Mathematik **XXVI**, 115–131 (1843).

macroscopic realm it is perfectly conceivable to "mark" and to identify individual particles, as a matter of principle, this is not possible for micro-particles such as electrons, protons, π mesons, or photons. This strict indistinguishability is emphasized by the Born interpretation of quantum wave functions from which one concludes that, in general, it is impossible to predict the outcome of a measurement for one single particle. The predictions of quantum mechanics are probabilities. They apply only to a large number of identically prepared particles. Wave functions or self-adjoint operators which single out, in some way or other, one particular particle in an ensemble of N identical ones, cannot be physically meaningful. We discuss these matters first for the example of two particles, $N = 2$.

Let $\Psi(\boldsymbol{x}^{(1)}, m_s^{(1)}; \boldsymbol{x}^{(2)}, m_s^{(2)})$ be an arbitrary two-particle wave function. Any single-particle observable, constructed according to the rules of Sect. 1.5, must be of the form

$$\mathcal{O} = \mathcal{O}\left(\frac{\hbar}{\mathrm{i}}\boldsymbol{\nabla}^{(1)}, \boldsymbol{x}^{(1)}\right) + \mathcal{O}\left(\frac{\hbar}{\mathrm{i}}\boldsymbol{\nabla}^{(2)}, \boldsymbol{x}^{(2)}\right)$$

where the second term differs from the first solely by the replacement. ($\boldsymbol{x}^{(1)} \longleftrightarrow \boldsymbol{x}^{(2)}$, $\boldsymbol{p}^{(1)} \longleftrightarrow \boldsymbol{p}^{(2)}$) and, if it contains spin degrees of freedom, by simultaneous exchange of the spin operators. In case of a two-body observable, for which the interaction term (4.62) provides an example, this operator must be invariant when the two particles are interchanged with all their attributes.

The construction just sketched is an acceptable rule for obtaining one-body observables which are symmetric in the two identical particles. Alternatively, one may introduce the permutation operator Π_{12} which acts as follows:

$$\Pi_{12}\Psi(\boldsymbol{x}^{(1)}, m_s^{(1)}; \boldsymbol{x}^{(2)}, m_s^{(2)}) = \Psi(\boldsymbol{x}^{(2)}, m_s^{(2)}; \boldsymbol{x}^{(1)}, m_s^{(1)})$$

and whose properties are

$$\Pi_{12}^2 = \mathbb{1}\,, \quad \Pi_{12}^\dagger = \Pi_{12}\,.$$

Its eigenvalues are $+1$ and -1. In the first case its eigenstates are symmetric under exchange, in the second case they are antisymmetric. A one-particle observable is obtained by the rule

$$\mathcal{O} = \mathcal{O}\left(\frac{\hbar}{\mathrm{i}}\boldsymbol{\nabla}^{(1)}, \boldsymbol{x}^{(1)}\right) + \Pi_{12}\,\mathcal{O}\left(\frac{\hbar}{\mathrm{i}}\boldsymbol{\nabla}^{(1)}, \boldsymbol{x}^{(1)}\right)\Pi_{12}^\dagger\,.$$

Suppose a pure state Ψ is being prepared by measuring an observable of this kind. The state is described by the projection operator P_Ψ such that this operator commutes with Π_{12}, $[\Pi_{12}, P_\Psi] = 0$. This means, in turn, that if Ψ is an eigenstate of P_Ψ, then so is $\Pi_{12}\Psi$. Thus,

$$\Pi_{12}\Psi = z\Psi \quad \text{with} \quad z = \pm 1\,.$$

The state prepared in this way must be either symmetric or antisymmetric under exchange of the two particles. As the Hamiltonian of the two-body system itself is symmetric when the particles are interchanged, it commutes with Π_{12}, $[H, \Pi_{12}] = 0$.

The same conclusion applies to the operator (3.27) of temporal evolution of the system,

$$[\Pi_{12}, U(t, t_0)] = \left[\Pi_{12}, \exp\left(-\frac{\mathrm{i}}{\hbar}H(t-t_0)\right)\right] = 0 .$$

The symmetry character with respect to interchange of particles is not changed by the evolution in time. An initially symmetric state will remain to be symmetric for all times, an antisymmetric state stays antisymmetric forever. States which are neither symmetric nor antisymmetric, cannot be physical states. Also, from a physical point of view, it is not meaningful to superpose symmetric and antisymmetric states. We illustrate these conclusions by a few examples.

Example 4.2

The Hamiltonian (4.59) of a system with two identical particles with spin s may contain an internal central force, described by a potential $U(r)$ with $r := |\boldsymbol{x}^{(2)} - \boldsymbol{x}^{(1)}|$, but is assumed to contain no external forces. If one separates the kinematics into center-of-mass and relative motion, in agreement with (4.60), the wave function that describes the center-of-mass, by definition, is *symmetric* when the particles are interchanged. The wave function for relative motion is written in a factorized form in spherical polar coordinates,

$$\psi_{\alpha\ell m}(\boldsymbol{r}) = R_\alpha(r) Y_{\ell m}(\theta, \phi) , \quad \boldsymbol{r} = \boldsymbol{x}^{(2)} - \boldsymbol{x}^{(1)} .$$

Finally, the spin wave functions are coupled to total spin $\boldsymbol{S} = \boldsymbol{s}^{(1)} + \boldsymbol{s}^{(2)}$, viz.

$$|SM\rangle = \sum_{m_1, m_2} (sm_1, sm_2|SM)\, |sm_1\rangle\, |sm_2\rangle .$$

In $\mathbb{R}^3$ the exchange of the two particles implies the replacements

$$\Pi_{12} : \quad r \longmapsto r , \quad \theta \longmapsto \pi - \theta , \quad \phi \longmapsto \phi + \pi \quad \bmod 2\pi .$$

Obviously, the action of this mapping is the same as the action of space reflection, Sect. 4.2.1. While the radial function remains unchanged, the spherical harmonic receives the sign $(-)^\ell$. Thus, one has

$$\Pi_{12} : \quad \psi_{\alpha\ell m}(\boldsymbol{r}) \longmapsto \psi_{\alpha\ell m}(-\boldsymbol{r}) = (-)^\ell \psi_{\alpha\ell m}(\boldsymbol{r}) . \tag{4.65}$$

The spin function, in turn, receives its sign from the relation (4.34), viz.

$$\Pi_{12} : \quad |SM\rangle \longmapsto (-)^{2s-S} |SM\rangle . \tag{4.66}$$

Consider then the two characteristic cases:

1. *Two particles with spin* $s = 1/2$: According to the selection rules (4.30) the total spin S can have only the values 1 and 0. The eigenstates $|SM\rangle$ are constructed by means of the ladder operators (4.8) and the relations (4.13), whose action in the spinor representation is given by

$$\mathbf{J}_+ \left|\frac{1}{2}, -\frac{1}{2}\right\rangle = \left|\frac{1}{2}, +\frac{1}{2}\right\rangle ,$$

$$\mathbf{J}_- \left|\frac{1}{2}, +\frac{1}{2}\right\rangle = \left|\frac{1}{2}, -\frac{1}{2}\right\rangle , \quad \mathbf{J}_\pm \left|\frac{1}{2}, \pm\frac{1}{2}\right\rangle = 0 .$$

For the case of the triplet representation their action is

$$\mathbf{J}_{\pm}\,|1,\mp 1\rangle = \sqrt{2}\,|1,0\rangle\ , \quad \mathbf{J}_{\pm}\,|1,0\rangle = \sqrt{2}\,|1,\pm 1\rangle\ .$$

Proceeding like in Sect. 4.1.6 and as sketched in Fig. 4.4, one starts from the two-body state with $S = M = 1, |1,1\rangle = |1/2,+1/2\rangle\,|1/2,+1/2\rangle$, and applies the lowering operator $\mathbf{J}_{-} = \mathbf{J}_{-}^{(1)} + \mathbf{J}_{-}^{(2)}$ to it, thus obtaining

$$\begin{aligned} |1,+1\rangle &= \left|\frac{1}{2},+\frac{1}{2}\right\rangle\left|\frac{1}{2},+\frac{1}{2}\right\rangle,\\ |1,\ \ 0\rangle &= \frac{1}{\sqrt{2}}\left(\left|\frac{1}{2},+\frac{1}{2}\right\rangle\left|\frac{1}{2},-\frac{1}{2}\right\rangle + \left|\frac{1}{2},-\frac{1}{2}\right\rangle\left|\frac{1}{2},+\frac{1}{2}\right\rangle\right),\\ |1,-1\rangle &= \left|\frac{1}{2},-\frac{1}{2}\right\rangle\left|\frac{1}{2},-\frac{1}{2}\right\rangle. \end{aligned} \tag{4.67}$$

The state which has $S = 0$ is given by a linear combination which is orthogonal to the state $|1,0\rangle$ in (4.67),

$$|0,0\rangle = \frac{1}{\sqrt{2}}\left(\left|\frac{1}{2},+\frac{1}{2}\right\rangle\left|\frac{1}{2},-\frac{1}{2}\right\rangle - \left|\frac{1}{2},-\frac{1}{2}\right\rangle\left|\frac{1}{2},+\frac{1}{2}\right\rangle\right). \tag{4.68}$$

The three states (4.67) are symmetric, the state (4.68) is antisymmetric under exchange of the two particles. This agrees with the more general rule (4.66) which yields the sign $(-)^{1-S}$.

The action of exchange on the orbital wave function and the spin function taken together, is seen to be

$$\Pi_{12}:\ \psi_{\alpha\ell m}(\boldsymbol{r})\left|\left(\frac{1}{2},\frac{1}{2}\right)SM\right\rangle \longmapsto (-)^{\ell+S-1}\,\psi_{\alpha\ell m}(\boldsymbol{r})\left|\left(\frac{1}{2},\frac{1}{2}\right)SM\right\rangle. \tag{4.69}$$

2. *Two particles with spin $s = 1$*: According to the rules (4.30) the total spin can take the values $S = 2, 1, 0$. One starts from the state $|2,+2\rangle = |1,+1\rangle\,|1,+1\rangle$ and constructs from it the entire multiplet with $S = 2$, by means of the ladder operator $\mathbf{J}_{-}$. The state $|1,+1\rangle$ is determined using its orthogonality to the state $|2,+1\rangle$). The remaining triplet states follow from this state, as before. Finally, the state $|0,0\rangle$ is obtained by using its orthogonality to $|2,0\rangle$ and to $|1,0\rangle$. In summary, one obtains for total spin $S = 2$:

$$\begin{aligned} |2,\pm 2\rangle &= |1,\pm 1\rangle\,|1,\pm 1\rangle\\ |2,\pm 1\rangle &= \frac{1}{\sqrt{2}}\left(|1,\pm 1\rangle\,|1,0\rangle + |1,0\rangle\,|1,\pm 1\rangle\right)\\ |2,\ \ 0\rangle &= \frac{1}{\sqrt{6}}\left(|1,1\rangle\,|1,-1\rangle + 2\,|1,0\rangle\,|1,0\rangle + |1,-1\rangle\,|1,1\rangle\right), \end{aligned} \tag{4.70}$$

for total spin $S = 1$:

$$\begin{aligned} |1,\pm 1\rangle &= \frac{1}{\sqrt{2}}\left(\pm\,|1,\pm 1\rangle\,|1,0\rangle \mp |1,0\rangle\,|1,\pm 1\rangle\right)\\ |1,\ \ 0\rangle &= \frac{1}{\sqrt{2}}\left(|1,1\rangle\,|1,-1\rangle - |1,-1\rangle\,|1,1\rangle\right), \end{aligned} \tag{4.71}$$

and for total spin $S = 0$:

$$|0,0\rangle = \frac{1}{\sqrt{3}} \left(|1,1\rangle\,|1,-1\rangle - |1,0\rangle\,|1,0\rangle + |1,-1\rangle\,|1,1\rangle \right) . \tag{4.72}$$

The states (4.70) and (4.72) are seen to be symmetric while the state (4.71) is antisymmetric, in agreement with the general rule (4.66). Regarding the symmetry of the total, orbital and spin, wave function one has

$$\Pi_{12}: \quad \psi_{\alpha\ell m}(\boldsymbol{r})\,|(1,1)SM\rangle \longmapsto (-)^{\ell+S}\,\psi_{\alpha\ell m}(\boldsymbol{r})\,|(1,1)SM\rangle . \tag{4.73}$$

Before turning to a discussion of the relation between the spin of identical particles and their statistics which fixes the allowed signs in (4.69) and in (4.73), we generalize the results of this section to more than two particles.

4.3.3 Extension to *N* Identical Particles

Given a system of N identical particles with spin s, the general analysis of the preceding section applies to any pair (i, j) of them. As a consequence, it seems obvious that every observable is symmetric in *all* particles, i.e. that it commutes with the interchanging operation Π_{ij} for all i and j. Only those states of the N-particle system can be physically meaningful which are either completely *symmetric*, or completely *antisymmetric* under all permutations. Let us explain these matters in more detail: Let Π be a permutation of the N particles,

$$\Pi: \quad (1,2,3,\ldots,N) \longmapsto \big(\Pi(1), \Pi(2), \Pi(3), \ldots, \Pi(N)\big) ,$$

and let $(-)^{\Pi}$ be its sign. Permutations are generated by interchanging immediate neighbours. They are called *even*, and their sign is *positive*, if the number of neighbour exchanges is *even*. They are *odd*, and have a *minus sign* if the number of neighbour exchanges is *odd*. For example, the permutation $(1,2,3,4) \mapsto (4,1,2,3)$ is odd because it needs three exchanges of neighbours to reach the second ordering from the first.

Let $\Psi(1; 2; 3; \ldots; N)$ be a solution of the Schrödinger equation for N identical particles, the particle number "i" being a short-hand notation for the coordinates and the spin quantum numbers of a particle. A *completely symmetric* wave function is generated by the rule

$$\Psi_{\mathrm{S}} = N_{\mathrm{S}} \sum_{\Pi} \Pi\,\Psi(1; 2; 3; \ldots; N) , \tag{4.74}$$

while a *completely antisymmetric* wave function is obtained by the prescription

$$\Psi_{\mathrm{A}} = N_{\mathrm{A}} \sum_{\Pi} (-)^{\Pi}\,\Pi\,\Psi(1; 2; 3; \ldots; N) . \tag{4.75}$$

In these formulae N_{S} and N_{A} are normalization factors which in both cases must be determined such that Ψ_{S} and Ψ_{A}, respectively, are normalized to 1.

4.3.4 Connection Between Spin and Statistics

Particles whose spin is *half-integer*, i.e. who carry spin $s = 1/2, 3/2, 5/2, \ldots$, are called *fermions*; Particles whose spin is *integer*, i.e. $s = 0, 1, 2, \ldots$, are called *bosons*.[10] Note that it makes no difference whether the particles are elementary building blocs of nature such as an electron or a quark, or composite particles such as a proton p, the charged and neutral pions $\pi^{\pm}$ and π^0, respectively, atoms or atomic nuclei, whose spin, in reality, is the resulting sum of the spins and orbital angular momenta of their constituents. We show some examples for fermions and bosons in Table 4.1.

There is a deep relation, for every quantum system of N identical particles, between their spin class, bosonic or fermionic, and the symmetry of its physical states under permutations of the particles:

Table 4.1 Properties of a few elementary or composite particles

	Fermions		
Particle	Symbol	Charge	Spin
Electron	e^-	-1	1/2
Positron	e^+	$+1$	1/2
Proton	p	1	1/2
Neutron	n	0	1/2
Bismuth	^{209}Bi	83	9/2
Muon	μ^-	-1	1/2
Antimuon	μ^+	$+1$	1/2
Electron-	ν_e	0	1/2
Neutrino	$\overline{\nu}_e$	0	1/2
Up-quark	u	$+2/3$	1/2
Down-quark	d	$-1/3$	1/2
Strange-quark	s	$-1/3$	1/2
	Bosons		
Particle	Symbol	Charge	Spin
Photon	γ	0	1
$W^{\pm}$-bosons	$W^{\pm}$	± 1	1
Z^0-boson	Z^0	0	1
Higgs boson	H	0	0
Helium nucleus	α	2	0
Pions	$\pi^{\pm}, \pi^0$	$\pm 1, 0$	0

Charges are in units of the elementary charge

[10] The names go back to two physicists of the 20th century: Enrico Fermi (1901–1954), Italo-American, and Satyendra Nath Bose (1894–1974), from India.

Exchange Symmetry of N-Fermion-/N-Boson-States:
For any permutation Π of the particles, the physical states of N *fermions* are multiplied by $(-)^{\Pi}$. Such states are of the type of (4.75). Indeed, with $\Pi \circ \Pi' =: \Pi''$, one has

$$\begin{aligned}\Pi\Psi &= N_{\mathrm{A}} \sum_{\Pi'} (-)^{\Pi'} \Pi \circ \Pi' \Psi(1; \ldots; N) \\ &= (-)^{\Pi} N_{\mathrm{A}} \sum_{\Pi''} (-)^{\Pi''} \Pi'' \Psi(1; \ldots; N) \\ &= (-)^{\Pi} \Psi .\end{aligned} \tag{4.76}$$

The physical states of N *bosons* are totally symmetric under any permutation of the particles. They are of the type of (4.74) so that one has

$$\Pi\Psi = \Psi . \tag{4.77}$$

Before turning to the foundations on which this fundamental rule rests, let us explain it in some more detail, and illustrate it by some simple examples. The first part of the rule which concerns fermions, for the case $N = 2$, implies that every state must be antisymmetric when the two particles are exchanged with all their attributes. For the Example 4.2 of Sect. 4.3.2 this means that in (4.69) only those values of S and ℓ are admissible whose sum is *even*. The spin singlet can only occur when $\ell = 0, 2, \ldots$, while the spin triplet requires $\ell = 1, 3, \ldots$.

The second part which concerns bosons, for $N = 2$, says that in (4.73), too, only $S + \ell = even$ is admissible. With $S = 0$ or $S = 2$ the orbital angular momentum ℓ must have even-integer values, for $S = 1$ it must have odd-integer values.

Of course, the fermionic case is of special interest because of the alternating signs. Suppose, for instance, the Hamiltonian to be the sum of N copies of a one-body Hamiltonian $H(n)$, $H = \sum_{n=1}^{N} H(n)$, whose stationary solutions $\varphi_{\alpha_k}(n)$ and eigenvalues E_{α_k} are known,

$$H(n)\varphi_{\alpha_k}(n) = E_{\alpha_k}\, \varphi_{\alpha_k}(n) .$$

The N-particle wave function

$$\Psi(1; 2; \cdots ; N) = \varphi_{\alpha_1}(1)\varphi_{\alpha_2}(2) \ldots \varphi_{\alpha_N}(N) ,$$

although it is an eigenfunction of H pertaining to the eigenvalue $E = \sum_{k=1}^{N} E_{\alpha_k}$, does not obey the symmetry rule (4.76). It will do so only if we distribute the N fermions onto the normed states $\varphi_{\alpha_1}, \varphi_{\alpha_2}, \ldots , \varphi_{\alpha_N}$ in all possible ways and provide every permutation with the sign that pertains to it. Thus, the correct antisymmetrized product wave function, normalized to 1 must be constructed as follows:

$$\Psi_{\mathrm{A}} = \frac{1}{\sqrt{N!}} \det \begin{pmatrix} \varphi_{\alpha_1}(1) & \varphi_{\alpha_2}(1) & \ldots & \varphi_{\alpha_N}(1) \\ \varphi_{\alpha_1}(2) & \varphi_{\alpha_2}(2) & \ldots & \varphi_{\alpha_N}(2) \\ \vdots & \vdots & \ddots & \vdots \\ \varphi_{\alpha_1}(N) & \varphi_{\alpha_2}(N) & \ldots & \varphi_{\alpha_N}(N) \end{pmatrix} . \tag{4.78}$$

This product state is especially remarkable by the fact that it vanishes whenever $\alpha_i = \alpha_k$ for $i, k \in (1, 2, \ldots, N)$, that is, whenever two of the one-particle states are the same. This is a manifestation of

Pauli's Exclusion Principle: In a system of identical fermions two (or more) particles can never be in the same one-particle state.
Expressed differently: One-particle states in a product state have occupation number 0 or 1. A given such state φ_{α_k} may be unoccupied, or may contain at most one fermion of a given species.

It is precisely this restriction which is the defining property of *Fermi-Dirac statistics*. There is an intrinsic connection between the half-integral spin class of identical particles, i.e. their property of having *half-integral spin*, and the *Fermi-Dirac statistics* they obey.

If one wishes to determine the ground state of H for our example, there is no other possibility than to take the determinant (4.78) of the first N (energetically) lowest states. More qualitatively speaking: one fills the first N states with identical fermions by putting exactly one particle into each of these one-particle states. This model is the basis for building the electronic shells of atoms, as well as for the shell model of nuclei with protons and neutrons. The next example illustrates this construction:

Example 4.3

In a system of N identical fermions with spin $1/2$ consider the one-particle potential given by

$$U(r) = \frac{1}{2} m\omega^2 r^2 - C\,\boldsymbol{\ell} \cdot \boldsymbol{s} - D\boldsymbol{\ell}^2 .$$

This was the first ansatz for the shell model of nuclei which was able to explain the shell closures in especially stable nuclei, the so-called *magic numbers*. Note that the potential is assumed to be the same for all particles. The parameters C and D are positive constants. Choosing a basis $|JM\rangle$, the orbital angular momentum and the spin being coupled to $\boldsymbol{j} = \boldsymbol{\ell} + \boldsymbol{s}$, and writing the spin-orbit coupling in the form

$$\boldsymbol{\ell} \cdot \boldsymbol{s} = \frac{1}{2}(\boldsymbol{j}^2 - \boldsymbol{\ell}^2 - \boldsymbol{s}^2) ,$$

one realizes that this operator is already diagonal. With the result (1.148) for the energies of the spherical oscillator, the eigenvalues of the one-particle Hamiltonian are given by the following expression,

$$E_{n\ell} = \hbar\omega\left(2n + \ell + \frac{3}{2}\right) - \frac{C}{2}\left[j(j+1) - \ell(\ell+1) - \frac{3}{4}\right] - D\ell(\ell+1) .$$

This spectrum of one-particle energies is drawn in Fig. 4.6. For every value of j there are $(2j+1)$ substates $|JM\rangle$. If one sets out to fill this potential with N identical

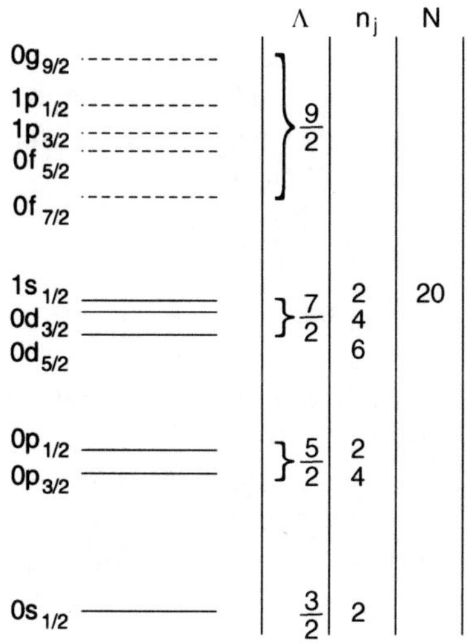

Fig. 4.6 The first eleven lowest one-particle levels in a simple shell model for nuclei. The number Λ is a short-hand notation for $\Lambda = 2n + \ell + 3/2$, n_j is the number of particles in the state $|n\ell j\rangle$, N is the total number of particles of the example. The *horizontal full lines* are the occupied levels, the *dashed* levels are empty

fermions such as to construct a product state with the lowest energy, one must start at the bottom of the spectrum and fill each state $|n\ell j\rangle$ with $2j+1$ particles. Figure 4.6 shows the example $N = 20$, where the states $0s_{1/2}$ to $1s_{1/2}$ are filled while all states higher than these remain unoccupied. The corresponding, properly antisymmetrized wave function is given by the determinant (4.78) which contains the first, lowest, one-particle states.

For bosons matters are completely different: If for a system of N bosons we assume a Hamiltonian, like above, which is the sum of N copies of some one-particle operator, $H = \sum_n H(n)$, then the rule (4.77) allows to place arbitrarily many particles into a given one-particle state $(E_{\alpha_k}, \varphi_{\alpha_k})$. This condition is in accord with *Bose-Einstein statistics*. The energetically lowest state is the one in which all N particles are put into the lowest one-particle state (E_0, φ_0). This phenomenon is called *Bose-Einstein condensation*, its existence at macroscopic scales was confirmed in recent years, in a series of beautiful experiments. Note, however, that the theoretical description cannot be as simple as it may seem here, because, for large numbers of particles, the mutual interaction of the particles cannot be neglected.

The connection between spin and statistics, developed here in an empirical and heuristic way, is the content of a theorem which goes back to M. Fierz and W. Pauli,

Spin-Statistics Theorem: All particles with *half-integer* spin obey *Fermi-Dirac statistics*, all particles with *integer* spin obey *Bose-Einstein statistics*.

Some Comments on the Spin-Statistics Theorem:

1. The two types of statistics are known from the quantum mechanical description of gases which consist of independently moving, identical bosons or fermions,

respectively, and which are in an equilibrium state under specific macroscopic conditions such as temperature, chemical potential, and total volume. The particles are distributed among the given one-particle states with energies ε_i and with occupation numbers n_i, so that for the total number N of particles and total energy E one has

$$\sum_i n_i = N, \quad \sum_i n_i \varepsilon_i = E .$$

In the case of bosons the occupation number, for every level i, can take any value between 0 and N. If $n_i \geq 2$ the $n_i!$ permutations of the bosons are indistinguishable. In the case of fermions, in turn, the occupation numbers n_i can take only the values 0 or 1.

2. For simplicity, let us consider the case of two bosons, or of two fermions. The symmetry relations (4.76) and (4.77), respectively, or, equivalently, the spin statistics theorem state that under exchange a two-body state fulfills the relation

$$\Psi(2, 1) = (-)^{2s} \Psi(1, 2) , \tag{4.79}$$

with s the spin of the particles. Note that in this operation the two particles are exchanged with all their attributes, position and spin. If the spin is integer the wave function is symmetric, if it is half-integer the wave function is antisymmetric.

It is a striking feature of the rotation group that a rotation of the frame of reference by the angle 2π, when applied to the spin function of a single particle, produces exactly this sign: For integer spin there is no sign change, while for half-integer spin there is a minus sign, cf. Sect. 4.1.4. Thus, rotating the spin of one of the particles by 2π, or, as an equivalent operation, rotating the spins of both particles about the same axis by the angle π, one obtains the same sign as in (4.79). Indeed, it turns out that in the proof of the spin-statistics theorem the symmetry character of a two-body function of identical particles eventually goes back to this sign. The proofs by Fierz and by Pauli hold in the framework of Lorentz covariant field theory and of what is called second quantization. The essential argument is as follows: One shows that it is only possible to construct a local, Lorentz covariant quantum field theory which respects all conditions imposed by causality (i.e. which guarantees propagation of physical actions with velocities smaller than or equal to the speed of light), if bosons obey Einstein-Bose statistics, fermions obey Fermi-Dirac statistics. We return to these matters in Part Two.

5 Applications of Quantum Mechanics

▶ **Introduction** Quantum mechanics provides the basis for most fields of modern physics and there are many well advanced methods of practical solution of specific and topical problems. These methods which may be perturbative or nonperturbative, often are specific to the various disciplines, and it would go far beyond the scope of a textbook to discuss them extensively and in due detail. For instance, atomic and molecular physics make extensive use of *variational calculus* and of *many-body techniques* the latter of which are of great importance also for the physics of condensed matter and for nuclear physics. Elementary particle physics, in turn, makes use of *covariant perturbation theory* as well as of various kinds of nonperturbative approaches. There are numerous methods to treat scattering off composite targets at low, intermediate, and high energies (optical potential, Green function techniques, eikonal approximation).

Exact solutions are often approximated by numerical procedures such as integration of differential equations, diagonalization of large matrices in truncated Hilbert spaces, discretization and simulation by means of Monte Carlo methods, etc., which are adapted for the problem one wishes to study. In this chapter we first sketch the possible application of quantum mechanics to information theory. We then discuss nonrelativistic perturbation theory in its time independent and its time dependent versions. Finally, we give an introduction to selected techniques for treating systems of many interacting fermions. Relativistic, Lorentz covariant perturbation theory will be dealt with in Part Two.

5.1 Correlated States and Quantum Information

The principles of quantum mechanics were formulated and laid down in Chap. 3. In particular, the description of quantum states by means of statistical operators or, equivalently, by density matrices was discussed in detail and illustrated by a number of instructive examples. So one might be tempted to say that there is little to be added to what we worked out in Chap. 3, from the point of view of basic principles, and all

F. Scheck, *Quantum Physics*, DOI: 10.1007/978-3-642-34563-0_5,

there is left to do is to develop practical methods for solving concrete problems of quantum mechanics which go beyond the few exactly solvable ones. Although the practical methods often are by no means simple and open up a wide field reaching far into modern research, the basic principles and the interpretation of quantum mechanics, after some further reflection, have perplexing consequences which often are different from expectations based on classical physics and which are testable in experiment. This is why we insert, as a first application, a discussion of nonlocalities in quantum mechanics, correlations, entangled states, as well as a short excursion to quantum information. All of these are topics of modern research and one should expect to see rapid progress forthcoming in the years to come.

5.1.1 Nonlocalities, Entanglement, and Correlations

The simplest quantum states of a system of N particles are the ones which are direct products of one-particle states such as

$$\Psi^{(0)}(1, 2, \ldots, N) = |\,\psi_1(1)\rangle\,|\,\psi_2(2)\,\rangle \cdots |\,\psi_N(N)\rangle \; . \tag{5.1}$$

For simplicity we number the states from 1 to N, irrespective, for the moment, of what they are dynamically and whether some of them are the same or not. The arguments $(\boldsymbol{x}_i, s_i)$ comprising the coordinates and spins, plus further attributes if the need arises, are also summarized by just writing "i" instead. The product state (5.1) is an element of the Hilbert space

$$\mathcal{H} = \mathcal{H}^{(1)} \otimes \mathcal{H}^{(2)} \otimes \cdots \mathcal{H}^{(N)} \equiv \bigotimes_{i=1}^{N} \mathcal{H}^{(i)} \, .$$

A product state of the kind of (5.1) is said to be *separable* with respect to the factors in $\mathcal{H}$. To illustrate the special nature of the state $\Psi^{(0)}$ let us define the *one-body density* in a more general N-body state Ψ by the expectation value

$$\varrho(\boldsymbol{x}) := \frac{1}{N} \langle \Psi |\sum_{i=1}^{N} \delta(\boldsymbol{x}_i - \boldsymbol{x})\,|\,\Psi\,\rangle \; . \tag{5.2}$$

Specializing to $\Psi = \Psi^{(0)}$ the density is just the sum of the one-body densities for everyone of the states in the product,

$$\varrho^{(0)}(\boldsymbol{x}) = \frac{1}{N}\sum_{i=1}^{N} |\psi_i(\boldsymbol{x})|^2 \, .$$

Its integral over the whole three-dimensional space gives 1,

$$\int \mathrm{d}^3x \; \varrho(\boldsymbol{x}) = 1 \, ,$$

if all single-particle wave functions are normalized to unity.

In most realistic situations where quantum mechanics is at work, the wave function of a many-body system is *not* of the simple product type (5.1). Any state which is not separable, is called *entangled state*.[1] Here are some arguments and examples that illustrate why this is so. As we will learn in Sect. 5.4 a product state such as $\Psi^{(0)}$ may be a useful starting approximation in analyzing a many-body system but as soon as there are interactions between the particles of the system, the eigenstates of their total Hamiltonian will be a coherent superposition of product states,

$$| \Psi \rangle = \sum_{n_1, n_2, \ldots, n_N} c_{n_1 n_2 \ldots n_N} \left| \psi_{n_1}(1) \right\rangle \left| \psi_{n_2}(2) \right\rangle \cdots \left| \psi_{n_N}(N) \right\rangle , \tag{5.3}$$

with complex coefficients $c_{n_1 n_2 \ldots n_N}$. States of this kind are entangled. Even though this is a pure state, in the terminology of quantum mechanics, Chap. 3, it correlates any chosen individual "i" with the quantum states of all others $j \neq i$. In fact, if one considers the particle i in isolation, by integrating out all other particles, one will find a density matrix whose square has a trace less than 1 and, hence, which describes a mixed state.

Even if there is no interaction between them, but if the particles are identical, there will be correlations due to the spin-statistics theorem. The determinant state (4.78) provides a good example for an entangled state whose entanglement is due to the Pauli principle. In order to work this out more explicitly we consider the example of two identical fermions:

Example 5.1 Two Identical Fermions

Consider two identical fermions with spin 1/2 to be placed into two orthogonal and normalized single-particle wave functions ψ_1 and ψ_2. If their spins are coupled to the triplet state $S = 1$ the total spin function (4.67) is symmetric. Hence their orbital wave function must be antisymmetric. Denoting the spatial coordinates of the particles by $\boldsymbol{x}$ and $\boldsymbol{y}$, and dropping the spin degrees of freedom, their orbital wave function must be

$$\Psi(\boldsymbol{x}, \boldsymbol{y}) = \frac{1}{\sqrt{2}} \left\{ \psi_1(\boldsymbol{x}) \psi_2(\boldsymbol{y}) - \psi_1(\boldsymbol{y}) \psi_2(\boldsymbol{x}) \right\} . \tag{5.4}$$

This is an entangled state which is correlated due to the Pauli principle. This correlation is made more explicit by calculating beyond the one-body density (5.2) the *two-body density* defined as follows

$$\varrho(\boldsymbol{x}, \boldsymbol{y}) := \frac{1}{N(N-1)} \langle \Psi | \sum_{n \neq m = 1}^{N} \delta(\boldsymbol{x}_n - \boldsymbol{x}) \delta(\boldsymbol{x}_m - \boldsymbol{x}) \, | \Psi \rangle . \tag{5.5}$$

The normalization factor is introduced for convenience: The number of ordered pairs $12, 13, \ldots, 1N, 23, \ldots, 2N$, etc. is $N(N-1)/2$; if the pairs in inverse order, $21, 31, \ldots, N1, \ldots$, are added, this number becomes $N(N-1)$. Calculating the

[1] The name was coined by E. Schrödinger who called them (in German) "verschränkt", i.e. entangled.

one- and two-body densities (5.2) and (5.5), for the example (5.4) one finds

$$\varrho(\boldsymbol{x}) = \frac{1}{4}\int \mathrm{d}^3x_1 \int \mathrm{d}^3x_2 \left[\psi_1^*(\boldsymbol{x_1})\psi_2^*(\boldsymbol{x_2}) - \psi_1^*(\boldsymbol{x_2})\psi_2^*(\boldsymbol{x_1})\right]$$
$$[\delta(\boldsymbol{x}_1 - \boldsymbol{x}) + \delta(\boldsymbol{x}_2 - \boldsymbol{x})]\,[\psi_1(\boldsymbol{x_1})\psi_2(\boldsymbol{x_2}) - \psi_1(\boldsymbol{x_2})\psi_2(\boldsymbol{x_1})]$$
$$= \frac{1}{2}\left\{ |\psi_1(\boldsymbol{x})|^2 + |\psi_2(\boldsymbol{x})|^2 \right\}, \tag{5.6}$$

i.e. the expected result. For the two-body density one finds

$$\varrho(\boldsymbol{x},\boldsymbol{y}) = \frac{1}{4}\int \mathrm{d}^3x_1 \int \mathrm{d}^3x_2 \left[\psi_1^*(\boldsymbol{x_1})\psi_2^*(\boldsymbol{x_2}) - \psi_1^*(\boldsymbol{x_2})\psi_2^*(\boldsymbol{x_1})\right]$$
$$\left[\delta(\boldsymbol{x}_1 - \boldsymbol{x})\delta(\boldsymbol{x}_2 - \boldsymbol{y}) + \delta(\boldsymbol{x}_2 - \boldsymbol{x})\delta(\boldsymbol{x}_1 - \boldsymbol{y})\right]$$
$$[\psi_1(\boldsymbol{x_1})\psi_2(\boldsymbol{x_2}) - \psi_1(\boldsymbol{x_2})\psi_2(\boldsymbol{x_1})]$$
$$= \frac{1}{2}\left\{|\psi_1(\boldsymbol{x})|^2\,|\psi_2(\boldsymbol{y})|^2 - \psi_1^*(\boldsymbol{x})\psi_2^*(\boldsymbol{y})\psi_1(\boldsymbol{y})\psi_2(\boldsymbol{x}) + (\boldsymbol{x} \leftrightarrow \boldsymbol{y})\right\}, \tag{5.7}$$

where the last term is obtained from the first two by exchanging the arguments $\boldsymbol{x}$ and $\boldsymbol{y}$.

In order to work out the correlations contained in (5.5) more clearly, it is useful to define the *two-body correlation function* $C(\boldsymbol{x},\boldsymbol{y})$ by the equation

$$N(N-1)\varrho(\boldsymbol{x},\boldsymbol{y}) = \left\{1 + C(\boldsymbol{x},\boldsymbol{y})\right\}N^2\varrho(\boldsymbol{x})\varrho(\boldsymbol{y}) . \tag{5.8}$$

In the example of $N = 2$ one finds from the results (5.6) and (5.7)

$$C(\boldsymbol{x},\boldsymbol{y}) = -\frac{\left|\psi_1^*(\boldsymbol{x})\psi_1(\boldsymbol{y}) + \psi_2^*(\boldsymbol{x})\psi_2(\boldsymbol{y})\right|^2}{\left(|\psi_1(\boldsymbol{x})|^2 + |\psi_2(\boldsymbol{x})|^2\right)\left(|\psi_1(\boldsymbol{y})|^2 + |\psi_2(\boldsymbol{y})|^2\right)} . \tag{5.9}$$

As the following discussion shows this result is quite instructive. For coinciding arguments the correlation function is seen to be equal to -1, $C(\boldsymbol{x},\boldsymbol{x}) = -1$, which says that $\varrho(\boldsymbol{x},\boldsymbol{x})$ vanishes and the probability to find the two identical fermions in the same position is equal to zero. In the other extreme, suppose the two wave functions ψ_1 and ψ_2 to be localized in different regions of space. Then as $\boldsymbol{x} \neq \boldsymbol{y}$, the correlation function $C(\boldsymbol{x},\boldsymbol{y})$ tends to zero, and the two-body density is approximately proportional to the product of the one-body densities.

Before concluding this example three further comments seem in order. First, one should note that we assumed ψ_1 and ψ_2 to be orthogonal. This need not necessarily be so. If these states are not orthogonal the one-body density is modified to

$$\varrho(\boldsymbol{x}) = \frac{1}{2}\left\{ |\psi_1(\boldsymbol{x})|^2 + |\psi_2(\boldsymbol{x})|^2 - 2\mathrm{Re}\left(\psi_1^*(\boldsymbol{x})\psi_2(\boldsymbol{x})\ \langle\psi_2|\,\psi_1\rangle\right\}, \tag{5.10a}$$

while the two-body density is as before

$$\varrho(\boldsymbol{x},\boldsymbol{y}) = \frac{1}{2}\left\{|\psi_1(\boldsymbol{x})|^2\,|\psi_2(\boldsymbol{y})|^2 - \psi_1^*(\boldsymbol{x})\psi_2^*(\boldsymbol{y})\psi_1(\boldsymbol{y})\psi_2(\boldsymbol{x}) + (\boldsymbol{x} \leftrightarrow \boldsymbol{y})\right\}. \tag{5.10b}$$

In either case, with orthogonal or nonorthogonal ψ_1 and ψ_2, one verifies that $\int \mathrm{d}^3y\,\varrho(\boldsymbol{x},\boldsymbol{y}) = \varrho(\boldsymbol{x})$. Second, if the orbital wave functions are not confined each to a finite domain, the presence of a second particle is felt under all circumstances. For example, assuming them to be plane waves,

$$\langle \boldsymbol{x} | \psi_1 \rangle = c\, e^{(i/\hbar)\boldsymbol{p}\cdot\boldsymbol{x}}, \quad \langle \boldsymbol{y} | \psi_2 \rangle = c\, e^{(i/\hbar\boldsymbol{q})\cdot\boldsymbol{y}} \text{ with } c = (2\pi\hbar)^{-3/2},$$

the two-body correlation function (5.8) is found to be

$$C(\boldsymbol{x}, \boldsymbol{y}) = -\frac{1}{2}\left\{1 + \cos\left((\boldsymbol{q} - \boldsymbol{p}) \cdot (\boldsymbol{x} - \boldsymbol{y})\right)\right\}.$$

It equals -1 both for $\boldsymbol{y} = \boldsymbol{x}$ and for $\boldsymbol{q} = \boldsymbol{p}$ but does not go to zero for large separations of the particles. Third, one should note that in the example (5.9) the particles are free, and the correlations are due exclusively to the Pauli exclusion principle. In a realistic picture, there will be interactions between the particles which will also cause two-body correlations. These might be called *dynamical correlations*.

In the further discussion, and for reasons to become clear below, let us stay with systems of only two particles. Assume the two particles a and b to be in a correlated, entangled state such as

$$|\Psi\rangle = \frac{1}{\sqrt{2}}\left\{|a:(+);b:(-)\rangle \pm |a:(-);b:(+)\rangle\right\}, \tag{5.11}$$

where $(+)$ and $(-)$ denote two eigenstates of a given observable. Examples for such states are the spin-singlet and the spin-triplet states, (4.68) and (4.67), of two fermions where the symbolic notation "$(\pm)$" is replaced by spin up and spin down, respectively along a given direction $\hat{\boldsymbol{n}}$ in space,

$$(+) \equiv \left(s_{\hat{\boldsymbol{n}}} = \frac{1}{2}\right), \qquad (-) \equiv \left(s_{\hat{\boldsymbol{n}}} = -\frac{1}{2}\right).$$

For instance, the decay $\pi^0 \to e^+e^-$, in the pion's rest system, would produce the electron and the positron with equal and opposite spatial momenta and in the spin-singlet state. Indeed, a by now well-known argument tells us that all partial waves of the plane wave describing the relative momentum have vanishing projection quantum number onto the direction of that momentum, $m_\ell = 0$. Therefore, the spin components of e^+ and e^- must sum to zero.[2] Another example is provided by the decay of the neutral pion into two photons, $\pi^0 \to \gamma\gamma$, which is its predominant decay mode. Here again the two photons move apart back to back, with equal and opposite momenta, but their spins are correlated due to the principle of conservation of total angular momentum.

Situations of the kind just described exhibit a typical property of quantum mechanics which escapes explanation in terms of *classical* statistical considerations. The individual states of the two particles are correlated due to conservation of momentum

---

[2] The decay $\pi^0 \to e^+e^-$, for dynamical reasons, is a rather rare bird and hence difficult to measure. It is very small as compared to the dominant decay into two photons, $\pi^0 \to \gamma\gamma$. The neutral pion has a heavier sister, called η, that has a better chance to decay into an e^+e^- or a $\mu^+\mu^-$ pair.

and of angular momentum, even though, in the course of time, they separate spatially. In the framework defined by Born's interpretation of the wave function there is no way to tell which of the particles in the final state will be in the (+) state or in the (−) state, respectively. On the other hand, as soon as the state of one of them is determined by a measurement, the state of the other is known instantaneously. In discussions of correlated, spatially separated states and in quantum information it is customary to install two imagined observers, Lady A, called Alice, and Sir B, called Bob, far away from each other, who set out each to measure her or his share of the correlated state. For example, Alice finds an electron with spin along a given direction $\hat{\boldsymbol{n}}$. The result of her measurement tells her that Bob who sits at the other end of an arrangement where the decaying pion was in the middle between them, must find a positron whose spin is antiparallel to $\hat{\boldsymbol{n}}$ – independently of whether or not he actually makes a measurement.

One might argue that Lady A and Sir B should rather do what the rules of quantum mechanics require them to do, that is, to repeat the same measurement very many times under identical conditions. Indeed, in doing so, they will both find the answers (+) and (−) with equal probabilities. However, nothing prevents them to do their measurements event by event, to record the answers in a long list, and eventually come together and compare their results, event by event. Note that there is no reason to worry about possible violations of causality in this comparison. If Alice wishes to communicate her result to Bob right after she has completed her measurement, she can do so only by sending him signals which travel (at most) at the speed of light.

These simple conclusions become even stranger if we ask Alice to measure the spin of the electron along any other direction $\hat{\boldsymbol{u}}$. She must find, and we know that for sure, spin up and spin down with equal probabilities. As soon as she finds e.g. spin up along $\hat{\boldsymbol{u}}$ she knows (and may tell us and Bob) that his particle is a positron which is spin down along $\hat{\boldsymbol{u}}$. If Bob does his measurement still with $\hat{\boldsymbol{n}}$ as his quantization axis, the spin down state along $\hat{\boldsymbol{u}}$ is a coherent linear superposition of spin up and spin down along $\hat{\boldsymbol{n}}$,

$$|\psi\rangle_B = \alpha_+ \left|+\tfrac{1}{2}\right\rangle_{\hat{n}} + \alpha_- \left|-\tfrac{1}{2}\right\rangle_{\hat{n}} , \quad \text{with} \tag{5.12}$$

$$\begin{pmatrix}\alpha_+\\ \alpha_-\end{pmatrix}_{\hat{n}} = \mathbf{D}^{(1/2)}(\phi,\theta,\psi)\begin{pmatrix}0\\1\end{pmatrix}_{\hat{u}} , \quad \text{and hence}$$

$$\alpha_+ = \sin(\theta/2)\,\mathrm{e}^{i(\psi-\phi)/2} , \quad \alpha_- = \cos(\theta/2)\,\mathrm{e}^{-i(\psi+\phi)/2} ,$$

where (ϕ, θ, ψ) are the Euler angles relating Bob's frame of reference to Alice's. This is strange because it is known that the spin operators $\boldsymbol{\sigma} \cdot \hat{\boldsymbol{n}}$ and $\boldsymbol{\sigma} \cdot \hat{\boldsymbol{u}}$ do not commute unless $\hat{\boldsymbol{u}} = \hat{\boldsymbol{n}}$.

Considerations of this type gave rise to a famous work by Einstein, Podolsky, and Rosen (EPR)[3] who introduced the notion of *element of physical reality* and who suggested that quantum mechanics was incomplete. The theory would be incomplete if there were one or more hidden variables such that what we actually observe would

[3] It is said to be the reference with the highest number of quotations in physics.

correspond to an average over some (unknown) distribution. An excellent description of the arguments of EPR, in an extended version developed by D. Bohm, as well as the theoretical and experimental resolution of the apparent paradox can be found, e.g. in [Basdevant and Dalibard (2002)] and in [Aharonov and Rohrlich (2005)].

In essence, one can derive inequalities for correlation functions which are different in quantum mechanics as such, as opposed to quantum mechanics supplemented by hidden variables. This is the content of inequalities discovered by J. Bell. Recent experiments which decided the issue in favour of plain quantum mechanics are also described in the literature quoted above.

The following example, although formally similar to Example 5.1, emphasizes the peculiar nature of correlated two- or more particle states.

Example 5.2 Nuclear isospin

Proton and neutron are baryons and, by convention, are assigned baryon number $B = 1$. Starting from the observation that they have almost the same mass, viz.

$$m_{\mathrm{p}} = 938.27\,\mathrm{MeV}\,, \quad m_{\mathrm{n}} = 939.56\,\mathrm{MeV}\,,$$

Heisenberg postulated that proton and neutron were in fact the same particle, appearing as two substates of a doublet. The simplest group providing this possibility being SU(2), it was postulated that nuclear interactions were approximately invariant under this group called *nuclear isospin* or *strong interaction isospin*. More precisely, the assumption is that proton and neutron are members of a doublet with isospin $I = 1/2$ such that

$$\mathrm{p} \equiv \left|I = \tfrac{1}{2}; I_3 = +\tfrac{1}{2}\right\rangle, \quad \mathrm{n} \equiv \left|I = \tfrac{1}{2}; I_3 = -\tfrac{1}{2}\right\rangle.$$

The electric charges (in units of the elementary charge) are obtained from the formula

$$Q(i) = \tfrac{1}{2}B(i) + I_3(i)\,, \quad i = \mathrm{p} \ \text{ or } \ \mathrm{n}\,.$$

In this picture of nuclear forces it was thought that the deviations from exact isospin invariance were due to electromagnetic interactions. It is then natural to classify nuclear ground and excited states by means of multiplets of SU(2). For example, a deuteron in the ground state must have its proton and neutron coupled to total isospin zero,

$$\begin{aligned} |I = 0, I_3 = 0\rangle = \frac{1}{\sqrt{2}}\Big\{&\left|I^{(1)} = \tfrac{1}{2}, I_3^{(1)} = +\tfrac{1}{2}; I^{(2)} = \tfrac{1}{2}, I_3^{(2)} = -\tfrac{1}{2}\right\rangle \\ &- \left|I^{(1)} = \tfrac{1}{2}, I_3^{(1)} = -\tfrac{1}{2}; I^{(2)} = \tfrac{1}{2}, I_3^{(2)} = +\tfrac{1}{2}\right\rangle\Big\}\,. \end{aligned} \tag{5.13}$$

Obviously, this state is of the same kind as the more general state (5.11). In a deuteron proton and neutron are bound, hence confined to a small domain in space, and it seems impossible to detect them spatially separated. However, it is perfectly possible to prepare a state of the kind of (5.13) in which proton and neutron move apart from each other. Return then to our two characters, Lady A and Sir B, and ask for instance Alice to identify the particle that moves towards her by measuring its electric charge. If she finds that particle to be a neutron, she will know immediately that Bob's particle is a proton, independently of whether he actually does measure the charge or not. If, however, she repeats her measurement a great number of times she will find, in the long run, an equal number of protons and neutrons.

Like in Example 5.1 Bob and Alice, when comparing their measurements afterwards, event by event, will be puzzled by the correlation they discover. Nevertheless there is no conflict with causality because, in order for Alice to tell Bob what she found in any particular event, she needs to send him signals which travel at most at the speed of light.

5.1.2 Entanglement, More General Considerations

The correlations and nonlocalities inherent to quantum mechanics can also be made visible in the statistical operator or density matrix formalism for states of two or more particles. Although it is difficult to formalize entanglement for completely general pure or mixed quantum states, states of two subsystems of a somewhat more general nature than the ones discussed above, can be analyzed as follows. Consider two systems A and B in a quantum state described by superpositions of a finite number of states, possibly more than 2. Their wave function has the form

$$| \Psi \rangle = \sum_{m=1}^{p} \sum_{n=1}^{q} c_{mn} \, | \phi_m \rangle_{\mathrm{A}} \, | \psi_n \rangle_{\mathrm{B}} \,, \tag{5.14}$$

where the states ϕ_m are part of an orthonormal basis of $\mathcal{H}_{\mathrm{A}}$, while the states ψ_n belong to an orthonormal basis of $\mathcal{H}_{\mathrm{B}}$.

The state (5.14) can be transformed to a diagonal form which comes closest to the two-state, two-particle wave function (5.11). For this purpose we study the $p \times q$–matrix $\mathbf{C} = \{c_{mn}\}$. Without restriction of generality we assume $p \leqslant q$. Write the matrix $\mathbf{C}$ as a product of three matrices as follows,

$$\mathbf{C} = \mathbf{U}^{\dagger}\mathbf{D}\mathbf{V} \,, \tag{5.15a}$$

$\mathbf{U}$ and $\mathbf{D}$ being $p \times p$-matrices, $\mathbf{V}$ a $p \times q$ matrix. If $\mathbf{V}$ has the property $\mathbf{V}\mathbf{V}^{\dagger} = \mathbb{1}_p$ then the product of $\mathbf{C}$ and its conjugate is

$$\mathbf{C}\mathbf{C}^{\dagger} = \mathbf{U}^{\dagger}\mathbf{D}\mathbf{V}\mathbf{V}^{\dagger}\mathbf{D}^{\dagger}\mathbf{U} = \mathbf{U}^{\dagger}\mathbf{D}\mathbf{D}^{\dagger}\mathbf{U} \,. \tag{5.15b}$$

Obviously, $\mathbf{U}$ can be chosen to be the unitary $p \times p$-matrix which diagonalizes the hermitian matrix $\mathbf{C}\mathbf{C}^{\dagger}$,

$$\mathbf{U}\left(\mathbf{C}\mathbf{C}^{\dagger}\right)\mathbf{U}^{\dagger} = \mathbf{D}\mathbf{D}^{\dagger}$$

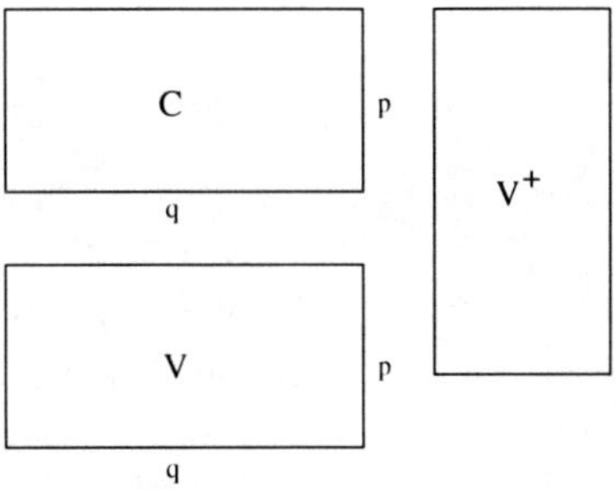

Fig. 5.1 The matrices C and V, in general, are *rectangular* having p lines and q columns

so that $\mathbf{D}\mathbf{D}^\dagger$ is a $p \times p$ diagonal matrix with real positive semidefinite entries,

$$\mathbf{D}\mathbf{D}^\dagger = \mathrm{diag}(w_1, w_2, \ldots, w_p)\,, \quad w_i \geqslant 0\,. \tag{5.15c}$$

This allows us to define the matrix $\mathbf{D}$ by the square root of $\mathbf{D}\mathbf{D}^\dagger$,

$$\mathbf{D} = \mathrm{diag}(\sqrt{w_1}, \ldots, \sqrt{w_p})\,. \tag{5.15d}$$

Regarding the other hermitian product of $\mathbf{C}$ and its conjugate one obtains

$$\mathbf{V}\left(\mathbf{C}^\dagger\mathbf{C}\right)\mathbf{V}^\dagger = \mathbf{D}^\dagger\mathbf{D} = \mathrm{diag}(w_1, \ldots, w_p, 0, \ldots, 0)\,. \tag{5.15e}$$

In the special case $p = q$ equation (5.15a) is the well-known diagonalization prescription of an arbitrary complex matrix by two unitaries (also called a bi-unitary transformation). The general case $p < q$ is illustrated by Fig. 5.1 which shows the $p \times q$ rectangular shape of $\mathbf{C}$, and the $p \times q$ rectangular $\mathbf{V}$. The product $\mathbf{V}\mathbf{V}^\dagger$ is the p-dimensional unit matrix, $\mathbf{V}\mathbf{V}^\dagger = \mathbb{1}_p$, while $\mathbf{V}^\dagger\mathbf{V}$ is the p-dimensional unit matrix embedded in a diagonal $q \times q$-matrix whose remaining elements are zero,

$$\mathbf{V}^\dagger\mathbf{V} = \begin{pmatrix} \mathbb{1}_p & 0 & \cdots \\ 0 & 0 & \cdots \\ \vdots & \vdots & \ddots \end{pmatrix}.$$

In case $\mathcal{H}_p$ is a subspace of $\mathcal{H}_q$ the matrix $\mathbf{V}^\dagger\mathbf{V}$ is seen to be the projector from $\mathcal{H}_q$ to $\mathcal{H}_p$. An example may help to illustrate these formulae.

Example 5.3 Two-Particle State with Coupled Spins

Let A be a particle with spin 1/2, B a particle with spin 1, and let the magnetic substates be numbered as usual,

$$A: \quad i = 1 : m_1 = +\frac{1}{2}\,, \quad i = 2 : m_1 = -\frac{1}{2}\,,$$
$$B: \quad k = 1 : m_2 = +1\,, \quad k = 2 : m_2 = 0\,, \quad k = 3 : m_2 = -1\,.$$

For a state such as (5.14) with general coefficients c_{ik}

$$|\Psi\rangle = \sum_{i=1}^{2}\sum_{k=1}^{3} c_{ik}\left|j_1 = \tfrac{1}{2}, i\right\rangle_A |j_2 = 1, k\rangle_B \tag{5.16}$$

one obtains

$$\mathbf{CC}^\dagger = \begin{pmatrix} A_{11} & A_{12} \\ A_{12}^* & A_{22} \end{pmatrix}, \quad \text{with}$$

$$A_{jj} = |c_{j1}|^2 + |c_{j2}|^2 + |c_{j3}|^2 , \; j = 1, 2 ,$$

$$A_{12} = c_{21}c_{11}^* + c_{22}c_{12}^* + c_{23}c_{13}^* .$$

The unitary $\mathbf{U}$ is obtained by diagonalizing this matrix whose eigenvalues are

$$\begin{matrix} w_1 \\ w_2 \end{matrix} = \frac{1}{2}(A_{11} + A_{22}) \pm \frac{1}{2}\sqrt{(A_{11} - A_{22})^2 + 4\,|A_{12}|^2} .$$

Suppose the state (5.16) is the one where the two spins are coupled to an eigenstate $|\,JM\,\rangle$ of total angular momentum, say with $M = 1/2$, then only c_{12} and c_{21} are different from zero and are equal to the appropriate Clebsch-Gordan coefficients, viz.

$$c_{12} = (\tfrac{1}{2}\,\tfrac{1}{2}, 1\,0|J\,\tfrac{1}{2}), \quad c_{21} = (\tfrac{1}{2}\,-\tfrac{1}{2}, 1\,+1|J\,\tfrac{1}{2}),$$

while $c_{11} = c_{22} = c_{33} = c_{13} = c_{23} = 0$. The eigenvalues of $(\mathbf{CC}^\dagger)$ are $w_1 = |c_{21}|^2$ and $w_2 = |c_{12}|^2$, the unitary matrix $\mathbf{U}$ is

$$\mathbf{U} = \begin{pmatrix} 0 & 1 \\ 1 & 0 \end{pmatrix},$$

its effect on the states of A being a renumbering of the first two of them. The matrix $\mathbf{V}$ is calculated by means of (5.15a),

$$\mathbf{V} = \mathbf{D}^{-1}\mathbf{UC} = \begin{pmatrix} c_{21}/\sqrt{w_1} & 0 & 0 \\ 0 & c_{12}/\sqrt{w_2} & 0 \end{pmatrix}.$$

One verifies that $\mathbf{VV}^\dagger = \text{diag}(1,1) \equiv \mathbb{1}_2$, while $\mathbf{V}^\dagger\mathbf{V} = \text{diag}(\mathbb{1}_2, 0)$.

In the example as well as in the general case (5.14) the effect of the unitary $\mathbf{U}$ in $\mathcal{H}_A$ is the passage from the original basis $|\,\phi_m\,\rangle_A$ to a new basis, say $|\,\eta_k\,\rangle_A$. Simultaneously, the basis $|\,\psi_n\,\rangle_B$ of $\mathcal{H}_B$ is projected by means of $\mathbf{V}$ onto another basis $|\,\chi_k\,\rangle_B$ such that the total state becomes

$$|\,\Psi\,\rangle = \sum_{k=1}^{r} \sqrt{w_k}\,|\,\eta_k\,\rangle_A\,|\,\chi_k\,\rangle_B , \tag{5.17}$$

$$\text{where} \quad r \leqslant \min\{p, q\} \quad \text{and} \quad \sum_{k=1}^{r} w_k = 1 . \tag{5.18}$$

In the example $r = 2$, the transformation from (5.14) to (5.17) is simply $|\,\eta_1\,\rangle = |\,\phi_2\,\rangle$, $|\,\eta_2\,\rangle = |\,\phi_1\,\rangle$ while the states $|\,\psi_n\,\rangle$ remain unchanged.

The procedure just described is due to E. Schmidt and is called the singular value decomposition of a rectangular matrix. The real positive-semidefinite w_k are called *Schmidt weights*, the number r of them which are different from zero is called the *Schmidt rank*.

The result (5.17) has some interesting consequences.

First, the state (5.17) as such is still a pure state. *It is entangled if and only if its Schmidt rank r is greater than 1.*

The second consequence concerns the statistical operator of the state (5.14) which by (3.35) is given by

$$W_{\mathrm{A,B}} = |\Psi\rangle\langle\Psi| = \sum_{k=1}^{r} w_k (|\eta_k\rangle\langle\eta_k|)_{\mathrm{A}} \otimes (|\chi_k\rangle\langle\chi_k|)_{\mathrm{B}} \,. \tag{5.19a}$$

If for some reason the subsystem B is not observed, one should take the trace over all states of that particle. This yields a reduced statistical operator which describes only particle A, viz.

$$W_{\mathrm{A}}^{\mathrm{red}} = \operatorname*{tr}_{\mathrm{B}} (W_{\mathrm{A,B}}) = \sum_{k=1}^{r} w_k (|\eta_k\rangle\langle\eta_k|)_{\mathrm{A}} \,. \tag{5.19b}$$

Likewise, if particle A is integrated over because it is not observed, then the operator (5.19a) reduces to

$$W_{\mathrm{B}}^{\mathrm{red}} = \operatorname*{tr}_{\mathrm{A}} (W_{\mathrm{A,B}}) = \textstyle\sum_{k=1}^{r} w_k (|\chi_k\rangle\langle\chi_k|)_{\mathrm{B}} \,. \tag{5.19c}$$

If the Schmidt rank is greater than or equal to 2 then either $W_{\mathrm{A}}^{\mathrm{red}}$, (5.19b), or, depending on the experimental set-up, the statistical operator $W_{\mathrm{B}}^{\mathrm{red}}$, (5.19c), are convex sums of projection operators and, hence, describe *mixed* states. This is remarkable: In most classical situations we are able to study subsystems in isolation from everything else ("the rest of the Universe" as Feynman put it) because physical phenomena are strictly local. For a planet taken in isolation it suffices to know the local force fields that act on it if one sets out to study its equation of motion. In contrast, entangled states contain nonlocalities which, upon integrating out partial subsystems, turn a pure state into a reduced mixed state.

Third, the analysis of (5.14) shows that a two-state subsystem A, $p = 2$, may be entangled with an arbitrarily large subsystem B, $q > 2$, the singular value series (5.17) will have at most two terms.

5.1.3 Classical and Quantum Bits

In the realm of classical physics information can be encoded in packages of simple yes and no, or "true" and "false" answers. Classical *bits*[4] take the values 1 (yes–true), and 0 (no–false), messages of any kind are represented by strings of bits. For example, in ASCII (American Standard Code for Information Interchange) the name Max Born is encoded as follows

[4] The word *bit* stands for **bin**ary dig**it**.

Letter	Coding	Letter	Coding
M	0100 1101	**B**	0100 0010
a	0110 0001	**o**	0110 1111
x	0111 1000	**r**	0111 0010
		n	0110 1110

Binary coding is well adapted to computing because the Boolean algebra of bits is easily implemented by electronics, for instance by assigning the charged and uncharged states of a capacitor the bit values 1 and 0, respectively. Every letter of the alphabet, every digit in the decimal system, as well as any other symbol that is used, can be represented as a string of bits. These strings are distinguishable classical macroscopic states. The information contained in a package of strings can be read without modifying it, it can be replicated, it can be transmitted from A to B, and, if the need arises, it can be processed on a computer. In particular, there is no obstruction of principle against replicating or *cloning* information contained in a package of strings.

In the realm of quantum physics it seems natural to define the analogue of the classical bit by means of two-dimensional quantum systems. Quantum systems with two relevant states are realized, e.g., by the two spin orientations of a particle with spin 1/2 whose orbital wave function is given, or by photons in a fixed dynamical state and their two polarization states, or by some atomic system for which two specific states can be isolated from the rest of its dynamics. The Hilbert space of such a two-state system is isomorphic to $\mathbb{C}^2$, and admits a basis $\{|+\rangle, |-\rangle\}$. An arbitrary element $|\psi\rangle \in \mathcal{H} \simeq \mathbb{C}^2$ is a linear combination of these basis states with complex coefficients. Taking into account the freedom of choosing phases of wave functions, the state $|\psi\rangle$ is seen to depend on two parameters, say θ and ϕ,

$$|\psi\rangle = \cos(\theta/2)\,|-\rangle + \mathrm{e}^{-i\phi}\sin(\theta/2)\,|+\rangle \,. \tag{5.20}$$

(This is formula (5.12) with $\psi = -\phi$, the third Euler angle being irrelevant.) The state (5.20) is a realization of what is called a *quantum bit*, or *qubit* in analogy to the classical bit. Consequently, a string of n qubits is an element of the Hilbert space $\mathcal{H}^{(n)} \simeq \otimes_n \mathbb{C}^2 = \mathbb{C}^{2n}$ whose natural basis is

$$\underbrace{|-\rangle \otimes \cdots \otimes |-\rangle}_{n-k} \otimes \underbrace{|+\rangle \otimes \cdots \otimes |+\rangle}_{k}\,, \qquad k = 0, 1, \ldots, n\,.$$

At first sight, as compared to the classical bit, a qubit seems to contain much more information because $|\psi\rangle$ describes an infinity of states all of which lie on the unit sphere S^2 parametrized by θ and ϕ. This is illustrated in Fig. 5.2: The space of pure states for a quantum bit is parametrized by the points on the sphere S^2. The north pole corresponds to the classical 0-bit, the south pole corresponds to the classical 1-bit. In the quantum world qubits are linear superpositions of these two states, the angle θ characterizing the ratio of the moduli of the two components, the angle ϕ their relative phase. Loosely speaking, an infinity of intermediate answers somewhere

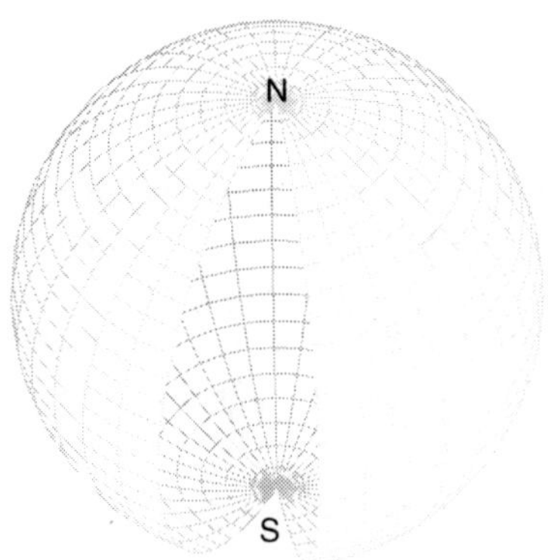

Fig. 5.2 The pure states of a quantum bit are points on the sphere S^2. The north pole N corresponds to the classical 0-bit, the south pole S to the classical 1-bit

between "no" and "yes" seem possible. However, because of the nonlocality of quantum mechanics and the intricacies of entanglement, matters are not so simple.[5] We illustrate this by a particularly striking property of qubits and strings of qubits. While classical information encoded in strings of bits, in principle, can be copied perfectly, this is not true for quantum information. One cannot make a duplicate of a qubit whose state is unknown, without perturbing the original in a uncontrollable way. This is the content of

The no-cloning theorem of Wootters and Zurek[6]

The argument is the following: Let $|\psi_1^{(o)}\rangle$ be an initial quantum state of which one tries to create a copy by means of some copying device. The copier initially is in a known "blanc" state $|\phi_0^{(c)}\rangle$ – comparable to the white paper in a copying machine. The cloning procedure should produce the transition

$$\left|\psi_1^{(\mathrm{orig})}\right\rangle \otimes \left|\phi_0^{(\mathrm{copy})}\right\rangle \longrightarrow \left|\psi_1^{(\mathrm{orig})}\right\rangle \otimes \left|\psi_1^{(\mathrm{orig})}\right\rangle , \tag{5.21}$$

that is, it should transform the initial known (but neutral) state $|\phi_0^{(\mathrm{copy})}\rangle$ into a perfect copy of the original $|\psi_1^{(\mathrm{orig})}\rangle$. Of course, the same cloning procedure applied to another state $|\psi_2^{(\mathrm{orig})}\rangle$, orthogonal to the first, should work in exactly the same way, viz.

$$\left|\psi_2^{(\mathrm{orig})}\right\rangle \otimes \left|\phi_0^{(\mathrm{copy})}\right\rangle \longrightarrow \left|\psi_2^{(\mathrm{orig})}\right\rangle \otimes \left|\psi_2^{(\mathrm{orig})}\right\rangle . \tag{5.22}$$

Consider then a third state

$$\left|\psi_3^{(\mathrm{orig})}\right\rangle = \frac{1}{\sqrt{2}}\left(c_1 \left|\psi_1^{(\mathrm{orig})}\right\rangle + c_2 \left|\psi_2^{(\mathrm{orig})}\right\rangle\right) \tag{5.23}$$

---

[5] An excellent introduction to the quantum aspects of information and computation is provided by the review article Galindo A., Martín-Delgado, M. A., Rev. Mod. Phys. **74** (2002) 2.

[6] W.K. Wootters, and W.H. Zurek, Nature **299** (1982) 802.

and create a copy of it by means of the same copying machine,

$$\left|\psi_3^{(\mathrm{orig})}\right\rangle \otimes \left|\phi_0^{(\mathrm{copy})}\right\rangle \longrightarrow \left|\psi_3^{(\mathrm{orig})}\right\rangle \otimes \left|\psi_3^{(\mathrm{orig})}\right\rangle . \tag{5.24}$$

Quantum theory tells us that the cloning transitions (5.21), (5.22), (5.24) are effected by the unitary evolution operator which is obtained from some Hamiltonian, and, therefore, cannot depend on the original that one wishes to copy. Thus, by the superposition principle, the result in copying the state (5.24) must be

$$\begin{aligned}&\left|\psi_3^{(\mathrm{orig})}\right\rangle \otimes \left|\phi_0^{(\mathrm{copy})}\right\rangle \\ &\longrightarrow \frac{1}{\sqrt{2}}\left(c_1\left|\psi_1^{(\mathrm{orig})}\right\rangle \otimes \left|\psi_1^{(\mathrm{orig})}\right\rangle + c_2\left|\psi_2^{(\mathrm{orig})}\right\rangle \otimes \left|\psi_2^{(\mathrm{orig})}\right\rangle\right) .\end{aligned} \tag{5.25}$$

This state is an entangled state and is definitely different from $|\psi_3^{(\mathrm{orig})}\rangle \otimes |\psi_3^{(\mathrm{orig})}\rangle$. No cloning is possible.

Duplication of a qubit is only possible if it is in a known state, say $|\psi_1\rangle = |-\rangle$ or $|\psi_2\rangle = |+\rangle$. In other terms, if the qubit is in an unknown superposition such as (5.20) which is to be transmitted from A to B, a spy may intercept it and analyze it but he or she will not remain unnoticed because of the impossibility to copy the information and send a perfect copy on to B.

A somewhat more general formulation of the no-cloning theorem goes as follows. We distinguish three Hilbert spaces, the space $\mathcal{H}^{(\mathrm{orig})}$ containing the state to be copied, the reservoir $\mathcal{H}^{(\mathrm{copy})}$ containing the initial "white" state onto which one tries to print a copy, and $\mathcal{H}^{(\mathrm{roU})}$ describing the "rest of the Universe", i.e. the environment. Let $\mathbf{U}$ be the unitary evolution operator which induces the transition

$$\mathbf{U}\left|\psi^{(\mathrm{orig})}\right\rangle \otimes \left|\phi_0^{(\mathrm{copy})}\right\rangle \otimes |\Omega_0\rangle \rightarrow \left|\psi^{(\mathrm{orig})}\right\rangle \otimes \left|\psi^{(\mathrm{orig})}\right\rangle \otimes |\Omega_\psi\rangle \tag{5.26}$$

for every state $|\psi^{(\mathrm{orig})}\rangle \in \mathcal{H}^{(\mathrm{orig})}$. We have allowed for the possibility that, while copying the original state, the environment moves from its initial state $|\Omega_0\rangle$ to some other, excited state $|\Omega_\psi\rangle$. While $\mathbf{U}$ must be independent of the state that we wish to copy, the final state of the environment may depend on the original. Compare then the action of $\mathbf{U}$ on two different originals,

$$\begin{aligned}&\mathbf{U}\left|\psi_i^{(\mathrm{orig})}\right\rangle \otimes \left|\phi_0^{(\mathrm{copy})}\right\rangle \otimes |\Omega_0\rangle \\ &\rightarrow \left|\psi_i^{(\mathrm{orig})}\right\rangle \otimes \left|\psi_i^{(\mathrm{orig})}\right\rangle \otimes |\Omega_{\psi_i}\rangle , \qquad i = a \text{ or } b ,\end{aligned} \tag{5.27}$$

and take the scalar product of this action on $|\psi_a^{(\mathrm{orig})}\rangle$ with its action on $|\psi_b^{(\mathrm{orig})}\rangle$. All states involved being normalized to 1, and $\mathbf{U}$ being unitary this scalar product is

$$\langle\psi_b^{(\mathrm{orig})}|\psi_a^{(\mathrm{orig})}\rangle = \langle\psi_b^{(\mathrm{orig})}|\psi_a^{(\mathrm{orig})}\rangle^2 \langle\Omega_{\psi_b}|\Omega_{\psi_a}\rangle . \tag{5.28}$$

The moduli of the transition amplitudes $\langle\psi_b^{(\mathrm{orig})}|\psi_a^{(\mathrm{orig})}\rangle$ and $\langle\Omega_{\psi_b}|\Omega_{\psi_a}\rangle$ are smaller than or equal to 1. Therefore, the result (5.28) tells us that the transition amplitude from a to b either vanishes, or is equal to 1,

$$\langle\psi_b^{(\mathrm{orig})}|\psi_a^{(\mathrm{orig})}\rangle = 0 \quad \text{or} \quad 1 . \tag{5.29}$$

While a *known* quantum state can be copied at will *it is impossible to create a copy of two different states which are not orthogonal to one another.*

Whether or not states may be copied *approximately* is a different matter. For this question and as well as questions regarding storage and retrieval of information in (future) quantum computers we refer to the literature quoted above.

In concluding this section it is instructive to recall the basic principles which cause the nonlocalities of the quantum world, entanglement, and, in particular, the impossibility to intercept and copy quantum messages. For that, note that we used (quantum) Hamiltonian theory in an essential way: the evolution operator which derives from a Hamiltonian, is *unitary*. The *superposition principle,* which is to say *linearity* of the Schrödinger equation, are essential, and so is the *probabilistic interpretation* of quantum mechanics. No new features of quantum mechanics were invoked, but, as the examples showed, quantum information is different from classical information and exhibits perplexing and perhaps surprising features.

5.2 Stationary Perturbation Theory

Suppose one wishes to solve an eigenvalue problem of the stationary Schrödinger equation $H\psi = E\psi$ that is expected to differ but little from some known, exactly solvable case, $H_0\psi^{(0)} = E_0\psi^{(0)}$. One assumes an expansion in terms of the difference $(H - H_0) =: H_1$, called the perturbation, solves the Schrödinger equation by iteration expecting the series one obtains to approach the exact solution. Depending on whether the unperturbed spectrum is nondegenerate or degenerate, this expansion is of a different nature. The results of this iterative solution, though of limited accuracy in practice, are instructive and can be well interpreted. Therefore, perturbation theory is of great value for estimates and for understanding the physics of a given problem.

5.2.1 Perturbation of a Nondegenerate Energy Spectrum

Let H_0 be a Hamiltonian whose energy spectrum and corresponding eigenfunctions are known. Assume the spectrum to be nondegenerate and, for the sake of simplicity, to be purely discrete. Thus, for every eigenvalue there is exactly one wave function,

$$H_0 \, |n\rangle = E_n^{(0)} \, |n\rangle \, . \qquad (5.30)$$

Consider then a system which is described by the Hamiltonian

$$H = H_0 + \varepsilon H_1 \qquad (5.31)$$

and which, in some sense, "lies close" to the unperturbed system described by H_0. The term εH_1 is interpreted as a perturbation of the eigenstates of H_0. The real and positive constant ε plays the role of an order parameter which is used for book-keeping but has no other purpose or physical significance. Its integer powers ε^n characterize successive perturbative corrections by their relative magnitude. When comparing coefficients in the perturbation series, ε will be seen to cancel out. Therefore, one may use the parameter ε for classifying perturbative contributions of successive orders, but may safely set it equal to 1 in the end.

The aim of perturbation theory is to determine approximately the eigenvalues E and the eigenfunctions ψ of the full Hamiltonian H, $H\psi = E\psi$, from the knowledge of the perturbation H_1, the wave functions $|n\rangle$, and the spectrum $\{E_n^{(0)}\}$. Inserting the decomposition (5.31) into the Schrödinger equation this means that one must solve the following equation:

$$(E - H_0)\psi = \varepsilon H_1 \psi \,. \tag{5.32}$$

We write the energy and the wave function as formal series in ε,

$$E = E^{(0)} + \varepsilon E^{(1)} + \varepsilon^2 E^{(2)} + \dots \,, \quad \text{and} \tag{5.33}$$

$$\psi = \sum_{m=0}^{\infty} c_m \,|m\rangle \quad \text{with} \quad c_m = c_m^{(0)} + \varepsilon c_m^{(1)} + \varepsilon^2 c_m^{(2)} + \dots \,. \tag{5.34}$$

Inserting the expansion (5.34) into (5.32) and taking the scalar product of this equation with the unperturbed state $\langle k|$, one obtains an algebraic system of equations,

$$(E - E_k^{(0)})c_k = \varepsilon \sum_{m=0}^{\infty} \langle k|H_1\,|m\rangle\, c_m \,. \tag{5.35}$$

Inserting then the expansions (5.33) for E, and (5.34) for c_m, this equation becomes

$$\begin{aligned} &[(E^{(0)} - E_k^{(0)}) + \varepsilon E^{(1)} + \varepsilon^2 E^{(2)} + \dots][c_k^{(0)} + \varepsilon c_k^{(1)} + \varepsilon^2 c_k^{(2)} + \dots] \\ &\quad = \sum_m \langle k|H_1\,|m\rangle\,[\varepsilon c_m^{(0)} + \varepsilon^2 c_m^{(1)} + \dots] \,. \end{aligned} \tag{5.36}$$

Note that the right-hand side carries one more power of ε than the left-hand side. The correction terms in (5.33) and (5.34) are obtained by identifying all terms of (5.36) which are multiplied by the same power of ε. We work this out more explicitly for the first three orders.

Order Zero $\mathcal{O}(\varepsilon^0)$: Equation (5.36) reduces to the equation $(E^{(0)} - E_k^{(0)})c_k^{(0)} = 0$. If we are interested in the perturbation of the state with quantum number n the solution is obvious;

$$E^{(0)} = E_n^{(0)} \,; \qquad c_n^{(0)} = 1 \,; \qquad c_m^{(0)} = 0 \quad \forall\, m \neq n \,. \tag{5.37}$$

First Order $\mathcal{O}(\varepsilon^1)$: The terms of first order in (5.36) yield the algebraic equation

$$(E_n^{(0)} - E_k^{(0)})c_k^{(1)} + E^{(1)} c_k^{(0)} = \langle k|H_1\,|n\rangle\, c_n^{(0)}$$

which is solved for $k = n$ and for $k \neq n$, respectively,

1. $k = n$:

$$\boxed{E^{(1)} = \langle n|H_1\,|n\rangle} \,. \tag{5.38}$$

To first order the displacement of the energy is given by the expectation value of the perturbation H_1.

2. $k \neq n$: Inserting the result (5.37) of order zero, i.e. $c_n^{(0)} = 1$, $c_k^{(0)} = 0$, one obtains

$$\boxed{c_k^{(1)} = \langle k|H_1\,|n\rangle\,/(E_n^{(0)} - E_k^{(0)})} \,. \tag{5.39}$$

The coefficient $c_n^{(1)}$, at first, seems to remain undetermined. Note, however, that in any case it must be chosen such that the wave function ψ is normalized to 1. To the order ε^1 this condition is met if

$$c_n^{(0)} + \varepsilon c_n^{(1)} = 1 + \varepsilon c_n^{(1)} = e^{ia\varepsilon} \approx 1 + a\varepsilon$$

with a real. The choice of a fixed value of a is seen to propagate through the whole perturbative series. It leads to a constant phase factor which multiplies ψ as a whole. Such a phase is not observable. Therefore, without restriction of generality one may set $a = 0$, i.e.

$$\boxed{c_n^{(1)} = 0} \,. \tag{5.37$'$}$$

Second Order $\mathcal{O}(\varepsilon^2)$: To this order (5.36) yields an equation that determines the correction of second order of the energy and of the expansion coefficients $c_k^{(2)}$,

$$(E_n^{(0)} - E_k^{(0)})c_k^{(2)} + E^{(1)}c_k^{(1)} + E^{(2)}c_k^{(0)} = \sum_m \langle k|H_1\,|m\rangle\, c_m^{(1)} \,.$$

Once more, one distinguishes the cases $k = n$ and $k \neq n$,

1. $k = n$: Making use of the results obtained in first order one has

$$\boxed{E^{(2)} = \sum_{m \neq n} |\langle n|H_1|m\rangle|^2\,/(E_n^{(0)} - E_m^{(0)})} \,. \tag{5.40}$$

2. $k \neq n$: The formulae for the expansion coefficients in second order are rather more complicated. One finds for $k \neq n$:

$$c_k^{(2)} = \sum_{m \neq n} \frac{\langle k|H_1\,|m\rangle\,\langle m|H_1\,|n\rangle}{(E_n^{(0)} - E_m^{(0)})(E_n^{(0)} - E_k^{(0)})} - \frac{\langle n|H_1\,|n\rangle\,\langle k|H_1\,|n\rangle}{(E_n^{(0)} - E_k^{(0)})^2} \,. \tag{5.41}$$

The corresponding coefficient with $k = n$ follows from the normalization condition for ψ,

$$c_n^{(2)} = -\frac{1}{2}\sum_{k \neq n} \frac{|\langle k|H_1\,|n\rangle|^2}{(E_n^{(0)} - E_k^{(0)})^2} \,. \tag{5.42}$$

The details of deriving (5.41) and (5.42) are left as an exercise.

Remarks

1. It is not difficult to guess the extension to a partially or completely continuous spectrum: In the expressions obtained above the sums over intermediate states are replaced by sums and/or integrals. It is important to keep in mind that one must always sum over a *complete* set of intermediate states. For example, in the case of the hydrogen atom we know that the bound states, by themselves, are not complete. Thus, it would be wrong to restrict the calculation of the perturbation of second order to these states only. The states in the continuum that we studied in Sect. 1.9.5 do contribute as well.
2. Perturbation theory of *first* order is applicable and quantitatively sufficient if the perturbation is small compared to typical energy differences of the unperturbed spectrum, that is, if
$$|\langle n|\varepsilon H_1\,|m\rangle| \ll \left|E_n^{(0)} - E_m^{(0)}\right| . \tag{5.43}$$
Obviously, the coefficients (5.39) measuring the admixtures are different from zero only if the operator H_1 connects the state $|n\rangle$ (this is the state whose perturbation we calculate) to the state $|k\rangle$. There may be obstructions from selection rules! Even in case the state $|k\rangle$ is admixed, the coefficient (5.39) is the smaller the further away the state lies on the energy scale.
3. The formula (5.40) for the energy shift to *second* order has a nice interpretation. The numerator
$$|\langle n|H_1\,|m\rangle|^2 = \langle n|H_1\,|m\rangle\,\langle m|H_1\,|n\rangle$$
may be understood as a transition from the state $|n\rangle$ to a *virtual* intermediate state $|m\rangle$, and the return from there to $|n\rangle$, or, symbolically, $n \to m \to n$. This transition is said to be virtual because initial and final states do not have the same energy. Such a transition violates the principle of energy conservation and, therefore, cannot be physical. Note, however, that all other selection rules such as angular momentum, parity, and the like, are respected.[7] Here again, the contribution is the smaller the further away the admixed state. The result (5.40) is particularly interesting in the case where $|n\rangle$ is the *ground state* of H_0. As now $E_n^{(0)}$ is the smallest eigenvalue of H_0 the formula (5.40) receives only negative contributions. One concludes: *In second order of perturbation theory the ground state is always lowered.*
4. In practice, the higher orders of perturbation theory are not very relevant. Indeed, as soon as the perturbation is too strong for first and second order to be adequate, it is advisable to determine the exact energy spectrum by other means – for instance, by diagonalization in a suitable basis of Hilbert space.

[7] For example, if $|n\rangle$ are states with definite momentum the transition amplitude $\langle m|H_1|n\rangle$ must fulfill the principle of momentum conservation. Nonrelativistic perturbation theory takes virtual intermediate states only out of their "energy shell". This will be different in relativistic, Lorentz covariant perturbation theory where virtual particles are taken out of the mass shell $p^2c^2 = E^2 - c^2\boldsymbol{p}^2 = m^2c^4$, both in their energy and their momentum.

5. The parameter ε of (5.31) is introduced for convenience and for the sake of clarity because it highlights the successive orders of perturbation theory. It does not appear in the results and, hence, may formally be set equal to 1 in the final formulae.
6. Nowadays the formulae (5.41) are rarely made use of in practice, in contrast to the equations (5.38)–(5.40) which are so important that one should remember them by heart. This is so because in practically all cases which require a higher accuracy one resorts to other, more direct methods of solution.
7. The calculation of the matrix elements $\langle n|H_1|m\rangle$ of the perturbation, taken between given eigenstates $|n\rangle$ and $|m\rangle$ of H_0, is an essential element in the formulae of perturbation theory. In many cases of practical relevance to molecular, atomic, or nuclear physics, the unperturbed Hamiltonian contains a central field, and its eigenfunctions can be taken to be products of radial functions $R(r)$ and of functions describing spin and orbital angular momentum. In those cases it is advisable to expand the perturbation in terms of *spherical tensor operators*, i.e. in terms of operators T^{κ}_{μ} which transform under rotations like an eigenstate of angular momentum with quantum numbers (κ, μ). In other terms, these are operators which transform by the matrices $\mathbf{D}^{(\kappa)}$. The advantage of this way of proceeding lies in the observation that matrix elements of the kind of
$$\langle j'm'|T^{\kappa}_{\mu}|jm\rangle$$
can be evaluated analytically by means of known techniques of the rotation group. These techniques are developed in Part Two. In particular, they allow to read off immediately important selection rules due to angular momentum and parity conservation. Furthermore, they reduce the calculation of the matrix elements to one-dimensional integrals over radial functions which may be done analytically or by means of standard numerical procedures.

5.2.2 Perturbation of a Spectrum with Degeneracy

As before, the eigenvalues and eigenfunctions of H_0 are assumed to be known. In contrast to the previous case, the eigenvalues may now be degenerate. If, for simplicity, we consider the case of a purely discrete spectrum, this means that the starting point is
$$H_0\,|\,n,\alpha\,\rangle = E^{(0)}_n\,|\,n,\alpha\,\rangle\,,\qquad \alpha = 1,2,\ldots,k_n\,. \tag{5.44}$$
The perturbed stationary Schrödinger equation $(H_0+\varepsilon H_1)\psi = E\psi$ is solved by means of the ansatz
$$E = E^{(0)} + \varepsilon E^{(1)} + \ldots\,,\qquad \psi = \psi^{(0)} + \varepsilon\psi^{(1)} + \ldots\,, \tag{5.45}$$
and by comparing terms of the same order in ε.

Order Zero $\mathcal{O}(\varepsilon^0)$: In this order one has
$$H_0\psi^{(0)} = E^{(0)}_n\psi^{(0)}\,.$$

While the energy remains at the value $E_n^{(0)}$, the corresponding eigenstate will be a linear combination of base functions $|n, \alpha\rangle, \alpha = 1, 2, \ldots, k_n$, which pertain to this eigenvalue of H_0 and which span the subspace $\mathcal{H}_n$,

$$E^0 = E_n^{(0)}, \qquad \psi^{(0)} = \sum_{\alpha=1}^{k_n} c_\alpha^{(n)} \,|n, \alpha\,\rangle\;. \tag{5.46}$$

The coefficients $c_\alpha^{(n)}$ are fixed in the next order.

First Order $\mathcal{O}(\varepsilon^1)$: To this order one obtains

$$H_0\psi^{(1)} + H_1\psi^{(0)} = E^{(0)}\psi^{(1)} + E^{(1)}\psi^{(0)}\,.$$

Taking the scalar product of this equation with $\langle n, \beta|$, and noting that

$$\left\langle n, \beta|H_0|\psi^{(1)}\right\rangle = E_n^{(0)}\left\langle n, \beta|\psi^{(1)}\right\rangle$$

(here H_0 is acting on the left-hand state!), and that $\langle n, \beta|n, \alpha\rangle = \delta_{\beta\alpha}$, one obtains a system of linear equations for the unknown coefficients, viz.

$$\sum_{\alpha=1}^{k_n} \left[\langle n, \beta|H_1\,|n, \alpha\,\rangle - E^{(1)}\delta_{\beta\alpha}\right] c_\alpha^{(n)} = 0\,. \tag{5.47}$$

Equation (5.47) has the form of a *secular equation* as it is known from celestial mechanics. The corrections of first order of the eigenvalues follow from the requirement that this system be soluable, i.e.

$$\det\left(\langle n, \beta|H_1\,|n, \alpha\,\rangle - E^{(1)}\mathbb{1}_{k_n\times k_n}\right) = 0\,, \tag{5.48}$$

Finally, the expansion coefficients $c_\alpha^{(n)}$ are calculated by solving (5.47) for each eigenvalue, one by one. The characteristic polynomial (5.48) has k_n real solutions. The corresponding eigenfunctions are k_n linear combinations of the states $|n, \alpha\rangle$ with fixed n. The degeneracy of the unperturbed problem may be lifted totally or partially.

Remarks

1. The restriction to a purely discrete spectrum is purely technical. Which of the matrix elements of H_1, within the subspace with fixed n, actually are different from zero depends on the selection rules that H_1 obeys.
2. The set of *all* eigenfunctions $|k, \alpha\rangle$ of H_0 provides a basis of the Hilbert space in which the problem is defined. If one starts from the more general ansatz

$$\psi = \sum_n \sum_{\beta=1}^{k_n} c_\beta^{(n)}\,|n, \beta\,\rangle$$

and replaces the operator H_1 by its matrix in this representation, then the analogue of (5.47) even yields the exact solution of the problem.

5.2.3 An Example: Stark Effect

A characteristic feature of the hydrogen spectrum is its ℓ-degeneracy: All bound states with the same principal quantum number n have the same energy, independently of the value $\ell = 0, 1, \ldots, n-1$ of the orbital angular momentum. So, for example, one has $E(2\mathrm{p}) = E(2\mathrm{s})$. This dynamical degeneracy which is a property of the $1/r$-potential acting here, is lifted in hydrogen-*like* atoms because the electrostatic potential no longer has the exact $1/r$ dependence. In order to understand this suppose that the nucleus of a hydrogen-like atom can be represented by a homogeneous charge distribution

$$\varrho(r) = \frac{3Ze}{4\pi R^3}\Theta(R-r)\,.$$

The potential which is felt by the electron and in which it moves then is

$$r \le R: \quad U(r) = -\frac{Ze^2}{R}\left[\frac{3}{2} - \frac{1}{2}\left(\frac{r}{R}\right)^2\right],$$

$$r > R: \quad U(r) = -\frac{Ze^2}{r}\,.$$

Making use of the formula (5.38) one easily estimates that both the 2s state and the 2p state experience upward shifts, as compared to the pure Coulomb potential. However, the shift of the 2s state is larger than the shift of the 2p state because the latter is driven out of the nucleus by its centrifugal tail and, therefore, feels less of the difference between the pure Coulomb and the hydrogen-like potentials. Thus, the former degeneracy is lifted.

Let H_0 be the Hamiltonian which describes the unperturbed hydrogen atom, or a hydrogen-like atom,

$$H_0 = -\frac{\hbar^2}{2m}\Delta + U(r)\,,$$

and let the perturbation be caused by a constant external electric field which acts along the 3 direction. This field couples to the dipole moment of the atom,

$$\boldsymbol{E} = E\hat{\boldsymbol{e}}_3\,, \qquad H_1 = -\boldsymbol{d}\cdot\boldsymbol{E} = -d_3 E\,. \tag{5.49}$$

The operator of the electric dipole moment reads $\boldsymbol{d} = -\,e\boldsymbol{x}$, or, upon transformation to a spherical basis, defined by $d_{\pm 1} = \mp(d_1 \pm i d_2)/\sqrt{2}$, $d_0 = d_3$,

$$d_\mu = -e\sqrt{\frac{4\pi}{3}}\, rY_{1\mu}\,. \tag{5.50}$$

Converting the field $\boldsymbol{E}$ to the same basis,

$$E_{+1} = -\frac{1}{\sqrt{2}}(E_1 + \mathrm{i}E_2)\,, \qquad E_{-1} = \frac{1}{\sqrt{2}}(E_1 - \mathrm{i}E_2)\,, \qquad E_0 = E_3\,,$$

the scalar product becomes

$$\boldsymbol{d}\cdot\boldsymbol{E} = \sum_\mu d_\mu^* E_\mu = \sum_\nu d_\nu E_\nu^*\,.$$

Provided the perturbation is small, and this is the case if the external field is small as compared to a typical internal field of the atom, $E \ll E_i \approx e/a_B^2 \approx 5 \times 10^9$ V/cm, the formalism developed above is applicable. We distinguish two cases:

1. *Hydrogen-like Atoms*: In these atoms the first order contribution vanishes because all matrix elements $\langle n\ell m|d_3|n\ell m\rangle$ are equal to zero by parity conservation: The matrix element $\int \mathrm{d}\Omega\, Y^*_{\ell m} Y_{10} Y_{\ell m}$ vanishes because Y_{10} is odd under space reflection while the product $|Y_{\ell m}|^2$ is even. Therefore, there is no Stark effect in hydrogen-like atoms which would be *linear* in the external field.
   A splitting of levels occurs only in second order perturbation theory. Indeed, by (5.40) one obtains:
   $$\Delta E^{(2)} = \sum_{n'\ell'm'} \frac{\langle n\ell m|\boldsymbol{d}\cdot\boldsymbol{E}\,|n'\ell'm'\rangle\langle n'\ell'm'|\boldsymbol{d}\cdot\boldsymbol{E}\,|n\ell m\rangle}{E_{n\ell}-E_{n'\ell'}}$$
   $$= \sum_{n'\ell'}\sum_{\mu\nu} E_\mu E^*_\nu \sum_{m'} \frac{\langle n\ell m|d_\nu\,|n'\ell'm'\rangle\langle n'\ell'm'|d^*_\mu\,|n\ell m\rangle}{E_{n\ell}-E_{n'\ell'}} .$$
   The numerator of this expression contains, among others, the product of the matrix elements
   $$\langle Y_{\ell m}|Y_{1\nu}\,|Y_{\ell'm'}\rangle \quad \text{and} \quad \langle Y_{\ell'm'}|Y_{1,-\mu}\,|Y_{\ell m}\rangle ,$$
   where use was made of the relation (1.117). As these are simultaneously different from zero only if the two selection rules $m'+\nu = m$ and $m-\mu = m'$ are fulfilled, one concludes that $\mu = \nu$. One then has
   $$\Delta E^{(2)} = \frac{1}{3}\boldsymbol{E}^2 \sum_{n'\ell'm'} \frac{\langle n\ell m|\boldsymbol{d}\,|n'\ell'm'\rangle\cdot\langle n'\ell'm'|\boldsymbol{d}\,|n\ell m\rangle}{E_{n\ell}-E_{n'\ell'}} \equiv -\boldsymbol{E}^2\frac{\alpha}{2} . \tag{5.51}$$
   This formula shows that the Stark effect in hydrogen-like atoms is *quadratic* in the external field. The constant of proportionality α is a characteristic property of the atom and is called *electric polarizability*.
2. *Hydrogen Atom:* We study the example of the subspace with principal quantum number $n = 2$. This space contains four orthogonal and normalized base states,
   $$|1\rangle \equiv |2\mathrm{s}, m=0\rangle , \qquad |2\rangle \equiv |2\mathrm{p}, m=0\rangle ,$$
   $$|3\rangle \equiv |2\mathrm{p}, m=+1\rangle , \qquad |4\rangle \equiv |2\mathrm{p}, m=-1\rangle ,$$
   all of which belong to the same eigenvalue $E^{(0)} = -\alpha^2 m_e c^2/8$. In this situation we must appeal to perturbation theory with degeneracy, that is, the wave function must be of the form
   $$\psi = \sum_{\alpha=1}^{4} c_\alpha\,|\alpha\rangle . \tag{5.52}$$
   All diagonal matrix elements of the perturbation H_1 vanish, cf. (5.49) and (5.50), $\langle\alpha|H_1|\alpha\rangle = 0$ (by parity conservation). Among the nondiagonal matrix elements only two are different from zero, and, in fact, are equal,
   $$\langle 1|H_1|2\rangle = \langle 2|H_1|1\rangle = -e\sqrt{\frac{4\pi}{3}}E\,\langle 2\mathrm{s}, m=0|rY_{10}\,|2\mathrm{p}, m=0\rangle .$$

This matrix element is calculated using eigenfunctions of the hydrogen atom. One finds the result

$$\Delta_{12} \equiv \langle 1|H_1|2\rangle = -3ea_B E\,. \tag{5.53}$$

The explicit form of the secular equation (5.48) for the total Hamiltonian $H = H_0 + H_1$, expressed in the basis $|1\rangle \dots |4\rangle$, is

$$\det\begin{pmatrix} E^{(0)} - E^{(1)} & \Delta_{12} & 0 & 0 \\ \Delta_{12} & E^{(0)} - E^{(1)} & 0 & 0 \\ 0 & 0 & E^{(0)} - E^{(1)} & 0 \\ 0 & 0 & 0 & E^{(0)} - E^{(1)} \end{pmatrix} = 0\,. \tag{5.54}$$

The four solutions are easily determined, they are

$$E^{(1)}_{1/2} = E^{(0)} \pm \Delta_{12}\,, \qquad E^{(1)}_3 = E^{(1)}_4 = E^{(0)}\,. \tag{5.55}$$

Thus, the original degeneracy is lifted only partially. The levels 1 and 2 are shifted by an amount which is *linear* in the modulus of the electric field. The corresponding eigenfunctions follow from the system of equations (5.47). One finds

$$\psi_1 = \frac{1}{\sqrt{2}}(|1\rangle + |2\rangle)\,, \qquad \psi_2 = \frac{1}{\sqrt{2}}(|1\rangle - |2\rangle)\,. \tag{5.56}$$

Note that the result (5.56) provides an example for the observation that two degenerate states are mixed strongly, in fact, completely, by a nondiagonal perturbation, independently of the strength of the matrix element Δ_{12}.

5.2.4 Two More Examples: Two-State System, Zeeman-Effect of Hyperfine Structure in Muonium

We discuss two more examples both of which correspond to specific physical situations and are of practical importance:

A Two-Level System with Variable Perturbation. Given a Hamiltonian $H = H_0 + H_1$ which has the following properties: Among others, H_0 has two stationary eigenstates $|\mathrm{n}_i(0)\rangle$, $i = 1, 2$, pertaining to the eigenvalues E_1 and E_2, respectively. In the subspace spanned by this basis, H_1 is represented by a matrix $\langle \mathrm{n}_i(0)|H_1|\mathrm{n}_j(0)\rangle = xAM_{ij}$ where A is a real number, and $\mathbf{M}$ a hermitian matrix whose trace is 1 and whose determinant is 0. The parameter x can be tuned from the value 0 to the value 1.

The properties $\det \mathbf{M} = 0$ and $\mathrm{tr}\,\mathbf{M} = 1$ imply that without loss of generality, this matrix can be written as

$$\mathbf{M} = \begin{pmatrix} \cos^2\alpha_0 & \cos\alpha_0 \sin\alpha_0 \\ \cos\alpha_0 \sin\alpha_0 & \sin^2\alpha_0 \end{pmatrix}.$$

In the basis $|\mathrm{n}_i(0)\rangle$ the Hamiltonian $H = H_0 + H_1$ has the explicit form

$$\mathbf{H} = \begin{pmatrix} E_1 + xA\cos^2\alpha_0 & xA\cos\alpha_0\sin\alpha_0 \\ xA\cos\alpha_0\sin\alpha_0 & E_2 + xA\sin^2\alpha_0 \end{pmatrix} .$$

Its eigenvalues follow from the quadratic equation

$$\lambda^2 - \lambda \,\mathrm{tr}\, \mathbf{M} + \det \mathbf{M} = \mathbf{0}$$

and, hence, are easily determined,

$$\lambda_{1/2} = \frac{1}{2}(\Sigma + xA)$$
$$\mp \frac{1}{2}\sqrt{(\Sigma - xA)^2 - 4(E_1E_2 + E_1xA\sin^2\alpha_0 + E_2xA\cos^2\alpha_0)} ,$$

where the abbreviation $\Sigma = E_1 + E_2$ is used. For the determination of the corresponding eigenstates let us make a digression which will turn out to be helpful for the interpretation of the system: Instead of the unperturbed basis $|\mathrm{n}_i(0)\rangle$ we use the basis in which $\mathbf{H_1}$ is diagonal and has the form $\mathbf{H}_1 = \mathrm{diag}(xA, 0)$. This new basis is given by

$$\begin{pmatrix} \nu_1 \\ \nu_2 \end{pmatrix} = \begin{pmatrix} \cos\alpha_0 & \sin\alpha_0 \\ -\sin\alpha_0 & \cos\alpha_0 \end{pmatrix} \begin{pmatrix} \mathrm{n}_1(0) \\ \mathrm{n}_2(0) \end{pmatrix} , \tag{5.57}$$

or, in a shorter notation, $|\nu_i\rangle = \sum_k V_{ik}(\alpha_0)|\mathrm{n}_k(0)\rangle$. Both bases, $|\nu_i\rangle$ and $|\mathrm{n}_k\rangle$, by their physical role, are distinguished bases: The states $|\mathrm{n}_i(0)\rangle$ are the eigenstates of the unperturbed Hamiltonian; $|\nu_1\rangle$ in turn is the state in which the perturbation xA acts, while in the state $|\nu_2\rangle$ it is not active.

The matrix $\mathbf{V}$ is unitary and, being real, is even orthogonal. In the new basis the Hamiltonian becomes $\widetilde{\mathbf{H}} = \mathbf{V}(\alpha_0)\mathbf{H}\mathbf{V}^T(\alpha_0)$. Using the abbreviations $\Sigma = E_1 + E_2$ (as before), and $\Delta := E_2 - E_1$ (new), and introducing trigonometric functions of the double angle, one obtains

$$\widetilde{\mathbf{H}} = \frac{1}{2}(\Sigma + xA)\begin{pmatrix} 1 & 0 \\ 0 & 1 \end{pmatrix} + \frac{1}{2}\begin{pmatrix} xA - \Delta\cos 2\alpha_0 & \Delta\sin 2\alpha_0 \\ \Delta\sin 2\alpha_0 & -xA + \Delta\cos 2\alpha_0 \end{pmatrix} . \tag{5.58}$$

Obviously, the eigenvalues are the ones we have given above. However, in the new representation they may be written in an alternative form:

$$\lambda_{1/2} = \frac{1}{2}\left\{(\Sigma + xA) \mp \sqrt{(\Delta\cos 2\alpha_0 - xA)^2 + \Delta^2\sin^2 2\alpha_0}\right\} . \tag{5.59}$$

It is useful to express the corresponding eigenstates in the basis

$$\left(|\nu_1\rangle, |\nu_2\rangle\right)^T .$$

They depend on the actual value of x. Therefore, we write them as

$$\left(|\mathrm{n}_1(x)\rangle, |\mathrm{n}_2(x)\rangle\right)^T$$

by inverting the formula (5.57), and by replacing α_0 by $\alpha(x)$. This x-dependent angle is determined from

$$\mathbf{V}^\dagger(\alpha(x))\,\widetilde{\mathbf{H}}\,\mathbf{V}(\alpha(x)) = \mathrm{diag}(\lambda_1, \lambda_2) .$$

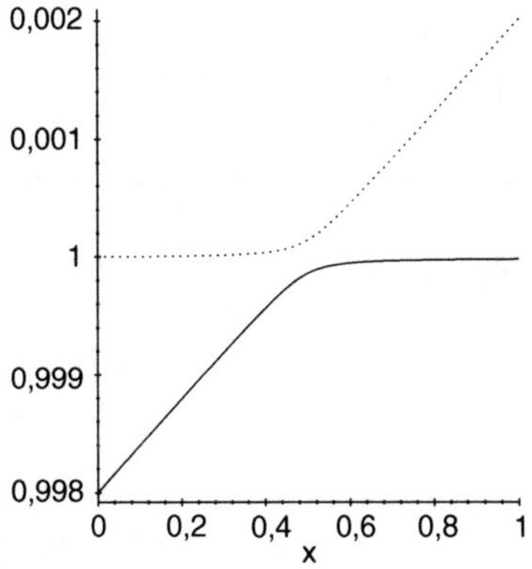

Fig. 5.3 Energy levels (5.59) of Example 5.1 with $\cos 2\alpha_0 = 0.99$ and $E_1 = 0.998$, $E_2 = 1$, $A = 0.002$ (arbitrary units). With increasing strength x of the perturbation the two eigenvalues initially move towards each other. However, instead of crossing at $x \approx \Delta \cos 2\alpha_0/a$, they move apart again. Meanwhile, the corresponding eigenstates practically exchange their roles

Using the representation (5.58) one shows that $\alpha(x)$, when expressed in terms of α_0, is given by:

$$\cos 2\alpha(x) = \frac{\Delta \cos 2\alpha_0 - xA}{\sqrt{(\Delta \cos 2\alpha_0 - xA)^2 + \Delta^2 \sin^2 2\alpha_0}} . \tag{5.60}$$

These results are beautifully interpretable. First, like in case 2 of Sect. 5.2.3, we note that if the unperturbed levels are degenerate, i.e. if $\Delta = 0$, then by (5.60), even a tiny perturbation xA leads to $\cos 2\alpha(x) = \pm 1$, that is, to maximal mixing of the states $|\nu_1\rangle$ and $|\nu_2\rangle$. If the unperturbed eigenvalues are different from each other we can take $E_2 > E_1$, without loss of generality, hence $\Delta > 0$, and $\alpha_0 < \pi/4$. The formulae (5.60) and (5.59) show that the perturbed system behaves differently, depending on whether A is positive or negative.

We discuss the more interesting case $A > 0$: For $x = 0$ we have $\lambda_i = E_i$, $\alpha(0) = \alpha_0$. If we let x grow from 0 to 1, and if we assume that $A > \Delta \cos 2\alpha_0$, then the eigenvalues $\lambda(x)$ follow the graphs shown in Fig. 5.3. They start at the unperturbed eigenvalues E_1 and E_2, respectively, and move towards each other until the point $x = \Delta \cos 2\alpha_0/A$, but there they do not cross. Rather, for increasing x beyond that point, they move apart again. At the same time the states $|\nu_i\rangle$ exchange their roles. This is seen as follows: Assume α_0 to be small, but the perturbative term A to be large as compared to $\Delta \cos 2\alpha_0$. At $x = 0$, according to (5.57), the basis is $|\mathrm{n}_1(0)\rangle \approx |\nu_1\rangle$, $|\mathrm{n}_2(0)\rangle \approx |\nu_2\rangle$. At the point $x = 1$ and with our assumption $A \gg \Delta \cos 2\alpha_0$ the formula (5.60) gives $\alpha(x = 1) \approx \pi/2$. Equation (5.57) now gives $|\mathrm{n}_2(0)\rangle \approx |\nu_1\rangle$, $|\mathrm{n}_1(0)\rangle \approx |\nu_2\rangle$. Thus, the two states have exchanged their physical content!

A possibly important application of this analysis is found in the dynamics of electron neutrinos $|\nu_e\rangle$ which are produced in the sun's interior, in a chain of reactions, which eventually leads to fusion of four hydrogen atoms into Helium ^{4}He (by the so-called pp-cycle). It seems plausible that the states ν_e and ν_μ which are created and annihilated in weak interactions, are not identical with eigenstates of mass, and that they are orthogonal mixtures of mass eigenstates.[8] In the application to this example H_0 is the mass operator with eigenvalues m_1^2 and m_2^2, $|\nu_1\rangle$ is identical with $|\nu_e\rangle$, $|\nu_2\rangle$ with $|\nu_\mu\rangle$. A ν_e created in the sun's interior feels a somewhat different interaction with matter in the sun than a ν_μ. In the example the term A is the additional interaction felt by ν_e, not by ν_μ, while the parameter x is proportional to the local density of the sun. If the relevant parameters fulfill the conditions described above this would mean that a ν_e which was created deep inside the sun, on its way to the sun's surface, would be "rotated" into a ν_μ.

Low energy ν_e's are detected on earth through inverse β-decay. Typical processes which were investigated are the transformation of Chlorine to Argon, or from Gallium to Germanium,

$$\nu_e + {}^{37}_{17}\mathrm{Cl} \longrightarrow {}^{37}_{18}\mathrm{Ar} + \mathrm{e}^- , \qquad \nu_e + {}^{71}_{31}\mathrm{Ga} \longrightarrow {}^{71}_{32}\mathrm{Ge} + \mathrm{e}^- ;$$

(the number in subscript gives the charge number of the element, the superscript shows the mass number of the isotope). In contrast, low energy muonic neutrinos ν_μ remain sterile, hence invisible because in the analogous processes $\nu_\mu + {}^{37}\mathrm{Cl}$ or $\nu_\mu + {}^{71}\mathrm{Ga}$ a muon μ^- would have to be created. The mass of the muon is approximately 200 times the value of the electron's mass, $m_\mu c^2 = 106\,\mathrm{MeV}$. The energy of neutrinos from the sun takes values no more than 7.2 MeV at most. The binding energies of the two nuclei involved differ by amounts on the order of a few MeV. Therefore, conservation of total energy forbids the creation of muons. This physically charming mechanism could help in explaining why terrestial experiments find fewer ν_e than predicted theoretically without mixing [Scheck (2012)].

Zeeman Effect of Hyperfine Structure in Muonium. Muonium is a perfect hydrogen-like atom consisting of a positive muon μ^+ and an electron e^-. As the muon has a mean-life of about 2.2×10^6 s and as this time is very long as compared to typical time scales of atomic transitions, these atoms can be produced copiously in experiment and can be analyzed in detail by means of high frequency techniques. The calculation of the interaction energy, as a function of the external homogeneous magnetic field, provides a simple, and physically instructive, application of quantum mechanical perturbation theory.

We choose the 3-direction along the homogeneous magnetic field, $\boldsymbol{B} = B\hat{\boldsymbol{e}}_3$. The Hamiltonian reads

$$H = H_0 - [\boldsymbol{\mu}(\mu) + \boldsymbol{\mu}(e)] \cdot \boldsymbol{B} - \frac{8\pi}{3}\boldsymbol{\mu}(\mu) \cdot \boldsymbol{\mu}(e)\,\delta(\boldsymbol{r}) .$$

Here H_0 is the unperturbed Hamiltonian (1.151) of the hydrogen atom if the reduced mass $\overline{m} = m_\mu m_e/(m_\mu + m_e)$ is inserted there. The second term describes the coupling

---

[8] In the weak interactions leptons always interact in pairs of a charged, electron-like partner, and an uncharged neutrino as well as their antiparticles. There are three families of such pairs: (e^-, ν_e), (μ^-, ν_μ), and (τ^-, ν_τ). For simplicity, we restrict the analysis of solar neutrinos to the first two of these families.

of the two magnetic moments to the exterior field, while the third term is the interaction of one of the magnetic moments with the magnetic field created by the other. The δ-distribution in the third term is derived either from classical electrodynamics (cf. [Jackson (1999)]), or from the Dirac equation in a nonrelativistic approximation. It says that the two magnetic moments couple only when the particles are at the same point in space. The operators $\boldsymbol{\mu}(\mu)$ and $\boldsymbol{\mu}(\mathrm{e})$ are as defined in (4.23). (Recall that the magnetic moment is defined by the eigenvalue of (4.23) for maximal m_s quantum number.) If

$$\mu_{\mathrm{B}}^{(i)} = \frac{|e|\,\hbar}{2m_i c}\,, \qquad i = \mathrm{e}, \mu\,,$$

denotes the Bohr magneton of the particle i, then the magnetic moment of the muon, and the corresponding operator are, respectively,

$$\mu(\mu) = g^{(\mu)}\mu_{\mathrm{B}}^{(\mu)}\frac{1}{2}\,, \quad \text{and} \quad \boldsymbol{\mu} = 2\mu(\mu)\boldsymbol{s}^{(\mu)}\,.$$

The muon's magnetic moment is positive, while the one of the electron is negative. Thus, the Hamiltonian reads

$$\begin{aligned} H = H_0 &- \left(- \left| g^{(\mathrm{e})} \right| \mu_{\mathrm{B}}^{(\mathrm{e})} \mathbf{s}_3^{(\mathrm{e})} + g^{(\mu)} \mu_{\mathrm{B}}^{(\mu)} \mathbf{s}_3^{(\mu)} \right) B \\ &+ \frac{16\pi}{3} \left| g^{(\mathrm{e})} \right| (\mu_{\mathrm{B}}^{(\mathrm{e})})^2 \frac{\mu(\mu)}{\mu_{\mathrm{B}}^{(\mathrm{e})}} \delta(\boldsymbol{r}) \left(\boldsymbol{s}^{(\mathrm{e})} \cdot \boldsymbol{s}^{(\mu)} \right). \end{aligned} \tag{5.61}$$

In transforming the third term we have introduced the ratio $\mu(\mu)/\mu_{\mathrm{B}}^{(\mathrm{e})}$ because the investigation of muonium allows to determine the magnetic moment of the muon (in units of the electronic Bohr magneton which is known) to high accuracy.

It is not difficult to estimate that typical matrix elements of the terms by which H and H_0 differ, are very small as compared to the differences of the eigenvalues of H_0. Therefore, the lowest order of perturbation theory is perfectly adequate.

We treat this system in the basis $|FM\rangle$ of eigenstates of total spin $\boldsymbol{F} = \boldsymbol{s}^{(\mathrm{e})} + \boldsymbol{s}^{(\mu)}$ which in this case can take the values $F = 1$ and $F = 0$.[9]

Taking $B = 0$, in a first step, the expectation value of H in the ground state of hydrogen, in the spin state $|FM\rangle$, is easily calculated by means of the formula

$$\boldsymbol{s}^{(\mathrm{e})} \cdot \boldsymbol{s}^{(\mu)} = \frac{1}{2}(\boldsymbol{F}^2 - \boldsymbol{s}^{(\mathrm{e})\,2} - \boldsymbol{s}^{(\mu)\,2}) :$$

$$\begin{aligned} \langle 1\mathrm{s}, FM|H\,|\,1\mathrm{s}, FM\,\rangle_{B=0} &= E_{1\mathrm{s}} + \frac{16\pi}{3} \left| g^{(\mathrm{e})} \right| (\mu_B^{(\mathrm{e})})^2 \frac{\mu(\mu)}{\mu_{\mathrm{B}}^{(\mathrm{e})}} \, |\psi_{1\mathrm{s}}(0)|^2 \\ &\times \frac{1}{2}\left(F(F+1) - \frac{3}{4} - \frac{3}{4} \right). \end{aligned}$$

[9] I have chosen the traditional notation of atomic physics. In atomic physics one studies the hyperfine structure and its Zeeman effect for an electron with total angular momentum $\boldsymbol{j} = \boldsymbol{\ell} + \boldsymbol{s}$ and for the nucleus with spin $\boldsymbol{I}$. The states of the combined system are classified by the resulting angular momentum $\boldsymbol{F} = \boldsymbol{j} + \boldsymbol{I}$.

The square of the wave function taken at relative position $\boldsymbol{r} = \boldsymbol{0}$ is

$$|\psi_{1s}(0)|^2 = |R_{1s}(0)Y_{00}|^2 = \frac{4}{a_B^3}\frac{1}{4\pi} = \frac{1}{\pi a_\infty^3}\left(1 + \frac{m_e}{m_\mu}\right)^{-3},$$

where we use the notation a_∞ for the Bohr radius (1.8) and make explicit the effect of the reduced mass. It is convenient to introduce here the Rydberg constant

$$Ry_\infty = \frac{\alpha^2 m_e c^2}{2hc},$$

so that

$$\langle 1s, FM|H|1s, FM\rangle_{B=0}$$
$$= E_{1s} + \frac{8}{3}\alpha^2 hcRy_\infty \frac{\mu(\mu)}{\mu_B^{(e)}}\left|g^{(e)}\right|\left(1 + \frac{m_e}{m_\mu}\right)^{-3}\frac{1}{2}\left[F(F+1) - \frac{3}{2}\right].$$

This formula yields the difference of the energies of the eigenstates with $F = 1$ and $F = 0$ at vanishing field $B = 0$, $\Delta E = E(1s, F = 1) - E(1s, F = 0)$ or, after division by h, the difference of the corresponding frequencies

$$\Delta\nu = \frac{1}{h}\Delta E = \frac{8}{3}\alpha^2 cRy_\infty \frac{\mu(\mu)}{\mu_B^{(e)}}\left|g^{(e)}\right|\left(1 + \frac{m_e}{m_\mu}\right)^{-3}. \tag{5.62}$$

If the magnetic field is switched on, the energies of the two states $|F = 1, M = \pm 1\rangle$ are seen to be

$$\langle 1, M = \pm 1|H - H_0|1, M = \pm 1\rangle \tag{$*$}$$
$$= \frac{1}{4}h\Delta\nu + \frac{1}{2}M\left(\left|g^{(e)}\right|\mu_B^{(e)} - g^{(\mu)}\mu_B^{(\mu)}\right)B.$$

In contrast, in the subspace with $M = 0$ spanned by $|10\rangle$ and $|00\rangle$, the operators $\mathbf{s}_3^{(i)}$ are not diagonal. Rows and columns being numbered by the base states $|10\rangle$ and $|00\rangle$, and introducing the abbreviation (5.62), the following matrix is to be diagonalized in this subspace

$$\mathbf{W} = \begin{pmatrix} E_{1s} + (1/4)\Delta E & W_{12} \\ W_{21} & E_{1s} - (3/4)\Delta E \end{pmatrix}.$$

The matrix element $W_{12} = \langle 10|\ldots|00\rangle$ receives contributions only from the second term in (5.61). Making use of the spin functions (4.67) and (4.68) one finds

$$W_{12} = W_{21} = \frac{1}{2}\left(g^{(\mu)}\mu_B^{(\mu)} + \left|g^{(e)}\right|\mu_B^{(e)}\right)B.$$

The eigenvalues are obtained from the characteristic polynomial

$$\lambda^2 - (\mathrm{tr}\,\mathbf{W})\lambda + (\det\mathbf{W}) = \mathbf{0}.$$

It is useful to introduce the dimensionless variable

$$x := \frac{1}{\Delta E}\left(g^{(\mu)}\mu_B^{(\mu)} + \left|g^{(e)}\right|\mu_B^{(e)}\right)B, \tag{5.63}$$

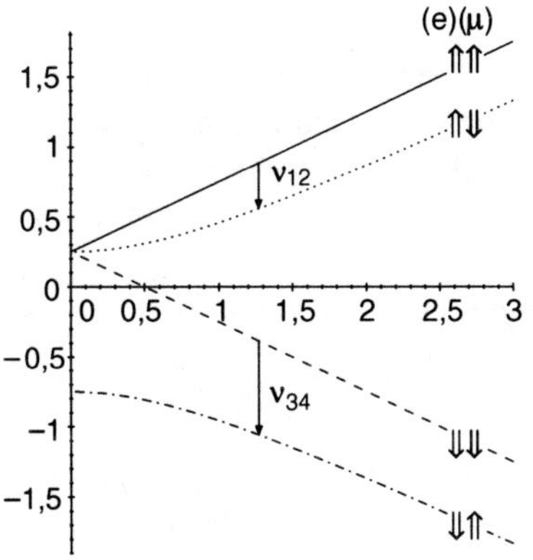

Fig. 5.4 Zeeman-effect of hyperfine structure in muonium, (5.64). We plot the dimensionless functions $y_{F,M} = [E(F,M) - E_{1s}]/\Delta E$, ΔE being the spacing between the hyperfine levels $F = 1$ and $F = 0$, as functions of the exterior magnetic field. This field appears in the dimensionless variable x, (5.63). For large values of x the upper two branches cross a second time (this cannot be seen in this figure, though, because of the large mass ratio m_μ/m_e)

so that the roots of this equation become

$$\lambda_{1/2} = E_{1s} - \frac{1}{4}\Delta E \pm \frac{1}{2}\Delta E\sqrt{1+x^2}\,.$$

The upper sign applies to the case $F = 1$, the lower sign to the case $F = 0$. Both cases are summarized by a factor $(-)^{F+1}$ in front of the square root.

Note that the results for the states $|1 \pm 1\rangle$ given above, may also be written as functions of the variable x: The following unified formula applies to all four eigenstates,[10]

$$E(F,M) = E_{1s} - \frac{1}{4}\Delta E - g^{(\mu)}\mu_B^{(\mu)}MB + (-)^{F+1}\frac{1}{2}\Delta E\sqrt{1+2Mx+x^2}\,. \quad (5.64)$$

This formula is of central importance in analyzing measurements of the transition frequencies in the ground state of muonium. The frequencies $[E(F,M) - E_{1s}]/h$ are drawn in Fig. 5.4 as functions of the applied magnetic field in the following way: One defines the functions $y_{F,M}(x) := [E(F,M) - E_{1s}]/\Delta E$ so that

$$y_{F,M}(x) = -\frac{1}{4} - \frac{g^{(\mu)}\mu_B^{(\mu)}}{g^{(\mu)}\mu_B^{(\mu)} + \left|g^{(e)}\right|\mu_B^{(e)}}Mx + (-)^{F+1}\frac{1}{2}\sqrt{1+2Mx+x^2}\,.$$

The frequencies proper as well as the transition frequencies between different substates (at fixed x) are obtained from Fig. 5.4 by multiplying with $\Delta\nu = \Delta E/\hbar$. For example, one easily confirms that the sum of the transition frequencies ν_{12} and ν_{34} yields the hyperfine interval $\Delta\nu$, for any value of the magnetic field. The physical

[10] Alternatively this problem can be solved in the basis $|s^{(e)}, m_1\rangle|s^{(\mu)}, m_2\rangle$ in which case one obtains this general formula. This was the original approach chosen by Breit and Rabi.

importance of this quantity lies in the fact that it is sensitive to radiative corrections characteristic for quantized electrodynamics, and to bound state effects. Its measurement allows for a test of the corresponding theoretical predictions. As frequencies can be measured to very high precision these tests are very sensitive. In particular, the hyperfine interval $\Delta\nu$ contains the magnetic moment of the positive muon and serves to determine this quantity.

We conclude this example with two comments:

Remarks

1. For very large values of the magnetic field, i.e. $x \gg 1$, the hyperfine interaction in (5.61) is negligible as compared to the interaction with the field. In this case the states may be classified again by the quantum numbers of the uncoupled basis of spin states. This limiting situation is alluded to by the arrows in the right part of Fig. 5.4 which show the alignment of the electron's and the muon's spins along the direction of $\boldsymbol{B}$, that is along the 3-axis.
   A closer look at this figure shows that, as it is drawn here, it cannot be correct: The magnetic moment of the electron is negative and, therefore, its interaction energy with the external field is *positive*, in accord with what the figure shows. However, the magnetic moment of the muon is positive and, hence, its interaction energy with the field is *negative*. Therefore, the state with $m_2 = +1/2$ must lie *below* the state with $m_2 = -1/2$. This is seen to be correct in the lower right part of the figure, but not in its upper right part! In fact, the upper two branches of the diagram intersect not only in $x = 0$ but also in the point $x_C \approx m_\mu/m_e$. Thus, by continuing the drawing to very large values of x, the state with $m_1 = +1/2$ and $m_2 = -1/2$, indeed, is found to lie higher in energy than the one with $(m_1 = +1/2, m_2 = +1/2)$.
2. The formula describing the Zeeman effect of hyperfine structure was derived by G. Breit and I. Rabi for s-states in Alkali atoms. As the derivation follows the same lines as our example, I quote here this more general case. Let $\mu(j)$ denote the atomic magnetic moment in the state with $j = 1/2$ (this is the sum of the orbital angular momentum and of the spin of the electron). The atom's nucleus carries angular momentum I and its magnetic moment is denoted by $\mu(I)$. The Hamiltonian then reads
   $$H = H_0 - \left(\frac{1}{j}\mu(j)\,\mathbf{j}_3 + \frac{1}{I}\mu(I)\mathbf{I}_3\right) B + f(r)\boldsymbol{j}\cdot\boldsymbol{I}\,.$$
   As before, the states with total angular momentum are denoted by $|FM\rangle$ (this is the sum of the angular momenta of the electron and the nucleus). The energy difference between the two hyperfine levels with $F = I \pm (1/2)$ now is
   $$\Delta E = E\left(F = I + \frac{1}{2}\right) - E\left(F = I - \frac{1}{2}\right) = \frac{1}{2}\langle f\rangle\,(2I+1)\,.$$
   The magnetic field is replaced by the analogous dimensionless variable
   $$x = \frac{1}{\Delta E}\left(-\frac{1}{j}\mu(j) + \frac{1}{I}\mu(I)\right)\,.$$

The formula of Breit and Rabi then reads

$$\frac{E - E_0}{\Delta E} = -\frac{1}{2(2I+1)} - \frac{\mu(I)/I}{\mu(j)/j + \mu(I)/I} Mx$$

$$\pm \sqrt{1 + 4M/(2I+1)x + x^2}\,, \qquad (5.65)$$

where the upper sign applies to $F = I + j$, the lower sign to $F = I - j$. The expectation value $\langle f \rangle$ is to be taken in the ground state whose unperturbed energy is E_0. The reader is invited to draw the examples of ^{6}Li ($I = 1$) and of ^{7}Li ($I = 3/2$) in a figure analogous to Fig. 5.4, and to interpret the figure.

5.3 Time Dependent Perturbation Theory and Transition Probabilities

As we just showed in Sects. 5.2.1 and 5.2.2 the perturbation H_1 of a system described by the Hamiltonian H_0 may lead to a displacement of the energies of the initial system, and to mixing of the corresponding wave functions. However, it may also cause the system to make a transition to another state, yet by respecting all conservation laws including the principle of energy conservation. Here is an example:

Suppose the unperturbed Hamiltonian contains two additive terms the first of which describes a hydrogen-like atom with its stable bound and continuum states, while the second contains the free radiation field and describes free electromagnetic waves in vacuum. Suppose further that a term is added to H_0 which describes the interaction of the electrons of the atom with the radiation field. In the initial state the atom is assumed to be in its ground state, the unpaired electron having the binding energy $E_0 = -B$. Furthermore, assume that there is a photon whose energy is sufficiently large so that $E_\gamma = \hbar\omega > B$. If the photon is absorbed, the electron is kicked out of its bound state and is taken to a state in the continuum with energy $E' = (E_\gamma - B) > 0$.

This raises several questions. Given an incoming beam of photons with energy E_γ, and a target consisting of many such atoms, how does one proceed in calculating the probability per unit of time, for the atom to make this transition, and how large is it? If the matrix elements of the interaction H_1 are small as compared to typical energy differences of the unperturbed system, is it possible to analyze the absorption process in a perturbation series?

5.3.1 Perturbative Expansion of Time Dependent Wave Function

The problem to be solved is the following: At time t_0 the system is in the state $\Psi(t_0) = |n_0\rangle$. For instance, the initial state consists of an atom in its ground state and a beam of photons with energy E_γ. Find the state $\Psi(t) = U(t, t_0)|n_0\rangle$ which, in the course of time, emanates from the initial state under the action of the perturbation H_1. For this purpose we expand the solutions of the time dependent Schrödinger equation

$$\mathrm{i}\hbar\dot{\Psi}(t) = H\Psi(t) = (H_0 + H_1)\Psi(t) \tag{5.66}$$

in terms of a basis of solutions of the stationary Schrödinger equation,

$$H_0\,|n\rangle = E_n\,|n\rangle\ ,$$

which contains the initial state $|n_0\rangle$. As we know from Sect. 1.8.1, Remark 3, stationary states have harmonic time dependence, viz.

$$|n\rangle: \quad \mathrm{e}^{-(\mathrm{i}/\hbar)E_n(t-t_0)}\,|n\rangle\ .$$

Written in a somewhat formal notation the ansatz for the time dependent solution must be

$$\Psi(t) = \sum\!\!\!\!\!\!\int c_n(t)\,\mathrm{e}^{-\mathrm{i}E_n(t-t_0)/\hbar}\,|n\rangle\ ,$$

where the hybrid symbol "sum/integral" hints at the fact that many, or all of these states lie in a continuum. Inserting this expansion and making use of the orthogonality of the base states $|n\rangle$ yields a system of coupled ordinary differential equations of first order for the time dependent coefficients. This system is seen to be of the form

$$\dot{c}_n(t) = -\frac{\mathrm{i}}{\hbar}\sum\!\!\!\!\!\!\int \langle n|H_1\,|m\rangle\ \mathrm{e}^{-\mathrm{i}\omega_{mn}t(t-t_0)}c_m(t)\,. \tag{5.67}$$

The energy differences that occur in these equations are replaced by the corresponding transition frequencies,

$$\omega_{mn} := (E_m - E_n)/\hbar\,.$$

The system of equations (5.67) must be solved with initial condition $c_n(t_0) = \delta_{nn_0}$.

If H_1 is "small", in a certain sense, the system of equations (5.67) can be solved by iteration, taking

$$c_n(t) = \sum_{\nu=0}^{\infty} c_n^{(\nu)}(t)\,.$$

The ν-th approximation is obtained by inserting the preceding one i.e. $c_n^{(\nu-1)}$, on the right-hand side of (5.67) and by integrating over time. (This ordering of terms becomes particularly transparent if one writes εH_1, instead of H_1, $\varepsilon^\nu c_n^{(\nu)}$ instead of $c_n^{(\nu)}$. As ε drops out of the final equations, one may safely take this parameter to 1 at the end.) Thus, one obtains

$$c_n^{(\nu)}(t) = -\frac{\mathrm{i}}{\hbar}\sum\!\!\!\!\!\!\int \int_{t_0}^{t} \mathrm{d}t_\nu\,\langle n|H_1(t_\nu)\,|m\rangle\ \mathrm{e}^{-\mathrm{i}\omega_{mn}(t_\nu-t_0)}c_m^{(\nu-1)}(t_\nu)\,. \tag{5.68}$$

In many cases only the first iteration is relevant. It reads

$$c_n^{(1)}(t) = -\frac{\mathrm{i}}{\hbar} \int\limits_{t_0}^{t} \mathrm{d}t_1 \langle n|H_1(t_1)\,|n_0\rangle\; \mathrm{e}^{\mathrm{i}\omega_{nn_0}(t_1-t_0)} \,. \tag{5.69}$$

The second approximation contains two integrations and reads

$$c_n^{(2)}(t) = \left(-\frac{\mathrm{i}}{\hbar}\right)^2 \sum\!\!\!\!\!\!\int \int\limits_{t_0}^{t} \mathrm{d}t_2 \int\limits_{t_0}^{t_2} \mathrm{d}t_1\; \langle n|H_1(t_2)\,|i\rangle\; \mathrm{e}^{\mathrm{i}\omega_{ni}(t_2-t_0)}$$
$$\times\; \langle \mathrm{i}|H_1(t_1)\,|n_0\rangle\; \mathrm{e}^{\mathrm{i}\omega_{in_0}(t_1-t_0)} \,. \tag{5.70}$$

It is relevant in those cases where the transition from the state $|n_0\rangle$ to the state $|n\rangle$ is impossible in first order because the matrix element $\langle n|H_1|n_0\rangle$ does not obey the selection rules and hence vanishes. We met an example for this situation towards the end of Sect. 4.2.1: If H_1 is supposed to describe an electric dipole transition and if we wish to calculate the transition 2s $\to$ 1s, the first order (5.69) vanishes. A transition of this kind which is forbidden in first order, will be possible in second order provided among the virtual intermediate states $|i\rangle$ there are one or more states whose transition matrix elements to $|n_0\rangle$ and to $|n\rangle$ are both different from zero. Such a process being of second order in the perturbation will be less frequent than a comparable process which is allowed in first order.

Without loss of generality the initial time t_0 may be taken to be the origin, i.e. we set $t_0 = 0$. Before turning to evaluate the first approximation (5.69) we note that, alternatively, one may use the interaction picture, introduced in Sect. 3.6, by replacing $H_1(t)$ by

$$H_1^{(\mathrm{int})}(t) = \mathrm{e}^{(\mathrm{i}/\hbar)H_0 t} H_1(t)\, \mathrm{e}^{-(\mathrm{i}/\hbar)H_0 t} \,.$$

In every matrix element of $H_1^{(\mathrm{int})}$ taken between eigenstates of H_0, the exponential containing H_0 acts both on the right and on the left states so that

$$\langle p|H_1^{(\mathrm{int})}(t)\,|q\rangle = \langle p|\, \mathrm{e}^{(\mathrm{i}/\hbar)H_0 t} H_1(t)\, \mathrm{e}^{-(\mathrm{i}/\hbar)H_0 t}\,|q\rangle$$
$$= \mathrm{e}^{\mathrm{i}\omega_{pq}t} \langle p|H_1(t)\,|q\rangle \,.$$

This means that in (5.70) the operator H_1 can be replaced by $H_1^{(\mathrm{int})}$, provided, at the same time, the exponential functions are replaced by 1.

In (5.70) the operator H_1 or $H_1^{(\mathrm{int})}$ acts once at time t_1, once at time t_2, but the integrations must be done such that the condition $t_2 > t_1$ is respected. In the product $H_1(t_2)H_1(t_1)$ the time arguments are ordered in increasing order from right to left. Note that

$$\int\limits_0^t \mathrm{d}t_2 \int\limits_0^{t_2} \mathrm{d}t_1 H_1(t_2)H_1(t_1) = \int\limits_0^t \mathrm{d}u_1 \int\limits_{u_1}^{t} \mathrm{d}u_2 H_1(u_2)H_1(u_1)$$

or, upon defining the *time-ordered product*,

$$\mathrm{P}\big(H_1(t_2)H_1(t_1)\big) := H_1(t_2)H_1(t_1)\Theta(t_2 - t_1) + H_1(t_1)H_1(t_2)\Theta(t_1 - t_2)\,, \tag{5.71}$$

the same double integral is equal to

$$\frac{1}{2!}\int_0^t \mathrm{d}t_2 \int_0^t \mathrm{d}t_1\, \mathrm{P}\big(H_1(t_2)H_1(t_1)\big)\,.$$

In the case considered here the original integral occurs just twice, hence the factor $1/2$. If one generalizes the definition (5.71) to a product of k operators which act at k times ordered from right to left, in increasing order, then the corresponding k-fold integral is equal to

$$\int_0^t \mathrm{d}t_k \int_0^{t_k} \mathrm{d}t_{k-1} \int_0^{t_{k-1}} \mathrm{d}t_{k-2} \cdots \int_0^{t_3} \mathrm{d}t_2 \int_0^{t_2} \mathrm{d}t_1\, A(t_k)\ldots Z(t_1)$$
$$= \frac{1}{n!}\int_0^t \mathrm{d}t_k \int_0^t \mathrm{d}t_{k-1} \cdots \int_0^t \mathrm{d}t_1\, \mathrm{P}\big(A(t_k)\ldots Z(t_1)\big)\,.$$

When applied to the second order expression (5.70) this contribution can be written in a much simplified and compact form by also inserting the completeness relation

$$\sum\!\!\!\!\!\!\!\int |i\rangle\,\langle i| = \mathbb{1}\,.$$

Thus, one obtains

$$c_n^{(2)}(t) = \left(-\frac{\mathrm{i}}{\hbar}\right)^2 \frac{1}{2!}\int_0^t \mathrm{d}t_2 \int_0^t \mathrm{d}t_1\, \langle n|\mathrm{P}\big(H_1^{(\mathrm{int})}(t_2)H_1^{(\mathrm{int})}(t_1)\big)\,|n_0\rangle\,. \tag{5.72}$$

It is obvious how to generalize this to the order k; the expression that is obtained in this way is called the *Dyson series.*[11]

5.3.2 First Order and Fermi's Golden Rule

We calculate the first iteration in more detail, using the abbreviation $\omega \equiv \omega_{nn_0}$ in (5.69) and choosing $t_0 = 0$ as before. In many cases H_1 does not depend on time explicitly. The integral over time in (5.69) can then be calculated directly. The

[11] After Freeman Dyson, mathematician and physicist, professor emeritus at the Institute for Advanced Studies, Princeton, USA.

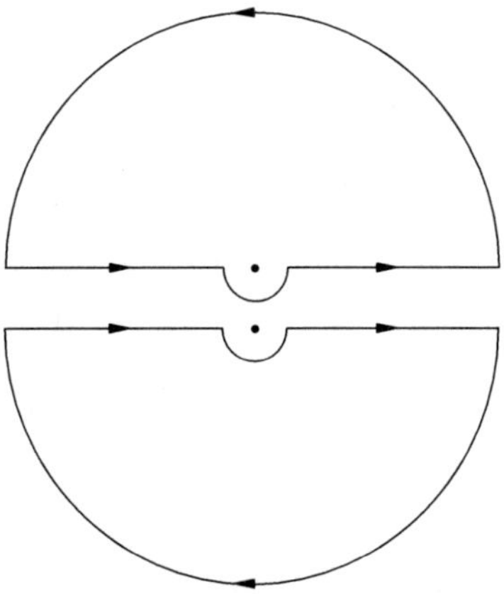

Fig. 5.5 Application of Cauchy's integral theorem in the calculation of the integral $\int_{-\infty}^{\infty} \mathrm{d}\omega g(\omega) I(t,\omega)$ with $g(\omega)$ a smooth function and $I(t,\omega)$ as defined in (5.74). The figure shows the complex ω-plane (twice). The path of integration $[-\infty, +\infty]$ is deformed at $\omega = 0$; it is supplemented by a semi-circle at infinity, once in the upper half-plane, once in the lower half-plane

expression for the probability, per unit of time, to pass from the state n_0 to the state n under the influence of H_1,

$$w(n_0 \to n) \approx \frac{1}{t} \left| c_n^{(1)}(t) \right|^2 , \tag{5.73}$$

contains the function of time and of transition frequency ω

$$\frac{1}{t} \left| \int_0^t \mathrm{d}t' \, \mathrm{e}^{\mathrm{i}\omega t'} \right|^2 = \frac{2(1 - \cos \omega t)}{t\omega^2} =: I(t, \omega) . \tag{5.74}$$

In the limit $t \to \infty$ this function becomes a well-known (tempered) distribution. In order to see this consider a smooth function of $g(\omega)$ which at infinity decays faster than any power, as well as the integral

$$\int_{-\infty}^{+\infty} \mathrm{d}\omega \, g(\omega) I(t, \omega)$$

$$= \lim_{\varepsilon \to 0} \int_{-\infty}^{+\infty} \mathrm{d}\omega \frac{g(\omega)}{\omega(\omega + \mathrm{i}\varepsilon)} \left\{ \frac{1}{t}(1 - \mathrm{e}^{\mathrm{i}\omega t}) + \frac{1}{t}(1 - \mathrm{e}^{-\mathrm{i}\omega t}) \right\} .$$

In this integral I have deformed the path of integration along the real axis of the complex ω plane in such a way that it avoids the singularity at the origin via the lower half-plane. As shown in Fig. 5.5, regarding the first term in curly brackets, this integration path is supplemented by the half circle at infinity in the upper half-plane. Regarding the second term in curly brackets, the contour is closed by the half circle at infinity in the lower half-plane. In this way it is guaranteed that in

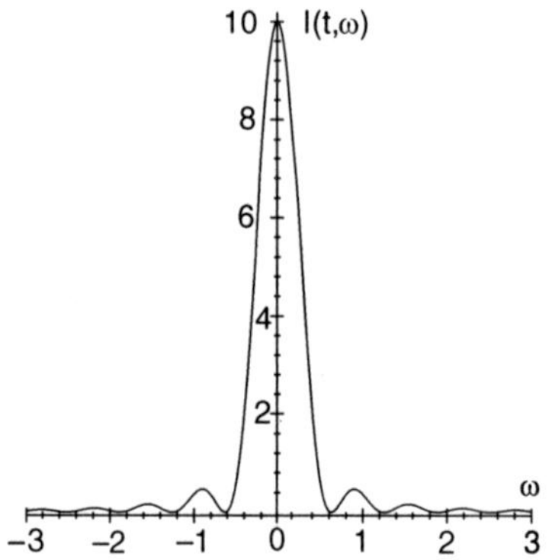

Fig. 5.6 The function $I(t, \omega)$ has a maximum at $\omega = 0$ which becomes the more pronounced the larger t is chosen. The figure shows $I(t, \omega)$ for $t = 10$ as a function of ω. In the limit $t \to \infty$ it tends to $2\pi\,\delta(\omega)$

either case the integrands vanish on these half circles. The integration running now over closed contours, one applies Cauchy's integral theorem which says that if the contour encloses the pole at $\omega = -\mathrm{i}\varepsilon$, the integral equals the residue of the integrand, multiplied by $2\pi\mathrm{i}$. This is what happens with the first term: The residue is

$$\left(\frac{1 - \mathrm{e}^{\mathrm{i}\omega t}}{\omega t}\right)_{\omega=0} = -\mathrm{i}\,,$$

and the contribution of the integral is $2\pi\mathrm{i}(-\mathrm{i}) = 2\pi$. The second term gives no contribution because the pole lies outside the contour. Thus, one finds that

$$\int\limits_{-\infty}^{+\infty} \mathrm{d}\omega\, g(\omega) I(t, \omega) = 2\pi\, g(0)$$

so that, when integrating over ω, the function $I(t, \omega)$ acts in the same way as the distribution $2\pi\,\delta(\omega)$,

$$I(t, \omega) \sim 2\pi\,\delta(\omega)\,. \tag{5.75}$$

Indeed, in the limit of large times, $t \to \infty$, the function (5.74) turns into the distribution $2\pi\,\delta(\omega)$. This is seen very clearly in Fig. 5.6 which shows $I(t, \omega)$ for a finite value of t yet large as compared to $1/\omega$. Note that the limit of large times corresponds to the experimental situation. As we explained previously, detection of the transition is done at a time which is practically infinite as compared to time scales which are characteristic for the microscopic process.

As a matter of application we consider the transition from a discrete state to a final state which lies in the continuum. Discrete states being eigenstates of energy, are given in the energy representation and are normalized to 1. In contrast, continuum states, as a rule, are defined in the momentum representation and are normalized to the δ-distribution. Thus, one has typically $\langle k|\, k' \rangle = 1/(g(k))\delta(k - k')$ where $g(k)$ is a real, positive-definite function. An example may clarify this. Using spherical polar coordinates in momentum space, $\boldsymbol{k} = (k, \theta_k, \phi_k)$, the eigenstates of momentum are normalized as follows,

$$\langle \boldsymbol{k}' | \boldsymbol{k} \rangle = \delta(\boldsymbol{k}' - \boldsymbol{k}) = \frac{1}{kk'} \delta(k' - k) \delta(\cos\theta_{k'} - \cos\theta_k) \delta(\phi_{k'} - \phi_k) .$$

Hence, in this case $g(k) = k^2$. Clearly, the states in the continuum must be normalized in the energy scale, not in the k-scale. The energy is a known function $E = E(k)$ of the modulus k of $\boldsymbol{k}$. The transformation that this requires is as follows. The projector onto a domain Δ of states $|k\rangle$ is given by

$$P_\Delta = \int\limits_\Delta \mathrm{d}k \; |k\rangle \, g(k) \langle k| .$$

Transforming to the energy scale then gives

$$P_\Delta = \int\limits_{\Delta(E)} \mathrm{d}E \; |k\rangle \, \varrho_k(E) \langle k| \quad \text{where} \quad \varrho_k(E) = g(k) \frac{\mathrm{d}k}{\mathrm{d}E} . \tag{5.76}$$

The function $\varrho_k(E)$ is called *level density* of the states $|k\rangle$ with energy $E(k)$.

Assembling the formulae (5.69), (5.73), (5.74), and (5.75) one obtains the transition probability per unit of time

$$w(n_0 \to k) \approx \frac{1}{\hbar^2} \int \mathrm{d}E \; \left| \langle k | H_1 | n_0 \rangle \right|^2 \varrho_k(E) 2\pi \delta(\omega) .$$

Finally, inserting the relation $\omega = (E - E_0)/\hbar$ from which $\delta(\omega) = \hbar \delta(E - E_0)$ follows, one obtains the final result

$$\boxed{w(n_0 \to k) \approx 2\pi \left| \langle k | H_1 | n_0 \rangle \right|^2 \varrho_k(E = E_0)/\hbar} . \tag{5.77}$$

This formula is also called

Fermi's Golden Rule: The transition probability per unit of time is proportional to the square of the matrix element between the initial and final states and to the level density of final states at the energy $E = E_0$.

Note that the principle of energy conservation is explicit in the result (5.77). All other conservation laws are hidden in the matrix element $\langle k | H_1 | n_0 \rangle$.

5.4 Stationary States of N Identical Fermions

Among the many-body problems that quantum mechanics deals with, systems consisting of a finite number of identical fermions are especially important for condensed matter physics, atomic physics, and nuclear physics. This is the principal reason why I restrict this section to systems of this kind and discuss the simplest methods for determining energies and wave functions of their ground states. By the same token this provides a basis on which the more specialized and refined procedures of many-body physics are built which are being used in these fields of physics.

5.4.1 Self Consistency and Hartree's Method

Let a system of N identical fermions be described by a Hamiltonian which, besides the kinetic energies, contains potential energies U_i as well as interaction potentials U_{ij} between pairs of particles,

$$H = \sum_{i=1}^{N}(T_i + U_i) + \sum_{i<j=1}^{N} U_{ij} \,. \tag{5.78}$$

In the description of the electronic shell of an atom the attractive terms U_i are determined by the electric field created by the atomic nucleus while U_{ij} is the repulsive Coulomb interaction of the electrons among themselves. Analogous situations are met in condensed matter physics: For example, in a lattice electrons move in an average periodic potential while experiencing the mutual Coulomb repulsion. In nuclear physics U_i represents an average potential which yields the single-particle spectra of protons or neutrons (shell model of nuclei), while U_{ik} is the effective residual interaction which is not taken into account in the average potential.

Assume, in a first step, that there are no single-particle potentials U_i. The simplest method to determine approximately the energy and the wave function of the ground state could consist in trying a product of initially unknown single-particle wave functions ψ_i,

$$\Psi(1, 2, \cdots, N) = \psi_1(\boldsymbol{x}^{(1)}, s^{(1)}) \cdot \ldots \cdot \psi_N(\boldsymbol{x}^{(1)}, s^{(1)})$$

which then are determined from the requirement that the expectation value $\langle \Psi | H | \Psi \rangle$ be a minimum under the subsidiary condition that Ψ remain normalized to 1, $\langle \Psi | \Psi \rangle = 1$.

One varies the product wave function by varying independently the single-particle functions, $\psi_i \mapsto \psi_i + \delta\psi_i$. The subsidiary condition is introduced by means of a Lagrange multiplier. Thus, the condition to be fulfilled is

$$\langle \delta\Psi | H | \Psi \rangle - \lambda \, \langle \delta\Psi | \Psi \rangle = 0 \,.$$

For the sake of clarity, the variation of the wave function number i is written $\eta_i = \delta\psi_i(\boldsymbol{x}^{(i)}, s^{(i)})$, and all spin arguments are suppressed. The above condition then reads

$$\int \mathrm{d}^3x^{(i)} \eta_i^* T_i \psi^{(i)} + \sum_{j \neq i} \int \mathrm{d}^3x^{(i)} \int \mathrm{d}^3x^{(j)} \eta_i^* \psi_j^*(\boldsymbol{x}^{(j)}) U_{ij} \psi_i(\boldsymbol{x}^{(i)}) \psi_j(\boldsymbol{x}^{(j)})$$
$$- \lambda \int \mathrm{d}^3x_i \eta_i^* \psi_i(\boldsymbol{x}^{(i)}) = 0 \,.$$

As the variations η_i are entirely arbitrary, this equation is fulfilled if the one-particle wave functions satisfy the following system of coupled equations

$$\left[T_i + \sum_{j \neq i} \int \mathrm{d}^3 x^{(j)} \psi_j^*(\boldsymbol{x}^{(j)}) U_{ij} \psi_j(\boldsymbol{x}^{(j)}) \right] \psi_i(\boldsymbol{x}^{(i)}) = \varepsilon_i \psi_i(\boldsymbol{x}^{(i)}) . \tag{5.79}$$

For fixed i this is seen to be a Schrödinger equation with a single-particle effective potential that is generated by all other particles $j \neq i$. One would try to solve this system of equations by an iterative procedure until the single-particle functions that it yields agree with the ones one uses to calculate the effective potential. If this is achieved one says that the solution is *self consistent.*

This first attempt is somewhat unfortunate for two reasons: First, the wave functions ψ_i and ψ_k belong to different potentials and, hence, are not orthogonal. Second, the product wave function Ψ is not antisymmetrized. In fact, it has no definite symmetry character at all and, hence, cannot satisfy the spin-statistics relation. Both difficulties can be avoided if, from the beginning, one assumes an antisymmetrized product of wave functions, that is, a Slater determinant (4.78) as the trial function. This modified procedure can be formulated in a particularly transparent way if one makes use of the method of second quantization.

5.4.2 The Method of Second Quantization

The idea of this method is simple: Instead of working with self-consistent one-particle wave functions (which are still to be determined!) in position space and in close analogy to the case of the harmonic oscillator (cf. Sect. 1.6) one introduces creation and annihilation operators $a_i^\dagger$ and a_i for particles in the state $|0, 0, \ldots, 0, i, 0, \ldots\rangle \equiv |\varphi_i(\boldsymbol{x}^{(i)})\rangle$ which act on a "vacuum", i.e. a state which contains no particles at all. A two-particle state where one particle is placed in the state φ_i, the other in the state φ_k, for example, then reads

$$a_i^\dagger a_k^\dagger \, |0, 0, \ldots\rangle .$$

The eigenvalues of the operators $N_i = a_i^\dagger a_i$ and $N_k = a_k^\dagger a_k$ are the number of particles in the state i and k, respectively. In order for such a two-particle state to be antisymmetric we must have $a_k^\dagger a_i^\dagger |0, 0, \ldots\rangle = -a_i^\dagger a_k^\dagger |0, 0, \ldots\rangle$ i.e. the two creation operators must *anti*commute. Indeed, with

$$\{ A \, , \, B \} := AB + BA \tag{5.80}$$

denoting the anticommutator, and postulating the commutation rules for creation and annihilation operators

$$\{a_i, a_k^\dagger\} = \delta_{ik} \, , \qquad \{a_i, a_k\} = 0 = \{a_i^\dagger, a_k^\dagger\} \, , \tag{5.81}$$

the product states describe antisymmetric product wave functions. This can be seen, on the one hand, by commuting any two particles. Indeed,

$$a_N^\dagger a_{N-1}^\dagger \ldots a_k^\dagger \ldots a_i^\dagger \ldots a_2^\dagger a_1^\dagger = (-)^\pi \, a_N^\dagger a_{N-1}^\dagger \ldots a_i^\dagger \ldots a_k^\dagger \ldots a_2^\dagger a_1^\dagger \, ,$$

where π is the specific permutation which maps $(1, 2, \ldots, i, \ldots, k, \ldots N)$ to $(1, 2, \ldots, k, \ldots, i, \ldots N)$. In this action one must take care that the particles are interchanged with all their attributes (position, spin, etc.).

On the other hand one shows by means of the relations (5.81) that every counting or number operator fulfills the relation

$$N_i(N_i - 1) = a_i^\dagger a_i(a_i^\dagger a_i - 1) = 0\,.$$

This implies that N_i has the only eigenvalues 0 and 1. Thus, the state i can either be empty, or be occupied by one single particle – in agreement with the Pauli principle.

The antisymmetrized product states that we just constructed are elements of what is called a *Fock space*.

Let $\mathcal{O}$ be a single-particle operator, $U(i,j)$ a two-body interaction, and define

$$\langle i|\mathcal{O}\,|\,k\,\rangle := \int \mathrm{d}^3x\,\varphi_i^*(\boldsymbol{x})\mathcal{O}\varphi_k(\boldsymbol{x})\,, \tag{5.82}$$

$$\langle ij|U\,|\,kl\,\rangle := \int \mathrm{d}^3x \int \mathrm{d}^3y\,\varphi_i^*(\boldsymbol{x})\varphi_j^*(\boldsymbol{y})U(\boldsymbol{x},\boldsymbol{y}) \tag{5.83}$$

$$[\varphi_k(\boldsymbol{x})\varphi_l(\boldsymbol{y}) - \varphi_k(\boldsymbol{y})\varphi_l(\boldsymbol{x})]\,,$$

where we have suppressed spin degrees of freedom, for the sake of clarity. (Note that the right-hand wave function in (5.83) is antisymmetric but is not normalized to 1.) With these definitions one now shows that in passing from the position space representation to a representation in Fock space, the one-body and two-body operators must be translated by the rules

$$\mathcal{O} \longmapsto \sum_{ik} \langle i|\mathcal{O}\,|\,k\,\rangle\, a_i^\dagger a_k\,, \quad \sum_{i<j} U(i,j) \longmapsto \frac{1}{4}\sum_{ij,kl} \langle ij|U\,|\,kl\,\rangle\, a_i^\dagger a_j^\dagger a_l a_k\,. \tag{5.84}$$

Note, in particular, the ordering of the last two indices in the second formula of (5.84)!

Denote by $|\Omega\rangle$ the vacuum state, that is, the state which contains no particles at all. As a matter of example, we prove the rule (5.84) for the case of two-body states

$$\Psi_a = a_m^\dagger a_n^\dagger\,|\,\Omega\,\rangle \quad \text{and} \quad \Psi_b = a_p^\dagger a_q^\dagger\,|\,\Omega\,\rangle\,,$$

the general case following from this. We have

$$\langle \Psi_b|\mathcal{O}\,|\,\Psi_a\,\rangle = \sum_{ij} \langle i|\mathcal{O}\,|\,j\rangle\, \langle \Omega\,|a_q a_p a_i^\dagger a_j a_m^\dagger a_n^\dagger\,|\,\Omega\,\rangle\,.$$

Any annihilation operator applied to the vacuum (to the right) gives zero, $a_i|\Omega\rangle = 0$. Likewise, every creation operator $a_i^\dagger$ applied to $\langle\Omega|$ (to the left) gives zero, $\langle\Omega\,|a_i^\dagger = 0$. Therefore, the strategy must be to shift the operator $a_i^\dagger$ to the left by means of the relations (5.81) until it hits $\langle\Omega|$, and, likewise, to let the operator a_j move to the right by exchanges of neighbours until it hits the vacuum state. This simple calculation shows that

$$\begin{aligned}\langle \Omega\,|a_q a_p a_i^\dagger a_j a_m^\dagger a_n^\dagger\,|\,\Omega\,\rangle &= \delta_{ip}\delta_{jm}\delta_{nq} + \delta_{iq}\delta_{jn}\delta_{mp} \\ &\quad - \delta_{iq}\delta_{jm}\delta_{pn} - \delta_{ip}\delta_{jn}\delta_{mq}\,.\end{aligned}$$

The states Ψ_a and Ψ_b can differ at most by one state. Clearly, we would have expected this result and the previous one if we had calculated the same matrix element in position space, by means of antisymmetrized wave functions:

$$\sum_{i=1}^{2} \langle \Psi_b | \mathcal{O}(i) \, | \Psi_a \rangle = \frac{1}{2} \int \mathrm{d}^3 x^{(1)} \int \mathrm{d}^3 x^{(2)} [\varphi_p^*(1)\varphi_q^*(2) - \varphi_p^*(2)\varphi_q^*(1)]$$
$$\times \sum_{i=1}^{2} \mathcal{O}(i) [\varphi_m(1)\varphi_n(2) - \varphi_m(2)\varphi_n(1)] \, .$$

Here we have used the abbreviations $i \equiv \boldsymbol{x}^{(i)}$ for the arguments. We note, furthermore, that it would have been sufficient to antisymmetrize only one of the two wave functions in these formulae. The reason for this is the fact that the operator is symmetric in the two particles.

The same remark also applies to the two-body matrix elements (5.83): As the operator is symmetric in all particles it is sufficient to choose only one of the states to be antisymmetric. Clearly, in this case, the states Ψ_a and Ψ_b can differ by no more than two single-particle states. Otherwise the matrix element vanishes. The proof of the second formula in (5.84) follows the same strategy as above: In the expectation value

$$\langle \Omega | a_n a_m (a_i^\dagger a_j^\dagger a_l a_k) a_p^\dagger a_q^\dagger \, | \Omega \rangle$$

the creation operators $a_i^\dagger$ and $a_j^\dagger$ are moved to the left, the annihilation operators a_l and a_k are moved to the right, by means of the canonical anticommutators (5.81), until they act on the vacuum. The result is to be compared with (5.83).

Consider the special case $|\Psi_a\rangle = |\Psi_b\rangle = a_p^\dagger a_q^\dagger |\Omega\rangle$. In this example one obtains the linear combination

$$\langle pq | U \, | pq \rangle - \langle pq | U \, | qp \rangle$$

of two-body matrix elements. The first of these is called the *direct interaction*, the second is called *exchange interaction.* If the one-particle wave functions do not overlap much then the exchange interaction is small as compared to the direct interaction. This is typical for atomic states of electrons. However, if they overlap strongly then exchange interaction and direct interaction are of the same order of magnitude. This situation occurs in the shell model of nuclei.

Remarks

1. The calculation of expectation values of monomials of creation and annihilation operators in the vacuum state $|\Omega\rangle$ is a combinatorial problem which one solves by means of a theorem of G.C. Wick, cf. Sect. 5.4.5.
2. There are always as many creation operators as there are annihilation operators so that the total number N of particles is the same in all states of the system as a whole. The action of an operator of the form $a_i^\dagger a_k$ may be visualized as the action of lifting a particle from the state k to the state i. The method of second quantization changes in no way the physical content of the theory. Its only

purpose is to simplify calculations and to guarantee that the states are properly antisymmetrized.

3. Unfortunately the remark about antisymmetry applies only to pure product states, but not to linear combinations of such states. For example, if single-particle states must be coupled to total angular momentum, then, in general, the coupled states must be antisymmetrized again.

5.4.3 The Hartree-Fock Equations

We start from the assumption that N identical fermions occupy N single-particle states such that their wave function Ψ is a determinant of the type of (4.78), or, upon using second quantization, an antisymmetric product state of the form

$$a_1^\dagger a_2^\dagger \cdots a_N^\dagger \,|\,\Omega\,\rangle\ .$$

The single-particle levels which are to be determined, are assumed to be the lowest in energy. In analogy to Hartree's method we vary the wave function Ψ, or, equivalently, its complex conjugate Ψ^*, by requiring that the variation of the energy of the ground state vanish, i.e. $\delta(\Psi, H\Psi) = 0$. Clearly, the variation of the single-particle states can only involve admixtures of states with $m > N$,

$$\psi_n \longmapsto \psi_n + \eta\psi_m\,, \qquad n \le N\,, \qquad m > N\,, \qquad \eta \ll 1\,.$$

The reason for this is that any admixture of an occupied state, $m < N$, leaves the Slater determinant (4.75) unchanged. Variation of, say, Ψ^* with $n \le N$ and $m > N$, means that we must have

$$\delta\Psi^* = \eta\langle\Psi|a_n^\dagger a_m : \qquad \eta(\Psi, a_n^\dagger a_m H\Psi) \stackrel{!}{=} 0\,.$$

Inserting the Hamiltonian H the variation yields the condition

$$\sum_{ij}\langle i|T\,|j\rangle\,(\Psi, a_n^\dagger a_m a_i^\dagger a_j\Psi) + \frac{1}{4}\sum_{ij,kl}\langle ij|U\,|kl\rangle\,(\Psi, a_n^\dagger a_m a_i^\dagger a_j^\dagger a_l a_k\Psi)$$
$$= 0\,.$$

The first term of this equation is different from zero only if $i = m$ and $j = n$. For the second term to be different from zero there are only the choices ($i = m$, $k = n$, $j = l$), ($j = m$, $l = n$, $i = k$), ($j = m$, $k = n$, $i = l$), and ($i = m$, $l = n$, $j = k$), of which the first two contribute with a positive sign, the latter two with a negative sign. In summary one obtains

$$\langle m|T\,|n\rangle + \frac{1}{4}\sum_{j=1}^{N}\big\{\langle mj|U\,|nj\rangle + \langle jm|U\,|jn\rangle$$
$$-\,\langle jm|U\,|nj\rangle - \langle mj|U\,|jn\rangle\big\} = 0\,.$$

Returning to the definition (5.83), one realizes that the last three terms are all equal to the second term so that the same equation comes in a shorter version and reads

$$\langle m|T\,|i\rangle + \sum_{j=1}^{N}\langle mj|U\,|ij\rangle = 0\,, \qquad i,j \le N\,, \quad m > N\,. \tag{5.85}$$

At this point it is useful to define the one-particle operator

$$H_{\text{s.c.}} := \sum_{m,n}\left(\langle m|T\,|n\rangle + \sum_{j=1}^{N}\langle mj|U\,|nj\rangle\right) a_m^{\dagger}a_n\,, \tag{5.86}$$

though *without* restriction on the indices m and n. Indeed, by relation (5.85) all matrix elements of the operator $H_{\text{s.c.}}$ relating an *occupied* to an *unoccupied* state are equal to zero. (The index "s.c." stands for *self consistent*, see below.)

The fact that all matrix elements $\langle p|H_{\text{s.c.}}|k\rangle$ with $p > N$ and $k < N$ vanish implies that one can diagonalize the Hamiltonian $H_{\text{s.c.}}$ separately in the space of all occupied states and in the space of the unoccupied states. Assuming this diagonalization to be carried out one is led to a new basis of single-particle states $|\alpha\rangle$ which pertain to the eigenvalues ε_α. In this *new* basis one obtains

$$\langle\sigma|T\,|\tau\rangle + \sum_{\alpha=1}^{N}\langle\alpha\sigma|U\,|\alpha\tau\rangle = \varepsilon_\sigma\,\delta_{\sigma\tau}\,. \tag{5.87}$$

Note that here either both states are occupied, i.e. $\sigma, \tau < N$, or both are unoccupied, i.e. $\sigma, \tau > N$.

Clearly, the ground state Ψ of the system must be the specific product state in which the N particles occupy the N lowest states of the new basis, i.e. the eigenstates of $H_{\text{s.c.}}$. Its energy is

$$\begin{aligned} E_0 = \langle\Psi|H_{\text{s.c.}}\,|\Psi\rangle &= \sum_{\sigma=1}^{N}\langle\sigma|T\,|\sigma\rangle + \frac{1}{2}\sum_{\sigma,\tau=1}^{N}\langle\sigma\tau|U\,|\sigma\tau\rangle \\ &= \sum_{\sigma=1}^{N}\varepsilon_\sigma - \frac{1}{2}\sum_{\sigma,\tau=1}^{N}\langle\sigma\tau|U\,|\sigma\tau\rangle\,. \end{aligned} \tag{5.88}$$

The Hamiltonian (5.86) is called the *Hartree-Fock operator*. The equation (5.87) are called *Hartree-Fock equations.*

It is instructive to rewrite these equations in coordinate space representation. In doing so, and recalling the definition of the two-body interaction (5.83), one sees that one would have to solve the following system of coupled Schrödinger equations

$$T\psi_\alpha(\boldsymbol{x}) + \mathcal{U}(\boldsymbol{x})\psi_\alpha(\boldsymbol{x}) - \int \mathrm{d}^3x'\,\mathcal{W}(\boldsymbol{x},\boldsymbol{x}')\psi_\alpha(\boldsymbol{x}') = \varepsilon_\alpha\psi_\alpha\,, \tag{5.89}$$

where the two potential terms are defined by

$$\mathcal{U}(\boldsymbol{x}) = \sum_{\sigma=1}^{N}\int \mathrm{d}^3x'\,\psi_\sigma^*(\boldsymbol{x}')U(\boldsymbol{x},\boldsymbol{x}')\psi_\sigma(\boldsymbol{x}')\,, \tag{5.90}$$

$$\mathcal{W}(\boldsymbol{x},\boldsymbol{x}') = \sum_{\sigma=1}^{N}\psi_\sigma^*(\boldsymbol{x}')U(\boldsymbol{x},\boldsymbol{x}')\psi_\sigma(\boldsymbol{x})\,. \tag{5.91}$$

Note that while the first of these is a local interaction the second is a nonlocal one.

Remarks

1. The two potentials contain the wave functions that one wishes to determine, and, a priori, it is not obvious how to tackle the system (5.89) of integro-differential equations. One might try an iterative procedure: Insert a set of trial wave functions $\psi_\alpha^{(0)}(\boldsymbol{x})$ in (5.90) and (5.91), and then solve these equations such as to obtain improved solutions $\psi_\alpha^{(1)}(\boldsymbol{x})$. These, in turn, must be inserted in the potential terms for a new round of solving the equations (5.89) yielding further improved solutions $\psi_\alpha^{(2)}(\boldsymbol{x})$. This iterative process will have to be continued until the point where the wave functions $\psi_\alpha^{(n-1)}(\boldsymbol{x})$ that are inserted in (5.90) and (5.91) practically coincide with the new solutions $\psi_\alpha^{(n)}(\boldsymbol{x})$ obtained from (5.89). A set of solutions which meets this condition, is said to be *self consistent*, hence the nomenclature introduced above.
2. In contrast to Hartree's method (5.79) described in Sect. 5.4.1 the term with $\sigma = \alpha$ no longer needs to be excluded because in this case the direct and the exchange terms cancel.
3. Two arbitrary solutions ψ_α and ψ_β of (5.89) which pertain to different eigenvalues, are orthogonal – in contrast to the Hartree equations (5.79) where they are not.
4. In practice one will not solve the system (5.89) of coupled equations (5.89) in coordinate space. Rather, one will try to reduce this sytem to a problem of diagonalizing finite dimensional matrices. Let $|a\rangle, |b\rangle, \dots$ be an arbitrary basis of Fock spaces (though selected from a practical point of view), and let $\psi_\alpha(\boldsymbol{x}) = \sum \varphi_a(\boldsymbol{x}) c_a^{(\alpha)}$ be the expansion of the unknown solution in this basis. Then the system (5.89) is equivalent to

$$\sum_a \left(\langle b|T\,|a\rangle + \sum_{d=1}^{N} \langle bd|U\,|ad\,\rangle \right) c_a^{(\alpha)} = \varepsilon_\alpha c_b^{(\alpha)} . \tag{5.92}$$

A system of algebraic equations of the kind of (5.92) must also be solved by iteration until the states $\{c_a^{(\alpha)}\}^T$ which are used as input, coincide, for all practical purposes, with the solutions of (5.92). Of course, this is nothing but a reformulation of the problem described above, in Remark 1. The advantage of this matrix representation is that in practice it might be sufficient to truncate the problem to a finite-dimensional subspace of states, the so-called model space. The system of equations (5.92) is then finite-dimensional and can be solved by repeated diagonalization of finite matrices.

The method just described yields at the same time the occupied and the unoccupied levels of the model space. It is difficult to judge the quality of the approximate solutions. Qualitatively speaking, in a favourable case one expects to find a single-particle spectrum of the kind sketched in Fig. 5.7 which has the distinctive feature that the highest occupied and the lowest unoccupied levels are separated by an energy gap Δ which is sizeably greater than typical level spacings in the occupied and in the unoccupied parts of the spectrum. This conjecture is supported by the observation

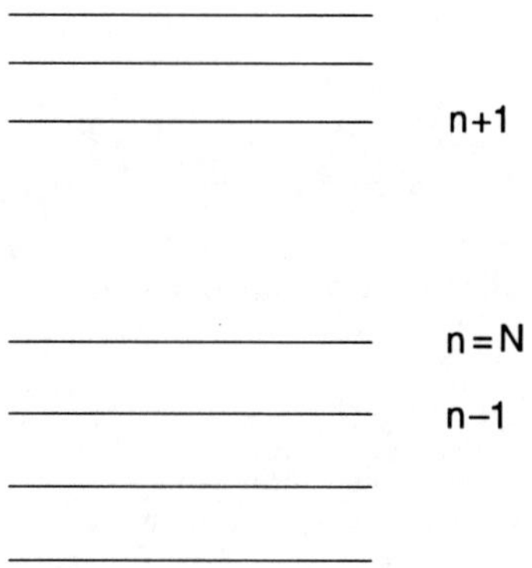

Fig. 5.7 The ground state of a system of N fermions as obtained by means of the Hartree-Fock method, presumably yields a realistic approximation if the highest occupied state $n = N$ is separated from the first unoccupied state $n + 1$ by an energy gap. (The spectrum shown here is purely fictitious)

that a single-particle excitation from $\alpha < N$ to $\beta > N$, in first and second order of perturbation theory, respectively, is suppressed by an energy denominator on the order of Δ or more.

5.4.4 Hartree-Fock Equations and Residual Interactions

The Hartree-Fock method as described in the preceding section, yields an approximate ground state described by an antisymmetric product wave function, that is, by a Slater determinant of single-particle states. For the sake of brevity, I call such a state a Fock state and write the symbol $|F\rangle$ for it. The true ground state of the N-fermion system whose energy might be lower than the one of the product state, will no longer be a product of single-particle wave functions. Instead it will be a superposition of different Fock states which differ from $|F\rangle$ by excitation(s) of one, or two, or more, particles into formerly unoccupied states of $|F\rangle$. Accordingly, these excited states are called one-particle, two-particle, etc. excitations.

In what follows I denote by $|\Psi\rangle$ the initially unknown, *true* ground state. Define then one-particle and two-particle densities, respectively, in the true ground state:

$$\varrho_{nm} := \langle \Psi | a_m^\dagger a_n | \Psi \rangle \ , \tag{5.93}$$

$$\varrho_{nmsr} := \langle \Psi | a_r^\dagger a_s^\dagger a_n a_m | \Psi \rangle \ . \tag{5.94}$$

The basis chosen in these definitions is arbitrary and need not be the basis of Hartree-Fock solutions. With a change of basis,

$$a_i'^\dagger = \sum_n \langle n | i \rangle a_n^\dagger \ , \qquad a_i' = \sum_n \langle n | i \rangle^* a_n = \sum_n \langle i | n \rangle a_n$$

and taking account of the completeness relation written in the form

$$\sum_n \langle i | n \rangle \langle n | j \rangle = \delta_{ij}$$

the trace of ϱ is seen to be invariant as it should.

One then shows: If the N-particle state is a Slater determinant $|F\rangle$ and if the densities in a state of this kind are denoted by superscripts (F), then $\operatorname{tr}\varrho^{(F)} = N$, $(\varrho^{(F)})^2 = \varrho^{(F)}$. For any one-particle operator $\mathcal{O}$ one has

$$\langle F|\mathcal{O}\,|F\rangle = \operatorname{tr}\left(\varrho^{(F)}\mathcal{O}\right).$$

The first and second assertions follow from the equation

$$\sum_n a_n\,|F\rangle\,\langle F|a_n^\dagger = 1\!\!1_{N\times N}\,,$$

which in turn is easy to prove. The sum over n is independent of the basis one chooses so that we may assume we transformed everything to the basis of Hartree-Fock states. Hence, the trace of the last expression is

$$\operatorname{tr}\sum_n a_n\,|F\rangle\,\langle F|a_n^\dagger = \langle F|\sum_{n=1}^{N} a_n^\dagger a_n\,|F\rangle = N\,.$$

If we now calculate the square ϱ^2, we find indeed

$$[(\varrho^{(F)})^2]_{mp} = \sum_n \langle F|a_m^\dagger a_n\,|F\rangle\,\langle F|a_n^\dagger a_p\,|F\rangle = \langle F|a_m^\dagger a_p\,|F\rangle = \varrho_{mp}^{(F)}\,.$$

The third relation is proved easily if one inserts the one-body operator in the notation of second quantization, viz.

$$\langle F|\mathcal{O}\,|F\rangle = \langle F|\sum_{i,j}\langle i|\mathcal{O}\,|j\rangle\,a_i^\dagger a_j\,|F\rangle = \operatorname{tr}\,(\varrho^{(F)}\mathcal{O})\,.$$

With the definitions (5.93) and (5.94) the expectation value of the Hamiltonian in the true ground state $|\Psi\rangle$ is given by

$$\langle\Psi|H\,|\Psi\rangle = \sum_{ij}\langle i|T\,|j\rangle\,\varrho_{ij} + \frac{1}{4}\sum_{ij,kl}\langle ij|U\,|kl\rangle\,\varrho_{lkji}\,.$$

If the N-particle state is a Slater determinant, $|\Psi\rangle \equiv |F\rangle$, then we have

$$\varrho_{lkji}^{(F)} = \varrho_{lj}^{(F)}\varrho_{ki}^{(F)} - \varrho_{li}^{(F)}\varrho_{kj}^{(F)}\,,$$

so that

$$\langle F|H\,|F\rangle = \sum_{ij}\langle i|T\,|j\rangle\,\varrho_{ij}^{(F)} + \frac{1}{2}\sum_{ij,kl}\langle ij|U\,|kl\rangle\,\varrho_{lj}^{(F)}\varrho_{ki}^{(F)}\,. \tag{5.95}$$

The Hartree-Fock method can equivalently be formulated in this representation by requiring that the Slater determinant $|F\rangle$ be chosen such that the expectation value of H is minimal. For that purpose choose a general variation to be

$$|F\rangle \longmapsto \left|F'\right\rangle = (1\!\!1 + \mathrm{i}\varepsilon\mathcal{O})\,|F\rangle$$

where $\mathcal{O}$ is an arbitrary self-adjoint one-particle operator and ε a positive infinitesimal. Then

$$\langle F'|H\left|F'\right\rangle = \langle F|H\,|F\rangle - \mathrm{i}\varepsilon\langle F|[\mathcal{O},H]\,|F\rangle$$

and the variational condition leads to the requirement

$$\langle F|a^\dagger_\alpha a_\beta H - H a^\dagger_\alpha a_\beta\,|F\rangle = 0\,. \tag{5.96}$$

The commutator which appears in this formula is calculated to be

$$[a^\dagger_\alpha a_\beta, H] = \sum_j \langle \beta|T\,|j\rangle\, a^\dagger_\alpha a_j - \sum_i \langle i|T\,|\alpha\rangle\, a^\dagger_i a_\beta$$
$$+\frac{1}{2}\left(\sum_{jkl}\langle \beta j|U\,|kl\rangle\, a^\dagger_\alpha a^\dagger_j a_l a_k - \sum_{ijl}\langle ij|U\,|\alpha l\rangle\, a^\dagger_i a^\dagger_j a_l a_\beta\right)\,.$$

The expectation value of this somewhat lengthy expression must be set equal to zero in the minimum of the energy. The following definition renders it somewhat more transparent and easier to interpret. Let

$$\langle m|T+\Gamma\,|n\rangle \equiv \langle m|T\,|n\rangle + \sum_{s,t\le N}\langle ms|U\,|nt\rangle\, \varrho^{(F)}_{ts} \tag{5.97}$$

be the Hartree-Fock Hamiltonian, the sum over s and over t covering only the occupied states. The condition (5.96) then says that

$$\sum_{n\le N}\langle \beta|T+\Gamma\,|n\rangle\, \varrho^{(F)}_{n\alpha} - \sum_{m\le N}\varrho^{(F)}_{\beta m}\langle m|T+\Gamma\,|\alpha\rangle = 0$$

must vanish. Equivalently, the condition to be imposed is

$$[T+\Gamma, \varrho^{(F)}] = 0 \quad\text{with}\quad (\varrho^{(F)})^2 = \varrho^{(F)}\,,\quad \operatorname{tr}\varrho^{(F)} = N\,. \tag{5.98}$$

We will return to these equations in Sect. 5.4.6.

5.4.5 Particle and Hole States, Normal Product and Wick's Theorem

The ground state of a system of N fermions which are not free but interact via the potential U with one another, can certainly not be described by a pure product wave function, i.e. a Slater determinant. Therefore, the energy formula as estimated by means of the Hartree-Fock equations, cannot be the final answer. Furthermore, the excited states of the system will not be the simple one particle-one hole excitations that follow from the Hartree-Fock solution. Rather, they will be linear combinations of such states, or may even contain excitations of more than one particle.

Many of the procedures that are used to treat many-body problems start from the idea that the Hartree-Fock ground state is to be interpreted as the "vacuum" on which to build a perturbative treatment of the true quantum states of the system. With this picture in mind it seems reasonable to assign to the perturbative ground state a special symbol of its own, viz.

$$|\,\Omega\,\rangle := a^\dagger_1 a^\dagger_2 \cdots a^\dagger_N\,|0\rangle\;,$$

and, more importantly, to define all creation and annihilation operators with reference to that state. As for $i > N$ a particle can be created, that is, the particle can be put into the state i, the definition and the action of the operators $a_i^\dagger$ and a_i remain unchanged. This is different for $i \leq N$, where one can only remove a particle from a formerly occupied state, or, equivalently, one can create a *hole* in the set of occupied states. If we wish to have a unified description of the two cases then it is mandatory to define new operators η_i all of which annihilate the state $|\Omega\rangle$, $\eta_i|\Omega\rangle = 0$. This goal is reached as follows:

Definition 5.1 Particle and Hole States

1. If $i > N$ let
$$\eta_i^\dagger := a_i^\dagger \,, \qquad \eta_i := a_i \,. \tag{5.99}$$
2. If $i \leq N$ let
$$\eta_i^\dagger := a_i \,, \qquad \eta_i := a_i^\dagger \,. \tag{5.100}$$

For all operators, $i > N$ and $i \leq N$, one then has
$$\eta_i \,|\, \Omega \,\rangle = 0 \,, \qquad \{\eta_i, \eta_j^\dagger\} = \delta_{ij} \,.$$

Remark

Note that this definition anticipates the extension of many-body theory to genuine quantum field theory. In the present context the action of $\eta_i^\dagger$ means no more than placing a particle in the state i, whenever $i > N$, or taking out a particle whenever $i \leq N$. All one-particle operators are of the form $\sum \mathcal{O}_{ik}\eta_i^\dagger \eta_k$ and, therefore, can do no more than to move a particle from one state to another. An analogous statement applies to two-particle operators which act in such a way that the total number of particles is conserved.

Both many-body quantum mechanics and quantum field theory make use of analytic, diagrammatic methods which, in essence, work in the spirit of perturbation theory as developed in Sect. 5.2. These methods are diagrammatic because to every term in the perturbative expansion one associates a diagram in which the interaction is represented by points (so-called *vertices*), while creation and annihilation of particles are represented by lines which emanate from vertices, or end in them. The rules of perturbation theory define the translation from diagrams to analytic expressions. The establishment of the rules and the analytic evaluation of a specific contribution in a given order of perturbation theory require some combinatorics for which there is an important theorem.

A set of creation and annihilation operators will be denoted, without distinction, by $A_i, A_j, \ldots$. For every product of a finite number of such operators one defines the normal product

Definition 5.2 Normal Product

A normal product of a given set of operators $A_i, A_j, \ldots, A_m$

$$:A_i A_j \cdots A_m: \,:= (-)^\pi A_x A_y \cdots A_z\,, \tag{5.101}$$

is obtained by dividing the set into two groups, one of which contains only creation operators while the other contains only annihilation operators. The second group as a whole is placed to the right of the first group, hence acting first. Furthermore, the normal product carries the sign of the permutation which takes the operators from their initial distribution to the normal ordering.

Normal products have the following properties:

$$\langle \Omega | :A_1 A_2 \cdots A_n: | \Omega \rangle = 0\,, \tag{5.102}$$

$$:A_1 A_2 \cdots A_n: = -:A_2 A_1 \cdots A_n:\,, \tag{5.103}$$

$$:(A_1 + A_2) A_3 \cdots A_n: = :A_1 A_3 \cdots A_n: + :A_2 A_3 \cdots A_n:\,. \tag{5.104}$$

Definition 5.3 Contraction

The contraction of two creation or annihilation operators is the difference of the original product and their normal product

$$\overbrace{A_1 A}_2 := A_1 A_2 - :A_1 A_2:\,. \tag{5.105}$$

If, in addition, the operators depend on time then one defines instead

$$\overbrace{A_1 A}_2 := T(A_1 A_2) - :A_1 A_2:\,, \tag{5.106}$$

where T denotes time ordering, times increasing from right to left.

The normal product is a real number and one has

$$\overbrace{A_1 A}_2 = \langle \Omega | A_1 A_2 | \Omega \rangle\,.$$

More generally, one talks about a contracted normal product if in a normal product one or more pairs of operators are already contracted. The following example where the operators A_1 and A_3, as well as A_4 and A_m, are contracted, clarifies the matter and shows how to manipulate expressions of this kind:

$$:\overbrace{A_1 A_2 A}_3\, \overbrace{A_4 \cdots A}_m A_n: = (-)^\pi\, \overbrace{A_1 A}_3\, \overbrace{A_4 A}_m :A_2 A_5 \cdots A_n:\,.$$

Here $(-)^\pi$ is the phase which is obtained by extracting the pairs of contracted operators from the product of the example.

The following theorem describes the combinatorics for the case of arbitrary operator products.

Theorem 5.1 Wick's Theorem

A given product of creation and annihilation operators is equal to the sum of all its contracted normal products, that is to say,

$$
\begin{aligned}
A_1 A_2 \cdots A_n = &:A_1 A_2 \cdots A_n: + :\overline{A_1 A}_2 A_3 \cdots A_n: + \dots \\
&+ :A_1 \overline{A_2 \cdots A}_m A_n: + :\overline{A_1 A}_2 \overline{A_3 \cdots A}_n: + \dots \\
&+ :\overline{A_1 \underline{A_2 A}}_3 \underline{\cdots A}_m A_n: + \dots .
\end{aligned}
\tag{5.107}
$$

The proof of this theorem is by induction and goes as follows. For $n = 2$ the formula (5.107) holds by the very definition (5.105) of the contraction of two operators,

$$
A_1 A_2 = :A_1 A_2: + \overline{A_1 A}_2 .
$$

Assume now the theorem to hold for some $n \geq 2$ and assume the product $A_1 \cdots A_n$ to be supplemented by one more operator B. If B is an annihilation operator and if it placed at the far right, then the product of B with any other operator is already in normal form. This means that the contraction of B with any other operator vanishes. Equation (5.107), assumed to hold for n, is multiplied by B from the right. Furthermore, B can be taken inside all normal products. As the contraction of B with any other operator vanishes, the formula (5.107) also holds for $n + 1$.

If B is a creation operator we place it at the far left and note that the arguments of the first case apply here as well.

Finally, if B is a polynomial in creation and annihilation operators one makes use of the properties (5.102)–(5.104) which hold for ordinary products, for normal products, as well as for contractions. This proves the theorem.

5.4.6 Application to the Hartree-Fock Ground State

We turn back to the creation and annihilation operators of the single-particle states of an arbitrary basis. The Definition 5.3 shows that the contraction of $a_i^\dagger$ with a_j is nothing but the one-body density. Therefore, for all one-body operators we have

$$
a_m^\dagger a_n = \varrho_{nm} + :a_m^\dagger a_n: .
$$

Regarding two-body operators such as the ones which appear in the Hamiltonian of the N-particle system, one applies Wick's theorem to them and thereby obtains

$$
\begin{aligned}
a_i^\dagger a_j^\dagger a_l a_k = &\varrho_{lj}\varrho_{ki} - \varrho_{li}\varrho_{kj} + \varrho_{lj}:a_i^\dagger a_k: + \varrho_{ki}:a_j^\dagger a_l: \\
&- \varrho_{li}:a_j^\dagger a_k: - \varrho_{kj}:a_i^\dagger a_l: + :a_i^\dagger a_j^\dagger a_l a_k: .
\end{aligned}
\tag{5.108}
$$

On the basis of these results the general Hamiltonian

$$
H = \sum_{ij} \langle i|T\,|j\rangle \, a_i^\dagger a_j + \frac{1}{4} \sum_{ijkl} \langle ij|U\,|kl\rangle \, a_i^\dagger a_j^\dagger a_l a_k
$$

is written in terms of the one-body density and of normal products, viz.

$$
\begin{aligned}
H = &\sum_{ij} \langle i|T\,|j\rangle \, (\varrho_{ij} + :a_i^\dagger a_j:) + \frac{1}{2} \sum_{ijkl} \langle ij|U\,|kl\rangle \, \varrho_{ki}\varrho_{lj} \\
&+ \sum_{ijkl} \langle ij|U\,|kl\rangle \, \varrho_{lj}:a_i^\dagger a_k: + \frac{1}{4} \sum_{ijkl} \langle ij|U\,|kl\rangle \, :a_i^\dagger a_j^\dagger a_l a_k: .
\end{aligned}
$$

In deriving this formula we made use of the antisymmetry of the two-body matrix element: Indeed, the first two terms on the right-hand side of (5.108) give the same contribution. Likewise, the four terms that follow all yield the same contribution.

Introducing the densities of the Hartree-Fock state, making use of the expression (5.95) and of the definition (5.97), one obtains

$$H = E_0 + \sum_{ij} \langle i|T + \Gamma\,|j\rangle :a_i^\dagger a_j: + \frac{1}{4}\sum_{ijkl} \langle ij|U\,|kl\rangle :a_i^\dagger a_j^\dagger a_l a_k: . \quad (5.109)$$

Note that this expression contains only normal products. It simplifies even further if the basis is chosen to be the Hartree-Fock basis of self-consistent single-particle states for which

$$\langle i|T + \Gamma\,|j\rangle = \varepsilon_i \delta_{ij}$$

holds, and if one takes the Hartree-Fock ground state to be the new vacuum state $|\Omega\rangle$. For *particle* states we have

$$:\eta_i^\dagger \eta_i: = a_i^\dagger a_i\,, \qquad i > N\,,$$

but for *hole* states we have

$$:\eta_j^\dagger \eta_j: = -:a_j^\dagger a_j:\,, \qquad j \le N\,.$$

In view of these relations it is useful to define

$$\widetilde{\varepsilon}_i := \varepsilon_i \quad \text{for} \quad i > N\,, \qquad \widetilde{\varepsilon}_i := -\varepsilon_i \quad \text{for} \quad i \le N\,. \quad (5.110)$$

Thus, in the basis of Hartree-Fock states the Hamiltonian reads

$$\boxed{H = E_0 + \sum_i \widetilde{\varepsilon}_i \eta_i^\dagger \eta_i + \frac{1}{4}\sum_{ijkl} \langle ij|U\,|kl\rangle :a_i^\dagger a_j^\dagger a_l a_k:}\,. \quad (5.111)$$

This result is the starting point for further analysis of the N-fermion system. Its physical interpretation is clear: The first term E_0 on the right-hand side is the energy of the ground state in Hartree-Fock approximation. The second term contains the single-particle energies which are obtained by solving the self-consistent Hartree-Fock equations (5.89) or (5.92). It describes the single-particle excitations of the system (which, however, are not eigenstates of the Hamiltonian). This part of the Hamiltonian could be termed the shell model approximation. The third term, finally, is what is called the *residual interaction*.

The residual interaction is dealt with by means of diagrammatic methods. In a first step one introduces the graphical representation of particle and hole states that is given in Fig. 5.8 in tabular form. All full lines describe particles or holes linked to a vertex which in turn symbolizes the two-particle interaction. A creation operator is drawn as a half-line going upwards from the vertex, an annihilation operator appears as a half-line going downwards from the vertex. For a *particle* the arrow points *towards* the vertex if it is annihilated, it points *away* from the vertex if it is created. For *hole* states analogous rules apply with the direction of the arrow reversed. Note that the direction of the arrows matters for the balance of the conserved quantities in the matrix elements.

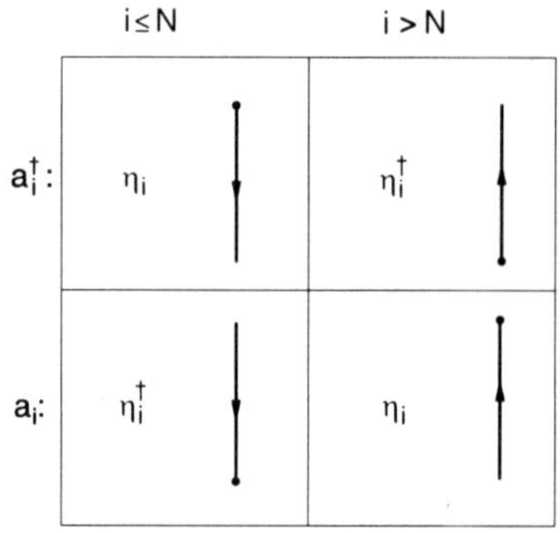

Fig. 5.8 Table summarizing the translation of creation and annihilation operators $a_i^\dagger$ and a_i, respectively, to analogous operators for holes when $i \leq N$, and for particles when $i > N$. The *arrows* define the rules of how to apply conservation laws, depending on whether the particle or hole state moves into the vertex, or away from it

The generic cases may be classified as follows:

1. $i, j, k, l > N$: All four indices pertain to unoccupied states. The residual interaction equals
$$+\frac{1}{4}\sum_{ijkl}\langle ij|U\,|kl\rangle\,\eta_i^\dagger\eta_j^\dagger\eta_l\eta_k$$
and is represented in Fig. 5.9a. Note that I have expanded the vertex somewhat such as to exhibit more clearly the direct and the exchange interactions which are contained in the matrix element. For example, if there is a contribution of second order whose analytic form is the product of two matrix elements between the ground state and a one particle-one hole excitation $|\Phi\rangle$
$$\langle\Omega|H\,|\,\Phi\,\rangle\,\langle\Phi|H\,|\,\Omega\,\rangle\ ,$$
its diagrammatic representation is the one of Fig. 5.9b.
2. $i, j, k, l \leq N$: All four states are occupied states. The residual interaction then reads
$$+\frac{1}{4}\sum_{ijkl}\langle ij|U\,|kl\rangle\,\eta_l^\dagger\eta_k^\dagger\eta_i\eta_j$$
and is visualized by the vertex of Fig. 5.10.
3. $i, k > N, j, l \leq N$: Two states are unoccupied, two are occupied, and the residual interaction is
$$-\frac{1}{4}\sum_{ijkl}\langle ij|U\,|kl\rangle\,\eta_i^\dagger\eta_l^\dagger\eta_j\eta_k\ .$$
This case is represented by the vertex in Fig. 5.11a. Except for the sign the particle-hole interaction equals the particle-particle or hole-hole interaction, respectively.

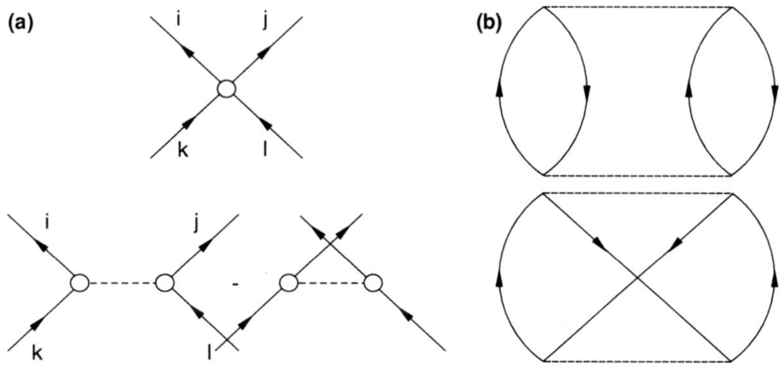

Fig. 5.9 **a** If in the residual interaction of (5.111) all four states are unoccupied states, then all four operators are *particle* operators, k and l are incoming, i and j are outgoing. The vertex is magnified such as to exhibit the direct contribution and the exchange term. **b** Diagrams describing the virtual transition from the ground state to an excited state and back, the excited state being a particle-hole excitation. Here too there is a direct term and an exchange contribution

Fig. 5.10 Representation analogous to the one of Fig. 5.9: Here all four states are occupied states, all four operators are *hole* operators. The orientation of the arrows follows the rules of Fig. 5.8

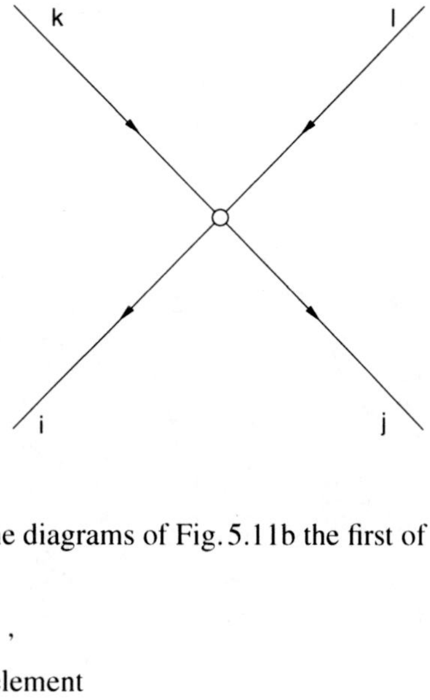

Terms of this class, for instance, allow for the diagrams of Fig. 5.11b the first of which corresponds to the matrix element

$$\langle ij|U\,|ij\rangle\ ,$$

while the second corresponds to the matrix element

$$\langle ij|U\,|ji\rangle\ .$$

Both of them describe dynamical correlations in the ground state of the system. We sketch the lines along which the analysis of the N-fermion system is pursued further: In principle the exact solutions of the Schrödinger equation may be obtained by diagonalization of the residual interaction in the Fock space of all single-particle states as obtained in the first step of the Hartree-Fock procedure. Of course, this is impossible in practice, and one has to resort to refined approximation methods. Such methods are known by the names of *Tamm-Dancoff method, time-dependent*

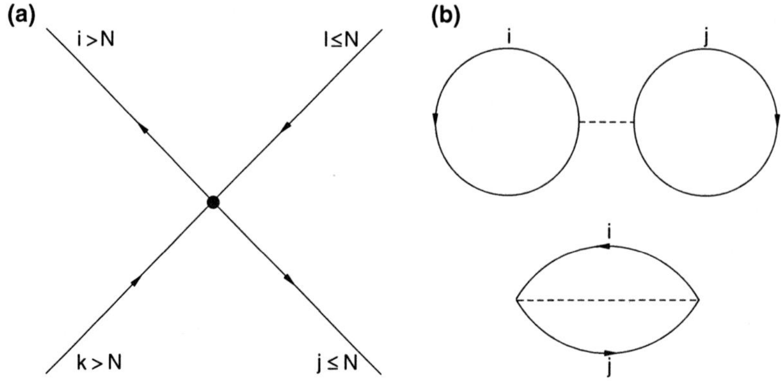

Fig. 5.11 **a** In this example i and k are unoccupied, j and l are occupied states; the orientation of the arrows follows the rules of Fig. 5.8. **b** For $k = i$ and $j = l$ there are two diagrams representing the particle-hole interaction

Hartree-Fock procedure, *Bogoliubov's method for pairing forces* etc. They go beyond perturbation theory in the sense that they sum certain classes of diagrams exactly. These methods which are standard techniques of the nonrelativistic N-body quantum theory, are the subject of more specialized monographs on many-body quantum theory, nuclear theory, or condensed matter theory.

Part II
From Symmetries in Quantum Physics to Electroweak Interactions

Symmetries and Symmetry Groups in Quantum Physics

6

▶ **Introduction** When one talks about discrete or continuous groups which are to describe symmetries of quantum systems, one must first identify the objects on which the elements of these groups are acting. In the case of the Galilei or the Lorentz groups an element $g \in G$ acts on points of space-time. One has the choice of interpreting a given element g as an *active* or as a *passive* transformation. The active interpretation is the right choice if one wants to compare, or to map onto each other, two identical physical processes which are located in different domains of space-time. In turn, the passive interpretation is the right one whenever one and the same physical process is to be studied from the viewpoint of two different frames of reference.

The discussion in Chap. 4 showed that transformations in space and/or time induce unitary (or antiunitary) transformations in Hilbert space. The action of a symmetry transformation $g \in G$, acting on points in space-time, entails the action $\mathbf{U}(g)$ on elements of Hilbert space. The cases quoted above are called *external symmetries*. An important example is provided by the rotation group $G = SO(3)$, interpreted as a group of passive transformations: Any of its elements $g \in G$, parametrized e.g. by three Eulerian angles (ϕ, θ, ψ), rotates the frame of reference about its origin such that any point whose coordinates are $(t, \boldsymbol{x})$ with respect to the original system, has coordinates $(t', \boldsymbol{x}')$ with respect to the new system, related to the former by

$$(t, \boldsymbol{x}) \underset{g}{\rightarrow} (t', \boldsymbol{x}') : \qquad t' = t , \quad \boldsymbol{x}' = \mathbf{R}(\phi, \theta, \psi)\boldsymbol{x} .$$

A wave function $\psi_{jm}(t, \boldsymbol{x})$ which is an eigenfunction of angular momentum and of its 3-component, is transformed by the unitary matrix $\mathbf{D}^{(j)*}(\phi, \theta, \psi)$, cf. Sect. 4.1.3.

There are symmetry actions in Hilbert space which have no relation to space-time and which concern only inner properties of a physical system. A well-known example is the (approximate) symmetry that relates the proton and the neutron, and which maps isometrically dynamical states of one to identical states of the other.

This raises a number of basic questions: Which among the physical states of quantum systems can be related by symmetry actions, which cannot? Is it true that in Hilbert space symmetry operations are always represented by unitary or

F. Scheck, *Quantum Physics*, DOI: 10.1007/978-3-642-34563-0_6,

antiunitary transformations? Why are continuous groups necessarily realized by unitary representations? We will answer the first question by establishing, in the first place, which states may appear as components in a coherent superposition and which cannot. The second question is the subject of an important theorem by E. Wigner.

We then continue, at a more advanced level, the analysis of the rotation group which is the prime example of a Lie group in quantum mechanics. The experience with the rotation group then provides some experience which helps understanding the role of further compact continuous groups describing inner symmetries. The last section treats the Poincaré group and some of its representations which are of central importance for the classification of particles by mass and spin.

6.1 Action of Symmetries and Wigner's Theorem

A simple example for the action of a possible symmetry is provided by the group $SO(2)$ of rotations about the origin in the plane $M = \mathbb{R}^2$. Interpreted as an active transformation $\mathbf{R}(\phi) \in SO(2)$ moves every point $\boldsymbol{x} \neq 0 \in M$ on the circle with radius $r = ||\boldsymbol{x}||$ as sketched in Fig. 6.1. The origin stays invariant (it is a fixed point of the action). A characteristic property of this group of transformations is the invariance of the norm and of the scalar product,

$$||\boldsymbol{x}|| = \sqrt{\boldsymbol{x}^2}\,, \qquad \langle \boldsymbol{x} | \boldsymbol{y} \rangle = \boldsymbol{x} \cdot \boldsymbol{y}\,.$$

If the Lagrangian function or the Hamiltonian function of a physical system is invariant under this group, then $SO(2)$ is a symmetry of the dynamics of this system.

In the Hilbert space of the physical states of a quantum system a representation $\mathbf{U}(g)$ of the group element g acts on unit rays

$$\{\psi^{(i)}\} = \left\{ e^{i\alpha} \psi^{(i)} \middle| \alpha \in \mathbb{R} \right\}$$

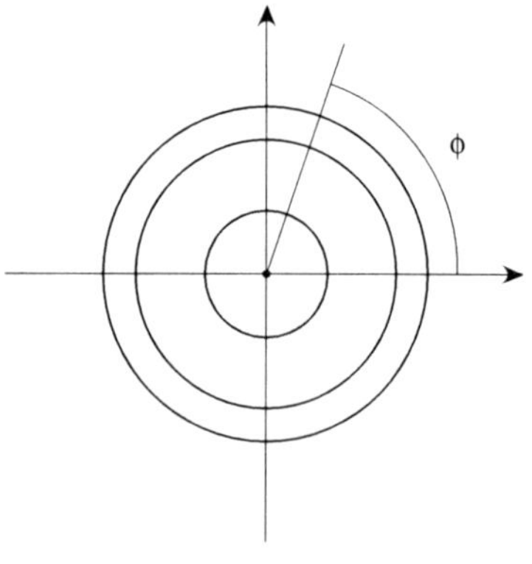

Fig. 6.1 When acting on the points of the plane $\mathbb{R}^2$ the rotation group $SO(2)$ moves them on *circles* whose center is the origin. The origin itself is a fixed point of this action

in such a way that with $\psi_U^{(i)} = \mathbf{U}(g)\psi^{(i)}$ and $\psi_U^{(k)} = \mathbf{U}(g)\psi^{(k)}$ one has the relation

$$\left|\left\langle\psi_U^{(i)}\middle|\psi_U^{(k)}\right\rangle\right| = \left|\left\langle\psi^{(i)}\middle|\psi^{(k)}\right\rangle\right|$$

for all transition amplitudes. Again, if g leaves invariant the Hamiltonian of the system then, with ψ a solution of the Schrödinger equation, $\mathbf{U}(g)\psi$ is also a solution. The group G describes a symmetry of the system.

6.1.1 Coherent Subspaces of Hilbert Space

As worked out in Part One the superposition principle is one of the most important rules of quantum mechanics. With any two solutions ψ_α and ψ_β of the Schrödinger equation every coherent linear combination $\lambda\,\psi_\alpha + \mu\,\psi_\beta$, with $\lambda,\ \mu \in \mathbb{C}$, is a solution. If ψ_α and ψ_β are pure states then this linear combination is a pure state, too, that is, a state which will exhibit interferences without any restriction. In turn, in order to define a pure state and to prepare it by means of measurements, one needs a set of observables all of which commute with each other. Our previous experience shows that a pure state for identically prepared electrons is identified if one of the following sets of four observables is chosen and if each observable has a definite value,

$$\begin{aligned} \{p_1, p_2, p_3, s_3\} \quad &\text{or} \quad \{x_1, x_2, x_3, s_3\} \\ &\text{or} \quad \left\{E = \boldsymbol{p}^2/(2m), \boldsymbol{\ell}^2, \ell_3, s_3\right\} . \end{aligned} \tag{6.1}$$

In the first case, for instance, the electrons are described in position space by a plane wave with momentum $\boldsymbol{p}$ and the spinor with eigenvalue m_s of the projection s_3 of spin onto the 3-axis,

$$\psi_{m_s}(t, \boldsymbol{x}) = \frac{1}{(2\pi\hbar)^{3/2}} \mathrm{e}^{-\mathrm{i}/\hbar(Et - \boldsymbol{p}\cdot\boldsymbol{x})}\ \chi_{m_s}, \quad \left(E = \frac{\boldsymbol{p}^2}{2m}\right) .$$

Obviously, this is a pure state. As there is no other information and as no further information is needed, the wave function $\psi_{m_s}(t, \boldsymbol{x})$ yields a *complete description* of the state of electrons with momentum $\boldsymbol{p}$ and spin projection m_s.

In the second and third examples (6.1) the wave functions in position space are given by, respectively,

$$\delta(\boldsymbol{x} - \boldsymbol{x}^{(0)})\ \chi_{m_s}, \quad \text{and} \quad \sqrt{\frac{2}{\pi\hbar}} k\, \mathrm{e}^{-i/\hbar Et}\ j_\ell(kr)\, Y_{\ell m_\ell}\ \chi_{m_s} .$$

In the second example the wave function describes an electron which is located at the point $\boldsymbol{x}^{(0)}$, while in the third example the electron has sharp values of ℓ and m_ℓ. In both cases the spin projection is fixed.

Nevertheless, this picture is incomplete, the description of quantum states, strictly speaking, must be supplemented by further statements. For example, what we just said about the sets of observables (6.1) applies equally well to the description of positrons. The eigenfunctions of the observables which pertain to different eigenvalues, may at any time be linearly combined. The linear combinations yield again pure states for electrons or positrons which are physically realizable. However, if one tries to mix a state $\psi_{m_s}^{(e)}(t, \boldsymbol{x})$ which describes electrons, with a state $\psi_{m_s}^{(\overline{e})}(t, \boldsymbol{x})$ describing positrons, then this linear combination is not a realistic physical state. The reason for this is that the electric charge is a strictly conserved quantity, with respect to all interactions, in conflict with the fact that the electron and the positron are eigenstates with different electric charges.

Similarly, a mixture of states $\psi_{m_s}^{(\overline{e})}$ of positrons with states $\psi_{m_s}^{(p)}$ of protons is not admissible. Indeed, although these particles carry the same spin and the same electric charge they differ by two further, charge-like quantum numbers, the *baryon number* B and the *lepton number* L. The positron has $(B = 0, L = -1)$,[1] but the proton has $(B = 1, L = 0)$. Both B and L, presumably, are additively conserved quantities.[2]

Likewise, it does not seem meaningful to mix states with *integer* angular momentum and states with *half-integer* angular momentum. The Bose nature of the first, and the Fermi nature of the second do not change under the temporal evolution of the dynamics. If one insisted on trying a superposition of a wave function Φ_S obeying Bose statistics, with a wave function χ_J obeying Fermi statistics, $\Psi_+ = \Phi_S + \chi_J$, with $2S = 2n$ and $2J = 2m + 1$, then this linear combination, after a complete rotation $\mathbf{D}(0, 2\pi, 0)$ of the frame of reference, would go over into $\Psi_- = \Phi_S - \chi_J$. To be indistinguishable, the total wave functions Ψ_+ and Ψ_- would have to belong to the same unit ray. This, however, is possible only when either the first or the second term vanishes identically.

For the sake of fixing the spin-statistics relationship of a wave function we introduce a *grading* $\Pi_S := (-)^{\partial}$ with $\partial(\Phi) = 0$ for every bosonic state, $\partial(\Psi) = 1$ for every fermionic state. Bosonic states belong to the value $\Pi_S(\Phi) = +1$, fermionic states belong to $\Pi_S(\Psi) = -1$. Electric charge Q, baryon number B, lepton number L, and, possibly, the three lepton numbers L_f, $f = e, \mu, \tau$, are *additively* conserved. Additive conservation of a quantity such as electric charge implies that in any reaction and in every decay process,

$$A + B \longrightarrow C_1 + C_2 + \ldots + C_m\,, \quad \text{or} \quad A \longrightarrow B_1 + B_2 + \ldots + B_n$$

the total charge of the initial state must equal the sum of charges in the final state, respectively,

$$Q(A) + Q(B) = \sum_{i=1}^{m} Q(C_i)\,, \quad \text{or} \quad Q(A) = \sum_{i=1}^{n} Q(B_i)\,.$$

[1] The story of lepton number is more complicated: The lepton number referred to in the text is the sum of individual lepton numbers for the three families of leptons which are known, $L = L_e + L_\mu + L_\tau$. For example, the electron has $L_e(e) = 1$, $L_\mu(e) = L_\tau(e) = 0$, while the muon μ^- has $L_e(\mu^-) = 0$, $L_\mu(\mu^-) = 1$, $L_\tau(\mu^-) = 0$. It seems that all three lepton numbers are conserved separately.

[2] This is certainly known to hold in a very good approximation as witnessed by the stability of the hydrogen atom, demonstrated, e.g., by the age of terrestrial oceans! Note, however, that there are theories in which B and L no longer are strictly conserved, while their difference $B - L$ still is.

The Fermi-Bose grading ∂ is also additively conserved, though modulo 2. This is equivalent to the statement that the spin-statistics characteristics Π_S is conserved multiplicatively. The operators Q, B, L, Π_S differ from operators such as those of the sets (6.1) by the fact that it is not admissible to mix eigenstates with different eigenvalues of Q or of B, etc. For example, a state containing a pion and a nucleon which has eigenvalue 0 of the charge operator Q, eigenvalue $+1$ of B, and eigenvalue 0 of L, may well be a linear combination of $\pi^- p$ and $\pi^0 n$, viz.

$$\Psi = c_1 \left|\pi^- p\right\rangle + c_2 \left|\pi^0 n\right\rangle, \quad c_1, c_2 \in \mathbb{C},$$

where the pion and the nucleon are in fixed dynamical orbital states which need not be specified any further. As usual, in a measurement of the individual charges, $|c_1|^2$ is the probability to find a π^- and a proton, while $|c_2|^2$ is the probability to find a neutral pion and a neutron. In contrast, it would not be physically meaningful to admix to this state components such as $c_3|\pi^+ n\rangle$, or $c_4|\pi^-\pi^0\rangle$. In the first case the total charge of the admixture would be $+1$, in the second case the total baryon number would be 0, both components differing from the original state.

These statements which derive from experiment, tell us that there are no observables which connect states with different eigenvalues of Q or B or another additively conserved quantity of this kind. In a two-body system such as a pion plus a proton, there may well be self-adjoint operators $\mathcal{O}_{12}$ which have the effect of raising the charge of the first particle by one unit while simultaneously lowering the charge of the second by one unit, such that the total charge remains unchanged. For instance, the scattering operator describing the charge exchanging reaction

$$\pi^- + p \longrightarrow \pi^0 + n,$$

does have this property. However, there can be no such operator which would change the sum of the charges.

Another example is provided by the process

$$e^+ + e^- \longrightarrow p + \overline{p},$$

in which an electron-positron pair is annihilated, and a proton-antiproton pair is created. The total lepton number as well as the total baryon number are equal to zero before and after the reaction, $L(e^-) + L(e^+) = 1 - 1$, $L(p) + L(\overline{p}) = 0 + 0$, $B(e^-) + B(e^+) = 0 + 0$, $B(p) + B(\overline{p}) = 1 - 1$, even though the quantum numbers of the individual particles do change.

In the light of these examples one realizes that the observables Q, B, L, Π_S, must differ in a qualitative sense from the observables contained in (6.1). Apparently, the Hilbert space decomposes into subspaces which are labelled by the eigenvalues of these operators, the superposition principle applies only to each of the subspaces separately but does not hold for states which are elements of *different* subspaces. This leads to the following definition:

Definition 6.1 Superselection Rule

A unit ray is said to be *physically admissible* if the projection operator P_ψ which projects onto this ray is an observable. If a self-adjoint operator which represents an absolutely conserved quantity, divides unit rays into two classes, the physically realizable and the physically inadmissible ones, one says that it defines a *superselection rule*.

Remarks

1. These restrictions on the validity of the superposition principle seem intuitively obvious and, surely, the reader would have resisted any temptation to mix states with different total electric charges, with well-defined phase relations. Yet, the problem of superselection rules was clarified in a systematic way only relatively late.[3] Expressed differently, one could make the following observation: If two unit rays which are described by the projection operators P_Ψ and P_Φ, are separated by a superselection rule, then for any observable one has $\langle\Psi|\mathcal{O}|\Phi\rangle = 0$. If one insisted on constructing the state $\lambda\Psi + \mu\Phi$ then the expectation value of $\mathcal{O}$ in this state would be $|\lambda|^2\langle\mathcal{O}\rangle_\Psi + |\mu|^2\langle\mathcal{O}\rangle_\Phi$. Therefore, no distinction would be possible from the *mixed* ensemble described by the statistical operator $W = w_\Psi P_\Psi + w_\Phi P_\Phi$, with $w_\Psi = |\lambda|^2$ and $w_\Phi = |\mu|^2$.
2. From a physical point of view it seems reasonable to assume that the observables Q, B, etc. which define the superselection rules of quantum theory, commute with one another. If this is indeed so then these operators may be chosen simultaneously diagonal. The Hilbert space thereby splits into orthogonal subspaces each of which is characterized by a set of definite eigenvalues of these operators. Any such subspace within which the superposition principle holds without restriction, is called a *coherent subspace*.
3. In the case assumed in the previous remark observables map coherent subspaces onto themselves. Systems which are different from each other, i.e. which correspond to different eigenvalues of the set of operators that defines the superselection rules, live in mutually orthogonal spaces and cannot interfere. In turn, self-adjoint operators which map a coherent subspace onto itself and which commute with all observables, are necessarily proportional to the identity on that space.

6.1.2 Wigner's Theorem

Let us assume that quantum mechanics contains only superselection rules which are compatible with one another. In other words, all strictly conserved observables that define superselection rules and, hence, whose eigenstates cannot be linearly combined, commute. The Hilbert space then decomposes into mutually orthogonal, coherent subspaces $\mathcal{H}_c$ which are characterized by the eigenvalues of these observables, that is to say, subspaces which pertain to definite values $c \equiv \{Q, B, L, \partial, \dots\}$ of electric charge, baryon number, lepton number, spin-statistics relation, etc.

[3] J.C. Wick, A.S. Wightman, and E.P. Wigner, Phys. Rev. **88**, 101 (1952).

Let unit rays temporarily be denoted by bold symbols, in an abbreviated notation,

$$\boldsymbol{\Psi}^{(i)} \equiv \left\{\psi^{(i)}\right\} = \left\{ e^{i\alpha}\psi^{(i)} \middle| \alpha \in \mathbb{R} \right\}, \quad \psi^{(i)} \in \mathcal{H}_c . \tag{6.2}$$

As explained in the introduction to this section a symmetry action $g \in G$ is a bijective mapping of unit rays

$$\boldsymbol{\Psi}^{(i)} \underset{g}{\longrightarrow} \boldsymbol{\Psi}_g^{(i)} ,$$

which is such that the modulus of the transition amplitudes is conserved,

$$\left|\left\langle \psi_g^{(i)} \middle| \psi_g^{(k)} \right\rangle\right| = \left|\left\langle \psi^{(i)} \middle| \psi^{(k)} \right\rangle\right| , \tag{6.3}$$
$$\psi^{(i)} \in \boldsymbol{\Psi}^{(i)} , \ \psi^{(k)} \in \boldsymbol{\Psi}^{(k)} , \psi_g^{(i)} \in \boldsymbol{\Psi}_g^{(i)} , \ \psi_g^{(k)} \in \boldsymbol{\Psi}_g^{(k)} .$$

In many cases, the images $\psi_g^{(i)}$ and $\psi_g^{(k)}$ are elements of the same coherent subspace $\mathcal{H}_c$ as their originals. Alternatively there are cases where they belong to the conjugate coherent subspace $\mathcal{H}_{\overline{c}}$ which differs from $\mathcal{H}_c$ by the property that all additively conserved quantum numbers are replaced by their opposites, i.e. $Q \to -Q$, $B \to -B$, etc. In particular, this is the case if the symmetry operation contains either *charge conjugation* **C**, or the product $\boldsymbol{\Theta} := \boldsymbol{\Pi}\mathbf{CT}$ of time reversal **T**, of **C**, and of space reflection $\boldsymbol{\Pi}$. By which kind of mapping of elements of Hilbert space the symmetry action is realized, is still an open question at this point. In general this mapping need neither be linear nor antilinear, that is, neither $(\lambda\psi)_g = \lambda\psi_g$ nor $(\lambda\psi)_g = \lambda^*\psi_g$, $\psi \equiv \psi^{(i)}$ or $\psi^{(k)}$, need hold. However, as only unit rays are physically distinguishable, phases of the states $\psi \in \boldsymbol{\Psi}$ being irrelevant, the remaining freedom can be utilized to establish the following important theorem:

Theorem 6.1 Unitarity–Antiunitarity of Symmetry Actions

A symmetry action $\boldsymbol{\Psi} \underset{g}{\longrightarrow} \boldsymbol{\Psi}_g$, $\boldsymbol{\Psi} \in \mathcal{H}_c$, $\boldsymbol{\Psi}_g \in \mathcal{H}_c$ or $\mathcal{H}_{\overline{c}}$, which has the property (6.3), can always be realized as a mapping

$$\boldsymbol{\Psi}_g = \mathbf{V}(g)\,\boldsymbol{\Psi} \tag{6.4}$$

which is additive and norm conserving, i.e. for which

$$\mathbf{V}(g)(\boldsymbol{\Psi}^{(i)} + \boldsymbol{\Psi}^{(k)}) = \mathbf{V}(g)\boldsymbol{\Psi}^{(i)} + \mathbf{V}(g)\boldsymbol{\Psi}^{(k)} \text{ and } \left\|\mathbf{V}(g)\boldsymbol{\Psi}^{(i)}\right\|^2 = \left\|\boldsymbol{\Psi}^{(i)}\right\|^2$$

holds. This mapping is fixed uniquely, up to a phase factor. It is either *unitary* or *antiunitary*.

Remarks

1. This theorem which was first proved by Wigner, ascertains that the symmetry G acts on elements of a given coherent subspace $\mathcal{H}_c$ by unitary *or* antiunitary transformations $\mathbf{V}(g)$. Thereby $\mathcal{H}_c$ is mapped onto itself, or, as in the case of **C** and of $\boldsymbol{\Theta}$, onto the conjugate coherent subspace $\mathcal{H}_{\overline{c}}$.

2. Which of the two possibilities allowed by the theorem, the unitary or the antiunitary realization, is the right one, depends on the dynamics of the system one is studying. In the case of time reversal **T** the Schrödinger equation stays form invariant only if **V** is antiunitary, cf. Part One, Sect. 4.2.2.
In turn, if G is a Lie group, then the action must be given by *unitary* transformations,
$$g \in G: \quad \mathbf{V}(g) \equiv \mathbf{U}(g) \quad \text{with} \quad \mathbf{U}^{\dagger}(g)\mathbf{U}(g) = \mathbb{1}\,.$$
The reason for this is that the product of g_1 and g_2 is realized by the action $\mathbf{U}(g_1)\mathbf{U}(g_2) = \mathbf{U}(g_1 g_2)$. In particular, every element $g \in G$ can be written as the square of another element g_0 so that by the group property
$$\mathbf{U}(g_0)\mathbf{U}(g_0) = \mathbf{U}(g_0^2) = \mathbf{U}(g)\,.$$
If one had chosen the antiunitary action, then the product $\mathbf{V}(g_0)\mathbf{V}(g_0)$ of the two antiunitary transformations would be *unitary*, cf. Theorem 4.1 in Sect. 4.2.2. This would contradict the initial choice.
3. Only unit rays, not their individual elements, are physically relevant. Therefore, the representation $\mathbf{U}(g)$ of an element $g \in G$, where G is a Lie group, initially, is only determined up to a phase factor. In particular, for the product of two elements one has
$$\mathbf{U}(g_1)\mathbf{U}(g_2) = e^{i\varphi(g_1,g_2)}\mathbf{U}(g_1 g_2)\,,$$
where the phase may depend on g_1 and g_2. A representation of this kind which is fixed up to a phase factor, is called *projective representation*. However, as there is the freedom to adjust the phase factor for every unitary representation $\mathbf{U}(g)$ appropriately, it is always possible, within each connected component of the group G, to convert the projective representation into an ordinary representation. In case the group G as a whole is simply connected, one can choose $e^{i\varphi(g)} = 1$. If it is twofold connected one can only achieve $e^{i\varphi(g)} = \pm 1$, that is to say, the representation may possibly have different signs in the individual connected components. (More generally, if G is n-fold connected then $e^{i\varphi(g)} = e^{i2\pi k/n}$, with $k = 0, 1, \dots, n-1$, can be achieved.)
4. Symmetry actions are important in relation with the dynamics of physical systems, i.e. with regard to their interactions. Consider a system whose dynamics is described by the Hamiltonian H. The time evolution of states $\psi(t, \boldsymbol{x})$ is given by
$$\psi(t, \boldsymbol{x}) = e^{-i/\hbar H t}\psi(0, \boldsymbol{x})$$
cf. Part One, Sect. 3.3.5. If the symmetry action does not reverse the direction of time the operator describing the time evolution must map the transformed states onto each other, too, viz.
$$\psi_g(t, \boldsymbol{x}) = e^{-i/\hbar H t}\psi_g(0, \boldsymbol{x})\,, \quad \psi_g(t, \boldsymbol{x}) = \mathbf{U}(g)\psi(t, \boldsymbol{x})\,. \tag{6.5}$$
The mapping $\mathbf{U}(g)$ must be time independent, linear, and unitary. In particular, we must have
$$\mathbf{U}^{-1}(g)e^{-i/\hbar H t}\mathbf{U}(g) = e^{-i/\hbar H t} \tag{6.6a}$$

and hence

$$[H, \mathbf{U}(g)] = 0 \,. \tag{6.6b}$$

An *anti*unitary realization $\mathbf{V}(g)$ of the symmetry would map states with positive energy onto states with negative energy. Furthermore, a mapping $\mathbf{U}(g)$ which does not commute with H is not a meaningful symmetry action because the Hamiltonian generates translations in time. The mapping $\psi \to \psi_g$ would not be independent of the frame of reference.

If the symmetry does reverse the arrow of time, then instead of (6.5) one has

$$\psi_g(t, \boldsymbol{x}) = \mathrm{e}^{+\mathrm{i}/\hbar H t} \psi_g(0, \boldsymbol{x}) \,, \quad \psi_g(t, \boldsymbol{x}) = \mathbf{V}(g)\psi(t, \boldsymbol{x}) \tag{6.7}$$

and in lieu of (6.6a)

$$\mathbf{V}^{-1}(g)\mathrm{e}^{\mathrm{i}/\hbar H t}\mathbf{V}(g) = \mathrm{e}^{-\mathrm{i}/\hbar H t} \,. \tag{6.8}$$

In this situation $\mathbf{V}$ must be antiunitary, i.e. it must take the form of a product $\mathbf{V} = \mathbf{K}^{(0)}\mathbf{U}$ of complex conjugation and a unitary transformation. With this choice one obtains $[H, \mathbf{V}] = 0$.

5. The reader will find a simple heuristic proof of Wigner's theorem in [Messiah (1964)], going back to Wigner's original proof. A more detailed and complete proof was given by V. Bargmann.[4]
6. In fact, Wigner's theorem follows from the fundamental theorem of projective geometry, suitably generalized to infinite dimension [see, e.g., Haag (1999)].

6.2 The Rotation Group (Part 2)

In this section we take up the rotation group and its specific role in quantum mechanics. First, we show that it is not the group of rotations in three real dimensions,

$$SO(3) = \left\{ \mathbf{R} \text{ real } 3 \times 3\text{-matrices} \mid \mathbf{R}^\dagger \mathbf{R} = \mathbb{1} \,, \det \mathbf{R} = 1 \right\} \tag{6.9}$$

but rather the *unimodular group in two complex dimensions*

$$SU(2) = \left\{ \mathbf{U} \text{ complex } 2 \times 2\text{-matrices} \mid \mathbf{U}^\dagger \mathbf{U} = \mathbb{1} \,, \det \mathbf{U} = 1 \right\} \tag{6.10}$$

which is relevant for the description of spin and angular momentum. We then study more systematically the unitary representations of $SU(2)$, derive from them the D-matrices. The latter are utilized to derive the Clebsch-Gordan series and further quantities related to Clebsch-Gordan coefficients.

[4] V. Bargmann, J. Math. Phys. **5**, 862 (1964).

6.2.1 Relationship between $SU(2)$ and $SO(3)$

The group $SU(2)$ is defined to be the group of unimodular unitary 2×2-matrices with complex entries. As one verifies easily every element $U \in SU(2)$ can be written in the form

$$\mathbf{U} = \begin{pmatrix} u_{11} \equiv u & u_{12} \equiv v \\ u_{21} = -v^* & u_{22} = u^* \end{pmatrix} \quad \text{with} \quad |u|^2 + |v|^2 = 1 \tag{6.11}$$

where u and v are complex numbers subject to the normalization condition as indicated. Indeed, one has

$$\mathbf{U}^\dagger = \begin{pmatrix} u^* & -v \\ v^* & u \end{pmatrix} \quad \text{and} \quad \mathbf{U}^\dagger\mathbf{U} = \begin{pmatrix} |u|^2 + |v|^2 & u^*v - vu^* \\ v^*u - uv^* & |v|^2 + |u|^2 \end{pmatrix} = \mathbb{1}_{2\times 2}\,.$$

Writing the complex entries u and v in terms of their real and imaginary parts, $u = x^1 + \mathrm{i}x^2$ and $v = x^3 + \mathrm{i}x^4$, the normalization condition becomes $\sum_{i=1}^4 x_i^2 = 1$. This shows that these parameters define points on the unit sphere S^3 in $\mathbb{R}^4$. Every element of $SU(2)$ is determined by four real numbers which are the coordinates of points on S^3. It is not difficult to convince oneself that this manifold is simply connected, that is to say, that there is only one class of closed loops which can be contracted to a point. To see this choose two open neighbourhoods $S^3 \setminus \{N\}$ and $S^3 \setminus \{S\}$ on S^3 which are mapped into two charts $\mathbb{R}^3$ by stereographic projection, once from the north pole N, once from the south pole S. The image of any closed loop on S^3 is a closed loop in one of the charts. This image in $\mathbb{R}^3$ can always be contracted to a point. As the mapping from the sphere to the charts is bijective and continuous in both directions, the same statement holds for the original loop. This proves that the group $SU(2)$ is a simply connected manifold.

Any element $\mathbf{U} \in SU(2)$ can also be written as an exponential series in a hermitian and traceless 2×2-matrix h,

$$\mathbf{U} = \exp\{\mathrm{i}h\} \qquad \text{with} \quad h^\dagger = h\,,\ \operatorname{tr} h = 0\,. \tag{6.12}$$

The three Pauli-matrices σ_i are themselves hermitian, have vanishing trace, and are linearly independent. On the other hand, a hermitian and traceless matrix h depends on only three real parameters. Therefore, any such matrix h can be written as a real linear combination of the Pauli matrices

$$h = \sum_{i=1}^{3} \alpha_i \sigma_i\,, \quad \alpha_i \in \mathbb{R}\,. \tag{6.13}$$

This decomposition which the reader is invited to prove as an exercise, is made plausible by the following reasoning: The matrix h is diagonalized by means of a unitary transformation. As it has trace zero its eigenvalues are equal and opposite. They are a and $-a$, so that the unitary matrix $\mathbf{U}$ which is transformed to diagonal form simultaneously, becomes $\mathbf{U}' = \operatorname{diag}(\mathrm{e}^{\mathrm{i}a}, \mathrm{e}^{-\mathrm{i}a})$, its determinant is equal to 1. Both the trace and the determinant are invariant under unitary transformations. Therefore, the original matrix $\mathbf{U}$ has the correct properties.

Interpreting the three real numbers as the components of a vector $\boldsymbol{\alpha} = (\alpha_1, \alpha_2, \alpha_3)$ and taking $\alpha = |\boldsymbol{\alpha}|$, the matrix $\mathbf{U}$ can be written as follows,

$$\mathbf{U} = \exp\left\{ \mathrm{i} \sum_{i=1}^{3} \alpha_i \sigma_i \right\} = \mathbb{1}_{2\times 2} \cos\alpha + \frac{\mathrm{i}}{\alpha} \boldsymbol{\alpha} \cdot \boldsymbol{\sigma} \sin\alpha$$
$$= \frac{1}{\alpha} \begin{pmatrix} \alpha\cos\alpha + \mathrm{i}\alpha_3 \sin\alpha & (\alpha_2 + \mathrm{i}\alpha_1)\sin\alpha \\ -(\alpha_2 - \mathrm{i}\alpha_1)\sin\alpha & \alpha\cos\alpha - \mathrm{i}\alpha_3\sin\alpha \end{pmatrix} . \tag{6.14}$$

This relation is verified in Exercise 6.3 by direct computation. The following simple argument makes it plausible. In the special case $\boldsymbol{\alpha} = (0, 0, \alpha)$ it follows from the property $\sigma_3^{2n} = \mathbb{1}_{2\times 2}$ and $\sigma_3^{2n+1} = \sigma_3$, and by writing the exponential series explicitly,

$$\exp\{\mathrm{i}\alpha\sigma_3\} = \sum_{n=0}^{\infty} \frac{\mathrm{i}^{2n}}{(2n)!} \alpha^{2n}\, \mathbb{1}_{2\times 2} + \mathrm{i} \sum_{m=0}^{\infty} \frac{\mathrm{i}^{2m}}{(2m+1)!} \alpha^{(2m+1)} \sigma_3$$
$$= \cos\alpha\, \mathbb{1}_{2\times 2} + \mathrm{i}\sin\alpha\, \sigma_3 .$$

The general case with $\boldsymbol{\alpha}$ pointing in an arbitrary direction, is reduced to this special case by first performing a rotation in $\mathbb{R}^3$ which takes $\boldsymbol{\alpha}$ to $\boldsymbol{\alpha}' = (0, 0, \alpha)$, inserting, in a second step, the exponential series, and, finally, by reversing the rotation. As the three Pauli matrices $\boldsymbol{\sigma} = (\sigma_1, \sigma_2, \sigma_3)$ transform in the same way[5] and since the scalar product is invariant, one obtains the result (6.14).

While the group $SU(2)$ is singly connected, the rotation group $SO(3)$ is known to be *doubly connected.* Two different geometric proofs of this statement are given, e.g., in Chap. 5 of [Scheck (2010)]. The two groups have the same Lie algebra. The relationship between $SU(2)$ and $SO(3)$ is summarized in the statement that $SU(2)$ is the *universal covering group* of $SO(3)$. What this means is clarified by an explicit construction:

Let $\boldsymbol{x}$ be some vector in $\mathbb{R}^3$, $\mathbf{X}$ a traceless, hermitian 2×2-matrix, which is built from the components of $\boldsymbol{x}$ and the three Pauli matrices $\boldsymbol{\sigma}$,

$$\boldsymbol{x} \longleftrightarrow \mathbf{X} := \boldsymbol{\sigma} \cdot \boldsymbol{x} \equiv x^1\sigma_1 + x^2\sigma_2 + x^3\sigma_3 = \begin{pmatrix} x^3 & x^1 - \mathrm{i}x^2 \\ x^1 + \mathrm{i}x^2 & -x^3 \end{pmatrix} . \tag{6.15}$$

What this correspondence tells us is that the space $\mathbb{R}^3$ and the space of hermitian and traceless 2×2-matrices are isomorphic.

One verifies that the determinant of this matrix, except for the sign, equals the squared norm of $\boldsymbol{x}$, $\det\mathbf{X} = -\boldsymbol{x}^2$. Furthermore, an arbitrary unitary transformation $\mathbf{U} \in SU(2)$, when applied to the hermitian matrix $\mathbf{X}$,

$$\mathbf{X}' = \mathbf{U}\,\mathbf{X}\,\mathbf{U}^\dagger , \tag{6.16}$$

yields another such matrix which by (6.15) is associated to a vector $\boldsymbol{x}'$. The determinant of $\mathbf{X}$ being invariant, $\det\mathbf{X}' = \det\mathbf{X}$, one concludes that $\boldsymbol{x}'$ has the same

[5] We come back to this when discussing spherical tensors in Sect. 6.2.5.

length as $\boldsymbol{x}$. Thus, the vector $\boldsymbol{x}'$ can differ from $\boldsymbol{x}$ only by a rotation, $\boldsymbol{x}' = \mathbf{R}\boldsymbol{x}$ with $\mathbf{R} \in SO(3)$. This provides a correspondence of rotations $\mathbf{R} \in SO(3)$ with elements $\mathbf{U} \in SU(2)$ that we need to explore in more detail.

To every $\mathbf{U} \in SU(2)$ there corresponds a unique $\mathbf{R} \in SO(3)$ such that with $\mathbf{X}' = \mathbf{U}\mathbf{X}\mathbf{U}^\dagger$ one has $\boldsymbol{x}' = \mathbf{R}\boldsymbol{x}$. Conversely, for every $\mathbf{R} \in SO(3)$ there is a $\mathbf{U}(\mathbf{R}) \in SU(2)$ such that with $\mathbf{X}' = \mathbf{U}\mathbf{X}\mathbf{U}^\dagger$ one again has $\boldsymbol{x}' = \mathbf{R}\boldsymbol{x}$; if $\mathbf{U}(\mathbf{R})$ is the pre-image of $\mathbf{R}$, then so is $-\mathbf{U}(\mathbf{R})$.

Remark

In the terminology of group theory the relation between $SU(2)$ and $SO(3)$ is described as follows. The group $H := \{\mathbb{1}, -\mathbb{1}\}$ is an invariant subgroup of the group $G = SU(2)$, that is, for all $g \in G$ one has

$$g\,H\,g^{-1} = H\,,$$

or, in somewhat more detail, to every $h_1 \in H$ and every $g \in G$ there is an element $h_2 \in H$ which fulfills the relation $h_1 g = g h_2$. If G possesses an invariant subgroup H then the cosets $\{gH\}$ form also a group. This group is called *factor group* and is denoted by G/H. That this is indeed so is confirmed by verifying the group axioms:

(i) The composition is defined to be multiplication of cosets. Indeed, one has

$$(g_1 H)(g_2 H) = g_1(H g_2)H = g_1(g_2 H)H = g_1 g_2 H\,.$$

The product of two cosets is again a coset.

(ii) Multiplication of cosets is associative.

(iii) The subgroup H as a whole takes the role of the unit element. Indeed, one calculates

$$H(gH) = (Hg)H = (gH)H = g(HH) = gH\,.$$

(iv) The inverse of gH is $g^{-1}H$.

Using the terminology of this remark the correspondence $\pm\mathbf{U} \to \mathbf{R}$ established above, shows that $SO(3)$ is isomorphic to the factor group which is obtained from the cosets of the invariant subgroup $\{\mathbb{1}, -\mathbb{1}\}$ of $SU(2)$,

$$SO(3) \cong SU(2)\,/\,\{\mathbb{1}, -\mathbb{1}\}\,. \tag{6.17}$$

Using Euler angles, for a specific example, every rotation in $\mathbb{R}^3$ can be written as a product $\mathbf{R}(\phi, \theta, \psi) = \mathbf{R}_{\overline{3}}(\psi)\mathbf{R}_\eta(\theta)\mathbf{R}_3(\phi)$. Therefore, it is sufficient to know the pre-images of a rotation about the 3-axis and about the 2-axis. These are, in the case of the 3-axis,

$$\pm\,\mathrm{e}^{\mathrm{i}(\phi/2)\sigma_3} = \pm\begin{pmatrix} \mathrm{e}^{\mathrm{i}\phi/2} & 0 \\ 0 & \mathrm{e}^{-\mathrm{i}\phi/2} \end{pmatrix},$$

(and likewise for the third rotation by the angle ψ), and

$$\pm\,\mathrm{e}^{\mathrm{i}(\theta/2)\sigma_2} = \pm\begin{pmatrix} \cos\theta/2 & \sin\theta/2 \\ -\sin\theta/2 & \cos\theta/2 \end{pmatrix}$$

in the case of the 2-axis. It is not difficult to verify that the formula (6.16) correctly describes the corresponding rotation. Let us test this for the second, somewhat more complicated case by calculating the product on the right-hand side of (6.16): With $\mathbf{U}(\theta) = \exp\{i(\theta/2)\sigma_2\}$ and introducing the abbreviations $x^+ = x^1 + ix^2$, $x^- = x^1 - ix^2$, one finds

$$\mathbf{U}(\theta)\mathbf{X}\mathbf{U}^\dagger(\theta) = \begin{pmatrix} \cos\theta/2 & \sin\theta/2 \\ -\sin\theta/2 & \cos\theta/2 \end{pmatrix} \begin{pmatrix} x^3 & x^- \\ x^+ & -x^3 \end{pmatrix} \begin{pmatrix} \cos\theta/2 & -\sin\theta/2 \\ \sin\theta/2 & \cos\theta/2 \end{pmatrix}$$

$$= \begin{pmatrix} x^3(\cos^2\frac{\theta}{2} - \sin^2\frac{\theta}{2}) & -2x^3\sin\frac{\theta}{2}\cos\frac{\theta}{2} \\ +(x^+ + x^-)\sin\frac{\theta}{2}\cos\frac{\theta}{2} & -x^+\sin^2\frac{\theta}{2} + x^-\cos^2\frac{\theta}{2} \\ -2x^3\sin\frac{\theta}{2}\cos\frac{\theta}{2} & x^3(\sin^2\frac{\theta}{2} - \cos^2\frac{\theta}{2}) \\ -x^-\sin^2\frac{\theta}{2} + x^+\cos^2\frac{\theta}{2} & -(x^+ + x^-)\sin\frac{\theta}{2}\cos\frac{\theta}{2} \end{pmatrix}$$

$$= \begin{pmatrix} x^3\cos\theta + x^1\sin\theta & -x^3\sin\theta + x^1\cos\theta - ix^2 \\ -x^3\sin\theta + x^1\cos\theta + ix^2 & -(x^3\cos\theta + x^1\sin\theta) \end{pmatrix}.$$

Solving for the cartesian components one recovers the familiar formulae

$$x'^1 = x^1\cos\theta - x^3\sin\theta$$
$$x'^2 = x^2$$
$$x'^3 = x^1\sin\theta + x^3\cos\theta\,.$$

Note, in particular, that the half angles in **U** in the product on the right-hand side of (6.16) turn into integer arguments by means of well-known addition theorems for trigonometric functions.

Combining the three rotations one obtains

$$\mathbf{U}(\mathbf{R}) = \exp\left\{i\left(\frac{\psi}{2}\right)\sigma_3\right\}\exp\left\{i\left(\frac{\theta}{2}\right)\sigma_2\right\}\exp\left\{i\left(\frac{\phi}{2}\right)\sigma_3\right\} = \begin{pmatrix} u & v \\ -v^* & u^* \end{pmatrix}, \tag{6.18}$$

where the complex entries u and v are

$$u = e^{i/2(\psi+\phi)}\cos\frac{\theta}{2}\,, \qquad v = e^{i/2(\psi-\phi)}\sin\frac{\theta}{2}\,, \tag{6.19}$$

thus recovering the general parametrization (6.11).

6.2.2 The Irreducible Unitary Representations of *SU*(2)

The group $SU(2)$ is the relevant Lie group for the description of angular momentum in quantum physics. This section constructs the irreducible unitary representations of this group by means of an explicit procedure. Let

$$\mathbf{U} = \begin{pmatrix} u & v \\ -v^* & u^* \end{pmatrix} \quad \text{with} \quad |u|^2 + |v|^2 = 1$$

be an arbitrary element of $SU(2)$. Let this element act on vectors $(c_1, c_2)^T$, i.e. on objects living in the space $\mathbb{C}^2$,

$$\begin{pmatrix} (c_1)_U \\ (c_2)_U \end{pmatrix} = \mathbf{U}(u, v) \begin{pmatrix} c_1 \\ c_2 \end{pmatrix} = \begin{pmatrix} u c_1 + v c_2 \\ -v^* c_1 + u^* c_2 \end{pmatrix} . \tag{6.20}$$

Furthermore, for every fixed value of j define the following set of homogeneous polynomials of degree j in the components c_1 and c_2,

$$f_m^{(j)} := \frac{(c_1)^{j+m}(c_2)^{j-m}}{\sqrt{(j+m)!(j-m)!}}, \quad j = 0, \frac{1}{2}, 1, \frac{3}{2}, 2, \ldots, \tag{6.21}$$
$$m = -j, -j+1, \ldots, +j .$$

Note that the exponents $j \pm m$ are integers if and only if j is either integer or half-integer, and if m takes one of the listed values. In these cases the definition (6.21) is meaningful. For example, for the first three values of j these polynomials are

$$j = 0: \quad f_0^{(0)} = 1 ,$$
$$j = \frac{1}{2}: \quad f_{1/2}^{(1/2)} = c_1 , \quad f_{-1/2}^{(1/2)} = c_2 ,$$
$$j = 1: \quad f_1^{(1)} = \frac{c_1^2}{\sqrt{2}}, \quad f_0^{(1)} = c_1 c_2 , \quad f_{-1}^{(1)} = \frac{c_2^2}{\sqrt{2}} .$$

The transformation $\mathbf{U} \in SU(2)$ maps the polynomials (6.21) with a fixed j onto polynomials of the same class. Therefore, these polynomials may serve as a basis for a $(2j+1)$-dimensional representation. The problem then is to find the $(2j+1) \times (2j+1)$-matrix $\mathbf{D}^{(j)}(u, v)$ which realizes this transformation.

Inserting the action (6.20) one finds, step by step,

$$\begin{aligned}
(f_m^{(j)})_U &= \frac{1}{\sqrt{(j+m)!(j-m)!}} (u c_1 + v c_2)^{j+m} (-v^* c_1 + u^* c_2)^{j-m} \\
&= \sum_{\mu,\nu=0} \frac{1}{\sqrt{(j+m)!(j-m)!}} \frac{(j+m)!}{\mu!(j+m-\mu)!} \frac{(j-m)!}{\nu!(j-m-\nu)!} . \\
&\quad \times (u c_1)^{j+m-\mu} (v c_2)^{\mu} (-v^* c_1)^{j-m-\nu} (u^* c_2)^{\nu} \\
&= \sum_{\mu,\nu=0} \frac{\sqrt{(j+m)!(j-m)!}}{(j+m-\mu)!\mu!(j-m-\nu)!\nu!} u^{j+m-\mu} u^{*\,\nu} v^{\mu} (-v^*)^{j-m-\nu} \\
&\quad \times c_1^{2j-\mu-\nu} c_2^{\mu+\nu} .
\end{aligned}$$

In the second step the binomial expansion in round parentheses was written out explicitly. The sums over μ and ν start at zero; they end whenever the argument of one of the factorials in the denominator becomes negative.[6] It remains to express

[6] Replace, for instance, $(j+m-\mu)! = \Gamma(j+m-\mu+1)$ and note that the Gamma function $\Gamma(z)$ has first-order poles in $z = 0, z = -1, z = -2, \ldots$. Hence, its reciprocal has zeroes in these points.

the right-hand side of the last equation in terms of the functions $f_m^{(j)}(c_1, c_2)$. This is achieved by setting $j - \mu - \nu =: m'$ because then the exponent of c_1 has the required form $2j - \mu - \nu = j + m'$, while the exponent of c_2 goes over into $\mu + \nu = j - m'$. As expected one finds the relation

$$(f_m^{(j)})_U = \sum_{m'} D_{mm'}^{(j)}(u, v)\, f_{m'}^{(j)} ,$$

where the matrix we wish to determine is obtained from the above calculation, viz.

$$D_{mm'}^{(j)}(u, v) = \sum_{\mu} \frac{\sqrt{(j+m)!(j-m)!(j+m')!(j-m')!}}{(j+m-\mu)!\mu!(j-m'-\mu)!(m'-m+\mu)!} \times u^{j+m-\mu}(u^*)^{j-m'-\mu} v^{\mu}(-v^*)^{m'-m+\mu} . \tag{6.22}$$

Before we analyze the formula (6.22) any further we collect the following statements:

(i) It is obvious that the polynomials $f_m^{(j)}$ are linearly independent.

(ii) Calculating $|f_m^{(j)}|^2$ and summing over m, we find

$$\sum_m \left| f_m^{(j)} \right|^2 = \sum_m \frac{|c_1^{j+m} c_2^{j-m}|^2}{(j+m)!(j-m)!} = \frac{1}{(2j)!}\left\{ |c_1|^2 + |c_2|^2 \right\}^{2j} = \sum_m \left| (f_m^{(j)})_U \right|^2 .$$

The squared norm $(f_m^{(j)}, f_m^{(j)}) = \sum_m |f_m^{(j)}|^2$ is invariant under unitary transformations. Note that this result justifies, a posteriori, the normalization factor in the definition (6.21). This implies that the representations (6.22) are *unitary*.

(iii) Finally, one shows that these representations are also *irreducible*, by means of the formula (6.22). This goes as follows:
The result (6.22) contains some special cases. For any m, but $m' = j$ fixed, the index μ must be zero so that

$$D_{m,m'=j}^{(j)}(u, v) = \sqrt{\frac{(2j)!}{(j+m)!(j-m)!}}\, u^{j+m}(-v^*)^{j-m} \tag{6.23}$$

On the other hand, taking $u = e^{i\alpha/2}$, $v = 0$, one has, for all values of m and m',

$$D_{mm'}^{(j)}(e^{i\alpha/2}, 0) = \delta_{mm'} e^{im\alpha} .$$

If a $(2j+1) \times (2j+1)$-matrix $\mathbf{A}$ commutes with $D_{mm'}^{(j)}(e^{i\alpha/2}, 0)$ it must be diagonal $\mathbf{A}_{mm'} = a_m \delta_{mm'}$. If $\mathbf{A}$ commutes with all transformations $\mathbf{D}^{(j)}(u, v)$ one has, in particular,

$$a_m\, D_{mj}^{(j)}(u, v) = D_{mj}^{(j)}\, a_j \quad \text{for all } m .$$

Equation (6.23) tells us that $D_{mj}^{(j)}$ is different from zero for all m and arbitrary arguments u, v. Thus, one concludes $a_m = a_j$ for all m. This means, in other

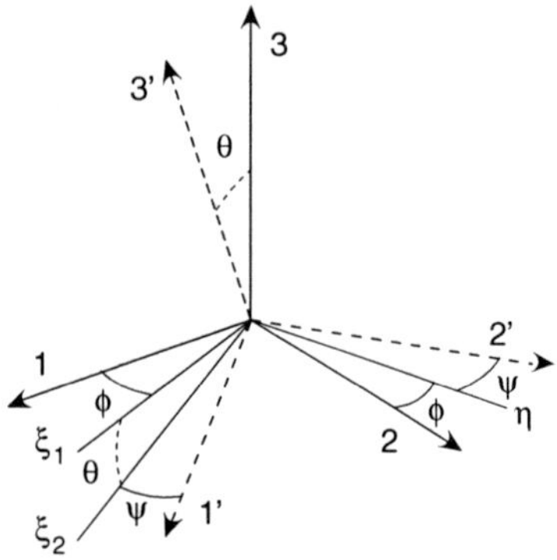

Fig. 6.2 Eulerian angles in a definition where the second rotation is about the intermediate position of the 2-axis

terms, that a matrix which commutes with $\mathbf{D}^{(j)}(u,v)$ for all u, v is a multiple of the unit matrix. This also means that the representations $\mathbf{D}^{(j)}(u,v)$ are irreducible (cf. Part One, Sect. 4.1.2.).

This is an important result: The expression (6.22) yields irreducible, unitary representations of $SU(2)$ as functions of Eulerian angles (ϕ, θ, ψ) in the definition of Fig. 6.2. The algebraic analysis of Sect. 4.1.2 tells us that these are *all* representations there are.

Replacing u and v as indicated in (6.19), and collecting the exponentials in the variables ϕ and ψ, one obtains

$$D^{(j)}_{mm'}(\psi,\theta,\phi) = \mathrm{e}^{\mathrm{i}m\psi}\, d^{(j)}_{mm'}(\theta)\, \mathrm{e}^{\mathrm{i}m'\phi}\,, \quad \text{where} \tag{6.24}$$

$$d^{(j)}_{mm'}(\theta) = \sum_{\mu} (-)^{m'-m+\mu} \frac{\sqrt{(j+m)!(j-m)!(j+m')!(j-m')!}}{(j+m-\mu)!\mu!(j-m'-\mu)!(m'-m+\mu)!} \times \left(\cos\frac{\theta}{2}\right)^{2j+m-m'-2\mu} \left(\sin\frac{\theta}{2}\right)^{m'-m+2\mu}.$$

This formula is written in a slightly different form by choosing $m' - m + 2\mu = m - m' + 2r$ that is, by replacing μ by $m - m' + r$. Noting that $(-)^{2m'-2m} = +1$, independently of whether the indices are integer or half-integer, the set of functions $\mathbf{d}^{(j)}(\theta)$ take their final form

$$d^{(j)}_{mm'}(\theta) = \sum_{r} (-)^{r} \frac{\sqrt{(j+m)!(j-m)!(j+m')!(j-m')!}}{(j+m'-r)!(m-m'+r)!(j-m-r)!r!} \times \left(\cos\frac{\theta}{2}\right)^{2j+m'-m-2r} \left(\sin\frac{\theta}{2}\right)^{m-m'+2r}. \tag{6.25}$$

This is the expression that we quoted in Sect. 4.1.4, without proof.

Remarks

1. **Symmetry relations:** Using the explicit formula (6.25) it is not difficult to prove the symmetry relations quoted in Part One. They are
$$d^{(j)}_{mm'}(\theta) = (-)^{m'-m} d^{(j)}_{m'm}(\theta)\,, \tag{6.26}$$
$$d^{(j)}_{-m',-m}(\theta) = (-)^{m'-m} d^{(j)}_{m'm}(\theta)\,, \tag{6.27}$$
$$d^{(j)}_{m',-m}(\theta) = (-)^{j-m'} d^{(j)}_{m'm}(\pi - \theta)\,. \tag{6.28}$$

2. **Haar measure and orthogonality:** Integration over Eulerian angles $\mathrm{d}\mathbf{R} = \varrho(\phi,\theta,\psi)\,\mathrm{d}\phi\,\mathrm{d}\theta\,\mathrm{d}\psi$ must be defined such that for any fixed rotation $\mathbf{R}_0$ and for $\mathbf{R}' = \mathbf{R}_0\mathbf{R}$ one has $\mathrm{d}\mathbf{R}' = \mathrm{d}(\mathbf{R}_0\mathbf{R}) = \mathrm{d}\mathbf{R}$. This is a consequence of the group property of $SU(2)$. The function $\varrho(\phi,\theta,\psi)$ follows from the Jacobian of the transformation from $\mathbf{R} \equiv (\phi,\theta,\psi)$ to $\mathbf{R}' \equiv (\phi',\theta',\psi')$. Indeed, a simple geometric construction shows that (cf. Exercise 6.4)
$$\frac{\partial(\phi,\theta,\psi)}{\partial(\phi',\theta',\psi')} = \frac{\sin\theta'}{\sin\theta}$$
from which follows the relation between $\mathrm{d}\mathbf{R}$ and $\mathrm{d}\mathbf{R}'$:
$$\mathrm{d}\mathbf{R} = \varrho(\phi,\theta,\psi)\frac{\partial(\phi,\theta,\psi)}{\partial(\phi',\theta',\psi')}\,\mathrm{d}\phi'\,\mathrm{d}\theta'\,\mathrm{d}\psi' = \frac{\varrho(\phi,\theta,\psi)}{\varrho(\phi',\theta',\psi')}\frac{\sin\theta'}{\sin\theta}\,\mathrm{d}\mathbf{R}'\,.$$
As a consequence, up to a multiplicative constant, $\varrho(\phi,\theta,\psi) = \sin\theta$, the choice $\mathrm{d}\mathbf{R} = \sin\theta\;\mathrm{d}\phi\,\mathrm{d}\theta\,\mathrm{d}\psi$ fulfills the requirement. This integral measure is called *Haar measure* for the group $SU(2)$.

   Let $\mathbf{X}$ be an arbitrary $(2j+1)\times(2j+1)$-matrix. Define the matrix
$$\mathbf{M} := \int \mathrm{d}\mathbf{R}\;\mathbf{D}^{(j')}(\mathbf{R})\mathbf{X}(\mathbf{D}^{(j)}(\mathbf{R}))^{-1}$$
where the integral is taken over the whole range of the Euler angles. Then, for all rotations $\mathbf{S}$, one obtains
$$\mathbf{M}\mathbf{D}^{(j)}(\mathbf{S}) = \mathbf{D}^{(j')}(\mathbf{S})\mathbf{M}\,. \tag{6.29}$$
This assertion is proven by using the group property of the D-functions and by making use of the integration measure derived above. We calculate the right-hand side
$$\begin{aligned}
&\mathbf{D}^{(j')}(\mathbf{S})\,\mathbf{M}\\
&= \int \mathrm{d}\mathbf{R}\;\mathbf{D}^{(j')}(\mathbf{S})\mathbf{D}^{(j')}(\mathbf{R})\;\mathbf{X}\;(\mathbf{D}^{(j)}(\mathbf{R}))^{-1}(\mathbf{D}^{(j)}(\mathbf{S}))^{-1}\;\mathbf{D}^{(j)}(\mathbf{S})\\
&= \left(\int \mathrm{d}\mathbf{R}\;\mathbf{D}^{(j')}(\mathbf{S}\mathbf{R})\;\mathbf{X}\;(\mathbf{D}^{(j)}(\mathbf{S}\mathbf{R}))^{-1}\right)\mathbf{D}^{(j)}(\mathbf{S}) = \mathbf{M}\mathbf{D}^{(j)}(\mathbf{S})\,.
\end{aligned}$$

   If $\mathbf{D}^{(j)}$ and $\mathbf{D}^{(j')}$ are not equivalent, then Schur's lemma tells us that $\mathbf{M}$ vanishes for every choice of $\mathbf{X}$. We make use of this freedom by choosing $X_{pq} = \delta_{pk}\delta_{lq}$, thus obtaining

$$M_{mn} = \int \mathrm{d}\mathbf{R}\; D^{(j')}_{mk}(\mathbf{R})(D^{(j)}(\mathbf{R}))^{-1}_{ln} = \int \mathrm{d}\mathbf{R}\; D^{(j)*}_{nl}(\mathbf{R}) D^{(j')}_{mk}(\mathbf{R}) = 0 .$$

Thus, the D-functions pertaining to different values of j are orthogonal.

In turn, if $\mathbf{D}^{(j)}$ and $\mathbf{D}^{(j')}$ are equivalent, that is, if $j' = j$, then Schur's lemma shows that $\mathbf{D}^{(j)} = \mathbf{D}^{(j')}$ and that $\mathbf{M}$ is a multiple of the unit matrix, $\mathbf{M} = c\,\mathbb{1}$. Written in more detail, we have

$$M_{mn} = \int \mathrm{d}\mathbf{R}\; D^{(j)}_{mk}(\mathbf{R}) D^{(j)}(\mathbf{R})^{-1}_{ln} = c\,\delta_{mn} .$$

The normalization constant c is obtained by calculating the trace of $\mathbf{M}$,

$$\begin{aligned}
\operatorname{tr}\mathbf{M} &= \sum_m \int \mathrm{d}\mathbf{R}\; (D^{(j)}(\mathbf{R}))^{-1}_{lm} D^{(j)}_{mk}(\mathbf{R}) = c\,(2j+1) \\
&= \delta_{lk} \int \mathrm{d}\mathbf{R} = \delta_{lk} \int_0^{2\pi} \mathrm{d}\psi \int_0^{\pi} \sin\theta\;\mathrm{d}\theta \int_0^{2\pi} \mathrm{d}\phi = 8\pi^2\,\delta_{lk} .
\end{aligned}$$

We thus obtain an important formula: *the orthogonality relation for D-functions*

$$\begin{aligned}
&\int_0^{2\pi} \mathrm{d}\psi \int_0^{\pi} \sin\theta\;\mathrm{d}\theta \int_0^{2\pi} \mathrm{d}\phi\; D^{(j')*}_{mm'}(\psi,\theta,\phi) D^{(j)}_{\mu\mu'}(\psi,\theta,\phi) \\
&\qquad = \frac{8\pi^2}{2j+1} \delta_{j'j}\delta_{m'\mu'}\delta_{m\mu} .
\end{aligned} \tag{6.30}$$

Note its similarity to the orthogonality relation of spherical harmonics.

Example 6.1

Let $\mathbf{R}$ be the rotation that takes the original 3-direction into the "i-direction". Let $\boldsymbol{b}^{(\lambda)}$ denote the eigenfunctions of the square of the orbital angular momentum, for the eigenvalue $\ell = 1$, and of the component ℓ_i. The relation between these functions and the eigenfunctions $\boldsymbol{a}^{(\lambda)}$ of ℓ_3, as well as the relation between the operators ℓ_i and ℓ_3 are given by, respectively,

$$\boldsymbol{b}^{(\lambda)} = \mathbf{D}^{(1)\dagger}(\mathbf{R})\boldsymbol{a}^{(\lambda)} , \quad \ell_i = \mathbf{D}^{(1)\dagger}(\mathbf{R})\,\ell_3\,\mathbf{D}^{(1)}(\mathbf{R}) .$$

From the general formula (6.25) one obtains

$$\begin{aligned}
\mathbf{d}^{(1)}(\theta) &= \begin{pmatrix} \cos^2\frac{\theta}{2} & \sqrt{2}\sin\frac{\theta}{2}\cos\frac{\theta}{2} & \sin^2\frac{\theta}{2} \\ -\sqrt{2}\sin\frac{\theta}{2}\cos\frac{\theta}{2} & \cos^2\frac{\theta}{2} - \sin^2\frac{\theta}{2} & \sqrt{2}\sin\frac{\theta}{2}\cos\frac{\theta}{2} \\ \sin^2\frac{\theta}{2} & -\sqrt{2}\sin\frac{\theta}{2}\cos\frac{\theta}{2} & \cos^2\frac{\theta}{2} \end{pmatrix} \\
&= \frac{1}{2}\begin{pmatrix} 1+\cos\theta & \sqrt{2}\sin\theta & 1-\cos\theta \\ -\sqrt{2}\sin\theta & 2\cos\theta & \sqrt{2}\sin\theta \\ 1-\cos\theta & -\sqrt{2}\sin\theta & 1+\cos\theta \end{pmatrix} .
\end{aligned}$$

Consider two special cases:
(a) $\ell_i \equiv \ell_1$: To obtain this choice one must choose $\mathbf{D}^{(1)}(0, \pi/2, 0)$. One then finds

$$\mathbf{D}^{(1)}(0, \frac{\pi}{2}, 0) = \frac{1}{2}\begin{pmatrix} 1 & \sqrt{2} & 1 \\ -\sqrt{2} & 0 & \sqrt{2} \\ 1 & -\sqrt{2} & 1 \end{pmatrix}, \qquad \ell_1 = \frac{1}{\sqrt{2}}\begin{pmatrix} 0 & 1 & 0 \\ 1 & 0 & 1 \\ 0 & 1 & 0 \end{pmatrix},$$

$$\boldsymbol{b}^{(1)} = \frac{1}{2}\begin{pmatrix} 1 \\ \sqrt{2} \\ 1 \end{pmatrix}, \quad \boldsymbol{b}^{(0)} = \frac{1}{2}\begin{pmatrix} -\sqrt{2} \\ 0 \\ \sqrt{2} \end{pmatrix}, \quad \boldsymbol{b}^{(-1)} = \frac{1}{2}\begin{pmatrix} 1 \\ -\sqrt{2} \\ 1 \end{pmatrix}.$$

This is the result obtained in Sect. 1.9.1, up to a phase factor (-1) in $\boldsymbol{b}^{(0)}$.
(b) $\ell_i \equiv \ell_2$: Here the choice must be $\mathbf{D}^{(1)}(\pi/2, \pi/2, 0)$. One finds

$$\mathbf{D}^{(1)}(\frac{\pi}{2}, \frac{\pi}{2}, 0) = \frac{1}{2}\begin{pmatrix} \mathrm{i} & \sqrt{2} & -\mathrm{i} \\ -\mathrm{i}\sqrt{2} & 0 & -\mathrm{i}\sqrt{2} \\ \mathrm{i} & -\sqrt{2} & -\mathrm{i} \end{pmatrix}, \quad \ell_2 = \frac{1}{\sqrt{2}}\begin{pmatrix} 0 & -\mathrm{i} & 0 \\ \mathrm{i} & 0 & -\mathrm{i} \\ 0 & \mathrm{i} & 0 \end{pmatrix},$$

$$\boldsymbol{b}^{(1)} = \frac{1}{2}\begin{pmatrix} -\mathrm{i} \\ \sqrt{2} \\ \mathrm{i} \end{pmatrix}, \quad \boldsymbol{b}^{(0)} = \frac{1}{2}\begin{pmatrix} \mathrm{i}\sqrt{2} \\ 0 \\ \mathrm{i}\sqrt{2} \end{pmatrix}, \quad \boldsymbol{b}^{(-1)} = \frac{1}{2}\begin{pmatrix} -\mathrm{i} \\ -\sqrt{2} \\ \mathrm{i} \end{pmatrix}.$$

Remarks

1. The unitary irreducible representations (6.22) are one-valued functions over the parameter manifold S^3 of the group $SU(2)$. These are the representations which we need for describing angular momentum and, more specifically, spin in quantum theory. Using (6.19) they are expressed in terms of Eulerian angles which, as we know, are defined in ordinary space $\mathbb{R}^3$. One then obtains the functions (6.24) and (6.25) which are *one*-valued functions for *integer* $j \equiv \ell$ but *two*-valued functions for *half-integer* values, $j = (2n+1)/2$.
   In particular, the relation between $SU(2)$ and $SO(3)$ noted in (6.17) says that the two elements $\mathbb{1}$ and $-\mathbb{1}$ of $SU(2)$ are mapped onto the identity in $SO(3)$. The element $-\mathbb{1} \in SU(2)$ is parametrized by $(u = -1, v = 0)$. (According to (6.18) and (6.19) this corresponds, e.g., to the choice $(\psi = 0, \theta = 0, \phi = 2\pi)$ of Euler angles.) Inserting this into (6.22) one obtains $D^{(j)}_{mm'}(u = -1, v = 0) = \delta_{mm'}(-1)^{2j}$. The action of the rotation $\mathbf{R}_3(\phi = 2\pi) = \mathbb{1}_{3\times 3}$ on a quantum state with angular momentum j is the identity only if j is an integer. In other terms, representations of $SU(2)$ are representations of $SO(3)$ if and only if j is an integer.
   This observation reveals a remarkable feature of physics. Indeed, nature could have restricted her choice to representations of $SO(3)$ only, that is, to representations with integer values of angular momentum. The existence of spin $1/2$ as well as of higher values of half-integer spin shows that, indeed, it is the group $SU(2)$, not $SO(3)$, which describes angular momentum in quantum theory.

2. The two groups $SU(2)$ and $SO(3)$ whose close relation was worked out above, have isomorphic Lie algebras. Indeed, the Lie algebra of $SU(2)$ is generated by the matrices $(\sigma_k/2)$ which obey the commutation relations

$$\left[\left(\frac{\sigma_i}{2}\right), \left(\frac{\sigma_j}{2}\right)\right] = \mathrm{i}\sum_k \varepsilon_{ijk}\left(\frac{\sigma_k}{2}\right) . \tag{6.31}$$

These commutators follow from the well-known formula

$$\sigma_i\sigma_j = \delta_{ij} + \mathrm{i}\sum_k \varepsilon_{ijk}\sigma_k , \tag{6.32}$$

that we first encountered in Sect. 4.1.7.
The Lie algebra of the rotation group $SO(3)$ is generated by the operators $\mathbf{J}_i$, $i = 1, 2, 3$, cf. Sect. 4.1.1. Their commutators are

$$\left[\mathbf{J}_i , \mathbf{J}_j\right] = \mathrm{i}\sum_k \varepsilon_{ijk}\,\mathbf{J}_k . \tag{6.33}$$

The isomorphism of the Lie algebras (6.31) and (6.33) means that *locally*, in the neighbourhood of the identity, the two groups are isomorphic. However, if we compare them *globally*, e.g. by studying continuous deformations of the group elements that lead back to the starting element, the two groups are seen to be different. The above analysis shows the relationship between them.
3. The D-functions satisfy a system of differential equations of first order in Euler angles from which the solutions $D^{(j)}_{\kappa m}(\phi, \theta, \psi)$ may be obtained, as an alternative. This system is obtained by considering the product of a finite rotation and an infinitesimal one in two different orderings, i.e.

$$\mathrm{e}^{\mathrm{i}\boldsymbol{\varepsilon}\cdot\boldsymbol{J}}\,\mathbf{D}(\mathbf{R}) = \mathbf{D}(\mathbf{R}_1) , \quad \text{and} \quad \mathbf{D}(\mathbf{R})\,\mathrm{e}^{\mathrm{i}\boldsymbol{\eta}\cdot\boldsymbol{J}} = \mathbf{D}(\mathbf{R}_2) .$$

One takes the differentials of the first equation with respect to the components ε_i, of the second with respect to η_i, in the limit $\boldsymbol{\varepsilon} = 0 = \boldsymbol{\eta}$ where the two rotations become the same, $\mathbf{R}_1 = \mathbf{R}_2$. We denote the Eulerian angles by the common symbol $\theta_m \equiv (\phi, \theta, \psi)$. In a first step one expresses the differentials $\partial/\partial\varepsilon_i$ and $\partial/\partial\eta_k$ in terms of differentials with respect to θ_m. The partial derivatives $\partial\theta_m/\partial\varepsilon_i$ and $\partial\theta_m/\partial\eta_k$ which are needed for this calculation, are best obtained from a good drawing.[7] Regarding the factors depending on ϕ and on ψ one obtains the equations

$$\frac{\partial D^{(j)}_{\kappa m}}{\partial\psi} = \mathrm{i}\,\kappa D^{(j)}_{\kappa m} , \quad \frac{\partial D^{(j)}_{\kappa m}}{\partial\phi} = \mathrm{i}\,m D^{(j)}_{\kappa m} , \tag{6.34}$$

thus confirming the decomposition (6.24). The matrix elements $d^{(j)}_{\kappa m}(\theta)$ are found to obey the system of coupled differential equations

$$\sqrt{j(j+1) - \kappa(\kappa \mp 1)}\,d^{(j)}_{\kappa\mp 1, m} = \frac{\kappa\cos\theta - m}{\sin\theta} d^{(j)}_{\kappa m} \pm \frac{\mathrm{d}d^{(j)}_{\kappa m}}{\mathrm{d}\theta} \tag{6.35}$$

[7] This calculation is found, e.g., in Appendix E of [Fano and Racah (1959)].

$$\sqrt{j(j+1)-m(m\pm 1)}d_{\kappa,m\pm 1} = \frac{\kappa - m\cos\theta}{\sin\theta} d^{(j)}_{\kappa m} \pm \frac{\mathrm{d}d^{(j)}_{\kappa m}}{\mathrm{d}\theta}\,. \tag{6.36}$$

This system is solved as follows: For $\kappa = m = j$ and with the upper sign, equation (6.36) gives

$$0 = j\frac{1-\cos\theta}{\sin\theta} d^{(j)}_{jj} + \frac{\mathrm{d}d^{(j)}_{jj}}{\mathrm{d}\theta}\,.$$

Its solution supplemented by the condition $d^{(j)}_{jj}(\theta = 0) = 1$ is easily seen to be

$$d^{(j)}_{jj}(\theta) = \left(\frac{1+\cos\theta}{2}\right)^j = \left(\cos\frac{\theta}{2}\right)^{2j}\,.$$

All other matrix elements $d^{(j)}_{\kappa m}(\theta)$ are obtained by recurrence, using (6.35) with the upper sign, and (6.36) with the lower sign.

The following comment is of interest for the example worked out right after these remarks. The first-order differential Eqs. (6.34), (6.35), and (6.36) can be combined to obtain a differential equation of *second* order for the D-functions. It reads[8]

$$-\left\{\frac{\partial^2}{\partial\theta^2} + \cot\theta\frac{\partial}{\partial\theta} + \frac{1}{\sin^2\theta}\left(\frac{\partial^2}{\partial\psi^2} + \frac{\partial^2}{\partial\phi^2} - 2\cos\theta\frac{\partial^2}{\partial\psi\,\partial\phi}\right)\right\}\mathbf{D}^{(j)}(\phi,\theta,\psi)$$
$$= j(j+1)\,\mathbf{D}^{(j)}(\phi,\theta,\psi)\,. \tag{6.37}$$

4. We already know that the D-functions span a system of orthogonal functions, cf. (6.30). Making use of the self-adjoint equations (6.34) and (6.37) one can show, in addition, that this system is also *complete* (see, e.g., [Fano and Racah (1959)]). Therefore, any square integrable function of Eulerian angles can be expanded in this basis. Furthermore, it is now easy to verify the formula (4.21) that we quoted in Part One.

$$Y_{\ell m}(\theta,\phi) = \sqrt{\frac{2\ell+1}{4\pi}} D^{(\ell)}_{om}(0,\theta,\phi) \tag{6.38}$$

We summarize this important result: The functions

$$\left\{\sqrt{\frac{2j+1}{8\pi^2}}\, D^{(j)}_{\kappa m}(\phi,\theta,\psi)\right\}, \quad j = 0, \frac{1}{2}, 1, \frac{3}{2}, \dots;$$
$$\kappa,\ m = -j, -j+1, \dots, +j \tag{6.39}$$

provide a complete system of orthogonal functions in the space of Eulerian angles. We also note that one makes frequent use of this fact in atomic and nuclear physics. For example, the calculation of angular correlations in decays of unstable states is simplified by using the properties of D-functions.

---

[8] Note, however, that it carries less information than the system of equations (6.34)–(6.36). This is similar to the wave equation of electromagnetic theory which contains less information than the system of Maxwell's equations.

Example 6.2 Symmetric Top

The free symmetric top of classical mechanics can be quantized in a canonical way by postulating Heisenberg commutation relations for the Euler angles (ϕ, θ, ψ) and their canonically conjugate momenta $(p_\phi, p_\theta, p_\psi)$,

$$\left[P_m \equiv p_{\theta_m}, Q^n \equiv \theta_n\right] = \frac{\hbar}{\mathrm{i}}\delta_{mn} .$$

In a position space representation where Eulerian angles are the coordinates, the momenta act by differentiation, viz.

$$p_\phi = -\mathrm{i}\hbar\frac{\partial}{\partial\phi}, \quad p_\theta = -\mathrm{i}\hbar\frac{\partial}{\partial\theta}, \quad p_\psi = -\mathrm{i}\hbar\frac{\partial}{\partial\psi} .$$

Writing partial differentials in an abbreviated notation, i.e. $\partial_\phi \equiv \partial/\partial\phi$ etc., and extracting, as before, a factor $\hbar$ in defining angular momenta, the components of angular momentum along cartesian, body-fixed axes read (cf., e.g., [Scheck (2010)], Eq. (3.89))

$$\begin{aligned}
\hbar\overline{L}_1 &= \frac{1}{\sin\theta}(p_\phi - p_\psi\cos\theta)\sin\psi + p_\theta\cos\psi \\
&= -\mathrm{i}\hbar\left[\frac{1}{\sin\theta}(\partial_\phi - \partial_\psi\cos\theta)\sin\psi + \partial_\theta\cos\psi\right], \\
\hbar\overline{L}_2 &= \frac{1}{\sin\theta}(p_\phi - p_\psi\cos\theta)\cos\psi - p_\theta\sin\psi \\
&= -\mathrm{i}\hbar\left[\frac{1}{\sin\theta}(\partial_\phi - \partial_\psi\cos\theta)\cos\psi - \partial_\theta\sin\psi\right], \qquad (6.40) \\
\hbar\overline{L}_3 &= p_\psi = -\mathrm{i}\hbar\partial_\psi .
\end{aligned}$$

The square of the angular momentum $\boldsymbol{L}^2$ is obtained from these expressions,

$$\begin{aligned}
\hbar^2\boldsymbol{L}^2 &= \hbar^2\left\{\overline{L}_1^2 + \overline{L}_2^2 + \overline{L}_3^2\right\} \\
&= -\hbar^2\left\{\frac{\partial^2}{\partial\theta^2} + \cot\theta\frac{\partial}{\partial\theta} + \frac{1}{\sin^2\theta}\left(\frac{\partial^2}{\partial\psi^2} + \frac{\partial^2}{\partial\phi^2} - 2\cos\theta\frac{\partial^2}{\partial\psi\,\partial\phi}\right)\right\} . \qquad (6.41)
\end{aligned}$$

The differential operator of second order that appears in (6.41) is seen to be the same as the one of the differential equation (6.37). Therefore, the D-functions with integer values $j \equiv L$ are eigenfunctions of the quantum top. We pursue this important observation by constructing a Hamiltonian for the symmetric top and by studying its energy spectrum.

The classical Hamiltonian describing the symmetric top, i.e. a top whose moments of inertia fulfill $I_1 = I_2 \neq I_3$,

$$H_{\text{class.}} = \frac{1}{2I_1\sin^2\theta}\left(p_\phi - p_\psi\cos\theta\right)^2 + \frac{1}{2I_1}p_\theta^2 + \frac{1}{2I_3}p_\psi^2$$

(cf. [Scheck (2010)], Eq. (3.90)), by the quantization procedure just described, is replaced by an operator in the variables (ϕ, θ, ψ) and their derivatives. The corresponding stationary Schrödinger equation is written in terms of the square of the angular momentum and its projection onto the symmetry axis. Making use of (6.40) one obtains

$$H\Psi(\phi,\theta,\psi) = \hbar^2 \left\{ \frac{\mathbf{L}^2 - \overline{L}_3^2}{2I_1} + \frac{\overline{L}_3^2}{2I_3} \right\} \Psi(\phi,\theta,\psi) = E\Psi(\phi,\theta,\psi) \,. \tag{6.42}$$

Comparison with the differential equation (6.37) shows that the eigenfunctions Ψ can be written in terms of D-functions with integer values of j. The precise relationship requires a further symmetry argument. Given the shape of the symmetric top there are different possibilities to choose the axes of the intrinsic, body-fixed system. Clearly, the eigenfunctions in (6.42) should not depend on any specific choice. It is not difficult to realize that every choice of the body-fixed axes can be reached by combining the following standard transformations: A rotation about the 1-axis by the angle π, a rotation about the 3-axis by the angle $\pi/2$, and by cyclic permutations of the three axes. Using the symmetry properties of the D-functions derived above, one sees that the eigenfunctions of the symmetric top, normalized to 1, are given by

$$\Psi(\phi,\theta,\psi) \equiv \Psi_{Lm\kappa}(\phi,\theta,\psi)$$
$$= \sqrt{\frac{2L+1}{16\pi^2(1+\delta_{\kappa 0})}} \left(D^{(L)}_{\kappa m}(\phi,\theta,\psi) + (-)^L D^{(L)}_{-\kappa m}(\phi,\theta,\psi) \right) . \tag{6.43}$$

The quantum number $L = 0, 1, 2, \ldots$ runs through zero and all positive integers, m and κ belong to the set of values known from the analysis of orbital angular momentum, i.e. $m, \kappa = -L, -L+1, \ldots, L$. The eigenvalues of the Hamiltonian (6.42) depend on the modulus of the angular momentum and on its projection onto the symmetry axis only. They are

$$E_{L\kappa} = \hbar^2 \left\{ \frac{L(L+1) - \kappa^2}{2I_1} + \frac{\kappa^2}{2I_3} \right\} . \tag{6.44}$$

This result plays an important role in the description of diatomic molecules and in the collective model of atomic nuclei which in their ground state exhibit a permanent deformation (deviation from spherical shape).[9] Strongly deformed nuclei are found in the group of Rare Earth and of Transuranium elements. They are either *prolate*, i.e. have the shape of a cigar, or are *oblate*, i.e. have a disc-like shape. Their third moment

[9] A. Bohr, Dan. Mat.-Fys. Medd. 26, No. 14 (1952), A. Bohr and B.R. Mottelson, Dan. Mat.-Fys. Medd. 27, No. 16 (1953) and 30, No. 1 (1955). The theoretical work of Aage Bohr, son of Niels Bohr, and of Ben Mottelson on the collective rotational and vibrational excitations of strongly deformed nuclei was awarded the Nobel prize 1975. They shared the prize with the experimentalist James Rainwater.

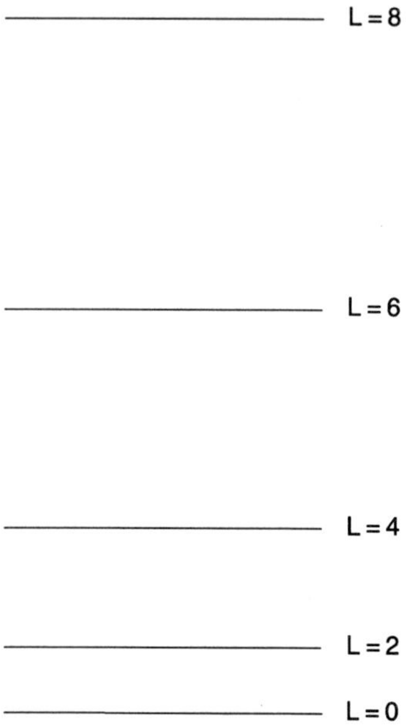

Fig. 6.3 Energy spectrum of a quantized symmetric top

of inertia I_3 seems to vanish or at least to be very small as compared to I_1, so that, classically speaking, these nuclei primarily rotate about axes which are perpendicular to the symmetry axis.

There is a pecularity when the projection of the angular momentum onto the symmetry axis vanishes. The eigenfunctions (6.43) with $\kappa = 0$ are different from zero only for L an *even integer* number. The eigenvalues (6.44) are

$$E_L = \hbar^2 \frac{L(L+1)}{2I_1}, \qquad L = 0, 2, 4, \ldots . \tag{6.45}$$

A typical rotational spectrum of this sort is shown in Fig. 6.3.

Strictly speaking the idea of a rigid body is incompatible with Heisenberg's uncertainty relation between position and momentum. Take as a model for the classical rigid body a set of mass points which are held fixed relative to each other by appropriately chosen forces. After quantization these mass points will exhibit small oscillations about their classical positions, in close analogy to the ground state of the harmonic oscillator, cf. Sect. 6.2.3. The problem posed by this remark was investigated and solved both for the case of the quantum mechanical three-body problem and for the more general case of the N-body system.[10]

[10] S. Flügge and A. Weiguny, Z. Physik **171**, 171 (1963), A. Weiguny, Z. Physik **186**, 227 (1965).

6.2.3 Addition of Angular Momenta and Clebsch-Gordan Coefficients

In this section we take up once more the question of coupling two or more angular momenta. We prove some properties of the Clebsch-Gordan series and indicate methods of calculating Clebsch-Gordan coefficients.

The Problem and Some Preliminaries: Representations of a compact Lie group G such as, e.g., $SU(n)$, $SO(m)$ etc. are also representations of the corresponding Lie algebra Lie(G) (often denoted by gothic letters such as $\mathfrak{g}$, $\mathfrak{su}(n)$, $\mathfrak{so}(m)$ etc.). Indeed, in Sect. 4.1.2 we derived the representations of $SU(2)$ from the properties of its Lie *algebra*, without making explicit use of the group itself. In doing so, however, some care is appropriate, as exemplified by the following remarks. (i) The representations of the Lie algebra of the group $G = U(1)$ have the spectrum $\mathbb{R}$, without any further restriction, while the group proper admits only integer eigenvalues $m \in \mathbb{Z}$ because the eigenfunctions $\exp\{im\phi\}$ must be one-valued. (ii) The comparison of $SU(2)$ with $SO(3)$ whose Lie algebras are isomorphic, showed that their Lie algebra correctly yields both the integer and the half-integer eigenvalues j, without distinction. In both cases the irreducible unitary representations characterized by the quantum numbers (j, m) are one-valued representations of the covering group $SU(2)$. However, with respect to $SO(3)$, only the representations with integer $j \equiv \ell$ are one-valued.

Let us start with a short summary of the problem posed by the addition of angular momenta: The product states $|j_1 m_1\rangle\,|j_2 m_2\rangle$ obtained from the base states $|j_1 m_1\rangle$ and $|j_2 m_2\rangle$ span a representation which, in contrast to its individual factors, is *reducible.* The factors span irreducible subspaces $\mathcal{H}^{(j_i)}$ whose dimensions are, respectively, $(2j_1 + 1)$ and $(2j_2 + 1)$. Under a rotation $\mathbf{R}$ in $\mathbb{R}^3$ their elements transform like $\mathbf{D}^{(j_1)}(\mathbf{R})$ and $\mathbf{D}^{(j_2)}(\mathbf{R})$, respectively. Thus, the (reducible) product space has dimension $(2j_1 + 1)(2j_2 + 1)$, its elements transform by the product $\mathbf{D}^{(j_1)}(\mathbf{R})\mathbf{D}^{(j_2)}(\mathbf{R})$. The problem to be solved is to expand the product state in terms of irreducible representations, that is, in a symbolic notation, to construct the series

$$j_1 \otimes j_2 = \sum \oplus J \tag{6.46}$$

In other terms and at the level of base states, the problem is to find the unitary transformation which maps product states to eigenstates of total angular momentum $J^2 = (j_1 + j_2)^2$ and of the 3-component $\mathbf{J}_3 = (\mathbf{j}_1)_3 + (\mathbf{j}_2)_3$,

$$|JM\rangle = \sum_{m_1, m_2} (j_1 m_1, j_2 m_2 | JM)\, |j_1 m_1\rangle\, |j_2 m_2\rangle\ . \tag{6.47}$$

Here $|JM\rangle$ denotes an eigenstate of J^2 and of $\mathbf{J}_3$ with eigenvalues $J(J+1)$ and M, respectively.

Are Products of Representations Themselves Representations? In a somewhat more general framework consider an algebra $\mathcal{A}$ which may, but need not, be a Lie algebra. Let ϱ_V and ϱ_W be two representations on vector spaces V and W, respectively, such that an element $a \in \mathcal{A}$ is mapped to $\varrho_V(a)$ in V, to $\varrho_W(a)$ in W.

The action of a on an element $v \in V$ and on an element $w \in W$ then is $\varrho_V(a)v$ and $\varrho_W(a)w$, respectively. In the case of matrix representations (this is the case we are considering here) this action is defined by the product of a matrix and a (column) vector. Considering now the set of all products vw, a *linear* action of a can only be given by the sum of tensor products

$$\varrho(a) = \varrho_V(a) \otimes \mathbb{1}_W + \mathbb{1}_V \otimes \varrho_W(a) \tag{6.48}$$

The question then is whether and under which conditions (6.48) is again a representation of the algebra. The product that is part of the definition of the algebra is denoted symbolically by $\bullet$. The ansatz yields a representation if and only if for the product $a \bullet b$ of any two elements $a, b \in \mathcal{A}$

$$\varrho_{V/W}(a \bullet b) = \varrho_{V/W}(a) \bullet \varrho_{V/W}(b) \quad \text{implies} \quad \varrho(a \bullet b) = \varrho(a) \bullet \varrho(b) .$$

In general, this does not hold: Indeed, by (6.48), one has

$$\begin{aligned} \varrho(a) \bullet \varrho(b) = \varrho_V(a) \bullet \varrho_V(b) \otimes \mathbb{1}_W + \mathbb{1}_V \otimes \varrho_W(a) \bullet \varrho_W(b) \\ + \varrho_V(a) \bullet \varrho_W(b) + \varrho_V(b) \bullet \varrho_W(a) . \end{aligned}$$

The first two terms yield the right answer $\varrho(a \bullet b)$ but the last two do not unless they are zero. If the algebra is a Lie algebra $(\mathcal{A}, \bullet) = (\mathfrak{g}, [\ ,\ ])$ whose product is the commutator, the unwanted terms cancel out, and (6.48) does yield a representation.

Comment on Notations: In the old mathematics literature and in much of the physics literature the reducible product representation whose base elements are $|j_1 m_1\rangle\, |j_2 m_2\rangle$, are denoted by the symbol $\times$ of the direct product. (Every base element of the first representation space V is multiplied with every base element of the other space W. Every matrix element $D^{(j_1)}_{\kappa_1 m_1}$ is multiplied with every matrix element $D^{(j_2)}_{\kappa_2 m_2}$.) In the modern mathematical literature this case is written as tensor product $V \otimes W$ with base elements $v \otimes w$, while the symbol $\times$ means that $V \times W$ is the set of elements (v, w) with two entries (which are not multiplied). Strictly speaking, the series (6.47) should be denoted as follows

$$|JM\rangle = \sum_{m_1, m_2} (j_1 m_1, j_2 m_2 | JM)\, |j_1 m_1\rangle \otimes |j_2 m_2\rangle .$$

Since at the level of the eigenstates there should not arise any confusion I stick to the simpler notation of (6.47), omitting the $\otimes$-sign. For operators, however, the correct notation of the tensor product as in (6.48) is essential.

Covariance and Contravariance: Base vectors such as $\boldsymbol{\varphi}^{(j)} \equiv \{\varphi^{(j)}_m\}$ and expansion coefficients $\boldsymbol{a}^{(j)} \equiv \{a^{(j)}_m\}$ representing a physical state transform by *contragredience*, cf. Sect. 4.1.3, that is to say, if the latter transform with respect to rotations by $\mathbf{D}^{(j)}$ then the transformation law for the basis is given by $\left(\mathbf{D}^{-1}\right)^T = \mathbf{D}^*$. Thus, in a short-hand notation we have

$$\boldsymbol{a}'^{(j)} = \mathbf{D}^{(j)} \boldsymbol{a}^{(j)} , \qquad \boldsymbol{\varphi}'^{(j)} = \mathbf{D}^{(j)*} \boldsymbol{\varphi}^{(j)} .$$

These transformation rules make sure that the physical state

$$\Psi = \sum_j \varphi'^{(j)} \cdot a'^{(j)} = \sum_j \varphi^{(j)} \mathbf{D}^{(j)\dagger} \mathbf{D}^{(j)} a^{(j)} = \sum_j \varphi^{(j)} \cdot a^{(j)}$$

does not depend on the choice of the basis.

Of course, the Clebsch-Gordan coefficients should be defined in such a way that they apply universally to $SU(2)$ and can be tabulated or calculated by means of standard computer routines. In order to fulfill this requirement one defines these coefficients in such a way that they apply to *cogredient* objects, i.e. to quantities that transform in the same way under rotations. In other terms, both factors must either transform by $\mathbf{D}$ or by $\mathbf{D}^*$.

For the sake of clarity we call the base elements *co*variant, the expansion coefficients *contra*variant. Let the Clebsch-Gordan coefficients be defined as in (6.47) so that they couple covariantly transforming objects such as $|j_1 m_1\rangle$ and $|j_2 m_2\rangle$. We note, however, that we will also have to deal with cases where the Clebsch-Gordan series is needed for contragredient objects, i.e. for the coupling of a covariant and a contravariant set. This will be the case, for example, when particle-hole states, or states consisting of a particle and an antiparticle, will have to be constructed which are eigenstates of total angular momentum. An example may help to clarify this problem.

Let the one-particle states of an atom be classified by j and m, the eigenvalues of angular momentum (vector sum of orbital angular momentum and of spin) and of its 3-component, respectively. Assume these states to be occupied by electrons up to the limit energy E_F. If one excites an electron from the occupied state $|j_1 m_1\rangle$ with energy $E_1 < E_F$ to a formerly unoccupied state $|j_2 m_2\rangle$ with energy $E_2 > E_F$ one obtains a particle-hole excitation whose angular momentum state is

$$\overline{|j_1 m_1\rangle}\, |j_2 m_2\rangle \ .$$

Its second factor is covariant but the first factor is contravariant. Therefore, the Clebsch-Gordan series needed to transform this product basis to a basis of total angular momentum cannot be applied directly. Indeed, one must first convert the "wrong" transformation behaviour of the hole state $\overline{|j_1 m_1\rangle}$ in such a way that both factors become cogredient, i.e. become covariant with respect to rotations. As we now show the solution to this problem is easy. We prove the following lemma:

Lemma 6.1

A covariantly transforming set $\boldsymbol{b}^{(j)} \equiv \{b_m^{(j)}, m = +j, \ldots, m = -j\}$, by the unitary transformation

$$\mathbf{U}_0 := \mathbf{D}^{(j)}(0, \pi, 0) \ , \tag{6.49}$$

is mapped to a contravariantly transforming set $\tilde{\boldsymbol{b}}^{(j)} \equiv \{\tilde{b}^{(j)m}, m = +j, \ldots, m = -j\}$. Conversely, the same transformation maps a contravariant set to a covariant set.

Proof

If the assertion is correct then $\tilde{\boldsymbol{b}}^{(j)} = \mathbf{U}_0 \boldsymbol{b}^{(j)}$. Let $\mathbf{U}$ be an arbitrary unitary transformation acting on $\boldsymbol{b}^{(j)}$,

$$\mathbf{U}\,\boldsymbol{b}^{(j)} = \boldsymbol{b}^{(j)\prime} .$$

Denote the action of the same transformation on $\tilde{\boldsymbol{b}}^{(j)}$ by $\widetilde{\mathbf{U}}$,

$$\tilde{\boldsymbol{b}}^{(j)\prime} = \widetilde{\mathbf{U}}\,\tilde{\boldsymbol{b}}^{(j)} .$$

If $\tilde{\boldsymbol{b}}^{(j)}$ is contragredient to $\boldsymbol{b}^{(j)}$ and if this is to hold for $\tilde{\boldsymbol{b}}^{(j)\prime}$ and $\boldsymbol{b}^{(j)\prime}$, too, then one must have

$$\widetilde{\mathbf{U}} = \left(\mathbf{U}^{-1}\right)^T = \mathbf{U}^* \quad \text{and} \quad \mathbf{U}^* = \mathbf{U}_0\,\mathbf{U}\,\mathbf{U}_0^{-1} .$$

Special cases are provided by the unitary transformations $\mathbf{U} = \mathbf{D}^{(j)}(\mathbf{R})$ induced by rotations $\mathbf{R}$ in ordinary space. In this case the above requirements read

$$\mathbf{U}_0\,\mathbf{D}^{(j)}(\mathbf{R})\mathbf{U}_0^{-1} = \mathbf{D}^{(j)*}(\mathbf{R}) \quad \text{for all} \quad \mathbf{R} \in SO(3) .$$

Writing $\mathbf{D}^{(j)}$ as an exponential series in the generators $\boldsymbol{J} = \{\mathbf{J}_1, \mathbf{J}_2, \mathbf{J}_3\}$ these conditions translate to

$$\mathbf{U}_0\,\mathbf{D}^{(j)}(\mathbf{R})\mathbf{U}_0^{-1} = \mathrm{e}^{\mathrm{i}\varphi \mathbf{U}_0 \boldsymbol{J} \mathbf{U}_0^{-1}} = \mathbf{D}^{(j)*}(\mathbf{R}) = \mathrm{e}^{-\mathrm{i}\varphi \boldsymbol{J}^*} , \text{ or} \tag{6.50}$$

$$\mathbf{U}_0\mathbf{J}_k + \mathbf{J}_k^*\mathbf{U}_0 = 0 , \quad k = 1, 2, 3 . \tag{6.51}$$

The latter condition (6.51), indeed, is fulfilled if use is made of the Condon-Shortley phase convention, cf. Sect. 4.1.3. In this convention $\mathbf{J}_1$ and $\mathbf{J}_3$ are real while $\mathbf{J}_2$ is pure imaginary: The unitary $\mathbf{U}_0$ being induced by a rotation about the 2-axis by the angle π, takes $\mathbf{J}_1$ to $-\mathbf{J}_1$, $\mathbf{J}_3$ to $-\mathbf{J}_3$, but leaves $\mathbf{J}_2$ unchanged. This shows that (6.51) is fulfilled and thus proves the assertion.

Within the Condon-Shortley phase convention the unitary map $\mathbf{U}_0$ takes a very simple form. Setting $\theta = \pi$ the formula (6.25), or, even simpler, the symmetry relation (6.28) yields the explicit form of this matrix, viz.

$$(U_0)_{m'm} \equiv D^{(j)}_{m'm}(0, \pi, 0) = (-)^{j-m'}\,\delta_{m',-m} . \tag{6.52}$$

Let us return once more to the example of a particle-hole excitation. The correct coupled state which is an eigenstate of total angular momentum and its 3-component, is found to be

$$|JM\rangle = \sum_{m_1 m_2} (-)^{j_1+m_1} (j_1, -m_1; j_2, m_2|JM)\,|j_1, -m_1\rangle\,|j_2 m_2\rangle .$$

A hole in a level with quantum numbers (j_1, m_1) behaves like a particle in the state $(j_1, -m_1)$ supplemented by the characteristic sign $(-)^{j_1+m_1}$.

Clebsch-Gordan Coefficients are Real: The transformation $\mathbf{U}_0$ has the canonical form (6.49) or (6.52) which applies to all irreducible representations. Its action on the left-hand side of (6.47) reads

$$(U_0^{(J)})_{M'M} = D^{(J)}_{M'M}(0, \pi, 0) ,$$

Its action on the right-hand side (6.47) is

$$(U_0)_{m'_1m'_2,m_1m_2} = D^{(j_1)}_{m'_1m_1}(0,\pi,0)D^{(j_2)}_{m'_2m_2}(0,\pi,0)\,.$$

Let $\mathbf{C}$ be an arbitrary unitary transformation. By the very definition of $\mathbf{U}_0$ we have, as a general rule, $\mathbf{C}^*\mathbf{U}_0 = \mathbf{U}_0\mathbf{C}$ or $\mathbf{C}^*\mathbf{U}_0\mathbf{C}^\dagger = \mathbf{U}_0$. The specific matrix which serves the purpose of the coupling (6.47), i.e. whose entries are the Clebsch-Gordan coefficients, is a special case of such a unitary transformation so that

$$\mathbf{C}^*\mathbf{U}_0\mathbf{C}^\dagger = \mathbf{U}_0^{(J)}\,.$$

In this particular case the matrices $\mathbf{U}_0^{(J)}$ and $\mathbf{U}_0$, in addition, are equivalent which is to say that they fulfill

$$\mathbf{C}\mathbf{U}_0\mathbf{C}^\dagger = \mathbf{U}_0^{(J)}\,.$$

These two relations are compatible only if $\mathbf{C}$ is real, hence orthogonal. Obviously, this result is relevant for many practical purposes and we summarize it here:

Theorem 6.2

In the framework of the Condon-Shortley phase convention for the generators of the rotation group the Clebsch-Gordan coefficients are *real*.

This result presents the following advantages:

(i) Two cogredient objects are coupled in the same way, independently of whether both are covariant or both contravariant;

(ii) The matrix $\mathbf{C}$ is orthogonal, which means that

$$\sum_{J,M} (j_1m_1, j_2m_2|JM)\left(JM|j_1m'_1, j_2m'_2\right) = \delta_{m_1m'_1}\delta_{m_2m'_2}\,;$$

(iii) The entries $(j_1m_1, j_2m_2|JM)$ can be calculated once and for ever. As they are real, they may be obtained by means of simple computer routines, or may be listed in tables.

6.2.4 Calculating Clebsch-Gordan Coefficients; The 3*j*-Symbols

The results of the preceding section provide a possibility of principle for calculating the Clebsch-Gordan coefficients. The following formula which is due to E. Wigner, should be immediately clear:

$$\begin{aligned}
&(j_1\mu_1, j_2\mu_2|j\mu)(j_1m_1, j_2m_2|jm) \\
&= \frac{2j+1}{8\pi^2} \\
&\times \int_0^{2\pi}\mathrm{d}\psi\int_0^{\pi}\sin\theta\,\mathrm{d}\theta\int_0^{2\pi}\mathrm{d}\phi\; D^{(j)*}_{\mu m}(\phi,\theta,\psi)D^{(j_1)}_{\mu_1m_1}(\phi,\theta,\psi)D^{(j_2)}_{\mu_2m_2}(\phi,\theta,\psi)\,.
\end{aligned} \tag{6.53}$$

Doing this integral, solving for the coefficients, and making use of the normalization condition

$$\sum_{m_1 m_2} (j_1 m_1, j_2 m_2 | jm)^2 = 1$$

and of the convention $(j_1, m_1 = j_1; j_2, m_2 = j - j_2 | j, m = j) \geq 0$, one obtains an explicit formula for the Clebsch-Gordan coefficients. This particular method, however, is somewhat cumbersome and I skip it as well as some more practicable alternatives for calculating the Clebsch-Gordan coefficients in the framework of the Lie algebra of $SU(2)$. They are worked out, e.g., in [Fano and Racah (1959)]. I merely quote the final result

$$(j_1 m_1, j_2 m_2 | j_3 m_3) = \delta_{m_1+m_2, m_3} \sqrt{2j_3+1}\, \Delta(j_1, j_2, j_3) \sum_r (-)^r$$
$$\times \frac{\sqrt{(j_1+m_1)!(j_1-m_1)!(j_2+m_2)!(j_2-m_2)!(j_3+m_3)!(j_3-m_3)!}}{r!(j_1+j_2-j_3-r)!(j_1-m_1-r)!(j_3-j_2+m_1+r)!}$$
$$\times \frac{1}{(j_2+m_2-r)!(j_3-j_1-m_2+r)!} \,. \tag{6.54}$$

This formula contains a symbol $\Delta(j_1, j_2, j_3)$ which is symmetric in all three angular momenta and which is defined by

$$\Delta(j_1, j_2, j_3) = \sqrt{\frac{(j_1+j_2-j_3)!(j_2+j_3-j_1)!(j_3+j_1-j_2)!}{(j_1+j_2+j_3+1)!}} \,. \tag{6.55}$$

This quantity is different from zero only if the arguments of the three factorials in the numerator are greater than or equal to zero, that is, if

$$\mathrm{Max}\ ((j_1 - j_2), (j_2 - j_1)) \leq j_3 \leq j_1 + j_2 \tag{6.56}$$

holds true. This is the *triangle rule for angular momenta* that we encountered in Sect. 4.1.6. The structure of the expression (6.54) is quite remarkable: It consists in a finite sum of terms each of which is the square root of a rational number. This fact is often used in their practical evaluation and in writing computer routines for Clebsch-Gordan coefficients.

The symmetry relations (4.34) and (4.35) cf. Sect. 4.1.5, are obtained directly from (6.54): Except for the Kronecker δ and for the denominator all factors are invariant under arbitrary permutations of the pairs (j_i, m_i). We denote the denominator temporarily by $N(j_1, m_1; j_2, m_2 | j)$. Replacing the summation index r by $r = j_1 + j_2 - j_3 - s$ the denominator becomes

$$\begin{aligned} &(j_1+j_2-j_3-s)!s!(j_3-j_2-m_1+s)!(j_1+m_1-s)! \\ &(j_3-j_1+m_2+s)!(j_2-m_2-s)! \\ &\equiv N(j_2, m_2; j_1, m_1 | j) \,. \end{aligned}$$

By the same replacement the sign factor in (6.54) goes over into $(-)^{j_1+j_2-j_3}(-)^{-s}$. As s is always an integer there follows the symmetry relation

$$(j_2m_2, j_1m_1|j_3m_3) = (-)^{j_1+j_2-j_3}(j_1m_1, j_2m_2|j_3m_3)\ . \tag{6.57}$$

Similarly, replacing r by $r = j_1 - m_1 - t$ the denominator becomes

$$\begin{aligned}&(j_1 - m_1 - t)!(j_2 - j_3 + m_1 + t)!\\ &t!((j_3 - j_2 + j_1 - t)!(j_2 + m_2 - j_1 + m_1 + t)!\\ &\times (j_3 - m_1 - m_2 - t)! \equiv N(j_1, m_1; j_3, -(m_1 + m_2)|j_2)\,,\end{aligned}$$

the sign factor becomes $(-)^{j_1-m_1}(-)^t$. As $m_3 = m_1 + m_2$ one obtains the second symmetry relation

$$(j_1m_1, j_2m_2|j_3m_3) = (-)^{j_1-m_1}\sqrt{\frac{(2j_3+1)}{(2j_2+1)}}\\ (j_1, m_1, j_3, -m_3|j_2, -m_2)\ . \tag{6.58}$$

In practice, the Clebsch-Gordan coefficients are used, in this form, whenever their normalization

$$\sum_{m_1m_2}(j_1m_1, j_2m_2|j_3m_3)\left(j_1m_1, j_2m_2|j_3'm_3'\right) = \delta_{j_3j_3'}\delta_{m_3m_3'}$$

matters. On the other hand, their symmetry properties and the selection rules which are coded in them, become more transparent and are easier to remember if one introduces the $3j$-*symbols* which are defined as follows

$$\begin{aligned}\begin{pmatrix} j_1 & j_2 & j_3\\ m_1 & m_2 & m_3\end{pmatrix} &= \delta_{m_1+m_2+m_3,0}\,(-)^{j_1-j_2-m_3}\Delta(j_1, j_2, j_3)\sum_r(-)^r\\ &\times\frac{\sqrt{(j_1+m_1)!(j_1-m_1)!(j_2+m_2)!(j_2-m_2)!(j_3+m_3)!(j_3-m_3)!}}{r!(j_1+j_2-j_3-r)!(j_1-m_1-r)!(j_3-j_2+m_1+r)!}\\ &\times\frac{1}{(j_2+m_2-r)!(j_3-j_1-m_2+r)!}\ .\end{aligned} \tag{6.59}$$

They are related to the Clebsch-Gordan coefficients by the formula

$$(j_1m_1, j_2m_2|j_3m_3) = (-)^{j_2-j_1-m_3}\sqrt{2j_3+1}\begin{pmatrix} j_1 & j_2 & j_3\\ m_1 & m_2 & -m_3\end{pmatrix}\ . \tag{6.60}$$

Their properties follow directly from the properties of Clebsch-Gordan coefficients. We summarize them here:

Properties of the 3*j*-Symbols

(i) The $3j$-symbols (6.59) are equal to zero whenever one of the selection rules

$$j_1 + j_2 + j_3 = n \,, \quad n \in \mathbb{N}_0 \,, \tag{6.61}$$

$$|j_1 - j_2| \le j_3 \le j_1 + j_2 \text{ (cyclic)} \,,$$

$$m_1 + m_2 + m_3 = 0 \,, \tag{6.62}$$

is not fulfilled.

(ii) They are invariant under cyclic permutations of the columns

$$\begin{pmatrix} j_1 & j_2 & j_3 \\ m_1 & m_2 & m_3 \end{pmatrix} = \begin{pmatrix} j_2 & j_3 & j_1 \\ m_2 & m_3 & m_1 \end{pmatrix} \quad \text{(cyclic)} \,. \tag{6.63}$$

In odd (or anticyclic) permutations of their columns they receive a sign factor which depends on the sum $j_1 + j_2 + j_3$,

$$\begin{pmatrix} j_2 & j_1 & j_3 \\ m_2 & m_1 & m_3 \end{pmatrix} = (-)^{j_1+j_2+j_3} \begin{pmatrix} j_1 & j_2 & j_3 \\ m_1 & m_2 & m_3 \end{pmatrix} \quad \text{(anticyclic)} \,. \tag{6.64}$$

(iii) The same factor appears when the signs of all three m_i are changed simultaneously,

$$\begin{pmatrix} j_1 & j_2 & j_3 \\ -m_1 & -m_2 & -m_3 \end{pmatrix} = (-)^{j_1+j_2+j_3} \begin{pmatrix} j_1 & j_2 & j_3 \\ m_1 & m_2 & m_3 \end{pmatrix} \,. \tag{6.65}$$

(iv) The orthogonality of the Clebsch-Gordan coefficients is inherited by the $3j$-symbols for which it reads

$$\sum_{m_1 m_2} \begin{pmatrix} j_1 & j_2 & j \\ m_1 & m_2 & m \end{pmatrix} \begin{pmatrix} j_1 & j_2 & j' \\ m_1 & m_2 & m' \end{pmatrix} = \frac{1}{2j+1} \delta_{jj'} \delta_{mm'} \,. \tag{6.66}$$

(v) Some special cases of relevance for atomic and nuclear physics are

$$\begin{aligned} \begin{pmatrix} j & j' & 0 \\ m & m' & 0 \end{pmatrix} &= \frac{(-)^{j-m}}{\sqrt{2j+1}} \delta_{jj'} \delta_{m,-m'} \,, \\ \begin{pmatrix} j & 1 & j \\ -m & 0 & m \end{pmatrix} &= \frac{(-)^{j-m} m}{\sqrt{j(2j+1)(j+1)}} \,, \\ \begin{pmatrix} j & 2 & j \\ -m & 0 & m \end{pmatrix} &= (-)^{j-m} \frac{3m^2 - j(j+1)}{\sqrt{(2j-1)j(2j+1)(j+1)(2j+3)}} \,. \end{aligned} \tag{6.67}$$

(vi) An interesting special case occurs when all three magnetic quantum numbers are zero. Denoting the three angular momenta (which in this case are integers!) by a, b, and c, one has

$$\begin{pmatrix} a & b & c \\ 0 & 0 & 0 \end{pmatrix} = 0 \quad \text{if } a+b+c = 2m+1 , \tag{6.68}$$

$$\begin{pmatrix} a & b & c \\ 0 & 0 & 0 \end{pmatrix} = \Delta(a,b,c) \frac{(-)^n n!}{(n-a)!(n-b)!(n-c)!} \quad \text{if } a+b+c = 2n . \tag{6.69}$$

(vii) While the definition of the Clebsch-Gordan coefficients for $SU(2)$ are generally accepted there are differing definitions for the $3j$-symbols in the literature. For example, the $\overline{V}$-coefficients of [Fano and Racah (1959)] are related to the $3j$-symbols (6.59) by

$$\begin{pmatrix} a & b & c \\ \alpha & \beta & \gamma \end{pmatrix} = (-)^{a+b+c}\, \overline{V} \begin{pmatrix} a & b & c \\ \alpha & \beta & \gamma \end{pmatrix} .$$

Other definitions and conventions may be found in [Fano and Racah (1959)].

6.2.5 Tensor Operators and Wigner–Eckart Theorem

A set of operators $\{T_\mu^{(\kappa)}\}$, $\kappa \in \mathbb{N}_0$, $\mu \in (-\kappa, -\kappa+1, \ldots, \kappa)$ which transforms contravariantly under rotations $\mathbb{R}^3$ (thus, by $\mathbf{D}^{(\kappa)}$), or else transforms covariantly (i.e. by $\mathbf{D}^{(\kappa)*}$), is called a *tensor operator of rank* κ.

Here is an example: The operator $|\boldsymbol{x}_1 - \boldsymbol{x}_2|^{-1}$ is invariant with respect to rotations. However, if it is expanded in terms of multipoles,

$$\frac{1}{|\boldsymbol{x}_1 - \boldsymbol{x}_2|} = \sum_{\ell=0}^{\infty} \frac{4\pi}{2\ell+1} \frac{r_<^\ell}{r_>^{\ell+1}} \sum_{m=-\ell}^{+\ell} Y_{\ell m}^*(\hat{\boldsymbol{x}}_1) Y_{\ell m}(\hat{\boldsymbol{x}}_2) , \tag{6.70}$$

then it appears to be the invariant product of two tensor operators of rank ℓ,

$$T_m^{(\ell)}(\boldsymbol{x}) = \sqrt{\frac{4\pi}{2\ell+1}}\, r^\lambda Y_{\ell m}(\hat{\boldsymbol{x}})$$

where $r = |\boldsymbol{x}|$ and $\lambda = \ell$ or $\lambda = -\ell - 1$.

A further example is provided by the three components of angular momentum. Written in the spherical basis, viz.

$$\mathbf{J}_{\pm 1} = \mp \frac{1}{\sqrt{2}} (\mathbf{J}_1 \pm \mathrm{i}\mathbf{J}_2) , \qquad \mathbf{J}_0 = \mathbf{J}_3 ,$$

they form a tensor operator of rank 1.

Tensor operators are always defined with reference to a frame in $\mathbb{R}^3$ and, therefore, they depend on $\boldsymbol{x}$, explicitly or implicitly. Thus, under a rotation a covariant tensor operator transforms according to the rule

$$T_\mu^{(\kappa)\prime}(\boldsymbol{x}') = \sum_{\nu=-\kappa}^{\kappa} D_{\mu\nu}^{(\kappa)*}(\phi, \theta, \psi)\, T_\nu^{(\kappa)}(\boldsymbol{x}) . \tag{6.71}$$

Matrix elements of an operator of this kind, taken between eigenstates of the square and of the 3-component of angular momentum, have an especially simple structure: Their dependence on all magnetic quantum numbers is solely contained in a sign and a $3j$-symbol. In other terms, all such matrix elements are proportional to one another, their ratios have universal values which do not depend on the specific nature of the operator. This is the content of an important theorem.

Theorem 6.3 Theorem of Wigner and Eckart

The matrix elements of a tensor operator between eigenstates of angular momentum are given by the universal formula

$$\langle JM|T_\mu^{(\kappa)}|J'M'\rangle = (-)^{J-M}\begin{pmatrix} J & \kappa & J' \\ -M & \mu & M' \end{pmatrix}\left(J\left\|T^{(\kappa)}\right\|J'\right). \tag{6.72}$$

The proportionality factor $\left(J||T^{(\kappa)}||J'\right)$, common to all of them, is independent of the magnetic quantum numbers. It is called the *reduced matrix element*.

Proof

The following proof is somewhat heuristic but has the advantage of emphasizing the physical content of the theorem [Fano and Racah (1959)]. It proceeds in two steps.

1. Consider first an irreducible tensor field $T_\mu^{(\kappa)}(\boldsymbol{x})$ which has the same form and the same functional dependence in every frame of reference. This is a tensor field for which

$$T_\mu^{(\kappa)}(\boldsymbol{x}') = \sum_\nu D_{\mu\nu}^{(\kappa)\,*} T_\nu^{(\kappa)}(\boldsymbol{x})$$

holds. Here, the argument $\boldsymbol{x}$ stands for possibly more than one single argument. An example for such a tensor field is given by the spherical harmonics $T_m^{(\ell)}(\boldsymbol{x}) \equiv Y_{\ell m}(\hat{\boldsymbol{x}})$ which, indeed, have the same explicit functional form in all coordinate systems which differ by rotations about the origin. Writing $\int \mathrm{d}^3x$ for the possibly multi-dimensional integral $\int \mathrm{d}^3x_1\, \mathrm{d}^3x_2 \cdots$, we have for all $\kappa \neq 0$

$$F_\mu^{(\kappa)} := \int \mathrm{d}^3x\; T_\mu^{(\kappa)}(\boldsymbol{x}) = 0\,. \tag{6.73}$$

This assertion is proved by rotating the coordinate system, $\boldsymbol{x} \mapsto \boldsymbol{x}' = \mathbf{R}\boldsymbol{x}$, and by calculating its effect on the constant tensor $F_\mu^{(\kappa)}$. One has

$$F_\mu^{(\kappa)} = \int \mathrm{d}^3x'\; T_\mu^{(\kappa)}(\boldsymbol{x}') = \sum_\nu D_{\mu\nu}^{(\kappa)\,*}(\mathbf{R}) \int \mathrm{d}^3x\; T_\nu^{(\kappa)}(\boldsymbol{x})$$
$$= \sum_\nu D_{\mu\nu}^{(\kappa)\,*}(\mathbf{R})\, F_\nu^{(\kappa)}\,.$$

The set $F_\mu^{(\kappa)}$ is a tensor whose $(2\kappa+1)$ components are the same in every frame of reference. As the matrices $\mathbf{D}^{(\kappa)}$ are irreducible, the components $F_\mu^{(\kappa)}$ must vanish for all $\kappa \neq 0$.

2. Making use of the Clebsch-Gordan series (6.47) and of its inverse, the matrix element $\langle JM|T^{(\kappa)}_{\mu}|J'M'\rangle$ is expressed, step by step, in terms of irreducible spherical tensors to which the above result applies. In order to simplify matters we use the symbolic notation $[T^{(\kappa)} \otimes S^{(\lambda)}]^{(\sigma)}_{\nu}$ for the coupling of two spherical tensors. Furthermore, the expression $(-)^{J-M}\psi^*_{JM}$ is written as $\phi_{J,-M}$, by defining $\phi_{JM} := (-)^{J+M}\psi^*_{J,-M}$. These conventions should help to understand the following calculation:

$$\begin{aligned}\langle JM|T^{(\kappa)}_{\mu}|J'M'\rangle &= \int \mathrm{d}^3x\, \psi^*_{JM}(\boldsymbol{x})\, T^{(\kappa)}_{\mu}(\boldsymbol{x})\, \psi_{J'M'}(\boldsymbol{x}) \\ &= (-)^{J-M}\int \mathrm{d}^3x\, \phi_{J,-M}\sum_{\lambda\sigma}(\kappa\mu, J'M'|\lambda\sigma)\left[T^{(\kappa)} \otimes \psi_{J'}\right]^{(\lambda)}_{\sigma} \\ &= (-)^{J-M}\sum_{\tau\omega}\sum_{\lambda\sigma}(\kappa\mu, J'M'|\lambda\sigma)(J,-M,\lambda\sigma|\tau\omega)\, F^{(\tau)}_{\omega}\,,\end{aligned}$$

where the abbreviation

$$F^{(\tau)}_{\omega} = \int \mathrm{d}^3x\, \left[\phi_J \otimes [T^{(\kappa)} \otimes \psi_{J'}]^{(\lambda)}\right]^{(\tau)}_{\omega}$$

is used. This constant tensor, by (6.73), is equal to zero unless $\tau = \omega = 0$. In this case, using $(J,-M,\lambda\sigma|00) = (-)^{J+M}/\sqrt{2J+1}\,\delta_{J\lambda}\delta_{M\sigma}$, the product of the two Clebsch-Gordan coefficients becomes

$$\begin{aligned}&\sum_{\lambda\sigma}(\kappa\mu, J'M'|\lambda\sigma)(J,-M,\lambda\sigma|00) \\ &= \frac{(-)^{J+M}}{\sqrt{2J+1}}(\kappa\mu, J'M'|JM) = (-)^{J+M}(-)^{J'-\kappa-M}\begin{pmatrix} J & \kappa & J' \\ -M & \mu & M' \end{pmatrix}.\end{aligned}$$

This is the formula given in (6.72) if the reduced matrix element is defined as follows

$$\left(J\left\|T^{(\kappa)}\right\|J'\right) := (-)^{J-\kappa+J'}\int \mathrm{d}^3x\, \left[\phi_J \otimes [T^{(\kappa)} \otimes \psi_{J'}]^{(\lambda)}\right]^{(0)}_{0}. \qquad (6.74)$$

This completes the proof of the theorem.

Remarks

1. The proof of the theorem shows that it holds only for operators whose matrix elements $\langle JM|T^{(\kappa)}_{\mu}|J'M'\rangle$ do not depend on the choice of the frame of reference. The interaction of a magnetic moment $\boldsymbol{\mu}$ with a fixed external magnetic field $\boldsymbol{B}$ provides a counter-example because this field singles out an invariant direction in space. In this case the theorem can only be applied to one of the two factors.

2. Although the reduced matrix element may be calculated from its definition (6.74), it is often simpler to calculate the full matrix element for one specific set of magnetic quantum numbers and to divide by the corresponding $3j$-symbol[11] and the phase factor. We give some examples below.
3. The states $|JM\rangle$ of a quantum system, in general, contain also radial functions. The integral over the radial variable(s) is absorbed in the reduced matrix element. Furthermore, $|JM\rangle$ may be a coupled state such as, for instance, an eigenstate of the sum of orbital angular momentum and spin, $[\ell \otimes s]^{(J)}$. Yet, the operator may refer to only one of the two parts. In these cases the calculation of the reduced matrix element may be a little more involved but can be done once and for ever, the results may be tabulated in a compendium.
4. The Wigner-Eckart theorem not only reduces the calculation of very many, viz. $(2J+1)(2\kappa+1)(2J'+1)$, matrix elements to a single one, it also yields and exhibits the selection rules, via the $3j$-symbol, that follow from conservation of angular momentum. Indeed, it is immediately clear that J, J', and κ must obey the triangle rule, and that the sum of the magnetic quantum numbers of the initial state and of the operator must equal the magnetic quantum number of the final state, $M = \mu + M'$. Further selection rules which follow from the conservation of parity, or which are due to specific properties of the radial functions, are contained in the reduced matrix element and, therefore, are less obvious.
5. In general, tensor operators are bosonic operators, that is, κ is an *integer* number. They are classified by representations of $SO(3)$, and they are tensor operators with respect to $SO(3)$ (not only with respect to $SU(2)$).

Example 6.3 Quadrupole Interaction in Deformed Nuclei

As long as the electron in an atom does not penetrate the nucleus, that is to say, if in (6.70) the variable $r_<$ is the radial variable of a proton in the nucleus, while $r_>$ is the one of the electron, the quadrupole piece in the electrostatic interaction between the electron and the proton is given by

$$U_{\text{E2}} = -\frac{4\pi \mathrm{e}^2}{5}\frac{r_n^2}{r_e^3}\sum_m Y^*_{2m}(\hat{\boldsymbol{x}}_e)Y_{2m}(\hat{\boldsymbol{x}}_n)\,.$$

The expectation value of this interaction in the ground state of the nucleus and a given atomic state of the electron, factorizes in the two components. We consider here only the nuclear part. Traditionally the static *quadrupole moment* is defined as follows

$$Q_0 := \sqrt{\frac{16\pi}{5}}e\int_0^\infty \mathrm{d}r \int \mathrm{d}\Omega\ \varrho_{JJ}(\boldsymbol{x})r^2 Y_{20}(\theta,\phi)\,,$$

[11] There exist tables of $3j$-symbols but they may also be calculated directly from (6.59).

where the proton density in the state with $M = J$

$$\varrho_{JJ}(\boldsymbol{x}) = \langle J, M = J | \sum_{n=1}^{Z} \delta(\boldsymbol{x} - \boldsymbol{x}_n) | J, M = J \rangle$$

has to be inserted. Written differently, the static quadrupole moment is

$$Q_0 = \sqrt{\frac{16\pi}{5}}\, e \langle J, J | \sum_{n=1}^{Z} r_n^2 Y_{20}(\hat{\boldsymbol{x}}_n) | J, J \rangle \,. \tag{6.75}$$

This quadrupole moment is called the *spectroscopic* quadrupole moment because it is visible in the quadrupole hyperfine structure of atomic spectra.

The more general matrix elements which appear in the interaction with the electron, are expressed in terms of the quadrupole moment by the Wigner-Eckart theorem,

$$\begin{aligned}\langle JM' | Q | JM \rangle &= \sqrt{\frac{16\pi}{5}}\, e \langle JM' \rangle \sum_n r_n^2 Y_{2\mu}(\hat{\boldsymbol{x}}_n) | JM \rangle \\ &= Q_0 (-)^{J-M'} \begin{pmatrix} J & 2 & J \\ -M' & \mu & M \end{pmatrix} \begin{pmatrix} J & 2 & J \\ -J & 0 & J \end{pmatrix}^{-1} .\end{aligned}$$

This result as well as the formula (6.75) for the quadrupole moment tell us that $(J, 2, J)$ must obey the triangle rule. Thus, one must have $J \geq 1$ and $M' = \mu + M$. Only those nuclei whose ground state has spin $J \geq 1$ can have a static, spectroscopic quadrupole moment.[12] The quadrupole hyperfine structure is proportional to expectation values in the states $|JM\rangle$. By the Wigner-Eckart theorem, and inserting the third of the formulae (6.67), one obtains

$$\langle JM | Q | JM \rangle = Q_0 \frac{3M^2 - J(J+1)}{J(2J-1)} \,.$$

Unfortunately, this formula does not yet allow to predict the spectroscopic quadrupole splitting. The reason is that even if the spin of the electron may be neglected, its orbital angular momentum ℓ and the nuclear spin J must be coupled to total angular momentum F. Now, in order to obtain the matrix element $\langle (\ell J)F, M_F | U_{\mathrm{E2}} | (\ell J)F, M_F \rangle$ from the Wigner-Eckart theorem one needs the corresponding reduced matrix element $((\ell J)F || U_{\mathrm{E2}} || (\ell J)F)$. In the next section we will show how the latter is obtained.

Example 6.4 Spin Operator in Spin-Orbit States

Let $j_+ := \ell + 1/2$ and $j_- := \ell - 1/2$. The aim of this example is to calculate the reduced matrix elements

[12] There are strongly deformed nuclei in nature whose ground state has spin zero. Their quadrupole moment, then called intrinsic quadrupole moment, can be made visible only in their excitations, i.e. the rotational spectra of the Example 6.2.

$$\left((\ell\frac{1}{2})j_\pm \,||\sigma||\, (\ell\frac{1}{2})j_\pm\right) =: (\pm\,||\sigma||\,\pm)\,,$$
$$\left((\ell\frac{1}{2})j_\pm \,||\sigma||\, (\ell\frac{1}{2})j_\mp\right) =: (\pm\,||\sigma||\,\mp)\,,$$

where we introduced abbreviations as indicated. Three special cases must be evaluated:

1. The state which has $m_\ell = \ell$ and $m_s = +1/2$ reads $|\ell, \ell\rangle\,|1/2, 1/2\rangle$, and one has
$$\langle j_+, m = j_+|\sigma_0\,|j_+, m = j_+\rangle = 1 = \begin{pmatrix} j_+ & 1 & j_+ \\ -j_+ & 0 & j_+ \end{pmatrix}(+||\sigma||+)\,.$$
Inserting here the second formula (6.67) and the value $j_+ = \ell + 1/2$ one obtains the first result:
$$(+\,||\sigma||\,+) = \sqrt{\frac{(2\ell+2)(2\ell+3)}{2\ell+1}}\,.$$
2. By the action of the lowering operator $\mathbf{J}_-$ onto $|j_+, j_+\rangle$ one finds the state
$$|j_+, j_+ - 1\rangle = \frac{1}{\sqrt{2\ell+1}}\left\{\sqrt{2\ell}\,|\ell, \ell-1\rangle\,|1/2, 1/2\rangle + |\ell, \ell\rangle\,|1/2, -1/2\rangle\right\}$$
and, from this, the state which is orthogonal to it,
$$|j_-, j_- = j_+ - 1\rangle$$
$$= \frac{1}{\sqrt{2\ell+1}}\left\{-\,|\ell, \ell-1\rangle\,|1/2, 1/2\rangle + \sqrt{2\ell}\,|\ell, \ell\rangle\,|1/2, -1/2\rangle\right\}\,.$$
The matrix element of σ_0 in the state $|j_-, j_-\rangle$ is then found to be
$$\langle j_-, j_-|\sigma_0\,|j_-, j_-\rangle = \frac{1}{\sqrt{2\ell+1}}(1-2\ell) = \begin{pmatrix} j_- & 1 & j_- \\ -j_- & 0 & j_- \end{pmatrix}(-\,||\sigma||\,-)\,.$$
Here too, one inserts the second formula (6.67) as well as $j_- = \ell - 1/2$ to obtain
$$(-\,||\sigma||\,-) = -\sqrt{2\ell(2\ell-1)}\,.$$
3. Using the states $|j_+, j_-\rangle$ and $|j_-, j_-\rangle$ constructed in 2. one calculates
$$\langle j_-, j_-|\sigma_0\,|j_+, j_-\rangle = -\frac{2\sqrt{2\ell}}{\sqrt{2\ell+1}} = \begin{pmatrix} j_- & 1 & j_+ \\ -j_- & 0 & j_- \end{pmatrix}(-\,||\sigma||\,+)\,.$$
The $3j$-symbol with $j_- = j_+ - 1$ which appears in this formula, has the explicit value $-1/\sqrt{j_+(2j_+ + 1)} = -1/\sqrt{(2\ell+1)(\ell+1)}$. Inserting this yields
$$(-\,||\sigma||\,+) = 2\sqrt{2\ell(\ell+1)} = (+\,||\sigma||\,-)\,.$$
Thus, all three reduced matrix elements are calculated.

6.2.6 *Intertwiner, 6 j- and 9 j-Symbols

This section contains some more advanced material and may be skipped in a first reading. If one does so one should just note the definitions of $6j$- and $9j$-symbols which are used in the next section.

In order to make optimal, practical use of the techniques of the rotation group and of the Wigner-Eckart theorem a last step is still missing in the analysis of coupled quantum states. As a rule, the eigenstates of total angular momentum are composed of more than two individual angular momenta which, furthermore, may be coupled in various orders. An example will help to elucidate this remark: Suppose Z electrons are placed in the bound states of a given attractive potential and assume the interaction between them to be small. The one-particle states then carry definite orbital angular momentum and spin quantum numbers. The states $|FM\rangle$ of the total system, for fixed eigenvalues of total angular momentum F^2 and its 3-component $\mathbf{F}_3$ can be constructed, for example, by coupling all orbital angular momenta $\ell_1, \ell_2, \ldots, \ell_Z$ to a total orbital angular momentum L, all spins to a total spin S, and, eventually, by coupling these to the state $|(LS)FM\rangle$. Alternatively, one may first couple the orbital angular momentum and the spin of each electron to a resulting angular momentum, $|(\ell_i, s_i)j_i, m_i\rangle$, and then, in a second step, couple $j_1, j_2, \ldots, j_Z$ to F and M such as to form a state $|(j_1, j_2, \ldots, j_Z)FM\rangle$. The first choice is called ℓs-coupling, the second is called jj-coupling. Obviously, other, mixed couplings are also possible.

It should be clear that these constructions lead to different but equivalent representations of the rotation group which, hence, are connected by unitary transformations. Therefore, the aim must be to construct these unitary transformations for the rotation group for a few especially relevant cases.

Representations of a Lie group G are also representations of its Lie algebra $\mathfrak{g}$, cf. Sect. 6.2.3. In the theory of Lie algebras transformations of this kind are called *intertwiner,* because, indeed, they twine or twist together different representations in the following sense: Given a representation ϱ_V on the vector space V and a representation ϱ_W on the vector space W. Denote the action of the element $a \in \mathfrak{g}$ of the Lie algebra by $R_V(a)$ when acting in V, by $R_W(a)$ in W. A linear map $\varphi : V \to W$ which commutes with the action of the Lie algebra,

$$\varphi \circ R_V(a) = R_W(a) \circ \varphi \qquad \text{for all} \quad a \in \mathfrak{g}\,,$$

is called an *intertwiner*. In the case of the rotation group these maps follow from the knowledge of the Clebsch-Gordan series and can be given explicitly. We describe two situations of use in practice:

Three Angular Momenta in Different Coupling Schemes: Suppose three angular momenta j_1, j_2, j_3 are coupled in two different ways, viz.

$$\Big[[j_1 \otimes j_2]^{(j_{12})} \otimes j_3\Big]^{(j)} =: A\,, \quad \Big[j_1 \otimes [j_2 \otimes j_3]^{(j_{23})}\Big]^{(j)} =: B\,,$$

where we use a symbolic notation that should be immediately clear. Written more explicitly a state of the first set reads

$$\begin{aligned}
&|(j_{12}, j_3) jm\rangle \\
&\quad = \sum_{m_1,m_2,m_3} (j_1 m_1, j_2 m_2 | j_{12} m_{12})(j_{12} m_{12}, j_3 m_3 | jm) |j_1 m_1\rangle |j_2 m_2\rangle |j_3 m_3\rangle \ ,
\end{aligned}$$

while a state of the second set which carries the same values of j and m reads

$$\begin{aligned}
&|(j_1, j_{23}) jm\rangle \\
&\quad = \sum_{m'_1,m'_2,m'_3} \big(j_1 m'_1, j_{23} m_{23} | jm\big)\big(j_2 m'_2, j_3 m'_3 | j_{23} m_{23}\big) \big|j_1 m'_1\big\rangle \big|j_2 m'_2\big\rangle \big|j_3 m'_3\big\rangle \ .
\end{aligned}$$

In the first case the selection rules for magnetic quantum numbers are $m_{12} = m_1 + m_2$ and $m = m_1 + m_2 + m_3$, in the second case they are $m_{23} = m'_2 + m'_3$ and $m = m'_1 + m'_2 + m'_3$. Both sets span the representation space characterized by j and m. The mapping from the first to the second basis

$$|\sigma\ jm\rangle = \sum_\tau C_{\sigma\tau}\, |\tau\ jm\rangle \ , \quad \text{with} \quad \sigma \equiv (j_1, j_{23}) \text{ and } \tau \equiv (j_{12}, j_3)$$

is given by the scalar products $C_{\sigma\tau} = \langle \tau\ jm | \sigma\ jm\rangle$ and is calculated from the decompositions given above. Replacing the Clebsch-Gordan coefficients by $3j$-symbols, by way of the relation (6.60), one finds

$$\begin{aligned}
C_{\sigma\tau} &\equiv \langle (j_{12}, j_3)\, jm | (j_1, j_{23})\, jm\rangle \\
&= \sqrt{2j_{12}+1}\,(2j+1)\,\sqrt{2j_{23}+1} \\
&\quad \times \sum_{m_1 m_2 m_3} (-)^{j_2-j_1-m_{12}} (-)^{j_3-j_{12}-m} (-)^{j_{23}-j_1-m} (-)^{j_3-j_2-m_{23}} \\
&\quad \times \begin{pmatrix} j_1 & j_2 & j_{12} \\ m_1 & m_2 & -m_{12} \end{pmatrix} \begin{pmatrix} j_{12} & j_3 & j \\ m_{12} & m_3 & -m \end{pmatrix} \\
&\quad \times \begin{pmatrix} j_1 & j_{23} & j \\ m_1 & m_{23} & -m \end{pmatrix} \begin{pmatrix} j_2 & j_3 & j_{23} \\ m_2 & m_3 & -m_{23} \end{pmatrix} .
\end{aligned}$$

Note that use was made of the selection rules for the m-quantum numbers as well as of the orthogonality of the states $|j_i m_i\rangle$.

Strictly speaking, we could have written the mapping matrix in the form $\langle \sigma\ j'm' | \tau\ jm\rangle$, with different values of the total angular momentum and its 3-component. Yet, both sets of states A and B are irreducible. Although they are eigenstates of different sets of commuting operators, all of them are eigenstates of $J^2 = \left(\sum J_i\right)^2$ and of $\mathbf{J}_3 = \sum \mathbf{J}_{i\,3}$. Therefore, the map from A to B is diagonal in j and m. An even more important observation is that the transformation matrix $\mathbf{C} = \{C_{\sigma\tau}\}$ does not depend on m at all. Indeed, under the action of rotations in $\mathbb{R}^3$, both the set A and the set B transform by the unitary $\mathbf{D}^{(j)}(\mathbf{R})$. This means that $\mathbf{C}$ commutes with $\mathbf{D}^{(j)}(\mathbf{R})$ for all $\mathbf{R}$ and, therefore, cannot depend on m. Thus, the expression obtained above is written more precisely

$$C_{\sigma\tau} = \langle (j_{12}, j_3)\, j | (j_1, j_{23})\, j\rangle \ ,$$

that is, without any reference to the magnetic quantum number m.

$$
\begin{Bmatrix} \bullet & \bullet & \bullet \\ & & \end{Bmatrix}
\quad
\begin{Bmatrix} \bullet & & \\ & \bullet & \bullet \end{Bmatrix}
$$

$$
\begin{Bmatrix} & \bullet & \\ \bullet & & \bullet \end{Bmatrix}
\quad
\begin{Bmatrix} & & \bullet \\ \bullet & \bullet & \end{Bmatrix}
$$

Fig. 6.4 The $6j$-symbol has a chance of being different from zero only if the four triples marked by *bullets* obey the triangle rule

The mapping from one coupling scheme to another is of universal nature and, therefore, is independent of the specific choice of the basis in the subspace with fixed value of j. Therefore, it seems appropriate to account for this universal property by a definition of its own. In the case at hand, with three angular momenta, one extracts the factor $\sqrt{2j_{12}+1}\sqrt{2j_{23}+1}$ as well as a sign factor which is symmetric in all j's,

$$
\begin{aligned}
&\langle (j_{12}, j_3)\, j | (j_1, j_{23})\, j \rangle \\
&\quad =: (-)^{j_1+j_2+j_3+j} \sqrt{2j_{12}+1}\sqrt{2j_{23}+1} \begin{Bmatrix} j_1 & j_2 & j_{12} \\ j_3 & j & j_{23} \end{Bmatrix} .
\end{aligned}
\tag{6.76}
$$

The symbol in curly brackets which is defined by this equation, is called $6j$-*symbol.*

The $6j$-symbols can be expressed as sums over products of $3j$-symbols, from the formulae just given. A more useful, explicit formula was derived by G. Racah. This formula which exhibits the selection rules and the symmetries of $6j$-symbols, reads

$$
\begin{aligned}
\begin{Bmatrix} j_1 & j_2 & j_3 \\ l_1 & l_2 & l_3 \end{Bmatrix} &= \Delta(j_1, j_2, j_3)\Delta(j_1, l_2, l_3)\Delta(l_1, j_2, l_3)\Delta(l_1, l_2, j_3) \\
&\quad \times \sum_r (-)^r (r+1)!\, [(r-(j_1+j_2+j_3))!\, (r-(j_1+l_2+l_3))! \\
&\quad \times (r-(l_1+j_2+l_3))!\, ((r-(l_1+l_2+j_3))! \\
&\quad \times (j_1+j_2+l_1+l_2-r)!(j_2+j_3+l_2+l_3-r)! \\
&\quad \times (j_3+j_1+l_3+l_1-r)!]^{-1} .
\end{aligned}
\tag{6.77}
$$

The sum over r contains only a finite number of contributions because the terms of the sum vanish as soon as one of the arguments in the factorials of the denominator is negative. The first four factors contain the triangle symbol (6.55) and guarantee that the four triples of angular momenta on which they depend obey the triangle rule (6.56). If one marks those entries of a $6j$-symbol by a "bullet" which are subject to this rule one obtains the simple scheme of Fig. 6.4.

This scheme and the explicit formula (6.77) yield the

Symmetry Relations of $6j$-Symbols:

(i) The $6j$-symbol is invariant under any permutation of its columns.

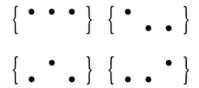

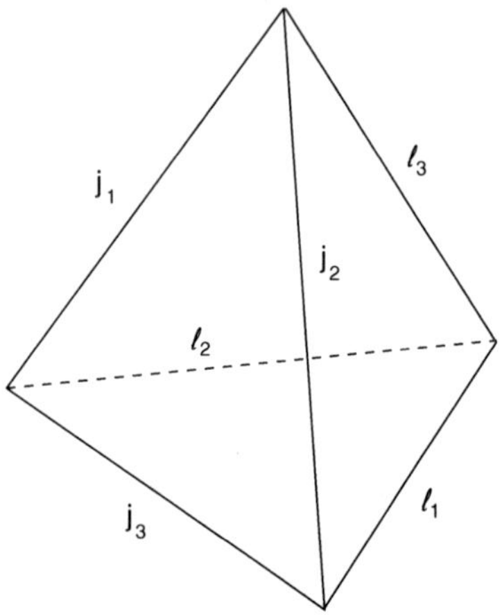

Fig. 6.5 Symmetries of the $6j$-symbols illustrated by a regular tetraeder: Opposite edges correspond to the columns of the symbol; the sides of each triangular surface fulfill the triangle rule

(ii) It stays invariant if in two of the columns the upper entries are exchanged with the lower entries, while leaving the third column unchanged. So, for example, one has

$$\begin{Bmatrix} j_1 & l_2 & l_3 \\ l_1 & j_2 & j_3 \end{Bmatrix} = \begin{Bmatrix} j_1 & j_2 & j_3 \\ l_1 & l_2 & l_3 \end{Bmatrix} .$$

(iii) The symmetry relations of $6j$-symbols with respect to permutations correspond to the symmetries of a regular tetrahedron under exchange of its edges. In Fig. 6.5 the edges are marked by the six angular momenta contained in the $6j$-symbol

$$\begin{Bmatrix} j_1 & j_2 & j_3 \\ l_1 & l_2 & l_3 \end{Bmatrix}$$

in such a way that each triangle is associated to one of the triples (j_1, j_2, j_3) etc., the three pairs of opposing edges corresponding to the columns of the $6j$-symbol.

(iv) Whenever one of the entries equals zero the symbol simplifies greatly. For example, one finds

$$\begin{Bmatrix} j_1 & j_2 & j_3 \\ l_1 & l_2 & 0 \end{Bmatrix} = (-)^{j_1+j_2+j_3} \delta_{j_1 l_2} \delta_{j_2 l_1} \delta(j_1, j_2, j_3) \frac{1}{\sqrt{(2j_1+1)(2j_2+1)}} ,$$

where the symbol $\delta(j_1, j_2, j_3)$ stands for the requirement that j_1, j_2, and j_3 fulfill the triangle rule (6.56). It is equal to 1 when this rule is fulfilled, and is zero in all other cases. (One should not confound it with the symbol $\Delta(j_1, j_2, j_3)$ of (6.55).)

(v) The transformation matrix $\mathbf{C}$ is unitary (or orthogonal, respectively),

$$\sum_{j_{23}} \langle (j_{12}, j_3)\, j | (j_1, j_{23})\, j \rangle \, \langle (j_1, j_{23})\, j | (j'_{12}, j_3)\, j \rangle = \delta_{j_{12} j'_{12}} .$$

Furthermore, by the group property it is associative,

$$\sum_{j_{13}} \langle (j_{12}, j_3)\, j | (j_{13}, j_2)\, j \rangle \langle (j_{13}, j_2)\, j | (j_1, j_{23})\, j \rangle$$
$$= \langle (j_{12}, j_3)\, j | (j_1, j_{23})\, j \rangle .$$

From this follow two relations for $6j$-symbols which read, in a somewhat simplified notation,

$$\sum_c (2c+1) \begin{Bmatrix} a & b & c \\ d & e & f \end{Bmatrix} \begin{Bmatrix} a & b & c \\ d & e & g \end{Bmatrix} = \delta_{fg} \delta(a, e, f) \delta(d, b, f) \frac{1}{2f+1} , \tag{6.78}$$

$$\sum_c (-)^{c+f+g} (2c+1) \begin{Bmatrix} a & b & c \\ d & e & f \end{Bmatrix} \begin{Bmatrix} a & b & c \\ e & d & g \end{Bmatrix} = \begin{Bmatrix} a & e & f \\ b & d & g \end{Bmatrix} . \tag{6.79}$$

The δ-symbol containing three arguments is equal to 1 if these obey the triangle rule (6.56), it vanishes in all other cases.

Three and Four Angular Momenta Coupled to Zero: The $3j$- and $6j$-symbols come into play also when one couples three or four angular momenta to total angular momentum zero. This observation turns out to be useful in applications of the Wigner-Eckart theorem. We first consider the case

$$\left[[j_1 \otimes j_2]^{(j_{12})} \otimes j_3 \right]^{(0)}_0$$

and note that we must have $j_{12} = j_3$. The triple product is easily evaluated

$$\left[[j_1 \otimes j_2]^{(j_{12})} \otimes j_3 \right]^{(0)}_0 = \sum_{m's} \frac{(-)^{j_3+m_3}}{\sqrt{2j_3+1}}$$
$$(j_3, -m_3 | j_1 m_1, j_2 m_2) | j_1 m_1 \rangle | j_2 m_2 \rangle \, | j_3 m_3 \rangle .$$

Comparing this with the other coupling which is possible

$$\left[j_1 \otimes [j_2 \otimes j_3]^{(j_{23})} \right]^{(0)}_0 = \sum_{m's} \frac{(-)^{j_1-m_1}}{\sqrt{2j_1+1}}$$
$$(j_1, -m_1 | j_2 m_2, j_3 m_3) | j_1 m_1 \rangle | j_2 m_2 \rangle \, | j_3 m_3 \rangle ,$$

the relation (6.58) shows the two right-hand sides to be equal. Thus, the triple product is associative and may be written in the simplified form $[j_1 \otimes j_2 \otimes j_3]^{(0)}_0$. Note, however, that in general it is not commutative. Indeed, one shows easily that

$$[j_3 \otimes j_1 \otimes j_2]^{(0)}_0 = (-)^{2j_3} [j_1 \otimes j_2 \otimes j_3]^{(0)}_0$$
$$[j_2 \otimes j_3 \otimes j_1]^{(0)}_0 = (-)^{2j_1} [j_1 \otimes j_2 \otimes j_3]^{(0)}_0 .$$

As the sum $j_1 + j_2 + j_3$ is always an integer, the sign will change only if the angular momentum in the middle position, by the permutation, turns from an integer to a half-integer, or vice versa. This implies that the quantity

$$(-)^{2j_2} [j_1 \otimes j_2 \otimes j_3]^{(0)}_0 ,$$

(thus multiplied by the phase factor $(-)^{2j_2}$ of the middle angular momentum) is invariant under cyclic permutations of its arguments, and obtains the phase factor $(-)^{j_1+j_2+j_3}$ under anticyclic permutations. These are precisely the sign rules of $3j$-symbols and, indeed, one finds

$$(-)^{2j_2}\,[j_1 \otimes j_2 \otimes j_3]^{(0)}_0 = (-)^{j_1+j_2+j_3} \begin{pmatrix} j_1 & j_2 & j_3 \\ m_1 & m_2 & m_3 \end{pmatrix} .$$

Thus, the $3j$-symbols describe the coupling of three angular momenta to total angular momentum zero.

As we now show, an analogous statement holds for the coupling of *four* angular momenta coupled to total angular momentum zero, involving $6j$-symbols instead of $3j$-symbols. We consider the following two couplings

$$\Big[[j_1 \otimes j_2]^{(j_{12})} \otimes [j_3 \otimes j_4]^{(j_{34})}\Big]^{(0)}_0 , \quad \Big[\Big[[j_1 \otimes j_2]^{(j_{12})} \otimes j_3\Big]^{(j_{123})} \otimes j_4\Big]^{(0)}_0 .$$

The two schemes are seen to be products of *three* angular momenta coupled to zero, j_{12}, j_3, and j_4. As shown above this product is associative. Therefore, the two products are the same. Of course, the same conclusion applies to the products of four where j_2 and j_3 are interchanged,

$$\Big[[j_1 \otimes j_3]^{(j_{12})} \otimes [j_2 \otimes j_4]^{(j_{34})}\Big]^{(0)}_0 = \Big[\Big[[j_1 \otimes j_3]^{(j_{13})} \otimes j_2\Big]^{(j_{132})} \otimes j_4\Big]^{(0)}_0 .$$

We conclude that there are only two genuinely different ways of coupling the product of four to zero (the schemes where j_1 and j_2 are interchanged and/or where j_3 and j_4 are interchanged, differ only by signs). We write them, in a shorter notation,

$$[j_{12} \otimes j_{34}]^{(0)}_0 \quad \text{and} \quad [j_{13} \otimes j_{24}]^{(0)}_0 .$$

The mapping that relates them reduces to $\langle (j_{12}, j_3)\, j_4 | (j_{13}, j_2)\, j_4 \rangle$, i.e. a transformation known from (i) above. With (6.76) one obtains

$$\begin{aligned} \langle (j_{12}, j_{34})\, 0 \,|\, (j_{13}, j_{24})\, 0 \rangle &= \langle (j_{12}, j_3)\, j_4 | (j_{13}, j_2)\, j_4 \rangle \\ &= \sqrt{(2j_{12}+1)(2j_{13}+1)} \\ &\qquad (-)^{j_{12}+j_{13}+j_2+j_3} \begin{Bmatrix} j_1 & j_2 & j_{12} \\ j_4 & j_3 & j_{13} \end{Bmatrix} . \end{aligned}$$

The $6j$-symbols come into play in the case of four angular momenta which are coupled to zero in different ways. We now turn to the determination of these mappings between different coupling schemes in cases where the total angular momentum is not zero.

Four Angular Momenta in Different Couplings: Clearly, it is always possible to couple neighbouring angular momenta in different orderings such as, e.g., $(j_2 \otimes j_1)^{(j_{12})}$ instead of $(j_1 \otimes j_2)^{(j_{12})}$. By the relation (6.57) this can cause at most a sign change $(-)^{j_1+j_2-j_{12}}$. For this reason we have not distinguished these simple exchanges of neighbours. When we now study recouplings of four angular momenta we will not keep track of the permutations of neighbours either.

The experience gained in the preceding paragraph shows that matters simplify considerably if one classifies the coupling schemes of irreducible products of four with respect to products of *five* of rank zero. As an example, consider the product $[j_{12} \otimes j_{34}]^{(j)}$. It corresponds to the product of five angular momenta $\left[[j_{12} \otimes j_{34}]^{(j)} \otimes j_5\right]^{(0)}$, coupled to zero, hence with $j = j_5$. As the product of *three* with rank zero is associative, it may be written in the simplified notation $\left[j_{12} \otimes j_{34} \otimes j_5\right]^{(0)}$. In analogy the product of four $\left[[j_{12} \otimes j_3]^{(j_{123})} \otimes j_4\right]^{(j_5)}$ corresponds to the scheme $\left[\left[[j_{12} \otimes j_3]^{(j_{123})} \otimes j_4\right]^{(j_5)} \otimes j_5\right]^{(0)}$ which, in turn, is equivalent to $[[j_{12} \otimes j_3] \otimes [j_4 \otimes j_5]]^{(0)}$, i.e. to the associative product of three angular momenta $[j_{12} \otimes j_3 \otimes j_{45}]^{(0)}$.

These examples show that any product of five which has rank zero can be written as a product of three composed of two pairs and a single angular momentum. The various coupling schemes differ only by the choice of pairs and by the ordering of its three factors. Except for exchange of neighbours this means that the intertwiners are classified in three types that we now analyze, one after the other.

(a) $\langle (j_{12} j_{34} j_5)\, 0 \,|(j_{12} j_3 j_{45})\, 0 \rangle$,

(b) $\langle (j_{12} j_{34} j_5)\, 0 \,|(j_1 j_{23} j_{45})\, 0 \rangle$,

(c) $\langle (j_{12} j_{34} j_5)\, 0 \,|(j_{13} j_{24} j_5)\, 0 \rangle$.

Type (a): As j_{12} remains unchanged, this map reduces to the recoupling of products of four with rank zero (that is to say, j_{12}, j_3, j_4, and j_5 coupled to zero), or to the equivalent mapping $\langle (j_{34} j_5) j_{12} | (j_3 j_{45}) j_{12} \rangle$ of products of three with rank j_{12}.
Type (b): These mappings are products of two mappings of type (a). This is seen best by means of an example:

$$\begin{aligned}\langle (j_{12} j_{34} j_5)\, 0 \,|(j_1 j_{23} j_{45})\, 0 \rangle &= \langle (j_{12} j_{34} j_5)\, 0 \,|(j_{12} j_3 j_{45})\, 0 \rangle \\ &\quad \langle (j_{12} j_3 j_{45})\, 0 \,|(j_1 j_{23} j_{45})\, 0 \rangle \\ &= \langle (j_{34} j_5) j_{12} | (j_3 j_{45}) j_{12} \rangle \\ &\quad \langle (j_{12} j_3) j_{45} | (j_1 j_{23}) j_{45} \rangle \; .\end{aligned}$$

Type (c): These mappings decompose into products of three recoupling transformations of products of three where one has to sum over the rank of the factor in the middle. For instance,

$$\begin{aligned}&\big\langle (j_{12} j_{34} j_5)\, 0 \,\big|(j_{13} j_{24} j_5)\, 0 \big\rangle \\ &= \sum_{j_{45}} \big\langle (j_{34} j_5) j_{12} \big| (j_3 j_{45}) j_{12} \big\rangle \big\langle (j_{12} j_3) j_{45} \big| (j_{13} j_2) j_{45} \big\rangle \big\langle (j_2 j_{45}) j_{13} \big| (j_{24} j_5) j_{13} \big\rangle \; .\end{aligned}$$

The left-hand side is equivalent to

$$\langle (j_{12} j_{34}) j | (j_{13} j_{24}) j \rangle \; , \tag{6.80}$$

hence a transformation which was analyzed in the first case dealt with above. Inserting its results and taking account of the fact that there j_{12} and j_3, or j_1 and j_{23}, respectively, were coupled to j, while here it is the pairs (j_{12}, j_{34}) and (j_{13}, j_{24}), respectively, which are coupled to j, it is immediately clear that (6.80) is given by a product of *six* $3j$-symbols (instead of four $3j$-symbols as there). Again, it seems appropriate to replace this product by a new definition, the *$9j$-symbols:*

$$\begin{Bmatrix} j_1 & j_2 & J_1 \\ j_3 & j_4 & J_2 \\ j_5 & j_6 & J \end{Bmatrix} := \sum_{m's} \begin{pmatrix} j_1 & j_2 & J_1 \\ m_1 & m_2 & M_1 \end{pmatrix} \begin{pmatrix} j_3 & j_4 & J_2 \\ m_3 & m_4 & M_2 \end{pmatrix}$$
$$\times \begin{pmatrix} j_5 & j_6 & J \\ m_5 & m_6 & M \end{pmatrix} \begin{pmatrix} j_1 & j_3 & j_5 \\ m_1 & m_3 & m_5 \end{pmatrix}$$
$$\times \begin{pmatrix} j_2 & j_4 & j_6 \\ m_1 & m_4 & m_6 \end{pmatrix} \begin{pmatrix} J_1 & J_2 & J \\ M_1 & M_2 & M \end{pmatrix} . \tag{6.81}$$

Equipped with the experience gained so far it is not difficult to derive the symmetries of $9j$-symbols. Let $\Sigma := \sum_1^6 j_i + J_1 + J_2 + J$ denote the sum of all nine angular momenta. Then one has the following

Symmetry Relations for $9j$-Symbols:

(i) A $9j$-symbol can be different from zero only if the three angular momenta in each row and in each column obey the triangle relation (6.56).

(ii) A $9j$-symbol is invariant under *cyclic* permutations of its columns as well as under *cyclic* permutations of its rows.

(iii) An *odd* permutation of its columns, or of its rows, multiplies the $9j$-symbol by the phase factor $(-)^{\Sigma}$. In particular, if two columns or two rows are equal, and if Σ is an odd integer, the $9j$-symbol vanishes.

(iv) If one of its entries is equal to zero then the $9j$-symbol reduces to a $6j$-symbol. For example, one has

$$\begin{Bmatrix} j_1 & j_2 & J \\ j_4 & j_3 & j \end{Bmatrix} = (-)^{j_2+j_3+j+J} \sqrt{(2J+1)(2j+1)} \begin{Bmatrix} j_1 & j_2 & J \\ j_3 & j_4 & J \\ j & j & 0 \end{Bmatrix} . \tag{6.82}$$

Very much like the $3j$- and $6j$-symbols the $9j$-symbols are universal objects of $SU(2)$, independent of any choice of basis, and, thus, may be tabulated as well. The practical use of these *intertwiner* mappings will be seen when deriving the relations for reduced matrix elements to which we turn in the next section.

6.2.7 Reduced Matrix Elements in Coupled States

In the framework of perturbation theory, or in calculating transition probabilities, one often encounters matrix elements of two-body interactions with coupled eigenstates of angular momentum. For example, the tensor operator $T^{(\kappa_1)}$, acting on particle "1", and the tensor operator $S^{(\kappa_2)}$, acting on particle "2", may appear coupled to a new tensor operator

$$M^{(\kappa)}(1,2) = \left[T^{(\kappa_1)}(1) \otimes S^{(\kappa_2)}(2) \right]^{(\kappa)} , \tag{6.83}$$

which reads, when written in components,

$$M_\mu^{(\kappa)}(1,2) = \sum_{\tau=-\kappa_1}^{+\kappa_1} \sum_{\sigma=-\kappa_2}^{+\kappa_2} (\kappa_1\tau, \kappa_2\sigma|\kappa\mu)\, T_\tau^{(\kappa_1)}(1)\, S_\sigma^{(\kappa_2)}(2)\,. \tag{6.84}$$

Applying the Wigner-Eckart theorem to matrix elements between coupled states of the type $|(j_1 j_2)J)\rangle$, at first, leads to reduced matrix elements such as

$$\left((j_1 j_2)J \,\|\, M^{(\kappa)}(1,2) \,\|\, (j_1' j_2')J'\right)\,.$$

Clearly, in order to express these as functions of reduced matrix elements of one-body operators

$$\left(j_1 \,\|\, T^{(\kappa_1)}(1) \,\|\, j_1'\right)\,, \quad \left(j_2 \,\|\, S^{(\kappa_2)}(2) \,\|\, j_2'\right)$$

which either are known, or may be easier to calculate in practice, one must "decouple" again the states and the operators. With this in mind one will not be surprised to discover that these relations involve the *intertwiner* transformations studied in the preceding section. The following formulae are of great practical importance. They all follow from the recoupling transformations studied above. For details of their derivation I refer to the literature, e.g., [de Shalit and Talmi (1963)].

One must distinguish two cases:

(a) If, as assumed above, the two operators act on two *different* systems, one has

$$\begin{aligned}\left((j_1 j_2)J \,\left\|\, M^{(\kappa)}(1,2) \,\right\|\, (j_1' j_2')J'\right) &= \sqrt{(2J+1)(2\kappa+1)(2J'+1)} \begin{Bmatrix} j_1 & j_2 & J \\ j_1' & j_2' & J' \\ \kappa_1 & \kappa_2 & \kappa \end{Bmatrix} \\ &\times \left(j_1 \,\left\|\, T^{(\kappa_1)}(1) \,\right\|\, j_1'\right) \left(j_2 \,\left\|\, S^{(\kappa_2)}(2) \,\right\|\, j_2'\right)\,. \end{aligned} \tag{6.85}$$

The reduced matrix elements of the individual operators $T^{(\kappa_1)}$, or $S^{(\kappa_2)}$, in coupled states are given by the formulae

$$\begin{aligned}\left((j_1 j_2)J \,\left\|\, T^{(\kappa_1)}(1) \,\right\|\, (j_1' j_2')J'\right) &= (-)^{j_1+j_2+J'+\kappa_1} \sqrt{(2J+1)(2J'+1)} \\ &\times \left(j_1 \,\left\|\, T^{(\kappa_1)}(1) \,\right\|\, j_1'\right) \begin{Bmatrix} j_1 & J & j_2 \\ J' & j_1' & \kappa_1 \end{Bmatrix} \delta_{j_2 j_2'}\,, \end{aligned} \tag{6.86}$$

$$\begin{aligned}\left((j_1 j_2)J \,\left\|\, S^{(\kappa_2)}(2) \,\right\|\, (j_1' j_2')J'\right) &= (-)^{j_1+j_2'+J+\kappa_2} \sqrt{(2J+1)(2J'+1)} \\ &\times \left(j_2 \,\left\|\, S^{(\kappa_2)}(2) \,\right\|\, j_2'\right) \begin{Bmatrix} j_2 & J & j_1 \\ J' & j_2' & \kappa_2 \end{Bmatrix} \delta_{j_1 j_1'}\,. \end{aligned} \tag{6.87}$$

An important special case is one where the two operators $T^{(\kappa_1)}$ and $S^{(\kappa_2)}$ are coupled to a scalar. In this case it is useful to replace the expression (6.83) by a *scalar product* which is defined as follows

$$\left(T^{(\kappa)} \cdot S^{(\kappa)}\right) := (-)^\kappa \sqrt{2\kappa+1} \left[T^{(\kappa)} \otimes S^{(\kappa)}\right]_0^{(0)}\,, \tag{6.88}$$

and which is equal to

$$= (-)^{\kappa}\sqrt{2\kappa+1}\sum_{\tau,\sigma}(\kappa\tau,\kappa\sigma|00)\,T^{(\kappa)}_{\tau}S^{(\kappa)}_{\sigma} = \sum_{\mu=-\kappa}^{+\kappa}(-)^{\mu}T^{(\kappa)}_{\mu}S^{(\kappa)}_{-\mu}\,.$$

The formula (6.85) then reduces to the special case

$$\Big((j_1 j_2)J\Big|\Big|\Big(T^{(\kappa)}\cdot S^{(\kappa)}\Big)\Big|\Big|(j_1' j_2')J'\Big) = (-)^{j_2+J+j_1'}\sqrt{2J+1}\,\delta_{JJ'}\begin{Bmatrix} j_1 & j_2 & J \\ j_2' & j_1' & \kappa \end{Bmatrix}$$
$$\times\Big(j_1\Big|\Big|T^{(\kappa)}\Big|\Big|j_1'\Big)\Big(j_2\Big|\Big|S^{(\kappa)}\Big|\Big|j_2'\Big)\,. \qquad (6.89)$$

In general, the angular momenta j_i and j_k' of the formulae (6.85)–(6.89) are accompanied by further quantum numbers α_i and α_k', respectively. In (6.85) and in (6.89) these just go through, down to the individual reduced matrix elements so that j_i is to be replaced by α_i, j_i (and likewise for the primed quantities). In formula (6.86), however, one has the additional condition $\alpha_2 = \alpha_2'$, while in formula (6.87) one must have $\alpha_1 = \alpha_1'$.

(b) If the two operators act on the *same* system, i.e. if one has

$$\Big[T^{(\kappa_1)}(i)\otimes S^{(\kappa_2)}(i)\Big]^{(\kappa)} = T^{(\kappa)}(i)\,, \qquad (6.90)$$

and if the angular momenta are accompanied by further quantum numbers α, then one has

$$\Big(\alpha j\Big|\Big|M^{(\kappa)}(i)\Big|\Big|\alpha' j'\Big) = (-)^{j+\kappa+j'}\sqrt{2\kappa+1}\sum_{\alpha'' j''}\Big(\alpha j\Big|\Big|T^{(\kappa_1)}(i)\Big|\Big|\alpha'' j''\Big)$$
$$\times\Big(\alpha'' j''\Big|\Big|S^{(\kappa_2)}(i)\Big|\Big|\alpha' j'\Big)\begin{Bmatrix}\kappa_1 & \kappa_2 & \kappa \\ j' & j & j''\end{Bmatrix}\,. \qquad (6.91)$$

The sum over j'' runs through all values which are allowed by the $6j$-symbol, that is, which are compatible with the triangle rules (κ_1, j, j'') and (j', κ_2, j'').

This section concludes with some formulae that the reader is invited to confirm, and which will be needed in many examples of practical relevance.

Special Cases:

– If the operator $T_0^{(\kappa)}$ is self-adjoint then we have

$$\Big(J\big|\big|T^{(\kappa)}\big|\big|J'\Big) = (-)^{J-J'}\Big(J'\big|\big|T^{(\kappa)}\big|\big|J\Big)^{*}\,. \qquad (6.92)$$

– As a consequence of the conventions adopted in the Wigner-Eckart theorem the reduced matrix element of the unity is not equal to 1 but is

$$\big(J\,||\,\mathbb{1}\,||\,J'\big) = \delta_{JJ'}\sqrt{2J+1}\,. \qquad (6.93)$$

– For the operators of the angular momentum itself one has

$$(J\,||\,\mathbf{J}\,||\,J) = \sqrt{J(J+1)(2J+1)}\,. \qquad (6.94)$$

Applying this to the spin operator of the electron the right-hand side is equal to $\sqrt{3/2}$. Thus, for the operator $\boldsymbol{\sigma} = 2\mathbf{s}$ one has $(1/2||\boldsymbol{\sigma}||1/2) = \sqrt{6}$.

– The reduced matrix elements of spherical harmonics taken between eigenstates of $\boldsymbol{\ell}^2$ are

$$\left(\ell || \mathbf{Y}_\lambda || \ell'\right) = \frac{(-)^\ell}{\sqrt{4\pi}} \sqrt{(2\ell+1)(2\lambda+1)(2\ell'+1)} \begin{pmatrix} \ell & \lambda & \ell' \\ 0 & 0 & 0 \end{pmatrix} . \tag{6.95}$$

– The reduced matrix element of spherical harmonics in states in which the orbital angular momentum and the spin are coupled to total angular momentum j is equal to

$$\left((\ell \tfrac{1}{2}) j || \mathbf{Y}_\lambda || (\ell' \tfrac{1}{2}) j'\right) = \frac{(-)^{j+1/2}}{\sqrt{4\pi}} \sqrt{(2j+1)(2\lambda+1)(2j'+1)} \times \begin{pmatrix} j & \lambda & j' \\ 1/2 & 0 & -1/2 \end{pmatrix} \frac{1+(-)^{\ell+\lambda+\ell'}}{2} . \tag{6.96}$$

Regarding the formulae (6.95) and (6.96) the following remark is appropriate: The last factor on the right-hand side of (6.96) reflects the selection rule due to parity. Indeed, it is equal to 1 if $(-)^{\ell'}(-)^\lambda = (-)^\ell$; it vanishes in all other cases. The same selection rule in (6.95) is hidden in the $3j$-symbol which according to (6.64) is equal to zero whenever $\ell + \lambda + \ell'$ is *odd*. If spin and orbital angular momentum are coupled in the ordering $(1/2\ \ell)j$ then (6.96) is modified by a phase that follows from the symmetry relation (6.57).

– Let $\mathbf{T}$ be an arbitrary *vector* operator (i.e. a tensor operator of rank 1). Its matrix elements between states pertaining to the *same* value of J are proportional to the corresponding matrix elements of the angular momentum operator,

$$\langle \alpha J M | \mathbf{T} \left| \alpha' J M' \right\rangle = \frac{\langle \alpha J M | \left(\mathbf{J} \cdot \mathbf{T}\right) \left| \alpha' J M \right\rangle}{J(J+1)} \langle J M | \mathbf{J} \left| J M' \right\rangle . \tag{6.97}$$

Note, however, that the operator $\mathbf{T}$ may also have nonvanishing matrix elements between states with different values J and J' – in contrast to the operator $\mathbf{J}$. The formula (6.97) only holds within the subspace with a given value of J.

6.2.8 Remarks on Compact Lie Groups and Internal Symmetries

The rotation group $SU(2)$ is not only of central importance for the description of spin. It also provides a prototype model which is helpful as a landmark in the study of other symmetry groups in quantum physics. Finite-dimensional Lie groups, and among them, more specifically, the *compact* Lie groups, are applied to nuclear and elementary particle physics in various interpretations. For example, the group $SU(2)$ is useful in describing the charge symmetry of proton, neutron, and composite states thereof. In this variant one talks about (strong interaction) *isospin*, having in mind the empirical symmetry observed in the interactions of nucleons. This symmetry has no relation to space and time but refers to an extension of Hilbert space whose basis is characterized by internal degrees of freedom. Symmetries of this kind are called *internal symmetries*.

Further compact Lie groups of relevance for physics are: the groups $SU(3)$, $SU(4)$, or, more generally, the unitary groups $SU(n)$ in n complex dimensions. Many of the techniques that one has learnt in studying the example of $SU(2)$, with due care and caution, can be generalized to higher groups. One studies irreducible unitary representations by first constructing a maximal set of commuting operators which are the analogues of $\boldsymbol{J}^2$ and $\mathbf{J}_3$, and whose eigenvalues serve to classify the representations. There exist Clebsch-Gordan series, that is, decompositions of the tensor product of two representations in terms of irreducible representations, and there exists the generalization of the Wigner-Eckart theorem. We sketch here an example which illustrates some similarities with the rotation group but also shows some essential differences.

Example 6.5 The Group $SU(3)$

The group $SU(3)$, the *unimodular group in three complex dimensions*, is defined as follows

$$SU(3) = \left\{ \mathbf{U} \text{ complex } 3 \times 3\text{-matrices} \mid \mathbf{U}^\dagger \mathbf{U} = \mathbb{1}\,,\ \det \mathbf{U} = 1 \right\}. \tag{6.98}$$

Simple counting shows that its elements depend on 8 unrestricted real parameters: Any complex 3×3-matrix initially depends on 9 complex, or 18 real entries. The condition $\mathbf{U}^\dagger \mathbf{U} = \mathbb{1}$, when written out explicitly, is seen to yield 3 real and 3 complex equations (diagonal and nondiagonal elements, respectively). In addition, the condition $\det \mathbf{U} = 1$ yields one more real equation (why is this only one condition?). In total, there remain $18 - (9 + 1) = 8$ parameters. As one is dealing with a *compact* group these parameters can be chosen to be generalized Euler angles. Therefore, they lie in intervals $(0, \pi)$ or $(0, 2\pi)$. (Note that more generally, the elements of $SU(n)$ depend on $n^2 - 1$ real parameters.) Thus, the corresponding Lie algebra $\mathfrak{su}(3)$ has 8 generators $\lambda_1, \dots, \lambda_8$, for which the following choice is a convenient one

$$\lambda_1 = \begin{pmatrix} 0 & 1 & 0 \\ 1 & 0 & 0 \\ 0 & 0 & 0 \end{pmatrix}, \quad \lambda_2 = \begin{pmatrix} 0 & -\mathrm{i} & 0 \\ \mathrm{i} & 0 & 0 \\ 0 & 0 & 0 \end{pmatrix}, \quad \lambda_3 = \begin{pmatrix} 1 & 0 & 0 \\ 0 & -1 & 0 \\ 0 & 0 & 0 \end{pmatrix},$$

$$\lambda_4 = \begin{pmatrix} 0 & 0 & 1 \\ 0 & 0 & 0 \\ 1 & 0 & 0 \end{pmatrix}, \quad \lambda_5 = \begin{pmatrix} 0 & 0 & -\mathrm{i} \\ 0 & 0 & 0 \\ \mathrm{i} & 0 & 0 \end{pmatrix}, \quad \lambda_6 = \begin{pmatrix} 0 & 0 & 0 \\ 0 & 0 & 1 \\ 0 & 1 & 0 \end{pmatrix}, \tag{6.99}$$

$$\lambda_7 = \begin{pmatrix} 0 & 0 & 0 \\ 0 & 0 & -\mathrm{i} \\ 0 & \mathrm{i} & 0 \end{pmatrix}, \quad \lambda_8 = \frac{1}{\sqrt{3}} \begin{pmatrix} 1 & 0 & 0 \\ 0 & 1 & 0 \\ 0 & 0 & -2 \end{pmatrix}.$$

(These matrices are called *Gell-Mann matrices*.[13])

[13] After Murray Gell-Mann who, along with Yuval Ne'eman, proposed the $SU(3)$ *flavour* classification of mesons and baryons, cf. Gell-Mann, M., Ne'eman, Y.: *The Eightfold Way* (Benjamin, New York 1964).

The matrices (6.99) are linearly independent, they all have trace zero. Every element of the Lie algebra $\mathfrak{su}(3)$, i.e. every traceless hermitian 3×3-matrix, can be expressed as a linear combination of these matrices. Obviously, they were constructed in analogy to the Pauli matrices which one recognizes as being embedded in λ_1, λ_2, and λ_3, and likewise in λ_4, λ_5, and $(\lambda_3 + \sqrt{3}\lambda_8)/2$, as well as in the group of three λ_6, λ_7, and $(-\lambda_3 + \sqrt{3}\lambda_8)/2$. Defining the generators in analogy to $SU(2)$ by

$$T_i := \frac{1}{2}\lambda_i\,, \quad i = 1, 2, \ldots, 8\,, \tag{6.100}$$

they are seen to have the following normalization and to fulfill the commutation rules

$$\mathrm{tr}\,(T_i T_k) = \frac{1}{2}\delta_{ik}\,, \quad [T_i, T_j] = \mathrm{i}\sum_k f_{ijk} T_k\,. \tag{6.101}$$

The constants f_{ijk} are the *structure constants of SU(3)*. The structure constants are antisymmetric in all three indices and have the values

ijk	123	147	156	246	257	345	367	458	678
f_{ijk}	1	1/2	−1/2	1/2	1/2	1/2	−1/2	$\sqrt{3}/2$	$\sqrt{3}/2$

(6.102)

All structure constants which do not appear explicity here, either are related to the ones in the table by even or odd permutations, or vanish.

As analogues of the operator $\boldsymbol{J}^2$ in $SU(2)$ the group $SU(3)$ contains *two* operators which commute with all generators. They are called *Casimir operators*. (They are not written down here.) The diagonal, mutually commuting generators λ_3 and λ_8, the analogues of $\mathbf{J}_3$, are contained in the set (6.99). In applications to particle physics, instead of λ_8, or T_8, one prefers to define

$$Y := \frac{2}{\sqrt{3}} T_8 = \begin{pmatrix} 1/3 & 0 & 0 \\ 0 & 1/3 & 0 \\ 0 & 0 & -2/3 \end{pmatrix}.$$

The fundamental representation is three-dimensional. It is denoted by **3** and contains the states[14]

$$\mathbf{3}\,:\,(T_3, Y) = \left(\frac{1}{2}, \frac{1}{3}\right), \quad \left(-\frac{1}{2}, \frac{1}{3}\right), \quad \left(0, -\frac{2}{3}\right).$$

Plotting these states in a diagram whose axes are T_3 and Y yields the triangle (u, d, s) shown in Fig. 6.6. A further fundamental representation has the quantum numbers

$$\overline{\mathbf{3}}\,:\,(T_3, Y) = \left(-\frac{1}{2}, -\frac{1}{3}\right), \quad \left(\frac{1}{2}, -\frac{1}{3}\right), \quad \left(0, \frac{2}{3}\right).$$

This is the triangle $(\overline{u}, \overline{d}, \overline{s})$ of Fig. 6.6. In contrast to $SU(2)$ where the doublet and its conjugate are equivalent, the representations **3** and $\overline{\mathbf{3}}$ are not equivalent. The first of them is said to be the triplet representation, the second is said to be the antitriplet representation of $SU(3)$.

[14] For example, these are interpreted as the quantum numbers of the *up-*, *down-*, and *strange*-quarks in the quark model of strongly interacting particles.

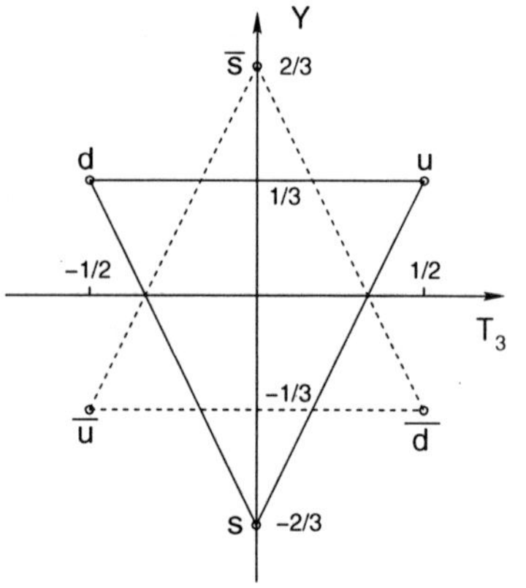

Fig. 6.6 The triplet representation **3** and the antitriplet $\overline{\mathbf{3}}$ of $SU(3)$, realized here by quark states (u, d, s) and antiquark states $(\overline{u}, \overline{d}, \overline{s})$, respectively

As for the rotation group all other unitary irreducible representations of $SU(3)$ can be obtained from the **3** and the $\overline{\mathbf{3}}$ (in the old literature this was called the spinor method). Without delving into the details I quote here the most relevant Clebsch-Gordan series for $SU(3)$. They read

$$\mathbf{3} \otimes \overline{\mathbf{3}} = \mathbf{1} \oplus \mathbf{8}\,, \ \mathbf{3} \otimes \mathbf{3} = \mathbf{3}_a \oplus \mathbf{6}_s\,, \ \mathbf{3} \otimes \mathbf{3} \otimes \mathbf{3} = \mathbf{1}_a \oplus \mathbf{8} \oplus \mathbf{8} \oplus \mathbf{10}_s\,. \quad (6.103)$$

The numbers printed in boldface give the dimension of the irreducible representation. Like in the rotation group the **1** is called the *singlet,* the **8** is new and is called the *octet,* and so is the **10** which is called the *decuplet* representations. The suffix "a" or "s" says that the corresponding representation is antisymmetric or symmetric, respectively, in the two factors of the left-hand side. The second Clebsch-Gordan series in (6.103), for instance, is analogous to the coupling of two spin-1/2 states in $SU(2)$ which yields the antisymmetric singlet (whose dimension is 1) and the symmetric triplet (with dimension 3). Thus, if one used the same notation as above, one would write $\mathbf{2} \otimes \mathbf{2} = \mathbf{1} \oplus \mathbf{3}$ for $SU(2)$.

The first and the third Clebsch-Gordan series of (6.103) are new. In particular, the third series shows that in decomposing a tensor product, a representation may occur more than once. In this example it is the octet which comes in twice on the right-hand side. This also means that in the analogue of the Wigner-Eckart theorem there can be more than one reduced matrix element with given quantum numbers with respect to $SU(3)$.

In elementary particle physics the octet, i.e. the adjoint representation of $SU(3)$, serves to classify some of the stable or quasi-stable, strongly interacting particles, the *mesons* and *baryons*. The decuplet **10** and its conjugate $\overline{\mathbf{10}}$ describe further multiplets of quasi-stable and unstable baryons. Thus, a further Clebsch-Gordan series of $SU(3)$ which is relevant for particle physics, is

$$\mathbf{8} \otimes \mathbf{8} \ = \ \mathbf{1} \oplus \mathbf{8} \oplus \mathbf{8} \oplus \mathbf{10} \oplus \overline{\mathbf{10}} \oplus \mathbf{27}\,. \quad (6.104)$$

In this case, too, the octet shows up twice on the right-hand side. The octet, the decuplet, and the antidecuplet are illustrated in Figs. 6.7, 6.8 and 6.9. The eigenvalues of

Fig. 6.7 Octet or adjoint representation **8** of $SU(3)$. It contains two doublets ($T = 1/2, T_3 = \pm 1/2$) with $Y = \pm 1$, one triplet ($T = 3, T_3 = +1, 0, -1$) with $Y = 0$, and one singlet ($T = 0, T_3 = 0$) with $Y = 0$

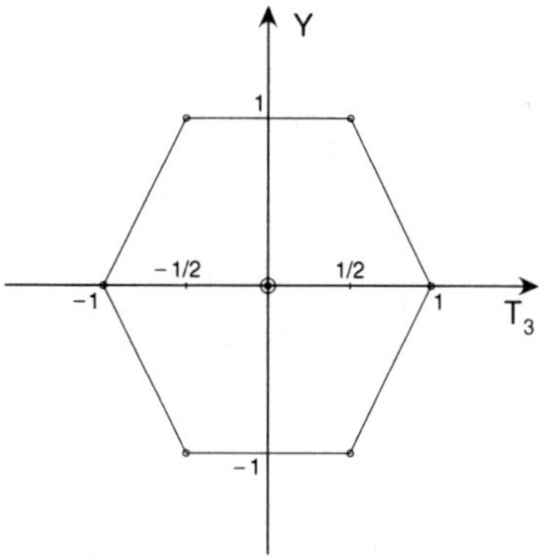

Fig. 6.8 The decuplet representation **10** of $SU(3)$. When resolved in terms of multiplets of the T-subgroup $SU(2)_T$ it contains one quartet with $Y = 1$, one triplet with $Y = 0$, one doublet with $Y = -1$, and one singlet with $Y = -2$

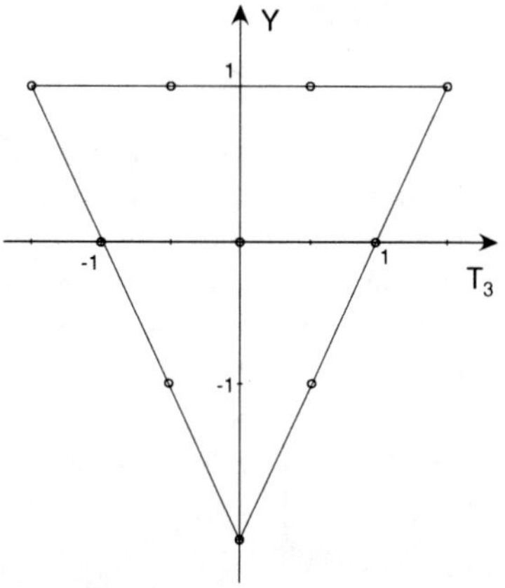

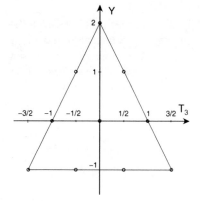

Fig. 6.9 The representation $\overline{\mathbf{10}}$ also contains one quartet, one triplet, one doublet, and one singlet with respect to the subgroup $SU_T(2)$. Like in the case of the representations **3** and $\overline{\mathbf{3}}$ it is not equivalent to the representation **10**

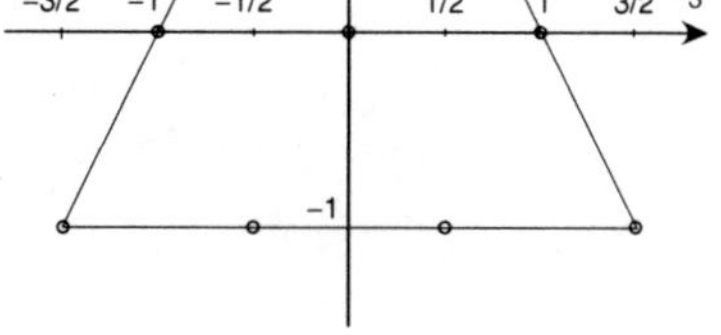

T_3 and of Y are read on the abscissa and on the ordinate, respectively. The quantum number (I) is obtained from the number $2I + 1$ of states in each horizontal line, with $I(I + 1)$ the eigenvalue of $\boldsymbol{I}^2 = T_1^2 + T_2^2 + T_3^2$.

Remark

Perhaps, among all Lie groups, the group $SU(2)$ is the most significant for physics. Thus, one should not be surprised that it plays a special role also from a mathematical point of view. Indeed, it has the distinctive property that the transformation $\mathbf{U}_0$, (6.49), exists in all its representations. Looking back at its matrix representation (6.52) one realizes that $\mathbf{U}_0$ is a symmetric bilinear form for *integer* j, an antisymmetric bilinear form for *half-integer* j. There is a theorem in the theory of compact Lie groups which rests on this condition and which says that all representations with integer angular momentum are *real*, in essence, that is to say they are unitarily equivalent to a real form, while all representations with half-integer eigenvalues are of quaternionic type.[15]

The first of these assertions is illustrated by Sect. 4.1.1 where we started from a real representation of rotations with $j = 1$. For half-integer values $j = (2n+1)/2$, in turn, the preceding paragraphs taught us that the conjugate spinor representation is equivalent to the original spinor representation – unlike the case of, say, $SU(3)$! Keeping in mind the special role of the transformation (6.49) for quantum physics may be helpful, for mathematically oriented readers, in illustrating the rather abstract notions used in the book by Bröcker und tom Dieck (1985) and in understanding better the proof of this theorem given there.

6.3 Lorentz- and Poincaré Groups

In any approach to (special-)relativistic quantum physics one must analyze the Lorentz and Poincaré groups in the light of quantum theory. These groups are non-compact Lie groups. Their representation theory is more involved than that of a compact group such as $SU(2)$ and it would take too much space to go into much detail here. Therefore, we concentrate on the most essential question, from a physical point of view, of how to describe quantum states with given masses and spins (particles, nuclei, atoms).

[15] Th. Bröcker und T. tom Dieck; *Representations of Compact Lie Groups*, Springer, 1985, Theorem (6.4).

6.3.1 The Generators of the Lorentz and Poincaré Groups

The elements of the Poincaré group contain two entries $(\boldsymbol{\Lambda}, a)$ the first of which, $\boldsymbol{\Lambda}$, is a real 4×4-matrix and belongs to the proper, orthochronous Lorentz group $L_+^\uparrow$. The second argument a is a constant four-vector which describes translations in space and time. With x and x' points in Minkowski space one has

$$x' = \boldsymbol{\Lambda} x + a\,, \quad \text{or, in components}\,, \quad x'^\mu = \Lambda^\mu{}_\nu x^\nu + a^\mu\,, \qquad (6.105)$$

where use is made of the sum convention which says that one should sum over any pair of equal and contragredient indices. That is to say, the first term on the right-hand side is to be supplemented by $\sum_{\nu=0}^{3}$.

The homogeneous part $\boldsymbol{\Lambda}$ of the transformation obeys the condition

$$\boldsymbol{\Lambda}^T \mathbf{g} \boldsymbol{\Lambda} = \mathbf{g}\,, \qquad (6.106)$$

with $\mathbf{g} = \mathrm{diag}(1, -1, -1, -1)$ denoting the metric tensor of Minkowskian spacetime. The decomposition theorem for proper orthochronous Lorentz transformations (see e.g. [Scheck (2010)]) tells us that every such transformation $\boldsymbol{\Lambda} \in L_+^\uparrow$ can be written uniquely as the product of a rotation and a boost. Furthermore, it was shown that these factors are expressed in terms of three generators each, i.e., in the real notation used in mechanics,

$$\boldsymbol{\Lambda} = \exp\left(-\boldsymbol{\varphi} \cdot \boldsymbol{J}\right) \exp\left(\lambda \hat{\boldsymbol{w}} \cdot \boldsymbol{K}\right)\,.$$

In quantum physics it is more convenient to use a hermitian form for the generators of rotations (cf. Sect. 4.1.1), and an antihermitian form for boosts, viz.

$$\widetilde{\mathbf{J}}_k := i\mathbf{J}_k\,, \qquad \widetilde{\mathbf{K}}_j := -i\mathbf{K}_j\,.$$

Inserting these definitions, but omitting the tilde for the sake of clarity, one has

$$\boldsymbol{\Lambda} = \exp\left(i\boldsymbol{\varphi} \cdot \boldsymbol{J}\right) \exp\left(i\lambda \hat{\boldsymbol{w}} \cdot \boldsymbol{K}\right) \qquad (6.107)$$

Using this convention, the generators in the defining representation are

$$\mathbf{J}_1 = \begin{pmatrix} 0 & 0 & 0 & 0 \\ 0 & 0 & 0 & 0 \\ 0 & 0 & 0 & -\mathrm{i} \\ 0 & 0 & \mathrm{i} & 0 \end{pmatrix}, \ \mathbf{J}_2 = \begin{pmatrix} 0 & 0 & 0 & 0 \\ 0 & 0 & 0 & \mathrm{i} \\ 0 & 0 & 0 & 0 \\ 0 & -\mathrm{i} & 0 & 0 \end{pmatrix}, \ \mathbf{J}_3 = \begin{pmatrix} 0 & 0 & 0 & 0 \\ 0 & 0 & -\mathrm{i} & 0 \\ 0 & \mathrm{i} & 0 & 0 \\ 0 & 0 & 0 & 0 \end{pmatrix}. \qquad (6.108)$$

$$\mathbf{K}_1 = \begin{pmatrix} 0 & -\mathrm{i} & 0 & 0 \\ -\mathrm{i} & 0 & 0 & 0 \\ 0 & 0 & 0 & 0 \\ 0 & 0 & 0 & 0 \end{pmatrix}, \ \mathbf{K}_2 = \begin{pmatrix} 0 & 0 & -\mathrm{i} & 0 \\ 0 & 0 & 0 & 0 \\ -\mathrm{i} & 0 & 0 & 0 \\ 0 & 0 & 0 & 0 \end{pmatrix}, \ \mathbf{K}_3 = \begin{pmatrix} 0 & 0 & 0 & -\mathrm{i} \\ 0 & 0 & 0 & 0 \\ 0 & 0 & 0 & 0 \\ -\mathrm{i} & 0 & 0 & 0 \end{pmatrix}. \qquad (6.109)$$

The commutation rules for these generators are given by

$$[\mathbf{J}_1, \mathbf{J}_2] = \mathrm{i}\mathbf{J}_3\,, \qquad [\mathbf{J}_1, \mathbf{K}_1] = 0\,,$$
$$[\mathbf{J}_1, \mathbf{K}_2] = \mathrm{i}\mathbf{K}_3\,, \qquad [\mathbf{K}_1, \mathbf{K}_2] = -\mathrm{i}\mathbf{J}_3\,, \qquad (6.110)$$

supplemented by the same commutators with the indices cyclically permuted. The first commutator is familiar from the analysis of the rotation group. The second says that a rotation about a given axis commutes with all boosts whose velocity points along that axis: Indeed, the rotation changes only components perpendicular to the axis of rotation but these are precisely the ones which remain unchanged under a special transformation. The third commutator expresses the fact that the triple $(\mathbf{K}_1, \mathbf{K}_2, \mathbf{K}_3) =: \boldsymbol{K}$ forms a vector operator. For an interpretation of the most interesting fourth commutator see Sect. 4.5.2 of [Scheck (2010)].

The inhomogeneous part of the Poincaré group can also be expressed in terms of generators if, instead of the components x^μ, one introduces *homogeneous coordinates* y^μ. These are obtained by supplementing the components (x^0, x^1, x^2, x^3) by a fifth, inert, component and by replacing the first four by $y^\mu := y^4 x^\mu$. A Poincaré transformation then takes the form

$$y'^\mu = \Lambda^\mu{}_\nu y^\nu + y^4 a^\mu \,, \; (\mu = 0, 1, 2, 3) \quad y'^4 = y^4 \,.$$

Let indices run from 0 to 4, that is to say, take

$$y'^M = \widetilde{\Lambda}^M{}_N y^N \,, \quad (M = 0, 1, 2, 3, 4) \,, \tag{6.111}$$

and write

$$\widetilde{\boldsymbol{\Lambda}} = \begin{pmatrix} \Lambda^\mu{}_\nu & a^\mu \\ \mathbf{0} & 1 \end{pmatrix} . \tag{6.112}$$

Then homogeneous Lorentz transformations and translations are included in a single scheme. A pure translation has the form (with $\Lambda = \mathbb{1}$)

$$\begin{pmatrix} y'^0 \\ y'^1 \\ y'^2 \\ y'^3 \\ y'^4 \end{pmatrix} = \begin{pmatrix} 1 & 0 & 0 & 0 & a^0 \\ 0 & 1 & 0 & 0 & a^1 \\ 0 & 0 & 1 & 0 & a^2 \\ 0 & 0 & 0 & 1 & a^3 \\ 0 & 0 & 0 & 0 & 1 \end{pmatrix} \begin{pmatrix} y^0 \\ y^1 \\ y^2 \\ y^3 \\ y^4 \end{pmatrix} .$$

Choosing the coefficients a^μ infinitesimally small and writing

$$y' \approx \left(\mathbb{1} + \mathrm{i} \sum_{\nu=0}^{3} a^\nu \mathbf{P}_\nu \right) y \,,$$

the generators $\mathbf{P}_\nu$ for translations in space and time are readily obtained: $\mathrm{i}\mathbf{P}_\nu$ are 5×5-matrices with a 1 in position ν of the last column and zeroes elsewhere.[16] For instance one has

$$\mathbf{P}_0 = -\mathrm{i} \begin{pmatrix} 0 & 0 & 0 & 0 & 1 \\ 0 & 0 & 0 & 0 & 0 \\ 0 & 0 & 0 & 0 & 0 \\ 0 & 0 & 0 & 0 & 0 \\ 0 & 0 & 0 & 0 & 0 \end{pmatrix} .$$

[16] Translations form Abelian groups, the generators $\mathbf{P}_\nu$ commute with one another, cf. (6.113). Thus, a finite translation can be written in the form $\exp\{\mathrm{i} \sum_\nu a^\nu \mathbf{P}_\nu\}$.

The commutation rules for these generators and the generators $\mathbf{J}_i$ und $\mathbf{K}_j$ are easily evaluated. Denoting as usual spacetime indices by Greek letters, and pure space indices referring to $\mathbb{R}^3$ by Latin letters, one obtains

$$\left[\mathbf{P}_\mu, \mathbf{P}_\nu\right] = 0\,, \quad \left[\mathbf{J}_i, \mathbf{P}_j\right] = \mathrm{i}\,\varepsilon_{ijk}\,\mathbf{P}_k\,, \quad [\mathbf{J}_i, \mathbf{P}_0] = 0\,,$$
$$\left[\mathbf{K}_i, \mathbf{P}_j\right] = -\mathrm{i}\,\delta_{ij}\,\mathbf{P}_0 \quad [\mathbf{K}_k, \mathbf{P}_0] = -\mathrm{i}\mathbf{P}_k\,.$$

While the generators $\mathbf{J}_i$ and $\mathbf{K}_j$ refer to three-dimensional space $\mathbb{R}^3$ only and, hence, do not have a simple transformation law with respect to boosts, the set of generators $\mathbf{P}_\mu$ form a four-vector with respect to $L_+^\uparrow$. Converting this covariant vector to a contravariant one,

$$\mathbf{P}^\lambda = g^{\lambda\mu}\mathbf{P}_\mu$$

these commutators become

$$\left[\mathbf{P}^\mu, \mathbf{P}^\nu\right] = 0 \tag{6.113}$$
$$\left[\mathbf{J}_i, \mathbf{P}^j\right] = \mathrm{i}\,\varepsilon_{ijk}\,\mathbf{P}^k\,, \qquad \left[\mathbf{J}_i, \mathbf{P}^0\right] = 0\,, \tag{6.114}$$
$$\left[\mathbf{K}_i, \mathbf{P}^j\right] = -\mathrm{i}\,g^{ij}\,\mathbf{P}^0\,, \qquad \left[\mathbf{K}_k, \mathbf{P}^0\right] = \mathrm{i}\,\mathbf{P}^k\,. \tag{6.115}$$

The first of these reflects the fact that all translations in space or time, being elements of Abelian groups, commute. The commutators (6.114) tell us that $\boldsymbol{P} = (\mathbf{P}^1, \mathbf{P}^2, \mathbf{P}^3)$ is a vector operator, and that the energy remains unchanged under rotations in $\mathbb{R}^3$. Regarding (6.115) the left-hand commutator says that $\mathbf{K}_i$ commutes with $\mathbf{P}^j$ as long as $i \neq j$, but that a boost along a given direction does not commute with a translation in the same direction. The right-hand commutator in (6.115), finally, tells that a "boosted" state has different energies before and after a special Lorentz transformation.

The components $\mathbf{P}_\mu$ may be contracted to obtain the invariant $\mathbf{P}^2 = \mathbf{P}_\mu\mathbf{P}^\mu = (\mathbf{P}_0)^2 - \boldsymbol{P}^2$ which commutes with all generators,

$$\left[\boldsymbol{P}^2, \mathbf{P}^\mu\right] = 0\,, \quad \left[\boldsymbol{P}^2, \mathbf{J}_i\right] = 0\,, \quad \left[\boldsymbol{P}^2, \mathbf{K}_i\right] = 0\,. \tag{6.116}$$

As we shall see, in the representations which are relevant for quantum physics, the operator $\boldsymbol{P}$ describes the momentum, while $c\mathbf{P}_0$ represents the energy. As a consequence, $\mathbf{P}^2/c^2$ will be the square m^2 of a mass.

This description of the group $L_+^\uparrow$ and of the translations has a technical disadvantage: While the generators $\mathbf{J}_i$, $\mathbf{K}_j$, $\mathbf{P}^i$ and $\mathbf{P}^0$ have obvious geometric and physical interpretations, their definition assumes a separation of spacetime into a physical position space $\mathbb{R}^3$ and a time $\mathbb{R}_t$ measurable in the laboratory. This is somehow in conflict with the idea of covariance because, in Special Relativity, what we call "space" and what we call "time" depends on the chosen class of reference systems. As is well-known, a boost mixes time and space coordinates and, therefore, changes the splitting of spacetime into $\mathbb{R}^3$ and $\mathbb{R}_t$. There is no such problem with the inhomogeneous part of the Poincaré group because any set $\mathbf{P}^\mu$ that one may choose transforms like a four-vector under $L_+^\uparrow$. In this sector covariance is explicit. Therefore, we may restrict our analysis to the homogeneous part of the Poincaré group, i.e. to the proper orthochronous Lorentz group, and construct a manifestly covariant form of its generators which should replace the set $(\boldsymbol{J}, \boldsymbol{K})$.

Suppose we write the (real) Lorentz transformation $\boldsymbol{\Lambda} \in L_+^\uparrow$ in the neighbourhood of the identity $\mathbb{1}$ as follows

$$\Lambda^\mu{}_\nu \approx \delta^\mu{}_\nu + \alpha^\mu{}_\nu\,.$$

The matrix of the coefficients for a boost along the 1-direction and for a rotation about the 3-axis, for example, then is, respectively,

$$\alpha^\mu{}_\nu = \begin{pmatrix} 0 & \varepsilon & 0 & 0 \\ \varepsilon & 0 & 0 & 0 \\ 0 & 0 & 0 & 0 \\ 0 & 0 & 0 & 0 \end{pmatrix}, \quad \text{and} \quad \alpha^\mu{}_\nu = \begin{pmatrix} 0 & 0 & 0 & 0 \\ 0 & 0 & \varepsilon & 0 \\ 0 & -\varepsilon & 0 & 0 \\ 0 & 0 & 0 & 0 \end{pmatrix}.$$

In either case the covariant tensor of rank two, defined by

$$\alpha_{\mu\nu} = g_{\mu\mu'}\alpha^{\mu'}{}_\nu\,,$$

is antisymmetric,

$$\alpha_{\mu\nu} + \alpha_{\nu\mu} = 0\,. \tag{6.117}$$

With $g = \text{diag}(1, -1, -1, -1)$ the two examples are, respectively,

$$\alpha_{\mu\nu} = \begin{pmatrix} 0 & \varepsilon & 0 & 0 \\ -\varepsilon & 0 & 0 & 0 \\ 0 & 0 & 0 & 0 \\ 0 & 0 & 0 & 0 \end{pmatrix}, \quad \text{and} \quad \alpha_{\mu\nu} = \begin{pmatrix} 0 & 0 & 0 & 0 \\ 0 & 0 & -\varepsilon & 0 \\ 0 & \varepsilon & 0 & 0 \\ 0 & 0 & 0 & 0 \end{pmatrix}.$$

Obviously, the real tensor $\alpha_{\mu\nu}$ is a covariant tensor with respect to the group $L_+^\uparrow$ which contains the parameters of the infinitesimal boost and the infinitesimal rotation, respectively. A new form of the generators behaving properly under $L_+^\uparrow$ is obtained if we introduce a set of 4×4-matrices $\mathbf{M}^{\mu\nu}$ which are also antisymmetric,

$$\mathbf{M}^{\mu\nu} + \mathbf{M}^{\nu\mu} = 0\,, \tag{6.118}$$

and which are defined such that infinitesimal transformations can be written in the form

$$\boldsymbol{\Lambda} \approx \mathbb{1} + \frac{\mathrm{i}}{2}\sum_{\mu,\nu=0}^{3} \alpha_{\mu\nu}\mathbf{M}^{\mu\nu} = \mathbb{1} + \mathrm{i}\sum_{\mu<\nu} \alpha_{\mu\nu}\mathbf{M}^{\mu\nu}\,. \tag{6.119}$$

Because of their antisymmetry (6.118) there are precisely 6 matrices of this kind, i.e. the same number as with the choice $(\boldsymbol{J}, \boldsymbol{K})$. The relation between the new and the previous form of the generators is easily established once a class of reference systems is chosen which fixes the splitting of spacetime into $\mathbb{R}^3$ and $\mathbb{R}_t$.

If we choose $\alpha_{01} = \varepsilon = -\alpha_{10}$, and all others equal to zero, we obtain

$$\Lambda \approx \mathbb{1} + \mathrm{i}\,\varepsilon\,\mathbf{K}_1 = \mathbb{1} + \mathrm{i}\,\varepsilon\,\mathbf{M}^{01}\,.$$

If we choose $\alpha_{12} = -\varepsilon = -\alpha_{21}$, then

$$\Lambda \approx \mathbb{1} + \mathrm{i}\,\varepsilon\,\mathbf{J}_3 = \mathbb{1} - \mathrm{i}\,\varepsilon\,\mathbf{M}^{12}\,.$$

Therefore, the relation between the two sets of generators is

$$\mathbf{K}_j = \mathbf{M}^{0j}\,, \qquad \mathbf{J}_3 = -\mathbf{M}^{12} \quad \text{(and cyclic permutations)}\,. \tag{6.120}$$

The commutators of the $\mathbf{M}^{\mu\nu}$ with the generators $\mathbf{P}^{\nu}$ and their mutal commutators are found to be (see Exercise 6.13)

$$\left[\mathbf{M}^{\mu\nu}, \mathbf{P}^{\sigma}\right] = -\mathrm{i}\left(\mathbf{P}^{\mu} g^{\nu\sigma} - \mathbf{P}^{\nu} g^{\mu\sigma}\right), \tag{6.121}$$

$$\left[\mathbf{M}^{\mu\nu}, \mathbf{M}^{\sigma\tau}\right] = \mathrm{i}\left(\mathbf{M}^{\mu\sigma} g^{\nu\tau} + \mathbf{M}^{\nu\tau} g^{\mu\sigma} - \mathbf{M}^{\mu\tau} g^{\nu\sigma} - \mathbf{M}^{\nu\sigma} g^{\mu\tau}\right). \tag{6.122}$$

As an easy exercise one confirms that the relations (6.121) agree with the commutators (6.114) and (6.115), and that (6.122) agrees with (6.110). Of course, in calculating the commutators with $\mathbf{P}^{\mu}$ one must again introduce homogeneous coordinates (6.111) and must write the Lorentz transformations (6.112) and their generators $\mathbf{M}^{\mu\nu}$ as 5×5-matrices by filling the fifth row and the fifth column with zeroes.

Remarks

1. When constructing unitary, reducible or irreducible, representations of the Poincaré group the generators $\mathbf{J}_i$ or $\mathbf{M}^{ij}$, as well as $\mathbf{P}^{\mu}$, are replaced by hermitian matrices or self-adjoint operators which obey the same commutation rules (6.113)–(6.114), (6.116), or (6.121) and (6.122), respectively. Strictly speaking we should denote them by new symbols such as $\mathbf{U}(\mathbf{J}_i)$, $\mathbf{U}(\mathbf{P}^{\mu})$. Note that, up to exceptions that will be mentioned explicitly, we will refrain from doing so and will use the same notations as before. A similar comment concerns the generators $\mathbf{K}_j$ and $\mathbf{M}^{\mu=0,i}$, respectively, which are antihermitian.
2. One might be surprised to find the generators $\mathbf{K}_j$ and $\mathbf{M}^{0i}$ to be antihermitian, not hermitian. Let us temporarily denote all generators, $\mathbf{J}_i$ and $\mathbf{K}_j$, by $\mathbf{T}_n$ without distinction and consider an infinitesimal Lorentz transformation generated by $\mathbf{T}_n$,

$$\Lambda \approx \mathbb{1} + \mathrm{i}\,\varepsilon\,\mathbf{T}_n\,,$$

cf. (6.107). We then have $\Lambda^T = (\mathbb{1}+\mathrm{i}\varepsilon\mathbf{T}_n)^T = \mathbb{1}-\mathrm{i}\varepsilon\mathbf{T}_n^{\dagger}$. The condition (6.106) yields the equation

$$\mathbf{g}\,\mathbf{T}_n - \mathbf{T}_n^{\dagger}\,\mathbf{g} = 0\,.$$

If the generator is a component of the angular momentum, then the metric $\mathbf{g}$ acts like the (negative) unit matrix and the condition reduces to $\mathbf{J}_i - \mathbf{J}_i^{\dagger} = 0$.

If, in turn, the generator is a component of $\boldsymbol{K}$, then, as one easily verifies, $\mathbf{g}\mathbf{K}_i + \mathbf{K}_i\mathbf{g} = 0$, which means that the above condition is fulfilled only if $\mathbf{K}_i^\dagger = -\mathbf{K}_i$.
Obviously, the same argument applies to the generators $\mathbf{M}^{ik}$ and $\mathbf{M}^{0i}$ which are nothing but different notations for the components of $\boldsymbol{J}$ and $\boldsymbol{K}$, respectively.

3. If one replaces the $\mathbf{J}_i$ and $\mathbf{K}_i$ by the linear combinations

$$\mathbf{A}_i := \frac{1}{2}(\mathbf{J}_i + \mathrm{i}\,\mathbf{K}_i)\,, \qquad \mathbf{B}_i := \frac{1}{2}(\mathbf{J}_i - \mathrm{i}\,\mathbf{K}_i)\,, \tag{6.123}$$

then a simple calculation shows that every $\mathbf{A}_i$ commutes with every $\mathbf{B}_j$ while the components of $\boldsymbol{A}$ as well as those of $\boldsymbol{B}$ fulfill the commutation relations of $\mathfrak{su}(2)$, the Lie algebra of $SU(2)$, viz.

$$\left[\mathbf{A}_i, \mathbf{A}_j\right] = \mathrm{i}\,\varepsilon_{ijk}\mathbf{A}_k\,, \quad \left[\mathbf{B}_i, \mathbf{B}_j\right] = \mathrm{i}\,\varepsilon_{ijk}\mathbf{B}_k\,, \quad \left[\mathbf{A}_i, \mathbf{B}_j\right] = 0\,. \tag{6.124}$$

This is an interesting result: On the one hand all generators $\boldsymbol{A}$ and $\boldsymbol{B}$ now are hermitian, in agreement with Wigner's theorem and with the Remark 2 in Sect. 6.1.2. On the other hand it shows that the proper orthochronous Lorentz group has the structure of a direct product

$$SU(2) \times SU(2)\,. \tag{6.125}$$

In particular, it possesses *two* inequivalent spinor representations,

$$\left(\frac{1}{2}, 0\right) \text{ and } \left(0, \frac{1}{2}\right), \quad \text{or} \quad (\mathbf{2}, \mathbf{1}) \text{ and } (\mathbf{1}, \mathbf{2})\,,$$

if instead of the angular momentum we give the dimension of the representations (cf. Sect. 6.2.8).
The action of space reflection $\boldsymbol{\Pi} = \mathrm{diag}(1, -1, -1, -1)$ leaves the generators $\mathbf{J}_i$ invariant, while the generators $\mathbf{K}_i$ change sign,

$$\boldsymbol{\Pi}\mathbf{J}_i\boldsymbol{\Pi}^{-1} = \mathbf{J}_i\,, \quad \boldsymbol{\Pi}\mathbf{K}_i\boldsymbol{\Pi}^{-1} = -\mathbf{K}_i\,.$$

This means that the generators $\mathbf{A}_i$ and $\mathbf{B}_i$ exchange their roles, and the two spinor representations are mapped onto each other. As both the group $L_+^\uparrow$ and space reflection (or parity) play important roles in quantum physics both kinds of spinors will be needed in the description of fermions with spin 1/2.

6.3.2 Energy-Momentum, Mass and Spin

The generators $\mathbf{P}^\mu$ and $\mathbf{M}^{\mu\nu}$ are used to construct the scalar

$$\mathbf{P}^2 = \mathbf{P}_\mu \mathbf{P}^\mu = (\mathbf{P}^0)^2 - \boldsymbol{P}^2\,, \tag{6.126}$$

having the physical dimension of $(mc)^2$, and the four-vector

$$\mathbf{W}_\sigma := \frac{1}{2}\,\varepsilon_{\mu\nu\lambda\sigma}\mathbf{M}^{\mu\nu}\mathbf{P}^\lambda\,. \tag{6.127}$$

This vector is called the *spin vector of Pauli und Lubanski.* Here $\varepsilon_{\mu\nu\lambda\sigma}$ is the totally antisymmetric Levi-Cività symbol in four dimensions, supplemented by the convention

$$\varepsilon_{0123} = +1 \tag{6.128}$$

It equals +1 if the indices are an *even* permutation of (0123), it equals −1 if they are an *odd* permutation of (0123).[17] (It should be clear that we use here the sum convention $a_\mu b^\mu \equiv \sum_{\mu=0}^{3} a_\mu b^\mu$.) If $\mu = 0$ then $\varepsilon_{0ijk} = \varepsilon_{ijk}$, i.e. it equals the corresponding antisymmetric symbol in dimension 3 (where cyclic permutations *are* even permutations).

If we choose a class of reference systems which single out the time coordinate then

$$\mathbf{W}_0 = \frac{1}{2}\varepsilon_{ijk}\varepsilon_{ijl}\mathbf{J}^l\mathbf{P}^k = \boldsymbol{J}\cdot\boldsymbol{P}\,. \tag{6.129}$$

From the definition (6.127) and by the antisymmetry of the ε-symbol one concludes

$$\mathbf{W}_\sigma\,\mathbf{P}^\sigma = 0\,.$$

Furthermore, one easily verifies the following commutators:

$$\left[\mathbf{W}_\sigma, \mathbf{P}^\mu\right] = 0\,, \quad \left[\mathbf{M}^{\mu\nu}, \mathbf{W}^\sigma\right] = -\mathrm{i}\left(\mathbf{W}^\mu g^{\nu\sigma} - \mathbf{W}^\nu g^{\mu\sigma}\right)\,. \tag{6.130}$$

The second commutator (6.130) follows from the observation that $\mathbf{W}^\sigma$ is a four-vector and, therefore, must have the same commutation rules with $\mathbf{M}^{\mu\nu}$ as $\mathbf{P}^\sigma$. The commutators of the components follow from this:

$$\begin{aligned}
[\mathbf{W}_\lambda, \mathbf{W}_\sigma] &= \frac{1}{2}\varepsilon_{\alpha\beta\gamma\lambda}\left\{\left[\mathbf{M}^{\alpha\beta}, \mathbf{W}_\sigma\right]\mathbf{P}^\gamma + \mathbf{M}^{\alpha\beta}\left[\mathbf{P}^\gamma, \mathbf{W}_\sigma\right]\right\}\\
&= -\mathrm{i}\varepsilon_{\alpha\beta\gamma\lambda}\left(\mathbf{W}^\alpha\delta^\beta_\sigma - \mathbf{W}^\beta\delta^\alpha_\sigma\right)\mathbf{P}^\gamma\\
&= -\mathrm{i}\varepsilon_{\lambda\sigma\alpha\gamma}\mathbf{W}^\alpha\mathbf{P}^\gamma\,.
\end{aligned} \tag{6.131}$$

Taking the square $\mathbf{W}^2 := \mathbf{W}_\sigma\mathbf{W}^\sigma$ yields another invariant which commutes with all $\mathbf{P}^\mu$ and with $\mathbf{M}^{\mu\nu}$. In total we obtain six operators which mutually commute and, thus, which may be used for classifying representations of the Poincaré group:

$$\mathbf{P}^\mu\ (\mu = 0, 1, 2, 3)\,,\ \mathbf{W}^2\,,\ \text{and one component}\ \mathbf{W}_\lambda\,. \tag{6.132}$$

Instead of the four components $\mathbf{P}^\mu$ one may alternatively use only three of them together with the square $\mathbf{P}^2$. Note, however, that only *one* component of $\mathbf{W}$ may

---

[17] Note that permutations are always defined by means of neighbour exchanges. An even permutation from (0123) to $(nm\ pq)$ is one that needs an even number of exchanges of neighbours. Unlike in dimension 3, cyclic permutations in dimension 4 are not even!
The convention (6.128) which is not adopted generally in the literature, implies that $\varepsilon^{0123} = -1$.

be contained in the set (6.132) because the components do not commute – in analogy to the discussion of the components of ordinary angular momentum. Clearly, the components $\mathbf{P}^\mu$ must be the operators of energy and momentum while angular momentum and/or spin must be hidden in $\mathbf{W}^2$ and $\mathbf{W}_\sigma$. Working out the precise relationship is the purpose of the section immediately following this one.

6.3.3 Physical Representations of the Poincaré Group

The description of elementary objects of microphysics such as atoms, nuclei, or elementary particles, builds upon a postulate which is due to E. Wigner. For simplicity we shall call any such object a particle. The postulate then reads as follows.

Postulate Classification of Particles

Particles are classified by eigenvalues of mass and spin, i.e. by the eigenvalues of $\mathbf{P}^2$ and of $\mathbf{W}^2$, the spin taking only integer or half-integer values. The dynamical states of free particles can be characterized by the eigenvalues of the four operators $\mathbf{P}^\mu$ of the energy and the momentum, and by one component of spin.

Write the eigenvalues of $\mathbf{P}^2$ as m^2c^2, where m is the invariant rest mass of the particle. Then $\mathbf{P}^0$ has eigenvalue E/c, the three-vector $\boldsymbol{P}$ has eigenvalues $\boldsymbol{p}$, with E and $\boldsymbol{p}$ the energy and the momentum, respectively, in a class of systems of reference in which the time axis is given. The eigenvalues fulfill the energy-momentum relation for a free particle,

$$(E/c)^2 + \boldsymbol{p}^2 = m^2c^2 .$$

Interestingly enough it turns out that one must distinguish the two possible alternatives of *massive* particles, $m \neq 0$, and of *massless* particles, $m = 0$, the analysis of their representations being rather different.

The Case $m \neq 0$: Every massive particle possesses a rest system. In other terms, if it is given in a state with momentum $\boldsymbol{p}$ one can always find a special Lorentz transformation which transforms to the particle's rest frame where its four-momentum is $(mc, \mathbf{0})^T$. In the rest system the eigenvalue of $\mathbf{W}_0$ is zero by (6.127) because $\mathbf{P}^\lambda$ contributes only for $\lambda = 0$, while the ε-tensor vanishes whenever two indices are equal. Regarding the spatial components, the definition (6.127), together with (6.120) and $p^0 = mc$ yield the relation $\mathbf{W}_i = mc\,\mathbf{J}^i$. From this one concludes that the commutators are

$$\left[\left(\frac{\mathbf{W}^i}{mc}\right), \left(\frac{\mathbf{W}^j}{mc}\right)\right] = \mathrm{i} \sum_{k=1}^{3} \varepsilon_{ijk} \left(\frac{\mathbf{W}^k}{mc}\right)$$

and, in particular, that $W^2/(mc)^2$ has eigenvalues $j(j+1)$. Let n be a spacelike unit vector which is perpendicular to p, i.e. which fulfills $n^2 = -1$ and $(n \cdot p) = 0$. In the rest system it must have the form $\overset{0}{n} = (0, \widehat{\boldsymbol{n}})^T$ so that $n^2 = -\widehat{\boldsymbol{n}}^2 = -1$ and

$$\frac{1}{mc}(\mathbf{W} \cdot n) = \boldsymbol{J} \cdot \widehat{\boldsymbol{n}} =: \mathbf{J}^{\widehat{n}} .$$

Denote the eigenvalue of this operator in the rest system by μ. As the (Lorentz-) scalar product $(\mathbf{W} \cdot n)$ is an invariant under all $\boldsymbol{\Lambda} \in L_+^\uparrow$, μ is its eigenvalue in *all* systems of reference. Of course, the set of values that this eigenvalue can take, is well-known to us from the study of the rotation group. It is

$$\mu = -j, -j+1, \ldots, j-1, j .$$

In summary, we note the following: The spin of a massive particle is described, in a Lorentz invariant manner, by the operators

$$\frac{1}{(mc)^2}\mathbf{W}_\sigma \mathbf{W}^\sigma = \frac{1}{(mc)^2}\mathbf{W}^2 \quad \text{and} \quad \frac{1}{mc}(\mathbf{W} \cdot n) . \tag{6.133}$$

Pictorially speaking this means that in order to measure the spin of a massive particle one must go to its momentaneous rest system and perform all kinds of rotations in $\mathbb{R}^3$. If its state responds by $\mathbf{D}^{(j)}$ then the particle carries spin j; the admissible values of the projection of the spin onto an arbitrary axis in $\mathbb{R}^3$ are the numbers $-j, -j+1, \ldots, j$ that are well-known from nonrelativistic quantum mechanics.

The Case $m = 0$: A massless particle has no rest system, in vacuum it moves always with the velocity of light and, for a massive observer, there is no causal way of "catching" the particle. Therefore, the analysis of the previous section cannot be applied to this case. This essential difference can also be seen from a group theoretical point of view:

A massive particle has *time*like four-momenta, $p = (E/c, \boldsymbol{p})^T$ with $p^2 > 0$, and, thus, can always be brought to rest. The maximal subgroup of the Lorentz group $L_+^\uparrow$ which leaves invariant its four-momentum $(mc, \mathbf{0})^T$, obviously, is the full rotation group.

A mass*less* particle has *light*like four-momenta, $p = (E/c = |\boldsymbol{p}|, \boldsymbol{p})^T$ with $p^2 = 0$. The maximal subgroup of $L_+^\uparrow$ which leaves this momentum invariant, is the one-parameter group of rotations about the direction $\widehat{\boldsymbol{p}}$ which, obviously, is smaller than the full rotation group. Therefore, it is not surprising that the spin of a massless particle is defined differently than for a massive particle, and has different properties.

Imagine that we are in a representation in which the massless particle takes an arbitrary but fixed eigenvalue p of $\mathbf{P}$. Our aim is to find the operators which describe the spin of the massless particle. In order to express the components of $\mathbf{W}_\lambda$ we introduce a system of base vectors

$$(n^{(0)}, n^{(1)}, n^{(2)}, n^{(3)})$$

the first of which, $n^{(0)}$, is timelike, the others being spacelike, and which are pairwise orthogonal, $(n^{(\alpha)} \cdot n^{(\beta)}) = 0$ for $\alpha \neq \beta$. Furthermore, they are taken to be normalized to ± 1, with $n^{(0)\,2} = 1$, $n^{(i)\,2} = -1$ $(i = 1, 2, 3)$, and $n^{(1)}$ and $n^{(2)}$ to be orthogonal to p,

$$(p \cdot n^{(1)}) = 0 = (p \cdot n^{(2)}) .$$

One verifies that $n^{(3)}$ is the following linear combination of p and of $n^{(0)}$

$$n^{(3)} = \frac{1}{(p \cdot n^{(0)})}\, p - n^{(0)} .$$

For example, if we chose the frame of reference such that $p = (q, 0, 0, q)^T$ then the base vectors could be

$$n^{(0)} = (1, 0, 0, 0)^T\,, \quad n^{(1)} = (0, 1, 0, 0)^T\,,$$
$$n^{(2)} = (0, 0, 1, 0)^T\,, \quad n^{(3)} = (0, 0, 0, 1)^T\,.$$

A base system of this kind is orthogonal and complete in the following sense

$$\left(n^{(\alpha)} \cdot n^{(\beta)}\right) = g^{\alpha\beta} \qquad \text{(orthogonality)} \tag{6.134}$$

$$n_\sigma^{(\alpha)} g_{\alpha\beta} n_\tau^{(\beta)} = g_{\sigma\tau} \qquad \text{(completeness)}\,, \tag{6.135}$$

(where we again make use of the sum convention).

Define new operators $\mathbf{J}^{(\sigma)}$ through the expression

$$\mathbf{W}_\lambda = \mathbf{J}^{(\sigma)} g_{\sigma\tau} n_\lambda^{(\tau)}\,. \tag{6.136}$$

From the orthogonality relation (6.134) one has

$$\mathbf{J}^{(\mu)} = \mathbf{W}_\lambda\, n^{(\mu)\,\lambda} \equiv (\mathbf{W} \cdot n^{(\mu)})\,.$$

One concludes from this formula that $\mathbf{J}^{(0)}$ and $\mathbf{J}^{(3)}$ are equal and opposite since, with $(\mathbf{W} \cdot p) = 0$, one has

$$\mathbf{J}^{(0)} + \mathbf{J}^{(3)} = (\mathbf{W} \cdot n^{(0)}) + \left(\mathbf{W} \cdot \left[\frac{1}{(p \cdot n^{(0)})}\, p - n^{(0)}\right]\right) = 0\,.$$

Thus, the (four-)square of the angular momentum is

$$\mathbf{J}^2 = \mathbf{J}^{(0)\,2} - \mathbf{J}^{(1)\,2} - \mathbf{J}^{(2)\,2} - \mathbf{J}^{(3)\,2} = -\left(\mathbf{J}^{(1)\,2} + \mathbf{J}^{(2)\,2}\right)\,.$$

In the next step one proves that the operators $\mathbf{J}^{(i)}$, $i = 1, 2, 3$, fulfill the following commutators

$$\left[\mathbf{J}^{(1)}, \mathbf{J}^{(2)}\right] = 0\,, \quad \left[\mathbf{J}^{(2)}, \mathbf{J}^{(3)}\right] = -\mathrm{i}\,(p \cdot n^{(0)})\,\mathbf{J}^{(1)}\,,$$
$$\left[\mathbf{J}^{(3)}, \mathbf{J}^{(1)}\right] = -\mathrm{i}\,(p \cdot n^{(0)})\,\mathbf{J}^{(2)}\,. \tag{6.137}$$

We begin with the first commutator and calculate

$$\begin{aligned}\left[\mathbf{J}^{(1)}, \mathbf{J}^{(2)}\right] &= n^{(1)\,\lambda} n^{(2)\sigma}\,[\mathbf{W}_\lambda, \mathbf{W}_\sigma] \\ &= -\mathrm{i}\,\varepsilon_{\lambda\sigma\alpha\beta}\, n^{(1)\,\lambda} n^{(2)\sigma}\, \mathbf{W}^\alpha\, p^\beta \\ &= -\mathrm{i}\,\varepsilon_{\lambda\sigma\alpha\beta}\, n^{(1)\,\lambda} n^{(2)\sigma} n^{(\nu)\alpha} g_{\mu\nu} \mathbf{J}^{(\mu)} p^\beta\,,\end{aligned}$$

where we used the commutator (6.131) and the definition (6.136). The index ν can only take the values 0 and 3, because otherwise the ε-tensor which is antisymmetric, is contracted with the product of two equal base vectors which is symmetric, thus giving zero. For the remaining values $\nu = 0$ and $\nu = 3$, using $\mathbf{J}^{(3)} = -\mathbf{J}^{(0)}$ and $g_{00} = 1 = -g_{33}$ one obtains

$$\left[\mathbf{J}^{(1)}, \mathbf{J}^{(2)}\right] = \mathrm{i}\,\varepsilon_{\lambda\sigma\alpha\beta}\, n^{(1)\,\lambda} n^{(2)\sigma} (n^{(3)\alpha} + n^{(0)\alpha})\, p^\beta \mathbf{J}^{(3)}\,.$$

However, as $(n^{(3)\alpha} + n^{(0)\alpha})$ is proportional to p^α and, hence, since the symmetric product $p^\alpha p^\beta$ is contracted with the antisymmetric ε-tensor the commutator is equal to zero.

Among the two remaining commutators we calculate the second, as an example,

$$\begin{aligned}\left[\mathbf{J}^{(2)}, \mathbf{J}^{(3)}\right] &= n^{(2)\lambda} n^{(3)\sigma} \left[\mathbf{W}_\lambda, \mathbf{W}_\sigma\right] \\ &= -\mathrm{i}\, \varepsilon_{\lambda\sigma\mu\nu}\, n^{(2)\lambda} n^{(3)\sigma} \mathbf{W}^\mu p^\nu \\ &= -\mathrm{i}\, \varepsilon_{\lambda\sigma\mu\nu}\, n^{(2)\lambda} n^{(3)\sigma} n^{(\beta)\mu} g_{\alpha\beta} \mathbf{J}^{(\alpha)} p^\nu .\end{aligned}$$

Inserting here $p = (p \cdot n^{(0)})(n^{(3)} + n^{(0)})$ one sees that only the second term can contribute. This also means that β can only take the value 1. Inserting this and commuting indices such that the base vectors appear in increasing order yields with $g_{11} = -1$

$$\left[\mathbf{J}^{(2)}, \mathbf{J}^{(3)}\right] = -\mathrm{i}\, \varepsilon_{\nu\mu\lambda\sigma} n^{(0)\nu} n^{(1)\mu} n^{(2)\lambda} n^{(3)\sigma} \,(p \cdot n^{(0)})\, \mathbf{J}^{(1)} .$$

The factor $\varepsilon_{\nu\mu\lambda\sigma} n^{(0)\nu} n^{(1)\mu} n^{(2)\lambda} n^{(3)\sigma}$ is seen to be the determinant of the 4×4-matrix constructed with the base vectors and is equal to 1. This proves the second commutator. The proof of the third commutator goes along the same lines.

Finally, if we set

$$\mathbf{S}^{(i)} := -\frac{1}{(p \cdot n^{(0)})} \mathbf{J}^{(i)} ,$$

then the commutators (6.137) go over into a form that is well-known from Euclidean geometry, viz.

$$\left[\mathbf{S}^{(1)}, \mathbf{S}^{(2)}\right] = 0 , \quad \left[\mathbf{S}^{(2)}, \mathbf{S}^{(3)}\right] = \mathrm{i}\, \mathbf{S}^{(1)} , \quad \left[\mathbf{S}^{(3)}, \mathbf{S}^{(1)}\right] = \mathrm{i}\, \mathbf{S}^{(2)} . \qquad (6.138)$$

This algebra is isomorphic to the Lie algebra of the Euclidean group in two real dimensions. $\mathbf{S}^{(1)}$ and $\mathbf{S}^{(2)}$ correspond to the generators for translations along the 1- and 2-directions, respectively, $\mathbf{S}^{(3)}$ corresponds to the generator of rotations around the origin, cf. Exercise 6.11. Much like for the rotation group this information is sufficient to construct the representations which are suitable candidates for classifying massless particles. It will turn out, however, that in addition to our experience with the rotation group we will need one further empirical information for that purpose.

It comes as a surprise that with (6.134) the operator

$$\mathbf{W}^2 = \mathbf{J}^2 = -\left((\mathbf{J}^{(1)})^2 + (\mathbf{J}^{(2)})^2\right)$$

can take any negative value. Apparently this operator is not quantized. As in nature there are no physical states with continuous spin, one postulates that the physical states of a massless particle pertain to the eigenvalue $w^2 = 0$ of the operator $\mathbf{W}^2 = \mathbf{W}_\sigma \mathbf{W}^\sigma$. The conditions following from this additional, empirical postulate

$$w^2 = 0 , \quad p^2 = 0 , \quad (w \cdot p) = 0$$

allow to draw an important conclusion: The four-vectors w and p must be *collinear*,

$$w = h\,p\,. \tag{6.139}$$

The factor of proportionality h is called the *helicity*. The word is derived from the notion of helix and refers to a sense of rotation with respect to some given direction. This sense of rotation is defined by the correlation between the spin alignment and the spatial momentum of the massless particle. This is best understood if one returns to a class of frames in which the time axis is fixed. According to (6.129) one then has $\mathbf{W}_0 = \boldsymbol{p} \cdot \boldsymbol{J}$. Denote the operator whose eigenvalues are h by the bold symbol $\mathbf{h}$. With (6.139) one also has $\mathbf{W}_0 = p_0\mathbf{h}$, where $p_0 = |\boldsymbol{p}|$, so that one concludes

$$\mathbf{h} = \boldsymbol{p} \cdot \boldsymbol{J}/|\boldsymbol{p}|\,. \tag{6.140}$$

The operator $\mathbf{h}$ describes the projection of the angular momentum onto the spatial momentum of the particle. In principle, it contains both orbital and spin angular momenta. However, since we know that the projection m_ℓ of the *orbital* angular momentum onto $\boldsymbol{p}$ always vanishes (cf. Sect. 1.9.3), it describes the projection of the *spin* onto the direction of flight. A massless particle having no rest system, this is the only possibility to define the spin of the particle. *One says that $s = |h|$ is the spin of the massless particle.*

What are the eigenvalues of the spin so defined? Is there an analogue to the magnetic quantum number? If $w = hp$, as imposed by the empirical condition above, the operators $\mathbf{J}^{(1)}$ and $\mathbf{J}^{(2)}$ have the eigenvalues 0. The eigenvalue of $\mathbf{J}^{(3)}$ is $-(p{\cdot}n^{(0)})h$ which means that the eigenvalue of $\mathbf{S}^{(3)}$ is h. To obtain more information it is instructive to study the action of space reflection onto the helicity operator (6.140). The operator $\boldsymbol{J}$ is invariant, while $\boldsymbol{p}$ changes sign. Therefore, helicity is what is called a *pseudoscalar*, i.e. a quantity which is invariant under rotations but changes sign under the action of space reflection. As a consequence of this observation, the spin s is seen to admit two polarization states, $+h$ and $-h$, which are related by space reflection. Finally, as $\mathbf{S}^{(3)}$ generates rotations about the direction of $\boldsymbol{p}$ and since wave functions of bosons must be one-valued, those of fermions must be two-valued, the quantum number s can take only integer and half-integer values, respectively,

$$s = 0, \frac{1}{2}, 1, \frac{3}{2}, 2, \ldots\,.$$

Remarks

1. It is interesting to note that helicity is an invariant under $L_+^\uparrow$ only for massless particles. Trying the same definition for a massive particle one realizes that (6.140) is no longer invariant under special Lorentz transformations (boosts).
2. The photon is the best known, most certainly massless particle. According to the definition given above its spin is 1. The two distinct orientations $h = +1$ and $h = -1$ are realizable in physics because the electromagnetic interaction

is not only invariant with respect to $L_+^\uparrow$ but also with respect to space reflection $\boldsymbol{\Pi}$. Classically, these two "magnetic" states correspond to plane waves with right- or left-circular polarization.

3. For many years one thought that the three neutrinos ν_e, ν_μ, and ν_τ, all of which carry spin 1/2, were strictly massless. Nowadays there is experimental evidence that all or some of them have nonvanishing masses, though very small as compared to the masses of their electrically charged partners e, μ, and τ. If they were strictly massless we would classify them by their helicity states. Neutrinos participate in the weak interactions about which we know that they are not invariant under $\boldsymbol{\Pi}$. Thus, it seems plausible, that only one of the two helicity states is physically realized.
4. The case of spin $s = 1/2$ is somewhat special insofar as both a massive, and a massless spin-1/2 fermion shows two orientations of the spin. This shows an essential difference from a particle with spin $s = 1$ (or any higher integer or half-integer spin). If the spin-1 particle is massive the spin has three physically possible orientations $m_s = 1$, $m_s = 0$, and $m_s = -1$. If it is massless there are only two physical states, $h = +1$ and $h = -1$. More generally, a particle with nonvanishing mass M and spin $s \geq 1$ possesses the $2s + 1$ substates well-known from the rotation group. If its mass is zero its helicity has only the eigenvalues $h = s$ and $h = -s$. As we shall see in our discussion of the Dirac equation the limit $mass \to 0$ is continuous in the case of spin-1/2 particles but is discontinuous for particles with $s \geq 1$.

6.3.4 Massive Single-Particle States and Poincaré Group

Though anticipating a little I wish to add one more topic to the representation theory of the preceding section: the explicit action of Poincaré transformations on single-particle states. In doing so I make use of second quantization that was introduced in Sect. 5.3.2 and to which I return in more detail in the next chapter.

The particle is assumed to have a nonvanishing mass M and to carry spin s. Any other quantum numbers which might be needed to characterize its state but which are irrelevant for this discussion, are summarized by the symbol α. The aim is to work out the action $\mathbf{U}(\boldsymbol{\Lambda}, a)$ of an arbitrary Poincaré transformation $(\boldsymbol{\Lambda}, a)$ on the state $|\alpha; s\, m; p\rangle$. For this purpose we make use of the following

Remark

In the present context it is important to keep in mind the difference between *passive* and *active* interpretations of symmetry transformations. Take the example of rotations in space $\mathbb{R}^3$. The passive interpretation assumes that the frame of reference is rotated while the physical object is kept unchanged; In the active interpretation it is that object which is rotated while the frame is kept fixed. Thus, an *active* rotation, when expressed in terms of Eulerian angles, has the form

$$\mathbf{R} = e^{i\mathbf{J}_3\psi} e^{i\mathbf{J}_2\theta} e^{i\mathbf{J}_3\phi} .$$

With the conventions introduced in Sect. 4.1.3 the D-matrices are given by

$$D^{(j)}_{m\mu}(\psi, \theta, \phi) = \langle jm | \mathbf{R} \, | j\mu \rangle \; , \tag{6.141}$$

in agreement with (6.18). Note that in Part One rotations were always interpreted as *passive* transformations. Furthermore, we noted that base states transform by

$$|jm\rangle' = \sum_{m'} D^{(j)*}_{mm'} \left| jm' \right\rangle .$$

A passive rotation is described by the inverse of the above, active one, viz.

$$\mathbf{R}^{-1} = \mathrm{e}^{-\mathrm{i}\mathbf{J}_3\phi} \mathrm{e}^{-\mathrm{i}\mathbf{J}_2\theta} \mathrm{e}^{-\mathrm{i}\mathbf{J}_3\psi}$$

Thus, the formula just given is equivalent to

$$|jm\rangle' \equiv \mathbf{U}(\mathbf{R}^{-1}) \, |jm\rangle = \sum_{m'} D^{(j)*}_{mm'} \left| jm' \right\rangle = \sum_{m'} \left(D^{(j)\dagger} \right)_{m'm} \left| jm' \right\rangle .$$

(Note that in the last expression summation is over the first index). Replacing the passive rotation by an active one, means replacing $\mathbf{R}^{-1}$ by $\mathbf{R}$, and $\mathbf{D}^{-1} = \mathbf{D}^{(j)\dagger}$ by $\mathbf{D}^{(j)}$. Therefore, one has

$$\mathbf{U}(\mathbf{R}) \, |jm\rangle = \sum_{m'} D^{(j)}_{m'm}(\psi, \theta, \phi) \left| jm' \right\rangle , \tag{6.142}$$

(summing again over the first index). A remark which is relevant for what follows is this: One should realize that the action of the Poincaré transformations is an active one. For instance, the action of $\mathbf{U}(\mathbf{L}(p))$ on a particle state really means actively "boosting" from the rest system to four-momentum p.

With this comment in mind we return to the construction of representations of Poincaré transformations. In the rest system of the particle the eigenstates of s^2 and of $\mathbf{s}_3$ are

$$\left| \alpha; s\, m; \overset{0}{p} \right\rangle \quad \text{with} \quad \overset{0}{p} = (Mc, 0, 0, 0)^T .$$

Writing proper orthochronous Poincaré transformations as (Λ, a), with $\Lambda \in L_+^\uparrow$, and a the four-vector of time and space translation, and denoting by $\mathbf{U}$ their representations in the space of single-particle states, an *active* rotation $\mathbf{R} \in SO(3)$ acts as follows:

$$\mathbf{U}\left(\mathbf{R}, 0\right) \left| \alpha; s\, m; \overset{0}{p} \right\rangle = \sum_{m'} D^{(s)}_{m'm}(\mathbf{R}) \left| \alpha; s\, m' \overset{0}{p} \right\rangle .$$

If that same state is boosted, by means of the special Lorentz transformation $\mathbf{L}(p)$ to a momentum p then

$$|\alpha; s\, m; p\rangle = \mathbf{U}\left(\mathbf{L}(p), 0\right) \left| \alpha; s\, m; \overset{0}{p} \right\rangle .$$

The result is a state of the massive particle carrying spin quantum numbers (s, m) and moving with four-momentum p. As one sees very clearly, the spin of a massive particle is defined with reference to its rest system.

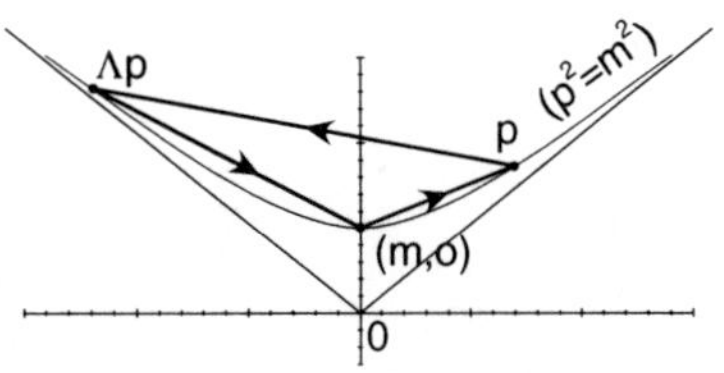

Fig. 6.10 The momentum p of a massive particle is reached by boosting it from the rest system as sketched in the figure. In the second step an arbitrary proper, orthochronous Lorentz transformation is applied to it, taking p to Λp. Finally, the transformed momentum is boosted back to the rest system

We now have all tools ready that are needed to answer the question: What is the action of an arbitrary Poincaré transformation $\mathbf{U}(\boldsymbol{\Lambda}, a)$ on the state $|\alpha; s\, m;\, p\rangle$?

Homogeneous Transformations ($\boldsymbol{a} = \mathbf{0}$): It is convenient to first consider the proper orthochronous transformations only and to set $a = 0$. In this case one has

$$\mathbf{U}(\boldsymbol{\Lambda}, 0)\,|\alpha; s\, m;\, p\rangle = \mathbf{U}(\boldsymbol{\Lambda}, 0)\mathbf{U}(\mathbf{L}(p), 0)\left|\alpha; s\, m;\, \overset{0}{p}\right\rangle .$$

defining $p_\Lambda = \boldsymbol{\Lambda} p$ and making use of the group property in the representation at stake, one has

$$\mathbf{U}(\boldsymbol{\Lambda}, 0)\mathbf{U}(\mathbf{L}(p), 0) = \mathbf{U}\left(\mathbf{L}(p_\Lambda)\right)\mathbf{U}\left(\mathbf{L}^{-1}(p_\Lambda)\boldsymbol{\Lambda}\mathbf{L}(p)\right) .$$

Applying this to the state $|\alpha; s\, m;\, p\rangle$ one obtains

$$\mathbf{U}(\boldsymbol{\Lambda}, 0)\,|\alpha; s\, m;\, p\rangle = \mathbf{U}\left(\mathbf{L}(p_\Lambda)\right)\mathbf{U}\left(\mathbf{L}^{-1}(p_\Lambda)\boldsymbol{\Lambda}\mathbf{L}(p)\right)\left|\alpha; s\, m;\, \overset{0}{p}\right\rangle .$$

A sketch may help to visualize the action of the product $\mathbf{L}^{-1}(p_\Lambda)\boldsymbol{\Lambda}\mathbf{L}(p)$ on a state in the rest system. The first (rightmost) factor $\mathbf{L}(p)$ boosts the particle from a state of rest $(Mc, 0, 0, 0)^T$ to a state with momentum p. Then, the Lorentz transformation $\boldsymbol{\Lambda}$ acts on this state by taking p to $p_\Lambda = \boldsymbol{\Lambda} p$. Finally, the inverse of $\mathbf{L}(p_\Lambda = \boldsymbol{\Lambda} p)$ takes the particle back to a state of rest. All three values of the momentum lie on the upper branch of the hyperboloid

$$(\overset{0}{p})^2 = p^2 = (\boldsymbol{\Lambda} p)^2 = (Mc)^2 .$$

As the particle is at rest both before and after the round trip sketched in Fig. 6.10 the product of the three Lorentz transformations must be a pure, active rotation[18]

$$\mathbf{L}^{-1}(p_\Lambda = \boldsymbol{\Lambda} p)\boldsymbol{\Lambda}\mathbf{L}(p) := \mathbf{R}_W . \tag{6.143}$$

This resulting rotation is called *Wigner rotation*.

[18] There should be no reason for confusion if I use the same notation $\mathbf{R}$ for rotations, in $\mathbb{R}^3$ as well as in $\mathbb{R}^4$. Of course, the correct notation in spacetime is $\mathcal{R} = \text{diag}\,(1, \mathbf{R})$.

Its action on the state in the rest system is, according to (6.142),

$$\mathbf{U}\Big(\mathbf{L}^{-1}(\boldsymbol{\Lambda} p)\boldsymbol{\Lambda}\mathbf{L}(p)\Big)\left|\alpha;\, s\, m;\, \overset{0}{p}\right\rangle = \sum_{m'} D^{(s)}_{m'm}(\mathbf{R}_W)\left|\alpha;\, s\, m';\, \overset{0}{p}\right\rangle .$$

Finally, the remaining transformation $\mathbf{U}\,(\mathbf{L}(p_\Lambda))$ can be shifted past the matrix elements of the rotation matrix $\mathbf{D}^{(s)}$, and one obtains the preliminary result

$$\mathbf{U}(\boldsymbol{\Lambda}, 0)\,|\alpha;\, s\, m;\, p\rangle = \sum_{m'} D^{(s)}_{m'm}(\mathbf{R}_W)\,|\alpha;\, s\, m';\, (\boldsymbol{\Lambda} p)\rangle\ .$$

This "shifting" needs a little reflection. The action of the Wigner rotation results in a linear combination of the states $|sm\rangle$ whose coefficients are elements of the corresponding D-matrix. The remaining special Lorentz transformation whose argument is p_Λ, does not change these coefficients. It just boosts the momentum to the value p_Λ.

Translations: We now include translations in space and time. By the group property one has

$$\mathbf{U}(\boldsymbol{\Lambda}, a) = \mathbf{U}(\mathbb{1}, a)\mathbf{U}(\boldsymbol{\Lambda}, 0)\ .$$

The operator $\exp\{\mathrm{i}/\hbar\, a_\mu \mathbf{P}^\mu\}$ is the unitary operator which describes the action of the translation $(\mathbb{1}, a)$, p^μ is the eigenvalue of $\mathbf{P}^\mu$, and $(a\cdot p)$ is the eigenvalue of $(a\cdot\mathbf{P})$. Therefore, one obtains

$$\mathbf{U}(\mathbb{1}, a)\,|\alpha;\, s\, m;\, p\rangle = \mathrm{e}^{(\mathrm{i}/\hbar)(a\cdot p)}\,|\alpha;\, s\, m;\, p\rangle\ .$$

Composing the two parts one obtains the desired result

$$\mathbf{U}(\boldsymbol{\Lambda}, a)\,|\alpha;\, s\, m;\, p\rangle = \mathrm{e}^{\mathrm{i}/\hbar(a\cdot p_\Lambda)} \sum_{m'} D^{(s)}_{m'm}\Big(\mathbf{L}^{-1}(p_\Lambda)\boldsymbol{\Lambda}\mathbf{L}(p)\Big)\,|\alpha;\, s\, m';\, p_\Lambda\rangle\ , \tag{6.144}$$

where the abbreviation $p_\Lambda = \boldsymbol{\Lambda} p$ was used.

When is the Representation (6.144) **Unitary?** It is obvious that the representation (6.144) is irreducible. But is it also unitary? In order to answer this question one must know the normalization of the single-particle states $|\alpha;\, s\, m;\, p\rangle$. The particle, being free, moves on its mass shell, that is to say, its four-momentum fulfills the condition $p^2 = (Mc)^2$, its time component p^0 is fixed by the modulus of the spatial momentum, $(p^0)^2 = (Mc)^2 + \boldsymbol{p}^2$. If the plane waves were normalized to a simple δ-distribution as in nonrelativistic quantum mechanics then this normalization would depend on the frame of reference and, hence, would not be very useful. However, one shows that the product of p^0 and $\delta(\boldsymbol{p}-\boldsymbol{p}')$ is invariant both under rotations and under special Lorentz transformations, hence, under the whole group $L_+^\uparrow$, see Exercise 6.12. This result suggests to introduce a $L_+^\uparrow$-invariant normalization, the *covariant normalization*. We define it as follows

$$\langle p'|p\rangle = 2p^0\,\delta(\boldsymbol{p}-\boldsymbol{p}')\,. \tag{6.145}$$

Written out in more detail for single-particle states it reads

$$\langle \alpha';s'm';p'|\alpha;sm;p\rangle = \delta_{\alpha'\alpha}\delta_{s's}\delta_{m'm}\,2\sqrt{(Mc)^2+\boldsymbol{p}^2}\,\delta(\boldsymbol{p}'-\boldsymbol{p})\,. \tag{6.146}$$

One now shows that, indeed, the representations (6.144) are unitary, i.e that in the space of single-particle states

$$\mathbf{U}^\dagger\mathbf{U} = 1\!\!1 = \mathbf{U}\mathbf{U}^\dagger\,.$$

For the sake of simplicity we drop the quantum numbers α, α' and use the following abbreviations for the Wigner rotations

$$\mathbf{R}_W = \mathbf{L}^{-1}(\boldsymbol{\Lambda}p)\boldsymbol{\Lambda}\mathbf{L}(p)\,, \qquad \mathbf{R}'_W = \mathbf{L}^{-1}(\boldsymbol{\Lambda}p')\boldsymbol{\Lambda}\mathbf{L}(p')\,.$$

We calculate the scalar product of two states constructed according to (6.144),

$$\langle s'm';p'|\mathbf{U}^\dagger(\Lambda,a)\mathbf{U}(\Lambda,a)|sm;p\rangle =$$
$$\sum_\mu\sum_{\mu'} D^{(s)*}_{\mu'm'}(\mathbf{R}'_W)D^{(s)}_{\mu m}(\mathbf{R}_W)\,\mathrm{e}^{-\mathrm{i}(p'_\Lambda\cdot a)}\mathrm{e}^{\mathrm{i}(p_\Lambda\cdot a)}\,\langle s'\mu';p'_\Lambda|s\mu;p_\Lambda\rangle\,.$$

The last factor in this expression is evaluated by means of (6.146) and is found to be equal to

$$\begin{aligned}\langle s'\mu';p'_\Lambda|s\mu;p_\Lambda\rangle &= \delta_{s's}\delta_{\mu'\mu}\,2\sqrt{(Mc)^2+\boldsymbol{p}^2_\Lambda}\,\delta(\boldsymbol{p}'_\Lambda-\boldsymbol{p}_\Lambda)\\ &= \delta_{s's}\delta_{\mu'\mu}\,2\sqrt{(Mc)^2+\boldsymbol{p}^2}\,\delta(\boldsymbol{p}'-\boldsymbol{p})\,.\end{aligned}$$

In the second step the invariance of $p^0\delta(\boldsymbol{p}'-\boldsymbol{p})$ was used. The Kronecker symbols contribute only for $s'=s$ and $\mu'=\mu$. The δ-distribution in the spatial momenta is different from zero only if the spatial momenta $\boldsymbol{p}'$ and $\boldsymbol{p}$ are equal. However, in this case also the two rotations are the same and one obtains

$$\begin{aligned}\sum_\mu D^{(s)*}_{\mu m'}(\mathbf{R}_W)D^{(s)}_{\mu m}(\mathbf{R}_W) &= \sum_\mu \left(D^{(s)\dagger}(\mathbf{R}_W)\right)_{m'\mu}\left(D^{(s)}(\mathbf{R}_W)\right)_{\mu m}\\ &= \delta_{m'm}\,.\end{aligned}$$

From these results one concludes

$$\langle s'm';p'|\mathbf{U}^\dagger(\Lambda,a)\mathbf{U}(\Lambda,a)\,|sm;p\rangle = \langle s'm';p'|sm;p\rangle\,. \tag{6.147}$$

This proves the unitarity of the representation (6.144).

The representation (6.144) with $s=1/2$ will be a good starting point for the construction of the force-free Dirac equation. We return to this topic in Chap. 9.

Quantized Fields and Their Interpretation 7

Introduction This chapter deals with the quantum theory of systems with an infinite number of degrees of freedom and provides elements of *quantum field theory*. A classical field such as the real scalar field, a Maxwell field, or some continuous mechanical system, when subject to the rules of quantum theory, turns into a field operator which can create or annihilate quanta of this field and which describes the kinematics and the spin properties of these quanta. This step allows to extend scattering theory to processes in which quanta, or particles, are created or annihilated, and, thus, which do not necessarily conserve particle numbers. Canonical quantization is based on a formalism making use of Lagrange densities which are constructed in view of a mild generalization of Hamilton's variational principle of point mechanics. Hence, it is not difficult to build in, or take care of, symmetries and invariances of the theory. In particular, creation and annihilation of particles by interaction terms which determine reactions and decay processes, will always be in accord with the selection rules of the theory.

Perhaps the simplest and most transparent approach to the theory of free quantized fields is *canonical quantization* as developed by Born, Heisenberg, and Jordan. It was obtained by its close analogy to the quantization of mechanical systems with a finite number of degrees of freedom. An alternative, more intuitive approach makes use of path integrals and was developed by Dirac and Feynman. In this chapter we concentrate mainly on the first of these because the canonical formalism is of great practical importance. Path integral quantization of fields is an important topic on its own. Sections 7.7 and 7.8 give an introduction to this topic, for more information one should consult the monographs listed in the bibliography.

F. Scheck, *Quantum Physics*, DOI: 10.1007/978-3-642-34563-0_7,

7.1 The Klein-Gordon Field

In a sense to be described more precisely the Klein-Gordon equation is an analogue of the force-free Schrödinger equation. It reads

$$\Box\phi + \kappa^2\phi = \frac{1}{c^2}\frac{\partial^2}{\partial t^2}\phi - \Delta\phi + \kappa^2\phi = 0\,, \quad \text{with} \quad \kappa = \frac{mc}{\hbar}\,. \tag{7.1}$$

The differential operator $\Box$ that it contains is the generalization of the Laplace operator to Minkowskian spacetime. Its explicit form in terms of the time and space coordinates of an arbitrarily chosen frame of reference, and the characteristic relative minus sign between the second time and space derivatives follow from the Lorentz invariant contraction of the partial derivatives $\partial/\partial x^\mu$ and $\partial/\partial x_\mu$. Resolved in time and space coordinates these are

$$\partial_\mu \equiv \frac{\partial}{\partial x^\mu} = \left(\frac{\partial}{\partial x^0}, \nabla\right)\,, \quad \partial^\mu \equiv \frac{\partial}{\partial x_\mu} = \left(\frac{\partial}{\partial x^0}, -\nabla\right)\,, \tag{7.2}$$

from which one obtains

$$\Box = \partial_\mu \partial^\mu = \frac{\partial^2}{(\partial x^0)^2} - \Delta = \frac{1}{c^2}\frac{\partial^2}{\partial t^2} - \Delta\,. \tag{7.3}$$

The constant in the second term of (7.1), except for a factor 2π, is the inverse of the Compton wave length of a particle with mass m,

$$\frac{1}{\kappa} = \frac{\lambda^{(m)}}{2\pi} = \frac{\hbar}{mc} = \frac{\hbar c}{mc^2}\,.$$

This means that processes in which this particle participates, are characterized by the length $(\hbar c)/(mc^2) = (197.33\ \text{MeV}/mc^2)$ fm. For instance, in the case of a charged π-meson which has the mass $m_{\pi^\pm} = 139.57$ MeV this length is equal to 1.41 fm.

Note that up to this point already two physical assertions are made: The Klein-Gordon equation is the relativistic analogue of the force-free Schrödinger equation, and $\lambda^{(m)}$ is a length which is relevant for physical processes. Before I continue with the main matter I wish to illustrate these two statements.

1. Try to solve the Klein-Gordon equation (7.1) by means of the following ansatz

   $$f_p(x) = f_p(t, \boldsymbol{x}) = \mathrm{e}^{-(\mathrm{i}/\hbar)p\cdot x} = \mathrm{e}^{-\mathrm{i}/\hbar(cp^0 t - \boldsymbol{p}\cdot\boldsymbol{x})}\,,$$

   where $p = (p^0, \boldsymbol{p})$ is an arbitrary four-vector endowed with the physical dimension of a momentum. Then from (7.1) there follows the condition

   $$p^2 = (p^0)^2 - \boldsymbol{p}^2 = (mc)^2\,. \tag{7.4}$$

   Thus, the fourth component p^0 is fixed in terms of the spatial momentum $\boldsymbol{p}$ and the mass m. The component p^0 which carries the physical dimension (energy$/c$) may be replaced by the energy $E = c\,p^0$ which then satisfies the known relativistic energy-momentum relation

   $$E = \sqrt{(c\boldsymbol{p})^2 + (mc^2)^2}\,.$$

This is a simple but important result: If one expands an arbitrary solution $\phi(x)$ in terms of plane waves $f_p(x)$, this expansion will contain only momenta which obey the condition (7.4). Every partial solution must lie on the *mass shell* of the particle with mass m. The main purpose of the Klein-Gordon equation is to guarantee that the free particle satisfies the relativistic energy-momentum relation. It is in this sense that, indeed, it is analogous to the Schrödinger equation without forces

$$\mathrm{i}\hbar \frac{\partial \psi(t, \boldsymbol{x})}{\partial t} = -\frac{\hbar^2}{2m} \boldsymbol{\Delta} \psi(t, \boldsymbol{x})$$

which, by the same ansatz $f_p(t, \boldsymbol{x})$, yields the nonrelativistic relation $E = \boldsymbol{p}^2/(2m)$ between energy and momentum. To this observation I wish to add the following comment:

Remark

A free particle which carries spin s is described by a set of fields which contain the information about that spin s and its projection quantum number s_3. Calling the fields of this set simply "components" of a single-particle field one concludes:

Every component of a field which describes a free particle with spin s must satisfy the Klein-Gordon equation.

The Klein-Gordon equation guarantees the correct relation between energy and momentum. However, by itself, it does not suffice to describe the spin content of a (multi-component) field. An example taken from electrodynamics may illustrate this. Both the electric and the magnetic field fulfill the wave equation in vacuum

$$\left(\frac{1}{c^2}\frac{\partial^2}{\partial t^2} - \boldsymbol{\Delta}\right) \boldsymbol{F}(t, \boldsymbol{x}) = 0\,,$$

which tells us, in the language of particle physics, that the photon is massless. The relative orientation of the magnetic and the electric fields which is characteristic for Maxwell theory, follow from Maxwell's equations proper which contain more information than the wave equation. Again, in the language of particle physics, only the full set of Maxwell's equations show that the photon carries helicity 1.

2. The Klein-Gordon equation follows from the Lagrange density

$$\begin{aligned}\mathcal{L}_{\mathrm{KG}}(\phi, \partial_\mu \phi) &= \hbar c \frac{1}{2}[\partial_\mu \phi(x) \partial^\mu \phi(x) - \kappa^2 \phi^2(x)] \\ &= \hbar c \frac{1}{2}[\partial_\mu \phi(x) g^{\mu\nu} \partial_\nu \phi(x) - \kappa^2 \phi^2(x)]\,. \end{aligned} \tag{7.5}$$

Taking the partial derivatives with respect to ϕ and to $\partial_\mu \phi$ and inserting into the Euler-Lagrange equation,

$$\frac{\partial \mathcal{L}_{\mathrm{KG}}}{\partial \phi} - \partial_\mu \frac{\partial \mathcal{L}_{\mathrm{KG}}}{\partial(\partial_\mu \phi)} = -\hbar c \left(\kappa^2 \phi + \partial_\mu \partial^\mu \phi\right) = 0\,,$$

indeed, yields the Klein-Gordon equation.

The Lagrange density (7.5) must have the physical dimension (energy/volume). With the factor as given here this is correct if the field $\phi(x)$ has the dimension $(\text{length})^{-1}$ – unlike solutions $\psi(t, \boldsymbol{x})$ of the Schrödinger equation whose physical dimension is $(\text{length})^{-3/2}$. What is the reason for this difference? I wish to give three answers to this question:

(a) Firstly, some reflection shows that the solutions $\phi(x)$ of the Klein-Gordon equation, unlike in nonrelativistic quantum mechanics, cannot be interpreted as being probability amplitudes and that, therefore, they must not necessarily have the same dimension as those of the Schrödinger equation. In contrast to the latter the Klein-Gordon wave functions cannot be probability amplitudes in the sense of Born's interpretation. A first reason for this is the fact that the Klein-Gordon equation is a differential equation of *second* order, not first order, in time. Besides the second initial condition that has to be imposed on solutions of the Klein-Gordon equation the proof of the conservation of probability in Sect. 1.4 no longer goes through. Another, deeper reason is that the Klein-Gordon field, like any other field in relativistic quantum theory, describes both particle and antiparticle degrees of freedom. Therefore, the Klein-Gordon equation can no longer be interpreted as a single-particle theory.

(b) As we will show in the next section, a meaningful, Lorentz covariant scalar product of normalizable solutions of the Klein-Gordon equation is given by

$$(\phi_1, \phi_2) = \mathrm{i} \int \mathrm{d}^3x \, \left\{\phi_1^*(x)\partial_0\phi_2(x) - \left(\partial_0\phi_1^*(x)\right)\phi_2(x)\right\}$$
$$(x^0 = \text{const}) \, .$$

It is reasonable to require scalar products to have no dimension. As ∂_0 carries dimension (1/length) the solutions $\phi_i(x)$ must have the dimension found above.

(c) On the other hand, in admitting also complex valued solutions of the Klein-Gordon equation, we will find that the electromagnetic four-current density is given by

$$j^\mu(x) = Q\, ec\, \mathrm{i}\, \phi^*(x) \overset{\leftrightarrow}{\partial^\mu} \phi(x)$$

where e is the elementary charge and Q a dimensionless positive or negative integer number. (The symbol for the derivative stands for the skew-symmetric partial derivative acting to the right and to the left. Its definition is repeated in (7.9) below.) If this expression is correct then, indeed, ϕ must have the dimension $(\text{length})^{-1}$.

To $\mathcal{L}_{\text{KG}}$ we add an interaction term $\mathcal{L}_{\text{int}} = -\hbar c\phi(x)\varrho(x)$ where $\varrho(x)$ represents an external source whose physical interpretation will soon become clear by way of an example. The Euler-Lagrange equation following from the new Lagrange density reads

$$\Box\phi + \kappa^2\phi = -\varrho(x) \, , \tag{7.6}$$

with $\varrho(x)$ taking the role of an external source for the Klein-Gordon field $\phi(x)$. As an example we consider a static, point-like source of strength g, i.e. $\varrho(x) = g\,\delta(\boldsymbol{x})$. For static solutions of (7.6), $\phi(x) = \phi(\boldsymbol{x})$, and with $\Box\psi(\boldsymbol{x}) = -\boldsymbol{\Delta}\psi(\boldsymbol{x})$ this differential equation goes over into

$$\left(\boldsymbol{\Delta} - \kappa^2\right)\phi(\boldsymbol{x}) = g\,\delta(\boldsymbol{x})\,. \tag{7.7}$$

This latter equation is solved best by first determining its Fourier transform: Using

$$\phi(\boldsymbol{x}) = \frac{1}{(2\pi)^{3/2}}\int \mathrm{d}^3k\ \mathrm{e}^{\mathrm{i}\boldsymbol{k}\cdot\boldsymbol{x}}\widetilde{\phi}(\boldsymbol{k})$$

the differential equation (7.7) is turned into an algebraic equation

$$\left(\boldsymbol{k}^2 + \kappa^2\right)\widetilde{\phi}(\boldsymbol{k}) = -g\frac{1}{(2\pi)^{3/2}}\,,$$

whose solution is obvious. The inverse Fourier transform is obtained by integration using spherical polar coordinates in $\mathbb{R}^3_k$. With $r = |\boldsymbol{x}|$ and $k = |\boldsymbol{k}|$ one has

$$\begin{aligned}\phi(\boldsymbol{x}) &= -\frac{g}{(2\pi)^3}\int \mathrm{d}^3k\ \frac{\mathrm{e}^{\mathrm{i}\boldsymbol{k}\cdot\boldsymbol{x}}}{\boldsymbol{k}^2+\kappa^2}\\ &= -2\pi\frac{g}{(2\pi)^3}\int_0^\infty k^2\mathrm{d}k\ \frac{\mathrm{e}^{\mathrm{i}kr}-\mathrm{e}^{-\mathrm{i}kr}}{(k^2+\kappa^2)\mathrm{i}kr} = -\frac{g}{4\pi}\frac{\mathrm{e}^{-\kappa r}}{r}\,.\end{aligned}$$

Assume the source to be a nucleon whose mass is large as compared to m and which is located at the point $\boldsymbol{x}_0$. In an arbitrary point $\boldsymbol{x}$ it creates the field

$$\phi^{(0)}(\boldsymbol{x}) = -\frac{g}{4\pi}\frac{\mathrm{e}^{-\kappa|\boldsymbol{x}-\boldsymbol{x}_0|}}{|\boldsymbol{x}-\boldsymbol{x}_0|}\,.$$

Consider now the Hamilton density of the interaction $\mathcal{H}_{\mathrm{int}} = -\mathcal{L}_{\mathrm{int}}$ as well as the interaction energy (obtained by integrating over the whole space), $H_{\mathrm{int}} = \int \mathrm{d}^3x\ \mathcal{H}_{\mathrm{int}}(x)$. Inserting $\varrho^{(1)}(\boldsymbol{x}) = g\,\delta(\boldsymbol{x}-\boldsymbol{x}_1)$ one obtains

$$H_{\mathrm{int}} = \hbar c\int \mathrm{d}^3x\ \phi^{(0)}(\boldsymbol{x})\varrho^{(1)}(\boldsymbol{x}) = -\hbar c\,\frac{g^2}{4\pi}\frac{\mathrm{e}^{-\kappa|\boldsymbol{x}_1-\boldsymbol{x}_0|}}{|\boldsymbol{x}_1-\boldsymbol{x}_0|}\,. \tag{7.8}$$

This energy which may be interpreted as the interaction energy of two (very heavy) nucleons, is called *Yukawa potential*.[1] The strength of this interaction is characterized by the parameter $g^2/(4\pi)$, its range by $1/\kappa = \lambda^{(m)}/(2\pi)$. In Fig. 7.1 it is represented schematically by full incoming and outgoing lines for the two nucleons, and by a dashed line joining two vertices for the π-meson.

---

[1] After H. Yukawa who in 1936 predicted the existence of π-mesons from an interpretation of nuclear forces by the exchange of mesons, and from the range of these forces.

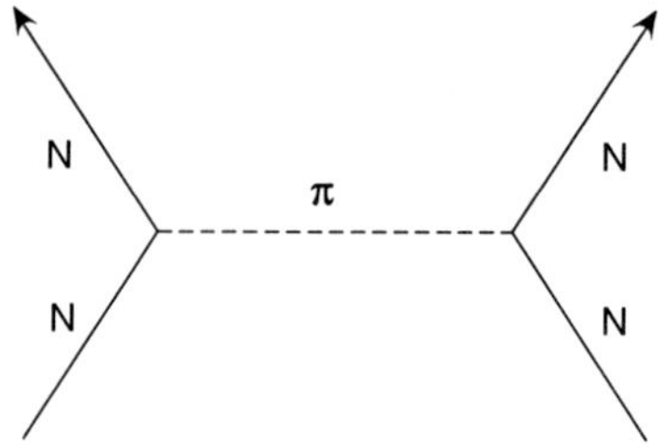

Fig. 7.1 A π-meson of mass m_π is exchanged between two nucleons of mass m_N. If $m_N \gg m_\pi$ holds true then the nucleons can be regarded as inert external sources in the Klein-Gordon equation

7.1.1 The Covariant Normalization

It is useful to introduce an abbreviation for partial derivatives acting to the right and to the left with a relative minus sign (see also (2.3) of Sect. 2.2),

$$f \overset{\leftrightarrow}{\partial}_\mu g := f \left(\frac{\partial g}{\partial x^\mu} \right) - \left(\frac{\partial f}{\partial x^\mu} \right) g \,. \tag{7.9}$$

Let $\phi_n(x)$ and $\phi_m(x)$ be square integrable solutions of the Klein-Gordon equation whose quantum numbers are denoted symbolically by n and m, respectively. Define a scalar product by the following integral with a fixed value of the time coordinate x^0

$$(\phi_n, \phi_m)|_{x^0} = \mathrm{i} \int \mathrm{d}^3x \; \phi_n^*(x) \overset{\leftrightarrow}{\partial}_0 \phi_m(x) \,. \tag{7.10}$$

As one easily confirms, the integral (7.10) has the properties of a scalar product: It is linear in the second argument, and antilinear in the first. It fulfills the relations

$$(\phi_n, \phi_m)^* = (\phi_m, \phi_n) \,, \qquad (\phi_m, \phi_m) \geq 0 \,,$$

and for fixed ϕ_0, and for all ϕ_n it vanishes if and only if ϕ_0 is the null element.

Although the integral (7.10) seems to assume the choice of a class of reference systems which define the time axis, the definition is Lorentz covariant. What really matters in this formula is the integration over a spacelike three-dimensional hypersurface in spacetime. A surface Σ of this kind is characterized by the property that any two arbitrarily chosen points $x \in \Sigma$ and $y \in \Sigma$ are *spacelike* with respect to each other. In our conventions this means that $(x - y)^2 < 0$. The three-dimensional hypersurface Σ_0 that one obtains by choosing a fixed section in time, $x^0 = \text{const.}$, is a specific example.

The property of a hypersurface to be spacelike is a Lorentz invariant feature. The covariance is proven if we succeed in showing that in (7.10) the section Σ_0 can be continuously deformed into any other spacelike hypersurface Σ without changing the value of the integral. This is shown as follows.

Let $n^\mu(x)$ be the local normal to the surface in the point $x \in \Sigma$ whose orientation is positive timelike, i.e. for which $n^2 = 1$, $n \in V_x^+$, with V_x^+ denoting the future light cone at x. In the case of Σ_0 this is $n^\mu = (1, 0, 0, 0)^T$, and the integral in (7.10) can be written as

$$\int_{\Sigma_0} \mathrm{d}\sigma \; n^\mu \phi_n^*(x) \overset{\leftrightarrow}{\partial}_\mu \phi_m(x) \,,$$

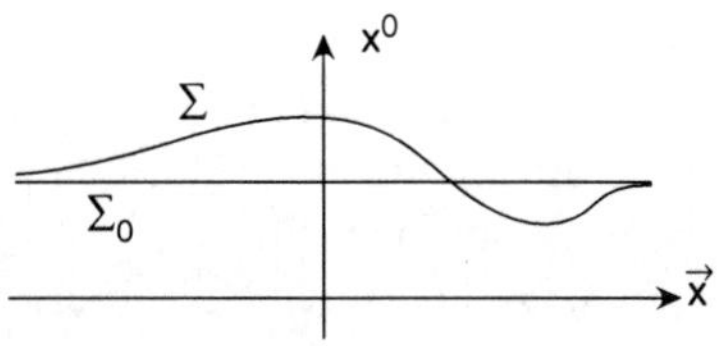

Fig. 7.2 A spacelike hypersurface, i.e. a three-dimensional submanifold of spacetime in which any two points are relatively spacelike, is obtained from the time section Σ_0 by local continuous deformation

where $\mathrm{d}\sigma$ is the three-dimensional volume element d^3x. Suppose that Σ_0 is locally and continuously deformed into another spacelike hypersurface Σ such as the one that we sketched in Fig. 7.2.[2] Consider then the difference of the integrals over Σ and over Σ_0. Gauss' theorem relates this difference to an integral over the four-dimensional volume enclosed by Σ and Σ_0,

$$\int_{\Sigma} \mathrm{d}\sigma\, n^\mu \phi_n^*(x) \overleftrightarrow{\partial_\mu} \phi_m(x) - \int_{\Sigma_0} \mathrm{d}\sigma\, n^\mu \phi_n^*(x) \overleftrightarrow{\partial_\mu} \phi_m(x)$$

$$= \int_{V(\Sigma-\Sigma_0)} \mathrm{d}^4x\, \partial^\mu \left(\phi_n^*(x) \overleftrightarrow{\partial_\mu} \phi_m(x) \right) .$$

The integrand of the volume integral on the right-hand side is easy to evaluate because ϕ_n and ϕ_m are solutions of the Klein-Gordon equation,

$$\begin{aligned} \partial^\mu \left(\phi_n^*(x) \overleftrightarrow{\partial_\mu} \phi_m(x) \right) &= (\partial^\mu \phi_n^*)(\partial_\mu \phi_m) + \phi_n^* \Box \phi_m - (\Box \phi_n^*)\phi_m \\ &\quad - (\partial_\mu \phi_n^*)(\partial^\mu \phi_m) \\ &= (\kappa^2 - \kappa^2)\phi_n^* \phi_m = 0 . \end{aligned}$$

This shows that the scalar product (7.10) has the same value for any spacelike hypersurface that approaches Σ_0 smoothly at infinity. This proves the assertion.

7.1.2 A Comment on Physical Units

In many parts of relativistic quantum theory and of quantum field theory one can spare much writing by using what are called *natural units*, instead of the customary physical units such as the SI-system, or the older cgs-system. They all have in common that Planck's constant $\hbar$ and the velocity of light c take the value 1.

$$\hbar = 1 , \qquad c = 1 . \tag{7.11}$$

[2] As the fields decrease sufficiently fast at infinity there is no loss of generality in assuming Σ to approach Σ_0 smoothly at spatial infinity.

With the following arguments in mind one will quickly get used to this class of conventions. Note, first, that the convention (7.11) does not completely fix physical units because, while fixing *relative* units of length, time, momentum, and energy, it leaves the absolute scale undetermined. Indeed, writing dimensionful physical quantities, when expressed in natural units, with a subscript "nat", and denoting the physical dimension of an observable A as usual by $[A]$, then one has, with $c = 1$

$$[\boldsymbol{x}_{\text{nat}}] = [t_{\text{nat}}] , \quad [\boldsymbol{p}_{\text{nat}}] = [E_{\text{nat}}] .$$

If, in addition, one chooses $\hbar = 1$ then energy and momentum have the inverse unit of length and time, viz.

$$[\boldsymbol{p}_{\text{nat}}] = [E_{\text{nat}}] = [\boldsymbol{x}_{\text{nat}}]^{-1} = [t_{\text{nat}}]^{-1} .$$

This also holds for any pair of canonically conjugate variables (q, p) whose product is known to have the dimension of an action. But action has no dimension in natural units. Thus, with $\hbar = 1$, one has

$$[p_{\text{nat}}] = [q_{\text{nat}}]^{-1} .$$

The missing scale is fixed by choosing a physical unit for the energy. This may be the MeV, that is, 10^6 eV, (where 1 eV is the energy that an elementary charge gains in passing through a potential difference of 1 V). It could also be the rest mass $mc^2 \mathrel{\widehat{=}} m_{\text{nat}}$ of some elementary particle. For example, in quantum electrodynamics which deals primarily with electrons and photons, it seems reasonable to introduce the rest mass of the electron $m_e c^2 = 0.511$ MeV as the scale for energies. Another example is the physics of π-mesons and nucleons at low energies where the rest mass of the charged pions $m_\pi c^2 = 139.57$ MeV is an obvious candidate for the energy scale.

If one does not work in a restricted domain of physics such as the ones just mentioned it is more sensible to stay with the unit eV and multiples thereof. Common notations for powers of ten of the electron Volt are

$$\begin{aligned} &1\,\text{meV} = 10^{-3}\,\text{eV} , \quad 1\,\text{keV} = 10^{3}\,\text{eV} , \quad 1\,\text{MeV} = 10^{6}\,\text{eV} , \\ &1\,\text{GeV} = 10^{9}\,\text{eV} , \quad 1\,\text{TeV} = 10^{12}\,\text{eV} . \end{aligned}$$

They are called, in the order given, *milli-*, *kilo-*, *mega-*, *giga-*, and *tera-electron Volt.*

If at the end of a calculation one wishes to return to conventional units the following formulae for lengths, times, and cross sections are helpful

$$\begin{aligned} l\,\text{fm} &\mathrel{\widehat{=}} \hbar c \cdot l_{\text{nat}}\ \text{MeV}^{-1} \\ t\,\text{s} &\mathrel{\widehat{=}} \frac{\hbar c}{c} \cdot t_{\text{nat}}\ \text{MeV}^{-1} \\ \sigma\ \text{fm}^2 &\mathrel{\widehat{=}} (\hbar c)^2 \cdot \sigma_{\text{nat}}\ \text{MeV}^{-2} , \end{aligned} \tag{7.12}$$

together with the numerical value

$$\hbar c = 197.3270\,\text{MeV fm} = 197.3270 \times 10^{-15}\,\text{MeV m} .$$

Masses are expressed in MeV, the relation between their notation in natural units and physical units is simply $m_{\text{nat}} \mathrel{\widehat{=}} mc^2$. Two examples may clarify the conversion:

A state which has the mean life $\tau = 8.4 \times 10^{-17}$ s (this is the mean life of neutral pions π^0) has an uncertainty in energy, or width, given by

$$\Gamma = \frac{\hbar c}{c}\frac{1}{\tau} = 7.836\,\text{eV}\,.$$

A cross section that was measured or calculated to be $\sigma_{\text{nat}} = 1\,\text{GeV}^{-2}$, when expressed in millibarn (1 barn $= 10^{-28}\,\text{m}^2$), is equal to

$$\sigma = \sigma_{\text{nat}} \cdot (\hbar c)^2 = 0.3894\,\text{mbarn}\,.$$

A further simplification that we shall make use of in the sequel, concerns the choice of units in Maxwell's equations of electrodynamics. Denoting as usual electric and magnetic fields by $(\boldsymbol{E}, \boldsymbol{H})$, respectively, displacement and magnetic induction fields by $(\boldsymbol{D}, \boldsymbol{B})$, respectively, then Maxwell's equations and the Lorentz force read, in any system of units,

$$\nabla \cdot \boldsymbol{B}(t, \boldsymbol{x}) = 0 \quad \nabla \times \boldsymbol{E}(t, \boldsymbol{x}) + f_1 \frac{\partial \boldsymbol{B}(t,\boldsymbol{x})}{\partial t} = 0 \tag{7.13}$$

$$\nabla \cdot \boldsymbol{D}(t, \boldsymbol{x}) = f_2\, \varrho(t, \boldsymbol{x}),\, \nabla \times \boldsymbol{H}(t, \boldsymbol{x}) - f_3 \frac{\partial \boldsymbol{D}(t,\boldsymbol{x})}{\partial t} = f_4\, \boldsymbol{j}(t, \boldsymbol{x})\,, \tag{7.14}$$

$$\boldsymbol{F}(t, \boldsymbol{x}) = e\left(\boldsymbol{E}(t, \boldsymbol{x}) + f_1\, \boldsymbol{v} \times \boldsymbol{B}(t, \boldsymbol{x})\right)\,. \tag{7.15}$$

In the vacuum the two types of fields are related by

$$\boldsymbol{D} = \varepsilon_0\, \boldsymbol{E}\,, \qquad \boldsymbol{B} = \mu_0\, \boldsymbol{H}\,. \tag{7.16}$$

The dielectric constant ε_0 and the magnetic permeability μ_0 of the vacuum are always chosen such that $f_1 = f_3$. The continuity equation

$$\frac{\partial \varrho(t, \boldsymbol{x})}{\partial t} + \nabla \cdot \boldsymbol{j}(t, \boldsymbol{x}) = 0\,,$$

which follows from the two inhomogeneous Maxwell equations (7.14), yields the relation $f_4 = f_2 f_3 = f_2 f_1$. Using the system of Gauss' units, for example, one has

$$f_1 = f_3 = \frac{1}{c}\,, \quad f_2 = 4\pi\,, \quad f_4 = f_1 f_2 = \frac{4\pi}{c}\,, \quad \varepsilon_0 = \mu_0 = 1\,.$$

This particularly transparent sytem of units simplifies even further if one is allowed to set $c = 1$, and if one rescales charge and current densities such that all factors 4π in (7.14) are made to disappear. This is achieved if the observables of Maxwell's theory are replaced by natural quantities defined as follows

$$\boldsymbol{E}_{\text{nat}} := \frac{1}{\sqrt{4\pi}} \boldsymbol{E}_{\text{Gauss}}\,, \quad \boldsymbol{B}_{\text{nat}} := \frac{1}{\sqrt{4\pi}} \boldsymbol{B}_{\text{Gauss}},$$

$$\varrho_{\text{nat}} := \sqrt{4\pi}\, \varrho_{\text{Gauss}}\,, \qquad \boldsymbol{j}_{\text{nat}} := \sqrt{4\pi}\, \boldsymbol{j}_{\text{Gauss}}\,. \tag{7.17}$$

With this convention and with $c = 1$ all constants f_i in (7.13)–(7.15) are now equal to 1. As a consequence of the choice (7.17) the field strength tensor $F_{\mu\nu}$ is rescaled by the same factor as the fields $\boldsymbol{E}, \ldots, \boldsymbol{D}$, and the Lagrange density of Maxwell theory takes the form

$$\mathcal{L} = -\frac{1}{4}(F_{\text{nat}})_{\mu\nu}(F_{\text{nat}})^{\mu\nu}$$

i.e. the customary factor $1/(16\pi)$ is replaced by $1/4$.

Together with the choice $\hbar = 1$ one sees that at the end of a calculation the square of the elementary charge in natural units must be replaced by Sommerfeld's fine structure constant following the rule

$$e_{\text{nat}}^2 = 4\pi\alpha\,, \qquad \text{where} \quad \alpha = 1/137.036\,. \tag{7.18}$$

In natural units the Bohr radius and the energies of bound states in hydrogen are, respectively,

$$(a_B)_{\text{nat}} = \frac{1}{\alpha m_{\text{nat}}} \mathrel{\widehat{=}} \frac{\hbar c}{\alpha m c^2}\,, \qquad (E_n)_{\text{nat}} = -\alpha^2 \frac{1}{2n^2} m_{\text{nat}} \mathrel{\widehat{=}} -\alpha^2 \frac{1}{2n^2} mc^2\,.$$

The Compton wave length of a particle with mass m reads $\lambda = h/(mc) \mathrel{\widehat{=}} 2\pi/m_{\text{nat}}$, the constant κ of the Klein-Gordon equation (7.1) is simply replaced by m_{nat}, and the Yukawa potential reads $\mathrm{e}^{-m_{\text{nat}} r}/r$.

Unless stated otherwise we will employ natural units in the sequel but will suppress the subscript "nat" from here on. Planck's constant $\hbar$, however, will be written explicitly if the transition to classical physics matters, or if the structure of a quantum theoretic expression is to be analyzed as a power series in $\hbar$.

7.1.3 Solutions of the Klein-Gordon Equation for Fixed Four-Momentum

Expressed in natural units the Lagrange density (7.5) and the Klein-Gordon equation (7.1) read

$$\mathcal{L}_{\text{KG}} = \frac{1}{2}[\partial_\mu \phi(x) \partial^\mu \phi(x) - m^2 \phi^2(x)]\,,$$
$$\Box \phi(x) + m^2 \phi(x) = 0\,.$$

The solutions with fixed values of the momentum $p = (E, \boldsymbol{p})^T$ that we considered above, take the form

$$f_p(x) = \frac{1}{(2\pi)^{3/2}} \mathrm{e}^{-\mathrm{i}p\cdot x}\,, \quad \text{with} \quad p^0 = E_p = \sqrt{m^2 + \boldsymbol{p}^2}\,. \tag{7.19}$$

These solutions are not square-integrable but can be normalized to δ-distributions. As one easily verifies they are orthogonal and normalized in the following sense. For fixed but arbitrarily chosen time x^0 one finds

$$\mathrm{i}\int \mathrm{d}^3x\; f_{p'}^*(x) \overleftrightarrow{\partial_0} f_p(x) = 2E_p\, \delta(\boldsymbol{p} - \boldsymbol{p}')\,, \tag{7.20}$$

$$\int \mathrm{d}^3x\; f_{p'}(x) \overleftrightarrow{\partial_0} f_p(x) = \int \mathrm{d}^3x\; f_{p'}^*(x) \overleftrightarrow{\partial_0} f_p^*(x) = 0\,. \tag{7.21}$$

The left-hand side of the formulae (7.20) and (7.21) is the generalization of the covariant scalar product (7.10) to solutions which are not normalizable in the ordinary sense. The right-hand side is identical with the covariant normalization (6.145) in momentum space that was introduced in the context of representations of the Poincaré group. Of course, the invariance of the scalar product proven in Sect. 7.1.1 shows that this normalization does not depend on the choice of the frame of reference. Nevertheless, it is instructive to clarify this fact directly, by working in momentum space. Indeed, taking the components of the four-momentum p to be unconstrained variables, that is, assuming no relation between them, it is clear that an integral $\int \mathrm{d}^4 p \; \mathcal{I}$ is Lorentz invariant provided the integrand $\mathcal{I}$ itself is an invariant. However, for any free physical state one must require p to lie on the mass shell $p^2 = m^2$ and its time component p^0 to be positive in every frame. (Free particles always have positive energy.) These conditions are fulfilled if the integral is modified by means of suitable constraints, viz. by replacing the integrand as follows,

$$\int \mathrm{d}^4 p \; \mathcal{I} \longmapsto \int \mathrm{d}^4 p \, \delta(p^2 - m^2)\Theta(p^0) \; \mathcal{I} .$$

The question then is whether the δ-distribution and the step function are compatible with the invariance. The first factor, $\delta(p^2 - m^2)$, is invariant while the second factor, $\Theta(p^0)$, taken in isolation, is not. For instance, if p were a spacelike vector then there would exist Lorentz transformations $\Lambda \in L_+^\uparrow$ which reverse the sign of the time component p^0. The product of the two factors, in turn, is invariant because the δ-distribution restricts the vector p to a hyperboloid whose two shells lie within the forward light cone (future), and within the backward light cone (past), respectively. A proper orthochronous Lorentz transformation leaves these two shells invariant, cf. Fig. 7.3, and, hence, cannot map a positive $p^0 > 0$ onto a negative $p^0 < 0$, or vice versa.

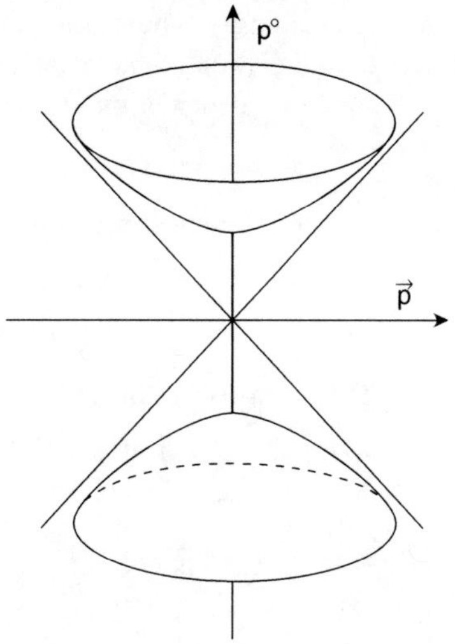

Fig. 7.3 The two shells of the mass hyperboloid $p^2 = m^2$ in a representation in momentum space. Only the upper shell corresponds to physically allowed states

If we choose a class of reference frames which single out the zero component, then, with $E_p = \sqrt{\boldsymbol{p}^2 + m^2}$, we use the well-known formula for the δ-distribution of a function of x, $\delta(f(x))$, (see Appendix A.1)

$$\delta(p^2 - m^2) = \delta\left((p^0)^2 - (\boldsymbol{p}^2 + m^2)\right) = \frac{1}{2p^0}\{\delta(p^0 - E_p) + \delta(p^0 + E_p)\}$$

and perform the integral over the variable p^0. The step function insures that only positive values of p^0 contribute and one obtains

$$\int \mathrm{d}^4 p\, \delta(p^2 - m^2)\Theta(p^0) \cdots = \int \frac{\mathrm{d}^3 p}{2E_p} \cdots .$$

This argument shows that $\mathrm{d}^3 p/(2E_p)$ is an invariant volume element on the mass shell $p^2 = m^2$ and that, indeed, the normalization of single-particle states must be as given in (6.145) and in (7.20).

Remarks

1. Clearly, only the factor E_p really matters. The factor 2 seems natural in view of the skewsymmetric derivative (7.20) but by no means it is compulsory. Indeed, some authors use normalization to $2E_p$ only for bosons but normalize to E_p/m in the case of fermions, where m is the rest mass of the fermion that is considered. In this book I choose normalization to $2E_p$ throughout, both for bosons and for fermions.
2. Using covariant normalization the volume element in momentum space is

$$\frac{\mathrm{d}^3 p}{2E_p} \quad \text{instead of} \quad \mathrm{d}^3 p .$$

As a consequence, the formulae for cross sections and Fermi's Golden Rule which depend on the density of states in the final states, must be modified correspondingly. The single-particle states with momentum p and spin quantum numbers (s, μ) that we denote by $|p; s, \mu\rangle$ then no longer are normalized to 1 (or to a δ-distribution whose factor is 1). Their covariant normalization reads

$$\left\langle p'; s, \mu' \middle| p; s, \mu \right\rangle = \delta_{\mu\mu'} 2E_p\, \delta(\boldsymbol{p} - \boldsymbol{p}') . \tag{7.22}$$

The completeness relation, in a formal notation, is modified to

$$\sum_\sigma \int \frac{\mathrm{d}^3 q}{2E_q} \,|q; s, \sigma\,\rangle\, \langle q; s, \sigma| = \mathbb{1} . \tag{7.23}$$

This is easily confirmed by the formal calculation

$$\sum_\sigma \int \frac{\mathrm{d}^3 q}{2E_q} \,|q; s, \sigma\,\rangle\, \langle q; s, \sigma|\, p; s, \mu\rangle$$
$$= \sum_\sigma \int \mathrm{d}^3 q \,|q; s, \sigma\,\rangle\, \delta_{\sigma\mu} \delta(\boldsymbol{q} - \boldsymbol{p}) = |\,p; s, \mu\,\rangle .$$

7.1.4 Quantization of the Real Klein-Gordon Field

We noted earlier that, in contrast to those of the Schrödinger equation, the solutions of the Klein-Gordon equation cannot be probability amplitudes. Born's interpretation of a normalizable solution $\phi(t, \boldsymbol{x})$ makes no physical sense. However, if we understand the solutions in the sense of second quantization, i.e. if we construct operator-valued solutions of the Klein-Gordon equation then we obtain quasi automatically a physically convincing interpretation.

Consider the field $\phi(x)$ as a genuine scalar with respect to Poincaré transformations, not as one of the components of some field with nonvanishing spin. In other terms, in the perspective of representation theory of the Poincaré group, cf. (6.144), it is meant to describe a particle with spin $s = 0$. We return to the Lagrange density (7.5) for a real field $\phi(x)$ and define the generalized, canonically conjugate momentum,

$$\pi(x) := \frac{\partial \mathcal{L}}{\partial(\partial_0 \phi)} = \partial^0 \phi(x) . \tag{7.24}$$

Obviously, this definition is inspired by classical canonical mechanics, the canonically conjugate pair (q, p) being replaced by $(\phi(x), \pi(x))$. One then constructs

$$\widetilde{\mathcal{H}} = \pi\, \partial_0 \phi - \mathcal{L} = \frac{1}{2}\left\{(\partial^0 \phi)(\partial_0 \phi) + (\nabla \phi)^2 + m^2 \phi^2\right\}$$

and, by Legendre transformation of this function, obtains the Hamilton density

$$\mathcal{H}(\phi, \pi) = \frac{1}{2}\left\{\pi^2(x) + (\nabla \phi)^2 + m^2 \phi^2\right\} . \tag{7.25}$$

In close analogy to the transition from classical to quantum point mechanics, summarized by the prescription for conversion of Poisson brackets to commutators,

$$\{p_i, q_k\} = \delta_{ik} \longmapsto \frac{\mathrm{i}}{\hbar}[p_i, q_k] = \delta_{ik} ,$$

canonical quantization of classical fields builds upon the following postulate:

Postulate 7.1

In the realm of quantum theory the real-valued functions $\phi(x)$ and $\pi(x)$ are replaced by operator-valued distributions $\Phi(x)$ and $\Pi(x)$, respectively, which in any frame of reference and for equal times, obey the commutator relations

$$\frac{\mathrm{i}}{\hbar}\left[\Pi(x), \Phi(x')\right]_{x^0 = x^{0\prime}} = \delta(\boldsymbol{x} - \boldsymbol{x}') \tag{7.26}$$

$$\left[\Phi(x), \Phi(x')\right]_{x^0 = x^{0\prime}} = 0 = \left[\Pi(x), \Pi(x')\right]_{x^0 = x^{0\prime}} . \tag{7.27}$$

Before working out the consequences of the quantization rules (7.26) with the aim of clarifying the physical roles of Φ and Π, I add a few further remarks:

Remarks

1. In this section and for the sake of clarity, operator-valued functions are denoted by Greek capital letters. Later on this convention will not be applied everywhere but it will always be clear from the context whether one is dealing with functions or with operators. It is common usage in quantum field theory to call complex functions $\phi(x)$ etc. *c-number valued*, while the operators $\Phi(x)$ which are obtained from them are said to *operator-valued*. Needless to add that units will mostly be such that $\hbar = 1$.
2. By imposing equal times in (7.26) it seems as if one preferred certain frames of reference while destroying Lorentz covariance. That this is not so will be shown explicitly below. The rule (7.26), together with the covariant normalization of single-particle states yields an invariant formalism. To witness, one could modify the quantization rule already at this point such that it stays manifestly covariant, by choosing an arbitrary spacelike hypersurface Σ instead of the hyperplane $x^0 = x'^0$.
3. This mode of quantization rests in an essential way on classical mechanics. Without the notion of Lagrange density, and the canonically conjugate momentum that follows from it, it would be difficult to guess the commutation rules.
4. In the example studied here the Legendre transform of the Lagrange density exists. However, in cases where this is not so, one must study quantization with constraints, a topic that we do not enter, for lack of space.

The Hamiltonian $H = \int \mathrm{d}^3x\ \mathcal{H}(x)$ is obtained from the Hamilton density by integration over the whole space, keeping the time coordinate fixed. One then calculates the commutators of H with Φ, with Π, and with arbitrary polynomials in these operators. For example, one has

$$[H, \Phi(x)] = \int\limits_{y^0=x^0} \mathrm{d}^3y\ [\mathcal{H}(y), \Phi(x)] = \frac{1}{2}\int \mathrm{d}^3y\ [\Pi^2(y), \Phi(x)]_{y^0=x^0}$$

$$= -\mathrm{i}\int \mathrm{d}^3y\ \Pi(y)_{y^0=x^0}\delta(\boldsymbol{y}-\boldsymbol{x}) = -\mathrm{i}\Pi(x) = -\mathrm{i}\partial_0\Phi(x)\,.$$

This calculation makes use of the analogue of the elementary manipulation

$$[p^2, q] = ppq - qpp = p(-\mathrm{i} + qp) - (\mathrm{i} + pq)p = -2\mathrm{i}p$$

of point (quantum) mechanics. In much the same way one finds

$$[H, \Pi(x)] = -\mathrm{i}\,\partial_0\Pi(x)$$

and, similarly, for a polynomial $F(x)$ in the variables Φ and Π

$$[H, F(x)] = -\mathrm{i}\,\partial_0 F(x)\,.$$

The analogy to the equations of motion

$$\frac{\mathrm{d}f}{\mathrm{d}t} = \{H, f\}\,, \qquad \text{and} \qquad \frac{\mathrm{d}F}{\mathrm{d}t} = \frac{\mathrm{i}}{\hbar}[H, F]$$

of classical mechanics (with Poisson brackets), and of quantum mechanics (with commutators), respectively, is obvious. As we are working in a Lorentz covariant framework (explicitly or implicitly), it seems plausible that H is the time-component of a set of four operators $P^\nu = (P^0 = H, P^i)$ which define the energy-momentum operator of the theory, and that the space components P^i are also given by space integrals of corresponding densities. Indeed, if one constructs the classical tensor field of energy and momentum

$$\mathcal{T}^{\mu\nu} = \frac{\partial \mathcal{L}}{\partial(\partial_\mu \Phi)} \partial^\nu \Phi - g^{\mu\nu} \mathcal{L} , \tag{7.28}$$

then one sees that $\partial_\mu \mathcal{T}^{\mu\nu} = 0$ holds for all solutions of the Euler-Lagrange equation, and that the Hamilton density is the time-time component of this tensor field, $\mathcal{H} = \mathcal{T}^{00}$. Its space components are the missing momentum densities. Making use of the Klein-Gordon equation these densities and their space integrals are calculated to be

$$\mathcal{P}^i = -\Pi(x) \, (\nabla \Phi)^i , \qquad \boldsymbol{P} = -\int \mathrm{d}^3 x \; \Pi(x) \nabla \Phi .$$

The commutator of the operator $\boldsymbol{P}$ with a polynomial $F(x)$ in Φ and Π is calculated in the same way as the one of H with F,

$$[\boldsymbol{P}, F(x)] = -\int \mathrm{d}^3 y \; [\Pi(y) \nabla \Phi(y), F(x)]_{y^0 = x^0} = \mathrm{i} \, \nabla F(x) .$$

By the definition (7.2) the commutators calculated above are summarized in a covariant notation

$$\left[P^\mu, F(x)\right] = -\mathrm{i} \, \partial^\mu F(x) . \tag{7.29}$$

These are *Heisenberg's equations of motion* in a Lorentz covariant form.

We note that the Heisenberg equations of motion can also be written in an integral form. Let x and y be arbitrarily chosen points of spacetime. Then the following *translation formula* holds

$$F(y) = \mathrm{e}^{\mathrm{i} P \cdot (y-x)} F(x) \mathrm{e}^{-\mathrm{i} P \cdot (y-x)} , \tag{7.30}$$

(see Exercise 7.2). Working out the Taylor expansion of F about the point $y = x$ one easily confirms that (7.29) follows from the translation formula (7.30). One of the prime uses of the translation formula is to reduce operator-valued functions, or distributions with an arbitrary argument y to the same quantities at a fixed reference point x such as, e.g., $x = 0$. We shall make frequent use of this possibility.

7.1.5 Normal Modes, Creation and Annihilation Operators

The aim of this section and the following one is to clarify the physical role of the operator $\Phi(x)$. A first idea that comes to mind is to expand the, to some extent, arbitrary operator Φ in terms of a complete system of solutions which carry simple, easily interpretable single-particle quantum numbers. A set of special relevance is given by the plane waves (7.19) which are known to describe the free motion of particles with mass m and given eigenvalues of momentum $\boldsymbol{p}$ on the mass shell $(p^0)^2 = \boldsymbol{p}^2 + m^2 = E_p^2$. We define

$$\Phi(x) = \int \frac{\mathrm{d}^3 p}{2E_p} \left[f_p(x) a(p) + f_p^*(x) a^\dagger(p) \right] , \tag{7.31}$$

where the integral contains the invariant volume element. Obviously, the operator nature of Φ in (7.31) is taken over by the coefficients $a(p)$ and $a^\dagger(p)$, the latter being the adjoint of $a(p)$. The specific combination of the terms in square brackets follows from the requirement that before quantization $\Phi(x)$ be a *real* field, and after quantization be a *self-adjoint* operator. The relations (7.20) and (7.21) allow to express the operators $a(p)$ and $a^\dagger(p)$ in terms of Φ. With x^0 being an arbitrary but fixed time one obtains

$$a(p) = \mathrm{i} \int \mathrm{d}^3 x \; f_p^*(x) \overleftrightarrow{\partial_0} \, \Phi(x) , \quad a^\dagger(p) = -\mathrm{i} \int \mathrm{d}^3 x \; f_p(x) \overleftrightarrow{\partial_0} \, \Phi(x) . \tag{7.32}$$

If $\Phi(x)$ fulfills the free Klein-Gordon equation then these operators are independent of time. We check this for $a(p)$:

$$\begin{aligned} \partial_0 a(p) &= \mathrm{i} \int \mathrm{d}^3 x \; \left[f_p^*(x) \partial_0^2 \Phi(x) - (\partial_0^2 f_p^*) \Phi(x) \right] \\ &= \mathrm{i} \int \mathrm{d}^3 x \; \left[f_p^*(x) \boldsymbol{\Delta} \Phi(x) - (\boldsymbol{\Delta} f_p^*(x)) \Phi(x) \right] = 0 . \end{aligned}$$

In the last step one transforms either the first or the second term by two partial integrations. There are no boundary contributions provided Φ decreases sufficiently fast at spatial infinity. (In case this assumption is not fulfilled replace $f_p(x)$ by localized wave packets.)

The commutators of the operators $a(p)$ and $a^\dagger(q)$ are obtained from (7.31), the formulae (7.32), and the orthogonality of the basis. Here is an example of this calculation:

$$\begin{aligned} \left[a(p), a^\dagger(q) \right] &= -\mathrm{i}^2 \int \mathrm{d}^3 x \int \mathrm{d}^3 y \; f_p^*(x) f_q(y) \frac{\overleftrightarrow{\partial}}{\partial x^0} \frac{\overleftrightarrow{\partial}}{\partial y^0} \left[\Phi(x), \Phi(y) \right] \\ &= \mathrm{i}^2 \int \mathrm{d}^3 x \int \mathrm{d}^3 y \; f_p^*(x) (\partial_0 f_q(y)) \left[\partial_0 \Phi(x), \Phi(y) \right] \\ &\quad + \mathrm{i}^2 \int \mathrm{d}^3 x \int \mathrm{d}^3 y \; (\partial_0 f_p^*(x)) f_q(y) \left[\Phi(x), \partial_0 \Phi(y) \right] . \end{aligned}$$

By the definition (7.24) and the postulate (7.26) one has

$$\left[\partial_0 \Phi(x), \Phi(y) \right] = -\mathrm{i} \delta(\boldsymbol{x} - \boldsymbol{y}) , \qquad \left[\Phi(x), \partial_0 \Phi(y) \right] = +\mathrm{i} \delta(\boldsymbol{x} - \boldsymbol{y}) .$$

Inserting this one sees that one of the integrations can be carried out, e.g., the integration over y, so that

$$\left[a(p), a^\dagger(q) \right] = \mathrm{i} \int \mathrm{d}^3 x \; f_p^*(x) \overleftrightarrow{\partial_0} \, f_q(x) = 2E_p \delta(\boldsymbol{p} - \boldsymbol{q}) .$$

The commutator $[a(p), a(q)]$ is calculated along the same lines, and is found to be zero. We summarize this important result:

$$\left[a(p), a^\dagger(q)\right] = 2E_p\delta(\boldsymbol{p}-\boldsymbol{q}), \tag{7.33}$$

$$[a(p), a(q)] = 0 = \left[a^\dagger(p), a^\dagger(q)\right]. \tag{7.34}$$

The analogy to the one-dimensional harmonic oscillator, Sect. 1.6, is obvious. However, in contrast to that case, there is an infinity of oscillators each of which is characterized by an eigenvalue of the momentum (and hence of energy). This conclusion is corroborated by a calculation of the Hamiltonian and of the operator of spatial momentum in terms of $a(p)$ and $a^\dagger(p)$.

The Hamilton density is given in (7.25) where $\Pi(x)$ is to be calculated by means of (7.24). Inserting the expansion (7.31) and making use of the orthogonality of the basis $f_p(x)$, one finds

$$H = \frac{1}{2}\int \frac{\mathrm{d}^3 p}{2E_p}\, E_p \left\{a^\dagger(p)a(p) + a(p)a^\dagger(p)\right\}. \tag{7.35}$$

In much the same way the operator of spatial momentum is represented in terms of the operators a and $a^\dagger$. One finds

$$\boldsymbol{P} = \frac{1}{2}\int \frac{\mathrm{d}^3 p}{2E_p}\, \boldsymbol{p} \left\{a^\dagger(p)a(p) + a(p)a^\dagger(p)\right\}. \tag{7.36}$$

Before we continue with a further analysis of these expressions we recapitulate the formalism of second quantization for bosons that we first discussed in Part One, Sect. 1.6.

For simplicity and for the sake of illustration, we replace the dependence of creation and annihilation operators on the momentum p by a discrete index, and assume that they fulfill the following commutators

$$[a_i, a_k^\dagger] = \delta_{ik}, \quad [a_i, a_k] = 0 = [a_i^\dagger, a_k^\dagger]. \tag{7.37}$$

These commutation rules are formally similar to the rules (7.33) and (7.34) except that their right-hand side is a finite c-number while the right-hand side of (7.33) is a distribution.

Define the self-adjoint operators

$$N_i := a_i^\dagger a_i, \qquad N := \sum_i N_i,$$

the first of which is called particle number operator of the kind i, while the second is called total particle number. Let their eigenstates be denoted by

$$|\,\nu_i\,\rangle, \quad \text{or} \quad |\,\nu_1, \nu_2, \ldots, \nu_i, \ldots\rangle$$

such that one has $N_i\,|\,\nu_i\,\rangle = \nu_i|\nu_i\rangle$ and $N\,|\nu_1, \nu_2, \ldots, \nu_i, \ldots\rangle = (\sum \nu_i)|\nu_1, \nu_2, \ldots, \nu_i, \ldots\langle$, and, hence, $N\,|\nu_1, \nu_2, \ldots, \nu_i, \ldots\rangle = \nu\,|\,\nu_1, \nu_2, \ldots, \nu_i, \ldots\rangle$ with $\nu = \sum_i \nu_i$. All states are assumed to be normalized to 1. First, one notes that

$$\langle\nu|N_i\,|\,\nu\rangle = \nu_i = \langle\nu|a_i^\dagger a_i\,|\,\nu\rangle = \|\, a_i\,|\,\nu\rangle\,\|^2 .$$

Hence, ν_i is positive or zero, $\nu_i \geq 0$. Furthermore, one confirms the commutators

$$[N_i, a_k] = -a_k \delta_{ik}\,, \quad [N, a_k] = -a_k\,,$$
$$[N_i, a_k^\dagger] = a_k^\dagger \delta_{ik}\,, \quad [N, a_k^\dagger] = a_k^\dagger\,.$$

One concludes that if $|\nu\rangle$ is an eigenstate of N_i and N then also $a_i|\nu\rangle$ and $a_i^\dagger|\nu\rangle$ are eigenstates of N_i and N, and belong to eigenvalues shifted by -1 and by $+1$, respectively. For example, one has

$$N_i\,(a_i\,|\,\nu\,\rangle) = a_i(N_i - 1)\,|\,\nu\,\rangle = (\nu_i - 1)(a_i\,|\,\nu\,\rangle)\,,$$
$$N_i\,(a_i^\dagger\,|\,\nu\,\rangle) = a_i^\dagger(N_i + 1)\,|\,\nu\,\rangle = (\nu_i + 1)(a_i^\dagger\,|\,\nu\,\rangle)\,.$$

As noted above, the eigenvalues must be positive or zero. Therefore, one necessarily has $\nu_i \in \mathbb{N}_0$ and thus also $\nu \in \mathbb{N}_0$. In summary, only the values

$$\nu = n\,, \quad \nu_i = n_i \qquad \text{with} \quad n = 0, 1, 2, \ldots, \; n_i = 0, 1, 2, \ldots \tag{7.38}$$

are admissible. In turn, if this were not so, the Hamiltonian (7.35) would have arbitrarily negative eigenvalues! At the same time this shows that there exists a state $|0\rangle$ pertaining to the smallest eigenvalue of N which has the property

$$a_i\;|\,0\,\rangle = 0 \qquad \text{for all} \quad i\,.$$

One easily convinces oneself that the operators $O_i := a_i a_i^\dagger$ and $O := \sum_i a_i a_i^\dagger$ have properties which are very similar to those of N_i and N. Their eigenvalues differ from the eigenvalues of N_i and N by one unit. Therefore, a Hamiltonian of the form $\sum_i E_i\{a_i^\dagger a_i + a_i a_i^\dagger\}/2$ differs from $H = \sum_i E_i a_i^\dagger a_i$ only by a constant which, however, is infinite. On the other hand, only *differences* of eigenvalues are observable and, hence, physically relevant. The constant, though infinite, is expected to drop out from all observable effects.

This physically irrelevant difficulty would have been avoided if, from the start, one had defined all operators in *normal order,* i.e. if one had replaced all monomials in creation and annihilation operators by the corresponding normal products in which all creation operators are grouped to the left of all annihilation operators. In fact, this is what we shall do in the sequel.

Thus, the Hamilton density (7.25) is replaced by the normal ordered expression

$$\mathcal{H}(\phi, \pi) = \frac{1}{2} : \left\{\pi^2(x) + (\nabla\phi)^2 + m^2\phi^2\right\} :, \tag{7.39}$$

and the momentum density is replaced by the normal product

$$\mathcal{P}^i = -: \Pi(x)\,(\nabla\Phi)^i :. \tag{7.40}$$

Repeating the calculation that led to the expressions (7.35) and (7.36) one now obtains

$$H = \int \frac{\mathrm{d}^3 p}{2E_p}\, E_p\, a^\dagger(p) a(p) = \int \frac{\mathrm{d}^3 p}{2E_p}\, E_p\, N(p)\,, \tag{7.41}$$

$$\boldsymbol{P} = \int \frac{\mathrm{d}^3 p}{2E_p}\, \boldsymbol{p}\, a^\dagger(p) a(p) = \int \frac{\mathrm{d}^3 p}{2E_p}\, \boldsymbol{p}\, N(p)\,, \tag{7.42}$$

where $N(p)$ is the number operator of particles with momentum p and corresponds to the operator N_i of the example.

The ground state $|0\rangle$ which is annihilated by all operators of type $a(p)$, i.e. for which $a(p)|0\rangle = 0$ for all p, is an eigenstate of H and of $\boldsymbol{P}$, all four eigenvalues being equal to zero. A state which is obtained from the ground state by application of the creation operator $a^\dagger(q)$, has energy E_q and spatial momentum $\boldsymbol{q}$. Indeed, one finds

$$H\left(a^\dagger(q)\,|\,0\rangle\right) = \int \frac{\mathrm{d}^3 p}{2E_p}\, E_p\, a^\dagger(p)a(p)a^\dagger(q)\,|\,0\rangle$$
$$= \int \frac{\mathrm{d}^3 p}{2E_p}\, E_p\, a^\dagger(p)(2E_q)\delta(\boldsymbol{p}-\boldsymbol{q})\,|\,0\rangle = E_q\left(a^\dagger(q)\,|\,0\rangle\right),$$

and, in a similar way, also

$$\boldsymbol{P}\left(a^\dagger(q)\,|\,0\rangle\right) = \boldsymbol{q}\left(a^\dagger(q)\,|\,0\rangle\right).$$

As the relation $E_q^2 - \boldsymbol{q}^2 = m^2$ holds true the state $a^\dagger(q)|0\rangle \equiv |q\rangle$ must be the free state of a single particle of mass m, energy E_q, and spatial momentum $\boldsymbol{q}$. This state is normalized covariantly since, using (7.33), one shows

$$\langle q' \mid q\rangle = \langle 0|a(q')a^\dagger(q)\,|\,0\rangle = 2E_q\,\delta(\boldsymbol{q}'-\boldsymbol{q}).$$

Letting two or more creation operators act on the ground state yields many-body states of the kind

$$\left(a^\dagger(p_1)\right)^{n_{p_1}}\left(a^\dagger(p_2)\right)^{n_{p_2}}\cdots|\,0\rangle,$$

which are invariant under any permutation of the creation operators but are not yet normalized. They are elements of a Hilbert space which is generated by the Hilbert space $\mathcal{H}_1$ of single-particle states, by taking products with total particle number N and taking the infinite sum over all N,

$$\mathcal{H} = \sum_{N=0}^{\infty} \oplus \underbrace{(\mathcal{H}_1 \otimes \mathcal{H}_1 \otimes \ldots \otimes \mathcal{H}_1)}_{N} = \sum_{N=0}^{\infty} \oplus \,(\mathcal{H}_1)^{\otimes N}.$$

Let $n(p)$ be the number of particles with energy-momentum p, and let the corresponding normalized state be denoted by $|n(p)\rangle$. The normalization of the states is obtained from the relations

$$a^\dagger(p)\,|\,n(p)\rangle = \sqrt{n(p)+1}\,|\,n(p)+1\rangle,$$
$$a(p)\,|\,n(p)\rangle = \sqrt{n(p)}\,|\,n(p)-1\rangle. \tag{7.43}$$

One finds

$$|\,n(p_1)n(p_2)\cdots\rangle = \prod_{p_i} \frac{1}{\sqrt{n(p_i)!}}\left(a^\dagger(p_i)\right)^{n(p_i)}|\,0\rangle. \tag{7.44}$$

Remarks

1. The physical interpretation of the operators $a^\dagger(p)$ and $a(p)$ in the expansion (7.31) should now be clear: The first of them creates a single particle which carries four-momentum p with $p^2 = m^2$. With respect to a given frame of reference this particle has the spatial momentum $\boldsymbol{p}$ and the energy $E_p = \sqrt{m^2 + \boldsymbol{p}^2}$. Its adjoint operator $a(p)$ annihilates a particle from the same kinematic state. The corresponding solutions of the Klein-Gordon equation are the plane waves (7.19) which are normalized as in (7.20) and (7.21). If instead we used the modified wave functions
$$\widetilde{\phi}(p) := \frac{1}{\sqrt{2E_p}} f_p \tag{7.45}$$
in momentum space and their Fourier transforms $\phi(x)$ in position space then they would be normalized like wave functions of Schrödinger theory,
$$(\phi', \phi) = \int\limits_{x^0=t_0} \mathrm{d}^3x \, \phi'^*(x)\phi(x) .$$
The complex function $\phi(x)$ obtained in this way, is to be interpreted as the probability amplitude for finding the particle at time t_0 in the position $\boldsymbol{x}$.
2. In this section we discuss the classical and quantum Klein-Gordon field, real or hermitian, respectively. Therefore, the single-particle states $|p\rangle$ carry no further attributes beyond energy and momentum. In the perspective of representation theory of the Poincaré group the single-particle states $|p\rangle = a^\dagger(p)|0\rangle$ obey the transformation rule (6.144) with $s = 0$.
If the particles have a nonvanishing spin, if they are electrically charged, or carry any other quantum property, the creation and annihilation operators must contain these quantum numbers. The spin states of massive particles are the irreducible representations (6.144) of the Poincaré group that we studied in Chap. 6. In the case of massless particles one must use the helicity representations of Sect. 6.3.3. Examples we shall study further below are the complex scalar field, the photon (Maxwell) field, and the Dirac field.
3. As the creation operators $a^\dagger(p_i)$ all commute, the state (7.44) is automatically *symmetric* under exchange of any two excitations. For example, in a two-body state one has
$$\left| p_j, p_k \right\rangle = a^\dagger(p_j)a^\dagger(p_k)\,|0\rangle = a^\dagger(p_k)a^\dagger(p_j)\,|0\rangle = \left| p_k, p_j \right\rangle .$$
The free particles whose states we constructed above, satisfy *Bose-Einstein statistics*.
4. The set of plane waves (7.19) provide but one of many possibilities for expanding the quantized field in terms of a complete and normalized basis. Depending on the specific physical situation another such base system may be more suitable than plane waves. Here is an example of great importance in atomic and nuclear physics: In transitions between bound states of an atom, or a nucleus which, besides their energy, are classified by angular momentum and parity,

photons are created or absorbed which are in states with good total angular momentum and definite parity. In such a situation it is suggestive to expand the photon field in terms of a complete, normalized system of eigenstates of angular momentum and parity, instead of plane waves. This requires introducing creation and annihilation operators for photons which carry these attributes. Suppose we had expanded the self-adjoint scalar field in terms of the orthogonal and appropriately normalized base system $\{\phi_n(\alpha)\}$. The commutation rules (7.33) and (7.34) are then replaced by

$$[a_m(\alpha), a_n^\dagger(\beta)] = (\phi_m(\alpha), \phi_n(\beta)) , \tag{7.46}$$

$$[a_m(\alpha), a_n(\beta)] = 0 , \tag{7.47}$$

the right-hand side of (7.46) containing the scalar product of the base functions with respect to which the creation and annihilation operators are defined. Whether this basis has a proper normalization, or whether the scalar product yields a distribution like in the example (7.33), (7.34), does not make any difference.

5. The general translation formula (7.30) is particularly useful for the analysis of single-particle matrix elements. Choose, for instance, $x = 0$, and consider matrix elements of a field operator between single-particle states with four momenta q and q', respectively. Then, from (7.30), one obtains

$$\langle q'|F(y)\,|q\rangle = \langle q'|e^{iP\cdot y}F(0)e^{-iP\cdot y}\,|q\rangle = e^{-i(q-q')\cdot y}\langle q'|F(0)\,|q\rangle . \tag{7.48}$$

In this formula P stands for the operator of four-momentum while q or q' are its eigenvalues in the two states. Thus, if one knows the matrix element $\langle q'|F(0)|q\rangle$ at the point $x = 0$ or any other point of spacetime, then (7.48) allows to "transport" it to any other point y. An example of practical use is provided by the divergence of a four-current density $J^\mu(y)$. The following calculation

$$\langle q'|\partial_\mu J^\mu(y)\,|q\rangle = \partial_\mu\langle q'|J^\mu(y)\,|q\rangle = -i(q-q')_\mu\langle q'|J^\mu(0)\,|q\rangle\, e^{-i(q-q')\cdot y}$$

leads to the important formula

$$\langle q'|\partial_\mu J^\mu(0)\,|q\rangle = -i(q-q')_\mu\langle q'|J^\mu(0)\,|q\rangle . \tag{7.49}$$

Note that it is $q - q'$, that is, the momentum transfer from the initial state $|q\rangle$ to the final state $|q'\rangle$ which appears on the right-hand side of (7.49). Why is this formula important?

In many situations it will not be possible to really calculate the matrix elements $\langle q'|J^\mu(0)|q\rangle$ because the operator $J^\mu(y)$ is not known explicitly. However, its one-particle matrix element $\langle q'|J^\mu(0)|q\rangle$, having a well-defined transformation behaviour under Lorentz transformations, parity, and time reversal, can be decomposed in terms of all covariants which can be constructed from the momenta (and, if the particle carries spin, from the spin functions) which are compatible with this transformation behaviour. Furthermore, if the divergence $\partial_\mu J^\mu(y)$ is known (for example, it may be zero, or may be proportional to some known scalar field), then (7.49) yields additional constraints for this decomposition. A simple but typical example is found in Exercise 7.4.

7.1.6 Commutator for Different Times, Propagator

The quantization rule (7.26), (7.27) which leads to the commutators (7.33), (7.34) is Lorentz covariant. This assertion is confirmed by a calculation of the commutator of the field $\Phi(x)$ with the same field $\Phi(y)$ taken at some other, arbitrary point of spacetime.

Inserting the expansion (7.31) at two stages of the calculation, and making use of the commutators (7.33) and (7.34) one finds step by step

$$[\Phi(x), \Phi(y)] = \frac{1}{(2\pi)^3} \int \frac{d^3q}{2E_q} \int \frac{d^3p}{2E_p} \left[\left(a(q)e^{-iqx} + a^\dagger(q)e^{iqx}\right), \left(a(p)e^{-ipx} + a^\dagger(p)e^{ipx}\right)\right]$$

$$= \frac{1}{(2\pi)^3} \int \frac{d^3q}{2E_q} \int \frac{d^3p}{2E_p} \left\{\left[a(q), a^\dagger(p)\right]e^{-iqx+ipy} + \left[a^\dagger(q), a(p)\right]e^{iqx-ipy}\right\}$$

$$= \frac{1}{(2\pi)^3} \int \frac{d^3q}{2E_q} \left\{e^{-iq(x-y)} - e^{iq(x-y)}\right\} .$$

The integral obtained on the right-hand side, strictly speaking, is a tempered distribution. As this is an important quantity that appears in various contexts in quantum field theory it is given a definition of its own. With the abbreviation $z = x - y$ one defines the distribution

$$\Delta_0(z; m) := -\frac{i}{(2\pi)^3} \int \frac{d^3q}{2E_q} \left(e^{-iqz} - e^{iqz}\right)\bigg|_{q^0=E_q} . \tag{7.50}$$

It is called the *causal distribution for mass m* and has the following properties:
(a) It satisfies the Klein-Gordon equation

$$\left(\Box + m^2\right) \Delta_0(z; m) = 0 , \tag{7.51}$$

(b) At $z^0 = 0$, i.e. for equal times $x^0 = y^0$ it vanishes for any z

$$\Delta_0(z^0 = 0, z; m) = 0 , \tag{7.52}$$

(c) Its partial derivative with respect to z^0, taken at $z^0 = 0$, equals Dirac's δ-distribution but for a sign,

$$\frac{\partial}{\partial z^0} \Delta_0(z; m)|_{z^0=0} = -\delta(z) , \tag{7.53}$$

(d) With respect to simultaneous reflection in space and time $z \mapsto -z$ it is *antisymmetric*,

$$\Delta_0(-z; m) = -\Delta_0(z; m)\,, \tag{7.54}$$

(e) It vanishes for all spacelike values of z, i.e. for all $z^2 < 0$,

$$\Delta_0(z; m) = 0 \quad \text{for} \quad z^2 < 0\,. \tag{7.55}$$

The properties (a)–(d) follow directly from the defining expression (7.50) and are easily verified. The proof of property (e) may be done by means of the following argument:

The distribution Δ_0 is Lorentz invariant. This is seen directly from the explicit representation (7.50). Alternatively, one may consult the equivalent representation

$$\Delta_0(z; m) = -\frac{\mathrm{i}}{(2\pi)^3} \int \mathrm{d}^4 q\ \mathrm{e}^{-\mathrm{i}qz} \delta(q^2 - m^2)\Big(\Theta(q^0) - \Theta(-q^0)\Big) \tag{7.56}$$

The distinction between positive and negative q^0 is invariant for timelike q. Indeed, the condition $q^2 = m^2$ which follows from the distribution $\delta(q^2 - m^2)$ shows that q is timelike. Therefore, the integrand is invariant, and so is the unrestricted volume element $\mathrm{d}^4 q$. If $z^2 < 0$ as required in (e) then one can always find a frame of reference in which the time coordinate vanishes, $z'^0 = 0$. Then, by (b) one has $\Delta_0 = 0$. Thus, by its Lorentz invariance, it vanishes for all spacelike values of z, i.e. in our convention for the Minkowski metric, for all $z^2 < 0$.

We note the result of the calculation performed above:

$$[\Phi(x), \Phi(y)] = \mathrm{i}\Delta_0(x - y; m)\,. \tag{7.57}$$

This result, among others, has two important properties: The commutator of $\Phi(x)$ and $\Phi(y)$ is a distribution and is Lorentz invariant. It vanishes for *spacelike* values of the difference $x - y$. This is an expression of the causality of the theory, or, as one says more precisely, of its *micro-causality* or *locality*. Indeed, the property (e) says that the field operators commute whenever their arguments are relatively spacelike, that is to say, x and y cannot communicate in a causal way.

We now turn to a quantity of central importance for the covariant version of perturbation theory: the particle propagator. Turning back to the expansion (7.31) with the plane waves (7.19) inserted there, one finds the field operator $\Phi(x)$ to be the sum

$$\Phi(x) = \Phi^{(+)}(x) + \Phi^{(-)}(x)$$

of a positive frequency part $\Phi^{(+)}$, containing all terms of the kind $a_p\,\mathrm{e}^{-\mathrm{i}p^0x^0}\mathrm{e}^{\mathrm{i}\boldsymbol{p}\cdot\boldsymbol{x}}$, and a negative frequency part $\Phi^{(-)}$, which contains the corresponding conjugate terms $a_p^\dagger\,\mathrm{e}^{+\mathrm{i}p^0x^0}\mathrm{e}^{-\mathrm{i}\boldsymbol{p}\cdot\boldsymbol{x}}$. The nomenclature "positive frequency" is chosen because the time dependence $\mathrm{e}^{-\mathrm{i}p^0x^0}$ has the familiar form $\mathrm{e}^{-\mathrm{i}/\hbar Et}$ of a state with positive energy of quantum mechanics.[3] If the field operator $\Phi(x)$ acts on the ground state, called

[3] It would seem unfortunate to talk about "positive energy term" because then its adjoint would refer to negative energies, instead of negative frequencies. Free particles (or antiparticles) with mass m always have *positive* energy.

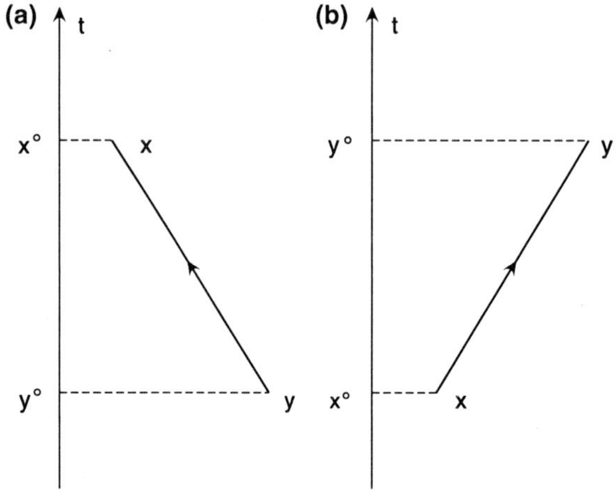

Fig. 7.4 The two terms of the time-ordered product (7.58) describe: **a** the creation of a particle at the point y and its annihilation at x which is later than y; **b** creation at x and subsequent annihilation at y, in the case $y^0 > x^0$

vacuum, to the right, $\Phi(x)|0\rangle$, then only the negative frequency part contributes, $\Phi|0\rangle = \Phi^{(-)}|0\rangle$. If it acts to the left then only the positive frequency part contributes, $\langle 0|\Phi = \langle 0|\Phi^{(+)}$. Thus, the expectation value of the product of two field operators is given by

$$\langle 0|\Phi(x)\Phi(y)\,|0\rangle = \langle 0|\Phi^{(+)}(x)\Phi^{(-)}(y)\,|0\rangle\ .$$

Assuming the time x^0 to be later than the time y^0, as sketched in Fig. 7.4a, this term would describe the creation of a particle at the point y of spacetime, and the annihilation of the same particle at the point x. From the point of view of nonrelativistic perturbation theory it seems plausible that a process of this kind is the intermediate state in a contribution of second order. If this is so there exists also the analogous process with $y^0 > x^0$ where the particle is created at x and is annihilated at y, cf. Fig. 7.4b.

The two contributions can be combined in an elegant way by ordering creation and annihilation terms in the form of a *time-ordered product* defined by

$$T\,\Phi(x)\Phi(y) := \Theta(x^0 - y^0)\Phi(x)\Phi(y) + \Theta(y^0 - x^0)\Phi(y)\Phi(x)\,. \qquad (7.58)$$

The expectation value of this product in the ground state (the vacuum state)

$$\begin{aligned}&\langle 0|T\,\Phi(x)\Phi(y)\,|0\rangle\\&= \langle 0|\Theta(x^0 - y^0)\Phi(x)\Phi(y) + \Theta(y^0 - x^0)\Phi(y)\Phi(x)\,|0\rangle\end{aligned} \qquad (7.59)$$

relates creation in one point with annihilation in some other, later, point of spacetime. This synthesis is called the *propagator* and takes a remarkable form:

$$\begin{aligned}\langle 0|T\,\Phi(x)\Phi(y)\,|0\rangle &= \frac{\mathrm{i}}{(2\pi)^4}\int \mathrm{d}^4k\; \mathrm{e}^{-\mathrm{i}k(x-y)}\frac{1}{k^2 - m^2 + \mathrm{i}\varepsilon}\\&\equiv P_F(x-y;m)\,.\end{aligned} \qquad (7.60)$$

Before analyzing this expression in detail let us prove (7.60). A starting point for the proof is provided by two integral representations of the Heaviside, or step, function. They are

$$\Theta(u) = \frac{1}{2\pi \mathrm{i}} \int_{-\infty}^{+\infty} \mathrm{d}\lambda \, \frac{\mathrm{e}^{\mathrm{i}\lambda u}}{\lambda - \mathrm{i}\varepsilon} \, , \quad \Theta(-u) = -\frac{1}{2\pi \mathrm{i}} \int_{-\infty}^{+\infty} \mathrm{d}\lambda \, \frac{\mathrm{e}^{\mathrm{i}\lambda u}}{\lambda + \mathrm{i}\varepsilon} \, . \tag{7.61}$$

Their proof rests on the Cauchy integral theorem and is deferred to the exercises (s. Exercise 7.5).

Performing a calculation analogous to the one at the beginning of this section one finds

$$\langle 0|\Phi(x)\Phi(y)\,|\,0\rangle = \frac{1}{(2\pi)^3} \int \frac{\mathrm{d}^3 q}{2E_q} \mathrm{e}^{-\mathrm{i}qz} \, , \quad \text{with} \quad z = x - y \, ,$$

where the momentum $\boldsymbol{q}$ is on the mass shell $q^2 = m^2$. Multiplying by the step function $\Theta(x^0 - y^0)$ and using the first of its integral representations (7.61) one has

$$\langle 0|\Phi(x)\Phi(y)\Theta(x^0 - y^0)\,|\,0\rangle = \frac{1}{\mathrm{i}(2\pi)^4} \int_{-\infty}^{+\infty} \mathrm{d}\lambda \, \frac{\mathrm{e}^{\mathrm{i}\lambda u}}{\lambda - \mathrm{i}\varepsilon} \int \frac{\mathrm{d}^3 q}{2E_q} \mathrm{e}^{-\mathrm{i}qz} \, .$$

The idea then is to write the integration over the three-momentum $\boldsymbol{q}$ and over the auxiliary variable λ in such a way that they turn into one common integral over an unrestricted four-momentum which no longer lies on the mass shell. In order to achieve this let

$$k^0 := E_q - \lambda \, , \qquad \boldsymbol{k} := \boldsymbol{q} \, .$$

Although the energy which pertains to the three-momentum $\boldsymbol{k}$ is the same as the one that pertains to $\boldsymbol{q}$, i.e. $E_k = \sqrt{\boldsymbol{k}^2 + m^2} = E_q$, the Lorentz invariant $(k^0)^2 - \boldsymbol{k}^2$ remains unrestricted. The integration over λ becomes integration over k^0,

$$\int_{-\infty}^{+\infty} \mathrm{d}\lambda \longmapsto - \int_{+\infty}^{-\infty} \mathrm{d}k^0 = + \int_{-\infty}^{+\infty} \mathrm{d}k^0 \, .$$

In summary, one now has to integrate over the four independent variables k^0, k^1, k^2, k^3. Inserting the definitions one obtains

$$\langle 0|\Phi(x)\Phi(y)\Theta(x^0 - y^0)\,|\,0\rangle = -\frac{i}{(2\pi)^4} \int \mathrm{d}^4 k \, \frac{\mathrm{e}^{-ikz}}{2E_k(E_k - k^0 - \mathrm{i}\varepsilon)} \, .$$

The second term of (7.59) which contains the other time ordering, is transformed in an analogous manner. One uses the second representation (7.61), sets now $k^0 := E_q + \lambda$, $\boldsymbol{k} = \boldsymbol{q}$, and replaces k by $-k$ in the integral that one obtains. One then finds

$$\langle 0|\Phi(y)\Phi(x)\Theta(y^0 - x^0)\,|\,0\rangle = -\frac{i}{(2\pi)^4} \int \mathrm{d}^4 k \, \frac{\mathrm{e}^{-ikz}}{2E_k(E_k + k^0 - \mathrm{i}\varepsilon)} \, .$$

The two integrals are combined, the sum of the integrands giving

$$\frac{1}{2E_k} \left(\frac{1}{E_k - k^0 - \mathrm{i}\varepsilon} + \frac{1}{E_k + k^0 - \mathrm{i}\varepsilon} \right) = \frac{1}{E_k^2 - (k^0)^2 - 2\mathrm{i}E_k\varepsilon} + \mathcal{O}(\varepsilon^2) \, .$$

The difference in the denominator of the right-hand side is $E_k^2 - (k^0)^2 = m^2 + \boldsymbol{k}^2 - (k^0)^2 = m^2 - k^2$. The infinitesimal quantity ε is no more than a prescription how to deform the path of integration in the complex λ-plane. The term $2E_k\varepsilon$ plays the same role as ε and may be replaced by the latter. Thus, the formula (7.60) is proven.[4]

Remarks

1. Besides the causal distribution Δ_0, (7.50), the Klein-Gordon equation has another distribution-valued solution linearly independent of the former. This solution can be written as follows:
$$\Delta_1(z; m) = \frac{1}{(2\pi)^3} \int \frac{\mathrm{d}^3 q}{2E_q} \left(\mathrm{e}^{-\mathrm{i}qz} + \mathrm{e}^{\mathrm{i}qz}\right)\Big|_{q^0=E_q} . \tag{7.62}$$
It is called *acausal distribution for mass m*, because for spacelike argument, $z^2 < 0$, it does not vanish. Indeed, and in contrast to the property (d) of Δ_0 it is *symmetric*, $\Delta_1(-z; m) = \Delta_1(z; m)$, and the proof given for (e) no longer goes through.
2. The time-ordered product (7.58) arranges the field operators Φ from right to left, with increasing time arguments, following the popular saying "the early bird catches the worm". Its expectation value (7.59) sums up two contributions which were distinct in nonrelativistic perturbation theory.
3. The propagator $P_F(x - y; m)$, (7.60), in position space, is given in the form of a Fourier transform. In momentum space and up to numerical factors, its representation is given by $1/(k^2 - m^2 + \mathrm{i}\varepsilon)$. It comes in whenever a spinless particle described by the field Φ is exchanged between two external lines. Consider, for example, the diagram of Fig. 7.1. It looks like a perturbative contribution of second order of the kind well-known from nonrelativistic perturbation theory. However, there is an essential difference: While in the latter the intermediate state violates only energy conservation but obeys all other selection rules, in the former the particle is taken off its mass shell $p^2 = m^2$. Its virtual four-momentum k which is the integration variable, does not lie on the hyperboloid for mass m. Neither energy conservation not momentum conservation are fulfilled.
4. The preceding remark should not be surprising in a Lorentz covariant theory where the splitting of a four-momentum into what is called energy and what is called momentum depends on the frame of reference. Therefore, it is very plausible that in covariant perturbation theory the propagator holds a central role and that the denominator $k^2 - m^2 + \mathrm{i}\varepsilon$ replaces the energy denominator $E - E_0$ of quantum mechanical perturbation theory. This observation will

[4] Strictly speaking, the integrals that one is dealing with are tempered distributions. However, all definitions for tempered distributions are chosen such that formal calculus follows the same rules as the calculus with functions and genuine integrals. This remark shows that the derivation of (7.60) as given here is correct.

receive further support in the next section which deals with the complex scalar field.

7.2 The Complex Klein-Gordon Field

Like in the Sects. 7.1.4–7.1.6 we stay with the description of free particles with spin zero, i.e. we interpret the classical field $\phi(x)$ and its quantized partner $\Phi(x)$ as before, but allow for $\phi = \phi_1 + \mathrm{i}\phi_2$ to be complex. The field operator $\Phi(x)$ no longer is self-adjoint. The classical Lagrange density which must still be real, then has the form

$$\mathcal{L}_{\mathrm{KG}}(\phi_i, \partial_\mu \phi_i) \equiv \mathcal{L}_{\mathrm{KG}}(\phi, \phi^*, \partial_\mu \phi, \partial_\mu \phi^*) \qquad (7.63)$$
$$= \partial_\mu \phi^*(x) \partial^\mu \phi(x) - m^2 \phi^*(x)\phi(x) .$$

One has the choice of either defining the two real fields ϕ_1 and ϕ_2, i.e. the real and imaginary parts of the complex field $\phi(x)$, to be the independent degrees of freedom, or, alternatively, to interpret the full field ϕ and the complex conjugate field ϕ^* as the independent variables. With regard to the physical interpretation the second choice is more transparent.

Both fields, ϕ and ϕ^*, satisfy the Klein-Gordon equation with mass m. The momentum canonically conjugate to ϕ is given by

$$\phi(x) \longleftrightarrow \pi(x) = \frac{\partial \mathcal{L}_{\mathrm{KG}}}{\partial(\partial_0 \phi)} = \partial^0 \phi^* ,$$

while for ϕ^* it reads

$$\phi^* \longleftrightarrow \pi^* = \partial^0 \phi .$$

The Hamilton density (7.25) is replaced by

$$\mathcal{H}(\phi, \phi^*, \pi, \pi^*) = \pi^*(x)\pi(x) + \nabla\phi^*(x) \cdot \nabla\phi(x) + m^2 \phi^*(x)\phi(x) . \qquad (7.64)$$

The quantization of this model is determined by the Postulate 7.1 according to which

$$\left[\Pi(x), \Phi(x')\right]_{x^0=x'^0} = -\mathrm{i}\,\delta(\boldsymbol{x} - \boldsymbol{x}') = \left[\Pi^*(x), \Phi^*(x')\right]_{x^0=x'^0} , \qquad (7.65)$$

all other commutators vanishing. One confirms that the commutators with the Hamiltonian $H = \int \mathrm{d}^3x\, \mathcal{H}(x)$ are

$$[H, \Phi(x)] = -\mathrm{i}\Pi^*(x) = -\mathrm{i}\partial^0 \Phi(x) , \quad [H, \Pi(x)] = -\mathrm{i}\partial^0 \Pi(x) , \quad \text{etc}$$

as expected.

As the field operator $\Phi(x)$ is not self-adjoint the expansion in terms of normal modes analogous to (7.31), contains two terms which are no longer adjoints of each other. One has

$$\Phi(x) = \int \frac{\mathrm{d}^3 p}{2E_p} \left[f_p(x) a(p) + f_p^*(x) b^\dagger(p) \right] \equiv \Phi^{(+)}(x) + \Phi^{(-)}(x) , \qquad (7.66)$$

its adjoint being given by

$$\Phi^\dagger(x) = \int \frac{d^3p}{2E_p} \left[f_p^*(x) a^\dagger(p) + f_p(x) b(p) \right] \equiv \Phi^{\dagger(-)}(x) + \Phi^{\dagger(+)}(x) . \quad (7.67)$$

Suppose one inserted the plane waves (7.19), and decomposed the field operators in positive and negative frequency parts, the first of which contains the operators $a(p)$ or $b(p)$, the second of which contains the operators $a^\dagger(p)$ or $b^\dagger(p)$, respectively. A calculation completely analogous to the case of the real field shows that the quantization rules (7.65) lead to the commutators

$$\left[a(p), a^\dagger(q)\right] = 2E_p \, \delta(\boldsymbol{p} - \boldsymbol{q}) = \left[b(p), b^\dagger(q)\right] , \quad (7.68)$$

$$[a(p), a(q)] = 0 = [b(p), b(q)] , \quad [a(p), b(q)] = 0 = \left[a(p), b^\dagger(q)\right] . \quad (7.69)$$

As before we define the quantized form of the Hamiltonian and of the momentum operator by normal products, cf. (7.39) and (7.40). Thus, it is not surprising to find expressions for the Hamiltonian and the momentum which are entirely analogous to (7.41) and (7.42), respectively. They are

$$H = \int \frac{d^3p}{2E_p} \, E_p \left[a^\dagger(p) a(p) + b^\dagger(p) b(p) \right] , \quad (7.70)$$

$$\boldsymbol{P} = \int \frac{d^3p}{2E_p} \, \boldsymbol{p} \left[a^\dagger(p) a(p) + b^\dagger(p) b(p) \right] . \quad (7.71)$$

These results suggest the same interpretation of creation operators $a^\dagger(q)$ and $b^\dagger(q)$ as in Sect. 7.1.5: when applied to the vacuum state both create one-particle states with momentum $\boldsymbol{q}$ and energy E_q of spinless bosons which satisfy Bose-Einstein statistics.

What is it then that renders the two types of particles different from each other? In order to answer this question we must study more carefully the invariances of the Lagrange density (7.63). As we know from classical field theory the fact that eigenstates of H can be classified by energy and momentum, lastly, is a consequence of the homogeneity of the Lagrange density in space and time, via the theorem of E. Noether. Obviously, the Lagrange densities of the real as well as the complex field have this property. However, the density (7.63) has a further, rather obvious, symmetry: If one multiplies the field ϕ by a complex number with modulus 1, hence its complex conjugate by the complex conjugate number, then $\mathcal{L}_{\text{KG}}$, (7.63), remains unchanged. A transformation of this sort,

$$\begin{aligned} \phi(x) &\longmapsto \phi'(x) = e^{i\alpha} \phi(x) , \\ \phi^*(x) &\longmapsto \phi'^*(x) = e^{-i\alpha} \phi^*(x) , \quad \alpha \in \mathbb{R} , \end{aligned} \quad (7.72)$$

is called *global gauge transformation*.[5] As it leaves the Lagrange density invariant it entails one further conservation principle. It will turn out that it is this conserved quantity that allows to distinguish the excitations generated by $a^\dagger$ and by $b^\dagger$.

[5] The term "gauging" is used here in a figurative sense but alludes to genuine gauging, or rescaling, of a (real) metric, $g_{\mu\nu}(x) \mapsto g'(x) = \exp\{\lambda(x)\} g_{\mu\nu}(x)$. In fact, this was the context within which Hermann Weyl invented the notion of gauge transformation in 1919.

Choosing the real parameter α infinitesimal, $|\alpha| \ll 1$, the result (7.72) is seen to be a variation of the degrees of freedom ϕ and ϕ^*

$$\phi' = \phi + \delta\phi, \; \phi'^* = \phi^* + \delta\phi^* , \quad \text{with} \quad \delta\phi = \mathrm{i}\alpha\phi, \; \delta\phi^* = -\mathrm{i}\alpha\phi^* ,$$

which leaves $\mathcal{L}_{\mathrm{KG}}$ invariant. Using $\delta(\partial_\mu\phi) = \partial_\mu(\delta\phi)$ one then has

$$\begin{aligned} 0 = \delta\mathcal{L} &= \frac{\partial\mathcal{L}}{\partial\phi}\delta\phi + \frac{\partial\mathcal{L}}{\partial(\partial_\mu\phi)}\partial_\mu(\delta\phi) + \text{h.c.} \\ &= \mathrm{i}\alpha\left\{\left(\frac{\partial}{\partial x^\mu}\frac{\partial\mathcal{L}}{\partial(\partial_\mu\phi)}\phi + \frac{\partial\mathcal{L}}{\partial(\partial_\mu\phi)}\partial_\mu\phi\right) - (\phi \leftrightarrow \phi^*)\right\} \\ &= \mathrm{i}\alpha\partial_\mu\left(\frac{\partial\mathcal{L}}{\partial(\partial_\mu\phi)}\phi - (\phi \leftrightarrow \phi^*)\right) . \end{aligned}$$

In the second step use is made of the equation of motion

$$\frac{\partial\mathcal{L}}{\partial\phi} = \frac{\partial}{\partial x^\mu}\frac{\partial\mathcal{L}}{\partial(\partial_\mu\phi)} .$$

It follows from this that

$$j^\mu(x) := -\mathrm{i}\left(\frac{\partial\mathcal{L}}{\partial(\partial_\mu\phi)}\phi - \frac{\partial\mathcal{L}}{\partial(\partial_\mu\phi^*)}\phi^*\right) \tag{7.73}$$

fulfills the continuity equation $\partial_\mu j^\mu(x) = 0$. Inserting the Lagrange density (7.63) one obtains in our example

$$j^\mu(x) = \mathrm{i}\,\phi^*(x)\,\overleftrightarrow{\partial^\mu}\,\phi(x) . \tag{7.74}$$

(Except for a numerical factor this is the form of the electromagnetic current density announced at the beginning of Sect. 7.1.) If the fields decrease sufficiently fast at spatial infinity then the continuity equation implies that the space integral of the time component

$$Q := \int \mathrm{d}^3x \; j^0(x) \tag{7.75}$$

is a conserved quantity, i.e. that $\partial^0 Q = 0$ holds.

At this point we replace the classical fields ϕ and ϕ^* by the field operators as constructed by means of the postulate 2.1. We also replace the classical current density (7.74) by the normal-ordered operator

$$J^\mu(x) = \mathrm{i} \,{:}\Phi^\dagger(x)\,\overleftrightarrow{\partial^\mu}\,\Phi(x){:} . \tag{7.76}$$

Inserting again the expansions (7.66) and (7.67) as well as the plane waves (7.19) one finds a simple, yet important result:

$$Q = \int \frac{\mathrm{d}^3p}{2E_p}\left[a^\dagger(p)a(p) - b^\dagger(p)b(p)\right] . \tag{7.77}$$

It tells us that the states $a^\dagger(q)|0\rangle$ and $b^\dagger(q)|0\rangle$ differ by the eigenvalue of Q: The first state carries the eigenvalue $+1$, the second carries the eigenvalue -1. This suggests to interpret Q as the operator which describes the electric charge. For that

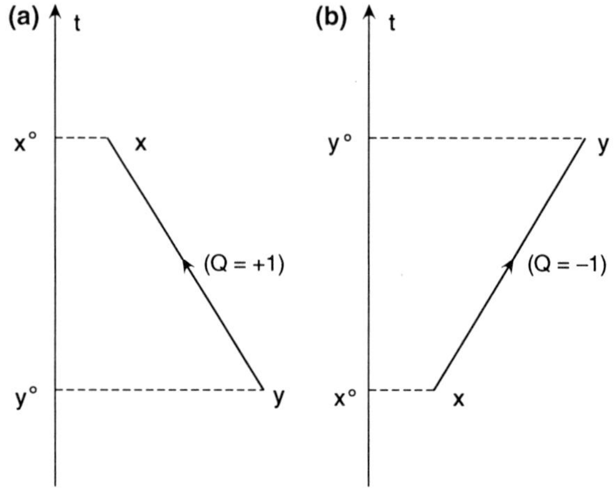

Fig. 7.5 Like in Fig. 7.4 the two terms of the time-ordered product describe once the propagation from y to x, once from x to y. Note, however, that if in the first case (**a**) it is a particle that propagates virtually, then in the second case (**b**) it is an antiparticle

it suffices to multiply it by the elementary charge. The two types of excitations, the ones generated by $a^\dagger$ and the ones generated by $b^\dagger$, pertain to the same value of mass and have the same kinematic properties, but carry equal and opposite electric charge. This is the first hint at the possibility of using a complex field to describe simultaneously particles and antiparticles.

By repeating the calculation that led to (7.57) in the real case one finds the slightly modified result

$$\left[\Phi(x), \Phi^\dagger(y)\right] = \mathrm{i}\Delta_0(x-y; m)\,. \tag{7.78}$$

The explicit form (7.60) of the propagator which holds for the complex field is of particular interest,

$$\langle 0|T\,\Phi(x)\Phi^\dagger(y)\,|0\rangle\ .$$

Depending on the time ordering of the points x and y, different positive and negative frequency parts contribute here:

$$x^0 > y^0 : \langle 0|\Phi^{(+)}(x)\Phi^{\dagger\,(-)}(y)\,|0\rangle$$
$$x^0 < y^0 : \langle 0|\Phi^{\dagger\,(+)}(y)\Phi^{(-)}(x)\,|0\rangle\ .$$

The first term is of the type $\langle 0|aa^\dagger|0\rangle$ and describes the propagation of a *particle* from y to x, the second term is of the type $\langle 0|bb^\dagger|0\rangle$ and describes the propagation of an *antiparticle* from x to y, as sketched in Fig. 7.5. The propagator describes the two virtual processes, creation and annihilation of a particle and of an antiparticle, respectively, without regard to the time ordering. Thus, the two contributions are contained in one single expression!

Of course, this interpretation of the propagator remains an academic one as long as the particle and the antiparticle are free particles which do not interact with any other objects. So far there are no physically testable predictions. The following example serves the purpose of filling this gap.

Example 7.1

The coupling of an electrically charged scalar field to the Maxwell field is obtained by the principle of *minimal coupling*, i.e. by the replacement

$$\partial_\mu \longmapsto \partial_\mu + \mathrm{i} q A_\mu \tag{7.79}$$

where $A_\mu(x)$ is a vector potential which yields the electric and magnetic Maxwell fields while q is a multiple of the elementary charge. Inserting this into the kinetic term of $\mathcal{L}_{\mathrm{KG}}$, (7.63) yields

$$\begin{aligned}\partial_\mu \Phi^\dagger \partial^\mu \Phi &\mapsto \left(\left(\partial_\mu - \mathrm{i} q A_\mu\right) \Phi^\dagger\right) \left(\partial^\mu + \mathrm{i} q A^\mu\right) \Phi \\ &= \partial_\mu \Phi^\dagger \partial^\mu \Phi - q J^\mu A_\mu + q^2 \Phi^\dagger \Phi A_\mu A^\mu ,\end{aligned} \tag{7.80}$$

with $J^\mu(x)$ the operator of current density defined in (7.76). Anticipating later developments we suppose that the field $A_\mu(x)$ is also quantized, according to the Postulate 7.1, and is written as a sum of positive and negative frequency parts, $A_\mu = A_\mu^{(+)} + A_\mu^{(-)}$. The first term contains annihilation operators $c_\lambda(k)$ (k denotes the momentum, λ stands for the spin) for photons with momentum k and polarization λ. The second term contains the corresponding creation operator $c_\lambda^\dagger(k)$ and creates photons with the same properties. All this is in complete analogy to the scalar field discussed previously. Resolving the second term on the right-hand side of (7.80) with respect to its content in creation and annihilation operators,

$$\begin{aligned}J^\mu A_\mu &= \left(\Phi^{\dagger(-)} + \Phi^{\dagger(+)}\right) \overleftrightarrow{\partial^\mu} \left(\Phi^{(-)} + \Phi^{(+)}\right) \left(A_\mu^{(+)} + A_\mu^{(-)}\right) \\ &\propto \left(a^\dagger + b\right) \left(b^\dagger + a\right) \left(c + c^\dagger\right) ,\end{aligned}$$

one realizes that it describes absorption, or emission of a photon on a scalar particle, say a π^+. To second order in the charge e this interaction term describes, for example, the elastic scattering process

$$\pi^+(q) + \gamma(k) \longrightarrow \pi^+(q') + \gamma(k') , \quad (q + k = q' + k') ,$$

in the way drawn in Fig. 7.6. This figure shows time running upwards, from bottom to top. Initially, there is an incoming state π^+, γ with four-momenta q and k, respectively. The final state contains the two outgoing particles with momenta q' and k', respectively. Furthermore, there are two different intermediate states: In Fig. 7.6a the intermediate state consists of the incoming and outgoing photons plus a virtual π^+ that carries the momentum $(q - k') = (q' - k)$, while the intermediate state in Fig. 7.6b contains the incoming and the outgoing π^+ as well as a virtual π^- with momentum $(k - q')$, but no photon. In every time section $t =$ const. the total charge is equal to $+1$. This example confirms very clearly that the propagator sums up the two virtual processes and does not distinguish the temporal order of interactions and vertices.

We will find a completely analogous situation when we study the interaction of *fermions* with photons with the exception that the current density has a somewhat different structure. Therefore, the interaction term of the example is a prototype which is characteristic for quantized electrodynamics.

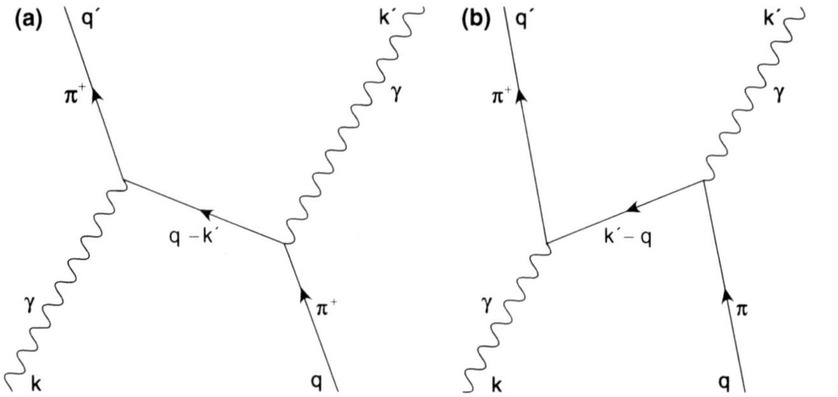

Fig. 7.6 Time flows from bottom to top. In **a** a virtual π^+ is exchanged, in **b** a virtual π^-, the antiparticle of π^+, is exchanged. Note that the roles of incoming and outgoing photons can be exchanged

Fig. 7.7 The contact interaction $\Phi^\dagger(x)\Phi(x)A_\mu(x)A^\mu(x)$ allows for the creation and/or annihilation of two photons and two π particles. Of course, this happens in such a way that electric charge is conserved

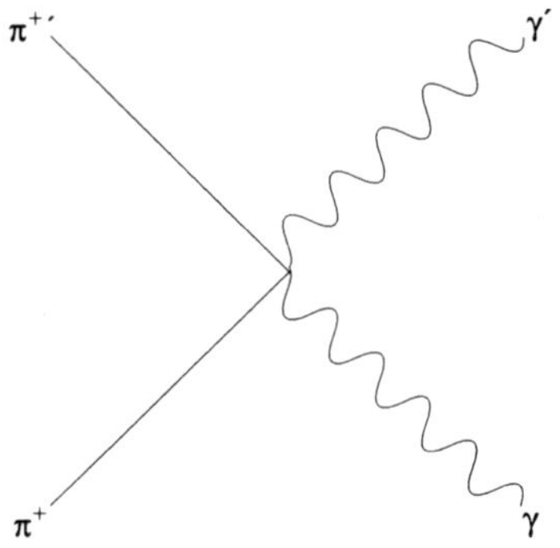

The third term on the right-hand side of (7.80) comes in only in the case of charged bosons. As it is bilinear both in the boson and the photon fields it describes a kind of contact interaction whereby two massive scalar particles and two photons are hooked to the same spacetime point, without the occurrence of a propagator. This contribution which is drawn in Fig. 7.7 is often called the *seagull term*.[6]

Remarks

1. The interpretation of $b^\dagger$-type excitations as the antiparticles of the $a^\dagger$-type excitations receives a better physical basis from the principle of minimal coupling: Indeed, the two vertices in Fig. 7.6b show the processes of pair annihilation and pair creation, respectively,

[6] Turning one's head to the right, and barring too much artistic scrutiny, one will recognize a seagull in Fig. 7.7.

$$\pi^+ + \pi^- \longrightarrow \gamma\,, \quad \gamma \longrightarrow \pi^+ + \pi^-\,,$$

which are only allowed if π^+ and π^- have equal and opposite charges. As they have the same mass they are antiparticles of one another.

2. The theory developed up to this point is completely symmetric in particles and antiparticles, the states $b^\dagger(q)|0\rangle$ and $a^\dagger(q)|0\rangle$ are kinematically identical. They differ only by the sign of their charge. However, this sign can be changed by modifying the definition of the electromagnetic current density (7.76), simply by replacing e by $-e$. Thus, we conclude:

What we call particle and what we call antiparticle is a matter of convention.

3. The arguments we gave with regard to the invariance of the Lagrange density under global gauge transformations (7.72) do not tell us whether the conserved "charge" (7.75) really is the electric charge. In fact, there is the possibility that the particles described by the field operator Φ, beyond the electric charge, carry further, additively conserved, quantum numbers. Whether or not this is the case is a question of the interactions and of their structure. Hints at additive charges often come from experiments testing conservation laws. For example, besides π-mesons, there exists a set of four K-mesons K^0, K^-, $\overline{K^0}$, and K^+, which carry the charges indicated in the superscript, but also carry a further quantum number S, called *strangeness*. The K^+ and the K^- are antiparticles of one another, and so are the K^0 and the $\overline{K^0}$. Assigning the value $S = -1$ to the K^- and the $\overline{K^0}$ (this is the conventional choice in particle physics), the K^+ and K^0 carry the value $S = 1$. This quantum number was discovered empirically. It is found to be additively conserved in strong and electromagnetic interactions. It can change, however, in weak interactions, and does so following a well-defined pattern. This implies, for example, that in a reaction with strong interaction

$$A + B \longrightarrow C + D + E\,,$$

in which particles with strangeness participate, the sum of the eigenvalues of S must be the same in the initial and the final states, $S(A) + S(B) = S(C) + S(D) + S(E)$. In decays such as

$$\text{(a) } \pi^+ \longrightarrow \pi^0 + e^+ + \nu_e \quad \text{(b) } K^+ \longrightarrow \pi^0 + e^+ + \nu_e\,,$$

which are caused by weak interactions, the strangeness remains unchanged, branch (a), or changes by one unit, branch (b), (the positron and the neutrino carry no strangeness).

7.3 The Quantized Maxwell Field

The Maxwell field is an obvious candidate for trying the principle of canonical quantization. As compared to the previous examples, the real and the complex Klein-Gordon fields, it exhibits two new, intimately related properties. The photon which is the field quantum of the quantized form of electrodynamics, has no mass and, hence, satisfies the Klein-Gordon equation for mass zero. As it is massless, in accord with Sect. 6.3.3, it is to be described by its helicity whose eigenvalues are $h = 1$ and $h = -1$. (A *massive* particle with spin $s = 1$ would possess *three* spin degrees of freedom!) In the classical, not yet quantized form of the theory these helicity states correspond to left- and right-circular polarization. This is equivalent to the statement that electromagnetic plane waves have only transverse but no longitudinal polarization.

7.3.1 Maxwell's Theory in the Lagrange Formalism

The essential dynamical variable of the free Maxwell field in vacuum is the tensor field of electromagnetic field strengths,

$$F^{\mu\nu}(x) = \begin{pmatrix} 0 & -E^1(x) & -E^2(x) & -E^3(x) \\ E^1(x) & 0 & -B^3(x) & B^2(x) \\ E^2(x) & B^3(x) & 0 & -B^1(x) \\ E^3(x) & -B^2(x) & B^1(x) & 0 \end{pmatrix} \tag{7.81}$$

which satisfy the homogeneous and inhomogeneous Maxwell equations, in natural units,

$$\varepsilon_{\mu\nu\sigma\tau}\partial^\nu F^{\sigma\tau}(x) = 0\,, \tag{7.82}$$

$$\partial_\mu F^{\mu\nu}(x) = j^\nu(x)\,. \tag{7.83}$$

We briefly confirm that this form of the equations which is manifestly Lorentz covariant, agrees with Eqs. (7.13) and (7.14) of Sect. 7.1.2. Given $F^{\mu\nu}$ the electric and magnetic fields are, respectively,

$$E^i(x) = -F^{0i}(x)\,, \qquad B^i(x) = -\frac{1}{2}\varepsilon_{ijk}F^{jk}(x)\,,$$

where ε_{ijk} is the antisymmetric Levi-Civita tensor in three dimensions and the sum convention is used. The homogeneous equations with $\mu = 0$, $F^{jk} = -\,\varepsilon_{jkl}B^l$, and $\varepsilon_{0ijk} = \varepsilon_{ijk}$[7] yield

$$\varepsilon_{ijk}\partial^i F^{jk} = -\varepsilon_{ijk}\varepsilon_{jkl}\partial^i B^l = 0\,,$$

All indices appearing in contragredient pairs are to be summed, Latin indices running from 1 to 3. As $\varepsilon_{ijk}\varepsilon_{jkl} = 2\delta_{il}$, the first homogeneous equation (7.13) follows.

Choosing $\mu = i$, precisely one of the other indices of $\varepsilon_{i\nu\sigma\tau}$ must be zero. For example, one has $\varepsilon_{i0jk} = -\varepsilon_{ijk}$ and $\varepsilon_{ij0k} = \varepsilon_{ijk}$. Therefore, one obtains

$$\begin{aligned} \varepsilon_{i0jk}\partial^0 F^{jk} + \varepsilon_{ij0k}\partial^j F^{ok} &= \varepsilon_{ijk}\varepsilon_{jkl}\partial^0 B^l + \varepsilon_{ijk}(-\partial_j)(-E^k) \\ &= 0 = \partial_0 B^i + (\nabla \times \boldsymbol{E})^i\,. \end{aligned}$$

This is seen to be the second of the homogeneous equations (7.13) with $f_1 = 1$.

[7] Note that we had chosen the convention $\varepsilon_{0123} = +1$.

It is even simpler to reproduce the inhomogeneous equations: From (7.83), inserting the operator (7.2) with $\nu=0$, one has

$$\partial_i F^{i0}(x) = j^0(x) \qquad \text{or} \quad \nabla \cdot \boldsymbol{E}(x) = \varrho(x)\,.$$

This is the first equation (7.14) with $\boldsymbol{D} = \boldsymbol{E}$ and $f_2 = 1$. Furthermore, for $\nu = i$, we have

$$\partial_0 F^{0i} + \partial_k F^{ki} = -\frac{\partial}{\partial x^0} E^i - \varepsilon_{kil} \partial_k B^l = j^i = -\left(\frac{\partial}{\partial x^0}\boldsymbol{E} - \nabla \times \boldsymbol{B}\right)^i ,$$

which is seen to be the second equation (7.14) with $f_3 = f_4 = 1$.

Expressing the field strength tensor field by means of a (four-)potential field $A^\mu(x)$,

$$F^{\mu\nu}(x) = \partial^\mu A^\nu(x) - \partial^\nu A^\mu(x)\,, \tag{7.84}$$

the homogeneous equations (7.82) are fulfilled automatically. The inhomogeneous Maxwell equations are shown to be the Euler-Lagrange equations of the following Lagrange density

$$\mathcal{L}(A^\alpha, \partial_\mu A^\alpha, j^\mu) := -\frac{1}{4} F_{\mu\nu}(x) F^{\mu\nu}(x) - j_\mu(x) A^\mu(x)\,. \tag{7.85}$$

This is confirmed by a little calculation. Writing the kinetic term as

$$-\frac{1}{4} F_{\alpha\beta} g^{\alpha\mu} g^{\beta\nu} F_{\mu\nu} = -\frac{1}{4}\left(\partial_\alpha A_\beta - \partial_\beta A_\alpha\right) g^{\alpha\mu} g^{\beta\nu} \left(\partial_\mu A_\nu - \partial_\nu A_\mu\right) ,$$

and noting that all indices must be summed, one realizes that the derivative terms $(\partial_\sigma A_\tau)$ appear in four places. Therefore, the derivative of the Lagrange density by this term is

$$\frac{\partial \mathcal{L}}{\partial(\partial_\sigma A_\tau)} = -\frac{1}{4}\left(4\partial^\sigma A^\tau - 4\partial^\tau A^\sigma\right) = -F^{\sigma\tau}\,.$$

The partial derivative by the potential itself is $\partial\mathcal{L}/\partial A_\sigma = -j^\sigma$, so that the Euler-Lagrange equations are

$$\frac{\partial \mathcal{L}}{\partial A_\tau} - \partial_\sigma \left(\frac{\partial \mathcal{L}}{\partial(\partial_\sigma A_\tau)}\right) = 0 = -j^\tau(x) + \partial_\sigma F^{\sigma\tau}(x)\,.$$

These are the inhomogeneous Maxwell equations.

Remarks

1. When the definition (7.84) is expressed in terms of electric and magnetic fields one obtains the familiar relations

$$E^i(x) = F^{i0}(x) = \partial^i A^0 - \partial^0 A^i = -\left(\nabla A^0 + \frac{\partial}{\partial x^0}\boldsymbol{A}\right)^i ,$$

$$B^i(x) = -\frac{1}{2}\varepsilon_{ijk} F^{jk} = -\frac{1}{2}\varepsilon_{ijk}\left(\partial^j A^k - \partial^k A^j\right) = (\nabla \times \boldsymbol{A})^i\,.$$

(Note that $\{\partial^j\} = -\{\partial_j\} = -\nabla$.)

2. The Lagrange density (7.85) provides an example for a classical field theory with more than one real field. In a symbolic, more general notation we can write the corresponding Lagrange density as $\mathcal{L}(\phi^{(i)}, \partial_\mu \phi^{(i)})$ where $i = 1, 2, \ldots, n$ serves to number the fields. In the case at hand there are four real fields, (A^0, A^1, A^2, A^3), which, in addition, have a well-defined transformation behaviour under Lorentz transformations. Every one of the fields $\phi^{(i)}$ fulfills an Euler-Lagrange equation.
3. The expression (7.28) for the energy-momentum field that was introduced in Sect. 7.1.4 translates as follows: One calculates the derivative

$$\begin{aligned}\partial^\nu \mathcal{L}(\phi^{(i)}, \partial_\mu \phi^{(i)}) &= \sum_i \left\{ \frac{\partial \mathcal{L}}{\partial \phi^{(i)}} \frac{\partial \phi^{(i)}}{\partial x_\nu} + \frac{\partial \mathcal{L}}{\partial(\partial_\mu \phi^{(i)})} \frac{\partial(\partial_\mu \phi^{(i)})}{\partial x_\nu} \right\} \\ &= \sum_i \left\{ \left(\partial_\mu \frac{\partial \mathcal{L}}{\partial(\partial_\mu \phi^{(i)})} + \frac{\partial \mathcal{L}}{\partial(\partial_\mu \phi^{(i)})} \partial_\mu \right) \partial^\nu \phi^{(i)} \right\} \\ &= \partial_\mu \left\{ \sum_i \frac{\partial \mathcal{L}}{\partial(\partial_\mu \phi^{(i)})} \partial^\nu \phi^{(i)} \right\} .\end{aligned}$$

In the second step $\partial \mathcal{L}/\partial \phi^{(i)}$ is replaced by means of the equation of motion. Alternatively, the same partial derivative can be written simply

$$\partial^\nu \mathcal{L}(\phi^{(i)}, \partial_\mu \phi^{(i)}) = g^{\nu\mu} \partial_\mu \mathcal{L}(\phi^{(i)}, \partial_\mu \phi^{(i)}) .$$

A conservation law is obtained by taking the difference of the two expressions

$$\partial_\mu \mathcal{T}^{\mu\nu}(x) = 0 \quad \text{with}$$

$$\mathcal{T}^{\mu\nu} = \sum_i \frac{\partial \mathcal{L}}{\partial(\partial_\mu \phi^{(i)})} \partial^\nu \phi^{(i)} - g^{\mu\nu} \mathcal{L} . \tag{7.86}$$

4. In the vacuum, i.e. without external sources, the two kinds of fields $\boldsymbol{E}(x)$ and $\boldsymbol{B}(x)$ obey the wave equation, i.e. the Klein-Gordon equation for mass $m = 0$. If one chooses the class of Lorenz gauges,[8] $\partial_\mu A^\mu(x) = 0$, then also $A^\mu(x)$ satisfies the wave equation. This implies that the quantized form of the theory will describe massless particles.
5. As it should, the Lagrange density (7.85) is invariant with respect to proper orthochronous Lorentz transformations. Thus, the corresponding equations of motion are Lorentz covariant. Furthermore, the Lagrange density is invariant under space reflection and time reversal. This property is a physical one because it can be tested in the interaction of the Maxwell fields with matter (which comes in by its current density j^μ).
6. In addition, the Lagrange density (7.85) is invariant under *local* gauge transformations

$$A_\mu(x) \longmapsto A'_\mu(x) = A_\mu(x) - \partial_\mu \chi(x) \tag{7.87}$$

[8] This condition was first found by Ludvig Valentin Lorenz (1829–1891), a Danish physicist, long before Hendrik Antoon Lorentz' times to whom this relation is often but erroneously attributed.

where $\chi(x)$ is a Lorentz scalar, differentiable function. This is obvious for the first term in (7.85) because $F^{\mu\nu}(x)$ does not change when (7.87) is applied to A_μ. The second term which describes the interaction with the current density $j^\mu(x)$ changes by the term $j_\mu(x)\partial^\mu\chi(x)$. Using partial integration this term is converted to $(\partial^\mu j_\mu(x))\chi(x)$ up to possible boundary terms at infinity. These boundary terms do not contribute if the current density vanishes sufficiently fast at infinity. As the current density satisfies the continuity equation the additional term equals zero.

7.3.2 Canonical Momenta, Hamilton- and Momentum Densities

The generalized momenta canonically conjugate to $A_\mu(x)$ are calculated in analogy to the prescription (7.24). One finds

$$\pi^0(x) = \frac{\partial \mathcal{L}}{\partial(\partial_0 A_0)} \equiv 0\,, \quad \pi^i(x) = \frac{\partial \mathcal{L}}{\partial(\partial_0 A_i)} = -F^{0i} = E^i\,. \tag{7.88}$$

The momentum conjugate to A_0 is identically zero, the momentum conjugate to A_i is seen to be the ith component of the electric field. One then calculates the Hamilton density and the momentum density as follows. With

$$F_{\mu\nu}F^{\mu\nu} = -F_{\mu\nu}F^{\nu\mu} = -\mathrm{tr}\left(\{F_{\alpha\nu}\}\{F^{\nu\beta}\}\right) = 2\left(\boldsymbol{B}^2 - \boldsymbol{E}^2\right),$$

with the spatial canonical momenta π^i, and with $E^i = -\partial A^i/\partial x^0 - \nabla^i A_0$ one obtains

$$\begin{aligned}\mathcal{H} = -\pi^i \frac{\partial A^i}{\partial x^0} - \mathcal{L} &= \boldsymbol{E}\cdot(\boldsymbol{E} + \nabla A_0) + \frac{1}{2}(\boldsymbol{B}^2 - \boldsymbol{E}^2) + j_\mu A^\mu \\ &= \frac{1}{2}(\boldsymbol{B}^2 + \boldsymbol{E}^2) + j_\mu A^\mu + \boldsymbol{E}\cdot\nabla A_0\,.\end{aligned}$$

The Hamilton function is obtained by integration over the whole space. The contribution of the last term of the result for $\mathcal{H}$ is calculated as follows,

$$\int \mathrm{d}^3x\ \boldsymbol{E}\cdot\nabla A_0 = -\int \mathrm{d}^3x\ (\nabla\cdot\boldsymbol{E})\,A_0 = -\int \mathrm{d}^3x\ j^0(x)A_0(x)\,.$$

Eventually, these equations yield

$$H = \int \mathrm{d}^3x\ \mathcal{H} = \int \mathrm{d}^3x\ \left\{\frac{1}{2}\left(\boldsymbol{B}^2 + \boldsymbol{E}^2\right) - \boldsymbol{j}\cdot\boldsymbol{A}\right\}. \tag{7.89}$$

The first term on the right-hand side is the well-known field energy, the second term is the interaction energy of the fields with the given electromagnetic currents.

The Hamilton density just calculated is nothing but the component $\mathcal{H} = \mathcal{T}^{00}$ of the energy-momentum tensor field (7.86). The momentum density is obtained from the latter and is found to be

$$\mathcal{P}^k = \mathcal{T}^{0k} = \sum_i \pi^{(i)}\partial^k\phi^{(i)} = \sum_{j=1}^{3} E^j\left(\frac{\partial A_j}{\partial x_k}\right) = \boldsymbol{E}\cdot\left(\frac{\partial \boldsymbol{A}}{\partial x^k}\right). \tag{7.90}$$

We will analyze this expression further below, in the framework of quantization.

7.3.3 Lorenz- and Transversal Gauges

As mentioned before, by choosing the class of Lorenz gauges

$$\partial_\mu A^\mu(x) = 0\,, \tag{7.91}$$

the inhomogeneous equations (7.83) yield the equation of motion

$$\Box A^\mu(x) = j^\mu(x)\,. \tag{7.92}$$

This is an inhomogeneous Klein-Gordon equation for mass zero. Its external source contains the current density of matter. Note that the condition (7.91) does not fix the gauge completely. Indeed, further gauge transformations (7.87) preserving the Lorenz condition (7.91) can be found. They must be generated by gauge functions $\chi(x)$ which are solutions of the homogeneous wave equation $\Box\chi(x) = 0$. Clearly, these additional gauge transformations leave the equation of motion (7.92) form invariant.

At this point there is a bifurcation into two rather different paths along which to pursue the analysis of the quantized radiation field: A manifestly covariant approach which raises interesting conceptual questions, or a frame-dependent approach which hides the covariance but underpins more clearly the physical content of the theory. Depending on which direction one chooses the same theory will exhibit rather different facets. As this is of great importance for the physical understanding we dwell for a while on this aspect in a couple of remarks which also serve to motivate the next steps of our analysis.

Remarks

1. In a strict sense, only the field strengths, i.e. the entries of the tensor field $F^{\mu\nu}$ are observable. Yet, after quantization, the auxiliary field A^μ seems to be the natural candidate for the description of the photon. However, matters do not really match. We already know that photons have strictly no mass and, as they are described classically by a vector field, presumably carry spin $s = 1$. The general analysis described in Sect. 6.3.3 says that massless particles have to be described by their helicity, not by the spin of nonrelativistic quantum theory. It also says that massless vector particles have *two*, not *three*, spin degrees of freedom, viz. the components $h = 1$ and $h = -1$. The vector field $A^\mu(x)$, initially, has four degrees of freedom – obviously too many. Indeed, an arbitrary vector field which transforms like a four-vector under Lorentz transformations, contains the spins $s = 1$ and $s = 0$, and it is plausible that the spin-0 part is contained in its (four-)divergence. It is precisely this part which is set equal to zero by the Lorenz condition (7.91). The three remaining degrees of freedom would be appropriate if the vector particle had a nonvanishing mass because then the spin would have the usual three independent orientations in space. In the case of the photon only two spin projections are physical so that there is still one degree of freedom too much. This conclusion is reflected in the freedom of gauge that remains after imposing the condition (7.91).

2. Maxwell's equations (7.82) and (7.83) as well as the condition (7.91) and the equation of motion (7.92) are manifestly Lorentz covariant. Qualitatively speaking, this means that these equations have a well-defined transformation behaviour under Lorentz transformations and that they are *form invariant*. Talking about gauges and gauge fixing there are two, formally different frameworks in which one can work. One may choose any further gauge transformation such that all relevant equations remain *manifestly covariant*. This approach has advantages when deriving general properties of the classical theory and its quantized partner. The prize that one has to pay is that one will have to work with unphysical degrees of freedom which disappear from the theory only in the final step of calculating observables.
   Alternatively, the remaining gauge freedom may be used to eliminate the redundant degree of freedom from the start. We will see, however, that in doing so one unavoidably breaks manifest covariance. The advantage of this approach is that it deals only with physical degrees of freedom and, at every step, is amenable to physical interpretation in a transparent manner. As a disadvantage it renders calculations in higher orders of perturbation theory cumbersome, if not impossible. Of course, all genuine observables such as, e.g., cross sections, come out the same, irrespective of the framework one has chosen.

As our first aim is to understand the physical meaning of quantized electromagnetism, we choose here the second alternative, that is, we abandon explicit covariance in favour of a formalism that makes use of physical fields only. We return to quantization in a manifestly covariant formulation in Sect. 7.5 below.

The freedom of gauging in

$$A_\mu \longmapsto A'_\mu = A_\mu - \partial_\mu \chi(x)$$

is chosen such as to render the spatial part of the vector field A'_μ divergenceless. Assume the original field A_μ had a nonvanishing spatial divergence $\boldsymbol{\nabla} \cdot \boldsymbol{A}$ and choose the gauge function to be

$$\chi(x) \equiv \chi(t, \boldsymbol{x}) = \frac{1}{4\pi} \int \mathrm{d}^3 y \, \frac{1}{|\boldsymbol{x} - \boldsymbol{y}|} \boldsymbol{\nabla}_y \cdot \boldsymbol{A}(t, \boldsymbol{y}) .$$

Then, with $\boldsymbol{\Delta}(1/|\boldsymbol{x} - \boldsymbol{y}|) = -4\pi\delta(\boldsymbol{x} - \boldsymbol{y})$ one has $\boldsymbol{\Delta}\chi = -\boldsymbol{\nabla} \cdot \boldsymbol{A}$ so that one concludes $\boldsymbol{\nabla} \cdot \boldsymbol{A}' = \boldsymbol{\nabla} \cdot \boldsymbol{A} + \boldsymbol{\Delta}\chi = 0$. From here on one remains within the class of gauge potentials A_μ which fulfill the condition

$$\boldsymbol{\nabla} \cdot \boldsymbol{A}(x) = 0 . \tag{7.93}$$

A gauge of this class is called a *transverse gauge,* or *Coulomb gauge*. Inserting this in the equation of motion (7.83) with $\nu = 0$ one obtains

$$\Box A^0(x) - \frac{\partial^2 A^0(x)}{\partial (x^0)^2} = -\boldsymbol{\Delta} A^0(x) = j^0(x) .$$

A solution of this differential equation (Poisson equation) is known from electrostatics, viz.

$$A^0(t, \boldsymbol{x}) = \frac{1}{4\pi} \int \mathrm{d}^3 y \, \frac{j^0(t, \boldsymbol{y})}{|\boldsymbol{x} - \boldsymbol{y}|} \, .$$

If one is concerned with a free Maxwell field, i.e. if the external sources j^μ are identically zero, the field A^0 can be made to vanish

$$A^0(x) \equiv 0 \, .$$

In order to achieve this choose the gauge transformation

$$A'_\mu = A_\mu - \partial_\mu \psi \quad \text{and} \quad \psi = \int_0^{x^0} \mathrm{d}t' \, A^0(t', \boldsymbol{x}) \, .$$

The Lorenz condition which led to the equation of motion (7.92) can always be imposed. However, transforming the field A^0 to zero (whose conjugate momentum vanishes identically) is possible only in the theory in vacuum, i.e. without external sources.

While the magnetic field is always transversal, i.e. fulfills the condition $\nabla \cdot \boldsymbol{B} = 0$, the electric field

$$\boldsymbol{E} = -\nabla A^0 - \partial_0 \boldsymbol{A} \equiv \boldsymbol{E}_\parallel + \boldsymbol{E}_\perp$$

has both a parallel and a transverse component,

$$\nabla \times \boldsymbol{E}_\parallel = 0 \, , \qquad \nabla \cdot \boldsymbol{E}_\perp = 0 \, .$$

They are defined by space or time derivatives of the vector potential $\boldsymbol{A}$ as given above.

In a Coulomb gauge the results obtained are used to transform the electric part of the field energy in (7.89) as follows. One has

$$\int \mathrm{d}^3 x \, \boldsymbol{E}^2 = \int \mathrm{d}^3 x \, (\boldsymbol{E}_\parallel^2 + \boldsymbol{E}_\perp^2) - 2 \int \mathrm{d}^3 x \, \nabla A^0 \cdot \boldsymbol{E}_\perp \, .$$

The second term does not contribute because, by partial integration, the nabla operator is shifted to $\boldsymbol{E}_\perp$. Furthermore, one has

$$\int \mathrm{d}^3 x \, \boldsymbol{E}_\parallel^2 = \int \mathrm{d}^3 x \, (\nabla A^0)^2 = \int \mathrm{d}^3 x \, \nabla \cdot (A^0 \nabla A^0) - \int \mathrm{d}^3 x \, A^0 \boldsymbol{\Delta} A^0 \, ,$$

the first term of which vanishes, while the second is rewritten by means of $\boldsymbol{\Delta} A^0 = -j^0$. Inserting the solution for A^0 given above, one obtains

$$\int \mathrm{d}^3 x \, \boldsymbol{E}_\parallel^2 = \int \mathrm{d}^3 x \int \mathrm{d}^3 y \, \frac{j^0(t, \boldsymbol{x}) j^0(t, \boldsymbol{y})}{4\pi \, |\boldsymbol{x} - \boldsymbol{y}|} \, .$$

With these results the Hamilton function (7.89) becomes

$$H = \int d^3x \left\{ \frac{1}{2} \left(\boldsymbol{B}^2 + \boldsymbol{E}_\perp^2 \right) - \boldsymbol{j} \cdot \boldsymbol{A} \right\} + \frac{1}{2} \int d^3x \int d^3y \, \frac{j^0(t, \boldsymbol{x}) j^0(t, \boldsymbol{y})}{4\pi |\boldsymbol{x} - \boldsymbol{y}|}$$
$$= \int d^3x \left\{ \frac{1}{2} \left((\nabla \times \boldsymbol{A})^2 + \left(\frac{\partial \boldsymbol{A}}{\partial x^0} \right)^2 \right) - \boldsymbol{j} \cdot \boldsymbol{A} \right\}$$
$$+ \frac{1}{2} \int d^3x \int d^3y \, \frac{j^0(t, \boldsymbol{x}) j^0(t, \boldsymbol{y})}{4\pi |\boldsymbol{x} - \boldsymbol{y}|} . \tag{7.94}$$

This result is remarkable by the fact that, besides the field energy of the transverse fields and the interaction between spatial current density and vector potential, the Hamilton function H contains an *instantaneous* Coulomb interaction (last term). Note that the form (7.94) of the Hamilton function is an exact result.

7.3.4 Quantization of the Maxwell Field

Let us accept that $A^\mu(x)$ is the field which is appropriate for the description of photons but, for the reasons described above, let us constrain it by gauge conditions such as (7.91) or (7.93). If we try to quantize this field by blindly applying the Postulate 7.1 we run into a contradiction: The time component $\pi^0(x)$ being identically zero, we would obtain the requirement

$$\left[\pi^i(x), A^j(y) \right]_{x^0=y^0} = -i \, \delta^{ij} \, \delta(\boldsymbol{x} - \boldsymbol{y}) = -i \, \delta^{ij} \frac{1}{(2\pi)^3} \int d^3k \, e^{i\boldsymbol{k}\cdot(\boldsymbol{x}-\boldsymbol{y})} . \tag{7.95}$$

This postulate is in conflict with the first inhomogeneous Maxwell equation (without external source). Differentiating (7.95) by $\boldsymbol{x}$, its left-hand side becomes (Gauss' law)

$$\nabla_x \left[\boldsymbol{E}^i(t, \boldsymbol{x}), A^j(t, \boldsymbol{y}) \right] = 0 ,$$

while its right-hand side yields a nonvanishing expression

$$-i \sum_{i=1}^{3} \partial_i \, \delta^{ij} \, \delta(\boldsymbol{x} - \boldsymbol{y}) = \frac{1}{(2\pi)^3} \int d^3k \, k^j e^{i\boldsymbol{k}\cdot(\boldsymbol{x}-\boldsymbol{y})} \neq 0 .$$

This contradiction can be resolved by a modification of the quantization rule for the Maxwell field. Assume

$$\left[\pi^i(x), A^j(y) \right]_{x^0=y^0} = \frac{i}{(2\pi)^3} \int d^3k \, e^{i\boldsymbol{k}\cdot(\boldsymbol{x}-\boldsymbol{y})} \left(\delta^{ij} - \frac{k^i k^j}{\boldsymbol{k}^2} \right) . \tag{7.96}$$

As we shall soon learn, the expression which appears on the right-hand side does not fall from heaven. Indeed, the specific linear combination in the integrand will be seen to be the sum over the physically allowed helicity states of the photon.

By direct analogy to the expansion (7.31) in terms of the base functions (7.19) we write

$$\boldsymbol{A}(t, \boldsymbol{x}) = \frac{1}{(2\pi)^{3/2}} \int \frac{d^3k}{2k^0} \sum_\lambda \boldsymbol{\varepsilon}_\lambda(\boldsymbol{k}) \left[c_\lambda(\boldsymbol{k}) e^{-ikx} + c_\lambda^\dagger(\boldsymbol{k}) e^{ikx} \right] , \tag{7.97}$$

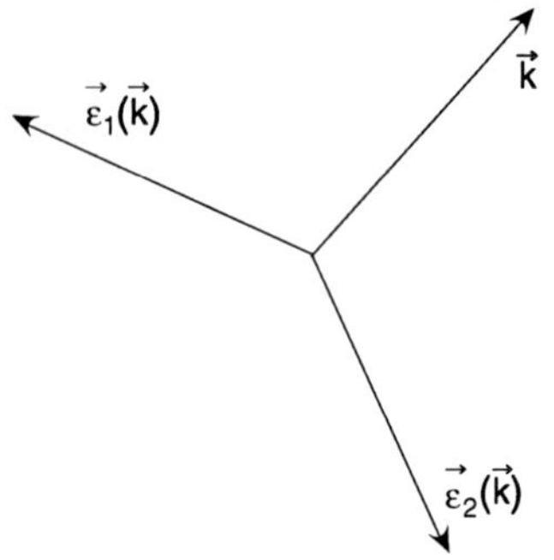

Fig. 7.8 The vector $\boldsymbol{k}$ defines the direction of propagation of the plane wave. Its polarization can only be perpendicular to this direction, that is, it must be a linear combination of $\boldsymbol{\varepsilon}_1$ and $\boldsymbol{\varepsilon}_2$

where, as usual, $kx = k^0x^0 - \boldsymbol{k}\cdot\boldsymbol{x}$, and where $k^0 = |\boldsymbol{k}| =: \omega_k$ is the (physical) energy of a photon with spatial momentum $\boldsymbol{k}$. The vector behaviour of $\boldsymbol{A}$ is carried over to the polarization directions $\boldsymbol{\varepsilon}_\lambda = (\varepsilon_\lambda^1, \varepsilon_\lambda^2, \varepsilon_\lambda^3)$ whose number is determined by the transversality condition. Using a Coulomb gauge, for instance, Eq. (7.93) and the ansatz (7.97) yield the condition

$$\boldsymbol{\varepsilon}_\lambda(\boldsymbol{k})\cdot\boldsymbol{k} = 0$$

whose significance is easily understood. The spatial vector $\boldsymbol{k}$ defines the direction in which a monochromatic wave with wave number $|\boldsymbol{k}|$ propagates. The polarization of this wave must be transversal to this direction, i.e. the index λ can take only two values, say, $\lambda = 1$ and $\lambda = 2$, as sketched in Fig. 7.8. For these one has

$$\boldsymbol{\varepsilon}_\lambda(\boldsymbol{k})\cdot\boldsymbol{\varepsilon}_{\lambda'}(\boldsymbol{k}) = \delta_{\lambda\lambda'}\,.$$

Defining, in addition to the two previous components, $\boldsymbol{\varepsilon}_3(\boldsymbol{k}) := \hat{\boldsymbol{k}} = \boldsymbol{k}/|\boldsymbol{k}|$, one also has

$$\sum_{\lambda=1}^{3}\varepsilon_\lambda^i(\boldsymbol{k})\varepsilon_\lambda^j(\boldsymbol{k}) = \delta^{ij}\,.$$

If one sums over the first two values only one obtains the important relation

$$\sum_{\lambda=1}^{2}\varepsilon_\lambda^i(\boldsymbol{k})\varepsilon_\lambda^j(\boldsymbol{k}) = \delta^{ij} - \frac{k^ik^j}{\boldsymbol{k}^2}\,.$$

This is the sum over spins that appears in the integrand of (7.96). In view of the explicit calculations that follow it is convenient to introduce an additional convention:

$$\boldsymbol{\varepsilon}_1(-\boldsymbol{k}) = -\boldsymbol{\varepsilon}_1(\boldsymbol{k})\,,\quad \boldsymbol{\varepsilon}_2(-\boldsymbol{k}) = +\boldsymbol{\varepsilon}_2(\boldsymbol{k})\,. \tag{7.98}$$

As a consequence of this convention, one has

$$\boldsymbol{\varepsilon}_\lambda(\boldsymbol{k})\cdot\boldsymbol{\varepsilon}_{\lambda'}(-\boldsymbol{k}) = (-)^\lambda\,\delta_{\lambda\lambda'}\,. \tag{7.99}$$

Furthermore, the corresponding spherical basis then satisfies a relation which is familiar from the theory of spherical tensor operators, viz.

$$\mp\frac{1}{\sqrt{2}}\left\{\boldsymbol{\varepsilon}_1(-\boldsymbol{k}) \pm \mathrm{i}\,\boldsymbol{\varepsilon}_2(-\boldsymbol{k})\right\} = \pm\frac{1}{\sqrt{2}}\left\{\boldsymbol{\varepsilon}_1(\boldsymbol{k}) \mp \mathrm{i}\,\boldsymbol{\varepsilon}_2(\boldsymbol{k})\right\}\,.$$

The operators $c_\lambda(\boldsymbol{k})$ and $c_\lambda^\dagger(\boldsymbol{k})$ are obtained from the expansion (7.97) in a manner which is completely analogous to the formulae (7.32). One finds

$$c_\lambda(\boldsymbol{k}) = \frac{\mathrm{i}}{(2\pi)^{3/2}} \int \mathrm{d}^3x\ \mathrm{e}^{\mathrm{i}kx}\ \overset{\leftrightarrow}{\partial_0}\ \boldsymbol{\varepsilon}_\lambda(\boldsymbol{k}) \cdot \boldsymbol{A}(x)\,, \tag{7.100}$$

$$c_\lambda^\dagger(\boldsymbol{k}) = -\frac{\mathrm{i}}{(2\pi)^{3/2}} \int \mathrm{d}^3x\ \mathrm{e}^{-\mathrm{i}kx}\ \overset{\leftrightarrow}{\partial_0}\ \boldsymbol{\varepsilon}_\lambda(\boldsymbol{k}) \cdot \boldsymbol{A}(x)\,. \tag{7.101}$$

These formulae are obtained as follows. By the choice $A^0 = 0$ one has $\pi^i = E^i = -\dot{A}^i$. Construct then

$$\omega_k \int \mathrm{d}^3x\ \mathrm{e}^{\mathrm{i}kx} \boldsymbol{\varepsilon}_\lambda \cdot \boldsymbol{A} = (2\pi)^{3/2} \frac{1}{2} \left(c_\lambda(\boldsymbol{k}) + (-)^\lambda c_\lambda^\dagger(-\boldsymbol{k}) \mathrm{e}^{2\mathrm{i}\omega_k x^0} \right),$$

$$\mathrm{i} \int \mathrm{d}^3x\ \mathrm{e}^{\mathrm{i}kx} \boldsymbol{\varepsilon}_\lambda \cdot \dot{\boldsymbol{A}} = (2\pi)^{3/2} \frac{1}{2} \left(c_\lambda(\boldsymbol{k}) - (-)^\lambda c_\lambda^\dagger(-\boldsymbol{k}) \mathrm{e}^{2\mathrm{i}\omega_k x^0} \right),$$

and take the sum of the two equations,

$$c_\lambda(\boldsymbol{k}) = \frac{1}{(2\pi)^{3/2}} \int \mathrm{d}^3x\ \mathrm{e}^{\mathrm{i}kx} \boldsymbol{\varepsilon}_\lambda(\boldsymbol{k}) \left(\omega_k \boldsymbol{A}(x) + \mathrm{i}\dot{\boldsymbol{A}}(x) \right).$$

Now, the expression within parentheses in the integrand equals i times the left-right derivative given in (7.100). The formula (7.101) is the hermitian conjugate of (7.100).

In the next step one uses these formulae to calculate the commutators for the operators $c_\lambda(\boldsymbol{k})$ and $c_\lambda^\dagger(\boldsymbol{k})$.

$$\begin{aligned}
\left[c_\lambda(\boldsymbol{k}), c_{\lambda'}^\dagger(\boldsymbol{k}') \right] &= \frac{1}{(2\pi)^3} \int \mathrm{d}^3x\ \mathrm{e}^{\mathrm{i}kx} \int \mathrm{d}^3y\ \ \mathrm{e}^{-\mathrm{i}k'y} \sum_{m,n=1}^{3} \varepsilon_\lambda^m(\boldsymbol{k})\, \varepsilon_{\lambda'}^n(\boldsymbol{k}') \\
&\quad \times \left[\left(\omega_k A^m(x) + \mathrm{i}\dot{A}^m(x) \right), \left(\omega_k' A^n(y) - \mathrm{i}\dot{A}^n(y) \right) \right]_{x^0=y^0} \\
&= -\frac{\mathrm{i}(\omega_k + \omega_k')}{(2\pi)^3} \int \mathrm{d}^3x\ \mathrm{e}^{\mathrm{i}kx} \int \mathrm{d}^3y\ \ \mathrm{e}^{-\mathrm{i}k'y} \\
&\quad \times \sum_{m,n=1}^{3} \varepsilon_\lambda^m(\boldsymbol{k})\, \varepsilon_{\lambda'}^n(\boldsymbol{k}') \left[\pi^m(x), A^n(y) \right]_{x^0=y^0}\,.
\end{aligned}$$

At this point one inserts the commutator (7.96)

$$\left[\pi^m(x), A^n(y) \right]_{x^0=y^0} = \frac{\mathrm{i}}{(2\pi)^3} \int \mathrm{d}^3q\ \mathrm{e}^{\mathrm{i}\boldsymbol{q}\cdot(\boldsymbol{x}-\boldsymbol{y})} \left(\delta^{mn} - \frac{q^m q^n}{\boldsymbol{q}^2} \right),$$

then takes the integral over x and over y, and makes use of the relations

$$\int \mathrm{d}^3q\ \delta(\boldsymbol{k} - \boldsymbol{q})\delta(\boldsymbol{q} - \boldsymbol{k}') = \delta(\boldsymbol{k} - \boldsymbol{k}')\,,$$

$$\sum_{m,n=1}^{3} \left(\delta^{mn} - \frac{k^m k^n}{\boldsymbol{k}^2} \right) \varepsilon_\lambda^m(\boldsymbol{k})\, \varepsilon_{\lambda'}^n(\boldsymbol{k}) = \boldsymbol{\varepsilon}_\lambda(\boldsymbol{k}) \cdot \boldsymbol{\varepsilon}_{\lambda'}(\boldsymbol{k}) = \delta_{\lambda\lambda'}\,.$$

The commutator between $c_\lambda(\boldsymbol{k})$ and $c_{\lambda'}(\boldsymbol{k}')$ is obtained in the same manner. In summary one obtains

$$\left[c_\lambda(\boldsymbol{k}), c^\dagger_{\lambda'}(\boldsymbol{k}')\right] = 2\omega_k\, \delta_{\lambda\lambda'}\, \delta(\boldsymbol{k} - \boldsymbol{k}')\,, \tag{7.102}$$

$$\left[c_\lambda(\boldsymbol{k}), c_{\lambda'}(\boldsymbol{k}')\right] = 0 = \left[c^\dagger_\lambda(\boldsymbol{k}), c^\dagger_{\lambda'}(\boldsymbol{k}')\right]\,. \tag{7.103}$$

On the basis of what we learned in studying the scalar field, the interpretation of these results seems clear: The operators $c^\dagger_\lambda(\boldsymbol{k})$ and $c_\lambda(\boldsymbol{k})$ are creation and annihilation operators, respectively, for one-photon states with momentum $\boldsymbol{k}$ and polarization λ. Their energy is $\omega_k = |\boldsymbol{k}|$. The one-particle states

$$|\,\boldsymbol{k}, \lambda\,\rangle = c^\dagger_\lambda(\boldsymbol{k})\,|\,0\,\rangle$$

are covariantly normalized, i.e.

$$\left\langle \boldsymbol{k}', \lambda' \,\middle|\, \boldsymbol{k}, \lambda \right\rangle = 2\omega_k\, \delta_{\lambda\lambda'}\, \delta(\boldsymbol{k} - \boldsymbol{k}')\,.$$

In order to consolidate this interpretation, however, we should calculate the Hamiltonian and the momentum operator for the free photon field.

7.3.5 Energy, Momentum, and Spin of Photons

We split the Hamilton function (7.94) into a purely field dependent term and the interaction terms with charge and current densities of matter, $H = H_0 + H_{\text{int}}$. After (canonical) quantization H_0 is replaced by the normal-ordered operator

$$H_0 = :\int \mathrm{d}^3x \left\{ \frac{1}{2}\left((\nabla \times \boldsymbol{A})^2 + \left(\frac{\partial \boldsymbol{A}}{\partial x^0}\right)^2 \right) \right\}:. \tag{7.104}$$

A little calculation using the expansion (7.97) yields the result (cf. Exercise 7.7)

$$H_0 = \sum_{\lambda=1}^{2} \int \frac{\mathrm{d}^3k}{2\omega_k}\, \omega_k\, c^\dagger_\lambda(\boldsymbol{k}) c_\lambda(\boldsymbol{k})\,. \tag{7.105}$$

The momentum operator is obtained from the formula (7.90)

$$\mathcal{P}^k = :\boldsymbol{E}\cdot\left(\frac{\partial \boldsymbol{A}}{\partial x^k}\right): = -:\left(\dot{\boldsymbol{A}} + \nabla A^0\right)\cdot\left(\frac{\partial \boldsymbol{A}}{\partial x^k}\right): = -:\dot{\boldsymbol{A}}\cdot\left(\frac{\partial \boldsymbol{A}}{\partial x^k}\right):,$$

(keeping track of the fact that A^0 was transformed to zero). Inserting (7.97) and integrating over all space one finds

$$\boldsymbol{P} = \sum_{\lambda=1}^{2} \int \frac{\mathrm{d}^3k}{2\omega_k}\, \boldsymbol{k}\, c^\dagger_\lambda(\boldsymbol{k}) c_\lambda(\boldsymbol{k})\,. \tag{7.106}$$

It is now easy to verify that the interpretation given above was correct, that is to say, that one has

$$H\,|\,\boldsymbol{k}, \lambda\,\rangle = \omega_k\,|\,\boldsymbol{k}, \lambda\,\rangle\,, \qquad \boldsymbol{P}\,|\,\boldsymbol{k}, \lambda\,\rangle = \boldsymbol{k}\,|\,\boldsymbol{k}, \lambda\,\rangle$$

with $\omega_k = |\boldsymbol{k}|$.

It remains to identify the spin carried by the single-particle states $|\boldsymbol{k}, \lambda\rangle$.

7.3.6 Helicity and Orbital Angular Momentum of Photons

The basis that we used in the expansion (7.97) of the quantized photon field is the complete system of (covariantly normalized) plane waves. If one talks about spin and orbital angular momentum one must keep in mind that although the plane wave contains all values of the orbital angular momentum, from $\ell = 0$ to $\ell = \infty$, the projection m_ℓ of any partial wave onto the direction $\hat{\boldsymbol{k}}$ of propagation equals zero, cf. Sect. 1.9.3. Furthermore, the general analysis of representations of the Poincaré group tells us that massless particles do not have spins in the sense of nonrelativistic quantum theory but are characterized by their helicity which takes only two values $h = \pm s$ (no matter what the value of $s \geqslant 1/2$ is). This was found to be a consequence of the observation that the relevant subgroup of the Lorentz group leaves invariant the spatial momentum $\boldsymbol{k}$ of the massless particle. What then could be more natural than to study the behaviour of its quantum states under rotations about the axis defined by $\boldsymbol{k}$?[9]

It is appropriate to replace the unit vectors $\boldsymbol{\varepsilon}_1$ and $\boldsymbol{\varepsilon}_2$ by the spherical basis

$$\boldsymbol{\zeta}_\pm := \mp\frac{1}{\sqrt{2}}\left(\boldsymbol{\varepsilon}_1 \pm \mathrm{i}\boldsymbol{\varepsilon}_2\right) . \tag{7.107}$$

If one now performs a rotation about the axis $\hat{\boldsymbol{k}}$ by the angle ϕ then $\boldsymbol{\zeta}_+$ and $\boldsymbol{\zeta}_-$ transform according to

$$\boldsymbol{\zeta}_\pm \longmapsto \boldsymbol{\zeta}'_\pm = \mathrm{e}^{\pm\mathrm{i}\phi}\,\boldsymbol{\zeta}_\pm .$$

The handedness of a particle with spin s was defined earlier by the projection of its spin onto the direction of the momentum, $h = \boldsymbol{s}\cdot\hat{\boldsymbol{k}}$. It is then clear that the photon carries the helicities $+1$ and -1. Thus, according to the definition given in Sect. 1.3.3, $s = |h|$, one concludes:

The photon carries spin 1.

Another basis which may be used in expanding the quantized photon field, is provided by the eigenfunctions of angular momentum (orbital angular momentum plus spin). For instance, this is the appropriate choice when one studies emission and absorption of photons in atomic or nuclear states which are classified by angular momentum and parity. For example, a photon emitted in the transition

$$\left|n', (\ell', s)j'\right\rangle \longrightarrow |n, (\ell, s)j\rangle$$

carries (total) angular momentum J which together with j and j' satisfies the triangle rule. Furthermore, the parity of the photon state is equal to the product of the parities of initial and final state. This gives rise to selection rules for transitions in atoms or nuclei.

---

[9] In the case of Einstein's equations for the gravitation field the analogous analysis shows that the graviton is massless and has spin/helicity 2.

Base states with definite angular momentum for photons are constructed as follows. Assuming harmonic time dependence

$$\boldsymbol{A}(t, \boldsymbol{x}) = \mathrm{e}^{-\mathrm{i}k^0 t} \boldsymbol{A}(\boldsymbol{x})$$

in the equation of motion (7.92) without external sources, and making use of the energy-momentum relation $k^0 = |\boldsymbol{k}| \equiv \kappa$, the vector field $\boldsymbol{A}(\boldsymbol{x})$ satisfies the differential equation

$$\left(\boldsymbol{\Delta} + \kappa^2\right) \boldsymbol{A}(\boldsymbol{x}) = 0 \,. \tag{7.108}$$

This is the well-known *Helmholtz equation.* The aim then is to find solutions of this equation which are eigenfunctions of angular momentum and have definite parity.

A very useful tool for this purpose is provided by the *vector spherical harmonics* which are obtained by coupling ordinary spherical harmonics and the elements of the spherical basis. Adding to the unit vectors (7.107) a third one

$$\boldsymbol{\zeta}_0 := \boldsymbol{\varepsilon}_3 \tag{7.109}$$

one obtains the spherical basis $\boldsymbol{\zeta}_\mu$, $\mu = 1, 0, -1$ whose properties are

$$\boldsymbol{\zeta}^*_\mu = (-)^\mu \boldsymbol{\zeta}_{-\mu} \,, \qquad \boldsymbol{\zeta}^*_\mu \cdot \boldsymbol{\zeta}_{\mu'} = \delta_{\mu\mu'} \,. \tag{7.110}$$

The vector spherical harmonics are defined by

$$\boldsymbol{T}_{J\ell M}(\theta, \phi) := \sum_{m,\mu} (\ell m, 1\mu \,|JM)\, Y_{\ell m}(\theta, \phi)\, \boldsymbol{\zeta}_\mu \,. \tag{7.111}$$

Obviously, they transform like irreducible spherical tensors of rank J, and, in addition, have vector character in $\mathbb{R}^3$. Their behaviour with respect to space reflection is

$$\boldsymbol{T}_{J\ell M}(\pi - \theta, \phi + \pi) = (-)^\ell\, \boldsymbol{T}_{J\ell M}(\theta, \phi) \; : \tag{7.112}$$

They fulfill the orthogonality relation

$$\int \mathrm{d}\Omega \, \left(\boldsymbol{T}^*_{J'\ell' M'} \cdot \boldsymbol{T}_{J\ell M}\right) = \delta_{J'J}\delta_{\ell'\ell}\delta_{M'M} \,. \tag{7.113}$$

The result (7.113) follows from the orthogonality of the $Y_{\ell m}$, from (7.110), and from the unitarity of the Clebsch-Gordan coefficients. Finally, one verifies that the set $\boldsymbol{T}_{J\ell M}(\theta, \phi)$ is a *complete* orthonormal system.

We return to the Helmholtz equation, assuming a factorized form for its solutions

$$\boldsymbol{A}_{J\ell M}(\boldsymbol{r}) = R_\ell(r)\, \boldsymbol{T}_{J\ell M}(\theta, \phi) \,.$$

The Laplace operator in spherical polar coordinates reads

$$\boldsymbol{\Delta} = \frac{1}{r^2}\frac{\partial}{\partial r}\left(r^2 \frac{\partial}{\partial r}\right) - \frac{\boldsymbol{\ell}^2}{r^2} \,,$$

cf. Sect. 1.9.2. Inserting this in (7.108) and introducing the dimensionless variable $\varrho := \kappa r$ one obtains a differential equation for the function $R_\ell(r)$ which depends on the radial variable only,

$$\frac{1}{\varrho^2}\frac{\mathrm{d}}{\mathrm{d}\varrho}\left(\varrho^2\frac{\mathrm{d}R_\ell(\varrho)}{\mathrm{d}\varrho}\right)-\frac{\ell(\ell+1)}{\varrho^2}R_\ell(\varrho)+R_\ell(\varrho)=0\,.$$

This is the differential equation for spherical Bessel and Hankel functions that we first met in Sect. 1.9.3. The solutions regular at the origin $r=0$ are the spherical Bessel functions $j_\ell(\kappa r)$. In turn, the spherical Hankel functions $h_\ell^{(\pm)}(\kappa r)$ are the solutions which oscillate at infinity like $\mathrm{e}^{\pm\mathrm{i}(\varrho-\ell(\pi/2))}/\varrho$. Both classes of solutions fulfill the relation

$$\int_0^\infty r^2\mathrm{d}r\; f_\ell(\kappa' r)f_\ell(\kappa r)=\frac{\pi}{2\kappa\kappa'}\delta(\kappa-\kappa')\,.$$

Let $R(r)$ be an arbitrary smooth function of r and let $\boldsymbol{F}$ be the vector field

$$\boldsymbol{F}(\boldsymbol{r}):=\boldsymbol{\nabla}\left(R(r)Y_{LM}(\theta,\phi)\right)\,.$$

The aim is to expand this vector field in terms of the basis $\boldsymbol{T}_{J\ell M}$. The nabla operator is a tensor operator of rank 1. Thus, expressing it in terms of the spherical basis and denoting its components by ∇_μ, one has (cf. Exercise 7.8)

$$\left[\ell_3,\nabla_\mu\right]=\mu\nabla_3\,,\qquad\left[\ell_\pm,\nabla_\mu\right]=\sqrt{2-\mu(\mu\pm1)}\;\nabla_{\mu\pm1}\,.$$

The same relation is written by means of $3j$-symbols if one uses the decomposition of $\boldsymbol{\ell}$ in the spherical basis. One has

$$\left[\ell_m,\nabla_\mu\right]=(-)^{m-\mu}\sqrt{6}\begin{pmatrix}1 & 1 & 1\\ m+\mu & -m & -\mu\end{pmatrix}\nabla_{\mu+m}\,.$$

It is then clear that the decomposition of $\boldsymbol{F}$ can only contain values of J which obey the triangle rule $\Delta(L,1,J)$ with L and 1. The parity of $\boldsymbol{F}$ is $(-)^{L+1}$. Because of the property (7.112), the expansion can contain only those base functions which have $\ell=L\pm1$. Finally, one finds that J takes the value L only.

The nabla operator can be decomposed into a radial and a tangent part,

$$\boldsymbol{\nabla}=\boldsymbol{x}\,(\boldsymbol{x}\cdot\boldsymbol{\nabla})-\boldsymbol{x}\times(\boldsymbol{x}\times\boldsymbol{\nabla})=\boldsymbol{x}\frac{\partial}{\partial r}-\frac{\mathrm{i}}{r}\,(\boldsymbol{x}\times\boldsymbol{\ell})\,.$$

These formulae and some angular momentum algebra are used to prove the so-called *gradient formula*

$$\begin{aligned}\boldsymbol{\nabla}\left(R(r)Y_{LM}(\theta,\phi)\right)=&-\sqrt{\frac{L+1}{2L+1}}\left(\frac{\mathrm{d}R(r)}{\mathrm{d}r}-\frac{L}{r}R(r)\right)\boldsymbol{T}_{LL+1M}\\&+\sqrt{\frac{L}{2L+1}}\left(\frac{\mathrm{d}R(r)}{\mathrm{d}r}+\frac{L+1}{r}R(r)\right)\boldsymbol{T}_{LL-1M}\,.\end{aligned}\tag{7.114}$$

The two terms which contain the derivative $R'(r)$ come from the radial part, the terms in $1/r$ come from the tangent part. The reason why they are grouped as done in (7.114) is that the spherical Bessel and Hankel functions fulfill the relations

$$\left(\frac{\mathrm{d}}{\mathrm{d}\varrho}-\frac{\ell}{\varrho}\right)R_\ell(\varrho)=R_{\ell+1}(\varrho)\,,\qquad\left(\frac{\mathrm{d}}{\mathrm{d}\varrho}+\frac{\ell+1}{\varrho}\right)R_\ell(\varrho)=R_{\ell-1}(\varrho)\,.$$

This renders the gradient formula quite transparent when applied to these special functions.

Two solutions of the Helmholtz equation with definite angular momentum and which are regular at $r = 0$, are

(i) the *electric multipole fields*

$$\boldsymbol{A}^{(E)}_{LM} = -\sqrt{\frac{L}{2L+1}} j_{L+1}(\kappa r)\boldsymbol{T}_{LL+1M} + \sqrt{\frac{L+1}{2L+1}} j_{L-1}(\kappa r)\boldsymbol{T}_{LL-1M} , \qquad (7.115)$$

(ii) the *magnetic multipole fields*

$$\boldsymbol{A}^{(M)}_{LM} = j_L(\kappa r)\boldsymbol{T}_{LLM} . \qquad (7.116)$$

One confirms that they have a number of properties of particular importance for applications:

1. Both types of fields satisfy the transverse gauge condition (7.93), i.e.
$$\nabla \cdot \boldsymbol{A}^{(E/M)}_{LM} = 0 ,$$
and, hence, are admissible solutions. Using these solutions in the analogue of the expansion (7.97) of the field operator $\boldsymbol{A}$ in terms of creation and annihilation operators, one obtains creation (annihilation) operators for photons in states with definite angular momentum and definite parity;
2. For $L = 0$ both solutions are identically zero. There are no transverse multipole fields with total angular momentum zero;
3. Applying space reflection, the field $\boldsymbol{A}^{(E)}_{LM}$ takes the sign $(-)^{L+1}$, while $\boldsymbol{A}^{(M)}_{LM}$ takes the sign $(-)^L$.

Applying these properties to the interaction $\boldsymbol{j} \cdot \boldsymbol{A}$, and noting that the current density $\boldsymbol{j}$ is odd under space reflection, one obtains selection rules for transitions between states with given angular momenta and parities. These selection rules are:

(a) The angular momenta J_i, J_f, and L of initial and final states, respectively, and of the multipole photon satisfy the triangle relation $\Delta(J_i, L, J_f)$. There are no monopole transitions $L = 0$;
(b) A photon of type $(L, (E))$ changes the parity by the factor $(-)^L$;
(c) A photon of type $(L, (M))$ changes the parity by the factor $(-)^{L+1}$.

A few examples for the interaction of the radiation field with hydrogen, or hydrogen-like atoms, may help to illustrate these results. The transition (2p $\rightarrow$ 1s) is a pure E1-transition, i.e. an electric dipole transition, because the selection rules fix $L = 1$ and parity $(-)$. In the nonrelativistic description no isolated photon can be emitted in the transition from 2s to 1s because there is no multipole field[10] with $L = 0$. A transition (3d $\rightarrow$ 1s) necessarily is an E2-transition. However, we meet here a peculiarity of atomic physics which is easy to understand and that I wish to explain briefly. The light emitted by ordinary atoms has wave lengths which are large as compared to typical dimensions of the atom, $\lambda \gg a_B$. This, in turns, means that

[10] The transition (2s $\rightarrow$ 1s) is possible through emission of *two* photons, or through a small relativistic term, unless it is induced by collisions with other atoms.

the argument κr of the spherical Bessel functions, over atomic distances, remains *small* as compared to 1, $(\kappa r) \ll 1$. The transition probabilities depend on the matrix elements of the fields (7.115) or (7.116) between initial and final states. The radial functions $R_{n\ell}$ and $R_{n'\ell'}$ in the radial part $\langle n\ell | j_\ell(\kappa r) | n'\ell' \rangle$ of the matrix elements, for increasing r, decrease to very small values while (κr) is still small. Therefore, the Bessel function can be approximated by its behaviour for small values of the argument,

$$j_\ell(\kappa r) \sim \frac{(\kappa r)^\ell}{(2\ell+1)!!} .$$

This shows that the radial matrix elements suppress higher values of ℓ in favour of smaller ones. Thus, an E2-transition such as the one between 3d and 1s in hydrogen is strongly suppressed. It is more favourable to make the detour via the 2p-state

$$3\mathrm{d} \longrightarrow 2\mathrm{p} \longrightarrow 1\mathrm{s} ,$$

through two E1-transitions.

In a muonic atom the relative length scales are less pronounced, as one easily confirms, so that E2-transitions such as (3d $\to$ 1s) become perfectly observable.

Likewise, in nuclear physics length scales are not so clearly ordered. Typical transitions have energies in the range of MeV. If one calculates κ from this and estimates the product κR, with R the radius of the nucleus, one will find cases where (κr) is comparable with 1. Indeed, nuclear spectroscopy reveals higher multipoles with sizeable intensities.

7.4 Interaction of the Quantum Maxwell Field with Matter

This section deals with the interaction of the radiation field with electrons in matter in a framework still akin to nonrelativistic perturbation theory. That is to say, while the creation and annihilation of photons is described by means of the field operator (7.97), the quantum states of the electron still are solutions of the Schrödinger equation, transition probabilities and cross sections are calculated like in nonrelativistic quantum theory. This somewhat old-fashioned treatment illustrates the physical interpretation in a particularly simple way. In Chap. 10 it will be replaced by the modern covariant perturbation theory.

The simplest case is the interaction of the radiation field with a system of N nonrelativistic electrons whose Hamiltonian has the form

$$H_0 = \sum_{i=1}^{N} \frac{\boldsymbol{p}^{(i)\,2}}{2m} + \sum_{i<j=1}^{N} U(i,j) .$$

The coupling to the photon field is derived from the rule (7.79) with $q = -e$ the (negative) elementary charge. Thus, the operator $\boldsymbol{p} = -\mathrm{i}\nabla$ is to be replaced by $\boldsymbol{p} - q\boldsymbol{A}$. The Hamiltonian then becomes

$$H = H_0 + \sum_{i=1}^{N} \left\{ -\frac{q}{2m} \left(\boldsymbol{p}^{(i)} \cdot \boldsymbol{A}(t, \boldsymbol{x}^{(i)}) + \boldsymbol{A}(t, \boldsymbol{x}^{(i)}) \cdot \boldsymbol{p}^{(i)} \right) + \frac{q^2}{2m} \boldsymbol{A}^2(t, \boldsymbol{x}^{(i)}) \right\}$$
$$\equiv H_0 + H_{\text{int}} \,. \tag{7.117}$$

Adding the interaction of the magnetic field with the magnetic moment of the electron, H is supplemented by the term

$$H_{\text{int}}^{\text{spin}} = -\sum_{i=1}^{N} \frac{q}{2m} \boldsymbol{\sigma}^{(i)} \cdot \left(\nabla \times \boldsymbol{A}(t, \boldsymbol{x}^{(i)}) \right) . \tag{7.118}$$

Upon insertion of the expansion (7.97) of the field operator $\boldsymbol{A}$ in these expressions, one sees that, to first order in $\mathcal{O}(q)$, the terms linear in $\boldsymbol{A}$ create or annihilate a single photon, while to second order in $\mathcal{O}(q)$, the quadratic term in $\boldsymbol{A}$ scatters a photon from an incoming to an outgoing state or else, creates or annihilates two photons.

7.4.1 Many-Photon States and Matrix Elements

The calculation of matrix elements of the field operator (7.97) needs a little more care than the simple example of Sect. 7.1.5 because multi-photon states, instead of a standard norm, have a distribution-valued normalization. We study a few cases which are needed for the calculations to be done below.

The normalization of a state containing one single photon with momentum $\boldsymbol{q}$ and polarization λ, follows from the commutator (7.102),

$$c_\lambda^\dagger(\boldsymbol{q}) \,|0\rangle \equiv \left| (\boldsymbol{q}, \lambda)^1 \right\rangle : \quad \left\langle (\boldsymbol{q}', \lambda')^1 \middle| (\boldsymbol{q}, \lambda)^1 \right\rangle = 2\omega_q \delta_{\lambda'\lambda} \delta(\boldsymbol{q}' - \boldsymbol{q}) \,.$$

Integrating and summing over the attributes of the left-hand state this implies

$$\sum_{\lambda'} \int \frac{\mathrm{d}^3 q'}{2\omega_q'} \left\langle (\boldsymbol{q}', \lambda')^1 \middle| (\boldsymbol{q}, \lambda)^1 \right\rangle = 1 \,.$$

A state containing two photons $c_\lambda^\dagger(\boldsymbol{q}) c_\mu^\dagger(\boldsymbol{p})|0\rangle$ which carry different momenta, has the squared norm

$$\langle 0| c_{\mu'}(\boldsymbol{p}') c_{\lambda'}(\boldsymbol{q}') c_\lambda^\dagger(\boldsymbol{q}) c_\mu^\dagger(\boldsymbol{p}) \,|0\rangle = 2\omega_q 2\omega_p \left\{ \delta_{\lambda'\lambda} \delta(\boldsymbol{q}' - \boldsymbol{q}) \delta_{\mu'\mu} \delta(\boldsymbol{p}' - \boldsymbol{p}) \right.$$
$$\left. + \, \delta_{\lambda'\mu} \delta(\boldsymbol{q}' - \boldsymbol{p}) \delta_{\mu'\lambda} \delta(\boldsymbol{p}' - \boldsymbol{q}) \right\} .$$

This is easily shown by applying the commutation rules (7.102) and (7.103) several times in such a way that all annihilation operators are shifted to the right, while all creation operators are moved to the left.[11] If the two momenta are equal, $\boldsymbol{p} = \boldsymbol{q}$, and if $\lambda = \mu$, then the right-hand side is equal to

$$2!\,(2\omega_q)^2 \delta_{\lambda'\lambda}\delta_{\mu'\lambda}\delta(\boldsymbol{q}' - \boldsymbol{q})\delta(\boldsymbol{p}' - \boldsymbol{q})\,.$$

In other terms, the state

$$\big|(\boldsymbol{q},\lambda)^2\big\rangle = \frac{1}{\sqrt{2!}}\left(c_\lambda^\dagger(\boldsymbol{q})\right)^2 |0\rangle$$

is correctly normalized provided some care is taken. One cannot calculate the squared norm of this state because the square of a δ-distribution is ill-defined. However, one can evaluate the scalar product

$$\frac{1}{\sqrt{2!}}\int \frac{\mathrm{d}^3 q_1}{2\omega_{q_1}} \int \frac{\mathrm{d}^3 q_2}{2\omega_{q_2}} \langle 0|c_\lambda(\boldsymbol{q}_1)c_\lambda(\boldsymbol{q}_2)\,\big|(\boldsymbol{q},\lambda)^2\big\rangle = 1\,.$$

A state of n photons with momentum $\boldsymbol{q}$ and polarization λ is given by

$$\big|(\boldsymbol{q},\lambda)^n\big\rangle = \frac{1}{\sqrt{n!}}\left(c_\lambda^\dagger(\boldsymbol{q})\right)^n |0\rangle\,. \tag{7.119}$$

It is normalized in the same sense as above, i.e. one has

$$\frac{1}{\sqrt{n!}}\int \frac{\mathrm{d}^3 q_1}{2\omega_{q_1}} \cdots \int \frac{\mathrm{d}^3 q_n}{2\omega_{q_n}} \langle 0|c_\lambda(\boldsymbol{q}_1)\cdots c_\lambda(\boldsymbol{q}_n)\,\big|(\boldsymbol{q},\lambda)^n\big\rangle = 1\,.$$

This multiple integral is precisely the one needed in the calculation of transition matrix elements. Indeed, recall some formulae of nonrelativistic perturbation theory: A given operator $\mathcal{O}$ acts on the normalized state (7.119). The question then is to which, possibly different, final states it can lead. In order to answer this question we calculate

$$\sum_{\lambda_1}\cdots\sum_{\lambda_n}\cdots\int \frac{\mathrm{d}^3 q_1}{2\omega_{q_1}} \cdots \int \frac{\mathrm{d}^3 q_n}{2\omega_{q_n}} \cdots \langle 0|c_{\lambda_1}(\boldsymbol{q}_1)\cdots c_{\lambda_n}(\boldsymbol{q}_n)\cdots\ \mathcal{O}\,\big|(\boldsymbol{q},\lambda)^n\big\rangle$$

and express the final state(s) by means of states normalized as in (7.119). So, for instance,

$$c_\lambda^\dagger(\boldsymbol{q})\,\big|(\boldsymbol{q},\lambda)^n\big\rangle = \sqrt{n+1}\,\big|(\boldsymbol{q},\lambda)^{n+1}\big\rangle,\quad c_\lambda(\boldsymbol{q})\,\big|(\boldsymbol{q},\lambda)^n\big\rangle = \sqrt{n}\,\big|(\boldsymbol{q},\lambda)^{n-1}\big\rangle.$$

These are the formulae (7.43), the corresponding transition amplitudes being

$$\big\langle(\boldsymbol{q},\lambda)^{n+1}\big|c_\lambda^\dagger(\boldsymbol{q})\,\big|(\boldsymbol{q},\lambda)^n\big\rangle = \sqrt{n+1},\ \big\langle(\boldsymbol{q},\lambda)^{n-1}\big|c_\lambda(\boldsymbol{q})\,\big|(\boldsymbol{q},\lambda)^n\big\rangle = \sqrt{n}\,. \tag{7.120}$$

These are the relevant formulae for emission and absorption of single photons, the topic to which we now turn.

7.4.2 Absorption and Emission of Single Photons

As a first example consider the absorption of a photon $\gamma(\boldsymbol{q},\lambda)$ on an initial state $|i\rangle$ which may be thought to be a bound atomic state, under the influence of the interaction (7.117). For simplicity, we assume that before the absorption only photons of the

[11] This is a detailed calculation which becomes more systematic by Wick's theorem, cf. Theorem 5.1.

same kind are present, if at all. We calculate the amplitude for the transition into a final state $|f\rangle$ which may be another bound state or a state in the continuum,

$$|A\rangle \longrightarrow |B\rangle\,, \qquad |A\rangle \equiv \left|i;(\boldsymbol{q},\lambda)^n\right\rangle\,,\ |B\rangle \equiv \left|f;(\boldsymbol{q},\lambda)^{n-1}\right\rangle.$$

At order $\mathcal{O}(q)$ and with $\boldsymbol{p}^{(j)} = -\mathrm{i}\boldsymbol{\nabla}_{(j)}$ one has

$$\langle B|H_{\text{int}}|A\rangle = -\frac{q}{2m}\frac{1}{(2\pi)^{3/2}}$$
$$\sum_\mu \int \frac{\mathrm{d}^3k}{2\omega_k} \sum_{j=1}^{N} \langle B|\boldsymbol{\varepsilon}_\mu(\boldsymbol{k}) \left(2\boldsymbol{p}^{(j)} - \boldsymbol{k}\right) c_\mu(k)\mathrm{e}^{-\mathrm{i}kx}\,|A\rangle\,.$$

The second term on the right-hand side does not contribute because of $\boldsymbol{\varepsilon}_\mu \cdot \boldsymbol{k} = 0$. The first term contains the second of the matrix elements (7.120) so that, upon insertion, one obtains

$$\langle B|H_{\text{int}}|A\rangle = -\frac{q}{m}\frac{1}{(2\pi)^{3/2}}\sqrt{n}\sum_{j=1}^{N}\langle f|\mathrm{e}^{\mathrm{i}\boldsymbol{q}\cdot\boldsymbol{x}^{(j)}}\boldsymbol{p}^{(j)}\cdot\boldsymbol{\varepsilon}_\lambda\,|i\rangle\,\mathrm{e}^{-\mathrm{i}q^0t}\,, \tag{7.121}$$

with n the number of photons before the absorption. The amplitude for the transition in which a photon $\gamma(\boldsymbol{q},\lambda)$ is emitted,

$$|A\rangle \longrightarrow |C\rangle\,, \qquad |A\rangle \equiv \left|i;(\boldsymbol{q},\lambda)^n\right\rangle,\ |C\rangle \equiv \left|f;(\boldsymbol{q},\lambda)^{n+1}\right\rangle,$$

is calculated in much the same way

$$\langle C|H_{\text{int}}|A\rangle = -\frac{q}{m}\frac{1}{(2\pi)^{3/2}}\sqrt{n+1}\sum_{j=1}^{N}\langle f|\mathrm{e}^{-\mathrm{i}\boldsymbol{q}\cdot\boldsymbol{x}^{(j)}}\boldsymbol{p}^{(j)}\cdot\boldsymbol{\varepsilon}_\lambda\,|i\rangle\,\mathrm{e}^{\mathrm{i}q^0t}\,. \tag{7.122}$$

These results are remarkable as such and warrant the following comment:

Remark

In the early times of quantum theory absorption and emission of photons was treated in the framework of a *semi-classical theory of radiation*. While in the classical, unquantized theory of radiation transition probabilities are proportional to the square of the *amplitude* $|\boldsymbol{A}|^2$, the quantized theory shows them to be proportional to n, the number of photons in the state $(\boldsymbol{q},\lambda)$. Intuitively one expects classical and quantum descriptions to give the same answers for large numbers of photons,[12] $n \gg 1$. The matrix element (7.121) for *absorption* is proportional to $\sqrt{n}$. Thus, if in the semi-classical theory one defines

$$\boldsymbol{A}_{\text{cl}}^{(\text{abs})} = \frac{1}{(2\pi)^{3/2}}\sqrt{n}\,\boldsymbol{\varepsilon}_\lambda \mathrm{e}^{-\mathrm{i}q\cdot x}\,,$$

[12] Presumably, no one would be tempted to treat the radiation emitted by an FM-radio station in the framework of quantum electrodynamics. As one easily estimates, the number of emitted photons per unit of volume is many orders of magnitude and, hence, very large as compared to 1.

then one obtains the correct answer, even for small n. This is no longer true when one deals with a process of *emission*. In this case one would have to define

$$\boldsymbol{A}_{\rm cl}^{(\rm emi)} = \frac{1}{(2\pi)^{3/2}}\sqrt{n+1}\,\boldsymbol{\varepsilon}_\lambda \mathrm{e}^{\mathrm{i}\boldsymbol{q}\cdot\boldsymbol{x}}$$

in order to get the correct quantum theoretic answer. However, this is not the conjugate of the absorption potential. Conversely, the complex conjugate of $\boldsymbol{A}_{\rm cl}^{(\rm abs)}$ would give the right answer only in the limit $n \gg 1$. This difference becomes a marked one for $n = 0$, i.e. in the case where there is no photon at all in the initial state. The classical description would have $\boldsymbol{A}_{\rm cl} = 0$ and there would be no emission of photons. The state $|A\rangle$ could not go over into the state $|C\rangle$ by emission of a photon. In contrast, the quantum theoretic formula (7.122) yields a nonvanishing probability even for $n = 0$, for the state $|A\rangle$ to emit a photon. This is the reason why the early, semi-classical theory needed to distinguish *spontaneous emission* from *induced emission*. The quantum theoretic result (7.122) contains both, no distinction between spontaneous and induced emission is necessary.

Example 7.2

We calculate the probability for E1-transitions between circular orbits (i.e. orbits with $\ell = n-1$) in hydrogen or in hydrogen-like atoms,

$$|n;\ell = n-1\rangle \longrightarrow \big|n' = n-1;\ell' = n'-1 = n-2\big\rangle + \gamma(\boldsymbol{k})$$

in the limit of long wave lengths, $|\boldsymbol{k}|r \ll 1$. The relevant matrix element is given in (7.122), the transition probability is calculated by the Golden Rule (5.77). In energy representation, and with covariant normalization the phase space factor in this formula reads

$$\varrho(\omega_k)\mathrm{d}\omega_k\mathrm{d}\Omega = \frac{\mathrm{d}^3k}{2\omega_k} = \frac{\omega_k^2\mathrm{d}\omega_k}{2\omega_k}\mathrm{d}\Omega\,,$$

the second form of which is expressed in spherical polar coordinates, with $\omega_k = |\boldsymbol{k}|$. As a result one has $\varrho(\omega_k) = \omega_k/2$. Thus, as there is only one electron that can change its state, the probability is

$$\frac{\mathrm{d}W}{\mathrm{d}\Omega} = 2\pi\frac{q^2}{2m^2(2\pi)^3}\frac{1}{2\ell+1}\sum_{m,m'}\left|\langle n';\ell' m'|\mathrm{e}^{-\mathrm{i}\boldsymbol{q}\cdot\boldsymbol{x}}\,\boldsymbol{p}\cdot\boldsymbol{\varepsilon}_\lambda\,|n;\ell m\rangle\right|^2\omega_k\mathrm{d}\Omega\,.$$

The following comment is in order: The energy of the photon now equals the difference of the binding energies of initial and final states,

$$\begin{aligned}\omega_k &= \Delta E = E_n - E_{n-1} = -\frac{1}{2}\left(\frac{1}{n^2} - \frac{1}{(n-1)^2}\right)(Z\alpha)^2 m\\ &= \frac{2n-1}{2n^2(n-1)^2}(Z\alpha)^2 m\,.\end{aligned}$$

As no measurement of the magnetic quantum number is made, one must take the incoherent sum over all m and m'. Furthermore, since all magnetic substates of the initial state are equally probable, one must divide by the statistical factor $(2\ell+1)$.

In the approximation of long wave lengths one has $(\kappa r) \ll 1$. Therefore, the exponential can be replaced by 1, $\exp\{-\mathrm{i}\boldsymbol{k}\cdot\boldsymbol{x}\} \approx 1$. The remaining matrix element of the momentum operator can be reduced to a matrix element of the position operator, provided the potential U commutes with $\boldsymbol{x}$. Using

$$\left[p^2, \boldsymbol{x}\right] = -2\mathrm{i}\,\boldsymbol{p}\,,$$

the operator $\boldsymbol{p}$ is replaced by the commutator of H_0 with $\boldsymbol{x}$ so that

$$\begin{aligned}\langle n';\ell' m'|\boldsymbol{p}\,|n;\ell m\rangle &= \mathrm{i}\,m\langle n';\ell' m'|[H_0,\boldsymbol{x}]\,|n;\ell m\rangle \\ &= \mathrm{i}\,m\,(E_{n'}-E_n)\,\langle n';\ell' m'|\boldsymbol{x}\,|n;\ell m\rangle \\ &= \mathrm{i}\,m\omega_k\,\langle n';\ell' m'|\boldsymbol{x}\,|n;\ell m\rangle\ .\end{aligned}$$

Before calculating the matrix element of $\boldsymbol{x}$ we evaluate the sum over the helicities of the emitted photon (provided these are not measured in the experiment) and do the integral over the polar angles. The matrix element $\langle n';\ell' m'|\boldsymbol{x}|n;\ell m\rangle$ is a real vector on $\mathbb{R}^3$ which is drawn in Fig. 7.9 together with $\hat{\boldsymbol{k}}$ and the two spin vectors $\boldsymbol{\varepsilon}_1$ and $\boldsymbol{\varepsilon}_2$. Define

$$\cos\theta_\lambda := \frac{\langle n';\ell' m'|\boldsymbol{x}|n;\ell m\rangle\cdot\boldsymbol{\varepsilon}_\lambda}{|\langle n';\ell' m'|\boldsymbol{x}|n;\ell m\rangle|}\ .$$

These functions are decomposed in terms of the spherical coordinates of the direction which is defined by the matrix element of $\boldsymbol{x}$,

$$\cos\theta_1 = \sin\theta\cos\phi\,,\qquad \cos\theta_2 = \sin\theta\sin\phi\,.$$

We then have

$$\begin{aligned}\frac{\mathrm{d}W}{\mathrm{d}\Omega} &= \frac{q^2\omega_k^3}{8\pi^2}\frac{1}{2\ell+1}\sum_{m,m'}\sum_{\lambda}|\langle n';\ell' m'|\,\boldsymbol{x}\,|n;\ell m\rangle|^2\cos^2\theta_\lambda \\ &= \frac{q^2\omega_k^3}{8\pi^2}\frac{1}{2\ell+1}\sum_{m,m'}|\langle n';\ell' m'|\,\boldsymbol{x}\,|n;\ell m\rangle|^2\sin^2\theta\,.\end{aligned}$$

Integrating over the solid angle one finds

$$\int \mathrm{d}\Omega\ \sin^2\theta = 2\pi\int \mathrm{d}(\cos\theta)\ (1-\cos^2\theta) = \frac{8\pi}{3}\,.$$

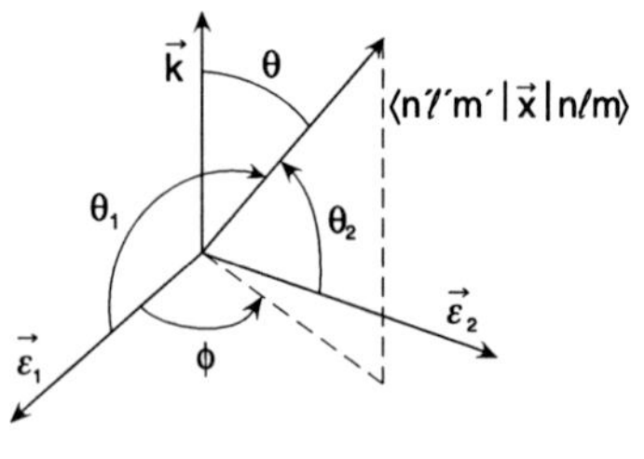

Fig. 7.9 The direction of propagation of the emitted photon, its two admissible polarizations, and the transition matrix element of the position operator, drawn as vectors on space $\mathbb{R}^3$

With $q^2 = e^2 = 4\pi\alpha$, cf. (7.18), the integrated probability is found to be

$$W = \frac{4}{3}\alpha\omega_k^3 \frac{1}{2\ell+1} \sum_{m,m'} |\langle n';\ell' m'| \boldsymbol{x} |n;\ell m\rangle|^2 . \tag{7.123}$$

The remaining matrix element is calculated in a general and transparent way by means of the Wigner-Eckart theorem and of a few formulae of angular momentum algebra. Writing the vector $\boldsymbol{x}$ in the spherical basis,

$$x_\mu = r\sqrt{\frac{4\pi}{3}}\, Y_{1\mu}(\theta,\phi) , \qquad \mu = 1, 0, -1 ,$$

the Wigner-Eckart theorem (6.72) yields the expression

$$\langle n';\ell' m'| r\sqrt{\frac{4\pi}{3}} Y_{1\mu} |n;\ell m\rangle$$
$$= \sqrt{\frac{4\pi}{3}} \langle n';\ell' | r | n;\ell\rangle (-)^{\ell'-m'} \begin{pmatrix} \ell' & 1 & \ell \\ -m' & \mu & m \end{pmatrix} \left(\ell' \| Y_1 \| \ell\right) .$$

In the sums over the absolute square the two 3j-symbols give a factor 1, after summing over m, m', and μ, and using the orthogonality relation (6.66). There remains

$$\frac{1}{2\ell+1} \sum_{m,m'} |\langle n';\ell' m'| \boldsymbol{x} | n;\ell m\rangle|^2 = \frac{1}{2\ell+1}\frac{4\pi}{3} \langle n';\ell' | r | n;\ell\rangle^2 \left(\ell' \| Y_1 \| \ell\right)^2 .$$

The reduced matrix element of Y_1 is given by the formula (6.95). Here it yields the value

$$\left(\ell' = \ell - 1 \| Y_1 \| \ell\right) = -\sqrt{\frac{3\ell}{4\pi}} .$$

At this point we specialize to $n' = n - 1$, $\ell' = \ell - 1$, and obtain

$$W = \frac{4}{3}\alpha\omega_k^3 \frac{\ell}{2\ell+1} \langle n-1;\ell-1| r |n;\ell\rangle^2 , \tag{7.124}$$

or, inserting all factors $\hbar$ and c,

$$W = \frac{4}{3}\alpha c \left(\frac{\Delta E}{\hbar c}\right)^3 \frac{\ell}{2\ell+1} \langle n-1;\ell-1| r |n;\ell\rangle^2 .$$

The radial matrix element, finally, is calculated using the radial functions of the hydrogen atom (1.155). In order to express everything in terms of the principal quantum number n of the initial state we replace ℓ by $\ell = n - 1$. We then find

$$\langle n-1; n-2| r |n; n-1\rangle = a_B \frac{2^{2n+1} n^{n+1} (n-1)^{n+2}}{(2n-1)^{2n}\sqrt{2(2n-1)(n-1)}} .$$

Collecting all partial results the final result is found to be

$$\Gamma \equiv \hbar W(E1; n \to n-1) = \frac{2^{4n} n^{2n-4} (n-1)^{2n-2}}{3(2n-1)^{4n-1}} \alpha^5 mc^2 Z^4 . \tag{7.125a}$$

Alternatively, with reference to the electron mass this may be written as follows

$$W(E1; n \to n-1) = \frac{\alpha^5 m_e c^2}{3\hbar} \frac{2^{4n} n^{2n-4}(n-1)^{2n-2}}{(2n-1)^{4n-1}} \frac{m}{m_e} Z^4$$
$$= 5.355 \times 10^9\,\text{s}^{-1} \frac{2^{4n} n^{2n-4}(n-1)^{2n-2}}{(2n-1)^{4n-1}} \frac{m}{m_e} Z^4 . \quad (7.125\text{b})$$

This result warrants a more detailed analysis and some comments:

Remarks

1. The transition probability per unit of time, when multiplied by $\hbar$, is the width of the (unstable) state (n, ℓ). Together with the analogous quantity of the final state, this defines the measurable *line width*. The line width is proportional to the mass m. To very good approximation this mass equals the mass of the electron. If the electron is replaced by some other, heavier charged particle such as the muon μ^- ($m_\mu/m_e = 206.77$), or the antiproton $\overline{p}$ ($m_{\overline{p}}/m_e = 1836$), the width increases essentially linearly with the mass. The time that the particle needs for the transition, correspondingly, decreases with $1/m$. Comparing heavy atoms with light atoms, i.e. large and small values of the nuclear charge number Z, one sees that transition times in heavy atoms are shorter by the factor $(Z_{\text{light}}/Z_{\text{heavy}})^4$.
2. In order to get a feeling for orders of magnitude, one should calculate the average transition time from the formula

$$\tau(E1; n \to n-1) = \frac{1}{W(E1; n \to n-1)}$$
$$\approx 1.867 \times 10^{-10} \frac{(2n-1)^{4n-1}}{2^{4n} n^{2n-4}(n-1)^{2n-2}} \frac{m_e}{m} Z^{-4}\,\text{s} .$$

   For (2p $\to$ 1s) and $m = m_e$ one finds $\tau(E1; 2\text{p} \to 1\text{s}) \approx 1.59 \times 10^{-9}/Z^4$ s, while for a muon this time is about two hundred times shorter, $\tau(E1; 2\text{p} \to 1\text{s}) \approx 7.71 \times 10^{-12}/Z^4$ s. These numbers can be used to estimate the total time that a muon captured with a high value of n needs to run through the whole cascade down to the 1s-state.
3. There is a correction to the formula (7.125a) which is negligible for electrons[13] but may become important for heavier particles. Its physical origin is easy to understand. When an atom makes a spontaneous E1-transition by emission of a photon, both charged partners, the electron (with charge $q = -e$) and the nucleus (charge Ze) contribute to the dipole transition. Indeed, both of them

[13] A notable exception where this correction cannot be neglected are precision measurements by means of high-frequency spectroscopy. Such measurements are done in view of detecting radiative corrections as predicted by quantum electrodynamics. They are sensitive to even very small effects.

move relative to the common center of mass. As a consequence of this remark the dipole operator must be replaced as follows

$$-e\boldsymbol{x} \longmapsto -e\left(1+\frac{(Z-1)m}{m+m_A}\right)\boldsymbol{x}\,, \tag{7.126}$$

where m_A is the mass of the nucleus (see Exercise 7.9).

4. Of course, one could have done the same calculation using the multipole fields (7.115) by taking the limit $(\kappa r) \ll 1$ there. Indeed, the result (7.124) is a special case of a general formula for Eλ-transitions between states with angular momenta (J_i, M_i) and (J_f, M_f), respectively, in the approximation of long wave lengths. This formula reads

$$W(E\lambda; n, J_i \to n', J_f) = 8\pi c\alpha \frac{\lambda+1}{\lambda(2\lambda+1)!!}\left(\frac{\Delta E}{\hbar c}\right)^{2\lambda+1} B(\mathrm{E}\lambda)\,, \tag{7.127}$$

with the reduced probability $B(\mathrm{E}\lambda)$ containing the radial matrix element and the matrix element of the spherical harmonics,

$$B(\mathrm{E}\lambda) = \frac{1}{2J_i+1}\sum_{M_i, M_f} \left|\langle J_f M_f \left| r^\lambda Y_{\lambda\mu} \right| J_i M_i\rangle\right|^2 . \tag{7.128}$$

The latter is again analyzed by means of the Wigner-Eckart theorem.

5. There are analogous formulae for magnetic transitions which go back to the interaction terms (7.117) and (7.118). Finally, it is not difficult to extend these results to cases where (κr) is no longer small as compared to 1. The essential difference is that the radial parts now contain the full spherical Bessel functions instead of the approximation $(\kappa r)^\ell/((2\ell+1)!!)$.

7.4.3 Rayleigh- and Thomson Scattering

Consider the scattering of a photon on an electron,

$$\gamma(\boldsymbol{k}, \lambda) + e(i) \longrightarrow \gamma(\boldsymbol{k}', \lambda') + e(f)$$

for small values of the momentum transfer $|\boldsymbol{k}-\boldsymbol{k}'|$ under the action of the interaction (7.117). For simplicity, the spin interaction (7.118) is neglected. This process is not only a nice application of the theory of the quantized radiation field in interaction with matter, it also plays a fundamental physical role:

(a) The special case of elastic scattering $\omega = \omega'$, $|f\rangle = |i\rangle$, called *Rayleigh scattering,* provides the basis for explaining the blue color of the sky during the day and its reddening in the evening.

(b) Furthermore, in the limit of very small momentum transfer one obtains the (classical) scattering cross section for *Thomson scattering,* which is proportional to e^4, where $q = (-e)$ is the *physical* charge of the electron. This is important because quantum electrodynamics distinguishes the *bare charge* of the electron from its *observable charge*. The starting parameter q_0 that appears in the

Lagrange density is modified by radiative corrections so that, at each order of perturbation theory, the physical, observable charge q must be identified anew. One says that the charge is *renormalized* from its bare value. Quantum electrodynamics has the special feature, as compared to other gauge theories, that it possesses a classical limit. Quite generally, renormalization refers to some energy scale μ which is a physical scale at which the charge $q(\mu)$ is measured. The limit of Thomson scattering is important because it tells us that the scale $\mu = 0$ is a natural one for quantum electrodynamics and because it yields the physical charge at this scale.

Encouraged by this motivation we apply – for the last time! – nonrelativistic perturbation theory.

The initial and final states of the electron are given the summary notation $|i\rangle$ and $|f\rangle$, respectively, its virtual intermediate states are written $|n\rangle$. The energy of the photon before and after the scattering is denoted by, respectively,

$$\omega = |\boldsymbol{k}|\,, \quad \text{and} \quad \omega' = |\boldsymbol{k}'|\,.$$

In *first* order of perturbation theory only the second term on the right-hand side of (7.117) can contribute because this is the only term that contains two field operators $\boldsymbol{A}$. One of these annihilates the incoming photon, the other creates the outgoing photon. In turn, the terms $\boldsymbol{p} \cdot \boldsymbol{A}$ and $\boldsymbol{A} \cdot \boldsymbol{p}$ are *linear* in $\boldsymbol{A}$ and, therefore, contribute only in second order. However, checking their factors one realizes that the two contributions obtained in first and second order perturbation theory, are proportional to $q^2 \equiv e^2$. Thus, expanding in terms of the physically relevant parameter $\alpha = e^2/(4\pi)$ both contribute at the same level: the amplitude is of order α, the cross section, hence, of order α^2. The two contributions are calculated as follows.

1. The term in $\boldsymbol{A} \cdot \boldsymbol{A}$:
   This is the "seagull" term that we met previously in Example 7.1 and in Fig. 7.7 and which we repeat in Fig. 7.10a. Keeping only terms which contain one creation

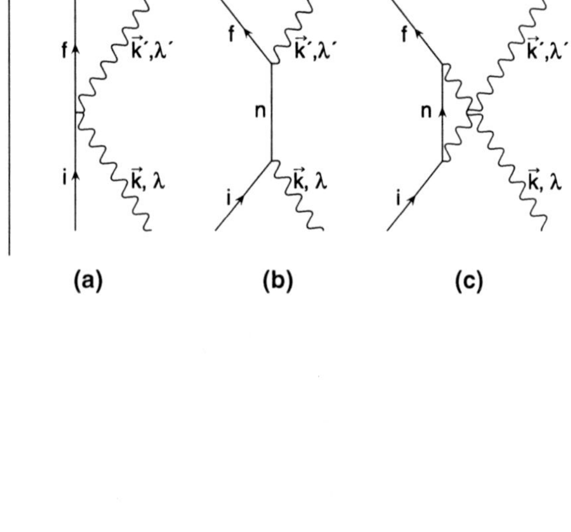

Fig. 7.10 **a** Time flows from bottom to top; a photon is absorbed, another one is emitted at the same point; **b** and **c** the contributions of second order

and one annihilation operator, and making use of (7.97), one has

$$
\begin{aligned}
\langle f; \boldsymbol{k}'\lambda' | \frac{e^2}{2m} \boldsymbol{A}\cdot\boldsymbol{A} \,|\, i; \boldsymbol{k}\lambda\rangle &\equiv c^{(1)}(t) \\
&= \frac{e^2}{2m}\frac{1}{(2\pi)^3}\int \frac{\mathrm{d}^3 q}{2\omega_q}\int \frac{\mathrm{d}^3 p}{2\omega_p}\sum_{\alpha,\beta} \boldsymbol{\varepsilon}_\alpha(\boldsymbol{q})\cdot\boldsymbol{\varepsilon}_\beta(\boldsymbol{p}) \\
&\quad\times \langle f; \boldsymbol{k}'\lambda' | c^\dagger_\alpha(\boldsymbol{q}) c_\beta(\boldsymbol{p}) \mathrm{e}^{\mathrm{i}(\omega_q-\omega_p)t}\mathrm{e}^{-\mathrm{i}(\boldsymbol{q}-\boldsymbol{p})\cdot\boldsymbol{x}} \\
&\quad + c_\alpha(\boldsymbol{q}) c^\dagger_\beta(\boldsymbol{p}) \mathrm{e}^{-\mathrm{i}(\omega_q-\omega_p)t}\mathrm{e}^{\mathrm{i}(\boldsymbol{q}-\boldsymbol{p})\cdot\boldsymbol{x}} \,|\, i; \boldsymbol{k}\lambda\rangle \\
&= \frac{e^2}{2m}\frac{1}{(2\pi)^3}\int \frac{\mathrm{d}^3 q}{2\omega_q}\int \frac{\mathrm{d}^3 p}{2\omega_p}\sum_{\alpha,\beta} \\
&\quad\times \left\{\delta(\boldsymbol{q}-\boldsymbol{k}')\delta_{\alpha\lambda'}\delta(\boldsymbol{p}-\boldsymbol{k})\delta_{\beta\lambda} + \delta(\boldsymbol{p}-\boldsymbol{k}')\delta_{\beta\lambda'}\delta(\boldsymbol{q}-\boldsymbol{k})\delta_{\alpha\lambda}\right\} \\
&\quad\times (2\omega_{k'})(2\omega_k)\mathrm{e}^{\mathrm{i}(\omega'-\omega)t}\langle f|\mathrm{e}^{-\mathrm{i}(\boldsymbol{k}'-\boldsymbol{k})\cdot\boldsymbol{x}}\,|\, i\rangle\, \boldsymbol{\varepsilon}_{\lambda'}(\boldsymbol{k}')\cdot\boldsymbol{\varepsilon}_\lambda(\boldsymbol{k})\,.
\end{aligned}
$$

The momentum transfer $(\boldsymbol{k}-\boldsymbol{k}')$ is chosen to be small so that the exponential function can be approximated, $\mathrm{e}^{-\mathrm{i}(\boldsymbol{k}'-\boldsymbol{k})\cdot\boldsymbol{x}} \approx 1$, thus simplifying this lengthy formula enormously. As $\langle f\,|\,i\rangle = \delta_{fi}$ one has

$$
\begin{aligned}
c^{(1)}(t) &= \left\langle f; \boldsymbol{k}'\lambda' \left| \frac{e^2}{2m}\boldsymbol{A}\cdot\boldsymbol{A}\right| i; \boldsymbol{k}\lambda\right\rangle \\
&= \frac{e^2}{m}\frac{1}{(2\pi)^3}\boldsymbol{\varepsilon}_{\lambda'}(\boldsymbol{k}')\cdot\boldsymbol{\varepsilon}_\lambda(\boldsymbol{k})\mathrm{e}^{\mathrm{i}(\omega'-\omega)t}\,\delta_{fi}\,.
\end{aligned}
$$

This approximation is realistic because the electron is bound in an atomic state. Assuming the momentum transfer to be small means that the wave length is large as compared to atomic dimensions.

2. The terms linear in $\boldsymbol{A}$:
   On the basis of the general formula (5.72) and in the same approximation of long wave lengths $\mathrm{e}^{-\mathrm{i}(\boldsymbol{k}'-\boldsymbol{k})\cdot\boldsymbol{x}} \approx 1$ as before, these terms are seen to contribute as follows

$$
\begin{aligned}
c^{(2)}(t) &= (-\mathrm{i})^2\left(\frac{e}{m}\right)^2\frac{1}{(2\pi)^3}\int_0^t \mathrm{d}t_2 \int_0^{t_2}\mathrm{d}t_1 \;\sum\!\!\!\!\!\!\int \\
&\quad\times\Big\{\langle f|\boldsymbol{p}\cdot\boldsymbol{\varepsilon}_{\lambda'}\,|n\rangle\,\mathrm{e}^{\mathrm{i}(E_f-E_n+\omega')t_2}\langle n|\boldsymbol{p}\cdot\boldsymbol{\varepsilon}_\lambda\,|i\rangle\,\mathrm{e}^{\mathrm{i}(E_n-E_i-\omega)t_1} \\
&\quad + \langle f|\boldsymbol{p}\cdot\boldsymbol{\varepsilon}_\lambda\,|n\rangle\,\mathrm{e}^{\mathrm{i}(E_f-E_n-\omega)t_2}\langle n|\boldsymbol{p}\cdot\boldsymbol{\varepsilon}_{\lambda'}\,|i\rangle\,\mathrm{e}^{\mathrm{i}(E_n-E_i+\omega')t_1}\Big\} \\
&= \mathrm{i}\left(\frac{e}{m}\right)^2\frac{1}{(2\pi)^3}\int_0^t \mathrm{d}t_2\,\mathrm{e}^{\mathrm{i}(E_f-E_i+\omega'-\omega)t_2} \\
&\quad\times \sum\!\!\!\!\!\!\int\left\{\frac{\langle f|\boldsymbol{p}\cdot\boldsymbol{\varepsilon}_{\lambda'}\,|n\rangle\,\langle n|\boldsymbol{p}\cdot\boldsymbol{\varepsilon}_\lambda\,|i\rangle}{E_n-E_i-\omega} + \frac{\langle f|\boldsymbol{p}\cdot\boldsymbol{\varepsilon}_\lambda\,|n\rangle\,\langle n|\boldsymbol{p}\cdot\boldsymbol{\varepsilon}_{\lambda'}\,|i\rangle}{E_n-E_i+\omega'}\right\}.
\end{aligned}
$$

The "sum/integral" symbol stands for the sum over discrete intermediate states, and the integral over states in the continuum, both cases being denoted symbolically by $|n\rangle$.

Both terms, $c^{(1)}(t)$ and $c^{(2)}(t)$, are proportional to $e^{i\widetilde{\omega}t}$ with $\widetilde{\omega} = E_f + \omega' - E_i - \omega$, with $E_f = E_i$ in the first case. The calculation of the transition probability per unit of time requires to evaluate the distribution

$$\frac{1}{t}\left|\int_0^t dt'\, e^{i\widetilde{\omega}t}\right|^2 \equiv I(t, \widetilde{\omega})$$

for t going to infinity. As shown in Sect. 5.2 this yields the distribution

$$2\pi\, \delta(\widetilde{\omega})\,.$$

Upon integration over the energy ω' of the outgoing photon this distribution guarantees the principle of energy conservation, $\widetilde{\omega} = 0$. Therefore, to order e^4, the transition probability is

$$\frac{dW}{d\Omega} = 2\pi \int d\omega' \left|c^{(1)} + c^{(2)}\right|^2 \delta(E_f + \omega' - E_i - \omega)\varrho(\omega')\,,$$

where, as before, $\varrho(\omega') = \omega'/2$ is the density of the photonic final states. The differential cross section is obtained from this by dividing by the incoming flux of photons, the flux factor being

$$\frac{2\omega}{(2\pi)^3}\,,$$

as a consequence of covariant normalization.[14] The cross section reads

$$\frac{d\sigma}{d\Omega} = \left(\frac{e^2}{4\pi m}\right)^2 \left(\frac{\omega'}{\omega}\right) \Bigg| \delta_{fi}\, \boldsymbol{\varepsilon}_{\lambda'}(\boldsymbol{k}') \cdot \boldsymbol{\varepsilon}_{\lambda}(\boldsymbol{k}) \\ - \frac{1}{m} \sum\!\!\!\!\!\!\int \left(\frac{\langle f|\boldsymbol{p}\cdot\boldsymbol{\varepsilon}_{\lambda'}|n\rangle \langle n|\boldsymbol{p}\cdot\boldsymbol{\varepsilon}_{\lambda}|i\rangle}{E_n - E_i - \omega} + \frac{\langle f|\boldsymbol{p}\cdot\boldsymbol{\varepsilon}_{\lambda}|n\rangle \langle n|\boldsymbol{p}\cdot\boldsymbol{\varepsilon}_{\lambda'}|i\rangle}{E_n - E_i + \omega'} \right) \Bigg|^2 . \tag{7.129}$$

This formula is due to Kramers and Heisenberg.

The diagrams of Fig. 7.10b, c allow for an interpretation of the last two terms on the right-hand side of (7.129): The denominator of the first of them is the difference of the energy of the intermediate state (electron in the state $|n\rangle$, no photon) and the energy of the initial state (electron in state $|i\rangle$, incoming photon), $E_n - (E_i + \omega)$. In the second

[14] This flux factor would be 1 if the wave function were simply $e^{i\boldsymbol{k}\cdot\boldsymbol{x}}$. If the covariantly normalized plane waves (7.19) are employed which have the property (7.20), then it is $2E_k/(2\pi)^3$.

term the intermediate state is composed of the electron in $|n\rangle$, and of the incoming and outgoing photons. The energy difference (intermediate state – initial state) is equal to

$$(E_n + \omega + \omega') - (E_i + \omega) = E_n - E_i + \omega' ,$$

in agreement with the formulae of perturbation theory in second order.

The following remarks serve to analyze this important formula and to illustrate its application in specific physical situations.

Remarks

1. The first of the two factors has physical dimension (length)^2. Adding the correct factors $\hbar$ and c, and taking account of the fact that e^2 is expressed in natural units, one has
$$r_0 := \frac{e^2}{4\pi m} \mathrel{\widehat{=}} \alpha \frac{\hbar}{mc} = 2.82\,\text{fm} = 2.82 \times 10^{-15}\,\text{m} . \tag{7.130}$$
In the earlier literature this length is often called the *classical electron radius*.
2. Rayleigh scattering, by definition, is *elastic* scattering for which $|f\rangle = |i\rangle$, $E_f = E_i$, and $\omega' = \omega$. In order to tailor the Kramers-Heisenberg formula (7.129) such that it fits this case, we make use of the completeness relation
$$\mathbb{1} = \sum\!\!\!\!\!\!\int |n\rangle \langle n|$$
in order to transform $|\ldots|^2$ such that it resemble the second term. Furthermore, we use the commutators
$$\left[x_i, p_j\right] = \mathrm{i}\, \delta_{ij} , \quad \left[\boldsymbol{p}^2, \boldsymbol{x}\right] = -2\mathrm{i}\, \boldsymbol{p} ,$$
to convert matrix elements of the momentum operator to matrix elements of the position operator and vice versa, cf. Example 7.2 above,
$$\langle b|\boldsymbol{p}\,|a\rangle = \mathrm{i}\; m(E_b - E_a)\langle b|\boldsymbol{x}\,|a\rangle .$$
This yields, in a first step,
$$\begin{aligned}\boldsymbol{\varepsilon}_{\lambda'} \cdot \boldsymbol{\varepsilon}_\lambda &= \frac{1}{\mathrm{i}} \sum\!\!\!\!\!\!\int \{\langle i|\boldsymbol{x}\cdot\boldsymbol{\varepsilon}_{\lambda'}\,|n\rangle\langle n|\boldsymbol{p}\cdot\boldsymbol{\varepsilon}_\lambda\,|i\rangle - \langle i|\boldsymbol{p}\cdot\boldsymbol{\varepsilon}_{\lambda'}\,|n\rangle\langle n|\boldsymbol{x}\cdot\boldsymbol{\varepsilon}_\lambda\,|i\rangle\} \\ &= \frac{2}{m} \sum\!\!\!\!\!\!\int \frac{1}{\omega_{ni}} \langle i|\boldsymbol{p}\cdot\boldsymbol{\varepsilon}_{\lambda'}\,|n\rangle\,\langle n|\boldsymbol{p}\cdot\boldsymbol{\varepsilon}_\lambda\,|i\rangle ,\end{aligned}$$
with $\omega_{ni} := E_n - E_i$. In a second step the two terms within the absolute bars on the right-hand side of (7.129) are combined as follows
$$\begin{aligned}&\delta_{fi}\, \boldsymbol{\varepsilon}_{\lambda'} \cdot \boldsymbol{\varepsilon}_\lambda - \frac{1}{m} \sum\!\!\!\!\!\!\int \left(\frac{\langle f|\boldsymbol{p}\cdot\boldsymbol{\varepsilon}_{\lambda'}\,|n\rangle\,\langle n|\boldsymbol{p}\cdot\boldsymbol{\varepsilon}_\lambda\,|i\rangle}{\omega_{ni} - \omega} + \left\{\omega \leftrightarrow -\omega, \lambda \leftrightarrow \lambda'\right\} \right) \\ &= -\frac{1}{m} \sum\!\!\!\!\!\!\int \left(\frac{\omega\,\langle f|\boldsymbol{p}\cdot\boldsymbol{\varepsilon}_{\lambda'}\,|n\rangle\,\langle n|\boldsymbol{p}\cdot\boldsymbol{\varepsilon}_\lambda\,|i\rangle}{\omega_{ni}\,(\omega_{ni} - \omega)} + \left\{\omega \leftrightarrow -\omega, \lambda \leftrightarrow \lambda'\right\} \right) .\end{aligned}$$

In the approximation assumed above one has $\omega \ll \omega_{ni}$ and, therefore,

$$\frac{\omega}{\omega_{ni}(\omega_{ni} \mp \omega)} \approx \frac{\omega}{\omega_{ni}^2}\left(1 \pm \frac{\omega}{\omega_{ni}}\right) .$$

The 1 in parentheses does not contribute, due to the completeness relation. Indeed, one has

$$\begin{aligned}
&\sum\!\!\!\!\!\!\int \frac{1}{\omega_{ni}^2}\left(\langle i|\boldsymbol{p}\cdot\boldsymbol{\varepsilon}_{\lambda'}|n\rangle\langle n|\boldsymbol{p}\cdot\boldsymbol{\varepsilon}_{\lambda}|i\rangle - \langle i|\boldsymbol{p}\cdot\boldsymbol{\varepsilon}_{\lambda}|n\rangle\langle n|\boldsymbol{p}\cdot\boldsymbol{\varepsilon}_{\lambda'}|i\rangle\right)\\
&= m^2 \sum\!\!\!\!\!\!\int \left(\langle i|\boldsymbol{x}\cdot\boldsymbol{\varepsilon}_{\lambda'}|n\rangle\langle n|\boldsymbol{x}\cdot\boldsymbol{\varepsilon}_{\lambda}|i\rangle - \langle i|\boldsymbol{x}\cdot\boldsymbol{\varepsilon}_{\lambda}|n\rangle\langle n|\boldsymbol{x}\cdot\boldsymbol{\varepsilon}_{\lambda'}|i\rangle\right)\\
&= m^2\langle i|\left[(\boldsymbol{x}\cdot\boldsymbol{\varepsilon}_{\lambda'}), (\boldsymbol{x}\cdot\boldsymbol{\varepsilon}_{\lambda})\right]|i\rangle = 0 .
\end{aligned}$$

The next, nonvanishing term of the expansion yields the final result:

$$\begin{aligned}
&\left.\frac{\mathrm{d}\sigma}{\mathrm{d}\Omega}\right|_{\text{Rayleigh}}\\
&= \left(\frac{r_0}{m}\right)^2 \omega^4 \left|\sum\!\!\!\!\!\!\int \frac{1}{\omega_{ni}^3}\left(\langle i|\boldsymbol{p}\cdot\boldsymbol{\varepsilon}_{\lambda'}|n\rangle\langle n|\boldsymbol{p}\cdot\boldsymbol{\varepsilon}_{\lambda}|i\rangle + \langle i|\boldsymbol{p}\cdot\boldsymbol{\varepsilon}_{\lambda}|n\rangle\langle n|\boldsymbol{p}\cdot\boldsymbol{\varepsilon}_{\lambda'}|i\rangle\right)\right|^2\\
&= (m r_0)^2 \omega^4 \left|\sum\!\!\!\!\!\!\int \frac{1}{\omega_{ni}}\left(\langle i|\boldsymbol{x}\cdot\boldsymbol{\varepsilon}_{\lambda'}|n\rangle\langle n|\boldsymbol{x}\cdot\boldsymbol{\varepsilon}_{\lambda}|i\rangle + \langle i|\boldsymbol{x}\cdot\boldsymbol{\varepsilon}_{\lambda}|n\rangle\langle n|\boldsymbol{x}\cdot\boldsymbol{\varepsilon}_{\lambda'}|i\rangle\right)\right|^2 .
\end{aligned} \tag{7.131}$$

This formula provides the basis for explaining the colours of the sky. The atmosphere contains uncoloured gases, in fact, gases whose typical transition frequencies ω_{ni} lie in the ultraviolet spectrum. When we talk about colours of the sky we have in mind wave lengths λ in the visible range which, as one knows, are much larger than those in the ultraviolet domain. Thus, the frequencies are $\omega \ll \omega_{ni}$, and, hence, the Rayleigh formula is applicable to the atmosphere. The essential result is that the cross section is inversely proportional to the fourth power of the wave length

$$\left.\frac{\mathrm{d}\sigma}{\mathrm{d}\Omega}\right|_{\text{Rayleigh}} \propto \frac{1}{\lambda^4} .$$

Light that has a long wave length is scattered less than light of short wave length. If we look up to the sky at noon, avoiding to look directly into the sun, we see primarily light scattered on molecules of the air which, by Rayleigh's formula, is predominantly of short wave length, i.e. *blue*. At dawn or at dusk, in turn, there is no harm in looking to the sun. Because of its tangent orbit the light of the sun traverses a much thicker layer of air than when the sun is at the zenith. On its way through the air mostly blue tones are scattered off the light's trajectory. Therefore, what we see is predominantly *red* or *reddish*. Recall that Eos, the goddess of the dawn, was characterized by Homer as *rhododactylos*, i.e. indeed *rosy*-fingered.

3. The other limiting case of elastic scattering applies when the wave length of the scattered light is very small as compared to the spatial size of the atom on which it is scattered, $\lambda \ll d$, i.e. if $\omega \gg \omega_{ni}$. In this case one has

$$\omega \gg \frac{\langle n|\boldsymbol{p}\,|\,i\,\rangle^2}{2m}\,,$$

which implies that the second term on the right-hand side of (7.129) is negligible as compared to the first term. The scattering amplitude is given almost exclusively by the "seagull" graph of Fig. 7.10a,

$$f \approx r_0 \delta_{fi} \boldsymbol{\varepsilon}_{\lambda'} \cdot \boldsymbol{\varepsilon}_{\lambda}\,.$$

The scattering cross section that follows from this amplitude describes Thomson scattering

$$\left(\frac{\mathrm{d}\sigma}{\mathrm{d}\Omega}\right)\bigg|_{\text{Thomson}} = r_0^2\, |\boldsymbol{\varepsilon}_{\lambda'} \cdot \boldsymbol{\varepsilon}_{\lambda}|^2\,. \tag{7.132}$$

As we emphasized previously this provides an experimental possibility to determine the physical electric charge, i.e. the fine structure constant $\alpha(0)$ at the reference scale $\mu = 0$.

4. There remains the question of how to describe elastic or inelastic scattering in situations where the light crosses one of the characteristic frequencies of the atom on which it is scattered, i.e. when $\omega \approx \omega_{ni}$. In such cases one talks about *resonance scattering* or *resonance fluorescence.* The formula (7.129) of Kramers and Heisenberg can no longer be correct because the second term on its right-hand side has a pole at every constellation where $\omega = E_n - E_i$. While the ground state $|i\rangle$ may safely be assumed to be stable, the excited state $|n\rangle$ exhibits a certain uncertainty of its energy, due to spontaneous emission, to Doppler effect, and, depending on the density of the scattering gas, to atomic collisions. If Γ_n is the resulting total width of the state $|n\rangle$ this means that the denominator of this term must be replaced as follows

$$E_n - E_i - \omega \longmapsto (E_n - \mathrm{i}\,\Gamma_n/2) - E_i - \omega\,. \tag{7.133}$$

The additional term which is completely negligible in the two limiting cases discussed above, provides a summary description of radiation damping. More about the phenomenon of resonance fluorescence and about other special cases of the Kramers-Heisenberg formula can be found in [Sakurai (1984)].

7.5 Covariant Quantization of the Maxwell Field

In this section we return to the remarks of Sect. 7.3.3 and work out the alternative of a manifestly covariant quantization of electrodynamics. The basic idea of this approach as well as its somewhat surprising properties are not difficult to understand.

As it provides the basis for covariant perturbation theory in quantum electrodynamics and for the technique of Feynman diagrams, one should not skip it. In turn, I marked by an asterisk some more formal developments which contribute to a deeper understanding but are not so important for practical matters. One can skip those in a first reading without risking to get stuck in the subsequent analysis.

7.5.1 Gauge Fixing and Quantization

The vanishing of the momentum canonically conjugate to $A_0(x)$ was the first difficulty into which one ran by naïve application of canonical quantization to the Maxwell field, cf. (7.88). This problem was circumvented in Sect. 7.3 by making use of the gauge freedom of the classical theory in a very special way: On the one hand one imposed the gauge condition (7.93), i.e. $\nabla \cdot \boldsymbol{A} = 0$, which filtered the transverse, physical degrees of freedom. On the other hand, the potential $A_0(x)$ was set to zero in vacuum, i.e. outside all sources, and this pathological degree of freedom disappeared from the theory altogether. The free theory and the interaction with matter then decomposed into one term that contains only transverse quantized fields, and the unquantized, instantaneous Coulomb interaction.

As the Lorenz condition (7.91) must be handled with care already at the classical level, it seems appropriate to impose it only after quantization, as a constraint on the physical states of the theory. One way of effecting this is not to assume from the start that the divergence $\partial_\mu A^\mu(x)$ vanishes but, rather, to introduce it in the Lagrange density like a scalar field that has no kinetic energy. That is to say, one replaces (7.85) by

$$\begin{aligned}\mathcal{L}(A^\alpha, \partial_\mu A^\alpha, j^\mu) := &-\frac{1}{4} F_{\mu\nu}(x) F^{\mu\nu}(x) - j_\mu(x) A^\mu(x) \\ &-\frac{1}{2}\Lambda \left(\partial_\mu A^\mu(x)\right)^2 .\end{aligned} \tag{7.134}$$

The factor Λ is taken to be real and positive, $\Lambda \in \mathbb{R}^+ \setminus \{0\}$. The equations of motion which follow from the Lagrange density (7.134) as its Euler-Lagrange equations, replace (7.92) by

$$\Box A^\mu(x) - (1-\Lambda)\partial^\mu \left(\partial_\nu A^\nu(x)\right) = j^\mu(x) . \tag{7.135}$$

There is an immediate and interesting consequence of this equation. Assuming $A_\mu(x)$ to be at least C^3 and taking account of the continuity equation $\partial_\mu j^\mu(x) = 0$, the divergence of the Eq. (7.135) yields

$$\Box \partial_\mu A^\mu - (1-\Lambda)\Box \left(\partial_\nu A^\nu(x)\right) = \Lambda \Box \left(\partial_\nu A^\nu(x)\right) = 0 . \tag{7.136}$$

Thus, the divergence $\partial_\nu A^\nu(x)$ satisfies the Klein-Gordon equation with mass zero. In spite of the fact that we are dealing with an interacting theory, $(\partial_\nu A^\nu(x))$ is a massless *free* scalar field.

Defining now the momenta π^{μ} which are canonically conjugate to the A_{μ}, none of them is identically zero,

$$A_{\mu}(x) \longleftrightarrow \pi^{\mu}(x) = -F^{0\mu} - \Lambda g^{\mu 0}\left(\partial_{\nu} A^{\nu}(x)\right)$$
$$= F^{\mu 0} - \Lambda g^{\mu 0}\left(\partial_{\nu} A^{\nu}(x)\right) . \tag{7.137}$$

According to the rules of canonical quantization, Postulate 7.1, one has

$$\left[A_{\mu}(x), \pi^{\nu}(y)\right]_{x^0=y^0} = \mathrm{i}\,\delta_{\mu}^{\ \nu}\delta(\boldsymbol{x}-\boldsymbol{y}) ,$$

or, converting the covariant index μ to a contravariant one by means of the metric tensor,

$$\left[A^{\mu}(x), \pi^{\nu}(y)\right]_{x^0=y^0} = \mathrm{i}\, g^{\mu\nu}\delta(\boldsymbol{x}-\boldsymbol{y}) . \tag{7.138}$$

To these one adds the commutators of the fields as well as of the momenta among themselves,

$$\left[A^{\mu}(x), A^{\nu}(y)\right]_{x^0=y^0} = 0 = \left[\pi^{\mu}(x), \pi^{\nu}(y)\right]_{x^0=y^0} . \tag{7.139}$$

The commutation rules for the fields and their time derivatives follow from the postulates (7.138) and (7.139). They read

$$\left[\dot{A}^{\mu}(x), \dot{A}^{k}(y)\right]_{x^0=y^0} = \mathrm{i}\, g^{\mu 0}\frac{\Lambda-1}{\Lambda}\partial_x^k \delta(\boldsymbol{x}-\boldsymbol{y}) ,$$
$$\left[\dot{A}^{\mu}(x), A^{\nu}(y)\right]_{x^0=y^0} = \mathrm{i}\, g^{\mu\nu}\left(1+\frac{1-\Lambda}{\Lambda}g_{\mu 0}\right)\delta(\boldsymbol{x}-\boldsymbol{y}) . \tag{7.140}$$

They are derived as follows: For simplicity, we drop temporarily all subscripts "$x^0 = y^0$". It follows from (7.139) that all spatial derivatives of the fields commute,

$$[\partial^i A^{\mu}, \partial^j A^{\nu}] = 0 .$$

One uses these commutators and the expressions obtained previously for π^0 and π^i, viz.

$$\pi^0 = -\Lambda(\dot{A}^0 - \sum_j \partial^j A^j) ,$$
$$\pi^i = F^{i0} = \partial^i A^0 - \partial^0 A^i \equiv \partial^i A^0 - \dot{A}^i ,$$

in evaluating a first group of commutators,

$$[A^0, \pi^0] = -\Lambda[A^0, \dot{A}^0] = \mathrm{i}\,\delta(\boldsymbol{x}-\boldsymbol{y}) , \quad [A^0, \pi^i] = -[A^0, \dot{A}^i] = 0 ,$$
$$[A^i, \pi^0] = -\Lambda[A^i, \dot{A}^0] = 0 , \quad [A^k, \pi^i] = -[A^k, \dot{A}^i] = \mathrm{i}\, g^{ik}\delta(\boldsymbol{x}-\boldsymbol{y}) .$$

The commutators between $\dot{A}^\mu$ and $\dot{A}^\nu$ are verified in a similar manner. Thus, the commutator $[\pi^0, \pi^i] = 0$ implies

$$\begin{aligned}
0 &= [\dot{A}^0 - \sum_j \partial^j A^j, \partial^i A^0 - \dot{A}^i] \\
&= [\dot{A}^0, \partial^i A^0] + \left[\sum_j \partial^j A^j, \dot{A}^i\right] - [\dot{A}^0, \dot{A}^i] \\
&= \left(\frac{1}{\Lambda} \mathrm{i}\partial_y^i + \mathrm{i}\sum_j \delta^{ij} \partial_x^j\right) \delta(\boldsymbol{x}-\boldsymbol{y}) - [\dot{A}^0, \dot{A}^i] \\
&= \mathrm{i}\frac{\Lambda - 1}{\Lambda} \partial_x^i \delta(\boldsymbol{x}-\boldsymbol{y}) - [\dot{A}^0, \dot{A}^i] .
\end{aligned}$$

In much the same way one proves the implication

$$[\pi^i, \pi^k] = 0 \quad \Longrightarrow \quad [\dot{A}^i, \dot{A}^k] = 0 .$$

Thus, all commutators (7.140) are proven.

The second group of the commutators (7.140) is particularly interesting: Indeed, they give

$$\begin{aligned}
\left[\dot{A}^0(x), A^0(y)\right]_{x^0=y^0} &= +\mathrm{i}\,\frac{1}{\Lambda}\delta(\boldsymbol{x}-\boldsymbol{y}) , \\
\left[\dot{A}^i(x), A^k(y)\right]_{x^0=y^0} &= -\mathrm{i}\,\delta^{ik}\,\delta(\boldsymbol{x}-\boldsymbol{y}) .
\end{aligned}$$

Comparing with the commutator (7.26) for scalar fields,

$$\left[\dot{\Phi}(x), \Phi(y)\right]_{x^0=y^0} = -\mathrm{i}\,\delta(\boldsymbol{x}-\boldsymbol{y}) ,$$

it seems as though the commutator between $\dot{A}^0$ and A^0 had the wrong sign or, expressed differently, as if the field and its conjugate momentum had exchanged their roles! We will clarify the implications of this sign below.

7.5.2 Normal Modes and One-Photon States

In expanding the four field operators A_μ in terms of plane waves, of creation and annihilation operators, four polarization directions $\varepsilon_\mu^{(\lambda)}(k)$, $\lambda = 0, 1, 2, 3$ are needed. In order to construct the latter we return to the analysis of massless particles in the framework of the Poincaré group and introduce the basis $n^{(\alpha)}$ that was defined in Chap. 6. Let

$$\begin{aligned}
&\varepsilon^{(0)}(k) \equiv t\,, \quad \text{with} \quad t^2 = 1, t^0 > 0; \\
&\varepsilon^{(1)}(k) = n^{(1)},\ \varepsilon^{(2)}(k) = n^{(2)}, \quad \text{with } \varepsilon^{(i)}(k)\cdot\varepsilon^{(j)}(k) = -\delta^{ij},\ i, j = 1, 2\,; \\
&\varepsilon^{(3)}(k) = \frac{1}{(k\cdot t)}k - t\,.
\end{aligned}$$

These base vectors fulfill the orthogonality and completeness relations (6.134) and (6.135), respectively, viz.

$$(\varepsilon^{(\lambda)} \cdot \varepsilon^{(\lambda')}) = g^{\lambda\lambda'}, \tag{7.141}$$

$$\varepsilon_\mu^{(\lambda)} g_{\lambda\lambda'} \varepsilon_\nu^{(\lambda')} = g_{\mu\nu} \,. \tag{7.142}$$

The polarizations $\varepsilon^{(1)}(k)$ and $\varepsilon^{(2)}(k)$ are orthogonal to k and t, $(\varepsilon^{(i)} \cdot t) = 0 = (\varepsilon^{(i)} \cdot k)$ for $i = 1, 2$, and correspond to the transverse polarizations of Sect. 7.3.4. The polarizations $\varepsilon^{(0)}$ and $\varepsilon^{(3)}$ are called *timelike* and *longitudinal*, respectively. Their sum is proportional to k,

$$\varepsilon_\mu^{(0)}(k) + \varepsilon_\mu^{(3)}(k) = \frac{1}{(k \cdot t)} k_\mu \,.$$

Regarding the field operators A_μ, one follows the examples (7.31) and (7.97) by assuming an expansion in terms of plane waves (or any other complete base system),

$$A_\mu(x) = \frac{1}{(2\pi)^{3/2}} \int \frac{\mathrm{d}^3k}{2\omega_k} \sum_{\lambda=0}^{3} \left\{ \varepsilon_\mu^{(\lambda)}(k) c^{(\lambda)}(k) \mathrm{e}^{-ikx} + \mathrm{h.c.} \right\}, \tag{7.143}$$

where h.c. stands for the hermitian conjugate of the first term. The negative frequency part contains the operators $c^{(\lambda)\dagger}(k)$ and the polarizations $\varepsilon_\mu^{(\lambda)*}$. (The complex conjugate signs matter if in $\mathbb{R}^3$ a spherical basis is used, instead of a cartesian basis.) We use covariant normalization in which case the volume element is $\mathrm{d}^3k/(2\omega_k)$ with $\omega_k = |\boldsymbol{k}|$ the energy of the photon. Inserting these formulae in the postulated commutators a by now well-known calculation yields

$$\left[c^{(\lambda)}(k), c^{(\lambda')\dagger}(k') \right] = -2\omega_k \, g^{\lambda\lambda'} \delta(\boldsymbol{k} - \boldsymbol{k}') \,, \tag{7.144}$$

$$\left[c^{(\lambda)}(k), c^{(\lambda')}(k') \right] = 0 \,. \tag{7.145}$$

The interpretation of the operators $c^{(\lambda)\dagger}(k)$ and $c^{(\lambda)}(k)$ is reached by formal application of the analysis of Sect. 7.1.5: The first of them creates a photon with momentum k and polarization λ, while the second annihilates a photon in that same state. Note, however, that while we rediscover the physical photons of Sect. 7.3 in the transverse modes $\lambda = 1$ and $\lambda = 2$, the *scalar photons* with $\lambda = 0$ and the *longitudinal photons* with $\lambda = 3$ are unphysical degrees of freedom. With $g^{ii} = -1$ the polarizations $\lambda = 1, 2, 3$ give the "right" sign on the right-hand side of (7.144), but for $\lambda = 0$ there is a minus sign.

The commutator of $A_\mu(x)$ and $A_\nu(y)$ for arbitrary spacetime points x and y is obtained from the commutation rules for the creation and annihilation operators. One finds

$$\left[A_\mu(x), A_\nu(y) \right] = -\mathrm{i}\, g_{\mu\nu} \Delta_0(x - y; m = 0) \,. \tag{7.146}$$

This result shows once more that the Lorenz condition is incompatible with the postulated quantization procedure. Indeed, from (7.146) one concludes

$$\left[\partial^\mu A_\mu(x), A_\nu(y) \right] = -\mathrm{i}\, \partial_\nu \Delta_0(x - y; m = 0) \neq 0 \,.$$

The commutator (7.146) is proportional to the causal distribution (7.56). It vanishes whenever x and y are spacelike relative to each other. For space indices $\mu = \nu = j$ the right-hand side of (7.146) has precisely the form of the equation (7.57) (including the correct sign) which holds for the case of the real scalar field. For $\mu = \nu = 0$ it has the opposite sign. The significance of this sign is seen best on an example: Let a scalar photon be in the state

$$\left|\lambda = 0, \widetilde{f}\right\rangle = \int \frac{d^3k}{2\omega_k} \widetilde{f}(k) c^{(0)\dagger}(k) \,|0\rangle \,,$$

with $\widetilde{f}(k)$ a L^2-integrable function. The squared norm of this state is

$$\begin{aligned}\left\langle\lambda = 0, \widetilde{f}\right|\lambda = 0, \widetilde{f}\left.\right\rangle &= \int \frac{d^3k}{2\omega_k} \int \frac{d^3k'}{2\omega_{k'}} \widetilde{f}^*(k')\widetilde{f}(k)\langle 0|c^{(0)}(k')c^{(0)\dagger}(k)\,|0\rangle \\ &= \int \frac{d^3k}{2\omega_k} \int \frac{d^3k'}{2\omega_{k'}} \widetilde{f}^*(k')\widetilde{f}(k)(-2\omega_k')\delta(\boldsymbol{k} - \boldsymbol{k}') \\ &= -\int \frac{d^3k}{2\omega_k} \,|\widetilde{f}(k)|^2\end{aligned}$$

and, hence, negative!

7.5.3 Lorenz Condition, Energy and Momentum of the Radiation Field

All of the peculiar properties noted above are related, directly or indirectly, to the Lorenz condition. This condition cannot be fulfilled as an operator condition but it should appear in the observable aspects of the theory, as a physical constraint. This raises the question whether the classical condition (7.91) can be replaced by some weaker requirement. This could be, for instance, the following

Postulate 7.2

Let $|\psi\rangle$ be a physically realizable state of the theory. The Lorenz condition is required to hold for the expectation value of any such state.

$$\langle\psi\,|\partial_\mu A^\mu(x)|\psi\rangle \overset{!}{=} 0\,. \tag{7.147}$$

Note that this is not equivalent to requiring the operator $\partial_\mu A^\mu(x)$ itself to vanish identically. We noted above that this requirement is too strong and leads to an inconsistency. However, the condition (7.147) is already satisfied if the positive frequency part of $\partial_\mu A^\mu(x)$ annihilates the state,

$$\left(\partial_\mu A^\mu(x)\right)^{(+)} |\,\psi\,\rangle = 0\,.$$

Indeed, if this is the case, then every expectation value vanishes,

$$\langle\psi\,|\partial_\mu A^\mu(x)\,|\,\psi\,\rangle = \langle\psi|\left\{\left(\partial_\mu A^\mu(x)\right)^{(-)} + \left(\partial_\mu A^\mu(x)\right)^{(+)}\right\}|\,\psi\,\rangle = 0\,.$$

This modified condition is meaningful: Even in the presence of interaction with matter the divergence fulfills the Klein-Gordon equation (7.136) of a massless, *free* scalar particle. Therefore, the division into positive and negative frequency parts is well-defined.

It is now easy to work out the consequences of the Postulate 7.2. As by assumption $\varepsilon^{(1)} \cdot k = 0 = \varepsilon^{(2)} \cdot k$, one has

$$\left(\partial_\mu A^\mu(x)\right)^{(+)} = -\mathrm{i}\frac{1}{(2\pi)^3}\int \frac{\mathrm{d}^3 k}{2\omega_k}\, \mathrm{e}^{-\mathrm{i}kx} \sum_{\lambda=0,3} c^{(\lambda)}(k) k_\mu\, \varepsilon^{(\lambda)\,\mu}(k)\,.$$

Therefore, the condition (7.147) is equivalent to

$$\left(k \cdot \varepsilon^{(0)}(k) c^{(0)}(k) + k \cdot \varepsilon^{(3)}(k) c^{(3)}(k)\right) |\, \psi\,\rangle = 0\,,$$

or, as $\varepsilon^{(3)} = k/(k \cdot t) - \varepsilon^{(0)}$ and $k^2 = 0$, to

$$\left[c^{(0)}(k) - c^{(3)}(k)\right] |\, \psi\,\rangle = 0\,. \tag{7.148}$$

This condition says that the unphysical longitudinal and scalar photons of the covariant theory are correlated in a specific manner. If at all, every physical state contains as many of either kind. The bonus is seen when calculating the Hamiltonian of the theory. By the established techniques that we need not repeat here, one obtains

$$H = -\sum_{\lambda,\lambda'=0}^{3} g^{\lambda\lambda'} \int \frac{\mathrm{d}^3 k}{2\omega_k} \omega_k c^{(\lambda)\dagger}(k) c^{(\lambda')}(k)\,,$$

$$= -\int \frac{\mathrm{d}^3 k}{2\omega_k} \omega_k c^{(0)\dagger}(k) c^{(0)}(k) + \sum_{\lambda=1}^{3} \int \frac{\mathrm{d}^3 k}{2\omega_k} \omega_k c^{(\lambda)\dagger}(k) c^{(\lambda)}(k)$$

i.e. an operator whose spectrum, a priori, is not positive. However, if one calculates its expectation value in an arbitrary physical state which fulfills the condition (7.147), this operator counts only the contributions of the physical transverse photons,

$$\langle \psi | H \,|\, \psi\,\rangle = \int \frac{\mathrm{d}^3 k}{2\omega_k}\, \omega_k \sum_{\lambda=1}^{2} \langle \psi | c^{(\lambda)\,\dagger}(k) c^{(\lambda)}(k) \,|\, \psi\,\rangle\,,$$

the negative contributions of the scalar photons are compensated by those of the longitudinal photons.

An analogous result is found for the operator of momentum,

$$\langle \psi | \boldsymbol{P} \,|\, \psi\,\rangle = \int \frac{\mathrm{d}^3 k}{2\omega_k}\, \boldsymbol{k} \sum_{\lambda=1}^{2} \langle \psi | c^{(\lambda)\,\dagger}(k) c^{(\lambda)}(k) \,|\, \psi\,\rangle\,.$$

The program of canonical quantization is concluded by the calculation of the propagator. One finds the result

$$\langle 0 | T A^\mu(x) A^\nu(y) | 0 \rangle = -\frac{\mathrm{i}}{(2\pi)^4} \int \mathrm{d}^4 k\, \frac{\mathrm{e}^{-\mathrm{i}k\cdot(x-y)}}{k^2 + \mathrm{i}\varepsilon} \left[g^{\mu\nu} + \frac{1-\Lambda}{\Lambda} \frac{k^\mu k^\nu}{k^2 + \mathrm{i}\varepsilon} \right]. \tag{7.149}$$

The denominator $(k^2 + i\varepsilon)$ is understood if one recalls that the photon has no mass. The additional terms in square brackets are new and possibly unfamiliar because they depend on the parameter Λ. As a comment which remains qualitative at this point, one might say this: Terms in momentum space which are multiplied by k_μ, become partial derivatives ∂_μ in position space. The interaction always contains the operator A_μ contracted with the electromagnetic current density j^μ. As the latter is conserved, it seems plausible that the terms depending on Λ do not contribute. Therefore, one has some freedom in choosing this parameter.

A look at the equations (7.140) shows that an obvious choice is $\Lambda = 1$. This gauge fixing is called *Feynman gauge*. In this gauge the propagator (7.149) has the form expected from rather general considerations. Similarly, the limit $\Lambda \to \infty$ is a special case because it yields an expression which comes very close to the propagator in the Coulomb gauge, cf. (7.96). This choice is called *Landau gauge*. The value $\Lambda = 0$ leads to ill-defined expressions (7.140), in accord with our previous experience which taught us that covariant quantization is incompatible with the Lorenz condition.

7.6 *The State Space of Quantum Electrodynamics

The results of the preceding section are intriguing. On the one hand, we found that the same theory which within the class of Coulomb gauges could be quantized consistently and in agreement with general principles, contains unphysical degrees of freedom in its manifestly covariant quantum form: the scalar and the longitudinal photons. On the other hand, the two species conspire through the relation (7.148) in such a way that physical states always contain equal numbers of them. This correlation and the negative squared norm of scalar photon states imply that the contributions of the scalar photons cancel those of the longitudinal photons. Hence, these degrees of freedom are not observable asymptotically.

This seems strange: One has the choice of formulating the same theory in a transverse gauge, at the expense of manifest Lorentz covariance, or in a manifestly covariant form. With the first choice there are only physical degrees of freedom, i.e. transverse photons and instantaneous Coulomb interactions, with the second choice one inherits two unphysical degrees of freedom. The results which can be tested by experiment, are the same in the two cases.

One can work out this equivalence and, once this is done, apply the manifestly covariant formalism in practice, without further scruples. However, one may feel uneasy in doing so because the underlying state space no longer is a Hilbert space. The aim of this section is to investigate this larger space as well as the formal framework of quantum electrodynamics.

7.6.1 *Field Operators and Maxwell's Equations

Upon quantization the components of the tensor field of electromagnetic field strengths $F^{\mu\nu}(x)$ become operator-valued, tempered distributions, i.e. taking

$$F^{\mu\nu}(g) = \int \mathrm{d}^4x \; F^{\mu\nu}(x) g(x), \, g \in \mathcal{S},$$

where $\mathcal{S}$ denotes the space of smooth, strongly decreasing functions, these operators $F^{\mu\nu}(g)$ are defined on a dense subset $\mathcal{D} \subset \mathcal{H}$. The vacuum Ω as well as the image of $\mathcal{D}$ by $F^{\mu\nu}(g)$ are assumed to lie in $\mathcal{D}$. Then the states

$$F^{\alpha\beta}(g_1) F^{\gamma\delta}(g_2) \cdots \Omega$$

are well-defined and lie in the same domain of Hilbert space. The field operators are taken to be Poincaré covariant, i.e. transform under $x \mapsto \Lambda x + a$ according

$$F^{\alpha\beta}(\Lambda x + a) = \Lambda^{\alpha}{}_{\mu} \Lambda^{\beta}{}_{\nu} \; \mathbf{U}(\Lambda, a) F^{\mu\nu}(x) \mathbf{U}^{-1}(\Lambda, a) .$$

Furthermore, we assume *micro-causality* to hold in the form

$$\left[F^{\mu\nu}(x), F^{\alpha\beta}(y) \right] = 0 \qquad \text{for} \quad (x - y)^2 < 0 ,$$

cf. (7.57). Note that here we consider only the electric and magnetic field operators, not the potentials which are no observables.

Even these modest constraints on physical degrees of freedom of the theory are hampered by difficulties. In order to see this consider the vacuum expectation value

$$\left(\Omega, F^{\mu\nu}(x) \Omega \right) =: F^{\mu\nu}(x) .$$

Under translations of the argument and using the invariance of the vacuum expectation value one has

$$\begin{aligned} F^{\mu\nu}(x + a) &= \left(\Omega, F^{\mu\nu}(x + a) \Omega \right) = \left(\Omega, \mathbf{U}(a) F^{\mu\nu}(x) \mathbf{U}^{-1}(a) \Omega \right) \\ &= F^{\mu\nu}(x) . \end{aligned}$$

The quantity $F^{\mu\nu}$ must be a tensor of rank two which is invariant under translations, $F^{\mu\nu}(x+a) = F^{\mu\nu}(x)$. This is possible only if it is proportional to $g^{\mu\nu}$. On the other hand, $F^{\mu\nu} = -F^{\nu\mu}$, and one concludes that $F^{\mu\nu}(x)$ vanishes identically.

Before discussing in which way the assumptions must be modified in order to avoid this catastrophe, let us show this: One could be tempted to try to work with the unobservable operators A^{μ}, instead of the observables $F^{\mu\nu}$, and to abandon the requirement of micro causality only for them, by not imposing $[A^{\mu}(x), A^{\nu}(y)]$ for all $(x - y)^2 < 0$. This attempt goes wrong as well. This is the content of a theorem of Strocchi.[15] Define the Fourier transforms of the fields and the test functions by

$$\begin{aligned} F^{\mu\nu}(p) &= \tfrac{1}{(2\pi)^{5/2}} \int \mathrm{d}^4x \; F^{\mu\nu}(x) \mathrm{e}^{\mathrm{i}px} , \quad A^{\mu}(p) = \tfrac{1}{(2\pi)^{5/2}} \int \mathrm{d}^4x \; A^{\mu}(x) \mathrm{e}^{\mathrm{i}px} , \\ g(p) &= \tfrac{1}{(2\pi)^{3/2}} \int \mathrm{d}^4x \; g(x) \mathrm{e}^{-\mathrm{i}px} , \qquad g(x) \in \mathcal{S}, \quad g(p) \in \mathcal{S} . \end{aligned}$$

[15] F. Strocchi, Phys. Rev. **162** (1967) 1429.

In case the Fourier transforms of $F^{\mu\nu}$ and A^μ exist, we have the Parseval equation

$$\int d^4p\ F^{\mu\nu}(p)g(p) = \int d^4x\ F^{\mu\nu}(x)g(x)$$

as well as the analogous equation for A^μ. The Maxwell equations (7.82) and (7.83) without external sources translate to momentum space as follows

$$\varepsilon_{\mu\nu\sigma\tau}p^\nu F^{\sigma\tau}(p) = 0\,, \tag{7.150}$$

$$p_\mu F^{\mu\nu}(p) = 0\,. \tag{7.151}$$

In momentum space the relation between A^μ and $F^{\mu\nu}$ reads

$$F^{\mu\nu}(p) = -\mathrm{i}\left(p^\mu A^\nu(p) - p^\nu A^\mu(p)\right)\,. \tag{7.152}$$

Theorem of Strocchi

Let the action of A^μ on the vacuum be well-defined, $A^\mu(g)\Omega$. Furthermore assume A^μ to transform covariantly with respect to Poincaré transformations, i.e.

$$A^\mu(\Lambda x + a) = \Lambda^\mu{}_\nu \mathbf{U}(\Lambda, a)A^\nu(x)\mathbf{U}^\dagger(\Lambda, a)\,.$$

Then neither $A^\mu(p)$ nor $F^{\mu\nu}(p)$ exist.

Proof

One first shows that the expectation value of the product of $A^\mu(p)$ and $A^\nu(q)$ can be written as follows

$$\begin{aligned}\langle\Omega|A^\mu(p)A^\nu(q)\,|\Omega\rangle &= \delta^{(4)}(p+q)\langle\Omega|A^\mu(0)A^\nu(q)\,|\Omega\rangle\,(2\pi)^{3/2}\\ &\equiv \delta^{(4)}(p+q)D^{\mu\nu}(q)\,.\end{aligned}$$

This is a consequence of covariance and of $\mathbf{U}^\dagger(a) = \mathbf{U}(-a)$. One has

$$\begin{aligned}\langle\Omega|A^\mu(p)A^\nu(q)|\Omega\rangle &= \frac{1}{(2\pi)^5}\int d^4x\int d^4y\, e^{-ipx}e^{iqy}\langle\Omega|A^\mu(x)A^\nu(y)|\Omega\rangle\\ &= \frac{1}{(2\pi)^5}\int d^4x\int d^4y\ e^{-ipx}e^{iqy}\\ &\quad\times\langle\Omega|\mathbf{U}(x)A^\mu(0)\mathbf{U}(y-x)A^\nu(0)\mathbf{U}(-y)\,|\Omega\rangle\,,\end{aligned}$$

Let $z := y - x$. As the vacuum is translation invariant, one has $\langle\Omega|\mathbf{U}(x) = \langle\Omega|$ and $\mathbf{U}(-y)|\Omega\rangle = \mathbf{U}^\dagger(z)|\Omega\rangle$ and, hence,

$$\begin{aligned}\langle\Omega|A^\mu(p)A^\nu(q)\,|\Omega\rangle &= \frac{1}{(2\pi)^5}\int d^4x\int d^4z\ e^{-i(p+q)x}e^{iqz}\\ &\quad\times\langle\Omega|A^\mu(0)A^\nu(z)\,|\Omega\rangle\\ &= (2\pi)^{3/2}\delta^{(4)}(p+q)\langle\Omega|A^\mu(0)A^\nu(q)\,|\Omega\rangle\,.\end{aligned}$$

The quantity $D^{\mu\nu}$ thus defined is seen to be a covariant distribution. As such it may be decomposed in terms of the two covariants that are available here, $g^{\mu\nu}$ and $q^\mu q^\nu$:

$$D^{\mu\nu}(q) = g^{\mu\nu}D_1(q) + q^\mu q^\nu D_2(q)\,,$$

where $D_1(q)$ and $D_2(q)$ must be Poincaré invariant distributions. Calculate then the expectation value

$$\langle \Omega | A^\mu(p) F^{\nu\sigma}(q) | \Omega \rangle = -\mathrm{i}\, \delta^{(4)}(p+q) \left\{ \left(q^\nu g^{\mu\sigma} - q^\sigma g^{\mu\nu} \right) D_1(q) + \left(q^\mu q^\nu q^\sigma - q^\nu q^\mu q^\sigma \right) D_2(q) \right\} .$$

The second term is zero. Contracting with q_σ and using the inhomogeneous Maxwell equations (7.151), one obtains

$$q_\sigma \langle \Omega | A^\mu(p) F^{\nu\sigma}(q) | \Omega \rangle = 0 = -\mathrm{i}\, \delta^{(4)}(p+q) \left(q^\nu q^\mu - q^2 g^{\mu\nu} \right) D_1(q) .$$

The resulting condition $\left(q^\nu q^\mu - q^2 g^{\mu\nu} \right) D_1(q) = 0$ allows conclusions about the support of the distribution $D_1(q)$. Firstly, it must be contained in the set of points for which $q^\nu q^\mu - q^2 g^{\mu\nu}$ vanishes,

$$\operatorname{supp} D_1 \subset \left\{ \left(q^\nu q^\mu - q^2 g^{\mu\nu} \right) = 0 \right\} .$$

Furthermore, it must be invariant under Poincaré transformations. These conditions are met by the point $q = 0$ only, that is, $\operatorname{supp} D_1 = \{0\}$. A distribution of this kind can only contain Dirac's δ-distribution (or invariant derivatives thereof). So let us try $D_1(q) = c\, \delta^{(4)}(q)$.[16] With this choice one has

$$\langle \Omega | A^\mu(p) F^{\nu\sigma}(q) | \Omega \rangle = \text{const.}\ \delta^{(4)}(p+q) \delta^{(4)}(q) \left(q^\nu g^{\mu\sigma} - q^\sigma g^{\mu\nu} \right) ,$$

from which one derives the expectation value

$$\langle \Omega | F^{\mu\sigma}(p) F^{\nu\tau}(q) | \Omega \rangle = \text{const.}\ \delta^{(4)}(p+q) \delta^{(4)}(q) \times \left(p^\mu q^\nu g^{\sigma\tau} - p^\sigma q^\nu g^{\mu\tau} - p^\mu q^\tau g^{\sigma\nu} + p^\sigma q^\tau g^{\mu\nu} \right) .$$

The expression in parentheses on the right-hand side vanishes at $p = -q$ and $q = 0$. Furthermore, as $F^{\mu\nu}$ is hermitian, we have $F^{\mu\nu}(-q) = F^{\mu\nu\,\dagger}(q)$. Thus, one concludes

$$\langle \Omega | F^{\mu\nu\,\dagger}(q) F^{\mu\nu}(q) | \Omega \rangle = \| F^{\mu\nu}(q) | \Omega \rangle \|^2 = 0 ,$$

so that the fields vanish altogether. This proves the theorem.

We already know a first way out of this dilemma: Making use of its gauge invariance the theory is reformulated in a way which hides its covariance – but, of course, without touching its physical content! – and works with its physical degrees of freedom only. In this version the theory is defined on a *genuine* Hilbert space $\mathcal{H}_{\text{phys.}}$ of physical states.

There is an other solution which leads to the same observable predictions while keeping the manifest covariance. In essence, it consists in embedding the Hilbert space in a larger space

$$\mathcal{H}_{\text{phys.}} \subset \mathcal{H} ,$$

which is no longer a Hilbert space. This is the topic of the next section.

---

[16] The ansatz $D_1 = c\, \Box_q \delta^{(4)}(q)$ is excluded as well. To see this calculate $\int \mathrm{d}^4 q\ (q^\mu q^\nu - q^2 g^{\mu\nu}) D_1 g(q) \propto g(0)$, which may be different from zero – in contradiction with the conclusion reached above.

7.6.2 *The Method of Gupta and Bleuler

Support of $F^{\mu\nu}(p)$: The support of $F^{\mu\nu}(p)$ is the light cone $p^2 = 0$. To see this, contract the homogeneous Maxwell equations (7.150)

$$p^\alpha F^{\beta\gamma}(p) + p^\beta F^{\gamma\alpha}(p) + p^\gamma F^{\alpha\beta}(p) = 0$$

with p_α and insert the inhomogeneous equations (7.151)

$$p_\alpha p^\alpha F^{\beta\gamma}(p) = -p^\beta p_\alpha F^{\gamma\alpha}(p) + p^\gamma p_\alpha F^{\beta\alpha}(p) = 0\,.$$

The result just obtained,

$$p^2 F^{\beta\gamma}(p) = 0\,, \tag{7.153}$$

says that the support of $F^{\beta\gamma}(p)$ is the set of points $p^2 = 0$, i.e. precisely the light cone. Splitting into contributions on the *positive* and on the *negative* light cone yields

$$F^{\mu\nu}(p) = \delta_+(p)\widehat{F}^{\mu\nu}(\boldsymbol{p}) + \delta_-(p)\widehat{F}^{\mu\nu}(-\boldsymbol{p}) \equiv F^{(+)\mu\nu}(p) + F^{(-)\mu\nu}(p)\,,$$

where $\delta_\pm$ stand for the two terms with positive and negative p^0 (cf. Exercise 7.10),

$$\delta_\pm(p) := \Theta(\pm p^0)\,\delta(p^2)\,.$$

Let P be the operator of four-momentum, p its eigenvalues. The translation formula (7.30), the definition of the Fourier transforms, and the fact that the vacuum carries no momentum are used to show that $F^{\mu\nu}(p)|\Omega\rangle$ is an eigenstate of P pertaining to the eigenvalue $-p$,

$$P\,F^{\mu\nu}(p)\,|\Omega\rangle = -pF^{\mu\nu}(p)\,|\Omega\rangle\,.$$

Inserting the decomposition $F^{\mu\nu} = F^{(+)\,\mu\nu} + F^{(-)\,\mu\nu}$ this result can hold only for the second term, or, in other terms, the positive frequency part annihilates the (perturbative) vacuum,

$$F^{(+)\mu\nu}(p)\,|\Omega\rangle = 0\,.$$

The commutator of $F^{\alpha\beta}(p)$ and $F^{\sigma\tau}(q)$ has the form[17]

$$\left[F^{\alpha\beta}(p), F^{\sigma\tau}(q)\right] = \delta^{(4)}(p+q)M^{\alpha\beta\sigma\tau}(q)\,(\delta_+(q) - \delta_-(q))\,,$$

where $M^{\alpha\beta\sigma\tau}$ is an abbreviation for the expression

$$M^{\alpha\beta\sigma\tau} = -g^{\alpha\sigma}q^\beta q^\tau + g^{\alpha\tau}q^\beta q^\sigma + g^{\beta\sigma}q^\alpha q^\tau - g^{\beta\tau}q^\alpha q^\sigma\,.$$

This tensor of rank four is fixed by its properties: It must be antisymmetric in the pair (α, β) and in the pair (σ, τ). In either factor $F^{\alpha\beta}$ or $F^{\sigma\tau}$ it satisfies the Maxwell equations. Furthermore, terms which contain the invariant tensor $\varepsilon_{\alpha\beta\sigma\tau}$ are excluded because they have the wrong behaviour with respect to the parity operation. Returning to spacetime this means that

$$\left[F^{\alpha\beta}(x), F^{\sigma\tau}(y)\right] = \mathrm{i}\,M^{\alpha\beta\sigma\tau}(\partial_y)\Delta_0(x-y; m=0)\,,$$

[17] It is not obvious that the commutator is indeed a c-number. This needs a detailed proof.

with Δ_0 the causal distribution (7.56) for mass zero. Thus, the commutator of the observable field strengths is *causal.*

Remark

The remaining inhomogeneous Maxwell equations (7.151) follow from the homogeneous equations (7.150), the support condition (7.153), and from *one* of the inhomogeneous equations such as, e.g., $p_\nu F^{0\nu}(p) = 0$.

State Space with Indefinite Metric: Let $\mathcal{H}$ be a space equipped with a nondegenerate scalar product (ψ_i, ψ_j), that is to say, in a short-hand notation,

$$(\psi_i, \psi_{i_0}) = 0 \quad \forall\, \psi_i \in \mathcal{H} \quad \Longrightarrow \quad \psi_{i_0} = 0\,.$$

However, the scalar product need not be positive-definite. Suppose potentials A^μ are introduced so that one has as in (7.152)

$$F^{\mu\nu}(p) = -\mathrm{i}\left(p^\mu A^\nu(p) - p^\nu A^\mu(p)\right)\,.$$

Gauge transformations translate to momentum space according to

$$A^\mu(p) \longmapsto A^\mu(p) + \mathrm{i}p^\mu \chi(p)\,.$$

The field operators are assumed to satisfy the requirements:

(a) The operators A^μ are covariant, i.e. the Poincaré transformation (Λ, a) is unitarily represented with reference to the scalar product of $\mathcal{H}$, and yields the known transformation behaviour of the A^μ;

(b) The fields $F^{\mu\nu}(p)$, as well as $A^\mu(p)$, have their support on the light cone. From $p^2 A^\mu(p) = 0$ one concludes

$$0 = p_\nu(p^\nu A^\mu(p)) = -\mathrm{i}p_\nu F^{\mu\nu}(p) + \mathrm{i}p^\mu\left(-\mathrm{i}p_\nu A^\nu(p)\right)\,.$$

Defining $-\mathrm{i}p_\nu A^\nu(p) =: B(p)$, the spacetime representation of the latter is the divergence $B(x) = \partial_\mu A^\mu(x)$. The equation $p_\nu F^{\mu\nu}(p) = p^\mu B(p)$ tells us that the inhomogeneous Maxwell equations (7.151) are restored provided

$$B(p)\mathcal{H}_{\text{phys.}} = 0 \tag{7.154}$$

holds true. This condition is the same as (7.147) which was formulated on spacetime.

The vacuum expectation value of the product of two operators A^μ is written as

$$\langle \Omega | A^\mu(p) A^\nu(q) | \Omega \rangle = \delta^{(4)}(p+q)\delta_-(q)\left(\alpha g^{\mu\nu} + \beta q^\mu q^\nu\right)\,.$$

The first parameter must have the value $\alpha = -1$ in order to yield the correct expectation value of the product of two $F^{\mu\nu}$. The second parameter β reflects the freedom of choosing gauges and remains undetermined. Inserting (7.152) one has

$$\begin{aligned}
\langle \Omega | F^{\mu\alpha}(p) F^{\nu\beta}(q) | \Omega \rangle = &-p^\mu q^\nu \langle \Omega | A^\alpha(p) A^\beta(q) | \Omega \rangle \\
&- p^\alpha q^\beta \langle \Omega | A^\mu(p) A^\nu(q) | \Omega \rangle \\
&+ p^\mu q^\beta \langle \Omega | A^\alpha(p) A^\nu(q) | \Omega \rangle \\
&+ p^\alpha q^\nu \langle \Omega | A^\mu(p) A^\beta(q) | \Omega \rangle\,.
\end{aligned}$$

Choose $\beta = 0$, thus fixing the gauge, so that

$$\langle \Omega | A^{\mu}(p) A^{\nu}(q) | \Omega \rangle = -\delta^{(4)}(p+q)\delta_{-}(q) g^{\mu\nu} .$$

This yields the expression for $\langle \Omega | F^{\mu\alpha}(p) F^{\nu\beta}(q) | \Omega \rangle$ given above. Also here, one realizes that $\langle \Omega | A^0(p) A^0(p) | \Omega \rangle$, while being a squared norm, is not positive.

The operator A^{μ} is decomposed in terms of positive and negative frequency parts, too,

$$A^{\mu}(p) = \delta_{+}(p)\widehat{A}^{\mu}(\boldsymbol{p}) + \delta_{-}(p)\widehat{A}^{\mu}(-\boldsymbol{p}) \equiv A^{(+)\mu} + A^{(-)\mu}$$

and then is quantized in the canonical way,

$$\left[\widehat{A}^{\mu}(\boldsymbol{p}), \widehat{A}^{\nu}(\boldsymbol{q})\right] = 0 , \tag{7.155}$$

$$\left[\widehat{A}^{\mu}(\boldsymbol{p}), \widehat{A}^{\nu\,\dagger}(\boldsymbol{q})\right] = -2\omega_p g^{\mu\nu}\, \delta(\boldsymbol{p}-\boldsymbol{q}) . \tag{7.156}$$

Decomposing the operators A^{μ} in terms of polarizations, as before,

$$A^{\mu}(p) = \sum_{\lambda=0}^{3} A^{(\lambda)}(p)\varepsilon^{(\lambda)\,\mu}(p)$$

and choosing the basis as we did above, one sees that the divergence is proportional to the difference of $A^{(0)}$ and $A^{(3)}$

$$B(p) = -\mathrm{i}p^0 \left(A^{(0)}(p) - A^{(3)}(p)\right) =: -\mathrm{i}p^0 \overline{B}(p) .$$

All components $F^{\mu\nu}$ except for F^{03} depend on the transverse operators $A^{(1)}$ and $A^{(2)}$ only. For the exceptional component one has

$$F^{03}(p) = -\mathrm{i}p^3 \left(A^{(3)}(p) - A^{(0)}(p)\right) = \mathrm{i}p^3 \overline{B}(p) .$$

The commutators of the transverse operators $\widehat{A}^{(1,2)}$ and $\widehat{\overline{B}}$ (these are the operators which remain after separating off δ_+ and δ_-) are all equal to zero,

$$\left[\widehat{\overline{B}}(\boldsymbol{p}), \widehat{A}^{(1,2)}(\boldsymbol{q})\right] = 0 = \left[\widehat{\overline{B}}(\boldsymbol{p}), \widehat{A}^{(1,2)\,\dagger}(\boldsymbol{q})\right],$$

$$\left[\widehat{\overline{B}}(\boldsymbol{p}), \widehat{\overline{B}}(\boldsymbol{q})\right] = 0 = \left[\widehat{\overline{B}}(\boldsymbol{p}), \widehat{\overline{B}}^{\dagger}(\boldsymbol{q})\right] .$$

Embedding of the Physical States: Let $\mathcal{H}' \subset \mathcal{H}$ be the subspace generated by successive application of the creation operators $\widehat{A}^{(1,2)\,\dagger}$ and $\widehat{\overline{B}}^{\dagger}$ on the state $|\Omega\rangle$. As the divergence is a *free* field, and keeping track of the commutators, one finds that $\widehat{\overline{B}}(\boldsymbol{p})\mathcal{H}' = 0$. Therefore, for any two elements $\phi, \psi \in \mathcal{H}'$ one has

$$\langle \phi | \widehat{\overline{B}} \,|\, \psi \rangle = 0 = \langle \phi | \widehat{\overline{B}}^{\dagger} \,|\, \psi \rangle \quad \text{i.e.} \quad \langle \phi | \overline{B} \,|\, \psi \rangle = 0 .$$

For example, this statement holds for $|\phi\rangle \in \mathcal{H}'$ and for $|\psi\rangle = \widehat{\overline{B}}^{\dagger} |\phi\rangle$, in which case

$$\langle \psi | \widehat{\overline{B}}^{\dagger} \,|\phi\rangle = \|\psi\|^2 = 0 .$$

Thus, there are states in $\mathcal{H}'$ whose norm is equal to zero. Call $\mathcal{H}_0$ the subspace spanned by these states,

$$\mathcal{H}_0 = \left\{ \psi \mid \; \|\psi\|^2 = 0 \right\} \subset \mathcal{H}' .$$

The space $\mathcal{H}'$ cannot yet be the physical state space because it contains states whose norm vanishes. On the other hand, if one restricted the states to the space $\mathcal{H}_\perp$ which is generated by the operators $\widehat{A}^{(1)\dagger}$ and $\widehat{A}^{(2)\dagger}$, then this choice would not suffice for two reasons: The space $\mathcal{H}_\perp$ is not Lorentz invariant; furthermore, there are field strengths $F^{\mu\nu}$ (these are observables!) which lead out of this space.

We are now well prepared for the last step that is needed to identify the physical states. Let $\mathcal{P}(F^{\mu\nu})$ be a polynomial in the field operators $F^{\mu\nu}$, and let $|\phi\rangle, |\psi\rangle \in \mathcal{H}'$ be two arbitrary states in $\mathcal{H}'$. Measurable quantities are always of the form

$$\langle\phi|\mathcal{P}(F^{\mu\nu})\,|\,\psi\,\rangle \,, \qquad |\,\psi\,\rangle\,, |\,\phi\,\rangle \in \mathcal{H}' .$$

For an arbitrary pair of elements of $\mathcal{H}_0$, $|\psi_0\rangle, |\phi_0\rangle \in \mathcal{H}_0$, one has

$$\begin{aligned}\langle\phi+\phi_0|\mathcal{P}(F^{\mu\nu})\,|\,\psi+\psi_0\,\rangle &= \langle\phi|\mathcal{P}(F^{\mu\nu})\,|\,\psi\,\rangle + \langle\phi_0|\mathcal{P}(F^{\mu\nu})\,|\,\psi+\psi_0\,\rangle \\ &\quad + \langle\phi|\mathcal{P}(F^{\mu\nu})\,|\,\psi_0\,\rangle \,.\end{aligned}$$

The second and the third term on the right-hand side are equal to zero. As an example, we show this for the second term: The result of the application of $\mathcal{P}(F^{\mu\nu})$ on $|\psi+\psi_0\rangle \in \mathcal{H}'$ is contained in $\mathcal{H}'$. Now, $|\phi_0\rangle$ is generated by applying $\widehat{\overline{B}}$ to an element $|\psi\rangle \in \mathcal{H}'$, so that $\langle\phi_0| = \langle\psi|\widehat{\overline{B}}$. However, $\widehat{\overline{B}}$ acting on any element of $\mathcal{H}'$ gives zero. Therefore, one sees that

$$\langle\phi+\phi_0|\mathcal{P}(F^{\mu\nu})\,|\,\psi+\psi_0\,\rangle = \langle\phi|\mathcal{P}(F^{\mu\nu})\,|\,\psi\,\rangle \,.$$

This allows for an important conclusion:

Physical Space of Quantum Electrodynamics: In $\mathcal{H}'$ define the equivalence classes of all states whose difference is in $\mathcal{H}_0$,

$$\mathcal{H}_{\text{phys.}} = \mathcal{H}'/\mathcal{H}_0 \,. \tag{7.157}$$

This is the physical space of the quantized Maxwell fields.

In conclusion, this clarifies the structure of the state space of quantum electrodynamics: When applying manifestly covariant quantization one must construct the theory in the space $\mathcal{H}$ which, although a linear space, is not a Hilbert space because of its indefinite scalar product. It is only at the end of a calculation that one restricts to the physical *in*- and *out*-states. That this is possible is due to the fact that scalar and longitudinal degrees of freedom always appear in the linear combination $A^{(0)}(p) - A^{(3)}(p)$. The contributions of the scalar and the longitudinal, unphysical, fields cancel. This method was developed first by S.N. Gupta and K. Bleuler.

7.7 Path Integrals and Quantization

The method of *path integrals*, as developed by Dirac and Feynman, provides an alternative to canonical quantization of point mechanics and of classical field theories. As far as nonrelativistic quantum theory is concerned this method is not of central importance. However, in covariant quantum field theory, as well as in some other domains of physics, it has become an important technique. Yet, its application to quantum field theory, mathematically speaking, is not well-founded and needs special care and some tricks. Therefore, it seems useful to explain the essential idea in the framework of nonrelativistic quantum mechanics first, and to illustrate it by means of simple examples.

We start with a few remarks and recall some classical notions.

7.7.1 The Action in Classical Mechanics

The canonical transformations which are generated by smooth functions of the *original* coordinates and the *new* momenta

$$\Big\{q, p, H(q, p, t)\Big\} \to S(q, P, t) \to \Big\{Q, P, \widetilde{H}(Q, P, t)\Big\} \tag{7.158}$$

play a special role: The Hamilton-Jacobi differential equation is derived by means of them, and so is the important class of infinitesimal canonical transformations. Using the notation $q = (q_1, q_2, \ldots, q_f)$, $P = (P_1.P_2, \ldots, P_f)$, where f is the number of degrees of freedom, the transformation formulae in phase space are

$$Q_i = \frac{\partial S}{\partial P_i}, \quad p_k = \frac{\partial S}{\partial q_k}, \quad \widetilde{H}(Q, P, t) = H + \frac{\partial S}{\partial t}. \tag{7.159}$$

Assume that the problem posed by the Hamilton-Jacobi differential equation

$$\widetilde{H}(Q, P, t) = H\left(q, p = \frac{\partial S}{\partial q}, t\right) + \frac{\partial S}{\partial t} = 0 \tag{7.160}$$

was already solved. The function S then depends on f coordinates q and on f integration parameters $\alpha = (\alpha_1, \ldots, \alpha_f)$ which replace the variables P. The new momenta and coordinates are given by

$$P_k = \alpha_k, \quad \text{and} \quad Q_i = \frac{\partial S(q, \alpha, t)}{\partial \alpha_i} = \text{const.} =: \beta_i,$$

respectively. Clearly, the second of these equations must be solved for $q = q(\alpha, \beta, t)$. Assume the Legendre transform relating the Hamiltonian function and the Lagrangian function to exist. Calculate the total time derivative of the generating function,

$$\frac{\mathrm{d}S}{\mathrm{d}t} = \frac{\partial S}{\partial t} + \sum_{i=1}^{f} \frac{\partial S}{\partial q_i}\dot{q}_i = \left[-H(q, p, t) + \sum_{i=1}^{f} p_i \dot{q}_i\right]_{p_i = -\partial H/q_i}.$$

The variables p that appear on the right-hand side are eliminated by expressing them in terms of q and $\dot{q}$. The expression in square brackets is seen to be the Lagrangian function evaluated along solutions of the equations of motion. After integration over the time variable t one obtains *Hamilton's principal function*

$$S(q,\alpha,t) = \int_{t_0}^{t} \mathrm{d}t'\, L(q,\dot{q},t')\,. \tag{7.161}$$

Note that the integrand is a *function* of solutions $q = q(t)$ of the equations of motion, and of the corresponding velocities $\dot{q}(t)$. The function (7.161) must not be mistaken for the *action functional*

$$I[q] = \int_{t_1}^{t_2} \mathrm{d}t\, L(q,\dot{q},t)$$

which is at the heart of Hamilton's variational principle. In the case of the latter q and $\dot{q}$ are independent variables, $I[q]$ is a functional of q, and not an ordinary function of $q(t)$ and its derivative.

For the sake of simplicity we shall call Hamilton's principal function simply the *action* in what follows.

As a simple example consider the generating function

$$S(\vec{q},\vec{\alpha},t) = \vec{\alpha}\cdot\vec{q} - \frac{1}{2m}\vec{\alpha}^2 t + c$$

which refers to force-free motion of a particle with mass m in three-dimensional space. The solutions to be inserted in S, read

$$\vec{q}(t) = \vec{\beta} + \frac{\vec{\alpha}}{m}t\,.$$

The function S and its derivative are, respectively,

$$S(\vec{q},\vec{\alpha},t) = \frac{\vec{\alpha}^2}{2m}t + \vec{\alpha}\cdot\vec{\beta} + c\,,$$

$$\frac{\mathrm{d}S(\vec{q},\vec{\alpha},t)}{\mathrm{d}t} = \vec{\alpha}\cdot\dot{\vec{q}} - \frac{1}{2m}\vec{\alpha}^2 = \frac{\vec{\alpha}^2}{2m}\,.$$

The Lagrangian function is $L = (1/2)m\dot{\vec{q}}^{\,2} = \vec{\alpha}^2/2m$ and, therefore, the action (7.161) is

$$S(\vec{q},\vec{\alpha},t) = \frac{\vec{\alpha}^2}{2m}(t - t_0)\,.$$

The interpretation of the constants of integration is obvious: $\vec{\alpha}$ is the momentum of the particle, $\vec{\beta}$ is its initial position in space. The action is the product of the kinetic energy and the time difference $(t - t_0)$.

7.7.2 The Action in Quantum Mechanics

In what follows the distinction between the self-adjoint operators and their eigenvalues is essential. Therefore, operators will temporarily be denoted by underlined symbols such as, for example, $\underline{p} = -\mathrm{i}\hbar\partial/\partial q$, while their eigenvalues are written as ordinary Latin letters. So, for example, one has $\underline{p}|p\rangle = p|p\rangle$. The only exception is the Hamilton operator that we continue to write as H, in order to avoid confusion. As a first example consider the position operator in Heisenberg representation, Eq. (3.47),

$$\underline{q}_t = \mathrm{e}^{(\mathrm{i}/\hbar)Ht}\underline{q}_0\mathrm{e}^{-(\mathrm{i}/\hbar)Ht} .$$

Its eigenvalue at time t_i is denoted q_i, i.e. $\underline{q}_{t_i}|q_i\rangle = q_i|q_i\rangle$. Consider then the following series of equations

$$\begin{aligned}
q_2 q_1\ \langle q_2|\,q_1\rangle &= \langle q_2|\underline{q}^{\dagger}_{t_2}\underline{q}_{t_1}\,|q_1\,\rangle \\
&= \langle q_2|\mathrm{e}^{(\mathrm{i}/\hbar)Ht_2}\underline{q}_0\mathrm{e}^{(\mathrm{i}/\hbar)H(t_1-t_2)}\underline{q}_0\mathrm{e}^{(-\mathrm{i}/\hbar)Ht_1}\,|q_1\,\rangle \\
&= \langle q_2(t_2)|\underline{q}_0\mathrm{e}^{(\mathrm{i}/\hbar)H(t_1-t_2)}\underline{q}_0\,|q_1(t_1)\rangle \\
&= q_2 q_1\langle q_2(t_2)|\mathrm{e}^{(\mathrm{i}/\hbar)H(t_1-t_2)}\,|q_1(t_1)\rangle\ .
\end{aligned}$$

Note, in particular, that the states in the first two lines are eigenstates in the Heisenberg picture whereas the last two lines contain eigenstates $|q(t)\rangle$ expressed in the Schrödinger picture. Thus, one obtains the relation

$$\langle q_2|\,q_1\rangle = \langle q_2(t_2)|\mathrm{e}^{-(\mathrm{i}/\hbar)H(t_2-t_1)}\,|q_1(t_1)\rangle\ , \tag{7.162}$$

which shows that the Hamilton operator boosts the sytem from the position $q_1(t_1)$ at time t_1 to the position $q_2(t_2)$ reached at time t_2.

This observation is not restricted to the position operator and its eigenfunctions. For example, let

$$H = \frac{\underline{\vec{p}}^2}{2m} + U(\underline{\vec{q}}) \tag{7.163}$$

be a given one-particle Hamilton operator, and let $|a(t_1)\rangle$ and $|b(t_2)\rangle$ be two solutions of the time-dependent Schrödinger equation $H\psi = (i/\hbar)\dot{\psi}$. These need not necessarily be stationary eigenstates of H. The transition amplitude from $|a(t_1)\rangle$ to $|b(t_2)\rangle$ which is mediated by the Hamilton operator, is given by the generalization of (7.162), viz.

$$\langle b|\,a\rangle = \langle b(t_2)|\mathrm{e}^{-(\mathrm{i}/\hbar)H(t_2-t_1)}\,|a(t_1)\rangle\ . \tag{7.164}$$

The essential idea of the method of path integrals is to decompose the "boost" of the state $|a(t_1)\rangle$ from time t_1 to time t_2 into very many small steps and to make use of the superposition principle of quantum mechanics. The quantum system which starts in a given initial configuration evolves into its final configuration by means of a weighted sum over all possible intermediate states. The principle of the method is best demonstrated by an example in one space dimension. For this purpose we study the Hamilton operator (7.163) restricted to one spatial variable q,

$$H = \frac{\underline{p}^2}{2m} + U(\underline{q})\,. \tag{7.165}$$

Denote by $|q\rangle$ and $|p\rangle$ the eigenstates of $\underline{q}$ and $\underline{p}$, respectively, and let these states be normalized in the sense of distributions, i.e.

$$\langle q'|q\rangle = \delta(q'-q)\,,\quad \langle p'|p\rangle = \delta(p'-p)\,,\quad \text{with} \tag{7.166a}$$

$$\langle q|p\rangle = \frac{1}{\sqrt{2\pi\hbar}}\mathrm{e}^{(\mathrm{i}/\hbar)pq}\,,\quad \langle p|q\rangle = \frac{1}{\sqrt{2\pi\hbar}}\mathrm{e}^{-(\mathrm{i}/\hbar)pq}\,. \tag{7.166b}$$

For an arbitrary Heisenberg state $|a\rangle$ and generalizing (7.166b) one defines the amplitudes

$$a(q) := \langle q|a\rangle\,,\qquad \tilde{a}(p) := \langle p|a\rangle \tag{7.166c}$$

as well as analogous amplitudes for $|b\rangle$.

As long as the time interval $t_2 - t_1 = \Delta t$ is very small the evolution operator (7.164) can be approximated as follows

$$\mathrm{e}^{-(\mathrm{i}/\hbar)H\Delta t} \simeq \mathrm{e}^{-(\mathrm{i}/\hbar)(\underline{p}^2/2m)\Delta t}\mathrm{e}^{-(\mathrm{i}/\hbar)U(\underline{q})\Delta t}\,. \tag{7.167}$$

In spite of the fact that, in general, e^{A+B} is not the same as $\mathrm{e}^A\mathrm{e}^B$, the correction terms to the approximation (7.167) are of order $(\Delta t)^2$ and, for reasons of consistency, must be neglected. To show this one makes use of the Campbell-Hausdorff formula

$$\mathrm{e}^A\mathrm{e}^B = \mathrm{e}^{C(A,B)} \quad \text{with}$$

$$C(A,B) = A + B + \frac{1}{2}[A,B] + \frac{1}{12}[[A,B],B] + \frac{1}{12}[[B,A],A] + \cdots\,. \tag{7.168}$$

Indeed, the next-to-leading term in (7.167) is proportional to

$$\left[\underline{p}^2, U(\underline{q})\right](\Delta t)^2$$

which must be neglected being of second order in Δt.

As an example we calculate the transition amplitude

$$\langle q_2(t+\Delta t)|\,q_1(t)\rangle \simeq \langle q_2|\mathrm{e}^{-(\mathrm{i}/\hbar)(\underline{p}^2/2m)\Delta t}\mathrm{e}^{-(\mathrm{i}/\hbar)U(\underline{q})\Delta t}\,|q_1\rangle$$

by inserting the completeness relation $\int \mathrm{d}^3p|p\rangle\langle p|$ in two positions, once to the left of the exponentials, once to their right. The potential energy $U(\underline{q})$ acts multiplicatively, therefore the operator $\underline{q}$ can be replaced by the eigenvalue q_1 if it acts to the right. Regarding the first exponential one has

$$\int_{-\infty}^{\infty}\mathrm{d}p\int_{-\infty}^{\infty}\mathrm{d}p'\;\langle q_2|\,p'\rangle\langle p'|\mathrm{e}^{-(\mathrm{i}/\hbar)(\underline{p}^2/2m)\Delta t}\,|p\rangle\,\langle p|\,q_1\rangle$$

$$= \frac{1}{2\pi\hbar}\int_{-\infty}^{\infty}\mathrm{d}p\int_{-\infty}^{\infty}\mathrm{d}p'\;\mathrm{e}^{-(\mathrm{i}/\hbar)(p^2/2m)\Delta t}\delta(p'-p)\mathrm{e}^{(\mathrm{i}/\hbar)p'q_2}\mathrm{e}^{-(\mathrm{i}/\hbar)pq_1}$$

$$= \frac{1}{2\pi\hbar}\int_{-\infty}^{\infty}\mathrm{d}p\;\mathrm{e}^{-(\mathrm{i}/\hbar)(p^2/2m)\Delta t}\mathrm{e}^{-(\mathrm{i}/\hbar)p(q_1-q_2)}$$

$$= \sqrt{\frac{m\mathrm{e}^{-\mathrm{i}\pi/2}}{2\pi\hbar\Delta t}}\mathrm{e}^{(\mathrm{i}/\hbar)m(q_1-q_2)^2/(2\Delta t)}\,.$$

In this calculation use was made of the relations (7.166b) and the formula (1.46) for the Gauss integral. The transition amplitude of the example then becomes

$$\langle q_2(t+\Delta t)|\,q_1(t)\rangle \simeq \sqrt{\frac{m\mathrm{e}^{-\mathrm{i}\pi/2}}{2\pi\hbar\Delta t}}\exp\frac{\mathrm{i}}{\hbar}\left\{\frac{m(q_2-q_1)^2}{2\Delta t}-\Delta t U(q_1)\right\}$$

$$\simeq \sqrt{\frac{m\mathrm{e}^{-\mathrm{i}\pi/2}}{2\pi\hbar\Delta t}}\exp\frac{\mathrm{i}}{\hbar}\left\{\frac{m(q_2-q_1)^2}{2\Delta t}-\Delta t\frac{U(q_1)+U(q_2)}{2}\right\}.$$

The expression in curly brackets is seen to be the product of the Lagrangian function, evaluated along a solution, and the time shift Δt. By definition this is the *action*. Indeed, by following the classical orbit one has $q(t') \simeq q_1 + ((t'-t)/\Delta t)(q_2-q_1)$ from which one concludes

$$S = \int\limits_{q_1=q(t)}^{q_2=q(t+\Delta t)} \mathrm{d}t'\; L(q,\dot{q}) \simeq \frac{m(q_2-q_1)^2}{2\Delta t} - \Delta t\frac{U(q_1)+U(q_2)}{2}.$$

Thus, the transition amplitude reads

$$\langle q_2(t+\Delta t)|\,q_1(t)\rangle \simeq \sqrt{\frac{m\mathrm{e}^{-\mathrm{i}\pi/2}}{2\pi\hbar\Delta t}}\,\mathrm{e}^{(\mathrm{i}/\hbar)S(q_2,q_1)}. \tag{7.169a}$$

At this point use is made of the superposition principle of quantum mechanics: The transition amplitude for a *finite* time interval is calculated by following the evolution of the system from the initial configuration $q_i(t_i)$ to its final configuration $q_f(t_f)$ by means of very many small steps of the kind of (7.169a). Using the notations

$$q_0 \equiv q_i,\; t_0 \equiv t_i,\; q_n \equiv q_f,\; t_n \equiv t_f \quad\text{and}\quad t_j = t_i + \frac{j}{n}(t_f - t_i)$$

the transition amplitudes are found to be

$$\langle q_f(t_f)|\,q_i(t_i)\rangle = \lim_{n\to\infty}\int\prod_{k=1}^{n-1}\mathrm{d}q_k\prod_{j=0}^{n-1}\langle q_{j+1}(t_{j+1})|\,q_j(t_j)\rangle$$

$$= \lim_{n\to\infty}\int\prod_{k=1}^{n-1}\mathrm{d}q_k\left(\frac{nm\mathrm{e}^{-\mathrm{i}\pi/2}}{2\pi\hbar\Delta(t_f-t_i)}\right)^{n/2}\exp\left\{\frac{\mathrm{i}}{\hbar}\int\limits_{t_i}^{t_f}\mathrm{d}t\; L(q,\dot{q})\right\}. \tag{7.169b}$$

In a somewhat symbolic notation this formula is written as follows

$$\langle q_f(t_f)|\,q_i(t_i)\rangle = \int \mathcal{D}[q]\exp\left\{\frac{\mathrm{i}}{\hbar}\int\limits_{t_i}^{t_f}\mathrm{d}t\; L(q,\dot{q})\right\}, \tag{7.169c}$$

where the integration "measure" $\mathcal{D}[q]$ is defined by the limit spelled out in (7.169b).

Remarks

1. It should be clear that only the initial and the final configurations are given. In the example these are the positions q_i and q_f, respectively. There is no information about which paths the system chooses between these boundary values, except for the case when $\hbar$ tends to zero.
2. In the limit $\hbar \to 0$ the exponential in the integrand oscillates rapidly and, therefore, gives a significant contribution only when the action S is nearly constant. This happens when S is stationary, i.e. when $q(t)$ is the physical orbit of classical mechanics which connects q_i and q_f. In this limit one returns to Hamilton's principle of classical mechanics.
3. As the superposition principle applies in quantum mechanics one can subdivide the path integral from "i" to "f" into two or more steps. For example, with $t_i < t_k < t_f$ one obtains
$$\int \mathcal{D}[q]\mathrm{e}^{(\mathrm{i}/\hbar)S_{fi}} = \int \mathrm{d}q(t) \int \mathcal{D}[q]\mathrm{e}^{(\mathrm{i}/\hbar)S_{fk}} \int \mathcal{D}[q]\mathrm{e}^{(\mathrm{i}/\hbar)S_{ki}} . \qquad (7.170)$$
For an infinitesimal change of the action one has
$$\delta \left\langle q_f(t_f) \middle| q_i(t_i) \right\rangle = \frac{\mathrm{i}}{\hbar} \int \mathcal{D}[q]\mathrm{e}^{(\mathrm{i}/\hbar)S_{fi}} \delta S_{fi} .$$
For example for $t \mapsto t + \delta t$ Eq. (7.159) gives $\delta S = -H\delta t$. From this one obtains
$$\delta \left\langle q(t+\delta t) \middle| q(t) \right\rangle = -\frac{\mathrm{i}}{\hbar} \left\langle q(t+\delta t) \middle| H \middle| q(t) \right\rangle \delta t . \qquad (7.171)$$
Obviously, this is the Schrödinger equation.
4. The generalization of the one-dimensional example worked out above, to three space dimensions is obvious and need not be worked out here.
5. More information on path integrals is found in several monographs [Feynman and Hibbs 1965], [Simon 1979], [Roepstorff 1992], [Reuter and Dittrich 2001]. There exist also tables of path integrals, cf. [Kleinert 1990], [Grosche and Steiner 1998].

7.7.3 Classical and Quantum Paths

Let t be an arbitrary intermediate time situated between the initial time t_i and the final time t_f,
$$t_i < t < t_f .$$
The product rule (7.170) for path integrals was seen to be a consequence of the superposition principle. This rule can also be applied to matrix elements of operators. If $A(t)$ is an operator in Heisenberg representation one has
$$\begin{aligned}\left\langle q_f(t_f) \middle| A(t) \middle| q_i(t_i) \right\rangle &= \left\langle q_f \middle| \mathrm{e}^{-(\mathrm{i}/\hbar)H(t_f-t)} A_0 \mathrm{e}^{-(\mathrm{i}/\hbar)H(t-t_i)} \middle| q_i \right\rangle \\ &= \int \mathrm{d}q' \int \mathrm{d}q'' \int \mathcal{D}[q]\, \mathrm{e}^{(\mathrm{i}/\hbar)S(q_f,t_f;q'',t)} \\ &\quad \times \left\langle q'' \middle| A_0 \middle| q' \right\rangle \int \mathcal{D}[q]\, \mathrm{e}^{(\mathrm{i}/\hbar)S(q',t;q_i,t)} . \end{aligned} \qquad (7.172)$$

This formula and the rule (7.170) can be applied to the position operator at the ordered times $t_1, t_2, \ldots, t_n$ all of which are assumed to lie in the interval (t_i, t_f),

$$t_i < t_1 < t_2 < \cdots < t_n < t_f \,.$$

One then obtains

$$\langle q_f(t_f) | \underline{q}(t_1)\underline{q}(t_2)\cdots\underline{q}(t_n) \,|\, q_i(t_i) \rangle$$

$$= \int \mathcal{D}[q] \, \big(q(t_1)q(t_2)\ldots q(t_n)\big) \exp\left\{ \frac{\mathrm{i}}{\hbar} \int_{t_i}^{t_f} \mathrm{d}t \; L(q,\dot{q}) \right\} . \qquad (7.173)$$

This result illustrates very nicely the nature of the path integral in the quantum world. If $\hbar$ were very small, or would even be sent to zero, then only the stationary action would contribute which, in turn, is only realized for the *classical* solution of the Euler-Lagrange equation which relates q_i with q_f. In the quantum case where $\hbar$ is a nonvanishing quantum of action, the particle is free to travel along all possible paths which connect q_i and q_f. However, the paths that the particle may choose are weighted by the exponential function $\exp\{(\mathrm{i}/\hbar)S_{fi}\}$ of the action. The idea of a particle trajectory is no longer meaningful. Yet, the quantum states still bear some properties of the corresponding classical dynamics.

These examples show that the method of path integrals shed light on another facet of quantum mechanics but is not of central importance for its formulation. Matters change, however, when one constructs quantum field theory. In spite of the observation that they are not mathematically well-defined, path integrals have become an important tool in this field and they are being used in many applications in classical and quantum field theories.

7.8 Path Integral for Field Theories

As we emphasized previously, field theory is characterized by nondenumerable, infinitely many degrees of freedom. The role of the finite number of degrees of freedom $q \equiv (q_1, q_2, \ldots, q_f)$ of a quantum mechanical system is taken over by *fields* $\phi(x)$ defined on spacetime. Intuitively, it seems plausible that path integrals might be generalized to field theories because many notions and general principles of canonical mechanics can be adapted to field theory, either by direct generalization or by analogy. Nevertheless, the comparison with formula (7.169c) shows that there appear certain difficulties not known in quantum mechanics. One of them is the question of how to generalize the integration measure $\mathcal{D}[q]$ to something like $\mathcal{D}[\phi]$ for fields.

In this section we start by deriving rules for differential calculus with functionals and develop a formal calculus for path integrals with fields. The chapter closes with the simplest example for a propagator.

7.8.1 The Functional Derivative

Let $\mathcal{S}(\mathbb{R}^n)$ be the space of test functions defined on a space $\mathbb{R}^n$. This is the linear space of all C^∞ functions on $\mathbb{R}^n$ which decrease faster at infinity than any inverse power of the distance to the origin, with all their derivatives having the same property. The corresponding dual space of continuous linear functionals, that is, the space of the tempered distributions on $\mathbb{R}^n$, is denoted by $\mathcal{S}'(\mathbb{R}^n)$.

The *functional derivative* of the distribution $\Delta \in \mathcal{S}'$ by a function $f \in \mathcal{S}$ is defined as follows:

$$\frac{\delta \Delta(f)}{\delta f(y)} := \lim_{\varepsilon \to 0} \frac{1}{\varepsilon} \left[\Delta\left(f(x) + \varepsilon\delta(x-y)\right) - \Delta\left(f(x)\right)\right] . \tag{7.174}$$

Here x and y are points in $\mathbb{R}^n$, $\delta(z)$ is Dirac's δ-distribution. A few examples will be helpful in illustrating this definition.

(i) Consider the distribution $\Delta(f) = \int \mathrm{d}^n x \; f(x)\delta(x)$. Using (7.174) one calculates

$$\begin{aligned} \frac{\delta \Delta(f)}{\delta f(y)} &= \lim_{\varepsilon \to 0} \frac{1}{\varepsilon} \left[\int \mathrm{d}^n x \; \left(f(x) + \varepsilon\delta(x-y)\right)\delta(x) - \int \mathrm{d}^n x \; f(x)\delta(x) \right] \\ &= \int \mathrm{d}^n x \; \delta(x-y)\delta(x) = \delta(y) . \end{aligned} \tag{7.175a}$$

(ii) As an even simpler example let Δ be the integral of the function f, i.e. $\Delta(f) = \int \mathrm{d}^n x \; f(x)$. Its functional derivative by f is found to be 1:

$$\frac{\delta \Delta(f)}{\delta f(y)} = \lim_{\varepsilon \to 0} \frac{1}{\varepsilon} \left[\int \mathrm{d}^n x \; \left(f(x) + \varepsilon\delta(x-y)\right) - \int \mathrm{d}^n x \; f(x) \right] = 1 . \tag{7.175b}$$

(iii) As is well known smooth functions may also be interpreted as distributions. So choosing $\Delta(f) = f$ one finds

$$\frac{\delta \Delta(f)}{\delta f(y)} = \lim_{\varepsilon \to 0} \frac{1}{\varepsilon} \left[f(x) + \varepsilon\delta(x-y) - f(x)\right] = \delta(x-y) . \tag{7.175c}$$

Examples (7.175b) and (7.175c) are helpful in deriving a general property of functional derivatives. Like in the second example choose $\Delta(f) = \int \mathrm{d}^n x \; f(x)\delta(x)$. Then, from (7.175c) one obtains

$$\frac{\delta \Delta(f)}{\delta f(y)} = 1 = \int \mathrm{d}^n x \; \delta(x-y) = \int \mathrm{d}^n x \; \frac{\delta f(x)}{\delta f(y)} .$$

Choosing now $\Delta(f) = \int \mathrm{d}^n x \; \Gamma(f(x))$ where Γ is a tempered distribution acting on the function f, the definition (7.174) yields

$$\begin{aligned} \frac{\delta \Delta(f)}{\delta f(y)} &= \lim_{\varepsilon \to 0} \frac{1}{\varepsilon} \int \mathrm{d}^n x \; \left[\Gamma\left(f(x) + \varepsilon\delta(x-y)\right) - \Gamma\left(f(x)\right)\right] \\ &= \int \mathrm{d}^n x \; \frac{\mathrm{d}\Gamma(f(x))}{\mathrm{d}f} \delta(x-y) = \left. \frac{\mathrm{d}\Gamma}{\mathrm{d}f} \right|_y . \end{aligned}$$

In turn, calculating the functional derivative of Γ by f one has

$$\frac{\delta\Gamma(f(x))}{\delta f(y)} = \left.\frac{\mathrm{d}\Gamma}{\mathrm{d}f}\right|_y \delta(x-y)\,.$$

Thus, one obtains the general formula

$$\frac{\delta\Delta(f)}{\delta f(y)} = \int \mathrm{d}^n x\, \frac{\delta\Gamma(f(x))}{\delta f(y)}\,. \tag{7.176}$$

We note that the functional derivative (7.174) satisfies the rules of differential calculus. In particular, the product rule and the chain rule are fulfilled.

7.8.2 Functional Power Series and Taylor Series

Define the following series

$$\begin{aligned}\Pi\,(f(x)) &= K_0(x) + \int \mathrm{d}^n x_1\; K_1(x,x_1)f(x_1)\\ &+ \int \mathrm{d}^n x_1 \int \mathrm{d}^n x_2\; K_2(x,x_1,x_2)f(x_1)f(x_2) + \cdots\,.\end{aligned} \tag{7.177a}$$

Making use of (7.174) one derives the following formulae for functional derivatives of $\Pi \in \mathcal{S}'$ by $f \in \mathcal{S}$:

$$\begin{aligned}&\Pi(0) = K_0(x)\,, \qquad \left.\frac{\delta\Pi(f(x))}{\delta f(y)}\right|_{f=0} = K_1(x,y)\\ &\left.\frac{\delta^2\Pi(f(x))}{\delta f(y_1)\delta f(y_2)}\right|_{f=0} = K_2(x,y_1,y_2) + K_2(x,y_2,y_1)\,.\end{aligned} \tag{7.177b}$$

A case of special interest for quantum field theory is one where the "coefficients" $K_m(x,y_1,y_2,\ldots,y_m)$ are totally symmetric in the variables $y_1,\ldots,y_m$. In this case one has

$$K_m(x,y_1,\ldots,y_m) = \frac{1}{m!}\left.\frac{\delta^m\Pi(f)}{\delta f(y_1)\cdots\delta f(y_m)}\right|_{f=0}\,. \tag{7.178}$$

To give an example let us calculate the vacuum expectation values of the time ordered product of m copies of the classical field $\phi(x)$, viz.

$$K_m(x,y_1,\ldots,y_m) = \langle 0|T\,\phi(y_1)\cdots\phi(y_m)\,|0\rangle\,.$$

These functions are independent of x. Consider then

$$\Pi(f) := \sum_{m=0}^{\infty}\frac{1}{m!}\int \mathrm{d}^n y_1\ldots\mathrm{d}^n y_m\;\langle 0|T\,\phi(y_1)\cdots\phi(y_m)\,|0\rangle\; f(y_1)\cdots f(y_m)\,.$$

The m-th functional derivative at $f=0$ is found to be

$$\left.\frac{\delta^m\Pi(f)}{\delta f(y_1)\cdots\delta f(y_m)}\right|_{f=0} = \langle 0|T\,\phi(y_1)\cdots\phi(y_m)\,|0\rangle\,. \tag{7.179}$$

The right-hand side of (7.179) shows the Green functions which belong to the field theory of a scalar field ϕ which is still classical. If one had a closed expression for $\Pi(f)$ this would allow to derive all Green functions $\langle 0|T\ \cdots|0\rangle$ by functional derivatives.

These considerations and the examples given show that it is possible to construct functional Taylor series. Let f and g be functions in $\mathcal{S}$, λ a real or complex variable. Then for a $\Gamma \in \mathcal{S}'$ one has

$$\Gamma(f+\lambda\, g) = \sum_{k=0}^{\infty} \frac{\lambda^k}{k!} \int \mathrm{d}^n x_1 \cdots \int \mathrm{d}^n x_k\ g(x_1)\cdots g(x_k) \frac{\delta^k \Gamma(f)}{\delta f(x_1)\cdots\delta f(x_k)} .$$

Evaluating this expansion at the point $\lambda = 1$, with $f = 0$, one obtains

$$\Gamma(f+\lambda\, g)|_{\lambda=1} = \sum_{k=0}^{\infty} \frac{1}{k!} \int \mathrm{d}^n x_1 \cdots \int \mathrm{d}^n x_k\ g(x_1)\cdots g(x_k) \left.\frac{\delta^k \Gamma(f)}{\delta f(x_1)\cdots\delta f(x_k)}\right|_{f=0} . \tag{7.180}$$

This is the analogue of an ordinary Taylor series in one variable.

7.8.3 Generating Functional

The true physical ground state of a quantum field theory can be very different from the vacuum of a theory without interaction. An example which illustrates this observation is provided by quantum electrodynamics in which *interacting* electron fields are different from *free* electron fields and where the ground state possesses nontrivial internal structure (cf. Chap. 10). A similar statement applies to nonrelativistic theories such as the theory of superconductivity or the theory of superfluidity: The ground states contain correlations which do not exist in the free theory of N electrons.

In defining the path integral a basic concept comes from investigating the transition of the ground state from an initial configuration at $t = -\infty$ to a final configuration at $t = +\infty$ under the action of an external interaction. The response of the system contains information on the structure of the theory and of its ground state. In fact, as we shall see, this response reflects the full dynamics of the system.

Formally the generalization of Feynman's path integral to infinitely many, non-denumerable degrees of freedom goes as follows: For each individual field, by analogy to (7.169c), one writes down the ansatz

$$\langle \phi_2 | \phi_1 \rangle = N \int_{\phi_1}^{\phi_2} \mathcal{D}[\phi] \exp\left\{ \mathrm{i} \int_1^2 \mathrm{d}^4 x\ \mathcal{L}(x) \right\} , \tag{7.181}$$

where ϕ_1 and ϕ_2 are two given field configurations and where the second integral

$$\int_1^2 \mathrm{d}^4x \equiv \int_{t_1}^{t_2} \mathrm{d}x_0 \int \mathrm{d}^3x$$

is to be understood in the sense that $\phi_i \equiv \phi(t_i, \vec{x})$, $i = 1, 2$. The factor N is a normalization factor which need not be specified at this point. For the sake of simplicity we consider here a single scalar field. We note that the generalization to cases of more than one kind of boson field, with or without spin, is straightforward. The application to *fermionic* fields, however, must be discussed separately.

Adding a coupling term to some external source j the expression (7.181) is replaced by

$$\langle\phi_2|\phi_1\rangle = N \int_{\phi_1}^{\phi_2} \mathcal{D}[\phi] \exp\left\{\mathrm{i} \int_1^2 \mathrm{d}^4x \; [\mathcal{L}(x) + \phi(x)j(x)]\right\} . \tag{7.182}$$

For example, in the case of a real scalar field the action reads

$$S = \int \mathrm{d}^4x \left[\frac{1}{2}\partial_\mu\phi\partial^\mu\phi - \frac{1}{2}m^2\phi^2 - V(\phi)\right] . \tag{7.183}$$

Here the term $V(\phi)$ denotes a possible self-coupling of the field, that is to say, a kind of potential. Define then

$$W[j] := N \int \mathcal{D}[\phi] \exp\left\{\mathrm{i} \int \mathrm{d}^4x \left[\frac{1}{2}\partial_\mu\phi\partial^\mu\phi - \frac{1}{2}m^2\phi^2 - V(\phi) + j\phi\right]\right\} . \tag{7.184}$$

This definition raises two difficulties which warrant some comments. In spite of the fact that it is mathematically not well-defined, the formal integration measure $\mathcal{D}[\phi]$ may be viewed qualitatively as an infinite product of elements $\mathrm{d}\phi_k$, with $\phi_k \equiv \phi(x_k)$, on a discretized version of spacetime. Furthermore, even if it were well-defined, an integral such as (7.184) would initially not be convergent. This can be remedied either by replacing the term m^2 by $(m^2 - \mathrm{i}\varepsilon)$, or by continuing the functional $W[j]$ to a space $\mathbb{R}^4$ to which one assigns a Euclidean structure. If one chooses the second alternative some preparation is needed.

Let vectors on a Euclidean four-dimensional space be denoted by x_E. Let $x_\mathrm{E} = (\vec{x}, x^4)^T$ and $x^4 = \mathrm{i}\, x^0$. Then one has

$$\mathrm{d}^4x = -\mathrm{i}\mathrm{d}^4x_\mathrm{E} , \quad \partial_\mu\phi\, \partial^\mu\phi = -\partial_\mu^\mathrm{E}\phi\, \partial_\mu^\mathrm{E}\phi,$$

$$x_\mathrm{E}^2 = \sum_{i=1}^4 (x^i)^2 = \vec{x}^2 + (x^4)^2 = \vec{x}^2 - (x^0)^2 = -x^2 .$$

The corresponding definitions in Euclidean momentum space are chosen such that $k^4x^4 = k^0x^0$ so that conventions for the signs in plane waves are preserved. With $k_\mathrm{E} = (\vec{k}, k^4)^T$ this is achieved by the choice $k_\mathrm{E}^4 = -\mathrm{i}\; k^0$. Indeed, this entails $\mathrm{d}^4k = \mathrm{i}\; \mathrm{d}^4k_\mathrm{E}$ and $k_\mathrm{E}^2 = -k^2$. Furthermore, one has

$$k \cdot x = k^0x^0 - \vec{k} \cdot \vec{x} = k^4x^4 - \vec{k}_\mathrm{E} \cdot \vec{x}_\mathrm{E} .$$

The expression (7.184) is replaced by

$$W[j]_{\mathrm{E}} := N_{\mathrm{E}} \int \mathcal{D}[\phi] \exp\left\{ -\int \mathrm{d}^4x \left[\frac{1}{2} \partial^{\mathrm{E}}_{\mu}\phi \partial^{\mathrm{E}}_{\mu}\phi + \frac{1}{2} m^2\phi^2 + V(\phi) - j\phi \right] \right\} . \tag{7.185}$$

This formula is the starting point for an actual calculation of a path integral. As such a calculation quickly becomes rather technical and fairly involved I restrict matters to some comments and to the essence of the method. Even though the integration measure is not well-defined the functional $W[j]_{\mathrm{E}}$ may be evaluated provided it has the formal structure of a Gaussian integral. An example from analysis may be helpful here. Let I be the integral in one real dimension

$$I := \int \mathrm{d}x \; \mathrm{e}^{-a(x)}$$

where $a(x)$ is a differentiable function, and let x_0 be a stationary point of this function. This means that $a(x)$ can be expanded around x_0,

$$a(x) \simeq a(x_0) + \frac{1}{2}(x - x_0)^2 a''(x_0) .$$

The integral then is approximately a Gaussian integral

$$I \simeq \mathrm{e}^{-a(x_0)} \int \mathrm{d}x \; \mathrm{e}^{-(x-x_0)^2/2} a''(x_0) .$$

In field theory the analogues of stationary values x_0 are the solutions ϕ_0 of the classical equations of motion. Expanding around such solutions leads to Gaussian integrals also in field theory.

7.8.4 An Example: Propagator of the Scalar Field

As an example we calculate $W^{(0)}[j]$ for a free scalar field, i.e. an action with vanishing potential $V(\phi) \equiv 0$,

$$W^{(0)}[j] := N \int \mathcal{D}[\phi] \exp\left\{ \mathrm{i} \int \mathrm{d}^4x \; \left[\tfrac{1}{2} \partial_{\mu}\phi \partial^{\mu}\phi - \tfrac{1}{2}(m^2 - \mathrm{i}\varepsilon)\phi^2 + \phi j \right] \right\} . \tag{7.186}$$

In this example, and for the sake of illustration, we enforce convergence by means of the first alternative described above. If Fourier transform in dimension four is defined by

$$\widetilde{F}(p) = \int \frac{\mathrm{d}^4x}{(2\pi)^2} \mathrm{e}^{-\mathrm{i}px} F(x) ,$$

then $\tilde{F}^*(p) = F(-p)$. The argument of the exponential function in (7.186) goes over into

$$\frac{\mathrm{i}}{2} \int \mathrm{d}^4p \, \{ \widetilde{\phi}'(p) \left[p^2 - m^2 + \mathrm{i}\varepsilon \right] \widetilde{\phi}'(-p) \\ - \widetilde{j}(p) \left[p^2 - m^2 + \mathrm{i}\varepsilon \right]^{-1} \widetilde{j}(-p) \} , \tag{7.187a}$$

with the abbreviation

$$\widetilde{\phi}'(p) = \widetilde{\phi}(p) + \frac{1}{p^2 - m^2 + \mathrm{i}\varepsilon}\widetilde{j}(p)\,. \tag{7.187b}$$

Spelled out more explicitly, this transformation of the integral comes about as follows:

$$\frac{\mathrm{i}}{2}\int \mathrm{d}^4x\left[\partial_\mu\phi\partial^\mu\phi - (m^2 - \mathrm{i}\varepsilon)\phi^2 + 2\phi j\right] = \frac{\mathrm{i}}{2(2\pi)^4}\int \mathrm{d}^4x$$

$$\iint \mathrm{d}^4p\mathrm{d}^4p'\mathrm{e}^{\mathrm{i}px}\mathrm{e}^{\mathrm{i}p'x}\left[(p^2 - m^2 + \mathrm{i}\varepsilon)\,\widetilde{\phi}(p)\widetilde{\phi}(p') + 2\widetilde{\phi}(p\widetilde{j}(p)')\right]$$

$$= \frac{\mathrm{i}}{2}\int \mathrm{d}^4p\left[\widetilde{\phi}(p)(p^2 - m^2 - \mathrm{i}\varepsilon)\widetilde{\phi}(-p) + \widetilde{\phi}(p)\widetilde{j}(-p) + \widetilde{\phi}(-p)\widetilde{j}(p)\right].$$

Inserting the abbreviation $\tilde{\phi}'(p)$ yields the expression shown in (7.187a).

The additional term in (7.187b) which is proportional to $\tilde{j}$ acts like a constant term in the functional integral. Therefore, one has $\mathcal{D}[\phi'] = \mathcal{D}[\phi]$ and the prime on the variable may be omitted. Thus, the functional to be calculated becomes

$$W^{(0)}[j] = N\exp\left\{-\frac{\mathrm{i}}{2}\int \mathrm{d}^4p\frac{\widetilde{j}(p)\widetilde{j}(-p)}{p^2 - m^2 + \mathrm{i}\varepsilon}\right\}$$

$$\times\int \mathcal{D}[\phi]\exp\left\{\frac{\mathrm{i}}{2}\int \mathrm{d}^4x\left[\partial_\mu\phi\partial^\mu\phi - (m^2 - \mathrm{i}\varepsilon)\phi^2\right]\right\}$$

$$\equiv W^{(0)}[0]\exp\left\{-\frac{\mathrm{i}}{2}\int \mathrm{d}^4p\frac{\widetilde{j}(p)\widetilde{j}(-p)}{p^2 - m^2 + \mathrm{i}\varepsilon}\right\}.$$

At this point one can return to position space by inverse Fourier transform. One obtains

$$-\frac{\mathrm{i}}{2}\int \mathrm{d}^4p\frac{\widetilde{j}(p)\widetilde{j}(-p)}{p^2 - m^2 + \mathrm{i}\varepsilon}$$

$$= -\frac{\mathrm{i}}{2(2\pi)^4}\int \mathrm{d}^4p\iint \mathrm{d}^4x\mathrm{d}^4y\;\mathrm{e}^{-\mathrm{i}p(x-y)}\frac{j(x)j(y)}{p^2 - m^2 + \mathrm{i}\varepsilon}$$

$$= -\frac{1}{2}\iint \mathrm{d}^4x\mathrm{d}^4y\;j(x)\Delta(x - y)j(y)\,.$$

The distribution $\Delta(x - y)$ denotes the propagator (7.60) obtained earlier, viz.

$$\Delta(x - y) = \frac{\mathrm{i}}{(2\pi)^4}\int \mathrm{d}^4p\;\mathrm{e}^{-\mathrm{i}p(x-y)}\frac{1}{p^2 - m^2 + \mathrm{i}\varepsilon} \tag{7.188}$$

$$\equiv P_F(x - y; m)\,. \tag{7.189}$$

We close this section with a few comments on these results.

Remarks

1. It is remarkable to rediscover the propagator $\Delta(x - y)$ in the formalism of path integrals even though the fields were not quantized explicitly. This is a strong hint at the fact that this formalism is a genuine alternative to canonical quantization.

2. It is instructive to calculate the functional derivatives of $W[j]$ by the external source j. What one finds is this: The first derivative as well as all higher *odd* derivatives vanish at $j = 0$. The *even* derivatives at $j = 0$ can all be expressed in terms of the propagator (7.188) or by products of this propagator. In close analogy to the series (7.180) the functional $W[j]$ is given by

$$W[j] = \sum_{n=0}^{\infty} \frac{\mathrm{i}^n}{n!} \int \mathrm{d}^4 x_1 \cdots \int \mathrm{d}^4 x_n \times j(x_1) \ldots j(x_n)\, G_n(1, \ldots, n) \,, \tag{7.190a}$$

$$\text{with} \quad G_n(1, \ldots, n) = \frac{1}{\mathrm{i}^n} \frac{\delta}{\delta j(x_1)} \cdots \frac{\delta}{\delta j(x_n)} W[j] \bigg|_{j=0} \,. \tag{7.190b}$$

So, for instance, in the example of a free scalar field one obtains the formulae

$$G_{2n+1}^{(0)} = 0 \,,$$
$$G_2^{(0)}(u_1, u_2) = P_F(u_1 - u_2) \,,$$
$$G_4^{(0)}(u_1, u_2, u_3, u_4) = P_F(u_1 - u_2) P_F(u_3 - u_4) + P_F(u_1 - u_3) P_F(u_2 - u_4) + P_F(u_2 - u_3) P_F(u_1 - u_4) \,.$$

3. Since all Green functions $G_{2n}^{(0)}$ are reducible to $G_2^{(0)}$ it is meaningful not to expand the functional $W[j]$ proper but rather the exponent in its exponential representation

$$W[j] = \mathrm{e}^{\mathrm{i} Z[j]} \,. \tag{7.191}$$

The series expansion

$$\mathrm{i} Z[j] = \sum_{n=0}^{\infty} \frac{\mathrm{i}^n}{n!} \int \mathrm{d}^4 x_1 \cdots \int \mathrm{d}^4 x_n \, G^{(c)}(1, \ldots, n) \tag{7.192}$$

leads to the modified Green functions, called *connected Green functions* $G^{(c)}$.

7.8.5 Complex Scalar Field and Path Integrals

It is not difficult to extend the example (7.186) to the case of a complex scalar field or the case of multi-component scalar fields. In the case of a complex scalar field ϕ and ϕ^* are independent variables. Likewise, to the exterior source j one must add its conjugate j^*. The functional $W[j, j^*]$ for the free complex scalar field then reads

$$W^{(0)}[j, j^*] := N \int \mathcal{D}[\phi] \int \mathcal{D}[\phi^*] \exp\left\{ \mathrm{i} \int \mathrm{d}^4 x \, \left[\tfrac{1}{2} \partial_\mu \phi^* \partial^\mu \phi - \tfrac{1}{2} (m^2 - \mathrm{i}\varepsilon) \phi^* \phi + \phi^* j + j^* \phi \right] \right\} \,. \tag{7.193}$$

The argument of the exponential function contains terms which are sesquilinear in ϕ and terms which are linear in ϕ or ϕ^*. A possible self-interaction $V(\phi)$ would also be a function of $(\phi^*\phi)$ like the mass term. All individual terms are real. Even if the complex field ϕ is replaced by a multi-component field $\Phi = (\phi_1, \dots, \phi_N)^T$ the structure of the functional (7.193) remains unchanged. Of course, in this case one has to introduce a corresponding number of external source terms which might be grouped in a column $J = (j_1, \dots, j_N)^T$ as well. The path integral to be calculated then has the form

$$I[J, J^*] = \int \mathcal{D}[\phi_1] \int \mathcal{D}[\phi_1^*] \cdots \int \mathcal{D}[\phi_N] \int \mathcal{D}[\phi_N^*]$$
$$\times \exp\{-\left(\Phi^\dagger \mathbf{C}\Phi - \Phi^\dagger J - J^\dagger \Phi\right)\}, \qquad (7.194)$$

where $\mathbf{C}$ is a hermitian $N \times N$-matrix which we assume to be nonsingular. All its eigenvalues are different from zero. Furthermore, for physical reasons, they must be positive. Note the identity

$$\Phi^\dagger \mathbf{C}\Phi - \Phi^\dagger J - J^\dagger \Phi = \left(\Phi - \mathbf{C}^{-1}J\right)^\dagger \mathbf{C}\left(\Phi - \mathbf{C}^{-1}J\right) - J^\dagger \mathbf{C}^{-1} J \, .$$

Thus, defining $\Psi = \Phi - \mathbf{C}^{-1}J$ and noting that the last term in this transformation is independent of the integration variables, one is led to calculate

$$I[J, J^*] = \int \mathcal{D}[\psi_1] \int \mathcal{D}[\psi_1^*] \cdots \int \mathcal{D}[\psi_N] \int \mathcal{D}[\psi_N^*]$$
$$\times \exp\{-\left(\Psi^\dagger \mathbf{C}\Psi\right)\} \, \mathrm{e}^{-J^\dagger \mathbf{C}^{-1} J} \, . \qquad (7.195)$$

Also this integral is a Gaussian one to which one associates a value by the following reasoning.

Let $z \in \mathbb{C}$ be a complex variable, and let $K(\alpha)$ be the integral

$$K(\alpha) = \int \mathrm{d}z \int \mathrm{d}z^* \, \mathrm{e}^{-\alpha |z|^2} \, .$$

One writes z in a polar decomposition, $z = r\mathrm{e}^{\mathrm{i}\phi}$, and calculates the modulus of the Jacobian of the transformation from (z, z^*) to (r, ϕ), to obtain

$$K(\alpha) = 2 \int_0^\infty \mathrm{d}r \int_0^{2\pi} \mathrm{d}\phi \, \mathrm{e}^{-\alpha r^2} = \frac{2\pi}{\alpha} \, .$$

In a next step let $Z = (z_1, \dots, z_N)^T$ be a column vector of complex variables z_1 to z_N and let $\mathbf{C}$ be a nonsingular hermitian matrix with positive eigenvalues γ_i, $i = 1, \dots, N$. Denote by $\mathbf{U}$ the unitary matrix which diagonalizes $\mathbf{C}$, viz.

$$\mathbf{C} = \mathbf{U}^\dagger \overset{0}{\mathbf{C}} \mathbf{U} \, .$$

Defining $\mathbf{U}Z =: W$ one has

$$\prod_{i=1}^N \int \frac{\mathrm{d}z_i}{2\pi} \frac{\mathrm{d}z_i^*}{2\pi} \, \mathrm{e}^{-Z^\dagger \mathbf{C} Z} = \prod_{i=1}^N \int \frac{\mathrm{d}w_i}{2\pi} \frac{\mathrm{d}w_i^*}{2\pi} \, \mathrm{e}^{-W^\dagger \overset{0}{\mathbf{C}} W}$$
$$= \frac{1}{\gamma_1 \gamma_2 \cdots \gamma_N} = \frac{1}{\det \mathbf{C}} \, .$$

These calculations show that to the symbolic integral (7.194) one can associate the following expression:

$$I[J, J^*] = \frac{1}{\det \mathbf{C}} e^{-J^\dagger \mathbf{C}^{-1} J} . \tag{7.196}$$

This expression contains the determinant $\det \mathbf{C}$ in its denominator. Of course, it exists only if none of the eigenvalues γ_i is equal to zero. We note in passing that this expression may alternatively serve as a starting point for the calculation of the Green functions by taking its functional derivatives by the source terms J and J^*.

As a final comment we note that $I[J, J^*]$ can be defined also in case the determinant of $\mathbf{C}$ vanishes, i.e. when one or several of its eigenvalues γ_k are equal to zero. In such a situation one modifies the "integration" measure by introducing δ-distributions $\delta(\gamma_k)$ for those eigenvalues.

Scattering Matrix and Observables in Scattering and Decays

8

Introduction As an interlude in the analysis of canonical field quantization, this section describes important concepts of scattering theory for Lorentz covariant quantum field theories that will be needed for the calculation of observables such as scattering cross sections and decay probabilities. The framework thus provided which rests on general, physically plausible assumptions, though fairly general is somewhat abstract. This is the reason why, in a first step, I turn back to nonrelativistic scattering theory, a topic that the reader is already familiar with. This theory is reformulated and formalized in such a way that the notions used in the covariant theory where particle creation and annihilation are allowed, are motivated by the analogy to the nonrelativistic case. As a bonus we will obtain formulae which allow to convert the (complex) transition amplitudes obtained from a perturbation series, into formulae for cross sections or decay widths.

8.1 Nonrelativistic Scattering Theory in an Operator Formalism

As a preparation to the definition of the scattering matrix, a notion of central importance in quantum field theory, we take up potential scattering on the basis of the Schrödinger equation. This leads to the definition of the T-matrix and, by the same token, yields the relation between the T-matrix and the scattering amplitude $f(\theta)$.

8.1.1 The Lippmann-Schwinger Equation

The stationary Schrödinger equation in position space has the form, using natural units,

$$(H_0 - (E - U(\boldsymbol{x}))\,\psi(\boldsymbol{x}) = 0\,, \quad \text{with} \quad H_0 = -\frac{1}{2m}\boldsymbol{\Delta} \tag{8.1}$$

F. Scheck, *Quantum Physics*, DOI: 10.1007/978-3-642-34563-0_8,

The solutions of the force-free equation, i.e. with $U(\boldsymbol{x}) = 0$, with energy $E = \boldsymbol{k}^2/2m$, and definite momentum, are the plane waves

$$\phi(\boldsymbol{k}, \boldsymbol{x}) = \mathrm{e}^{\mathrm{i}\boldsymbol{k}\cdot\boldsymbol{x}} .$$

The way they are normalized here, they are orthogonal and complete in the following sense,

$$\int \mathrm{d}^3x\, \phi^*(\boldsymbol{k}', \boldsymbol{x})\phi(\boldsymbol{k}, \boldsymbol{x}) = (2\pi)^3\delta(\boldsymbol{k}' - \boldsymbol{k}), \tag{8.2}$$

$$\int \mathrm{d}^3k\, \phi^*(\boldsymbol{k}, \boldsymbol{x}')\phi(\boldsymbol{k}, \boldsymbol{x}) = (2\pi)^3\delta(\boldsymbol{x}' - \boldsymbol{x}) . \tag{8.3}$$

The scattering solutions of the full Schrödinger equation which contain an *out*going or an *in*coming spherical wave $\mathrm{e}^{\pm\mathrm{i}\kappa r}/r$ are denoted by $\psi_{\text{out}}(\boldsymbol{k}, \boldsymbol{x})$ and $\psi_{\text{in}}(\boldsymbol{k}, \boldsymbol{x})$, respectively. In these expressions $\kappa = |\boldsymbol{k}|$ is the modulus of the spatial momentum long *before* or long *after* the scattering, respectively. In this notation the integral equation equivalent to the Schrödinger equation reads, cf. (2.26),

$$\psi_{\text{out/in}}(\boldsymbol{k}, \boldsymbol{x}) = \phi(\boldsymbol{k}, \boldsymbol{x}) + \int \mathrm{d}^3x'\, G_0^{\text{out/in}}(\boldsymbol{x}, \boldsymbol{x}')U(\boldsymbol{x}')\psi_{\text{out/in}}(\boldsymbol{k}, \boldsymbol{x}') . \tag{8.4}$$

Note that the definition of the Green function is somewhat modified, viz.

$$G_0^{\text{out/in}}(\boldsymbol{x}, \boldsymbol{x}') = -\frac{m}{2\pi}\frac{\mathrm{e}^{\pm\mathrm{i}\kappa|\boldsymbol{x}-\boldsymbol{x}'|}}{|\boldsymbol{x} - \boldsymbol{x}'|} \tag{8.5}$$

The choice of the factor in front has no other purpose than to free the second term in (8.4) from all factors that differ from 1. This is useful when one writes this equation in a way independent of the representation. The two signs in (8.5) express the different asymptotics of ψ: As we noted in Sect. 2.4 the plus sign asymptotically yields an outgoing spherical wave, while the minus sign yields an incoming spherical wave. As an equivalent representation, the Green functions can be written as Fourier integrals, viz.

$$G_0^{\text{out/in}}(\boldsymbol{x}, \boldsymbol{x}') = \frac{2m}{(2\pi)^3}\int \mathrm{d}^3k'\, \frac{\mathrm{e}^{\mathrm{i}\boldsymbol{k}'\cdot(\boldsymbol{x}-\boldsymbol{x}')}}{\kappa^2 - \kappa'^2 \pm \mathrm{i}\varepsilon} .$$

We prove this formula for the positive sign, for the sake of an example. It is useful to introduce spherical polar coordinates for the integration variable,

$$\mathrm{d}^3k' = \kappa'^2\, \mathrm{d}\kappa' \sin\theta_k\, \mathrm{d}\phi_k \equiv \kappa'^2\, \mathrm{d}\kappa'\, \mathrm{d}\Omega$$

and to take the 3-axis in the direction of $\boldsymbol{x} - \boldsymbol{x}'$. Write $|\boldsymbol{x} - \boldsymbol{x}'| = \varrho$. Then one has

$$\begin{aligned}\int \mathrm{d}^3k'\, \frac{\mathrm{e}^{\mathrm{i}\boldsymbol{k}\cdot(\boldsymbol{x}-\boldsymbol{x}')}}{\kappa^2 - \kappa'^2 + \mathrm{i}\varepsilon} &= 2\pi\int_0^\infty \kappa'^2\, \mathrm{d}\kappa'\, \frac{\mathrm{e}^{\mathrm{i}\kappa'\varrho} - \mathrm{e}^{-\mathrm{i}\kappa'\varrho}}{(\kappa^2 - \kappa'^2 + \mathrm{i}\varepsilon)\mathrm{i}\kappa'\varrho} \\ &= -\pi\int_{-\infty}^\infty \mathrm{d}\kappa'\, \frac{(\mathrm{e}^{\mathrm{i}\kappa'\varrho} - \mathrm{e}^{-\mathrm{i}\kappa'\varrho})\kappa'}{(\kappa' - \kappa - \mathrm{i}\varepsilon)(\kappa' + \kappa + \mathrm{i}\varepsilon)\mathrm{i}\varrho} .\end{aligned}$$

Note that κ is positive. Apply now Cauchy's integral theorem in the following way. The integral over $\exp\{i\kappa'\varrho\}$ is closed by a semi-circle in the upper half of the complex κ'-plane. Thus, one obtains the residue of the pole at $\kappa' = \kappa + i\varepsilon$, multiplied by $2\pi i$. The integral over the second term with $\exp\{-i\kappa'\varrho\}$ is closed by means of a semi-circle in the lower half-plane and yields the residue of the pole at $\kappa' = -\kappa - i\varepsilon$, multiplied by $2\pi i$. Taking care of the orientation of the contour integrals one sees that one obtains the same contribution as in the first case. Thus, the sum of the two contributions is

$$\int d^3k' \, \frac{e^{ik\cdot(x-x')}}{\kappa^2 - \kappa'^2 + i\varepsilon} = -2\pi^2 \frac{e^{i\kappa\varrho}}{\varrho} \, ;$$

which, upon insertion, yields the result given above.

The integral equation (8.4) contains the same information as (8.1), but it is supplemented by the given asymptotics of the scattering wave. In what follows our aim is to write this equation as an operator equation, in a form independent of the representation. The operator which is the inverse of the operator $(E - H_0)$ is denoted by $(E - H_0)^{-1}$. Its eigenfunctions are the plane waves, its eigenvalues follow from

$$(E - H_0)^{-1}\phi(\boldsymbol{k}', \boldsymbol{x}) = \frac{1}{E - E'}\phi(\boldsymbol{k}', \boldsymbol{x}) = \frac{2m}{k^2 - k'^2}\phi(\boldsymbol{k}', \boldsymbol{x}) \, ,$$

provided the energies are chosen such that $E' \neq E$ and $E' > 0$. Using this operator the differential equation (8.1) and the integral equation (8.4) can also be rewritten:

$$\psi(\boldsymbol{k}, \boldsymbol{x}) = (E - H_0)^{-1} U(\boldsymbol{x})\psi(\boldsymbol{k}, \boldsymbol{x}) \, ,$$

$$\psi_{\text{out/in}}(\boldsymbol{k}, \boldsymbol{x}) = \phi(\boldsymbol{k}, \boldsymbol{x}) + (E - H_0 \pm i\varepsilon)^{-1} U(\boldsymbol{x})\psi_{\text{out/in}}(\boldsymbol{k}, \boldsymbol{x}) \, .$$

The second of these equations deserves a comment. The functions (8.5) are Green functions for the operator $(E - H_0 \pm i\varepsilon)^{-1}$ in the following sense: Write the integral equation (8.4) as

$$\psi_{\text{out/in}}(\boldsymbol{k}, \boldsymbol{x}) = \phi(\boldsymbol{k}, \boldsymbol{x}) + \int d^3x' \, (E - H_0 \pm i\varepsilon)^{-1} \, \delta(\boldsymbol{x}' - \boldsymbol{x}) U(\boldsymbol{x}')\psi_{\text{out/in}}(\boldsymbol{k}, \boldsymbol{x}') \, ,$$

insert here the completeness relation (8.3), and note that the functions ϕ are eigenfunctions of the inverse operator,

$$(E - H_0 \pm i\varepsilon)^{-1} \, \phi(\boldsymbol{k}', \boldsymbol{x}) = \left(E - E' \pm i\varepsilon\right)^{-1} \phi(\boldsymbol{k}', \boldsymbol{x}) \, .$$

This then leads back to the original form (8.4).

Even though, so far, all equations were written in position space they can be formulated in a form that is independent of this specific representation. Denote the operators (8.5) by G_0^{out} and G_0^{in} when written in representation-free form, and write the plane waves as $|\phi_{\boldsymbol{k}}\rangle$, and the scattering states as $|\psi_{\text{out/in}}\rangle$. The relation to the representation in position space is recovered by means of

$$\langle \boldsymbol{x} | G_0^{\text{out/in}} | \boldsymbol{x}' \rangle = G_0^{\text{out/in}}(\boldsymbol{x}, \boldsymbol{x}') \, ,$$

$$\langle \boldsymbol{x} \mid \psi_{\text{out/in}} \rangle = \psi_{\text{out/in}}(\boldsymbol{k}, \boldsymbol{x}) \, , \quad \langle \boldsymbol{x} \mid \phi_{\boldsymbol{k}} \rangle = e^{i\boldsymbol{k}\cdot\boldsymbol{x}} \, .$$

Thus, in position space the action of these Green operators on states $|\psi\rangle$ is given by the integral

$$\langle \boldsymbol{x}|G_0^{\text{out/in}}\left|\psi_{\text{out/in}}\right\rangle = \int \mathrm{d}^3x' \langle \boldsymbol{x}|G_0^{\text{out/in}}\left|\boldsymbol{x}'\right\rangle \left\langle \boldsymbol{x}'\right| \psi_{\text{out/in}}\rangle$$
$$= \int \mathrm{d}^3x'\ G_0^{\text{out/in}}(\boldsymbol{x},\boldsymbol{x}')\psi_{\text{out/in}}(\boldsymbol{k},\boldsymbol{x}')\,.$$

The operators G_0^{out} and G_0^{in} are adjoint of one another, $(G_0^{\text{out}})^\dagger = G_0^{\text{in}}$. Defining

$$G_0^{\text{out/in}} = (E - H_0 \pm \mathrm{i}\varepsilon)^{-1}\,, \tag{8.6}$$

the formula (8.4) takes the abstract form

$$\left|\psi^{\text{out/in}}\right\rangle = |\phi_{\boldsymbol{k}}\rangle + G_0^{\text{out/in}}\,U\left|\psi^{\text{out/in}}\right\rangle\,. \tag{8.7}$$

This equation is called the *Lippmann-Schwinger equation.*

Instead of the Green functions $G_0^{\text{out/in}}$ which belong to the operators $(E-H_0\pm\mathrm{i}\varepsilon)^{-1}$ one may as well use the Green functions which pertain to $(E - H \pm \mathrm{i}\varepsilon)^{-1}$ with $H = H_0 + U$. They are

$$G^{\text{out/in}} := (E - H \pm \mathrm{i}\varepsilon)^{-1} \tag{8.8}$$

The relation between G and G_0 follows from the operator identity

$$A^{-1} - B^{-1} = B^{-1}(B - A)A^{-1}\,,$$

by inserting in a first time ($A = E - H \pm \mathrm{i}\varepsilon$, $B = E - H_0 \pm \mathrm{i}\varepsilon$) such as to obtain the relation

$$G^{\text{out/in}} = G_0^{\text{out/in}} + G_0^{\text{out/in}}UG^{\text{out/in}}\,, \tag{8.9}$$

in a second time ($A = E - H_0 \pm \mathrm{i}\varepsilon$,$B = E - H \pm \mathrm{i}\varepsilon$) to obtain the relation

$$G_0^{\text{out/in}} = G^{\text{out/in}} - G^{\text{out/in}}UG_0^{\text{out/in}}\,. \tag{8.10}$$

If one inserts the second of these into the Lippmann-Schwinger equation one obtains another form of this equation which is relevant for many applications, viz.

$$\left|\psi^{\text{out/in}}\right\rangle = |\phi_{\boldsymbol{k}}\rangle + G^{\text{out/in}}\,U\,|\phi_{\boldsymbol{k}}\rangle\,. \tag{8.11}$$

This form of the Lippmann-Schwinger equation differs from (8.7) by the second term on the right-hand side where G_0 is replaced by G and where the free solution replaces the full scattering solution.

8.1.2 *T*-Matrix and Scattering Amplitude

In the second form (8.11) of the Lippmann-Schwinger equation the scattering state $|\psi^{\text{out/in}}\rangle$ is generated by the action of one of the two operators

$$\mathbf{\Omega}^{\text{out/in}} := \mathbb{1} + G^{\text{out/in}}\, U \tag{8.12}$$

on the force-free solution $|\phi_{\boldsymbol{k}}\rangle$, $|\psi^{\text{out/in}}\rangle = \mathbf{\Omega}^{\text{out/in}}|\phi_{\boldsymbol{k}}\rangle$. The operators $\mathbf{\Omega}^{\text{out}}$ and $\mathbf{\Omega}^{\text{in}}$ are called *Møller operators*.[1]

In natural units, i.e. with $\hbar = 1$, and using the "bracket" notation the scattering amplitude (2.27) reads

$$f(\theta, \phi) = -\frac{m}{2\pi}\langle\phi_{\boldsymbol{k}'}|U\,|\psi^{\text{out}}\rangle\,.$$

It is determined by the matrix element

$$T_{\boldsymbol{k}'\boldsymbol{k}} := \langle\phi_{\boldsymbol{k}'}|U\,|\psi^{\text{out}}\rangle = \langle\phi_{\boldsymbol{k}'}|U\,\mathbf{\Omega}^{\text{out}}\,|\phi_{\boldsymbol{k}}\rangle\,. \tag{8.13}$$

Thus, the relation between the scattering amplitude and the matrix element $T_{\boldsymbol{k}'\boldsymbol{k}}$ reads

$$f(\theta, \phi) = -\frac{m}{2\pi}T_{\boldsymbol{k}'\boldsymbol{k}}\,. \tag{8.14}$$

This suggests to define the corresponding *T*-operator by

$$\mathbf{T} := U\,\mathbf{\Omega}^{\text{out}}\,. \tag{8.15}$$

In the present context this operator describes purely elastic scattering, but, more importantly for our purposes, it can be generalized to inelastic scattering and to relativistic, Lorentz covariant scattering theory. In particular, the relation (8.14) between the scattering amplitude whose square yields the cross section, and the matrix element of the operator **T** between initial and final states, remains the same. We will make use of this relation, among others, in the derivation of the optical theorem.

Let us dwell for a while upon the subject of nonrelativistic scattering theory, and derive an integral equation for **T** itself, as well as a third form of the Lippmann-Schwinger equation which contains **T**. One has

$$\begin{aligned}\mathbf{T} &= U\mathbf{\Omega}^{\text{out}} = U\left(\mathbb{1} + G^{\text{out}}U\right)\\ &= U + UG_0^{\text{out}}U\left(\mathbb{1} + G^{\text{out}}U\right)\\ &= U + UG_0^{\text{out}}\,\mathbf{T}\,.\end{aligned}$$

In the first line of this calculation the definition (8.15) of **T** and the definition (8.12) of $\mathbf{\Omega}^{\text{out}}$ are inserted. In the second line use is made of the relation (8.9) for G^{out}. The integral equation for the **T**-matrix that I repeat here,

$$\mathbf{T} = U + UG_0^{\text{out}}\,\mathbf{T} \tag{8.16}$$

is determined by the free Green function G_0^{out} and the potential U.

[1] They were introduced by Christian Møller, Danish physicist (1904–1980).

The scattering solution on the right-hand side of the Lippmann-Schwinger equation (8.7) is expressed in terms of the Møller operator $\mathbf{\Omega}^{\text{out}}$, $|\psi^{\text{out/in}}\rangle = \mathbf{\Omega}^{\text{out/in}}|\phi_{\boldsymbol{k}}\rangle$. Recalling the definition (8.15) of the operator $\mathbf{T}$, a third form of the Lippmann-Schwinger equation is obtained:

$$|\psi^{\text{out}}\rangle = |\phi_{\boldsymbol{k}}\rangle + G_0^{\text{out}}\,\mathbf{T}\,|\phi_{\boldsymbol{k}}\rangle\ . \tag{8.17}$$

This equation is a good starting point, still in the framework of potential theory, when one wishes to study multiple scattering of a projectile on a target composed of many scattering centers. As an adequate description of this subject is beyond the scope of this book I refer to the literature on multiple scattering.

8.2 Covariant Scattering Theory

The analysis presented in the preceding section was mostly of a somewhat formal character and we did not go into a more rigorous mathematical treatment of the relevant operators and of the integral equations they satisfy. An analogous comment applies to the Lorentz covariant scattering theory to be dealt with here: We do not dwell on a deeper mathematical analysis and justification. Rather we concentrate on down-to-earth results which are of the greatest use in practical perturbative quantum field theory. Indeed, the outcome will be formulae which allow to relate matrix elements of the scattering matrix with expressions for cross sections and decay widths, both of which are measurable in realistic experiments.

8.2.1 Assumptions and Conventions

As a rule, the processes to be described in quantum field theory concern transitions between stable or quasi-stable states which are prepared *long before*, and/or are detected *long after* the interaction proper. We will assume these asymptotic *in*- and *out*-states to be free, i.e. noninteracting one- or many-body states which are elements of Hilbert spaces of the type

$$\mathcal{H} = \sum_{N=0}^{\infty} \oplus\,(\mathcal{H}_1)^{\otimes N}\ ,$$

(cf. Sect. 7.1.5). The states will be denoted by their occupation numbers. This representation in "second quantization" takes account of the possibility of particle annihilation and creation, in accord, of course, with the selection rules of the theory. Examples are the reactions

$$e^+ + e^- \longrightarrow e^+ + e^-\ ,\quad e^+ + e^- \longrightarrow \gamma + \gamma\ ,$$
$$e^+ + e^- \longrightarrow e^+ + e^- + \gamma\ ,$$

as well as the decay processes

$$\mu^- \longrightarrow e^- + \nu_\mu + \overline{\nu_e}\,, \quad n \longrightarrow p + e^- + \overline{\nu_e}\,.$$

We shall consistently make use of covariant normalization (7.22),

$$\langle p'; s, \mu' | p; s, \mu \rangle = 2E_p \delta_{\mu'\mu} \delta(\boldsymbol{p} - \boldsymbol{p}')\,, \quad E_p = \sqrt{m^2 + \boldsymbol{p}^2}\,,$$

irrespectively of whether we deal with bosons or fermions. As consequences of this convention, integration over momenta is done with invariant volume element $\mathrm{d}^3p/(2E_p)$, the completeness relation takes the form (7.23), the number of particles with momentum $\boldsymbol{p}$ per unit of volume is $2E_p/(2\pi)^3$.

8.2.2 *S*-Matrix and Optical Theorem

Although scattering theory is described using an essentially stationary picture it is useful to distinguish the incoming states, as produced by a preparation measurement, from the final states which are measured in detectors. In a realistic situation the initial state will either be a one-particle state, in the case of a decay process, or a two-particle state, consisting of the projectile and the target. The final state, in turn, can contain many free particles, typically,

$$A \text{ (unstable)} \longrightarrow C + D + E + \ldots\,,$$
$$A + B \longrightarrow C + D + E + F + \ldots\,.$$

The initial state A or $A + B$ will be considered to be an *in*-state, the final state $C + D + E + \ldots$ to be an *out*-state, even in cases where both are represented by stationary plane waves, and not by wave packets. This description is summarized schematically as follows:

Let $\{|\psi_\alpha^{\text{in}}\rangle\}$ be an orthonormal system of *in*-states, $\{|\psi_\alpha^{\text{out}}\rangle\}$ an analogous system of *out*-states, both of which are assumed to be complete. What is the probability amplitude for finding a specific final state in a given initial state? The answer to this question is obtained by expanding the given *in*-state in the basis of the *out*-states. The probability amplitude is given by the (in general complex) matrix element

$$S_{\beta\alpha} = \left\langle \psi_\beta^{\text{out}} \middle| \psi_\alpha^{\text{in}} \right\rangle\,, \quad \text{with } |\psi_\alpha^{\text{in}}\rangle = \sum_{\beta'} S_{\beta'\alpha} |\psi_{\beta'}^{\text{out}}\rangle\,.$$

If, as assumed here, the two asymptotic base systems are complete, and denoting initial and final states by i and f, respectively, then the matrix $\mathbf{S} = \{S_{fi}\}$ is unitary,

$$\mathbf{S}^\dagger\,\mathbf{S} = \mathbb{1} = \mathbf{S}\,\mathbf{S}^\dagger\,. \tag{8.18}$$

The assumption stated above is called *asymptotic completeness.* This property is essential in proving the unitarity of the S-matrix.

In a theory which contains no interaction at all the S-matrix is diagonal and, with an appropriate choice of phases, it is equal to the identity, $\mathbf{S} = \mathbb{1}$. The physical interpretation of this case is simple: the particles just pass by one another without any mutual influence. Every prepared *in*-state is found unchanged in the set of the outgoing states. However, in reality there are interactions between the particles of the theory. It is then meaningful to subtract the trivial part $\mathbf{S}^{(0)} = \mathbb{1}$ from the S-matrix. This subtraction yields the *reaction matrix*

$$R_{fi} := S_{fi} - \delta_{fi}\,, \quad \text{or} \quad \mathbf{R} = \mathbf{S} - \mathbb{1}\,. \tag{8.19}$$

As we shall see in Chap. 10 covariant perturbation theory, as exemplified by the method of Feynman rules, yields precisely the R-matrix. Note that the diagonal term in (8.19) occurs only if initial and final states are identical. Independently of which theory one is dealing with, the R-matrix must always be proportional to a four-dimensional δ-distribution which guarantees conservation of the total four-momentum, $P^{(f)} = p^{(C)} + p^{(D)} + p^{(E)} + \cdots = P^{(i)}$, where $P^{(i)} = p^{(A)}$ for a decay of particle A, $P^{(i)} = p^{(A)} + p^{(B)}$ in a reaction $A + B \to C + D + E + \ldots$. It is useful to extract this δ-distribution from $\mathbf{R}$ by a further definition. Extracting also a factor $\mathrm{i}\,(2\pi)^4$ (which is a matter of convention) one has

$$R_{fi} =: \mathrm{i}\,(2\pi)^4 \delta(P^{(i)} - P^{(f)})\, T_{fi}\,. \tag{8.20}$$

The matrix $\mathbf{T}$ which is defined by this equation is called the *scattering matrix*. The next section will show that the formulae for decay widths, for cross sections, and other observables are functions of entries of this matrix. The intrinsic logic of these definitions may be summarized as follows:

- The S-matrix $\mathbf{S}$ tells us whether the theory describes interactions at all, by checking whether or not it is different from the identity. A theory which yields the S-matrix $\mathbf{S} = \mathbb{1}$ is called *trivial;*
- Rules of covariant perturbation theory such as the Feynman rules for quantum electrodynamics, yield the matrix $\mathbf{R}$, not $\mathbf{T}$. This is so because at each vertex energy and momentum are conserved so that $\mathbf{R}$ can be different from zero only if the total initial momentum $P^{(i)}$ equals the total final momentum $P^{(f)}$;
- The decay width $\Gamma(i \to X)$ of an unstable particle, and the cross section for a reaction $A + B \to X$ are proportional to $|T_{Xi}|^2$ and to $|T_{X,(A+B)}|^2$, respectively, but not to $|R_{fi}|^2$ which would be ill-defined.

The unitarity of the S-Matrix implies the relation

$$\mathbf{R} + \mathbf{R}^\dagger = -\mathbf{R}^\dagger\, \mathbf{R} \tag{8.21}$$

for the R-matrix. This, in turn, means that the scattering matrix satisfies the relation

$$\mathrm{i}\,\left(\mathbf{T}^\dagger - \mathbf{T}\right) = (2\pi)^4 \mathbf{T}^\dagger\, \mathbf{T}\,.$$

This third form of the unitarity relation is to be understood as exemplified by a typical transition $i \to f$ where it reads

$$\mathrm{i}\,\left(T^\dagger_{fi} - T_{fi}\right) = (2\pi)^4 \sum\!\!\!\!\!\!\int T^\dagger_{fn} T_{ni}\, \delta^{(4)}\left(P^{(n)} - P^{(i)}\right)\,. \tag{8.22}$$

Here the sum/integral contains all intermediate states which are compatible with the selection rules and the conservation laws of the theory. Before illustrating this important relation by the example of elastic scattering of two particles, and before deriving the optical theorem from the unitarity relation, a few comments seem in order.

Remarks

1. It is important to realize, very much like initial and final states $|i\rangle$ and $|f\rangle$ are on their mass shells, all intermediate states $|n\rangle$ in (8.22) are physical states, i.e. lie on their respective mass shells. This is unlike perturbation theory where intermediate states are *virtual* states, i.e. states which are not on their mass shell.
2. In passing from (8.21) to (8.22), as well as in the calculation of $|R_{fi}|^2$ further down, one is faced with the square of the δ-distribution for energy-momentum conservation which is not defined. This is not a real problem but rather an artefact due to the fact that one described scattering states by plane waves. A more careful analysis shows that a description by means of localized wave packets does not run into this difficulty and, yet, yields the same results for observables.
3. A more subtle problem is the following. In the definition of the S-matrix it is assumed that all particles participating in a given reaction are massive. Only in this case is the vacuum separated by a finite gap in energy from the hyperboloid of the smallest mass involved. An example is sketched in Fig. 8.1. This assumption is called the *spectrum condition.* Matters change fundamentally as soon as physical particles without mass are involved. (In fact, this is the rule rather than the exception.) For example, in quantum electrodynamics every electron is accompanied by a flight of photons with very small energy so that, strictly speaking, there are no isolated one-electron states. Through its coupling to the radiation field the electron becomes what is sometimes called an *infraparticle.* In this situation the S-matrix is not well-defined in the way described above. The problem is of physical, not only technical nature. Indeed, realistic detectors can never define or measure states with arbitrarily sharp energy. Rather, they always have a certain finite resolution. Therefore, such detectors cannot separate the state of a single electron without any photons from a state of the same electron in which there are in addition *soft* photons, i.e. photons with very small energy. There exists a modified definition of the S-matrix which copes with this empirical situation,[2] but it is not very tractable and, hence, not very useful in practice.

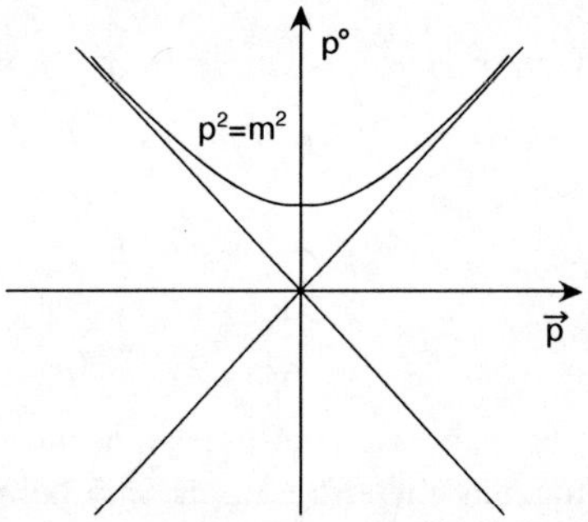

Fig. 8.1 Surface of physical four-momenta of a massive particle. The vacuum ($p^0 = 0, \boldsymbol{p} = 0$) is separated in energy from the hyperboloid $p^2 = m^2$ of the particle with the smallest mass

[2] O. Steinmann, Acta Physica Austriaca, Suppl. XI, (1973) 167; Fortschr. Physik **22** (1974) 367; *Scattering of Infraparticles*.

Concrete calculations in the framework of perturbation theory usually apply an intuitive method which is somewhat of a recipe but yields correct answers. One calculates the process at stake (cross section, decay width, or the like) as if there were isolated electrons, muons, etc., at a given order $\mathcal{O}(\alpha^n)$ in the fine structure constant α, and notices that the expressions thus obtained contain divergencies, called *infrared divergencies*. In a second step, one calculates the closely related process in which besides the electron (muon), one or more soft photons are emitted, to the same order in α. This process also contains an infrared divergence. One then integrates the second process over all momenta of the emitted photon(s) which are in the range permitted by the experimental resolution. Adding the results, the bare process and the accompanying radiative process, one finds that their infrared divergencies cancel. The realistically measurable observable comes out perfectly finite.

Of course, these divergencies must first be regularized, that is, must be replaced by finite expressions in a consistent way so that well-defined results are obtained. This is often done by temporarily assigning a finite mass m_γ to the photon, and to take the limit $m_\gamma \to 0$ at the end. As the infrared divergence is due to the masslessness of the photon, it is avoided when the photon is given a mass.

4. In spite of the problem discussed above the concept of the S-matrix is extremely fruitful. It serves to formalize the principle that the theory should focus on those objects which are observable. In the case of scattering the microscopic interaction region proper is not accessible. The physical information is contained in the relation between the preparation of the quantum system long before the scattering, far from the interaction region, and the detection of the scattering products, a long time after the scattering, and again at asymptotic distances. Thus, the reader will not be surprised to learn that this concept was invented by W. Heisenberg.

In Chap. 2 we derived the optical theorem in the framework of potential theory. In a step which goes far beyond this, one proves this theorem in its most general form, from the unitarity of the S-matrix, (8.18). We start from (8.22) and analyze first *elastic* scattering of two particles,

$$A + B \longrightarrow A + B : \quad p + q = p' + q' ,$$

with $p^2 = p'^2 = m_A^2$, $q^2 = q'^2 = m_B^2$. When written out in more detail, (8.22) reads

$$\begin{aligned} &\mathrm{i}\left[T^*(p,q|p',q') - T(p',q'|p,q)\right] \\ &\quad = (2\pi)^4 \sum_{\text{spins}} \int \frac{\mathrm{d}^3p''}{2E_{p''}} \int \frac{\mathrm{d}^3q''}{2E_{q''}} \delta^{(4)}(p+q-p''-q'') \\ &\qquad \times T^*(p'',q''|p',q')T(p'',q''|p,q) . \end{aligned}$$

This expression assumes that the intermediate states are reached by elastic scattering and, therefore, holds only below the first inelastic threshold. For example, it is applicable to the process $\pi^+ + p \to \pi^+ + p$ in the range of energies in the center-of-mass system which remain below the mass of the resonance N^{*++}. In contrast, it is not applicable to the process $\pi^- + p \to \pi^- + p$ because here, even at very low energies, charge exchange $\pi^- + p \to \pi^0 + n$ is kinematically possible.

It is useful to explore the unitarity relation in the center-of-mass system because it takes a particularly simple form in that system and because the comparison to the corresponding nonrelativistic situation becomes especially transparent. As in this frame of reference we have $\boldsymbol{p} + \boldsymbol{q} = 0 = \boldsymbol{p}' + \boldsymbol{q}'$, the spatial momenta are chosen as follows

$$(\boldsymbol{p} = \boldsymbol{\kappa}\,,\, \boldsymbol{q} = -\boldsymbol{\kappa})\,, \quad (\boldsymbol{p}' = \boldsymbol{\kappa}'\,,\, \boldsymbol{q}' = -\boldsymbol{\kappa}')\,, \quad |\boldsymbol{\kappa}| = |\boldsymbol{\kappa}'| \equiv \kappa\,.$$

The integration variables $\boldsymbol{p}''$ and $\boldsymbol{q}''$ are replaced by

$$\boldsymbol{Q} := \frac{1}{2}(\boldsymbol{q}'' - \boldsymbol{p}'')\,, \quad \boldsymbol{P} := \boldsymbol{p}'' + \boldsymbol{q}''$$

the Jacobian being $\partial(\boldsymbol{Q}, \boldsymbol{P})/\partial(\boldsymbol{q}'', \boldsymbol{p}'') = 1$ and, hence, $\mathrm{d}^3p''\,\mathrm{d}^3q'' = \mathrm{d}^3Q\,\mathrm{d}^3P$. Carrying out the integral $\int \mathrm{d}^3P$, one obtains $\boldsymbol{p}'' + \boldsymbol{q}'' = 0$ or $\boldsymbol{Q} = \boldsymbol{q}''$. Introduce polar coordinates for $\boldsymbol{Q}$,

$$\mathrm{d}^3Q = x^2\,\mathrm{d}x\,\mathrm{d}\Omega_Q\,,$$

and calculate the following integrals

$$\int_0^\infty x^2\,\mathrm{d}x \int \mathrm{d}\Omega_Q \, \frac{T^* T}{4\sqrt{x^2 + m_A^2}\sqrt{x^2 + m_B^2}} \\ \times \delta^{(1)}\left(\sqrt{x^2 + m_A^2} + \sqrt{x^2 + m_B^2} - \sqrt{\kappa^2 + m_A^2} - \sqrt{\kappa^2 + m_B^2}\right)\,.$$

The integral over the modulus x of $\boldsymbol{Q}$ is done by means of the well-known formula

$$\int \mathrm{d}x\, f(x)\delta(g(x)) = \sum_i \frac{1}{|g'(x_i)|} f(x_i)\,, \quad x_i : \text{simple zeroes of } g(x)$$

(see Appendix A.1). One has

$$g = \sqrt{x^2 + m_A^2} + \sqrt{x^2 + m_B^2} - \sqrt{\kappa^2 + m_A^2} - \sqrt{\kappa^2 + m_B^2}$$

and, setting $x_i = \kappa$, one obtains

$$g'\big|_{x_i} = \left|\frac{x}{\sqrt{x^2 + m_A^2}} + \frac{x}{\sqrt{x^2 + m_B^2}}\right|_{x_i = \kappa} = \frac{\kappa}{E_A E_B}(E_A + E_B)\,.$$

Upon insertion of the inverse of this expression, the product $E_A E_B$ cancels out. The denominator contains the sum $E_A + E_B =: E$, i.e. the total energy in the center-of-mass system. Inserting into the unitarity relation yields

$$2\,\mathrm{Im}\,T(\boldsymbol{\kappa}', \boldsymbol{\kappa}) = (2\pi)^4 \frac{\kappa}{4E} \sum_{\text{spins}} \int \mathrm{d}\Omega_{\kappa''} T^*(\boldsymbol{\kappa}'', \boldsymbol{\kappa}') T(\boldsymbol{\kappa}'', \boldsymbol{\kappa})\,.$$

When this formula is applied to forward scattering, $\boldsymbol{\kappa}' = \boldsymbol{\kappa}$, one obtains the optical theorem in its application to purely elastic scattering: The imaginary part of the forward scattering amplitude is proportional to the total elastic cross section which in this case equals the total cross section.

In the special case of potential scattering that we treated in Chap. 2 the relation between scattering amplitude and cross section was given by

$$\mathrm{Im} f(E, \theta = 0) = \frac{\kappa}{4\pi}\sigma_{\text{elastic}} = \frac{\kappa}{4\pi}\int \mathrm{d}\Omega_{\kappa''}\, \left|f(\kappa'', \kappa)\right|^2 .$$

Comparison to the result obtained above yields an important formula relating the scattering amplitude $f(E, \theta)$ and an element of the T-matrix,

$$f(\kappa', \kappa) \equiv f(E, \theta) = \left(\frac{8\pi^5}{E}\right) T(\kappa', \kappa) . \tag{8.23}$$

In the next step we allow for inelastic channels to be open, i.e. extend the result to energies which are above the first inelastic threshold. The unitarity relation then reads

$$\begin{aligned} \mathrm{i}\left[T^*(p, q|p', q') - T(p', q'|p, q)\right] &= (2\pi)^4 \sum_{\text{spins}} \sum_n \int \frac{\mathrm{d}^3 k_1}{2E_1} \frac{\mathrm{d}^3 k_2}{2E_2} \cdots \frac{\mathrm{d}^3 k_n}{2E_n} \\ &\times \delta^{(4)}(p + q - k_1 - k_2 \ldots - k_n) \\ &\times T^*(k_1, k_2, \ldots, k_n|p', q') T(k_1, k_2, \ldots, k_n|p, q) , \end{aligned}$$

where sums and integrals are taken over all intermediate states which respect energy-momentum conservation and which are allowed by the selection rules. In the forward direction ($p' = p, q' = q$), writing "n" as a short-hand for the n-particle intermediate state, the relation becomes

$$\mathrm{Im}\, T(p, q|p, q) = \frac{2\kappa E}{(2\pi)^6} \sum_n \sigma(p + q \to n) = \frac{2\kappa E}{(2\pi)^6} \sigma_{\text{tot}} . \tag{8.24}$$

We need to comment on this result in a little more detail. It should intuitively be clear that the multiple integral of the squared T-amplitude is proportional to the partial integrated cross section $\sigma(p + q \to n)$. What is missing, however, is the correct factor which normalizes the cross section to the incoming flux. In the next section we will show that $\sigma(p + q \to n)$, correctly normalized, is given by the formula:

$$\begin{aligned} \sigma(p + q \to n) &= \frac{(2\pi)^{10}}{4\kappa E} \int \frac{\mathrm{d}^3 k_1}{2E_1} \frac{\mathrm{d}^3 k_2}{2E_2} \cdots \frac{\mathrm{d}^3 k_n}{2E_n} \\ &\times \sum_{\text{spins}} \delta(p + q - k_1 - \ldots - k_n) \left|T(k_1, \ldots, k_n|p, q)\right|^2 . \end{aligned}$$

This was the result that we inserted above. Finally, replacing the T-amplitude by the scattering amplitude, by means of (8.23), one obtains the final form of the *optical theorem*

$$\operatorname{Im} f(E, \theta = 0) = \left(\frac{\kappa}{4\pi}\right) \sigma_{\text{tot}}(E)\,. \tag{8.25}$$

Remarks

1. Note that the left-hand side of the optical theorem contains the amplitude for *elastic* scattering in the forward direction, while the right-hand side contains the *total* cross section for all final states which are allowed at the given center-of-mass energy E. This theorem provides a powerful relation, indeed. For instance, it tells us that the imaginary part of the elastic forward scattering amplitude is always positive.
2. While the optical theorem for potential scattering seems to follow from the Schrödinger equation and its specific properties, the derivation given here shows that it rests on more fundamental principles. The theorem is a direct consequence of the unitarity of the S-matrix. This property, in turn, is proven from asymptotic completeness.
3. The optical theorem is of great importance for experimental physics. The total cross section at the energy E, in principle, is easy to measure. One needs no more than to measure the transmission of a beam through the target.
4. Returning to the earlier form (8.24) of the optical theorem, one sees that both the total cross section, and the element of the T-matrix are Lorentz invariants. Regarding the factor κE one notices that E is nothing but the square root of the invariant $s := (p+q)^2$, while for κ one derives the following invariant expression

$$\begin{aligned}\kappa &= \frac{1}{2\sqrt{s}}\sqrt{(s - m_A^2 - m_B^2)^2 - 4m_A^2 m_B^2} \\ &= \frac{1}{2\sqrt{s}}\sqrt{\left(s - (m_A + m_B)^2\right)\left(s - (m_A - m_B)^2\right)}\,,\end{aligned} \tag{8.26}$$

(see Exercise 8.1). The Lorentz invariance of T, in turn, is a consequence of covariant normalization.

8.2.3 Cross Sections for two Particles

Assume for the moment that the spatial momenta of the two incoming particles of the reaction

$$A\,(p) + B\,(q) \longrightarrow 1\,(k_1) + 2\,(k_2) + \ldots + n\,(k_n)$$

are *collinear*, i.e. that $\boldsymbol{p}$ and $\boldsymbol{q}$ are parallel. In experiment this assumption holds true if one of the particles (called the *target*) is at rest while the other particle (called the *projectile*) moves towards it, or if the two particles move towards each other with equal and opposite momenta. In the first case, one deals with the kinematics of the laboratory system, in the second case with the kinematics of the center-of-mass system. Of course, any other *colliding beam* constellation is also allowed in which the particle's momenta are collinear (for example, have opposite directions) but do not have equal magnitude.

The assumption of collinearity is no real restriction because, on the one hand, the cross section is an invariant, and, on the other hand, because two momenta which are arbitrarily oriented in space can always be made collinear, by means of a Special Lorentz transformation. Thus, one will proceed as follows: In a first step one constructs the cross section in a collinear situation as described above. One then expresses the result exclusively in terms of Lorentz invariant variables. Finally, depending on the specific experimental situation, one expresses the cross section in the appropriate system of reference.

It looks as though the cross section $\mathrm{d}\sigma(i \to f)$ were proportional to $|R_{fi}|^2$, with R_{fi} as defined in (8.19). However, blindly inserting the definition (8.20) of the T-matrix leads to an ill-defined expression: The square of the distribution $\delta(P^{(i)} - P^{(f)})$, with $P^{(i)} = p + q$ and $P^{(f)} = k_1 + k_2 \ldots + k_n$ is not defined. This problem has its origin in our describing scattering like a stationary process and in using plane waves both for the *in*-states and for the *out*-states. Plane waves are present "everywhere" in space and with the same density. If, instead, one describes the scattering process by means of incoming wave packets for the target and for the projectile, and takes the limit of plane waves only at the end, then one can show that this is equivalent to the prescription

$$\left|R_{fi}\right|^2 \longmapsto (2\pi)^4 \delta(P^{(i)} - P^{(f)}) \left|T_{fi}\right|^2 . \tag{8.27}$$

All other parts in the formulae for cross sections are the same as before, when plane waves are used instead of wave packets. Note that a similar comment applies to decay widths.

As noted earlier, the covariant normalization implies that there are $2E_{A/B}/(2\pi)^3$ particles of the species A or B in the unit volume, respectively. Given their relative velocity $\boldsymbol{v}_{AB}$ the flux factor by which we must divide, is given by

$$\frac{2E_A}{(2\pi)^3} \frac{2E_B}{(2\pi)^3} |\boldsymbol{v}_{AB}| . \tag{8.28}$$

Collecting the formulae (8.27) and (8.28) as well as the density of final states $\mathrm{d}^3 k_j/(2E_j)$ of particle number j, the (multiple) differential cross section reads

$$\begin{aligned} &\mathrm{d}^{3n}\sigma_{fi}(A + B \to 1 + 2 + \ldots + n) \\ &\quad = \frac{(2\pi)^{10} \delta(P^{(i)} - P^{(f)})}{2E_A 2E_B |\boldsymbol{v}_{AB}|} \left|T_{fi}\right|^2 \prod_{j=1}^{n} \frac{\mathrm{d}^3 k_j}{2E_j} . \end{aligned} \tag{8.29}$$

Although, as such, it does not yet represent anything measurable, this is a master formula because it can be adapted to any possible experimental set-up with two incoming particles. Indeed, for this purpose it suffices to integrate over all momenta which either are fixed by the kinematics or which are not measured in the specific experiment. If the incoming and/or outgoing particles have nonvanishing spins and if these are not measured, then one must average incoherently over the spin orientations of the incoming particles(s) and sum (again incoherently) over those of "i".

We analyze first the flux factor in the denominator of (8.29). Assuming, as before, $\boldsymbol{p}$ and $\boldsymbol{q}$ to be collinear, one has

$$E_A\,E_B|\boldsymbol{v}_{AB}| = \sqrt{(p\cdot q)^2 - p^2q^2} = \sqrt{(p\cdot q)^2 - m_A^2 m_B^2}\,.$$

This is easily shown: With $pq = E_A E_B - \boldsymbol{p}\cdot\boldsymbol{q} = E_A E_B \mp |\boldsymbol{p}||\boldsymbol{q}|$, $\boldsymbol{v}_A = \boldsymbol{p}/E_A$, and $\boldsymbol{v}_B = \boldsymbol{q}/E_B$ as well as $m_A^2 = E_A^2 - \boldsymbol{p}^2$ and $m_B^2 = E_B^2 - \boldsymbol{q}^2$, one has

$$\begin{aligned}(p\cdot q)^2 &= m_A^2 m_B^2 + E_A^2 E_B^2\left[\boldsymbol{v}_A^2 + \boldsymbol{v}_B^2 \mp 2|\boldsymbol{v}_A||\boldsymbol{v}_B|\right]\\ &= m_A^2 m_B^2 + E_A^2 E_B^2\left(\boldsymbol{v}_A - \boldsymbol{v}_B\right)^2\ ;\end{aligned}$$

Here, collinearity was used in two places. This proves the assertion

$$\sqrt{(p\cdot q)^2 - m_A^2 m_B^2} = E_A E_B|\boldsymbol{v}_A - \boldsymbol{v}_B| \equiv E_A E_B|\boldsymbol{v}_{AB}|\,.$$

Obviously, this flux factor is a Lorentz scalar. In the earlier literature it was often called the *Møller-factor.* By introducing the Lorentz invariant variable $s := (p+q)^2$ it can be expressed in terms of s and of the masses m_A and m_B, or, alternatively, in terms of the modulus κ of the spatial momentum, (8.26), in the center-of-mass system,

$$E_A E_B|\boldsymbol{v}_{AB}| = \frac{1}{2}\sqrt{\left(s - m_A^2 - m_B^2\right)^2 - 4m_A^2 m_B^2} = \kappa\sqrt{s}\,. \tag{8.30}$$

Both forms are useful for practical applications.

Example 8.1

Two spinless charged particles are scattered elastically due to their electromagnetic interaction. The target (particle "2") is assumed to be a spinless atomic nucleus which is characterized by its form factor $F^{(K)}(Q^2)$. The projectile is taken to be an electron (particle "1") which has no inner structure and whose spin is neglected, for simplicity. In this case the formula (8.29) reads

$$\mathrm{d}^6\sigma = \frac{(2\pi)^{10}\delta^{(4)}(p+q-k_1-k_2)}{4\kappa\sqrt{s}}\left|T_{fi}\right|^2\frac{\mathrm{d}^3k_1}{2E_1}\frac{\mathrm{d}^3k_2}{2E_2}\,.$$

We calculate the differential cross section $\mathrm{d}\sigma/\,\mathrm{d}\Omega^*$ for scattering of the electron in the center-of-mass system. Integration over the spatial momentum $\boldsymbol{k}_1$ yields

$$\frac{\mathrm{d}^3\sigma}{\mathrm{d}\Omega^*\,\mathrm{d}\kappa'} = \frac{(2\pi)^{10}}{16\kappa\sqrt{s}E_1E_2}\left|T_{fi}\right|^2\delta^{(1)}(W - E_1 - E_2)\kappa'^2\,,$$

with $W = \sqrt{m_A^2 + \kappa'^2} + \sqrt{m_B^2 + \kappa'^2}$. The quantity κ' denotes the modulus of the spatial momentum in the final state. By the remaining δ-distribution it contributes only for $\kappa' = \kappa$. Integration over κ', or, equivalently, over W, using

$$\mathrm{d}\kappa' = \frac{\mathrm{d}\kappa'}{\mathrm{d}W}\,\mathrm{d}W = \frac{E_1E_2}{\kappa'(E_1+E_2)}\,\mathrm{d}W\,,$$

with $\kappa' = \kappa$ and $E_1 + E_2 = W = \sqrt{s}$, yields

$$\frac{\mathrm{d}^2\sigma}{\mathrm{d}\Omega^*} = \frac{1}{16s}(2\pi)^{10}\left|T_{fi}\right|^2\,.$$

Note that here we use a somewhat pedantic notation: As $\mathrm{d}\Omega^* = \mathrm{d}(\cos\theta^*)\,\mathrm{d}\phi^*$ the expression given above, strictly speaking, yields a doubly differential cross section. It is common practice to write it simply as

$$\frac{\mathrm{d}\sigma}{\mathrm{d}\Omega^*},$$

independently of whether the azimuth ϕ is relevant or whether one integrates over this angle. Of course we rush to take up that convention!

The Feynman rules of quantum electrodynamics that will be derived in Chap. 10, yield the following expression for the R-matrix element of Fig. 8.2 for the process studied here:

$$R_{fi} = \mathrm{i}^3(-e^2)(2\pi)^4 \int \mathrm{d}^4Q\, \delta(p - k_1 - Q)\delta(q + Q - k_2)$$
$$\times \langle k_1 | j_\mu(0) | p \rangle \frac{-g^{\mu\nu}}{Q^2} \langle k_2 | j_\nu(0) | q \rangle \,. \tag{8.31}$$

In this formula one recognizes the photon propagator (7.149) in the Feynman gauge, and the electromagnetic current operator $j_\mu(x)$, noting that the elementary charge $+e$ (for the nucleus) and $-e$ (for the electron) are factored. At each vertex there is a four-dimensional δ-distribution which serves to balance the four-momenta. The momentum Q of the virtual photon is integrated over. This latter integral fixes the integrand at the value $Q = p - k_1$. As one has

$$\delta(p - k_1 - Q)\delta(q + Q - k_2) = \delta(Q - (p - k_1))\delta((p + q) - (k_1 + k_2))$$

there remains the δ-distribution for energy-momentum conservation, as expected, that was extracted explicitly from the T-matrix in the definition (8.20). The numerical factors follow from the perturbation series and from those multiplying the propagators, cf. Chap. 10.

The matrix elements of the electromagnetic current density can be analyzed further. The current density behaves like a Lorentz vector. As the nucleus of the example has no spin, there remain only the four-vectors q and k_2, or, alternatively, their sum and their difference, that can be used as covariants in an expansion of the matrix

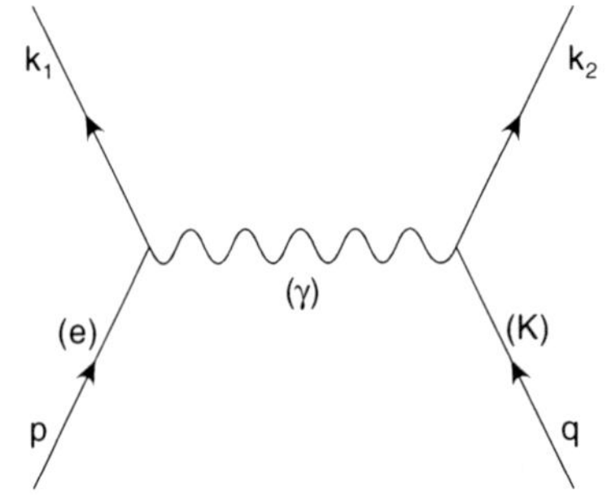

Fig. 8.2 A (spinless) electron is scattered elastically on a spinless nucleus, by exchange of a virtual photon

element. Making use of the translation formula (7.48) one writes

$$\langle k_2 | j_\mu(0) | q \rangle \equiv \frac{1}{(2\pi)^3} \left\{ F^{(K)}(Q^2)(q+k_2)_\mu + G^{(K)}(Q^2)(q-k_2)_\mu \right\},$$
$$Q = q - k_2 ,$$

so that the vector nature of the matrix elements is taken up by $(q+k_2)_\mu$ and by $(q-k_2)_\mu$, while the form factors $F^{(K)}(Q^2)$ and $G^{(K)}(Q^2)$ are Lorentz scalar functions. These form factors can only depend on the variable Q^2 because any other invariant kinematic variable such as $(q+k_2)^2$, or $q \cdot k_2$ is linearly dependent on Q^2, q^2, and k_2^2, the latter two of which are constant, $q^2 = k_2^2 = m_B^2$. (The dependence of form factors on the mass of the particle is never written explicitly.)

Recalling the formula (7.49) for the divergence $\partial^\mu j_\mu(x) = 0$, one sees that the second form factor must vanish identically, $G^{(K)}(Q^2) \equiv 0$. Concerning the first of them, $F^{(K)}(Q^2)$, one can determine its value at $Q^2 = 0$. Using (7.48) one has

$$\langle k_2 | j_\mu(x) | q \rangle = \frac{1}{(2\pi)^3} F^{(K)}(Q^2)\, \mathrm{e}^{-\mathrm{i}(q-k_2)x} (q+k_2)_\mu .$$

Since $j_\mu(x)$ is conserved, the space integral of its time component is a constant. In the present case, after having extracted the elementary charge, this is the nuclear charge number Z. Thus, with covariant normalization,

$$\langle k_2 | \int \mathrm{d}^3x\, j_0(x) | q \rangle = Z \langle k_2 | q \rangle = Z\, 2E_q \delta(\boldsymbol{q} - \boldsymbol{k}_2) .$$

Furthermore, the integral on the left-hand side can be calculated directly, making use of the translation formula and the covariant decomposition of the matrix element,

$$\begin{aligned}\langle k_2 | \int \mathrm{d}^3x\, j_0(x) | q \rangle &= \int \mathrm{d}^3x\; \mathrm{e}^{-\mathrm{i}(q-k_2)x} \langle k_2 | j_0(0) | q \rangle \\ &= (2\pi)^3 \delta(\boldsymbol{q} - \boldsymbol{k}_2) \frac{1}{(2\pi)^3} F^{(K)}(Q^2 = 0)\, 2E_q .\end{aligned}$$

Comparing the two formulae yields the result $F^{(K)}(0) = Z$. Note, finally, that $F^{(K)}(Q^2)$ is the electric form factor of a spinless nucleus that we studied in Sect. 2.4.2 in a more general framework.[3]

The same kinematic analysis also applies to the electron, with q replaced by p, and k_2 by k_1. As the charge $(-e)$ was already extracted one obtains $F^{(e)}(0) = 1$. As long as we are allowed to treat the electron as a genuine point particle the electric form factor has the same value 1 at all momentum transfers, $F^{(e)}(Q^2) = 1$.

The T-matrix element follows from the definition (8.20). With the above results it is equal to

$$T_{fi} = -\frac{Ze^2}{(2\pi)^6} \frac{F^{(K)}(Q^2)}{Q^2} \left((p+k_1) \cdot (q+k_2) \right) .$$

---

[3] It is not obvious that the Lorentz scalar form factor $F(Q^2)$ can be identified with the form factor in nonrelativistic scattering theory. In fact, this identification is correct because here we work in the center-of-mass system where $\boldsymbol{p} = -\boldsymbol{q}$ and $\boldsymbol{k}_1 = -\boldsymbol{k}_2$. In the limit of the momenta going to zero one lands directly in the nonrelativistic kinematic configuration.

At this point it is useful to replace all scalar products by the Lorentz invariant variables

$$s := (p+q)^2 = (k_1+k_2)^2 \,, \qquad t := (p-k_1)^2 = (k_2-q)^2 \tag{8.32}$$

and to express the cross section in a form which is independent of the frame of reference. That is to say, instead of $\mathrm{d}\sigma/\,\mathrm{d}\Omega^*$ one calculates the expression $\mathrm{d}\sigma/\,\mathrm{d}t$. In the center-of-mass one has $t = -2\kappa^2(1-\cos\theta^*)$. Integration over the azimuth of the volume element $\mathrm{d}\Omega^* = \mathrm{d}(\cos\theta^*)\,\mathrm{d}\phi^*$ is trivial, noting that $|T_{fi}|^2$ is independent of it. One finds

$$\frac{\mathrm{d}\sigma}{\mathrm{d}(\cos\theta^*)} = \frac{\mathrm{d}\sigma}{\mathrm{d}t}\frac{\mathrm{d}t}{\mathrm{d}(\cos\theta^*)} = 2\kappa^2\frac{\mathrm{d}\sigma}{\mathrm{d}t}$$

and inserts here the invariant (8.30) for κ^2. In a similar way one calculates

$$(p+k_1)\cdot(q+k_2) = 2(s-m_A^2-m_B^2)+t\,.$$

Finally, one inserts $e^2 = 4\pi\alpha$ and obtains the final result

$$\frac{\mathrm{d}\sigma}{\mathrm{d}t} = \pi\frac{(Z\alpha)^2|F^{(K)}(t)|^2}{t^2}\frac{\left[2(s-m_A^2-m_B^2)+t\right]^2}{(s-m_A^2-m_B^2)^2-4m_A^2m_B^2}\,. \tag{8.33}$$

This invariant expression can now be evaluated in every frame of reference which is either given or is singled out by the experimental arrangement. For example, consider the laboratory system and assume the rest mass $m_A \equiv m_e$ of the electron to be small as compared to its energy and, of course, in comparison with the mass m_B of the nucleus. In this situation, and with $q = (m_B, \mathbf{0})^T$, $p = (E, \boldsymbol{p})^T$, $k_1 = (E', \boldsymbol{k}_1)^T$, and $|\boldsymbol{p}| \approx E$, $|\boldsymbol{k}_1| \approx E'$, one has

$$s \approx m_B^2 + 2m_BE\,,$$

$$t \approx -2EE'(1-\cos\theta) = -2E^2\frac{1-\cos\theta}{1+(E/m_B)(1-\cos\theta)}\,,$$

$$\frac{\mathrm{d}t}{\mathrm{d}(\cos\theta)} \approx \frac{2E^2}{(1+(E/m_B)(1-\cos\theta))^2}\,.$$

The relationship between E' and E that was inserted here, viz.

$$E' = \frac{E}{1+(E/m_B)(1-\cos\theta)}\,,$$

follows from the energy balance with $m_A \approx 0$. One now calculates

$$\frac{\mathrm{d}\sigma}{\mathrm{d}\Omega} = \frac{1}{2\pi}\frac{\mathrm{d}t}{\mathrm{d}\cos\theta}\left(\frac{\mathrm{d}\sigma}{\mathrm{d}t}\right)_{\mathrm{Lab}}\,.$$

The result is found to be

$$\frac{\mathrm{d}\sigma}{\mathrm{d}\Omega} = \frac{(Z\alpha)^2|F^{(k)}(Q^2)|^2}{E^2(1-\cos\theta)^2}\frac{[1+(E/2m_B)(1-\cos\theta)]^2}{[1+(E/m_B)(1-\cos\theta)]^2}\,.$$

Disregarding terms in (E/m_B) (these are typical recoil terms) one recognizes the well-known Rutherford cross section

$$\left(\frac{\mathrm{d}\sigma}{\mathrm{d}\Omega}\right)_{\text{Rutherford}} = \frac{(Z\alpha)^2}{E^2(1-\cos\theta)^2} = \left(\frac{Z\alpha}{2E}\right)^2 \frac{1}{\sin^4(\theta/2)},$$

multiplied by the square of the electric form factor of the nucleus. The cross section (8.33), when multiplied by the square of the charge $Z = 2$ and by the form factor of the α-particle, applies, for instance, to elastic scattering of α-particles on a spinless nucleus.

8.2.4 Decay Widths of Unstable Particles

Matrix elements of the T-matrix serve also to describe the decay of a single unstable particle into n particles with four-momenta as indicated here,

$$A\ (q) \longrightarrow 1\ (k_1) + 2\ (k_2) + \ldots + n\ (k_n)$$

provided the decay is kinematically allowed and fulfills all selection rules. The $3n$-fold differential decay rate reads

$$\mathrm{d}^{3n}\Gamma_{fi} = (2\pi)^4\delta(k_1 + k_2 + \ldots + k_n - q)\frac{(2\pi)^3}{2E_q}|T_{fi}|^2 \prod_{i=1}^{n} \frac{\mathrm{d}^3 k_i}{2E_i}. \tag{8.34}$$

As before, and as the need arises, one must sum over the spin orientations of the particles of the final state, and/or, if the unstable particle carries nonvanishing spin, one must average over its spin orientations. In either case it is the experimental set-up that must be consulted.

Perhaps, the structure of the formula (8.34) needs no further explanation: it contains a δ-distribution which takes care of energy and momentum conservation. One divides by $2E_q/(2\pi)^3$ because one considers the decay rate for one unstable particle per volume element. Finally, every particle of the final state has its phase space density $\mathrm{d}^3k_i/(2E_i)$.

Remarks

1. Like (8.29) the formula (8.34) does not represent an observable as such. In order to obtain observable quantities one must first integrate or sum over all those variables which are fixed by the conservation laws and, hence, are redundant.
2. Although the master formula (8.34) is correct its justification is only a qualitative one. More rigorously, the kinematic state $|q; s\rangle$ of an unstable particle cannot be an *in-* or an *out*-state because the particle decays after a finite mean lifetime. However, it is perfectly meaningful to talk about a *quasi-stable particle* if the uncertainty of the energy which is caused by the decay probability, is small as compared with the rest mass, $\Gamma \ll m_A$.

3. In certain applications it is useful to analyze more closely the physical dimensions of the quantities contained in (8.34). As Γ_{fi} has dimension (energy), the δ-distribution has dimension $(\text{energy})^{-4}$, and $\mathrm{d}^3k_i/(2E_i)$ has dimension $(\text{energy})^2$, one concludes that

$$\left[|T_{fi}|^2\right] = (\text{energy})^{2(3-n)} .$$

Thus, in two-body decays $|T_{fi}|^2$ has dimension $(\text{energy})^2$, while in three-body decays the same quantity has no dimension.

Example 8.2 Two-body Decay

With respect to the rest system of the decaying particle, $q = (M, \mathbf{0})^T$, the two particles in the final state have fixed energies and equal and opposite three-momenta,

$$k_1 = (E_1, \boldsymbol{\kappa}) , \qquad k_2 = (E_2, -\boldsymbol{\kappa}) ,$$

where the energies E_i and the modulus $\kappa := |\boldsymbol{\kappa}|$ fulfill the relations

$$E_1 + E_2 = M , \quad E_1 = \sqrt{m_1^2 + \kappa^2} , \quad E_2 = \sqrt{m_2^2 + \kappa^2} . \tag{8.35}$$

Upon integration over $\boldsymbol{k}_2$ one obtains

$$\mathrm{d}^3\Gamma(A \to 1+2) = \frac{(2\pi)^7}{8ME_1E_2} |T(A \to 1+2)|^2 \delta(E_1 + E_2 - M)\, \mathrm{d}^3k_1 .$$

One then converts to spherical polar coordinates $\mathrm{d}^3k_1 = \kappa^2\, \mathrm{d}\kappa\, \mathrm{d}\Omega^*$, and notes that $E_1^2 = m_1^2 + \kappa^2$ implies the relation $\kappa\, \mathrm{d}\kappa = E_1\, \mathrm{d}E_1$, so that the integral over κ becomes an integral over E_1. As E_2 depends on κ, or, what is equivalent, on E_1, we also need the derivative of the argument of the δ-distribution, evaluated at its simple zero. This derivative reads

$$\frac{\mathrm{d}}{\mathrm{d}E_1}(E_1 + E_2 - M) = 1 + \frac{\mathrm{d}E_2}{\mathrm{d}\kappa}\frac{\mathrm{d}\kappa}{\mathrm{d}E_1} = \frac{E_1 + E_2}{E_2} = \frac{M}{E_2} .$$

Inserting this and integrating over the azimuth (note that the decay probability cannot depend on this angle) one finds

$$\mathrm{d}\Gamma(A \to 1+2) = \frac{(2\pi)^8\kappa}{8M^2} |T(A \to 1+2)|^2\, \mathrm{d}(\cos\theta^*) . \tag{8.36}$$

What does the angle θ^* refer to? If the unstable particle A has no spin, or if it carries a nonvanishing spin but is unpolarized, then the angle θ^* is not a physically relevant one. The decay then is isotropic, that is, all directions along which the particles of the final state may fly, are equally probable. In these cases one integrates over θ^*, thus obtaining the following formula for the total decay probability,

$$\Gamma(A \to 1+2) = \frac{(2\pi)^8\kappa}{4M^2} |T(A \to 1+2)|^2 . \tag{8.37}$$

This formula, too, is a Lorentz invariant. In order to exhibit this property more clearly one expresses the modulus of the spatial momentum in the center-of-mass system by the three masses: The energy balance (8.35) yields

$$\kappa = \frac{1}{2M}\sqrt{(M^2 - m_1^2 - m_2^2)^2 - 4m_1^2 m_2^2}\,. \tag{8.38}$$

In a situation where the particle A carries spin s and is polarized along the 3-direction, the angle θ^* is the angle between the expectation value of s_3 and the spatial momentum of the decay products. The differential decay probability describes the correlation between the spin of A and the direction of the momentum $\hat{\boldsymbol{\kappa}}$. It then contains physically relevant information.[4]

The decay $\pi^0 \to \gamma + \gamma$ may serve as an example. In this process the neutral pion decays into two photons by electromagnetic interaction, although this happens via possibly complicated hadronic intermediate states. The pion is a spinless particle, the photons have spin 1 (helicity). By means of simple arguments involving the conservation laws relevant for this decay, one shows that the T-matrix element must have the form,[5]

$$T(\pi^0 \to \gamma + \gamma) = \frac{e^2}{(2\pi)^{9/2}} \frac{2F}{M} \varepsilon_{\alpha\beta\sigma\tau} \varepsilon_1^\alpha \varepsilon_2^\beta k_1^\sigma k_2^\tau$$

where $M \equiv m_{\pi^0}$, where ε_i denotes the polarization of the photon "i", k_i its momentum. As before, F is a Lorentz scalar form factor which in the present case must be constant because, according to (8.38), the momentum has the fixed value $\kappa = M/2$. The factors 2 and $1/M$ are introduced by convention, the latter, in particular, with the intention to render the form factor dimensionless. This form factor F parametrizes our ignorance about the possible virtual intermediate states in the transition of the π^0 to the two photons. In calculating the absolute square of the amplitude, summing over the polarizations of the photons,

$$\sum_{\lambda_1=1}^{2} \sum_{\lambda_2=1}^{2} |\varepsilon_{\mu\nu\sigma\tau} \varepsilon_1^\mu \varepsilon_2^\nu k_1^\sigma k_2^\tau|^2\,,$$

the sums over the spins are replaced by

$$\sum_{\lambda=1}^{2} \varepsilon_\alpha^{(\lambda)} \varepsilon_\beta^{(\lambda)} \longmapsto - \sum_{\lambda,\lambda'=0}^{4} \varepsilon_\alpha^{(\lambda)} g_{\lambda\lambda'} \varepsilon_\beta^{(\lambda')}$$

which, by (7.142) is equal to $-g_{\alpha\beta}$. All terms of the difference of these two expressions, when contracted with the totally antisymmetric symbol $\varepsilon_{\alpha\beta\gamma\delta}$, give zero. Thus,

---

[4] If the initial state is a pure eigenstate of parity then a nonvanishing spin-momentum correlation is a signal that the interaction that is responsible for the decay, does not conserve parity, see also Sect. 4.2.1.

[5] A detailed analysis can be found, e. g., in [Scheck (2012)], Sect. 4.2.

summing the absolute square over the spins yields a result proportional to

$$\varepsilon_{\mu\nu\sigma\tau}\varepsilon^{\mu\nu\alpha\beta}k_1^\sigma k_{1\alpha}k_2^\tau k_{2\beta} = -2\left(\delta_\sigma^\alpha\delta_\tau^\beta - \delta_\sigma^\beta\delta_\tau^\alpha\right)k_1^\sigma k_{1\alpha}k_2^\tau k_{2\beta}$$
$$= 2(k_1\cdot k_2)^2 = \frac{1}{2M^4}\,.$$

Inserting these intermediate results, taking care of the replacement $e^2 = 4\pi\alpha$ and of the fact that the decay rate must be divided by 2! because the two photons are indistinguishable, one finds

$$\Gamma(\pi^0 \to \gamma + \gamma) = \pi\alpha^2|F|^2 M \quad (M \equiv m_{\pi^0})\,. \tag{8.39}$$

The decay width was determined by experiment (this is the same as the decay rate when natural units are used) and was found to have the value $\Gamma_{\text{exp}} = 7.85\,\text{eV}$. From this number one extracts the value $F = 1.86 \times 10^{-2}$ for the form factor.

Example 8.3 Three-Body Decay

Consider the decay of a particle A into three particles,

$$A \longrightarrow 1\,(k_1) + 2\,(k_2) + 3\,(k_3)\,.$$

Going to the rest system of A and integrating (8.34) over one of the three momenta, say $\boldsymbol{k}_3$, one has

$$\mathrm{d}^6\Gamma = \frac{(2\pi)^7\kappa_1\kappa_2}{16ME_3}|T(A\to 1+2+3)|^2$$
$$\delta(E_1+E_2+E_3-M)\,\mathrm{d}E_1\,\mathrm{d}E_2\,\mathrm{d}\Omega_1\,\mathrm{d}\Omega_2\,,$$

with

$$E_3 = \sqrt{m_3^2 + (\boldsymbol{k}_1+\boldsymbol{k}_2)^2}\,, \qquad \kappa_i := |\boldsymbol{k}_i|\,,\ i = 1,2\,.$$

In a next step one may integrate $\mathrm{d}\Omega_1$, and by choosing $\boldsymbol{k}_1$ as the direction of reference, by using axial symmetry about this direction, one may also integrate over the azimuth of the remaining solid angle. One then obtains

$$\mathrm{d}^3\Gamma = \frac{(2\pi)^9\kappa_1\kappa_2}{8ME_3}|T(A\to 1+2+3)|^2$$
$$\times\,\delta(E_1+E_2+E_3-M)\,\mathrm{d}E_1\,\mathrm{d}E_2\,\mathrm{d}(\cos\theta)\,,$$

where now one must insert

$$E_3 = \sqrt{m_3^2 + \kappa_1^2 + \kappa_2^2 + 2\kappa_1\kappa_2\cos\theta}\,.$$

At this point it is useful to replace the variables $(E_1, E_2, \cos\theta)$ by (E_1, E_2, E_3), the Jacobian of this transformation being

$$\frac{\partial(E_1, E_2, \cos\theta)}{\partial(E_1, E_2, E_3)} = \frac{E_3}{\kappa_1\kappa_2}\,.$$

In a last step one replaces E_2 and E_3 by their sum $W := E_2 + E_3$ and their difference $w := E_2 - E_3$, noting that $\partial(w, W)/\partial(E_2, E_3) = 2$, and integrates over W. This yields the double differential decay rate

$$\mathrm{d}^2\Gamma = \frac{(2\pi)^9}{16M}|T(A \to 1+2+3)|^2 \,\mathrm{d}E_1\,\mathrm{d}w \tag{8.40}$$

which is a genuine observable. The total decay rate is obtained by integration over the kinematic ranges of E_1 and w.

Remarks

1. In the previous examples one may have noticed that every external particle yields a factor $(2\pi)^{-3/2}$ in the T-matrix element, this factor going back to our choice of normalization of one-particle states. Indeed, this factor is characteristic for external particle lines and will be contained in one of the Feynman rules of perturbation theory. By this rule $|T(A \to 1+2)|^2$ receives the factor $(2\pi)^{-9}$, while $|T(A \to 1+2+3)|^2$ receives the factor $(2\pi)^{-12}$.
2. In contrast to two-body decays, the particles in the final state of a three-body decay have no fixed energy (with respect to the rest system of the decaying particle). From a historical point of view, the β-decay of atomic nuclei provides an important case. Examples of three-body decays are the triton decay

$$^3H \longrightarrow {}^3\,\mathrm{He} + e^- + \overline{\nu_e}\,,$$

or the decay of a negatively charged muon,

$$\mu^- \longrightarrow e^- + \nu_\mu + \overline{\nu_e}\,,$$

whose neutrinos are not easily detected. In both cases the electron has a continuous energy spectrum ranging from a minimal value to the maximally allowed energy. For instance, for muon decay one has

$$m_e \le E_e \le \frac{m_\mu^2 + m_e^2}{2m_\mu}\,.$$

On the basis of this observation (and assuming the principle of energy-momentum conservation) Wolfgang Pauli postulated the existence of the electron neutrino ν_e.
3. Like cross sections differential decay rates may be expressed in terms of Lorentz invariant quantities, keeping in mind that $|T|^2$, by itself, is invariant. This presents the advantage that they may then be tailored to fit any given frame of reference. In particular, one can calculate the kinematic and dynamic characteristics of decay in flight.

8.3 Comment on the Scattering of Wave Packets

The physically realistic *in*-states which are prepared in experiment are best modeled by wave packets which are localized in the neighbourhood of the spatial momenta $\boldsymbol{p}$ and $\boldsymbol{q}$, respectively. In a two-particle reaction $A + B \to 1 + 2 + \ldots + n$ the initial state then has the form

$$|A, B\rangle_{\text{in}} = \int \frac{\mathrm{d}^3 p'}{2E_{p'}} \int \frac{\mathrm{d}^3 q'}{2E_{q'}} \, \widetilde{\varphi}_A(\boldsymbol{p}') \, \widetilde{\varphi}_B(\boldsymbol{q}') \, |\boldsymbol{p}', \boldsymbol{q}'\rangle \, ,$$

where the wave packets are covariantly normalized to 1,

$$\int \frac{\mathrm{d}^3 p'}{2E_{p'}} \, |\widetilde{\varphi}_A(\boldsymbol{p}')|^2 = 1 \, ,$$

(and similarly for the second particle). The Fourier components are chosen such that the wave packets are centered in $\boldsymbol{p}$ and in $\boldsymbol{q}$, respectively. If one defines the Fourier transform by

$$\varphi(\boldsymbol{x}) = \frac{1}{(2\pi)^{3/2}} \int \frac{\mathrm{d}^3 p}{\sqrt{2E_p}} \, \mathrm{e}^{\mathrm{i}px} \widetilde{\varphi}(\boldsymbol{p}) \, ,$$

then the absolute square of $\varphi(\boldsymbol{x})$ is normalized to 1,

$$\int \mathrm{d}^3 x \, |\varphi(\boldsymbol{x})|^2 = 1 \, .$$

With T denoting the T-matrix, as before, the *out*-state which contains n particles, is characterized by the function

$$\phi(\boldsymbol{k}_1, \ldots, \boldsymbol{k}_n) = \mathrm{i}(2\pi)^4 \int \frac{\mathrm{d}^3 p'}{2E_{p'}} \int \frac{\mathrm{d}^3 q'}{2E_{q'}} \, \delta(P_f - p' - q')$$
$$\times \, \langle \boldsymbol{k}_1, \ldots, \boldsymbol{k}_n | T \, |\boldsymbol{p}', \boldsymbol{q}'\rangle \, \widetilde{\varphi}_A(\boldsymbol{p}') \, \widetilde{\varphi}_B(\boldsymbol{q}') \, .$$

These formulae are sufficient for a realistic calculation of the transition probability that is not hampered by the spurious problem of having to take the square of a δ-distribution. For lack of space we do not work out this case here (see, e. g., [Goldberger and Watson (1964)]). We just note that it yields precisely the result obtained in (8.29) that was at the basis of the examples of the preceding section.

Remarks

1. Although the details of this calculation (that we skipped) are somewhat cumbersome, they show that the problem of the "squared δ-distribution" was an artefact of the unrealistic choice of basis, viz. of plane waves. If one works with properly prepared wave packets from the start the problem does not occur.
2. All one-particle states, independently of whether they are contained in the *in*- or *out*-states, or in one of the intermediate states, are on their physical mass shells. Therefore, it makes no difference whether such states are denoted by their four-momentum, that is by $|p\rangle$ etc., or by their spatial momentum, that is by $|\boldsymbol{p}\rangle$ etc. As the reader sees I made use of this freedom on various occasions.

Particles with Spin 1/2 and the Dirac Equation

9

Introduction In order to identify the spin of a massive particle one must go to its rest system, perform rotations of the frame of reference, and study the transformation behaviour of one-particle states. This prescription was one of the essential results of Chap. 6. Furthermore, the spin $1/2$ (electrons, protons, other fermions) is described by the fundamental representation of the group $SU(2)$. The eigenstates of the observables s^2 and s_3 transform by the D-matrix $\mathbf{D}^{(1/2)}(\mathbf{R})$ which is a *two*-valued function on $\mathbb{R}^3$. If one-fermion states with momentum p have the form $|\alpha; 1/2, m; p\rangle$ where α summarizes all attributes other than the spin, its projection, and the momentum, then the formula (6.144) expresses the action of an arbitrary Poincaré transformation on this state. In the *active* interpretation, and with $\hbar = 1$, it reads

$$\begin{aligned} &\mathbf{U}(\boldsymbol{\Lambda}, a)\big|\alpha; \frac{1}{2}, m; p\big\rangle \\ &\quad = \mathrm{e}^{\mathrm{i}a\cdot(\boldsymbol{\Lambda} p)} \sum_{m'} D^{(1/2)}_{m'm}\left(\mathbf{L}^{-1}(\boldsymbol{\Lambda} p)\boldsymbol{\Lambda}\mathbf{L}(p)\right)\big|\alpha; \frac{1}{2}, m'; (\boldsymbol{\Lambda} p)\big\rangle . \end{aligned} \tag{9.1}$$

Here $\boldsymbol{\Lambda} \in L_+^\uparrow$ is a proper orthochronous Lorentz transformation, and a is the constant four-vector of translation in time and space. The origin of this somewhat ascetic formula (9.1) was explained in Sect. 6.3.4. For this reason a short summary may be sufficient at this point: The first factor $\exp\{\mathrm{i}a \cdot (\boldsymbol{\Lambda} p)\}$ represents the action of the translation. The second factor contains the product

$$\mathbf{L}^{-1}(\boldsymbol{\Lambda} p)\boldsymbol{\Lambda}\mathbf{L}(p) =: \mathbf{R}_W , \tag{9.2}$$

i.e. a three-step round-trip on the mass shell $p^2 = M^2$, namely, (1) from the rest system to the point p, effected by the boost $\mathbf{L}(p)$, (2) from there to the point $\boldsymbol{\Lambda} p$, by the action of $\boldsymbol{\Lambda}$, (3) and, finally, the return to the rest system. The product of the three moves is a rotation in the rest system, called Wigner rotation. The 2×2-matrix $\mathbf{D}^{(1/2)}$ is unitary and has determinant 1. It is an element of $SU(2)$, $\mathbf{D}^{(1/2)} \in SU(2)$.

What follows is the attempt to take apart, so to speak, the argument of this D-function, that is, to write the latter as a product of three 2×2-matrices,

F. Scheck, *Quantum Physics*, DOI: 10.1007/978-3-642-34563-0_9,

$$\mathbf{D}^{(1/2)}\left(\mathbf{L}^{-1}(\mathbf{\Lambda}p)\mathbf{\Lambda}\mathbf{L}(p)\right) = \mathbf{A}\left(\mathbf{L}^{-1}(\mathbf{\Lambda}p)\right)\mathbf{A}(\mathbf{\Lambda})\,\mathbf{A}(\mathbf{L}(p))\,. \qquad (9.3)$$

It turns out that, in general, these matrices no longer belong to $SU(2)$ but that they are elements of $SL(2, \mathbb{C})$, the special linear group in two complex dimensions. As a consequence, the spin states in a Lorentz covariant quantum theory are no longer to be found (only) in representations of $SU(2)$, but, rather, in those of the larger group $SL(2, \mathbb{C})$. In contrast to $SU(2)$ the group $SL(2, \mathbb{C})$ contains *two* inequivalent spinor representations. Although these transform alike under rotations $\mathbf{R} \in L_+^\uparrow$, they have different transformation behaviour with respect to Special Lorentz transformations $\mathbf{L} \in L_+^\uparrow$.

The two inequivalent spinor representations are related to each other by the operation of space reflection (parity). Since space reflection is known to be an important symmetry operation in the quantum theory of elementary particles it will become clear that the description of massive spin-1/2 particles involves *both types* of spinor representations. This explains why solutions of the Dirac equation have four components, not two, even though they describe particles with spin 1/2.

On the basis of this experience it is not difficult to construct a linear equation in momentum space which describes force-free fermions. Transformed back to position space, this yields a linear differential equation of first order in space and time coordinates which, in a sense to be analyzed more closely, is an analogue of the Schrödinger equation without external forces. This equation which was discovered by Dirac, supplemented by the interaction with the Maxwell radiation field, is then studied in detail. Perhaps the most important result will be that the Dirac equation cannot be interpreted as a relativistic wave equation of one-particle theory (or, at least, only in a rather limited range of applications). This is its main difference from the Schrödinger equation which has a perfectly consistent interpretation in terms of nonrelativistic quantum mechanics and Born's interpretation of the wave function.

It is only in the framework of field quantization that the Dirac equation has a consistent and physically convincing interpretation. One is led, in a rather natural way, to quantize the Dirac field following the rules of the Klein-Gordon and the Maxwell fields and, thus, to introduce creation and annihilation operators for fermions and their antiparticles. This will provide the second building block for the fully Lorentz covariant form of quantum electrodynamics.

9.1 Relationship Between $SL(2, \mathbb{C})$ and $L_+^\uparrow$

It is no accident that the title of this section sounds alike the one of Sect. 6.2.1. Indeed, the analysis that follows here is a direct generalization of the correspondence (6.17) that established the relationship between $SU(2)$ (which is a subgroup of $SL(2, \mathbb{C})$) and the rotation group $SO(3)$ (which, in turn, is a subgroup of $L_+^\uparrow$).

Every hermitian 2×2-matrix whose trace vanishes, can be written as a linear combination of the three Pauli matrices, with real coefficients. If we add to the set of the Pauli matrices a fourth one,

$$\sigma^{(0)} := \begin{pmatrix} 1 & 0 \\ 0 & 1 \end{pmatrix}, \tag{9.4}$$

i.e. the unit matrix, then *every* hermitian 2×2-matrix (with arbitrary trace) can be written as a real linear combination of

$$\{\sigma_\mu\} := \left(\sigma^{(0)}, \boldsymbol{\sigma}\right), \quad \left(\sigma_\mu^\dagger = \sigma_\mu\right), \tag{9.5}$$

where $\boldsymbol{\sigma} = \left(\sigma^{(1)}, \sigma^{(2)}, \sigma^{(3)}\right)$ are the usual Pauli-matrices,

$$\sigma^{(1)} = \begin{pmatrix} 0 & 1 \\ 1 & 0 \end{pmatrix}, \quad \sigma^{(2)} = \begin{pmatrix} 0 & -\mathrm{i} \\ \mathrm{i} & 0 \end{pmatrix}, \quad \sigma^{(3)} = \begin{pmatrix} 1 & 0 \\ 0 & -1 \end{pmatrix}. \tag{9.6}$$

Note that we changed the notation of the Pauli matrices slightly, writing $\sigma^{(i)}$ instead of σ_i, for $i = 1, 2, 3$, in order to emphasize that, together with the unit matrix $\sigma^{(0)}$, the object (9.5) now carries a Lorentz index. This will become clear from the following construction.

Let x be an arbitrary covariant vector on Minkowski spacetime, and $x = \left(x^0, \boldsymbol{x}\right)^T$ its representation in a given frame of reference. To this vector let there correspond a 2×2-matrix $\mathbf{X}$ constructed from the components of x and the matrices σ_μ,

$$x \longleftrightarrow \mathbf{X} := \sigma_\mu x^\mu = \begin{pmatrix} x^0 + x^3 & x^1 - \mathrm{i}x^2 \\ x^1 + \mathrm{i}x^2 & x^0 - x^3 \end{pmatrix}. \tag{9.7}$$

This matrix is hermitian but, unlike linear combinations of Pauli matrices only, is no longer traceless. As one easily sees the (Lorentz invariant) squared norm of x equals the determinant of $\mathbf{X}$,

$$x^2 = \det \mathbf{X} = (x^0)^2 - (x^3)^2 - (x^1 - \mathrm{i}x^2)(x^1 + \mathrm{i}x^2) = (x^0)^2 - \mathbf{x}^2 .$$

The correspondence (9.7) shows that the space of the vectors x and the set $\mathfrak{H}(2)$ of hermitian 2×2-matrices are isomorphic.

Perform then a transformation $\mathbf{\Lambda} \in L_+^\uparrow$ on x,

$$x \longmapsto x' = \mathbf{\Lambda} x$$

and consider the hermitian matrix $\mathbf{X}'$ that corresponds to x' by the definition (9.7). There must exist a nonsingular matrix $\mathbf{A}(\mathbf{\Lambda})$ such that one has

$$\mathbf{X} \longmapsto \mathbf{X}' = \mathbf{A}(\mathbf{\Lambda})\, \mathbf{X}\, \mathbf{A}^\dagger(\mathbf{\Lambda}). \tag{9.8}$$

What are the properties of this matrix $\mathbf{A}(\mathbf{\Lambda})$? On the one hand one must have $x'^2 = x^2$, i.e. $\det \mathbf{X}' = \det \mathbf{X}$ from which one concludes that the determinant of $\mathbf{A}$ must have the modulus 1, $|\det \mathbf{A}| = 1$. On the other hand, every $\mathbf{\Lambda} \in L_+^\uparrow$ can be deformed continuously into the identity $\mathbb{1} \in L_+^\uparrow$. The partner $\mathbf{A}(\mathbf{\Lambda})$ of $\mathbf{\Lambda}$ must follow this deformation and must go over into $\pm \mathbb{1}_{2\times 2}$. Both, $+\mathbb{1}$ and $-\mathbb{1}$, in even dimensions, have determinant $+1$. These two observations suggest the natural condition

$$\det \mathbf{A} = 1 . \tag{9.9}$$

The complex 2×2-matrices whose determinant has the value 1, form a group which is called *special linear group*, or *unimodular group*

$$SL(2, \mathbb{C}) := \{\mathbf{A} \in M_2(\mathbb{C}) | \det \mathbf{A} = 1\} \, . \tag{9.10}$$

The similarity of this definiton to the case of $SU(2)$ is striking. Indeed, returning to the definition (6.10), one sees that it differs from (9.10) only by the requirement of unitarity of the elements. Obviously, $SU(2)$ is a subgroup of $SL(2, \mathbb{C})$. This is in perfect accord with the statement that the rotation group $SO(3)$ is a subgroup of the proper orthochronous Lorentz group $L_+^\uparrow$. Indeed, the relation between the proper orthochronous Lorentz group $L_+^\uparrow$ and the special linear group $SL(2, \mathbb{C})$ is closely analogous to the relation between the rotation group $SO(3)$ and the unimodular group $SU(2)$.

Given a matrix $\mathbf{A}(\mathbf{\Lambda})$ that fulfills the condition (9.9) and maps x to x' according to (9.8), then, obviously, the matrix $-\mathbf{A}(\mathbf{\Lambda})$ yields the same mapping. In other terms, the factor group $SL(2, \mathbb{C})/\{\mathbb{1}, -\mathbb{1}\}$ is isomorphic to the proper orthochronous Lorentz group,

$$L_+^\uparrow \cong SL(2, \mathbb{C})/\{\mathbb{1}, -\mathbb{1}\} \, . \tag{9.11}$$

This is a direct generalization of the isomorphism (6.17) discussed in Sect. 6.2.1.

At this point recall the decomposition theorem for proper orthochronous Lorentz transformations (see, e.g., [Scheck (2010)], Chap. 4): Every $\mathbf{\Lambda} \in L_+^\uparrow$ can be written in a unique way as the product

$$\mathbf{\Lambda} = \mathbf{L}(\boldsymbol{v})\, \mathbf{R}(\boldsymbol{\theta})$$

of a rotation and a Special Lorentz transformation where $\boldsymbol{\theta}$ stands for the set of three angles of rotation, and $\boldsymbol{v}$ is a three-velocity. The originals $\pm\mathbf{A}(\mathbf{R})$ of the rotation $\mathbf{R}$ are given by (6.18),

$$\mathbf{A}(\mathbf{R}) = \exp\left\{\mathrm{i}\frac{1}{2}\boldsymbol{\sigma} \cdot \boldsymbol{\theta}\right\} . \tag{9.12}$$

For Special Lorentz transformations the correspondence looks as follows. Let

$$\boldsymbol{v} = \hat{\boldsymbol{v}} \tanh \lambda \, , \quad \text{and} \quad \boldsymbol{\zeta} = \lambda \hat{\boldsymbol{v}}$$

($\lambda = \tanh^{-1} |\boldsymbol{v}|$ is the *rapidity* parameter). One then has

$$\mathbf{L}(\boldsymbol{v}) \longleftrightarrow \pm\mathbf{A}(\mathbf{L}(\boldsymbol{v}))\, , \quad \mathbf{A}(\mathbf{L}(\boldsymbol{v})) = \exp\left\{\frac{1}{2}\boldsymbol{\sigma} \cdot \boldsymbol{\zeta}\right\} . \tag{9.13}$$

This is proven as follows. By an appropriate rotation the 3-direction can always be made to coincide with the direction of $\boldsymbol{v}$. In this case $\mathbf{A}(\mathbf{L}) = \mathrm{diag}\left(\mathrm{e}^{\lambda/2}, \mathrm{e}^{-\lambda/2}\right)$ and

$$\begin{aligned}
\mathbf{X}' = \mathbf{A}\mathbf{X}\mathbf{A}^\dagger &= \begin{pmatrix} \mathrm{e}^{\lambda/2} & 0 \\ 0 & \mathrm{e}^{-\lambda/2} \end{pmatrix} \begin{pmatrix} x^0 + x^3 & x^1 - ix^2 \\ x^1 + ix^2 & x^0 - x^3 \end{pmatrix} \begin{pmatrix} \mathrm{e}^{\lambda/2} & 0 \\ 0 & \mathrm{e}^{-\lambda/2} \end{pmatrix} \\
&= \begin{pmatrix} \mathrm{e}^{\lambda}(x^0 + x^3) & x^1 - \mathrm{i}x^2 \\ x^1 + \mathrm{i}x^2 & \mathrm{e}^{-\lambda}(x^0 - x^3) \end{pmatrix} .
\end{aligned}$$

Indeed, solving for the cartesian components one finds

$$x'^0 = x^0 \cosh\lambda + x^3 \sinh\lambda\,, \; x'^1 = x^1\,, \; x'^2 = x^2\,,$$
$$x'^3 = x^0 \sinh\lambda + x^3 \cosh\lambda\,,$$

i.e. the well-known formulae for a Special Lorentz transformation along the 3-axis.

Remarks

1. The correspondence between $L_+^\uparrow$ and $SL(2, \mathbb{C})$ is established explicitly by
$$\Lambda \longleftrightarrow \pm\mathbf{A} = \pm\exp\{\frac{1}{2}\boldsymbol{\sigma}\cdot\boldsymbol{\zeta}\}\exp\{\mathrm{i}\frac{1}{2}\boldsymbol{\sigma}\cdot\boldsymbol{\theta}\} \qquad (9.14)$$
with six real parameters $\boldsymbol{\theta}$ and $\boldsymbol{\zeta}$. It is important to note that the exponential series which represents a Special Lorentz transformation contains no factor i. As one easily verifies, this part can also be written as
$$\mathbf{A}(\mathbf{L}(\boldsymbol{v})) = \mathbb{1}\cosh(\lambda/2) + \boldsymbol{\sigma}\cdot\hat{\boldsymbol{v}}\sinh(\lambda/2)\,.$$
2. While the matrix $\mathbf{A}(\mathbf{R})$ is *unitary* the matrix $\mathbf{A}(\mathbf{L}(\boldsymbol{v}))$ is *hermitian*, its determinant is equal to 1,
$$\mathbf{A}(\mathbf{R})\mathbf{A}^\dagger(\mathbf{R}) = \mathbb{1}_{2\times 2}\,, \quad \mathbf{A}(\mathbf{L}(\boldsymbol{v})) = \mathbf{A}^\dagger(\mathbf{L}(\boldsymbol{v}))\,, \quad \det\mathbf{A}(\mathbf{L}(\boldsymbol{v})) = 1\,.$$

This result is a reflection of the decomposition theorem for Lorentz transformations mentioned above. It tells us that every element $\mathbf{A} \in SL(2, \mathbb{C})$ can be decomposed into a unitary matrix and a hermitian matrix with determinant 1. One notices the analogy to the decomposition of complex numbers into phase factor and modulus.

9.1.1 Representations with Spin 1/2

The formula (9.14) yields a representation of $SL(2, \mathbb{C})$ in a two-dimensional space. In this representation the corresponding Lie algebra is generated by the elements

$$\mathbf{J}_i = \frac{1}{2}\sigma^{(i)}\,, \quad \mathbf{K}_j = \frac{\mathrm{i}}{2}\sigma^{(j)}\,. \qquad (9.15)$$

One confirms that their commutators are precisely the ones given in (6.110), viz.

$$\left[\mathbf{J}_i, \mathbf{J}_k\right] = i\,\varepsilon_{ikl}\mathbf{J}_l\,, \quad \left[\mathbf{J}_i, \mathbf{K}_k\right] = i\,\varepsilon_{ikl}\mathbf{K}_l\,, \quad \left[\mathbf{K}_i, \mathbf{K}_k\right] = -\mathrm{i}\,\varepsilon_{ikl}\mathbf{J}_l\,. \qquad (9.16)$$

Define the following constant element of $SL(2, \mathbb{C})$

$$\varepsilon := \mathrm{i}\sigma^{(2)} = \begin{pmatrix} 0 & 1 \\ -1 & 0 \end{pmatrix} = \exp\{\mathrm{i}(\pi/2)\sigma^{(2)}\}\,. \qquad (9.17)$$

One verifies that it has the properties $\varepsilon^{-1} = \varepsilon^T = -\varepsilon$. What is striking about this definition is that ε is the same as the transformation $\mathbf{U}_0$ for $j = 1/2$ that was introduced in Sect. 6.2.3

$$\varepsilon = \mathbf{U}_0^{(1/2)} = \mathbf{D}^{(1/2)}(0, \pi, 0)\,,$$

and which is known to map covariant quantities to contravariant ones, and vice versa. The relation (6.51) then shows immediately that

$$\varepsilon\,\sigma^*_\mu\,\varepsilon^{-1} = \left(\sigma^{(0)}, -\boldsymbol{\sigma}\right) =: \hat{\sigma}_\mu \tag{9.18}$$

holds true, the star denoting complex conjugation (like in Chap. 6). It is useful, as was done here, to denote this set of matrices by the new symbol $\hat{\sigma}_\mu$.

Applying the relation (9.18) to rotations $\mathbf{\Lambda} = \mathbf{R}$, i.e. to $\mathbf{A}(\mathbf{R})$ one has

$$\varepsilon\,\mathbf{A}^*(\mathbf{R})\,\varepsilon^{-1} = \mathbf{A}(\mathbf{R})\,.$$

The matrices $\mathbf{A}$ and $\mathbf{A}^*$ are related by an equivalence relation, they are unitarily equivalent. In other terms, if we considered the subgroup of rotations, then there would be only *one* spinor representation. Matters are different for Special Lorentz transformations $\mathbf{\Lambda} = \mathbf{L}(\boldsymbol{v})$. Indeed, with $\mathbf{\Lambda} = \mathbf{L}(\boldsymbol{v})\mathbf{R}(\boldsymbol{\theta})$ one finds

$$\begin{aligned}\varepsilon\,\mathbf{A}^*(\mathbf{LR})\,\varepsilon^{-1} &= \varepsilon\;\mathrm{e}^{(-\mathrm{i}/2)\boldsymbol{\sigma}^*\cdot\boldsymbol{\theta}}\;\mathrm{e}^{(1/2)\boldsymbol{\sigma}^*\cdot\boldsymbol{\zeta}}\;\varepsilon^{-1}\\ &= \varepsilon\;\mathrm{e}^{(-\mathrm{i}/2)\boldsymbol{\sigma}^*\cdot\boldsymbol{\theta}}\;(\varepsilon^{-1}\varepsilon)\;\mathrm{e}^{(1/2)\boldsymbol{\sigma}^*\cdot\boldsymbol{\zeta}}\;\varepsilon^{-1}\\ &= \mathrm{e}^{(\mathrm{i}/2)\boldsymbol{\sigma}\cdot\boldsymbol{\theta}}\;\mathrm{e}^{-(1/2)\boldsymbol{\sigma}\cdot\boldsymbol{\zeta}} \equiv \widehat{\mathbf{A}}\,.\end{aligned}$$

Comparing this with (9.14) one realizes that here one is dealing with a spinor representation in which the generators are represented by

$$\mathbf{J}'_\mathrm{i} = \frac{1}{2}\sigma^{(i)}, \qquad \mathbf{K}'_j = -\frac{\mathrm{i}}{2}\sigma^{(j)}, \tag{9.19}$$

and which differs from (9.15) by the sign of the generators for Special Lorentz transformations. It is easy to check that the generators (9.19) obey the commutation rules (9.16). This spinor representation is not equivalent to the representation (9.15). (They are equivalent only when the Lorentz group is restricted to its subgroup of rotations.)

Of particular interest is the observation that the transition from (9.15) to (9.19), and vice versa, is effected by space reflection: Indeed, under this mapping $\boldsymbol{K}$ behaves like a vector, i.e. changes its sign, while $\boldsymbol{J}$ transforms like an axial vector, i.e. does not change sign. Thus, the two inequivalent spinor representations of the proper orthochronous Lorentz group are related by the operation of space reflection.

Summarizing, we note that $\mathbf{A}(\boldsymbol{\Lambda})$ and $\widehat{\mathbf{A}}(\boldsymbol{\Lambda})$ yield spinor representations which are not equivalent. They take the role of the unique spinor representation $\mathbf{D}^{(1/2)}(\mathbf{R})$ of the rotation group in nonrelativistic quantum theory. Very much like for the rotation group, the transition to contragredience, i.e. the mapping relating covariant and contravariant quantities, is effected by $\mathbf{U}_0 = \mathbf{D}^{(1/2)}(0, \pi, 0) = \varepsilon$. Expressed in formulae, this is

$$\left(\mathbf{A}^T\right)^{-1} = \varepsilon\,\mathbf{A}\,\varepsilon^T, \quad \left(\widehat{\mathbf{A}}^T\right)^{-1} = \varepsilon\,\widehat{\mathbf{A}}\,\varepsilon^T, \quad \left(\varepsilon^T = \varepsilon^{-1}\right). \tag{9.20}$$

We shall make use of these relations below.

9.1.2 *Dirac Equation in Momentum Space

I have marked this section by an asterisk because it addresses, in the first place, the more theoretically minded reader. Its primary aim is to derive the force-free Dirac equation, starting from representation theory of the Lorentz group and, specifically, from its two nonequivalent spinor representations. Of course, one may skip some of this material and go directly to (9.30), (9.31) and (9.32) which define Dirac spinors and the linear equations they satisfy, respectively.

The first problem posed above is solved by now: the D-matrix in (9.1) can indeed be taken apart as sketched in (9.3). However, the factors of this decomposition, in general, are no longer unitary. With this in mind we return to the analysis of the spinor representations of Sect. 6.3.4 elaborating the case $s = 1/2$ in more detail. In contrast to Sect. 6.3.4 we use here the *passive* interpretation of Poincaré and Lorentz transformations. That is to say, the spin-1/2 fermion in its quantum state $|s, 1/2; p\rangle$ is fixed, while it is the frame of reference which is transformed. For example, if $\mathbf{L}(p)$ is a Special Lorentz transformation with momentum p then

$$\mathbf{U}(\mathbf{L}(p))|s, m; p\rangle = \big|s, m; \overset{0}{p}\big\rangle$$

equals the spinor in the rest system. If one wishes to visualize this transformation one might say that the frame of reference in which the fermion had momentum p is replaced by another one which moves along with the particle. With this passive interpretation the action of the arbitrary Lorentz transformation $\boldsymbol{\Lambda}$ on the fermion state is given by (in a somewhat simplified notation)

$$\begin{aligned}
\mathbf{U}(\boldsymbol{\Lambda}^{-1})|s, m; p\rangle &= \mathbf{U}(\boldsymbol{\Lambda}^{-1})\mathbf{U}(\mathbf{L}_p^{-1})\mathbf{U}(\mathbf{L}_p)|s, m; p\rangle \\
&= \mathbf{U}(\boldsymbol{\Lambda}^{-1})\mathbf{U}(\mathbf{L}_p^{-1})|s, m; \overset{0}{p}\rangle \\
&= \mathbf{U}(\mathbf{L}_{\Lambda p}^{-1})\mathbf{U}\left(\mathbf{L}_{\Lambda p}\boldsymbol{\Lambda}^{-1}\mathbf{L}_p^{-1}\right)|s, m; \overset{0}{p}\rangle \\
&= \mathbf{U}\left(\mathbf{L}_{\Lambda p}\boldsymbol{\Lambda}^{-1}\mathbf{L}_p^{-1}\right)|s, m; \Lambda p\rangle .
\end{aligned}$$

With the conventions used in this book the representation of rotations fulfill the relations

$$\mathbf{U}(\mathbf{R}^{-1}) = \mathbf{D}^*(\mathbf{R}^{-1}) = \left(\mathbf{D}^{-1}(\mathbf{R})\right)^* = \mathbf{D}^T(\mathbf{R}) .$$

Note that for spinors the matrix $\mathbf{U}$ is either $\mathbf{A}$ or $\widehat{\mathbf{A}}$, depending on which of the two representations is considered. Thus, with the decomposition (9.3), $\mathbf{D}(\mathbf{R}_W) = \mathbf{A}(\mathbf{L}_{\Lambda p}^{-1})\mathbf{A}(\boldsymbol{\Lambda})\mathbf{A}(\mathbf{L}_p)$ for the Wigner rotation we obtain for the first spinor representation

$$\mathbf{D}^T(\mathbf{R}) = \mathbf{A}^T(\mathbf{L}_p)\mathbf{A}^T(\boldsymbol{\Lambda})\mathbf{A}^T(\mathbf{L}_{\Lambda p}^{-1}) .$$

$$\mathbf{U}(\boldsymbol{\Lambda}^{-1})|1/2, m; p\rangle = \sum_{m'}\left(\mathbf{A}^T(\mathbf{L}_p)\mathbf{A}^T(\boldsymbol{\Lambda})\mathbf{A}^T(\mathbf{L}_{\Lambda p}^{-1})\right)_{mm'}|1/2, m'; \Lambda p\rangle . \quad (9.21)$$

The corresponding equation for the other representation looks alike, with all $\mathbf{A}$ replaced by $\widehat{\mathbf{A}}$. Taking the inverse of the equations (9.20), multiplying by ε^T from the left, and by ε from the right, one obtains

$$\varepsilon^T \mathbf{A}^T \varepsilon = \mathbf{A}^{-1}, \qquad \varepsilon^T \widehat{\mathbf{A}}^T \varepsilon = \widehat{\mathbf{A}}^{-1}.$$

One multiplies (9.21) by ε^T from the left and inserts the identity $\varepsilon\,\varepsilon^T = \mathbb{1}$ in three places. In this way, the transposed matrices are replaced by the original matrices, their arguments are replaced by their inverse. This yields

$$\begin{aligned} \mathbf{U}(\Lambda^{-1}) &\sum_{m''} \left(\varepsilon^T\right)_{mm''} \left|1/2, m''; p\right\rangle \\ &= \sum_{m'} \left(\mathbf{A}(\mathbf{L}_p^{-1})\mathbf{A}(\Lambda^{-1})\mathbf{A}(\mathbf{L}_{\Lambda p})\,\varepsilon^T\right)_{mm'} \left|1/2, m'; \Lambda p\right\rangle, \end{aligned} \tag{9.22}$$

as well as the analogous equation where all matrices $\mathbf{A}$ are replaced by $\widehat{\mathbf{A}}$.

Although, at first, it may look complicated, this transformation rule is quite remarkable. Indeed, it suffices to define

$$\widetilde{\phi}_m(p) := \sum_{m'} \left(\mathbf{A}(\mathbf{L}_p)\,\varepsilon^T\right)_{mm'} \left|1/2, m'; p\right\rangle \tag{9.23}$$

$$\widetilde{\chi}_m(p) := \sum_{m'} \left(\widehat{\mathbf{A}}(\mathbf{L}_p)\,\varepsilon^T\right)_{mm'} \left|1/2, m'; p\right\rangle \tag{9.24}$$

to obtain spinors which are covariant with respect to Lorentz transformations. To see this, multiply (9.22) with $\mathbf{A}(\mathbf{L}_p)$ from the left and insert the definitions (9.23) and (9.24). Then one obtains (suppressing the projection quantum numbers and the summation indices)

$$\mathbf{U}(\Lambda^{-1})\,\widetilde{\phi}(p) = \mathbf{A}(\Lambda^{-1})\,\widetilde{\phi}(\Lambda p) \tag{9.25}$$

$$\mathbf{U}(\Lambda^{-1})\,\widetilde{\chi}(p) = \widehat{\mathbf{A}}(\Lambda^{-1})\,\widetilde{\chi}(\Lambda p)\,. \tag{9.26}$$

In this way we have constructed spinors which transform covariantly under Lorentz transformations. Furthermore, one sees that $\widetilde{\phi}$ and $\widetilde{\chi}$ are linearly dependent. As a consequence, there exists an equation which relates the two types of covariant spinor fields and which is easily derived: Eliminate the factor $\varepsilon^T\left|1/2, m; p\right\rangle$ from (9.24) and (9.23) to obtain

$$\widetilde{\chi}(p) = \widehat{\mathbf{A}}(\mathbf{L}_p)\mathbf{A}^{-1}(\mathbf{L}_p)\widetilde{\phi}(p)$$

Without loss of generality the 3-axis of $\mathbb{R}^3$ may be chosen along the direction of the space part $\boldsymbol{p}$ of p. One then has

$$\mathbf{A}^{-1}\left(\mathbf{L}_p\right) = \begin{pmatrix} e^{-\lambda/2} & 0 \\ 0 & e^{\lambda/2} \end{pmatrix},$$

where λ denotes the *rapidity* which is related by some well-known equations to the relativistic γ-factor and the modulus of the three-velocity,

$$\cosh\lambda = \gamma = \frac{p^0}{m}, \quad \sinh\lambda = \gamma v = \frac{p^0}{m}\frac{|\boldsymbol{p}|}{p^0} = \frac{|\boldsymbol{p}|}{m},$$

or to the energy $p^0 = \sqrt{m^2 + \boldsymbol{p}^2}$, the spatial momentum $\boldsymbol{p}$, and the mass m of the particle. One calculates

$$\begin{aligned}\widehat{\mathbf{A}}\left(\mathbf{L}_p\right) &= \varepsilon\, \mathbf{A}^*\left(\mathbf{L}_p\right) \varepsilon^{-1} \\ &= \begin{pmatrix} 0 & 1 \\ -1 & 0 \end{pmatrix} \begin{pmatrix} \mathrm{e}^{\lambda/2} & 0 \\ 0 & \mathrm{e}^{-\lambda/2} \end{pmatrix} \begin{pmatrix} 0 & -1 \\ 1 & 0 \end{pmatrix} \\ &= \begin{pmatrix} \mathrm{e}^{-\lambda/2} & 0 \\ 0 & \mathrm{e}^{\lambda/2} \end{pmatrix} .\end{aligned}$$

Using these results one concludes

$$\begin{aligned}\widehat{\mathbf{A}}\left(\mathbf{L}_p\right) \mathbf{A}^{-1}\left(\mathbf{L}_p\right) &= \begin{pmatrix} \mathrm{e}^{-\lambda} & 0 \\ 0 & \mathrm{e}^{\lambda} \end{pmatrix} \\ &= \frac{1}{2}\left\{\begin{pmatrix} \mathrm{e}^{\lambda}+\mathrm{e}^{-\lambda} & 0 \\ 0 & \mathrm{e}^{\lambda}+\mathrm{e}^{-\lambda} \end{pmatrix}\right. \\ &\qquad \left. - \begin{pmatrix} \mathrm{e}^{\lambda}-\mathrm{e}^{-\lambda} & 0 \\ 0 & -\left(\mathrm{e}^{\lambda}-\mathrm{e}^{-\lambda}\right) \end{pmatrix}\right\} \\ &= \mathbb{1}_2 \cosh\lambda - \sigma^{(3)} \sinh\lambda = \frac{1}{m}\left(p^0\sigma^{(0)} - p^3\sigma^{(3)}\right) \\ &\longrightarrow \frac{1}{m}\left(p^0\sigma^{(0)} - \boldsymbol{p}\cdot\boldsymbol{\sigma}\right) .\end{aligned}$$

In the last line the result was generalized in an obvious way to the case where the momentum $\boldsymbol{p}$ points in an arbitrary direction. Thus, one obtains

$$m\widetilde{\chi}(p) = \left(p^0\sigma^{(0)} - \boldsymbol{p}\cdot\boldsymbol{\sigma}\right)\widetilde{\phi}(p) .$$

Making use of the relation $(p^0\sigma^{(0)} + \boldsymbol{p}\cdot\boldsymbol{\sigma})(p^0\sigma^{(0)} - \boldsymbol{p}\cdot\boldsymbol{\sigma}) = p^2\,\mathbb{1}_2 = m^2\,\mathbb{1}_2$ one concludes, without further calculation, that

$$m\widetilde{\phi}(p) = \left(p^0\sigma^{(0)} + \boldsymbol{p}\cdot\boldsymbol{\sigma}\right)\widetilde{\chi}(p)$$

must also hold. These two relations can be written in a more transparent form by making use of the definitions (9.5) and (9.18),

$$\left\{\sigma_\mu\right\} = \left(\sigma^{(0)}, \boldsymbol{\sigma}\right) , \quad \left\{\widehat{\sigma}_\mu\right\} = \left(\sigma^{(0)}, -\boldsymbol{\sigma}\right) ,$$

as well as of the corresponding contravariant versions that are obtained by the action of the metric,

$$\left\{\sigma^\mu\right\} = \left(\sigma^{(0)}, -\boldsymbol{\sigma}\right) , \quad \left\{\widehat{\sigma}^\mu\right\} = \left(\sigma^{(0)}, \boldsymbol{\sigma}\right) .$$

The above relations then read

$$\begin{aligned} m\widetilde{\chi}(p) &= p^\mu \widehat{\sigma}_\mu\, \widetilde{\phi}(p) = p_\mu \widehat{\sigma}^\mu\, \widetilde{\phi}(p) , \\ m\widetilde{\phi}(p) &= p^\mu \sigma_\mu\, \widetilde{\chi}(p) = p_\mu \sigma^\mu\, \widetilde{\chi}(p) . \end{aligned} \tag{9.27}$$

An even more compact form of the equations for $\widetilde{\phi}(p)$ and $\widetilde{\chi}(p)$ is obtained by means of the following definitions. Let

$$\gamma^{\mu} := \begin{pmatrix} 0 & \sigma^{\mu} \\ \widehat{\sigma}^{\mu} & 0 \end{pmatrix}, \tag{9.28}$$

or, when spelled out in components,[1]

$$\gamma^{0} := \begin{pmatrix} 0 & \mathbb{1}_2 \\ \mathbb{1}_2 & 0 \end{pmatrix}, \quad \gamma^{i} := \begin{pmatrix} 0 & -\sigma^{(i)} \\ \sigma^{(i)} & 0 \end{pmatrix}. \tag{9.29}$$

(Note that the entries are themselves blocs of 2×2-matrices.) Finally, the two spinors $\widetilde{\phi}(p)$ and $\widetilde{\chi}(p)$ are put together to form a spinor with four components

$$u(\boldsymbol{p}) := \begin{pmatrix} \widetilde{\phi}(p) \\ \widetilde{\chi}(p) \end{pmatrix}. \tag{9.30}$$

The two equations which were obtained above and which relate the two types of two-spinors then take the compact form

$$\left(\gamma^{\mu} p_{\mu} - m \mathbb{1}_{4\times 4}\right) u(\boldsymbol{p}) = 0. \tag{9.31}$$

Before continuing on this, it is useful to collect some properties of the matrices γ^{μ} which are defined by (9.28). Take the product $\sigma^{\mu}\widehat{\sigma}^{\nu}$, add to it the same product with the indices interchanged. By the properties of the Pauli matrices this gives zero whenever $\mu \neq \nu$. For $\mu = \nu = 0$ one obtains twice the unit matrix, while for $\mu = \nu = i$ one obtains minus twice the unit matrix. These cases are summarized in the formula

$$\sigma^{\mu}\widehat{\sigma}^{\nu} + \sigma^{\nu}\widehat{\sigma}^{\mu} = 2g^{\mu\nu}\mathbb{1}_{2\times 2}.$$

When transcribed to the γ-matrices one obtains an important relation

$$\gamma^{\mu}\gamma^{\nu} + \gamma^{\nu}\gamma^{\mu} = 2g^{\mu\nu}\mathbb{1}_{4\times 4}. \tag{9.32}$$

Thus, the square of γ^0 equals the positive 4×4 unit matrix, the square of γ^i, $i = 1, 2, 3$, equals the negative unit matrix. As defined in (9.28) the matrix γ^0 is hermitian, while the three matrices γ^i are antihermitian. As γ^0, by (9.32), anticommutes with γ^i, these relations are

$$\left(\gamma^0\right)^{\dagger} = \gamma^0, \quad \left(\gamma^i\right)^{\dagger} = -\gamma^i, \quad \text{but} \quad \gamma^0\left(\gamma^i\right)^{\dagger}\gamma^0 = \gamma^i. \tag{9.33}$$

Contracting the relation (9.32) with p_μ and with p_ν, one finds

$$\frac{1}{2} p_\mu p_\nu \left(\gamma^{\mu}\gamma^{\nu} + \gamma^{\nu}\gamma^{\mu}\right) = p^2 \mathbb{1}_{4\times 4} = m^2 \mathbb{1}_{4\times 4}.$$

[1] We often write the unit matrix in dimension n in the shorter notation $\mathbb{1}_n$, instead of $\mathbb{1}_{n\times n}$, or, in case the dimension is obvious from the context, just $\mathbb{1}$.

Thus, the operator $\gamma^\mu p_\mu$ has the eigenvalues $(+m)$ and $(-m)$, both of which have multiplicity 2. From this one concludes that there exists another four-component spinor $v(\boldsymbol{p})$ which is linearly independent of $u(\boldsymbol{p})$ and which satisfies the equation

$$\left(\gamma^\mu p_\mu + m \mathbb{1}_{4\times 4}\right) v(\boldsymbol{p}) = 0\,. \tag{9.34}$$

The set of two equations (9.31) and (9.34) constitute the *Dirac equation in momentum space*.

Remarks

1. The action of translations on the spinors (9.23) and (9.24) is the same as the one on the original states $|s, m;\, p\rangle$. This is the reason why we restricted our discussion to the special case $a = 0$. It is obvious from the construction given above that the Dirac equation is *covariant*. Thus, there is no need to verify this property, neither here nor in position space representation.
2. Besides the four matrices γ^μ one defines one further 4×4-matrix as follows

$$\gamma_5 := \mathrm{i}\gamma^0\gamma^1\gamma^2\gamma^3 = \begin{pmatrix} \mathbb{1}_2 & 0 \\ 0 & -\mathbb{1}_2 \end{pmatrix}. \tag{9.35}$$

   This matrix is hermitian. As one easily verifies, it anticommutes with all four γ^μ,

$$\gamma_5\gamma^\mu + \gamma^\mu\gamma_5 = 0\,. \tag{9.36}$$

   Note that there is no difference between γ_5 and γ^5, the subscript 5 being no Lorentz index.
3. The matrices γ^μ and γ_5, endowed with the product (9.32), generate an algebra that belongs to the class of *Clifford algebras*. We wish to sketch this important notion by means of the example we are studying here. For a more systematic study we refer to the literature.[2] The Minkowski space M^4, understood as a four-dimensional $\mathbb{R}$-vector space, admits the scalar product $(p, q) = p^0q^0 - \boldsymbol{p}\cdot\boldsymbol{q}$. A cartesian basis $\{\hat{e}_\mu\}$ of this vector space fulfills the relations

$$(\hat{e}_\mu, \hat{e}_\nu) = g_{\mu\nu}\,,$$

   that is, one has $(\hat{e}_0, \hat{e}_0) = 1$, $(\hat{e}_i, \hat{e}_i) = -1$, and $(\hat{e}_\mu, \hat{e}_\nu) = 0$ whenever $\mu \neq \nu$. One defines a new product $p\,q$ of vectors $p, q \in M^4$ which is associative and, with respect to addition, is distributive. It satisfies the condition[3]

$$q\,p + p\,q = 2(p, q)\,. \tag{9.37}$$

---

[2] Coquereaux, R., *Lectures on Clifford Algebras* (Trieste 1982); Choquet-Bruhat, Y., DeWitt-Morette, C., Dillard-Bleick, M., Analysis, Manifolds, and Physics (North-Holland Publ. Amsterdam 1982).

[3] There should be no confusion if for once we write the Minkowski product somewhat more pedantically as $p^0q^0 - \boldsymbol{p}\cdot\boldsymbol{q} = (p, q)$.

The resulting algebra of all (finite) sums and products is called *Clifford algebra* over M^4. It is denoted by Cl (M^4). When expressed in the basis one has

$$\hat{e}_\mu \hat{e}_\nu + \hat{e}_\nu \hat{e}_\mu = 2g_{\mu\nu}\,, \quad \hat{e}_\nu \hat{e}_\mu = -\hat{e}_\mu \hat{e}_\nu \text{ for } \mu \neq \nu, \quad q^2 = (q, q)\,.$$

The Clifford algebra is also a vector space whose basis can be chosen as follows:

$$\left\{1, \hat{e}_\alpha, \hat{e}_\alpha \hat{e}_\beta (\alpha < \beta), \hat{e}_\alpha \hat{e}_\beta \hat{e}_\gamma (\alpha < \beta < \gamma), \hat{e}_0 \hat{e}_1 \hat{e}_2 \hat{e}_3\right\}\,.$$

This explicit form of the basis allows to determine the dimension of this space. One has[4]

$$1 + 4 + 6 + 4 + 1 = 2^4\,.$$

The Dirac γ-matrices, independently of how they are represented, satisfy these relations, the product being the ordinary matrix product. They generate a basis of $\mathrm{Cl}(M^4)$ which often is chosen to be:

$$\begin{aligned} &\Gamma_S = 1\!\!1_4\,, \qquad \Gamma_P = \mathrm{i}\gamma_5\,, \\ &\Gamma_V^\alpha = \gamma^\alpha\,, \qquad \Gamma_A^\alpha = \gamma^\alpha \gamma_5\,, \\ &\Gamma_T^{[\alpha,\beta]} = \frac{\mathrm{i}}{2\sqrt{2}} \left(\gamma^\alpha \gamma^\beta - \gamma^\beta \gamma^\alpha\right)\,. \end{aligned} \tag{9.38}$$

The letters in the subscript are abbreviations for Scalar, Pseudoscalar, Vector, Axial vector, and Tensor covariants, respectively, and refer to different types of couplings of fermions to spin-1 bosons.

Furthermore, the Clifford algebra $\mathrm{Cl}(M^4)$ is a *graded* algebra, its elements can be classified into *even* and *odd* elements, depending on whether they *commute* with γ_5, or *anticommute* with that matrix. With $[A, B] = AB - BA$ denoting the commutator, $\{A, B\} = AB + BA$ the anticommutator, the basis given in (9.38) has the properties

$$\begin{aligned} &\left[\Gamma_S\,, \gamma_5\right] = 0\,, \; \left[\Gamma_P\,, \gamma_5\right] = 0, \; \left[\Gamma_T\,, \gamma_5\right] = 0\,, \\ &\{\Gamma_V, \gamma_5\} = 0\,, \; \{\Gamma_A, \gamma_5\} = 0\,. \end{aligned}$$

Thus Γ_S, Γ_P, and Γ_T are even, while Γ_V and Γ_A are odd. The matrix γ_5 takes the role of a grading automorphism of the algebra. These seemingly abstract properties are important for physics. For instance, the couplings of fermions to gauge bosons in the gauge theories of electromagnetic, weak, and strong interactions are always of vector or axial vector type. All Yukawa couplings, i.e. couplings to bosons with spin zero, are of scalar or pseudoscalar type.

---

[4] In the general case the real Clifford algebra $\mathrm{Cl}(V^n)$ over the n-dimensional real vector space V^n has the dimension

$$\sum_{k=0}^{n'} \binom{n}{k} = 2^n\,,$$

thus, in the present case, the dimension is 16.

4. Consider the linear combinations
$$P_+ = \frac{1}{2}\left(1\!\!1 + \gamma_5\right), \quad P_- = \frac{1}{2}\left(1\!\!1 - \gamma_5\right). \tag{9.39}$$
As one easily verifies, these are projection operators. Indeed, they obey the rules
$$P_+^2 = P_+ , \quad P_-^2 = P_- , \quad P_+ P_- = 0 = P_- P_+ , \quad P_+ + P_- = 1\!\!1 .$$
They project the four-spinors $u(\boldsymbol{p})$ and $v(\mathbf{p})$ to their upper and lower two components, respectively. The physical significance of these projections will become clear once we will have obtained explicit solutions of the Dirac equation.
5. As the γ-matrices often are contracted with four-vectors it is customary to use a new symbol, called the *slash*, viz.
$$\not p \equiv \gamma^\mu p_\mu .$$
We shall make use of this useful abbreviation on many occasions below.
6. Using the definition (9.18) one verifies that $v(\boldsymbol{p})$, too, may be expressed in terms of the spinors defined in (9.23) and (9.24),
$$v(\boldsymbol{p}) = \begin{pmatrix} \varepsilon \widetilde{\chi}^*(p) \\ -\varepsilon \widetilde{\phi}^*(p) \end{pmatrix} .$$
As we will see further below this relation is an expression of charge conjugation, i.e. of the mapping of fermion states to antifermion states.
7. The representation (9.28), or (9.29), of the γ-matrices is what may be termed a *natural representation* because it was obtained from the spinor representations of the Lorentz group. It is also called *high-energy representation* because of the fact that the two two-component equations for $\widetilde{\phi}(p)$ and for $\widetilde{\chi}(p)$ decouple completely in the limit $E_p \equiv p^0 \gg m$. Without changing the content of the Dirac equation in any respect, one can always subject $u(\boldsymbol{p})$ and $v(\boldsymbol{p})$ to a nonsingular linear transformation $\mathbf{S}$ which reshuffles its components
$$u(\boldsymbol{p}) \longmapsto u'(\boldsymbol{p}) = \mathbf{S}\, u(\boldsymbol{p}) , \quad v(\boldsymbol{p}) \longmapsto v'(\boldsymbol{p}) = \mathbf{S}\, v(\boldsymbol{p}) ,$$
and, at the same time, introduce transformed γ-matrices
$$\gamma'^\mu = \mathbf{S} \gamma^\mu \mathbf{S}^{-1}$$
such that the Dirac equation (9.31) or (9.34) remains form invariant. Obviously, the relation (9.32) remains unchanged. As an example choose
$$\mathbf{S} = \frac{1}{\sqrt{2}} \begin{pmatrix} 1\!\!1_2 & 1\!\!1_2 \\ 1\!\!1_2 & -1\!\!1_2 \end{pmatrix} = \mathbf{S}^{-1} , \tag{9.40}$$
where $1\!\!1_2$ denotes the 2×2 unit matrix. One obtains
$$\gamma'^0 = \begin{pmatrix} 1\!\!1_2 & 0 \\ 0 & -1\!\!1_2 \end{pmatrix} , \quad \gamma'^i = \begin{pmatrix} 0 & \sigma^{(i)} \\ -\sigma^{(i)} & 0 \end{pmatrix} , \quad \gamma_5' = \begin{pmatrix} 0 & 1\!\!1_2 \\ 1\!\!1_2 & 0 \end{pmatrix} . \tag{9.41}$$
This representation is called the *standard representation*. It is particularly useful in cases where one studies weakly relativistic motion and where one

wishes to expand in terms of v/c. Atomic and nuclear physics make frequent use of this representation. As an exercise, the reader is invited to verify the properties (9.33) for this representation. The reason for this is that (9.40) is not only nonsingular but also unitary.

All representations which are generated from the natural representation by a unitary transformation **S** *have the specific property*

$$\gamma^0 \left(\gamma^\alpha\right)^\dagger \gamma^0 = \gamma^\alpha \,. \tag{9.42}$$

We have suppressed the prime on the γ-matrices because this property concerns a whole class of representations.

8. As an example for the preceding remark let us consider the class of *Majorana representations* whose defining property is that all γ-matrices are pure imaginary,

$$\left(\gamma^{(\mathrm{M})\,\alpha}\right)^* = -\gamma^{(\mathrm{M})\,\alpha} \,. \tag{9.43}$$

Given any of the representations constructed in Remark 7 above, choose the transformation

$$\mathbf{S}^{(\mathrm{M})} := \frac{1}{\sqrt{2}}\gamma^0(\mathbb{1} + \gamma^2)$$

which is seen to be unitary by means of the formulae

$$(\mathbb{1} + \gamma^{2\,\dagger})\gamma^{0\,\dagger} = (\mathbb{1} - \gamma^2)\gamma^0 = \gamma^0(\mathbb{1} + \gamma^2) \,.$$

Starting from the natural, or high-energy, representation (9.28) one has

$$\mathbf{S}^{(\mathrm{M})}_{\mathrm{HE}} = \frac{1}{\sqrt{2}}\begin{pmatrix} \sigma^{(2)} & \mathbb{1}_2 \\ \mathbb{1}_2 & -\sigma^{(2)} \end{pmatrix} .$$

The corresponding Majorana representation is found to be

$$\gamma^{(\mathrm{M})\,0}_{\mathrm{HE}} = \begin{pmatrix} \sigma^{(2)} & 0 \\ 0 & -\sigma^{(2)} \end{pmatrix} , \quad \gamma^{(\mathrm{M})\,1}_{\mathrm{HE}} = \mathrm{i}\begin{pmatrix} \sigma^{(3)} & 0 \\ 0 & \sigma^{(3)} \end{pmatrix} ,$$
$$\gamma^{(\mathrm{M})\,2}_{\mathrm{HE}} = \begin{pmatrix} 0 & \sigma^{(2)} \\ -\sigma^{(2)} & 0 \end{pmatrix} , \quad \gamma^{(\mathrm{M})\,3}_{\mathrm{HE}} = -\mathrm{i}\begin{pmatrix} \sigma^{(1)} & 0 \\ 0 & \sigma^{(1)} \end{pmatrix} ,$$
$$\gamma^{(\mathrm{M})}_{\mathrm{HE}\,5} = \begin{pmatrix} 0 & \sigma^{(2)} \\ \sigma^{(2)} & 0 \end{pmatrix} . \tag{9.44}$$

In turn, starting from the standard representation (9.41) one has

$$\mathbf{S}^{(\mathrm{M})}_{\mathrm{St}} = \frac{1}{\sqrt{2}}\begin{pmatrix} \mathbb{1}_2 & \sigma^{(2)} \\ \sigma^{(2)} & -\mathbb{1}_2 \end{pmatrix} .$$

The corresponding Majorana representation reads

$$\gamma_{\rm St}^{(\rm M)\,0} = \begin{pmatrix} 0 & \sigma^{(2)} \\ \sigma^{(2)} & 0 \end{pmatrix}, \quad \gamma_{\rm St}^{(\rm M)\,1} = \mathrm{i}\begin{pmatrix} \sigma^{(3)} & 0 \\ 0 & \sigma^{(3)} \end{pmatrix},$$
$$\gamma_{\rm St}^{(\rm M)\,2} = \begin{pmatrix} 0 & -\sigma^{(2)} \\ \sigma^{(2)} & 0 \end{pmatrix}, \quad \gamma_{\rm St}^{(\rm M)\,3} = -\mathrm{i}\begin{pmatrix} \sigma^{(1)} & 0 \\ 0 & \sigma^{(1)} \end{pmatrix},$$
$$\gamma_{\rm St\,5}^{(\rm M)} = \begin{pmatrix} \sigma^{(2)} & 0 \\ 0 & -\sigma^{(2)} \end{pmatrix}. \tag{9.45}$$

Note that in accord with the Condon-Shortley phase convention the Pauli matrix $\sigma^{(1)}$ is real positive, $\sigma^{(2)}$ is pure imaginary, and $\sigma^{(3)}$ is real and diagonal. Therefore, all five γ-matrices of the representations (9.44) and (9.45), indeed, are pure imaginary.

9.1.3 Solutions of the Dirac Equation in Momentum Space

After having aquired a certain virtuosity in converting different representations of the γ-matrices into one another, we choose one of them and determine explicit solutions for fixed four-momentum p. As a first test we verify that the two linear systems of equations (9.31) and (9.34) are soluable. Indeed, the determinants vanish,

$$\det(\not{p} - m\,\mathbb{1}) = \det(\not{p} + m\,\mathbb{1}) = \left((p^0)^2\boldsymbol{p}^2 m^2\right)^2 = (p^2 - m^2)^2 = 0\,,$$

provided the momentum is on its mass shell.

The simplest approach consists in constructing solutions in the particle's rest system, using the standard representation. The standard representation (9.41) has the distinctive property that in the rest frame $p = (m, \boldsymbol{0})^T$ the spinor $u_{\rm St}(\boldsymbol{p})$ takes the form $(\chi^{(1)}, 0, 0)^T$ with $\chi^{(1)}$ a two component spinor as we know it from nonrelativistic quantum mechanics. This is shown by evaluating the Dirac equation (9.31) with an ansatz $(\chi^{(1)}, \chi^{(2)})^T$ at $\boldsymbol{p} = 0$, $\chi^{(i)}$ denoting two-component spinors,

$$m\begin{pmatrix} 0 & 0 \\ 0 & \mathbb{1}_2 \end{pmatrix}\begin{pmatrix} \chi^{(1)} \\ \chi^{(2)} \end{pmatrix} = 0\,.$$

It is obvious that $\chi^{(2)}$ must vanish identically, $\chi^{(2)} \equiv 0$, while $\chi^{(1)} \equiv \chi$ is the spin function of a nonrelativistic particle.

Similarly, (9.34) is first solved for $v(\boldsymbol{0}) = (\phi^{(1)}, \phi^{(2)})^T$, in which case this equation reduces to

$$m\begin{pmatrix} \mathbb{1}_2 & 0 \\ 0 & 0 \end{pmatrix}\begin{pmatrix} \phi^{(1)} \\ \phi^{(2)} \end{pmatrix} = 0\,.$$

In this case $\phi^{(1)}$ must vanish identically, $\phi^{(1)} \equiv 0$, while $\phi^{(2)} \equiv \phi$ is a nonrelativistic spinor.

In a second step the solutions $u(\boldsymbol{0}) = (\chi, \boldsymbol{0})^T$ and $v(\boldsymbol{0}) = (\boldsymbol{0}, \phi)^T$ are boosted to arbitrary three-momentum $\boldsymbol{p}$ by means of the following trick. One makes use of the relations

$$(\not{p} \pm m\,\mathbb{1})(\not{p} \mp m\,\mathbb{1}) = p^2 - m^2 = 0\,,$$

in order to verify that the spinors

$$u(\boldsymbol{p}) = N\,(\not{p} + m\,\mathbb{1})\,u(\boldsymbol{0})$$
$$v(\boldsymbol{p}) = -N\,(\not{p} - m\,\mathbb{1})\,v(\boldsymbol{0})$$

satisfy the equations (9.31) and (9.34), respectively. Except for the normalization constant N which can be chosen at will, these solutions are unique. The minus sign in the expression for $v(\boldsymbol{p})$ is chosen for convenience because, upon insertion of the γ-matrices (9.41) in

$$(\not{p} \pm m\,\mathbb{1}_4) = (\gamma^0 p^0 - \gamma^i p^i \pm m\,\mathbb{1}_4) = \begin{pmatrix} (p^0 \pm m)\,\mathbb{1}_2 & -\boldsymbol{\sigma}\cdot\boldsymbol{p} \\ \boldsymbol{\sigma}\cdot\boldsymbol{p} & -(p^0 \mp m\,\mathbb{1}_2) \end{pmatrix},$$

the solutions take the explicit form

$$u(\boldsymbol{p}) = N \begin{pmatrix} (p^0 + m)\chi \\ \boldsymbol{\sigma}\cdot\boldsymbol{p}\,\chi \end{pmatrix} \tag{9.46}$$

$$v(\boldsymbol{p}) = N \begin{pmatrix} \boldsymbol{\sigma}\cdot\boldsymbol{p}\phi \\ (p^0 + m)\,\phi \end{pmatrix}. \tag{9.47}$$

The normalization factor N must be chosen such that both solutions are normalized covariantly,

$$u^\dagger(\boldsymbol{p})u(\boldsymbol{p}) = 2p^0 = v^\dagger(\boldsymbol{p})v(\boldsymbol{p})\,. \tag{9.48}$$

We demonstrate this calculation on the example of the solutions $u(\boldsymbol{p})$. Let the spinor χ be normalized to 1, $\chi^\dagger\chi = 1$. Then

$$u^\dagger(\boldsymbol{p})u(\boldsymbol{p}) = N^2\left(\chi^\dagger(p^0 + m)\;\; \chi^\dagger\boldsymbol{\sigma}\cdot\boldsymbol{p}\right)\begin{pmatrix} (p^0 + m)\chi \\ \boldsymbol{\sigma}\cdot\boldsymbol{p}\,\chi \end{pmatrix}$$
$$= N^2\{(p^0 + m)^2 + \boldsymbol{p}^2\} = 2p^0(p^0 + m)N^2\,.$$

Thus, the condition (9.48) fixes the normalization factor to be

$$N = \frac{1}{\sqrt{p^0 + m}}\,. \tag{9.49}$$

We conclude this section with some supplements and remarks.

Remarks

1. Of course, the spinors $\chi^{(i)}$ carry a magnetic quantum number for which one often writes $r = 1$ when $m_s = +1/2$, $r = 2$ when $m_s = -1/2$. Hence, the Dirac spinors should be written more explicitly $u^{(r)}(\boldsymbol{p})$ and $v^{(s)}(\boldsymbol{p})$, $r, s = 1, 2$, and the normalization condition (9.48) is to be supplemented by the orthogonality in this quantum number, viz.

$$u^{(r)\dagger}(\boldsymbol{p})u^{(s)}(\boldsymbol{p}) = 2p^0\delta_{rs} = v^{(r)\dagger}(\boldsymbol{p})v^{(s)}(\boldsymbol{p})\,. \tag{9.50}$$

In some calculations to follow below another ortogonality relation will be needed. It reads

$$u^{(r)\dagger}(\boldsymbol{p})v^{(s)}(-\boldsymbol{p}) = 0 \tag{9.51}$$

and is easily verified from the explicit solutions.

2. The case of massless particles is also interesting for the following reason: In contrast to spins higher than $1/2$, the limit $m \to 0$ can be taken directly in the solutions (9.46) and (9.47). This is so because the number of states of a spin-$1/2$ particle with $m \neq 0$ is *equal* to the number of helicity states of a spin-$1/2$ particle without mass. This is different for the case of particles with spin $s \geq 1$ where the massive particle has $(2s+1)$ possible orientations of its spin while the massless particle has no more than the two helicity states $h = \pm s$, cf. Sect. 6.3.3, case $m = 0$. Taking $m = 0$ in the solutions (9.46), one has (9.47) with $p^0 = |\boldsymbol{p}|$ and with

$$h = \frac{1}{2}\frac{\boldsymbol{\sigma}\cdot\boldsymbol{p}}{p^0},$$

$$u(\boldsymbol{p}, m=0) = \sqrt{p^0}\begin{pmatrix}\chi \\ 2h\,\chi\end{pmatrix}$$

$$v(\boldsymbol{p}, m=0) = \sqrt{p^0}\begin{pmatrix}2h\,\phi \\ \phi\end{pmatrix}.$$

Turning to the natural, or high-energy representation the first of these spinors becomes

$$u_{\mathrm{HE}}(\boldsymbol{p}, m=0) = \sqrt{\frac{p^0}{2}}\begin{pmatrix}(\mathbb{1}_2 + 2h)\chi \\ (\mathbb{1}_2 - 2h)\chi\end{pmatrix}.$$

For the choice of the eigenvalue $h = +1/2$ the lower component vanishes, for the choice $h = -1/2$ the upper component vanishes.[5] Analogous statements apply to the spinors $v(\boldsymbol{p}, m=0)$. This is in accord with the original form of the Dirac equation (9.27): The two equations in (9.27) decouple if m is set to zero. Massive particles can come very close to this remarkably simple limit: In any kinematic situation in which the energy is large as compared to the rest mass, $p^0 \gg m$, the solutions can be taken to be eigenstates of helicity, in very good approximation. The natural representation (9.28) of the γ-matrices is best adapted to this situation because the two equations (9.27) (nearly) decouple. This justifies, a posteriori, why this representation is also called high-energy representation.

3. As we shall learn in Sect. 9.2.3 the spinors $u(\boldsymbol{p})$ and $v(\boldsymbol{p})$ are to be associated to pairs of spin-$1/2$ particles which are antiparticles of one another. Note, however, that the correct identification of the spin states requires some care. The Majorana representation (9.45) is helpful in clarifying this relation. Applying the transformation $\mathbf{S}_{\mathrm{St}}^{(\mathrm{M})}$ to the solutions (9.46) and (9.47) one obtains the expressions

[5] The helicity of massless particles with spin $1/2$ is often defined as

$$h' := \boldsymbol{\sigma}\cdot\boldsymbol{p}/|\boldsymbol{p}|,$$

hence, differing from h by a factor of 2. Its eigenvalues then are ± 1.

$$u_{\text{St}}^{(\text{M})}(\boldsymbol{p}) = \frac{N}{\sqrt{2}}\Big((p^0+m+\sigma^{(2)}\boldsymbol{\sigma}\cdot\boldsymbol{p})\chi\,,\,((p^0+m)\sigma^{(2)}-\boldsymbol{\sigma}\cdot\boldsymbol{p})\chi\Big)^T\,,$$

$$v_{\text{St}}^{(\text{M})}(\boldsymbol{p}) = \frac{N}{\sqrt{2}}\Big(((p^0+m)\sigma^{(2)}+\boldsymbol{\sigma}\cdot\boldsymbol{p})\phi\,,\,(-(p^0+m)+\sigma^{(2)}\boldsymbol{\sigma}\cdot\boldsymbol{p})\phi\Big)^T\,.$$

(We write $p^0 + m$ for $(p^0 + m)\mathbb{1}_2$, for the sake of simplicity.) We now wish to show that, in essence, v is obtained from the complex conjugate of u. For this purpose take $u^*(\boldsymbol{p})$ and use $(\sigma^{(2)})^2 = \mathbb{1}_2$ as well as the relation

$$\sigma^{(2)}\boldsymbol{\sigma}\sigma^{(2)} = -\boldsymbol{\sigma}^*\,.$$

This is obtained from (9.18) which, in turn, is closely related to the transformation $\mathbf{U}_0$ that was analyzed in Sect. 6.2.3. One finds

$$\begin{aligned} u_{\text{St}}^{(\text{M})*}(\boldsymbol{p}) &= \frac{N}{\sqrt{2}}\Big(((p^0+m)\sigma^{(2)}+\boldsymbol{\sigma}\cdot\boldsymbol{p})(\sigma^{(2)}\chi^*)\,,\\ &\qquad (-(p^0+m)+\sigma^{(2)}\boldsymbol{\sigma}\cdot\boldsymbol{p})(\sigma^{(2)}\chi^*)\Big)^T\\ &= \frac{-iN}{\sqrt{2}}\Big(((p^0+m)\sigma^{(2)}+\boldsymbol{\sigma}\cdot\boldsymbol{p})(\varepsilon\chi)^*\,,\\ &\qquad (-(p^0+m)+\sigma^{(2)}\boldsymbol{\sigma}\cdot\boldsymbol{p})(\varepsilon\chi)^*\Big)^T\,, \end{aligned}$$

where, in the second step, the definition (9.17) was inserted. Except for the factor $(-\mathrm{i})$ this is precisely the solution $v(\boldsymbol{p})$ if the two-component spinor ϕ is replaced by $(\varepsilon\chi)^*$. Thus, if $u(\boldsymbol{p})$ describes the spin state of a *particle* which reduces to the spinor χ in the rest frame, then its image $u^*(\boldsymbol{p})$ describes the corresponding *antiparticle*. Its spinor in the rest frame is $(\varepsilon\chi)^*$. In other terms, writing both the momentum $\boldsymbol{p}$ and the two-component spinor in the argument, then for every Majorana representation one has

$$u_{\text{St}}^{(\text{M})*}(\boldsymbol{p},\chi) = v_{\text{St}}^{(\text{M})}(\boldsymbol{p},-i\varepsilon\chi^*)\,,\quad v_{\text{St}}^{(\text{M})*}(\boldsymbol{p},\phi) = u_{\text{St}}^{(\text{M})}(\boldsymbol{p},i\varepsilon\phi^*)\,. \tag{9.52}$$

4. We turn to the interesting question of how to obtain invariants (with respect to the Lorentz group) from the spinors $u(\boldsymbol{p})$ and $v(\boldsymbol{p})$. We give the answer first in the high-energy representation (9.28) of the γ-matrices, and translate them to other representations later. The normalization condition (9.48) shows that products such as $u^\dagger u$ or $v^\dagger v$ are proportional to the energy. This is to say that such products transform like the time component of a four-vector. Returning to the two-component spinors $\widetilde{\phi}(p)$ and $\widetilde{\chi}(p)$ one sees that both $\widetilde{\chi}^\dagger\widetilde{\phi}$ and $\widetilde{\phi}^\dagger\widetilde{\chi}$ are invariant under all $\boldsymbol{\Lambda} \in L_+^\uparrow$. For instance, by means of (9.25) and (9.26) one has

$$\widetilde{\chi}'^\dagger\widetilde{\phi}' = \widetilde{\chi}^\dagger\,\widehat{\mathbf{A}}^\dagger(\boldsymbol{\Lambda}^{-1})\mathbf{A}(\boldsymbol{\Lambda}^{-1})\,\widetilde{\phi} = \widetilde{\chi}^\dagger\,\left(\varepsilon\mathbf{A}^T\varepsilon^{-1}\right)\mathbf{A}\,\widetilde{\phi}\,.$$

Here the hermitian conjugate of the definition $\widehat{\mathbf{A}} = \varepsilon\mathbf{A}^*\varepsilon^{-1}$ was inserted. Finally, the transposed of the relation (9.20) together with $\varepsilon^T = -\varepsilon$ shows that $\varepsilon\mathbf{A}^T\varepsilon^{-1} = \mathbf{A}^{-1}$. This proves the invariance $\widetilde{\chi}'^\dagger\widetilde{\phi}' = \widetilde{\chi}^\dagger\widetilde{\phi}$. The invariance of $\widetilde{\phi}^\dagger\widetilde{\chi}$ is shown in an analogous way. Applying now the explicit representations

(9.29) and (9.35) one realizes that the two independent invariants above can be expressed in terms of $u(p)$ and its hermitian conjugate (or, alternatively, in terms of $v(p)$ and $v^\dagger(p)$) as follows:

$$\widetilde{\chi}^\dagger \widetilde{\phi} = \frac{1}{2}\left(u^\dagger(p)\gamma^0 u(p) + u^\dagger(p)\gamma^0\gamma_5 u(p)\right),$$

$$\widetilde{\phi}^\dagger \widetilde{\chi} = \frac{1}{2}\left(u^\dagger(p)\gamma^0 u(p) - u^\dagger(p)\gamma^0\gamma_5 u(p)\right).$$

Consider, in addition, space reflection and time reversal: These discrete operations interchange the spinors $\widetilde{\phi}$ and $\widetilde{\chi}$. This implies that $u^\dagger(p)\gamma^0 u(p)$ is invariant under *all* Lorentz transformations and, hence, is a genuine Lorentz-scalar, while $u^\dagger(p)\gamma^0\gamma_5 u(p)$ changes sign under space reflection and, hence is a Lorentz pseudoscalar.
This observation motivates a special notation which is quite useful in the calculus of Dirac spinors: The product of $u^\dagger$ or of $v^\dagger$ with γ^0, respectively, is denoted as follows

$$\overline{u(p)} := u^\dagger(p)\gamma^0, \qquad \overline{v(p)} := v^\dagger(p)\gamma^0. \tag{9.53}$$

How does this definition translate to other representations of the γ-matrices? The answer is easy to verify: The definition (9.53) remains meaningful, that is to say, $\overline{u'(p)}u'(p)$ and $\overline{v'(p)}v'(p)$ are Lorentz scalars, if the substitution matrix $\mathbf{S}$ is *unitary*.
Note that this definition of the "over-bar" is a generally accepted one in relativistic quantum theory. This is the reason why in the physics literature complex conjugation is denoted by an asterisk and not by the overbar which is customary in the mathematical literature.
As an exercise one calculates the products $\overline{u(p)}u(p)$ and $\overline{v(p)}v(p)$ for the solutions (9.46) and (9.47). One finds

$$\overline{u^{(r)}(p)}u^{(s)}(p) = 2m\,\delta_{rs}, \quad \overline{v^{(r)}(p)}v^{(s)}(p) = -2m\,\delta_{rs}. \tag{9.54}$$

Indeed, these are genuine invariants.

9.1.4 Dirac Equation in Spacetime and Lagrange Density

It is not difficult to translate the Dirac equation (9.31) and (9.34) to Minkowski spacetime. The solutions

$$\left\{u^{(r)}(\boldsymbol{p})\,\mathrm{e}^{-\mathrm{i}px},\; v^{(r)}(\boldsymbol{p})\,\mathrm{e}^{\mathrm{i}px}\right\}, \quad r = 1,2, \quad \boldsymbol{p} \in \mathbb{R}^3,$$

form a complete base system with respect to the spin degrees of freedom and to the momenta. Therefore, a Dirac spinor $\psi(x)$ which depends on space and time, can be expanded in terms of this basis, viz.

$$\psi(x) = \sum_{r=1}^{2}\int \frac{\mathrm{d}^3 p}{2p^0}\left\{a^{(r)}(\boldsymbol{p})u^{(r)}(\boldsymbol{p})\,\mathrm{e}^{-\mathrm{i}px} + b^{(r)*}(\boldsymbol{p})v^{(r)}(\boldsymbol{p})\,\mathrm{e}^{\mathrm{i}px}\right\}. \tag{9.55}$$

The coefficients $a^{(r)}(\boldsymbol{p})$ and $b^{(r)}(\boldsymbol{p})$ are complex numbers which depend on the spin orientation and on the momentum, px denotes the usual product in Minkowski space, $px = p^0x^0 - \boldsymbol{p}\cdot\boldsymbol{x}$, with $p^0 = \sqrt{m^2+\boldsymbol{p}^2}$ the energy. As $u^{(r)}(\boldsymbol{p})$ and $v^{(r)}(\boldsymbol{p})$ are four-component spinors, also $\psi(x)$ is a four-component Dirac spinor, $\psi(x) = (\psi_1(x), \psi_2(x), \psi_3(x), \psi_4(x))^T$.

In the product $m\,\psi(x)$ replace $mu(\boldsymbol{p})$ and $mv(\boldsymbol{p})$ by means of the equations (9.31) and (9.34), respectively, then make use of the identities

$$\not{p}\,\mathrm{e}^{-\mathrm{i}px} = \mathrm{i}\gamma^\mu\partial_\mu\,\mathrm{e}^{-\mathrm{i}px}\,, \quad -\not{p}\,\mathrm{e}^{\mathrm{i}px} = \mathrm{i}\gamma^\mu\partial_\mu\,\mathrm{e}^{\mathrm{i}px}\,,$$

and extract $\gamma^\mu\partial_\mu$ from the integral. On the right-hand side this yields the linear combination $\mathrm{i}\gamma^\mu\partial_\mu\psi(x)$ of first derivatives of the function $\psi(x)$. Thus, one obtains the *Dirac equation in spacetime*

$$\left(\mathrm{i}\gamma^\mu\partial_\mu - m\,\mathbb{1}\right)\psi(x) = 0\,. \tag{9.56}$$

It is equivalent to and replaces the two equations (9.31) and (9.34) in momentum space. By using the "slash"-notation and suppressing the explicit 4×4 unit matrix the Dirac equation takes the elegant and compact form

$$(i\not{\partial} - m)\,\psi(x) = 0\,.$$

Like in (9.53) one defines the adjoint spinor

$$\overline{\psi(x)} := \psi^\dagger(x)\gamma^0$$

and derives the adjoint form of the Dirac equation as follows. Multiply the hermitian conjugate of the Dirac equation (9.56) by γ^0 from the right and insert $(\gamma^0)^2 = \mathbb{1}_4$ at the appropriate position, to obtain

$$\left(\psi^\dagger(x)\gamma^0\right)\gamma^0\left(\mathrm{i}\gamma^{\mu\,\dagger}\partial_\mu - m\right)\gamma^0 = 0\,.$$

Use then the property (9.42) to obtain

$$\overline{\psi(x)}\left(\mathrm{i}\gamma^\mu\overleftarrow{\partial}_\mu + m\,\mathbb{1}\right) = 0 \quad\text{or}\quad \overline{\psi(x)}\left(\mathrm{i}\overleftarrow{\not{\partial}} + m\,\mathbb{1}\right) = 0\,. \tag{9.57}$$

(The arrow indicates that the derivative acts on the function on the left of the operator, i.e. $\overline{\psi}\overleftarrow{\not{\partial}} = (\partial_\mu\overline{\psi})\gamma^\mu$, not to be confused with the right-left derivative $f\overleftrightarrow{\partial}g = f\partial g - (\partial f)g$!).

It is not difficult to test the general condition which says that every component of the field $\psi(x)$ obeys the Klein-Gordon equation (cf. the remark in Sect. 7.1): If the operator $\left(i\not{\partial} + m\right)$ acts from the left on the Dirac equation (9.56), then with

$$\left(\mathrm{i}\not{\partial} + m\,\mathbb{1}\right)\left(\mathrm{i}\not{\partial} - m\,\mathbb{1}\right) = \left(-\partial_\mu\partial^\mu - m^2\right)\mathbb{1}$$

one obtains the Klein-Gordon operator $\Box + m^2$ and the differential equation

$$\left(\Box + m^2\right)\psi(x) = 0\,.$$

Indeed, this equation holds for every component $\psi_\alpha(x)$, $\alpha = 1, 2, 3, 4$.

A Lagrange density for the Dirac field is constructed along the following lines. We know that $\overline{\psi}\psi$ is a Lorentz invariant. Furthermore, one shows that $\overline{\psi}\gamma^\mu\psi$ transforms like a contravariant Lorentz vector (see Exercise 9.2) so that $\overline{\psi}\gamma^\mu \overleftrightarrow{\partial_\mu} \psi$ is a Lorentz scalar. The following invariant complies with all requirements that a Lagrange density must fulfill,

$$\mathcal{L}_\mathrm{D} = \overline{\psi(x)}\Big(\frac{\mathrm{i}}{2}\gamma^\mu \overleftrightarrow{\partial_\mu} - m\mathbb{1}\Big)\psi(x)\,. \tag{9.58}$$

Indeed, with this choice of factors and relative sign it yields the correct equations of motion. In analogy to the case of the scalar field one does not vary real and imaginary parts of the field but, rather, the field ψ itself and its adjoint $\overline{\psi}$. If one does so, one finds e.g.

$$\frac{\partial \mathcal{L}}{\partial \overline{\psi}_\alpha} = \Big(\frac{\mathrm{i}}{2}\gamma^\mu\partial_\mu\psi - m\psi\Big)_\alpha\,, \quad \frac{\partial \mathcal{L}}{\partial(\partial_\mu\overline{\psi}_\alpha)} = \Big(-\frac{\mathrm{i}}{2}\gamma^\mu\psi\Big)_\alpha\,.$$

When these are inserted in the Euler-Lagrange equation one obtains

$$\frac{\partial \mathcal{L}}{\partial \overline{\psi}_\alpha} - \partial_\mu\left(\frac{\partial \mathcal{L}}{\partial(\partial_\mu\overline{\psi}_\alpha)}\right) = \big(\mathrm{i}\gamma^\mu\partial_\mu\psi - m\psi\big)_\alpha = 0\,, \quad \alpha = 1,2,3,4\,.$$

This is the Dirac equation (9.56). If, in turn, one varies the field ψ, then one finds the adjoint Dirac equation (9.57).

From a physical point of view the free Dirac equation is not very informative: True, it guarantees the correct dispersion relation for energy and momentum, it is Lorentz covariant, and describes the spin content covariantly. But, as long as no interaction is introduced, it does not predict anything that is observable. This we can change very easily, however, by introducing the coupling of the Dirac field to electromagnetic fields. At the same time this modification prepares the first step in the interpretation of the Dirac spinor ψ.

We apply the *minimal substitution rule* as explained in Sect. 7.2, i.e. replace the derivatives of the Dirac spinors as follows

$$\partial_\mu\psi \longmapsto (\partial_\mu + \mathrm{i}qA_\mu)\psi\,, \quad \partial_\mu\overline{\psi} \longmapsto (\partial_\mu - \mathrm{i}qA_\mu)\overline{\psi}\,, \tag{9.59}$$

Applying these rules to the Lagrange density (9.58) and adding the Lagrange density of the free Maxwell field, one obtains

$$\mathcal{L} = \mathcal{L}_\mathrm{D} + \mathcal{L}_\gamma - q\,\overline{\psi(x)}\gamma^\mu\psi(x)A_\mu(x)\,, \quad \text{with } \mathcal{L}_\gamma = -\frac{1}{4}F_{\mu\nu}F^\mu\,. \tag{9.60}$$

One recognizes the close relationship to the more general form (7.85) of Lagrange densities. Furthermore, one obtains an expression for the electromagnetic current density of the particle described by the Dirac equation:

$$j^\mu(x) = q\,\overline{\psi(x)}\gamma^\mu\psi(x)\,. \tag{9.61}$$

The Lagrange density (9.60) with $q = -\,|e|$, i.e. the negative elementary charge, provides the basis for quantum electrodynamics with electrons and photons.

The Dirac equation, by the principle of minimal substitution, becomes

$$\big(\mathrm{i}\partial\!\!\!/ - q A\!\!\!/ - m\mathbb{1}\big)\psi(x) = 0\,. \tag{9.62}$$

The equation now contains the interaction with external electromagnetic potentials and, therefore, makes predictions about the scattering of photons on charged fermions which are observable.

For every solution $\psi(x)$ of this equation there is a Dirac field $\psi_C(x)$ which satisfies the equation of motion

$$\left(\mathrm{i}\not{\partial} + q\not{A} - m\,\mathbb{1}\right)\psi_C(x) = 0\,. \tag{9.63}$$

Note that this differential equation differs from (9.62) by the sign of the charge. In a Majorana representation the equation (9.63) is nearly obvious: It suffices to take the complex conjugate of the equation (9.62), to make use of the property $\gamma^{(M)\mu\,*} = -\gamma^{(M)\mu}$ and to choose $\psi_C(x) = c\psi(x)$ with $c \in \mathbb{C}$.

Already this simple example shows that the Dirac theory for fermions with electric charge q predicts the existence of a partner which has the same mass but opposite charge. To every fermion there is an antiparticle: for the electron the positron, for the muon μ^- the antimuon μ^+, for the proton p the antiproton $\overline{p}$, etc. We shall see, after having quantized the Dirac field, that the theory not only predicts the existence of antiparticles, but, in addition, is completely symmetric in both, the particle and its partner. Therefore, it is a matter of convention what we call a *particle* and what we call an *antiparticle.*

Starting from the high-energy representation (9.28) the spinor

$$\psi_C(x) = \mathrm{i}\gamma^2\psi^*(x) = \mathrm{i}\gamma^2\gamma^0\left(\overline{\psi(x)}\right)^T \tag{9.64}$$

is shown to be a solution of (9.63). Indeed, if one takes the complex conjugate of (9.62), inserts the identity $(\mathrm{i}\gamma^2)^2 = \mathbb{1}$ in front of ψ^*, and multiplies the equation obtained in this way by $(\mathrm{i}\gamma^2)$ from the left, then one has

$$(\mathrm{i}\gamma^2)\left(-\mathrm{i}\gamma^{\mu\,*}\partial_\mu - q\gamma^{\mu\,*}A_\mu - m\,\mathbb{1}\right)(\mathrm{i}\gamma^2)(\mathrm{i}\gamma^2\psi^*(x)) = 0\,.$$

By the definition (9.64) and using the relation

$$(\mathrm{i}\gamma^2)\gamma^{\mu\,*}(\mathrm{i}\gamma^2) = -\gamma^\mu\,, \tag{9.65}$$

the equation (9.63) follows. We note that (9.65) holds in any representation of the γ-matrices which is obtained from the high-energy representation by a transformation $\mathbf{S}$ which, in addition to be unitary, is also real (that is to say which is orthogonal).[6] An example is provided by the standard representation (9.41). The mapping from ψ to ψ_C and its inverse is called *charge conjugation.*

Since the adjoint spinor fields $\overline{\psi}$ seem to play a special role it often is convenient to make use of the second form of (9.64). The operator of charge conjugation then contains the matrix $\mathbf{C} = i\gamma^2\gamma^0$ whose properties are

$$\mathbf{C} = \mathrm{i}\gamma^2\gamma^0 = -\mathbf{C}^{-1} = -\mathbf{C}^\dagger = -\mathbf{C}^T\,. \tag{9.66}$$

Note that twofold application of charge conjugation leads back to the original spinor,

$$(\psi_C)_C = \mathrm{i}\gamma^2(\mathrm{i}\gamma^2\psi^*)^* = \gamma^2\gamma^{2\,*}\psi = \psi\,.$$

[6] This property no longer holds after transforming to the corresponding Majorana representations. This is the reason why in such representations ψ_C is not given by (9.64).

Furthermore, there are situations where the action of charge conjugation on a one-particle state brings to light, in addition, a phase η_C which is characteristic for the particle one is describing. But even if this happens one can arrange for the square of this phase to be 1, $\eta_C^2 = 1$, without restriction.

If a scalar quantum field equals its hermitian adjoint, $\phi^\dagger(x) = \phi(x)$, then it describes an electrically neutral boson with spin zero which is identical with its antiparticle. For a spin-1/2 particle the analogous condition is

$$\psi_C(x) = \pm\psi(x)\,. \tag{9.67}$$

A Dirac field which has this property describes what is called a *Majorana particle.* Majorana particles are identical with their antiparticle and carry no electric charge or, for that matter, any other additively conserved quantum number. Whether or not there are such particles in nature is still an open issue.

A similar analysis of the action of time reversal **T** on the free Dirac equation shows that for every solution there is a time-reversed spinor field ψ_T which is also a solution. In the representation (9.28) as well as in any other representation obtained from it by an orthogonal transformation **S** one has

$$\psi_T(x) = \mathbf{T}\big(\overline{\psi(\mathbf{T}x)}\big)^T\,, \quad \mathbf{T} = \mathrm{i}\gamma_5\gamma^2\,, \quad \mathbf{T}x = (-x^0, \boldsymbol{x})^T\,. \tag{9.68}$$

The following properties of **T** are verified by direct calculation,

$$\mathbf{T} = -\mathbf{T}^{-1} = -\mathbf{T}^\dagger = -\mathbf{T}^T\,. \tag{9.69}$$

From a physical point of view the combined symmetry operations $\mathbf{C}\boldsymbol{\Pi}$ and $\mathbf{C}\boldsymbol{\Pi}\mathbf{T} = \boldsymbol{\Theta}$ are particularly important. Note that we described these symmetries previously in the context of Wigner's theorem in Sect. 6.1.2.

9.2 Quantization of the Dirac Field

The investigation of spinor representations of $SL(2,\mathbb{C})$ led us in an elegant and transparent way to the free Dirac equation which describes the relativistic states of motion of a fermion as well as its spin degrees of freedom in a covariant form. The equation of motion also gave a first hint at the fact that to every spin-1/2 particle there corresponds an antiparticle. Without having "asked" her, the Dirac equation yields a second class of solutions which seem to have negative energies. This peculiar result can be emphasized by reformulating the Dirac equation in a form which is analogous to the Schrödinger equation. For this purpose consider the Dirac equation on spacetime. Upon multiplication of (9.56) by γ^0 from the left one obtains

$$\mathrm{i}\frac{\partial}{\partial t}\psi(t,\boldsymbol{x}) = H\psi(t,\boldsymbol{x})\,, \text{ with } H = -\mathrm{i}\sum_{i=1}^{3}(\gamma^0\gamma^i)\nabla^i + m\gamma^0\,. \tag{9.70}$$

In the standard representation (9.41) there is a conventional notation which is generally accepted since the times of Dirac. One defines

$$\beta := \gamma^0_{\mathrm{St}} = \begin{pmatrix} \mathbb{1}_2 & 0 \\ 0 & -\mathbb{1}_2 \end{pmatrix}\,, \quad \boldsymbol{\alpha} := \beta\gamma^i_{\mathrm{St}} = \begin{pmatrix} 0 & \boldsymbol{\sigma} \\ \boldsymbol{\sigma} & 0 \end{pmatrix}\,, \tag{9.71}$$

so that (9.70) can also be written as

$$\mathrm{i}\frac{\partial}{\partial t}\,\psi(t,\boldsymbol{x}) = \left(-\mathrm{i}\,\boldsymbol{\alpha}\cdot\nabla + m\beta\right)\psi(t,\boldsymbol{x})\,. \tag{9.72}$$

This differential equation is called the *Hamiltonian form of the Dirac equation*. The fact that this equation admits negative, arbitrarily large energies, expressed more concisely, means that the "Hamiltonian" $H = -\mathrm{i}\,\boldsymbol{\alpha}\cdot\nabla + m\beta$ has a spectrum which is unbounded from below. In the perspective of quantum mechanics this is a very peculiar, if not inacceptable result.

This is another example for a problem purely of *physics*: The equation of motion was derived from general and deep mathematical principles. It has all important properties following from very general considerations. But it does not disclose how it is to be interpreted.

Yet, already at this level, we have at least two hints which will lead to the correct interpretation and, hence, to a solution of apparent contradictions. On the one hand, the considerations given above and the Hamiltonian formulation (9.72) fit with a one-particle theory, in close analogy to the Schrödinger equation for a single particle in nonrelativistic motion. On the other hand, charge conjugation hints at the antiparticle which comes along with the particle. The key for solving the puzzle is here: The Dirac equation, with or without interactions, cannot be a theory for the isolated single particle. Rather, by its very nature, it must contain features of a many-body theory. This remark suggests to subject the Dirac field to the rules of canonical quantization, that is, to interpret the field ψ as an operator which creates and annihilates particles and antiparticles. As we will see all pecularities then disappear and the apparent contradictions are resolved.

9.2.1 Quantization of Majorana Fields

We take as our starting point the Dirac equation in spacetime and use the natural representation (9.28) of the γ-matrices. Like in momentum space, cf. (9.30), the spinor field $\psi(x)$ is written in terms of two-component spinor fields $\phi(x)$ and $\chi(x)$,

$$\psi(x) = \begin{pmatrix}\phi(x)\\ \chi(x)\end{pmatrix}, \tag{9.73}$$

whose transformation behaviour under Lorentz transformations is given by the analogues to (9.25) and (9.26). Here again, the Dirac equation is a comprehensive notation for the algebraic two-component equations (9.27) which upon translation to spacetime, become coupled differential equations,

$$\mathrm{i}\,\sigma^{\mu}\partial_{\mu}\chi(x) = m\phi(x) \tag{9.74}$$

$$\mathrm{i}\,\widehat{\sigma}^{\mu}\partial_{\mu}\phi(x) = m\chi(x)\,. \tag{9.75}$$

In this context we recall the definition of the σ-matrices:

$$\left\{\sigma^{\mu}\right\} = \left(\sigma^{(0)}, -\boldsymbol{\sigma}\right), \quad \left\{\widehat{\sigma}^{\mu}\right\} = \left(\sigma^{(0)}, \boldsymbol{\sigma}\right).$$

As Hermann Weyl had discovered this form of the Dirac equation for the case $m = 0$, before Dirac's work, it seems appropriate to call these equations the *Weyl-Dirac equations*.

In the natural representation (9.28) one has

$$\mathrm{i}\gamma^2 = \begin{pmatrix} 0 & -\mathrm{i}\sigma^{(2)} \\ \mathrm{i}\sigma^{(2)} & 0 \end{pmatrix} = \begin{pmatrix} 0 & -\varepsilon \\ \varepsilon & 0 \end{pmatrix},$$

charge conjugation acting on $\psi(x)$ yields

$$\psi_{\mathrm{C}}(x) = \begin{pmatrix} 0 & -\varepsilon \\ \varepsilon & 0 \end{pmatrix} \psi^*(x) = \begin{pmatrix} -\varepsilon\chi^*(x) \\ \varepsilon\phi^*(x) \end{pmatrix}.$$

Therefore, when we consider a Majorana field with the choice of sign $\psi_{\mathrm{C}}(x) = \psi(x)$ we have

$$\phi(x) = -\varepsilon\chi^*(x), \quad \chi(x) = \varepsilon\phi^*(x).$$

The equations of motion (9.74) and (9.75) are replaced by a single one for which one may choose either

$$\mathrm{i}\sigma^\mu\partial_\mu\chi(x) = -m\varepsilon\chi^*(x) \quad \text{or} \quad \mathrm{i}\widehat{\sigma}^\mu\partial_\mu\phi(x) = m\varepsilon\phi^*(x),$$

the two being equivalent. A Lagrange density which yields these equations as its Euler-Lagrange equations, can be chosen to be

$$\mathcal{L}_{\mathrm{M}} = \phi^\dagger(x)\frac{\mathrm{i}}{2}\widehat{\sigma}^\mu \overleftrightarrow{\partial_\mu}\phi(x) + \frac{m}{2}\Big(\phi^T(x)\varepsilon\phi(x) - \phi^\dagger(x)\varepsilon\phi^*(x)\Big).$$

This is easily confirmed. Indeed, one has

$$\frac{\partial\mathcal{L}_{\mathrm{M}}}{\partial\phi^*} = \frac{\mathrm{i}}{2}\widehat{\sigma}^\mu\partial_\mu\phi - m\varepsilon\phi^*, \qquad \frac{\partial\mathcal{L}_{\mathrm{M}}}{\partial(\partial_\mu\phi^*)} = -\frac{\mathrm{i}}{2}\widehat{\sigma}^\mu\phi.$$

Subtracting the divergence $\partial_\mu(\cdots)$ of the second equation from the first, one obtains

$$\frac{\partial\mathcal{L}_{\mathrm{M}}}{\partial\phi^*} - \partial_\mu\left(\frac{\partial\mathcal{L}_{\mathrm{M}}}{\partial(\partial_\mu\phi^*)}\right) = \mathrm{i}\widehat{\sigma}^\mu\partial_\mu\phi - m\varepsilon\phi^* = 0.$$

The Lagrange density serves to define the momentum canonically conjugate to $\phi(x)$,

$$\pi(x) := \frac{\partial\mathcal{L}_{\mathrm{M}}}{\partial(\partial^0\phi)} = \frac{\mathrm{i}}{2}\phi^*\widehat{\sigma}^{(0)} = \frac{\mathrm{i}}{2}\phi^*. \tag{9.76}$$

One draws two conclusions from these calculations: Firstly, the Majorana field has only one degree of freedom $\phi(x)$ (of course, being a two-component spinor, it contains the components ϕ_1 and ϕ_2), all other fields are related to ϕ. Secondly, the canonically conjugate momentum equals ϕ^* which, in turn, can be expressed in terms of ϕ by means of the Weyl-Dirac equation.

In the framework of canonical quantization we expect the commutator of ϕ and π to be a c-number, i.e. not an operator. Thus, commutators or anticommutators of two spinor operators A and B have the form

$$\big[A_\alpha, B_\beta\big]_\pm = A_\alpha B_\beta \pm B_\beta A_\alpha = c_{\alpha\beta}$$

with $\alpha, \beta = 1, 2$ or $\alpha, \beta = 1, 2, 3, 4$, depending on whether they act on two-spinors or on four-spinors. If we write this commutator while suppressing the components, i.e. in the form $[A, B]_{\pm}$, then the result is a 2×2- or 4×4-matrix, respectively, whose entries are c-numbers.

In the example of the Majorana field it is sufficient to discuss the commutator (or anticommutator) of the field $\phi(x)$ at the point x with $\phi(y)$ at the point y of spacetime. All other (anti)commutators are obtained from this one by charge conjugation, or by the action of the equation of motion. Thus, let us try the quantization rule

$$[\phi(x), \phi(y)] = \mathbf{t}\, \Delta(x - y; m) \tag{9.77}$$

and determine the c-number valued matrix on the right-hand side. Based on the experience of Chap. 7 one expects this ansatz to be Lorentz covariant and to satisfy micro-causality. The left-hand side contains two operators which are classified by the spinor representation of the first kind of $L_+^\uparrow$, their transformation behaviour under $\mathbf{\Lambda} \in L_+^\uparrow$ is well-known. If the quantization rule (9.77) is to be covariant with respect to Lorentz transformations, the 2×2-matrix $\mathbf{t}$ must be an invariant tensor with respect to $SL(2, \mathbb{C})$. Furthermore, if the rule is supposed to be (micro)causal then $\Delta(x - y; m)$ must be the causal distribution $\Delta_0(x - y; m)$ for the mass m that we studied in Sect. 7.1.6.

The tensor ε (cf. (9.20)) is the only invariant tensor of $SL(2, \mathbb{C})$. Therefore, the right-hand side must be proportional to the product $\varepsilon\, \Delta_0(x - y; m)$. Now, upon exchange of the two operators $\phi(x)$ and $\phi(y)$ in (9.77) the tensor ε is replaced by $\varepsilon^T = -\varepsilon$ while $(x - y)$ is replaced by $(y - x)$. The causal distribution Δ_0 being antisymmetric, cf. (7.54), the product of ε and $\Delta_0(x - y; m)$ is *symmetric* under exchange of the two fields. As a consequence the ansatz (9.77) contains a contradiction and cannot be maintained: Indeed, for $\phi(x) \longleftrightarrow \phi(y)$ the left-hand side is antisymmetric but the right-hand side is symmetric.

Is this state of affairs a catastrophe? Are there modifications of the quantization rule which resolve the contradiction?

If one insists on a *commutator* then there is no other choice than to replace Δ_0 by $\Delta_1(x - y; m)$, (7.62), which is symmetric under exchange of x and y. However, one would then inherit a new, rather unwanted property: The commutator of $\phi(x)$ with $\phi(y)$, for *space*-like distances of x and y would not vanish. In other terms, the field $\phi(x)$ would disturb the field $\phi(y)$ even in constellations where x and y are not causally related. At this microscopic level one might be tempted to ignore this deficiency for a while and to continue along the lines prescribed by canonical quantization. If one does so the spinor field is expanded in terms of creation and annihilation operators for particles with spin 1/2, as in the previous cases. However, one discovers a Hamiltonian whose spectrum is unbounded from below and one does not get rid of the problem of "negative energies."[7]

---

[7] Many textbooks follow this historical path and find a Hamiltonian which contains terms of the kind $\sum_i E_i(a_i^\dagger a_i - b_i b_i^\dagger)$ where $a_i^\dagger$ creates particles, $b_i^\dagger$ creates antiparticles in the states "i". Then one postulates that these operators fulfill anticommutation rules, not commutators, such that the

The only physically meaningful alternative is to replace the commutator on the left-hand side of (9.77) by the *anticommutator*. Then, indeed, both sides are *symmetric* under exchange, and the right-hand side satisfies the requirement of micro-causality. This modified ansatz is consistent and physically acceptable. Thus, using the notation $\{\cdot, \cdot\}$ for the anticommutator, $\{A, B\} = AB + BA$, we postulate

$$\{\phi(x), \phi(y)\} = c^{(\mathrm{M})}\varepsilon\, \Delta_0(x - y; m)\,, \tag{9.78}$$

where $c^{(\mathrm{M})}$ is a complex number that has to be determined. For this purpose let the operator $i\widehat{\sigma}^{\mu}\partial^{x}_{\mu}$ act on (9.78) from the left and insert the Weyl-Dirac equation for $\phi(x)$. This yields

$$\left\{\phi^*(x), \phi(y)\right\} = c^{(\mathrm{M})}\frac{\mathrm{i}}{m}\,\varepsilon^{-1}\widehat{\sigma}^{\mu}\varepsilon\,\partial^{x}_{\mu}\Delta_0(x - y; m)\,.$$

Choosing the two time arguments to be equal, $x^0 = y^0$, the spatial derivatives of the distribution Δ_0 are equal to zero while the time derivative, according to (7.53), is given by

$$\partial^{x}_{0}\Delta_0(x - y; m)\big|_{x^0=y^0} = -\delta(\boldsymbol{x} - \boldsymbol{y})\,.$$

Inserting this one obtains

$$\left\{\phi^*(x), \phi(y)\right\}\big|_{x^0=y^0} = -c^{(\mathrm{M})}\frac{\mathrm{i}}{m}\,\mathbb{1}_2\delta(\boldsymbol{x} - \boldsymbol{y})\,.$$

The left-hand side is hermitian and positive. Therefore, $c^{(\mathrm{M})}$ must lie on the upper part of the imaginary half-axis. Thus, we choose

$$c^{(\mathrm{M})} = \mathrm{i}\,m\,. \tag{9.79}$$

At this point we leave the quantization of the Majorana field. The quantization rule (9.78) with $c^{(\mathrm{M})} = im$ leads to a consistent and physically satisfactory theory. We show this for the more general case of the unconstrained Dirac field which contains the Majorana field as a special case.

9.2.2 Quantization of Dirac Fields

In contrast to the case of Majorana fields, the two-spinors ϕ and χ in a Dirac field are independent degrees of freedom. Nevertheless, the arguments against quantization by *commutators* are very similar to the ones given for Majorana fields. We start by constructing a Lagrange density and derive the canonical momenta such as to identify the relevant commutators or anticommutators. A possible choice is

---

(Footnote 7 continued)
expression for the energy is transformed to $\sum_i E_i(a_i^\dagger a_i + b_i^\dagger b_i)$ which assigns positive energies both to particles and to antiparticles. Regarding the result, this is correct and, furthermore, it also yields the Pauli exclusion principle. However, this method is unsatisfactory because one cures the symptoms instead of modifying the theory at its roots.

$$\mathcal{L}_D = \frac{i}{2}\left(\phi^\dagger(x)\widehat{\sigma}^\mu \overleftrightarrow{\partial_\mu}\,\phi(x) + \chi^\dagger(x)\sigma^\mu \overleftrightarrow{\partial_\mu}\,\chi(x)\right)$$
$$- m\left(\chi^\dagger(x)\phi(x) + \phi^\dagger(x)\chi(x)\right) .$$

One easily verifies that the Euler-Lagrange equations which pertain to this Lagrange density are identical with the equations of motion (9.74) and (9.75).

The momenta canonically conjugate to ϕ and χ are seen to be (again written in two-component form)

$$\pi_\phi(x) = \frac{i}{2}\phi^*(x) , \qquad \pi_\chi(x) = \frac{i}{2}\chi^*(x) .$$

The right-hand sides are related to χ^* and ϕ^*, respectively, via the Weyl-Dirac equations. Therefore, it suffices to consider the commutator, or anticommutator, of $\phi(x)$ and $\chi^\dagger(y)$. Like in the preceding section we first try the commutator, i.e.

$$\left[\phi(x), \chi^\dagger(y)\right] = c\,\Delta_0(x - y; m) .$$

Application of charge conjugation to the left-hand side of this equation yields

$$\left[-\varepsilon\chi^*(x), \varepsilon\phi(y)\right] = -\varepsilon\left[\chi^*(x), \phi(y)\right]\varepsilon^{-1} = c\Delta_0(y - x; m)$$
$$= -c\Delta_0(x - y; m) .$$

Note that in the second step the above ansatz was inserted, while in the third step the antisymmetry of Δ_0 was used. As the right-hand side of the ansatz remains unchanged, there is a contradiction unless $c = 0$.

Here again, there are two possibilities of resolving the contradiction: Either one insists on the *commutator* and replaces the causal distribution Δ_0 by $\Delta_1(x - y; m)$, but then runs into the same difficulties as with commutators for Majorana fields. Or one replaces commutators by *anticommutators*. With this second choice all inconsistencies disappear and the theory has a physically meaningful interpretation. This idea is worked out in what follows.

On the basis of these considerations it seems natural to postulate

$$\left\{\phi(x), \chi^\dagger(y)\right\} = i\,m\,\Delta_0(x - y; m) . \tag{9.80}$$

The constant is chosen in accord with the Majorana case which is contained in (9.80) by the choice $\chi^* = \varepsilon\phi$.

We now show that the four-component field operator $\psi(x)$ and its adjoint $\psi^\dagger(y)$ satisfy a causal anticommutator for arbitrary spacetime points x and y which reads

$$\left\{\psi(x), \psi^\dagger(y)\right\} = i\left(m\gamma^0 + i\gamma^\mu\gamma^0\partial_\mu^x\right)\Delta_0(x - y; m) . \tag{9.81}$$

Proof

When expressed in terms of the two-component spinors of (9.73) the 4×4-matrix (9.81) can be written in terms of four 2×2-blocs of anticommutators,

$$\left\{\psi(x), \psi^\dagger(y)\right\} = \begin{pmatrix} \left\{\phi(x), \phi^\dagger(y)\right\} & \left\{\phi(x), \chi^\dagger(y)\right\} \\ \left\{\chi(x), \phi^\dagger(y)\right\} & \left\{\chi(x), \chi^\dagger(y)\right\} \end{pmatrix} .$$

The entries are 2 × 2-matrices all of which can be reduced to the ansatz (9.80). This is obvious for the nondiagonal blocs while for the blocs in the diagonal one makes use of the Weyl-Dirac equations. Thus, by (9.74) one has

$$\left\{\phi(x), \phi^\dagger(y)\right\} = \frac{\mathrm{i}}{m}\partial^x_\mu\left\{\sigma^\mu\chi(x), \phi^\dagger(y)\right\} = -\sigma^\mu\partial^x_\mu\Delta_0(x-y;m)\,.$$

Likewise, by (9.75) one obtains

$$\left\{\chi(x), \chi^\dagger(y)\right\} = \frac{\mathrm{i}}{m}\partial^x_\mu\left\{\widehat{\sigma}^\mu\phi(x), \chi^\dagger(y)\right\} = -\widehat{\sigma}^\mu\partial^x_\mu\Delta_0(x-y;m)\,.$$

As a result one finds

$$\left\{\psi(x), \psi^\dagger(y)\right\} = \mathrm{i}\begin{pmatrix} \mathrm{i}\sigma^\mu\partial^x_\mu & m\mathbb{1}_2 \\ m\mathbb{1}_2 & \mathrm{i}\widehat{\sigma}^\mu\partial^x_\mu \end{pmatrix}\Delta_0(x-y;m)\,.$$

Upon rewriting this in terms of the matrices γ^0 and γ^μ in the representation (9.28) this is seen to be precisely the form asserted in (9.81). ◁

In order to obtain still another form of the anticommutator of the Dirac field and its adjoint we write the anticommutator (9.81) more explicity in components. It then reads

$$\left\{\psi_\alpha(x), \psi^\dagger_\lambda(y)\right\} = \mathrm{i}\left(m\gamma^0 + \mathrm{i}\gamma^\mu\gamma^0\partial^x_\mu\right)_{\alpha\lambda}\Delta_0(x-y;m)\,,$$
$$\alpha, \lambda = 1, \ldots 4\,.$$

We multiply this result from the right by $(\gamma^0)_{\lambda\beta}$ and take the sum over λ. This yields

$$\left\{\psi_\alpha(x), \overline{\psi(y)}_\beta\right\} = \mathrm{i}\left(m\mathbb{1}_4 + \mathrm{i}\gamma^\mu\partial^x_\mu\right)_{\alpha\beta}\Delta_0(x-y;m)\,,$$

or, in a more compact notation,

$$\left\{\psi(x), \overline{\psi(y)}\right\} = \mathrm{i}\left(m\mathbb{1} + \mathrm{i}\gamma^\mu\partial^x_\mu\right)\Delta_0(x-y;m)\,. \tag{9.82}$$

This quantization rule has a number of important implications that we wish to work out next and which will be useful in clarifying their physical interpretation.

Choosing equal times, $x^0 = y^0$, and making use of the well-known properties of the distribution Δ_0, one finds

$$\left.\left\{\psi_\alpha(x), \psi^\dagger_\beta(y)\right\}\right|_{x^0=y^0} = \delta_{\alpha\beta}\delta(\boldsymbol{x}-\boldsymbol{y})\,. \tag{9.83}$$

This formula is understood better if one recalls that by (9.58) the momentum canonically conjugate to ψ is

$$\pi_\psi = \frac{\partial\mathcal{L}_\mathrm{D}}{\partial(\partial_0\psi)} = \frac{\mathrm{i}}{2}\psi^\dagger\,.$$

The formal analogy to the quantization of the Klein-Gordon field is obvious and, in fact, suggests to decompose the operators $\psi(x)$ and $\overline{\psi(x)}$ in terms of normal modes. As we know the solutions with given momentum and spin orientation, the following expansion in terms of creation and annihilation operators seems appropriate

$$\psi_\alpha(x) = \frac{1}{(2\pi)^{3/2}} \sum_{r=1}^{2} \int \frac{\mathrm{d}^3 p}{2E_p} \left[a^{(r)}(\boldsymbol{p})\, u_\alpha^{(r)}(\boldsymbol{p})\, \mathrm{e}^{-\mathrm{i}px} + b^{(r)\dagger}(\boldsymbol{p})\, v_\alpha^{(r)}(\boldsymbol{p})\, \mathrm{e}^{\mathrm{i}px} \right] . \tag{9.84}$$

Taking the hermitian conjugate of this expansion, and multiplying by γ^0 from the right, one has

$$\overline{\psi(x)_\alpha} = \frac{1}{(2\pi)^{3/2}} \sum_{r=1}^{2} \int \frac{\mathrm{d}^3 p}{2E_p} \left[a^{(r)\dagger}(\boldsymbol{p})\, \overline{u_\alpha^{(r)}(\boldsymbol{p})}\, \mathrm{e}^{-\mathrm{i}px} + b^{(r)}(\boldsymbol{p})\, \overline{v_\alpha^{(r)}(\boldsymbol{p})}\, \mathrm{e}^{\mathrm{i}px} \right] . \tag{9.85}$$

The spinor nature $\alpha = 1, \ldots, 4$ of these field operators is taken over by the solutions $u^{(r)}(\boldsymbol{p})$ etc. while their operator nature is transferred to $a^{(r)}(\boldsymbol{p})$, $b^{(r)}(\boldsymbol{p})$, and their adjoints. One integrates over all three-momenta using the invariant volume element $\mathrm{d}^3 p/(2E_p)$, $E_p = \sqrt{\boldsymbol{p}^2 + m^2}$, and sums over the two orientations of the spin. These inverse formulae which allow to calculate the operators $a^{(r)}(\boldsymbol{p})$ and $b^{(r)}(\boldsymbol{p})$ from ψ and $\overline{\psi}$, read

$$a^{(s)}(\boldsymbol{p}) = \frac{1}{(2\pi)^{3/2}} \int \mathrm{d}^3 x\ \mathrm{e}^{\mathrm{i}px} \sum_{\alpha=1}^{4} u_\alpha^{(s)\dagger}(\boldsymbol{p}) \psi_\alpha(x) \tag{9.86}$$

$$b^{(s)}(\boldsymbol{p}) = \frac{1}{(2\pi)^{3/2}} \int \mathrm{d}^3 x\ \mathrm{e}^{\mathrm{i}px} \sum_{\alpha,\beta=1}^{4} \overline{\psi_\alpha(x)} \gamma^0_{\alpha\beta} v_\beta^{(s)}(\boldsymbol{p}) , \quad s = 1, 2 . \tag{9.87}$$

Here the relations (9.50) and (9.51) were used. Calculating now the anticommutators of these operators and their adjoints, one obtains a remarkably simple result:

$$\left\{ a^{(r)}(\boldsymbol{p}), a^{(s)\dagger}(\boldsymbol{q}) \right\} = 2E_p \delta_{rs} \delta(\boldsymbol{p} - \boldsymbol{q}) = \left\{ b^{(r)}(\boldsymbol{p}), b^{(s)\dagger}(\boldsymbol{q}) \right\} , \tag{9.88}$$

$$\left\{ a^{(r)}(\boldsymbol{p}), a^{(s)}(\boldsymbol{q}) \right\} = 0 = \left\{ b^{(r)}(\boldsymbol{p}), b^{(s)}(\boldsymbol{q}) \right\} , \tag{9.89}$$

$$\left\{ a^{(r)}(\boldsymbol{p}), b^{(s)}(\boldsymbol{q}) \right\} = 0 = \left\{ a^{(r)}(\boldsymbol{p}), b^{(s)\dagger}(\boldsymbol{q}) \right\} . \tag{9.90}$$

The interpretation of these operators follows from the experience with bosonic fields gained in Chap. 7 as well as from the nonrelativistic systems of fermions that we studied in Sect. 5.3.2. Thus one concludes:

(i) The operators $a^{(r)\dagger}(\boldsymbol{q})$ and $b^{(r)\dagger}(\boldsymbol{p})$ create one-particle states both of which carry momentum $\boldsymbol{p}$ and spin orientation r,

$$a^{(r)\dagger}(\boldsymbol{p})|0\rangle = |(a);\, \boldsymbol{p}, r\rangle ,$$
$$b^{(r)\dagger}(\boldsymbol{p})|0\rangle = |(b);\, \boldsymbol{p}, r\rangle .$$

These states are normalized covariantly, i.e. one has

$$\langle \boldsymbol{q}, s|\boldsymbol{p}, r\rangle = 2E_p \delta_{rs} \delta(\boldsymbol{q} - \boldsymbol{p}) .$$

(ii) The operators $a^{(r)}(\boldsymbol{p})$ and $b^{(r)}(\boldsymbol{p})$ are the corresponding annihilation operators.

(iii) The occupation numbers of such states can only take the values 0 or 1. States containing two different creation operators such as

$$a^{(r)\dagger}(\boldsymbol{p})a^{(s)\dagger}(\boldsymbol{q})|0\rangle$$

are antisymmetric upon exchange $(r, \boldsymbol{p}) \leftrightarrow (s, \boldsymbol{q})$. The quantized Dirac field describes two species of particles which obey the Pauli principle and which have identical kinematic and spin degrees of freedom. It only remains to find out by which additional property they differ.

9.2.3 Electric Charge, Energy, and Momentum

In Sect. 9.1.4 we introduced the coupling of the Dirac field to the electromagnetic field by means of the rule (9.59) of minimal coupling. This allowed to identify the current density (9.61). The factor q in front is chosen to be 1, a choice that for applications is equivalent to having all charges come in as multiples of the elementary charge. One confirms that $j^\mu = \overline{\psi(x)}\gamma^\mu\psi(x)$ is conserved by making use of the equations of motion (9.56) and (9.57). Indeed, one has

$$\partial_\mu j^\mu(x) = \left(\partial_\mu\overline{\psi(x)}\gamma^\mu\right)\psi(x) + \overline{\psi(x)}\left(\gamma^\mu\partial_\mu\psi(x)\right) = \mathrm{i}m(1-1)\overline{\psi}\psi = 0\,.$$

Under the assumption that the fields vanish at infinity sufficiently rapidly this continuity equation implies that the *charge operator*

$$Q := \int \mathrm{d}^3x\; j^0(x) \tag{9.91}$$

is a constant of the motion, i.e. $\mathrm{d}Q/\,\mathrm{d}t = 0$. After quantization and upon replacement of the classical current density by the normal-ordered operator

$$j^\mu(x) := :\overline{\psi(x)}\gamma^\mu\psi(x):\,, \tag{9.92}$$

one discovers an expression for Q in terms of number operators for particles and antiparticles which is familiar from Sect. 2.2, viz.

$$Q = \sum_{r=1}^{2}\int\frac{\mathrm{d}^3p}{2E_p}\left[a^{(r)\dagger}(\boldsymbol{p})a^{(r)}(\boldsymbol{p}) - b^{(r)\dagger}(\boldsymbol{p})b^{(r)}(\boldsymbol{p})\right]. \tag{9.93}$$

It confirms the conjecture made in Sect. 9.1.4: The particles which are created by $a^\dagger$, and the particles created by $b^\dagger$, differ by the sign of their electric charge. Before continuing on this interpretation let us make up for the calculation that leads to (9.93). Inserting the expansion (9.84) as well as its hermitian conjugate, one finds

$$Q =: \int \mathrm{d}^3x\;\psi^\dagger(x)\psi(x): = \sum_{r=1}^{2}\sum_{s=1}^{2}\int\frac{\mathrm{d}^3p}{2E_p}\int\frac{\mathrm{d}^3q}{2E_q}$$

$$:\Big(\delta(\boldsymbol{q}-\boldsymbol{p})a^{(r)\dagger}(\boldsymbol{p})a^{(s)}(\boldsymbol{q})u^{(r)\dagger}(\boldsymbol{p})u^{(s)}(\boldsymbol{q})$$

$$+\delta(\boldsymbol{p}+\boldsymbol{q})a^{(r)\dagger}(\boldsymbol{p})b^{(s)\dagger}(\boldsymbol{q})u^{(r)\dagger}(\boldsymbol{p})v^{(s)}(\boldsymbol{q})$$
$$+\delta(\boldsymbol{p}+\boldsymbol{q})b^{(r)}(\boldsymbol{p})a^{(s)}(\boldsymbol{q})v^{(r)\dagger}(\boldsymbol{p})u^{(s)}(\boldsymbol{q})$$
$$+\delta(\boldsymbol{p}-\boldsymbol{q})b^{(r)}(\boldsymbol{p})b^{(s)\dagger}(\boldsymbol{q})v^{(r)\dagger}(\boldsymbol{p})v^{(s)}(\boldsymbol{q})\Big):,$$

where, so far, only the integration over $\mathbb{R}^3$ was performed. By the δ-distributions the spatial momenta in the first and in the fourth term must be equal, $\boldsymbol{p}=\boldsymbol{q}$, and, hence, also $E_p = E_q$. One of the two integrations then collapses while the other yields the orthogonality relation (9.50). Regarding the second and third terms, one concludes that $\boldsymbol{q}=-\boldsymbol{p}$ (but again $E_q = E_p$), so that one of the integrations is dropped while the other, by (9.51), gives zero. Thus, only the first and the fourth terms contribute. When taking the normal order, finally, the first term remains unchanged because the creation operator is already on the left of the annihilation operator. In the fourth term, however, the creation operator must be shifted to the left, past the annihilation operator. This yields a minus sign and proves the expression (9.93) for the charge operator. The energy-momentum tensor field (7.86) is calculated from the Lagrange density (9.58). In particular, the energy density is the time-time component of this tensor field and is found to be

$$\mathcal{H}(x)=\mathcal{T}^{00}(x)=:\overline{\psi(x)}\big(-\frac{\mathrm{i}}{2}\boldsymbol{\gamma}\cdot\overset{\leftrightarrow}{\nabla}+m\mathbb{1}\big)\psi(x):$$
$$=:\psi^\dagger(x)\big(-\frac{\mathrm{i}}{2}\boldsymbol{\alpha}\cdot\overset{\leftrightarrow}{\nabla}+m\beta\big)\psi(x):,$$

where, in the second step, the representation (9.71) was used. Integrating over the whole space the derivative can be shifted entirely to the right by means of partial integration,

$$H=\int \mathrm{d}^3x\; \mathcal{H}(x)=:\int \mathrm{d}^3x\; \psi^\dagger(x)\big(-\mathrm{i}\boldsymbol{\alpha}\cdot\nabla+m\beta\big)\psi(x):,$$

thus obtaining an expression which is akin to (9.72).

The momentum density is given by the time-space components of $\mathcal{T}^{\mu\nu}$,

$$\mathcal{T}^{0k}(x)=\frac{\mathrm{i}}{2}:\overline{\psi(x)}\gamma^0\overset{\leftrightarrow}{\partial^k}\psi(x):=-\frac{\mathrm{i}}{2}:\big(\psi^\dagger(x)\overset{\leftrightarrow}{\nabla}\psi(x)\big)^k:.$$

As above, upon integrating over $\mathbb{R}^3$ the derivatives can be shifted to the factor on the right such that one obtains

$$\boldsymbol{P}=-\mathrm{i}:\int \mathrm{d}^3x\; \psi^\dagger(x)\nabla\psi(x):.$$

By a calculation which is practically the same as for the charge operator, energy and momentum are expressed in terms of creation and annihilation operators, as follows:

$$(H,\boldsymbol{P})=\sum_{r=1}^{2}\int\frac{\mathrm{d}^3p}{2E_p}(E_p,\boldsymbol{p})\big[a^{(r)\dagger}(\boldsymbol{p})a^{(r)}(\boldsymbol{p})+b^{(r)\dagger}(\boldsymbol{p})b^{(r)}(\boldsymbol{p})\big]. \tag{9.94}$$

This concludes the interpretation of the quantum version of the Dirac field. In summary, we found:

One-Particle States of the Dirac Field: Let $|0\rangle$ denote the vacuum state, i.e. the state which is annihilated by all operators $a^{(s)}(\boldsymbol{q})$ and $b^{(s)}(\boldsymbol{q})$,

$$a^{(s)}(\boldsymbol{q})|0\rangle = 0 = b^{(s)}(\boldsymbol{q})|0\rangle \,.$$

The one-particle states created from the vacuum,

$$a^{(r)\dagger}(\boldsymbol{p})|0\rangle \quad \text{and} \quad b^{(r)\dagger}(\boldsymbol{p})|0\rangle$$

are eigenstates of the operators $\left(H, \boldsymbol{P}\right)$ with eigenvalues $E_p = \sqrt{\boldsymbol{p}^2 + m^2}$ and $\boldsymbol{p}$, respectively, and with spin orientation r (the same for both). Furthermore, they are eigenstates of the charge operator Q pertaining to the eigenvalues $+1$ and -1, respectively.
If one of them is called *particle* state then the other describes the *antiparticle* state. The occupation number of a state with eigenvalues $(E_p, \boldsymbol{p}, r, q = \pm 1)$ can only be 1 or 0. States containing several fermions with distinct quantum numbers are antisymmetric under permutations.

9.3 Dirac Fields and Interactions

The fundamental interactions, i.e. the weak, electromagnetic, and strong interactions, all have vector or axial vector character. This is to say, the basic vertices schematically have the structure

$$\overline{\psi(x)}\Gamma^\mu \psi(x) A_\mu(x)\,, \quad \text{with} \quad \Gamma^\mu = \gamma^\mu \quad \text{or} = \gamma^\mu \gamma_5 \,.$$

This structure entails correlations of spins and helicities of the participating incoming or outgoing particles which are characteristic for the interaction. For this reason we work out in more detail the description of the spin of fermions and then proceed to an analysis of vertices in fundamental interactions.

9.3.1 Spin and Spin Density Matrix

In nonrelativistic quantum mechanics a beam of identical spin-1/2 particles with momentum $\boldsymbol{p}$ which is partially or fully polarized, is described by a density matrix ϱ. Let $\hat{\boldsymbol{n}}$ be the direction of the polarization, let w_+ be the weight of the state which is polarized in the positive direction, w_- the weight of the state polarized in the negative direction. As worked out in Sect. 4.1.8 one has

$$\varrho = \frac{1}{2}\left(\mathbb{1} + \boldsymbol{\zeta} \cdot \boldsymbol{\sigma}\right), \quad \text{with} \quad \boldsymbol{\zeta} = (w_+ - w_-)\,\widehat{\boldsymbol{n}} \,.$$

The density matrix is the convex sum of two terms

$$\varrho = w_+|\widehat{\boldsymbol{n}}, +\rangle\langle\widehat{\boldsymbol{n}}, +| \; + \; w_-|\widehat{\boldsymbol{n}}, -\rangle\langle\widehat{\boldsymbol{n}}, -|, \; w_+ + w_- = 1 \,.$$

The expectation value of spin is calculated as usual, viz.

$$\langle \boldsymbol{s} \rangle = \frac{1}{2}\boldsymbol{\zeta}\,.$$

This expectation value which, as we know, is a classical observable, is easily generalized to relativistic kinematics. One just has to boost the axial vector $(0, \boldsymbol{\zeta})$ constructed in the rest system to a system in which the particle has momentum $\boldsymbol{p}$, by means of a Special Lorentz transformation. One obtains

$$s = \mathbf{L}_p\,\left(0, \boldsymbol{\zeta}\right)^T = \left(\frac{\boldsymbol{p}\cdot\boldsymbol{\zeta}}{m},\ \boldsymbol{\zeta} + \frac{\boldsymbol{p}\cdot\boldsymbol{\zeta}}{m(E_p+m)}\boldsymbol{p}\right)^T\,. \tag{9.95}$$

This four-vector is orthogonal to the four-momentum p. Up to the sign its square equals $\boldsymbol{\zeta}^2$ and, hence, has a value between -1 and 0,

$$s\cdot p = 0\,,\quad s^2 = -\boldsymbol{\zeta}^2\,,\quad -1 \le s^2 \le 0\,.$$

Hence, the degree of polarization is

$$P = \sqrt{-s^2} = \frac{w_+ - w_-}{w_+ + w_-} = w_+ - w_-\,.$$

The next goal is to construct spin density matrices for particles and antiparticles which generalize the nonrelativistic density matrix given above. For this purpose consider spinors $u(\boldsymbol{p})$ and $v(\boldsymbol{p})$ pertaining to momentum p and to mass m describing states that are *fully polarized* along the direction $\hat{\boldsymbol{n}}$. Clearly, with given $\hat{\boldsymbol{n}}$, there is a vector s, like in (9.95), but here with full polarization. In order to avoid any confusion we assign to it a new symbol,

$$n = \mathbf{L}_p\,\left(0, \widehat{\boldsymbol{n}}\right)^T = \left(\frac{\boldsymbol{p}\cdot\widehat{\boldsymbol{n}}}{m},\ \widehat{\boldsymbol{n}} + \frac{\boldsymbol{p}\cdot\widehat{\boldsymbol{n}}}{m(E_p+m)}\boldsymbol{p}\right)^T\,. \tag{9.96}$$

Analogously to s it fulfills $n^2 = -1$ and $n\cdot p = 0$. If we were in the rest system ($\boldsymbol{p} = \boldsymbol{0}$) this would mean that, in the standard representation and given the solutions (9.46) and (9.47), we would have

$$u(\boldsymbol{0})\overline{u(\boldsymbol{0})} = 2m\begin{pmatrix} 1/2\left(\mathbb{1}_2 + \boldsymbol{\sigma}\cdot\widehat{\boldsymbol{n}}\right) & 0 \\ 0 & 0\end{pmatrix},$$

$$v(\boldsymbol{0})\overline{v(\boldsymbol{0})} = 2m\begin{pmatrix} 0 & 0 \\ 0 & 1/2\left(\mathbb{1}_2 - \boldsymbol{\sigma}\cdot\widehat{\boldsymbol{n}}\right)\end{pmatrix}.$$

One shows that for arbitrary values of the momentum these expressions are replaced by the following covariant formulae:

$$u(\boldsymbol{p})\overline{u(\boldsymbol{p})} = \frac{1}{2}\left(\not{p} + m\,\mathbb{1}\right)\left(\mathbb{1} + \gamma_5\not{n}\right), \tag{9.97}$$

$$v(\boldsymbol{p})\overline{v(\boldsymbol{p})} = \frac{1}{2}\left(\not{p} - m\,\mathbb{1}\right)\left(\mathbb{1} + \gamma_5\not{n}\right). \tag{9.98}$$

(Note that on the left-hand side of (9.97) and of (9.98) a column vector is multiplied by a row vector so that the result is a 4×4-matrix. This differs, e.g., from (9.54) where a row is multiplied by a column, giving the scalar product.) The proof of these formulae is the content of Exercise 9.4 but we note that it is largely analogous to the derivation of (9.81). Taking the sum over the two directions of polarization, i.e. the sum of these expressions for $+\widehat{\boldsymbol{n}}$ and for $-\widehat{\boldsymbol{n}}$, they simplify and become

$$\sum_{\text{Spin}} u(\boldsymbol{p})\overline{u(\boldsymbol{p})} = \left(\not{p} + m\,\mathbb{1}\right), \tag{9.99}$$

$$\sum_{\text{Spin}} v(\boldsymbol{p})\overline{v(\boldsymbol{p})} = \left(\not{p} - m\,\mathbb{1}\right). \tag{9.100}$$

The results (9.97) and (9.98) which appear also in the calculation of traces of squared amplitudes, allow to identify some important projection operators. The operators

$$\Omega_\pm := \pm\frac{1}{2m}\left(\not{p} \pm m\,\mathbb{1}\right) \tag{9.101}$$

are projection operators. Indeed, one finds

$$\Omega_+ + \Omega_- = \mathbb{1},$$

$$\Omega_\pm^2 = \frac{1}{4m^2}\Big((p^2+m^2)\,\mathbb{1} \pm 2m\not{p}\Big) = \pm\frac{1}{2m}\left(\not{p} \pm m\,\mathbb{1}\right) = \Omega_\pm .$$

Obviously, Ω_+ projects onto the positive frequency part for given momentum $\boldsymbol{p}$, while Ω_- projects onto the negative frequency part. Furthermore, the operator defined by

$$\Pi_{\hat{n}} := \frac{1}{2}\left(\mathbb{1} + \gamma_5\not{n}\right), \tag{9.102}$$

is the projection operator which projects onto the spin state polarized in the positive $\hat{\boldsymbol{n}}$-direction. Indeed, one has

$$\Pi_{\hat{n}}^2 = \frac{1}{4}\left(\mathbb{1} + 2\gamma_5\not{n} + \gamma_5\not{n}\gamma_5\not{n}\right) = \frac{1}{4}\left(\mathbb{1} + 2\gamma_5\not{n} - \mathbb{1}\,n^2\right) = \frac{1}{2}\left(\mathbb{1} + \gamma_5\not{n}\right) = \Pi_{\hat{n}} .$$

Here use was made of the fact that γ_5 and $\not{n}$ anticommute and that $n^2 = -1$. In the special case $\boldsymbol{p} = 0$ this operator goes over into the correct nonrelativistic result

$$\Pi_{\hat{n}}|_{\boldsymbol{p}=0} = \frac{1}{2}\begin{pmatrix} \mathbb{1}_2 + \boldsymbol{\sigma}\cdot\widehat{\boldsymbol{n}} & 0 \\ 0 & \mathbb{1}_2 - \boldsymbol{\sigma}\cdot\widehat{\boldsymbol{n}} \end{pmatrix}.$$

One shows that the projection operator $\Pi_{\hat{n}}$ commutes both with Ω_+ as well as with Ω_-. We verify one example:

$$\begin{aligned}\left[\Pi_{\hat{n}}, \Omega_+\right] &= \frac{1}{4m}\left(\gamma_5\not{n}\not{p} - \not{p}\gamma_5\not{n}\right) = \frac{1}{4m}\gamma_5\left(\not{n}\not{p} + \not{p}\not{n}\right) \\ &= \frac{1}{4m}\left(\not{n}\not{p} + 2p\cdot n - \not{n}\not{p}\right) = 0 .\end{aligned}$$

In a similar way one verifies that $\left[\Pi_{\hat{n}}, \Omega_-\right] = 0$ holds true.

The two types of projection operators can be combined such as to obtain covariant density matrices for particles and for antiparticles, respectively. For a system of *particles* with momentum $\boldsymbol{p}$ which carry partial polarization $\boldsymbol{\zeta}$ one multiplies the projector onto positive frequencies by the spin projector which is obtained by replacing $\hat{\boldsymbol{n}}$ by $\boldsymbol{\zeta}$ (complete polarization), i.e. with n replaced by s,

$$\varrho^{(+)} = 2m\,\Omega_{+}\frac{1}{2}\,(\mathbb{1} + \gamma_5 \not{s}) = \frac{1}{2}\,(\not{p} + m\,\mathbb{1})\,(\mathbb{1} + \gamma_5 \not{s})\ . \tag{9.103}$$

This operator has the following property

$$\operatorname{tr}\varrho^{(+)} = \frac{1}{2}\,(m\operatorname{tr}\mathbb{1}_4 + \operatorname{tr}\not{p} + m\operatorname{tr}(\gamma_5\not{s}) + \operatorname{tr}(\not{p}\gamma_5\not{s})) = 2m\,.$$

Note that here only the first term contributes, the matrices of the second, third, and fourth term having trace zero. Taking the square of (9.103) one obtains

$$\begin{aligned}\varrho^{(+)\,2} &= \frac{1}{4}(\not{p} + m\,\mathbb{1})(\mathbb{1} + \gamma_5\not{s})(\not{p} + m\,\mathbb{1})(\mathbb{1} + \gamma_5\not{s})\\ &= \frac{1}{4}(\not{p} + m\,\mathbb{1})^2(\mathbb{1} + \gamma_5\not{s})^2 = \frac{1}{4}\big(2m^2\,\mathbb{1} + 2m\not{p}\big)\big(\mathbb{1} + 2\gamma_5\not{s} - s^2\,\mathbb{1}\big)\\ &= m\big(\not{p} + m\,\mathbb{1}\big)\left(\frac{1}{2}\big(1 + \zeta^2\big)\,\mathbb{1} + \gamma_5\not{s}\right) = 4m^2\Omega_{+}\frac{1}{2}\left(\frac{1}{2}(1 + \zeta^2)\,\mathbb{1} + \gamma_5\not{s}\right),\end{aligned}$$

i.e. an expression which is equal to $2m\varrho^{(+)}$ precisely if $|\boldsymbol{\zeta}| = 1$, i.e. if there is maximal polarization. In turn, in the general case one has

$$\operatorname{tr}\left(\frac{\varrho^{(+)\,2}}{(2m)^2}\right) = \frac{1}{2}(1 + \zeta^2) \le \operatorname{tr}\left(\frac{\varrho^{(+)}}{2m}\right) = 1\,.$$

The matrix $\varrho^{(+)}$ is not hermitian but it fulfills the relation

$$\gamma^0\,\varrho^{(+)\dagger}\,\gamma^0 = \varrho^{(+)}\,,$$

that we know from a similar property of the γ-matrices. From the point of view of physics, this relation is perfectly acceptable because $\varrho^{(+)\dagger}$ describes the parity-mirror state of $\varrho^{(+)}$, i.e. the state which has $\boldsymbol{p} \mapsto -\boldsymbol{p}$, but $\boldsymbol{\zeta} \mapsto \boldsymbol{\zeta}$ (s. Exercise 9.5). On the other hand, if one multiplies $\varrho^{(+)}$ by γ^0 from the left then the product $\gamma^0\varrho^{(+)}$ is hermitian.

In a completely analogous manner one shows that the density matrix

$$\varrho^{(-)} = -2m\,\Omega_{-}\frac{1}{2}\,(\mathbb{1} + \gamma_5\not{s}) = \frac{1}{2}\,(\not{p} - m\,\mathbb{1})\,(\mathbb{1} + \gamma_5\not{s})\ . \tag{9.104}$$

describes *antiparticles* with momentum $\boldsymbol{p}$ and polarization $\boldsymbol{\zeta}$.

The two operators (9.103) and (9.104), when multiplied by γ^0 from the left, can be combined in a single definition, viz.

$$P^{(\pm)} := \gamma^0\varrho^{(\pm)} = \frac{1}{2}\,(E\,\mathbb{1} - \boldsymbol{p}\cdot\boldsymbol{\alpha} \pm m\beta)\,(\mathbb{1} + \gamma_5\not{s})\ , \tag{9.105}$$

where the standard representation and the definition (9.71) were used on the right-hand side. These matrices are also acceptable density matrices for particles and antiparticles, respectively. They are hermitian,

$$P^{(\pm)\dagger} = P^{(\pm)}\,,$$

and their trace is

$$\operatorname{tr} P^{(\pm)} = 2E\,,$$

reflecting the covariant normalization.

Example 9.1 Extreme Relativistic Motion

Extreme relativistic motion, $E \gg m$, is of particular interest. It occurs in typical experimental set-ups with very fast electrons, or with nearly massless neutrinos. Let the 3-axis point along the direction of the spatial momentum $\boldsymbol{p}$ of the particle and suppose the polarization vector $\boldsymbol{\zeta}$ of (9.95) is decomposed in terms of its longitudinal component ζ_l (along $\boldsymbol{p}$), and two transverse components ζ_t^1 and ζ_t^2 (perpendicular to $\boldsymbol{p}$). The density matrices (9.103) and (9.104) then go over into

$$\varrho^{(\pm)} \approx \frac{1}{2} \not{p} \left\{ \mathbb{1} - \gamma_5 \left(\pm \zeta_l \mathbb{1} + \zeta_t^1 \gamma^1 + \zeta_t^2 \gamma^2 \right) \right\}. \tag{9.106}$$

This is shown as follows. With $E \gg m$, with $\{p^\mu\} \approx E(1, 0, 0, 1)$, and inserting (9.95) one has

$$\varrho^{(\pm)} \approx \frac{1}{2} \left(E(\gamma^0 - \gamma^3) \pm m \mathbb{1} \right) \left(\mathbb{1} + \gamma_5 \left[\frac{E}{m} \zeta_l (\gamma^0 - \gamma^3) - \zeta_t^1 \gamma^1 - \zeta^2 \gamma^2 \right] \right).$$

The potentially dangerous term, i.e. the one which is multiplied by E^2/m, is zero because $\gamma_5(\gamma^0 - \gamma^3) = -(\gamma^0 - \gamma^3)\gamma_5$ and because the square

$$\left(\gamma^0 - \gamma^3\right)^2 = \left(\gamma^0\right)^2 + \left(\gamma^3\right)^2 - \left\{\gamma^0, \gamma^3\right\}$$

is equal to zero. The only terms that remain in the limit $m \to 0$, are

$$\varrho^{(\pm)} \approx \frac{1}{2} E \left(\gamma^0 - \gamma^3\right) \left\{ \mathbb{1} + \gamma_5 \left[\mp \zeta_l - \zeta_t^1 \gamma^1 - \zeta_t^2 \gamma^2 \right] \right\}.$$

The factor in front of the right-hand side is nothing but $\not{p}$ because $\boldsymbol{p}$ points in the direction of the 3-axis. This proves the formula (9.106).

The hermitian form of the density matrix whose trace is normalized to $2E$ is obtained in much the same way. It is equal to

$$P^{(\pm)} \approx \frac{1}{2} \gamma^0 \not{p} \left\{ \mathbb{1} - \gamma_5 \left(\pm \zeta_l \mathbb{1} + \zeta_t^1 \gamma^1 + \zeta_t^2 \gamma^2 \right) \right\}.$$

These density matrices describe electrons and positrons, respectively, at very high energies which are polarized completely or partially, along an arbitrary direction. Consider as an example a beam of electrons with momentum $\boldsymbol{p} = |\boldsymbol{p}|\hat{\boldsymbol{e}}_3 \approx E\hat{\boldsymbol{e}}_3$ which has the polarization $P = w_+ - w_-$ along the 3-direction. The spin state of this beam is described by the density matrix

$$\varrho^{(+)} \approx \frac{1}{2} \not{p} \left\{ \mathbb{1} - \left(w_+ - w_- \right) \gamma_5 \right\}. \tag{9.107}$$

A beam of positrons which was prepared under identical conditions, has the density matrix

$$\varrho^{(-)} \approx \frac{1}{2} \not{p} \left\{ \mathbb{1} + \left(w_+ - w_- \right) \gamma_5 \right\}. \tag{9.108}$$

Application to Neutrinos: We have the following empirical information on the three neutrinos ν_e, ν_μ, and ν_τ that are produced or annihilated in weak interactions: (i) In comparison to their charged partners e^-, μ^-, and τ^- they have very small masses; (ii) They exist only in spin states which are fully polarized in a direction opposite to their spatial momentum. This is to say that they are described by (9.107) with $(w_+ = 0, w_- = 1)$. (This is equivalent to having $\zeta_l = -1$, $\zeta_t^1 = 0 = \zeta_t^2$.) If their mass were exactly equal to zero then these would be eigenstates of helicity with eigenvalue -1. States of this kind, independently of whether they are massive or massless, are called *lefthanded*, or *left chiral*.

Their antiparticles $\overline{\nu_e}$, $\overline{\nu_\mu}$, and $\overline{\nu_\tau}$, in turn, appear only in spin states which are completely aligned along the positive $\boldsymbol{p}$-direction, or, using the terminology just introduced, which are *righthanded*, or *right chiral*. Thus, they are described by (9.108) with $(w_+ = 1, w_- = 0)$.

This is an important result of physics: The spin properties of both neutrinos and antineutrinos are described by the density matrix

$$\varrho^{(\nu)} = \frac{1}{2} \not{p} \left(\mathbb{1} + \gamma_5 \right) . \tag{9.109}$$

The reason why physical neutrinos are lefthanded, and antineutrinos are righthanded, must be found in the dynamics of weak interactions.

Let us consider for a while neutrinos to be strictly massless. In a world with massless particles the symmetry operations, within $L_+^\uparrow$, which leave the momentum $\boldsymbol{p}$ invariant, are the rotations about the direction of $\boldsymbol{p}$ and the reflections on planes which contain this vector. While these rotations do not change the helicity, the reflections turn positive helicity into negative helicity and vice versa. For example, positive and negative helicity of photons are described by the polarization vectors

$$\hat{\boldsymbol{e}}_+ = -\frac{1}{\sqrt{2}} \left(\hat{\boldsymbol{e}}_1 + \mathrm{i} \hat{\boldsymbol{e}}_2 \right) , \quad \text{and} \quad \hat{\boldsymbol{e}}_- = \frac{1}{\sqrt{2}} \left(\hat{\boldsymbol{e}}_1 - \mathrm{i} \hat{\boldsymbol{e}}_2 \right) ,$$

respectively, if their momentum points along the 3-direction. Rotations about the 3-axis do not change them, but a reflection on, e.g., the (1, 3)-plane interchanges $\hat{\boldsymbol{e}}_+$ and $\hat{\boldsymbol{e}}_-$. In other terms, the reflection changes the orientation of the frame of reference unless the positive 3-direction is simultaneously replaced by the negative 3-direction. The latter operation again exchanges the helicities. Therefore, this argument applies also to fermions.

Furthermore, one knows that electromagnetic interactions are invariant under rotations as well as under space reflection, and, in particular, under reflections on planes containing the momentum $\boldsymbol{p}$. Therefore, it is not surprising that photons of either helicity are physical photons. In contrast to electromagnetism, the weak interactions are not invariant under space reflection. To the contrary, in weak interactions mediated by what are called *charged currents* one finds *maximal* parity violation. An example is provided by the decay

$$\mu^- \longrightarrow e^- + \overline{\nu_e} + \nu_\mu$$

which involves the charged current linking e^- with $\overline{\nu_e}$, and μ^- with ν_μ, and where maximal spin-momentum correlations are observed. Therefore, the reflections described above cannot be admissible symmetry operations in weak interactions. The two theoretically expected helicities of neutrinos or of antineutrinos are not dynamically degenerate. Thus, it is not really surprising to find only one of each pair: neutrinos with positive helicity, and antineutrinos with negative helicities completely decouple from all observable physical processes.

These considerations may be supplemented by a further remark. If parity $\boldsymbol{\Pi}$ is not a valid symmetry in weak interactions, then these interactions cannot be invariant under charge conjugation either. Indeed, as one easily shows, the charge conjugate density matrix of (9.109) is equal to

$$\varrho_{\mathrm{C}}^{(\nu)}(m=0) = \frac{1}{2}\not{p}\left(\mathbb{1} - \gamma_5\right).$$

This density matrix would describe *righthanded* neutrinos and *lefthanded* antineutrinos which are not found in nature. In turn, the combined operation $\boldsymbol{\Pi}\mathbf{C}$ leaves the density matrix (9.109) unchanged. Indeed, the *combination* of space reflection and charge conjugation is an admissible symmetry of weak interactions.

9.3.2 The Fermion-Antifermion Propagator

In the framework of perturbative quantum field theory the propagator for Dirac fields is perhaps the most important quantity. It combines the exchange of particles and antiparticles in a Lorentz invariant way and is obtained in much the same way as the propagator (7.60) for the Klein-Gordon field. Denoting the positive and negative frequency parts of the Dirac field (9.84) and of the adjoint field (9.85) by superscripts (+) and (−), respectively, the vacuum expectation value of the time-ordered product of two fields in the points x and y of spacetime is equal to

$$\begin{aligned}&\langle 0|T\,\psi(x)_\alpha\overline{\psi(y)}_\beta|0\rangle\\&= \langle 0|\psi(x)_\alpha^{(+)}\overline{\psi(y)}_\beta^{(-)}\Theta(x^0-y^0)0\rangle - \langle 0\overline{\psi(y)}_\beta^{(+)}\,\psi(x)_\alpha^{(-)}\Theta(y^0-x^0)|0\rangle\,.\end{aligned}$$

Note that every positive frequency term acting on the vacuum to the right, as well as any negative frequency term acting on the vacuum state to the left, give zero. Furthermore, the fields of the second term are interchanged. As we are dealing with fermions this explains the minus sign. Before actually calculating the two terms let us give a physical interpretation of them: In the first term a *particle* (i.e., in the notation of creation and annihilation operators, a particle of the kind "a") is created in the world point y and is annihilated in the world point x, the time coordinate x^0 being later than the time y^0. In the second term in which y^0 is later than x^0, an *antiparticle* (i.e. a particle of the kind "b") is created in the point x, and is annihilated in the point y. Needless to add that these terms must contribute to physical amplitudes in such a way that at every vertex the conservation laws of the theory are respected.

Inserting the expansions (9.84) and (9.85) one finds in a first step

$$\langle 0|\psi(x)_\alpha \overline{\psi(y)}_\beta|0\rangle \Theta(x^0-y^0) = \sum_{r,s}\int \frac{\mathrm{d}^3 p}{2E_p}\int \frac{\mathrm{d}^3 q}{2E_q}$$

$$\frac{1}{(2\pi)^3}\,\mathrm{e}^{-\{ipx}\,\mathrm{e}^{iqy}\Theta(x^0-y^0)u^{(r)}(\boldsymbol{p})_\alpha\overline{u^{(s)}(\boldsymbol{q})}_\beta\langle 0|a^{(r)}(\boldsymbol{p})a^{(s)\dagger}(\boldsymbol{q})|0\rangle\,.$$

The vacuum expectation value on the right-hand side yields $2E_p\delta_{rs}\delta(\boldsymbol{p}-\boldsymbol{q})$ so that the integration over $\boldsymbol{q}$, and the sum over s reduce the integrand to $\boldsymbol{q}=\boldsymbol{p}$ and $s=r$. Furthermore, according to (9.99) the product of the two spinors in momentum space yields

$$\sum_r u^{(r)}(\boldsymbol{p})_\alpha\overline{u^{(r)}(\boldsymbol{p})}_\beta = \left(\not{p}+m\,\mathbb{1}\right)_{\alpha\beta}\,.$$

Using the first of the integral representations (7.61) for the step function, and defining $k^0 := E_p-\lambda$, $\boldsymbol{k} := \boldsymbol{p}$ (whereby also $E_p = E_k$), one obtains

$$\langle 0|\psi(x)_\alpha\overline{\psi(y)}_\beta|0\rangle\Theta(x^0-y^0)$$
$$= -\frac{\mathrm{i}}{(2\pi)^4}\int \mathrm{d}^4k\;\mathrm{e}^{-\mathrm{i}k(x-y)}\frac{\left(E_k\gamma^0-\boldsymbol{k}\cdot\boldsymbol{\gamma}+m\,\mathbb{1}\right)_{\alpha\beta}}{2E_k\left(E_k-k^0-\mathrm{i}\varepsilon\right)}\,.$$

The second term of the time-orderd product is calculated in the same way as the first, noting that here only the vacuum expectation value

$$\langle 0|b^{(r)}(\boldsymbol{p})b^{(s)\dagger}(\boldsymbol{q})|0\rangle = 2E_p\,\delta_{rs}\,\delta(\boldsymbol{p}-\boldsymbol{q})$$

contributes so that the spin sum over v-type spinors (9.100)

$$\sum_r \overline{v^{(r)}(\boldsymbol{p})}_\beta v^{(r)}(\boldsymbol{p})_\alpha = \left(\not{p}-m\,\mathbb{1}\right)_{\alpha\beta}$$

has to be inserted. One now uses the second integral representation (7.61) of the step function, and defines $k^0 := E_p+\lambda$, and $\boldsymbol{k} := \boldsymbol{p}$. Upon replacing the integration variable k by $-k$, one obtains

$$\langle 0|\overline{\psi(y)}_\beta\psi(x)_\alpha|0\rangle\Theta(y^0-x^0)$$
$$= \frac{\mathrm{i}}{(2\pi)^4}\int \mathrm{d}^4k\;\mathrm{e}^{-\mathrm{i}k(x-y)}\frac{\left(E_k\gamma^0+\boldsymbol{k}\cdot\boldsymbol{\gamma}-m\,\mathbb{1}\right)_{\alpha\beta}}{2E_k\left(-E_k-k^0+\mathrm{i}\varepsilon\right)}\,.$$

Like in the case of the scalar field the sum of the two contributions eventually yields a simple and covariant expression, viz.

$$\langle 0|T\,\psi(x)_\alpha\overline{\psi(y)}_\beta|0\rangle = -\frac{\mathrm{i}}{(2\pi)^4}\int \mathrm{d}^4k\;\mathrm{e}^{-\mathrm{i}k(x-y)}$$
$$\times\frac{1}{2E_k}\left[\frac{\left(E_k\gamma^0-\boldsymbol{k}\cdot\boldsymbol{\gamma}+m\,\mathbb{1}\right)_{\alpha\beta}}{E_k-k^0-\mathrm{i}\varepsilon}-\frac{\left(E_k\gamma^0+\boldsymbol{k}\cdot\boldsymbol{\gamma}-m\,\mathbb{1}\right)_{\alpha\beta}}{E_k+k^0-\mathrm{i}\varepsilon}\right]$$
$$= \frac{\mathrm{i}}{(2\pi)^4}\int \mathrm{d}^4k\;\mathrm{e}^{-\mathrm{i}k(x-y)}\frac{\left(\not{k}+m\,\mathbb{1}\right)_{\alpha\beta}}{k^2-m^2+\mathrm{i}\varepsilon}\,.$$

Dropping the spinor indices we note the final result:

$$\langle 0|T\,\psi(x)\overline{\psi(y)}|0\rangle = \frac{\mathrm{i}}{(2\pi)^4}\int \mathrm{d}^4k\; \mathrm{e}^{-\mathrm{i}k(x-y)}\frac{(\not{k}+m\,\mathbb{1})}{k^2-m^2+\mathrm{i}\varepsilon}$$
$$\equiv -\frac{1}{2}S_F(x-y;m)\,. \tag{9.110}$$

The symbol S_F is a standard abbreviation for the fermion propagator. The structure of the propagators (7.60), (7.149), and (9.110) are very similar. The numerator of the integrand contains the spin sum which is characteristic for the type of particle one is dealing with (scalar particle, vector particle, or spin-1/2 fermion). The denominator always equals [(virtual momentum)^2 − (mass)^2], and, by the term iε, gives a prescription of how to deform the path of integration for k in its complex plane.

9.3.3 Traces of Products of γ-Matrices

In calculating cross sections or decay probabilities for processes in which fermions are born or absorbed, one must calculate absolute squares of amplitudes containing spinors $u(p)$, $v(q)$, etc. Calculations of this kind become economic and elegant if one makes use of so-called *trace techniques*. A simple example may illustrate the problem at stake.

Example 9.2

Suppose the process sketched in Fig. 9.1, i.e. a process wich describes an incoming fermion with momentum q and the same fermion in the outgoing state with momentum p, is described by the scattering amplitude

$$T = Q_\mu\,\overline{u(p)}\Gamma^\mu u(q) \equiv Q_\mu \sum_{\beta,\sigma=1}^{4} \overline{u(p)}_\beta \Gamma^\mu_{\beta\sigma} u_\sigma(q)\,,$$

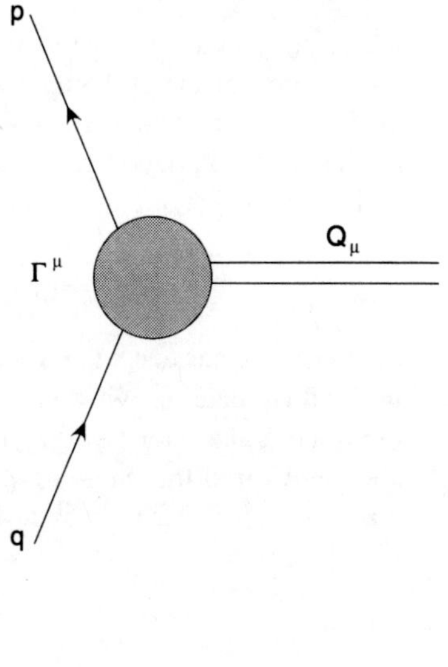

Fig. 9.1 A fermion with momentum q comes in, it interacts with other particles whose influence is described implicitly by the quantity Γ^μ, and then leaves with momentum p

where Q_μ is a real four-vector (which, in general, depends on the momenta of further particles), and where

$$\Gamma^\mu = a\gamma^\mu + b\gamma^\mu\gamma_5 \equiv a\Gamma^\mu_V + b\Gamma^\mu_A$$

is an arbitrary linear combination of base elements (9.38) of the Clifford algebra $\mathrm{Cl}(M^4)$. In analogy to the summation convention for Lorentz indices it is useful to agree on summing over pairs of repeated spinor indices, without writing the summation symbol explicitly. When taking the absolute square of the amplitude T, one must calculate

$$\begin{aligned}|T|^2 &= Q_\mu Q_\nu \left(\overline{u(p)}\Gamma^\mu u(q)\right)\left(\overline{u(p)}\Gamma^\nu u(q)\right)^* \\ &= Q_\mu Q_\nu \overline{u(p)}_\beta \left(\Gamma^\mu\right)_{\beta\sigma} u_\sigma(q)\left(u^\dagger(q)\gamma^0\right)_\tau \left(\gamma^0\left(\Gamma^\nu\right)^\dagger\gamma^0\right)_{\tau\alpha} u_\alpha(p) \\ &= Q_\mu Q_\nu u_\alpha(p)\overline{u(p)}_\beta \left(\Gamma^\mu\right)_{\beta\sigma} u_\sigma(q)\overline{u(q)}_\tau \left(\gamma^0\left(\Gamma^\nu\right)^\dagger\gamma^0\right)_{\tau\alpha}\end{aligned}$$

In the first step we have made use of

$$\begin{aligned}\left(\overline{u(p)}\Gamma^\nu u(q)\right)^* &= \left(u^\dagger(p)\gamma^0\Gamma^\nu u(q)\right)^* \\ &= u^\dagger(q)\left(\Gamma^\nu\right)^\dagger\left(\gamma^0\right)^\dagger u(p), \quad \left(\gamma^0\right)^\dagger = \gamma^0\end{aligned}$$

and, in the second step, have inserted $\left(\gamma^0\right)^2 = \mathbb{1}_4$. Inspection of the last line shows that in summing over the (Dirac-)spinor indices α, β, σ, and τ one must, in fact, calculate the trace of the product of the following 4×4-matrices,

$$u(p)\overline{u(p)}\,, \quad \Gamma^\mu\,, \quad u(q)\overline{u(q)}\,, \quad \gamma^0\left(\Gamma^\nu\right)^\dagger\gamma^0\,,$$

the first and third of which are given by the formula (9.97). Thus, we have to calculate

$$\begin{aligned}|T|^2 = Q_\mu Q_\nu \frac{1}{4}\mathrm{tr}\,\Big\{&(\not{p} + m\,\mathbb{1})(\mathbb{1} + \gamma_5\not{n}_f) \\ &\Gamma^\mu(\not{q} + m\,\mathbb{1})(\mathbb{1} + \gamma_5\not{n}_i)\gamma^0\left(\Gamma^\nu\right)^\dagger\gamma^0\Big\}\,,\end{aligned}$$

where n_i characterizes the polarization of the incoming state, and n_f the polarization of the outgoing state. Of course, it is a matter of the experimental set-up whether or not these polarizations must be kept track of. If the incoming particle is unpolarized one must take the average over the two spin orientations n_i with equal weights. This is equivalent to replacing the fourth and the fifth factors as follows

$$\frac{1}{2}(\not{q} + m\,\mathbb{1})(\mathbb{1} + \gamma_5\not{n}_i) \longrightarrow (\not{q} + m\,\mathbb{1})\,.$$

Likewise one has to decide whether the experiment is sensitive to the polarization in the final state, or whether the outgoing fermion is counted independently of its spin orientation. Regarding the first alternative, the quantity $|T|^2$ yields probabilities to find the outgoing particle with its spin pointing in the direction of $+\hat{\boldsymbol{n}}_f$, or of $-\hat{\boldsymbol{n}}_f$ from which the polarization can be calculated. In the second case where the spin is not detected, one must replace the product of the first and second factors according to the rule

$$\frac{1}{2}(\not{p} + m\,\mathbb{1})(\mathbb{1} + \gamma_5 \not{n}_f) \longrightarrow (\not{p} + m\,\mathbb{1})\,.$$

We also note that the elements $\Gamma_V^\mu = \gamma^\mu$ and $\Gamma_A^\mu = \gamma^\mu\gamma_5$, by the property (9.42) of the γ-matrices, have the same property, i.e.

$$\gamma^0 \left(\Gamma_{V/A}^\mu\right)^\dagger \gamma^0 = \Gamma_{V/A}^\mu\,. \tag{9.111}$$

Thus, in the simplest case, i.e. when the incoming particle is unpolarized and the spin of the outgoing particle is not detected, one has to calculate

$$\sum_{\text{spins}} |T|^2 = Q_\mu Q_\nu \text{tr}\,\left\{(\not{p} + m\,\mathbb{1})\Gamma^\mu(\not{q} + m\,\mathbb{1})\Gamma^\nu\right\}\,. \tag{9.112}$$

In any event, with or without the spin degrees of freedom, one eventually will have to calculate a trace over products of γ-matrices. There are general formulae for such traces. We derive and summarize them next.

The starting point is obtained from the relation (4.37) which yields the symmetric sum of two Pauli matrices

$$\sigma^{(i)}\sigma^{(j)} + \sigma^{(j)}\sigma^{(i)} = 2\delta_{ij}\,\mathbb{1}\,.$$

With our previous definitions $\{\sigma^\mu\} = (\mathbb{1}_2, -\boldsymbol{\sigma})$ and $\{\widehat{\sigma}^\mu\} = (\mathbb{1}_2, \boldsymbol{\sigma})$, cf. Sect. 9.1.2, this implies the relation

$$\sigma^\mu\widehat{\sigma}^\nu + \sigma^\nu\widehat{\sigma}^\mu = 2g^{\mu\nu}\,\mathbb{1}_2\,.$$

When reformulated in terms of γ-matrices (9.28) it is seen to be equivalent to (9.32), viz.

$$\gamma^\mu\gamma^\nu + \gamma^\nu\gamma^\mu = 2g^{\mu\nu}\,\mathbb{1}_4\,.$$

All formulae for traces follow from this relation and from the representation (9.28). However, as the trace is invariant under cyclic permutations of its factors, these formulae hold in *all* representations. Indeed, the transformation matrix $\mathbf{S}$ in the formula

$$\gamma'^\mu\gamma'^\nu \cdots \gamma'^\tau = \mathbf{S}\gamma^\mu\gamma^\nu \cdots \gamma^\tau\mathbf{S}^{-1}$$

cancels out in the trace.

The following rules apply to traces of products of γ-matrices:

(i) The trace of the unit matrix is 4, the trace of γ_5 vanishes,

$$\text{tr}\;\mathbb{1}_4 = 4\,, \qquad \text{tr}\;\gamma_5 = 0\,, \tag{9.113}$$

(ii) The trace of the product of an *odd* number of γ-matrices vanishes

$$\text{tr}\;\underbrace{\left\{\gamma^\alpha\gamma^\beta \cdots \gamma^\tau\right\}}_{2k+1} = 0\,. \tag{9.114}$$

This is shown by the following consideration: γ^μ in (9.28) is an *odd*, bloc-like matrix in the sense of having nonzero entries only in the off-diagonal blocs. In turn, the product of two γ-matrices is *even*, that is, has nonzero entries in the diagonal blocs only. As a consequence, all products with an even number of factors are *even*, all products with an odd number of factors are *odd* and, hence, have trace zero.

(iii) Generally, the trace of products with an even number $2n$ of factors is reduced to traces of $2n-2$ factors by multiple application of the anticommutator (9.32) and by the use of the cyclicity of the trace. The formulae for two, four, and six factors read

$$\operatorname{tr}\{\gamma^\alpha\gamma^\beta\} = 4g^{\alpha\beta}\,, \tag{9.115}$$

$$\operatorname{tr}\{\gamma^\alpha\gamma^\beta\gamma^\sigma\gamma^\tau\} = 4\big(g^{\alpha\beta}g^{\sigma\tau} - g^{\alpha\sigma}g^{\beta\tau} + g^{\alpha\tau}g^{\beta\sigma}\big)\,, \tag{9.116}$$

$$\begin{aligned}\operatorname{tr}\{\gamma^\alpha\gamma^\beta\gamma^\mu\gamma^\nu\gamma^\sigma\gamma^\tau\} &= g^{\alpha\beta}\operatorname{tr}\{\gamma^\mu\gamma^\nu\gamma^\sigma\gamma^\tau\}\\ &\quad - g^{\alpha\mu}\operatorname{tr}\{\gamma^\beta\gamma^\nu\gamma^\sigma\gamma^\tau\} + g^{\alpha\nu}\operatorname{tr}\{\gamma^\beta\gamma^\mu\gamma^\sigma\gamma^\tau\}\\ &\quad - g^{\alpha\sigma}\operatorname{tr}\{\gamma^\beta\gamma^\mu\gamma^\nu\gamma^\tau\} + g^{\alpha\tau}\operatorname{tr}\{\gamma^\beta\gamma^\mu\gamma^\nu\gamma^\sigma\}\,, \end{aligned}\tag{9.117}$$

As an example verify (9.116) in detail: One has

$$\begin{aligned}\frac{1}{4}\operatorname{tr}\{\gamma^\alpha\gamma^\beta\gamma^\sigma\gamma^\tau\} &= 2g^{\alpha\beta}\frac{1}{4}\operatorname{tr}\{\gamma^\sigma\gamma^\tau\} - \frac{1}{4}\operatorname{tr}\{\gamma^\beta\gamma^\alpha\gamma^\sigma\gamma^\tau\}\\ &= 2g^{\alpha\beta}g^{\sigma\tau} - 2g^{\alpha\sigma}\frac{1}{4}\operatorname{tr}\{\gamma^\beta\gamma^\tau\} + \frac{1}{4}\operatorname{tr}\{\gamma^\beta\gamma^\sigma\gamma^\alpha\gamma^\tau\}\\ &= 2g^{\alpha\beta}g^{\sigma\tau} - 2g^{\alpha\sigma}g^{\beta\tau} + 2g^{\alpha\tau}\frac{1}{4}\operatorname{tr}\{\gamma^\beta\gamma^\sigma\} - \frac{1}{4}\operatorname{tr}\{\gamma^\beta\gamma^\sigma\gamma^\tau\gamma^\alpha\}\\ &= 2g^{\alpha\beta}g^{\sigma\tau} - 2g^{\alpha\sigma}g^{\beta\tau} + 2g^{\alpha\tau}g^{\beta\sigma} - \frac{1}{4}\operatorname{tr}\{\gamma^\beta\gamma^\sigma\gamma^\tau\gamma^\alpha\}\,.\end{aligned}$$

The last trace is the same as the one on the left-hand side so that (9.116) follows.

(iv) Traces containing γ_5: If γ_5 is multiplied with one, two, or three γ-matrices then the product has trace zero

$$\operatorname{tr}\{\gamma_5\gamma^\alpha\} = 0,\quad \operatorname{tr}\{\gamma_5\gamma^\alpha\gamma^\beta\} = 0,\quad \operatorname{tr}\{\gamma_5\gamma^\alpha\gamma^\beta\gamma^\sigma\} = 0\,. \tag{9.118}$$

The matrix γ_5, (9.35), is proportional to the product of all four γ-matrices γ^μ, $\mu = 0, 1, 2, 3$. Therefore, the first and third relations (9.118) follow from the rule (9.114). In the second relation α and β either are equal in which case the second of (9.113) applies, or they are different in which case both are contained in γ_5. By exchanges of neighbours one shifts them until their square appears which is plus or minus the unit matrix.

There remains the product of two γ-matrices whose Lorentz indices necessarily are different. By the rule (9.115) the trace of this product vanishes. It is only in a product with four γ-matrices that there is a result different from zero. It reads

$$\operatorname{tr}\{\gamma_5\gamma^\alpha\gamma^\beta\gamma^\sigma\gamma^\tau\} = 4\mathrm{i}\varepsilon^{\alpha\beta\sigma\tau}\,. \tag{9.119}$$

One might be puzzled by the observation that the result of rule (9.119) is pure imaginary, even though one is calculating the absolute square of an amplitude, hence, a real quantity. In fact, this trace neither occurs in isolation, nor in an expression where it is multiplied by one of the real expressions (9.115)–(9.117). It does occur, however, when two terms of the kind (9.119) are multiplied with one another. In practice, the first statement follows from the fact that two of the indices of the ε-symbol are contracted with the same kinematic variable. The product of a symmetric tensor with the antisymmetric symbol ε gives zero.

When two terms of the kind of (9.119) are multiplied and are contracted partially or completely, a useful formula is the following

$$\varepsilon^{\alpha\beta\sigma\tau}\varepsilon_{\alpha\beta\mu\nu} = -2\left\{\delta^{\sigma}{}_{\mu}\delta^{\tau}{}_{\nu} - \delta^{\sigma}{}_{\nu}\delta^{\tau}{}_{\mu}\right\}. \qquad (9.120)$$

For instance, let us return to the example (9.112) and choose $\Gamma^{\mu} = \gamma^{\mu}(\mathbb{1} - \gamma_5)$, for the sake of illustration. Then one has

$$\begin{aligned}\sum_{\text{Spins}} |T|^2 &= Q_{\mu} Q_{\nu} \text{tr}\left\{(\not p + m\,\mathbb{1})\gamma^{\mu}(\mathbb{1} - \gamma_5)(\not q + m\,\mathbb{1})\gamma^{\nu}(\mathbb{1} - \gamma_5)\right\} \\ &= Q_{\mu} Q_{\nu} 2\,\text{tr}\left\{(\not p + m\,\mathbb{1})\gamma^{\mu}\not q\gamma^{\nu}(\mathbb{1} - \gamma_5)\right\} \\ &= 2 Q_{\mu} Q_{\nu}\left[\text{tr}\left\{\not p\gamma^{\mu}\not q\gamma^{\nu}\right\} - \text{tr}\left\{\gamma_5\not p\gamma^{\mu}\not q\gamma^{\nu}\right\}\right] \\ &= 8 Q_{\mu} Q_{\nu}\left[p^{\mu} q^{\nu} - g^{\mu\nu} p\cdot q + q^{\mu} p^{\nu} - \mathrm{i}\varepsilon^{\alpha\mu\beta\nu} p_{\alpha} q_{\beta}\right]\end{aligned}$$

Here one made use of $(\mathbb{1} - \gamma_5)^2 = 2(\mathbb{1} - \gamma_5)$ and $(\mathbb{1} + \gamma_5)(\mathbb{1} - \gamma_5) = 0$. Only the first three terms in square brackets contribute and are real. The fourth term which is pure imaginary, gives zero.

Example 9.3

Assume the scattering amplitude is obtained from the exchange of a photon, with propagator (7.149), between two charged spin-1/2-fermions. In the Feynman gauge the amplitude reads

$$T = c\,\overline{u^{(2)}(p_2')}\gamma^{\mu} u^{(2)}(p_2)\frac{-g_{\mu\sigma}}{Q^2 + \mathrm{i}\varepsilon}\overline{u^{(1)}(p_1')}\gamma^{\sigma} u^{(1)}(p_1),$$

with $Q = p_1 - p_1' = p_2' - p_2$ and c a constant of no particular relevance at this point. Taking the absolute value of the square of the amplitude, averaged and summed over the spins in the initial and final states, respectively, one obtains an expression of the form

$$\begin{aligned}&\left[p_2'^{\mu} p_2^{\nu} + (m_2^2 - p_2'\cdot p_2) g^{\mu\nu} + p_2^{\mu} p_2'^{\nu}\right] g_{\mu\sigma} g_{\nu\tau} \\ &\left[p_1'^{\sigma} p_1^{\tau} + (m_1^2 - p_1'\cdot p_1) g^{\sigma\tau} + p_1^{\sigma} p_1'^{\tau}\right].\end{aligned}$$

Though in this example the contractions can be done directly, it may be simpler to first simplify the traces. For example, the second factor in square brackets was obtained from

$$\frac{1}{4}\text{tr}\left\{(\not p_1' + m_1\,\mathbb{1})\gamma^{\sigma}(\not p_1 + m_1\,\mathbb{1})\gamma^{\tau}\right\}.$$

Considering the term proportional to $g^{\mu\nu}$ in the first factor and noting that

$$g^{\mu\nu} g_{\mu\sigma} g_{\nu\tau} = g_{\sigma\tau}$$

one sees that in the term

$$g_{\sigma\tau} \frac{1}{4} \mathrm{tr} \left\{ (\not{p}_1' + m_1 \, 1\!\!1) \gamma^\sigma (\not{p}_1 + m_1 \, 1\!\!1) \gamma^\tau \right\}$$

one has to sum the indices σ and τ. The remark is that there are formulae which reduce the products by two factors and, hence, simplify matters. These formulae read

$$g_{\mu\nu} \gamma^\mu \gamma^\nu = 4 \, 1\!\!1 , \quad g_{\mu\nu} \gamma^\mu \not{p} \gamma^\nu = -2 \not{p} \tag{9.121}$$
$$g_{\mu\nu} \gamma^\mu \not{p} \not{q} \gamma^\nu = 4 \, p \cdot q , \quad g_{\mu\nu} \gamma^\mu \not{p} \not{q} \not{r} \gamma^\nu = -2 \not{r} \not{q} \not{p} \tag{9.122}$$
$$g_{\mu\nu} \gamma^\mu \not{p} \not{q} \not{r} \not{s} \gamma^\nu = 2 \left(\not{s} \not{p} \not{q} \not{r} + \not{r} \not{q} \not{p} \not{s} \right) \tag{9.123}$$

Again, these relations are proved by means of the basic anticommutator (9.32).

Remarks

1. It is certainly a good exercise to work out "by hand" the traces of the somewhat schematic examples of this section as well as the lowest-order processes to be discussed in the next chapter, by making use of the rules given above. Nowadays, however, there are advanced algebraic program packages which allow to evaluate traces of products of γ-matrices on computers. One will prefer to use those in all cases where the products contain many factors and/or if one wants to make sure that the results are free of errors, of signs or other.
2. As an alternative, the trace techniques can be formulated for two-component spinors, the products of γ-matrices being replaced by products of the kind

$$\sigma^\alpha \widehat{\sigma}^\beta \sigma^\gamma \widehat{\sigma}^\delta \ldots$$

in which σ-matrices and $\widehat{\sigma}$-matrices alternate.[8] In fact, in situations where the masses may be neglected, calculations are often much simpler than with γ-matrices.

9.3.4 Chiral States and Their Couplings to Spin-1 Particles

In this section we define chiral fermion fields, we explain their physical role, and show in which way they occur in interaction terms of electromagnetic, weak, and strong interactions.

[8] For concrete and realistic examples see. e.g., A. Kersch und F. Scheck, Nucl. Phys. B **263**, 475, 1986.

Definition 9.1 Chiral Fields

Let $\psi(x)$ be a quantized Dirac spinor which is a solution of the Dirac equation. The fields which are generated from it by the action of the projection operators (9.39),

$$\big(\psi(x)\big)_{\mathrm{R}} := P_+\psi(x) = \frac{1}{2}\big(\mathbb{1} + \gamma_5\big)\psi(x)\,, \tag{9.124}$$

$$\big(\psi(x)\big)_{\mathrm{L}} := P_-\psi(x) = \frac{1}{2}\big(\mathbb{1} - \gamma_5\big)\psi(x)\,, \tag{9.125}$$

are called *right chiral* and *left chiral fields*.

The physical role of chiral fields will become clear through the following remarks. We return to the natural, or high-energy, representation (9.28) and remind the reader of the explicit form (9.35) of γ_5. Inspection of the spinors (9.30) in momentum space shows that ψ_{R} is identical with the two-component spinor $\widetilde{\phi}(p)$, while ψ_{L} is identical with the spinor $\widetilde{\chi}(p)$, both of which fulfill the Weyl equations (9.27). For high energies, $E \gg m$, or in cases where the mass is exactly zero, these equations decouple and become[9]

$$p_\mu\widehat{\sigma}^\mu\widetilde{\phi}(p, m=0) = 0\,, \qquad p_\mu\sigma^\mu\widetilde{\chi}(p, m=0) = 0\,. \tag{9.126}$$

Recalling the definition of σ^μ and of $\widehat{\sigma}^\mu$ as well as of $\{p_\mu\} = (E, -\boldsymbol{p})$ with $|\boldsymbol{p}| = E$, one sees that $\widetilde{\phi}(p, m = 0)$ and $\widetilde{\chi}(p, m = 0)$ are eigenstates of the helicity with eigenvalues $+1/2$ and $-1/2$, respectively. In the limit $m \to 0$ the chiral states go over into eigenstates of helicity. The analogous statements apply to the spinor $v(p)$, i.e. to eigenstates of momentum of antiparticles.

The relation between the chirality of a particle and the polarization along the direction of its spatial momentum can be worked out further. We do this by means of an example that also clarifies the general case.

Example 9.4 Density Matrix for Left Chiral Field

The density matrix $\varrho^{(+)}$, (9.103), was obtained from the matrix $u(p)\overline{u(p)}$ of (9.97). Calculating in the same way the density matrix for a particle in a left chiral state, i.e.

$$u_{\mathrm{L}}(p) = \frac{1}{2}(\mathbb{1} - \gamma_5)u(p)$$

one obtains the corresponding density matrix

$$\varrho_{\mathrm{L}}^{(+)} = u_{\mathrm{L}}(p)\overline{u_{\mathrm{L}}(p)} = \frac{1}{4}\big(\mathbb{1} - \gamma_5\big)\big(\not{p} - m\not{t}\big) = P_-\big(\not{p} - m\not{t}\big)P_+\,. \tag{9.127}$$

[9] These are the original equations that H. Weyl had proposed before the discovery of the Dirac equation. Initially, Weyl's equations were critized and rejected because they are not invariant under space reflection. Their role for physics was recognized only much later, after the discovery of parity violation in the weak interactions in the year 1956.

The proof of this formula is not difficult. With

$$\overline{u_{\rm L}(p)} \equiv u_{\rm L}^\dagger(p)\gamma^0 = \frac{1}{2}u^\dagger(p)(\mathbb{1} - \gamma_5)\gamma^0 = \frac{1}{2}\overline{u(p)}(\mathbb{1} + \gamma_5)\,,$$

inserting and working out the matrix (9.97), one finds

$$u_{\rm L}(p)\overline{u_{\rm L}(p)} = \frac{1}{2}(\mathbb{1} - \gamma_5)\frac{1}{2}\left(\not{p} + \not{p}\gamma_5\not{n} + m\,\mathbb{1} + m\gamma_5\not{n}\right)\frac{1}{2}(\mathbb{1} + \gamma_5)$$
$$= \frac{1}{2}(\mathbb{1} - \gamma_5)\frac{1}{2}\left(\not{p} - m\not{n} + (\mathbb{1} + \gamma_5)m\not{n}\right)\frac{1}{2}(\mathbb{1} + \gamma_5)\,.$$

Here the term $m\not{n}$ was added and subtracted such as to obtain a factor $(\mathbb{1} + \gamma_5)$ on the left of $\not{n}$. When multiplied with the overall left factor $(\mathbb{1} - \gamma_5)$ this gives zero. As γ_5 anticommutes with $\not{p}$ and $\not{n}$, the overall factor $(\mathbb{1} + \gamma_5)$ on the right can be shifted to the left. This proves the formula (9.127).

In order to interpret the result (9.127), consider the polarization along the spatial momentum $\boldsymbol{p}$, choosing the 3-direction parallel to this momentum,

$$\{p^\mu\} = (E, 0, 0, p)\,, \quad (p \equiv |\boldsymbol{p}|)\,, \quad \widehat{\boldsymbol{n}} = \pm\widehat{\boldsymbol{e}}_3\,.$$

Using the expression (9.96) for the four-vector n one has

$$m\,\{n^\mu\} = \left(\pm p, 0, 0, \pm(m + \frac{p^2}{E+m})\right)^T = \pm\left(p, 0, 0, E\right)^T\,,$$

where $p^2 = E^2 - m^2 = (E+m)(E-m)$ was inserted. Thus, the density matrix is

$$\varrho_{\rm L}^{(+)} = \frac{1}{4}\left(\mathbb{1} - \gamma_5\right)\left(\gamma^0 + \gamma^3\right)(E \mp p)\,. \tag{9.128}$$

This result tells us that the relative probability to find the spin oriented along the positive or the negative direction of spatial momentum, with $\beta = p/E$, is given by

$$R := \frac{w(h = +1/2)}{w(h = -1/2)} = \frac{E - p}{E + p} = \frac{1 - \beta}{1 + \beta}\,. \tag{9.129}$$

If the velocity of the fermion is close or equal to the speed of light then this ratio is approximately equal to

$$R = \frac{1 - \beta^2}{1 + \beta^2} \approx \frac{1}{4\gamma^2} = \frac{m^2}{4E^2}\,,$$

in agreement with our earlier statement that $u_{\rm L}(p)$ with $m = 0$ describes a state with helicity $-1/2$. In other terms, if a massive fermion is created in a left-chiral state with high energy, $E \gg m$, then the state with the spin oriented parallel to the momentum is suppressed by this ratio as compared to the state in which the spin is antiparallel to the momentum.

An analogous result holds for the spinors $P_- v(p)$ which describe antiparticles. Note, however, that this spinor represents *right* chiral states because one has

$$P_-\left(\not{p} + m\,\not{n}\right)P_+ =: \varrho_{\rm R}^{(-)}\,. \tag{9.130}$$

Its interpretation is taken over from the previous case by exchanging parallel and antiparallel spin orientations. For example, a massless antiparticle which was created in this chiral state has helicity $+1/2$. Thus, we summarize the correspondence of the projection operators P_+ and P_- to the chirality as follows:

$$P_- v(p) \equiv v_{\mathrm{R}}(p)\,, \qquad P_+ v(p) \equiv v_{\mathrm{L}}(p)\,. \tag{9.131}$$

The operator P_+ projects onto *right* chiral particle, and onto *left* chiral antiparticle states, while the operator P_- projects onto *left* chiral particle, and onto *right* chiral antiparticle states.

Vector and Axial Vector Couplings: After having clarified the physical interpretation of the chiral fields (9.124) and (9.125) we consider the typical interaction vertex of the Example 9.2

$$\overline{\psi(x)}\Gamma^\mu \psi(x) \qquad \text{with} \qquad \Gamma^\mu = \gamma^\mu\big(a\,\mathbb{1} + b\gamma_5\big)\,. \tag{9.132}$$

The projection operators being idempotent, and using $\gamma^\mu\gamma_5 = -\gamma_5\gamma^\mu$ one has

$$\begin{aligned}\Gamma^\mu &= \gamma^\mu\big(a\,\mathbb{1} + b\gamma_5\big) = (a+b)\gamma^\mu P_+^2 + (a-b)\gamma^\mu P_-^2 \\ &= (a+b)P_-\gamma^\mu P_+ + (a-b)P_+\gamma^\mu P_-\,.\end{aligned} \tag{9.133}$$

Inspection of the expansions (9.84), and (9.85) in terms of creation and annihilation operators shows that

(a1) an *in*coming particle with momentum p and spin projection r is represented by a spinor $u^{(r)}(p)$,

(a2) an *out*going particle with (p', r') by $\overline{u^{(r')}(p')}$,

(b1) an *in*coming antiparticle with (q, s) by $\overline{v^{(s)}(p)}$,

(b2) an *out*going antiparticle with (q', s') by $v^{(s')}(p')$.

Depending on the choice of the external particles, this means that the vertex $\overline{\psi(x)}\Gamma^\mu\psi(x)$ is replaced by the vertices in momentum space

$$\overline{u^{(r')}(p')}\Gamma^\mu u^{(r)}(p),\ \overline{v^{(s)}(q)}\Gamma^\mu v^{(s')}(q'),\ \overline{u^{(r)}(p)}\Gamma^\mu v^{(s)}(q)\,,$$
$$\overline{v^{(s)}(q)}\Gamma^\mu u^{(r)}(p)\,.$$

The first term stands for the scattering of a particle from (r, p) to (r', p'), the second term stands for the scattering of an antiparticle from (s, q) to (s', q'). The third term describes the creation of a particle-antiparticle pair, the fourth the annihilation of such a pair. Note that in identifying the chirality it does not matter whether the two spinors refer to the same particle or to two different particles. In the latter case one would have to mark the spinors by a particle index, as we did in Example 9.3. Recalling the relationships proved above,

$$P_{+/-}u^{(r)}(p) = u^{(r)}_{\mathrm{R/L}}(p)\,, \quad \overline{u^{(r)}(p)}P_{+/-} = \overline{u^{(r)}_{\mathrm{L/R}}(p)}\,,$$
$$P_{+/-}v^{(s)}(q) = v^{(s)}_{\mathrm{L/R}}(q),\ \overline{v^{(s)}(q)}P_{+/-} = \overline{v^{(s)}_{\mathrm{R/L}}(q)}$$

one reads off an important selection rule for interactions which couple to the vertex $\overline{\psi^{(k)}(x)}\Gamma^\mu\psi^{(i)}(x)$: (The indices i and k denote two, possibly different, species of fermions. They are omitted when they refer to the same particle.)

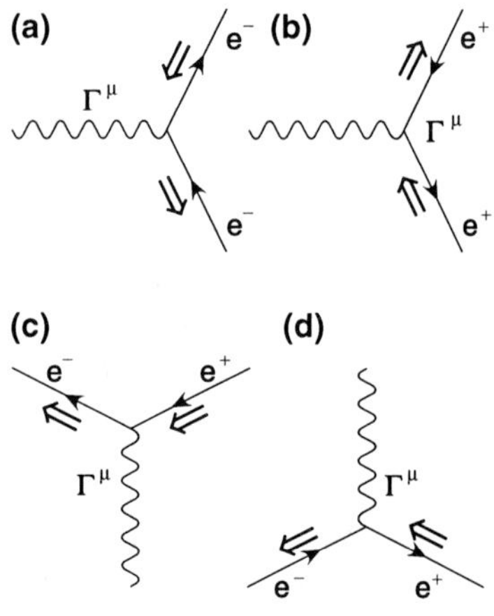

Fig. 9.2 **a**–**d** An arbitrary mixture of vector and axial vector couplings correlates the chiralities of the external particles as drawn in this figure: example **a** shows the scattering of a left chiral electron, **b** the scattering of a right chiral positron, **c** shows the creation of an electron-positron pair, **d** the annihilation of an electron-positron pair

Selection Rule for Vector/Axial Vector Coupling: An interaction in which two fermions couple through the operator

$$\overline{\psi^{(k)}(x)}\gamma^{\mu}\big(a\,\mathbb{1}+b\gamma_5\big),\,\psi^{(i)}(x) \tag{9.134}$$

conserves chirality if one the particles is incoming, the other is outgoing. If a particle-antiparticle pair is created or annihilated then this happens with opposite chitalities, (see Fig. 9.2).

Remarks

It is a standard convention in quantum electrodynamics to mark fermion lines with arrows along the flow of *negative* charge. If in Fig. 9.2 one follows these arrows then this selection rule says that at every vertex chirality is conserved (relative to the direction of arrows), irrespective of whether one is dealing with scattering, with pair creation, or pair annihilation.

To the schematic examples of Fig. 9.2, we add three realistic cases taken from electromagnetic and weak interactions:

Example 9.5 Electromagnetic Pair Creation

In lowest order perturbation theory the process $e^- + e^+ \to \mu^- + \mu^+$ is mediated by the exchange of a single virtual photon as shown in Fig. 9.3. The coupling of charged fermions to photons being given by (9.61) we must take $a = 1, b = 0$ in (9.132). The expression (9.133) then shows that left- and right-chiral states couple with the same strength. When an unpolarized beam of positrons collides

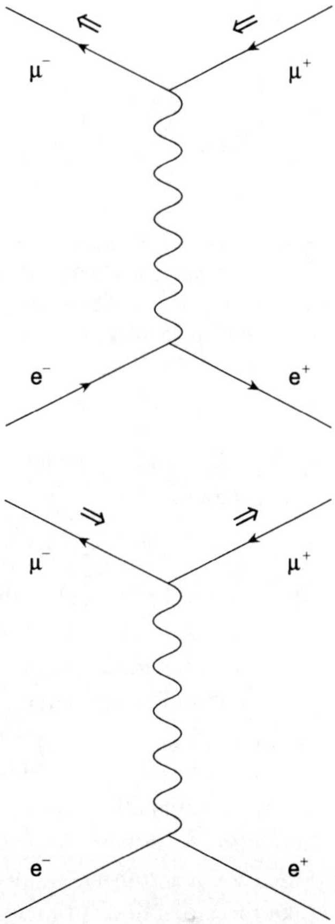

Fig. 9.3 Creation of a $\mu^+\mu^-$-pair from unpolarized beams of electrons and positrons

with an unpolarized beam of electrons the two possible constellations of chiralities in the final state are created with the same weights. Therefore, the outgoing beams of μ^- and μ^+ are also unpolarized. In turn, if in the same experiment one succeeds in accepting only those μ^+ which have their spin oriented along the momentum, then one knows that the accompanying μ^- is lefthanded. Indeed, in practice one uses a variant of this process in which the muons are replaced by τ-leptons, i.e. the process $e^- + e^+ \to \tau^- + \tau^+$, to produce beams of polarized τ-leptons.

Example 9.6 The Decays $\pi^0 \to e^+e^-$, $\eta \to \mu^+\mu^-$, $\eta \to e^+e^-$

The selection rules due to chirality have an interesting application in the decays of spin-0 mesons into two leptons that we work out in this and the next example. The three pions as well as the η-meson have spin zero. They are unstable, strongly interacting hadrons whose masses are, in MeV and in units of the electron mass, $m_{\pi^0} = 134.98\,\text{MeV} = 264.14\,m_e$, $m_\eta = 547.3\,\text{MeV} = 1071.0\,m_e$. When a meson in its rest frame decays into two particles then these move back-to-back

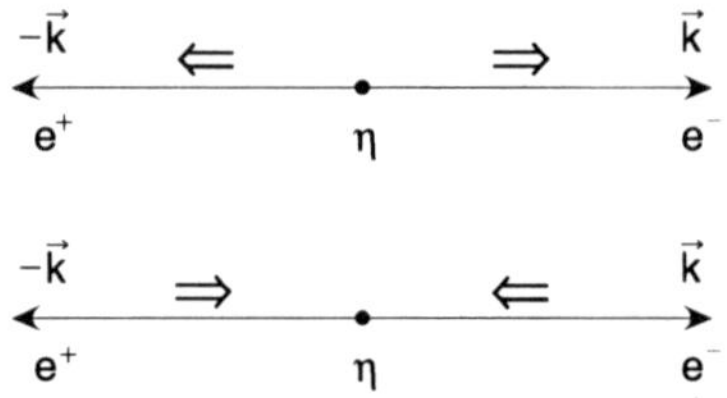

Fig. 9.4 If a meson with spin zero decays into two fermions the spins must point in opposite directions (conservation of angular momentum). However, as the example shows a particle-antiparticle pair, every vector or axial vector coupling favours parallel orientation of the spins, in conflict with conservation of angular momentum. If the masses of the fermions vanish then this decay is not possible

with equal and opposite three-momenta. For instance, in the case of the η, the kinematics is

$$q = (m_\eta, 0, 0, 0) = (E_1, \boldsymbol{k}) + (E_2, -\boldsymbol{k}), \quad E_i = \sqrt{m_i^2 + \boldsymbol{k}^2}\,.$$

It is convenient to choose the 3-axis along the direction of $\boldsymbol{k}$. In the decay process the squared total angular momentum $\boldsymbol{J}^2$ and its projection $\mathbf{J}_3$ are conserved. The angular momentum $\boldsymbol{J}$, in the final state, is the sum of the relative orbital angular momentum $\boldsymbol{\ell}$ of the two particles and of their spins. The magnetic quantum numbers fulfill $J_3 = m_\ell + m_s^{(1)} + m_s^{(2)}$. As the decaying meson is at rest and has spin zero, one has $J_3 = 0$. Although the relative motion of the two particles in the final state contains all values of the angular momentum ℓ, the projection m_ℓ onto the direction $\boldsymbol{k}$ vanishes for every partial wave. This was shown in Sect. 1.9.3. Thus, the two quantum numbers $m_s^{(1)}$ and $m_s^{(2)}$ whose sum must be zero, are nothing but the projections of their spins onto the spatial momentum. Therefore, there is a physically interesting conflict between the conservation of angular momentum and the selection rule (9.134) for chiralities: Angular momentum conservation implies that the spins are oriented in either of the two constellations sketched in Fig. 9.4. As the meson carries no lepton number, the fermions in the final state are antiparticles of one another. The interaction is of the type of (9.132), more specifically, in the present example we have $a = 1$ and $b = 0$. This interaction favours the production of the pair with *opposite* chiralities. Therefore, the constellations shown in Fig. 9.4 which are imposed by angular momentum conservation, are suppressed by the factor $(m_i/E)^2$. Indeed, the decay $\pi^0 \to e^+e^-$ is extremely rare. The decay of the η-meson into a $\mu^+\mu^-$-pair is approximately $(m_\mu/m_e)^2$ times more frequent than the decay into an e^+e^--pair, in spite of the fact that the available phase space is smaller. If the masses of the fermions in the final state were exactly zero the decay would be strictly forbidden.

Example 9.7 Weak Interaction with Charged Currents

In the weak interaction which is mediated by exchange of virtual charged $W^\pm$-bosons, the coupling to pairs of fermions such as (e^-, ν_e), (μ^-, ν_μ), or (τ^-, ν_τ), is of the type of (9.132) with $a = 1$ and $b = -1$. More explicitly, the vertices on which physical amplitudes depend, are

$$\overline{\psi^{(\nu_e)}(x)}\gamma^\mu\big(\mathbb{1} - \gamma_5\big)\psi^{(e)}(x) \quad \text{and} \quad \overline{\psi^{(e)}(x)}\gamma^\mu\big(\mathbb{1} - \gamma_5\big)\psi^{(\nu_e)}(x)$$

with analogous expressions for the other two lepton families. As here the two particles are created or annihilated with different electric charges, one says that the interaction takes place via *charged currents*. Of course, the overall electric charge must be conserved in the scattering or the decay process. For instance, this requirement is fulfilled in the decays

$$\pi^+ \to \mu^+ + \nu_\mu, \pi^+ \to e^+ + \nu_e, \; \pi^- \to \mu^- + \overline{\nu_\mu}, \; \pi^- \to e^- + \overline{\nu_e},$$

For neutrinos which are (nearly or) exactly massless, chirality is (practically) identical with helicity. Therefore, neutrinos are always produced lefthanded, antineutrinos are produced righthanded. As the decaying meson has spin zero the same arguments concerning conservation of angular momentum apply here, as in the previous example. Thus, in the decay of the π^+ the charged partner is created with longitudinal polarization -1, while in the decay of the π^- it is created with longitudinal polarization $+1$, in conflict with the selection rule for vector/axial vector coupling which favours the opposite polarizations. Without knowing the details of the dynamics these simple arguments suffice to predict that the probabilities for the decays $\pi^+ \to e^+\nu_e$ and $\pi^+ \to \mu^+\nu_\mu$ have the approximate ratio

$$\frac{\Gamma(\pi^+ \to e^+\nu_e)}{\Gamma(\pi^+ \to \mu^+\nu_\mu)} \approx \frac{m_e^2}{m_\mu^2} = \left(\frac{0.511}{105.658}\right)^2 = 2.34 \times 10^{-5}.$$

Though of the right order of magnitude, this is not quite correct as yet. On the one hand, comparing to the general formula (8.37) there is a factor $\kappa = (M^2 - m^2)/(2M)$ with $M = m_\pi$ and $m = m_e$ or $m = m_\mu$. On the other hand, the decay probability contains the absolute square of the T-matrix element. The two factors taken together yield a factor proportional to $(1 - m^2/M^2)^2$, so that one expects the ratio

$$\frac{\Gamma(\pi^+ \to e^+\nu_e)}{\Gamma(\pi^+ \to \mu^+\nu_\mu)} \approx \frac{m_e^2}{m_\mu^2}\left(\frac{1 - (m_e/m_\pi)^2}{1 - (m_\mu/m_\pi)^2}\right)^2 = 1.28 \times 10^{-4}. \qquad (9.135)$$

Indeed, this estimate lies very close to the experimental result which is 1.23×10^{-4}. In spite of the fact that the electronic decay offers the larger phase space it is strongly suppressed as compared to the muonic decay. If both the ν_e and the electron were strictly massless the decay $\pi^+ \to e^+\nu_e$ would not be possible.

Like the electromagnetic interaction the strong interactions are characterized by pure vector couplings. In the weak interaction via charged currents the $W^\pm$-bosons couple by the term (9.132) with $a = -b = 1$, in the weak interaction via neutral currents the Z^0-boson couples by similar terms, but with different coefficients a and b which depend on the external fermions at the vertex. In all of these interactions chiral fields play an important dynamical role. In particular, the selection rules worked out above apply to the basic vertices.

Scalar and Pseudoscalar Couplings: We conclude this section with a few remarks on those cases whose vertices contain sesquilinear terms of the kind

$$\overline{\psi(x)}\Gamma\psi(x) \qquad \text{with} \qquad \Gamma = c\,\mathbb{1} + \mathrm{i}\,d\gamma_5\,. \tag{9.136}$$

For example, the mass terms in the Lagrange density (9.58) are of this type, with $c = 1$ and $d = 0$, and so are the Yukawa couplings to *scalar* or to *pseudoscalar* (spin-0) particles whose vertices have, in the first case, $(c = 1, d = 0)$, in the second case, $(c = 0, d = 1)$. As for vector and axial vector couplings there are selection rules for chiralities but these are different from, i.e. opposite to those of the previous case.

If one decomposes the quantum fields (9.84) and (9.85) into their right chiral and left chiral components, $\psi = P_+\psi + P_-\psi$ and $\overline{\psi} = \overline{\psi}P_- + \overline{\psi}P_+$, then, by the analysis carried out above, one has

$$P_{+/-}\psi(x)$$
$$= \frac{1}{(2\pi)^{3/2}}\sum_{r=1}^{2}\int\frac{\mathrm{d}^3p}{2E_p}\left[a^{(r)}(\boldsymbol{p})\,u^{(r)}_{R/L}(\boldsymbol{p})\,\mathrm{e}^{-\mathrm{i}px} + b^{(r)\dagger}(\boldsymbol{p})\,v^{(r)}_{L/R}(\boldsymbol{p})\,\mathrm{e}^{\mathrm{i}px}\right],$$

$$\overline{\psi(x)}P_{+/-}$$
$$= \frac{1}{(2\pi)^{3/2}}\sum_{r=1}^{2}\int\frac{\mathrm{d}^3p}{2E_p}\left[a^{(r)\dagger}(\boldsymbol{p})\,\overline{u^{(r)}_{L/R}(\boldsymbol{p})}\,\mathrm{e}^{-\mathrm{i}px} + b^{(r)}(\boldsymbol{p})\,\overline{v^{(r)}_{R/L}(\boldsymbol{p})}\,\mathrm{e}^{\mathrm{i}px}\right].$$

As the two terms in (9.136) commute with γ_5 there remain only two nonvanishing terms which read

$$\overline{\psi(x)}\Gamma\psi(x) = \overline{\psi(x)}P_+\Gamma P_+\psi(x) + \overline{\psi(x)}P_-\Gamma P_-\psi(x)\,.$$

From these expressions and in comparison with the case of (9.132), discussed above, one extracts the following rule:

Selection Rule for Scalar/Pseudoscalar Coupling: An interaction to which two fermions couple by the operator

$$\overline{\psi^{(k)}(x)}\big(c\,\mathbb{1} + \mathrm{i}\,d\gamma_5\big)\psi^{(i)}(x)\,, \tag{9.137}$$

changes the chirality when one of the particles is incoming, the other is outgoing. If a particle-antiparticle pair is created or annihilated then this happens with equal helicities.

This rule is illustrated in Fig. 9.5 by some typical vertices. If one agrees on marking fermion lines by arrows in the direction of the flux of negative charge, or, what amounts to the same, if

- for *incoming particles* the arrow points *towards* the vertex,
- for *out*going *particles* the arrow points *away* from the vertex,

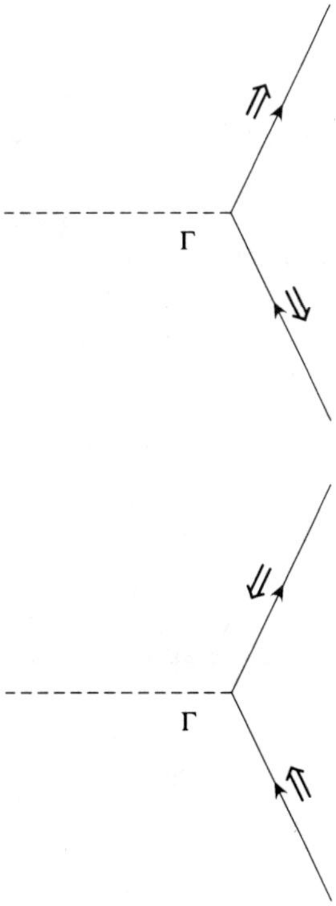

Fig. 9.5 A coupling which contains an arbitrary linear combination of the operators $1\!\!1$ and γ_5, correlates the chiralities of external lines as shown here. The *dashed lines* symbolize a spin-0 particle which may be a scalar or a pseudoscalar. If the fermions are strictly massless then the *arrows* represent helicities

– for *in*coming *antiparticles* the arrow points *away* from the vertex,
– for *out*going *antiparticles* the arrow points *towards* the vertex,
then the rule is: the chirality always changes at vertices with scalar or pseudoscalar coupling.

To illustrate this case suppose the interaction in Example 9.7 contained the operator $c\,1\!\!1 + 1\!\!1\gamma_5$, instead of $\gamma^\mu(a\,1\!\!1 - b\gamma_5)$. In this case the branching ratio (9.135) would not be suppressed by the factor $(m_e/m_\mu)^2$ and, hence, would have the value 5.49, thus more than four orders of magnitude larger than what one finds in experiment!

9.4 When is the Dirac Equation a One-Particle Theory?

After having demonstrated that the Dirac equation in its interpretation differs markedly from the Schrödinger equation and that, strictly speaking, it can only be understood in its quantized form as a theory of particles and antiparticles, one might be surprised if, in spite of this experience, in this section we attempt to use it as a one-particle theory. There are good reasons for this which refer to concrete physical situations.

In many problems of atomic and molecular physics where binding energies are generally small as compared to rest masses and where relativistic corrections of the results of Schrödinger theory are still small, it is perfectly legitimate to neglect degrees of freedom related to antiparticles. For instance, it is reasonable to evaluate effects of genuine radiative corrections (these are effects of higher order in α which are predicted by quantum electrodynamics) with reference to the hydrogen spectrum as obtained from the one-particle Dirac equation into which the Coulomb potential is inserted. Similarly, the Dirac equation with external potential is well suited for the analysis of hydrogen-like atoms in which the electron is replaced by a μ^-, or an antiproton $\overline{p}$. Formally speaking, this approach is one in which one restricts the spectrum of the Dirac operator with a potential to positive energies[10] and in which the degrees of freedom involving virtual electron-positron pairs are neglected. In this approach one must keep in mind that a bound state with binding energy B has the total energy $E = m - B$ which stays positive as long as $B \ll m$.

9.4.1 Separation of the Dirac Equation in Polar Coordinates

There is an elegant method, due to Dirac, which allows to reduce the Dirac equation to a differential equation in the radial variable. One starts from the Hamiltonian form (9.72) with an external, spherically symmetric Coulomb potential $U(r)$. Since $U(r) = q\Phi(r)$ is proportional to the time component of $A^\mu = (\Phi(r), \mathbf{0})$ which was introduced into (9.62) by the principle of minimal coupling, the potential $U(r)$ appears multiplied by γ^0. In passing to the form (9.70) of the Dirac equation an additional factor γ^0 comes in so that, by $(\gamma^0)^2 = \mathbb{1}_4$, the potential in (9.72) is multiplied by the unit matrix $\mathbb{1}_4$.[11] If one restricts solutions to stationary states with positive energy,

$$\Psi(t, \boldsymbol{x}) = \mathrm{e}^{-\mathrm{i}Et}\psi(\boldsymbol{x}),$$

then the following differential equation has to be solved:

$$E\psi(\boldsymbol{x}) = \Big(-\mathrm{i}\boldsymbol{\alpha}\cdot\nabla + U(r)\mathbb{1} + m\beta\Big)\psi(\boldsymbol{x}), \tag{9.138}$$

the matrices $\boldsymbol{\alpha}$ and β being defined in (9.71). Besides these matrices one defines three more matrices,

$$\boldsymbol{S} := (\gamma_5)_{\mathrm{St}}\boldsymbol{\alpha} = \begin{pmatrix} \boldsymbol{\sigma} & 0 \\ 0 & \boldsymbol{\sigma} \end{pmatrix}. \tag{9.139}$$

The gradient operator is decomposed into radial part and orbital angular momentum by

$$\nabla = \widehat{\boldsymbol{r}}(\widehat{\boldsymbol{r}}\cdot\nabla) - \widehat{\boldsymbol{r}}\times(\widehat{\boldsymbol{r}}\times\nabla) = \widehat{\boldsymbol{r}}(\widehat{\boldsymbol{r}}\cdot\nabla) - \mathrm{i}\frac{1}{r}\widehat{\boldsymbol{r}}\times\boldsymbol{\ell}\,.$$

---

[10] see e.g. G.E. Brown and D.G. Ravenhall, Proc. Roy. Soc. London A **208**, 552, 1952.

[11] If, in turn, one had a potential that is a genuine Lorentz *scalar* then this would be multiplied by β, and, hence, would be added to the mass m. Such scalar potentials are used in the description of strong interactions in, e.g., antiprotonic atoms.

The scalar product of $\widehat{\boldsymbol{r}}$ and ∇ (taken in this ordering!) equals $\partial/\partial r$, and one has

$$\boldsymbol{\alpha} \cdot \nabla = \boldsymbol{\alpha} \cdot \widehat{\boldsymbol{r}} \frac{\partial}{\partial r} - \mathrm{i} \frac{1}{r} \boldsymbol{\alpha} \cdot (\widehat{\boldsymbol{r}} \times \boldsymbol{\ell}) .$$

The second term is transformed by means of a well-known relation for Pauli matrices, see (4.37). For any two vectors $\boldsymbol{a}$ and $\boldsymbol{b}$ one has the relation

$$(\boldsymbol{\sigma} \cdot \boldsymbol{a})(\boldsymbol{\sigma} \cdot \boldsymbol{b}) = \boldsymbol{a} \cdot \boldsymbol{b} + \mathrm{i} \boldsymbol{\sigma} \cdot (\boldsymbol{a} \times \boldsymbol{b}) . \tag{9.140}$$

From this formula, and with $\boldsymbol{\alpha} = \gamma_5 \boldsymbol{S}$ and $\hat{\boldsymbol{r}} \cdot \boldsymbol{\ell} = 0$ one obtains

$$\mathrm{i} \boldsymbol{\alpha} \cdot (\widehat{\boldsymbol{r}} \times \boldsymbol{\ell}) = \mathrm{i} \gamma_5 \boldsymbol{S} \cdot (\widehat{\boldsymbol{r}} \times \boldsymbol{\ell}) = \gamma_5 (\boldsymbol{S} \cdot \widehat{\boldsymbol{r}})(\boldsymbol{S} \cdot \boldsymbol{\ell}) \quad \text{and}$$

$$\boldsymbol{\alpha} \cdot \nabla = \gamma_5 (\boldsymbol{S} \cdot \widehat{\boldsymbol{r}}) \left(\frac{\partial}{\partial r} - \frac{1}{r} (\boldsymbol{S} \cdot \boldsymbol{\ell}) \right) .$$

The trick is to introduce an operator which contains the whole information on the orbital angular momentum $\boldsymbol{\ell}$, the spin $\boldsymbol{s} = \boldsymbol{\sigma}/2$, and the total angular momentum $\boldsymbol{j} = \boldsymbol{\ell} + \boldsymbol{s}$. This goal is achieved by Dirac's operator K which is defined as follows:

$$K := \beta \left(\boldsymbol{S} \cdot \boldsymbol{\ell} + \mathbb{1}_4 \right) \equiv \begin{pmatrix} K^{(0)} & 0 \\ 0 & -K^{(0)} \end{pmatrix} . \tag{9.141}$$

Properties of Dirac's Operator:

1. The operator $K^{(0)} = \boldsymbol{\sigma} \cdot \boldsymbol{\ell} + \mathbb{1}_2$ which is contained in Dirac's operator K, can be expressed by the squares of all angular momenta involved. With $\boldsymbol{j} = \boldsymbol{\ell} + \boldsymbol{s}$ one has

$$K^{(0)} = \boldsymbol{\sigma} \cdot \boldsymbol{\ell} + \mathbb{1} = 2\boldsymbol{s} \cdot \boldsymbol{\ell} + \mathbb{1} = \boldsymbol{j}^2 - \boldsymbol{\ell}^2 - \boldsymbol{s}^2 + \mathbb{1} .$$

   Obviously, its eigenfunctions are the coupled spin-orbit states

$$|j\ell m\rangle = \sum_{m_\ell, m_s} (\ell, m_\ell; 1/2, m_s | jm) |\ell m_\ell\rangle |1/2, m_s\rangle$$

   where $|\ell m_\ell\rangle$ are spherical harmonics, and $|1/2, m_s\rangle$ are eigenstates of spin. Denoting its eigenvalues by $-\kappa$, i.e. $K^{(0)}|j\ell m\rangle = -\kappa |j\ell m\rangle$, these are equal to

$$\kappa = -j(j+1) + \ell(\ell+1) - \frac{1}{4} .$$

   As one sees, these eigenvalues take positive as well as negative values.
2. The square of the operator $K^{(0)}$ is obtained by means of the formula (9.140) which yields for $\boldsymbol{a} = \boldsymbol{b} = \boldsymbol{\ell}$ and with $\boldsymbol{\ell} \times \boldsymbol{\ell} = \mathrm{i}\boldsymbol{\ell}$

$$\left(\boldsymbol{\sigma} \cdot \boldsymbol{\ell} \right)^2 = \boldsymbol{\ell}^2 - \boldsymbol{\sigma} \cdot \boldsymbol{\ell} .$$

   Thus, the square of $K^{(0)}$ is found to be

$$\left(K^{(0)} \right)^2 = \boldsymbol{\ell}^2 + \boldsymbol{\sigma} \cdot \boldsymbol{\ell} + \mathbb{1} = \boldsymbol{j}^2 - \boldsymbol{s}^2 + \mathbb{1} .$$

   Inserting the eigenvalues one sees that the square of κ is related to j in a unique way, viz.

$$\kappa^2 = j(j+1) - \frac{3}{4} + 1 = \left(j + \frac{1}{2} \right)^2 .$$

Thus, the rules for calculating the eigenvalues ℓ and j from a given quantum number κ are

$$j = |\kappa| - \frac{1}{2}, \tag{9.142}$$

$$\text{for} \quad \kappa > 0 \quad \text{one has} \quad \ell = \kappa, \tag{9.143}$$

$$\text{for} \quad \kappa < 0 \quad \text{one has} \quad \ell = -\kappa - 1. \tag{9.144}$$

In other terms, once a positive or negative integer κ is given, the values of j and of ℓ are fixed. In particular, the sign of κ defines the coupling of spin and orbital angular momenta: If $\kappa > 0$, then $j = \ell - 1/2$, if $\kappa < 0$, then $j = \ell + 1/2$. Therefore, it is appropriate to denote the eigenstates of $K^{(0)}$ by $|\kappa, m\rangle$, instead of $|j\ell m\rangle$.

3. Thus, the eigenvalue and eigenfunctions of Dirac's operator K, (9.141), are as follows

$$K \begin{pmatrix} |\kappa, m\rangle \\ |-\kappa, m\rangle \end{pmatrix} = -\kappa \begin{pmatrix} |\kappa, m\rangle \\ |-\kappa, m\rangle \end{pmatrix}. \tag{9.145}$$

The number κ takes the set of values

$$\kappa = \pm 1, \pm 2, \pm 3, \ldots. \tag{9.146}$$

Introducing the operator K into the stationary Dirac equation (9.138) one obtains

$$E\psi(\boldsymbol{x}) = H\psi(\boldsymbol{x})$$
$$\text{with} \quad H = -\mathrm{i}\gamma_5 \boldsymbol{S} \cdot \widehat{\boldsymbol{r}} \left(\frac{\partial}{\partial r}\mathbb{1} + \frac{1}{r}\mathbb{1} - \beta \frac{1}{r} K \right) + U(r)\, \mathbb{1} + \beta m. \tag{9.147}$$

We show next that the operators H and K commute. For this purpose it suffices to calculate the commutator of $\gamma_5 \boldsymbol{S} \cdot \widehat{\boldsymbol{r}} = \boldsymbol{\alpha} \cdot \widehat{\boldsymbol{r}}$ with $K = \mathrm{diag}\big((\boldsymbol{\sigma} \cdot \boldsymbol{\ell} + \mathbb{1}), -(\boldsymbol{\sigma} \cdot \boldsymbol{\ell} + \mathbb{1})\big)$. We have

$$\left[\begin{pmatrix} 0 & \boldsymbol{\sigma} \cdot \widehat{\boldsymbol{r}} \\ \boldsymbol{\sigma} \cdot \widehat{\boldsymbol{r}} & 0 \end{pmatrix}, \begin{pmatrix} \boldsymbol{\sigma} \cdot \boldsymbol{\ell} + \mathbb{1} & 0 \\ 0 & -(\boldsymbol{\sigma} \cdot \boldsymbol{\ell} + \mathbb{1}) \end{pmatrix} \right] \equiv \begin{pmatrix} 0 & -A \\ A & 0 \end{pmatrix},$$
$$\text{where} \quad A = (\boldsymbol{\sigma} \cdot \widehat{\boldsymbol{r}})(\boldsymbol{\sigma} \cdot \boldsymbol{\ell} + \mathbb{1}) + (\boldsymbol{\sigma} \cdot \boldsymbol{\ell} + \mathbb{1})(\boldsymbol{\sigma} \cdot \widehat{\boldsymbol{r}}).$$

Making again use of the relation (9.140) and noting that $\widehat{\boldsymbol{r}} \cdot \boldsymbol{\ell} = 0$, one obtains

$$(\boldsymbol{\sigma} \cdot \widehat{\boldsymbol{r}})(\boldsymbol{\sigma} \cdot \boldsymbol{\ell}) + (\boldsymbol{\sigma} \cdot \boldsymbol{\ell})(\boldsymbol{\sigma} \cdot \widehat{\boldsymbol{r}}) = \frac{\mathrm{i}}{r} \boldsymbol{\sigma} \cdot \big(\boldsymbol{r} \times \boldsymbol{\ell} + \boldsymbol{\ell} \times \boldsymbol{r} \big),$$

where we took account of the fact that $r = |\boldsymbol{x}|$ and $\boldsymbol{\ell}$ commute. Note that the sum in brackets on the right-hand side is not equal to zero because $\boldsymbol{\ell}$ does not commute with $\boldsymbol{r}$. Rather, recalling the commutator $[\ell_i, x_j] = \mathrm{i}\varepsilon_{ijk} x_k$ (see Sect. 1.9.1), one has

$$\boldsymbol{\ell} \times \boldsymbol{r} = -\boldsymbol{r} \times \boldsymbol{\ell} + 2\mathrm{i}\boldsymbol{r}.$$

Inserting this formula one sees that the auxiliary operator A vanishes. Thus, the commutator of H and K equals zero.

As $[H, K] = 0$ there exist common eigenfunctions which can be chosen to be products of radial functions and spin-orbit states. One makes the ansatz

$$\psi_{\kappa,m}(\boldsymbol{x}) = \begin{pmatrix} g_\kappa(r) \; |\kappa, m\rangle \\ \mathrm{i}\, f_\kappa(r) \, |-\kappa, m\rangle \end{pmatrix} . \tag{9.148}$$

The factor i is a matter of convention. Its introduction is useful because the coupled differential equations for the radial functions $g_\kappa(r)$ and $f_\kappa(r)$ then are real. In order to derive these equations one must calculate the action of $\boldsymbol{\sigma} \cdot \widehat{\boldsymbol{r}}$ on the functions $|\kappa, m\rangle$. One finds

$$(\boldsymbol{\sigma} \cdot \widehat{\boldsymbol{r}})|\kappa, m\rangle = -|-\kappa, m\rangle . \tag{9.149}$$

The proof of this equation goes as follows: The operator $(\boldsymbol{\sigma} \cdot \widehat{\boldsymbol{r}})$ is a tensor operator of rank zero, and it is odd with respect to space reflection. Therefore, $(\boldsymbol{\sigma} \cdot \widehat{\boldsymbol{r}})|\kappa, m\rangle$ must be proportional to $|-\kappa, m\rangle$. Only this state has the same values of j and of m, and, upon comparison of (9.143) and of (9.144), differs by its parity from the state $|\kappa, m\rangle$,

$$(\boldsymbol{\sigma} \cdot \widehat{\boldsymbol{r}})|\kappa, m\rangle = \alpha|-\kappa, m\rangle .$$

From (9.140) one sees that $(\boldsymbol{\sigma} \cdot \widehat{\boldsymbol{r}})^2 = 1$, the norm of the state $(\boldsymbol{\sigma} \cdot \widehat{\boldsymbol{r}})|\kappa, m\rangle$ is equal to 1,

$$\langle\kappa, m|(\boldsymbol{\sigma} \cdot \widehat{\boldsymbol{r}})^2|\kappa, m\rangle = 1 = |\alpha|^2 .$$

In order to determine the coefficient α whose absolute value is 1, it suffices to evaluate the relation between $|\kappa, m\rangle$ and $|-\kappa, m\rangle$ in a special case. We choose $\widehat{\boldsymbol{r}} = \widehat{\boldsymbol{e}}_3$ and $\theta = 0$. Then one has

$$Y_{\ell m_\ell}(0, \phi) = \sqrt{\frac{2\ell + 1}{4\pi}} \delta_{m_\ell 0} .$$

Inserting the definition of the coupled state $|\kappa, m\rangle$, with $m_s = m$, one finds the condition

$$2m\sqrt{2\ell + 1}\big(\ell, 0; 1/2, m|jm\big) = \alpha\sqrt{2\bar{\ell} + 1}\big(\bar{\ell}, 0; 1/2, m|jm\big) ,$$

where the quantum numbers take the values $\ell = \kappa$, $\bar{\ell} = \ell - 1$, $j = \ell - 1/2$ if κ is positive, while they take the values $\ell = (-\kappa) - 1$, $\bar{\ell} = \ell + 1$ and $j = \ell + 1/2$ if κ is negative. The numerical values of the Clebsch-Gordan coefficients $\big(\ell, m_\ell; 1/2, m_s|jm\big)$ are given in the table (note that they can be calculated directly, cf. Exercise 9.7),

$j \backslash m_s$	$\frac{1}{2}$	$-\frac{1}{2}$
$\ell + \frac{1}{2}$	$\sqrt{\frac{\ell+1/2+m}{2\ell+1}}$	$\sqrt{\frac{\ell+1/2-m}{2\ell+1}}$
$\ell - \frac{1}{2}$	$-\sqrt{\frac{\ell+1/2-m}{2\ell+1}}$	$\sqrt{\frac{\ell+1/2+m}{2\ell+1}}$

One now takes $m_\ell = 0$, $m_s = m = \pm 1/2$ and confirms that for $\kappa > 0$ as well as for $\kappa < 0$ one obtains the same value $\alpha = -1$. This proves the relation (9.149).

Insert now (9.148) into (9.147) and use the relation (9.149), to obtain a system of two coupled, homogeneous, and real differential equations for the radial functions, viz.

$$f'_\kappa(r) = \frac{\kappa - 1}{r} f_\kappa(r) - \big(E - U(r) - m\big) g_\kappa(r) , \tag{9.150}$$

$$g'_\kappa(r) = -\frac{\kappa + 1}{r} g_\kappa(r) + \big(E - U(r) + m\big) f_\kappa(r) . \tag{9.151}$$

Two limiting cases are of special interest for physics: The limit of energies which are very *large* as compared to the rest mass, and the limit of energies which are very *close* to the rest mass.

(i) If $E \gg m$ so that the mass term in (9.150), (9.151) can be neglected, the radial functions fulfill the symmetry relations

$$g_{-\kappa}(r) = f_\kappa(r) , \quad f_{-\kappa}(r) = -g_\kappa(r) . \tag{9.152}$$

The upper and the lower components in (9.148) are of the same order of magnitude. This limit is relevant, for example, when one studies the scattering of electrons or muons on nuclei at energies which are large as compared to their rest mass.

(ii) In weakly bound atoms the binding energy B in $E = m - B$ is small as compared to m, i.e. $|E - m| \ll m$. As the potential is small, too, the second term on the right-hand side of (9.150) is suppressed by a factor $B/(2m)$ as compared to the second term on the right-hand side of (9.151). One concludes that $f_\kappa(r)$ must be sizeably smaller than $g_\kappa(r)$. Indeed, in the nonrelativistic limit of the hydrogen atom the "large" component $g_\kappa(r)$ tends to the corresponding radial function of the Schrödinger equation while the "small" component $f_\kappa(r)$ tends to zero.

9.4.2 Hydrogen-like Atoms from the Dirac Equation

In this section it is shown that the radial equations (9.150), (9.151), with the potential energy of point-like charges

$$U(r) = -\frac{Z\alpha}{r}$$

can be solved exactly. In particular, the case of bound states, i.e. of states with energy $E < m$, is of special interest.

For very large values of the radial variable r the potential $U(r)$ and the terms $(\kappa \pm 1)/r$ may be neglected. Then, from (9.150) one obtains approximately

$$f'_\kappa(r) \approx -(E - m) g_\kappa(r) = (m - E) g_\kappa(r) .$$

We differentiate (9.151), neglect all terms of the order $\mathcal{O}(1/r)$ and insert the approximate equation for $f'_\kappa(r)$, thus obtaining

$$g''_\kappa(r) \approx (m^2 - E^2) g_\kappa(r) .$$

As we are interested in bound states we have $E < m$, the difference $m^2 - E^2$ is real and positive. Thus, we define

$$\lambda := \sqrt{m^2 - E^2} \tag{9.153}$$

and note that the asymptotics of $g_\kappa(r)$ must be $g_\kappa(r) \sim e^{-\lambda r}$. The asymptotic form of $f_\kappa(r)$ is the same but is multiplied by the factor $\sqrt{(m-E)/(m+E)}$. Like in the nonrelativistic case it is useful to extract a factor $1/r$. Therefore, one substitutes as follows

$$g_\kappa(r) = \frac{1}{r}\, e^{-\lambda r} \sqrt{m+E}\, \big(u(r) + v(r)\big)\,,$$
$$f_\kappa(r) = \frac{1}{r}\, e^{-\lambda r} \sqrt{m-E}\, \big(u(r) - v(r)\big)\,.$$

Again in analogy to the case of the Schrödinger equation one defines a dimensionless variable

$$\varrho := 2\lambda r\,. \tag{9.154}$$

Passing from (g_κ, f_κ) to the functions u and v the latter are found to satisfy the system of differential equations

$$\frac{du}{d\varrho} = \left(1 - \frac{Z\alpha E}{\lambda\varrho}\right) u(\varrho) - \left(\frac{\kappa}{\varrho} + \frac{Z\alpha m}{\lambda\varrho}\right) v(\varrho)\,, \tag{9.155}$$

$$\frac{dv}{d\varrho} = \left(-\frac{\kappa}{\varrho} + \frac{Z\alpha m}{\lambda\varrho}\right) u(\varrho) + \frac{Z\alpha E}{\lambda\varrho} v(\varrho)\,. \tag{9.156}$$

To find out the behaviour of the solutions in the neighbourhood of $r = 0$ is not as simple as in the case of the Schrödinger equation where this behaviour was determined by the centrifugal term $\ell(\ell+1)/r^2$. One possibility could be to eliminate the "small" component $f_\kappa(r)$ and to derive a differential equation of *second* order for the "large" component $g_\kappa(r)$. One would then find that this second-order differential equation contains the potential $U(r)$ as well its square. The term $U^2(r) = (Z\alpha)^2/r^2$ competes with the centrifugal term and, therefore, the behaviour of the solutions at $r = 0$ is modified as compared to the nonrelativistic case.

There is a somewhat simpler approach, however. Let

$$u(\varrho) = \varrho^\gamma\, \phi(\varrho)\,, \qquad v(\varrho) = \varrho^\gamma\, \chi(\varrho) \quad \text{with} \quad \phi(0) \neq 0,\ \chi(0) \neq 0\,.$$

The exponent γ is determined from the system (9.155) and (9.156). At the point $\varrho = 0$ this system of differential equations yields a linear algebraic system for $\phi(0)$ and $\chi(0)$ which reads

$$\gamma\, \phi(0) = -\frac{Z\alpha E}{\lambda}\phi(0) - \left(\kappa + \frac{Z\alpha m}{\lambda}\right) \chi(0)$$
$$\gamma\, \chi(0) = \left(-\kappa + \frac{Z\alpha m}{\lambda}\right) \phi(0) + \frac{Z\alpha E}{\lambda}\chi(0)\,.$$

This is a homogeneous system which has a solution different from zero if and only if its determinant equals zero,

$$\det\begin{pmatrix}\gamma + Z\alpha E/\lambda & \kappa + Z\alpha m/\lambda \\ \kappa - Z\alpha m/\lambda & \gamma - Z\alpha E/\lambda\end{pmatrix} = 0\,,$$

that is to say, if $\gamma^2 - \kappa^2 + (Z\alpha)^2 = 0$. Of the two solutions of this equation only the positive square root

$$\gamma = \sqrt{\kappa^2 - (Z\alpha)^2}\,. \tag{9.157}$$

is admissible for the solutions to be regular at $r = 0$. Factoring out the behaviour of the radial functions at $r = 0$, the differential equations (9.155) and (9.156) yield a system of differential equations for $\phi(\varrho)$ and $\chi(\varrho)$ which read

$$\varrho\frac{\mathrm{d}\phi}{\mathrm{d}\varrho} = \left(\varrho - \left(\gamma + \frac{Z\alpha E}{\lambda}\right)\right)\phi(\varrho) - \left(\kappa + \frac{Z\alpha m}{\lambda}\right)\chi(\varrho) \tag{9.158}$$

$$\varrho\frac{\mathrm{d}\chi}{\mathrm{d}\varrho} = -\left(\kappa - \frac{Z\alpha m}{\lambda}\right)\phi(\varrho) - \left(\gamma - \frac{Z\alpha E}{\lambda}\right)\chi(\varrho)\,. \tag{9.159}$$

At this point it is instructive to pause for a while and to consider the state of the analysis from a physical perspective. This also allows to develop a strategy for what remains to be done. The functions $\phi(\varrho)$ and $\chi(\varrho)$ are the remnants of the unknown functions $f_\kappa(r)$ and $g_\kappa(r)$, respectively, after having extracted their asymptotics $\sim \mathrm{e}^{-\lambda r}$ as well as their behaviour near the origin $\sim r^\gamma$. By the obvious analogy to the corresponding nonrelativistic problem one expects $\phi(\varrho)$ and $\chi(\varrho)$ to be simple polynomials if $f_\kappa(r)$ and $g_\kappa(r)$ are to be L^2-normed eigenfunctions of H. That this is indeed true will be seen from the following calculation which, at the same time, also yields the eigenvalues of the energy and the eigenfunctions. For the sake of simplifying the notation a little, we temporarily set $Z\alpha E/\lambda =: \sigma$ and $Z\alpha m/\lambda =: \tau$. Differentiating, e.g., the differential equation (9.159) by ϱ, one obtains

$$\chi' + \varrho\chi'' = -(\kappa - \tau)\phi' - (\gamma - \sigma)\chi'\,.$$

The derivative ϕ' is taken from (9.158). One eliminates ϕ by making use once more of (9.159). These replacements yield

$$\varrho\chi'' + (2\gamma + 1 - \varrho)\chi' = \chi\left((\gamma - \sigma) + \frac{1}{\varrho}(-\gamma^2 + \sigma^2 + \kappa^2 - \tau^2)\right).$$

However, since $-\gamma^2 + \kappa^2 = (Z\alpha)^2$ and with $\sigma^2 - \tau^2 = -(Z\alpha)^2$ the second term on the right-hand side vanishes. There remains a differential equation of *second* order for the function $\chi(\varrho)$ alone which reads

$$\varrho\chi''(\varrho) + (2\gamma + 1 - \varrho)\chi'(\varrho) - \left(\gamma - \frac{Z\alpha E}{\lambda}\right)\chi(\varrho) = 0\,. \tag{9.160}$$

The equation (9.160) is nothing but Kummer's differential equation (cf. Appendix A.2.2)

$$zw''(z) + (c - z)w'(z) - a\,w(z) = 0\,,$$

whose solution regular at $z = 0$ is the well-known confluent hypergeometric function, $w(z) = {}_1F_1(a; c; z)$. Inserting the parameters (which are real here)

$$a \equiv \gamma - \frac{Z\alpha E}{\lambda}\,, \qquad c \equiv 2\gamma + 1\,.$$

the solution reads, up to a normalization factor,

$$\chi(\varrho) = {}_1F_1\big(\gamma - Z\alpha E/\lambda; 2\gamma + 1; \varrho\big)\,. \tag{9.161}$$

The other radial function $\phi(\varrho)$ is obtained in the same way by deriving the analogous differential equation of second order that it satisfies. One finds Kummer's differential equation also in this case. Alternatively and more simply, $\phi(\varrho)$ is obtained from (9.159) and the relation

$$z\ {}_1F_1'(a; c; z) + a\ {}_1F_1(a; c; z) = a\ {}_1F_1(a + 1; c; z)$$

for the derivative of the confluent hypergeometric function. This calculation yields

$$\phi(\varrho) = \frac{Z\alpha E/\lambda - \gamma}{\kappa - Z\alpha m/\lambda}\,{}_1F_1\big(\gamma + 1 - Z\alpha E/\lambda; 2\gamma + 1; \varrho\big)\,. \tag{9.162}$$

From this point on the arguments are the same as in Schrödinger theory for central field problems. The asymptotics of the function ${}_1F_1$

$${}_1F_1\big(a; c; \varrho\big) \sim \frac{\Gamma(c)}{\Gamma(c-a)}(-\varrho)^{-a} + \frac{\Gamma(c)}{\Gamma(a)}\,\mathrm{e}^{\varrho}\varrho^{a-c}$$

would ruin the exponential decay found above,

$$f_\kappa(r)\,,\ g_\kappa(r) \sim \mathrm{e}^{-\lambda r} = \mathrm{e}^{-\varrho/2}$$

if the second term could not be made to vanish. Simple inspection shows that this is possible if and only if a is equal to zero, or to a negative integer, i.e. if

$$\frac{Z\alpha E}{\lambda} - \gamma = n'\,, \quad n' \in \mathbb{N}_0\,, \quad \text{with } \lambda = \sqrt{m^2 - E^2}\,. \tag{9.163}$$

If this condition is fulfilled then, as expected, the function $\chi(\varrho)$ is a polynomial in ϱ. If $n' \geq 1$, then this is also true for the function $\phi(\varrho)$. The case $n' = 0$ is exceptional and must be analyzed separately. The function ${}_1F_1\big(1; 2\gamma + 1; \varrho\big)$ in (9.162) is no longer a polynomial and, as a consequence, grows too strongly at asymptotic values. At the same time the numerator of the overall factor on the right-hand side of (9.162) is equal to zero. In this case the function $\phi(\varrho)$ vanishes provided the *denominator* of the overall factor is different from zero. This is precisely what one has to check:

If $\gamma = Z\alpha E/\lambda$, then

$$\kappa^2 = \gamma^2 + (Z\alpha)^2 = (Z\alpha)^2\frac{\lambda^2 + E^2}{\lambda^2} = \Big(Z\alpha\frac{m}{\lambda}\Big)^2$$

and, therefore, $\kappa = \pm Z\alpha m/\lambda$. Choosing the positive value the denominator of the factor vanishes, for the negative value it does not. This implies a condition:
For $n' = 0$ only the negative value of κ is admissible.

One easily sees that the principal quantum number n of the nonrelativistic hydrogen atom is related to the quantum number n' by

$$n = n' + |\kappa| , \qquad n \in \mathbb{N} . \tag{9.164}$$

The condition (9.163) for the energy yields a formula for the eigenvalues of the Hamiltonian H,

$$E_{n|\kappa|} = m \left\{ 1 + \left(\frac{Z\alpha}{n - |\kappa| + \sqrt{\kappa^2 - (Z\alpha)^2}} \right)^2 \right\}^{-1/2} . \tag{9.165}$$

Here the quantum numbers n, κ, and $j = |\kappa| - 1/2$ take the values

$$\begin{aligned} & n \in \mathbb{N}; \\ & \kappa = \pm 1, \pm 2, \ldots, -n; \\ & j = \frac{1}{2}, \frac{3}{2}, \ldots, n - \frac{1}{2} . \end{aligned} \tag{9.166}$$

It is then easy to go back to the original radial functions $f_\kappa(r)$ and $g_\kappa(r)$, and, by making use of well-known formulae for integrals with exponentials, powers, and confluent hypergeometric functions, to normalize them according to

$$\int_0^\infty r^2 \, \mathrm{d}r \left(f_\kappa^2(r) + g_\kappa^2(r) \right) = 1 .$$

One finds the result

$$\begin{aligned} g_{n\kappa}(r) = {} & 2\lambda N(n,\kappa)\sqrt{m+E}\, \varrho^{\gamma-1}\, \mathrm{e}^{-\varrho/2} \\ & \Big\{ -(n - |\kappa|)\, {}_1F_1\left(-n + |\kappa| + 1; 2\gamma + 1; \varrho\right) \\ & + \left(\frac{Z\alpha m}{\lambda} - \kappa \right) {}_1F_1\left(-n + |\kappa|; 2\gamma + 1; \varrho\right) \Big\} , \end{aligned} \tag{9.167}$$

$$\begin{aligned} f_{n\kappa}(r) = {} & -2\lambda N(n,\kappa)\sqrt{m-E}\, \varrho^{\gamma-1}\, \mathrm{e}^{-\varrho/2} \\ & \Big\{ (n - |\kappa|)\, {}_1F_1\left(-n + |\kappa| + 1; 2\gamma + 1; \varrho\right) \\ & + \left(\frac{Z\alpha m}{\lambda} - \kappa \right) {}_1F_1\left(-n + |\kappa|; 2\gamma + 1; \varrho\right) \Big\} , \end{aligned} \tag{9.168}$$

the normalization constant being $N(n, \kappa)$ being given by

$$N(n,\kappa) = \frac{\lambda}{m} \frac{1}{\Gamma(2\gamma+1)} \left\{ \frac{\Gamma(2\gamma + n - |\kappa| + 1)}{2Z\alpha(Z\alpha m/\lambda - \kappa)\Gamma(n - |\kappa| + 1)} \right\}^{1/2} . \tag{9.169}$$

The radial variable and the parameters are

$$\begin{aligned} & \varrho = 2\lambda r , \\ & \lambda = \sqrt{m^2 - E_{n|\kappa|}^2} = \frac{Z\alpha m}{\sqrt{n^2 - 2(n - |\kappa|)(|\kappa| - \gamma)}} , \\ & \gamma = \sqrt{\kappa^2 - (Z\alpha)^2} . \end{aligned} \tag{9.170}$$

In summary, the problem of bound states in the hydrogen atom and in hydrogen-like atoms is solved completely. The states with positive energy which are in the continuum, are derived by the same procedure, see Remark 4 below.

Remarks

1. The binding energies (9.165) depend on the principal quantum number n, on the absolute value of κ, but not on its sign. This means that states with the same angular momentum j which belong to different pairs of orbital angular momenta $(\ell, \bar{\ell})$, are degenerate in energy. For example, the state with $(n = 2, \kappa = -1, j = 1/2)$ contains the orbital angular momenta $\ell = 0$ in the large component of (9.148) and $\bar{\ell} = 1$ in the small component. It corresponds to the nonrelativistic $2s_{1/2}$-state. In turn, the state with $(n = 2, \kappa = +1, j = 1/2)$ contains the orbital angular momenta $\ell = 1$ and $\bar{\ell} = 0$, and corresponds to the nonrelativistic $2p_{1/2}$-state.
   The state $(n = 2, \kappa = -2, j = 3/2)$ which is the analogue of the $2p_{3/2}$-state, has an energy higher than that of the "$2p_{1/2}$"-state. As $\kappa = +2$ is excluded, this state is nondegenerate. I have put the spectroscopic notation in quotation marks because the relativistic bound states are no longer eigenfunctions of orbital angular momentum. Indeed, they each contain two values, ℓ and $\bar{\ell} = \ell \pm 1$. However, as the component with the radial function $g_{n\kappa}(r)$ is large as compared to its partner $f_{n\kappa}(r)$, the value ℓ pertaining to the large component may be used to denote the state. One uses the nonrelativistic notation but should keep in mind what it stands for.
   In Fig. 9.6 we sketch the spectrum (9.165). Although the very strong degeneracy of the nonrelativistic spectrum is lifted partially, there remains the degeneracy of those levels which have the same principal quantum number and the same total angular momentum $j = |\kappa| - 1/2$.

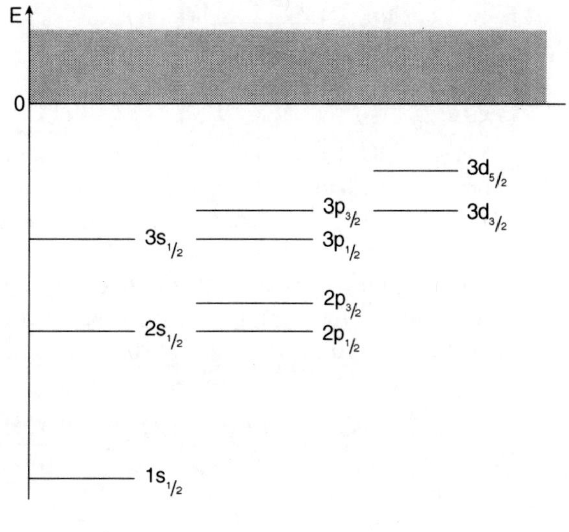

Fig. 9.6 Spectrum of binding energies in a relativistic hydrogen-like atom (not to scale). States with equal principal quantum numbers and equal angular momenta j stay degenerate

2. The weakly relativistic limit is attained by an expansion in terms of powers of $(Z\alpha)$. For example, for the energy (9.165) one finds
$$E_{n|\kappa|} \simeq m \left\{1 - \frac{(Z\alpha)^2}{2n^2} - \frac{(Z\alpha)^4}{2n^4}\left(\frac{n}{|\kappa|} - \frac{3}{4}\right)\right\} . \tag{9.171}$$
The first term in curly brackets is the rest mass, the second term yields the binding energy which is known from the nonrelativistic theory. The third term represents the first relativistic correction and is remarkable by the fact that it depends on $Z\alpha$ and on the angular momentum j, but is independent of the mass. Therefore, in muonic atoms and in electronic atoms the relativistic corrections to $E_{n|\kappa|}/m$ are the same. Keeping the principal quantum number n fixed, these corrections are seen to be larger for small values of j than for large values.
In the same limit the "small" component $f_{n\kappa}(r)$ tends to zero, the "large" component goes over into the corresponding nonrelativistic wave function. More precisely: For $\kappa > 0$ the component $g_{n\kappa}(r)$ goes over into the eigenfunction $y_{n\ell}(r)/r$ with $\ell = \kappa$. For $\kappa' < 0$ it goes over into $y_{n\bar{\ell}}(r)/r$ with $\bar{\ell} = |\kappa'| - 1$. In comparison of $\kappa = \ell$ and of $\kappa' = -(\ell + 1)$, this corresponds to the statement that the states with $(n, j = \ell - 1/2)$ and with $(n, j = \ell + 1/2)$, in this limit, have the same radial wave function.
3. Another case of practical importance is the one where the charge density $\varrho(r)$ is still spherically symmetric but is no longer singular in the origin. This happens when the nucleus of the atom has a finite spatial extension in which case its charge density can be expanded in a Taylor series around $r = 0$,
$$\varrho(r) = \varrho^{(0)} + \varrho^{(1)} r + \frac{1}{2!}\varrho^{(2)} r^2 + \mathcal{O}(r^3) . \tag{9.172}$$
The corresponding potential is then also regular at $r = 0$,
$$U(r) = -4\pi Z\alpha^2 \left\{ \int_0^\infty r' \, \mathrm{d}r' \, \varrho(r') - \frac{1}{6}\varrho^{(0)} r^2 \right.$$
$$\left. - \frac{1}{12}\varrho^{(1)} r^3 - \frac{1}{40}\varrho^{(2)} r^4 + \mathcal{O}(r^5) \right\} . \tag{9.173}$$
This potential has the shape of a parabola near the origin and, thus, is no longer competing with the centrifugal terms in the differential equations for $f(r)$ and $g(r)$. Therefore, the behaviour of these functions at $r = 0$ no longer is r^γ with γ as given in (9.157), but depends only on κ.
Following the example of (9.160) one may derive differential equations (of second order) for $f(r)$ and for $g(r)$ separately, by working on the sytem of equations (9.150), (9.151). If one does so one finds that these new equations contain the centrifugal terms, respectively,
$$\frac{\kappa(\kappa - 1)}{r^2} \quad \text{and} \quad \frac{\kappa(\kappa + 1)}{r^2} .$$

It is then appropriate to set

$$f(r) = r^{\alpha} \sum_{n=0}^{\infty} a_n r^n \,, \quad g(r) = r^{\beta} \sum_{n=0}^{\infty} b_n r^n \,, \qquad (a_0 \neq 0 \,, \, b_0 \neq 0) \,.$$

As a result one finds that the characteristic exponents α and β fulfill the equations

$$\alpha(\alpha+1) = \kappa(\kappa-1) \,, \quad \beta(\beta+1) = \kappa(\kappa+1) \,.$$

For each sign of κ one selects those solutions of the pairs $\alpha = (\kappa-1, -\kappa)$ and $\beta = (\kappa, -\kappa-1)$ which are positive or zero. This means, in more detail, that the function $g(r)$ starts like r^{κ} when $\kappa > 0$, and like $r^{-\kappa-1}$ when $\kappa < 0$. Likewise for $f(r)$: If $\kappa > 0$ it behaves like $r^{\kappa-1}$, if $\kappa < 0$ like $r^{-\kappa}$.
Thus, extracting their respective limiting forms near the origin,

$$\kappa > 0 \begin{cases} g_\kappa(r) = r^{\kappa} G_\kappa(r) \,, \\ f_\kappa(r) = r^{\kappa-1} F_\kappa(r), \end{cases} \tag{9.174}$$

$$\kappa < 0 \begin{cases} g_\kappa(r) = r^{-\kappa-1} F_\kappa(r) \,, \\ f_\kappa(r) = -r^{-\kappa} G_\kappa(r), \end{cases} \tag{9.175}$$

the radial functions $F_\kappa(r)$ and $G_\kappa(r)$ are seen to satisfy the following system of differential equations of first order

$$\begin{aligned} \frac{\mathrm{d}F_\kappa(r)}{\mathrm{d}r} &= \big(U(r) - E + m \,\mathrm{sign}\, \kappa\big) r G_\kappa(r) \\ \frac{\mathrm{d}G_\kappa(r)}{\mathrm{d}r} &= -\frac{1}{r} \left\{ \big(U(r) - E - m \,\mathrm{sign}\, \kappa\big) F_\kappa(r) + \big(2|\kappa|+1\big) G_\kappa(r) \right\} . \end{aligned} \tag{9.176}$$

This system must be solved with initial conditions

$$F_\kappa(0) = a_0 \,, \quad G_\kappa(0) = -\frac{U(0) - E - m \,\mathrm{sign}\, \kappa}{2|\kappa|+1} a_0 \,, \tag{9.177}$$

$$F'_\kappa(0) = 0 \quad G'_\kappa(0) = 0 \,. \tag{9.178}$$

A system of differential equations of the kind of (9.176) is well suited for numerical integration, in those cases where no analytical solution is found. The reader will find some hints at practical methods of integration and of determining eigenvalues of bound states, e.g., in [Scheck (2012), Sect. 2.6.3]. Note that these procedures are relevant for the analysis of muonic atoms.

4. There is no particular problem in solving exactly the radial equations (9.150) and (9.151) for a Coulomb-Potential $U(r) \propto 1/r$ and for *positive* energy, cf. [Scheck (2012)]. Scattering states in the field of a pointlike charge are needed in phase shift analyses of scattering of electrons on potentials with infinite range. We note, however, that unlike the Schrödinger equation the Dirac equation does not yield the classical Rutherford cross section.

9.5 Path Integrals with Fermionic Fields

At first sight the path integral for fermionic fields formally looks similar to the path integral for bosonic fields. By analogy to (7.193) and to (7.183) the action for a Dirac field and its adjoint reads

$$S = \int \mathrm{d}^4x \; \mathcal{L}(\psi, \eta, x) \quad \text{with} \tag{9.179a}$$

$$\mathcal{L} = \overline{\psi}\left(\frac{\mathrm{i}}{2}\overset{\leftrightarrow}{\partial}_{\mu}\gamma^{\mu} - m\right)\psi + \mathcal{L}_{\mathrm{W}} + \overline{\eta}\psi + \overline{\psi}\eta \,. \tag{9.179b}$$

In this Lagrange density the term η takes the role of an external source. The Euler-Lagrange equations that follow from the Lagrange density (9.179b) yield the Dirac equation

$$(\mathrm{i}\partial\!\!\!/ - m)\,\psi = \eta \tag{9.180}$$

and its hermitian conjugate, up to possible additional terms which come from the interaction Lagrange density $\mathcal{L}_W$. This implies that η is a four-component spinor field too, whose adjoint is defined in analogy to $\bar{\psi}$, that is,

$$\overline{\eta} = \eta^{\dagger}\gamma^{0} \,.$$

It is less obvious how to make sense out of an integration $\int \mathcal{D}\psi$ over a Dirac field since neither ψ nor η are bosonic fields. Even though the Dirac field is not quantized explicitly in this formalism, both ψ and the source η must be Grassmann-valued fields. What this means is best explained by means of a simple one-dimensional example.

Consider an n-dimensional real vector space V, and let $\Lambda^*(V)$ be the algebra of exterior forms on V. The exterior, or Cartan, derivative d is an antiderivation of degree 1,

$$\mathrm{d} \;:\; \Lambda^k(V) \longrightarrow \Lambda^{k+1}(V)$$

named "anti" because it fulfills a graded Leibniz rule. With $\omega \in \Lambda^k$ and $\chi \in \Lambda^l$ the exterior derivative of the product $\omega \wedge \chi$ is given by

$$\mathrm{d}\,(\omega \wedge \chi) = (\,\mathrm{d}\omega) \wedge \chi + (-)^k \omega \wedge (\,\mathrm{d}\chi) \,, \tag{9.181}$$

with the characteristic sign factor determined by the degree of the first form which is commuted with the derivative.

More generally, one calls an operation of this kind an *antiderivation of degree r* if it has the following properties:

(i) It maps Λ^k onto Λ^{k+r},

$$T : \Lambda^k(V) \longrightarrow \Lambda^{k+r}(V) \,. \tag{9.182a}$$

(ii) It acts linearly, that is, with $\omega, \chi \in \Lambda^*$ and $\lambda, \mu \in \mathbb{C}$ one has

$$T\,(\lambda\omega + \mu\chi) = \lambda T\omega + \mu T\chi \,. \tag{9.182b}$$

(iii) It fulfills the graded Leibniz rule (9.181)

$$T(\omega \wedge \chi) = (T\omega) \wedge \chi + (-)^k \omega \wedge (T\chi) , \tag{9.182c}$$

where ω is a k-form.

As a special case we consider antiderivations θ_i of degree (-1). These act between the spaces $\Lambda^k(V)$ as follows

$$\Lambda^n(V) \to_\theta \to \Lambda^{n-1}(V) \to_\theta \ldots \to \Lambda^1(V) \to_\theta \to \Lambda^0(V) .$$

Let $\{\omega_k\}$ be a set of base one-forms in the space Λ^1. The set which is dual to it is the set of antiderivations θ_i, $i = 1, 2, \ldots, n$, and is defined by their action on zero-forms and on one-forms by

$$f \in \Lambda^0 : \quad \theta_i f = 0; \quad \omega_k \in \Lambda^1 : \quad \theta_i \omega_k = \delta_{ik} . \tag{9.183}$$

An arbitrary smooth p-form $\chi \in \Lambda^p$ is expanded in terms of the basis $\omega_{k_1} \wedge \omega_{k_2} \wedge \ldots \wedge \omega_{k_p}$. The action of the antiderivation θ_i on χ follows from its action on the base forms of Λ^p,

$$\theta_i \left(\omega_{k_1} \wedge \omega_{k_2} \wedge \ldots \wedge \omega_{k_p}\right) = \delta_{ik_1} \omega_{k_2} \wedge \ldots \wedge \omega_{k_p}$$
$$- \delta_{ik_2} \omega_{k_1} \wedge \omega_{k_3} \wedge \ldots \wedge \omega_{k_p} + \cdots + (-)^{p-1} \delta_{ik_p} \omega_{k_1} \wedge \omega_{k_2} \wedge \ldots \wedge \omega_{k_{p-1}} .$$

The one-forms θ_i may act in series such as $\theta_{i_2} \circ \theta_{i_1}$ or $\theta_{i_3} \circ \theta_{i_2} \circ \theta_{i_1}$ up to

$$\theta_n \circ \theta_{n-1} \circ \cdots \circ \theta_1 \equiv \int \mathrm{d}\omega . \tag{9.184}$$

One verifies easily that the so-defined operator $\int \mathrm{d}\theta$ gives zero when applied to any p-form whose degree p is smaller than n. When it is applied to the (unique) base n-form it yields the value 1,

$$\int \mathrm{d}\omega \, \omega = \quad \text{for} \quad \omega \in \Lambda^p \quad \text{if} \quad p < n ,$$
$$\int \mathrm{d}\omega \, (\omega_1 \wedge \cdots \wedge \omega_n) = 1 .$$

As one sees the operation $\int \mathrm{d}\theta$ has the properties of an integral. This opens up a line that one can explore further. For example, given a linear combination of exterior forms such as

$$F(\omega) = c_0 + \sum_{i=1}^{n} c_i \, \omega_i + \sum_{i<j} c_{ij} \, \omega_i \wedge \omega_j + \ldots + c_{1\ldots n} \, \omega_1 \wedge \cdots \wedge \omega_n$$

its integral (9.184) yields the coefficient multiplying the form with the highest degree,

$$\int \mathrm{d}\omega \, F(\omega) = c_{1\ldots n} .$$

How does this integral behave under a change of basis and under a translation of its variables?

Under a change of basis for the one-forms in Λ^1

$$\omega_i = \sum_{k=1}^{n} a_{ik} \chi_k \,, \quad \text{or, in compact notation,} \quad \omega = \mathbf{A}\chi \,. \tag{9.185}$$

also the antiderivations θ_i are transformed and are replaced by other antiderivations γ_k. Defining the integral like in (9.184), one has

$$\omega_1 \wedge \omega_2 \wedge \cdots \wedge \omega_n = (\det \mathbf{A})\, \chi_1 \wedge \chi_2 \wedge \cdots \wedge \chi_n$$

and one obtains the integral

$$\int \mathrm{d}\chi \ (\omega_1 \wedge \omega_2 \wedge \cdots \wedge \omega_n) = \det \mathbf{A} \,.$$

This yields the following equations,

$$c_{12\ldots n} \int \mathrm{d}\chi \ (\omega_1 \wedge \cdots \wedge \omega_n) = \int \mathrm{d}\chi \ F(\omega) = \int \mathrm{d}\chi \ F(\mathbf{A}\chi)$$
$$= (\det \mathbf{A})\, c_{12\ldots n} \int \mathrm{d}\chi \ \chi_1 \wedge \cdots \wedge \chi_n = (\det \mathbf{A}) \int \mathrm{d}\chi \cdot F(\chi)$$

These equalities show that the change of basis has the following consequence for the integral

$$\int \mathrm{d}\chi \ F(\mathbf{A}\chi) = (\det \mathbf{A}) \int \mathrm{d}\omega \ F(\omega) \,. \tag{9.186}$$

This result is quite remarkable when compared to the analogous transformation formula for ordinary integrals. For such integrals, with $x = \mathbf{A}y$ and $\partial x/\partial y = \det \mathbf{A}$ one would have found

$$\int \mathrm{d}y \ f(\mathbf{A}y) = (\det \mathbf{A})^{-1} \int \mathrm{d}x \ f(x) \,.$$

The factor mutipling the integral in (9.186) is the inverse of the Jacobi determinant that occurs in ordinary integrals over commuting variables.

A translation of Grassmannian variables can be handled as follows. One embeds the vector space V in a larger vector space W whose dimension is $\dim W = N$, with $N > n$. Consider the exterior algebra $\Lambda^*(W)$ on this larger space. Construct a basis of one-forms for $\Lambda^1(W)$ from the previous one for $\Lambda^1(V)$, supplemented by $(N-n)$ additional base forms,

$$(\omega_1, \omega_2, \ldots, \omega_n, \chi_{n+1}, \ldots \chi_N) \,.$$

If ω is an arbitrary exterior form represented in terms of the ω_i, and χ a form expanded in terms of the χ_k, then one has

$$\theta_n \circ \theta_{n-1} \circ \cdots \circ \theta_1 \ (\omega + \chi) = \theta_n \circ \theta_{n-1} \circ \cdots \circ \theta_1 \ \omega \,.$$

There follows the generalized translation formula

$$\int \mathrm{d}\omega \ F(\omega + \chi) = \int \mathrm{d}\omega \ F(\omega) \,. \tag{9.187}$$

This finite dimensional example suggests a possibility to assign a well-defined result to a formal integral of the type

$$I := \int \mathcal{D}\overline{\psi}\,\mathcal{D}\psi\,\exp\left\{\sum_{r,s}\overline{\psi}_r C_{rs}\psi_s + \sum_r\left(\overline{\psi}_r\eta_r + \overline{\eta}_r\psi_r\right)\right\}. \tag{9.188}$$

For this purpose consider first a simpler Gaussian integral:

$$\begin{aligned}\int \mathrm{d}\bar{\omega}\,\mathrm{d}\omega\,\exp\left\{\sum \lambda_i\bar{\omega}_i\omega_i\right\} &= \int \mathrm{d}\bar{\omega}\,\mathrm{d}\omega\,\exp\{\lambda_1\bar{\omega}_1\omega_1\}\cdots\exp\{\lambda_n\bar{\omega}_n\omega_n\}\\ &= \int \mathrm{d}\bar{\omega}\,\mathrm{d}\omega\,(1+\lambda_1\bar{\omega}_1\omega_1)(1+\lambda_2\bar{\omega}_2\omega_2)\cdots(1+\lambda_n\bar{\omega}_n\omega_n)\\ &= \prod_{i=1}^{n}\lambda_i\,.\end{aligned}$$

The integral (9.188) is seen to be of this type, the matrix $\mathbf{C}$ having the property

$$\gamma^0\mathbf{C}^\dagger\gamma^0 = \mathbf{C}\,.$$

Using a matrix notation one calculates

$$\begin{aligned}\overline{(\psi+\mathbf{C}^{-1}\eta)}\mathbf{C}\left(\psi+\mathbf{C}^{-1}\eta\right) &= \left(\overline{\psi}+\eta^\dagger(\mathbf{C}^{-1})^\dagger\gamma^0\right)\mathbf{C}\left(\psi+\mathbf{C}^{-1}\eta\right)\\ &= \left(\overline{\psi}+\overline{\eta}\mathbf{C}^{-1}\right)\mathbf{C}\left(\psi+\mathbf{C}^{-1}\eta\right) = \overline{\psi}\mathbf{C}\psi+\overline{\eta}\psi+\overline{\psi}\eta+\overline{\eta}\mathbf{C}^{-1}\eta\,.\end{aligned}$$

Except for the last term which is independent of ψ this is seen to be the argument of the exponential function in (9.188). Making use of the integration formula (9.186) and of the translation formula (9.187), and integrating over the shifted variable $(\psi+\mathbf{C}^{-1}\eta)$ and its adjoint, one obtains the result

$$I = (\det\mathbf{C})\,\exp\{-\overline{\eta}\mathbf{C}^{-1}\eta\}\,. \tag{9.189}$$

Comparison with the formula (7.196) which holds for bosonic fields shows that the difference is only in the position of the determinant: In the bosonic case it appears in the denominator, while in the fermionic case it appears in the numerator.

In concluding this section we study the generating functional for a free fermionic field. Let

$$F\,[\overline{\eta},\eta] := \frac{W^{(0)}[\overline{\eta},\eta]}{W^{(0)}[0,0]}\,, \tag{9.190}$$

where the functional $W^{(0)}[\overline{\eta},\eta]$ is given by the formal expression

$$\begin{aligned}&W^{(0)}[\overline{\eta},\eta]\\ &= \int \mathcal{D}\overline{\psi}\,\mathcal{D}\psi\,\exp\left\{\mathrm{i}\int \mathrm{d}^4x\,\left[\overline{\psi(x)}(\mathrm{i}\not{\partial}-m)\psi(x)+\overline{\eta(x)}\psi(x)+\overline{\psi(x)}\eta(x)\right]\right\}.\end{aligned}$$

On the basis of the example worked out above and adapting the calculation in Sect. 2.8.4 to the present case one finds the following result for this integral

$$W^{(0)}[\overline{\eta},\eta] = \det(\mathrm{i}\not{\partial}-m)\exp\left\{\frac{1}{2}\int \mathrm{d}^4x\int \mathrm{d}^4y\,\overline{\eta(x)}S_F(x-y)\eta(y)\right\}. \tag{9.191}$$

Here too, it is remarkable that even though the Dirac field is Grassmann-valued, it was not (canonically) quantized, and, yet, one rediscovers the Feynman propagator (9.110). Note, finally, that all higher Green functions are obtained by means of functional derivatives of the functional $F[\overline{\eta},\eta]$.

10 Elements of Quantum Electrodynamics and Weak Interactions

▶ **Introduction** Quantum field theory in its application to electroweak and strong interactions has two rather different facets: A pragmatic, empirical one, and an algebraic, systematical one. The pragmatic approach consists in a set of rules and formal calculational procedures which are extremely successful in their application to concrete physical processes, but rest on mathematically shaky ground. The mathematically rigorous approach, in turn, is technically difficult and not very useful, from a practical point of view, for reaching results which can be compared with phenomenology. Generally speaking, quantum field theory quickly becomes rather technical if one wants to understand it in some depth, and goes far beyond the scope of a textbook such as this one. We refer to the many excellent monographs on this topic some of which are listed in the bibliography.

This chapter introduces the most important calculational techniques of quantum electrodynamics and illustrates its impressive practical successes when confronted with experiment. Furthermore, it is known that quantum electrodynamics is only one aspect of what is called the *standard model of electroweak and strong interactions.* A detailed description of this model and of its quantization would go far beyond the size of this book. Nonetheless, this chapter concludes with some remarks on the standard model and several examples of weak interactions in tree approximation.

10.1 *S*-Matrix and Perturbation Series

We understand realistic quantum field theories (realistic in the sense of describing physically interesting processes) almost exclusively in a perturbative framework. One starts from a well-defined and solvable theory such as quantum field theory of free, noninteracting fields, or some exact limit such as the theory of electrons in external fields, and constructs the physically relevant quantities from the interaction terms by expansion in terms of "small" parameters. The expansion parameters can be a set of coupling constants, or some momenta, or masses which are typical for the processes one is studying. For example, scattering processes with electrons and

F. Scheck, *Quantum Physics*, DOI: 10.1007/978-3-642-34563-0_10,

photons are expanded in terms of Sommerfeld's fine structure constant α, while for radiative corrections in bound states of atoms with charge number Z the expansion parameters are $(Z\alpha)$ and the average momentum of the bound charged lepton.The *spirit* of perturbation theory, to a large extent, is the same as in nonrelativistic quantum mechanics, and results can often be interpreted in very similar terms. Regarding *techniques* and specific difficulties, however, there are important differences. Techniques are different because modern perturbation theory of quantum field theory is constructed such that at every stage it is manifestly covariant.[1] The basic and very specific difficulties of quantum field theory that do not occur in nonrelativistic perturbation theory are due to the fact that every field theory describes systems with an infinite number of degrees of freedom. A consequence, among several others, is that the Heisenberg or Schrödinger pictures on the one hand, and the interaction picture on the other (cf. Sect. 3.6), no longer are unitarily equivalent so that calculations of the S-matrix, without a more refined analysis, have no more than heuristic value.

The starting point is provided by a formal expansion of the S-matrix in terms of products of the interaction $H_1^{(\text{int})}$ in the interaction picture that we know from time dependent perturbation theory,

$$\mathbf{S} = \mathbb{1} + \sum_{n=1}^{\infty} \frac{(-\mathrm{i})^n}{n!} \int\limits_{-\infty}^{+\infty} \mathrm{d}t_1 \int\limits_{-\infty}^{+\infty} \mathrm{d}t_2 \cdots \int\limits_{-\infty}^{+\infty} \mathrm{d}t_n$$
$$\times\, T\left(H_1^{(\text{int})}(t_1) H_1^{(\text{int})}(t_2) \ldots H_1^{(\text{int})}(t_n)\right). \tag{10.1}$$

The symbol T denotes time-ordering, with the time arguments increasing from right to left. The idea is that the Hamiltonian of the theory can be decomposed into a free, noninteracting term H_0 whose eigenstates are known, and an interaction term which in some sense is small. An example is provided by quantum electrodynamics with electrons which at the classical level is described by the Lagrange density (9.60) with minimal coupling. In this case we have

$$\mathcal{L} = \mathcal{L}_\gamma + \mathcal{L}_D + \mathcal{L}_1\,, \qquad \text{with} \tag{10.2}$$

$$\mathcal{L}_\gamma = -\frac{1}{4} {:}F_{\mu\nu}F^{\mu\nu}{:}\,, \tag{10.3}$$

$$\mathcal{L}_D = {:}\overline{\psi(x)}\left(\frac{1}{2}\mathrm{i}\gamma^\mu \overset{\leftrightarrow}{\partial}_\mu - m_e\mathbb{1}\right)\psi(x){:}\,, \tag{10.4}$$

$$\mathcal{L}_1 = -e\,{:}\overline{\psi(x)}\gamma^\mu\psi(x)A_\mu(x){:}\,, \quad (e = -\left|e\right|)\,. \tag{10.5}$$

The Hamilton density is obtained from this by the rule explained in Sect. 7.1.4, i.e. by constructing the canonically conjugate field momenta, and the function

$$\widetilde{\mathcal{H}} = \sum_i \pi^i \partial_0 \phi^i - \mathcal{L}\,,$$

[1] In the early times the pioneers of quantum field theory constructed the perturbation series in the same way as in quantum mechanics. Examples are the calculation of vacuum polarization by Uehling, and the analysis of the Lamb shift by Bethe. (E.A. Uehling, Phys. Rev. **8**, 55, 1935; H.A. Bethe, Phys. Rev. **72**, 339, 1947). Thus the adjective "modern" for the covariant formulation.

and by a subsequent Legendre transformation which eventually yields the Hamilton density $\mathcal{H}(\phi^i, \pi^k)$. In the example above this is very simple: As $\mathcal{L}_1$ contains no derivatives one obtains $\mathcal{H}_1 = -\mathcal{L}_1$ and

$$H_1 = \int \mathrm{d}^3x \; \mathcal{H}_1(x) = e \int \mathrm{d}^3x \; {:}\overline{\psi(x)}\gamma^\mu \psi(x) A_\mu(x){:} \; .$$

Upon inserting this in (10.1) one obtains a Lorentz covariant series

$$\mathbf{S} = \mathbb{1} + \sum_{n=1}^{\infty} \frac{(-\mathrm{i})^n}{n!} \int \mathrm{d}^4x_1 \int \mathrm{d}^4x_2 \cdots \int \mathrm{d}^4x_n \\ \times T\Big(\mathcal{H}_1(x_1)\mathcal{H}_1(x_2) \ldots \mathcal{H}_1(x_n)\Big) \tag{10.6}$$

in increasing monomials in the Hamilton density of the interaction.[2] This series is called *Dyson series.*

Before entering the more detailed analysis of this perturbative series let us consider the general problem posed here, (though still quoting the example of quantum electrodynamics with electrons). The equations of motion that follow from the Lagrange density (10.2) are

$$\big(\mathrm{i}\gamma^\mu \partial_\mu - m\mathbb{1}\big)\psi(x) = e \,{:}\gamma^\nu \psi(x) A_\nu(x){:} \, , \tag{10.7}$$

$$\Box A^\mu(x) = e \,{:}\overline{\psi(x)}\gamma^\mu \psi(x){:} \, . \tag{10.8}$$

This is a system of coupled differential equations for the fermionic field operator $\psi(x)$ and the bosonic operator $A^\mu(x)$. Suppose, for a moment, this system were exactly solvable and we knew its solutions, say $\boldsymbol{\psi}$ and $\mathbf{A}^\mu$. Most certainly, these would turn out to be very complicated operators, rather different from the free fields ψ and A^μ that are solutions of the force-free Dirac equation and of the Maxwell equations without external sources, respectively. For instance, the operator $\overline{\psi}$ would do much more than create a free electron. The states it would create, though carrying the quantum numbers of a single electron, would contain an arbitrary number of photons. Likewise, $\mathbf{A}^\mu$ would create states which contain e^+e^--pairs besides the isolated photon. We would neither know the structure of the ground state of the exact theory nor in which respect it differs from the perturbative vacuum of the theory without interaction. On the basis of our experience with ground states of systems of a finite number N of fermions (see Sect. 5.3) we presume that the ground state of full quantum electrodynamics is a highly correlated state that differs markedly from the perturbative vacuum.

Perturbation theory circumvents this inextricable problem by starting from *free fields* and the *perturbative vacuum*, and by constructing the effects of mutual interactions by a systematic expansion. However, one has to struggle with problems of technical nature and of interpretation that are unknown in the nonrelativistic N-body system. Furthermore, several steps of this heuristic approach are of a formal nature

---

[2] Here and in what follows we omit the superscript "(int)" that we used above to denote the interaction picture.

and require much care and physical intuition in working out unique results that can be tested by comparison with experiment.

We now turn to the derivation of the Dyson series (10.1). One makes use of the operator $\mathbf{U}(t, t_0)$ which describes the temporal evolution in the interaction picture and which is a solution of the Schrödinger equation with Hamiltonian H_1, see Sect. 3.3.5,

$$\mathrm{i}\dot{\mathbf{U}}(t, t_0) = H_1 \mathbf{U}(t, t_0) .$$

Given the boundary condition $\mathbf{U}(t_0, t_0) = \mathbb{1}$ this operator satisfies the integral equation

$$\mathbf{U}(t, t_0) = \mathbb{1} - \mathrm{i} \int_{t_0}^{t} \mathrm{d}t' \, H_1(t') \mathbf{U}(t', t_0) , \tag{10.9}$$

which is solved by iteration, very much like in nonrelativistic quantum theory. The formal solution reads

$$\mathbf{U}(t, t_0) = \mathbb{1} - \mathrm{i} \int_{t_0}^{t} \mathrm{d}\tau_1 \, H_1(\tau_1) \left[\mathbb{1} - \mathrm{i} \int_{t_0}^{\tau_1} \mathrm{d}\tau_2 \, H_1(\tau_2) \mathbf{U}(\tau_2, t_0) \right] = \ldots$$

$$= \mathbb{1} + \sum_{n=1}^{\infty} (-\mathrm{i})^n \int_{t_0}^{t} \mathrm{d}\tau_1 \int_{t_0}^{\tau_1} \mathrm{d}\tau_2 \cdots \int_{t_0}^{\tau_{n-1}} \mathrm{d}\tau_n \, H_1(\tau_1) H_1(\tau_2) \ldots H_1(\tau_n) .$$

The limits of integration tell us that the time arguments are ordered from right to left, in increasing order, $\tau_1 \geq \tau_2 \geq \ldots \geq \tau_{n-1} \geq \tau_n$. This takes account of the fact that $H_1(t)$ and $H_1(t')$, taken at different times, possibly do not commute. All integrations can be extended to the interval (t_0, t) if one introduces a time ordering in the monomials of the integrand and divides by $n!$. We show this for the example $n = 2$. In this case one has

$$\int_{t_0}^{t} \mathrm{d}\tau_1 \int_{t_0}^{\tau_1} \mathrm{d}\tau_2 \, H_1(\tau_1) H_1(\tau_2) = \int_{t_0}^{t} \mathrm{d}\sigma_2 \int_{\sigma_2}^{t} \mathrm{d}\sigma_1 \, H_1(\sigma_1) H_1(\sigma_2) .$$

If one replaces the integrands by the time-ordered product

$$T\big(H_1(\tau_1) H_1(\tau_2)\big)$$
$$= H_1(\tau_1) H_1(\tau_2) \Theta(\tau_1 - \tau_2) + H_1(\tau_2) H_1(\tau_1) \Theta(\tau_2 - \tau_1) ,$$

and takes the two integrations over the whole interval from t_0 to t, then one obtains twice the double integral just above it. Thus, one can replace this integral by

$$\frac{1}{2!} \int_{t_0}^{t} \mathrm{d}\tau_1 \int_{t_0}^{t} \mathrm{d}\tau_2 \, T\big(H_1(\tau_1) H_1(\tau_2)\big) ,$$

which yields the same value. This argument is readily generalized to the n-fold product in which case the multiple integral must be divided by $n!$.

The S-matrix (10.1) is obtained by the simultaneous limit ($t_0 \to -\infty, t \to +\infty$),

$$\mathbf{S} = \lim_{t_0 \to -\infty} \lim_{t \to +\infty} \mathbf{U}(t, t_0) .$$

As expected on general grounds, it contains a diagonal term $\mathbb{1}$ which stands for "no scattering", and a series in the interaction H_I and, hence, in creation and annihilation operators of the quantum fields which are contained in the interaction H_I. Thus, when calculating matrix elements S_{fi} of the S-matrix between a given initial state $|\, i\,\rangle$ and a selected final state $|\, f\,\rangle$ the essential question will be how many particles are contained in these states. In this way the number of creation operators and the number of annihilation operators that are needed are fixed, and, thus, the minimal order n of perturbation theory is defined at which the process will happen.

In evaluating specific matrix elements S_{fi} by means of the Dyson series (10.6) one makes use of Wick's theorem that we proved earlier, in Sect. 5.4.5. We do not repeat this theorem here but illustrate its application by a number of explicit examples that are worked out further down.

10.1.1 Tools of Quantum Electrodynamics with Leptons

Quantum electrodynamics with electrons, muons, and τ-leptons is a beautiful theory. It is not only *the* model theory for all renormalizable quantum gauge theories, per se, it is also amazingly successful. In spite of its lacking complete mathematical rigour, its predictions for radiative corrections are well-defined and agree with all precision experiments known today. In this section we collect the tools which, via the general series expansion (10.6), define formal rules for covariant perturbation theory. This provides the basis for the calculation and the discussion of some important examples.

The defining Lagrange density reads

$$\mathcal{L} = \mathcal{L}_\gamma + \sum_{f=e,\mu,\tau} \mathcal{L}_D^{(f)} + \mathcal{L}_\mathrm{I} , \tag{10.10}$$

where the symbol f is an abbreviation for the electron-positron field, or the $\mu^-\mu^+$ field, or the $\tau^-\tau^+$ field. The Lagrange density $\mathcal{L}_\gamma$ of the free photon field is given by (10.3), while

$$\mathcal{L}_D^{(f)} = :\overline{\psi^{(f)}(x)}\big(\frac{1}{2}\mathrm{i}\gamma^\mu \overset{\leftrightarrow}{\partial}_\mu - m_f \mathbb{1}\big)\psi^{(f)}(x): , \tag{10.11}$$

$$\mathcal{L}_\mathrm{I} = -e \sum_f :\overline{\psi^{(f)}(x)}\gamma^\mu \psi^{(f)}(x) A_\mu(x): , \quad (e = -\,|e|) . \tag{10.12}$$

The interactions of any one of the charged leptons, e, μ, or τ, with the radiation field are identically the same. Any differences in the observables can only be caused by the differences in their masses which are

$$m_e = 0.511\,\mathrm{MeV} , \ m_\mu = 105.66\ \mathrm{MeV} , \ m_\tau = 1777\ \mathrm{MeV} . \tag{10.13}$$

This empirical observation which holds as well for the weak interactions of leptons, is called *lepton universality*. Their neutral partners $\nu_f = (\nu_e, \nu_\mu, \nu_\tau)$ do not couple to the Maxwell field because they carry no electric charge. Charged leptons have no direct interactions among each other. They talk to each other via the radiation field, and via their weak interaction with $W^{\pm}$- and Z-bosons. Uncharged leptons (the neutrinos) interact only through intermediate $W^{\pm}$- and Z-bosons.

Inserting the expansion (7.143) for the photon field, and the expansions (9.84) and (9.85) for the fermion field and its adjoint, respectively, the characteristic interaction term of quantum electrodynamics

$$\mathcal{L}_1 = -e{:}\overline{\psi^{(f)}(x)}\gamma^{\mu}\psi^{(f)}(x)A_{\mu}(x){:}$$

is seen to have the following structure in terms of creation and annihilation operators $a^{(f)}$, $b^{(f)}$, $a^{(f)\dagger}$, $b^{(f)\dagger}$, for leptons of the family f, and c, $c^{\dagger}$ for the photon (suppressing spin degrees of freedom)

$$\begin{aligned}\mathcal{L}_1 \sim -\, e&\Big\{\dots\overline{u_f(q)}\,a^{(f)\dagger}(q) + \dots\overline{v_f(q)}\,b^{(f)}(q)\Big\}\gamma^{\mu}\varepsilon_{\mu}(k)\\ &\times\Big\{\dots c(k) + \dots c^{\dagger}(k)\Big\}\Big\{\dots u_f(p)a^{(f)}(p) + \dots v_f(p)b^{(f)\dagger}(p)\Big\}.\end{aligned}$$

What matters here is the number of creation and annihilation operators contained in a product of field operators. In the present case there are always *two* fermionic but only *one* photonic creation or annihilation operators. Thus, in quantum electrodynamics there is one basic vertex, linking a single photon line to two fermion/antifermion lines. Furthermore, the terms in curly brackets represent the decompositions into positive and negative frequency parts,

$$\psi = \psi^{(+)} + \psi^{(-)}, \quad \overline{\psi} = \overline{\psi}^{(+)} + \overline{\psi}^{(-)}, \quad A^{\mu} = \left(A^{\mu}\right)^{(+)} + \left(A^{\mu}\right)^{(-)}.$$

The following formulae apply to *external* fermions, incoming or outgoing:

$$\begin{aligned}\langle 0|\psi^{(f)}(x)\,|\,f^{-}(p,r)\rangle &= \langle 0|\left(\psi^{(f)}(x)\right)^{(+)}|\,f^{-}(p,r)\rangle\\ &= \frac{1}{(2\pi)^{3/2}}u_f^{(r)}(p)\;\mathrm{e}^{-\mathrm{i}p\cdot x}, \qquad (10.14)\end{aligned}$$

$$\begin{aligned}\langle f^{+}(p,s)|\psi^{(f)}(x)\,|0\rangle &= \langle f^{+}(p,s)|\left(\psi^{(f)}(x)\right)^{(-)}|0\rangle\\ &= \frac{1}{(2\pi)^{3/2}}v_f^{(s)}(p)\;\mathrm{e}^{\mathrm{i}p\cdot x}, \qquad (10.15)\end{aligned}$$

$$\begin{aligned}\langle f^{-}(p,r)|\overline{\psi^{(f)}(x)}\,|0\rangle &= \langle f^{-}(p,r)|\left(\overline{\psi^{(f)}(x)}\right)^{(-)}|0\rangle\\ &= \frac{1}{(2\pi)^{3/2}}\overline{u_f^{(r)}(p)}\;\mathrm{e}^{\mathrm{i}p\cdot x}, \qquad (10.16)\end{aligned}$$

$$\begin{aligned}\langle 0|\overline{\psi^{(f)}(x)}\,|\,f^{+}(p,s)\rangle &= \langle 0|\left(\overline{\psi^{(f)}(x)}\right)^{(+)}|\,f^{+}(p,s)\rangle\\ &= \frac{1}{(2\pi)^{3/2}}\overline{v_f^{(s)}(p)}\;\mathrm{e}^{-\mathrm{i}p\cdot x}. \qquad (10.17)\end{aligned}$$

Likewise, for incoming or outgoing *external* photons we have, respectively,

$$\langle 0|A_{\mu}(x)\,|k,\lambda\rangle = \frac{1}{(2\pi)^{3/2}}\,\varepsilon_{\mu}^{(\lambda)}(k)\;\mathrm{e}^{-\mathrm{i}k\cdot x}, \qquad (10.18)$$

$$\langle k,\lambda|A_{\mu}(x)\,|0\rangle = \frac{1}{(2\pi)^{3/2}}\,\varepsilon_{\mu}^{(\lambda)}(k)\;\mathrm{e}^{\mathrm{i}k\cdot x}. \qquad (10.19)$$

Note that these must be *physical* photons which means that only the two *transverse* states $\lambda = 1$ and $\lambda = 2$ (or linear combinations thereof) are allowed.

When integrating over the momenta the exponentials in these formulae yield δ-distributions for energy-momentum conservation at every vertex. Furthermore, already at this point we take note of a factor $1/(2\pi)^{3/2}$ for every external particle, as well as of the fermion spinors and the polarizations, respectively, which are as follows.

(f1) $u_f^{(r)}(p)$ for every *incoming* f^- with momentum p and spin orientation r,

(f2) $v_f^{(s)}(p)$ for every *outgoing* f^+ with momentum/spin (p, s),

(f3) $\overline{u_f^{(r)}}(p)$ for every *outgoing* f^- with momentum/spin (p, r),

(f4) $\overline{v_f^{(s)}}(p)$ for every *incoming* f^+ with momentum/spin (p, s), and

(ph) $\varepsilon_\mu^{(\lambda)}(k)$ for an incoming or outgoing photon with momentum k and polarization λ.

An internal, i.e. virtual photon which is exchanged between two vertices, is represented by the propagator (7.149) in Feynman gauge

$$\langle 0|T\, A^\mu(x)A^\nu(y)\,|0\rangle = \frac{\mathrm{i}}{(2\pi)^4}\int \mathrm{d}^4k\ \mathrm{e}^{-\mathrm{i}k\cdot(x-y)}\frac{-g^{\mu\nu}}{k^2+\mathrm{i}\varepsilon}\,. \tag{10.20}$$

This will be reflected by a simple rule in momentum space: Every internal photon line translates into the factor $-g^{\mu\nu}/(k^2 + \mathrm{i}\varepsilon)$.

A virtual lepton f which propagates from $x \in M^4$ to $y \in M^4$ is represented by the propagator (9.110)

$$\langle 0|T\,\psi^{(f)}(x)\overline{\psi^{(f)}(y)}|0\rangle = \frac{\mathrm{i}}{(2\pi)^4}\int \mathrm{d}^4p\ \mathrm{e}^{-\mathrm{i}p\cdot(x-y)}\frac{(\not{p}+m_f\mathbb{1})}{p^2-m_f^2+\mathrm{i}\varepsilon}\,. \tag{10.21}$$

Thus, in momentum space representation an internal fermion line comes with the factor

$$\frac{(\not{p} + m_f\mathbb{1})}{p^2 - m_f^2 + \mathrm{i}\varepsilon}\,.$$

In summary, the series (10.6) and the above formulae provide the essential tools which allow to formulate general rules for constructing amplitudes describing specific processes of quantum electrodynamics.

There is an obvious question which is much more subtle to answer: What are the values of the masses m_f in (10.11) and of the coupling constant e, or α, in (10.12) that should be chosen? Indeed, it turns out that if we insert here the physical masses m_f as they are known from experiment, and the physical charge e as measured in Thomson scattering, cf. Sect. 7.4.3, these initial values are modified by contributions in higher orders of perturbation theory. Therefore, we should expect, in the course of the development, the Lagrange density (10.10) to be supplemented by counterterms, in order to take account of these corrections. If, for example, $m_f^{(0)}$ are the original, *bare* masses, this means that (10.11) must be replaced by

$$\mathcal{L}_D^{(f)} + \delta m_f{:}\overline{\psi^{(f)}(x)}\psi^{(f)}(x){:}\,, \quad \text{where} \quad \delta m_f = m_f - m_f^{(0)}$$

is the difference of the measured mass m_f and the starting parameter $m_f^{(0)}$. This phenomenon of renormalization of masses and coupling constants is not new and occurs also in nonrelativistic quantum theory of N particles. However, there is an essential difference: In the quantum N-body system the differences between the physical quantities and the parameters of the defining theory are *finite,* while in a renormalizable quantum field theory they are usually *infinite*. In other terms, since the physical values are finite, this means that the bare parameters must be infinitely large. In fact, also the wave functions of the interacting theory, obtained from perturbation theory, differ from those of the free theory by infinitely large factors. These phenomena are called *mass renormalization*, *charge renormalization*, and *wave function renormalization,* respectively. They are the cause of a filigree of mathematical subtleties of renormalizable quantum field theory and they imply aspects of the theory which are not easy to visualize.

10.1.2 Feynman Rules for Quantum Electrodynamics with Charged Leptons

In this section we first formulate the Feynman rules for perturbative quantum electrodynamics in the form of a list of instructions on how to translate diagrams into formulae. As is shown subsequently, the functions, the factors, and the signs that have to be inserted, either follow from the general principles described in Chaps. 7 and 8, or from the set of tools described above.

Rules for Amplitudes of Quantum Electrodynamics:

(R0) The Aim: The rules yield the second term of the symbolic decomposition of the S-matrix into the diagonal part $\mathbb{1}$ and the reaction matrix $\mathbf{R}$,

$$\mathbf{S} = \mathbb{1} + \mathbf{R}$$

for the transition from the initial state "i" to the final state "f" that one wishes to calculate, viz.

$$S_{fi} = \delta_{fi} + R_{fi} \,. \tag{10.22}$$

The T-matrix whose elements are needed for the calculation of cross sections or decay widths, cf. Sects. 8.2.3 and 8.2.4, is then obtained by extracting the factor $\mathrm{i}(2\pi)^4\delta(P_f - P_i)$ from R_{fi} where P_i and P_f denote the sum of all four-momenta in the initial and final states, respectively.

(R1) Diagrams: For a given process $A + B \to C + D + \cdots$ one draws all *connected* diagrams of order n. External and internal lepton lines are provided with arrows such that the direction of the arrows coincides with the direction of flow of *negative* charge. The rules that we formulated in Sect. 9.3.4 are equivalent to this: One defines who should be termed *particle* (traditionally, in quantum electrodynamics these are the e^-, the μ^-, and the τ^-) and thus defines the corresponding antiparticle. For a *particle* the arrow points towards the vertex if it is incoming, and away from it when it is outgoing. For an

antiparticle the arrow leaves the vertex when it is incoming, and enters the vertex when it is outgoing.
The factors of the following rules (R2) to (R4) are to be written down *from right to left*, by following the direction of the arrow as defined above.

(R2) External Lines: An incoming lepton f^- is represented by the spinor $u_f^{(r)}(p)$, an outgoing lepton by the spinor $\overline{u_f^{(r)}}(p)$, whose arguments are the given momenta and spin orientations. An antiparticle f^+ is represented by $\overline{v_f^{(r)}}(p)$ if it is an incoming one, and by $v_f^{(r)}(p)$ if it is outgoing.
Every incoming or outgoing photon yields the real function $\varepsilon_\mu^{(\lambda)}(k)$, with $\lambda = 1, 2$. The index μ is to be contracted with the factor γ^μ at the vertex to which the photon is hooked (see next rule).

(R3) Vertices: At every vertex insert $e\gamma^\mu$ as well as a δ-distribution for the four-momenta attached to this vertex, such that energy and momentum are conserved.
For this balance one must first clarify the flow of the momenta (see next rule).

(R4) Internal Lines: Every internal lepton line is represented by the propagator in momentum space

$$\frac{(\not{p} + m_f \mathbb{1})}{p^2 - m_f^2 + \mathrm{i}\varepsilon} .$$

The direction of the virtual momentum p follows the arrows, that is, it is the same as that of the flow of negative charge.
Every internal photon line stands for the propagator in momentum space

$$\frac{-g_{\mu\nu}}{k^2 + \mathrm{i}\varepsilon}$$

(provided the Feynman gauge is chosen). The indices μ and ν refer to the two vertices on which the propagator hinges. They are to be contracted with the matrices γ^μ and γ^ν, respectively, that are there by rule (R3).

(R5) Integrations: The analytical expression obtained from the previous rules must be integrated over the momenta of all *internal* lines. Note that the spin sums as well as the sum over lepton and antilepton contributions are contained in the propagators. The integrations over the internal momenta leaves us with a δ-distribution for the difference $P_i - P_f$ as an overall factor. This factor drops out when passing from the R-matrix to the T-matrix.

(R6) Signs: (a) The amplitude T_{fi} receives a factor $(-)^\Pi$ where Π denotes the permutation of the leptons of the same species in the final state.

(b) With L the number of closed lepton loops contained in the diagram, the amplitude T_{fi} receives the overall factor $(-)^L$.

(R7) ***C*-Invariance of Maxwell Theory**: A closed loop of virtual lepton lines to which an *odd* number of photons are attached, gives no contribution.

(R8) Numerical Factors: The matrix element R_{fi} with, in total, l_a external lepton lines and b_a external photon lines obtains the numerical factor

$$\left((2\pi)^{-3/2}\right)^{l_a+b_a} .$$

In other terms, every external particle yields a factor $(2\pi)^{-3/2}$, independently of whether it is a fermion or a photon.
Let l_i denote the number of internal lepton lines, b_i the number of internal boson lines, and, as before, let n be the order. Then R_{fi} receives the additional factor

$$\mathrm{i}^{n+l_i+b_i}\,(2\pi)^{4(n-l_i-b_i)} .$$

Note that these factors hold for the matrix element R_{fi}. The matrix element T_{fi} which represents the physical amplitude (scattering or decay) follows from R_{fi} by separating the numerical factor $\mathrm{i}\,(2\pi)^4$ and the distribution $\delta(P_i - P_f)$.

Rule (R1) talks about connected diagrams. These are diagrams which are not composed of two or more disjoint parts. This restriction is plausible because otherwise there would be contributions to the reaction matrix where particles would pass by one another without interacting. Rule (R2) and the first numerical factor in (R8) are consequences of the decompositions (7.143) and (9.84), (9.85) of the quantized fields in terms of normal modes, as well as of the conventions we have chosen. The second numerical factor in (R8) stems from the following contributions
- $(-\mathrm{i})^n$ from the Dyson series (10.6),
- the sign $(-)^n$ from the sign of the interaction term (10.12),
- a factor i from every lepton propagator (10.21) and every photon propagator (10.20), so, in total, $\mathrm{i}^{l_i+b_i}$,
- a factor $(2\pi)^{-4(l_i+b_i)}$ which stems from the propagators as well, and a factor $(2\pi)^4$ from every integration $\int \mathrm{d}^4x$ with exponentials, i.e., in total, $(2\pi)^{4n}$.

Regarding rule (R3) on vertices: Every vertex links three internal or external lines, two of which are fermionic, the third being bosonic. Both the propagators and the field operators proper yield exponentials whose argument is x times the balance of the four-momenta at the vertex. Upon integration over x one obtains 4π times a δ-distribution for every vertex. The integration over the internal momenta, in turn, is done without restriction.

The rule (R6a) is obvious because fermion fields of the same kind anticommute. The rules (R6b) and (R7) need a more detailed analysis: Consider the current density

$$j^\alpha(x) = :\overline{\psi(x)}\gamma^\alpha\psi(x):$$

of a given, fixed lepton species "f". Since in a closed loop only one and the same lepton runs through, we omit the superscript f. As one realizes easily the contribution of any such closed loop contains the factor

$$\langle 0|T\; j^{\mu_1}(x_1) j^{\mu_2}(x_2) \dots j^{\mu_m}(x_m)\,|0\rangle\;. \tag{10.23}$$

In order to reduce this expression to a product of propagators (10.21) one must reorder the field operators within the vacuum expectation value until they appear in the order $\psi\overline{\psi}\psi\overline{\psi}\dots$. This procedure always needs an *odd* number of permutations. For the sake of simplicity we consider an example:

$$\begin{aligned}
&\langle 0|T\;\overline{\psi(x)}_\alpha (\gamma^\mu)_{\alpha\beta}\psi_\beta(x)\overline{\psi(y)}_\sigma(\gamma^\nu)_{\sigma\tau}\psi_\tau(y)\,|0\rangle \\
&\quad = -(\gamma^\mu)_{\alpha\beta}(\gamma^\nu)_{\sigma\tau}\langle 0|T\;\psi_\beta(x)\overline{\psi(y)}_\sigma\psi_\tau(y)\overline{\psi(x)}_\alpha\,|0\rangle \\
&\quad = -(\gamma^\mu)_{\alpha\beta}(\gamma^\nu)_{\sigma\tau}\langle 0|T\;\psi_\beta(x)\overline{\psi(y)}_\sigma\,|0\rangle\,\langle 0|T\;\psi_\tau(y)\overline{\psi(x)}_\alpha\,|0\rangle
\end{aligned}$$

Thus, for every closed loop there is a factor -1.
If m photon lines are attached to the loop, that is, if it contains m vertices, then rule (R7) says that this number must be even. This is a consequence of the invariance of the interaction with respect to charge conjugation. Indeed, applying **C** to (10.23), and taking account of the fact that the current operator is odd under charge conjugation, $\mathbf{C}^{-1} j^\mu(x)\mathbf{C} = -j^\mu(x)$, one obtains

$$\begin{aligned}
&\langle 0|T\; j^{\mu_1}(x_1) j^{\mu_2}(x_2) \dots j^{\mu_m}(x_m)\,|0\rangle \\
&\quad = \langle 0|\mathbf{C}^{-1}\mathbf{C}T\; j^{\mu_1}(x_1) j^{\mu_2}(x_2) \dots j^{\mu_m}(x_m)\mathbf{C}^{-1}\mathbf{C}\,|0\rangle \\
&\quad = (-)^m\langle 0|T\; j^{\mu_1}(x_1) j^{\mu_2}(x_2) \dots j^{\mu_m}(x_m)\,|0\rangle\;.
\end{aligned}$$

This contribution vanishes if the number of vertices m is odd. Fig. 10.1a shows an example for which the amplitude vanishes due to (R7), while Fig. 10.1b shows another example which gives a nonvanishing contribution.

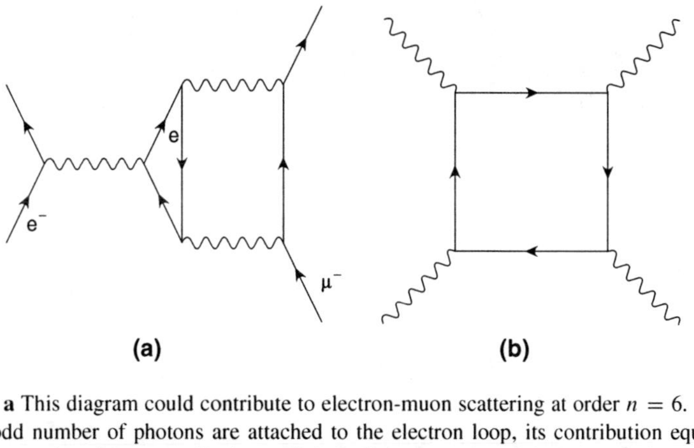

Fig. 10.1 **a** This diagram could contribute to electron-muon scattering at order $n = 6$. However, since an odd number of photons are attached to the electron loop, its contribution equals zero. **b** This diagram describes a contribution to light by light scattering of the lowest order, $n = 4$. The contribution does not vanish because four photons couple to the loop

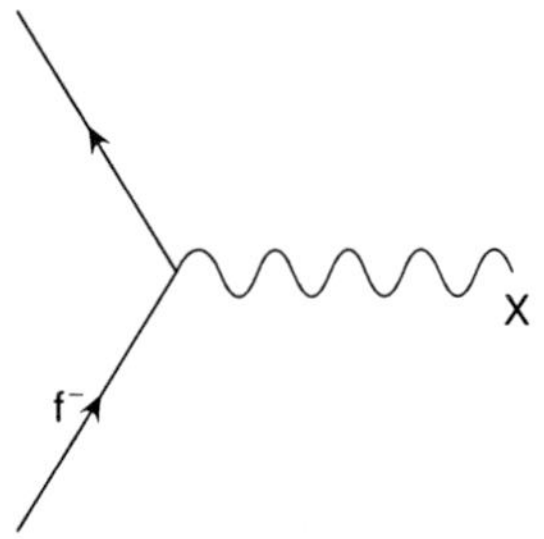

Fig. 10.2 A virtual photon is being exchanged between the external potential, symbolized by X, and the scattering lepton f^-

The Feynman rules are supplemented by a further rule which allows to calculate scattering on an external Coulomb potential. This potential is a model for a heavy charged particle, an atomic nucleus for instance, on which the electron is scattered. In a process of this kind the energy must be conserved, but the spatial momentum need not be conserved. The heavy partner can absorb, or deliver, an arbitrary amount of momentum, without changing its state of rest in any appreciable manner. A patient scattering partner of this sort is denoted by an X, as sketched in Fig. 10.2.

The rule reads as follows

(R9) Scattering on an External Potential: The charged lepton and the virtual photon are to be treated like in the rules (R1) to (R8), with the exception of (R3). At the (only) vertex write a one-dimensional δ-distribution for the energy (replacing the four-dimensional one of rule (R3)). Furthermore, the factor $e\gamma^\mu$ must be replaced by

$$\delta_{\mu 0} \frac{Ze}{(2\pi)^3} \frac{\widetilde{\varrho}(\boldsymbol{k})}{\boldsymbol{k}^2}, \tag{10.24}$$

where $\widetilde{\varrho}(\boldsymbol{k})$ is the form factor which corresponds to the charge density $\varrho(\boldsymbol{x})$,

$$\widetilde{\varrho}(\boldsymbol{k}) = \int \mathrm{d}^3 x \; \mathrm{e}^{-\mathrm{i}\boldsymbol{k}\cdot\boldsymbol{x}} \varrho(\boldsymbol{x})$$

the charge density being normalized to 1, $\int \mathrm{d}^3 x \; \varrho(\boldsymbol{x}) = 1$.

The proof of this rule is the content of Exercise 10.1.

10.1.3 Some Processes in Tree Approximation

In this section we illustrate the Feynman rules of quantum electrodynamics by the processes

$$
\begin{aligned}
&(a): \quad e^\mp + \gamma \longrightarrow e^\mp + \gamma\,,\\
&(b): \quad e^- + e^- \longrightarrow e^- + e^-\,,\\
&(c): \quad e^- + e^+ \longrightarrow e^- + e^+\,,\\
&(d): \quad e^- + e^+ \longrightarrow \gamma + \gamma\,,\\
&(e): \quad e^- + e^+ \longrightarrow \mu^- + \mu^+\,,\\
&(f): \quad e^- + e^+ \longrightarrow \tau^- + \tau^+
\end{aligned}
$$

which we calculate to lowest order, i.e. $n = 2$. In this order of perturbation theory there are no internal loops yet. Therefore, the corresponding diagrams look like the fir-trees that we used to draw when we were children. Indeed, diagrams of this sort are called *tree diagrams.*

The process (a) is the *Compton effect* of the electron or positron, (b) is called *Møller scattering,* (c) is called *Bhabha scattering,* all three of them referring to well-known physicists of the twentieth century. The process (d) is called *pair annihilation in flight,* while (e) and (f) are pair creation processes from colliding electron and positron beams. The latter series continues and includes $e^-e^+ \to p\bar{p}$, $e^-e^+ \to q\bar{q}$ (where q denotes a quark), etc.

The reactions (c), (e), and (f) are of particular relevance for experiments on e^+e^--colliders because, on the one hand, they serve to test quantum electrodynamics and the radiative corrections it predicts at high energies, and, on the other hand, they are reference reactions for the creation of hadrons by e^-e^+ pair annihilation. Regarding the second perspective, one measures the ratio

$$R = \frac{\mathrm{d}\sigma(e^-e^+ \to q\bar{q})}{\mathrm{d}\sigma(e^-e^+ \to \mu^-\mu^+)}$$

of quark-antiquark and of muon-antimuon creation in order to investigate hadronic physics.

Bhabha scattering (c) as well as pair annihilation often serve as analyzing reactions for the polarization of the positron: One scatters the incoming positrons, e.g. on a polarized iron foil and uses the spin dependence of the cross sections for determining their polarization. Likewise, the spin dependence of Compton scattering (a) is used to measure the polarization of photons. This shows that the reactions (a)–(f) yield far more than topics of academic exercises!

Before analyzing these reactions in more detail let us work out some of their general properties as well as possible relations between them. For this purpose we consider the slightly more general case of a two-body reaction $A + B \to C + D$, the masses and four-momenta being as indicated in parantheses,

$$A\,(m_1, p_1) + B\,(m_2, p_2) \longrightarrow C\,(M_1, q_1) + D\,(M_2, q_2)\,. \tag{10.25}$$

Define the Lorentz invariant kinematic variables

$$s := (p_1 + p_2)^2 = (q_1 + q_2)^2\,, \tag{10.26}$$

$$t := (p_1 - q_1)^2 = (p_2 - q_2)^2\,, \tag{10.27}$$

$$u := (p_1 - q_2)^2 = (q_1 - p_2)^2\,, \tag{10.28}$$

where use was made of the relation $p_1 + p_2 = q_1 + q_2$ for energy-momentum conservation. These variables which are called *Mandelstam variables* are not linearly independent. Indeed, one has

$$s + t + u = m_1^2 + m_2^2 + M_1^2 + M_2^2\,, \tag{10.29}$$

or, in words, their sum equals the sum of the squares of the four external masses. This is easily verified by direct calculation: With

$$s = m_1^2 + m_2^2 + 2p_1 \cdot p_2 ,$$
$$t = m_1^2 + M_1^2 - 2p_1 \cdot q_1 ,$$
$$u = m_1^2 + M_2^2 - 2p_1 \cdot q_2$$

and $p_2 - q_1 - q_2 = -p_1$ one obtains

$$s + t + u = m_1^2 + m_2^2 + M_1^2 + M_2^2 + 2m_1^2 - 2p_1^2 = m_1^2 + m_2^2 + M_1^2 + M_2^2 .$$

The physical meaning of s and t is understood most easily by evaluating them in the center-of-mass system. Knowing that $\boldsymbol{p}_1 = -\boldsymbol{p}_2$ one sees that s is the square of the total energy

$$s = (E_{p_1} + E_{p_2})^2 .$$

The variable t describes the transfer of momentum and, therefore, is a function of the scattering angle in the center-of-mass system. We study two special cases which may be sufficient for the examples (a)–(f):

(i) Pairs of equal masses in the initial and final states, i.e. $m_1 = m_2 \equiv m$ and $M_1 = M_2 \equiv M$: Let κ and κ' denote the modulus of the spatial momentum before and after the scattering, respectively. One then has

$$s = 4(m^2 + \kappa^2) = 4(M^2 + \kappa'^2) , \tag{10.30}$$

$$\begin{aligned} t &= m^2 + M^2 - 2\sqrt{(\kappa^2 + m^2)(\kappa'^2 + M^2)} + 2\kappa\kappa' \cos\theta \\ &= m^2 + M^2 - \frac{s}{2} + \frac{1}{2}\sqrt{(s - 4m^2)(s - 4M^2)} \cos\theta \end{aligned} \tag{10.31}$$

$$u = m^2 + M^2 - \frac{s}{2} - \frac{1}{2}\sqrt{(s - 4m^2)(s - 4M^2)} \cos\theta . \tag{10.32}$$

(ii) Elastic scattering, $m_1 = M_1 \equiv m$, and $m_2 = M_2 \equiv M$:
As we are dealing with elastic scattering we now have $\kappa = \kappa'$. A short calculation yields the formulae

$$\kappa = \frac{1}{2\sqrt{s}} \sqrt{\left(s - (m + M)^2\right)\left(s - (m - M)^2\right)} = \kappa' , \tag{10.33}$$

$$s = m^2 + M^2 + 2\kappa^2 + 2\sqrt{(\kappa^2 + m^2)(\kappa^2 + M^2)} , \tag{10.34}$$

$$t = -2\kappa^2(1 - \cos\theta) . \tag{10.35}$$

The scattering process (10.25) which is sketched in Fig. 10.3a, can be analyzed further in the light of possible symmetries, without disentangling yet the interaction region in terms of perturbation theory. For example, the particle D can be replaced by its antiparticle $\overline{D}$ by requiring that it should be incoming, instead of outgoing. At the same time A is replaced by $\overline{A}$ and is taken to be outgoing. This means that to the process (10.25) one associates the process

$$B + \overline{D} \longrightarrow C + \overline{A} ,$$

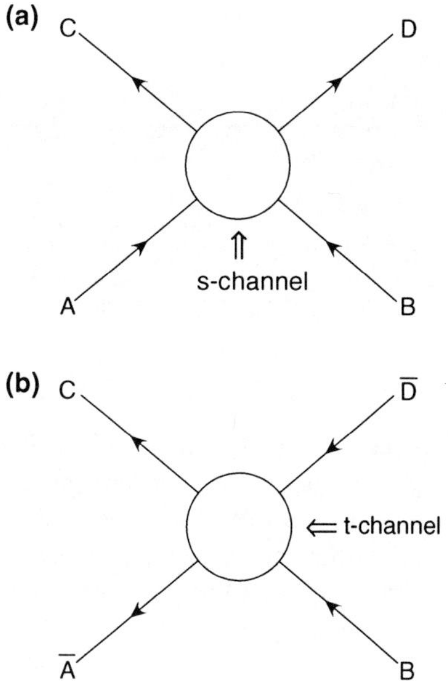

Fig. 10.3 **a** shows the reaction $A + B \to C + D$ in the s-channel; **b** shows the reaction $B + \overline{D} \to C + \overline{A}$, i.e. the t-channel related to the first reaction. Here, the *arrows* at the outer lines do not obey Feynman rules but refer to incoming and outgoing particles, respectively. The *bubble* in the interaction area stands for all diagrams contributing to this process

as shown in Fig. 10.3b. Formally, the amplitude for this process, Fig. 10.3b, is obtained from the amplitude for (10.25), Fig. 10.3a, by continuing q_2 to $-q_2$, and p_1 to $-p_1$. The variable t of the first process turns into the s-variable of the second, while the s-variable of the first becomes the physical t-variable of the second process. For this reason the original reaction is called the *s-channel reaction*, while the second which is obtained from the first by continuation of kinematic variables, is called the *t-channel reaction.* On the analytical side this means the following: Suppose one knows the amplitude of the s-channel reaction (10.25) as a function of the variables s and t. Then the amplitude for the corresponding t-channel reaction $B + \overline{D} \to C + \overline{A}$ is obtained by exchanging s and t. More importantly, however, this crossing needs analytic continuation of the amplitude to the correct physical domains of the variables.

The t-channel reaction can also be viewed in a different orientation of the momenta, that is to say, as the process $A + \overline{C} \to \overline{B} + D$ which is the charge conjugate, inverse reaction of $B + \overline{D} \to C + \overline{A}$. This operation which is called *crossing*, is especially interesting in those cases where crossing the external lines yields again the same process. An example is provided by Bhabha scattering

$$e^- + e^+ \longrightarrow e^- + e^+ ,$$

whose t-channel partner coincides with the original reaction. As a consequence, the absolute square of the scattering amplitude, when expressed in terms of the invariant variables s and t, must be symmetric under the exchange $s \leftrightarrow t$.

Note that here we are talking about the corresponding *relative* antiparticles, and, therefore, that it might be better to talk about the *charge conjugate* partners. Charge

Table 10.1 Some typical processes of the leptonic quantum electrodynamics with their crossed channels

s-channel	t-channel	u-channel	Symmetry
$e^- e^+ \to \gamma\gamma$	$e^\pm \gamma \to e^\pm \gamma$	$e^- \gamma \to e^- \gamma$	$t \leftrightarrow u$
$e^- e^- \to e^- e^-$	$e^- e^+ \to e^- e^+$	$e^- e^+ \to e^- e^+$	$t \leftrightarrow u$
$e^+ e^- \to \mu^+ \mu^-$	$e^- \mu^+ \to e^- \mu^+$	$e^- \mu^- \to e^- \mu^-$	
$e^- e^+ \to e^- e^+$	$e^- e^+ \to e^- e^+$	$e^- e^- \to e^- e^-$	$s \leftrightarrow t$

conjugation changes the signs of all additively conserved quantum numbers, including in particular electric charge. Therefore, the selection rules for the t-channel reaction are fulfilled precisely if they are fulfilled for the original reaction.

Returning to the general reaction (10.25), there is one more partner of it: let the particle A and the antiparticle $\overline{D}$ of D be incoming, the particles C and $\overline{B}$ be outgoing, viz.

$$A + \overline{D} \longrightarrow C + \overline{B} .$$

The process obtained in this way is called the u-channel reaction associated to (10.25). In this case it is the variables s and u which exchange their roles, while t keeps its role. Consider the example of Møller scattering (b), $e^- + e^- \to e^- + e^-$. Both the associated t-channel and u-channel reactions yield the process $e^- + e^+ \to e^- + e^+$. Therefore, the absolute square of the amplitude for Møller scattering must by symmetric under interchange of t with u.

The scattering processes that we study in this section are listed in the table above, together with their t- and u-channel partners and their possible symmetries in the variables s, t, and u.

The $t - u$ symmetry in the first two rows of Table 10.1, and the $s - t$ symmetry in the last row are marked by boxes. These symmetries are useful in short-cutting calculations. Crossing and the analytic continuation that goes with it are useful also in cases where there is no such symmetry because, starting from one out of a set of three associated cross sections, they allow to deduce the other two.

Compton Scattering on Electrons and Positrons: To lowest order, $n = 2$, scattering of a photon on an electron is described by the two tree diagrams of Fig. 10.4 which, in terms of physics, say this: In the Fig. 10.4a the incoming photon was absorbed before the outgoing photon is emitted, while in the Fig. 10.4b the outgoing photon is emitted before the incoming one was absorbed. The rules (R1)–(R8) of the preceding section are easy to implement here. To order $n = 2$ the only connected diagrams are the ones sketched in Fig. 10.4. The integration over the internal momentum of the virtual electron/positron line yields two δ-distributions. These are rewritten such that one obtains the expected distribution $\delta(P_i - P_f)$ of energy-momentum conservation, and, at the same time fixes the momentum in the propagator. Consider the example of the Fig. 10.4a, for the sake of illustration. Denoting the momentum of the virtual

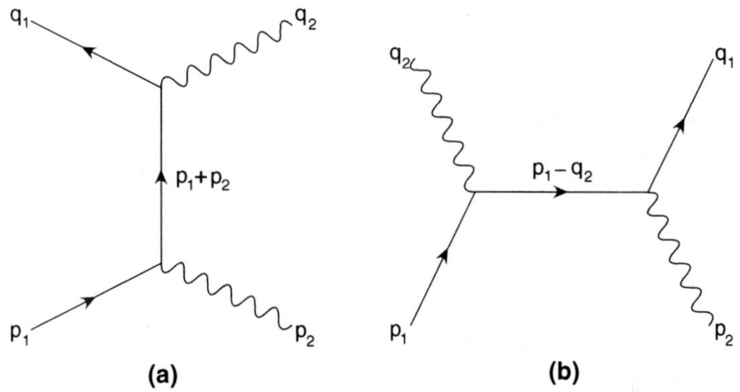

Fig. 10.4 **a** The incoming photon is swallowed by the incoming electron which then runs on as a electron-positron propagator with non-physical momentum $(p_1 + p_2)$ until it finally disgorges the outgoing photon and returns to its mass shell. **b** The incoming electron first emits the outgoing photon and then absorbs the incoming photon

line by Q, rule (R3) requires the factor

$$\delta(p_1 + p_2 - Q)\delta(Q - p_2 - q_2)$$
$$= \delta\big(Q - (q_1 + q_2)\big)\delta\big((p_1 + p_2) - (q_1 + q_2)\big) .$$

Thus, it yields the expected δ-distribution for energy and momentum conservation and it determines Q to be $Q = p_1 + p_2$. The second diagram is worked out in the same manner. This means that we can construct the T-matrix element directly and need not go via the function R_{fi}.

We use a short-hand notation for the polarizations of the incoming and outgoing photons, $\varepsilon' \equiv \varepsilon_\mu^{(\lambda')}(q_2)$ and $\varepsilon \equiv \varepsilon_\nu^{(\lambda)}(p_2)$, suppress the spin orientations of the in- and out-states of the electron, and write the electron mass as $m_1 \equiv m$. The T-amplitude then reads

$$T(e^-\gamma \to e^-\gamma)$$
$$= -\frac{e^2}{(2\pi)^6}\overline{u(q_1)}\left(\not{\varepsilon}'\frac{(\not{p}_1+\not{p}_2)+m\mathbb{1}}{(p_1+p_2)^2-m^2}\not{\varepsilon} + \not{\varepsilon}\frac{(\not{p}_1 - \not{q}_2) + m\mathbb{1}}{(p_1 - q_2)^2 - m^2}\not{\varepsilon}'\right) u(p_1)$$
$$= -\frac{e^2}{(2\pi)^6}\overline{u(q_1)}\left(\not{\varepsilon}'\frac{(\not{p}_1 + \not{p}_2) + m\mathbb{1}}{2p_1 \cdot p_2}\not{\varepsilon} - \not{\varepsilon}\frac{(\not{p}_1 - \not{q}_2) + m\mathbb{1}}{2p_1 \cdot q_2}\not{\varepsilon}'\right) u(p_1) . \quad (10.36)$$

In the second step we made use of the mass shell conditions $p_1^2 = m^2 = q_1^2$ and $p_2^2 = 0 = q_2^2$. Before actually calculating the cross section from this amplitude I wish to add a few remarks:

Remarks

1. The invariance under gauge transformations in position space, $A_\mu(x) \mapsto A'_\mu(x) = A_\mu - \partial_\mu \chi(x)$, when translated to momentum space (i.e. by Fourier

transform), reads

$$\widetilde{A}_\mu(k) \longmapsto \widetilde{A}'_\mu = \widetilde{A}_\mu(k) + c\, k_\mu \,,$$

where c is a number whose value is of no importance here (it is proportional to $\widetilde{\chi}(k)$). This is equivalent to the rule that one should make the replacement

$$\varepsilon_\mu \longmapsto \varepsilon_\mu + c\, k_\mu \tag{10.37}$$

in all amplitudes with external photons. If upon replacement of ε_μ by k_μ the amplitude yields a term that vanishes, then the gauge invariance of one's calculation is verified.

Let us perform this simple test for the example of $\not\varepsilon \mapsto \not p_2$ in the amplitude (10.36): Replacing $\not\varepsilon$ by $\not p_2$ in the first term its numerator yields $\not p_1 \not p_2 + \not p_2 \not p_1 = 2 p_1 \cdot p_2$ and, using $\not p_2 \not p_2 = p_2^2 = 0$,

$$(\not p_1 + \not p_2 + m\mathbb{1})\not p_2 = 2 p_1 \cdot p_2 - \not p_2 \not p_1 + m \not p_2 \,.$$

Applying then $\not p_1$ to $u(p_1)$ one obtains $\not p_1 u(p_1) = m u(p_1)$, so that the last two terms on the right-hand side cancel when applied to $u(p_1)$. Thus, one obtains

$$\overline{u(q_1)} \not\varepsilon' \frac{\not p_1 + \not p_2 + m\mathbb{1}}{2 p_1 \cdot p_2} \not p_2 u(p_1) = \overline{u(q_1)} \not\varepsilon'\, u(p_1) \,.$$

Regarding the second term, the replacement $\not\varepsilon \to \not p_2$ turns the numerator into

$$\not p_2 (\not p_1 - \not q_2 + m\mathbb{1}) = \not p_2 (\not q_1 - \not p_2 + m\mathbb{1}) = 2 p_2 \cdot q_1 - \not q_1 \not p_2 + m \not p_2 \,.$$

Here the energy-momentum balance was used in the form $p_1 - q_2 = q_1 - p_2$ from whose square one obtains $p_1 \cdot q_2 = p_2 \cdot q_1$. Letting $\not q_1$ act on $\overline{u(q_1)}$ to the left, and making use of the Dirac equation $\overline{u(q_1)} \not q_1 = m \overline{u(q_1)}$, yields

$$\overline{u(q_1)} \left(-\not p_2 \frac{(\not p_1 - \not q_2) + m\mathbb{1}}{2 p_1 \cdot q_2} \not\varepsilon' \right) u(p_1) = -\overline{u(q_1)} \not\varepsilon'\, u(p_1) \,.$$

Indeed, the sum of the two contributions gives zero.

The reader is invited to work out the second test of gauge invariance where $\not\varepsilon'$ is replaced by $\not q_2$ and where she or he should find a vanishing result, too.

2. The test performed in the previous example refers to a general property. For this reason terms in which $\varepsilon_\mu^{(\lambda)}(k)$ is replaced by k_μ are called *gauge terms*. As the process that we are studying here contains only external, and hence transverse photons the amplitude (10.36) contains only the polarizations $\lambda = 1$ and $\lambda = 2$. Nevertheless, in the calculation of $|T|^2$ and of the sums over spins one may include all four polarizations of the photons, i.e. the unphysical degrees of freedom $\varepsilon^{(0)}$ and $\varepsilon^{(3)}$, as well. Indeed, it will turn out that the latter yield gauge terms of the kind just described whose contribution vanishes.
3. Compton scattering on the positron is treated in complete analogy to the example above. The two diagrams of order $n = 2$ are drawn in Fig. 10.5. Following

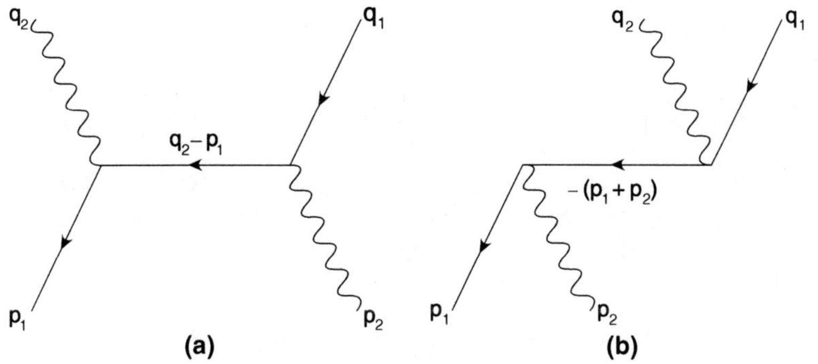

Fig. 10.5 Compton scattering on the positron: instructive as an example for a process in tree approximation, though experimentally quite exotic

the rules (R1)–(R8) they are transcribed into a scattering amplitude which reads

$$
\begin{aligned}
&T(e^+\gamma \to e^+\gamma) \\
&= -\frac{e^2}{(2\pi)^6}\overline{v(p_1)} \\
&\left(\not{\varepsilon}'\frac{(\not{q}_2-\not{p}_1)+m\mathbb{1}}{(p_1-q_2)^2-m^2}\not{\varepsilon}+\not{\varepsilon}\frac{-(\not{p}_1+\not{p}_2)+m\mathbb{1}}{(p_1+p_2)^2-m^2}\not{\varepsilon}'\right)v(q_1) \\
&= -\frac{e^2}{(2\pi)^6}\overline{v(p_1)} \\
&\left(-\not{\varepsilon}'\frac{(\not{q}_2-\not{p}_1)+m\mathbb{1}}{2p_1\cdot q_2}\not{\varepsilon}+\not{\varepsilon}\frac{-(\not{p}_1+\not{p}_2)+m\mathbb{1}}{2p_1\cdot p_2}\not{\varepsilon}'\right)v(q_1)\,, \qquad (10.38)
\end{aligned}
$$

where, again, $p_2^2 = 0 = q_2^2$ and $p_1^2 = m^2 = q_1^2$ were inserted. It is a simple exercise to perform the analogous test of gauge invariance.

We continue with the calculation of the differential cross section

$$
\begin{aligned}
&\mathrm{d}\sigma(e^-\gamma \to e^-\gamma) = (2\pi)^{10}\delta(q_1+q_2-p_1-p_2) \\
&\frac{1}{4E_{p_1}E_{p_2}\,|\boldsymbol{v}|}\frac{1}{4}\sum_{r,s}\sum_{\lambda,\lambda'}\left|T(e^-\gamma \to e^-\gamma)\right|^2\frac{\mathrm{d}^3q_1}{2E_{q_1}}\frac{\mathrm{d}^3q_2}{2E_{q_2}}\,. \qquad (10.39)
\end{aligned}
$$

The first factor right after the δ-distribution is due to the flux factor (8.28), the second factor is due to the averaging over the polarizations of the electron and the photon (there are two for each of them). With (10.33), $m_1 = m$, and $M = 0$ the modulus of the spatial momentum is $\kappa = (s-m^2)/(2\sqrt{s})$, and from (8.30) the flux factor becomes

$$
4E_{p_1}E_{p_2}\,|\boldsymbol{v}| = 4\kappa\sqrt{s} = 2(s-m^2)\,.
$$

Evaluation of the invariant squared momentum transfer t by means of (10.35) yields

$$t = -\frac{\left(s - m^2\right)^2}{2s}\left(1 - \cos\theta\right),$$

so that $\mathrm{d}(\cos\theta)$ can be expressed by $\mathrm{d}t$. Since the cross section, when averaged over spins, cannot depend on ϕ one integrates over this angle. Thus,

$$2\pi\,\mathrm{d}(\cos\theta) = 2\pi\frac{\mathrm{d}(\cos\theta)}{\mathrm{d}t}\mathrm{d}t = 2\pi\frac{2s}{(s - m^2)^2}\,\mathrm{d}t,$$

and the differential cross $\mathrm{d}\sigma/\mathrm{d}\Omega$ can be converted to the invariant cross section $\mathrm{d}\sigma/\mathrm{d}t$ using

$$\frac{\mathrm{d}\sigma}{\mathrm{d}t} = \frac{2s}{(s - m^2)^2}\int_0^{2\pi}\mathrm{d}\phi\,\frac{\mathrm{d}\sigma}{\mathrm{d}\Omega}. \tag{10.40}$$

The remaining procedure is now well defined: one calculates $\mathrm{d}\sigma/\mathrm{d}\Omega$ in the center-of-mass system, expresses the result as a function of the invariants s and t, and inserts in (10.40). The resulting expression $\mathrm{d}\sigma/\mathrm{d}t$ is Lorentz invariant as a whole and, therefore, may later be evaluated in any frame of reference which is defined by an experimental set-up.

The differential cross section in the center-of-mass system is obtained by integration of the general formula (10.39) over the momentum $\boldsymbol{q}_1$ of the outgoing electron and over the modulus κ' of the momentum of the outgoing photon. The integration $\int \mathrm{d}^3 q_1 \ldots$ is neutralized by the three spatial δ-distributions so that, using spherical polar coordinates for q_2, $\mathrm{d}^3 q_2 = \kappa'^2 \mathrm{d}\kappa'\,\mathrm{d}\Omega$, one obtains the intermediate result

$$\frac{\mathrm{d}\sigma}{\mathrm{d}\Omega} = (2\pi)^{10}\frac{1}{2(s - m^2)}$$
$$\times \int_0^\infty \kappa'^2 \mathrm{d}\kappa'\,\frac{1}{(2E_{q_1}2\kappa')}\,\frac{1}{4}\sum_{\text{Spins}}|T|^2\,\delta^{(1)}(E_{q_1} + \kappa' - \sqrt{s}).$$

Here $E_{q_2} = \left|\boldsymbol{q}_2\right| = \kappa'$ was inserted and we made use of the fact that the sum of the energies in the initial state equals $\sqrt{s}$ (in the center-of-mass system). The energy of the electron in the final state is $E_{q_1} = \sqrt{\kappa'^2 + m^2}$ and there remains the integral

$$J \equiv \int_0^\infty \frac{\kappa' \mathrm{d}\kappa'}{\sqrt{\kappa'^2 + m^2}}\delta^{(1)}(\sqrt{\kappa'^2 + m^2} + \kappa' - \sqrt{s})\sum |T|^2.$$

Of course, J receives contributions only when

$$\kappa' = \frac{s - m^2}{2\sqrt{s}},$$

and we have to deal with an integral of the type $\int \mathrm{d}x \; f(x)\delta\big(g(x)\big)$ which one obtains from the formula (cf. Appendix A.1)

$$\delta\big(g(x)\big) = \sum_i \frac{1}{|g'(x_i)|}\delta(x - x_i)$$

where the sum runs over simple zeroes only. One obtains

$$J = \sum_{\text{Spins}} |T|^2 \frac{s - m^2}{2s} .$$

Collecting all factors, inserting (10.40), and taking into account that in natural units $e^2 = 4\pi\alpha$, one finds

$$\frac{\mathrm{d}\sigma}{\mathrm{d}t} = \frac{\alpha^2\pi}{(s - m^2)^2} \frac{1}{4} \sum_{\text{Spins}} \left| \frac{(2\pi)^6}{e^2} T \right|^2 .$$

Finally, one calculates the spin sums of $|T|^2$. We start from (10.36), and insert $s = (p_1 + p_2)^2$, $u = (p_1 - q_2)^2$ in the numerators, thus obtaining

$$M \equiv -\frac{(2\pi)^6}{e^2} T = \varepsilon_\mu^{(\lambda')}(q_2)\, \overline{u^{(s)}(q_1)} Q^{\mu\nu} u^{(r)}(p_1)\, \varepsilon_\nu^{(\lambda)}(p_2) \quad \text{with}$$

$$Q^{\mu\nu} = \gamma^\mu \frac{\not{p}_1 + \not{p}_2 + m\mathbb{1}}{s - m^2}\gamma^\nu + \gamma^\nu \frac{\not{p}_1 - \not{q}_2 + m\mathbb{1}}{u - m^2}\gamma^\mu .$$

As noted above, in principle, the sums over λ and over λ' should comprise only the values 1 and 2. However, the following argument shows that one may as well sum over all four polarizations, including the longitudinal and timelike polarizations, without modifying the value of the cross section. As an example, we consider the incoming photon, the case of the outgoing photon being very similar. The relation (7.142) reads

$$\varepsilon_\mu^{(\lambda)}(k)\; g_{\lambda\bar{\lambda}}\; \varepsilon_\nu^{(\bar{\lambda})}(k) = g_{\mu\nu} .$$

Using the formulae for $\varepsilon_\mu^{(0)}(k)$ and $\varepsilon_\mu^{(3)}(k)$ that were given in Sect. 7.5.2 above, one has

$$\sum_{\lambda=1}^{2} \varepsilon_\mu^{(\lambda)}(k)\varepsilon_\nu^{(\lambda)}(k) = -g_{\mu\nu} + \varepsilon_\mu^{(0)}(k)\varepsilon_\nu^{(0)}(k) - \varepsilon_\mu^{(3)}(k)\varepsilon_\mu^{(3)}(k)$$

$$= -g_{\mu\nu} + \frac{1}{k \cdot t}\big(k_\mu t_\nu + t_\mu k_\nu - k_\mu k_\nu/(k \cdot t)\big) .$$

The second term on the right-hand side contains three gauge terms, i.e. terms which are proportional to either k_μ or k_ν or to both. Therefore, in the sums over the polarizations of the incoming as well as the outgoing photon the following replacements do not change the result:

$$\sum_{\lambda=1}^{2} \varepsilon_\mu^{(\lambda)}(p_2)\varepsilon_\nu^{(\lambda)}(p_2) \longmapsto -g_{\mu\nu} , \quad \sum_{\lambda'=1}^{2} \varepsilon_\sigma^{(\lambda')}(q_2)\varepsilon_\tau^{(\lambda')}(q_2) \longmapsto -g_{\sigma\tau} . \qquad (10.41)$$

This is a reflection of the general observation made earlier in Sect. 7.5.3: the contributions of timelike and longitudinal photons cancel in the observables.

The sums over the spin orientations of the electron are performed by means of the trace techniques. This yields

$$\frac{1}{4}\sum_{\text{Spins}}|M|^2 = \frac{1}{4}\,\text{tr}\,\left\{(\not{q}_1 + m\mathbb{1})Q^{\sigma\tau}(\not{p}_1 + m\mathbb{1})\widetilde{Q}_{\sigma\tau}\right\}$$

$$= \frac{1}{4}\,\text{tr}\,\left\{(\not{q}_1 + m\mathbb{1})Q^{\sigma\tau}(\not{p}_1 + m\mathbb{1})Q_{\tau\sigma}\right\}$$

where the relation

$$\widetilde{Q}_{\sigma\tau} \equiv \gamma^0\left(Q_{\sigma\tau}\right)^{\dagger}\gamma^0 = Q_{\tau\sigma}$$

was inserted (note the position of the indices!). This relation is easily verified.

Of course, at this point one must replace $Q^{\mu\nu}$ by its complete expression, then work out the traces by means of the formulae of Sect. 9.3.3, and write the result in terms of the variables s, t, and u. Note, however, that one may halve the effort by making use of the symmetry $s \leftrightarrow u$ noted previously. In view of this symmetry we write

$$\frac{1}{4}\sum_{\text{spins}}|M|^2 = a(s,u) + b(s,u) + a(u,s) + b(u,s)\,,$$

where the functions $a(s,u)$ and $b(s,u)$ are given by

$$a(s,u) = \frac{1}{4}\frac{1}{(s-m^2)^2}$$
$$\times\,\text{tr}\,\left\{(\not{q}_1+m\mathbb{1})\gamma^{\mu}(\not{p}_1+\not{p}_2+m\mathbb{1})\gamma^{\nu}(\not{p}_1+m\mathbb{1})\gamma_{\nu}(\not{p}_1+\not{p}_2+m\mathbb{1})\gamma_{\mu}\right\},$$

$$b(s,u) = \frac{1}{4}\frac{1}{(s-m^2)(u-m^2)}$$
$$\times\,\text{tr}\,\left\{(\not{q}_1+m\mathbb{1})\gamma^{\mu}(\not{p}_1+\not{p}_2+m\mathbb{1})\gamma^{\nu}(\not{p}_1+m\mathbb{1})\gamma_{\mu}(\not{p}_1-\not{q}_2+m\mathbb{1})\gamma_{\nu}\right\}.$$

One then uses the formulae of Sect. 9.3.3 and, specifically, the summed expressions (9.121)–(9.123), and converts the scalar products to the variables s and u according to

$$p_1\cdot p_2 = q_1\cdot q_2 = \frac{1}{2}(s-m^2)\,,\qquad p_1\cdot q_1 = m^2 - \frac{1}{2}t = \frac{1}{2}(s+u)\,,$$
$$p_1\cdot q_2 = q_1\cdot p_2 = \frac{1}{2}(m^2-u)\,,\qquad p_2\cdot q_2 = -\frac{1}{2}t = \frac{1}{2}(s+u)-m^2\,.$$

In this way one obtains

$$a(s,u) = \frac{2}{(s-m^2)^2}\left\{4m^4 - (s-m^2)(u-m^2) + 2m^2(s-m^2)\right\},$$

$$b(s,u) = \frac{2m^2}{(s-m^2)(u-m^2)}\left\{4m^2 + (s-m^2) + (u-m^2)\right\}.$$

Finally, the invariant cross section is found to be

$$\frac{d\sigma}{dt} = \frac{8\pi\alpha^2}{(s-m^2)^2}\left\{\left(\frac{m^2}{s-m^2}+\frac{m^2}{u-m^2}\right)^2 + \frac{m^2}{s-m^2}+\frac{m^2}{u-m^2}-\frac{1}{4}\left(\frac{s-m^2}{u-m^2}+\frac{u-m^2}{s-m^2}\right)\right\}. \quad (10.42)$$

This result whose $s \leftrightarrow u$-symmetry is obvious, is Lorentz invariant: $d\sigma$ is a physical quantity and, hence, cannot depend on the frame of reference one has chosen, while the variables s, t, and u, by their very definition, are Lorentz scalars.

Remarks

1. Noting that the $i\varepsilon$ term in the denominator of the electron-positron propagator is irrelevant for tree diagrams, we dropped this prescription from the start. However, as soon as the contribution of a diagram contains integrations over genuine internal loops this rule is important.
   Although it certainly is instructive to calculate traces of the kind encountered here analytically and by means of paper and pencil, one nowadays makes use of algebraic program packages which allow to perform complicated traces in an efficient way.
2. The evaluation of the expression (10.42) in the laboratory system (where the electron in the initial state is at rest) is of special interest for experimental purposes. Let the energies of the photon before and after the scattering be denoted by ω and by ω', respectively, let the scattering angle be θ_L (this is the angle between the incoming and the outgoing photons). With $p_1 = (m, \mathbf{0})^T$ one has

$$\begin{aligned} s &= (p_1+p_2)^2 = m^2 + 2m\omega\,, \\ t &= (p_1-q_1)^2 = (q_2-p_2)^2 = -2\omega\omega'(1-\cos\theta_L)\,, \\ u &= (p_1-q_2)^2 = m^2 - 2m\omega' \\ &= 2m^2 - s - t = m^2 - 2m\omega + 2\omega\omega'(1-\cos\theta_L)\,. \end{aligned}$$

   The above equations for u yield a relation between ω' and ω which reads

$$\omega' = \frac{m\omega}{m+\omega(1-\cos\theta_L)}\,. \quad (10.43)$$

   In order to derive the cross section in the laboratory system from the invariant (10.42) we need the derivative of t by $\cos\theta_L$. One has

$$t = -2\omega\omega'(1-\cos\theta_L) = -2m\omega^2\frac{1-\cos\theta_L}{m+\omega(1-\cos\theta_L)}\,,$$

   from which one concludes $dt/d\cos\theta_L = 2\omega'^2$. Thus,

$$\frac{d\sigma}{d\Omega_L} = \frac{d\sigma}{dt}\frac{dt}{d\Omega_L} = \frac{d\sigma}{dt}\frac{\omega'^2}{\pi}\,.$$

Inserting the result (10.42) one obtains

$$\frac{\mathrm{d}\sigma}{\mathrm{d}\Omega_{\mathrm{L}}} = \frac{1}{2}\left(\frac{\alpha}{m}\right)^2\left(\frac{\omega'}{\omega}\right)^2\left\{\frac{\omega}{\omega'} + \frac{\omega'}{\omega} - \sin^2\theta_{\mathrm{L}}\right\}. \tag{10.44}$$

3. In calculating the cross Sects. (10.42) and (10.44) it was assumed that the incoming particles are unpolarized, and that the polarization of the outgoing ones is not measured. It is not difficult to generalize (10.44) to the case where the incoming photon is polarized and where the polarization of the outgoing photon is determined. One finds

$$\left(\frac{\mathrm{d}\sigma}{\mathrm{d}\Omega_{\mathrm{L}}}\right)^{\gamma\text{ pol.}} = \frac{1}{4}\left(\frac{\alpha}{m}\right)^2\left(\frac{\omega'}{\omega}\right)^2\left\{\frac{\omega}{\omega'} + \frac{\omega'}{\omega} + 4\left(\boldsymbol{\varepsilon}'\cdot\boldsymbol{\varepsilon}\right)^2 - 2\right\}. \tag{10.45}$$

This formula is due to O. Klein and Y. Nishina. In the limit of small energies of the photon one has $\omega' \approx \omega \ll m$ and the formula (10.45) goes over into the cross section for Thomson scattering

$$\left(\frac{\mathrm{d}\sigma}{\mathrm{d}\Omega_{\mathrm{L}}}\right)_{\text{Thomson}} = \left(\frac{\alpha}{m}\right)^2\left(\boldsymbol{\varepsilon}'\cdot\boldsymbol{\varepsilon}\right)^2$$

that we obtained in the framework of the semi-classical theory of Chap. 7.
4. Of course, one may as well take the initial electron to be polarized and/or consider the possibility of discriminating the polarization of the electron in the final state. The calculation of the traces which makes use of the formulae (9.97) (and of (9.98) in the case of positrons) is not much more involved than the unpolarized case, and, hence, may be done by hand. Nevertheless, this might be a good exercise for trying one of the algebraic program packages on a computer.
5. The invariant notation of $\sum|T|^2$ in terms of the variables (s, t, u) is particularly useful because one may obtain the processes of the crossed channels by analytic continuation to the corresponding kinematic domains. For example, in the case of Compton scattering there is not only the symmetry between the s- and the u-channels. Its t-channel partner $e^+e^- \to \gamma\gamma$ is of special interest, too. Pair annihilation in flight of electrons whose polarization is known, and of polarized positrons is often used to measure the polarization of the positron. The formulae for cross sections with polarized lepton partners are found in the literature.[3]

Bhabha Scattering $e^+e^- \to e^+e^-$: This example is the first which contains the photon propagator. To lowest order, $n = 2$, the relevant diagrams are those of Fig. 10.6. The virtual photon may be exchanged between the electron and the positron but it may also be created by pair annihilation and disappear by pair creation. Electrons and positrons occur only in external lines and, hence, are on their mass shell. The photon appears in an internal line only and, hence, is not on its mass shell. As before,

[3] Cf. L.A. Page, Phys. Rev. **106**, 394, 1957; V.N. Baier, Enrico Fermi School "Physics with intersecting storage rings", Academic Press, 1971.

one writes down the T-amplitude, avoiding the detour via the R-matrix. Using the same notations as in Fig. 10.6 it is seen to be

$$T(e^+e^- \to e^+e^-) = -\frac{e^2}{(2\pi)^6}\left\{\overline{u(q_-)}\gamma^\mu u(p_-)\frac{-g_{\mu\nu}}{(p_- - q_-)^2}\overline{v(p_+)}\gamma^\nu v(q_+)\right.$$
$$\left. + \overline{v(p_+)}\gamma^\mu u(p_-)\frac{-g_{\mu\nu}}{(p_- + p_+)^2}\overline{u(q_-)}\gamma^\nu v(q_+)\right\} . \quad (10.46)$$

The invariant scalar products contained in the denominators are

$$(p_- - q_-)^2 = t \,, \qquad (p_- + p_+)^2 = s \,.$$

We skip the calculation of the invariant cross section because it is very similar to the preceding example. One finds

$$\frac{\mathrm{d}\sigma}{\mathrm{d}t} = \frac{2\pi\alpha^2}{s(s-4m^2)}\left\{\frac{1}{s^2}\Big((t-2m^2)^2 + (u-2m^2)^2 + 4m^2 s\Big)\right.$$
$$\left. + \frac{1}{t^2}\Big((s-2m^2)^2 + (u-2m^2)^2 + 4m^2 t\Big) + \frac{2}{st}(u-2m^2)(u-6m^2)\right\} . \quad (10.47)$$

The $s \leftrightarrow t$ symmetry of this formula is obvious, thus confirming our earlier and more general argument. As in the previous case, the evaluation of the cross section is easily done in either the center-of-mass, or the laboratory systems. One just needs to insert the representations of s, t, and u in the respective frame of reference.

Pair Creation of Leptons and Quarks: There exist collider rings which allow to focus a positron beam onto an electron beam, both of well-defined energies. Besides elastic scattering, $e^+e^- \to e^+e^-$, one observes various production processes in which the e^+e^- pair disappears, while states are produced which carry the (additive) quantum numbers of the vacuum. Among these, the leptonic production reactions

$$e^+ + e^- \longrightarrow \mu^+ + \mu^- \quad \text{and} \quad e^+ + e^- \longrightarrow \tau^+ + \tau^- \,, \qquad (10.48)$$

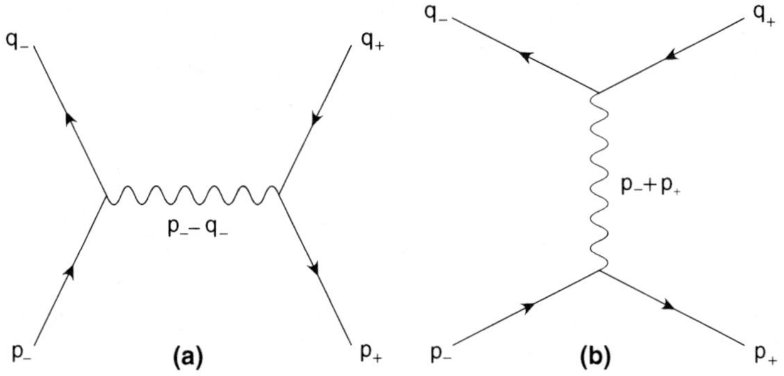

Fig. 10.6 Tree diagrams for Bhabha scattering; **a** exchange of a virtual photon between electron and positron; **b** electron and positron annihilate into a virtual photon which subsequently passes into a e^+e^- pair

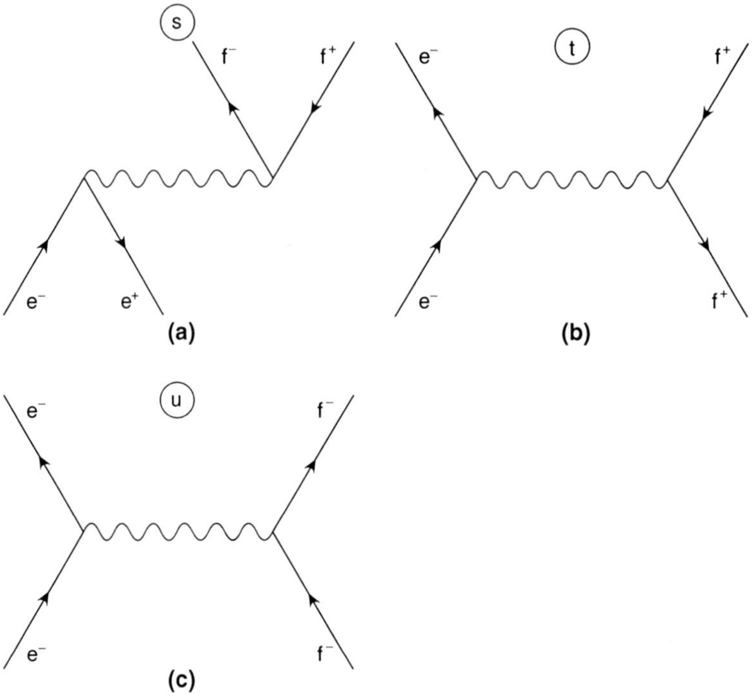

Fig. 10.7 a–c If $e^+e^- \to f^+f^-$ is the s-channel, then the associated t- and u-channel reactions are $e^-f^+ \to e^-f^+$ and $e^-f^- \to e^-f^-$, respectively. The amplitude for process (**c**) differs from the one for process (**b**) only in the sign of the right vertex' charge

as well as the creation of quark-antiquark pairs

$$e^+ + e^- \longrightarrow \overline{q} + q\,, \qquad q = u, d, s, c, b, t\,, \tag{10.49}$$

are of special importance. Here, the symbols in (10.49) stand for the *up*, *down*, *strange*, *charm*, *bottom*, and *top* quarks (in the order of increasing masses). Since the quarks never occur as free particles, the $q\overline{q}$-states "hadronize" into more complicated *out*-states of physical hadrons. Thus, in the case of quarks, we calculate no more than the first step of what really happens in experiment.

Suppose, as seen from the laboratory system, the positron and electron beams are focussed collinearly onto each other and have the same energy. Then the experiment as recorded in the laboratory, takes place in the kinematics of the center-of-mass system.[4]

Consider first pair creation $e^+e^- \to f^+f^-$ into a lepton-antilepton pair in which f is not an electron. Then, to order $n = 2$ only the diagram (a) of Fig. 10.7 contributes

[4] There are also asymmetric colliders in which the colliding beams do not have the same energy. Furthermore, the beams may cross at an angle that differs from 180°.

(this is the same as diagram (b) of Fig. 10.6) with the electron of the final state replaced by f^-, the positron by f^+. Using the same notation for the momenta as in Fig. 10.6, writing $m_e \equiv m$ and $m_f \equiv M$, and inserting $s = (p_- + p_+)^2$, the T-amplitude reads

$$T(e^+e^- \to f^+f^-) = \\ -\frac{e^2}{(2\pi)^6}\left(\overline{v_e(p_+)}\gamma^\mu u(p_-)\right)\frac{g_{\mu\nu}}{s}\left(\overline{u_f(q_-)}\gamma^\nu v_f(q_+)\right). \tag{10.50}$$

The flux factor (8.30) for this case is, using (10.30),

$$E_{e^+}E_{e^-}\,|\boldsymbol{v}| = \kappa\sqrt{s} = \frac{1}{2}\sqrt{s(s-4m^2)}\,.$$

After integration of the general expression of Sect. 8.2.3 over $\boldsymbol{q}_+$, and using $\kappa' \equiv |\boldsymbol{q}_+|$, there remains to calculate

$$\int \mathrm{d}q_+^3\,\mathrm{d}\sigma = \frac{(2\pi)^{10}}{8\sqrt{s(s-4m^2)}}\mathrm{d}\Omega \\ \times \int_0^\infty \frac{\kappa'^2\mathrm{d}\kappa'}{\kappa'^2+M^2}\delta(\sqrt{s}-2\sqrt{\kappa'^2+M^2})\frac{1}{4}\sum_{\text{Spins}}|T|^2\,.$$

The argument of the δ-distribution has its only, single zero at $\kappa' = \sqrt{s-4M^2}/2$. Using the well-known rule for the evaluation of this distribution, one finds

$$\int \mathrm{d}^3q_+\,\mathrm{d}\sigma = \frac{\sqrt{s-4M^2}}{s\sqrt{s-4m^2}}\mathrm{d}\Omega\,\frac{(2\pi)^{10}}{64}\sum_{\text{Spins}}|T|^2\,.$$

The trace rules which should be familiar by now, yield the absolute square of the amplitude, averaged and summed over the spin orientations in the initial and final states, respectively,

$$\sum_{\text{Spins}}\left|\frac{(2\pi)^6}{e^2}T(e^+e^- \to f^+f^-)\right|^2 \\ = \frac{1}{s^2}\mathrm{tr}\,\left\{(\not p_+ - m\mathbb{1})\gamma^\mu(\not p_- + m\mathbb{1})\gamma^\sigma\right\}\,\mathrm{tr}\,\left\{(\not q_- + M\mathbb{1})\gamma_\mu(\not q_+ - M\mathbb{1})\gamma_\sigma\right\} \\ = \frac{16}{s^2}\left\{p_+^\mu p_-^\sigma - (p_+\cdot p_- + m^2)g^{\mu\sigma} + p_+^\sigma p_-^\mu\right\} \\ \times\left\{q_{-\,\mu}q_{+\,\sigma} - (q_-\cdot q_+ + M^2)g_{\mu\sigma} + q_{-\,\sigma}q_{+\,\mu}\right\} \\ = \frac{16}{s^2}\Big\{2p_-\cdot q_-\,p_+\cdot q_+ + 2p_-\cdot q_+\,p_+\cdot q_- \\ + 2M^2p_-\cdot p_+ + 2m^2q_-\cdot q_+ + 4m^2M^2\Big\} \\ = \frac{16}{s^2}\Big\{4m^2M^2 + M^2(s-2m^2) + m^2(s-2M^2) \\ + \frac{1}{2}(m^2+M^2-t)^2 + \frac{1}{2}(m^2+M^2-u)^2\Big\}\,.$$

This expression is remarkable: Indeed, it is seen to be invariant under the exchange $m \leftrightarrow M$ as well as under $t \leftrightarrow u$. The former symmetry becomes obvious if one recalls that the reaction may take place in both directions and that there is invariance under time reversal. The latter symmetry follows from Fig. 10.7 and from the remark made above which said that the amplitudes for $e^- f^- \to e^- f^-$ and for $e^- f^+ \to e^- f^+$ differ only by the sign of the charge at the $\overline{f} f \gamma$-vertex.

It is then not difficult to express the differential cross section in the center-of-mass system in terms of s and of $z := \cos\theta$. One finds, with $e^2 = 4\pi\alpha$,

$$\frac{\mathrm{d}\sigma}{\mathrm{d}\Omega} = \frac{\alpha^2\sqrt{s-4M^2}}{4s\sqrt{s}} \left\{1 + z^2 + \frac{4(m^2+M^2)}{s}(1-z^2) + \frac{16m^2M^2}{s^2}z^2\right\} . \qquad (10.51)$$

In practice one will often have $M^2 \gg m^2$ and $s \gg m^2$ so that the terms in the electron mass may be neglected. Furthermore, it is useful to introduce the β-factor of the created particle f,

$$\beta^{(f)} = \frac{|\boldsymbol{q}_-|}{E_{q_-}} = \frac{|\boldsymbol{q}_+|}{E_{q_+}} = \frac{\kappa'}{\sqrt{s}/2} = \frac{\sqrt{s-4M^2}}{\sqrt{s}} , \qquad (M \equiv m_f) ,$$

Then one has $4M^2/s = 1 - \beta^{(f)2}$ and obtains

$$\frac{\mathrm{d}\sigma}{\mathrm{d}\Omega} \approx \frac{\alpha^2}{4s}\beta^{(f)} \left\{(1+\cos^2\theta) + (1-\beta^{(f)2})\sin^2\theta\right\} .$$

Let us calculate the cross section integrated over all angles, in this approximation, i.e.

$$\sigma(e^+e^- \to f^+f^-) = \int \mathrm{d}\Omega \, \frac{\mathrm{d}\sigma}{\mathrm{d}\Omega} \approx \frac{4\pi\alpha^2}{3s}\left\{1 + \frac{1}{2}\left(1-\beta^{(f)2}\right)\right\}\beta^{(f)}$$
$$= \frac{4\pi\alpha^2}{3s}\frac{(s+2M^2)\sqrt{s-4M^2}}{s^{3/2}} . \qquad (10.52)$$

This integrated cross section is plotted in Fig. 10.8, as a function of the square of the center-of-mass energy s.

An application for e^+e^--colliders is provided by the search for heavy quark-antiquark pairs at high energies in which the integrated production cross Sect. (10.52) is normalized to the cross section for a $\mu^+\mu^-$ pair. If s is large enough such that m_μ^2 can be neglected, too, $s \gg m_\mu^2$, one has

$$\sigma(e^+e^- \to \mu^+\mu^-) \approx \frac{4\pi\alpha^2}{3s} . \qquad (10.53)$$

Assume now that there exists a series of particles with masses M_i, and charges q_i (in units of the elementary charge), with $M_i \gg m_\mu$, then the total cross section normalized to the muon-antimuon cross section (10.53), is equal to

$$\frac{\sigma\left(e^+e^- \to \sum(f^+f^-)\right)}{\sigma(e^+e^- \to \mu^+\mu^-)} \approx 1 + \sum_{i=1\,(i\neq\mu)}^{N} q_i^2 \frac{(s+2M_i^2)\sqrt{s-4M_i^2}}{s^{3/2}}$$
$$= 1 + \sum_{i=1\,(i\neq\mu)}^{N} q_i^2 \sqrt{1 - \frac{12M_i^4}{s^2} - \frac{16M_i^6}{s^3}} . \qquad (10.54)$$

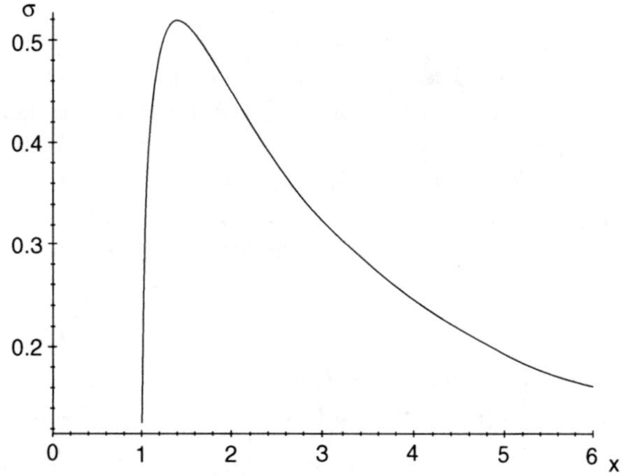

Fig. 10.8 The integrated cross section (10.52) for the creation of a $f^+ f^-$ pair out of $e^+ e^-$ annihilation in units of $\pi\alpha^2/3M^2/3$ as a function of the dimensionless variable $s/(4M^2)$

The sum over the new species of increasing mass terminates where $4M_{N+1}^2 \geq s$, i.e. as soon as the center-of-mass energy no longer allows to create the $f^+ f^-$-pair. In fact, it is very easy, at least in principle, to discover new fundamental particles: At every threshold for the production of a new pair $f_i^+ f_i^-$ the ratio (10.54) increases by a step whose height is proportional to the squared charge of these particles.

10.2 Radiative Corrections, Regularization, and Renormalization

As soon as one goes beyond the realm of tree diagrams and studies Feynman diagrams containing closed fermion loops, one encounters serious mathematical difficulties which are not so easy to repair. In fact, a complete and satisfactory treatment would go beyond the scope of this book and one should rather consult an advanced course on quantum field theory or monographs on this field. On the other hand, the analysis of higher orders of perturbation theory reveals new phenomena which are physically interesting in their own right and which raise rather basic issues. For anyone interested in fundamental physics, these phenomena should be part of his or her general physical culture. This section provides some insight into the difficulties to master, as well as an impression of the predictions of quantum electrodynamics for radiative corrections which can be tested in experiment.

10.2.1 Self-Energy of Electrons to Order $\mathcal{O}(e^2)$

The diagram of Fig. 10.9 shows a virtual process whereby an electron emits and reabsorbs a photon in such a way that both the electron and the photon of the closed loop are not on their respective mass shells. (As we know, this description is not quite correct. The fermionic intermediate state is represented by the electron-positron propagator and, hence, the diagram stands for two virtual processes, one containing a virtual electron, and one a virtual positron.) Independently of whether the lines with momentum p and q are external or internal lines, the electron in this diagram only "talks to itself". Thus, this process describes a *self-energy* of order e^2 whose prime effect is to change the mass m of the electron.

A short-hand notation for the fermionic propagator in momentum space is

$$S_F(p) := \frac{\not{p} + m\mathbb{1}}{p^2 - m^2 + i\varepsilon}. \tag{10.55}$$

Replacing the *internal* fermion line in the left part of Fig. 10.10 by the sum shown in the right part of Fig. 10.10 this implies the substitution

$$S_F(p) \longmapsto S_F(p) + S_F(p)\Sigma^{(0)}(p)S_F(p),$$

where $\Sigma^{(0)}(p)$ is given by the integral over the loop,

$$\Sigma^{(0)}(p) = -i\frac{e^2}{(2\pi)^4}\int d^4k\, \gamma^\mu S_F(p-k)\gamma_\mu \frac{1}{k^2 + i\varepsilon}.$$

This expression must be handled with some care because the integral is seen to have two distinct problems. On the one hand, it runs into difficulties for $k^2 \to 0$, i.e. for *small* values of the momentum k where the virtual photon is "soft". On the other hand, for *large* values of k it behaves like $\int d^4k/k^3$. The first problem is called the *infrared problem*, in analogy to the spectrum of visible light, and is due to the fact that the photon has no mass. One may temporarily repair it by assigning a small but finite mass m_γ to the photon, and by collecting *all* processes of a given order,

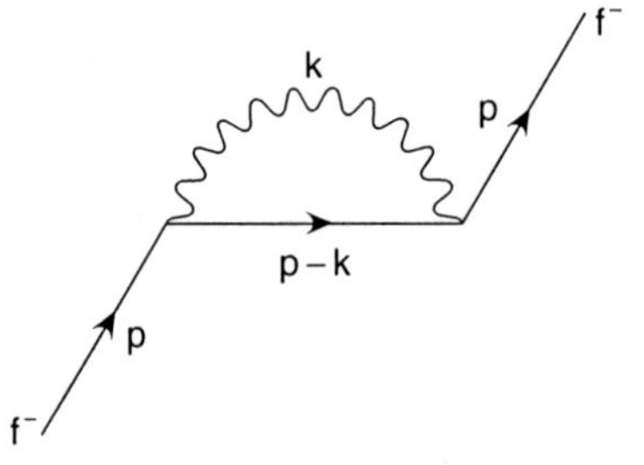

Fig. 10.9 An external or internal fermion line with momentum p is modified by a closed loop in which a virtual photon and a virtual electron or positron are circulating

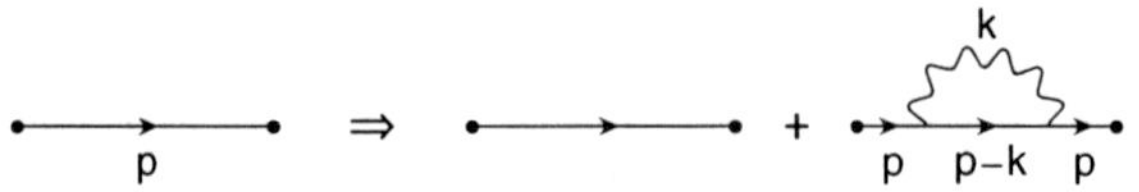

Fig. 10.10 An internal fermion line is replaced as stated by the sum of two diagrams. The vertices between which the fermion line proceeds are denoted by *dots*

hoping that the result stays finite when m_γ is sent to zero in the final result. Thus, one substitutes

$$\Sigma^{(0)} \mapsto \Sigma(p) = -\mathrm{i}\frac{e^2}{(2\pi)^4}\int \mathrm{d}^4k\, \gamma^\mu S_F(p-k)\gamma_\mu \frac{1}{k^2 - m_\gamma^2 + \mathrm{i}\varepsilon}\,.$$

The divergence at *large* values of k which presumably is a logarithmic divergence, is more serious. As this problem occurs in the range of "hard" photons, again by analogy to visible light, one talks about an *ultraviolet divergence.*

The first topic of this section is a more detailed analysis of the quantity $\Sigma(p)$ which will serve to justify the conjecture that the integral indeed diverges logarithmically, and, in spite of this, to find ways of giving it a well-defined physical meaning. From here on the purely calculational aspects may appear somewhat dry and technical. Therefore, some simple but tedious intermediate calculations are relegated to appendices so that one can concentrate on the main ideas of the program.

We start with the example of $\Sigma(p)$ which can be rewritten as follows, cf. Appendix A.3:

$$\Sigma(p) = \frac{\alpha}{2\pi}\int_0^1 \mathrm{d}z\, \left(2m\mathbb{1} - \not{p}(1-z)\right) I(p, m, m_\gamma)\,, \tag{10.56}$$

where $I(p, m, m_\gamma)$ is given by

$$I(p, m, m_\gamma) := \int_0^\infty \frac{\mathrm{d}\lambda}{\lambda}\, \exp\left\{\mathrm{i}\lambda\Big(p^2 z(1-z) - m^2 z - m_\gamma^2(1-z) + \mathrm{i}\varepsilon\Big)\right\}\,. \tag{10.57}$$

Equation (10.57) shows very clearly where to identify the ultraviolet problem: The integral over λ is logarithmically divergent.

Facing this difficulty one should not capitulate. Rather, one should realize that with this isolated contribution of perturbation theory one moves even further away from what is really measurable. If we correct internal or external fermion lines by this term then there are a number of other corrections which occur at the same order in α, and that we must also take into account, for reasons of consistency. Furthermore, there remains the question mentioned above: Which of the parameters are physical (masses and charges) and how are they related to the parameters of the Lagrange density?

A mathematically rigorous treatment should consist in answering these questions in a single step and to reformulate the theory in such a way that there are no infinities anywhere, (assuming the theory to be renormalizable). An approach of this kind is cumbersome, mathematically challenging, and not very practical. Therefore, in a more heuristic treatment, one tries to assign a finite value to divergent expressions such as (10.56) by applying to them a procedure called *regularization.* In other terms, one designs a procedure which replaces $\Sigma(p)$ as well as other divergent results of perturbation theory by finite integrals. As we shall see, this amounts to split $\Sigma(p)$ into finite and infinite parts.

Obviously, such a procedure can only be meaningful if it meets the following conditions:

(a) The Lorentz covariance of the theory as well as the invariance of its observables under gauge transformations are not destroyed;
(b) Physical, measurable predictions of the theory do not depend on the specific choice of regularization of divergent quantities;
(c) The mathematically ill-defined terms in the perturbation series which depend on the method of regularization, can be absorbed by a redefinition of the parameters and, hence, remain unobservable.

There are various methods of regularization in quantum electrodynamics and other quantized gauge theories. Among them, a method of great practical importance is *dimensional regularization.* It consists in an analytic continuation of Lorentz scalar integrals in the spacetime dimension n, considered as a complex variable, such that the integrals become convergent and hence well-defined. A simple example may help to illustrate this method. Suppose we wish to make sense of the integral

$$J = \int \mathrm{d}^\nu p\, g(p^2)\,, \qquad \nu \in \mathbb{C}\,,$$

for complex values of the variable ν, knowing that it is well-defined for some real integer dimension $\nu = n$ of a vector space $\mathbb{R}^n$, with p a vector on $\mathbb{R}^n$ and with p^2 its squared norm. For integer dimension one can always introduce spherical polar coordinates and integrate over the entire sphere S^{n-1} such as to reduce the integral to a one-dimensional integral over the modulus $\kappa = |p|$, viz.

$$\mathrm{d}^n p = \kappa^{n-1}\mathrm{d}\kappa\, \mathrm{d}\phi \prod_{k=1}^{n-2} \sin^k \theta_k\, \mathrm{d}\theta_k\,.$$

The integral over the surface of the unit sphere S^{n-1} is $\int \mathrm{d}\Omega = 2\pi^{n/2}/\Gamma(n/2)$, and, therefore, one has

$$J = \frac{2\pi^{n/2}}{\Gamma(n/2)} \int_0^\infty \kappa^{n-1}\mathrm{d}\kappa\; g(\kappa^2)\,.$$

Assume, furthermore, that there is a domain in the complex ν-plane which contains the point $\nu = n$ and where the integral is convergent. Then the above formula yields an analytic continuation in the complex ν-plane, away from $\nu = n$. This provides the possibility to continue J to the physical dimension $\nu = 4$ and, if it is divergent, to diagnose the nature of its singularity as well as to separate the singular terms in the sense of the splitting described above.

There is another regularization procedure due to Pauli and Villars[5] which, though perhaps less practical for phenomenology, is useful for investigations of principle because it respects Lorentz covariance as well as gauge invariance at every step. It consists in introducing auxiliary particles with large masses M_i whose number and possibly unphysical couplings C_i are chosen such that the sum of all contributions, i.e. of physical and unphysical particles, becomes convergent. At the end of a calculation

[5] W. Pauli and F. Villars, Rev. Mod. Phys. **21**, 434, 1949.

one takes the limit $M_i \to \infty$, thus obtaining an additive splitting of the contribution one is studying into divergent parts (in the limit $M_i \to \infty$), and finite terms which are independent of the masses M_i and coupling constants C_i.

We illustrate the Pauli-Villars regularization by the example (10.56). In this example it is sufficient to introduce one auxiliary particle with mass M and an imaginary coupling $\mathrm{i}e$. This construction amounts to replace (10.57) as follows

$$I(p, m, m_\gamma) \longmapsto I(p, m, m_\gamma) - I(p, m, M) .$$

The critical integral I is replaced by a convergent one because

$$\int_0^\infty \frac{\mathrm{d}x}{x} \left(\mathrm{e}^{\mathrm{i}ax} - \mathrm{e}^{\mathrm{i}bx} \right) = \ln\left(\frac{b}{a}\right) .$$

The expression (10.56) is replaced by a finite, regularized quantity that we denote by $\Sigma^{\mathrm{reg}}(p)$,

$$\Sigma^{\mathrm{reg}}(p) = \frac{\alpha}{2\pi} \int_0^1 \mathrm{d}z \left(2m\mathbb{1} - \not{p}(1-z)\right) \times \ln\left(\frac{M^2(1-z)}{m^2 z + m_\gamma^2(1-z) - p^2 z(1-z) - \mathrm{i}\varepsilon} \right) .$$

The terms $p^2 z(1-z)$ and $m^2 z$ of the numerator are neglected as compared to $M^2(1-z)$ because M is assumed to be large anyway. Splitting the logarithm according to $\ln(b/a) = \ln(b/c) + \ln(c/a)$ the term Σ^{reg} can be rewritten

$$\begin{aligned}\Sigma^{\mathrm{reg}}(p) =& \frac{\alpha}{2\pi} \int_0^1 \mathrm{d}z \left(2m\mathbb{1} - \not{p}(1-z)\right) \ln\left(\frac{M^2(1-z)}{m^2 z^2 + m_\gamma^2(1-z)} \right) \\ &+ \frac{\alpha}{2\pi} \int_0^1 \mathrm{d}z \left(2m\mathbb{1} - \not{p}(1-z)\right) \\ &\times \ln\left(\frac{m^2 z^2 + m_\gamma^2(1-z)}{m^2 z + m_\gamma^2(1-z) - p^2 z(1-z) - \mathrm{i}\varepsilon} \right) .\end{aligned}$$

In the first term the limit $m_\gamma \to 0$ is harmless so that the term $m_\gamma^2(1-z)$ of the denominator can be dropped. Furthermore, the $\mathrm{i}\varepsilon$ prescription in the denominator of the second term is of no relevance. In the first term one writes

$$\ln\left(\frac{M^2(1-z)}{m^2 z^2}\right) = \ln\left(\frac{M^2}{m^2}\right) + \ln\left(\frac{1-z}{z^2}\right)$$

and notes that the second term of these yields a finite contribution which can be neglected for large values of M^2. Thus,

$$\int_0^1 \mathrm{d}z\big(2m\mathbb{1} - \not{p}(1-z)\big)\ln\left(\frac{M^2(1-z)}{m^2z^2}\right)$$

$$\approx \ln\left(\frac{M^2}{m^2}\right)\left(\frac{3}{2}m\mathbb{1} - \frac{1}{2}(\not{p} - m\mathbb{1})\right).$$

This yields

$$\Sigma^{\text{reg}}(p) \approx \frac{3\alpha}{4\pi}m\ln\left(\frac{M^2}{m^2}\right)\mathbb{1} - \frac{\alpha}{4\pi}\ln\left(\frac{M^2}{m^2}\right)(\not{p} - m\mathbb{1})$$
$$+\frac{\alpha}{2\pi}\int_0^1 \mathrm{d}z\big(2m\mathbb{1} - \not{p}(1-z)\big)$$
$$\times \ln\left(\frac{m^2z^2 + m_\gamma^2(1-z)}{m^2z + m_\gamma^2(1-z) - p^2z(1-z)}\right).$$

The regularized expression then takes the form

$$\Sigma^{\text{reg}}(p) \equiv A\mathbb{1} + B\big(\not{p} - m\mathbb{1}\big) + C(p), \tag{10.58}$$

$$A = \frac{3\alpha}{2\pi}m\ln\left(\frac{M}{m}\right), \quad B = -\frac{\alpha}{4\pi}m\ln\left(\frac{M^2}{m^2}\right). \tag{10.59}$$

The constants A and B depend on the auxiliary mass M, and both tend to infinity logarithmically when the limit $M \to \infty$ is taken. The function $C(p)$ is finite but vanishes whenever the fermion is on its mass shell $p^2 = m^2$. Thus, this part poses no problem. What is the role of the unphysical divergent quantities A and B?

In what follows we show that the term A disappears from the theory altogether provided one identifies properly the physical mass of the fermion. This is equivalent to saying that the mass parameter in the original Lagrange density is renormalized at the given order of perturbation theory. Regarding the term B, matters are different. This term must be discussed in connection with other radiative corrections of the same order, hoping that it might cancel against other divergent contributions.

10.2.2 Renormalization of the Fermion Mass

A simple but somewhat lengthy calculation shows that the regularized expression (10.58) can be written as follows (cf. Appendix A.4),

$$\Sigma^{\text{reg}}(p) = A\,\mathbb{1} + \big(\not{p} - m\,\mathbb{1}\big)\left[B + \Sigma^{\text{finite}}(p)\right], \tag{10.60}$$

where

$$\Sigma^{\text{finite}}(p)\,\mathbb{1} = \Sigma_a(p^2)\,\mathbb{1} + \left\{\frac{\not p + m\,\mathbb{1}}{p^2 - m^2} - 2m\,\mathbb{1}\,\frac{\partial}{\partial p^2}\bigg|_{p^2=m^2}\right\}\Sigma_b(p^2)\,,$$

$$\Sigma_a(p^2) \approx \frac{\alpha}{4\pi}\left(1 - \frac{m^2}{p^2}\right)\left\{1 + \left(1 + \frac{m^2}{p^2}\right)\Lambda(p^2)\right\},$$

$$\Sigma_b(p^2) \approx \frac{\alpha}{4\pi} m\left(1 - \frac{m^2}{p^2}\right)\left\{1 - \left(3 - \frac{m^2}{p^2}\right)\Lambda(p^2)\right\},$$

$$\Lambda(p^2) = -p^2\int_0^1 \mathrm{d}z\,\frac{1-z}{m^2(1-z) + m_\gamma^2 z - p^2 z(1-z)}\,.$$

The Lagrange density (10.11) from which the perturbative series was obtained, contained the bare, uncorrected mass parameters $m_f^{(0)}$. These would be identical with the measurable masses of the leptons if we could neglect all corrections of higher order. However, if corrections of the kind discussed above are included then the free Lagrange density should contain the *physical* masses m_f. This amounts to add to $\mathcal{L}$ a term of the kind

$$\mathcal{L}_M = \sum_f \delta m_f :\overline{\psi^{(f)}(x)}\psi^{(f)}(x):\,,\quad \text{with } \delta m_f = m_f - m_f^{(0)} \tag{10.61}$$

which contains the differences of the physical and the bare masses,

$$\mathcal{L} \longmapsto \mathcal{L}' = \mathcal{L} + \mathcal{L}_M\,.$$

Indeed, a free single-particle state will not be scattered only if the free Lagrange density contains the physical mass, so that the S-matrix is

$$\langle q|\mathbf{S}\,|\,p\rangle = \langle q|\,p\rangle = 2E_p\,\delta(\boldsymbol{q} - \boldsymbol{p})\,. \tag{10.62}$$

As m_f is corrected in every order of perturbation theory this means that one modifies the interaction picture which is used for the Dyson series, order by order.

The Feynman rules for the new term $\mathcal{L}_M$ are easily formulated. A diagram describing the action of this term onto a fermion line of the kind f is drawn as shown in Fig. 10.11, by means of an X on the external or internal fermion line. (As before, this holds for each of the three charged leptons. We discuss the case of the electron and drop the index f on the mass and on the field.) The additional term $\mathcal{L}_M$ contributes already at first order. For the example of an electron line one has

$$\begin{aligned}\langle q|\mathbf{R}^{(1)}\,|\,p\rangle &= \mathrm{i}\delta m\langle q|\int \mathrm{d}^4x\,:\overline{\psi^{(f)}(x)}\psi^{(f)}(x):|\,p\rangle \\ &= \mathrm{i}\frac{\delta m}{(2\pi)^3}(2\pi)^4\delta(\boldsymbol{q} - \boldsymbol{p})\overline{u(q)}u(p)\,.\end{aligned}$$

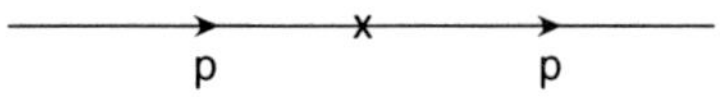

Fig. 10.11 The perturbation term $\mathcal{L}_M = \sum_f \delta m_f \colon \overline{\psi^{(f)}(x)}\psi^{(f)}(x)\colon$ does not contain a propagator, it creates and annihilates fermions at the same point of space-time and is illustrated by X

As we know, the diagram shown in Fig. 10.9 contributes in second order,

$$\begin{aligned}\langle q|\mathbf{R}^{(2)}\,|\,p\rangle &= -\frac{e^2}{(2\pi)^3}\delta(\boldsymbol{q}-\boldsymbol{p})\overline{u(q)}\\ &\quad\int \mathrm{d}^4k\;\gamma^\mu\frac{\not p - \not k + m\,\mathbb{1}}{(p-k)^2-m^2+\mathrm{i}\varepsilon}\gamma_\mu\frac{1}{k^2+\mathrm{i}\varepsilon}u(p)\\ &= -\mathrm{i}\frac{1}{(2\pi)^3}(2\pi)^4\delta(\boldsymbol{q}-\boldsymbol{p})\overline{u(q)}\Sigma(p)u(p)\,.\end{aligned}$$

The condition (10.62) implies that the sum of these two contributions be equal to zero. Inserting the regularized expression (10.60) and noting that $(\not p - m\,\mathbb{1})u(p)=0$, one concludes

$$\delta m = A \approx \frac{3\alpha m}{2\pi}\ln\left(\frac{M}{m}\right)\,. \tag{10.63}$$

This important result which is due to V. Weisskopf, in the light of the remark made at the beginning, has the following interpretation: Provided one introduces the physical mass of the electron into the equations of motion, right from the start, (and likewise for the other leptons), the constant A does not appear explicitly. It is hidden by the *renormalization* of the mass, $m_f^{(0)} \mapsto m_f$. Note, however, that in the limit $M \to \infty$ the mass is renormalized by an amount which is infinite.

In what follows we take the mass M of the auxiliary particle to be finite. The limit $M \to \infty$ will be taken only at the end of a calculation, and, of course, with due care.

The additional term $\mathcal{L}_M$ contributes also to every internal line as sketched in Fig. 10.12. When translated into formulae, and using the notation (10.55) one has

$$\begin{aligned}S_F(p) \longmapsto\; & S_F(p) + S_F(p)\Sigma(p)S_F(p) - S_F(p)\delta m S_F(p)\\ &= S_F(p) + S_F(p)\left[(\not p - m\,\mathbb{1})\left(B+\Sigma^{\text{finite}}(p)\right)\right]S_F(p)\\ &= S_F(p) + \left(B+\Sigma^{\text{finite}}(p)\right)S_F(p)\,.\end{aligned}$$

One should be aware that here one is working to order $\mathcal{O}(e^2)$. In the regularized version of the theory in which all contributions are finite, the following are equivalent:

$$\begin{aligned}&S_F(p) + \left(B+\Sigma^{\text{finite}}(p)\right)S_F(p)\\ &\quad\approx (1+B)\left(1+\Sigma^{\text{finite}}(p)\right)S_F(p) \approx (1+B)\frac{S_F(p)}{1-\Sigma^{\text{finite}}(p)}\,.\end{aligned}$$

Fig. 10.12 An internal fermion line is changed not only by self-energy but also by an additional term (10.61) in the Lagrange density

They differ by terms of higher order than $\mathcal{O}(e^2)$ which are neglected, for reasons of consistency. Note that the propagator (10.55) now contains the *physical* mass.

There remains the constant B which is also divergent in the limit $M \to \infty$. Keeping in mind that a propagator always links two vertices each of which means multiplication by an electric charge, one might wish to absorb the factor $(1 + B)$ by a renormalization of the charge. However, we will see that gauge invariance connects this term to some other, divergent correction on the vertex in such a way that these contributions cancel.

Traditionally in quantum electrodynamics the multiplicative quantity $(1 + B)$ as well as all contributions that higher than second orders add to it, is denoted by a symbol of its own, viz.

$$Z_2 := 1 + B + \mathcal{O}(e^4)\,. \tag{10.64}$$

A factor of this kind which is divergent in perturbation theory and of which there are three in quantum electrodynamics, is called *renormalization constant.*

Knowing how an internal fermion line is modified by the self-energy there remains the question whether and, if yes, how external fermion lines are to be modified. There are two alternatives in answering this question:

(1) Either one does not apply any radiative correction to external lines and uses consistently the physical masses. Then there are no further corrections, or,

(2) One corrects external lines in the way described above, and, for every such line, obtains the factor

$$\left(Z_2\right)^{1/2} = \sqrt{1 + B + \mathcal{O}(e^2)} \approx 1 + \frac{1}{2}B\,.$$

This implies that if one takes account of all diagrams which correct the external lines, one must divide the result by a factor $Z_2^{1/2}$ for every external line.

10.2.3 Scattering on an External Potential

The radiative corrections of order $\mathcal{O}(e^2)$ to the scattering of a charged lepton on external electromagnetic fields provide particularly instructive examples for the physical effects of quantum electrodynamics. These include scattering in an external static magnetic field, on a heavy point-like charge Ze, or on an atomic nucleus whose charge density has a finite spatial extension.

The g-Factors of Electron, Muon, and τ-Lepton

First, we show that the g-factor of the charged leptons has the natural value $g_{\text{nat}} = 2$ in Dirac theory without radiative corrections. Let q be the electric charge of the lepton, i.e. $q = -|e|$ for (e^-, μ^-, τ^-), $q = +|e|$ for (e^+, μ^+, τ^+). The interaction with an external four-potential is given by (10.12), but now with a static potential $A_\mu(x)$. As this four-potential is *classical* it can be taken out of the normal ordering,

$$\mathcal{H}_1 = \mathcal{L}_1 = q \sum_f :\overline{\psi^{(f)}(x)}\gamma^\mu \psi^{(f)}(x):\, A_\mu^{\text{class}}(\boldsymbol{x})\,.$$

In order to determine the magnetic moment that goes with the spin, we study scattering on a stationary magnetic field $\boldsymbol{B} = \nabla \times \boldsymbol{A}(\boldsymbol{x})$ in a kinematics where the spatial momenta before and after the scattering are equal and opposite,

$$\langle \boldsymbol{p}' | \boldsymbol{j}(x=0) \,|\, \boldsymbol{p} \rangle \cdot \boldsymbol{A}(\boldsymbol{p} - \boldsymbol{p}') , \quad \boldsymbol{p}' = -\boldsymbol{p} .$$

This kinematics presents the advantage that the limit $|\boldsymbol{p}| \to 0$ leads into the rest system of the particle, thus allowing for a comparison with the corresponding nonrelativistic expression. One uses the standard representation (9.41) which is adapted to the nonrelativistic limit, as well as the relation (9.140). One then has

$$\begin{aligned} \langle -\boldsymbol{p} | j^k(0) \,|\, \boldsymbol{p} \rangle &= \frac{1}{(2\pi)^3} q\, \overline{u(-\boldsymbol{p})} \gamma^k u(\boldsymbol{p}) \\ &= \frac{q}{(2\pi)^3} u^\dagger(-\boldsymbol{p}) \begin{pmatrix} 0 & \sigma^{(k)} \\ \sigma^{(k)} & 0 \end{pmatrix} u(\boldsymbol{p}) = \frac{2q}{(2\pi)^3} \mathrm{i} \varepsilon^{klm} p^l \left(\chi^\dagger \sigma^{(m)} \chi \right) . \end{aligned}$$

Thus, the scattering amplitude for magnetic scattering in the backward direction reads

$$T = \mathrm{i} \frac{2q}{(2\pi)^3} \varepsilon^{klm} p^l \left(\chi^\dagger \sigma^{(m)} \chi \right) A^k(2\boldsymbol{p}) . \tag{10.65}$$

This should be compared to the corresponding scattering amplitude in Schrödinger theory which is given by the matrix element

$$T^{\text{nonrel.}} = \langle -\boldsymbol{p} | H_1^{\text{nonrel.}} \,|\, \boldsymbol{p} \rangle$$

the interaction being given by

$$H_1^{\text{nonrel.}} = -\boldsymbol{\mu} \cdot \boldsymbol{B} = -g \frac{q}{2m} \boldsymbol{s} \cdot \boldsymbol{B} , \quad B^r = \varepsilon^{rst} \partial_s A^t(\boldsymbol{x}) ,$$

(cf. (4.22) and (4.23)) while the wave functions are

$$|\boldsymbol{p}\rangle = \sqrt{\frac{2m}{(2\pi)^3}}\, \mathrm{e}^{\mathrm{i}\boldsymbol{p}\cdot\boldsymbol{x}} \chi .$$

The normalization of the wave function is chosen in accord with the covariant normalization of $\langle \boldsymbol{p}' | \, \boldsymbol{p} \rangle$ in the nonrelativistic limit. The partial derivative acting on $\boldsymbol{A}(\boldsymbol{x})$ is shifted to the wave function and one obtains

$$T^{\text{nonrel.}} = \mathrm{i} \frac{g\, q}{(2\pi)^3} \varepsilon^{rst} p^s \left(\chi^\dagger \sigma^{(t)} \chi \right) A^r(2\boldsymbol{p}) . \tag{10.66}$$

Comparison of the amplitudes (10.65) and (10.66) yields the value of the g-factor:

$$g_{\text{nat}} = 2 . \tag{10.67}$$

This value is called the *natural* value of the g-factor.

Remarks

1. One often reads, or hears someone say, that $g_{\text{nat}} = 2$ is a consequence of the principle of minimal substitution (9.59). Unfortunately this is not quite correct, unless one postulates the Lagrange density (4.58) to be the *true and only* Lagrange density for fermions. It is well-known that by adding a sufficiently smooth divergence to the Lagrange density

$$\mathcal{L}_D \longmapsto \mathcal{L}'_D = \mathcal{L}_D + \partial_\mu M^\mu(x)\,,$$

the Euler-Lagrange equations, hence the Dirac equation and its adjoint, remain unchanged. For instance, one may choose

$$M^\mu(x) = -\mathrm{i}\sum_f \frac{a^{(f)}}{8m}\overline{\psi^{(f)}(x)}\sigma^{\mu\nu}\overset{\leftrightarrow}{\partial}_\nu\psi^{(f)}(x)\,,\quad\text{with} \tag{10.68}$$

$$\sigma^{\mu\nu} = \frac{\mathrm{i}}{2}\left(\gamma^\mu\gamma^\nu - \gamma^\nu\gamma^\mu\right) \tag{10.69}$$

where $a^{(f)}$ are real parameters. One applies minimal coupling to the modified Lagrange density and adds $\mathcal{L}_\gamma$ to it. This yields the theory (10.11), (10.12), to which a new interaction term is added,

$$\mathcal{L}_\mathrm{P} = q\sum_f \frac{a^{(f)}}{4m_f}\overline{\psi^{(f)}(x)}\sigma^{\mu\nu}\psi^{(f)}(x)\,F_{\mu\nu}(x)\,. \tag{10.70}$$

Working out this term shows that the g-factor is modified to

$$g' = g_{\text{nat}}\left(1 + a^{(f)}\right). \tag{10.71}$$

This term was introduced by Pauli. Of course, as long as there is no deeper reason, it seems unnatural to introduce this term ad hoc but, clearly, it cannot be excluded either.

2. Accepting $g_{\text{nat}} = 2$ to be the natural value for a lepton it is suggestive to call any deviation from it an *anomaly of the g-factor.*

$$a^{(f)} := \frac{1}{2}\left(g^{(f)} - g^{(f)}_{\text{nat}}\right) = \frac{1}{2}\left(g^{(f)} - 2\right). \tag{10.72}$$

Thus, in the case of leptons, we have $g^{(e)}_{\text{nat}} = g^{(\mu)}_{\text{nat}} = g^{(\tau)}_{\text{nat}} = 2$. The anomalies which are caused by radiative corrections will not be the same for the three charged leptons, due to the differences in their masses.

3. Charged and uncharged fermions which besides the electromagnetic interaction are also subject to strong interactions, have g-factors which strongly deviate from their natural value 2. For example, in the case of the proton and the neutron one finds

$$\boldsymbol{\mu}_{p/n} = g^{(p/n)}\frac{|e|}{2m_p}\boldsymbol{s}\,,\quad\text{with} \tag{10.73}$$

$$g^{(p)} = 5.585695\,,$$

$$g^{(n)} = -3.826086\,.$$

The origin of these numbers must be found in the properties of the strong interaction. Even though we may not be able to calculate their absolute values there are models built on what is called the additive quark model which allow to derive relations between the magnetic moments, or g-factors, of different strongly interacting fermions.[6]

4. When a negatively charged fermion is bound in a hydrogen-like state the anomaly modifies the formula (4.25) for the spin-orbit coupling so that it becomes

$$U_{\ell s}(r) = \frac{1}{2m_f^2}\left(1 + 2a^{(f)}\right)\frac{1}{r}\frac{\mathrm{d}U(r)}{\mathrm{d}r}\boldsymbol{\ell}\cdot\boldsymbol{s}\,. \tag{10.74}$$

This effect can serve to determine the anomaly from the atomic fine structure. An example is provided by the Σ^--baryon whose mass is $m(\Sigma^-) = 1197.45\,\text{MeV}$, its spin being 1/2, and which is sufficiently long-lived such that it can be captured in hydrogen-like atomic orbits. From the measured fine structure in Σ^--atoms the magnetic moment was found to be

$$\mu(\Sigma^-) = -1.160 \pm 0.025 \times \frac{|e|}{2m_p}\,. \tag{10.75}$$

(Note that it is expressed in units of the Bohr magneton of the proton.)

Electric and Magnetic Form Factors of Spin-1/2 Particles

Point-like fermions which have no more than the natural g-factor (10.67), and hence the natural magnetic moment $q/(2m)$, are but idealizations which do not occur in nature. This is evident for the strongly interacting nucleons and other baryons, but even in the case of leptons which are much closer to this idealization, one discovers deviations from the point-like particle as soon as one's measurements are accurate enough. An intuitive understanding is as follows: A fermion subject to interactions with photons, or other particles, will always be surrounded by such particles, though virtual, and, hence, will exhibit nontrivial form factors.

Through its interactions a fermion acquires an internal structure which changes its electromagnetic properties. It is useful to decompose the one-particle matrix elements of the electromagnetic current operator in terms of Lorentz covariants and of invariant form factors. One takes out the elementary charge as an overall factor, as usual, and makes use of the translation formula (7.30), to shift the current operator $j^\mu(x)$ to the origin $x = 0$. The form factors of a given spin 1/2 fermion are defined by the following decomposition in terms of covariants:

$$\begin{aligned}\langle q\,|\,j^\mu(0)\,|\,p\rangle &= \frac{1}{(2\pi)^3}\\ &\times \overline{u(q)}\left\{\gamma^\mu F_1(Q^2) - \frac{\mathrm{i}}{2m}\sigma^{\mu\nu}Q_\nu F_2(Q^2) - \frac{1}{2m}Q^\mu F_3(Q^2)\right\}u(p)\,. \end{aligned}\tag{10.76}$$

[6] For some early work in which relations of this kind were derived, see H. Rubinstein, R. Socolov, and F. Scheck, Phys. Rev. **154** (1967) 1608.

Here $Q = p - q$ denotes the momentum transfer, the factors $F_i(Q^2)$ are Lorentz scalar functions. The denominators $2m$ are introduced in order to give the same physical dimension to all form factors, while the factor i in front of the second term renders all form factors real if $j^\mu(x)$ is hermitian, cf. Exercise 10.2. It is not difficult to show that the third form factor vanishes identically if the current operator is conserved, i.e. if $\partial_\mu j^\mu(x) = 0$. Therefore, in the case of the electromagnetic current density one has to deal with two form factors only, F_1 and F_2. There physical significance is uncovered as follows:

Using the Dirac equation once for $u(p)$, and once for $\overline{u(q)}$, i.e.

$$\not p u(p) = m u(p)\,, \qquad \overline{u(q)}\not q = m\overline{u(q)}\,,$$

one calculates the matrix element $\mathrm{i}\sigma^{\mu\nu}(p_\nu - q_\nu)$ between these spinors,

$$\begin{aligned}
&\overline{u(q)}\mathrm{i}\sigma^{\mu\nu}(p_\nu - q_\nu)u(p)\\
&= -\frac{1}{2}\overline{u(q)}\left[\gamma^\mu \not p - \not p\gamma^\mu - \gamma^\mu \not q + \not q \gamma^\mu\right]u(p)\\
&= -\frac{1}{2}\overline{u(q)}\left[2\gamma^\mu \not p - 2p^\mu + 2\not q\gamma^\mu - 2q^\mu\right]u(p)\\
&= -2m\overline{u(q)}\gamma^\mu u(p) + \left(p+q\right)^\mu \overline{u(q)}u(p)\,,
\end{aligned}$$

This identity which holds for spinors in momentum space on the mass shell, is called the *Gordon identity*. It reads

$$\left(p+q\right)^\mu \overline{u(q)}u(p) = 2m\overline{u(q)}\gamma^\mu u(p) + \overline{u(q)}\mathrm{i}\sigma^{\mu\nu}(p_\nu - q_\nu)u(p)\,. \tag{10.77}$$

Upon inserting this identity in (10.76) one obtains the equivalent form

$$\langle q | j^\mu(0) | p\rangle = \frac{1}{(2\pi)^3}\overline{u(q)}\left\{(F_1 + F_2)\gamma^\mu - \frac{1}{2m}(p^\mu + q^\mu)F_2\right\}u(p)\,. \tag{10.78}$$

Starting from the decomposition (10.78) one now considers specific kinematic situations and individual components of the current operator, in close analogy to the analysis of the preceding section.

1. In the case of the *electric form factor* consider the 0-component of the current density, and its matrix element between states with equal and opposite spatial momenta

$$\langle \boldsymbol{q} = -\boldsymbol{p} | j^0(0) | \boldsymbol{p}\rangle = \frac{1}{(2\pi)^3}\overline{u(-\boldsymbol{p})}\left\{(F_1 + F_2)\gamma^0 - \frac{E_p}{m}F_2\right\}u(\boldsymbol{p})\,.$$

In the standard representation (9.41) of the γ-matrices we have

$$\overline{u(-\boldsymbol{p})}\gamma^0 u(\boldsymbol{p}) = u^\dagger(-\boldsymbol{p})u(\boldsymbol{p}) = 2m\,, \quad \overline{u(-\boldsymbol{p})}u(\boldsymbol{p}) = 2E_p\,.$$

The momentum transfer is $Q = p - q = (0, 2\boldsymbol{p})^T$ and, hence, the corresponding invariant is $t = (q-p)^2 = -4\boldsymbol{p}^2 = -4E_p^2 + 4m^2$. Thus, we have

$$\begin{aligned}
\langle -\boldsymbol{p} | j^0(0) | \boldsymbol{p}\rangle &= \frac{1}{(2\pi)^3}\left\{2m\left(F_1 + F_2\right) - \frac{2E_p^2}{m}F_2\right\}\\
&= \frac{2m}{(2\pi)^3}\left\{F_1 + \frac{t}{4m^2}F_2\right\}.
\end{aligned}$$

This specific linear combination of the two form factors is called *electric form factor*. It is denoted by

$$G_{\rm E}(t) := F_1(t) + \frac{t}{4m^2} F_2(t) \,. \tag{10.79}$$

One verifies that the value of F_1, and hence also of $G_{\rm E}$ at $t = 0$ equals the charge of the fermion in units of the elementary charge, i.e. for f^- it equals -1. In order to show this calculate first the matrix element of the charge operator,

$$\langle q| \int \mathrm{d}^3x \; j^0(x) \,|\, p\rangle = (-1) \; \langle q|\, p\rangle = (-1)\, 2E_p\, \delta(\boldsymbol{p} - \boldsymbol{q}) \,.$$

Alternatively, using the general decomposition (10.76) as well as the translation formula, one has

$$\int \mathrm{d}^3x \; \langle q| j^0(x) \,|\, p\rangle = (2\pi)^3 \delta(\boldsymbol{p} - \boldsymbol{q}) \langle q| j^0(0) \,|\, p\rangle$$
$$= \delta(\boldsymbol{p} - \boldsymbol{q}) F_1(0) u^\dagger(\boldsymbol{q}) u(\boldsymbol{p}) = F_1(0)\, 2E_p\, \delta(\boldsymbol{p} - \boldsymbol{q}) \,.$$

The comparison of the two calculations does indeed yield $F_1(0) = -1$.

2. The *magnetic form factor* is isolated by examining the spatial components of the current density,

$$\langle -\boldsymbol{p}| j^k(0) \,|\, \boldsymbol{p}\rangle = \frac{1}{(2\pi)^3} \Big(F_1 + F_2 \Big) \overline{u(-\boldsymbol{p})} \gamma^k u(\boldsymbol{p}) \,.$$

The product on the right-hand side is worked out using the standard representation,

$$\overline{u(-\boldsymbol{p})} \gamma^k u(\boldsymbol{p}) = 2\varepsilon_{klm} p^l \chi^\dagger \sigma^{(m)} \chi \,,$$

so that one concludes

$$\langle -\boldsymbol{p}| j^k(0) \,|\, \boldsymbol{p}\rangle = \frac{1}{(2\pi)^3} \Big(F_1 + F_2 \Big) \varepsilon_{klm} Q^l \chi^\dagger \sigma^{(m)} \chi \,.$$

This result suggests to define the sum of F_1 and F_2 to be the *magnetic form factor*,

$$G_{\rm M}(t) := F_1(t) + F_2(t) \,. \tag{10.80}$$

At the point $t = 0$ it yields the magnetic moment of the particle (in units of the corresponding magneton $e/(2m)$), where $F_1(0)$ is the natural magnetic moment while $F_2(0)$ is the anomalous magnetic moment.

Corrections to Scattering on an External Potential

We study the scattering of a lepton f on an external point-like charge Ze, and generalize then to the case where this charge is smeared out over a local domain in space. This is a model for the electrostatic potential created by an atomic nucleus of finite extension.

Translating the tree diagram (A) of Fig. 10.13 into an amplitude, rule (R9) yields

$$M = 2\pi e_0 \mathrm{i} \overline{u(q)} \gamma^\mu u(p) \widetilde{A}_\mu\big((\boldsymbol{p} - \boldsymbol{q})^2\big) \,, \quad \text{with} \tag{10.81}$$

$$\widetilde{A}_\mu\big((\boldsymbol{p} - \boldsymbol{q})^2\big) = \frac{Ze_0}{(2\pi)^3} \delta_{\mu 0} \frac{1}{(\boldsymbol{p} - \boldsymbol{q})^2} \,.$$

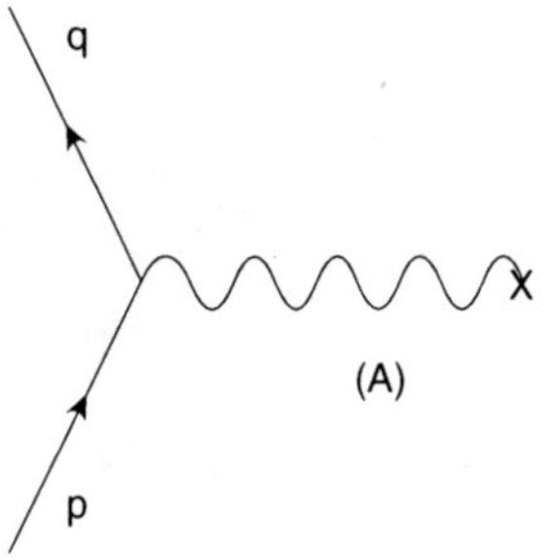

Fig. 10.13 Scattering on an external potential Ze/r

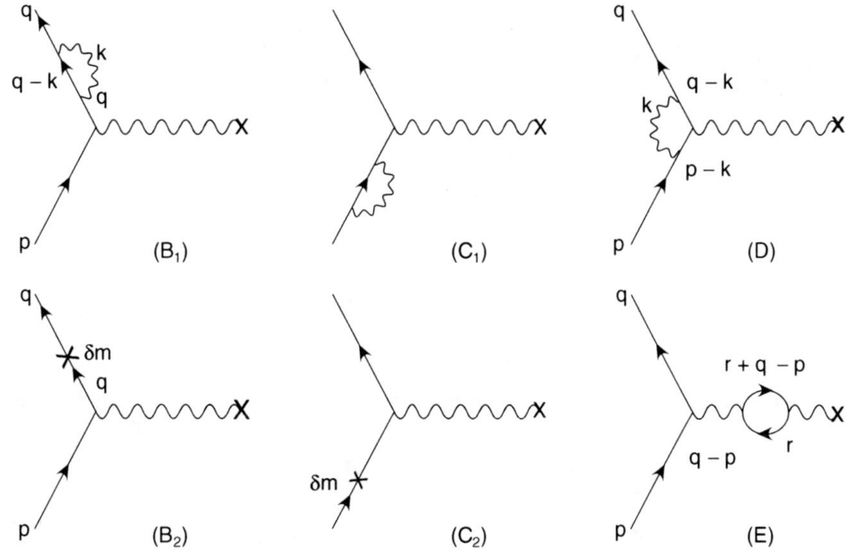

Fig. 10.14 Radiative corrections for scattering on an external potential to order e^2. Diagrams (B1), (C1), (B2), and (C2) illustrate the self-energy, (D) is the vertex correction, E the vacuum polarization

Figure 10.14 shows the connected diagrams which correct this amplitude to order $\mathcal{O}(e^2)$. When translated into formulae by means of Feynman rules, these yield the following terms:

Diagrams (B1) and **(B2)**:

$$
\begin{aligned}
B_{12} &= B_1 + B_2 \\
&= -\,2\pi\, \mathrm{i} e_0 \overline{u(q)} \\
&\quad \times \left\{ \delta m\, \mathbb{1} + \frac{\mathrm{i} e_0^2}{(2\pi)^4} \int \frac{\mathrm{d}^4 k}{k^2 + \mathrm{i}\varepsilon}\, \gamma^\lambda S_F(q-k)\gamma_\lambda \right\} S_F(q)\gamma^\mu u(p)\widetilde{A}_\mu\,;
\end{aligned}
$$

Diagram (C1) and **(C2)**:

$$\begin{aligned}C_{12} &= C_1 + C_2\\ &= -\,2\pi \mathrm{i} e_0 \overline{u(q)}\gamma^{\mu} S_F(p)\\ &\quad\times \left\{\delta m\,\mathbb{1} + \frac{\mathrm{i}e_0^2}{(2\pi)^4}\int \frac{\mathrm{d}^4k}{k^2+\mathrm{i}\varepsilon}\gamma^{\lambda} S_F(p-k)\gamma_{\lambda}\right\} u(p)\widetilde{A}_{\mu}\,;\end{aligned}$$

Vertex correction (D):

$$\begin{aligned}D = &-\,2\pi \mathrm{i} e_0 \overline{u(q)}\\ &\times \left\{\frac{\mathrm{i}e_0^2}{(2\pi)^4}\int \frac{\mathrm{d}^4k}{k^2+\mathrm{i}\varepsilon}\gamma^{\lambda} S_F(q-k)\gamma^{\mu} S_F(p-k)\gamma_{\lambda}\right\} u(p)\widetilde{A}_{\mu}\,;\end{aligned}$$

Vacuum polarization (E):

$$\begin{aligned}E &= 2\pi \mathrm{i} e_0 \overline{u(q)}\gamma^{\nu} u(p)\frac{1}{(q-p)^2+\mathrm{i}\varepsilon}\\ &\quad\times \frac{\mathrm{i}e_0^2}{(2\pi)^4}\sum_f \int \mathrm{d}^4r\ \mathrm{tr}\ \left\{\gamma_{\nu} S_F^{(f)}(r)\gamma^{\mu} S_F^{(f)}(r+q-p)\right\}\widetilde{A}_{\mu}\,.\end{aligned}$$

In the expression (E) the photon dissociates into a virtual f^+f^- pair which subsequently annihilates again into a photon. Independently of what the external lepton line is, the dominant contribution comes from virtual e^+e^- pairs, due to the smallness of the electron mass. The occurrence of the trace over the expression in curly brackets is easily understood if one strips off the left and right photon lines and calculates the relevant vacuum loop,

$$\langle 0|T\left(\overline{\psi^{(f)}(x)}\gamma^{\mu}\psi^{(f)}(x)\right)\left(\overline{\psi^{(f)}(y)}\gamma^{\nu}\psi^{(f)}(y)\right)|0\rangle\,.$$

As a special case of Wick's theorem and using the abbreviation

$$-\frac{1}{2}S_F(z) = \frac{\mathrm{i}}{(2\pi)^4}\int \mathrm{d}^4r\ \mathrm{e}^{-\mathrm{i}r\cdot z}\frac{(\not{r} + m_f\,\mathbb{1})}{r^2 - m_f^2 + \mathrm{i}\varepsilon}\,.$$

one obtains (summing, as before, over repeated indices)

$$\begin{aligned}&\langle 0|T\left(\overline{\psi^{(f)}(x)}\gamma^{\mu}\psi^{(f)}(x)\right)\left(\overline{\psi^{(f)}(y)}\gamma^{\nu}\psi^{(f)}(y)\right)|0\rangle\\ &= -\left(\gamma^{\mu}\right)_{\alpha\beta}\left(\gamma^{\nu}\right)_{\sigma\tau}\langle 0|T\,\psi_{\beta}^{(f)}(x)\overline{\psi_{\sigma}^{(f)}(y)}\,|0\rangle\,\langle 0|T\,\psi_{\tau}^{(f)}(y)\overline{\psi_{\alpha}^{(f)}(x)}\,|0\rangle\\ &= -\frac{1}{4}\left(\gamma^{\mu}\right)_{\alpha\beta}\left(S_F(x-y)\right)_{\beta\sigma}\left(\gamma^{\nu}\right)_{\sigma\tau}\left(S_F(y-x)\right)_{\tau\alpha}\\ &\equiv -\frac{1}{4}\mathrm{tr}\ \left\{\gamma^{\mu} S_F(x-y)\gamma^{\nu} S_F(y-x)\right\}\,.\end{aligned}$$

All diagrams of Fig. 10.14 contain two external fermion lines. Since we corrected these lines by self-energies and mass terms, all contributions must be multiplied by the factor Z_2^{-1}.

The sum of the uncorrected diagram (A) and of the terms B_{12} and C_{12} yields

$$
\begin{aligned}
(A) &+ B_{12} + C_{12} \\
&= 2\pi \mathrm{i} e_0 Z_2^{-1} \widetilde{A}_\mu \overline{u(q)} \left\{ \gamma^\mu - [(\delta m - A)\mathbb{1} - (\not{q} - m\,\mathbb{1})B]\, S_F(q)\gamma^\mu \right. \\
&\left. -\gamma^\mu S_F(p)\, [(\delta m - A)\mathbb{1} - B(\not{p} - m\,\mathbb{1})] \right\} u(p)\,.
\end{aligned}
$$

The finite terms Σ^{finite} do not contribute because they vanish on the mass shell $p^2 = m^2 = q^2$. Furthermore, one has $A = \delta m$ and $(\not{q} - m\,\mathbb{1})S_F(q) = \mathbb{1}$, and, of course, also $S_F(p)(\not{p} - m\,\mathbb{1}) = \mathbb{1}$. Finally, the emerging factor $1 + 2B$, in the given order, is transformed as follows

$$
1 + 2B \approx (1 + B)^2 \approx Z_2^2\,.
$$

Inserting these results one finds a very simple result for the sum

$$
(A) + B_{12} + C_{12} = 2\pi \mathrm{i} e_0 Z_2 \overline{u(q)} \gamma^\mu u(p) \widetilde{A}_\mu\,. \tag{10.82}
$$

The vertex correction of diagram (D) as well as the vacuum polarization in diagram (E) yield results whose physics is so interesting that we discuss them in two separate paragraphs.

10.2.4 Vertex Correction and Anomalous Magnetic Moment

The diagram (D) suffers from an infrared divergence and is ill-defined as long as the photon is massless. However, as the addition of further diagrams of the same order will be seen to cancel this divergence, one can avoid the problem by assigning a small but finite mass m_γ to the photon. Using the variables defined in Fig. 10.14 one has

$$
\begin{aligned}
D &= -2\pi \mathrm{i} e_0 Z_2^{-1} \frac{\mathrm{i} e_0^2}{(2\pi)^4} \\
&\int \frac{\mathrm{d}^4 k}{k^2 - m_\gamma^2 + \mathrm{i}\varepsilon} \overline{u(q)} \gamma^\lambda S_F(q - k) \gamma^\mu S_F(p - k) \gamma_\lambda u(p) \widetilde{A}_\mu\,.
\end{aligned}
$$

The expression between the two momentum space spinors in the numerator can be transformed by commuting $\not{q}$ to the left, and $\not{p}$ to the right, until one can use the Dirac equations $(\not{p} - m\,\mathbb{1})u(p) = 0$ and $\overline{u(q)}(\not{q} - m\,\mathbb{1}) = 0$, respectively. Furthermore, the sum over λ is done by means of the formulae of Sect. 9.3.3, Eqs. (9.121)–(9.123). In doing so one finds

$$
\begin{aligned}
&\overline{u(q)} \gamma^\lambda (\not{q} - \not{k} + m\,\mathbb{1}) \gamma^\mu (\not{p} - \not{k} + m\,\mathbb{1}) \gamma_\lambda u(p) \\
&= \overline{u(q)} \{ -2\not{k} \gamma^\mu \not{k} - 4mk^\mu + 4(q + p)^\mu \not{k} \\
&\quad + 4(q \cdot p - q \cdot k - p \cdot k)\gamma^\mu \} u(p)\,.
\end{aligned}
$$

As $q^2 = m^2 = p^2$, the product of the denominators of the three propagators, barring the $\mathrm{i}\varepsilon$'s, becomes

$$
(k^2 - m_\gamma^2)(k^2 - 2k \cdot p)(k^2 - 2k \cdot q)\,.
$$

It is not difficult to prove the following integral representations for terms of the type $1/(ab)$ and $1/(a^2b)$,

$$\frac{1}{ab} = \int_0^1 \mathrm{d}x \, \frac{1}{\left[ax + b(1-x)\right]^2}, \tag{10.83}$$

$$\frac{1}{a^2 b} = \int_0^1 \mathrm{d}x \, \frac{2x}{\left[ax + b(1-x)\right]^3}. \tag{10.84}$$

The first of these is proved directly while the second follows from the first by differentiation by the parameter a. Using (10.83) one has

$$\frac{1}{k^2 - m_\gamma^2} \frac{1}{(k^2 - 2k \cdot q)(k^2 - 2k \cdot p)} = \frac{1}{k^2 - m_\gamma^2} \int_0^1 \mathrm{d}x \, \frac{1}{\left[k^2 - 2k \cdot r(x)\right]^2},$$

where the abbreviation $r(x) = px + q(1-x)$ was introduced. Applying then (10.84) with $b = k^2 - m_\gamma^2$ and $a = [\ldots]$, the same expression is equal to

$$= \int_0^1 2y \mathrm{d}y \int_0^1 \mathrm{d}x \, \frac{1}{\left[\left(k^2 - 2k \cdot r(x)\right) y + \left(k^2 - m_\gamma^2\right)(1-y)\right]^3}$$

$$= \int_0^1 2y \mathrm{d}y \int_0^1 \mathrm{d}x \, \frac{1}{\left[(k - yr(x))^2 - r^2(x) y^2 - m_\gamma^2 (1-y)\right]^3}.$$

One inserts the transformed numerator and uses the integral representation for the product of the denominators, defines then the new integration variable $v := k - yr(x)$, and obtains eventually

$$\begin{aligned} D = & -2\pi \mathrm{i} e_0 Z_2^{-1} \frac{\mathrm{i} e_0^2}{(2\pi)^4} \int_0^1 2y \mathrm{d}y \int_0^1 \mathrm{d}x \int \mathrm{d}^4 v \\ & \times \frac{1}{\left[v^2 - y^2 r^2(x) - m_\gamma^2 (1-y)\right]^3} \overline{u(q)} \left\{ -2\not{v} \gamma^\mu \not{v} - 2y \not{r} \gamma^\mu \not{v} - 2y \not{v} \gamma^\mu \not{r} \right. \\ & -2y^2 \not{r} \gamma^\mu \not{r} - 4m v^\mu \mathbb{1} - 4m y r^\mu \mathbb{1} + 4(p+q)^\mu (\not{v} + y \not{r}) \\ & \left. + 4\gamma^\mu (p \cdot q - q \cdot v - yq \cdot r - p \cdot v - yp \cdot r) \right\} u(p) \widetilde{A}_\mu . \end{aligned}$$

The denominator depends on the square v^2 only. Therefore, all terms of the numerator which are odd in v give no contribution. The terms which are bilinear in v yield integrals of the form

$$\int \mathrm{d}^4 v \, \frac{v^\alpha v^\beta}{\left(v^2 - \Lambda^2\right)^3} = \frac{1}{4} g^{\alpha\beta} \int \mathrm{d}^4 v \, \frac{v^2}{\left(v^2 - \Lambda^2\right)^3},$$

Making use of these simplifications as well as of the summation formula $\gamma_\alpha \gamma^\mu \gamma^\alpha = -2\gamma^\mu$, the term D is seen to be proportional to

$$\int \mathrm{d}^4 v \, \frac{1}{(v^2 - \Lambda^2)^3} \times \overline{u(q)} \left\{ v^2 \gamma^\mu - 2y^2 \not{r}(x) \gamma^\mu \not{r}(x) - 4myr^\mu(x) \mathbb{1} + 4(p+q)^\mu y \not{r}(x) + 4\gamma^\mu \left[p \cdot q - y(p+q) \cdot r(x) \right] \right\} u(p) \, .$$

Inserting here $r(x)$ and, after some reordering, one obtains

$$D = -2\pi \mathrm{i} e_0 Z_2^{-1} \frac{\mathrm{i} e_0^2}{(2\pi)^4} \int_0^1 2y \mathrm{d}y \int_0^1 \mathrm{d}x \int \frac{\mathrm{d}^4 v}{(v^2 - \Lambda^2)^3} \times \overline{u(q)} \left\{ a(x,y) \gamma^\mu + b(x,y) p^\mu + c(x,y) q^\mu \right\} u(p) \widetilde{A}_\mu \, ,$$

where the four functions $\Lambda^2(x, y)$, $a(x, y)$, $b(x, y)$, and $c(x, y)$ are given by

$$\begin{aligned}
\Lambda^2(x,y) &= y^2 \left[m^2 x^2 + m^2 (1-x)^2 + 2p \cdot q \, x(1-x) \right] + m_\gamma^2 (1-y) \, , \\
a(x,y) &= v^2 + 4p \cdot q \, \left(1 - y + y^2 x(1-x)\right) \\
&\quad + 2m^2 y^2 (1 - 2x + 2x^2) - 4m^2 y \, , \\
b(x,y) &= 4my(1 - x - xy) \, , \\
c(x,y) &= 4my(x - y + xy) \, .
\end{aligned}$$

The first of these, $\Lambda^2(x, y)$, is invariant under the exchange $x \leftrightarrow (1-x)$. Thus, when integrating over x from 0 to 1 one can replace $b(x, y)$ as well as $c(x, y)$ by their average,

$$\int_0^1 \mathrm{d}x \, \cdots \left[b(x,y) p^\mu + c(x,y) q^\mu \right] = \int_0^1 \mathrm{d}x \, \cdots \frac{1}{2} \left[b(x,y) + c(x,y) \right] \left[p^\mu + q^\mu \right] .$$

This average depends on y only because $[b(x, y) + c(x, y)]/2 = 2my(1 - y)$, so that

$$D = -2\pi \mathrm{i} e_0 Z_2^{-1} \frac{\mathrm{i} e_0^2}{(2\pi)^4} \int_0^1 2y \mathrm{d}y \int_0^1 \mathrm{d}x \int \frac{\mathrm{d}^4 v}{\left(v^2 - y^2 r^2(x) - m_\gamma^2 (1-y)\right)^3} \times \overline{u(q)} \left\{ a(x,y) \gamma^\mu + 2my(1-y)(p+q)^\mu \right\} u(p) \widetilde{A}_\mu \, .$$

There is a last transformation of this result which allows to exhibit more clearly its physical significance. Indeed, if one succeeded in replacing the Lorentz covariants $\overline{u(q)}(p^\mu + q^\mu)u(p)$ by $\overline{u(q)}\gamma^\mu u(p)$ and $\overline{u(q)}\sigma^{\mu\nu}(p - q)_\nu u(p)$, with $\sigma^{\mu\nu}$ as defined in (10.69), then the effective magnetic moment of the fermion could be identified by means of the interaction term (10.70). For this purpose insert the Gordon decomposition (10.77) thus obtaining

$$D = -2\pi \mathrm{i} e_0 Z_2^{-1} \frac{\mathrm{i} e_0^2}{(2\pi)^4} \int_0^1 2y \mathrm{d}y \int_0^1 \mathrm{d}x$$

$$\int \frac{\mathrm{d}^4 v}{\left(v^2 - m^2 y^2 + y^2 x(1-x)(p-q)^2 - m_\gamma^2 (1-y)\right)^3}$$

$$\times \overline{u(q)} \Big\{ \big[a(x,y) + 4m^2 y(1-y)\big] \gamma^\mu$$

$$+ \mathrm{i}\sigma^{\mu\nu} (p-q)_\nu 2my(1-y) \Big\} u(p) \widetilde{A}_\mu \,.$$

In this way the term D takes a simple form which allows to calculate the magnetic moment of the f^-, including its radiative corrections,

$$D \equiv 2\pi \mathrm{i} e_0 Z_2^{-1} \overline{u(q)}$$

$$\left\{ F_1\big((p-q)^2\big)\gamma^\mu - \frac{\mathrm{i}}{2m} \sigma^{\mu\nu}(p-q)_\nu F_2\big((p-q)^2\big) \right\} u(p) \widetilde{A}_\mu \,.$$

The form factors F_1 and F_2 are defined by the above integral representation. Before working them out we note a physically important result: the lepton f^- of the defining theory (10.11) and (10.12) had the point charge -1 and the natural g-factor (10.67), but had no inner structure in the sense that its form factors were constant, $F_1^{(0)}(t) = 1$, $F_2^{(0)}(t) \equiv 0$, or $G_E^{(0)}(t) = G_M^{(0)}(t) = 1$. The radiative corrections modify both form factors in a nontrivial way and the lepton obtains some inner structure.

Unfortunately, it needs some more work to reach the eventual goal of this analysis. The form factor $F_1\big((p-q)^2\big)$ defined above is still given by a divergent integral. The technical reason for this is that the function $a(x, y)$ contains a term v^2 which ruins the integration $\int \mathrm{d}^4 v$. Therefore, here too, one should first regularize and test whether the observable parts can separated from the divergent ones. Without making a specific choice here, we assume the integral yielding the form factor F_1 to be regularized in some way or other, such that the following calculations become well-defined, but we do not specify the regularized expressions. Instead of $F_1(t)$ we write

$$F_1(t) = F_1(0) + [F_1(t) - F_1(0)] \,, \quad t = (p-q)^2 \,,$$

and note that the term in square brackets is convergent and is independent of the regularization procedure. Therefore, at this point, we need to examine only the potentially dangerous term $F_1(0)$. Collecting the diagrams (A) to (D) one has

$$(A) + B_{12} + C_{12} + D = 2\pi \mathrm{i} e_0 \left[Z_2 + Z_2^{-1} F_1(0) \right] \overline{u(q)} \gamma^\mu u(p) \widetilde{A}_\mu$$

$$+ 2\pi \mathrm{i} e_0 Z_2^{-1} \overline{u(q)} \left\{ [F_1(t) - F_1(0)]\, \gamma^\mu - \frac{\mathrm{i}}{2m} \sigma^{\mu\nu}(p-q)_\nu F_2(t) \right\} u(p) \widetilde{A}_\mu \,. \tag{10.85}$$

It is striking that after summing all contributions of the same order, the form factor $F_1(0)$ appears multiplied with the inverse of the renormalization constant Z_2. Indeed, this is the rescue! One proves the following identity (see Appendix A.5)

$$Z_2 + F_1(0) = 1 . \tag{10.86}$$

It says that the term B, (10.59) and (10.64), which is also divergent, after summing up all contributions of the same order of perturbation theory, cancels against the vertex correction $F_1(0)$. At the order we are analyzing here one concludes from this identity

$$Z_2 + Z_2^{-1} F_1(0) \approx Z_2 + \big(1 + F_1(0)\big) F_1(0) \approx Z_2 + F_1(0) = 1 .$$

For consistency, all form factors in the second term of (10.85) must be replaced as follows

$$Z_2^{-1} F_i \approx \big(1 + F_1(0)\big) F_i \approx F_i .$$

At the order to which we calculated the radiative corrections here, one finds a *convergent* result,

$$(A) + B_{12} + C_{12} + D = 2\pi \mathrm{i} e_0$$
$$\times \overline{u(q)} \left\{ \gamma^\mu \big(1 + F_1(t) - F_1(0)\big) - \frac{\mathrm{i}}{2m} \sigma^{\mu\nu} (p-q)_\nu F_2(t) \right\} u(p) \widetilde{A}_\mu , \tag{10.87}$$

that we now analyze with respect to its physical content.

Remarks

1. These calculations are well-defined as long as all potentially divergent quantities are regularized and, hence, finite. Of course, omitting in our examples all terms of order $\mathcal{O}(e_0^4)$, for consistency, is only meaningful in the framework of the regularized theory.
2. The result (10.87) which represents an observable scattering amplitude (but for the factor e_0 that must be discussed separately) contains no divergence at all. Although we have not shown this, it is plausible that the result is unique and does not depend on the chosen regularization scheme. Divergent contributions either are absorbed by renormalization, i.e. by transformation to the physical mass (and physical charge), or cancel like in the example (10.86).
3. Up to this point, our analysis is incomplete because the amplitude (10.87) still contains the bare charge e_0. The next section will show that e_0, by itself, is divergent but, when vacuum polarization (E) is added, it is replaced by the *physical* charge e through renormalization. With this argument in mind and subject to this proviso we insert the physical, hence finite charge e in (10.87).

Obviously, we are eager to learn about the physics contained in the perfectly finite physical result (10.87). As a first observation one sees that the f^- no longer is a point-like particle because the radiative corrections endow it with an electric form factor (10.79),

$$G_{\mathrm{E}}^{(f)}(t) = 1 + \big(F_1(t) - F_1(0)\big) + \frac{t}{4m^2} F_2(t) . \tag{10.88}$$

Likewise, the magnetic form factor $G_{\mathrm{M}}^{(f)}(t)$, (10.80), no longer is identically equal to 1. Both $G_{\mathrm{E}}^{(f)}(t)$, and $G_{\mathrm{M}}^{(f)}(t)$ show that the interaction with the radiation field provides the lepton f^- with a nontrivial internal structure. In particular it receives an anomalous magnetic moment that follows from (10.87). Replacing e_0 by e, with the above proviso, one has

$$F_2(t=0) = \frac{\mathrm{i}e^2}{(2\pi)^4} 4m^2 \int_0^1 2y\mathrm{d}y \int_0^1 \mathrm{d}x \int \mathrm{d}^4 v \, \frac{y(1-y)}{\left(v^2 - m^2y^2 + \mathrm{i}\varepsilon\right)^3} \, .$$

The integral over v yields (cf. Exercise 10.3)

$$\int \mathrm{d}^4 v \, \frac{1}{\left(v^2 - \Lambda^2 + \mathrm{i}\varepsilon\right)^3} = -\frac{\mathrm{i}\pi^2}{2\Lambda^2} \, ,$$

so that

$$F_2(0) = \frac{\mathrm{i}e^2}{(2\pi)^4} 4m^2 \frac{-\mathrm{i}\pi^2}{2m^2} \int_0^1 2y\mathrm{d}y \int_0^1 \mathrm{d}x \, \frac{1-y}{y} = \frac{e^2}{8\pi^2} \, .$$

Finally, inserting $e^2 = 4\pi\alpha$ one obtains

$$F_2(0) \equiv a^{(f)} = \frac{\alpha}{2\pi} \, . \qquad (10.89)$$

This is a classical result of quantum electrodynamics which was obtained by J. Schwinger[7] and which belongs to the great successes of quantum electrodynamics.

Remarks

1. Schwinger's result (10.89) is remarkable by the fact that the anomaly, in this order of perturbation theory, is independent of the nature of the lepton. The value $\alpha/(2\pi)$ is the same for the electron, the muon and the τ-lepton. This

Table 10.2 Anomaly of the magnetic moment of leptons. In the last column, $a_n^{(e)}$ denotes the contribution of order n to the anomaly of the electron

	$a^{(e)}$	$a^{(\mu)}$
Experiment	$1159{,}652186(4) \times 10^{-6}$	$1165{,}9208(6) \times 10^{-6}$
2. order	$\frac{\alpha}{2\pi}$	$\frac{\alpha}{2\pi}$
4. order	$-0.328478965\left(\frac{\alpha}{\pi}\right)^2$	$\approx a_4^{(e)} + \left(\frac{\alpha}{\pi}\right)^2$
6. order	$1.181\left(\frac{\alpha}{\pi}\right)^3$	$\approx a_6^{(e)} + 20\left(\frac{\alpha}{\pi}\right)^3$
Hadronic correction	≈ 0	$67(9) \times 10^{-9}$
Theory (total)	$1159.652359(282) \times 10^{-6}$	$1165.91628(77) \times 10^{-6}$

[7] J. Schwinger, Phys. Rev, **73**, 416 (1948).

no longer holds for higher orders, even in pure quantum electrodynamics, as can see from the results of Table 10.2. Furthermore, in higher orders there can also be virtual hadrons in the diagrams of perturbation theory. For the electron, hadronic corrections are still small and, in fact, below present experimental accuracies. However, for muons they are not so small and contribute by amounts exceeding the present experimental error bars. The experimental and theoretical uncertainties of Table 10.2 which are given in parantheses, refer to the last two digits of an entry. For example 1.2345(14) stands for the result 1.2345 ± 0.0014. One sees that the corrections predicted by pure quantum electrodynamics alone are different for muons and for electrons from the fourth order on up.[8]

2. The anomaly proper, i.e. the deviation of the magnetic moment from the corresponding Bohr magneton,

$$a^{(f)} = \frac{\mu^{(f)}}{e/(2m_f)} - 1 = \frac{1}{2}\left(g^{(f)} - 2\right)$$

can be measured *directly* by means of a simple physical effect. The principle that we describe here for the muon, is as follows: One injects the μ^- into the field of a magnetic bottle whose central magnetic field is homogeneous. The muon then follows a spiral orbit whose cyclotron frequency is[9]

$$\omega_c = \frac{eB}{m_\mu \gamma} .$$

Here, B is the strength of the magnetic field, and γ the relativistic factor $\gamma = 1/\sqrt{1-\beta^2}$ of the particle. The magnetic moment performs a precession about the B-field with angular velocity

$$\omega_s = \left(1 + \gamma a^{(\mu)}\right)\frac{eB}{m_\mu \gamma} .$$

As one sees the spin precession would be synchronous with the orbital motion if the anomaly were exactly zero. In this case the magnetic moment, and, hence, the spin, would always point in the same direction after a large number of revolutions. If the anomaly does not vanish then the spin precession is out of tune with the orbital spiral. The difference of the two circular frequencies is proportional to $a^{(\mu)}$,

$$\omega_a = \frac{e}{m_\mu} a^{(\mu)} B .$$

---

[8] References to the original theoretical and experimental work are found in the latest edition of the *Review of Particle Properties*, J. Beringer et al. (Particle Data Group), Phys. Rev. D86, 01001 (2012). See also their internet site pdg.lbl.gov, where pdg stands for particle data group.

[9] These formulae can be found, e.g., in the handbook *Muon Physics*, Vols. I–III, V.W. Hughes and C.S. Wu (eds.), Academic Press 1977.

With a constant field and after the time interval T the spin has moved by the angle

$$\theta = \frac{e}{m_\mu} a^{(\mu)} B T \,.$$

If one knows B and T and if one can measure θ, one obtains the anomaly without having measured the full magnetic moment. The spin orientation of the muon, in turn, is obtained from its decay distribution, $\mu^- \to e^- + \nu_\mu + \overline{\nu_e}$ which is asymmetric with respect to the spin direction (by parity violation of the weak interaction).

3. The theory of leptons and their interactions is invariant under the combined discrete transformation

$$\boldsymbol{\Theta} = \boldsymbol{\Pi}\mathbf{CT}$$

(from right to left: time reversal, charge conjugation, and space reflection). A consequence of this invariance is that the magnetic moments of f^- and of f^+ are equal and opposite. This was tested experimentally, the result being

$$\frac{g^{(e^+)} - g^{(e^-)}}{\langle g^{(e)} \rangle} = \left(-0.5 \pm 2.1\right) \times 10^{-12} \tag{10.90}$$

for the electron and the positron, and

$$\frac{g^{(\mu^+)} - g^{(\mu^-)}}{\langle g^{(\mu)} \rangle} = \left(-2.6 \pm 1.6\right) \times 10^{-8} \tag{10.91}$$

for the muon and its antiparticle. Here, $\langle g^{(f)} \rangle$ denotes the average.

10.2.5 Vacuum Polarization

When one takes account of the contribution (E), Fig. 10.14 the external potential is modified, $\widetilde{A}_\mu \mapsto \widetilde{A}'_\mu = \widetilde{A}_\mu + \Delta \widetilde{A}_\mu$. Defining $Q := p - q$ one has

$$\Delta \widetilde{A}_\mu = \frac{1}{Q^2} \Pi_{\mu\nu}(Q) \widetilde{A}^\nu(Q) \tag{10.92}$$

$$\Pi^{\mu\nu}(Q) = \frac{\mathrm{i}e_0^2}{(2\pi)^4} \sum_f \int \mathrm{d}^4 r \;\mathrm{tr}\, \left\{\gamma^\mu S_F^{(f)}(r) \gamma^\nu S_F^{(f)}(r - Q)\right\} . \tag{10.93}$$

As a first step, it is useful to investigate the tensor integral $\Pi^{\mu\nu}(Q)$ in the light of gauge invariance thus checking whether there are restrictions that might help to simplify calculations. A gauge transformation in x-space, $A_\mu(x) \mapsto A'_\mu(x) = A_\mu(x) - \partial_\mu \chi(x)$, is equivalent to the replacement in momentum space

$$\widetilde{A}_\mu(Q) \longmapsto \widetilde{A}'_\mu(Q) = \widetilde{A}_\mu(Q) + \widetilde{\chi}(Q) Q_\mu \,.$$

If the expression (10.92) must remain unchanged, one must have

$$Q_\mu \Pi^{\mu\nu}(Q) = 0 = \Pi^{\mu\nu}(Q) Q_\nu \,.$$

On the other hand, $\Pi^{\mu\nu}$ is a contravariant tensor field of rank two. As it depends only on Q the only tensors in terms of which $\Pi^{\mu\nu}$ may be decomposed, are $Q^\mu Q^\nu$ and the (inverse) metric $g^{\mu\nu}$, so that the decomposition in terms of covariants must have the form

$$\Pi^{\mu\nu}(Q) = \Pi(Q^2)Q^\mu Q^\nu + \Phi(Q^2)g^{\mu\nu} ,$$

with $\Pi(Q^2)$ and $\Phi(Q^2)$ Lorentz-scalar functions. Gauge invariance yields the condition $Q^2\Pi(Q^2) + \Phi(Q^2) = 0$ so that

$$\Pi^{\mu\nu}(Q^2) = \left(Q^\mu Q^\nu - Q^2 g^{\mu\nu}\right)\Pi(Q^2) . \tag{10.94}$$

These considerations rest on a somewhat uncertain basis, so far, because $\Pi^{\mu\nu}(Q)$ is given by an integral of the type $\int d^4r\ (1/r^2)$ which diverges. Obviously, one must first regularize divergent integrals. As an example, we apply the method of Pauli and Villars here, too. Denoting the contribution of a given lepton by $\Pi^{\mu\nu}(Q, m_f^2)$, this means that we substitute

$$\Pi^{\mu\nu}(Q, m_f^2) \longmapsto (\Pi^{\text{reg}})_f^{\mu\nu}(Q) = \Pi^{\mu\nu}(Q, m_f^2) + c_f \Pi^{\mu\nu}(Q, M_f^2) . \tag{10.95}$$

The new term depends on some large mass M_f, and the factor c_f is chosen so that the modified expression is well-defined and finite. "Well-defined" refers to two aspects: The regularization procedure should not violate gauge invariance. Furthermore, the finite part of regularized vacuum polarization must be fixed in a physically consistent manner. The first requirement is guaranteed by the Pauli-Villars procedure. The second is related to the question of how and at which energy scale the physical charge is defined.

A more detailed analysis of the expression (10.95) is found in Appendix A.6. This analysis shows that it is sufficient to choose $c_f = -1$. After regularization and a few more transformations as explained in the appendix the result becomes relatively simple. One finds

$$(\Pi_f^{\text{reg}})^{\mu\nu}(Q) = \frac{\alpha_0}{3\pi}\left(Q^\mu Q^\nu - Q^2 g^{\mu\nu}\right)$$
$$\left\{ \ln\left(\frac{M_f}{m_f}\right)^2 - 6\int_0^1 dz\, z(1-z)\ln\left(1 - \frac{Q^2}{m_f^2 - i\varepsilon} z(1-z)\right)\right\} , \tag{10.96}$$

where α_0 still contains the bare charge. The result (10.96) fulfills the condition (10.94) of gauge invariance. It contains two terms the first of which diverges logarithmically as $M_f \to \infty$ but, as we will show below, is absorbed by charge renormalization. The second term is finite and is fixed in such a way that it vanishes at the point $Q^2 = 0$. This term implies observable effects at $Q^2 \neq 0$ whose physics is remarkable.

Renormalization of the Electric Charge

To second order the tree diagram for Møller scattering, $e^- + e^- \to e^- + e^-$, is modified as sketched in Fig. 10.15. When translated to formulae this yields[10]

$$\frac{-g^{\mu\nu}}{Q^2} \longmapsto \frac{-g^{\mu\nu}}{Q^2} - \frac{g^{\mu\alpha}}{Q^2}\sum_f (\Pi_f^{\text{reg}})_{\alpha\beta}(Q)\frac{g^{\beta\nu}}{Q^2} , \tag{10.97}$$

---

[10] The minus sign of the second term is a consequence of rule (R6).

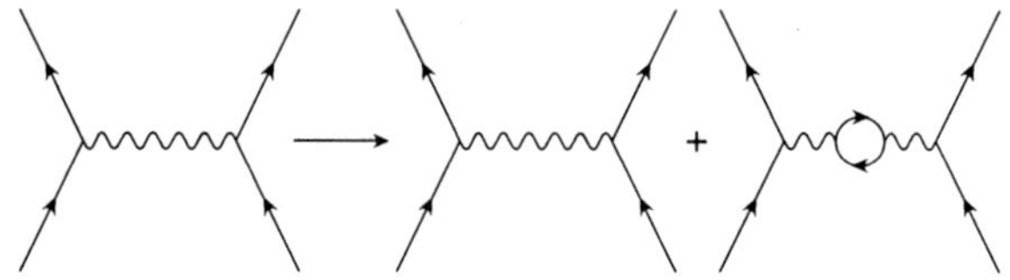

Fig. 10.15 A loop integral with virtual $f^- f^+$-pairs is added to Møller scattering

where now $Q = p_1 - q_1 = q_2 - p_2$. The term proportional to $Q^\mu Q^\nu$ of (10.96) does not contribute because, when "sandwiched" between the spinors of the external particles, it gives zero,

$$\overline{u(q_1)} \not{Q} u(p_1) = \overline{u(q)} (\not{p}_1 - \not{q}_1) u(p_1) = 0 = \overline{u(q_2)} \not{Q} u(p_2) .$$

Thus, the sum of the diagrams of Fig. 10.15 is proportional to $g^{\mu\nu}$,

$$-\frac{g^{\mu\nu}}{Q^2} \left\{ 1 - \frac{\alpha_0}{3\pi} \sum_f \ln \left(\frac{M_f}{m_f} \right)^2 \right.$$
$$\left. + \frac{2\alpha_0}{\pi} \sum_f \int_0^1 \mathrm{d}z\, z(1-z) \ln\Big(1 - \frac{Q^2}{m_f^2 - \mathrm{i}\varepsilon} z(1-z) \Big) \right\} .$$

This means that in the limit $Q^2 \to 0$ the photon propagator is modified multiplicatively by the factor

$$Z_3 \approx 1 - \sum_f \frac{\alpha_0}{3\pi} \ln \left(\frac{M_f}{m_f} \right)^2 . \tag{10.98}$$

This will be similar also in higher orders. Therefore, the renormalization factor is given a symbol and is generally denoted by Z_3. Note, however, that the expression on the right-hand side holds only in the order considered here, hence the $\approx$-sign.

As one verifies easily, the same factor also appears in scattering on an external potential, or in any other process with an internal photon line. This leads to the following conjecture: As the squared bare charge e_0^2 always appears multiplied by the renormalization constant Z_3, the product $e_0 Z_3^{1/2} =: e$ should be interpreted as the *physical charge*. In our case where we calculate the corrections to second order, for reasons of consistency, we replace α_0 by α, thus obtaining the presription

$$e_0^2 \longmapsto e^2 := Z_3 e_0^2 \approx e_0^2 \Big\{ 1 - \frac{\alpha}{3\pi} \sum_f \ln \left(\frac{M_f}{m_f} \right)^2 \Big\} . \tag{10.99}$$

The first step is the definition, the second yields the explicit expression in second order of perturbation theory.

Remarks

1. The physical charge of an electron can be determined by Thomson scattering, that is, at $Q^2 = 0$, cf. Sect. 7.4.3. The charge that is measured is related to the bare charge by

$$e = e_0 \sqrt{1 - \frac{\alpha}{3\pi} \sum_f \ln(\frac{M_f}{m_f})^2 + \mathcal{O}(\alpha^2)}\,.$$

Thus, once one has applied renormalization, there remain only finite radiative corrections which lead to well-defined, measurable effects.

2. The way we carried out renormalization of the charge singles out the scale $Q^2 = 0$. This is in agreement with the fact that the classical limit of Thomson scattering (by means of which the physical charge is obtained) is reached at $Q^2 = 0$. This specific point is called the *renormalization scale.* In quantum electrodynamics this scale is a physical one because in the limit of small energies one reaches the classical theory. Nevertheless, this is not the only possible choice. Below, we show that one may as well define α at some other scale whenever this seems adequate for physical reasons. In any case and with any choice of the scale, the perturbation series will tend to the true answer order by order. However, individual contributions will have different magnitudes depending on the choice of scale.
3. Renormalization is a phenomenon which is not only mathematically subtle but also largely counter-intuitive regarding its physics. According to the formula (10.99) the bare charge is *larger* than the physical charge. In particular, if the masses of the auxiliary particles are sent to infinity, the bare charge becomes infinite. In a regularized version of the theory all quantities are finite. In fact, for values of M_f which are still finite, the effects are not dramatic: For instance, choosing $M_f = 10\,\text{GeV}$, the fine structure constant α changes by about 3 %.
4. It remains an open question how to proceed when the photon belongs to an external line. Formally, the modification drawn in Fig. 10.16 yields

$$\frac{1}{k^2}\left(\Pi_f^{\text{reg}}\right)^{\mu\nu}(k)\varepsilon_\nu(k)$$
$$= -\frac{k^2}{k^2}\left\{\left(-\frac{\alpha}{3\pi}\right)\sum_f \ln\left(\frac{M_f}{m_f}\right)^2 + \mathcal{O}(k^2)\right\}\varepsilon^\mu(k)\,,$$

where we used $\varepsilon_\mu(k)k^\mu = 0$. As $k^2 = 0$ this result seems to remain undetermined because it yields 0/0. This problem can be avoided if one thinks of the photon as being emitted by a source very far away so that it slips off its mass shell but slightly. There must then be a factor Z_3, as before, which splits into a factor $Z_3^{1/2}$ for the charge at the vertex of Fig. 10.16, and the same factor for the far-away source. This, in turn, means that one should insert the renormalized, physical charges at the vertex as well as at the source.
Very much like in the case of the renormalization constant Z_2, (10.64), there are two alternatives:

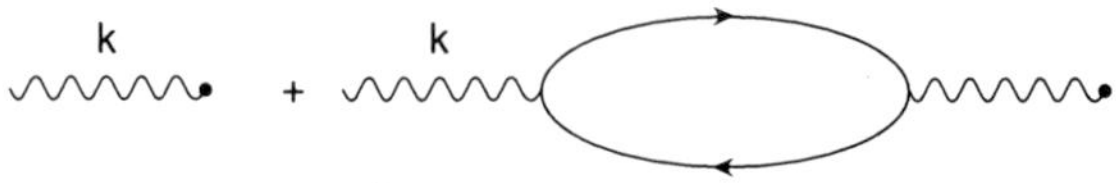

Fig. 10.16 Correction of an external photon line by vacuum polarization

(a) Either one does not correct external photon lines, and inserts the physical (renormalized) charge at all vertices, or,
(b) One applies the appropriate radiative corrections also to external photon lines, and divides the result by $Z_3^{1/2}$ for every such external line.

Observable Effects of Vacuum Polarization
In second order perturbation theory the regularized expression (10.96) takes the form

$$(\Pi^{\text{reg}})^{\mu\nu}(Q) = \big(Q^\mu Q^\nu - Q^2 g^{\mu\nu}\big)\big(C + \Pi^{\text{finite}}(Q^2)\big)\,, \tag{10.100}$$

the first term of which, $C = \alpha/(3\pi)\sum_f \ln\big(M_f/m_f\big)^2$, is absorbed by renormalizing the charge, while the second, finite term

$$\Pi^{\text{finite}}(Q^2) = -\frac{2\alpha}{\pi}\sum_f \int_0^1 \mathrm{d}z\; z(1-z)\ln\Big(\frac{m_f^2 - Q^2 z(1-z)}{m_f^2 - \mathrm{i}\varepsilon}\Big)$$

is defined by the condition that it vanish at $Q^2 = 0$. This term which is uniquely defined by this condition, by its Lorentz invariance, and by its gauge invariance, is discussed in this section.

By substitution of the variable $z := (1-x)/2$ and by means of a partial integration which liberates the logarithm, it becomes

$$\Pi^{\text{finite}}(Q^2) = \frac{\alpha}{\pi}Q^2\sum_f \int_0^1 \mathrm{d}x\; \frac{x^2(1-x^2/3)}{4m_f^2 - Q^2(1-x^2) - \mathrm{i}\varepsilon}\,.$$

One more substitution $s := 4m_f^2/(1-x^2)$ for which

$$\mathrm{d}s = \frac{s^2}{2m_f^2}x\,\mathrm{d}x\,, \quad x^2 = 1 - \frac{4m_f^2}{s}\,,$$

turns it into

$$\Pi^{\text{finite}}(Q^2) = \frac{\alpha}{3\pi}Q^2\sum_f \int_{4m_f^2}^{\infty} \mathrm{d}s\; \frac{\big(1+2m_f^2/s\big)\sqrt{1-4m_f^2/s}}{s(s-Q^2-\mathrm{i}\varepsilon)}\,. \tag{10.101}$$

This formula is interesting also from a function theoretic point of view because our finite term is represented by an integral over the cut from $(2m_f)^2$ to infinity. The virtual fermion pair f^+f^- contributes only if the variable s is larger than this threshold value $(2m_f)^2$.[11] In order to visualize better this finite part we use Fourier transformation to translate it to x-space. Furthermore, we return to our example of scattering on an external potential. Barring the radiative corrections $F_1(t) - F_1(0)$ and $F_2(t)$ of (10.87), and carrying out charge renormalization, one obtains

$$\begin{aligned} M(Q) &\equiv 2\pi \mathrm{i} e \overline{u(q)} \gamma^\mu u(p) \\ &\quad \times \left\{ g_{\mu\nu} + \frac{1}{Q^2}(Q_\mu Q_\nu - Q^2 g_{\mu\nu}) \Pi^{\text{finite}}(Q^2) \right\} \widetilde{A}^\nu(Q^2) \\ &= 2\pi \mathrm{i} e \overline{u(q)} \gamma^\mu u(p) \left\{ 1 - \Pi^{\text{finite}}(Q^2) \right\} \widetilde{A}_\mu(Q^2) . \end{aligned}$$

Here one again used $\overline{u(q)} \gamma^\mu u(p) Q_\mu = 0$. After charge renormalization the external potential becomes

$$\widetilde{A}_\mu\big((\boldsymbol{p}-\boldsymbol{q})^2\big) = \frac{Ze}{(2\pi)^3} \delta_{\mu 0} \frac{1}{(\boldsymbol{p}-\boldsymbol{q})^2} .$$

Regarding the remaining formulae one notes that $E_p = E_q$ and, hence, $Q^2 = -\boldsymbol{Q}^2$, so that

$$\Pi^{\text{finite}}(Q^2) = -\frac{\alpha}{3\pi} \boldsymbol{Q}^2 \sum_f \int_{4m_f^2}^{\infty} \mathrm{d}s \, \frac{\left(1 + 2m_f^2/s\right)\sqrt{1 - 4m_f^2/s}}{s(s + \boldsymbol{Q}^2 - \mathrm{i}\varepsilon)} .$$

We assume the external fermion to be either an electron or a muon, and the potential to be generated by an atomic nucleus with total charge Ze. Let

$$\widetilde{U}(\boldsymbol{Q}^2) := -Ze^2 \{1 - \Pi^{\text{finite}}(Q^2 = -\boldsymbol{Q}^2)\} \frac{1}{(2\pi)^3} \frac{1}{\boldsymbol{Q}^2} .$$

Upon transformation to position space one then has

$$\begin{aligned} U(\boldsymbol{x}) &= \frac{-Ze^2}{(2\pi)^3} \int \mathrm{d}^3 Q \; \mathrm{e}^{\mathrm{i}\boldsymbol{Q}\cdot\boldsymbol{x}} \\ &\quad \times \left\{ \frac{1}{\boldsymbol{Q}^2} + \frac{\alpha}{3\pi} \sum_f \int_{4m_f^2}^{\infty} \mathrm{d}s \, \frac{\left(1 + 2m_f^2/s\right)\sqrt{1 - 4m_f^2/s}}{s(s + \boldsymbol{Q}^2 - \mathrm{i}\varepsilon)} \right\} . \end{aligned}$$

The two integrals on $\boldsymbol{Q}$ are calculated by means of the formula

$$\frac{1}{(2\pi)^3} \int \mathrm{d}^3 Q \, \frac{\mathrm{e}^{\mathrm{i}\boldsymbol{Q}\cdot\boldsymbol{x}}}{\boldsymbol{Q}^2 + A^2} = \frac{\mathrm{e}^{-A|\boldsymbol{x}|}}{4\pi\,|\boldsymbol{x}|} , \tag{10.102}$$

---

[11] In fact, this expression may alternatively be derived from what is called a dispersion relation, that is, by means of Cauchy's integral theorem. In this approach it is the real part of a complex amplitude whose imaginary part describes $f^+ f^-$ creation, see, e.g., G. Källén, Handbuch der Physik, *Quantenelektrodynamik*, Vol. 1, Springer Berlin-Göttingen-Heidelberg, 1958.

and are reduced to a one-dimensional integral over the variable s, respectively. In particular, the second term in curly brackets yields

$$\frac{-Ze^2}{4\pi}\frac{\alpha}{3\pi}\sum_f \int_{4m_f^2}^{\infty} \mathrm{d}s\, \frac{1}{s}\Big(1+\frac{2m_f^2}{s}\Big)\sqrt{1-4m_f^2/s}\,\frac{\mathrm{e}^{-r\sqrt{s}}}{r}\,, \quad (r = |\boldsymbol{x}|)\,,$$

or, upon substitution by $y^2 := s/(4m_f^2)$, with $\mathrm{d}s = 8m_f^2 y\mathrm{d}y$,

$$= \frac{-Ze^2}{4\pi}\frac{2\alpha}{3\pi}\sum_f \int_1^{\infty} \mathrm{d}y\, \frac{1}{y^2}\Big(1+\frac{1}{2y^2}\Big)\sqrt{y^2-1}\,\frac{\mathrm{e}^{-2m_f r y}}{r}\,.$$

Thus, the total potential energy, corrected to order e^2, is equal to

$$U(r) = -\frac{Ze^2}{4\pi r}\left\{1+\sum_f \frac{2\alpha}{3\pi}\int_1^{\infty} \mathrm{d}y\;\mathrm{e}^{-2m_f r y}\Big(1+\frac{1}{2y^2}\Big)\frac{\sqrt{y^2-1}}{y^2}\right\}. \qquad (10.103)$$

The product $m_f r$ which appears in the argument of the exponential, supplemented by the correct factors $\hbar$ and c, is nothing but the ratio of the radial variable to the reduced Compton wave length of the virtual lepton $f, ƛ_C = \lambda_C/(2\pi)$,

$$m_f r \mathrel{\widehat{=}} \frac{r}{\hbar/(m_f c)}\,.$$

This means that the contribution of a given lepton species f to the vacuum polarization potential has a spatial range which is of the order of its Compton wave length $ƛ_C^{(f)}$. In the case of virtual electrons the relevant parameter is $ƛ_C^{(e)} = 386$ fm, for muons it is $ƛ_C^{(\mu)} = 1.87$ fm, and for τ-leptons it is $ƛ_C^{(\tau)} = 0.11$ fm. These numbers show that the by far largest contribution must be due to virtual e^+e^--pairs. Furthermore, since $ƛ_C^{(e)}$ is still small as compared to typical dimensions of ordinary, electronic atoms, one will not be surprised to learn that this contribution to the Lamb shift is numerically small: Vacuum polarization contributes about -17 MHz to the Lamb shift whose total magnitude is 1060 MHz.

Matters are different in muonic atoms which are smaller than electronic atoms by the ratio $m_e/m_\mu \approx 1/207$ and whose spatial extension is comparable to or smaller than $ƛ_C^{(e)}$. As a result, vacuum polarization is the dominant contribution to the Lamb shift of muonic atoms.

The limiting cases $m_f r \gg 1$ and $m_f r \ll 1$ can be further estimated. If $m_f r \gg 1$, that is, if $r \gg ƛ_C^{(e)}$, the integrand in the vacuum polarization potential is large only close to the lower boundary. Close to $y = 1$ one approximates

$$\Big(1+\frac{1}{2y^2}\Big)\frac{\sqrt{y^2-1}}{y^2} \approx \frac{3}{2}\sqrt{2(y-1)}$$

and uses the substitution $u := y - 1$ to obtain

$$\int_1^\infty dy\ e^{-2m_e r y}\left(1+\frac{1}{2y^2}\right)\frac{\sqrt{y^2-1}}{y^2}$$

$$\approx e^{-2m_e r}\int_0^\infty du\ e^{-2m_e r u}\frac{3}{2}\sqrt{2u} = \frac{3\sqrt{\pi}}{8}\frac{e^{-2m_e r}}{(m_e r)^{3/2}}\,.$$

The effective potential energy then is

$$U(r) \approx -\frac{Ze^2}{4\pi r}\left\{1+\frac{\alpha}{4\sqrt{\pi}}\frac{e^{-2m_e r}}{(m_e r)^{3/2}}\right\},\quad r \gg ƛ_C^{(e)}\,. \tag{10.104}$$

In this approximation the vacuum polarization which is induced by virtual electron-positron pairs, decreases with r more strongly than a Yukawa potential.

The other limiting situation $r \ll ƛ_C^{(e)}$ can also be worked out approximately. I quote the result without proof. It reads

$$U(r) \approx -\frac{Ze^2}{4\pi r}\left\{1-\frac{2\alpha}{3\sqrt{\pi}}\left[\ln(m_e r)+C_E+\frac{5}{6}\right]\right\},\quad r \ll ƛ_C^{(e)}\,, \tag{10.105}$$

where $C_E = 0.577216$ is Euler's constant.

The induced Charge Density of Vacuum Polarization

The result (10.103) has a striking physical property: The second term in the curly brackets is positive for all values of r and, hence, *amplifies* the unperturbed Coulomb potential. Indeed, by the action of vacuum polarization, the $1s$-state in a muonic lead atom is bound more strongly than in a pure Coulomb potential. This seems to contradict our intuition: Imagine vacuum polarization to be generated by virtual e^+e^--pairs which move within the range of the Compton wave length $ƛ_C^{(e)}$ of the electron. The virtual electrons should be attracted by the positive charge Ze while the virtual positrons should be repelled in such a way that the integral over the induced polarization charge density is equal to zero. But then the original charge $+Ze$ should be distributed over a larger domain in space, being smeared out over a range characterized by the Compton wave length. Such an effect always causes a *weakening* of binding. A smeared out charge distribution of total charge Ze binds less strongly than the same charge concentrated in a point (or in some smaller domain of space). This contradicts the observation made above!

In fact, the situation is even stranger. Given the potential, one can make use of Poisson's equation to derive the corresponding charge density. One finds that the latter has everywhere the same sign, for all $r \neq 0$. How can this be reconciled with the fact that the integral over the induced polarization density must be zero?

In order to answer this question consider the potential

$$\Phi^{\text{pol}}(r) = \frac{Ze}{4\pi r}\frac{2\alpha}{3\pi}\int_1^\infty dy\ e^{-2mry}\left(1+\frac{1}{2y^2}\right)\frac{\sqrt{y^2-1}}{y^2}\,.$$

(Note that this is the second term on the right-hand side of (10.103), taken for one of the leptons f, after extraction of a factor $(-e)$.) Defining $\xi := 2mr$ we study the dimensionless function

$$\varphi^{\text{pol}}(\xi) := \lim_{M\to\infty} \int_1^\infty \mathrm{d}y\ \mathrm{e}^{-(m/M)y}\ \mathrm{e}^{-\xi y}\Big(1+\frac{1}{2y^2}\Big)\frac{\sqrt{y^2-1}}{y^2}\,. \tag{10.106}$$

Here, the first exponential which enforces convergence, is introduced for a technical reason which will become clear below, and does not alter our argument. Use Poisson's equation to calculate the density,

$$\varrho^{\text{pol}}(\xi) = -\boldsymbol{\Delta}_\xi \varphi^{\text{pol}}(\xi)\,.$$

By the well-known formula

$$\big(\boldsymbol{\Delta} - \kappa^2\big)\frac{\mathrm{e}^{-\kappa r}}{r} = -4\pi\,\delta(\boldsymbol{r})$$

and with $\kappa = 2my$ one finds

$$\varrho^{\text{pol}}(\xi) = \lim_{M\to\infty} \left\{ 4\pi\,\delta(\boldsymbol{x}) \int_1^\infty \mathrm{d}y\ \mathrm{e}^{-(m/M)y}\Big(1+\frac{1}{2y^2}\Big)\frac{\sqrt{y^2-1}}{y^2} \right.$$
$$\left. -\frac{4m^2}{r}\int_1^\infty \mathrm{d}y\ \mathrm{e}^{-(m/M)y}\ \mathrm{e}^{-\xi y}\Big(1+\frac{1}{2y^2}\Big)\sqrt{y^2-1} \right\}\,. \tag{10.107}$$

It is clear that the convergence factor is needed only in the first term which otherwise is logarithmically divergent. It plays no role in the second term.

The strange behaviour of the polarization density (10.107) is easily understood. As long as r is not zero only the second term on the right-hand side contributes. It has the same sign for all $r \neq 0$. In the origin, $r = 0$, the density has two singularities, the δ-distribution of the first term multiplied by the logarithmically divergent factor,

$$4\pi\ \delta(\boldsymbol{x})\ \ln\Big(\frac{m}{M}\Big)\,,$$

as well as a pole of third order, $1/r^3$, in the second term on the right-hand side of (10.107). Indeed, calculating the charge density in the same approximation as (10.105), one obtains its leading term proportional to $1/r^3$. Nevertheless, the integral over the polarization density is equal to zero, as it should. In order to see this, calculate its integral using the formula

$$\int_0^\infty r^2\mathrm{d}r\ \frac{\mathrm{e}^{-\kappa r}}{r} = \frac{1}{\kappa^2}$$

(with $\kappa = 2my$, as before) thus finding

$$\int d^3x\, \varrho^{\text{pol}} = \lim_{M\to\infty} \left\{ \int_1^\infty dy\ e^{-(m/M)y}\left(1+\frac{1}{2y^2}\right)\frac{\sqrt{y^2-1}}{y^2} - \int_1^\infty dy\ e^{-(m/M)y}\left(1+\frac{1}{2y^2}\right)\frac{\sqrt{y^2-1}}{y^2} \right\} = 0\,.$$

In some sense this result is typical for quantum electrodynamics. The potential, being an observable, is well-defined but the polarization density which is not directly observable, is quite singular. Thus, in trying to interpret the induced charge density (10.107) one may say this: A test particle which passes very far from the given positive charge (that is, which scatters with small momentum transfers), sees essentially only its *physical* charge. If it approaches the charge, i.e. if it scatters with larger momentum transfers, then it sees more and more of its *bare* charge which, as we know, is larger than the physical charge. Note, however, that the bare charge, strictly speaking, is infinitely large. This phenomenon limits our ability to visualize vacuum polarization in a simple physical picture.

Remarks

1. The phenomenon of the running coupling constant:
   As a consequence of vacuum polarization the photon propagator is modified by the insertion of loops of virtual $f\bar{f}$-pairs according to
   $$-\frac{g^{\mu\nu}}{q^2+i\varepsilon} - \frac{g^{\mu\alpha}}{q^2+i\varepsilon}\Pi^{\text{reg}}_{\alpha\beta}(q)\frac{g^{\beta\nu}}{q^2+i\varepsilon} \equiv D'^{\mu\nu}(q)\,.$$
   This repeats the replacement (10.97), provided with the correct $i\varepsilon$-rules in the denominators. Like in (10.94) we make use of gauge invariance in decomposing $\Pi_{\alpha\beta}$ in terms of covariants. Extracting the factor e_0^2, we have
   $$\Pi^{\text{reg}}_{\alpha\beta}(q) = e_0^2\left[q_\alpha q_\beta - q^2 g_{\alpha\beta}\right]\pi(q^2)\,.$$
   (The Lorentz scalar $\pi(q^2)$ differs from $\Pi(q^2)$ of (10.94) by the factor e_0^2. For simplicity we omit the notation "reg".) Eq. (10.96) shows that
   $$\pi(q^2) = \frac{1}{12\pi^2}\sum_f \ln\left(\frac{M_f}{m_f}\right)^2 + \pi^{\text{finite}}(q^2)\,, \quad \text{with}$$
   $$\pi^{\text{finite}}(q^2) = -\frac{1}{2\pi^2}\sum_f \int_0^1 dz\ z(z-1)\ln\left(1-\frac{q^2 z(1-z)}{m_f^2 - i\varepsilon}\right).$$
   Thus, except for gauge terms whose contribution to every physical diagram vanishes, one has
   $$D'^{\mu\nu}(q) = -\left[g^{\mu\nu} - \frac{1}{q^2}q^\mu q^\nu\right]\frac{d'(q^2)}{q^2+i\varepsilon} \quad (\,+\text{ gauge terms }),$$
   where $d'(q^2) = 1 - e_0^2\pi(q^2)$.

Let μ^2 be the square of some arbitrary but fixed momentum. Define a new Lorentz scalar function $\pi^{\text{finite}}(q^2, \mu^2)$ by

$$\pi^{\text{finite}}(q^2, \mu^2) := \pi(q^2) - \pi(\mu^2) = \pi^{\text{finite}}(q^2) - \pi^{\text{finite}}(\mu^2) . \tag{10.108}$$

The quantity μ^2 is said to be the *renormalization point.* The function $\pi^{\text{finite}}(q^2, \mu^2)$ is free of any logarithmic divergence and, hence, is finite. Inserting this definition into $d'(q^2)$ and noting that we are working in second order perturbation theory, one finds

$$\begin{aligned} d'(q^2) &= 1 - e_0^2\pi(\mu^2) - e_0^2\pi^{\text{finite}}(q^2, \mu^2) \\ &\approx \left(1 - e_0^2\pi(\mu^2)\right)\left(1 - e_0^2\pi^{\text{finite}}(q^2, \mu^2)\right) . \end{aligned}$$

The first factor is denoted by a new symbol, viz.

$$Z(\mu^2) := 1 - e_0^2\pi(\mu^2) , \tag{10.109}$$

Taking account of the order at which we are calculating, the function $d'(q^2)$ can be written in the following way,

$$d'(q^2) = Z(\mu^2)\Big[1 - e_0^2 Z(\mu^2)\,\pi^{\text{finite}}(q^2, \mu^2)\Big] .$$

As we are using natural units, we have $\alpha_0 = e_0^2/(4\pi)$, and $\alpha^2 = e^2/(4\pi)$ for the physical coupling constant. If we define the product

$$\alpha_0 Z(\mu^2) =: \alpha(\mu^2) \tag{10.110}$$

to be the *physical coupling at the scale* μ^2, then the product

$$\alpha_0 d'(q^2) = \alpha(\mu^2)\Big[1 - 4\pi\alpha(\mu^2)\pi^{\text{finite}}(q^2, \mu^2)\Big] \tag{10.111}$$

is independent of μ^2.

One evaluates the result (10.110) both at $q^2 = \mu^2$, and at $q^2 = 0$, taking into account that $\pi^{\text{finite}}(q^2 = \mu^2, \mu^2)$ is equal to zero, and obtains

$$\alpha(\mu^2) = \alpha(0)\Big[1 - 4\pi\alpha(0)\pi^{\text{finite}}(q^2, 0)\Big] \approx \frac{\alpha(0)}{1 + 4\pi\alpha(0)\pi^{\text{finite}}(q^2, 0)} ,$$

or, upon inserting the integral representation of π^{finite},

$$\frac{\alpha(\mu^2)}{\alpha(0)} = 1 + \frac{2\alpha(0)}{\pi}\sum_f \int_0^1 dz\, z(z-1)\ln\left(1 - \frac{q^2 z(1-z)}{m_f^2 - i\varepsilon}\right) . \tag{10.112}$$

This formula is particularly interesting because it can be interpreted in two different ways:

(a) If one interprets $\alpha \equiv \alpha(0) \approx 1/137$ as being *the* coupling constant of quantum electrodynamics (this is the customary interpretation in quantum electrodynamics with light fermions), then (10.112) represents the modification of the photon propagator by the effect of vacuum polarization. The formula

$$\left[1 - 4\pi\alpha\pi^{\text{finite}}(q^2, 0)\right] = d^{\text{ren}}_{\text{QED}}(q^2)$$

is the radiatively corrected (to second order), renormalized expression for $d'(q^2)$.

(b) Alternatively, the formula (10.112) may be interpreted in the sense that it yields the dependence of the physical coupling on the scale μ^2. The numerical value of Sommerfeld's fine structure constant then depends on the scale μ^2 at which the physical elementary charge is determined. So, in fact, α is no longer constant, and becomes what is called a *running coupling constant* whose value depends on the renormalization scale μ^2.

As we emphasized previously, the renormalization point $\mu^2 = 0$ in quantum electrodynamics is not arbitrary. This choice is natural because it is singled out by Thomson scattering which fixes the physical charge of the electron by the limit $q^2 \to 0$ of photon-electron scattering.[12] Alternatively it is possible to define the electric charge, or α, at some other point $\mu^2 \neq 0$, that is, to apply the renormalization procedure at this scale. For example, in electroweak interactions a natural scale is defined by the squared masses of $W^\pm$ or of Z^0, in which case one obtains, e.g., $\alpha(m_Z^2) \approx 1/128$. An observable A which is calculated by perturbation theory, by choosing two different renormalization points, is then obtained from two different series expansions,

$$A \approx A_0 + A_2\alpha(0) + A_4\alpha^2(0)\,,$$

$$A \approx A_0' + A_2'\alpha(\mu^2) + A_4'\alpha^2(\mu^2)\,,$$

with different expansion coefficients. Although either series should approach the "true" answer, provided they are convergent or semi-convergent, there may be better, or worse, choices of the renormalization scale. Presumably, by choosing a scale which is defined by the physical situation at stake, one may truncate this series at a lower order than with some other choice far away from this natural scale.

While for quantum electrodynamics there is such a natural choice which is $\mu^2 = 0$, this is not true in other, nonabelian gauge theories such as the theory of electroweak interactions or the theory of strong interactions.

2. Where has Z_1 gone?

The reader will have noticed that there are renormalization constants such as Z_2, (10.64), correcting external fermion lines, as well as Z_3, (10.99), which multiplies the electric charge, but that there is no Z_1. Indeed, Z_2 which is called fermionic wave function renormalization, relates the renormalized fermion propagator to the bare propagator, Z_3, the wave function renormalization for the photon, relates the renormalized photon propagator to the bare one, while Z_1 contains the correction of the vertex part. One shows that the renormalization constants Z_1 and Z_2 of quantum electrodynamics are equal, $Z_1 = Z_2$. In second order perturbation theory, this result is contained in the relation (10.86). This equality that holds to all orders is the reason why we did not introduce Z_1 explicitly.

[12] This statement is but one of a series of low-energy theorems which say, qualitatively, that photon scattering on a fermion at small momentum transfers is determined by static electromagnetic properties of the fermion such as charge and magnetic moment.

3. Comments on the proof of renormalizability of quantum electrodynamics: The preceding sections demonstrated methods of *regularization* and the principle of *renormalization* on some examples in second order of perturbation theory. In this framework it was shown that, after having identified the physical masses and charges, all observables are free of divergences and are predicted without ambiguity. This intuitively appealing but mathematically incomplete procedure can be extended to *all* orders of perturbation theory. This is the content of the renormalization program which holds for quantum electrodynamics and for other, renormalizable gauge theories. In the case of quantum electrodynamics there are three essential objects whose ultraviolet divergence must be analyzed. These are: The exact fermion propagator[13]

$$\mathrm{i}S_F(p) = \int \mathrm{d}^4x \; \mathrm{e}^{\mathrm{i}p\cdot x} \langle 0|T\; \psi(x)\overline{\psi}(0)\,|0\rangle \; ; \tag{10.113}$$

The photon propagator, treated to all orders,

$$\mathrm{i}D^{\mu\nu}(q) = \int \mathrm{d}^4x \; \mathrm{e}^{\mathrm{i}q\cdot x} \langle 0|T\; A^\mu(x)A^\nu(0)\,|0\rangle \; ; \tag{10.114}$$

And the complete vertex part

$$\begin{aligned} &S_F(p)\Gamma_\mu(p,p')S_F(p') \\ &= -\int \mathrm{d}^4x \int \mathrm{d}^4y \; \mathrm{e}^{\mathrm{i}p\cdot x}\, \mathrm{e}^{-\mathrm{i}p'\cdot y} \langle 0|T\; \psi(x) j_\mu(0)\overline{\psi(y)}\,|0\rangle \; . \end{aligned} \tag{10.115}$$

These three fundamental quantities are sketched in Fig. 10.17, the shaded loops representing the sum of all Feynman diagrams of all finite orders n. If we were dealing with *free* fields then we would have

$$\begin{aligned} S_f^{\text{free}}(p) &= \frac{\not{p} + m_0\,\mathbb{1}}{p^2 - m_0^2 + \mathrm{i}\varepsilon} \,, \\ D^{\mu\nu}_{\text{free}}(q) &= -\frac{g^{\mu\nu}}{q^2 + \mathrm{i}\varepsilon} \,, \\ \Gamma_\mu^{\text{free}} &= \gamma_\mu \,. \end{aligned}$$

In analogy to the second order of perturbation theory one proves the following assertions for any arbitrary, finite order: Provided one adjusts the mass correction term $\delta m = m - m_0$ in such a way that on the mass-shell one has $\delta m - \Sigma(p) = 0$ for $\not{p} = m\,\mathbb{1}$, one obtains

$$S_F(p) \longrightarrow \frac{Z_2}{\not{p} - m} \quad \text{for} \quad \not{p} \to m\,\mathbb{1} \,.$$

Likewise $D^{\mu\nu}(q)$, for $q^2 \to 0$, has the limit

$$D^{\mu\nu}(q) \longrightarrow -Z_3 \frac{g^{\mu\nu}}{q^2} \,,$$

[13] We write it for a given lepton species, suppressing the index f.

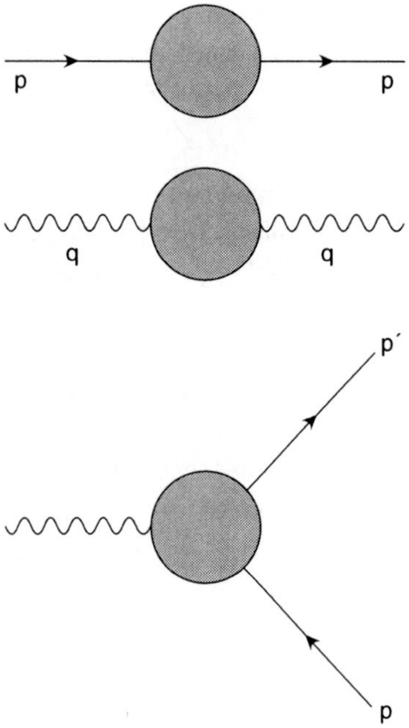

Fig. 10.17 Fermion propagator, photon propagator and vertex part in quantum electrodynamics to order n. The hatched areas hide all diagrams of the considered order that contribute to these quantities

the vertex at $p = p'$ and for $\not{p} \to m\,\mathbb{1}$ goes over into

$$\Gamma_\mu(p,p) \longrightarrow Z_1^{-1}\gamma_\mu\,.$$

The *renormalized* propagators, the *renormalized* vertex function, and the *renormalized,* physical charge are defined as follows:

$$S_F(p) = Z_2 S_F^{\text{ren}}(p)\,,$$

$$D^{\mu\nu}(q) = Z_3 D^{\mu\nu}_{\text{ren}}(q)\,,$$

$$\Gamma_\mu(p,p') = Z_1^{-1}\Gamma_\mu^{\text{ren}}(p,p')\,,$$

$$e_0 = \frac{Z_1}{Z_2\sqrt{Z_3}}e_{\text{ren}}\,. \tag{10.116}$$

One writes the renormalized functions defined in this way in terms of the physical mass and the physical charge and finds that they are *finite* for all p and p'. All terms which depend on the method of regularization (that is to say, all divergences of the theory) are contained in the three renormalization constants Z_1, Z_2, and Z_3. Furthermore, a more detailed analysis of the diagrams of order n shows that all factors Z_i cancel multiplicatively, and the final result is well-defined and finite. On says that self-energy and vertex parts are multiplicatively renormalizable.

4. Universality of electric charge:
As mentioned above one can prove that Z_1 and Z_2 are equal to arbitrary finite order of perturbation theory,

$$Z_1 = Z_2 . \tag{10.117}$$

This important relation follows from an identity that was proven by Ward and Takahashi. We quote this relation here but refer to Appendix A.7 for a somewhat formal proof,

$$(p - p')^\mu \Gamma_\mu(p, p') = \left(S_F(p)\right)^{-1} - \left(S_F(p')\right)^{-1} . \tag{10.118}$$

The relation (10.117) follows from this identity in the limit of $\not{p}'$ tending to $m\,\mathbb{1}$. Equation (10.117) has two important consequences: Renormalizability of the vertex function implies renormalizability of the self-energy, and vice versa. Furthermore, with $Z_1 = Z_2$ the definition (10.116) simplifies and becomes

$$e_{\text{ren}} = \sqrt{Z_3} e_0 . \tag{10.119}$$

The renormalization of the electric charge depends only on the photon propagator but not on the properties of the fermion that couples to the photon. In this sense renormalization of the charge is universal.
5. The S-matrix in presence of infraparticles:
Recall that the existence of isolated, free one-particle states was an essential assumption in defining the S-matrix. In quantum electrodynamics there are no such states. As the photon is strictly massless, an electron may be surrounded by a cloud of very soft photons, i.e. of photons whose energy is arbitrarily small. In other terms, the mass shell $p^2 = m^2$ of an electron never is isolated. Rather, it is the boundary of a continuum of energy-momentum surfaces on which it is accompanied by one, two, or more photons of very small energy. A particle which moves embedded in a photon cloud, is called an *infraparticle*, by allusion to the infrared divergences which are related to the masslessness of photons, too. As a consequence, strictly speaking, there is no S-matrix in quantum electrodynamics. But then, why is it that this ill-defined concept allows to derive correct formulae for cross sections even when one goes beyond the level of tree diagrams? The answer to this question which was discussed many times in the development of quantum electrodynamics, must be sought in a more careful analysis of realistic experimental arrangements by asking how beams are preparated and how they are detected. Arguing heuristically, a given process which is studied in an experiment, must be integrated over the *resolution* of the detector(s). Furthermore, in its theoretical description all those processes must be added, in which additional soft photons are emitted within the experimental resolution. In a symbolic representation this amounts to calculate the cross section

$$\mathrm{d}\sigma^{(n)}(A + B \to C + D)$$

radiatively corrected up to order $\mathcal{O}(\alpha^n)$, and to add to it

$$+ \int_{\Delta} \mathrm{d}\sigma^{(n-k)}(A + B \to C + D + \underbrace{\gamma + \gamma + \cdots + \gamma}_{k}),$$

radiatively corrected to order $\mathcal{O}(\alpha^{n-k})$.

In practical calculations one often introduces a ficticious finite photon mass. One calculates the cross sections at order n as indicated above, and studies the result in the limit of vanishing photon mass. As an alternative, one analyzes the amplitudes with dimensional regularization, noting that observables, after integration and addition as described above, come out perfectly finite. We note also that there is a mathematically rigorous approach to quantum electrodynamics which does not make use of the standard notion of S-matrix [Steinmann (2000)].

10.3 Epilogue: Quantum Electrodynamics in the Framework of Electroweak Interactions

The so-called *minimal standard model of electroweak interactions* combines weak interactions and quantum electrodynamics in a gauge theory which is constructed from the group $SU(2) \times U(1)$. The construction of this model requires some preparations and a sizeable technical apparatus. For this reason I restrict this epilogue to a number of general comments and some examples, and refer to more specialized monographs on this topic.

We begin with a list of particles which participate in the interactions described by the minimal standard model:

- Four bosons carrying spin 1:
  - the photon γ which is massless,
  - the Z^0 which is electrically neutral, and whose mass has the value $m_Z = 91.187 \pm 0.002\,\text{GeV}$,
  - the W-bosons, W^+ and W^-, which carry electric charges $+1$ and -1, respectively, and whose mass has the value $m_W = 80.385 \pm 0.015$ GeV;

  While γ and Z^0 coincide with their own antiparticles, the W^+ and W^- are antiparticles of one another.
- A boson with spin 0, electrically neutral, whose mass is expected to lie in the range 110–160 GeV. There is actually a strong candidate for this boson discovered at the Large Hadron Collider with mass about 125 GeV.
- Six leptons ordered in three pairs: (e^-, ν_e), (μ^-, ν_μ), (τ^-, ν_τ) as well as their antiparticles,
- Eighteen quarks which come in three pairs, too, as well as three species ("colours"): (u, d), (c, s), (t, b), and their antiparticles whose characteristics are summarized in Table 10.3 below those of the leptons.

The first five particles, i.e. the bosons with spin 1 and spin 0, respectively, play a double role: They can be created and detected as free, on-shell particles, but they

Table 10.3 Characteristics of leptons and quarks

Particle	Charge $\times \lvert e \rvert$	Mass (MeV)	Lepton numbers	Baryon number
e^-	-1	0.511	(1,0,0)	0
ν_e	0	$\simeq 0$	(1,0,0)	0
μ^-	-1	105.66	(0,1,0)	0
ν_μ	0	$\simeq 0$	(0,1,0)	0
τ^-	-1	1777.03	(0,0,1)	0
ν_τ	0	$\simeq 0$	(0,0,1)	0
u	$+2/3$	1.5–3	(0,0,0)	1/3
d	$-1/3$	3–7	(0,0,0)	1/3
c	$+2/3$	1000–1600	(0,0,0)	1/3
s	$-1/3$	100–300	(0,0,0)	1/3
t	$+2/3$	174200 (3300)	(0,0,0)	1/3
b	$-1/3$	4100–4500	(0,0,0)	1/3

also mediate interactions between the fermions, the real building blocs of matter. Following the pattern of the basic vertices of the standard model, the bosons can be exchanged between pairs of fermions.

Some, if not all of the neutrinos have nonvanishing masses which, however, are much smaller than those of their charged partners. Every lepton family comes with its own lepton family number, L_e, L_μ, and L_τ, each of which is additively conserved in electroweak interactions. Their values (L_e, L_μ, L_τ) are shown in the fourth column of Table 10.3.

Quarks can never occur in free asymptotic one-particle states and, therefore, their masses cannot be determined like those of leptons. Yet, there are mass parameters in the quark propagators which occur in a perturbative treatment of quantum chromodynamics. Although we do not go into this topic the hint given here might explain why quark masses are known only within error bars that are difficult to quantify.

The table shows only the *flavour* quantum numbers. These are the quantum numbers which are relevant for the selection rules of electromagnetic and weak interactions. Of each quark there are three exact copies which differ by one further quantum number, called *colour*. Colour plays a role in quantum chromodynamics, the theory of strong interactions while electroweak interactions do not see this degree of freedom.

10.3.1 Weak Interactions with Charged Currents

The minimal standard model contains different, partially unified interactions. Among these and as a matter of example, we consider the weak interactions which are responsible for nuclear β-decay and analogous processes in elementary particle physics.

Denote by $W_\mu(x)$ the quantum field which is constructed following the rules of canonical quantization. This field is linear in annihilation operators for positively charged W-bosons, and in creation operators for negatively charged W-bosons. Denote further by

$$f(x) \equiv \psi^{(f)}(x)$$

the quantum Dirac field describing a charged lepton of the species f, and by

$$\nu_f(x) \equiv \psi^{(\nu_f)}(x)$$

the quantum Dirac field of the corresponding neutrino. A typical interaction term of the electroweak model then reads

$$\mathcal{L}_{\mathrm{CC}} = -\frac{e}{2\sqrt{2}} \left\{ \sum_f \overline{\nu_f(x)}\gamma^\mu (\mathbb{1}-\gamma_5) f(x) W_\mu(x) + \overline{f(x)}\gamma^\mu (\mathbb{1}-\gamma_5)\nu_f(x) W^\dagger_\mu(x) \right\} . \tag{10.120}$$

This interaction has some remarkable properties:

(i) The factor in front of the curly brackets contains the electric elementary charge and a numerical factor which follows from the group structure of the model. The electromagnetic and the weak interactions are characterized by the same coupling constant. This, at first, seems surprising when one compares the phenomenology of electromagnetic physics of nuclei, atoms, and matter, with the rather different phenomenology of nuclear β-decay.

(ii) Like the Maxwell field the quantum field $W_\mu(x)$ is a vector field, though with a nonvanishing mass (in contrast to the photon). Its coupling to the leptons of a given family relates the charged to the neutral partner such that $\mathcal{L}_{\mathrm{CC}}$, as a whole, is electrically neutral. The leptonic factor of the interaction is a direct analogue of the electromagnetic current. As it changes the charge, by converting, e.g., a e^- into a ν_e, it is called the *charged current*. This also explains the notation $\mathcal{L}_{\mathrm{CC}}$ where "CC" stands for charged current.

(iii) The leptonic part is a coherent mixture of a *vector* current

$$\overline{\psi^{(1)}(x)}\gamma^\mu \psi^{(2)}(x) ,$$

whose structure is the same as the one of the electromagnetic current density, except for the fact that it relates two different fermion fields, as well as an *axial vector* current,

$$\overline{\psi^{(1)}(x)}\gamma^\mu \gamma_5 \psi^{(2)}(x) ,$$

which is new. As we remarked earlier, cf. Sect. 9.3.1, the mixture of vector and axial vector currents means that there may be parity-odd observables. Indeed, if the vector current transforms by a Lorentz transformation $\mathbf{\Lambda}$, then $\overline{\psi^{(1)}(x)}\gamma^\mu \gamma_5 \psi^{(2)}(x)$ transforms by $(\det \mathbf{\Lambda})\mathbf{\Lambda}$. This is to say that they transform in the same way with respect to $\mathbf{\Lambda} \in L_+^\uparrow$, but if one applies a space reflection in addition, there is a relative minus sign. The coefficients of the two terms being equal in magnitude, parity violation is even maximal! In other terms, using chiral fields (9.125), the W-boson couples only to left-chiral fields. This observation implies the specific selection rules for the coupling (10.120) that we derived and discussed in Sect. 9.3.4.

On the basis of the experience gained in Chap. 7 it is not difficult to see that the covariant propagator of W-bosons has the momentum space representation

$$\frac{-g^{\mu\nu} + q^\mu q^\nu / m_W^2}{q^2 - m_W^2 + i\varepsilon} . \tag{10.121}$$

Thus, it is obvious how the Feynman rules have to be generalized if this new interaction is part of the model. However, there is a pecularity: If we study processes such as muon decay, $\mu^- \to e^- + \overline{\nu_e} + \nu_\mu$ or nuclear β-decay, the momentum transfers are very small as compared to the mass of the W-boson. Therefore, one may safely neglect the second term of the numerator of (10.121) as well as the first term of its denominator. As a consequence, the exchange of a virtual W between lepton pairs $(1, 2)$ and $(3, 4)$, such as, e.g., the pairs (μ^-, ν_μ) and (e^-, ν_e), may be approximately described by the *effective* interaction

$$\begin{aligned}\mathcal{L}_{\text{eff}} \approx \frac{e^2}{8m_W^2} &\left(\overline{\psi^{(1)}(x)}\gamma^\mu(\mathbb{1} - \gamma_5)\psi^{(2)}(x)\right)\left(\overline{\psi^{(4)}(x)}\gamma_\mu(\mathbb{1} - \gamma_5)\psi^{(3)}(x)\right)\\ &+ \text{h. c.},\end{aligned} \tag{10.122}$$

from which the W-propagator has disappeared. (The addition "h. c." is a hint at the second term which must be the hermitian conjugate of the first.) This effective interaction relates four fermionic quantum fields at the same spacetime point and is called a *contact interaction.* A similar form of weak interaction was introduced by E. Fermi in the 1930's for the description of the β-decay of nuclei. As he chose to denote the common factor by $G/\sqrt{2}$, the quantity

$$G := \sqrt{2}\,\frac{e^2}{8m_W^2} \mathrel{\widehat{=}} \frac{\pi\alpha}{\sqrt{2}\,m_W^2} , \tag{10.123}$$

still today, is called *Fermi's constant.* This constant carries a dimension, $[G] = [E^{-2}]$, because it contains the factor $1/m_W^2$, as a remnant of the W-propagator. The large value of the mass of the W also explains the very large difference in the coupling strengths of electromagnetic versus weak interactions that is observed in atomic and nuclear physics, as well as in particle physics at low energies.

Although I do not go into this, I add here that the basic vertex which describes the coupling of fermions to the electrically neutral field $Z_\mu(x)$ of the Z^0-boson, contains a specific linear combination of the electromagnetic current and of the neutral partner of the charged current, viz.

$$\begin{aligned}\mathcal{L}_{\text{NC}} = \Big\{\alpha \Big[&\sum_f \overline{\nu_f(x)}\gamma^\mu(\mathbb{1} - \gamma_5)\nu_f(x)\\ &+ \sum_f \overline{f(x)}\gamma^\mu(\mathbb{1} - \gamma_5)f(x)\Big] + \beta\, j^\mu_{\text{e.m.}}(x)\Big\} Z_\mu(x) .\end{aligned} \tag{10.124}$$

These neutral current interactions (hence the index "NC" of the Lagrange density) again contain neutrinos only in their lefthanded realization. However, in contrast to the previous case of CC-interactions, this does not hold for their charged partners. The vector current $j^{\mu}_{\text{e.m.}}(x)$, for instance, has equal amounts of lefthanded and righthanded fermion fields. Therefore, observables that are due to this interaction may contain terms which are odd under space reflection, but parity violation will no longer be maximal.

Remarks

1. Regarding quarks the standard model of electroweak interactions fixes their vertices in a unique way, too. However, there are two complications which must be discussed separately. Firstly, the flavour eigenstates of Table 10.3 do not coincide exactly with the states that couple to weak vertices. Rather, at weak vertices with charged currents, unitary mixtures of these states are created or annihilated. Secondly, the matrix elements of the weak currents are modified by strong interactions in such a way that, though they are simple to parametrize, they are difficult to actually compute.
2. To a large extent, the quantization of the minimal standard model follows the model of quantum electrodynamics and leads to a generalized set of Feynman rules for computing amplitudes in the framework of covariant perturbation theory. This enlarged theory into which quantum electrodynamics is embedded, was proven to be renormalizable so that radiative corrections are predicted and can be tested in experiment.

10.3.2 Purely Leptonic Processes and Muon Decay

There are a few processes of weak interactions in which only leptons participate and which can be measured within a reasonable experimental effort. Among these there are:

– The decay of the negative or the positive muon

$$\mu^- \to e^- + \overline{\nu_e} + \nu_\mu\,, \quad \left(\mu^+ \to e^+ + \nu_e + \overline{\nu_\mu}\right); \tag{10.125}$$

– The analogous leptonic decays of $\tau^\pm$, viz.

$$\tau^- \to e^- + \overline{\nu_e} + \nu_\tau\,, \quad \left(\tau^+ \to e^+ + \nu_e + \overline{\nu_\tau}\right), \tag{10.126}$$

$$\tau^- \to \mu^- + \overline{\nu_\mu} + \nu_\tau\,, \quad \left(\tau^+ \to \mu^+ + \nu_\mu + \overline{\nu_\tau}\right); \tag{10.127}$$

– The so-called inverse muon decay

$$\nu_\mu + e^- \longrightarrow \mu^- + \nu_e\,, \tag{10.128}$$

where an incoming beam of muonic neutrinos hits electrons in a target and where the muon of the final state is detected;

– Elastic scattering of electronic or muonic neutrinos on electrons

$$\nu_e + e^- \rightarrow \nu_e + e^- , \qquad \overline{\nu_e} + e^- \rightarrow \overline{\nu_e} + e^- , \tag{10.129}$$

$$\nu_\mu + e^- \rightarrow \nu_\mu + e^- , \qquad \overline{\nu_\mu} + e^- \rightarrow \overline{\nu_\mu} + e^- . \tag{10.130}$$

– The decays of the Z^0 in lepton-antilepton pairs

$$Z^0 \rightarrow f + \overline{f} , \tag{10.131}$$

where f is a charged lepton or a neutrino.

These decays and reactions allow to study the weak interactions in a neat setting, without having to correct for effects of interactions other than electroweak. The purely electroweak radiative corrections generally are small but, depending on the accuracy of measurements, often must be taken into account.

Exercise 10.6 invites the reader to decide, on the basis of the vertices defined by (10.120) and by (10.124), which of the processes listed above are due to the charged current, or to the neutral current, or to both. Besides these purely weak processes there are others to which the electromagnetic and the weak interactions contribute simultaneously and, hence, in which they can interfere. Examples are the pair creation reactions (10.48),

$$e^+ + e^- \rightarrow \mu^+ + \mu^- , \quad e^+ + e^- \rightarrow \tau^+ + \tau^-$$

that we calculated in Sect. 10.1.3(c), in the framework of pure quantum electrodynamics. As one easily realizes, the diagrams in which the photon is replaced by a Z^0, also contribute. At energies which are small as compared to the mass of the Z^0, these contributions are suppressed by the factor $1/m_Z^2$. However, as soon as one reaches the neighbourhood of the Z^0-pole the corrections become large and, in fact, comparable to the electromagnetic amplitudes.

A classical example for a purely leptonic process is provided by the decay of the muon into an electron and two neutrinos,

$$\mu^- (q) \rightarrow e^- (p) + \overline{\nu_e} (k_1) + \nu_\mu (k_2) ,$$

the four-momenta being indicated in parentheses. Assume the muon, before its decay, to be at rest with respect to the frame of the observer, i.e. take $q = (m_\mu, \mathbf{0})^T$. When one does not, or cannot, observe the two neutrinos, the electron has a continuous energy spectrum whose lower and upper ends are easy to determine: In the kinematic situation where the neutrinos move back-to-back with equal and opposite spatial momenta, the electron has momentum $\boldsymbol{p} = 0$ and its energy is $E_p = m_e$. When the neutrinos move in the same direction, the electron moves in the opposite direction and has spatial momentum $\boldsymbol{p} = \boldsymbol{p}_{\max}$, while the neutrinos have momenta $-x\boldsymbol{p}_{\max}$ and $-(1-x)\boldsymbol{p}_{\max}$, respectively, with x a number between 0 and 1. Neglecting the masses of the neutrinos the sum of their energies is equal to the modulus of $\boldsymbol{p}_{\max}$ (in natural units). The principle of energy conservation then yields $E_p + |\boldsymbol{p}| = m_\mu$ from which one obtains $(E_p - m_\mu)^2 = E_p^2 - m_e^2$. Thus, the entire kinematic range of the electron is

$$m_e \leq E_p \leq W , \qquad \text{with} \quad W = \frac{m_\mu^2 + m_e^2}{2m_\mu} . \tag{10.132}$$

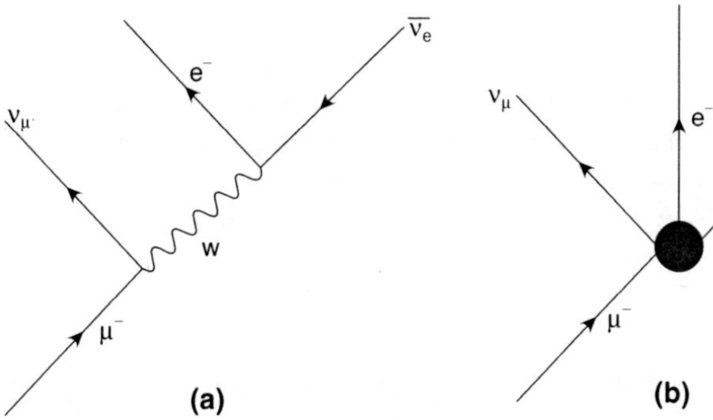

Fig. 10.18 **a** Muon decay in tree approximation is generated by exchange of a W-boson. **b** Due to $q^2 \ll m_W^2$ the diagram shrinks to a contact interaction with four fermions interacting with each other at the same point of space-time

Let the difference of the four-momenta q (of the muon) and p (of the electron) be denoted by $\Sigma := q - p$. Energy and momentum conservation tells us that Σ must also be $\Sigma = k_1 + k_2$. The general formula (8.34) yields the expression

$$\frac{\mathrm{d}^3\Gamma}{\mathrm{d}p^3} = \frac{(2\pi)^7}{4m_\mu E_p} \int \frac{\mathrm{d}^3k_1}{2E_1} \int \frac{\mathrm{d}^3k_2}{2E_2}\, \delta(\Sigma - k_1 - k_2) \sum_{\text{spin}} \left|T(\mu^- \to e^- \overline{\nu_e} \nu_\mu)\right|^2 ,$$

where we allow for the muon to be polarized but where the polarization of the electron is not measured. The effective contact interaction (10.122) provides an excellent approximation for the calculation of the amplitude and its absolute square. In order to see this, the reader should compare the squared momentum transfer $q^2 = m_\mu^2$ to the square of the W-mass: $m_\mu^2/m_W^2 \simeq 2 \times 10^{-6}$. Thus, the diagram containing the exchange of a W simplifies as shown schematically in Fig. 10.18.

The calculation of $|T|^2$ is done by means of the trace rules of Sect. 9.3.3. Integrating then over k_1 and k_2 one notices that the integration is symmetric in these momenta so that any *anti*symmetric terms may be omitted. All symmetric parts can be reduced to the following two integrals

$$\int \frac{\mathrm{d}^3k_1}{2E_1} \int \frac{\mathrm{d}^3k_2}{2E_2}\, (k_1 \cdot k_2)\delta(\Sigma - k_1 - k_2) = \frac{\pi}{4}\Sigma^2 \equiv I_0 , \tag{10.133}$$

$$\int \frac{\mathrm{d}^3k_1}{2E_1} \int \frac{\mathrm{d}^3k_2}{2E_2}\, \left(k_1^\mu k_2^\nu - (k_1 \cdot k_2)g^{\mu\nu} + k_2^\mu k_1^\nu\right) \times \delta(\Sigma - k_1 - k_2) = \frac{\pi}{6}\left(\Sigma^\mu \Sigma^\nu - \Sigma^2 g^{\mu\nu}\right) \equiv J^{\mu\nu} . \tag{10.134}$$

A way to calculate these integrals could be, for instance:

(a) As $\Sigma = k_1 + k_2$ and $k_1^2 = 0 = k_2^2$, one has $\left(k_1 \cdot k_2\right) = \Sigma^2/2$. The integral over the space components of k_2 is done by means of the spatial part of the δ-distribution. The remaining integral over $\boldsymbol{k}_1$ is best done in a frame of reference in which the spatial part of Σ vanishes, $\boldsymbol{\Sigma} = \boldsymbol{0}$. Then, indeed, one has $\boldsymbol{k}_2 = -\boldsymbol{k}_1$ and $E_2 = E_1$. Using spherical polar coordinates for $\boldsymbol{k}_1$, i.e. $\mathrm{d}^3 k_1 = E_1^2 \mathrm{d}E_1 \mathrm{d}\Omega_1$, one finds

$$I_0 = \frac{\Sigma^2}{2}\pi \int_0^\infty \mathrm{d}E_1 \; \delta(\Sigma^0 - 2E_1) = \frac{\pi}{4}\Sigma^2 .$$

(b) It is useful to decompose the second integral in terms of Lorentz covariants, that is,

$$J^{\mu\nu} = A\, \Sigma^\mu \Sigma^\nu + B\, \Sigma^2 g^{\mu\nu}$$

and to compute the coefficients A and B from the contractions

$$g_{\mu\nu} J^{\mu\nu} = (A + 4B)\Sigma^2 , \quad \Sigma_\mu \Sigma_\nu J^{\mu\nu} = (A + B)\Sigma^4 .$$

The first contraction is proportional to I_0, $g_{\mu\nu} J^{\mu\nu} = -2I_0$, and yields the relation $(A+4B)\Sigma^2 = -\pi\, \Sigma^2/2$. In the second expression the integrand is contracted with $\Sigma = k_1 + k_2$ in both indices, giving a vanishing result. Therefore, one has $B = -A$. Solving for A one obtains

$$A = \frac{\pi}{6} = -B .$$

The spin expectation value of the muon at rest is $s^{(\mu)} = (0, \widehat{\boldsymbol{n}})$ if we assume that the muon is completely polarized in the indicated direction. (If this is not so, replace the unit vector $\widehat{\boldsymbol{n}}$ by a vector $\boldsymbol{\zeta}$ whose modulus equals the degree of polarization $P^{(\mu)}$ of the muon.) Let the angle between the spatial momentum of the electron and the polarization of the muon be denoted by θ, i.e.

$$\boldsymbol{p} \cdot \widehat{\boldsymbol{n}} = \sqrt{E_p^2 - m_e^2} \cos\theta .$$

The energy of the electron is expressed by the ratio to the maximum (10.132),

$$x := \frac{E_p}{W} . \tag{10.135}$$

The range of this dimensionless variable is then

$$x_0 \le x \le 1 , \qquad \text{with} \quad x_0 = \frac{m_e}{W} .$$

Unless one looks very closely to the lower end of the energy spectrum the mass of the electron quickly becomes negligible as the energy E_p grows. Choosing $m_e \approx 0$, the kinematics simplifies to

$$W \approx \frac{m_\mu}{2} , \qquad 0 \le x \le 1 . \tag{10.136}$$

A calculation whose method should be familiar by now, and using this approximation, one finds the differential decay rate to be

$$\frac{\mathrm{d}^2 \Gamma(x, \theta)}{\mathrm{d}x\, \mathrm{d}\theta} \approx \frac{m_\mu^5 G^2}{192\pi^3} x^2 \left\{ (3 - 2x) + \cos\theta (1 - 2x) \right\} . \tag{10.137}$$

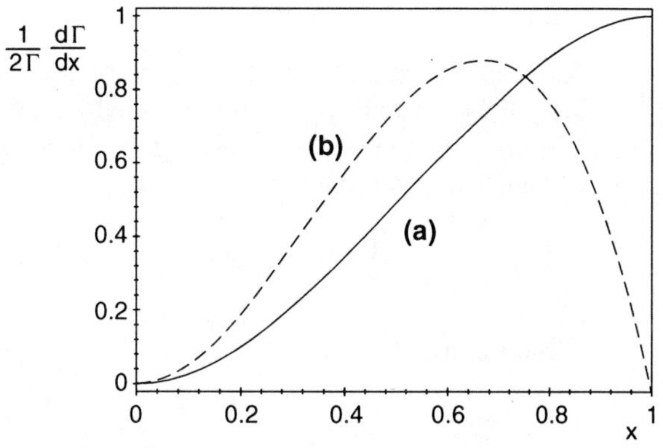

Fig. 10.19 **a** Energy spectrum of an electron during muon decay as a function of the dimensionless variable x, neither of the neutrinos being observed. **b** Spectrum of $\overline{\nu_e}$, electron and ν_μ are integrated out

This formula contains an isotropic term which yields the characteristic spectrum shown in Fig. 10.19, the so-called Michel spectrum,[14] and a term which describes the correlation between the spin expectation value of the muon and the momentum of the electron. As there is no such correlation in the initial state, its presence is an indication and a quantitative measure for parity violation. In case the muon is only partially polarized, this term is multiplied by $P^{(\mu)}$. Integration over the angle of emission θ as well as over the energy spectrum yields the decay width

$$\Gamma^{(0)}(\mu^- \to e^- \overline{\nu_e}\, \nu_\mu) \approx \int_{-1}^{+1} \mathrm{d}\cos\theta \int_0^1 \mathrm{d}x \frac{\mathrm{d}^2\Gamma(x,\theta)}{\mathrm{d}x\,\mathrm{d}\theta} = \frac{m_\mu^5 G^2}{192\pi^3}\,. \tag{10.138}$$

If all terms proportional to the electron mass are taken into account, and if one calculates the next-to-leading term in the expansion of the W-propagator in terms of m_μ^2/m_W^2, then (10.138) is replaced by

$$\Gamma(\mu^- \to e^- \overline{\nu_e}\, \nu_\mu) = \frac{m_\mu^5 G^2}{192\pi^3}\left\{1 - 8\frac{m_e^2}{m_\mu^2} + \mathcal{O}\left(\frac{m_e^3}{m_\mu^3}\right)\right\}\left\{1 + \frac{3m_\mu^2}{5m_W^2}\right\} \tag{10.139}$$

(cf. Exercise 10.7). This result is modified further by radiative corrections among which those from pure quantum electrodynamics are the most important. To lowest order in α these corrections amount to multiply the rate (10.139) by the factor

$$C_{\text{Photon}} = \left\{1 + \frac{\alpha}{8\pi}(25 - 4\pi^2)\right\}, \tag{10.140}$$

(see, e.g., [Scheck (2012)]).

[14] After Louis Michel, French theoretician (1923–1999) who derived the most general form of this spectrum at a time where the interaction responsible for the decay was not well known.

Remarks

1. Inserting numbers one sees that the "photonic" radiative corrections are of the order 0.44 % while the corrections due to the finite electron mass and to the W-propagator are practically negligible. Most important for practical purposes and applications is the fact that the muon lifetime is known to high accuracy, viz.
$$\tau^{(\mu)} = (2.1969811 \pm 0.0000022) \times 10^{-6}\text{s} \tag{10.141}$$
and, therefore, may serve as a reference information on the magnitude of Fermi's constant. From the number (10.141) one obtains
$$G = (1.1663787 \pm 0.0000006) \times 10^{-5}\ \text{GeV}^{-2}, \tag{10.142}$$
after having converted the result to the unit GeV, following common practice in particle physics.
2. The form of the energy spectrum of Fig. 10.19a is qualitatively well understood on the basis of the selection rule (9.134) for chirality that we derived in Sect. 9.3.4. It suffices to consider the two extreme cases of maximal and minimal energy of the electron, respectively, i.e. $x = 1$ and $x = x_0$:
   (a) At $x = 1$ the two neutrinos move in the same direction, cf. Fig. 10.20a, their spins are antiparallel because one of them is a particle, the other is an antiparticle. The electron which is created lefthanded by the interaction, moves in the opposite direction. Taking this direction to be the 3-axis one sees that $\mathbf{J}_3$, the projection of the total angular momentum, for the configuration drawn in Fig. 10.20a, is conserved. (Recall that the projection of the orbital angular momentum vanishes!) The electron takes over the chirality of the decaying muon. This simple analysis not only explains why the electron is preferentially emitted with large energy but also illustrates the angular distribution (10.137). Inserting there $x = 1$ the distribution has its maximum at $\theta = \pi$.
   (b) At $x = x_0$ the electron just stays where the muon was as rest before it decayed. In the neighbourhood of this point it is the phase space that closes as $x \to x_0$ and causes the differential decay rate to tend to zero.
3. If, instead of the electron spectrum, one wishes to calculate the energy spectrum of the ν_μ, by integrating over all momenta of the e^- and the $\overline{\nu_e}$, then one concludes from Fig. 10.20a that this spectrum must be the same as the one of the electron in Fig. 10.19. In turn, if one integrates over the momenta of e^- and of ν_μ, and determines the spectrum of $\overline{\nu_e}$, then the scheme in Fig. 10.20b readily shows that the antineutrino cannot be emitted with maximal energy. Close to threshold, on the other hand, the rate tends to zero anyway because of the closing phase space. Indeed, if one does this calculation one finds the following distribution, up to small terms in the electron mass,
$$\frac{m_\mu^5 G^2}{192\pi^3}\, g(x), \quad \text{with} \quad g(x) = 6x^2(1-x),$$

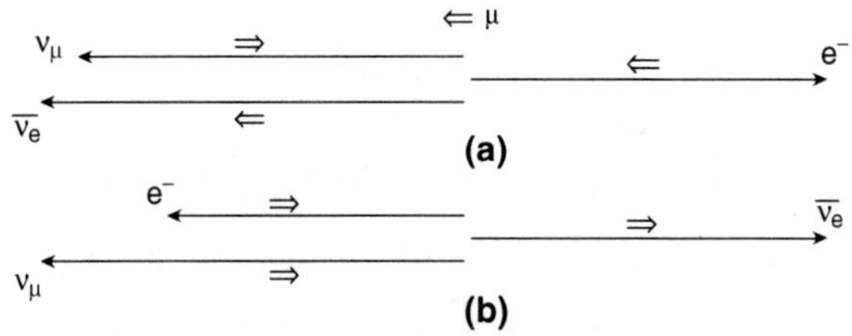

Fig. 10.20 **a** In the rest system of the decaying muon, the electron runs into the opposite direction to the direction of flight of the two neutrinos at maximal energy; conservation of projection of the total angular momentum is fulfilled. **b** If, however, $\overline{\nu_e}$ has maximum energy, electron and ν_μ have to fly into the opposite direction. Since both are *left-handed*, whereas $\overline{\nu_e}$ is *right-handed*, there is no possibility of conserving $\mathbf{J}_3$

which vanishes at $x = 0$ as well as at $x = 1$. This function is shown in Fig. 10.19, in comparison to the function $f(x) = x^2(3 - 2x)$, the isotropic part of (10.137).

4. When one integrates the doubly differential decay rate (10.137) over the energy only and uses the integrals

$$\int_0^1 \mathrm{d}x\, x^2(3 - 2x) = \frac{1}{2}, \quad \int_0^1 \mathrm{d}x\, x^2(1 - 2x) = -\frac{1}{6}$$

one finds the asymmetry of the decay, averaged over energy,

$$\frac{1}{\Gamma}\frac{\mathrm{d}\Gamma(\cos\theta)}{\mathrm{d}\cos\theta} = \frac{1}{2}\left(1 - \frac{1}{3}\cos\theta\right). \qquad (10.143)$$

This formula is important in practice. Indeed, recall that all terms in $\cos\theta$, both here and in (10.137), must be multiplied by $P^{(\mu)}$ whenever the muon is partially polarized only. The formula (10.143) then serves as an analyzer for the polarization of the muon. Of course, by lepton universality, all these results also apply to the leptonic decays (10.126) and (10.127) of the τ-leptons.

10.3.3 Two Simple Semi-leptonic Processes

To conclude we study two simple processes which involve both leptons and hadrons, and which belong to the class of what are called *semi-leptonic processes*. The first example takes up the decay $\pi^- \to \mu^- + \overline{\nu_\mu}$ whose charge-conjugate form was discussed in Sect. 9.3.4, Example 9.7, in comparison with the analogous electronic decay mode $\pi^- \to e^- + \overline{\nu_e}$,

$$\pi^-(q) \longrightarrow \mu^-(p) + \overline{\nu_\mu}(k). \qquad (10.144)$$

In the quark model a π^- is constructed from an antiquark $\overline{u}$ (electric charge $-2/3$), and a d-Quark (electric charge $-1/3$),

$$\pi^- \sim \left(\overline{u}d\right)_{\text{bound}}.$$

In pion decay, too, the liberated energy as well as all spatial momenta of the particles in the final state are small (as compared to m_W) so that we may appeal to the effective interaction (10.122) by inserting

$$\psi^{(1)}(x) = \psi^{(\mu)}(x) \equiv \mu(x)\,, \qquad \psi^{(2)}(x) = \psi^{(\nu_\mu)}(x) \equiv \nu_\mu(x)\,,$$
$$\psi^{(3)}(x) = \psi^{(d)}(x) \equiv d(x)\,, \qquad \psi^{(4)}(x) = \psi^{(u)}(x) \equiv u(x)\,,$$

(using a simplified notation like in (10.120)). However, the quark-antiquark pair is in a bound state of strong interactions that can hardly be computed so that we cannot do more than to parametrize the corresponding matrix element of the charged current. As one shows, the pseudoscalar nature of the pion implies that in the transition from a π^- into the vacuum only the axial vector charged current can contribute. With q denoting the momentum of the decaying pion, one has

$$\begin{aligned}&\langle 0|\overline{u(x)}\gamma^\mu(\mathbb{1}-\gamma_5)d(x)\,\big|\pi^-(q)\big\rangle \\ &\quad = -\langle 0|\overline{u(x)}\gamma^\mu\gamma_5 d(x)\,\big|\pi^-(q)\big\rangle = -\frac{\mathrm{i}}{(2\pi)^{3/2}}\widetilde{f}_\pi q^\mu\,. \end{aligned} \tag{10.145}$$

The form factor $\widetilde{f}_\pi$ which is defined by this ansatz, in principle, could also depend on q^2. However, since $q^2 = m_\pi^2$ has a fixed value in pion decay, this is a constant.[15] This parameter is called the *pion decay constant.* Using this parametrization, denoting by p the momentum of the muon, and by k the momentum of the antineutrino, the amplitude describing this decay reads

$$\begin{aligned} T(\pi^- \to \mu^-\overline{\nu_\mu}) &= \frac{-\mathrm{i}}{\sqrt{2}(2\pi)^{9/2}}\widetilde{f}_\pi G\,\overline{u^{(\mu)}(p)}\not{q}(\mathbb{1}-\gamma_5)v^{(\nu_\mu)}(k) \\ &= \frac{-\mathrm{i}}{\sqrt{2}(2\pi)^{9/2}}\widetilde{f}_\pi G\, m_\mu\overline{u^{(\mu)}(p)}(\mathbb{1}-\gamma_5)v^{(\nu_\mu)}(k)\,. \end{aligned}$$

In the second step $\not{q} = \not{p} + \not{k}$ was inserted and the Dirac equation was used for the muon and the antineutrino. The mass of the neutrino was neglected. Using the formula (8.37) and making use of the trace techniques a short calculation yields the decay width

$$\Gamma(\pi^- \to \mu^-\overline{\nu_\mu}) = \frac{G^2\widetilde{f}_\pi^2 m_\pi}{8\pi}\, m_\mu^2\left(1 - m_\mu^2/m_\pi^2\right)^2. \tag{10.146}$$

Its similarity to the formula (9.135) is remarkable. Here as well there the factor m_μ^2 is due to the conflict between conservation of angular momentum and the selection rule (9.134), while the factor in parantheses reflects the available phase space.

[15] This remark is relevant for situations where the pion is not on its mass shell and where q^2 is a genuine variable.

Remarks

1. The form factor $\widetilde{f}_\pi$ which in this case is constant, parametrizes our ignorance of what really happens inside a pion. We insert the measured lifetime of the pion,
$$\tau^{(\pi)} = 2.6033 \pm 0.0005 \times 10^{-8}\ \text{s}\,. \tag{10.147}$$
The formula (10.146) then yields approximately $\widetilde{f}_\pi \simeq 135$ MeV, i.e. a value in the neighbourhood of the mass of the pion.
2. The same analysis and the analogous calculation also apply to the decay channel $\pi^- \to e^- + \overline{\nu_e}$, one only has to replace m_μ by m_e. This confirms the ratio of the decay widths of the electronic and the muonic final state that we had given, without proof, in (9.135) of Sect. 9.3.4.
3. The result of the first remark above may seem disappointing because, apparently, we have not done much more than replace one experimental number (the lifetime) by another (the form factor). This deceptive impression vanishes immediately when we show that another decay such as the semi-leptonic τ-decay $\tau^+ \to \pi^+ + \overline{\nu_\tau}$ can now be predicted. Imagine the decay $\pi^- \to \tau^- + \overline{\nu_\tau}$ (which, of course, does not exist, for kinematic reasons) to be "crossed", that is to say, suppose the incoming π^- is replaced by an outgoing π^+, while the outgoing τ^- is replaced by an incoming τ^+, thus yielding the (kinematically allowed) process
$$\tau^+(p) \longrightarrow \pi^+(q) + \overline{\nu_\tau}(k) \tag{10.148}$$
whose T-matrix amplitude is given as follows:
$$\begin{aligned} T(\tau^+ \to \pi^+ \overline{\nu_\tau}) &= \frac{-\mathrm{i}}{\sqrt{2}(2\pi)^{9/2}} \widetilde{f}_\pi G\, \overline{v^{(\nu_\tau)}(k)}\, \not{q} \left(\mathbb{1} - \gamma_5\right) v^{(\tau)}(p) \\ &= \frac{-\mathrm{i}}{\sqrt{2}(2\pi)^{9/2}} \widetilde{f}_\pi G\, m_\tau \overline{v^{(\nu_\tau)}(k)} \left(\mathbb{1} + \gamma_5\right) v^{(\tau)}(p)\,. \end{aligned}$$
One should note that in the second step the energy-momentum balance $p = q + k$, i.e. $\not{q} = \not{p} - \not{k}$ was used and that $\not{p}$ was moved past γ_5 so that it acts on the spinor on the right-hand side (hence the minus sign). One then calculates the decay width for this channel, in close analogy to what we did above, and one finds
$$\Gamma(\tau^+ \to \pi^+ \overline{\nu_\tau}) = \frac{G^2 \widetilde{f}_\pi^2 m_\tau^3}{16\pi} \left(1 - m_\pi^2/m_\tau^2\right)^2. \tag{10.149}$$
In fact, one can do more than that: Assume that the decaying τ^+ was polarized. Due to parity violation in the weak interaction the angular distribution of the emitted pion, relative to the expectation value of the spin of τ^+, is not isotropic. Denoting by θ the angle between $\boldsymbol{s}^{(\tau)}$ and $\boldsymbol{q}$, one finds
$$\frac{1}{\Gamma} \frac{\mathrm{d}\Gamma(\tau^+ \to \pi^+ \overline{\nu_\tau})}{\mathrm{d}(\cos\theta)} = \frac{1}{2}\left(1 - \cos\theta\right). \tag{10.150}$$

This angular distribution may serve as an analyzer for the polarization of the τ^+. For example, in e^+e^--storage rings τ^+ and τ^- are produced in pairs by electromagnetic interaction. The selection rule (9.134) says that the chiralities of the τ^+ and of its partner τ^- are correlated. Thus, if one has determined the polarization of τ^+, by means of the asymmetry (10.150), then one knows the polarization of the accompanying negatively charged τ-lepton.

Appendix

A.1 Dirac's $\delta(x)$ and Tempered Distributions

Dirac's δ-distribution belongs to the class of *generalized functions*, more precisely, to the class of *tempered distributions*. In a physics perspective, the essential ideas which are at the root of this mathematical concept, can be understood by means of the following example.

Imagine a sequence of charge distributions ϱ_n all of which have total charge 1 and which are such that with increasing n they approach the *point* charge. For example, consider the functions

$$\varrho_n(\boldsymbol{r}) = \left(\frac{n}{\pi}\right)^{3/2} \mathrm{e}^{-nr^2} \tag{A.1}$$

whose normalization is equal to 1 for all n, as is easily verified,

$$\int \mathrm{d}^3x\, \varrho_n(\boldsymbol{r}) = 4\pi \int_0^\infty r^2 \mathrm{d}r\, \varrho_n(r) = 1\,.$$

Figure A.1 shows three examples of the charge distribution (A.1). In the limit $n \to \infty$ this becomes a rather singular graph: $\varrho_n(\boldsymbol{r})$ is essentially zero for all $\boldsymbol{r} \neq 0$, and takes a value approaching infinity in the origin $\boldsymbol{r} = 0$. While the object which is obtained by this procedure,

$$\lim_{n\to\infty} \varrho_n(\boldsymbol{r}) =: \delta(\boldsymbol{r})\,,$$

can no longer be a function in the usual sense, it remains true that the integral of ϱ_n multiplied by an arbitray continuous and bounded function $f(\boldsymbol{x})$ exists and, in the limit, takes the value

$$\lim_{n\to\infty} \int \mathrm{d}^3x\, \varrho_n(\boldsymbol{r}) f(\boldsymbol{x}) = f(0)\,.$$

This suggests to replace the ill-defined objects $\lim_{n\to\infty} \varrho_n(\boldsymbol{x})$ by the linear *functionals*

$$\delta_n(f) := \int \mathrm{d}^3x\, \varrho_n(\boldsymbol{r}) f(\boldsymbol{x})$$

F. Scheck, *Quantum Physics*, DOI: 10.1007/978-3-642-34563-0,

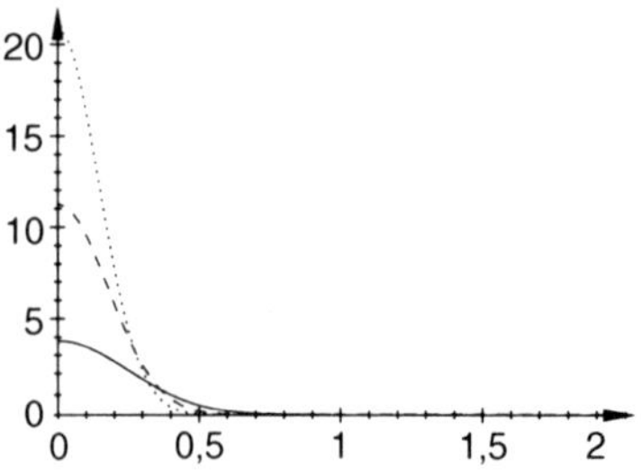

Fig. A.1 A family of functions approximating Dirac's δ-distribution $\delta(r)$. The figure shows the function (A.1) with $n = 8$, $n = 16$, and $n = 24$

which remain well-defined in this limit, and which yield the value

$$\delta(f) = f(0) \; . \tag{A.2}$$

One continues to use a symbolic integral of the kind

$$\int \mathrm{d}^3x \, \delta(\boldsymbol{x}) f(\boldsymbol{x}) \equiv \delta(f) := f(0)$$

but should note that this is not an integral in the sense of Lebesgue integrals. The expression (A.2) defines a linear and continuous functional over a linear function space. In other terms, if f is an element of the function space, then δ is an element of the dual space. Thus, the δ-distribution, like all other distributions, is defined with reference to a certain function space.

One proceeds as follows: One first defines the properties of the function space on the basis of what one needs, and then defines functionals which act on elements of this space and provide mappings to the real line $\mathbb{R}$. The guiding principle, qualitatively speaking, is to choose properties and definitions in such a manner that the calculus of distributions is formally the same as for ordinary functions.

We continue with the example: If the functions $f(\boldsymbol{x})$ possess continuous and bounded first derivatives then there exists the limit

$$\lim_{n\to\infty} \int \mathrm{d}^3x \, \frac{\partial \varrho_n(\boldsymbol{x})}{\partial x^i} f(\boldsymbol{x}) = -\lim_{n\to\infty} \int \mathrm{d}^3x \, \varrho_n(\boldsymbol{x}) \frac{\partial f(\boldsymbol{x})}{\partial x^i}$$
$$= -\delta\left(\frac{\partial f}{\partial x^i}\right) = -\frac{\partial f}{\partial x^i}(0) \; .$$

This allows to define another distribution which associates to every function f its negative derivative at 0,

$$\delta_{,i} \equiv \frac{\partial \delta}{\partial x^i} \; : \; f \longmapsto -\frac{\partial f}{\partial x^i}(0) \; . \tag{A.3}$$

This definition can be extended to higher derivatives under the condition that the corresponding higher derivatives of the functions f are continuous and bounded.

A.1.1 Test Functions and Tempered Distributions

While in the definition of the distribution (A.2) it was sufficient to require the functions on which it acts to be continuous and bounded, the distribution (A.3) is defined only on functions which, in addition, have continuous and bounded first derivatives. In this sense (A.3) is more singular than (A.2). More generally, the more singular a distribution, the more regular the functions must be onto which it acts.

The aim is to define distributions such that they are genuine generalized functions, that is to say, that they admit the same calculus as with ordinary functions. Therefore, it is useful to define all distributions with respect to one and the same function space and to choose this space sufficiently regular such that for every distribution all its derivatives of finite order are well defined, cf. the example (A.3). These requirements motivate the

Definition A.1 Space of Test Functions

The *space of test functions* $\mathcal{S}(\mathbb{R}^3)$ is the linear space of all C^∞-functions on $\mathbb{R}^3$ which, together with their finite derivatives, tend to zero for $|\boldsymbol{x}| \to \infty$ faster than any inverse power $|\boldsymbol{x}|^{-n}$.

A simple example is provided by the functions

$$f(\boldsymbol{x}) = P_k(\boldsymbol{x}) \mathrm{e}^{-|\boldsymbol{x}|^2} ,$$

with P_k polynomials. They are infinitely differentiable, the functions as well as their derivatives remain bounded, and, as the argument goes to infinity, they decrease faster than any inverse power of $|\boldsymbol{x}|$.

For the study of sequences of functions and the definition of convergence criteria a norm on the space of test functions is needed. The notation is simplified by using multi-indices and introducing some abbreviations,

$$k \equiv (k_1, k_2, k_3) , \qquad k_i \in \mathbb{N}_0 , \qquad \sum k \equiv k_1 + k_2 + k_3 ,$$

$$x^k \equiv (x^1)^{k_1} (x^2)^{k_2} (x^3)^{k_3} , \qquad D^k \equiv \frac{\partial^{k_1}}{\partial (x^1)^{k_1}} \frac{\partial^{k_2}}{\partial (x^2)^{k_2}} \frac{\partial^{k_3}}{\partial (x^3)^{k_3}} .$$

To every pair of naturals $p, q \in \mathbb{N}$ one associates the norm on $\mathcal{S}$

$$\|f\|_{p,q} := \sup_{\sum k \le p, \sum l \le q} \; \sup_{\boldsymbol{x} \in \mathbb{R}^3} \left| (1 + |\boldsymbol{x}|^2)^{\sum k} D^l f(\boldsymbol{x}) \right| . \tag{A.4}$$

One then says that a sequence f_n of elements of the space $\mathcal{S}$ converges and approaches $f \in \mathcal{S}$ if

$$\lim_{n \to \infty} \|f_n - f\|_{p,q} = 0 \quad \text{for all} \quad p, q \in \mathbb{N} . \tag{A.5}$$

Let T be a linear functional, $T : \mathcal{S} \to \mathbb{R}$, on the function space $\mathcal{S}$. The functional T is said to be *continuous* if for every null sequence one has $T(f_n) \to 0$. By the linearity of the functional one then has

$$T(f_n) \longrightarrow T(f) \quad \text{for} \quad f_n \longrightarrow f , \qquad f_n, f \in \mathcal{S} .$$

These concepts allow to define precisely what a distribution should be:

Definition A.2 Tempered Distribution

A tempered distribution is a continuous linear functional on the space $\mathcal{S}$ of rapidly decreasing functions.

As a necessary and sufficient condition for T to be continuous it must be possible to find a pair of naturals such that

$$|T(f)| \leq c \, \|f\|_{p,q} \quad \text{for all} \quad f \in \mathcal{S}, \tag{A.6}$$

where c is a positive constant. Here are two examples:

- Dirac's δ-distribution $\delta(f) := f(0)$ is continuous because it satisfies the inequality

$$|\delta(f)| \leq \|f\|_{0,0} \cdot$$

- Its first derivative by x^i which is defined by

$$\delta_{,i}(f) := -\frac{\partial f}{\partial x^i}(0) \,,$$

is continuous, too, because of

$$|\delta_{,i}(f)| \leq \|f\|_{0,1} \cdot$$

A.1.2 Functions as Distributions

One easily realizes that functions which do not increase faster than polynomials at infinity, fit into the definition of tempered distributions. Therefore and in this sense, the genuine distributions are indeed generalized functions. Let $T(\boldsymbol{x})$ be a continuous function for which there exists a suitable $m \in \mathbb{N}$ such that

$$(1 + |\boldsymbol{x}|^2)^{-m} |T(\boldsymbol{x})| \leq C \quad \text{with} \quad C \in \mathbb{R} \,.$$

If this function is used to define

$$T(f) := \int \mathrm{d}^3 x \, T(\boldsymbol{x}) f(\boldsymbol{x}) \,, \qquad f \in \mathcal{S} \,,$$

then $T(f)$ is a linear and continuous functional on $\mathcal{S}$ and one has

$$|T(f)| \leq \|f\|_{p,0} \int \mathrm{d}^3 x \; (1 + |\boldsymbol{x}|^2)^{-p} |T(\boldsymbol{x})| \,,$$

provided one chooses $p > m + 2$. Thus, the mapping $f \longmapsto T(f)$ is a distribution and is uniquely defined by the integral given above. Indeed, if $T(f) = 0$ for all $f \in \mathcal{S}$, then $T(\boldsymbol{x}) \equiv 0$. In this sense tempered distributions are generalized functions. This relationship also justifies the symbolic notation

$$T(f) \equiv \text{„} \int \mathrm{d}^3 x \, T(\boldsymbol{x}) f(\boldsymbol{x}) \text{“}$$

even though, in the case of genuine distributions, the integral is not defined.

The set of all tempered distributions spans a linear space which is the dual of the function space $\mathcal{S}$. For this reason it is denoted by $\mathcal{S}'$. All definitions of and operations with distributions are chosen such that, for genuine functions, they take the familiar form. The following example illustrates this statement. Let $T(\boldsymbol{x})$ be a function on $\mathbb{R}^3$. Under a Galilei transformation

$$\boldsymbol{x} \longmapsto \boldsymbol{x}' = \mathbf{R}\boldsymbol{x} + \boldsymbol{a}$$

and for every element $f \in \mathcal{S}$, we have the familiar transformation formula

$$\int \mathrm{d}^3x\, T(\mathbf{R}\boldsymbol{x} + \boldsymbol{a}) f(\boldsymbol{x}) = \int \mathrm{d}^3x'\, T(\boldsymbol{x}') f\left(\mathbf{R}^{-1}(\boldsymbol{x}' - \boldsymbol{a})\right)/|\det \mathbf{R}| .$$

By analogy to this example one *defines* the transformed distribution $T_{(\mathbf{R},\boldsymbol{a})}$ of a distribution T by

$$T_{(\mathbf{R},\boldsymbol{a})}(f) := T\left(f_{(\mathbf{R},\boldsymbol{a})}(\boldsymbol{x})\right) \quad \text{with } f_{(\mathbf{R},\boldsymbol{a})}(\boldsymbol{x}) := \frac{1}{|\det \mathbf{R}|} f\left(\mathbf{R}^{-1}(\boldsymbol{x}-\boldsymbol{a})\right) . \tag{A.7}$$

A.1.3 Support of a Distribution

The example of Dirac's δ-distribution shows that it is more difficult to characterize the support of a distribution than for ordinary functions. The functional $\delta(f)$ that we formally write as an integral

$$\int \mathrm{d}^3x\, \delta(\boldsymbol{x}) f(\boldsymbol{x}) = f(0) ,$$

yields a contribution from $f(\boldsymbol{x})$ at the point $\boldsymbol{x} = 0$, while for any function whose support does not contain the point $\boldsymbol{x} = 0$, it yields the value zero. Therefore, one expects the support of δ in $\mathbb{R}^3$ to be the origin, supp $\delta = \{0\}$.

In the case of test functions the support is defined as usual, i.e.

$$\operatorname{supp} f := \left\{\boldsymbol{x} \in \mathbb{R}^3 \middle| f(\boldsymbol{x}) \neq 0\right\} .$$

The complement of supp f is the largest open set in $\mathbb{R}^3$ on which the function f vanishes. In the case of a distribution T one says that it vanishes on an open set $\mathcal{O}$ if for all $f \in \mathcal{S}$ whose support is contained in $\mathcal{O}$, $T(f)$ equals zero

$$T(f) = 0 \quad \text{for all} \quad f \in \mathcal{S} \quad \text{with} \quad \operatorname{supp} f \subset \mathcal{O} .$$

In the special case of T a continuous function this means that $T(\boldsymbol{x})$ vanishes everywhere on $\mathcal{O}$. With these considerations in mind one defines the support of a distribution as follows:

Definition A.3 Support of a Distribution

The support of a distribution T is the complement of the largest open set on which T is zero.

Thus, it is meaningful to state that two distributions coincide on an *open* set, but it is not meaningful to claim that they are equal in individual points.

A.1.4 Derivatives of Tempered Distributions

Given a differentiable function $T = T(\boldsymbol{x})$ which grows no faster than a polynomial, one has, using partial integration,

$$\int \mathrm{d}^3x \left(\frac{\partial T(\boldsymbol{x})}{\partial x^k}\right) f(\boldsymbol{x}) = -\int \mathrm{d}^3x\, T(\boldsymbol{x}) \left(\frac{\partial f(\boldsymbol{x})}{\partial x^k}\right) .$$

Given a tempered distribution T, one defines its partial derivative with respect to x^k by the rule

$$T_{,k} \equiv \frac{\partial T}{\partial x^k}(f) := -T \left(\frac{\partial f(\boldsymbol{x})}{\partial x^k}\right) . \tag{A.8}$$

This defines again a linear continuous functional because

$$f \longmapsto -\frac{\partial f}{\partial x^k}$$

is a continuous mapping $\mathcal{S} \longrightarrow \mathcal{S}$ and because one has

$$\left\| \frac{\partial f}{\partial x^k} \right\|_{p,q} \leq \| f \|_{p,q+1} \cdot$$

From this one concludes that a tempered distribution is infinitely differentiable, and that its derivatives are again tempered distributions. Another conclusion one draws is this: If two distributions are equal on an open set, then also their derivatives are equal on the same set.

A.1.5 Examples of Distributions

We consider three simple examples and applications:

Example A.1 Dirac's δ-distribution in one dimension

According to the rules contained in (A.7) one has

$$\int \mathrm{d}x\, \delta(x-a) f(x) = \int \mathrm{d}x'\, \delta(x') f(x'+a) = f(a) , \tag{A.9}$$

$$\int \mathrm{d}x\, \delta(\Lambda x) f(x) = \frac{1}{\Lambda} f(0) , \quad (\Lambda > 0). \tag{A.10}$$

With $g(x)$ a continuous, bounded function which has *simple* zeroes in the points x_i, there follows the rule

$$\delta\big(g(x)\big) = \sum_i \frac{1}{|g'(x_i)|} \delta(x - x_i) . \tag{A.11}$$

This formula, too, follows from the definition (A.7) and from (A.9) by linearization of $g(x)$ in the neighbourhood of every simple zero x_i, viz.

$$g(x) \approx g'(x_i)(x - x_i) + \mathcal{O}[(x - x_i)^2] .$$

The rule (A.10) shows that, in general, $\delta(x)$ carries a physical dimension: This dimension is the inverse of the dimension of the argument.

Consider the product $\delta(x-a)\delta(y-x)$. Evaluating the first factor on the test functions $f(x)$, the second factor on the test functions $f(y)$, yields the same result as if one had evaluated $\delta(y-a)$. Therefore, one may write formally

$$\delta(x-a)\delta(y-x) = \delta(y-a)\,.$$

In turn, the square of a δ-distribution, $\delta^2(x)$, is not defined.

Example A.2 The step function

The Heaviside or step function

$$\Theta(x) = \begin{cases} 1 & \text{for} \quad x > 0\,, \\ 0 & \text{for} \quad x \le 0\,, \end{cases}$$

when interpreted as a distribution, has the derivative $\Theta' = \delta$. This follows from the rule (A.8). Indeed, one has

$$\Theta'(f) = -\Theta(f') = -\int_0^\infty \mathrm{d}x\, f'(x) = f(0)\,.$$

Example A.3 Point charge and Green function

The function

$$G(z) = -\frac{1}{4\pi}\frac{1}{|z|}$$

with $z \in \mathbb{R}^3$, is interpreted as a distribution. If so, its second derivatives fulfill the differential equation

$$\Delta G(z) = \delta(z)\,. \tag{A.12}$$

This says, with this interpretation, that for all test functions $f \in \mathcal{S}$ one finds

$$\Delta G(f) = f(0)\,. \tag{A.13}$$

This is proved by means of Green's second theorem,

$$\int_V \mathrm{d}^3x\,(\Phi\Delta\Psi - \Psi\Delta\Phi) = \int_{\partial V} \mathrm{d}^2\sigma \left(\Phi\frac{\partial\Psi}{\partial n} - \frac{\partial\Phi}{\partial n}\Psi\right),$$

where ∂V denotes the smooth surface enclosing the space volume V. Choosing V to be the volume obtained by cutting out a sphere of radius ε around the origin in the space $\mathbb{R}^3$, one obtains for the choice $\Phi = 1/|\boldsymbol{x}|$ and $\Psi = f \in \mathcal{S}$,

$$\int_{|\boldsymbol{x}|\ge\varepsilon} \mathrm{d}^3x\,\frac{1}{|\boldsymbol{x}|}\Delta f(\boldsymbol{x}) = \int_{|\boldsymbol{x}|\ge\varepsilon} \mathrm{d}^3x\, f(\boldsymbol{x})\Delta\left(\frac{1}{|\boldsymbol{x}|}\right)$$
$$- \int_{|\boldsymbol{x}|=\varepsilon} \mathrm{d}^2\sigma\,\frac{\partial f}{\partial r}\frac{1}{r} + \int_{|\boldsymbol{x}|=\varepsilon} \mathrm{d}^2\sigma\, f\frac{\partial}{\partial r}\left(\frac{1}{r}\right).$$

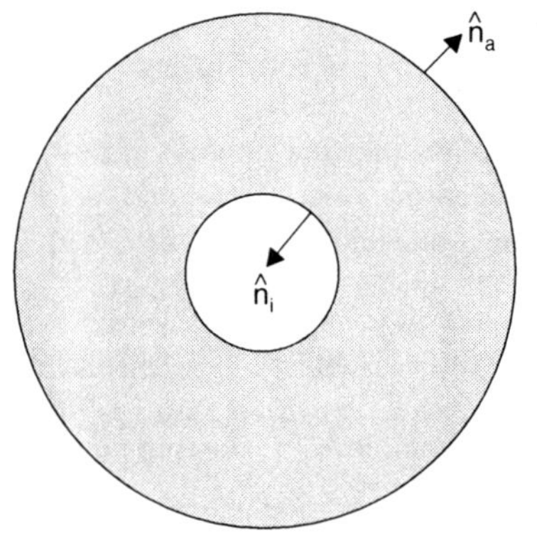

Fig. A.2 A volume which is encircled by two concentrical spheres. The radius of the outer sphere is taken to infinity, the radius of the inner sphere tends to zero

In these formulae $d^2\sigma = r^2 d\Omega = r^2 \sin\theta d\theta d\phi$. The volume which is the domain of integration, is sketched in Fig. A.2: The outer sphere (which, in fact, should be at infinity) has the normal $\hat{\boldsymbol{n}}_a$ pointing away from the origin, the inner sphere has the normal $\hat{\boldsymbol{n}}_i$ pointing towards the origin (hence the minus sign in the formula). The first term on the right-hand side equals zero because $\Delta(1/|\boldsymbol{x}|)$ vanishes everywhere outside the origin. The second term tends to zero as $\varepsilon \to 0$ because $d^2\sigma/r$ is proportional to $r = \varepsilon$ and the derivative of f remains bounded. The third term is different from zero. In the limit $\varepsilon \to 0$ it yields the contribution $-4\pi f(0)$. This proves the formula (A.13).

Remarks

1. Inserting in (A.12) the argument $\boldsymbol{z} = \boldsymbol{x} - \boldsymbol{y}$ yields a better known form of the differential equation for the Green function $G(\boldsymbol{x} - \boldsymbol{y})$:

$$\Delta G(\boldsymbol{x} - \boldsymbol{y}) = \delta(\boldsymbol{x} - \boldsymbol{y}) . \tag{A.14}$$

2. Similar considerations apply to the differential equation

$$(\Delta + k^2) G(k, \boldsymbol{x} - \boldsymbol{y}) = \delta(\boldsymbol{x} - \boldsymbol{y}) , \tag{A.15}$$

the Green function now being the one obtained in constructing the Born approximation in Chap. 2, Sect. 2.4,

$$G(k, \boldsymbol{x} - \boldsymbol{y}) = -\frac{1}{4\pi} \frac{1}{|\boldsymbol{x} - \boldsymbol{y}|} \left(a e^{ik|\boldsymbol{x}-\boldsymbol{y}|} + (1 - a) e^{-ik|\boldsymbol{x}-\boldsymbol{y}|} \right) . \tag{A.16}$$

3. Reversing the sign of the term k^2 in (A.15) and taking $\boldsymbol{z} = \boldsymbol{x} - \boldsymbol{y}$, one obtains the differential equation

$$(\Delta - \mu^2) G^{(\mu)}(\boldsymbol{z}) = \delta(\boldsymbol{z}) . \tag{A.17}$$

The Green function $G^{(\mu)}(z)$ is relevant for the description of the Yukawa potential, cf. Sect. 7.1. If no special boundary condition is imposed it is given by

$$G^{(\mu)}(z) = -\frac{1}{4\pi}\frac{e^{-\mu|z|}}{|z|} . \tag{A.18}$$

This formula is best proven by first converting (A.17) to an algebraic equation, by means of Fourier transformation. This algebraic equation is easily solved. In a second step, its solution is transformed back to coordinate space.

A.2 Gamma Function and Hypergeometric Functions

The properties of some special functions which are collected here, are relevant for the physical systems studied in Chap. 1. Also, they illustrate the richness of the theory of special functions and of the beautiful function theoretic methods that one uses in deriving the results quoted here. A nearly complete survey is given in the handbook [Abramowitz and Stegun (1965)] which also contains many hints to the literature on this topic. The tables [Gradshteyn and Ryzhik (1965)] are particularly useful for practical purposes. Among the many monographs on special functions we quote the Bateman Manuscript Project [Erdélyi et al. (1953)], as well as the classic [Whittaker, Watson (1958)].

A.2.1 The Gamma Function

The Gamma function $\Gamma(z)$ is the analytic continuation of the factorial $n!$ which, as a function, is defined only in the points $z = 0, 1, 2, \ldots$. A good starting point is provided by Euler's integral

$$\Gamma(z) = \int_0^\infty dt\, t^{z-1} e^{-t} . \tag{A.19}$$

One confirms that, indeed, for integer argument one has

$$\Gamma(n+1) = \int_0^\infty dt\, t^n e^{-t} = n! .$$

The integral (A.19) converges only for complex values of z whose real part is greater than zero. Therefore, this cannot yet be the analytic continuation to the entire z-plane. However, by splitting the integral into two integrals over the interval $[0, 1]$ and over the interval $[1, \infty)$, respectively, the first of these can be evaluated as follows

$$\int_0^1 dt\, t^{z-1} e^{-t} = \sum_{k=0}^\infty \frac{(-)^k}{k!} \int_0^1 dt\, t^{k+z-1} = \sum_{k=0}^\infty \frac{(-)^k}{k!} \frac{1}{z+k} .$$

The integral over $[1, \infty)$, in turn, converges for all z, and no restriction to the right half of the complex plane is needed. Thus, the full analytic continuation reads

$$\Gamma(z) = \sum_{k=0}^{\infty} \frac{(-)^k}{k!} \frac{1}{z+k} + \int_1^{\infty} \mathrm{d}t \, t^{z-1} \mathrm{e}^{-t} \,. \tag{A.20}$$

This shows the following:

The Gamma function is a meromorphic function which has poles of first order in the points $z = -k, k \in \mathbb{N}_0$. The residua in these points are given by $(-)^k/k!$.

There is an important functional equation which is obtained from the integral (A.19) by partial integration. It reads

$$\boxed{\Gamma(z+1) = z\Gamma(z)} \,, \tag{A.21}$$

and it holds for all $z \in \mathbb{C}$.

Another integral representation valid in the whole z-plane, is provided by *Hankel's contour integral* in the complex t-plane,

$$\frac{1}{\Gamma(z)} = \frac{\mathrm{i}}{2\pi} \int_C \mathrm{d}t \, (-t)^{-z} \mathrm{e}^{-t} \,, \tag{A.22}$$

the path of integration being as shown in Fig. A.3, with $|z| < \infty$.

Consider the integral over the same contour C cut at some large, positive value R of the real part of t,

$$\int_C \mathrm{d}t \, (-t)^{z-1} \mathrm{e}^{-t} \,.$$

In the limit $R \to \infty$ the integral representation (A.22) yields

$$\frac{2\pi}{\mathrm{i}} \frac{1}{\Gamma(1-z)} \,.$$

The same integral can be calculated directly and can be replaced by a line integral along the positive real axis of the t-plane. Above the real axis we have

$$\arg(-t) = -\pi \quad \text{and therefore} \quad (-t)^{z-1} = \mathrm{e}^{-\mathrm{i}\pi(z-1)} t^{z-1} \,,$$

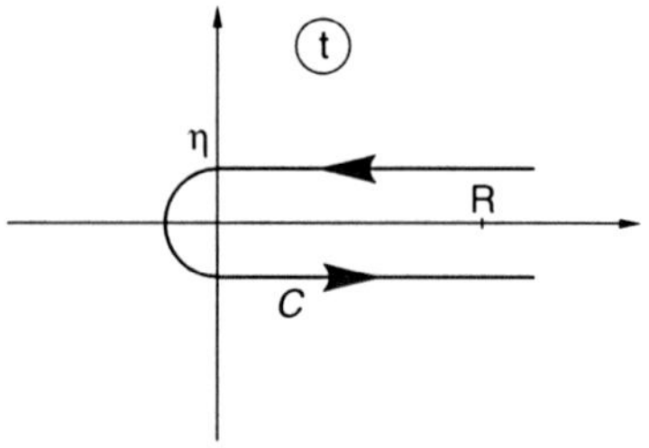

Fig. A.3 Path of integration in the complex t-plane in Hankel's contour integral of the Gamma function

while below the real axis we have

$$\arg(-t) = \pi \quad \text{and therefore} \quad (-t)^{z-1} = e^{+i\pi(z-1)} t^{z-1} .$$

The integral over the contour $\mathcal{C}$ decomposes into a line integral $\int_R^\eta = -\int_\eta^R$ parallel to the real axis, an integral over the half-circle with radius η which encloses the origin in the left half-plane, and another line integral extending from η to R in the lower half-plane. The integral over the half-circle is given by

$$i\eta^z \int_{-\pi}^{+\pi} d\phi \, e^{iz\phi + \eta(\cos\phi + i\sin\phi)} .$$

It tends to zero as $\eta \to 0$. The two line integrals can be combined and yield

$$-2i \sin(\pi z) \int_0^R dt \, t^{z-1} e^{-t} .$$

If R is sent to infinity, then, by (A.19), the right-hand side yields $\Gamma(z)$, up to a numerical factor. Comparison of the two results yields a relation that is valid on the whole complex z-plane, viz.

$$\boxed{\Gamma(z)\Gamma(1-z) = \frac{\pi}{\sin(\pi z)}} . \tag{A.23}$$

Some special values of the Gamma function with real argument are

$$\Gamma(n+1) = n! , \qquad \Gamma(2) = \Gamma(1) = 1 , \qquad \Gamma\left(\frac{1}{2}\right) = \sqrt{\pi} .$$

(The last of these follows from (A.19) by substituting $t = u^2$ and using the well-known Gauss integral.)

Of greatest practical importance are the asymptotic series for $\ln \Gamma(z)$ and for $\Gamma(z)$, the latter of which follows from the former. They are, in the first case,

$$\ln \Gamma(z) \sim \left(z - \frac{1}{2}\right) \ln z - z + \frac{1}{2} \ln(2\pi) + \sum_{m=1}^{\infty} \frac{B_{2m}}{2m(2m-1)z^{2m-1}} . \tag{A.24}$$

This series holds for $z \to \infty$ and $|\arg z| < \pi$, B_{2m} being the Bernoulli numbers. In the second case, the asymptotic series reads

$$\Gamma(z) \sim e^{-z} z^{z-1/2} \sqrt{2\pi} \times \left(1 + \frac{1}{12z} + \frac{1}{288z^2} - \frac{139}{51840z^3} - \frac{571}{2488320z^4} + \ldots\right) . \tag{A.25}$$

It holds for $z \to \infty$, and $|\arg z| < \pi$. Both formulae are used in numerical evaluations of $\Gamma(z)$ or $\ln \Gamma(z)$. For example, in the case of the Gamma function, one makes use of the functional equation (A.21) for mapping a given positive argument onto a positive, asymptotically large argument. A negative argument, in turn, is first mapped to a positive one by the mirror formula (A.23).

A.2.2 Hypergeometric Functions

In what follows one uses *Pochhammer's symbol* which is defined by

$$(x)_n := x(x+1)(x+2)\cdots(x+n-1) = \frac{\Gamma(x+n)}{\Gamma(x)}\,. \qquad \text{(A.26)}$$

In particular, one has $(x)_0 = 1$. The hypergeometric functions are denoted by the symbol ${}_mF_n(a_1, a_2, \dots a_m; c_1, c_2, \dots c_n; z)$ where m is the number of arguments which appear in the numerators, n is the number of arguments appearing in the denominators of the terms in the series representation of ${}_mF_n$. The real or complex variable is denoted by z.

The Hypergeometric Series. The hypergeometric series whose name emphasizes its similarity to the well-known geometric series $\sum x^n$, is defined as follows,

$$\begin{aligned} {}_2F_1(a,b;c;z) &= 1 + \frac{ab}{c}z + \frac{a(a+1)b(b+1)}{c(c+1)}\frac{z^2}{2!} + \dots \\ &\equiv \sum_{n=0}^{\infty} \frac{(a)_n(b)_n}{(c)_n}\frac{z^n}{n!} = \frac{\Gamma(c)}{\Gamma(a)\Gamma(b)} \sum_{n=0}^{\infty} \frac{\Gamma(a+n)\Gamma(b+n)}{\Gamma(c+n)}\frac{z^n}{n!}\,. \end{aligned} \qquad \text{(A.27)}$$

This series is absolutely and uniformly convergent in the interior of the circle $|z| = 1$. It is a solution of Gauss' differential equation

$$z(z-1)w''(z) + [(a+b+1)z - c]w'(z) + ab\,w(z) = 0\,. \qquad \text{(A.28)}$$

The differential equation (A.28) is of Fuchsian type, its first-order poles are situated in

$$z_1 = 0\,, \qquad z_2 = 1\,, \qquad z_3 = \infty\,.$$

Some special cases of hypergeometric series are the following

$$\begin{aligned} {}_2F_1(1,1;2;z) &= -\frac{1}{z}\ln(1-z)\,, \\ {}_2F_1(a,b;b;z) &= (1-z)^{-a}\,, \\ {}_2F_1(-\ell,\ell+1;1;z) &= P_\ell(1-2z) \qquad \text{(Legendre polynomials)}\,. \end{aligned}$$

The Confluent Hypergeometric Function. By the substitution $v = z_0 z$ the second singularity of the differential equation (A.28) is moved from 1 to the point z_0 on the positive real axis. This amounts to replace z by v/z_0 in the series (A.27). The series obtained in this way converges absolutely and uniformly for all arguments v for which $|v| < z_0$. As a special case, choose the parameter $b = z_0$, thus obtaining the series

$${}_2F_1(a, z_0; c; v/z_0) = \frac{\Gamma(c)}{\Gamma(a)} \sum_{n=0}^{\infty} \frac{\Gamma(a+n)}{\Gamma(c+n)}\frac{v^n}{n!}\left(\frac{\Gamma(z_0+n)}{\Gamma(z_0)z_0^n}\right)\,. \qquad \text{(A.29)}$$

It satisfies the differential equation (A.28) with $b = z_0$, i.e. it is a solution of the equation

$$v\left(1-\frac{v}{z_0}\right)w''(v)+\left[c-v\left(1+\frac{1+a}{z_0}\right)\right]w'(v)-aw(v)=0\,.$$

In this latter equation we let z_0 go to infinity, $z_0 \to \infty$. By this procedure one obtains a new differential equation, known as *Kummer's differential equation*,

$$vw''(v)+(c-v)w'(v)-aw(v)=0\,. \tag{A.30}$$

If it is possible to take the limit $z_0 \to \infty$ in every term of the series (A.29), one by one, then for all finite values of n, one has

$$\lim_{z_0\to\infty}\left(\frac{\Gamma(z_0+n)}{\Gamma(z_0)z_0^n}\right)=1$$

so that one obtains a series expansion for ${}_1F_1(a;c;z)$:

$$\lim_{z_0\to\infty}{}_2F_1(a,z_0;c;z/z_0)={}_1F_1(a;c;z)\,,$$

whereby

$${}_1F_1(a;c;z)=\frac{\Gamma(c)}{\Gamma(a)}\sum_{n=0}^{\infty}\frac{\Gamma(a+n)}{\Gamma(c+n)}\frac{z^n}{n!}=1+\frac{a}{c}z+\frac{a(a+1)}{c(c+1)}\frac{z^2}{2!}+\ldots \tag{A.31}$$

This function is called the *confluent hypergeometric function*. It takes its name from the fact that it is obtained by merging the poles at $z_2 = 1$ and at $z_3 = \infty$ of the hypergeometric function ${}_2F_1$ (confluent = "flowing together"). While the point $z_1 = 0$ continues to be a pole of ${}_1F_1$, the other two poles, by the merging process, turn into an essential singularity at infinity (up to exceptions such as the one where ${}_2F_1$ is a polynomial). The solution ${}_1F_1$ is regular at $z = 0$ but becomes singular at $z = \infty$. Standard criteria for convergence show that the series (A.31) converges for every finite argument. Thus, in a function theoretic sense, the series defines an *entire* function. In fact, these properties are plausible if one notes that the following well-known functions are contained in the definition,

$${}_1F_1(a;a;z)=e^{-z}\,,\qquad {}_1F_1(a=-n;c;z)=P_n(c,z)$$

where P_n is a polynomial of degree n.

We note without proof that Hermite polynomials, Laguerre polynomials, as well as Bessel functions, are special cases of (A.31), cf. [Abramowitz and Stegun (1965)], Chap. 13.

Some relations of special importance for practical purposes are

$${}_1F_1(a;c;z)=e^{z}\,{}_1F_1(c-a;c;z)\qquad\text{(Kummer's relation)}\,, \tag{A.32}$$

$$\frac{d^n}{dz^n}\,{}_1F_1(a;c;z)=\frac{(a)_n}{(c)_n}\,{}_1F_1(a+n;c+n;z)\,. \tag{A.33}$$

Integral Representations and Asymptotics. One proceeds in several steps. In a first step one shows that the transformation $w(z) = z^{1-c}v(z)$ transforms Kummer's differential equation (A.30) into

$$z^{1-c}[zv''(z)+(2-c-z)v'(z)-(1+a-c)v(z)]=0\,.$$

This equation is of the same type. This is seen by defining $a' = 1 + a - c$ and $c' = 2 - c$. This means that if ${}_1F_1(a;c;z)$ is a solution of (A.30) then so is

$$z^{1-c}\,{}_1F_1(1 + a - c;2 - c;z)\,, \tag{A.34}$$

unless c takes one of the values $0, -1, -2, \ldots$. Except for $c = 1$ the two solutions are linearly independent. If $c = 1$ they coincide.

Let $w(z)$ be a solution of (A.30). The problem is to find an analytic function $f(t)$ and a suitable contour $\mathcal{C}_0$ in the complex t-plane such that one has

$$w(z) = \frac{1}{2\pi \mathrm{i}} \int\limits_{\mathcal{C}_0} \mathrm{d}t \mathrm{e}^{tz} f(t)\,. \tag{A.35}$$

The derivatives of $w(z)$ being obtained by differentiation of the integrand, the differential equation yields the condition

$$\frac{1}{2\pi \mathrm{i}} \int\limits_{\mathcal{C}_0} \mathrm{d}t \mathrm{e}^{tz} [zt^2 f(t) + (c - z)tf(t) - af(t)] = 0\,.$$

By the relation $z\mathrm{e}^{tz} = \mathrm{d}/\mathrm{d}t\,(\mathrm{e}^{tz})$ and using partial integration this condition becomes

$$\int\limits_{\mathcal{C}_0} \mathrm{d}t \frac{\mathrm{d}}{\mathrm{d}t}[\mathrm{e}^{tz} t(t-1) f(t)] + \int\limits_{\mathcal{C}_0} \mathrm{d}t \mathrm{e}^{tz} \left[-\frac{\mathrm{d}}{\mathrm{d}t}\big[t(t-1)f(t)\big] + (ct - a)f(t) \right] = 0\,. \tag{A.36}$$

Sufficient conditions for (A.36) to be true are seen to be

$$\int\limits_{\mathcal{C}_0} \mathrm{d}t \frac{\mathrm{d}}{\mathrm{d}t}[\mathrm{e}^{tz} t(t-1) f(t)] = 0\,, \tag{A.37}$$

$$-\frac{\mathrm{d}}{\mathrm{d}t}[t(t-1)f(t)] + (ct - a)f(t) = 0\,. \tag{A.38}$$

The second of these is converted to a differential equation of first order

$$\frac{f'}{f} = \frac{a-1}{t} + \frac{c-a-1}{t-1}$$

for which a particular integral reads

$$f(t) = t^{a-1}(t-1)^{c-a-1}\,. \tag{A.39}$$

Thus, we obtain an integral representation (A.35) which reads

$$w(z) = \frac{1}{2\pi \mathrm{i}} \int_{\mathcal{C}_0} \mathrm{d}t \mathrm{e}^{tz} t^{a-1} (t-1)^{c-a-1} , \tag{A.40}$$

with (A.37) as a subsidiary condition. That is, we must have

$$\int_{\mathcal{C}_0} \mathrm{d}t \frac{\mathrm{d}}{\mathrm{d}t} [\mathrm{e}^{tz} t^a (t-1)^{c-a}] = 0;. \tag{A.41}$$

An integral representation for the second solution (A.34) is obtained in much the same way, with $a' - 1 = a - c$ and $c' - a' - 1 = -a$, and possibly a modified contour $\mathcal{C}_1$,

$$w(z) = \frac{1}{2\pi \mathrm{i}} z^{1-c} \int_{\mathcal{C}_1} \mathrm{d}t \mathrm{e}^{tz} t^{a-c} (t-1)^{-a} , \tag{A.42}$$

provided the corresponding condition (A.41) is fulfilled, i.e. that

$$\int_{\mathcal{C}_0} \mathrm{d}t \frac{\mathrm{d}}{\mathrm{d}t} [\mathrm{e}^{tz} t^{a-c+1} (t-z)^{-a+1}] = 0 \tag{A.43}$$

holds. The function (A.42) can be transformed further by taking the factor $z^{1-c} = z^{a-c} z^{-a} z$ inside the integral and by substituting the integration variable by $\tau := tz$:

$$w(z) = \frac{1}{2\pi \mathrm{i}} \int_{\mathcal{C}_1} \mathrm{d}\tau \mathrm{e}^{\tau} \tau^{a-c} (\tau - z)^{-a} .$$

The condition (A.43) then becomes

$$\int_{\mathcal{C}_1} \mathrm{d}\tau \frac{\mathrm{d}}{\mathrm{d}\tau} [\mathrm{e}^{\tau} \tau^{a-c+1} (\tau - z)^{-a+1}] = 0 . \tag{A.44}$$

We choose the contour $\mathcal{C}_1$ in such a way that it encloses the points z and 0, and the left real half-axis, as sketched in Fig. A.4.

The next step consists in showing that

$${}_1F_1(a;c;z) = \frac{\Gamma(c)}{2\pi \mathrm{i}} \int_{\mathcal{C}} \mathrm{d}\tau \mathrm{e}^{\tau} \tau^{a-c} (\tau - z)^{-a} \tag{A.45}$$

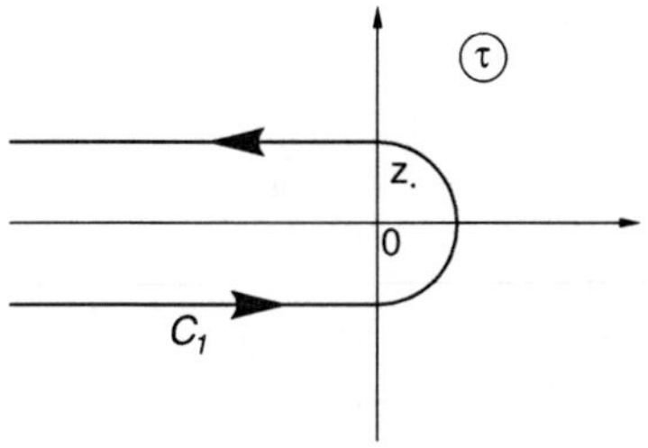

Fig. A.4 Path of integration in the complex τ-plane, including points 0 and z, in the integral representation of the confluent hypergeometric function

if the path of integration is chosen such that for all points z on the contour $\mathcal{C}$ the inequality

$$\left|\frac{z}{\tau}\right| \leq c < 1$$

is fulfilled. In this case on expands $(\tau - z)^{-a}$ in terms of powers of z/τ so that in (A.45) one obtains

$${}_1F_1(a;c;z) = \Gamma(c) \sum_{n=0}^{\infty} \binom{-a}{n} (-z)^n \frac{1}{2\pi \mathrm{i}} \int_{\mathcal{C}} \mathrm{d}\tau \mathrm{e}^{\tau} \tau^{-c-n} .$$

The right-hand side of this equation contains an integral that we know from the formula (A.22). Inserting this, one confirms the assertion as follows

$${}_1F_1(a;c;z) = \Gamma(c) \sum_{n=0}^{\infty} \binom{-a}{n} (-z)^n \frac{1}{\Gamma(c+n)} = \sum_{n=0}^{\infty} \frac{(a)_n}{(c)_n} \frac{z^n}{n!} .$$

In a further step one uses the integral representation (A.45) for deriving an asymptotic representation for ${}_1F_1$. For this purpose let

$$\Phi_k(a, c, z) = \frac{\Gamma(c)}{2\pi \mathrm{i}} \int_{\mathcal{C}_k} \mathrm{d}\tau \mathrm{e}^{\tau} \tau^{a-c} (\tau - z)^{-a} , \qquad k = 1, 2 ,$$

where the paths of integration $\mathcal{C}_1$ and $\mathcal{C}_2$ are the ones shown in Fig. A.5. Substituting $\tau' = \tau - z$ one shows that

$$\Phi_2(a, c, z) = \mathrm{e}^z \Phi_1(c - a, c, -z)$$

and, therefore,

$${}_1F_1(a;c;z) = \Phi_1(a, c, z) + \mathrm{e}^z \Phi_1(c - a, c, -z) .$$

One then expands Φ_1 in terms of τ/z, and obtains an asymptotic expansion for Φ_1 and, thus, also for ${}_1F_1$. The latter reads

$$\begin{aligned} {}_1F_1(a;c;z) \sim & \frac{\Gamma(c)}{\Gamma(c-a)} e^{\pm \mathrm{i}\pi a} z^{-a} \sum_{n=0}^{N} (a)_n (1 + a - c)_n \frac{(-z)^{-n}}{n!} \\ & + \frac{\Gamma(c)}{\Gamma(a)} \mathrm{e}^z z^{a-c} \sum_{n=0}^{M} (c - a)_n (1 - a)_n \frac{z^{-n}}{n!} . \end{aligned} \tag{A.46}$$

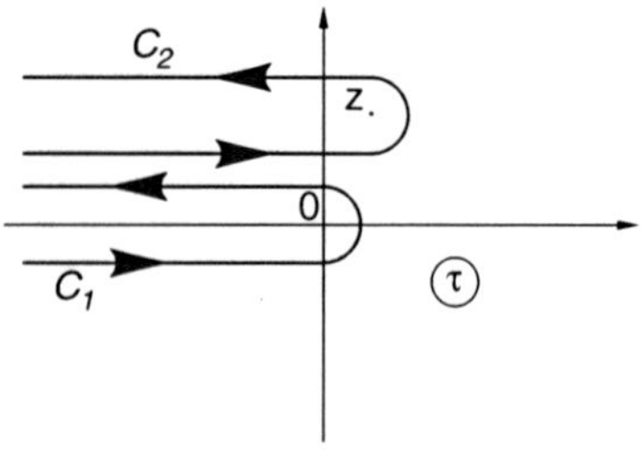

Fig. A.5 Split of the path of Fig. A.4 into two paths of integration which allows a development in τ/z thereby providing an asymptotic representation of the confluent hypergeometric function

This is an asymptotic series which holds for $|z| \to \infty$, with fixed values of a and c, and which is of the order $\mathcal{O}(|z|^{-N-1})$ and $\mathcal{O}(|z|^{-M-1})$, respectively. The sign in the first term must be chosen in accord with the following rule

$$\mathrm{e}^{+\mathrm{i}\pi a} \quad \text{holds for} \quad -\frac{\pi}{2} < \arg z < \frac{3\pi}{2}$$

$$\mathrm{e}^{-\mathrm{i}\pi a} \quad \text{holds for} \quad -\frac{3\pi}{2} < \arg z < -\frac{\pi}{2}\,.$$

In the discussion of bound states, in the context of the Schrödinger equation, we repeatedly made use of the fact that the second term in (A.46) which grows exponentially, is equal to zero if and only if a is a negative integer or zero. It was this condition that led to the quantization of the eigenvalues.

A.3 Self-Energy of the Electron

To second order $\mathcal{O}(e^2)$ and introducing a small mass m_γ for the photon, the Feynman rules yield the following modification of the electron propagator

$$S_F(p) \longmapsto S_F(p) + S_F(p)\Sigma(p)S_F(p)\,,$$

$$\Sigma(p) = -\mathrm{i}\frac{e^2}{(2\pi)^4}\int \mathrm{d}^4k\; \gamma^\mu S_F(p-k)\gamma_\mu \frac{1}{k^2 - m_\gamma^2 + \mathrm{i}\varepsilon}\,.$$

We give here some intermediate steps that lead to the expression (10.56). We use twice the formula

$$\frac{\mathrm{i}(\not{q} + m\,\mathbb{1})}{q^2 - m^2 + \mathrm{i}\varepsilon} = (\not{q} + m\,\mathbb{1})\int_0^\infty \mathrm{d}z\; e^{\mathrm{i}z}(q^2 - m^2 + \mathrm{i}\varepsilon)$$

with the choices $q = p - k$, and $q = k$, respectively. Dropping the common factor in front, we obtain

$$\Sigma(p) = \mathrm{i}\frac{e^2}{(2\pi)^4}\int_0^\infty \mathrm{d}z_1 \int_0^\infty \mathrm{d}z_2 \int \mathrm{d}^4k$$

$$\gamma^\mu\big(\not{p} - \not{k} + m\,\mathbb{1}\big)\gamma_\mu e^{\mathrm{i}z_1[(p-k)^2 - m^2 + \mathrm{i}\varepsilon]} e^{\mathrm{i}z_2[k^2 - m_\gamma^2 + \mathrm{i}\varepsilon]}\,.$$

By the formulae of Sect. 9.3.3, $\gamma^\mu\gamma_\mu = 4$ and $\gamma^\mu\not{q}\gamma_\mu = -2\not{q}$. Replacing the momentum k by the new variable

$$x := k - \frac{pz_1}{z_1 + z_2} = k - p + \frac{pz_2}{z_1 + z_2}$$

one has

$$\gamma^\mu\big(\not{p} - \not{k} + m\,\mathbb{1}\big)\gamma_\mu = 4m\,\mathbb{1} + 2\not{x} - 2\frac{z_2}{z_1 + z_2}\not{p}\,,$$

$$z_1(p-k)^2 + z_2 k^2 = (z_1+z_2)x^2 + \frac{z_1 z_2}{z_1+z_2} p^2 .$$

The x-integration needs three formulae which are not difficult to prove. They read

$$\int \frac{d^4x}{(2\pi)^4} \begin{Bmatrix} 1 \\ x^\mu \\ x^\mu x^\nu \end{Bmatrix} e^{ix^2(z_1+z_2)} = \frac{-i}{16\pi^2(z_1+z_2)^2} \begin{Bmatrix} 1 \\ 0 \\ ig^{\mu\nu}/2(z_1+z_2) \end{Bmatrix} \qquad (A.47)$$

The quantity $\Sigma(p)$ then becomes

$$\Sigma(p) = \frac{\alpha}{2\pi} \int_0^\infty dz_1 \int_0^\infty dz_2 \left(2m\,\mathbb{1} - \frac{z_2}{z_1+z_2}\not{p}\right) \frac{1}{(z_1+z_2)^2}$$
$$\times \exp\left[i\left(p^2 \frac{z_1 z_2}{z_1+z_2} - m^2 z_1 - m_\gamma^2 z_2 + i\varepsilon\right)\right] .$$

One inserts the identity

$$1 = \int_0^\infty \frac{d\lambda}{\lambda} \delta\left(1 - \frac{z_1+z_2}{\lambda}\right)$$

in the integrand and replaces the variables z_1 and z_2 by, respectively,

$$z = \frac{z_1}{\lambda}, \qquad \zeta = \frac{z_2}{\lambda} .$$

The δ-distribution enforces the condition $z + \zeta = 1$. As the ranges of the original variables are $z_1 \in [0, \infty]$, and $z_2 \in [0, \infty]$, the new variables z and ζ vary between 0 and 1, $z, \zeta \in [0, 1]$. Thus, after having done the integral over ζ there remains

$$\Sigma(p) = \frac{\alpha}{2\pi} \int_0^1 dz \, \left(2m\,\mathbb{1} - (1-z)\not{p}\right)$$
$$\int_0^\infty \frac{d\lambda}{\lambda} \exp\left[i\lambda\left(p^2 z(1-z) - m^2 z - m_\gamma^2(1-z) + i\varepsilon\right)\right] .$$

This is the expression (10.56) with (10.57), which is analyzed further in Sect. 10.2.1.

A.4 Renormalization of the Fermion Mass

This is a summary of some intermediate steps that lead from the expression (10.58) for the regularized self-energy to the form (10.60). There are two integrals to be calculated,

$$J_1 := \int_0^1 dz \, \ln\left(\frac{m^2 z^2 + m_\gamma^2(1-z)}{m^2 z + m_\gamma^2(1-z) - p^2 z(1-z)}\right)$$

$$= \int_0^1 dz \, \ln\left(1 + (m^2 - p^2)\frac{z(z-1)}{(m^2 - p^2)z + p^2 z^2 + m_\gamma^2(1-z)}\right),$$

$$J_2 := \int_0^1 dz \, (z-1) \ln\left(\frac{m^2 z^2 + m_\gamma^2(1-z)}{m^2 z + m_\gamma^2(1-z) - p^2 z(1-z)}\right).$$

In the first of them one uses partial integration according to

$$\int_0^1 dz \, \ln u(z) = z \ln u(z)|_0^1 - \int_0^1 dz \, z \frac{u'(z)}{u(z)}$$

to obtain

$$J_1 = (p^2 - m^2) \int_0^1 dz \, \frac{z\left[(m^2 - m_\gamma^2)z^2 + 2m_\gamma^2 z - m_\gamma^2\right]}{\left[(m^2 - p^2)z + p^2 z^2 + m_\gamma^2(1-z)\right]\left[m^2 z^2 + m_\gamma^2(1-z)\right]}$$

$$\approx (p^2 - m^2) \int_0^1 dz \, \frac{z}{(m^2 - p^2)z + p^2 z^2 + m_\gamma^2(1-z)}.$$

Here we used the fact that m_γ is very small, by assumption. In this case, the second factor in the numerator of the integrand cancels approximately against the second factor in its denominator. Substitution of the integration variable by $y = 1 - z$ then gives

$$J_1 \approx (p^2 - m^2) \int_0^1 dy \, \frac{1-y}{N(y)} \quad \text{with}$$

$$N(y) = m^2(1-y) + m_\gamma^2 y - p^2 y(1-y).$$

The integral J_2 is obtained by partial integration as well, by noting that the factor $(z-1) = u'(z)$ is the derivative of $u(z) = z(z/2 - 1)$.

$$J_2 = (p^2 - m^2) \int_0^1 dz \, z\left(\frac{z}{2} - 1\right)$$

$$\frac{-m_\gamma^2(1-z)^2 + m^2 z^2}{\left[m^2 z + m_\gamma^2(1-z) - p^2 z(1-z)\right]\left[m^2 z^2 + m_\gamma^2(1-z)\right]}$$

$$\approx (p^2 - m^2) \int_0^1 dz \, z\left(\frac{z}{2} - 1\right)\frac{1}{m^2 z + m_\gamma^2(1-z) - p^2 z(1-z)}.$$

Here again, the smallness of m_γ was used. Introducing the variable $y = 1 - z$ one finds

$$J_2 \approx -\frac{1}{2}(p^2 - m^2) \int_0^1 dy \, \frac{1-y^2}{N(y)}.$$

The integrand is split into two terms,

$$\frac{1-y^2}{N(y)} = \frac{1-y}{N(y)} + \frac{y(1-y)}{N(y)}$$

the second of which is written

$$\frac{y(1-y)}{N(y)} = \frac{1}{p^2}\left[-1 + \frac{m^2(1-y)+m_\gamma^2 y}{N(y)}\right]$$
$$\approx -\frac{1}{p^2} + \frac{m^2}{p^2}\frac{1-y}{N(y)}\,.$$

This shows that J_2 is expressed by a constant and by the same integral as J_1,

$$J_2 \approx -\frac{1}{2}(p^2-m^2)\left\{\left(1+\frac{m^2}{p^2}\right)\int_0^1 \mathrm{d}y\,\frac{1-y}{N(y)} - \frac{1}{p^2}\right\}.$$

Up to a factor $(-p^2)$ the integral is identical with the function $\Lambda(p^2)$ that was defined in Sect. 10.2.2,

$$\Lambda(p^2) = -p^2\int_0^1 \mathrm{d}y\,\frac{1-y}{N(y)}\,.$$

Thus, the expression

$$C(p) = \frac{\alpha}{2\pi}\int_0^1 \mathrm{d}z\,\left[2m\,\mathbb{1} - \not{p}(1-z)\right]\ln\left(\frac{m^2z^2+m_\gamma^2(1-z)}{m^2z+m_\gamma^2(1-z)-p^2z(1-z)}\right)$$

can be expressed in terms of this integral, or, equivalently, the function $\Lambda(p^2)$. There is no problem in showing that

$$C(p) \approx \frac{\alpha}{4\pi}(p^2-m^2)\left\{\frac{m}{p^2}\,\mathbb{1} - \left(\frac{1}{p^2}\Lambda(p^2)\right)m\left(3-\frac{m^2}{p^2}\right)\mathbb{1}\right.$$
$$\left. + (\not{p} - m\,\mathbb{1})\left(\frac{1}{p^2} + \left(1+\frac{m^2}{p^2}\right)\right)\left(\frac{1}{p^2}\Lambda(p^2)\right)\right\}$$

Decomposing according to

$$C(p) \equiv \Sigma_b(p)\,\mathbb{1} + (\not{p} - m\,\mathbb{1})\Sigma_a(p)\,,$$

the functions defined in this way, are given by

$$\Sigma_a(p^2) \approx \frac{\alpha}{4\pi}\left(1-\frac{m^2}{p^2}\right)\left\{1+\left(1+\frac{m^2}{p^2}\right)\Lambda(p^2)\right\},$$

$$\Sigma_b(p^2) \approx \frac{\alpha}{4\pi}m\left(1-\frac{m^2}{p^2}\right)\left\{1-\left(3-\frac{m^2}{p^2}\right)\Lambda(p^2)\right\},$$

$$\Lambda(p^2) = -p^2\int_0^1 \mathrm{d}z\,\frac{1-z}{m^2(1-z)+m_\gamma^2 z - p^2 z(1-z)}\,.$$

In the next step, the regularized quantity $\Sigma^{\text{reg}}(p)$ should be written in the form of (10.60)

$$\Sigma^{\text{reg}}(p) = A\,\mathbb{1} + (\not p - m\,\mathbb{1})\left[B + \Sigma^{\text{finite}}(p^2)\right].$$

The term Σ_a has already the right factor $\not p - m\,\mathbb{1}$ and need not be transformed any further. The term $\Sigma_b(p^2)$ could formally be obtained as follows. Making use of the identity $(\not p - m\,\mathbb{1})(\not p + m\,\mathbb{1}) = (p^2 - m^2)\,\mathbb{1}$ the term $C(p)$ is written

$$C(p) = (\not p - m\,\mathbb{1})\Big(\Sigma_a(p^2)\,\mathbb{1} + (\not p + m\,\mathbb{1})\frac{1}{p^2 - m^2}\Sigma_b(p^2)\Big). \tag{A.48}$$

However, this is not quite correct yet because the function $\Sigma^{\text{finite}}(p^2)$ must be constructed such that it vanishes on the mass shell $p^2 = m^2$. Although it is true that the two functions $\Sigma_a(p^2)$ and $\Sigma_b(p^2)$ are equal to zero at $p^2 = m^2$, the ratio $\Sigma_b(p^2)/(p^2 - m^2)$ yields $0/0$ and, hence, remains undetermined. This is repaired as follows. The function Σ_b is expanded in the neighbourhood of the mass shell $p^2 = m^2$,

$$\Sigma_b(p^2)\,\mathbb{1} \approx \mathbb{1}\,\Sigma_b(p^2)\Big|_{p^2=m^2} + 2m(\not p - m\,\mathbb{1})\frac{\partial \Sigma_b(p^2)}{\partial p^2}\Big|_{p^2=m^2},$$

where use was made of the relation

$$(p^2 - m^2)\,\mathbb{1} = (\not p + m\,\mathbb{1})(\not p - m\,\mathbb{1}) \approx 2m(\not p - m\,\mathbb{1}).$$

This shows that one should subtract an additional term $2m\partial\Sigma_b/\partial p^2$ in the big brackets of (A.48), taken at $p^2 = m^2$. If one does so the condition $\Sigma^{\text{finite}}(p^2 = m^2) = 0$ is fulfilled and one obtains the result quoted in Sect. 10.2.2,

$$\Sigma^{\text{finite}}\,\mathbb{1} = \Sigma_a(p^2)\,\mathbb{1} + (\not p + m\,\mathbb{1})\frac{1}{p^2 - m^2}\Sigma_b(p^2) - 2m\,\mathbb{1}\,\frac{\partial \Sigma_b}{\partial p^2}\Big|_{p^2=m^2} \tag{A.49}$$

Note that here again a condition imposed by physics was used: The propagator should contain the physical mass.

A.5 Proof of the Identity (10.86)

Let Σ(p) be the (regularized) self-energy (10.60) (suppressing the notation "reg", for the sake of clarity). Consider

$$\overline{u(q)}\frac{\partial \Sigma(p)}{\partial p_\mu}u(p)$$

at the argument $q = p$ and note that only the term B of (10.60) contributes, i.e. that

$$\overline{u(q)}\frac{\partial \Sigma(p)}{\partial p_\mu}u(p)\Big|_{q=p} = B\overline{u(p)}\gamma^\mu u(p).$$

On the other hand, if one writes down the original integral representation for Σ (without regard to regularization) one has

$$\overline{u(p)}\frac{\partial \Sigma(p)}{\partial p_\mu}u(p) = -\frac{\mathrm{i}e_0^2}{(2\pi)^4} \int \mathrm{d}^4k\, \overline{u(p)}\gamma^\lambda \left(\frac{\partial}{\partial p_\mu}S_F(p-k)\right)\gamma_\lambda \frac{1}{k^2+\mathrm{i}\varepsilon}u(p)\,.$$

Use now the identity which is easy to prove,

$$\frac{\partial}{\partial p_\mu}S_F(p) = -S_F(p)\gamma^\mu S_F(p)$$

to obtain

$$\overline{u(p)}\frac{\partial \Sigma(p)}{\partial p_\mu}u(p) = \frac{\mathrm{i}e_0^2}{(2\pi)^4}\int \frac{\mathrm{d}^4k}{k^2+\mathrm{i}\varepsilon}\overline{u(p)}\gamma^\lambda S_F(p-k)\gamma^\mu S_F(p-k)\gamma_\lambda u(p)$$

This expression is compared with the term D for $q = p$, that is, for momentum transfer equal to zero,

$$\begin{aligned} D(q=p) &= -2\pi \mathrm{i}e_0 Z_2^{-1}\frac{\mathrm{i}e_0^2}{(2\pi)^4} \int \frac{\mathrm{d}^4k}{k^2+\mathrm{i}\varepsilon}\overline{u(p)}\gamma^\lambda S_F(p-k)\gamma_\mu S_F(p-k)\gamma_\lambda u(p)\widetilde{A}_\mu \\ &= 2\pi \mathrm{i}e_0 Z_2^{-1}\overline{u(p)}\gamma^\mu u(p)F_1(0)\widetilde{A}_\mu\,. \end{aligned}$$

Comparing the two formulae yields

$$B\overline{u(p)}\gamma^\mu u(p) = -F_1(0)\overline{u(p)}\gamma^\mu u(p)\,,$$

and hence

$$F_1(0) + B + 1 = 1 \equiv F_1(0) + Z_2\,. \tag{A.50}$$

The proof as given here is purely formal, for the sake of brevity. It is fairly obvious that it works also for the regularized and infrared safe modifications of the integrals. As a more important remark, this identity holds to all finite orders (see Appendix A.7). Its physics content is that the internal structure of a particle is irrelevant when it interacts with photons in the limit of very small momentum transfer.

A.6 Analysis of Vacuum Polarization

We use the integral representation of the propagator

$$\frac{\mathrm{i}}{p^2-m^2+\mathrm{i}\varepsilon} = \int_0^\infty \mathrm{d}z\, e^{\mathrm{i}z(p^2-m^2+\mathrm{i}\varepsilon)}\,.$$

Writing m_i in lieu of $m_i \equiv m_f$ and $m_i \equiv M_f$, respectively, in (10.95), one has

$$\Pi^{\mu\nu}(Q, m_i^2) = -\frac{\mathrm{i}e_0^2}{(2\pi)^4} \int \mathrm{d}^4 k \int_0^\infty \mathrm{d}z_1 \int_0^\infty \mathrm{d}z_2$$
$$\mathrm{tr}\left\{\gamma^\mu(\not{k} - m_i)\gamma^\nu(\not{k} - \not{Q} - m_i)\right\}$$
$$\exp\Big[\mathrm{i}z_1(k^2 - m_i^2 + \mathrm{i}\varepsilon) + \mathrm{i}z_2\big((k-Q)^2 - m_i^2 + \mathrm{i}\varepsilon\big)\Big].$$

The trace is evaluated by means of the rules derived and collected in Sect. 9.3.3,

$$\mathrm{tr}\left\{\gamma^\mu(\not{k} - m_i)\gamma^\nu(\not{k} - \not{Q} - m_i)\right\}$$
$$= 4\big[k^\mu(k-Q)^\nu + k^\nu(k-Q)^\mu - g^{\mu\nu}(k^2 - k\cdot Q - m_i^2)\big].$$

The argument of the exponential function is transformed by substitution of the variable $k \mapsto x$. With

$$x := k - \frac{z_2}{z_1+z_2}Q = k - Q + \frac{z_1}{z_1+z_2}Q \quad \text{one obtains}$$
$$k^2 z_1 + (k-Q)^2 z_2 = (z_1+z_2)x^2 + \frac{z_1 z_2}{z_1+z_2}Q^2 .$$

One uses the integrals (A.47) of Appendix A.3 and obtains

$$\left(\Pi^{\mathrm{reg}}\right)^{\mu\nu}(Q) = -\frac{\alpha_0}{\pi}\sum_i c_i \int_0^\infty \mathrm{d}z_1 \int_0^\infty \mathrm{d}z_2$$
$$\times \frac{1}{(z_1+z_2)^2}\exp\left\{\frac{z_1 z_2}{z_1+z_2}Q^2 - (m_i^2 - \mathrm{i}\varepsilon)(z_1+z_2)\right\}$$
$$\times \left\{2(g^{\mu\nu}Q^2 - Q^\mu Q^\nu)\frac{z_1 z_2}{(z_1+z_2)^2}\right.$$
$$\left. + g^{\mu\nu}\left(-\frac{\mathrm{i}}{z_1+z_2} - \frac{z_1 z_2}{(z_1+z_2)^2}Q^2 + m_i^2\right)\right\} . \tag{A.51}$$

In deriving this formula $\alpha_0 = e_0^2/(4\pi)$ was inserted, and the expression resulting from the trace was transformed as follows

$$\big[k^\mu(k-Q)^\nu + k^\nu(k-Q)^\mu - g^{\mu\nu}(k^2 - k\cdot Q - m_i^2)\big]$$
$$= \left(x^\mu + \frac{z_2}{z_1+z_2}Q^\mu\right)\left(x^\nu - \frac{z_1}{z_1+z_2}Q^\nu\right) + \big((\mu \leftrightarrow \nu)\big)$$
$$- g^{\mu\nu}\left[x^2 + \frac{z_2 - z_1}{z_1+z_2}(x\cdot Q) - \frac{z_1 z_2}{(z_1+z_2)^2}Q^2 - m_i^2\right].$$

The intermediate result (A.51) shows that the first term in the curly brackets has the form imposed by gauge invariance but the term proportional to $g^{\mu\nu}$ does not. However, one shows that this second term vanishes (provided, of course, the masses and coupling constants (m_i, c_i) are chosen such that the integrals are convergent). Let

$$I := \int_0^\infty dz_1 \int_0^\infty dz_2 \, \frac{1}{(z_1+z_2)^2} \times \sum_i c_i \Big[m_i^2 - \frac{i}{z_1+z_2} - \frac{z_1 z_2}{(z_1+z_2)^2} Q^2\Big] \exp\Big\{i\Big[\frac{z_1 z_2}{z_1+z_2} Q^2 - (m_i^2 - i\varepsilon)(z_1+z_2)\Big]\Big\}$$

A trick consists in rescaling the two integration variables by a real factor λ and to consider the dependence of the integrals on λ. With $z_1 = \lambda x_1$ and $z_2 = \lambda x_2$ one has

$$I = \int_0^\infty dx_1 \int_0^\infty dx_2 \, \frac{1}{(x_1+x_2)^2} \times \sum_i c_i \Big[m_i^2 - \frac{i}{\lambda(x_1+x_2)} - \frac{x_1 x_2}{(x_1+x_2)^2} Q^2\Big] \times \exp\Big\{i\lambda\Big[\frac{x_1 x_2}{x_1+x_2} Q^2 - (m_i^2 - i\varepsilon)(x_1+x_2)\Big]\Big\} \equiv i\lambda \frac{\partial}{\partial \lambda} J(\lambda)\,,$$

the integral expression on the right-hand side being given by

$$J(\lambda) = \int_0^\infty dx_1 \int_0^\infty dx_2 \, \frac{1}{\lambda(x_1+x_2)^3} \sum c_i \exp\Big\{i\lambda\Big[\frac{x_1 x_2}{x_1+x_2} Q^2 - (m_i^2 - i\varepsilon)(x_1+x_2)\Big]\Big\}\,.$$

This integral $J(\lambda)$ is transformed to the original integration variable, by means of $x_1 = z_1/\lambda$ and $x_2 = z_2/\lambda$. One then sees that $J(\lambda)$, in fact, is independent of λ. Hence, one has indeed

$$I = i\lambda \frac{\partial J}{\partial \lambda} = 0\,.$$

This shows that the term which does not respect gauge invariance, does not contribute.

The first term which is gauge invariant, is transformed in a manner very similar to the self-energy, by inserting the factor

$$1 = \int_0^\infty \frac{d\lambda}{\lambda} \, \delta\Big(1 - \frac{z_1+z_2}{\lambda}\Big)$$

and by substituting $z := z_1/\lambda$ and $\zeta := z_2/\lambda$. One then obtains

$$\left(\Pi^{\mathrm{reg}}\right)^{\mu\nu}(Q) = \frac{2\alpha_0}{\pi}\left(Q^\mu Q^\nu - Q^2 g^{\mu\nu}\right)\int_0^\infty \mathrm{d}z \int_0^\infty \mathrm{d}\zeta \times z\zeta\,\delta(1-z-\zeta)\int_0^\infty \frac{\mathrm{d}\lambda}{\lambda} \sum_i c_i \exp\left\{\mathrm{i}\lambda\left[\frac{z\zeta}{z+\zeta}Q^2 - (m_i^2 - \mathrm{i}\varepsilon)(z+\zeta)\right]\right\} \tag{A.52}$$

$$= \frac{2\alpha_0}{\pi}\left(Q^\mu Q^\nu - Q^2 g^{\mu\nu}\right)\int_0^1 \mathrm{d}z\, z(1-z) \times \int_0^\infty \frac{\mathrm{d}\lambda}{\lambda}\sum_i c_i \exp\{\mathrm{i}\lambda[z(1-z)Q^2 - m_i^2 + \mathrm{i}\varepsilon]\} \tag{A.53}$$

Obviously, it is sufficient to introduce one auxiliary mass only: Choosing ($c_1 = 1$, $m_1 = m_f$) and ($c_2 = -1, m_2 = M_f$) (with $M_f \gg m_f$), one has

$$\left(\Pi^{\mathrm{reg}}\right)^{\mu\nu}(Q) = \Pi^{\mu\nu}(Q, m_f^2) - \Pi^{\mu\nu}(Q, M_f^2) \approx \frac{2\alpha_0}{\pi}\left(Q^\mu Q^\nu - Q^2 g^{\mu\nu}\right) \int_0^1 \mathrm{d}z(1-z)z\ln\left(\frac{M_f^2}{m_f^2 - \mathrm{i}\varepsilon - Q^2 z(1-z)}\right). \tag{A.54}$$

Here we used the integral formula

$$\int_0^\infty \frac{\mathrm{d}\lambda}{\lambda}\left(e^{\mathrm{i}a\lambda} - e^{\mathrm{i}b\lambda}\right) = \ln\left(\frac{b}{a}\right).$$

With $\int_0^1 \mathrm{d}z\, z(1-z) = 1/6$ and with

$$\ln\left(\frac{M_f^2}{m_f^2 - \mathrm{i}\varepsilon - Q^2 z(1-z)}\right) = \ln\left(\frac{M_f^2}{m_f^2}\right) - \ln\left(1 - \frac{Q^2 z(1-z)}{m_f^2 - \mathrm{i}\varepsilon}\right)$$

there follows the expression (10.96) given in the main text.

A.7 Ward-Takahashi Identity

The aim of this section is to show that the renormalization constants Z_1 and Z_2 of quantum electrodynamics are equal. The calculation that follows is purely formal because we do not ascertain that the integrals that are involved really exist. However, at the price of more writing, it applies as well to the regularized expressions.

We start from the expression (10.115), multiplied by the difference of the fermion momenta,

$$V \equiv (q-p)^{\mu} S_F(q)\Gamma_{\mu}(q,p)S_F(p)$$
$$= -(q-p)^{\mu}\int \mathrm{d}^4x \int \mathrm{d}^4y\, e^{\mathrm{i}q\cdot x}e^{-\mathrm{i}p\cdot y}\langle 0|T\,\psi(x)\,j_{\mu}(0)\overline{\psi(y)}|0\rangle\,. \qquad (A.55)$$

The vacuum state is invariant under translations. Therefore, the arguments of the time-ordered product can be shifted by the vector $-x$,

$$\langle 0|T\,\psi(x)\,j_{\mu}(0)\overline{\psi(y)}|0\rangle = \langle 0|T\,\psi(0)\,j_{\mu}(-x)\overline{\psi(y-x)}|0\rangle$$

without changing its expectation value. Substitution by $u := -x$ and $v := y-x$, allows to transform the term V, (A.55), as follows

$$V = -\int \mathrm{d}^4u \int \mathrm{d}^4v\,(q-p)^{\mu}e^{\mathrm{i}(p-q)\cdot u}e^{-\mathrm{i}p\cdot v}\langle 0|T\,\psi(0)\,j_{\mu}(u)\overline{\psi(v)}|0\rangle$$
$$= -\mathrm{i}\int \mathrm{d}^4u \int \mathrm{d}^4v\left(\frac{\partial}{\partial u_{\mu}}e^{\mathrm{i}(p-q)\cdot u}\right)e^{-\mathrm{i}p\cdot v}\langle 0|T\,\psi(0)\,j_{\mu}(u)\overline{\psi(v)}|0\rangle$$
$$= \mathrm{i}\int \mathrm{d}^4u \int \mathrm{d}^4v\,e^{\mathrm{i}(p-q)\cdot u}e^{-\mathrm{i}p\cdot v}\frac{\partial}{\partial u_{\mu}}\langle 0|T\,\psi(0)\,j_{\mu}(u)\overline{\psi(v)}|0\rangle\,.$$

In calculating the derivative of the vacuum expectation value by u_{μ} the product rule yields the term $\partial^{\mu}j_{\mu}(u)$ which vanishes by current conservation. The derivative by the time component u_0 yields nonvanishing contributions which stem from the step functions in the time-ordered product,

$$T\,\psi(0)j_{\mu}(u)\overline{\psi(v)} = \psi(0)j_{\mu}(u)\overline{\psi(v)}\Theta(-u_0)\Theta(u_0-v_0)$$
$$+\,\psi(0)\overline{\psi(v)}j_{\mu}(u)\Theta(-v_0)\Theta(v_0-u_0)$$
$$+\,j_{\mu}(u)\overline{\psi(v)}\psi(0)\Theta(u_0-v_0)\Theta(v_0)+\cdots\,,$$

$$\frac{\partial}{\partial u_0}\Theta(\pm u_0 - x) = \pm\,\delta(u_0 \mp x)\,,\quad \text{with}\ \ x=0\,, x=v_0 \text{ or } x=-v_0\,.$$

Working out these derivatives, V is found to be

$$V = \mathrm{i}\int \mathrm{d}^4u \int \mathrm{d}^4v\,e^{\mathrm{i}(p-q)\cdot u}e^{-\mathrm{i}p\cdot v}$$
$$\times\left\{\delta(u_0)\langle 0|T\left[j_0(u),\psi(0)\right]\overline{\psi(v)}|0\rangle\right.$$
$$\left.+\,\delta(u_0-v_0)\langle 0|T\,\psi(0)\left[j_0(u)\overline{\psi(v)}\right]|0\rangle\right\}\,.$$

The commutators at equal times are calculated from the canonical commutation relations (9.83),

$$\left[j_0(u),\psi(0)\right]\delta(u_0) = -\psi(u)\,\delta(u)\,, \qquad (A.56)$$

$$\left[j_0(u),\overline{\psi(v)}\right]\delta(u_0-v_0) = -\overline{\psi(v)}\,\delta(u-v)\,. \qquad (A.57)$$

One inserts the commutators (A.56) and (A.57) and makes use of the translation invariance of the vacuum expectation values, and obtains

$$V = S_F(p) - S_F(q),$$

that is, V is equal to the difference of the fermion propagator (10.113) for the momenta p and q, respectively. Multiplication of (A.55) by the inverse $\left(S_F(q)\right)^{-1}$ of the propagator from the left, and by the inverse $\left(S_F(p)\right)^{-1}$ from the right, yields

$$(q-p)^\mu \Gamma_\mu(q,p) = \left(S_F(q)\right)^{-1} - \left(S_F(p)\right)^{-1}. \tag{A.58}$$

This is one of the *Ward-Takahashi identities*. It plays a decisive role in the renormalization proof of quantum electrodynamics.

The identity (A.58) provides the key for proving the equality of Z_1 and Z_2. One takes the limit $\not{p} \to m\,\mathbb{1}$ where one has

$$\not{p} \to m\,\mathbb{1}: \quad \lim \left(S_F(p)\right)^{-1} = Z_2^{-1}\left(\not{p} - m\,\mathbb{1}\right), \quad \text{i.e.} \quad \left(S_F(p)\right)^{-1} = 0$$

while (A.58) yields

$$\left(S_F(q)\right)^{-1} = (q-p)^\mu \Gamma_\mu(q,p)\Big| \qquad \text{at}\, \not{p} \to m\,\mathbb{1}.$$

In the neighbourhood of the mass shell for q, that is close to $\not{q} = m\,\mathbb{1}$ one concludes

$$Z_2^{-1}\left(\not{q} - m\,\mathbb{1}\right) \approx (q-p)^\mu \Gamma_\mu(q,p)\Big| \approx Z_1^{-1}\left(\not{q} - m\,\mathbb{1}\right), \qquad \not{p} \to m\,\mathbb{1}.$$

This, in turn, implies that the first two renormalization constants are equal, $Z_1 = Z_2$.

A.8 Some Physical Constants and Units

Quantity	Symbol	Value	Dimension
Energy unit	1 eV	$1.602176565(35) \cdot 10^{-19}$	J
Unit of mass	$1\ \mathrm{eV}/c^2$	$1.782661845(39) \cdot 10^{-36}$	kg
Surface	1 barn	10^{-28}	m^2
Speed of light	c	299,792,458	$\mathrm{m\,s}^{-1}$
Planck's constant	h	$6.62606957(29) \cdot 10^{-34}$	J s
$h/(2\pi)$	$\hbar$	$6.58211928(15) \cdot 10^{-22}$	MeV s
Conversion factor	$\hbar c$	197.3269718(44)	MeV fm
Conversion factor	$(\hbar c)^2$	0.389379338(17)	GeV^2 mbarn
Elementary charge	e	$1.602176565(35) \cdot 10^{-19}$	C
Mass of electron	m_e	0.510998928(11)	MeV/c^2
Mass of muon	m_μ	105.6583715(35)	MeV/c^2
Mass of τ-lepton	m_τ	1776.82(16)	MeV/c^2
Mass of proton	m_p	938.272046(21)	MeV/c^2
Mass of neutron	m_n	939.565379(21)	MeV/c^2
n−p mass difference	$m_\mathrm{n} - m_\mathrm{p}$	1.2933322(4)	MeV/c^2
Mass of pion $\pi^\pm$	m_π	139.57018(35)	MeV/c^2
Fine structure constant	$\alpha = e^2/(\hbar c)$	1/137.035999074(44)	(none)
Rydberg energy	$hcR_\infty = m_\mathrm{e}c^2\alpha^2/2$	13.60569253(30)	eV
Bohr radius	$a_\infty = \hbar c/(\alpha m_\mathrm{e} c^2)$	$0.52917721092(17) \cdot 10^{-10}$	m
Bohr magneton	$\mu_\mathrm{B}^{(\mathrm{e})} = e\hbar/(2m_\mathrm{e}c)$	$5.7883818066(38) \cdot 10^{-11}$	$\mathrm{MeV\,T}^{-1}$
Nuclear magneton	$\mu_\mathrm{B}^{(\mathrm{p})} = e\hbar/(2m_\mathrm{p}c)$	$3.1524512605(22) \cdot 10^{-14}$	$\mathrm{MeV\,T}^{-1}$
Gravitation constant	G	$6.70837(80) \cdot 10^{-39}$	$\hbar c\,(\mathrm{GeV}/c^2)^{-2}$
Avogadro number	N_A	$6.02214129(27) \cdot 10^{23}$	mol^{-1}
Boltzmann constant	k	$8.6173324(78) \cdot 10^{-5}$	$\mathrm{eV\,K}^{-1}$
Fermi constant	$G_\mathrm{F}/(\hbar c)^3$	$1.1663787(6) \cdot 10^{-5}$	GeV^{-2}

Historical Notes

Without striving for completeness, I give here some biographical notes on scientists who made important contributions to the development of quantum mechanics and of quantum field theory. There is an extensive literature on the history of quantum mechanics, but much less so for quantum field theory. I recommend to turn primarily to original literature which often gives a more spontaneous and vivid impression of the developments, even though it may be somewhat subjective. For example, the letters of Max Born and Albert Einstein to which Max Born (as the editor) joined some of Pauli's letters concerning the debate on the interpretation of quantum mechanics, are very useful if one wants to understand better Einstein's attitude towards that theory. Regarding the early history of quantum field theory I strongly recommend the essais and lectures by Res Jost [Jost (1995)] who himself was one of the great actors in quantum field theory. (Note that most of his historical contributions are in German.)

The participation of renowned physicists in the development of nuclear weapons, and, in particular, in the "Manhattan project", is described and discussed in a large number of publications. Somewhat less well known is the early history of the application of nuclear fission and, in particular, of its French component under the leadership of Frédéric Joliot-Curie. For an excellent account I refer to [Weart (1979)]. One must know the scientific state of knowledge and the political circumstances of the time in some depth if one wants to understand the involvement of so many leading people in this dramatic story.

Among the popular scientific writings I recommend mostly the books written by leading scientists such as Freeman Dyson, Werner Heisenberg, Abraham Pais, Sylvan Schweber and a few others.

Bohr, Niels Hendrik David: ∗ 7 October 1885 in Copenhagen, † 18 November 1962 in Copenhagen. Nobel price 1922 for his work on the structure of atoms and on the radiation they emit.

Bohr turned to theoretical physics very early. His doctoral thesis dealt with a problem in the electron theory of metals. His stay, in 1912, at Rutherford's institute in Manchester gave the impetus and the basis for his work on the structure of atoms

F. Scheck, *Quantum Physics*, DOI: 10.1007/978-3-642-34563-0,

and of atomic transitions, this being the time when fundamental discoveries on radioactivity and atomic structure were made at this laboratory, while Planck's and Einstein's quantum hypothesis slowly became accepted knowledge. One should realize that the proof of the extreme smallness of atomic nuclei, by means of scattering experiments with α-particles (1911, E. Rutherford, H. Geiger and E. Marsden), was a great and important step in the understanding of the structure of matter. This was the scenario in which Bohr created what we now call "the Bohr model of the atom." Bohr's central role in the interpretation of the true quantum mechanics that was discovered in the early 1920s, can hardly be overestimated. The "Copenhagen interpretation" as we call it still today, is essentially his work. Note, however, that the Copenhagen school of thought was often criticized, perhaps one of the reasons being that it needs many words and, yet, does not explain completely the strange properties of the quantum world.

In the 1930s N. Bohr developed an important aspect of nuclei as systems showing collective excitations, and, in particular, the idea of the *compound nucleus* both of which were essential for the understanding of nuclear reactions. This model of a heavy nucleus as a quantum system with collective excitations (quantized droplet, quantized rigid rotator) which was further developed by Aage Bohr (son of Niels Bohr) and Ben Mottelson, and others, is in contrast to the shell model of nuclei. Both, in fact complementary, descriptions of nuclei were recognized by Nobel prices (1975, Aage Bohr and Ben Mottelson, together with James Rainwater), (1963, Maria Goeppert-Mayer and J. Hans D. Jensen, together with E. Wigner), respectively.

The lively and instructive book on Niels Bohr by Abraham Pais is highly recommended [Pais (1991)].

Born, Max: $*$ 11 December 1882 in Breslau, † 5 January 1970 in Bad Pyrmont. Received the Nobel price in 1954 (!) for his fundamental contributions to the development of quantum mechanics, and, in particular, for the statistical interpretation of the wave function.

Max Born's vita, his scientific career which embraces an unusually large spectrum of research topics, as well as his many interests and activities beyond his science, are described in his autobiography [Born (1975)] that I strongly recommend. Max Born was already a well established theoretical physicist who had made his reputation by his work and a book on the dynamics of crystal lattices when the fruitful Göttingen years (1921–1926) brought to light quantum mechanics in a breathtaking development. The list of students, assistants and collaborators who studied and worked with him during those years, comprises almost all the great names in quantum theory: Pauli, Heisenberg, Jordan, Dirac, Fermi, and many more. Born's contributions to the understanding of Heisenberg's matrix mechanics, to the interpretation of the Schrödinger wave function, and to many applications of quantum mechanics, were of central and decisive importance, even though the brilliance of the very young people around him seemed to outshine his achievements. This may be the reason why the Nobel price was awarded to him much later than to Heisenberg (1932), Pauli (1945), and Dirac (1933).

Max Born's life, among many other interesting aspects, is an impressive testimony of the irreparable destruction of cultural life in Germany, during the years of the Nazi dictatorship from 1933 until 1945. He was raised in a bourgeois German-Jewish family, imprinted by the conservative liberalism of the 1848 revolution, deeply rooted in German culture and tradition. Max Born studied in Wroclaw (then Breslau) (1901–1903), in Heidelberg (summer of 1902), Zurich (summer of 1903), and, for most of the time, in Göttingen (1903–1906) where eminent mathematicians such as Felix Klein, David Hilbert (whose private assistant he was), and Hermann Minkowski were his teachers. After a first military service and some time on an assistantship in the physical laboratory of Lummer in Breslau, he returned to Göttingen in 1908, following an invitation by Minkowski who had taken interest in Born's work on the theory of relativity. The collaboration in this field that they had planned, could not be realized because of Minkowski's untimely death during the winter of 1909. After having obtained habilitation in 1909 Born worked mostly on lattice dynamics. On a recommendation by Planck, Born obtained a professorship (Extraordinarius) at the university of Berlin. In 1919 he received a call to Frankfurt on the Main, changing positions with Max von Laue who moved to Berlin, and in 1921 was appointed full professor in Göttingen, succeeding Peter Debye. With the beginning of the Nazi dictatorship in 1933 Max Born was forbidden to teach and was put on unlimited leave, while continuing to receive his professor's salary. It was in 1938 only that he was deprived of his German citizenship and his belongings were confiscated. While working as a lecturer in Cambridge from 1933 until 1935, a planned call to the Indian Institute of Science, Bangalore, that had been initiated by Chandrasekhar Venkata Raman, did not realize because one of Raman's colleagues felt that "... a second-rate foreigner who was expelled from his country, is not good enough for us ...". From 1936 until his retirement in 1953 Born was professor in Edinburgh. In 1953 Max Born and his wife Hedwig returned to Germany. He was one of the signatories of the Göttingen manifesto against nuclear armament (1957). Born relentlessly spoke against participation of scientists in military projects such as the development of atomic bombs. In his autobiography Born remarks that his opposition may have been the cause for a certain reserve in his later relations with Robert Oppenheimer who never invited him to Princeton.

Max Born and Albert Einstein held a life-long close friendship which lasted until Einstein's death in 1955. The letters they exchanged during the years from 1916 until 1955, [Born (1969)], not only give an impression of the debates on quantum mechanics but also give instructive information on the disparate characters of the two friends, as well as on their very different relations to Germany which had been the prime cultural ground for both of them.

De Broglie, Prince Louis-Victor Pierre Raymond: ∗ 15 August 1892 in Dieppe, † 19 March 1987 in Paris. Received the Nobel price 1929 for his discovery of the wave nature of the electron.

If one wishes to get an impression of Prince Louis-Victor de Broglie's world one should read Marcel Proust's *A la Recherche du Temps Perdu* (Bibliothèque de la Pléiade (1954)). One will then not be surprised to learn that, originally, he wanted to become a diplomat and, for that purpose, studied history at the Sorbonne in Paris. Besides that, he also started studies of physics at age 18 but hesitated to enter research in that field. His great shot, the dual nature of the electron, was contained in his thesis (1924) and provided the basis for Schrödinger's later developments of wave mechanics. De Broglie became professor at the Institut Henri Poincaré, Paris, in 1928 and taught there until his retirement in 1962.

Dirac, Paul Adrien Maurice: $*$ 8 August 1902 in Bristol, † 20 October 1984 in Tallahassee, Florida. Was awarded the Nobel price in 1933, together with Erwin Schrödinger, for the discovery of "new productive forms of atomic theory".

Dirac's father who was Swiss, from Monthey (Valais), emigrated to England in 1888 where he worked as a teacher for French. Dirac who had intended to study mathematics, first became an electrical engineer at the university of Bristol. He had the impression that as a mathematician he would have to become a school teacher, a profession he wanted to avoid. He went on, eventually, to study mathematics, first in Bristol (1921–1923), and then at St. Johns College in Cambridge from 1923 to 1926, when he was awarded a Ph.D. on the basis of his doctoral thesis whose title was "Quantum Mechanics". This work was the result of Dirac's intense studies of matrix mechanics as developed by the Göttingen group shortly before. There followed visits to Niels Bohr in Copenhagen, and to Max Born in Göttingen (1927). His discovery of the relativistic equation for particles with spin-1/2 was made in the year 1928. Aged 28 only, Dirac became fellow of the Royal Society in 1930, as an early recognition of his outstanding achievement. In 1932 Dirac was appointed to the Lucasian Chair of mathematics at the university of Cambridge, a chair he held for 37 years, and that was previously held by Isaac Newton in the eighteenth century. After retirement Dirac moved to United States, where he did research at Florida State University until his death. Since 1995 a plaque in Westminster Abbey commemorates Paul Dirac, in the same place where Isaac Newton was buried.

Besides relativistic quantum theory of fermions Dirac's name is associated with a number of topics which have played an important role in physics ever since, and which were taken up again in various contexts, among them the theory of magnetic monopoles, the description of Lagrangian systems with boundary conditions, and his Lagrangian quantum theory as the basis for the path integral method whose richness and usefulness was worked out later by Feynman and others.

Dyson, Freeman: $*$ 1923, British-American mathematician and theoretical physicist. Professor Emeritus of the School of Natural Sciences, Institute for Advanced Studies in Princeton. Besides many other achievements, Freeman Dyson made important contributions to the development of quantum electrodynamics which are described in [Schweber (1994)].

By reconciling Feynman's pragmatic approach to quantum electrodynamics with Schwinger's mathematical techniques, he succeeded to resolve and to render transparent the filigree of quantum field theory. He surely would have deserved a Nobel price.

Ehrenfest, Paul: $*$ 18 January 1880 in Vienna, † 25 September 1933 in Leiden. Ehrenfest studied first in Vienna. He obtained his PhD in 1904, with L. Boltzmann as supervisor of his doctoral thesis. He continued his studies in Göttingen where Klein, Hilbert, Minkowski, and Carathéodory were among his teachers. After a stay at St. Petersburg (from 1907 on) he accepted a professorship at the university of Leiden, Netherlands, in 1912. Einstein who was a close friend of his, said about him that Ehrenfest had been the best teacher in our field he had ever known. Paul Ehrenfest committed suicide in 1933.

Einstein, Albert: ∗ 14 March 1879 in Ulm (Germany), † 18 April 1955 in Princeton, N.J. (USA). German-Swiss physicist, naturalized in the USA in 1940. Received the Nobel price 1921 for his achievements in theoretical physics and, in particular, for the discovery of the photoelectric effect in 1905. Einstein's contributions to quantum theory are numerous, though often underestimated because of his critical attitude regarding its interpretation. For instance, the hypothesis of elementary quanta which was the key to Planck's explanation of the black body radiation in 1900, became more generally accepted only after Einstein's 1905 work. His work with Podolsky and Rosen [Einstein, Podolsky, Rosen (1935)] on correlated, entangled states, is instrumental for modern work on quantum information. This work has been much discussed until this day, it belongs to the publications with the largest number of citations in physics. Last not least, in recent years Bose-Einstein condensation at macroscopic scales has become a standard experimental technique with many applications in fundamental research.

Fermi, Enrico: ∗ 29 September 1901 in Rome, † 29 November 1954 in Chicago. Italo-American physicist. Received the Nobel price 1938, not for his discovery of the statistics of spin-1/2 particles (which obey the Pauli principle), nor for his early theory of nuclear β-decay, but for his investigations of artificial radioactivity and of nuclear reactions induced by slow neutrons. From 1927 until 1938 Fermi was professor of theoretical physics at the university of Rome. In 1938 he emigrated to United States, escaping the Mussolini dictatorship in Italy. From 1938 until 1942 Fermi was professor of physics at Columbia university, New York. He conducted a number of nowadays classical experiments leading to the first operating nuclear reactor and the first controlled nuclear chain reaction. Fermi played a central role in the Manhattan Project. From 1946 until his death, he was professor at the Institute for Nuclear Studies of the university of Chicago.

Feynman, Richard Phillips: ∗ 11 May 1918 in New York, † 15 February 1988 in Pasadena. Together with Julian Schwinger and Sin-Itiro Tomonaga he was awarded the Nobel price 1965, for his work on quantum electrodynamics and on the deep consequences of this theory for elementary particle physics. He obtained his PhD in 1942 with J.A. Wheeler at Princeton University. From 1945 until 1950 he was professor of theoretical physics at Cornell University, Ithaka N.Y., and from 1950 on at the California Institute of Technology, Pasadena, California. Feynman made many contributions to particle physics, notably in the field of weak interactions (hypothesis of the conservation of the weak vector current), and to the parton hypothesis. The Feynman lectures on physics belong to the best known and most popular textbooks in physics.

Fock, Vladimir Aleksandrovich: ∗ 22 December 1898 in St. Petersburg, † 27 December 1974 in St. Petersburg. Russian theoretician. Made important contributions to quantum theory, among others the Hartree-Fock method in many-body physics, and the concept of Fock space. Fock also pointed out the close relationship between local gauge transformations in electrodynamics and local phases of Schrödinger wave functions (see Exercise 1.7 and its solution).

Heisenberg, Werner Karl: ∗ 5 December 1901 in Würzburg, † 1 February 1976 in Munich. Received the Nobel price 1932 for the development of quantum mechanics and for important applications of it.

Heisenberg studied in Munich, where W. Wien, A. Sommerfeld, A. Pringsheim and others were his academic teachers. During the winter term of 1922/23 he studied with Max Born, David Hilbert and others in Göttingen. After obtaining his PhD with Sommerfeld in 1923, again in Munich, he became assistant with Born at the university of Göttingen where he was granted the venia legendi in 1924. Max Born describes the near-to-catastrophe during Heisenberg's PhD exam where he had almost failed in experimental physics with Willy Wien on a question about the resolution of optical instruments. Heisenberg returned to Göttingen somewhat anxious whether professor Born would still accept him as assistant. He conscientiously worked on filling up the gaps in his knowledge. As is well known, he later invoked the resolution of the microscope as an illustration of the uncertainty relations.

There followed visits to Niels Bohr in Copenhagen, where he was appointed lecturer at the university in 1926. That same year, at the early age of 26 years, Heisenberg became full professor at the university of Leipzig. From 1941 on he was professor at the university of Berlin and director of the Kaiser-Wilhelm Institute (which became the Max-Planck Institute after the war). At the end of the war he was detained in England for a while, together with a group of other German scientists, because of his participation in the "Uran-Verein". He returned to Göttingen in 1946, as professor of theoretical physics and director of the newly founded Max-Planck Institute. Starting in 1958, after the Max-Planck Institute had moved from Göttingen to Munich, Heisenberg eventually became professor at the university of Munich, and continued acting as a director of the Max-Planck Institute for Physics and Astrophysics until 1970.

To some extent like Max Born's life, Heisenberg's vita may teach us a lot about the spirit and the atmosphere of a period of history which is extremely difficult to understand from our present perspective, and which was also the time of the dramatic transition of physics from its peaceful academic realm, with only few actors, to the public arena of nuclear energy and its dreadful applications. Heisenberg who was raised in the idealistic atmosphere of Jugendbewegung, had many interests beyond physics. Like Born he was a well versed pianist. His somewhat undecided, if not ambiguous, attitude towards the political authorities during the Nazi period is described in a fair and well-balanced manner by D. Cassidy [Cassidy (1992)].

Heisenberg's contributions to quantum mechanics and to quantum field theory are numerous and decisive. Besides matrix mechanics and the uncertainty relation there are many more subjects related to his name. Among them the theory of ferromagnetism, the concept of field quantization, the nuclear isospin, and the concept of scattering matrix. Regarding the latter his somewhat generous realization of the idea gave rise to the later rigorous investigations by R. Jost and his school in Zurich.

Hilbert, David: * 23 January 1862 in Königsberg, † 14 February 1943 in Göttingen. One of the greatest mathematicians of the twentieth century. Mathematical physics in general, general relativity and quantum mechanics, in particular, owe many important stimuli to David Hilbert.

Jordan, Pascual Ernst: * 18 October 1902 in Hannover, † 31 July 1980 in Hamburg. Studied and received his PhD (1924) in Göttingen with Max Born. Habilitation in 1926 at the university of Göttingen, and at the university of Hamburg. Starting 1928 he became professor at the university in Rostock, and in 1944 at the university of Berlin. From 1947 on Jordan was guest professor at the university in Hamburg. Pascual Jordan had an important share in the development of the Göttingen version of quantum mechanics and in the idea of field quantization, a share that is larger than is generally known. Max Born in his autobiography gives him proper credit.

Jost, Res: * 10 January 1918 in Berne, † 3 October 1990 in Zurich. Studied in Berne and at the university of Zurich where he received his PhD under the guidance of Gregor Wentzel in 1946, on a topic in meson theory. After a stay in Copenhagen, Jost became assistant of Pauli at the ETH in Zurich (Swiss Federal Institute of Technology). He followed Pauli on a sabbatical in Princeton in 1949 but stayed on until 1955, on a five-year membership at the Institute for Advanced Studies. In 1955 he was appointed Extraordinarius at the ETH, and in 1959 full professor, succeeding Pauli who had died in 1958.

Jost made important contributions to scattering theory, to quantum electrodynamics, to axiomatic quantum field theory, and other fields of theoretical physics. He belongs to the founding fathers of axiomatic field theory and was an eminent authority in this branch of theoretical physics. In his later years he turned to the early history of quantum field theory on which he left a number of essais and lectures [Jost (1995)].

Majorana, Ettore: * 5 August 1906 in Catania (Sicily), † 26 March 1938, presumably in the Tyrrhenian Sea. Italian (or, more precisely, Sicilian) theoretical physicist who was strongly influenced by Fermi, Heisenberg, and others. From 1937 until his myterious vanishing from the post ship on its way from Naples to Catania, he was professor at the university of Naples. He was an ingenuous young man, barely younger than Fermi, working on Dirac theory, on nuclear forces, and on nuclear physics more generally. Many of his contributions to theoretical physics are contained in unpublished calculations and notes. It seems that he was deeply scared by the consequences of nuclear conversion which he had realized almost immediately. His all too short life is beautifully described in the book by Leonardo Sciascia [Sciasca (1975)]. Did he commit suicide, or did he withdraw behind the walls of a Carthusian monastry because he did not want to be involved in the dreadful consequences of the nuclear physics known to him?

Noether, Emmy Amalie: $*$ 23 March 1882 in Erlangen, † 14 April 1935 in the United States.

Undoubtedly one of the great scientists in the mathematics of the twentieth century. Her seminal work *Invariante Variationsprobleme* of 1918 contains two theorems now bearing her name, which until this day provide keys to important parts of theoretical physics (mechanics, classical field theory, local gauge invariance, etc.). Barely any other mathematical publication has had such a profound and lasting impact on the physics of the twentieth century.

Emmy Noether grew up during the narrow-minded, man-dominated Second German Empire when, as a matter of principle, university careers were closed for women. During the Weimar republic when things had changed, Emmy Noether was granted habilitation in 1919 and was nominated "ausserplanmässige Professorin" (a purely honorary title) at Göttingen university. However, she never obtained a professorship in mathematics, in spite of the fact that her high qualification was never doubted, that she had the strongest support of David Hilbert, and enjoyed the highest esteem among her collaborators and students. Later, at the beginning of the Nazi regime she was deprived of her venia legendi. Even in the United States where she had emigrated in 1933, she never had a better position than a guest professorship at Bryn Mawr Women's College in Pennsylvania.

Pauli, Wolfgang Ernst: $*$ 25 April 1900 in Vienna, † 15 December 1958 in Zurich. Received the Nobel price 1945 for his discovery of the exclusion principle.

Although unusual in physics the young Wolfgang Pauli may justly be termed an infant prodigy. He obtained his doctoral degree in 1921, with Arnold Sommerfeld in Munich. He was barely twenty years old when he wrote a review article on General Relativity which still today is a standard reference. He spent a year each, as an assistant of Born in Göttingen, and with Bohr in Copenhagen. He was lecturer at the university of Hamburg from 1923 until 1928 when he received a call from the ETH Zurich. In 1940 he was appointed full professor of theoretical physics at Princeton University but returned to Zurich after World War II where he taught and worked until his death.

Pauli made many pathbreaking discoveries in quantum mechanics, quantum field theory, and particle physics. To quote just a few: Besides the exclusion principle that bears his name, he postulated the existence of a neutrino participating in nuclear β-decay, i.e., using modern terminology, the ν_e, and he proved the theorem on the correlation of spin and statistics, building on earlier work by M. Fierz.

Planck, Max Karl Ernst Ludwig: $*$ 23 April 1858 in Kiel, † 4 October 1947 in Göttingen. He received the Nobel price 1918 for his discovery of the energy quanta.

Max Planck studied in Munich and in Berlin where G. Kirchhoff and H. Helmholtz were among his teachers. He received his PhD in Munich in 1879, was then Privatdozent in Munich from 1880 until 1885, and Extraordinarius in Kiel until 1889. He was appointed to the chair formerly held by Gustav Kirchhoff, at the university of Berlin, teaching and doing research there until his retirement in 1926. His scientific work focussed on thermodynamics and, in particular, on the entropy in irreversible processes. His monumental work which after many attempts, led him to the correct description of black-body radiation, belongs to these extremely fruitful years in Berlin. It seems as though, for quite a while, his colleagues did not take seriously Planck's discovery, with one notable exception: Albert Einstein whose publication *Über einen die Erzeugung und Verwandlung des Lichtes betreffenden heuristischen Gesichtspunkt* of 1905 marked the beginning of quantum theory proper.

Max Planck's life was marked by several personal tragedies: His first wife, Marie Merck, died in 1909. Planck survived all four children of his first marriage. His son Karl died in 1916 in the battle of Verdun, his son Erwin was hung in January 1945 for having participated in the July 1944 upraisal against Hitler. Like many other families living in Berlin he lost his house and belongings in a bombardment of the city.

Max Planck was a courageous man who openly opposed some of the political decisions during the Nazi dictatorship. For instance he intervened at the highest level in favour of Fritz Haber who had been expelled from the country – unfortunately without success.

Schrödinger, Erwin Rudolf Josef Alexander: $*$ 12 August 1887 in Vienna, †4 January 1961 in Vienna. Schrödinger received the Nobel price 1933, together with P. Dirac, for his contributions to the quantum theory of the atoms.

Schrödinger studied in Vienna from 1906 until 1910, among others with F. Hasenöhrl. During World War I he was an artillery officer, from 1920 on assistant of Max Wien in Vienna. He held a first professorship (Extraordinarius) in Stuttgart, a full professorship in Wroclaw (then Breslau), and, at the university of Zurich, succeeding Max von Laue. In 1927 he was appointed to a chair at the university of Berlin, as successor of Max Planck. He left Germany in 1933, as a sign of protest against the new regime. After a number of stays in Oxford, Graz (until the annexation of Austria), and Princeton, he was appointed to the newly founded Institute for Advanced Studies in Dublin, whose director he became. He worked there until retirement in 1955. Besides de Broglie, Schrödinger was one of the most prominent critics of the dual nature of matter, i.e. of the particle-wave dualism, trying to interpret quantum mechanics as a pure wave theory.

Schwinger, Julian Seymour: $*$ 12 February 1918 in New York, † 16 August 1994 in Los Angeles. Received the Nobel price 1965, together with Tomonaga and Feynman, for his path-breaking contributions to quantum electrodynamics. Schwinger studied at City College of New York and at Columbia University, N.Y. He was assistant of R. Oppenheimer at the University of California in Berkeley (1939–1941). After research activities in relation with war efforts at the Radiation Laboratory of MIT in Cambridge, USA, he held a professorship at Harvard University from 1945 until 1972. From 1972 until his death in 1994 he was professor at the University of California in Los Angeles.

Sommerfeld, Arnold Johannes Wilhelm: $*$ 5 December 1868 in Königsberg, † 26 April 1951 in München.

Sommerfeld studied mathematics at the university of Königsberg, where, at that time, Hilbert, Hurwitz, and Lindemann belonged to his academic teachers. In 1893 he became assistant of Felix Klein in Göttingen. In 1897 he was appointed as professor for mathematics at Bergbauakademie (mining academy) in Clausthal, in 1900 professor of mechanics at the university of Aachen, and in 1906, finally, professor for theoretical physics at the university of Munich. Sommerfeld wrote important articles on partial differential equations and collaborated with Felix Klein on Klein's monumental work on the theory of spinning tops. Sommerfeld's work on atomic spectra and on quantum mechanics are from his first years in Munich: The elliptic orbits in Bohr's atomic model, the magnetic quantum number, etc. His work on the electron theory of metals was particularly important.

Sommerfeld founded a famous school of theoretical physics where many important physicists were trained. This unique and well-known school, too, was destroyed by the barbarism of the dictatorship after 1933. Arnold Sommerfeld authored a number of textbooks some of which are still being used today. His book on quantum theory, *Atombau und Spektrallinien* has had a lasting influence on the generation of his students, very much like Heitler's later book on the quantum theory of radiation. Sommerfeld, too, would have deserved the Nobel price for some of his achievements.

Tomonaga, Sin-Itiro: $*$ 31 March 1906 in Tokyo, † 8 August 1979 in Tokyo. Honored by the Nobel price 1965, together with Feynman and Schwinger, for his work on quantum electrodynamics. He studied in Tokyo with H. Yukawa and Y. Nishina, as well as in Leipzig with Heisenberg (1937–1939). A thesis on a subject from nuclear physics, on which he worked while in Leipzig, was accepted for his PhD in Tokyo. From 1941 on Tomonaga was professor of physics at Tokyo university. He developed a covariant formulation of quantum electrodynamics during the years 1941–1943 and, thus, was the first of the three. Due to the political circumstances his publications came to be known in the Western world only in 1947, at a time where both Feynman and Schwinger had independently developed their own solutions of the same subject.

Wigner, Eugene Paul: $*$ 17 November 1902 in Budapest, † 1 January 1995 in Princeton, New Jersey. Received the Nobel price 1963, together with Maria Goeppert-Mayer and J. Hans D. Jensen, but not for the nuclear shell model (like the latter two), but for his contributions to the theory of nuclei and of elementary particles, notably his discovery and application of fundamental symmetry principles.

He studied chemistry at the Technical Highschool in Berlin where he also obtained an engineer's degree. Already during the Berlin years 1928–1930 Wigner worked on applications of group theory to quantum mechanics, besides his teaching as a chemical engineer. During the years 1930–1933 Wigner spent some time of each year at Princeton. After Hitler came to power Wigner lost his position in Berlin and emigrated to the United States. As of 1938 he was professor of mathematical physics at Princeton University. Wigner, too, belonged to the group of highly talented physicists who collaborated on the Manhattan Project.

Exercises, Hints, and Selected Solutions

Exercises: Chap. 1

1.1 Determine the wave lengths of photons which are emitted in ($n = 2 \to n = 1$), ($n = 3 \to n = 2$), and ($n = 4 \to n = 3$) transitions in hydrogen. Situate them relative to the visible spectrum.
Hint: $\lambda = 2\pi \hbar c/\Delta E$.

1.2 Consider the following hydrogenlike atoms: (e^+e^-) (positronium), (μ^+e^-) (muonium), ($^4\mathrm{He}\mu^-$) (muonic Helium), ($\bar{p}p$) (antiprotonic hydrogen), ($^{12}\mathrm{C}\pi^-$) (pionic carbon), and ($p\,\Omega^-$). Calculate the reduced masses, the Bohr radii, and the transition energies $3 \to 2$ and $2 \to 1$ in eV as well as the corresponding wave lengths.
Hints: $m_{\mathrm{He}} = 2(m_{\mathrm{p}} + m_{\mathrm{n}})c^2 - 24\,\mathrm{MeV}$, $m_\Omega = 1672\,\mathrm{MeV}$.

1.3 Determine the de Broglie wave length for

1. an electron with velocity $v = \alpha c$, an electron with momentum $|\boldsymbol{p}| = 200\mathrm{MeV}/c$,
2. the earth on its orbit ($m_{\mathrm{E}} = 5.98 \times 10^{24}$ kg, $v = 29.8$ km/s).
3. How would you choose the energy, or the momentum, of a neutron for its wave length to be 10^5 fm?

1.4 Let ψ_1 and ψ_2 be two different solutions of the time-dependent Schrödinger equation for the same potential. Assume both of them to decrease sufficiently fast at infinity. Show that the transition current density $\varrho_{21} := \psi_1^*(t, \boldsymbol{x})\psi_2(t, \boldsymbol{x})$ fulfills a continuity equation.

1.5 By way of the correspondence

$$E \longleftrightarrow \mathrm{i}\hbar\frac{\partial}{\partial t}, \qquad \boldsymbol{p} \longleftrightarrow \frac{\hbar}{\mathrm{i}}\boldsymbol{\nabla}$$

the relativistic energy-momentum relation

$$E^2 = c^2\boldsymbol{p}^2 + (mc^2)^2$$

F. Scheck, *Quantum Physics*, DOI: 10.1007/978-3-642-34563-0,

yields a wave equation. This differential equation is called Klein-Gordon equation.

1. In which respect does this equation differ from the Schrödinger equation? What is its difference to the wave equation

$$\left(\frac{1}{c^2}\frac{\partial^2}{\partial t^2} - \Delta\right)\Phi(t,\boldsymbol{x}) = 0\,?$$

2. A wave function $\Phi(t,\boldsymbol{x})$ which satisfies the Klein-Gordon equation, is assumed to have the form

$$\Phi(t,\boldsymbol{x}) = \exp\left(-\frac{\mathrm{i}}{\hbar}mc^2 t\right)\psi(t,\boldsymbol{x})\,.$$

In which approximation does the Schrödinger equation follow from the Klein-Gordon equation?

1.6 Show: The Hermite polynomials $H_n(u)$, for which the coefficient of the highest power of u is $a_n = 2^n$ (see Sect. 1.6), fulfill the relations

$$\left(2u - \frac{\mathrm{d}}{\mathrm{d}u}\right)H_n(u) = H_{n+1}(u)\,,$$

$$\frac{\mathrm{d}}{\mathrm{d}u}H_n(u) = 2n\,H_{n-1}(u)\,,$$

$$H_{n+1}(u) = 2u\,H_n(u) - 2n\,H_{n-1}(u)\,.$$

1.7 A charged particle (mass m, electric charge e) and its interaction with external electric and magnetic fields $(\boldsymbol{E}, \boldsymbol{B})$, respectively, is described by the time-dependent Schrödinger equation (1.51) with the Hamiltonian (1.50). A smooth gauge transformation

$$\begin{aligned} \boldsymbol{A}(t,\boldsymbol{x}) &\mapsto \boldsymbol{A}'(t,\boldsymbol{x}) = \boldsymbol{A}(t,\boldsymbol{x}) + \nabla\chi(t,\boldsymbol{x})\,,\\ \Phi(t,\boldsymbol{x}) &\mapsto \Phi'(t,\boldsymbol{x}) = \Phi(t,\boldsymbol{x}) - \frac{1}{c}\frac{\partial\chi(t,\boldsymbol{x})}{\partial t} \end{aligned} \qquad (1)$$

leaves the fields invariant but not the Schrödinger equation. As a trial, let

$$\psi'(t,\boldsymbol{x}) = \exp[\mathrm{i}\eta(t,\boldsymbol{x})]\psi(t,\boldsymbol{x}) \qquad (2)$$

and show that one can find a real function $\eta(t,\boldsymbol{x})$ such that the Schrödinger equation remains form invariant, i.e. such that ψ' is a solution of the differential equation (1.51) with the transformed potentials.

Solution: With H as given in (1.50) and with the potentials (1) the transformed Hamiltonian is

$$H' = \frac{1}{2m}\left(\boldsymbol{p} - \frac{\mathrm{e}}{c}\mathbf{A}'(t,\boldsymbol{x})\right)^2 + e\Phi'(t,\boldsymbol{x})\,.$$

For the ansatz (2) we have

$$\dot\psi' = \mathrm{e}^{\mathrm{i}\eta}\dot\psi + \mathrm{i}\mathrm{e}^{\mathrm{i}\eta}\psi\frac{\partial\eta}{\partial t}\,, \qquad \nabla\psi' = \mathrm{e}^{\mathrm{i}\eta}\nabla\psi + \mathrm{i}\mathrm{e}^{\mathrm{i}\eta}\psi\nabla\eta\,,$$

$$\Delta\psi' = \mathrm{e}^{\mathrm{i}\eta}[\Delta\psi + 2\mathrm{i}(\nabla\eta)\cdot(\nabla\psi) - \psi(\nabla\eta)^2 + \mathrm{i}\psi\Delta\eta]\,.$$

This is to be compared to the Hamiltonian H' with the potentials given by (1),

$$H' = H - \frac{e}{2mc}\left[\left(\boldsymbol{p} - \frac{e}{c}\boldsymbol{A}\right)\cdot(\nabla\chi) + (\nabla\chi)\cdot\left(\boldsymbol{p} - \frac{e}{c}\boldsymbol{A}\right) - \frac{e}{c}(\nabla\chi)^2\right] - \frac{e}{c}\frac{\partial\chi}{\partial t}\,.$$

Note that $\boldsymbol{p} = \hbar\nabla/\mathrm{i}$ acts on all factors to the right of it, while the differential operator in $(\nabla\chi)$ acts on χ only. Therefore, we obtain

$$H' = H - \frac{e}{2mc}\left[\frac{\hbar}{\mathrm{i}}(\Delta\chi) + 2\frac{\hbar}{\mathrm{i}}(\nabla\chi)\cdot\nabla - 2\frac{\mathrm{e}}{c}\mathbf{A}\cdot(\nabla\chi) - \frac{e}{c}(\nabla\chi)^2\right] - \frac{e}{c}\frac{\partial\chi}{\partial t}\,.$$

The action of H' onto ψ' can now be computed,

$$H'\psi' = \mathrm{e}^{\mathrm{i}\eta}\Bigg[H\psi - \mathrm{i}\frac{\hbar^2}{m}(\nabla\eta)\cdot(\nabla\psi) - \mathrm{i}\frac{\hbar^2}{2m}\psi(\Delta\eta) + \frac{\hbar^2}{2m}\psi(\nabla\eta)^2 - \frac{\mathrm{e}\hbar}{mc}\boldsymbol{A}\cdot(\nabla\eta)\psi + \mathrm{i}\frac{\mathrm{e}\hbar}{2mc}\psi(\Delta\chi) + \frac{e^2}{mc^2}\boldsymbol{A}\cdot(\nabla\chi)\psi + \frac{e^2}{2mc^2}(\nabla\chi)^2\psi + \mathrm{i}\frac{\mathrm{e}\hbar}{mc}(\nabla\chi)\cdot(\nabla\psi) - \frac{\mathrm{e}\hbar}{mc}(\nabla\chi)\cdot(\nabla\eta)\psi - \frac{e}{c}\frac{\partial\chi}{\partial t}\psi\Bigg]\,.$$

Comparing this expression with $\mathrm{i}\hbar\dot{\psi}'$, with $\dot{\psi}'$ as calculated above, one sees that the terms within the square brackets should cancel, except for the first and the last ones. This happens if one chooses

$$\eta(t,\boldsymbol{x}) = \frac{e}{\hbar c}\,\chi(t,\boldsymbol{x})\,. \tag{3}$$

The last term is proportional to the additional term in $\dot{\psi}'$ so that, indeed,

$$\mathrm{i}\hbar\dot{\psi}' = H'\psi'\,.$$

The wave function is multiplied by a phase factor $\exp[\mathrm{i}\eta(t,\boldsymbol{x})]$ which, in general, is not constant, but depends on time and space. The function $\eta(t,\boldsymbol{x})$ is proportional to the gauge function $\chi(t,\boldsymbol{x})$.

1.8 Let $A, B, C, \ldots$ be operators all of which have the same domain of definition.

1. Prove the formulae

$$[A, BC] = [A, B]\,C + B\,[A, C]\,,$$

$$[A, B^n] = \sum_{r=0}^{n-1} B^r\,[A, B]\,B^{n-r-1}\,, \tag{4}$$

(use induction for the second one), as well as the Jacobi identity

$$[A, [B, C]] + [B, [C, A]] + [C, [A, B]] = 0\,. \tag{5}$$

2. Work out the following applications of these formulae: Let $F(\boldsymbol{q},\boldsymbol{p})$ be a dynamical quantity which depends polynomially on $\boldsymbol{q} = (q^1, q^2, q^3)$ and on $\boldsymbol{p} = (p_1, p_2, p_3)$. Show

$$[p_i, F(\boldsymbol{q},\boldsymbol{p})] = \frac{\hbar}{\mathrm{i}} \frac{\partial F(\boldsymbol{q},\boldsymbol{p})}{\partial q^i}, \qquad [q^i, F(\boldsymbol{q},\boldsymbol{p})] = \mathrm{i}\hbar \frac{\partial F(\boldsymbol{q},\boldsymbol{p})}{\partial p_i}.$$

3. Let $\ell_i = \sum_{jk} \varepsilon_{ijk} x^j p_k$, $(i = 1, 2, 3)$, be the operator which describes the i-th component of the orbital angular momentum of a particle. Calculate

$$[\ell_i, p_k], \qquad [\ell_i, \boldsymbol{p}^2], \qquad [\ell_i, \boldsymbol{x}^2]$$

and from this $[\ell_i, H]$ for $H = \boldsymbol{p}^2/(2m) + U(\boldsymbol{x}^2)$.

1.9 The Heisenberg uncertainty relations can be written in a rather general form: Let A and B be two self-adjoint operators representing observables. Their uncertainties (or standard deviations) in the state ψ fulfill the inequality

$$(\Delta A)_\psi \, (\Delta B)_\psi \ge \frac{1}{2} \left| \langle [A, B] \rangle_\psi \right| . \tag{6}$$

1. Prove this inequality by considering the operators $\Delta_A := A - \langle A \rangle_\psi$ and $\Delta_B := B - \langle B \rangle_\psi$, and calculating the norm of the state $(\Delta_A + ix\Delta_B)\psi$ with $x \in \mathbb{R}$.
2. Study the examples $(A = p_k, B = x^l)$ and $(A = \ell_2, B = \ell_3)$.

Solution: 1. The following calculation leads to the result (6).

$$\begin{aligned}
\|(\Delta_A + \mathrm{i}x\Delta_B)\psi\|^2 &= \|\Delta_A \psi\|^2 + x^2 \, \|\Delta_B \psi\|^2 \\
&\quad + \mathrm{i}x\big[\langle \Delta_A \psi, \Delta_B \psi \rangle - \langle \Delta_B \psi, \Delta_A \psi | \big] \\
&= \left\langle \Delta_A^2 \right\rangle_\psi + x^2 \left\langle \Delta_B^2 \right\rangle_\psi + \mathrm{i}x \, \langle [\Delta_A, \Delta_B] \rangle_\psi \\
&= \left\langle \Delta_A^2 \right\rangle_\psi + x^2 \left\langle \Delta_B^2 \right\rangle_\psi + \mathrm{i}x \, \langle [A, B] \rangle_\psi \ge 0 .
\end{aligned}$$

Here we used that with A and B self-adjoint, also Δ_A and Δ_B are self-adjoint. Furthermore $[\Delta_A, \Delta_B] = [A, B]$. The expectation value of the commutator of two self-adjoint operators is pure imaginary. Therefore, the last equation is correct only if the determinant of the quadratic equation $x^2 \langle \Delta_B^2 \rangle_\psi + \mathrm{i}x \, \langle [A, B] \rangle_\psi + \langle \Delta_A^2 \rangle_\psi = 0$ is smaller than or equal to zero,

$$\left(\mathrm{i} \, \langle [A, B] \rangle_\psi \right)^2 - 4(\Delta A)^2 (\Delta B)^2 \le 0 .$$

This is the asserted inequality (6).

2. In the first example one obtains $(\Delta p_k)(\Delta x^l) \ge \hbar/2$, in the second example one obtains $(\Delta \ell_2)(\Delta \ell_3) \ge |\langle \ell_1 \rangle_\psi|/2$.

1.10 Show the following

1. A state for which all three components of orbital angular momentum can be determined simultaneously, necessarily has $\ell = 0$.
2. In every eigenstate of $\boldsymbol{\ell}^2$ and of ℓ_3 the expectation values of x^1, x^2, p_1, p_2, ℓ_1, and ℓ_2 vanish.

Solution: 1. If all three components have sharp values simultaneously then the standard deviations are zero, $(\Delta\ell_i) = 0$. This means that $\langle \ell_i^2 \rangle = \langle \ell_i \rangle^2$ for all i. Since

$$(\Delta\ell_1)(\Delta\ell_2) \geq \frac{1}{2}|\langle[\ell_1, \ell_2]\rangle| = \frac{1}{2}|\langle\ell_3\rangle|$$

with its cyclic permutations, one has $\langle \ell_i^2 \rangle = \langle \ell_i \rangle^2 = 0$ for all i, hence $\langle \boldsymbol{\ell}^2 \rangle = 0$, and $\ell = 0$.

2. The expectation values of the first four operators in the state $|\ell m\rangle$ vanish because they have odd parity. The expectation values of ℓ_1 and of ℓ_2 are equal to zero because the operators $\ell_\pm$ raise and lower the m-quantum number, respectively, and because the states $|\ell m \pm 1\rangle$ and $|\ell m\rangle$ are orthogonal.

1.11 Given the operator $\mathcal{O} := \boldsymbol{x} \cdot \boldsymbol{p}$ where $\boldsymbol{x}$ and $\boldsymbol{p}$ are, respectively, the position and the momentum operators of a particle with mass m which is described by the Hamiltonian H.

1. Write down the Heisenberg equation of motion for $\mathcal{O}$ and show that the expectation value of the time derivative $\mathrm{d}\mathcal{O}/\mathrm{d}t$ in every eigenstate of H is equal to zero.
2. Calculate the commutator $[H, \mathcal{O}]$ for the case $H = \boldsymbol{p}^2/(2m) + U(\boldsymbol{x})$.
3. From this follows the virial theorem

$$2\,\langle T \rangle_\psi = \langle \boldsymbol{x} \cdot \nabla U \rangle_\psi \tag{7}$$

for the expectation values in stationary eigenstates of H. Apply this theorem to the spherical oscillator and to the hydrogen atom.

Solution: Heisenberg's equation of motion

$$\mathrm{d}(\boldsymbol{x} \cdot \boldsymbol{p})/\mathrm{d}t = \mathrm{i}[H, (\boldsymbol{x} \cdot \boldsymbol{p})]/\hbar\,,$$

evaluated in the eigenstate ψ of H, $H\psi = E\psi$, yields

$$\frac{\mathrm{d}}{\mathrm{d}t}\,\langle(\boldsymbol{x} \cdot \boldsymbol{p})\rangle_\psi = \frac{\mathrm{i}}{\hbar}\,\langle[H, (\boldsymbol{x} \cdot \boldsymbol{p})]\rangle_\psi \equiv \frac{\mathrm{i}}{\hbar}\langle\psi|[H, (\boldsymbol{x} \cdot \boldsymbol{p})]\,|\psi\rangle = 0\,,$$

because H acts to the left in the first term of the commutator, to the right in the second, in both cases answering by the same eigenvalue. The commutator is calculated by means of the formulae

$$[\boldsymbol{p}^2, (\boldsymbol{x} \cdot \boldsymbol{p})] = [\boldsymbol{p}^2, \boldsymbol{x}] \cdot \boldsymbol{p}\,,$$
$$[\boldsymbol{p}^2, x^k] = \hbar^2[\Delta, x^k] = -2\hbar^2\partial_k = -2\mathrm{i}\hbar p_k\,,$$
$$[U, (\boldsymbol{x} \cdot \boldsymbol{p})] = [U, \boldsymbol{p}] \cdot \boldsymbol{x} = -\frac{\hbar}{\mathrm{i}}\,(\nabla U) \cdot \boldsymbol{x}$$

Collecting all contributions, one obtains

$$\frac{\mathrm{i}}{\hbar}[H, (\boldsymbol{x} \cdot \boldsymbol{p})] = 2T - \boldsymbol{x} \cdot \nabla U\,.$$

In every stationary eigenstate of H the expectation value of this equation vanishes. With U a central potential of the form $U(r) = ar^\alpha$, one has $\boldsymbol{x} \cdot \nabla U = \alpha U(r)$ and, therefore, $2\,\langle T \rangle_\psi = \alpha\,\langle U \rangle_\psi$. In the case of the spherical oscillator ($a > 0, \alpha = 2$)

one obtains $\langle T\rangle_\psi = \langle U\rangle_\psi = E_{n\ell}/2$; in the case of bound states of the hydrogen atom ($a < 0, \alpha = -1$) one concludes $\langle T\rangle_\psi = -\langle U\rangle_\psi/2$, $\langle U\rangle_\psi = 2E_n$, $\langle T\rangle_\psi = -E_n$. These are the same results as in classical mechanics.

1.12 The nucleus of oxygen ^{16}O contains 8 protons. Two of these are assumed to be in the 0s-state, and 6 in the 0p-state of the spherical oscillator – in accord with the Pauli principle, see also Sect. 4.3.4, Example 4.3 and Fig. 4.6. The mean square charge radius of the nucleus is defined by

$$\left\langle r^2\right\rangle := \frac{1}{Z}\sum_{i=1}^{Z}\left\langle r^2\right\rangle_i\,, \tag{8}$$

where the index i characterizes the states (in our case, 0s and 0p, respectively). From electron scattering on ^{16}O and from the spectroscopy of muonic oxygen one knows that

$$\left[\left\langle r^2\right\rangle_{^{16}\mathrm{O}}\right]^{1/2} = 2.71 \pm 0.02\ \mathrm{fm}\,.$$

Calculate the oscillator parameter b, (1.143), and the energy $\hbar\omega$.
Solution: The potential reads $U(r) = m\omega^2 r^2/2$. Using the result of Exercise 1.11 one obtains $\left\langle r^2\right\rangle_{n\ell} = 2\langle U\rangle_{n\ell}/(m\omega^2) = E_{n\ell}/m\omega^2$, and from these

$$\left\langle r^2\right\rangle_{0\mathrm{s}} = \frac{3}{2}\frac{\hbar\omega}{m\omega^2} = \frac{3}{2}b^2\,,\qquad \left\langle r^2\right\rangle_{0\mathrm{p}} = \frac{5}{2}\frac{\hbar\omega}{m\omega^2} = \frac{5}{2}b^2\,.$$

Thus, one finds

$$\left\langle r^2\right\rangle_{^{16}\mathrm{O}} = \frac{1}{8}\left(2\times\frac{3}{2}b^2 + 6\times\frac{5}{2}b^2\right) = \frac{9}{4}b^2\,.$$

Inserting the experimental value given above one obtains $b = 1.81 \pm 0.01$ fm.

1.13 With p_r as defined in (1.123), show that

$$\hbar^2\boldsymbol{\ell}^2 = r^2(\boldsymbol{p}^2 - p_r^2)\,.$$

This provides an alternative derivation of (1.124).
Hints: Write $\hbar^2\boldsymbol{\ell}^2 = (\boldsymbol{x}\times\boldsymbol{p})\cdot(\boldsymbol{x}\times\boldsymbol{p})$ and use the identity

$$\sum_{i=1}^{3}\varepsilon_{ijk}\varepsilon_{ipq} = \delta_{jp}\delta_{kq} - \delta_{jq}\delta_{kp}\,. \tag{9}$$

Solution: 1. Making use of the formula (9) one shows

$$\hbar^2\boldsymbol{\ell}^2 = \sum_{jk}(x^j p_k x^j p_k - x^j p_k x^k p_j)\,.$$

In the first term on the right-hand side the second and third factors are replaced by $p_k x^j = x^j p_k - \mathrm{i}\hbar\delta_{jk}$. Similarly, in the second term the third and fourth factors are replaced by $x^k p_j = p_j x^k + \mathrm{i}\hbar\delta_{jk}$. This yields

$$\begin{aligned}\hbar^2\boldsymbol{\ell}^2 &= \boldsymbol{x}^2\boldsymbol{p}^2 - \mathrm{i}\hbar(\boldsymbol{x}\cdot\boldsymbol{p}) - (\boldsymbol{x}\cdot\boldsymbol{p})(\boldsymbol{p}\cdot\boldsymbol{x}) - \mathrm{i}\hbar(\boldsymbol{x}\cdot\boldsymbol{p})\\ &= r^2\boldsymbol{p}^2 - (\boldsymbol{x}\cdot\boldsymbol{p})^2 + \mathrm{i}\hbar(\boldsymbol{x}\cdot\boldsymbol{p})\,,\end{aligned}$$

where in the last step, one used $(\boldsymbol{p}\cdot\boldsymbol{x}) = (\boldsymbol{x}\cdot\boldsymbol{p}) - 3\mathrm{i}\hbar$.

2. Calculate then

$$(\boldsymbol{x}\cdot\boldsymbol{p}) = \frac{\hbar}{\mathrm{i}}\sum x^k \frac{\partial}{\partial x^k} = \frac{\hbar}{\mathrm{i}}\frac{\boldsymbol{x}^2}{r}\frac{\partial}{\partial r} = \frac{\hbar}{\mathrm{i}} r \frac{\partial}{\partial r} = r p_r + \mathrm{i}\hbar\,.$$

This is used to calculate

$$\begin{aligned}(\boldsymbol{x}\cdot\boldsymbol{p})^2 - \mathrm{i}\hbar(\boldsymbol{x}\cdot\boldsymbol{p}) &= [(\boldsymbol{x}\cdot\boldsymbol{p}) - \mathrm{i}\hbar](\boldsymbol{x}\cdot\boldsymbol{p}) = r p_r (r p_r + \mathrm{i}\hbar)\\ &= r p_r r p_r + \mathrm{i}\hbar r p_r = r^2 p_r^2\,.\end{aligned}$$

In the last step $p_r r = r p_r - \mathrm{i}\hbar$ was inserted. This proves the asserted decomposition.

1.14 The *classical* Hamiltonian function which is the ananlogue of the decomposition (1.125) is known to be

$$H_{\mathrm{cl}} = \frac{p_r^2}{2m} + \frac{\ell^2}{2mr^2} + U(r)$$

with $p_r = m\dot{r}$ and $\ell = |\boldsymbol{\ell}_{\mathrm{cl}}|$. Choose $\boldsymbol{\ell}_{\mathrm{cl}} = \ell\hat{\boldsymbol{e}}_3$: The classical orbits then lie in the $(1,2)$-plane. The corresponding quantum mechanical motion is no longer confined to this plane. Why is this so? For which quantum numbers does the quantum solution approximate the classical motion in the $(1,2)$-plane?

1.15 To show: A quantum system all of whose excitations are degenerate in energy with the ground state ψ_0, is *frozen*.
Hints: Use the Heisenberg equations of motion to show that the expectation values $\left\langle(\dot{x}^j)^2\right\rangle_{\psi_0}$ are equal to zero.

1.16 Calculate the following expectation values in the hydrogen atom: $\langle r\rangle_{n\ell}$, $\langle r^2\rangle_{n\ell}$, $\langle 1/r\rangle_{n\ell}$.
Solution:

$$\begin{aligned}\langle r\rangle_{n\ell} &= \frac{1}{2}a_{\mathrm{B}}[3n^2 - \ell(\ell+1)]\,,\\ \left\langle r^2\right\rangle_{n\ell} &= \frac{1}{2}n^2 a_{\mathrm{B}}^2[5n^2 + 1 - 3\ell(\ell+1)]\,,\\ \left\langle\frac{1}{r}\right\rangle_{n\ell} &= \frac{1}{n^2 a_{\mathrm{B}}}\,.\end{aligned}$$

1.17 Calculate the standard deviation (Δr) in the bound states of hydrogen which have maximal orbital angular momentum ℓ (these are called circular orbits).
Solution: Using the results of the preceding exercise one has

$$\begin{aligned}\left\langle r^2\right\rangle_{n,n-1} &= \frac{1}{2}n^2(n+1)(2n+1)a_{\mathrm{B}}^2\,,\\ \langle r\rangle_{n,n-1} &= \frac{1}{2}n(2n+1)a_{\mathrm{B}}\,.\end{aligned}$$

These are used to calculate the square of the standard deviation as well as the standard deviation proper,

$$(\Delta r)_{n,n-1} = \frac{1}{2} n \sqrt{2n+1}\, a_B \,.$$

Note that the terms proportional to n^4 cancel in the result for $(\Delta r)^2$. If one had restricted $\langle r^2 \rangle_{n,n-1}$ and $\langle r \rangle^2_{n,n-1}$ to this order one would have found the standard deviation to be zero.

1.18 Multipole interactions in hydrogenlike atoms depend on matrix elements of the operators $1/r^{\lambda+1} Y_{\lambda\mu}$ taken between bound states $R_{n\ell}(r) Y_{\ell m}$ and $R_{n\ell'} Y_{\ell' m}$. In some cases, besides the selection rules which are due to the angular integrals, the radial matrix element vanishes, too. Show that for hydrogen wave functions one obtains

$$\int_0^\infty r^2 \mathrm{d}r\, R_{n\ell}(r) \frac{1}{r^{\lambda+1}} R_{n\ell'}(r) = 0 \quad \text{for} \quad \lambda = |\ell - \ell'| \,.$$

Hints: For a physically relevant example and the proof see T.O. Ericson, F. Scheck, Nucl. Phys. B19 (1970) 450. You should be aware that there are various conventions for Laguerre polynomials: In the convention of Sect. 1.9.5 which is used in many textbooks on quantum mechanics, the associated Laguerre polynomials are denoted by L^σ_μ. Denoting the convention of, say, [Gradshteyn and Ryzhik (1965)] by $\widetilde{L}^z_k$, the relationship reads

$$L^\sigma_\mu(x) = \mu! \widetilde{L}^\sigma_{\mu-\sigma}(x) \,.$$

Exercises: Chap. 2

2.1 Consider an attractive, spherically symmetric square well potential

$$U(r) = -U_0 \Theta(R - r)$$

whose depth is $U_0 > 0$ and whose range is R, see Fig. 1.

1. Determine the solutions with $\ell = 0$ (s-waves) for positive energy E (scattering solutions).
2. In the case $-U_0 < E < 0$ there can be bound states. Discuss qualitatively the position of the eigenvalues of s-states.

Fig. 1

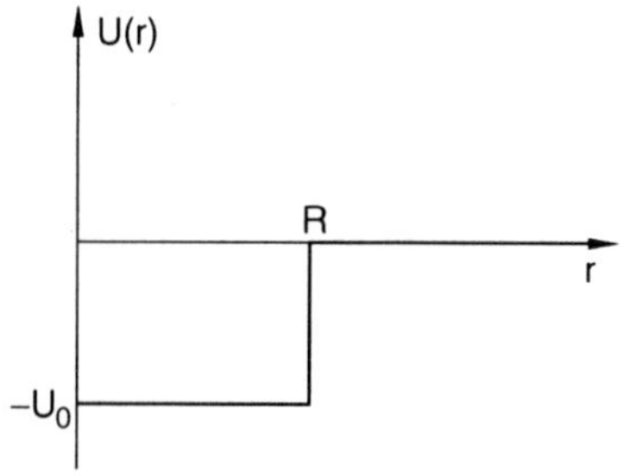

Hints: When $E > 0$ and for $r > R$ use the ansatz $\sin(kr + \delta)$ and determine δ. For $E < 0$ one must have $\mathrm{i}\delta \to \infty$.

Solution: With $\ell = 0$ and $R(r) = u(r)/r$ the differential equation for the function $u(r)$ becomes

$$u''(r) + \frac{2m(E - U(r))}{\hbar^2} u(r) = 0 .$$

1. $E > 0$ (cf. Sect. 2.5.5, Example 2.4): Outside the well, $r > R$, choose $u^{(a)}(r) = \sin(kr + \delta)$ with $k^2 = 2mE/\hbar^2$. Inside the well, $r \leq R$, choose $u^{(i)}(r) = \sin(\kappa r)$ where

$$\kappa^2 = \frac{2m(E + U_0)}{\hbar^2} \equiv k^2 + K^2 , \qquad K^2 = \frac{2m}{\hbar^2} U_0 .$$

The boundary condition at the point $r = R$ reads

$$\left. \frac{u^{(i)\prime}(r)}{u^{(i)}(r)} \right|_{r=R} = \left. \frac{u^{(a)\prime}(r)}{u^{(a)}(r)} \right|_{r=R} .$$

Working this out yields the scattering phase, the scattering length, and the effective range as given in Sect. 2.5.5.

2. $U_0 < E < 0$: Use $E = -B$ with $B > 0$, $\gamma^2 := 2mB/\hbar^2$, from which $\kappa^2 = K^2 - \gamma^2$. An analogous ansatz for the radial function yields

$$u^{(a)}(r) = \sin(\mathrm{i}\gamma r + \delta) = \frac{1}{2\mathrm{i}} \left(\mathrm{e}^{\mathrm{i}\delta} \mathrm{e}^{-\gamma r} - \mathrm{e}^{-\mathrm{i}\delta} \mathrm{e}^{+\gamma r} \right) .$$

There is a bound state only if the exponentially increasing term does not contribute, that is, if $\mathrm{i}\delta \to +\infty$, or, equivalently, if $\cot\delta = \mathrm{i}\coth(\mathrm{i}\delta) \to \mathrm{i}$. The continuity condition at $r = R$ yields the implicit equation

$$\kappa \cot(\kappa R) = -\gamma .$$

Let $x := \kappa R$, $x_0 = \sqrt{2mU_0} R/\hbar$, and $\gamma/\kappa = \sqrt{x_0^2 - x^2}/x$. The bound states are determined from the points of intersection of the curves $y_1(x) = \cot x$ and $y_2(x) = -\sqrt{x_0^2 - x^2}/x$. The graphs of these functions (Fig. 2) show that they intersect in n points (the moduli of which decrease from left to right) when x_0 lies in the interval

$$(2n - 1)\frac{\pi}{2} \leq x_0 < (2n + 1)\frac{\pi}{2} .$$

Fig. 2

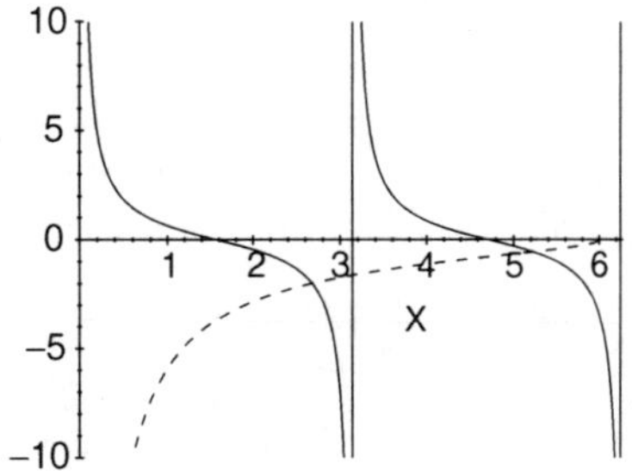

2.2 For the example of the Yukawa potential $U_Y(r) = g\mathrm{e}^{-\mu r}/r$, with $\mu = Mc/\hbar$, and M the mass of the exchanged particle, the scattering amplitude in first Born approximation is given by (2.31). Calculate the integrated cross section

$$\sigma_Y = \int \mathrm{d}\Omega \frac{\mathrm{d}\sigma}{\mathrm{d}\Omega}.$$

Discuss the limit $M \to 0$.

2.3 Given the relation (2.33) between the potential $U(\boldsymbol{x})$ and the density $\varrho(\boldsymbol{x})$, and making use of the limit $\mu \to 0$, show that

$$\begin{aligned} f(\theta) &= -\frac{2m}{4\pi\hbar^2}\int \mathrm{d}^3x\, \mathrm{e}^{\mathrm{i}\boldsymbol{q}\cdot\boldsymbol{x}} U(r) \\ &= -\frac{2mg}{\hbar^2 q^2}\int \mathrm{d}^3x\, \mathrm{e}^{\mathrm{i}\boldsymbol{q}\cdot\boldsymbol{x}} \varrho(\boldsymbol{x}) = -\frac{2mg}{\hbar^2 q^2} F(\boldsymbol{q}), \end{aligned}$$

where $F(\boldsymbol{q})$ is the form factor. After performing this limit $U(\boldsymbol{x})$ becomes the Coulomb potential, $\varrho(\boldsymbol{x})$ becomes the charge density.

Calculate the form factor for the examples

$$\text{(a):}\quad \varrho(\boldsymbol{x}) = \delta(\boldsymbol{x});$$

$$\text{(b):}\quad \varrho(r) = \frac{3}{4\pi R^3}\Theta(R-r).$$

2.4 Let the model (b) of Exercise 2.3 describe a nucleus with charge number Z.

1. Calculate the form factor $F(q^2)$. Determine the mean-square radius, first from the form factor, then directly from the density $\varrho(r)$.
2. If the electron has an energy E which is large as compared to its rest mass (times c^2) then one has, approximately, $E \approx \hbar ck$. Choose $R = 4.933$ fm and $E = 200$ MeV. Sketch the form factor as a function of q.

Solution: The form factor reads

$$F(q^2) = 3\left(\frac{\sin(qR)}{(qR)^3} - \frac{\cos(qR)}{(qR)^2}\right).$$

The mean square radius, calculated either from the Fourier transform of $\varrho(r)$, or by differentiation from $F(q^2)$, is

$$\left\langle r^2\right\rangle = -6\frac{\mathrm{d}F(q^2)}{\mathrm{d}q^2} = \frac{3}{5}R^2.$$

2.5 As a somewhat more extensive exercise, to be done on a PC, integrate numerically the radial Schrödinger equation for the model (b) of Exercise 2.3 with $Z = 8$, following the method described in Sect. 2.3.1. Determine the scattering phases $\delta_\ell(E)$ for various energies of the electron.

2.6 Suppose that in a partial wave analysis of elastic scattering one had

$$\delta_\ell(k) = \arcsin\left\{\left(\frac{k}{b\sqrt{2\ell+1}}\right)^{2\ell+1} \exp\left[-\frac{1}{2}\left(\frac{k^2}{b^2} - (2\ell+1)\right)\right]\right\}.$$

Discuss the behaviour of the amplitude $a_\ell(k)$ in the complex plane, as well as the behaviour of $\sigma_{\text{el}}(k)$ as a function of k.

2.7 Assume the charge distribution of a deformed nucleus to be

$$\varrho(\boldsymbol{x}) = \varrho_0(r) + \varrho_2(r) Y_{20}(\hat{\boldsymbol{x}}) .$$

1. Show: The calculation of the form factor can be reduced to integrals over the radial variable r.
2. Calculate the form factor and the differential cross section (in first Born approximation) for the example

$$\varrho_0(r) = \left(\frac{3}{2\pi R^2}\right)^{3/2} \exp\left(-\frac{3}{2}\frac{r^2}{R^2}\right) , \qquad \varrho_2(r) = \frac{N}{R^2}\delta(r-R) .$$

Hints:

$$j_2(x) = \frac{3\sin x}{x^3} - \frac{3\cos x}{x^2} - \frac{\sin x}{x} , \quad Y_{20}(\theta) = \sqrt{\frac{5}{16\pi}}(3\cos^2\theta - 1) .$$

Solution: 1. Insert the expansion (1.136) of the plane wave into the formula (2.35) for the form factor and use the orthogonality of spherical harmonics, to obtain

$$F(\boldsymbol{q}) = 4\pi \left[\int_0^\infty r^2 \mathrm{d}r\, j_0(qr)\varrho_0(r) - \int_0^\infty r^2 \mathrm{d}r\, j_2(qr)\varrho_2(r) Y_{20}(\hat{\boldsymbol{q}}) \right] .$$

2. The first integral is calculated by means of the formula

$$\int_0^\infty x \mathrm{d}x\, \mathrm{e}^{-ax^2} \sin(bx) = -\frac{\mathrm{d}}{\mathrm{d}b}\int_0^\infty \mathrm{d}x\, \mathrm{e}^{-ax^2}\cos(bx) = \frac{b}{4a}\sqrt{\frac{\pi}{a}}\mathrm{e}^{-b^2/(4a)}$$

and yields $\exp(-q^2R^2/6)$. The second integral yields the integrand at $r = R$. In total, the form factor is seen to be

$$F(\boldsymbol{q}) = \mathrm{e}^{-(1/6)q^2R^2} - 4\pi N j_2(qR) Y_{20}(\hat{\boldsymbol{q}}) .$$

The differential cross section is computed by inserting the expressions for $j_2(qR)$ and for $Y_{20}(\hat{\boldsymbol{q}})$ given above.

Exercises: Chap. 3

3.1 The Pauli matrices read

$$\sigma_1 = \begin{pmatrix} 0 & 1 \\ 1 & 0 \end{pmatrix} , \quad \sigma_2 = \begin{pmatrix} 0 & -\mathrm{i} \\ \mathrm{i} & 0 \end{pmatrix} , \quad \sigma_3 = \begin{pmatrix} 1 & 0 \\ 0 & -1 \end{pmatrix} . \tag{10}$$

Verify or show, respectively:

1. The matrices σ_i are hermitian als well as unitary. Every hermitian 2×2-matrix M can be written as a linear combination of the Pauli matrices and the unit matrix $\mathbb{1}_{2\times 2}$ with real coefficients. If M has trace zero then it is a linear combination of Pauli matrices only.

2. Using the short-hand notation $\boldsymbol{\sigma} \equiv (\sigma_1, \sigma_2, \sigma_3)$ the Pauli matrices obey the relations

$$\sigma_j \sigma_k = \delta_{jk} + \mathrm{i} \sum_{l=1}^{3} \varepsilon_{jkl} \sigma_l ,$$

$$(\boldsymbol{\sigma} \cdot \boldsymbol{a})(\boldsymbol{\sigma} \cdot \boldsymbol{b}) = (\boldsymbol{a} \cdot \boldsymbol{b})\, \mathbb{1}_{2\times 2} + \mathrm{i}\boldsymbol{\sigma} \cdot (\boldsymbol{a} \times \boldsymbol{b}) .$$

3. With $\omega = |\boldsymbol{\omega}|$ and $\hat{\boldsymbol{\omega}} = \boldsymbol{\omega}/\omega$ one has

$$\exp(\mathrm{i}\boldsymbol{\omega} \cdot \boldsymbol{\sigma}) = \mathbb{1} \cos\omega + \mathrm{i}\hat{\boldsymbol{\omega}} \cdot \boldsymbol{\sigma} \sin\omega .$$

4. Evaluate $(\boldsymbol{\sigma} \cdot \boldsymbol{\ell})(\boldsymbol{\sigma} \cdot \boldsymbol{\ell})$ keeping track of the fact that the components of $\boldsymbol{\ell}$ do not commute.

3.2 Show: Every unitary 2×2-matrix U can be written as an exponential series in $(\mathrm{i}H)$, where $H = H^{\dagger}$, i.e. $U = \exp(\mathrm{i}H)$.

3.3 Let A and B be operators defined on the same domain.

1. If they commute with their commutator, i.e. $[A, [A, B]] = 0 = [B, [A, B]]$, then, formally,

$$e^{A+B} = e^A e^B e^{-(1/2)[A,B]} . \tag{11}$$

2. Show that

$$e^A B e^{-A} = B + [A, B] + \frac{1}{2!}[A, [A, B]] + \dots . \tag{12}$$

Hints:

1. Define $F(x) := e^{x(A+B)} e^{-xB} e^{-xA}$ with x real, and show that $F(x)$ obeys the differential equation $F'(x) = -x\,[A, B]\,F(x)$.
2. Define

$$G(x) := e^{xA} B e^{-xA} = \sum_{n=0}^{\infty} \frac{x^n}{n!} G^{(n)}(0)$$

with $x \in \mathbb{R}$ and calculate the first and second derivatives of $G(x)$.

Solution: 1. With $F(x)$ as given, one calculates the derivative by x without shifting A or B past one another,

$$\begin{aligned} F'(x) &= e^{x(A+B)}[A e^{-xB} e^{-xA} - e^{-xB} A e^{-xA}] \\ &= e^{x(A+B)}[A - e^{-xB} A e^{xB}] e^{-xB} e^{-xA} \\ &= e^{x(A+B)} x\,[B, A] e^{-xB} e^{-xA} = -x\,[A, B] F(x) . \end{aligned}$$

This differential equation is easily solved: $F(x) = F(0) \exp(-(x^2/2) \times [A, B])$ with the initial condition $F(0) = \mathbb{1}$.

2. One has $G(0) = B$, $G'(0) = [A, G]$, $G''(0) = [A, G'] = [A, [A, G]]$.

3.4 With q and p two canonically conjugate variables, Q and P the associated self-adjoint operators on Hilbert space, define one-parameter groups of unitary operators

$$U_r^{(q)} := \mathrm{e}^{-(\mathrm{i}/\hbar)rQ}\,, \qquad U_s^{(p)} := \mathrm{e}^{-(\mathrm{i}/\hbar)sP}\,,$$

with real r and s.

1. Q is defined on all $f \in \mathcal{H}$ for which the limit
$$\lim_{r\to 0} \frac{\mathrm{i}\hbar}{r}\left[U_r^{(q)} - 1\!\!1\right] f = Qf$$
exists. (An analogous statement applies to P).
2. Postulate Weyl's commutation relation
$$U_r^{(q)} U_s^{(p)} = \mathrm{e}^{-(\mathrm{i}/\hbar)rs}\, U_s^{(p)} U_r^{(q)}\,. \tag{13}$$
Show that this is equivalent to $[P, Q] = (\hbar/\mathrm{i})\,1\!\!1$.

3.5 Let A and B be hermitian matrices. Under which assumption is their product AB hermitian? Show that their commutator $C := [A, B]$ is antihermitian.

3.6 At $t = 0$ and using the creation and annihilation operators (1.74)–(1.75), the coherent states of the example Sect. 1.8.2 can be written in the form

$$|\psi_z\rangle = \mathrm{e}^{-|z|^2/2}\mathrm{e}^{za^\dagger}\,|0\rangle$$

where $z \equiv z(0) = r\mathrm{e}^{-\mathrm{i}\phi(0)}$.

1. Show that
$$\mathrm{e}^{-za^\dagger} a \mathrm{e}^{za^\dagger} = a + z\,.$$
2. Show that ψ_z is an eigenstate of the annihilation operator a and calculate the eigenvalue.
3. If $F(a)$,$G(a^\dagger)$, $:\!H(a^\dagger, a)\!:$ are polynomials in their arguments what are the expectation values of these operators in the state ψ?
4. Any two states ψ_z and ψ_w with $w \neq z$, are not orthogonal. Calculate the transition matrix element.

Solution: 1. By the formula (12) and with $[a, a^\dagger] = 1$ one has

$$\mathrm{e}^{-za^\dagger}\, a\, \mathrm{e}^{za^\dagger} = a - z[a^\dagger, a] = a + z\,.$$

2. This result is used to show that
$$\begin{aligned} a|\psi_z\rangle &= \mathrm{e}^{-|z|^2/2}\mathrm{e}^{za^\dagger}(\mathrm{e}^{-za^\dagger} a \mathrm{e}^{za^\dagger})|0\rangle \\ &= \mathrm{e}^{-|z|^2/2}\mathrm{e}^{za^\dagger}(a + z)|0\rangle \\ &= z\,|\psi_z\rangle\,, \end{aligned}$$
where $a|0\rangle = 0$ was inserted.
3. Let $F(a)$ be a polynomial in a. By the previous result, one obtains the relation
$$\langle\psi_z|F(a)|\psi_z\rangle = F(z)$$
For a polynomial $G(a^\dagger)$ one shifts $a^\dagger$ as well as all powers of $a^\dagger$ to the "bra"-state, thus obtaining
$$\langle\psi_z|G(a^\dagger)|\psi_z\rangle = \langle[G^*(a)\psi_z]|\psi_z\rangle = \langle\psi_z|[G^*(a)\psi_z]\rangle^* = G(z^*)\,.$$
In a mixed polynomial all $a^\dagger$ are positioned to the left of all a, because of the prescription of normal ordering. Therefore, the above formulae imply
$$\langle\psi_z|\!:\!H(a^\dagger, a)\!:\!|\psi_z\rangle = H(z^*, z)\,.$$

4. The above results can be extended to exponential series. Thus, the overlap of two different states is calculated to be

$$\begin{aligned}\langle\psi_z|\,\psi_w\rangle &= \mathrm{e}^{-|w|^2/2}\langle\psi_z|\mathrm{e}^{wa^\dagger}|0\rangle\\ &= \mathrm{e}^{-|w|^2/2+|z|^2/2}\langle\psi_z|\mathrm{e}^{(w-z)a^\dagger}|\psi_z\rangle\\ &= \mathrm{e}^{-|w|^2/2+|z|^2/2+(w-z)z^*} = \mathrm{e}^{-(1/2)(|w|^2+|z|^2-2wz^*)}\,.\end{aligned}$$

The square of the modulus of this matrix element is $|\langle\psi_z|\,\psi_w\rangle|^2 = \exp(-|z-w|^2)$.

3.7 In a two-dimensional Hilbert space with basis $(|1\rangle, |2\rangle)^T$ the following matrices are given

$$A=\frac{1}{2}\begin{pmatrix}1&1\\1&1\end{pmatrix},\qquad B=\begin{pmatrix}0&1\\0&1\end{pmatrix},\qquad C=\begin{pmatrix}0&0\\0&\mathrm{i}\end{pmatrix},$$

$$D=\frac{1}{4}\begin{pmatrix}1&1\\1&3\end{pmatrix},\qquad E=\begin{pmatrix}0&0\\0&1\end{pmatrix},\qquad F=\frac{1}{3}\begin{pmatrix}1&0\\0&2\end{pmatrix}.$$

Which among these can be density matrices, which cannot? Which of them describe pure states, which describe mixed ensembles?

3.8 Solve the integral equation (3.30) by iteration for the given initial condition and express $U(t,t_0)$ by integrals over time-ordered products of the Hamiltonian.

3.9 The Hamiltonian of a physical system is assumed to have a purely discrete spectrum, $H|n\rangle = E_n|n\rangle$, $n\in\mathbb{N}_0$.

1. Prove: For every self-adjoint operator $\mathcal{O}$ which is defined on the states $|n\rangle$,

$$S := \sum_{n=0}^{\infty}(E_n-E_0)\big|\langle n|\mathcal{O}|0\rangle\big|^2 = \frac{1}{2}\langle 0|\,[\mathcal{O},[H,\mathcal{O}]]|0\rangle\,.$$

2. Calculate the quantity S for the example $\mathcal{O}=x$ and the Hamiltonian $H = p^2/(2m)+U(x)$ (in one dimension).

Exercises: Chap. 4

4.1 1. Construct the density matrix for the linear combination of eigenstates of $\boldsymbol{s}$ and s_3

$$|\chi\rangle = \cos\alpha\left|\frac{1}{2},\frac{1}{2}\right\rangle + \sin\alpha\,\mathrm{e}^{\mathrm{i}\beta}\left|\frac{1}{2},-\frac{1}{2}\right\rangle\,.$$

Calculate the expectation values of the observables

$$P(\theta,\phi) = \sin\theta\cos\phi\, s_1 + \sin\theta\sin\phi\, s_2 + \cos\theta\, s_3\,.$$

For which values of θ and ϕ is the polarization equal to 1?

2. Let the statistical operator

$$W = \cos^2\alpha\, P_{+1/2} + \sin^2\alpha\, P_{-1/2}$$

be assigned to a mixed ensemble. What is the magnitude of the polarization in the direction $\hat{\boldsymbol{n}}$ which has polar angles (θ,ϕ)?

4.2 In the subspace spanned by the eigenfunctions $|1\ m\rangle$ of $\boldsymbol{\ell}^2$ and of ℓ_3 the following matrix is given,

$$\varrho = \begin{pmatrix} u & v & -u \\ -v & 2u & v \\ -u & -v & u \end{pmatrix} \quad \text{with} \quad u = \frac{1}{4}\,, \quad v = \frac{\mathrm{i}}{2\sqrt{2}}\,.$$

Verify that this may be a density matrix for a state with orbital angular momentum 1. Determine the nature of the state, pure or mixed, and determine the eigenvalues of ϱ.

4.3 Assume a beam of particles with spin 1/2 to be characterized by its polarization $\boldsymbol{P} = \langle \boldsymbol{s} \rangle$ such that its first two cartesian components are $P_1 = 0.5$ and $P_2 = 0$. What is the maximal or minimal absolute value of the third component P_3? Can this be a pure state? If yes, what is its density matrix?

4.4 Show: Every linear combination of spin-1/2 states of the kind

$$a_+ \left| \frac{1}{2}, \frac{1}{2} \right\rangle + a_- \left| \frac{1}{2}, -\frac{1}{2} \right\rangle$$

is fully polarized. Determine the direction of polarization.

4.5 Polarized beams

1. A beam of protons is assumed to have fixed momentum $\boldsymbol{p}$ and to be 30 % polarized in the positive 3-direction. Construct the statistical operator which describes this beam.
2. Find out whether the density matrix

$$\varrho = \begin{pmatrix} \cos^2\theta/2 & \sin\theta/2\cos\theta/2 \\ \sin\theta/2\cos\theta/2 & \sin^2\theta/2 \end{pmatrix}$$

describes a pure state or a mixed ensemble. Calculate the expectation value of the observable

$$\mathcal{O} = \cos\alpha\, s_3 + \sin\alpha\, s_1$$

in this state. By an appropriate choice of the parameter α one can identify this state.

4.6 The partial waves for a particle with spin 1/2 can be classified by the total angular momentum $\boldsymbol{j}^2$ for whose eigenvalues $j(j+1)$, with fixed ℓ, j takes the values $j_1 = \ell + 1/2$ and $j_2 = \ell - 1/2$. Show that the operators

$$\Pi_{j_1} := \frac{1}{2\ell+1}(\ell + 1 + \boldsymbol{\sigma}\cdot\boldsymbol{\ell})\,, \qquad \Pi_{j_2} := \frac{1}{2\ell+1}(\ell - \boldsymbol{\sigma}\cdot\boldsymbol{\ell}) \tag{14}$$

are projection operators and that they project onto states with sharp j.
Solution: One has to show:

$$\Pi_{j_1} + \Pi_{j_2} = 1\,, \quad \Pi_{j_1}\Pi_{j_2} = 0 = \Pi_{j_2}\Pi_{j_1}\,, \quad \Pi_{j_k}^2 = \Pi_{j_k}\,, \quad k = 1, 2\,.$$

The first property is obvious. In proving the second property one uses

$$(\boldsymbol{\sigma}\cdot\boldsymbol{\ell})(\boldsymbol{\sigma}\cdot\boldsymbol{\ell}) = \boldsymbol{\ell}^2 - (\boldsymbol{\sigma}\cdot\boldsymbol{\ell})\,,$$

(cf. Exercise 3.1), to calculate

$$\Pi_{j_1}\Pi_{j_2} = \frac{1}{(2\ell+1)^2}[\ell(\ell+1) - (\boldsymbol{\sigma}\cdot\boldsymbol{\ell}) - \boldsymbol{\ell}^2 + (\boldsymbol{\sigma}\cdot\boldsymbol{\ell})]\,,$$

which, indeed, gives zero when applied to a state with sharp ℓ. The product of factors taken in the inverse order gives zero, too. It remains to verify that Π_{j_1} projects onto $j_1 = \ell + 1/2$, while Π_{j_2} projects onto $j_2 = \ell - 1/2$. In order to show this, replace the denominators in (14) as follows

$$2\ell + 1 = j_1(j_1+1) - j_2(j_2+1)$$

and use the formula $2\boldsymbol{s}\cdot\boldsymbol{\ell} = \boldsymbol{j}^2 - \boldsymbol{\ell}^2 - \boldsymbol{s}^2$ with $\boldsymbol{s} = \boldsymbol{\sigma}/2$, in order to replace $\boldsymbol{\sigma}\cdot\boldsymbol{\ell}$ acting on eigenstates of $\boldsymbol{j}^2$, by its eigenvalues $j(j+1) - \ell(\ell+1) - 3/4$. One obtains

$$\Pi_{j_1} = \frac{j(j+1) - j_2(j_2+1)}{j_1(j_1+1) - j_2(j_2+1)}\,, \qquad \Pi_{j_2} = \frac{j_1(j_1+1) - j(j+1)}{j_1(j_1+1) - j_2(j_2+1)}\,,$$

which confirm the assertion.

4.7 In the space of spinors and using the operators (14) the partial wave series of the elastic scattering amplitude of a particle with spin 1/2 has the form

$$\hat{f}(k,\theta) = \sum_{\ell=0}^{\infty}(2\ell+1)\{f_{j_1}\Pi_{j_1} + f_{j_2}\Pi_{j_2}\}P_\ell(\cos\theta)\,. \tag{15}$$

The scattering amplitude proper which must be a scalar, is obtained by evaluating $\hat{f}$ between Pauli spinors χ_p and χ_q. Show that $\hat{f}$ can be transformed as follows

$$\begin{aligned}\hat{f}(k,\theta) &= \sum_{\ell=0}^{\infty}\{\ell f_{j_2}(k) + (\ell+1)f_{j_1}(k)\}P_\ell(\cos\theta)\\ &\quad - \mathrm{i}\boldsymbol{\sigma}\cdot\left(\hat{\boldsymbol{k}}'\times\hat{\boldsymbol{k}}\right)\sum_{\ell=0}^{\infty}\{f_{j_1}(k) - f_{j_2}(k)\}P'_\ell(\cos\theta)\,.\end{aligned} \tag{16}$$

In the case where the amplitudes do not depend on spin, this expansion reduces to the one for spinless particles.

Solution: The amplitude must be evaluated in momentum space. Thus, one has

$$(\boldsymbol{\sigma}\cdot\boldsymbol{\ell})P_\ell(\hat{\boldsymbol{k}}'\cdot\hat{\boldsymbol{k}}) = \boldsymbol{\sigma}\cdot(-\mathrm{i}\hat{\boldsymbol{k}}'\times\nabla_{k'})P_\ell(\hat{\boldsymbol{k}}'\cdot\hat{\boldsymbol{k}}) = -\mathrm{i}\boldsymbol{\sigma}\cdot\left(\hat{\boldsymbol{k}}'\times\hat{\boldsymbol{k}}\right)P'_\ell(\hat{\boldsymbol{k}}'\cdot\hat{\boldsymbol{k}})\,.$$

4.8 The optical theorem in the examples of Exercises 4.6 and 4.7 derives from the unitarity relation

$$\operatorname{Im} f(\boldsymbol{k}',\boldsymbol{k}) = \frac{k}{4\pi}\sum_{\text{spins}}\int \mathrm{d}\Omega_{k''}\, f^*(\boldsymbol{k}',\boldsymbol{k}'')f(\boldsymbol{k}'',\boldsymbol{k})\,. \tag{17}$$

Show that every partial wave amplitude f_{j_i}, on its own, satisfies the unitarity relation

$$\operatorname{Im} f_{j_i}(k) = k\left|f_{j_i}(k)\right|^2, \qquad i = 1,2\,. \tag{18}$$

This implies that it can be expressed by means of scattering phases.

Solution: One evaluates $\hat{f}$ between two Pauli spinors and uses the addition theorem (1.121) of spherical harmonics. The integration over $\mathrm{d}\Omega_{k''}$ is simplified by the orthogonality of spherical harmonics. One obtains

$$4\pi \left\langle \chi_p \right| Y^*_{\ell m}(\hat{\boldsymbol{k}}') \left\{ \operatorname{Im} f_{j_1} \Pi_{j_1} + \operatorname{Im} f_{j_2} \Pi_{j_2} \right\} Y_{\ell m}(\hat{\boldsymbol{k}}) | \chi_q \rangle$$
$$= 4\pi k \langle \chi_p | Y^*_{\ell m}(\hat{\boldsymbol{k}}') [f^*_{j_1} \Pi_{j_1} + f^*_{j_2} \Pi_{j_2}][f_{j_1} \Pi_{j_1} + f_{j_2} \Pi_{j_2}] Y_{\ell m}(\hat{\boldsymbol{k}}) \left| \chi_q \right\rangle .$$

By the properties of the projection operators proven in Exercise 4.6, and by comparing coefficients, one obtains the relations (18). These, in turn, imply that one can write

$$f_{j_k} = \frac{1}{2ik}(\mathrm{e}^{2\mathrm{i}\delta_{j_k}(k)} - 1) = \frac{1}{k}\mathrm{e}^{\mathrm{i}\delta_{j_k}(k)} \sin \delta_{j_k}(k) .$$

4.9 In the subspace pertaining to the eigenvalue 2 of the operator $\boldsymbol{\ell}^2$, i.e. with $\ell = 1$, construct the 3×3-matrix $\langle 1m' | \ell_2 | 1m \rangle$ and show that

$$\exp(-\mathrm{i}\alpha \ell_2) = \mathbb{1} - \mathrm{i} \sin \alpha \, \ell_2 - (1 - \cos \alpha) \, \ell_2^2 .$$

4.10 Show: A beam of identical particles whose polarization $\boldsymbol{P}$ has a given modulus P, can be represented in the form

$$\varrho = \frac{1-P}{2} \mathbb{1} + P \mathbf{D}^{(1/2)}(\mathbf{R}) \begin{pmatrix} 1 & 0 \\ 0 & 0 \end{pmatrix} \mathbf{D}^{(1/2)\dagger}(\mathbf{R}) .$$

In which direction does the vector $\boldsymbol{P}$ point?

Exercises: Chap. 5

5.1 To the Hamiltonian of the hydrogen atom add the perturbation $H' = \hbar^2 C/(2mr^2)$ with positive constant C.

1. Determine the exact eigenvalues of $H + H'$ by adding H' to the centrifugal term. Is the degeneracy (2p–2s) lifted?
2. Calculate the displacement of the energy of the ground state in first order perturbation theory. Compare with the exact result.

5.2 Let the charge density of the lead nucleus ($Z = 82$) be described by a homogeneous distribution with radius $R = 6.5$ fm,

$$\varrho(r) = \frac{3Ze}{4\pi R^3} \Theta(R - r) .$$

Within first order of perturbation theory calculate the difference of the binding energies of the states 1s, 2s, and 2p, as compared to the pure Coulomb potential.

5.3 Compare the state 1s in a lead atom ($Z = 82$) with the corresponding 1s-state of *muonic* lead. How, approximately, does the potential at the position of the muon change when the electronic 1s-state of the host atom is occupied by one electron? How does the potential at the position of the electron change due to the presence of the muon?

5.4 Consider stationary perturbation theory for the case of a given eigenstate $|n\rangle$ of H_0. In first order one must have $c_n^{(1)} = \mathrm{e}^{\mathrm{i}\alpha} - 1$ with $\alpha \in \mathbb{R}$. Verify to second order included, that the results of the perturbation series remain unaffected by the choice of the free parameter α.
Hints: Show: The energy shifts are independent of α; in second order the resulting wave function

$$\psi \approx (1 + c_n^{(1)})|n\rangle + \sum_{k \neq n} c_k^{(1)} |k\rangle \,,$$

as a whole, can be multiplied by $\mathrm{e}^{-\mathrm{i}\alpha}$.

5.5 Since the nucleus of the host atom has a finite extension, the lowest bound states of the muonic atom are shifted as compared to the pure Coulomb potential. Interpret the difference

$$\Delta U(r) = -Ze^2 \left(\int \mathrm{d}^3 x' \, \frac{\varrho(r')}{|\boldsymbol{x} - \boldsymbol{x}|'} - \frac{1}{r} \right)$$

as a perturbation added to the Hamiltonian H_0 of the muonic atom in the field of a pointlike nucleus with charge Z. Calculate the shift of the muonic 1s-level to first order. In the case of light nuclei, show that it is proportional to the mean-square radius of the nucleus.

5.6 In a muonic atom we wish to compare two potentials U_1 and U_2 which differ in the nuclear interior only. Show: Expanding the muonic density according to $|\psi|^2 \approx a + br + cr^2$, the binding energies differ by

$$\Delta E \approx Ze^2 \frac{2\pi}{3} a \left(\Delta\left\langle r^2 \right\rangle + \frac{b}{2a} \Delta\left\langle r^3 \right\rangle + \frac{3c}{10a} \Delta\left\langle r^4 \right\rangle \right) .$$

In this expression $\Delta \left\langle r^\alpha \right\rangle$ denotes the change of the nuclear moment α. Derive an approximate expression for the energy difference $E(2\mathrm{p}) - E(2\mathrm{s})$.
Solution: By applying partial integration twice the difference $U_1 - U_2$ of the potentials is replaced by $\Delta(U_1 - U_2)$, so that one has

$$\begin{aligned} I &:= \int_0^\infty r^2 \mathrm{d}r \, (a + br + cr^2)(U_1 - U_2) \\ &= \int_0^\infty r^2 \mathrm{d}r \left(\frac{a}{6} r^2 + \frac{b}{12} r^3 + \frac{c}{20} r^4 \right) \Delta(U_1 - U_2) \,. \end{aligned}$$

Inserting now the Poisson equation, $\Delta(U_1 - U_2) = -4\pi Ze(\varrho_1 - \varrho_2)$, one obtains

$$\begin{aligned} \Delta E &= \mathrm{e} \int \mathrm{d}^3 x \, |\psi|^2 (U_1 - U_2) \\ &\approx -4\pi Ze^2 \left(\frac{a}{6} \Delta\left\langle r^2 \right\rangle + \frac{b}{12} \Delta\left\langle r^3 \right\rangle + \frac{c}{20} \Delta\left\langle r^4 \right\rangle \right) . \end{aligned}$$

The expansion coefficients are determined from the explicit form of the wave functions of the 2s- and 2p-states, respectively,

$$a^{(2s)} = \frac{1}{2a_B^3}\,, \qquad b^{(2s)} = -\frac{1}{a_B^4}\,, \qquad c^{(2s)} = \frac{7}{8a_B^5}\,;$$

$$a^{(2p)} = 0\,, \qquad b^{(2p)} = 0\,, \qquad c^{(2p)} = \frac{1}{12a_B^5}\,.$$

5.7 The fine structure in an atom with electric potential $U(r)$ is caused by the operator

$$U_{FS} = \frac{\hbar^2}{2m^2c^2}\frac{1}{r}\frac{\mathrm{d}U(r)}{\mathrm{d}r}\,\boldsymbol{\ell}\cdot\boldsymbol{s}\,. \tag{19}$$

Calculate the fine structure splitting for circular orbits in hydrogenlike atoms, in first order of perturbation theory.
Solution: The identity $2\boldsymbol{\ell}\cdot\boldsymbol{s} = \boldsymbol{j}^2 - \boldsymbol{\ell}^2 - \boldsymbol{s}^2$ yields the expectation value

$$\langle\boldsymbol{\ell}\cdot\boldsymbol{s}\rangle_{n\ell} = \frac{1}{2}\left[j(j+1) - \ell(\ell+1) - \frac{3}{4}\right].$$

For circular orbits insert here $\ell = n-1$. The radial matrix element of $1/r^3$ is calculated to be

$$\left\langle\frac{1}{r^3}\right\rangle_{n,n-1} = \frac{2}{a_B^3}\left[n^4(2n-1)(n-1)\right]^{-1}.$$

Thus, to first order of perturbation theory one obtains

$$\Delta E = \frac{(Z\alpha)^4}{2n^4(n-1)}\,mc^2\,.$$

Exercises: Chap. 6

6.1 In which way are the wave functions of the electron normalized in the first and the third example of Sect. 6.1.1?

6.2 Consider Klein's group $\{\mathbb{1}, \mathbf{P}, \mathbf{T}, \mathbf{PT}\}$, i.e. the group which contains the identity of the Lorentz group, space reflection, time reversal, and the product of space reflection and time reversal, in the light of Remark 2 following Wigner's theorem (Theorem 6.1). State in which way the elements of this group are represented in Hilbert space.

6.3 Confirm the relation (6.14) by direct evaluation of the exponential series.

6.4 Haar measure for $SU(2)$: Show that the weight function $\varrho(\phi,\theta,\psi)$ in the volume element $\mathrm{d}\mathbf{R} = \varrho(\phi,\theta,\psi)\,\mathrm{d}\phi\,\mathrm{d}\theta\,\mathrm{d}\psi$ of the space of Eulerian angles equals $\sin\theta$.

Fig. 3

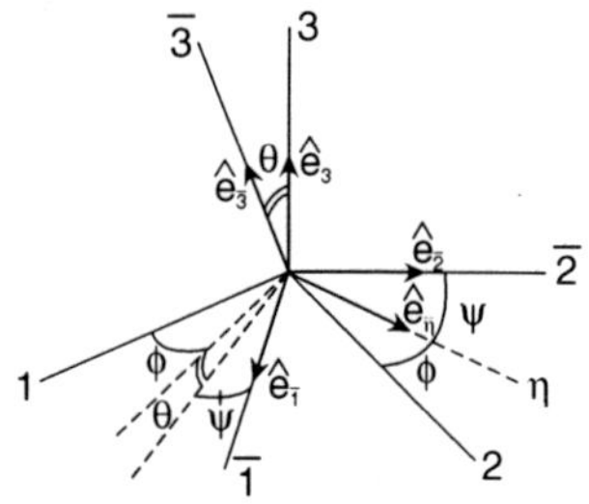

Solution: From Fig. 3 one reads off the relation (using notations as in the figure)

$$\hat{\boldsymbol{e}}_3\,\mathrm{d}\phi + \hat{\boldsymbol{e}}_\eta\,\mathrm{d}\theta + \hat{\boldsymbol{e}}_{\bar{3}}\,\mathrm{d}\psi = \hat{\boldsymbol{e}}_{\bar{1}}\,\mathrm{d}x'^{1} + \hat{\boldsymbol{e}}_{\bar{2}}\,\mathrm{d}x'^{2} + \hat{\boldsymbol{e}}_{\bar{3}}\,\mathrm{d}x'^{3}\,.$$

Multiply this equation from the left by $\hat{\boldsymbol{e}}_{\bar{1}}$, then by $\hat{\boldsymbol{e}}_{\bar{2}}$, and finally by $\hat{\boldsymbol{e}}_{\bar{3}}$, insert the scalar products of the unit vectors, as taken from the figure. This yields

$$\begin{aligned}
&\left(\hat{\boldsymbol{e}}_{\bar{1}}, \hat{\boldsymbol{e}}_3\right) = -\sin\theta\cos\psi\,, && \left(\hat{\boldsymbol{e}}_{\bar{1}}, \hat{\boldsymbol{e}}_\eta\right) = \sin\psi\,, && \left(\hat{\boldsymbol{e}}_{\bar{1}}, \hat{\boldsymbol{e}}_{\bar{3}}\right) = 0\,,\\
&\left(\hat{\boldsymbol{e}}_{\bar{2}}, \hat{\boldsymbol{e}}_3\right) = \sin\theta\sin\psi\,, && \left(\hat{\boldsymbol{e}}_{\bar{2}}, \hat{\boldsymbol{e}}_\eta\right) = \cos\psi\,, && \left(\hat{\boldsymbol{e}}_{\bar{2}}, \hat{\boldsymbol{e}}_{\bar{3}}\right) = 0\,,\\
&\left(\hat{\boldsymbol{e}}_{\bar{3}}, \hat{\boldsymbol{e}}_3\right) = \cos\theta\,, && \left(\hat{\boldsymbol{e}}_{\bar{3}}, \hat{\boldsymbol{e}}_\eta\right) = 0\,, && \left(\hat{\boldsymbol{e}}_{\bar{3}}, \hat{\boldsymbol{e}}_{\bar{3}}\right) = 1\,.
\end{aligned}$$

From these one obtains the relations

$$\begin{aligned}
\mathrm{d}x'^{1} &= -\sin\theta\cos\psi\,\mathrm{d}\phi + \sin\psi\,\mathrm{d}\theta\,,\\
\mathrm{d}x'^{2} &= \sin\theta\sin\psi\,\mathrm{d}\phi + \cos\psi\,\mathrm{d}\theta\,,\\
\mathrm{d}x'^{3} &= \cos\theta\,\mathrm{d}\phi + \mathrm{d}\psi\,.
\end{aligned} \tag{20}$$

Equation (20) yields the Jacobi determinant for $\left(\phi, \theta, \psi\right) \mapsto \left(x'^{1}, x'^{2}, x'^{3}\right)$:

$$\frac{\partial\left(x'^{1}, x'^{2}, x'^{3}\right)}{\partial\left(\phi, \theta, \psi\right)} = \det\begin{pmatrix} -\sin\theta\cos\psi & \sin\psi & 0\\ \sin\theta\sin\psi & \cos\psi & 0\\ \cos\theta & 0 & 1\end{pmatrix} = -\sin\theta\,.$$

Thus, the Jacobi determinant for the transformation relating $\mathbf{R} = \left(\phi, \theta, \psi\right)$ to $\mathbf{R}' = \left(\phi', \theta', \psi'\right)$ is given by

$$\frac{\partial\left(\phi, \theta, \psi\right)}{\partial\left(\phi', \theta', \psi'\right)} = \frac{\partial\left(\phi, \theta, \psi\right)}{\partial\left(x'^{1}, x'^{2}, x'^{3}\right)}\,\frac{\partial\left(x'^{1}, x'^{2}, x'^{3}\right)}{\partial\left(\phi', \theta', \psi'\right)} = \frac{\sin\theta'}{\sin\theta}\,. \tag{21}$$

Conclusion: Up to a constant that may be chosen to be 1, one obtains $\varrho(\phi, \theta, \psi) = \sin\theta$.

6.5 Denote the 2×2 unit matrix and the three Pauli matrices as follows $\{\sigma_\mu\} = \left(\mathbb{1}, \sigma^{(i)}\right)$, respectively. Let

$$\mathbf{A} := \sum_{0}^{3} a_\mu \sigma_\mu\,, \quad a_\mu \in \mathbb{R}\,.$$

By means of the known relations for Pauli matrices prove the relation:

$$\sum_{i=1}^{3} \sigma_i \mathbf{A} \sigma_i = 2\big(\text{tr}\,\mathbf{A}\big)\,\mathbb{1} - \mathbf{A}\,.$$

Show that the two $SU(2)$-partners of a given element $\mathbf{R} \in SO(3)$ are given by

$$\mathbf{U} = \pm \frac{1}{2\sqrt{1 + \text{tr}\,\mathbf{R}}}\Big(\mathbb{1} + \sum_{i,k} R_{ik}\sigma_i\sigma_k\Big)\,. \tag{22}$$

Solution: By means of the known anticommutator

$$\sigma_i\sigma_j + \sigma_j\sigma_i = 2\delta_{ij}$$

and the relation $\big(\sigma_i\big)^2 = \mathbb{1}$ that follows from it, one calculates

$$\sum_{i=1}^{3} \sigma_i \mathbf{A}\sigma_i = 3a_0\,\mathbb{1} + \sum_{j,i=1}^{3} a_j\sigma_i\sigma_j\sigma_i = 3a_0\,\mathbb{1} - \sum_{j=1}^{3} a_j\sigma_j = 4a_0\,\mathbb{1} - \sum_{\mu=0}^{3} a_\mu\sigma_\mu\,.$$

From this relation and from $\text{tr}\,\mathbf{A} = 2a_0$ one has

$$\sum_{i=1}^{3} \sigma_i \mathbf{A}\sigma_i = 4a_0\,\mathbb{1} - \mathbf{A} = 2\big(\text{tr}\,\mathbf{A}\big)\,\mathbb{1} - \mathbf{A}\,.$$

The transformation formula (6.16) reads more explicitly

$$\sum_i x'^i\sigma_i = \mathbf{U}\sum_j x^j\sigma_j\mathbf{U}^\dagger = R_{ik}\sigma_i x^k\,.$$

Comparing the coefficients of x^k in the second and third expression one concludes $\mathbf{U}\sigma_k\mathbf{U}^\dagger = \sum_i R_{ik}\sigma_i$. This is multiplied by σ_k from the right, and the sum is taken over k from 1 to 3, to obtain

$$\sum_{i,k} R_{ik}\sigma_i\sigma_k = \mathbf{U}\sum_k \sigma_k\mathbf{U}^\dagger\sigma_k = \mathbf{U}\Big\{2\big(\text{tr}\,\mathbf{U}\big)\,\mathbb{1} - \mathbf{U}^\dagger\Big\} = 2\big(\text{tr}\,\mathbf{U}\big)\mathbf{U} - \mathbb{1}\,.$$

This proves the formula, upon using the relation $\text{tr}\,\mathbf{U} = \sqrt{1 + \text{tr}\,\mathbf{R}}$ which follows from the same equation and from $\text{tr}\,\big(\sigma_i\sigma_k\big) = 2\delta_{ik}$.

6.6 The wave functions of the quantized, symmetric top are given by (6.43). Making use of the symmetry properties of the D-functions, show that the multiplicity in the choice of axes for the body-fixed system leads indeed to these functions.

6.7 Write the formula (6.70) as the coupling of two equal angular momenta to total angular momentum zero.

6.8 Prove the commutators of the generators $\mathbf{J}_3$ and $\mathbf{J}_\pm$ with spherical tensor operators of rank κ,

$$\Big[\mathbf{J}_3, T^{(\kappa)}_\mu\Big] = \mu\,T^{(\kappa)}_\mu\,, \qquad \Big[\mathbf{J}_\pm, T^{(\kappa)}_\mu\Big] = \sqrt{\kappa(\kappa+1) - \mu(\mu \pm 1)}\,T^{(\kappa)}_{\mu\pm1}\,.$$

Hint: Notice the relation

$$\left\langle \phi'_{jm'} \middle| \phi_{jm} \right\rangle = D^{(j)}_{m'm} = \langle jm' | \mathbf{D}^{(j)} | jm \rangle \,,$$

which follows from the definition of the D-matrices and the transformation behaviour of base functions with respect to rotations. Consider then the transformation

$$\mathbf{D}^{(\kappa)}\mathbf{T}^{(\kappa)}\mathbf{D}^{(\kappa)\,-1} = e^{\mathrm{i}\varepsilon\hat{\mathbf{e}}\cdot\boldsymbol{J}}\mathbf{T}^{(\kappa)}e^{-\mathrm{i}\varepsilon\hat{\mathbf{e}}\cdot\boldsymbol{J}}$$

for small ε, where $\hat{\boldsymbol{e}}$ is an arbitrary unit vector.

6.9 Derive the orthogonality relations for the $6j$- as well as for $9j$-symbols.

6.10 Prove the formula (6.97) which relates matrix elements of vector operators between states with equal values of J to matrix elements of the angular momentum operators.

6.11 Derive the commutators of the generators of the Euclidian group in two dimensions.
Hint: This group consists of rotations about the origin, and of translations along the 1- and the 2-axis. Therefore, it is useful to introduce homogeneous coordinates. Compare to (6.138).

6.12 Show that the product of the time component p^0 of the energy-momentum vector of a particle and the distribution $\delta(\boldsymbol{p} - \boldsymbol{p}')$ is invariant under all $\boldsymbol{\Lambda} \in L_+^\uparrow$. Do you see a relation to the consequences of covariant normalization?
Hints: According to the decomposition theorem every $\boldsymbol{\Lambda} \in L_+^\uparrow$ can be represented as the product of a rotation and a Special Lorentz transformation. Demonstrate the asserted invariance separately for rotations and for "boosts" along the 3-axis. In the second case one has

$$p'^0 = \gamma p^0 + \gamma\beta p^3 \,, \quad p'^3 = \gamma\beta p^0 + \gamma p^3 \,.$$

6.13 Calculate the commutators of the generators $\mathbf{M}^{\mu\nu}$ with the generators $\mathbf{P}^\nu$, as well as among themselves, i.e. verify the formulae (6.121) and (6.122).

6.14 Let $\mathbf{U}$, $\mathbf{V}$ and $\mathbf{T}$ be elements of $SU(2)$, and write the entries of $\mathbf{V}$ and of $\mathbf{T}$ as follows

$$V_{11} = y_1 + \mathrm{i}y_2 \quad V_{12} = y_3 + \mathrm{i}y_4 \qquad T_{11} = z_1 + \mathrm{i}z_2 \quad T_{12} = z_3 + \mathrm{i}z_4 \,.$$

Show that the points $y := (y_1, y_2, y_3, y_4)^T$ and $z := (z_1, z_2, z_3, z_4)^T$ lie on S^3, the unit sphere in $\mathbb{R}^4$. Show: the action of $\mathbf{U} : \mathbf{V} \mapsto \mathbf{T}$ is equivalent to a proper rotation $z = \mathbf{A}y$ on the sphere S^3, i.e. $\mathbf{A}^T\mathbf{A} = 1\!\!1_{4\times4}$ and $\det\mathbf{A} = 1$.
Hint: Notice that every $\mathbf{U} \in SU(2)$ can be represented by

$$\mathbf{U} = \begin{pmatrix} u & v \\ -v^* & u^* \end{pmatrix} , \quad \text{with } |u|^2 + |v|^2 = 1 \,.$$

6.15 Let (x, y, z) be cartesian coordinates in $\mathbb{R}^3$. Write xy, xz, and $x^2 - y^2$ in terms of components of a spherical tensor operator. The expectation value

$$e\langle\alpha; j, m = j|3z^2 - r^2|\alpha; j, m = j\rangle \equiv Q$$

is the quadrupole moment. Calculate the matrix elements

$$e\left\langle\alpha;\, jm'\middle|x^2 - y^2\middle|\alpha;\, j, m = j\right\rangle$$

for $m' = j, j-1, j-2, \ldots$. Express these in terms of Q and of Clebsch-Gordan coefficients.
Hint: You need the explicit form of the spherical harmonics with $\ell = 2$, viz.

$$Y_{20} = \sqrt{\frac{5}{16\pi}}\left(3\cos^2\theta - 1\right)$$

$$Y_{21} = -\sqrt{\frac{15}{8\pi}}\sin\theta\cos\theta\, e^{\mathrm{i}\phi}$$

$$Y_{22} = \frac{1}{4}\sqrt{\frac{15}{2\pi}}\sin^2\theta\, e^{2\mathrm{i}\phi}\,,$$

and you should use the symmetry relation $Y_{\ell,-m} = (-)^m Y^*_{\ell m}$.

6.16 Let $|(\ell s) j, m_j\rangle$ be the state of an unpaired proton or neutron in the shell model of nuclei. In a state of this kind the magnetic moment is defined by

$$\mu_j := g_j \langle j, m = j|\mathbf{J}_3|j, m = j\rangle \mu_B\,, \quad \text{with } \mu_B = \frac{e\hbar}{2M_N c} \widehat{=} \frac{e}{2M_N}\,.$$

More generally, the corresponding operator is the sum of the magnetic moments which are due to the orbital motion and to the spins,

$$\boldsymbol{\mu} = \mu_B\left\{g_\ell \boldsymbol{\ell} + g_s \boldsymbol{s}\right\},$$

with $g_s^{(p)} = 5.58$ and $g_s^{(n)} = -3.82$. The expectation value of this operator, by Exercise 6.10, is proportional to the expectation value of $\boldsymbol{J}$, that is to say,

$$g_j\,\langle jm|\boldsymbol{J}|jm\rangle = \langle jm|\left\{g_\ell \boldsymbol{\ell} + g_s \boldsymbol{s}\right\}|jm\rangle\,.$$

Calculate g_j as a function of g_ℓ and of g_s.
Comment and Hint: The shell model of nuclei assumes that the magnetic moment of the nucleus is given by the magnetic moment of the last proton or neutron which finds no partner in the state $|j, -m\rangle$.
(Example: The nucleus ^{209}Bi: A single proton sits in the state $1\mathrm{h}_{9/2}$, i.e. $\ell = 5$ and $j = 9/2$, above a closed shell of 82 protons and another closed shell of 126 neutrons.)
Replace the scalar products $\boldsymbol{J}\cdot\boldsymbol{\ell}$ and $\boldsymbol{J}\cdot\boldsymbol{s}$ by linear combinations of $\boldsymbol{J}^2$, $\boldsymbol{\ell}^2$, and $\boldsymbol{s}^2$.

6.17 To show: every element of $\mathfrak{su}(3)$, that is, every traceless hermitian 3×3-matrix, can be written as a linear combination of the Gell-Mann matrices (6.99). The group $SU(3)$ contains three subgroups $SU(2)$ that one can read off from the generators (6.99). Determine the generators of the Lie algebras of these subgroups as linear combinations of λ_k, $k = 1, 2, \ldots, 8$. Mark representations of these subgroups in Figs. 6.6, 6.7, 6.8, and 6.9.

6.18 Determine the structure constants f_{ijk} of $SU(3)$ by calculating the commutators of the generators λ_k. Consider then the anticommutators of the generators and show that

$$\{\lambda_i, \lambda_j\} = \frac{4}{3}\delta_{ij}\,1\!\!1 + 2\sum_{k=1}^{8} d_{ijk}\lambda_k\,.$$

Calculate the coefficients d_{ijk} (which are symmetric in their indices).
Solution: One finds

$$d_{118} = d_{228} = d_{338} = \frac{1}{\sqrt{3}} = -d_{888}\,,$$

$$d_{448} = d_{558} = d_{668} = d_{778} = -\frac{1}{2\sqrt{3}}\,,$$

$$d_{146} = d_{157} = d_{256} = d_{344} = d_{355} = \frac{1}{2} = -d_{247} = -d_{366} = -d_{377}\,.$$

(If the reader is used to algebraic program packages such as MATHEMATICA or MAPLE this is a good exercise for matrix calculations.)

6.19 Denote the generators of $SU(3)$ by $T_i = \lambda_i/2$, $i = 1,\ldots,8$. The three subgroups $SU(2)$ come out more clearly by the definitions

$$I_3 = T_3\,, \qquad I_\pm = T_1 \pm \mathrm{i}T_2\,,$$

$$U_3 = -\frac{1}{2}T_3 + \frac{\sqrt{3}}{2}T_8\,, \quad U_\pm = T_6 \pm \mathrm{i}T_7\,,$$

$$V_3 = \frac{1}{2}T_3 + \frac{\sqrt{3}}{2}T_8\,, \qquad V_\pm = T_4 \pm \mathrm{i}T_5\,.$$

Work out the commutators of these generators among themselves and with $Y := 2T_8/\sqrt{3}$. Identify the action of the ladder operators in (T_3, Y)-diagrams of representations of the group.
Use these commutators to show: The boundary of an irreducible representation can nowhere be concave. The corners of the boundary are occupied each by one state only.

Exercises: Chap. 7

7.1 Show that

$$\int \mathrm{d}^3x\; f(x)\,\overleftrightarrow{\partial_0}\,g(x)$$

is Lorentz invariant, if $f(x)$ and $g(x)$ are Lorentz scalar functions.

7.2 Prove the translation formula (7.30) by making use of the commutators (7.29).

7.3 Starting from the Hamiltonian density of the Klein-Gordon field (7.25) derive the Hamiltonian (7.35) expressed in terms of creation and annihilation operators. Carry out the analogous calculation which leads to the expression (7.36) for the total momentum.

7.4 Consider the matrix element of a vector current v_μ between the state of a charged pion π^+ with mass m and momentum q, and the state of a neutral pion π^0 with mass m_0 and momentum q',

$$\left\langle \pi^0(q') \left| v_\mu(0) \right| \pi^+(q) \right\rangle . \tag{23}$$

This matrix element is responsible for β-decay of pions,

$$\pi^+(q) \longrightarrow \pi^0(q') + e^+(p) + \nu_e(k) . \tag{24}$$

This decay is kinematically allowed because $m - m_0 = 4.6$ MeV, the charged pion being heavier than the neutral one. Work out a decomposition of the matrix element (23) in terms of Lorentz covariants and invariant form factors, and determine relations between the form factors for the case where the current $v_\mu(x)$ is conserved.

Solution: As in (23) the left and right single-particle states have the same behaviour under parity, the matrix element, as a whole, must be a Lorentz vector. Pions have no spin. Hence, the only covariants are q_μ and q'_μ, or, alternatively, their sum and difference,

$$P_\mu := q_\mu + q'_\mu , \qquad Q_\mu := q_\mu - q'_\mu ,$$

so that we may write

$$\left\langle \pi^0(q') \left| v_\mu(0) \right| \pi^+(q) \right\rangle = \frac{1}{(2\pi)^3} \left\{ P_\mu f_+(Q^2) + Q_\mu f_-(Q^2) \right\} . \tag{25}$$

The choice of the factor in front is a matter of convention. We choose it in agreement with the rule that every external particle should contribute a factor $1/(2\pi)^{3/2}$. The form factors depend on invariants only, hence on $q^2, q'^2, q \cdot q'$, or the corresponding scalar products containing P and Q. As both the π^+ and the π^0 are on their mass shell, only one of these scalar products is a genuine variable. Choose this variable to be Q^2. Furthermore, in pion-β-decay (24) the interval of variation of this variable is small as compared to the pion mass,

$$m_e^2 \le Q^2 \le (m - m_0)^2$$

(it is often approximated by $Q^2 = 0$). If the current is conserved, i.e. if $\partial^\mu v_\mu(x) = 0$, then one obtains the condition

$$\left(m^2 - m_0^2\right) f_+(Q^2) + Q^2 f_-(Q^2) = 0 . \tag{26}$$

The form factor $f_-(Q^2)$ is very small as compared to $f_+(Q^2)$; in the limit $m = m_0$ it vanishes identically.

7.5 By means of Cauchy's theorem of residua prove the integral representations (7.61) for $\Theta(\pm u)$.

7.6 An electrically neutral and a positively charged scalar fields are assumed to span the doublet representation of some $SU(2)$ (internal symmetry),

$$\boldsymbol{\Phi}(x) = \begin{pmatrix} \phi^{(+)}(x) \\ \phi^{(0)}(x) \end{pmatrix} , \qquad \left(t = \frac{1}{2}\right) . \tag{27}$$

Let the Lagrange density containing a self-interaction be:

$$\mathcal{L} = \frac{1}{2}\left(\partial_\mu \boldsymbol{\Phi}^\dagger, \partial^\mu \boldsymbol{\Phi}\right) - \frac{1}{2}\kappa\left(\boldsymbol{\Phi}^\dagger(x), \boldsymbol{\Phi}(x)\right) - \frac{\lambda}{4}\left(\boldsymbol{\Phi}^\dagger(x), \boldsymbol{\Phi}(x)\right)^2 . \tag{28}$$

The coupling parameter λ is assumed to be real and positive, the parameter κ real, positive or negative. The brackets $(\cdot, \cdot)$ symbolize coupling to $t = 0$ so that this $SU(2)$ leaves the Lagrange density (28) invariant.
Determine the position of the minimum of the energy for $\kappa > 0$, as well as for $\kappa < 0$.
What is the interpretation of κ in the first case? If in the process of quantization one wishes to start from the state with lowest energy, how should one proceed in the case of negative κ?
Does the ground state of the quantized theory with $\kappa < 0$ still possess the full $SU(2)$-symmetry?
Solution: If $\kappa > 0$ define $\kappa \equiv m^2$. Both fields are massive and have the same mass m. If $\kappa < 0$ the energy has degenerate minima at

$$\left(\boldsymbol{\Phi}^\dagger, \boldsymbol{\Phi}\right) = \frac{-\kappa}{\lambda} \equiv v^2 .$$

Each one of these minima is a possible ground state. In everyone of them $\boldsymbol{\Phi}(x)$ develops a nonvanishing expectation value $\langle \Omega | \boldsymbol{\Phi}(x) | \Omega \rangle$. As the vacuum is electrically neutral, this expectation value can only be due to the neutral field, $\langle \Omega | \phi^{(0)}(x) | \Omega \rangle$. The dynamical field that describes the neutral partner, must be of the form

$$\theta^{(0)}(x) := \phi^{(0)}(x) - \langle \Omega | \phi^{(0)}(x) \, | \Omega \rangle \; .$$

It can be quantized in a standard manner. The ground state no longer has the full $SU(2)$ symmetry.
Comment: More about this and on the relevance of this example for particle physics, see, e.g, [Scheck (2012)].

7.7 Calculate the Hamiltonian (7.104) for the quantized radiation field in terms of creation and annihilation operators for transverse photons, i.e. prove the expansion (7.105).

7.8 Express the components of the operator ∇ in spherical coordinates. Calculate the commutators of the nabla operator with the generators ℓ_3 and $\ell_\pm$, i.e. prove

$$\left[\ell_3, \nabla_\mu\right] = \mu \nabla_3 , \qquad \left[\ell_\pm, \nabla_\mu\right] = \sqrt{2 - \mu(\mu \pm 1)}\; \nabla_{\mu\pm 1} .$$

7.9 When an electron, or a muon, or a heavy meson (such as π^- or K^-), bound in an atom, makes an $E1$-transition, both the lepton (or meson) and the nucleus move while the center-of-mass remains approximately at rest. (The recoil of the emitted photon being neglected.) Show that this effect implies the replacement (7.126) in the dipole operator.
Solution: Denote by Z the center-of-mass of the isolated nucleus, by S the center-of-mass of the total system, nucleus plus bound particle. Denote by m and m_A

Fig. 4

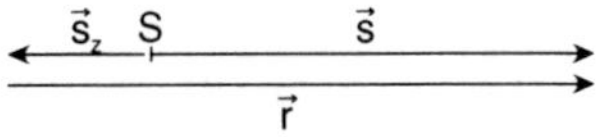

the masses of the particle and the nucleus, respectively. As the nucleus carries the charge $+Ze$, the dipole operator reads

$$\boldsymbol{D} = -e\,\boldsymbol{s} + Ze\boldsymbol{s}_Z\,,$$

with $\boldsymbol{s}$ the coordinate of the particle, $\boldsymbol{s}_Z$ the coordinate of the nucleus with respect to the common center-of-mass system. Note that atomic wave functions are always calculated in terms of $\boldsymbol{r}$, i.e. in terms of the coordinates of the particle with respect to the point Z. One has

$$m\boldsymbol{s} + m_A\boldsymbol{s}_Z = 0\,,$$

and, therefore, (see Fig. 4)

$$\boldsymbol{r} = \boldsymbol{s} - \boldsymbol{s}_Z = -\frac{m + m_A}{m}\boldsymbol{s}_Z\,.$$

One writes $\boldsymbol{s}$ and $\boldsymbol{s}_Z$ in terms of $\boldsymbol{r}$,

$$\boldsymbol{s} = \frac{m_A}{m + m_A}\boldsymbol{r}\,, \quad \boldsymbol{s}_Z = -\frac{m}{m + m_A}\boldsymbol{r}\,,$$

and inserts them in $\boldsymbol{D}$. One obtains

$$\boldsymbol{D} = -e\left(1 + \frac{(Z-1)m}{m + m_A}\right)\boldsymbol{r}\,. \tag{29}$$

This correction is negligible for electrons. In muonic atoms it is still small but it can be important in π^-- and K^--atoms.

7.10 Show that the definition of the distributions

$$\delta_\pm(p) := \Theta(\pm p^0)\,\delta(p^2)$$

is Lorentz invariant. (They are used in identifying the support of the field tensor $F^{\mu\nu}(p)$ in Sect. 7.6.2).

Exercises: Chap. 8

8.1 In a two-body reaction $A+B \to C+D+\cdots$ let κ denote the modulus of the spatial momentum in the center-of-mass system. Show that κ can be expressed in terms of m_A^2, m_B^2, and the variable $s = (p+q)^2$ (cf. (8.26)).

8.2 For a nonrelativistic particle with spin 1/2 construct the most general, rotationally invariant, but not necessarily parity invariant, potential with vanishing range. Show that the amplitude for scattering on this potential, in first Born approximation, only contains s- and p-waves.

8.3 The kinematics of the decay $\pi^+ \to \mu^+\nu_\mu$ allows to obtain a bound on the mass of the muon neutrino. Suppose that in the reference frame in which the pion was at rest before it decayed, you have measured $|\boldsymbol{p}|^{(\mu)} = 29.788 \pm 0.001$ MeV. Furthermore, one knows

$$m(\pi^+) = 139.57018 \pm 0.00035 \text{ MeV},$$
$$m(\mu) = 105.6583568 \pm 0.0000052 \text{ MeV}.$$

Use these data to draw a conclusion on $m(\nu_\mu)$.

8.4 Using the kinematics of the three-body decay

$$^3\text{He} \longrightarrow {}^3\text{He} + e^- + \overline{\nu_e}$$

at the upper end of the spectrum of the electron, one extracts information on the mass of the $\overline{\nu_e}$. Discuss the shape of this spectrum based on the following assumptions: In the neighbourhood of the upper end of the spectrum the squared decay amplitude $\sum |T|^2$ is constant. Both, the mother and the daughter nuclei are very heavy, so that the daughter nucleus takes over any momentum transfer but no energy.

Exercises: Chap. 9

9.1 Calculate the pseudoscalar products

$$\overline{u_i^{(r)}(p)}\gamma_5 u_k^{(s)}(p), \quad \overline{v_i^{(r)}(p)}\gamma_5 v_k^{(s)}(p)$$

using the solutions (9.46) and (9.47). Note that the spinors may pertain to two different particles i and k.

9.2 Prove that $\overline{\psi(x)}\gamma^\mu\psi(x)$ behaves like a Lorentz vector.

9.3 Using two-component spinors construct a Lagrange density for the Dirac field in the high-energy representation.
Solution: In position space the four-component Dirac spinor is composed of two-component spinor fields $\phi(x)$ and $\chi(x)$,

$$\psi(x) = \begin{pmatrix} \phi(x) \\ \chi(x) \end{pmatrix}.$$

The Lagrange density must be Lorentz scalar and hermitian, and it must yield the equations of motion (9.74) and (9.75). One verifies that

$$\mathcal{L}_\text{D} = \frac{\text{i}}{2}\left[\phi^*(x)\widehat{\sigma}^\mu \overleftrightarrow{\partial_\mu}\phi(x) + \chi^*(x)\sigma^\mu \overleftrightarrow{\partial_\mu}\chi(x)\right]$$
$$- m\left[\phi^*(x)\chi(x) + \chi^*(x)\phi(x)\right] \qquad (30)$$

fulfills these conditions.

9.4 Prove the formulae (9.97) and (9.98).

Solution: In the rest system and using the standard representation one has

$$u(\mathbf{0})\overline{u(\mathbf{0})} = 2m \begin{pmatrix} 1/2(\mathbb{1}_2 + \boldsymbol{\sigma}\cdot\widehat{\boldsymbol{n}}) & 0 \\ 0 & 0 \end{pmatrix}$$
$$= 2m\frac{1}{2}(\mathbb{1} + \gamma^0)\frac{1}{2}(\mathbb{1} - \gamma_5\widehat{\boldsymbol{n}}\cdot\boldsymbol{\gamma}),$$
$$v(\mathbf{0})\overline{v(\mathbf{0})} = 2m \begin{pmatrix} 0 & 0 \\ 0 & 1/2(\mathbb{1}_2 - \boldsymbol{\sigma}\cdot\widehat{\boldsymbol{n}}) \end{pmatrix}$$
$$= 2m\frac{1}{2}(\mathbb{1} - \gamma^0)\frac{1}{2}(\mathbb{1} - \gamma_5\widehat{\boldsymbol{n}}\cdot\boldsymbol{\gamma}).$$

Calculate then the sum and the difference $u(\boldsymbol{p})\overline{u(\boldsymbol{p})} \pm v(\boldsymbol{p})\overline{v(\boldsymbol{p})}$, by inserting the "boosted" solutions $u(\boldsymbol{p}) = N(\not{p} + m\,\mathbb{1})\,u(\mathbf{0})$ and $v(\boldsymbol{p}) = -N(\not{p} - m\,\mathbb{1})v(\mathbf{0})$. For example, one has

$$u(\boldsymbol{p})\overline{u(\boldsymbol{p})} + v(\boldsymbol{p})\overline{v(\boldsymbol{p})}$$
$$= \frac{1}{4(E_p+m)}\Big\{(\not{p} + m\,\mathbb{1})(\mathbb{1}+\gamma^0)(\mathbb{1} - \gamma_5 n^i\gamma^i)(\not{p} + m\,\mathbb{1})$$
$$- (\not{p} - m\,\mathbb{1})(\mathbb{1}-\gamma^0)(\mathbb{1} - \gamma_5 n^i\gamma^i)(\not{p} - m\,\mathbb{1})\Big\}$$
$$= \frac{1}{4(E_p+m)}\Big\{\Big[4m\not{p} + 2\not{p}\gamma^0\not{p} + 2m^2\gamma^0\Big]$$
$$- \gamma_5 n^i\Big[2m(\gamma^i\not{p} - \not{p}\gamma^i) - 2m^2\gamma^0\gamma^i + 2\not{p}\gamma^0\gamma^i\not{p}\Big]\Big\}.$$

Then, in each term, shift $\not{p}$ to the left by making use of the commutation rules of the γ-matrices and of the equation $\not{p}\not{p} = m^2$. One obtains, in detail,

$$u(\boldsymbol{p})\overline{u(\boldsymbol{p})} + v(\boldsymbol{p})\overline{v(\boldsymbol{p})}$$
$$= \Big\{\not{p} - \gamma_5 n^i\Big[-\not{p}\gamma^i + \not{p}\gamma^0\frac{p^i}{E_p+m} + \frac{m}{E_p+m}p^i\Big]\Big\}$$
$$= \not{p}\Big\{\mathbb{1} + \gamma_5\Big[-n^i\gamma^i + n^i p^i\Big(\frac{1}{E_p+m}\gamma^0 + \frac{1}{m(E_p+m)}\not{p}\Big)\Big]\Big\}$$
$$= \not{p}\Big\{\mathbb{1} + \gamma_5\Big[\frac{\widehat{\boldsymbol{n}}\cdot\boldsymbol{p}}{m}\gamma^0 - \Big(n^i + \frac{\widehat{\boldsymbol{n}}\cdot\boldsymbol{p}}{m(E_p+m)}p^i\Big)\gamma^i\Big]\Big\}$$
$$= \not{p}\big(\mathbb{1} + \gamma_5\not{n}\big),$$

where n is the four-vector (9.96). In much the same way one calculates

$$u(\boldsymbol{p})\overline{u(\boldsymbol{p})} - v(\boldsymbol{p})\overline{v(\boldsymbol{p})} = m\big(\mathbb{1} + \gamma_5\not{n}\big).$$

Taking the sum and the difference of these results yields the asserted equations.

9.5 To show: The density matrix $\varrho^{(+)\dagger}$ describes a state which is the parity mirror of the state described by $\varrho^{(+)}$.

9.6 Prove the formulae

$$u_{L/R}(p)\overline{u_{R/L}(p)} = \frac{1}{2} P_{\mp} \left(m \,\mathbb{1} - \gamma_5 \not{p} \not{n}\right).$$

Calculate the sum

$$u_L(p)\overline{u_L(p)} + u_R(p)\overline{u_R(p)} + u_L(p)\overline{u_R(p)} + u_R(p)\overline{u_L(p)}$$

and compare to a result from Exercise 9.4.

9.7 Compute the Clebsch-Gordan coefficients $\left(\ell, m_\ell; 1/2, m_s | jm\right)$ by constructing explicitly the states $|j, m\rangle$ from the highest state $|j = \ell + 1/2, m = \ell + 1/2\rangle$ (see also Sect. 4.1.6).

9.8 Let space reflection $x = (t, \boldsymbol{x}) \mapsto x' = (t, -\boldsymbol{x})$ be applied to the Dirac equation (9.56). Show that if $\psi(x)$ is a solution of the Dirac equation, then also

$$\psi_{\Pi}(x) = \gamma^0 \psi(x')$$

is a solution.

The following two exercises give a hint as to when the Dirac equation is applicable and when it fails as a single-particle equation, the failures being due to the admixture of states with "negative energy."

9.9 A fermion in one dimension is scattered by a step potential $U(x) = U_0 \Theta(x)$, see Fig. 5. Before the scattering it comes in from $x < 0$ with momentum p and spin orientation $m_s = +1/2$.

Write down, in the standard representation, the incoming wave, the reflected wave, and the wave that goes through towards positive x. Use continuity of the solution at $x = 0$ to fix the free parameters in the three waves (up to a common normalization). Calculate the current going through, and the reflected current, in relation to the incoming current. Discuss the results for the case $(E - U_0)^2 < m^2$ and for the case $(E - U_0)^2 > m^2$, $U_0 > E + m$.

Comment: The paradoxical results that one obtains in the second case, are called *Klein's paradox*.

9.10 In close analogy to (9.55) one constructs a wave packet $\psi(t, \boldsymbol{x})$ from the complete system of free solutions with positive and negative frequency. Calculate the spatial current J^i for this wave packet. Estimate the frequencies of the oscillations in the mixed, positive and negative frequency terms.

Comment: These rapid oscillations are called *Zitterbewegung*.

9.11 We consider bound states of an electron in the Coulomb potential for weakly relativistic motion. Show: For $\kappa < 0$, that is, for $\ell = -\kappa - 1$, we

Fig. 5

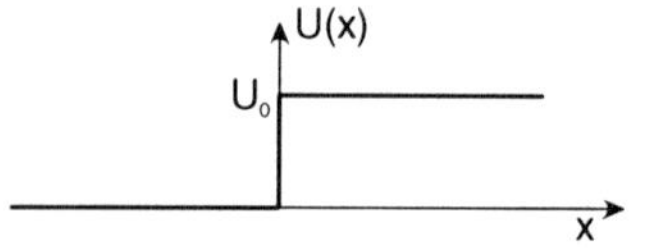

have $rg_{n\kappa}(r) \approx y_{n\ell}(r)$, where $y_{n\ell}(r)$ is the radial function of the nonrelativistic hydrogen atom, (cf. (1.155)).
Remark: The result shows that in the nonrelativistic limit

$$\left(n, \kappa, j = -\kappa - \frac{1}{2} = \ell + \frac{1}{2}\right) \quad \text{and}$$

$$\left(n, \kappa' = -\kappa - 1, j = \kappa' - \frac{1}{2} = \ell - \frac{1}{2}\right)$$

have the same radial function.

Exercises: Chap. 10

10.1 Work out the rule (R9) in detail.

10.2 Assume the current density $j^\mu(x)$ to be self-adjoint and conserved, $\partial_\mu j^\mu(x) = 0$. Then, in the decomposition in terms of covariants, only the form factors $F_1(Q^2)$ and $F_2(Q^2)$ are different from zero. Show that the form factors are *real*.
Solution: The current density being self-adjoint we have

$$\langle p| j_\mu^\dagger(0) |q\rangle = \langle q| j_\mu(0) |p\rangle^* = \langle p| j_\mu(0) |q\rangle \ . \tag{31}$$

The first equal sign is nothing but the definition of the adjoint operator while the second reflects the assumption. Inserting the decomposition (10.76) the middle term of (31) yields

$$\left\{ u^\dagger(q)\gamma_0 \left[\gamma_\mu F_1(Q^2) - \frac{\mathrm{i}}{2m}\sigma_{\mu\nu}(p-q)^\nu F_2(Q^2) \right] u(p) \right\}^*$$
$$= u^\dagger(p) \left[\left(\gamma_0\gamma_\mu\right)^\dagger F_1^*(Q^2) + \frac{\mathrm{i}}{2m}\left(\gamma_0\sigma_{\mu\nu}\right)^\dagger (p-q)^\nu F_2^*(Q^2) \right] u(q)$$
$$= \overline{u(p)} \left\{ \gamma_0\gamma_\mu^\dagger\gamma_0\, F_1^*(Q^2) - \frac{\mathrm{i}}{2m}\gamma_0\sigma_{\mu\nu}^\dagger\gamma_0 (q-p)^\nu F_2^*(Q^2) \right\} u(q) \, .$$

In the last step we inserted $\left(\gamma_0\right)^2 = \mathbb{1}$ between $u^\dagger(p)$ and the expression to the right of it. From the properties of the γ-matries we have

$$\gamma_0\gamma_\mu^\dagger\gamma_0 = \gamma_\mu \, , \qquad \gamma_0\sigma_{\mu\nu}^\dagger\gamma_0 = \sigma_{\mu\nu} \, .$$

Hence, the second equal sign in (31) yields

$$F_1^*(Q^2) = F_1(Q^2) \, , \qquad F_2^*(Q^2) = F_2(Q^2) \, . \tag{32}$$

Both form factors are indeed real.

10.3 Prove the integral that we used in calculating the anomalous magnetic moment,

$$I := \int \mathrm{d}^4 v \, \frac{1}{\left(v^2 - \Lambda^2 + \mathrm{i}\varepsilon\right)^3} = -\frac{\mathrm{i}\pi^2}{2\Lambda^2} \, .$$

Fig. 6

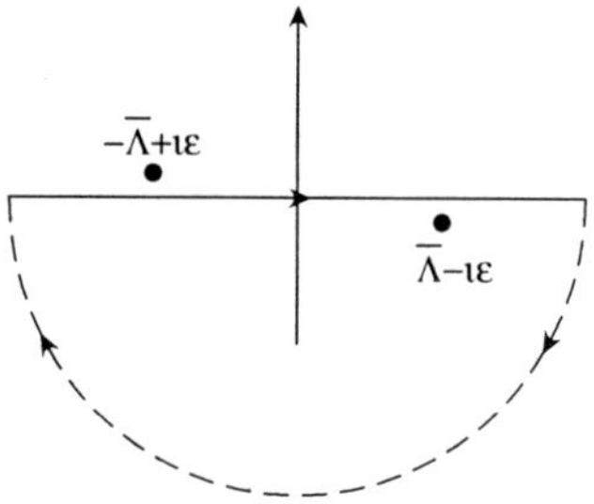

Solution: Splitting into integrals over v_0 and over $\boldsymbol{v}$, the latter is formulated in terms of spherical polar coordinates of $\mathbb{R}^3$. With $r \equiv |\boldsymbol{v}|$ one has

$$I = \int_{-\infty}^{+\infty} \mathrm{d}v^0 \int_0^{\infty} \mathrm{d}r\, r^2 \int \mathrm{d}\Omega\, \frac{1}{\left(v_0^2 - \overline{\Lambda}^2 + \mathrm{i}\varepsilon\right)^3}$$

$$= 4\pi \int_0^{\infty} \mathrm{d}r\, r^2 \int_{-\infty}^{+\infty} \mathrm{d}v^0 \frac{1}{\left(v_0^2 - \overline{\Lambda}^2 + \mathrm{i}\varepsilon\right)^3},$$

where $\overline{\Lambda}^2 := r^2 + \Lambda^2$. The denominator may equivalently be written as

$$v_0^2 - (\overline{\Lambda}^2 - \mathrm{i}\varepsilon) \simeq \left(v_0 - (\overline{\Lambda} - \mathrm{i}\varepsilon)\right)\left(v_0 + (\overline{\Lambda} - \mathrm{i}\varepsilon)\right).$$

The positions of the singularities of the integrand in the complex v_0-plane are as shown in Fig. 6. One closes the path of integration by a semi-circle at infinity, as sketched in the figure, and uses Cauchy's integral theorem in the form

$$\oint \mathrm{d}\zeta\, \frac{f(\zeta)}{(\zeta - z)^3} = \pi \mathrm{i} f''(z),$$

(It follows from the theorem of residua by twofold derivative in z). One obtains

$$I = -4\pi^2 \mathrm{i} \int_0^{\infty} \mathrm{d}r\, r^2 \frac{12}{(2\overline{\Lambda})^5} = -\frac{3}{2}\pi^2 \mathrm{i} \int_0^{\infty} \mathrm{d}r\, \frac{r^2}{\left(r^2 + \Lambda^2\right)^{5/2}}.$$

The remaining integral is done in an elementary way, by partial integration with $u'(r) = r/(r^2 + \Lambda^2)^{5/2}$ and $v(r) = r$. The result is

$$\int_0^{\infty} \mathrm{d}r\, \frac{r^2}{\left(r^2 + \Lambda^2\right)^{5/2}} = \frac{1}{3\Lambda^2},$$

from which the assertion follows directly.

10.4 At order $\mathcal{O}(\alpha)$ the differential cross sections in the center-of-mass system for the following processes are symmetric about 90°:

$$e^- + e^- \longrightarrow e^- + e^- \quad e^+ + e^- \longrightarrow \gamma + \gamma$$
$$e^+ + e^- \longrightarrow \mu^+ + \mu^- e^+ + e^- \longrightarrow \tau^+ + \tau^-.$$

Why is this so? Why does this not hold for $e^+ + e^- \to e^+ + e^-$?

10.5 Prove the integral formulae (A.11) which are needed in the analysis of the self-energy.
Hint: Since $x^2 = (x^0)^2 - \boldsymbol{x}^2$, the nonvanishing integrals factorize in integrals over each coordinate separately. In every one-dimensional integral the path of integration can be rotated by 45° in the complex plane. The integral then becomes a Gauss integral.

10.6 Consider the processes (10.125)–(10.131), sketch tree diagrams for these reactions, and decide to which of them the weak *charged* current contributes, to which the weak *neutral* current contributes, and to which both of them contribute.

10.7 Compute the decay rate of μ-decay taking into account the $W^{\pm}$-propagator. Show that the rate is modified as shown (10.139).

References

Mechanics and Electrodynamics

Landau, L. D., Lifschitz, E. M.: *The Classical Theory of Fields* (Pergamon Press, Oxford 1987)

Jackson, J. D.: *Classical Electrodynamics* (John Wiley, New York 1999)

For the sake of reference to specific topics in mechanics (further books on mechanics see therein)

Scheck, F.: *Mechanics: From Newton's Laws to Deterministic Chaos* (Springer-Verlag, Heidelberg 2010)

Classics on Quantum Theory

Condon, E. U., Shortley, E. H.: *The Theory of Atomic Spectra* (Cambridge University Press, London 1957)

Dirac, P. A. M.: *The Principles of Quantum Mechanics* (Oxford Science Publications, Clarendon Press Oxford 1996)

Heisenberg, W.: *Physikalische Prinzipien der Quantentheorie* (BI, Mannheim 1958)

Heitler, W.: *The Quantum Theory of Radiation* (Oxford University Press, Oxford 1953)

Von Neumann, J.: *Mathematical Foundations of Quantum Mechanics* (Princeton University Press, German original: Springer-Verlag, Berlin 1932)

Pauli, W.: *General Principles of Quantum Mechanics* (Springer-Verlag, Heidelberg 1980)

Wigner, E. P.: *Group Theory and its Applications to the Quantum Mechanics of Atomic Spectra* (Academic Press, New York 1959)

Selected Textbooks on Quantum Mechanics

Basdevant, J.-L., Dalibard, J.: *Quantum Mechanics* (Springer-Verlag, Berlin, Heidelberg 2002)

Basdevant, J.-L., Dalibard, J.: *The Quantum Mechanics Solver* (Springer-Verlag, Ecole Polytechnique, Heidelberg 2000)

Feynman, R. P., Hibbs, A. R.: *Quantum Mechanics and Path Integrals* (McGraw Hill 1965)

Gottfried, K., Yan, T. M.: *Quantum Mechanics: Fundamentals* (Springer-Verlag, New York, Berlin 2003)

Landau, L. D., Lifschitz, E. M.: *Quantum Mechanics* (Pergamon 1977)

Merzbacher, E.: *Quantum Mechanics* (John Wiley and Sons, New York 1997)

Sakurai, J. J.: *Modern Quantum Mechanics* (Addison-Wesley, Reading, Mass. 1994)

F. Scheck, *Quantum Physics*, DOI: 10.1007/978-3-642-34563-0,

Schiff, L. I.: *Quantum Mechanics* (McGraw-Hill 1968)

Schwinger, J.: *Quantum Mechanics, Symbolism of Atomic Measurements* (Springer-Verlag, New York, Heidelberg 2001)

Thirring, W.: *A Course in Mathematical Physics, Volume 3: Quantum Mechanics of Atoms and Molecules* (Springer, Berlin, Heidelberg 1981)

Extensive Monographs on Quantum Mechanics and Fields

Most of these might be to extensive for a first tour through quantum mechanics. All of them are suitable, however, for further study and as references for specific topics

Cohen-Tannoudji, C., Diu, B., Laloë, F.: *Mécanique Quantique I + II* (Hermann, Paris 1977)

Galindo, A., Pascual, P.: *Quantum mechanics I + II* (Springer-Verlag, Berlin Heidelberg New York 1990)

Itzykson, Cl., Zuber, J.-B.: *Quantum Field Theory* (McGraw-Hill, New York 1980)

Messiah, A.: *Quantum Mechanics 1 + 2* (North Holland Publ., Amsterdam 1964)

Zinn-Justin, J.: *Quantum Field Theory and Critical Phenomena* (Clarendon Press, Oxford 1994)

Problems of Quantum Mechanics on a PC

Feagin, J. M.: *Quantum Methods with Mathematica* (Springer, Berlin, Heidelberg 1994)

Horbatsch, M.: *Quantum Mechanics using Maple* (Springer, Berlin, Heidelberg 1995)

Selected Books on Relativistic Quantum Theory and Quantum Field Theory

Bethe, H. A., Jackiw, R. W.: *Intermediate Quantum Mechanics* (Benjamin Cummings, Menlo Park 1986)

Bethe, H. A., Salpeter, E. E.: *Quantum Mechanics of One and Two Electron Atoms* (Springer-Verlag, Berlin 1957)

Bjorkén, J. D., Drell, S. D.: *Relativistic Quantum Mechanics* (McGraw Hill, New York 1964)

Bjorkén, J. D., Drell, S. D.: *Relativistic Quantum Fields* (McGraw Hill, New York 1965)

Bogoliubov, N. N., Shirkov, D. V.: *Introduction to the Theory of Quantized Fields* (Interscience, New York 1959)

Collins, J. C.: *Renormalization* (Cambridge University Press, Cambridge 1998)

Dittrich, W., Reuter, M.: *Classical and Quantum Dynamics* (Springer, Heidelberg 2001)

Gasiorowicz, S.: *Elementary Particle Physics* (John Wiley & Sons, New York 1966)

Grosche, Ch., Steiner, F.: *Handbook of Feynman Path Integrals* (Springer, Heidelberg 1998)

Jauch, J. M., Rohrlich, F.: *The Theory of Photons and Electrons: The Relativistic Quantum Field Theory of Charged Particles with Spin one-half* (Springer-Verlag, New York 1976)

Jost, R.: *The General Theory of Quantized Fields* (Am. Mathematical Soc., Providence 1965)

Källen, G.: *Quantum Electrodynamics* (Springer-Verlag, Berlin 1972)

Kleinert, H.: *Path Integrals in Quantum Mechanics, Statistics, and Polymer Physics* (World Scientific, Singapore 1990)

Le Bellac, M.: *Quantum and Statistical Field Theory* (Clarendon Press, Oxford 1991)

Ramond, P.: *Field Theory – a Modern Primer* (Benjamin/Cummings Publ. Co., Reading 1981)

Sakurai, J. J.: *Advanced Quantum Mechanics* (Benjamin/Cummings 1984)

Salmhofer, M.: *Renormalization* (Springer-Verlag, Heidelberg 1999)

Scheck, F.: *Electroweak and Strong Interactions - Phenomenology, Concepts, Models* (Springer-Verlag, Heidelberg 2012)

Schweber, S. S.: *An Introduction to Quantum Field Theory* (Row, Peterson & Co., Evanston 1961)

Streater, R. F., Wightman, A. S.: *PCT, Spin & Statistics, and All That* (Benjamin, New York 1964)

Weinberg, S.: *The Quantum Theory of Fields I + II* (Cambridge University Press, Cambridge 1995, 1996)

The Rotation Group in Quantum Mechanics

Edmonds, A. R.: *Angular Momentum in Quantum Mechanics* (Princeton University Press, Princeton 1957)

Fano, U., Racah, G.: *Irreducible Tensorial Sets* (Academic Press, New York 1959)

Rose, M. E.: *Elementary Theory of Angular Momentum* (Wiley, New York 1957)

Rotenberg, M., Bivins, R., Metropolis, N., Wooten, J. K.: *The 3j- and 6j-Symbols* (Technical Press MIT, Boston 1959)

de Shalit, A., Talmi, I.: *Nuclear Shell Theory* (Academic Press, New York 1963)

Quantum Scattering Theory

Calogero, F.: *Variable Phase Approach to Potential Scattering* (Academic Press, New York 1967)

Goldberger, M. L., Watson, K. W.: *Collision Theory* (Wiley, New York 1964)

Newton, R. G.: *Scattering Theory of Waves and Particles* (McGraw Hill, 1966)

Selected Monographs on Quantized Gauge Theories, and Algebraic Quantum Field Theory

These references are meant primarily for those who are, or want to become, theoreticians. There are many more books on non-Abelian gauge theories whose list would be too long to be included here

Bailin, D., Love, A.: *Introduction to Gauge Field Theory* (Institute of Physics Publishing, 1993)

Bogoliubov, N. N., Logunov, A. A., Oksak, A. I., Todorov, I. T.: *General Principles of Quantum Field Theory* (Kluwer, 1990)

Cheng, T. P., Li, L. F.: *Gauge Theory of Elementary Particle Physics* (Oxford University Press, Oxford 1984)

Deligne, P. et al. (Eds.): *Quantum Fields and Strings: A Course for Mathematicians* (Am. Math. Soc., Inst. for Advanced Study, Providence 2000)

Faddeev, L. D., Slavnov, A. A.: *Gauge Fields: An Introduction to Quantum Theory* (Addison-Wesley, Reading 1991)

Haag, R.: *Local Quantum Physics* (Springer-Verlag, Heidelberg 1999)

Piguet, O., Sorella, S. P.: *Algebraic Renormalization* (Lecture Notes in Physics, Springer-Verlag 1995)

Steinmann, O.: *Perturbative Quantum Electrodynamics and Axiomatic Field Theory* (Springer-Verlag, Heidelberg 2000)

Handbooks, Special Functions

Abramowitz, M., Stegun, I. A.: *Handbook of Mathematical Functions* (Dover, New York 1965)

Erdélyi, A., Magnus, W., Oberhettinger, F., Tricomi, F. G.: *Higher Transcendental Functions, The Bateman Manuscript Project* (McGraw-Hill, New York 1953)

Gradshteyn, I. S., Ryzhik, I. M.: *Table of Integrals, Series and Products* (Academic Press, New York and London 1965)

Whittaker, E. T., Watson, G. N.: *A Course of Modern Analysis* (Cambridge University Press, London 1958)

Selected Books on the Mathematics of Quantum Theory

Some of these books are relatively elementary, others go deep into the theory of Lie groups and Lie algebras, and refer to present-day research in theoretical physics

de Azcárraga, J. A., Izquierdo, J. M.: *Lie Groups, Lie Algebras, Cohomology and some Applications in Physics* (Cambridge University Press, Cambridge 1995)

Blanchard, Ph., Brüning, E.: *Mathematical Methods in Physics* (Birkhäuser, Boston, Basel, Berlin 2003)

Bremermann, H.: *Distributions, Complex variables, and Fourier transforms* (Addison-Wesley, Reading, Mass. 1965)

Fuchs, J., Schweigert, Ch.: *Symmetries, Lie Algebras and Representations* (Cambridge University Press, Cambridge 1997)

Hamermesh, M.: *Group Theory and Its Applications to Physical Problems* (Addison-Wesley, Reading, Mass. 1962)

Lawson, H. B., Michelson, M.-L.: *Spin geometry* (Princeton University Press, Princeton 1989)

Mackey, G. W.: *The theory of group representations* (Univ. of Chicago Press, 1976) (Chicago lectures in mathematics)

O'Raifertaigh, L.: *Group Structure of Gauge Theories* (Cambridge Monographs on Mathematical Physics, Cambridge 1986)

Richtmyer, R. D.: *Principles of Advanced Mathematical Physics* (Springer-Verlag, New York 1981)

Sternberg, S.: *Group Theory and Physics* (Cambridge University Press, Cambridge 1994)

About the Interpretation of Quantum Theory

Aharonov, Y., Rohrlich, D.: *Quantum Paradoxes - Quantum Theory for the Perplexed* (Wiley-VCH Verlag, Weinheim 2005)

d'Espagnat, B.: *Conceptual Foundations of Quantum Mechanics* (Addison-Wesley, Redwood City 1989)

Omnès, R.: *The Interpretation of Quantum Mechanics* (Princeton University Press, Princeton 1994)

Selleri, F.: *Quantum Paradoxes and Physical Reality* (Kluwer, 1990)

References pertaining to the Historical Notes

Born, M.: *Mein Leben* (Nymphenburger Verlagshandlung, München 1975)

Born, M. (Hrsg.) : *Albert Einstein – Max Born, Briefwechsel 1916–1955* (Nymphenburger Verlagshandlung, München 1969)

Cassidy, D. C.: *Uncertainty: The Life and Science of Werner Heisenberg* (Freeman & Co, 1992)

Einstein, A., Podolsky, B., Rosen, N.: Phys. Rev. **47** (1935) 777

Jost, R.: *Das Märchen vom elfenbeinernen Turm* (Springer-Verlag, Heidelberg 1995)

Heisenberg, W.: *Der Teil und das Ganze* (Piper Verlag, München 1969)

Mehra, J.: *The Conceptual Completion and the Extensions of Quantum Mechanics: 1932–1941* (Springer-Verlag, New York 2001)

Pais, A.: *Niels Bohr's Times* (Clarendon Press, Oxford 1991)

Pais, A.: *"Subtle is the Lord ...", The Science and the Life of Albert Einstein* (Clarendon Press, Oxford 1982)

Sciaccia, L.: *La Scomparsa di Majorana* (Einaudi, Torino 1975)

Stern, F.: *Einstein's German World* (Princeton University Press 1999)

Weart, S. R.: *Scientists in Power* (Harvard University Press, Cambridge 1979)

Interesting information on Nobel price winners who contributed to the development of quantum mechanics and quantum field theory (curricula vitae, Nobel lectures, etc.) can also be found on the internet: http://www.nobel.se/physics/laureates/index.html

Index

F. Scheck, *Quantum Physics*, DOI: 10.1007/978-3-642-34563-0,

M

N

O

P

T

U

About the Author

Florian A. Scheck Born 1936 in Berlin, son of Gustav O. Scheck, flutist. Studied physics at the University of Freiburg, Germany. Diploma in Physics 1962, Ph.D. in Theoretical Physics 1964. Guest Scientist at the Weizmann Institute of Science, Rehovoth, Israel, 1964–1966. Research Assistant at the University of Heidelberg, 1966–1968. Research Fellow at CERN, Geneva, 1968–1970. From 1970 until 1976 Head of Theory Group at the Swiss Institute of Nuclear Research (PSI), Lecturer and Titular Professor at ETH Zurich. From 1976–2005 Professor of Theoretical Physics at Johannes Gutenberg University, Mainz. Professor emeritus since 2005. Principal field of activity: Theoretical particle physics.

F. Scheck, *Quantum Physics*, DOI: 10.1007/978-3-642-34563-0,

CPSIA information can be obtained at www.ICGtesting.com
Printed in the USA
LVOW07*1813210713

343877LV00016B/736/P

9 783642 345623